Lectures on ACCELERATOR PHYSICS

Lectures on
Accelerator Physics

Alexander Wu Chao
SLAC National Accelerator Laboratory, USA

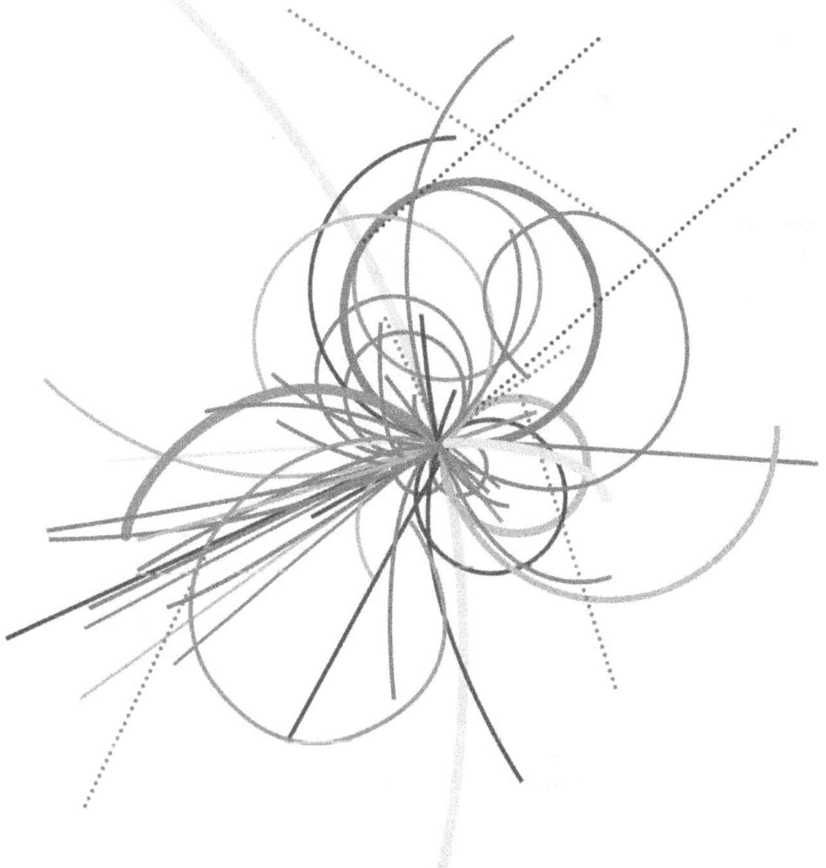

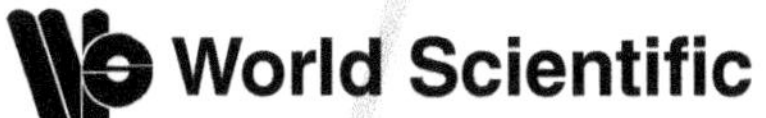

EW JERSEY • LONDON • SINGAPORE • BEIJING • SHANGHAI • HONG KONG • TAIPEI • CHENNAI • TOKYO

Published by

World Scientific Publishing Co. Pte. Ltd.
5 Toh Tuck Link, Singapore 596224
USA office: 27 Warren Street, Suite 401-402, Hackensack, NJ 07601
UK office: 57 Shelton Street, Covent Garden, London WC2H 9HE

Library of Congress Cataloging-in-Publication Data
Names: Chao, Alexander Wu, author.
Title: Lectures on accelerator physics / Alexander Wu Chao,
SLAC National Accelerator Laboratory, USA.
Description: New Jersey : World Scientific, [2020] |
Includes bibliographical references and index.
Identifiers: LCCN 2020038644 | ISBN 9789811226731 (hardcover) |
ISBN 9789811227967 (paperback) | ISBN 9789811226748 (ebook)
Subjects: LCSH: Particle accelerators. | Particle accelerators--Mathematical models.
Classification: LCC QC787.P3 C475 2020 | DDC 539.7/3--dc23
LC record available at https://lccn.loc.gov/2020038644

British Library Cataloguing-in-Publication Data
A catalogue record for this book is available from the British Library.

For any available supplementary material, please visit
https://www.worldscientific.com/worldscibooks/10.1142/12004#t=suppl

Preface

This book aims to provide the intended readers a basic textbook of accelerator physics, and hopefully to offer a view of accelerator physics as a rich and fascinating subject to learn. This is an academically oriented textbook, based on practical experience and tailored to the hands-on need of readers who want to gain basic knowledge in this field. It is my wish that this material helps in that regard, and it would be a great honor to inspire students eager to learn.

Accelerators are the tools driving many research and industrial advances. It has been driving the frontiers of physics in a major way during the past seven decades. This textbook, however, does not intend to give a historical review of these advances but to illustrate the physical principles behind these wonderful tools that we call accelerators. It introduces the inner working of its physical principles, often with intuitive pictures and simple mathematical models. Pure formalisms are avoided as much as possible. It is hoped that the readers would enjoy the fascinating physics behind this tool.

The style is informal at an introductory level. To serve as a textbook, references are listed only on the more established original literature instead of the constantly changing research frontiers. There are homework problems scattered in the text. Sometimes solutions are given. However, the purpose of these homework problems and solutions is to encourage thinking. A thoughtful answer is considered more worthy than a correct, standard answer.

I have warm thanks to express towards many individuals. I first of all thank the students. Their inquisitive questions helped sharpen up the contents. In addition, many of my colleagues helped me in occasions too numerous to list. Let me take this opportunity to thank Shyh-Yuan Lee of Indiana University, Phillip Morton, Karl Bane, Gennady Stupakov, Min-Huey Wang, Zhirong Huang and Juhao Wu of SLAC National Accelerator Laboratory, Guangrui Li, Tenghui Rui, Xiujie Deng of Tsinghua University, Rui Li of Jefferson Lab, Etienne Forest of KEK, and Helmut Wiedemann of Stanford University. Their generosity of helping and technical advice are much appreciated. A similarity of the title of these lectures to that of the classic *Lectures on Physics* is only accidental. The cover design is reproduction courtesy of the Superconducting Super Collider Central Design Group, 1986.

Alexander Chao
Stanford, February 2020

Prerequisites

- classical mechanics, including relativity
- classical electrodynamics
- mathematical physics

All prerequisites are at a senior undergraduate level. In addition, proficiency in algebra is a requirement.

References

These lectures are prepared hopefully in a self-contained manner. However, with its limited scope, several references are considered excellent accompanying and further readings.

- E.D. Courant and H.S. Snyder, Theory of the Alternating-Gradient Synchrotron, Ann. Phys. 3, 1 (1958).
- Matthew Sands, The Physics of Electron Storage Rings, An Introduction, SLAC-121 (1970).
- J.D. Jackson, Classical Electrodynamics, 3rd Ed., John Wiley (1999).
- S.Y. Lee, Accelerator Physics, 3rd Edition, World Scientific (2011).
- Helmut Wiedemann, Particle Accelerator physics, 4th Edition, Springer (2015).
- D.A. Edwards and M.J. Syphers, An Introduction to the Physics of High Energy Accelerators, John Wiley (1993).
- Handbook of Accelerator Physics and Engineering, Eds. A.W. Chao, K.H. Mess, M. Tigner and F. Zimmermann, 2nd Edition, World Scientific (2013).
- Kwang-Je Kim, Zhirong Huang, and Ryan Lindberg, Synchrotron Radiation and Free-Electron Lasers, Cambridge Univ. Press (2017).
- Alexander Wu Chao, Physics of Collective Beam Instabilities in High Energy Accelerators, John Wiley (1993).
- K.Y. Ng, Physics of Intensity Dependent Beam Instabilities, World Scientific (2006).

Contents

Chapter 1

Introduction

1.1 Basics

Accelerator physics is a branch of physics that studies charged particle beams. It is sometimes also called by a broader term of beam physics, but accelerator physics focuses more on the physics of beams specifically as encountered in accelerators.

In this section of Chapter 1, we shall first introduce a few basic skills needed in our lectures. In the remaining of Chapter 1, we introduce several topics of a more general nature. More specific topics are left for the later chapters to develop.

1.1.1 Maxwell equations

We start this section with Maxwell equations because of the fundamental role they play in accelerator physics. In the SI (Système International, MKSA) unit system, they read[1]

$$\begin{aligned} \nabla \cdot \vec{B} &= 0\,, \\ \nabla \cdot \vec{D} &= \rho\,, \\ \nabla \times \vec{H} - \frac{\partial \vec{D}}{\partial t} &= \vec{J}\,, \\ \nabla \times \vec{E} + \frac{\partial \vec{B}}{\partial t} &= 0\,, \end{aligned} \tag{1.1}$$

where we have adopted the conventional notations such as $\vec{E}, \vec{B}, \vec{D} = \epsilon\vec{E}, \vec{H} = \frac{\vec{B}}{\mu}$, $\rho, \vec{J}$ and c, and will not elaborate on their definitions here.

[1] Textbooks on electrodynamics; J.D. Jackson, Classical Electrodynamics, 3rd Edition, John Wiley (1999).

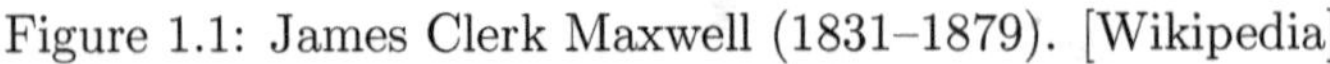
Figure 1.1: James Clerk Maxwell (1831–1879). [Wikipedia]

The first member of Eq. (1.1) states the nonexistence of magnetic monopoles. The second is Gauss law. The third is Ampere's law, and the fourth is Faraday's law. There will be short discussions of Gauss law after Eq. (1.31) and again in Sec. 8.1.1. The first and the fourth equations basically apply in vacuum; they are expressed in terms of $\vec{E}, \vec{B}$. The other two equations apply in material; they involve $\vec{D}, \vec{H}$.

Equation (1.1) is not complete unless we give also the boundary conditions for the fields. When boundaries are present, the steady state boundary conditions between two media are

$$\begin{aligned} (\vec{D}_2 - \vec{D}_1) \cdot \hat{n} &= \rho_{\text{surface}}\,, \\ (\vec{B}_2 - \vec{B}_1) \cdot \hat{n} &= 0\,, \\ (\vec{E}_2 - \vec{E}_1) \times \hat{n} &= \vec{0}\,, \\ (\vec{H}_2 - \vec{H}_1) \times \hat{n} &= -\vec{J}_{\text{surface}}\,, \end{aligned} \tag{1.2}$$

where ρ_{surface} is the surface charge density, $\vec{J}_{\text{surface}}$ is the surface current on the boundary, $\hat{n}$ is the unit vector normal to the boundary and points into medium 2.

Lorentz force Equations (1.1) and (1.2) are still not complete unless we tell what the fields do to the particle's motion. That is provided by the equation of motion with Lorentz force,[2]

$$\frac{d\vec{p}}{dt} = \vec{F} = q\vec{E} + q\vec{v} \times \vec{B}\,. \tag{1.3}$$

[2]Lorentz force is also written as $\vec{F} = q(\vec{E} + \vec{v} \times \vec{B})$ with q factored out, but that is slightly misleading. The electric force is proportional to the charge q. The magnetic force, on the other hand, does not react to the charge q, it only reacts to the current $q\vec{v}$. Factoring out q from the current $q\vec{v}$ is not wrong, but misleading.

where again we adopt the conventional notations of $\vec{v}, \vec{\beta} = \frac{\vec{v}}{c}, \gamma = \frac{1}{\sqrt{1-|\vec{\beta}|^2}}, \vec{p} = m\gamma\vec{v}$. To denote the charge of a particle, we use q, unless it is for a single electron or a single proton, then we use e. The particle sees $\vec{E}$ and $\vec{B}$ via the Lorentz force; it does not react to $\vec{D}$ or $\vec{H}$.

Given the source terms $\rho, \vec{J}$, we use Eqs. (1.1) and (1.2) to calculate the electric and magnetic fields. Given the fields, we calculate the Lorentz force delivered to the charged particles in the accelerator beam.

Application to accelerator physics Needless to say, applications of Maxwell equations to accelerators permeate throughout these lectures. We only make a few comments here of a more overview nature. Detailed discussions are left to the later chapters.

A short summary and overview of the Maxwell framework will be given in Fig. 5.10 when we discuss the electromagnetics of a radio-frequency cavity. A brief mention of the evolution from Maxwell equations to modern synchrotron radiation facilities is given at the beginning of Chapter 6.

An important notion of *prescribed motion* will be discussed in the chapter of synchrotron radiation, Sec. 6.1.2. At that time, we will point out the Maxwell equations have hidden an underlying geometric foundation.

In Chapter 8 of collective effects, we will stress the perturbative nature of our applications of the Maxwell equations in accelerator physics. Our perturbative applications typically stops at a first iteration. The only exception to this rule so far has been the case of free electron lasers, for which a second iteration is required.

Role in theoretical physics Maxwell equations obviously are extremely powerful and useful for accelerator physics. However, we may add a short mention of the pivotal role they played and may continue to play in the development and evolution of theoretical physics.

As mentioned, none of the four Maxwell equations really belonged to Maxwell. When first formulated in 1883, these equations were meant to be not much more than phenomenological. Maxwell summarized and extended four laws as known at the time in a particularly elegant manner. He probably did not realize how self-contained and how rich and deep his equations were destined to become. He realized the wave nature, but entrusted the aether rest frame to propagate his waves. He broke the Galileo invariance but did not explain how to remedy it.

Observe the following (and Homework 1.2):

1. It took 4 years after Maxwell equations to realize that they implied E&M waves, and that light is in fact a form of Maxwell's waves.
2. It took 29 years to realize that they implied synchrotron radiation, and then another 36 years to observe it in a synchrotron — and only accidentally!

3. It was the first of the unified theories to be realized, unifying electric and magnetic forces.
4. It was a deep puzzle that the Maxwell equations are not invariant under Galilean transformation, but is invariant under the then-mysterious Lorentz transformation.
5. Its Lorentz invariance laid the foundation leading to Einstein's special theory of relativity.
6. When quantum mechanics was developed, the attempt to quantize the Maxwell equations led to the quantum electrodynamics, the first golden triumph of quantum mechanics.
7. Maxwell equations possess a well-known gauge invariance in terms of the electromagnetic potentials. A generalization of this property led to the Yang-Mills theory, which in turn triggered the second of the unified theories, the electroweak theory unifying E&M and weak forces.
8. Gauge theories was applied to the strong force as well. Then a combination of the gauge electroweak and gauge strong theories led to the standard model of particle physics.
9. Maxwell equations have repeatedly served as the confident prototype leading to new revolutions in theoretical physics. Could there be more hidden treasures they still contain?

Homework 1.1 An electron going through the gap powered by a 1.5-V battery gains 1.5 eV energy. By stringing in series 10^9 batteries, therefore, the electron will gain 1.5 GeV. What's wrong with this picture?

Solution To shorten the total length, each battery needs to be short. But there is a limit on how short each battery can be. To avoid break-down, the voltage gradient is limited to about $G = 10^5$ V/m. This means each battery is 15 microns long and the whole accelerator is 15 km. The system must not see any intermediate grounding in its 1.5 km length. You may also estimate the electrostatic field energy that must be stored in the electric dipole system assuming free space.

Homework 1.2 We said in the text that Maxwell equations are one of the most fundamental, complete, self-contained theories in physics. This homework intends to illustrate their self-contained nature by a few minimal examples.

(a) Show that the framework of Maxwell equations (1.1) will not make sense unless the source terms $\vec{J}$ and ρ obeys

$$\nabla \cdot \vec{J} + \frac{\partial \rho}{\partial t} = 0 . \tag{1.4}$$

Indeed, this equation is called the continuity equation. This observation actually says the Maxwell equations have hidden in them the fact of charge conservation.

The equation of continuity (1.4) involves $\vec{D}$ and $\vec{H}$, and not the free space $\vec{E}$ and $\vec{B}$. Its validity holds rigorously inside any material regardless of the material properties. Charge conservation holds in any material according to Maxwell equations.

(b) Maxwell equations obey energy conservation. Show that, at least for the case in a medium with ϵ, μ constant in time and uniform in space,

$$\frac{\partial}{\partial t}\left(\frac{1}{2}\vec{E}\cdot\vec{D}+\frac{1}{2}\vec{B}\cdot\vec{H}\right)+\nabla\cdot(\vec{E}\times\vec{H}) = -\vec{E}\cdot\vec{J}.$$

This is basically a continuity equation for energy. The quantity $\vec{S}=\vec{E}\times\vec{H}$ is the energy flow Poynting vector. The quantity $\frac{1}{2}(\vec{E}\cdot\vec{D}+\vec{B}\cdot\vec{H})$ is the field energy density per unit volume.

(c) The text also mentioned other profound properties of the Maxwell equations. Hidden in them are the amazing Lorentz invariance and the gauge invariance. We shall not dwell in them here, referring them to other electrodynamics textbooks.

Homework 1.3 It has been suggested that one can consider an accelerator-based airplane (or ship) propeller. One might conceive an engine using a linear accelerator. A high intensity beam is accelerated to high energy and then expelled from the tail of the airplane. The airplane is propelled by the reciprocal force.

Perhaps we consider an electron induction linear accelerator. Let the beam current be $I = 100$ A and let the electrons be accelerated to $E = 100$ GeV. Calculate the propelling force delivered to the airplane.

Solution The propelling force is given by the reciprocal momentum received by the airplane per unit time. We obtain $F = \frac{IE}{ec} = 3.3\times10^4$ Newtons. If the airplane has a mass of 10^4 kg (including its induction accelerator), then it receives an acceleration of 3.3 m/s^2. The airplane becomes supersonic in less than 2 minutes!

One practical issue might be that, in order for the airplane not to accumulate residual charge, there needs to be two accelerators, one for positively-charged and one for negatively-charged particles. The positive side is that then the sound barrier will be broken in 1 minute.

Homework 1.4 It has been a dream that some kind of accelerator-based fusion reactor can be conceived. As one simplest example, consider an accelerator collider with a deuteron beam colliding with a tritium beam. How far is this from a conceivable fusion reactor? To maximize the fusion reaction, perhaps we choose a center-of-mass energy around 160 keV, where there is a peak resonance for $dt\to\alpha n$ production with cross-section $\sigma = 5\times10^{-28}$ m^2. Each event generates 17.6 MeV energy. We assume this energy can somehow be captured. Sketch a design of such a dream collider. The reason a tritium ion beam is used instead of a stationary tritium gas target is to avoid the presence of the tritium electrons — therefore, bremstrahlung.

Solution A possible set of parameters: Deuteron energy = 96 keV, tritium energy = 64 keV. The other parameters (for both beams) could be: volume density = 2.2×10^{26} m^{-3}, bunch length = 10 m, transverse radius = 0.12 μm, number of particle per bunch = 10^{14}. Then the collision of these two bunches generates about 150 J. The two original bunches have kinetic energy of 2.5 J. If the collision rate is 10 MHz, then the fusion power is 1.5 GW. However, the beam-beam interaction will be very strong. The dream has still some way to go.

1.1.2 Dimensionality

All physical quantities have dimensions. Dimensionality is intrinsically connected to all physical quantities. Pure numbers, say 5, are abstract objects without dimensionality, but they alone do not have physical meaning.

In physics, a number, e.g. 5, must be followed by its dimensionality to have meaning — for example, 5 apples. Now this quantity has the dimensionality of "apple". Indeed, you may add 5 apples with another 5 apples to obtain a total of 10 apples.

Suppose what you have are 5 apples and 5 oranges. This time you may not add them together because they have different dimensionalities. On the other hand, if you call them both "fruits", then they do have the same dimensionality now and you have a total of 10 fruits.

But if you have 5 green apples and 5 red apples, you are no longer allowed to add them again because in this case the dimensionalities become "green apples" and "red apples". Dimensionalities are not absolute. They depend on the context.

Sometimes in accelerator physics we have a quantity N, the number of particles. It is represented by a pure number, but again, they have in fact intrinsically a dimensionality of "electrons", or "protons", and they cannot be added together, and any expression such as $N_{\text{electron}} + N_{\text{proton}}$ is meaningless and will never occur in any physical analysis.

We will encounter countless number of expressions and equations in our lectures. All physical quantities must be defined of their dimensionalities. An equation can only be meaningful when its dimensionalities on both sides are balanced exactly.

Let us consider another example. An angle θ is measured in radians. It has no explicit units. However, being unitless does not mean it has no dimensionality. Any quantity with physical meaning will have its dimensionality, even if it is not assigned a unit. An angle has intrinsically the dimensionality of "angle" even though it may be measured in radians which is mislabeled to be "dimensionless". A wavelength λ and its corresponding wave number k are related by $k = \frac{2\pi}{\lambda}$. The dimensionality of k is inverse length. Dimensionality of λ is length times radian. In the same token, the quantity $\lambda\!\!\!^{-} = \frac{\lambda}{2\pi}$ has the dimensionality of length after λ is divided by 2π radians.

Similarly, we will learn a quantity called the tune ν, which is a ratio of two frequencies, and therefore again unitless. But it has its own dimensionality of "tune". You will never find an expression like $(\theta + \nu)$. Although they both are unitless, they have different dimensionalities. There is no physical process that would mix them to yield $(\theta + \nu)$. By the same token, you will not easily see expressions like $\pi + 1$ — you cannot add π angle to one apple.[3]

Angular frequency is often given in units of Hertz. Its dimensionality of "cycles per unit time", which is not to be considered the same as "s^{-1}" — the dimensionality of cycles is missing. Although they appear to use the same unit, they have different physical meanings, and therefore different dimensionalities.

Lorentz force and relativity An alert question might arise concerning the Lorentz force equation (1.3). Here we have two terms contributing to the force, one electric and one magnetic. Since electric and magnetic phenomena are supposedly two separate pieces of physics, the two terms have intrinsically different dimensionalities. So how could they be added together?

Hidden behind the expression (1.3), however, is Maxwell equations that demonstrate $\vec{E}$ and $\vec{B}$ are intrinsically linked. With Maxwell equations, they in fact belong to the same piece of physics. Furthermore, special theory of relativity tells us that they Lorentz transform into each other, and the transformation depends on the velocity of the observer (the particle is the observer). The appearance of the velocity $\vec{v}$ in (1.3) reflects this velocity dependence.

It is instructive to appreciate the fact that the mere appearance of the expression (1.3) has hidden in it the necessities of Maxwell equations (linking $\vec{E}$ and $\vec{B}$) and the theory of relativity (linking to $\vec{v}$) and they all appear in such a self-contained and self-consistent manner. See Homework 1.5.

Hill's equation Let us give one more illustration. Later we will encounter a Hill's equation which reads

$$\frac{d^2}{ds^2}x(s) = -K(s)x(s)\,, \tag{1.5}$$

where x is the horizontal displacement of a particle relative to the design orbit in an accelerator, $K(s)$ is a focusing function, s is a longitudinal distance along the design orbit.

The quantities x and s both represent distances. In this context, they both have the dimensionality of length, and one may even choose the same unit of meters for both of them. In that case, $K(s)$ has the dimensionality of $1/(\text{length})^2$ and its unit is m^{-2}.

[3]Admittedly, something like $\pi + 1$ can be concocted to occur, e.g. by asking for the total length of a storage ring plus a linac, but that is hardly a physical process.

However, x and s are really two different physical quantities, x referring to a horizontal displacement, s referring to a longitudinal coordinate distance. In this context, their respective dimensionalities are "horizontal length" and "longitudinal length", and they may even have different units, say, mm for x and inches for s. The two sides of Eq. (1.5) then would have a balance of their dimensionalities,

$$\begin{aligned}
[x] &= \text{horizontal length}\,, \\
[s] &= \text{longitudinal length}\,, \\
[K(s)] &= (\text{longitudinal length})^{-2}\,, \\
[\text{both sides of Eq. (1.5)}] &= \frac{\text{horizontal length}}{(\text{longitudinal length})^2}\,.
\end{aligned}$$

In terms of units, they are also balanced, as both sides have the units of

$$\frac{\text{mm}}{\text{in}^2}\,.$$

If we simplify the context and consider both x and s simply as lengths, then they are still balanced with both sides having the dimensionality of $(\text{length})^{-1}$ and unit of $(\text{meter})^{-1}$.

Homework 1.5 The text mentioned Lorentz transformation on the Lorentz force equation. Let us look into it in this homework by finding the Lorentz force seen by the particle in its rest frame, where the electric dimensionality and the magnetic dimensionality merge into one and the magnetic dimensionality disappears.

Solution The quantities $\vec{E}, \vec{B}, \vec{v}$ refer to the fields and the particle velocity in the laboratory frame. In the rest frame of the particle, i.e. in a reference frame that moves with velocity $\vec{v}$ relative to the laboratory frame, the fields are given by

$$\begin{aligned}
\vec{E}_0 &= \gamma(\vec{E} + \vec{v} \times \vec{B}) - \frac{\gamma^2}{\gamma+1}(\vec{v}\cdot\vec{E})\frac{\vec{v}}{c^2}\,, \\
\vec{B}_0 &= \gamma\left(\vec{B} - \frac{\vec{v}}{c^2} \times \vec{E}\right) - \frac{\gamma^2}{\gamma+1}(\vec{v}\cdot\vec{B})\frac{\vec{v}}{c^2}\,.
\end{aligned}$$

A particle in its rest frame sees no force from the magnetic field. It sees a force[4]

$$\vec{F}_0 = q\vec{E}_0\,.$$

Let us now make sure this solution of $\vec{F}_0$ is also what we get by making a Lorentz transformation on the force $\vec{F}$. The force transforms from the laboratory

[4]The rest mass, by its definition, does not change by Lorentz transformation. Note the less obvious fact that the charge q also is Lorentz invariant.

frame to the rest frame according to

$$\vec{F_0} = \gamma\vec{F} - \frac{\gamma^2}{\gamma+1}(\vec{v}\cdot\vec{F})\frac{\vec{v}}{c^2}\,.$$

Substituting $\vec{F} = q\vec{E} + q\vec{v}\times\vec{B}$, we indeed confirm $\vec{F_0} = q\vec{E}_0$.

Incidentally, the above mathematical expressions might look confusing. A visual clarification might be obtained by considering the case when $\vec{v}$ is along the $\hat{x}$ direction. We then have

$$F_x = F_{0x}, \qquad F_y = \frac{F_{0y}}{\gamma}, \qquad F_z = \frac{F_{0z}}{\gamma}\,.$$

1.1.3 Units

We mentioned that dimensionality is a context-dependent concept. Depending on the context, five green apples can have the dimensionality of fruit, or apple, or green apple. Units are then determined after the context of dimensionality has been chosen. In principle, one can have an infinite degrees of freedom in choosing the basic units. Also as emphasized, all dimensionalities, and therefore the units, must balance in all expressions and all equations.

However, it is not necessary and indeed impossible to keep all the dimensionalities or units. We often ignore most of the dimensionalities and the corresponding units. We choose and keep, by convention, only selectively a few of the more "fundamental" dimensionalities and the corresponding units, and simply ignore all the other dimensionalities assuming they are balanced. This is of course a drastic simplification. The customarily chosen dimensions to keep are mass, length, and time. All other dimensions (including our unfortunate "green apple") are dropped and those quantities are represented as pure numbers. Dimensionality balancing therefore is applied only to M, L, and T. Sometimes, we also keep temperature as another dimensionality K. When doing so, temperature unit must of course be balanced as well.

With this convention, all physical quantities then have reduced dimensionality represented in M, L, T. Distance has dimension of L. Velocity has LT^{-1}, etc. Table 1.1 shows the dimensionalities and the associated units in the MKSA system of some often used physical quantities.

In Table 1.1, there are various other units introduced. For example, pressure has the unit of pascal. In this convention, one pascal is just one newton/meter^2, or one kilogram/(meter·second^2). Similarly, one coulomb is the same as one $\sqrt{\text{kilogram}\cdot\text{meter}}$. It should be said, however, these are results of a shorthand convention based on the reduced dimensionalities. A discussion of pushing for even more shorthand abbreviations is mentioned in Sec. 1.1.4

In Table 1.2, we list the values of a few fundamental constants in the MKSA unit system.

Table 1.1: The MKSA units and dimensionalities of some physical quantities.

Quantity	MKSA unit	Dimensionality
Length ℓ	meter	L
Area A	m^2	L^2
Volume V	m^3	L^3
Mass M	kilogram	M
Density ρ	$\mathrm{kg/m}^3$	ML^{-3}
Time t	second	T
Speed v	m/s	LT^{-1}
Force F	newton	MLT^{-2}
Pressure P	pascal	$\mathrm{ML}^{-1}\mathrm{T}^{-2}$
Energy E	joule	$\mathrm{ML}^2\mathrm{T}^{-2}$
Momentum P		MLT^{-1}
Power P	watt	$\mathrm{ML}^2\mathrm{T}^{-3}$
Charge q	coulomb	$\mathrm{M}^{1/2}\mathrm{L}^{1/2}$
Current I	ampere	$\mathrm{M}^{1/2}\mathrm{L}^{1/2}\mathrm{T}^{-1}$
Potential V	volt	$\mathrm{M}^{1/2}\mathrm{L}^{3/2}\mathrm{T}^{-2}$
Elec. Field E	V/m	$\mathrm{M}^{1/2}\mathrm{L}^{1/2}\mathrm{T}^{-2}$
Mag. Field H	A-turn/m	$\mathrm{M}^{1/2}\mathrm{L}^{-1/2}\mathrm{T}^{-1}$
Displacement D	$\mathrm{C/m}^2$	$\mathrm{M}^{1/2}\mathrm{L}^{-3/2}$
Mag. flux Φ	weber	$\mathrm{M}^{1/2}\mathrm{L}^{3/2}\mathrm{T}^{-1}$
Flux density B	Tesla	$\mathrm{M}^{1/2}\mathrm{L}^{-1/2}\mathrm{T}^{-1}$
Conductivity σ_c	1/(ohm·m)	$\mathrm{L}^{-2}\mathrm{T}$
Resistivity ρ_r	ohm-meter	$\mathrm{L}^2\mathrm{T}^{-1}$
Resistance R	ohm	LT^{-1}
Capacitance C	farad	$\mathrm{L}^{-1}\mathrm{T}^2$
Inductance L	henry	L
Temperature T	kelvin	K

Numerical calculation The use of units in physics calculations is one of the most confusing things for the beginners. A trial-and-error technique, sometimes adopted by the students, should not be the way to do calculations. All numerical calculations must be done carrying units, and the units must all balance. Below we suggest one fool-proof way to do this.

For an illustration, let us take example to calculate the magnetic field B of a wire carrying a current I at a distance r from the wire,

$$B = \frac{\mu_0 I}{2\pi r}.$$

In this equation, the dimensionalities of the various quantities are, from Table 1.1,

$$[B] = \mathrm{M}^{1/2}\mathrm{L}^{-1/2}\mathrm{T}^{-1},$$

Table 1.2: Numerical values of some fundamental constants in the MKSA unit system. Quantities underlined are exact values without error bars.

Quantity	Value
speed of light c	$\underline{2.99792458}\ \times 10^{8}$ m s^{-1}
permeability of vacuum μ_0	$\underline{4\pi} \times 10^{-7}$ Henry m^{-1}
permittivity of vacuum $\epsilon_0 = \frac{1}{\mu_0 c^2}$	$8.854187817\ \times 10^{-12}$ Farad m^{-1}
electronic charge e	$\underline{1.602176634}\ \times 10^{-19}$ C
Planck constant h	$\underline{6.62607015}\ \times 10^{-34}$ J s
Boltzmann constant k_B	$\underline{1.380649}\ \times 10^{-23}$ J K^{-1}
gravitational constant G	$6.67428\ \times 10^{-11}$ Newton m^{2}kg^{-2}
Avogadro number N_A	$\underline{6.02214076}\ \times 10^{23}$
electron mass m_e	$9.1093822\ \times 10^{-31}$ kg
proton mass m_p	$1.6726216\ \times 10^{-27}$ kg
rest mass energy of electron $m_e c^2$	0.51099891 MeV
rest mass energy of proton $m_p c^2$	938.27201 MeV
fine structure constant $\alpha_F = \frac{e^2}{4\pi\epsilon_0 \hbar c}$	1/137.0359997
impedance of vacuum $Z_0 = \sqrt{\frac{\mu_0}{\epsilon_0}} = \mu_0 c = \frac{1}{\epsilon_0 c}$	376.7303 Ω
classical radius of electron $r_e = \frac{e^2}{4\pi\epsilon_0 m_e c^2}$	$2.8179403\ \times 10^{-15}$ m
classical radius of proton $r_p = \frac{e^2}{4\pi\epsilon_0 m_p c^2}$	$1.534698\ \times 10^{-18}$ m
electron Compton wavelength $\lambda_e = \frac{h}{m_e c}$	$2.4263106\ \times 10^{-12}$ m

$$
\begin{aligned}
[\mu_0] &= \text{Henry/L} &= L/L &= 1\,, \\
[I] &= \text{Ampere} &= \mathrm{M}^{1/2}\mathrm{L}^{1/2}\mathrm{T}^{-1}\,, \\
[r] &= L\,.
\end{aligned}
$$

Multiplying out the dimensionalities, the right-hand side yields $\mathrm{M}^{1/2}\mathrm{L}^{-1/2}\mathrm{T}^{-1}$, which balances with the left hand side as it should.

Once (and only when) the dimensionalities balance, we are ready to do numerical calculations. At this point, we must express all quantities in terms of the standard unit of each of the quantities involved according to Table 1.1. Once done, we then can confidently plug in numerical values of the quantities and trust the final result to be correct. No trials and errors. The final result is in the standard MKSA unit as well. For example, the magnetic field at 0.01 m away from a wire carrying 1 Ampere will be 2×10^{-5} Tesla, about half the magnetic field on earth surface.

Nature's choice of units One may ask the question what might Mother Nature choose as her fundamental dimensionalities and her fundamental units? The hint given to us is that Mother Nature seems to have chosen the three most fundamental constants to be the speed of light in vacuum c, the Planck constant

h, and gravitational constant G. She seems to have chosen speed, action, and gravity to be her focus. If so, it says that the fundamental dimensionalities we have chosen — mass, length, and time — are not what she chooses.

We then ask the next question. Suppose we insist to choose mass, length, and time to be the fundamental dimensionalities, what should we choose to be their units? To address that question, we need to use c, h, G to construct the three fundamental constants M_0, L_0, T_0, which are of the dimension of mass, length and time, respectively, for the use by humans (physicists). The result is

$$\begin{aligned} M_0 &= \sqrt{\frac{ch}{G}} = 5.45552 \times 10^{-8}\ \mathrm{kg}\,, \\ L_0 &= \sqrt{\frac{hG}{c^3}} = 4.05134 \times 10^{-35}\ \mathrm{m}\,, \\ T_0 &= \sqrt{\frac{hG}{c^5}} = 1.35138 \times 10^{-43}\ \mathrm{s}\,. \end{aligned}$$

Incidentally, we use h instead of $\hbar$ in these fundamental units because $\hbar = \frac{h}{2\pi}$ contains an additional dimensionality of rad^{-1} which messes up the dimensionality balance.

We note that our fundamental units of kg, m, and sec are many orders of magnitude away from these fundamental units. Perhaps this is a reflection that we are clueless in choosing our basic dimensionalities, and that even for these poor choice of basic dimensionalities, we are way off in choosing the appropriate magnitudes of the units for them.

Homework 1.6 Calculate the force between two stationary electrons separated by 1 Angström ($= 10^{-10}$ m).

Solution The force is

$$F = \frac{e^2}{4\pi\epsilon_0 r^2}\,.$$

Remember no trial-and-errors, check the balance of dimensionalities first. The answer is 2.3×10^{-8} Newton.

Homework 1.7 Electrons are considered point charges. However, due to historical reasons traced back to the 19th century before the advent of quantum mechanics, a quantity called "classical radius" of an electron was introduced. Today, it has lost all its original physical meanings, but the name stayed. Calculate the numerical value of the classical radii r_e and r_p listed in Table 1.2 using the values given in the same table of the other input quantities.

Homework 1.8

(a) In the SI system, often one sees the combination $\frac{1}{4\pi\epsilon_0}$. Show that this quantity can be cast in unit of $\Omega\,\mathrm{m\,s^{-1}}$,

$$\frac{1}{4\pi\epsilon_0} = 8.9875518 \times 10^9\ \Omega\,\mathrm{m\,s^{-1}}\,.$$

(b) Another often used quantity has a designated notation Z_0. It is called the impedance of the vacuum (see Table 1.2).[5] It is given by $Z_0 = \sqrt{\frac{\mu_0}{\epsilon_0}}$. Show that

$$Z_0 = \sqrt{\frac{\mu_0}{\epsilon_0}} = 376.730313\,\Omega\,.$$

Homework 1.9 Continuing the discussion on Nature's choice of units in the text, assuming three fundamental constants c, h, G, find a fundamental unit for the force F_0.

Solution

$$F_0 = \frac{c^4}{G} = 1.21026 \times 10^{44} \text{ Newton}\,.$$

This huge force does not involve h; it involves gravity but cannot have a quantum origin.

If we define a fundamental unit for energy, then we would invoke quantum effects, with $E_0 = F_0 L_0 = 4.90317 \times 10^9$ J.

Homework 1.10 Similar to the text when we chose c, h, G as the fundamental constants, but this time replace the gravitational constant G by the electric charge e. Since e can have both signs, the fundamental constant has better be e^2. Can we construct the three fundamental constants M_0, L_0, T_0, in terms of c, h, e^2?

Solution There is no solution because the three constants c, h, e^2 are redundant. Nature probably will not choose them as her most fundamental constants. In particular, $\frac{e^2 c}{h}$ is dimensionless in the MKSA unit system,

$$\frac{e^2 c}{h} = 11614.1\,.$$

In fact, it is just the fine structure constant α_F divided by $2\pi \times 10^{-7}$.

Homework 1.11 The volume density of field energy in electric and magnetic fields in free space are respectively

$$u_E = \frac{\epsilon_0 E^2}{2}, \qquad u_B = \frac{B^2}{2\mu_0}\,.$$

(a) Calculate the field energy density in a field $E = 1$ MV/m and in a field $B = 1$ Tesla. Which has a higher energy density?

(b) If we want the fields' volume densities to be equal, what would the ratio E/B be? What is the electric field when it has the same field energy density as a 1-Tesla magnetic field?

[5] Yes, the vacuum has an impedance, and it is a very large one; we will discuss it in Chapter 8.

Solution

(a) $u_E = 4.4$ J/m^3, $u_B = \frac{1}{8\pi} \times 10^7$ J/m^3.

(b) $\frac{\epsilon_0 E^2}{2} = \frac{B^2}{2\mu_0}$ means $\frac{E}{B} = \frac{1}{\sqrt{\epsilon_0\mu_0}} = c$.

The ratio of electric field to its equal-energy-density magnetic field is simply c. A 1-Tesla magnetic field, for example, is equivalent to a large 300 MV/m of electric field. Magnetic fields are much heavier than electric fields. Keep this comment in mind; we will return to it at a later point.

1.1.4 Shorthand expressions

Since, as mentioned, all expressions and equations must be dimensionally balanced, one can take advantage of it to simplify expressions and equations by shorthand abbreviations.

Let us take the equation $E = mc^2$ for example. The LHS has energy, with reduced dimensionality of $\mathrm{ML^2T^{-2}}$. The RHS contains a factor of mass m with dimensionality of M and factor of c^2 with dimensionality of $\mathrm{L^2T^{-2}}$. Dimensionalities are balanced as required. Without a risk of losing the validity of the equation, we can now omitted the factor c^2 on the RHS by setting the convention $c = 1$. The equation then reads $E = m$. As absurd as it might seem, there will actually be no loss of information. To recover the original equation before setting $c = 1$, we let $E = mc^\alpha$, then by balancing the dimensionalities, we immediately know that $\alpha = 2$.

As discussed earlier, we have by convention chosen only three fundamental dimensionalities, M, L, and T. In doing so, we have artificially reduced all the other dimensionalities to pure numbers — five electrons and five protons are both represented by a pure number five. So in a sense we have already been doing a great deal of shorthand abbreviations all along. In principle, we can do more if we so choose.

Since there are three remaining fundamental dimensions, we can in principle choose up to three additional abbreviations for shorthand. For example, we may take the convention that $c = 1$ and $\hbar = 1$ without the risk of losing information. The fine structure constant, a dimensionless quantity,

$$\alpha_F = \frac{e^2}{4\pi\epsilon_0\hbar c}$$

then reads in a shorthand $\alpha_F = \frac{e^2}{4\pi\epsilon_0}$. Again, by a dimensional analysis, if necessary, the original formula can be easily recovered.

The origin of this flexibility is related to the fact that one can choose the measuring units of M, L, T arbitrarily. They do not have to be meter, kilogram and seconds. As long as we fix our choices consistently, all physical quantities can be expressed properly. This degree of freedom in choosing the measuring units is reflected in the freedom of setting $c = 1$ and/or $\hbar = 1$. Throughout our discussions, however, we shall not apply any of these abbreviations.

Homework 1.12 In the text, we gave an example of fine structure constant $\alpha_F = \frac{e^2}{4\pi\epsilon_0}$ with the shorthand convention of $c = 1$ and $\hbar = 1$. Recover the full expression by dimensional analysis.

Solution Let the full expression be

$$\alpha_F = \frac{e^2}{4\pi\epsilon_0} c^\alpha \hbar^\beta .$$

Use dimensionality analysis to show that $\alpha = \beta = -1$.

Homework 1.13 A star becomes a blackhole when light cannot escape from its surface. Let M be the star's mass. Then it becomes a blackhole when its radius is smaller than the Schwarzschild radius

$$R_s = \frac{2GM}{c^2} ,$$

where G be the gravitational constant. When conventions $G = 1$ and $c = 1$ are chosen, we have a shorthand formula $R_S = 2M$. By a dimensionality analysis, recover the full expression from the shorthand expression.

1.1.5 Dimensional analysis

The previous application of shorthand abbreviations has the only benefit to make expressions look shorter. No actual real contents are generated. In this section, we discuss a much deeper application of dimensional analysis, yielding physical consequences.

Consider a physical system for which we know from basic physical grounds that there are three basic quantities that will play a role, and we are now looking for an expression of a fourth quantity in terms of the three basic quantities. Dimensional analysis would allow us to guess an expression of the fourth quantity.

It should be emphasized that, as surprising as it may be, this application yields physical results without any knowledge of the underlying dynamics or physical laws connecting these quantities.

Pendulum Take the example of a pendulum with mass m and string length ℓ, in a gravitational acceleration constant g ($g = 9.8$ ms^{-2} assuming the pendulum is on earth). The three basic quantities are m, ℓ, g. By a dimensional analysis, the oscillation period of the pendulum T can be found without any knowledge or Newton's law.

To do this explicitly, let

$$T = m^\alpha \ell^\beta g^\gamma .$$

By a dimensional analysis, it is easy to show that $\alpha = 0, \beta = \frac{1}{2}, \gamma = -\frac{1}{2}$. Therefore,

$$T = \sqrt{\frac{\ell}{g}} .$$

In particular, the conclusion that the pendulum period is independent of the pendulum mass would have been obtained immediately.

The only thing missing will be a numerical factor, which most likely will be of the order of unity, and in this pendulum case, we have lost a factor of 2π. Had we kept angle as a fundamental dimension, in fact, even the factor π can be obtained.

This application of dimensional analysis is much more powerful than the application just for a shorthand abbreviation mentioned in Sec. 1.1.4. In cases when the system is simple but the dynamics is involved or even unknown, it can be a powerful way to get to the answer.

Event rate We have considered the dimensional analysis with only three fundamental dimensionalities M, L, T above. If we had included more dimensionalities, we could analyze the result when more than three basic quantities are involved.

Consider the case of calculating the event rate of producing Higgs particles by colliding electrons and positrons. Let the electron beam have N_e electrons per bunch and the positron beam have N_p positrons per bunch, and let the bunches collide at a frequency f. Let the two beams have a cross-sectional transverse area A at the collision point.

The Higgs event rate will depend on N_e, N_p, f, A. It will also depend on another quantity called the collision cross-section Σ. The physical meaning of Σ is as follows. In order to produce a Higgs, an individual electron and an individual positron must pass by each other within a short distance b. If one electron and one positron collide with impact parameter less than b, then one Higgs is produced. The cross-section Σ is then defined by πb^2. In the MKSA system, it has the dimension of L^2, but it actually has the dimensionality

$$[\Sigma] = \mathrm{L}^2 \cdot \mathrm{Higgs} \cdot \mathrm{electron}^{-1} \cdot \mathrm{positron}^{-1}\,.$$

The other quantities have dimensionalities

$$[N_e] = \mathrm{electron}, \quad [N_p] = \mathrm{positron}, \quad [f] = \mathrm{T}^{-1}, \quad [A] = \mathrm{L}^2\,.$$

The quantity of interest is event rate of Higgs particles. It has the dimensionality

$$[\dot{H}] \;=\; \mathrm{Higgs} \cdot \mathrm{T}^{-1}\,.$$

The only way a dimensionality is balanced is

$$\dot{H} \;=\; \frac{N_e N_p f \Sigma}{A}\,.$$

Again, the only thing that could be missing from this equation is a numerical factor of the order of unity.

Homework 1.14 Consider two stars with equal mass M_0 circulating around each other at a distance R apart under their mutual gravitational attraction. Without invoking the Newton's gravitational law, find an expression of the period of revolution, T_0, of these two stars around each other.

Solution $T_0 = \sqrt{\frac{R^3}{GM_0}}$. We are very close to deriving Kepler's third law here without studying Kepler's volumes of planetary data.

1.1.6 Definition of second, meter, kilogram

This degree of freedom of choosing units also has been converted to define the units we use. The units of second, meter, and kilogram are very cleverly now *defined* through three fundamental constants by an international effort. As a result, when measurements of these fundamental constants become more accurate in the future, the meaning of these units becomes more accurate automatically through their definitions. In principle, we could choose any three fundamental constants for this purpose, but international convention has chosen them to be the speed of light c, the Planck constant h, and the atomic property of cesium atom,

- The second is the duration of exactly 9192631770 periods of the radiation corresponding to the transition between the two hyperfine levels of the ground state of the Cs^{133} atom.
- Speed of light is defined to be exactly 299792458 m/s.
- The Planck constant is defined to be exactly $6.62607015 \times 10^{-34}$ Joule-second.

Since these are definitions, the numbers listed here have no error bars.

This approach of defining the units has been taken and approved by the international community. For reference, Table 1.3 gives the 2019 revised International System of Units (SI) definitions. Both the pre- and post-2019 definitions are given.

1.1.7 Factorization

In later chapters, we will address the subject of circular accelerators in great length. As a simplest circular accelerator, one considers a charged particle circulating in a uniform magnetic field. Let the particle have charge q, momentum P_0, and let the magnetic field be B_0. Using the Lorentz force (1.3) to balance the centrifugal force, we find that the radius ρ of the circular trajectory is determined by

$$B_0\rho = \frac{P_0}{q}, \quad \text{or} \quad B_0\rho \,[\text{T-m}] = 3.3356\, P_0 \,[\text{GeV}/c]. \tag{1.6}$$

Table 1.3: The evolution of the International System of Units (SI). The base units became derived from fixed values of the fundamental constants seen in the energy equivalence relations $E = mc^2 = hf = eV = kT$ and other invariants of nature as of May 20, 2019. [Courtesy Leon Chao, 2019]

Base Unit	Definition (Pre-2019)	Fund. Const.	Definition (Post-2019)
s	The second is the duration of 9 192 631 770 periods of the radiation corresponding to the transition between the two hyperfine levels of the ground state of the cesium-133 atom.	$\Delta\nu_{Cs}$	The second is the SI unit of time. It is defined by taking the fixed numerical value of the cesium frequency $\Delta\nu_{Cs}$, the unperturbed ground-state hyperfine transition frequency of the cesium-133 atom, to be 9 192 631 770 when expressed in the unit Hz.
m	The meter is the length of the path travelled by light in vacuum during a time interval of 1/299 792 458 of a second.	c	The meter is the SI unit of length. It is defined by taking the fixed numerical value of the speed of light in vacuum c to be 299 792 458 when expressed in the unit $\mathrm{m\,s^{-1}}$, where the second is defined in terms of $\Delta\nu_{Cs}$.
kg	The kilogram is the unit of mass; it is equal to the mass of the international prototype of the kilogram.	h	The kilogram is the SI unit of mass. It is defined by taking the fixed numerical value of the Planck constant h to be $6.626\,070\,15 \times 10^{-34}$ when expressed in the unit J s, which is equal to kg $\mathrm{m^2\ s^{-1}}$, where the meter and the second are defined in terms of c and $\Delta\nu_{Cs}$.
A	The ampere is that constant current which, if maintained in two straight parallel conductors of infinite length, of negligible circular cross-section, and placed 1 m apart in vacuum, would produce between these conductors a force equal to 2×10^{-7} newton per meter of length.	e	The ampere is the SI unit of electric current. It is defined by taking the fixed numerical value of the elementary charge e to be $1.602\,176\,634 \times 10^{-19}$ when expressed in the unit C, which is equal to A s, where the second is defined in terms of $\Delta\nu_{Cs}$.

K	The kelvin, unit of thermodynamic temperature, is 1/273.16 of the thermodynamic temperature of the triple point of water.	k	The kelvin is the SI unit of thermodynamic temperature. It is defined by taking the fixed numerical value of the Boltzmann constant k to be $1.380\,649 \times 10^{-23}$ when expressed in the unit $\mathrm{J\,K^{-1}}$, which is equal to $\mathrm{kg\,m^2\,s^{-2}\,K^{-1}}$, where the kilogram, meter and second are defined in terms of h, c and $\Delta\nu_{Cs}$.
mol	The mole is the amount of substance of a system that contains as many elementary entities as there are atoms in 0.012 kilogram of carbon-12. When the mole is used, the elementary entities must be specified and may be atoms, molecules, ions, electrons, other particles, or specified groups of such particles.	N_A	The mole is the SI unit of amount of substance. One mole contains exactly $6.022\,140\,76 \times 10^{23}$ elementary entities. This number is the fixed numerical value of the Avogadro constant, N_A, when expressed in the unit $\mathrm{mol^{-1}}$ and is called the Avogadro number. The amount of substance of a system is a measure of the number of specified elementary entities. An elementary entity may be an atom, a molecule, an ion, an electron, or any other particle or specified group of particles.
cd	The candela is the luminous intensity, in a given direction, of a source that emits monochromatic radiation of frequency 540×10^{12} Hz and that has a radiant intensity in that direction of 1/683 watt per steradian.	K_{cd}	The candela is the SI unit of luminous intensity in a given direction. It is defined by taking the fixed numerical value of the luminous efficacy of monochromatic radiation of frequency 540×10^{12} Hz, K_{cd}, to be 683 when expressed in the unit $\mathrm{lm\,W^{-1}}$, which is equal to $\mathrm{cd\,sr\,W^{-1}}$, or $\mathrm{cd\,sr\,kg^{-1}\,m^{-2}\,s^3}$, where the kilogram, meter and second are defined in terms of h, c and $\Delta\nu_{Cs}$.

We will use Eq. (1.6) to illustrate the concept of factorization.

As will be seen repeatedly later, it is a common practice in accelerator physics analysis to define quantities that depends either purely on properties of the accelerator hardware or purely on properties of the charged particle beam. So quantities we define are either accelerator quantities or beam quantities. An accelerator quantity should not change when beam properties are changed. A beam quantity ideally should not change when the accelerator hardware is changed. We try to avoid defining quantities that have mixed dependences.

Accelerator designer's job is to find the necessary connections between these two sets of quantities. The connection is then expressed by an equation like Eq. (1.6). Magnetic rigidity, $B_0\rho$, is an accelerator quantity, while q and P_0 are beam quantities. Equation (1.6) connects them. We shall refer to this phenomenon of connecting purely-accelerator and purely-beam quantities as "factorization". Equation (1.6) is factorized.

An accelerator designer determines the accelerator parameters to meet the required beam parameters. Accelerator users determines the achievable beam parameters given the accelerator parameters. Both will use factorized equations such as (1.6).

In passing, we have said that any definition of a new physical quantity should try to obey this factorization rule. A quantity defined with a mixed nature is most likely not going to be very useful. We will return to this subject of factorization a few times in later lectures, but let us mention a few prime examples up front.

Courant-Snyder formalism is factorized We will discuss in Sec. 2.4 a Courant-Snyder formalism in some depth. The reason this formalism is so useful is because it factorizes the particle dynamics by the form

$$X_{\text{out}} = MX_{\text{in}}, \qquad X = \begin{bmatrix} x \\ x' \end{bmatrix}, \tag{1.7}$$

where X is a pure beam quantity and the transfer matrix M is a pure accelerator quantity. When we design the accelerator, we only work on M. Conversely, after we designed the accelerator and fixed M, the same M applies to all particles. Equation (1.7) is factorized. This is very practical, and therein is hidden the usefulness of the Courant-Snyder formalism.

Lie algebra formalism is factorized We will not discuss an important technique called Lie algebra in these lectures.[6] However, it is worth pointing out that the reason Lie algebra is so useful for accelerator physics is the fact that the particle dynamics has the form

$$X_{\text{out}} = e^{-:H:}X_{\text{in}},$$

where again, X is a beam quantity while the Lie map $e^{-:H:}$ is a purely accelerator quantity.

[6] Interested readers may refer to A. Dragt, AIP Conf. Proc. 87 (1982); A. Dragt et al., Ann. Rev. Nucl. Part. Sci. 38, 455 (1988); A. Dragt, Lie Methods for Nonlinear Dynamics with Applications to Accelerator Physics (2010), www.physics.umd.edu/dsat/; E. Forest, Beam Dynamics: A New Attitude and Framework, Harwood Academic (1998).

Impedance is a factorized quantity Still another example is found in the case of collective instabilities. We will discuss in Sec. 8.2 a quantity called the impedance. The reason impedance is so useful is again because it is defined purely by the accelerator with no reference to the beam. The perturbation delivered to a particle is given by the expression

$$V = IZ,$$

where the impedance Z is an accelerator quantity and I is a beam quantity.

That these concepts turn out to be so theoretically elegant and practically useful is attributed to the fact that they factorize.

Homework 1.15

(a) Derive Eq. (1.6), including the numerical value of 3.3356 (five significant figures).

(b) Show that the *cyclotron frequency* is given by

$$\omega_0 = \frac{qB_0}{\gamma m}. \tag{1.8}$$

Solution

$$\frac{1\times 10^9}{299792458} \approx 3.335641.$$

Homework 1.16 We discussed in some depth of the concept of dimensionalities in Sec. 1.1.2. We argued on a rather general ground then that there is no physical processes that give expressions like $\theta+\nu$, or quantities like $(\pi+1)$. Is there a possibility to find expressions like $A+B$, where A is a pure accelerator quantity and B is a pure beam quantity?

Homework 1.17 We derived the cyclotron revolution frequency in Eq. (1.8). When a particle is accelerated by a cyclotron, it is accelerated by a radio-frequency (RF) field once per revolution. The RF frequency is synchronized so that the particle gets an acceleration voltage each time it completes one revolution. Consider a particle that starts with a nonrelativistic energy $mc^2\gamma_0$ with $\gamma_0 \approx 1$. The synchronization requires the RF frequency to be

$$\omega_{\rm rf} = \omega_0 = \frac{qB_0}{\gamma_0 m} \approx \frac{qB_0}{m}.$$

This synchronization condition is somewhat miraculous. As the beam is accelerated, it gains momentum, and as a result its circumference increases. At the same time, its velocity also increases. Their ratio, the revolution frequency, however, remains fixed. This means once the RF frequency is synchronized initially, it will stay synchronized as the beam is accelerated turn after turn —

as long as the beam stays nonrelativistic. This was such a miracle coincidence that Lawrence was awarded Nobel prize in 1939.

(a) Show this coincidental synchronization explicitly.

(b) As acceleration continues, the beam begins to become relativistic. Its revolution frequency falls behind $\omega_{\rm rf}$. The acceleration process will cease. Give a quantitative estimate of the maximum beam energy that a Lawrence cyclotron can reach.

Solution (b) The revolution period is

$$T = \frac{2\pi m\gamma}{qB_0} = \frac{2\pi}{\omega_{\rm rf}}\frac{\gamma}{\gamma_0}.$$

During acceleration, in the n-th turn, the particle will lag behind the RF phase by

$$\Delta\phi(n) = \omega_{\rm rf}T - 2\pi = 2\pi\left(\frac{\gamma}{\gamma_0} - 1\right).$$

Let the acceleration be such that in the n-th turn, we have $\gamma(n) = \gamma_0(1 + gn)$, where g is the acceleration gradient. The acceleration ceases when the accumulated phase lag approaches π, or

$$\Delta\phi_{\rm tot} = \int_0^M dn\,\Delta\phi(n) = 2\pi\int_0^M dn\,gn = \pi g M^2 < \pi,$$

where M is the number of revolutions when the acceleration ceases. (Acceleration actually stops when the particle is 90° out of phase relative to the RF, but we can start the acceleration with some initial phase lead.)

The maximum number of revolution is

$$M = \frac{1}{\sqrt{g}}.$$

At the end of acceleration, the particle energy is $mc^2\gamma_0(1+\sqrt{g})$.

Consider the acceleration of protons at rest with $\gamma_0 = 1$. With 1 MV acceleration per turn, $g \approx 10^{-3}$. The acceleration ceases in 30 turns, and the kinetic energy gain is about 30 MeV.

One lesson learned from the cyclotron experience is the importance of synchronism conditions for sustained acceleration in particle accelerators. We will encounter the synchronization issue in our later lectures.

1.2 Design a first storage ring

How to design a magnet device in which a beam of charged particles can circulate in it and be stored indefinitely?

1.2.1 Uniform dipole

The simplest answer has been given in the previous section, i.e. a device that provides a uniform magnetic field. This can be achieved by making a continuous donut-shaped magnet (Fig. 1.2). A particle with momentum P_0, charge q makes circular motion in this magnet device. The accelerator designer's equation has been given in Eq. (1.6).

Figure 1.2: A donut magnet providing uniform magnetic field to circulating particles. The upper panel shows a cutaway 3-D configuration of the magnet. The lower left panel shows the cross-section of the magnet. The lower right panel shows the dynamics of the circulating particle with Lorentz force.

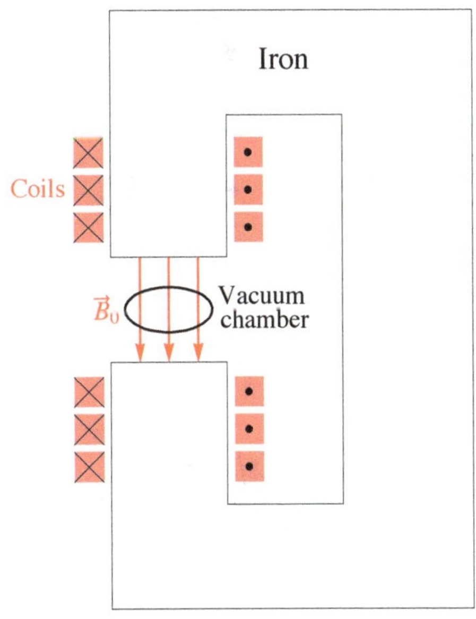

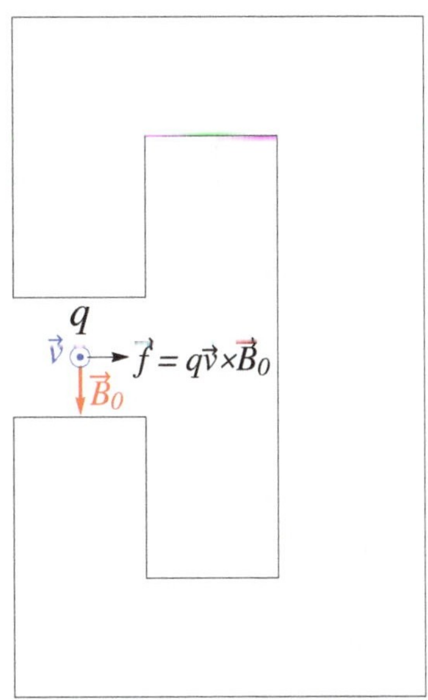

The quantity $B_0\rho$ is called the *magnetic rigidity.* Note that it is the designed relativistic *momentum*, not energy, that determines the magnetic rigidity. This donut-shaped magnet acts as a *bending magnet* or sometimes called a *dipole* magnet. The magnet device, that stores the beam, is called a *storage ring.*

Homework 1.18 The Superconducting Super Collider (SSC) was designed to be a 20-TeV proton storage ring. An SSC beam is supposed to contain 17000 bunches, each with 1×10^{10} protons.

(a) What is the stored energy (in units of Joules) in the circulating beam?

(b) SSC circumference is $C = 83$ km, with a dipole magnetic field of 6.6 T, what fraction of circumference has to be filled by dipoles?

Solution

(a) 545 MJ. This beam can easily burn big holes in the vacuum chamber and neighboring magnets if missteered.

(b) With $E = 20$ TeV, one obtains $B\rho = 66700$ T-m, therefore $\rho = 10.1$ km, and $\frac{2\pi\rho}{C} = 76\%$.

Homework 1.19 Consider an "iron-dominated" magnet (Sec. 1.3.6) whose cross-section is shown in Fig. 1.3 with gap size $2G = 20$ cm. To produce a magnetic field of 1 Tesla, how much current must be provided in the magnet coils?

Figure 1.3: The cross-section of an iron-dominated dipole magnet.

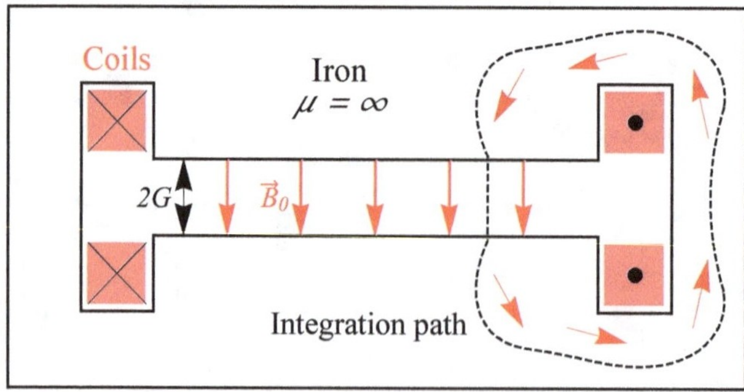

Solution

$$\frac{1}{\mu}\nabla \times \vec{B} = \vec{J} \qquad \Longrightarrow \qquad \text{Stokes theorem} \quad \frac{2GB_0}{\mu_0} + \int_{\text{iron}} \frac{\vec{B}\cdot\vec{ds}}{\mu} = I_{\text{tot}}\,.$$

With $\mu_{\text{iron}} = \infty$, we have

$$B_0 = \frac{\mu_0 I_{\text{tot}}}{2G}\,.$$

The required $I_{\rm tot}$ is therefore given by

$$I_{\rm tot}[{\rm A}] = \frac{B_0[{\rm G}]\, G[{\rm cm}]}{0.2\,\pi}\,.$$

For $B_0 = 1$ Tesla $= 10^4$ Gauss, and $G = 10$ cm, we require $I_{\rm tot} = 150$ kA. This hefty current is carried by current coils cooled by water flowing in its hollow center, and packed with multi-turn windings.

1.2.2 A uniform-dipole storage ring is unstable

But have we really succeeded in designing a storage ring? let us examine it more closely.

Stability principle The key point to note is that condition (1.6) holds for an *ideal* particle. The ideal particle indeed will be stored in the ring indefinitely according to the recipe. But for a realistic accelerator to work, it must satisfy a very basic *stability principle*,

> A stable storage ring must also store nonideal particles with "slight deviations" from the ideal conditions, i.e. the accelerator must have a finite acceptance around the ideal condition. Stability of the ideal condition alone does not make a stable accelerator.

In other words, we must be able to store a *beam* of particles, not just a single idealized particle. Let us examine the uniform-dipole design against this basic stability principle.

Coordinate system The ideal particle's trajectory is a circle. It circulates around the ring indefinitely. The coordinate system we use to describe this accelerator is shown in Fig. 1.4. We refer to x as the horizontal, y as the vertical and z as the longitudinal coordinate. In this coordinate system, the uniform magnetic field is designated as

$$\vec{B} = B_0 \hat{y}\,.$$

The coordinate system is chosen such that

- $(\hat{x}, \hat{y}, \hat{z})$ are right-handed.
- $\hat{x}$ points "outward".
- $\hat{y}$ points "upward".
- $\vec{P}_0 \parallel \hat{z}$.
- $\vec{B} \parallel \hat{y}$ for charge $q > 0$. When $q < 0$, we will choose $B_0 < 0$, keeping $\vec{P}_0 \parallel \hat{z}$.
- Lorentz force points "inward" to balance the centrifugal force as needed.
- For a planar accelerator, $\hat{y}$ is a constant, while $\hat{x}$ and $\hat{z}$ rotate around the accelerator.

Figure 1.4: Coordinate system for a circular accelerator.

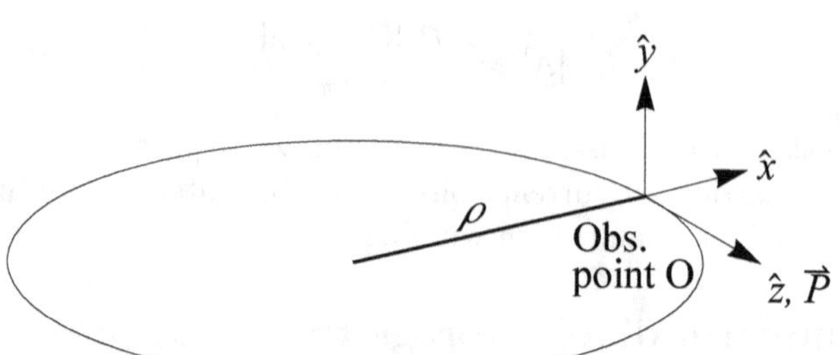

There are six "slight deviations" Consider observing the particles at position O. What "slight deviations" must we consider in the stability principle just mentioned?

The answer is that in a 3-D space (that we live in), there are six kinds of slight deviations to be considered. To describe particle motion in a 3-D space, we need a 6-D *phase space.* Each of the deviations then corresponds to one of the degrees of freedom in the phase space. Phase space is a big subject to be discussed in the next section. In the 3-D space, the six initial slight deviations to be considered are $x_0, x_0', \Delta P, z_0, y_0, y_0'$. Stability principle requires confined motion when there are small deviations on each and every one of the phase space dimensions. This is examined by the six cases shown in Fig. 1.5.

From Fig. 1.5, one observes that out of the six deviations, five are stable, i.e. the particle's motion is confined and the particle does not migrate to infinity and get lost. Deviation y_0' however is unstable. To have a stable storage ring, all six initial deviations must end up with stable motion. Five out of six is not good enough! This storage ring we designed with uniform magnetic field is therefore *unstable.* A beam injected into this storage ring will all eventually leak out in the y-direction.

Homework 1.20 The analysis of Fig. 1.5 has ignored the very weak gravitational force. What is the consequence if gravity is added to the consideration? We assume gravity is in the y-direction pulling particles downward. The same question can be asked later after vertical focusing is introduced.

1.2.3 Phase space

Here we make an important detour to introduce the concept of phase space before continuing on. The subject of phase space permeates throughout these lectures, but here we only introduce it briefly.

Phase space is a critical concept in dynamics. We are familiar with the 3-D real space (x, y, z). Galileo and Newton described their dynamics completely in the real space. The concept of phase space was developed only after a long

Figure 1.5: Six types of "small deviations" need to be considered in the 6-D phase space. The particle motion must be stable with all six kinds of initial deviations $x_0, x'_0, \Delta P, z_0, y_0, y'_0$. The case studied here is for a uniform dipole storage ring. The solid circles are the ideal design trajectory. The trajectories of the initially deviated particle are shown as dashed curves. It is clear that all cases are stable except the case with initial deviation in y'_0.

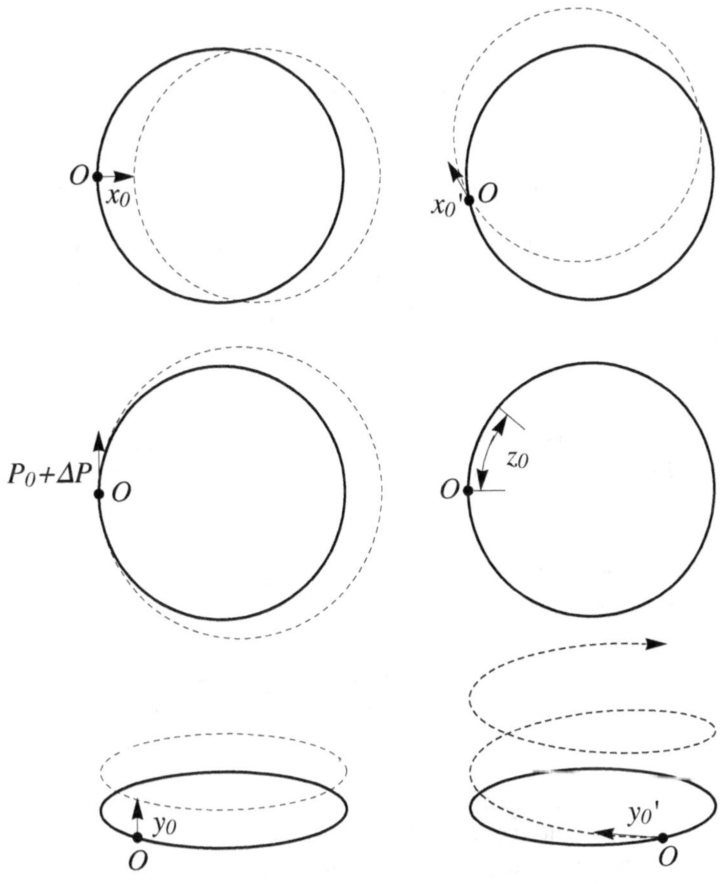

evolution, and is not to be taken for granted. On the surface, all that is done is to extend the 3-D real space to 6-D by including also the three momentum components, but hidden underneath is the accumulation of tremendous insights.

We introduced the phase space when we discussed the stability principle in the previous section. In general, living in an n-D world, we need to consider $2n$ kinds of slight deviations in the stability principle. Why *twice* of n? Because of the fundamental property of natural laws that all dynamics involve *second order* differential equations, led by Newton's second law $m\ddot{\vec{x}} = \vec{F}$ (or Hamilton equations if you so prefer). Had Newton's equation required a third order differential equation, then we will need to consider $3n$ deviations and $3n$-D phase space.

The extension from the 3-D real space to the 6-D phase space is an abstract construction. Newton's second order differential equation (one equation) is factorized into Hamilton's coupled first order differential equations (two equations, coupled). In the process, x splits into x and p, and there emerge the Hamiltonian and the phase space. It perhaps can be anticipated that by this artificial and abstract expansion, the phase space will no longer be fully filled as in the real space. There will be some degree of emptiness in the phase space compared with the real space, and this emptiness shall be rather severe because after all, we extended 3-D to 6-D, so basically up to half of the degrees of freedom in the 6-D phase space could potentially be empty.

How to describe and even to quantify this degree of emptiness in phase space? It turns out that most amazingly, there exists an elegant and concise condition to do exactly that, and in an if-and-only-if manner. This condition is called the *symplecticity condition*, to which we will devote Sec. 2.3. As we will explain then, the 6-D phase space we live in is $\frac{21}{36}$ filled in its available degrees of freedom.

In the present context, we are mentioning the phase space because each of its degrees of freedom corresponds to one possible small deviation that we must design our storage ring to confine stably within the accelerator.

Liouville theorem Another feature of phase space that has deep consequences is the Liouville theorem. Consider a certain volume of phase space. This piece of phase space moves and evolves in time — together with particles enclosed within it, if any. As it moves and evolves, its shape may change, but Liouville theorem says that its volume will be conserved, and particles enclosed will not escape and no foreign particles can sneak in.

The Liouville theorem has profound consequences. It states that all those intricate subtle dynamical effects in phase space, as predicted by the Hamiltonian beam dynamics, are rigorously preserved, and lasting an indefinitely long time. Predictions by single-particle dynamics, including intricate effects down to fine details, actually happen without dilution. These intricacies occur only in the 6-D phase space, while mostly obscured in the 3-D real space.

Liouville theorem does not specify how large or small the conserved phase space volume must be. In case we make only minute manipulations on the beam's phase space distribution, those minute effects will be forever kept. This leads to the possibility of "microbunching" a beam in accelerators. One application already realized of this microbunching possibility is the free electron laser. Some other phase space manipulations — we call them phase space gymnastics — will be the subject of Secs. 5.5 and 5.6. As accelerator physics and technology develops further, microbunching, phase space manipulation and phase space gymnastics are expected to play an increasingly prominent role. All that thanks to the Liouville theorem.

1.2.4 Focusing

Horizontal focusing There is another point important to notice in Fig. 1.5, and that is the fact that in a uniform dipole magnetic field, the x-motion is automatically focused.

The x-motion of a nonideal particle has automatically a focusing action imbedded in a uniform dipole magnetic field. To see that, we first note that the coordinates x and y of a particle are defined relative to the ideal circle designed for the ideal particle. If we plot x as a function of s, the distance along the path of the ideal circle, then we see that the case in the upper left panel of Fig. 1.5 represents an oscillatory motion of x relative to the ideal circle. In fact, its corresponding $x(s)$ is a cosine-like function with initial value $x(0) = x_0$. Similarly, the $x(s)$ corresponding to the upper-right panel is a sine-like function with initial value $x'(0) = x'_0$. These observations indicate that the x-motion in a uniform dipole field has a natural focusing yielding an oscillatory behavior. Furthermore, the resulting x-motion is such that the oscillation wavelength is equal to the ring circumference. The particle's x motion executes one complete oscillation every revolution around the ring.

Vertical focusing But now although the x-motion is focused and therefore naturally stable, the y-motion is not and, as mentioned before, is unstable. How to make this ring vertically stable? To stabilize this ring, one considers the following very clever trick: introduce a magnetic field component $B_x \propto -y$, i.e. now let

$$\vec{B} \;=\; B_0\hat{y} + Gy\hat{x}\,, \tag{1.9}$$

with a field gradient $G < 0$. A particle moving in the $+\hat{z}$ direction will then experience an additional force $q\vec{v} \times (Gy\hat{x}) \propto Gy\hat{y}$. The particle with a vertical displacement $y \neq 0$ is kicked always in the y direction and is always kicked toward $y = 0$ if $G < 0$. The particle's y-motion is therefore *focused* if $G < 0$. This focusing force will stabilize the storage ring.

Magnetic field (1.9) can be provided by tilting the pole faces in the uniform-field dipole donut magnet (pole face opens up toward outward x) as shown in Fig. 1.6. By analyzing Fig. 1.6, one observes that there is a $B_x \propto -y$ component. In particular, $B_x = 0$ when $y = 0$, i.e. the magnetic field is purely vertical in the midplane.

This magnet has a *dipole* component $B_0\hat{y}$. It now also acquires a *quadrupole* component $Gy\hat{x}$. This is now a *combined function* magnet. With $G < 0$, it is vertically focusing.

Weak-focusing storage ring We now have a storage ring using a combined-function magnet that is vertically focusing. But one must pause a little bit here;

Figure 1.6: Introducing vertical focusing to a uniform-field dipole by opening the magnet pole faces.

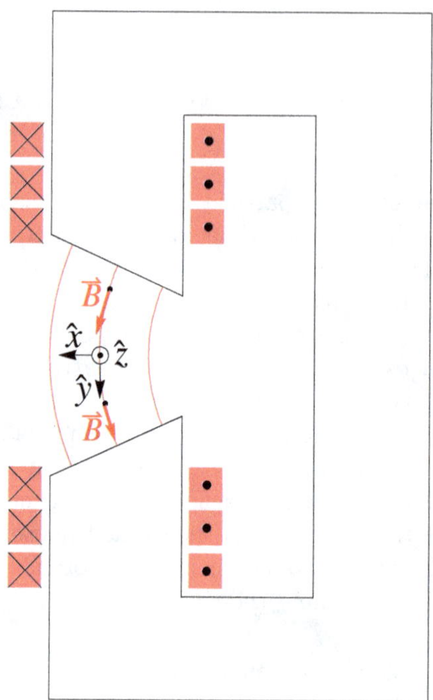

Eq. (1.9), our basic idea, unfortunately, is flawed. It does not satisfy the static Maxwell equation in free space $\nabla \times \vec{B} = 0$. The correct magnetic field that satisfies Maxwell equation also *necessarily* contains a component $B_y \propto x$,

$$\vec{B} \;=\; B_0\hat{y} + G(y\hat{x} + x\hat{y})\,, \tag{1.10}$$

where the *field gradient* can be written as

$$G \;=\; \frac{\partial B_y}{\partial x} \;=\; \frac{\partial B_x}{\partial y}\,.$$

Perhaps we summarize by stating that if you have a B_x that depends on y, then you must have also a B_y that depends on x, and vice versa.

So the magnet shown in Fig. 1.6 actually provides the field (1.10), not (1.9). The field can be decomposed as dipole + quadrupole contributions as sketched in Fig. 1.7.

The storage ring now has a stable y-motion because it contains $B_x = Gy$ with $G < 0$. However, the extra, by-necessity term $B_y = Gx$ is now causing a problem. If $G < 0$, it *defocuses* the x-direction. This is very unfortunate. Our

Figure 1.7: Field of a combined-function magnet is a superposition of a dipole contribution and a quadrupole contribution. The case shown corresponds to $B_0 > 0$, $G < 0$.

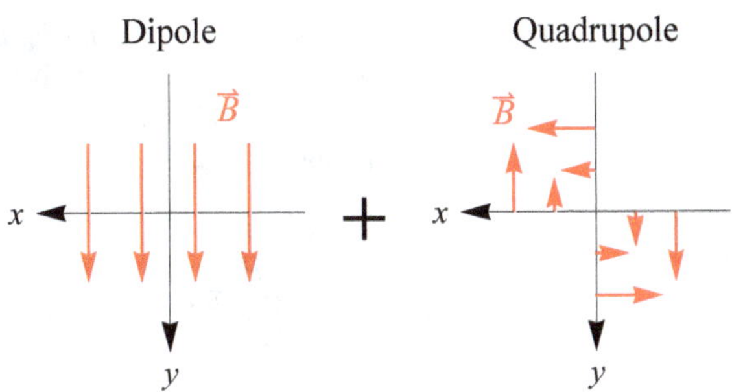

combined-function magnet is focusing in y and defocusing in x if $G < 0$ and the opposite if $G > 0$.

Fortunately, as mentioned before, the x-motion has a natural focusing from the *dipole* component. A little bit of defocusing from G can be tolerated if G is weak enough. We therefore put all these together, and have designed a stable storage ring, i.e. we use a combined-function configuration with a slightly negative gradient G. But since G must be weak, what we have designed is a *weak focusing* accelerator.

This weak focusing ring uses combined-function magnets. We shall see later that one can also design a *strong focusing* ring using *separated-function* magnets. This will be the subject of Chapter 2.

Pole face angle Consider a combined-function magnet with tilt pole faces. Let $g(x)$ as a function of x be the pole face gap. Using Ampere's law (with $\mu = \infty$ in the iron), we obtain approximately $B_y(x) \propto \frac{1}{g(x)}$ (Homework 1.19). Let $B_y(x) = B_0 \frac{g(0)}{g(x)}$, then the *field index* is defined,

$$n = -\frac{\rho}{B_0}\frac{\partial B_y}{\partial x} = \rho g(0)\frac{g'(x)}{g^2(x)} \approx \frac{\rho}{g(0)}g'(0)\,. \tag{1.11}$$

The pole face tilt angle is $g'(0)$.

Field index is an accelerator quantity. It has nothing to do with the charged particle beam. It is an important parameter describing the weak-focusing storage ring.

We already know that to provide stability of this weak-focusing ring, $g'(0)$ must be positive (and therefore $n > 0$). We also know that $g'(0)$ must not be too large, otherwise the natural focusing in the uniform bending field will not be able to stabilize the horizontal motion. In Chapter 2, we will analyze the beam dynamics of the weak-focusing storage ring more systematically and show that for a weak-focusing ring to be stable, n must lie between 0 and 1. This means the pole face tilt angle $g'(0) \approx n\frac{g(0)}{\rho}$ must lie between 0 and $\frac{g(0)}{\rho}$. We note that $\frac{g(0)}{\rho}$, the ratio of the magnet gap to the bending radius, is a small number. This means the needed pole face tilt angle is very small. Furthermore, a slight change in this small pole face tilt angle can changes the value of n substantially, which in turn seriously affects the beam dynamics. It follows that beam dynamics is very sensitive to magnet details — see Homework 2.4.

In terms of the field gradient G, we have

$$G = \left.\frac{\partial B_y}{\partial x}\right|_0 = -\frac{B_0}{\rho}n\,.$$

The condition $0 < n < 1$ means[7]

$$0 > G > -\frac{B_0}{\rho}\,.$$

Homework 1.21

(a) Check Fig. 1.7 for the coordinate system and the field lines drawn. Convince yourself that there is a $B_x \propto -y$.

(b) Figure 1.7 corresponds to the case when its dipole field $B_0 > 0$ and its quadrupole strength $G < 0$. There are three other types of combined-function magnets with other polarity combinations of B_0 and G. Draw their corresponding field patterns. Remember to pay attention to the directions of the coil currents.

Homework 1.22 Convince yourself that (a) the magnet configuration of Fig. 1.7 gives field (1.10), (b) it is also consistent with the tilted magnet pole face as shown in Fig. 1.6, and (c) the quadrupole component of the magnet is focusing in y and defocusing in x.

Homework 1.23 Consider a planet moving in a circular trajectory around a sun in its $\frac{1}{r^2}$ gravitational field. Is this planet motion stable? Consider small perturbations in the 6-D phase space to draw your conclusion.

1.3 2-D magnet

There are basically two types of magnet designs. One is *iron-dominated*, and the other is *current-dominated*. The former uses iron pole face to shape the magnetic field, while the latter uses little or no iron and sometimes uses superconducting

[7]It will be shown later in homework 6.50 that if the weak focusing ring is used to store electrons, the field index is further limited to $0 < n < \frac{3}{4}$. In that case, we have the condition $0 > G > -\frac{3}{4}\frac{B_0}{\rho}$.

wires to carry the large currents. Because the iron saturates when the magnetic field reaches 2 Tesla or so, iron-dominated magnets typically have maximum pole tip field not much beyond 1 Tesla, while the superconducting current-dominated magnets typically reach 4-10 Tesla. A brief comparison might look like Table 1.4. Note that 1 Tesla is not much higher than what can be reached by permanent magnets. In principle, many of the electromagnets can be replaced by permanent magnets, if not because of the requirements on field quality and magnet strength tunability.

Table 1.4: A brief comparison between iron-dominated and current-dominated magnets.

	Operation	Quality requirements	Max. field
Iron magnet, iron dominated	room temperature	pole face shaping	~2 T
Superconducting magnet, current dominated	cryogenic temperature	wire positioning	~10 T

Iron magnets were mentioned in Secs. 1.2.1 and 1.2.4 in passing so far. In this section we shall consider current-dominated magnets, and only on their 2-D designs.

1.3.1 Field from current filament

Consider a 2-D magnet, infinitely long in the z-direction and with its x-y cross-section uniform along the z-axis. To analyze current-dominated 2-D magnet designs like this, we first consider a filament infinitely long along the z-direction and carrying a current I_0 (positive I_0 means current going out of the board) as shown in Fig. 1.8.

By Ampere's law, and with help from Fig. 1.8, the magnetic field generated by this current filament is

$$B_x = -\frac{\mu_0 I_0}{2\pi r}\sin\theta, \qquad B_y = \frac{\mu_0 I_0}{2\pi r}\cos\theta\,. \tag{1.12}$$

We now introduce a trick very useful for 2-D magnetostatics called the Beth representation.[8] Instead of describing the field vectorially with x- and y-components, we describe it using complex numbers. Let

$$z = x + iy\,,$$

[8]R.A. Beth, J. Appl. Phys. 37, 2568 (1966); 38, 4689 (1967).

Figure 1.8: Magnetic field generated by a single current filament.

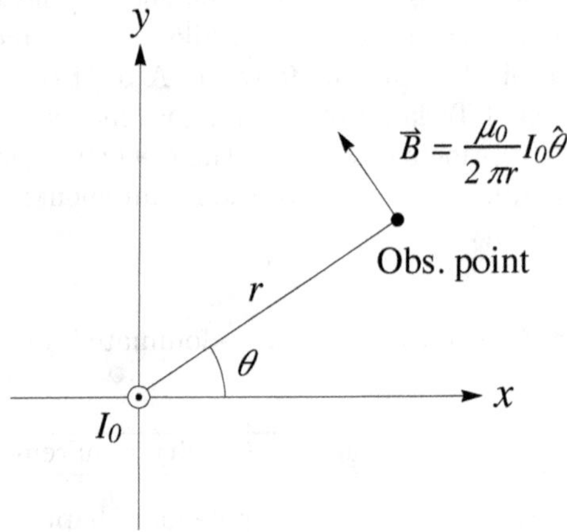

then

$$\begin{aligned} B_y + iB_x &= \frac{\mu_0 I_0}{2\pi r}(\cos\theta - i\sin\theta) \\ &= \frac{\mu_0 I_0}{2\pi r(\cos\theta + i\sin\theta)} = \frac{\mu_0 I_0}{2\pi(x+iy)} = \frac{\mu_0 I_0}{2\pi z}\,. \end{aligned} \tag{1.13}$$

The amplitude of the complex magnetic field $B_y + iB_x$ and the amplitude of the vector $\vec{B}$ are both equal to $\sqrt{B_x^2 + B_y^2}$.

Note that we have defined $z = x + iy$, where x part is real and y part is imaginary. But the complex magnetic field has the x part imaginary and y part real. This is one peculiarity of magnetic field — it is an *axial* vector. Magnetic field is a "skewed" object. Its origin requires taking the obscure curl operation on another vector, and its obscure action on the particle motion is to push the particle sideways.

One more thing to note is the $\frac{1}{z}$ dependence of the complex magnetic field due to a single current filament. This dependence will acquire further significance in the following discussions.

We now shift the location of the current filament from the origin to a location $z_0 = x_0 + iy_0$. The complex magnetic field observed at z is then given by

$$B_y + iB_x = \frac{\mu_0 I_0}{2\pi(z - z_0)}\,. \tag{1.14}$$

Equation (1.14) holds even when the wire has a finite radius, provided that the wire has a round cross-section and that the field is taken outside of the wire. This can be derived easily using the Ampere's law.

Homework 1.24 Verify Eq. (1.14) starting with first principle. A simple replacement of z by $z - z_0$ does not constitute a proof. Equation (1.14) is the basic formula used to calculate the magnetic field of a general 2-D current-dominated magnet design.

1.3.2 $\cos\theta$ magnet

Thin-sheet $\cos\theta$ magnet Having established the basic formula Eq. (1.14), we now consider the design of a *cosine-theta magnet*. The purpose of $\cos\theta$ magnets is to provide a uniform magnetic field to be used as dipoles. Consider a cylindrical infinitely-thin sheet of current distribution as shown in Fig. 1.9, with a $\cos\theta$ surface current flowing on the sheet.

Figure 1.9: Cross-section of a $\cos\theta$ dipole magnet.

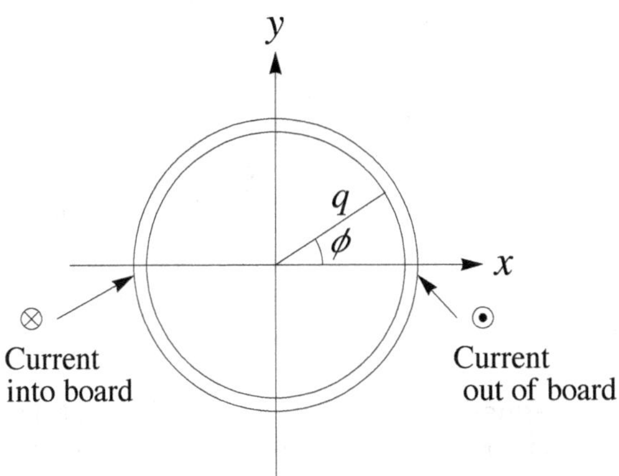

Let the current sheet carry a current density distribution

$$J(q, \phi) = \frac{I_0}{2a}\delta(q - a)\cos\phi\,,$$

where a is the current-carrying cylinder radius. The right half of the sheet ($\cos\phi > 0$) carries current out of the board. The left half ($\cos\phi < 0$) carries current into the board. There are no currents at the north and the south poles. This is called a $\cos\theta$ magnet for an obvious reason. The total current carried in each half sheet is given by

$$\int_0^\infty q dq \int_{-\pi/2}^{\pi/2} d\phi\, J(q, \phi) = \frac{I_0}{2}\int_{-\pi/2}^{\pi/2} d\phi\, \cos\phi = I_0\,.$$

We then obtain the complex magnetic field at position (x, y) by superposition,

$$\begin{aligned} B_y + iB_x &= \frac{\mu_0}{2\pi}\int_0^\infty q dq \int_0^{2\pi} d\phi \, \frac{J(q,\phi)}{x + iy - q\cos\phi - iq\sin\phi} \\ &= \frac{\mu_0 I_0}{4}\begin{cases} -\frac{1}{a} & \text{inside cylinder,} \\ \frac{a}{z^2} & \text{outside cylinder,} \end{cases} \end{aligned}$$

where the last step involves a straightforward integration (see Homework 1.33 for details of two tricks to do so). The real part of the result gives B_y; the imaginary part gives B_x. We have thus obtained the magnetic field of a pure thin-sheet "$\cos\theta$ magnet". The field inside the cylinder is uniform and is in the negative y-direction. The field outside the cylinder has a $\frac{1}{z^2}$ dependence, and has both x- and y-components. There is a discontinuity at $r = a$.

We may compute the magnetic field energy stored per unit length of the magnet as

$$\begin{aligned} \mathcal{E} &= \frac{1}{2\mu_0}\int_0^\infty r\,dr \int_0^{2\pi} d\phi\, |\vec{B}|^2 \\ &= \frac{1}{2\mu_0}\left(\frac{\mu_0 I_0}{4a}\right)^2 \int_0^{2\pi} d\phi \left[\int_0^a r\,dr + \int_a^\infty r\,\frac{a^4}{r^4}\,dr\right] \\ &= \frac{\pi\mu_0}{16\,a^2} I_0^2 \left[\frac{a^2}{2} + \frac{a^2}{2}\right] = \frac{\pi\mu_0}{16} I_0^2 = \pi a^2 \frac{B_0^2}{\mu_0}. \end{aligned}$$

The two terms in the square brackets correspond to the field energy stored inside and outside of the magnet cylinder respectively. Exactly half of the total field energy is stored in each.

Thick-coil $\cos\theta$ magnet One can extend the above calculation to a thick-coil $\cos\theta$ magnet. Let the coil with current density $J\cos\theta$ occupy the region between $r = a_1$ to $r = a_2$. Let us also wrap the coil from the outside with iron with radius $b > a_2$, as sketched in Fig. 1.10. Let us approximate the iron to have $\mu = \infty$, i.e. ignoring saturation.

The uniform magnetic field inside the magnet $(r < a_1)$ can be found (Homework 1.28) to be

$$B_y(r < a_1) = \frac{\mu_0 J}{2}(a_2 - a_1)\left[1 + \frac{a_1^2 + a_1 a_2 + a_2^2}{3\,b^2}\right], \tag{1.15}$$

where $J\cos\theta$ is the current area density in the coil. The second term in the square brackets comes from the iron. One sees that the iron enhances the field, and when $b \approx a_1 \approx a_2$, the enhancement approaches 100%. In practice, saturation is severe for high field magnets, and the enhancement factor is much less, around 20-30% for the SSC, for example. In that case, the wrapping iron is more for the purpose of mechanical support of the coil package which experiences great Lorentz repulsion forces (Homeworks 1.26 and 1.47).

Figure 1.10: Cross-section of a thick-coil $\cos\theta$ magnet. Iron region is shaded gray. The coil is indicated in red.

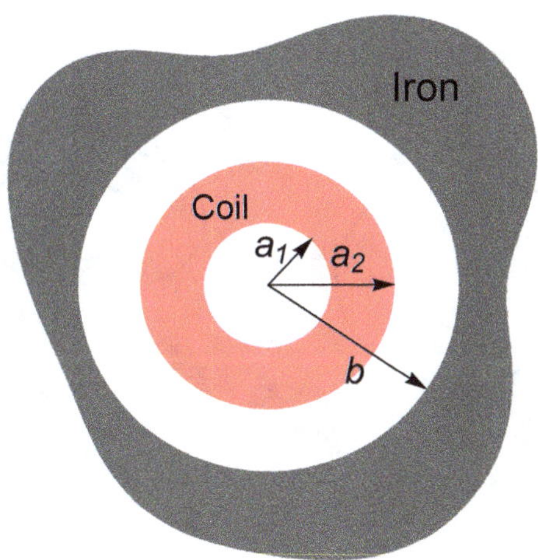

Figure 1.10 also gives a thick-coil quadrupole if the current density is $J\cos 2\theta$. Its field gradient is given by ($r < a_1$)

$$G = \frac{\mu_0 J}{2}\left[\ln\frac{a_2}{a_1} + \frac{a_2^4 - a_1^4}{4b^4}\right] . \tag{1.16}$$

Again, the second terms is contribution from the iron.

Homework 1.25

(a) Plot the field pattern for a thin-shell $\cos\theta$ magnet for both inside and outside the magnet.

(b) Repeat for a $\cos 2\theta$ magnet.

Solution Sketched in Fig. 1.11.

Homework 1.26 Take the SSC magnet design for example. Representing it by a thin $\cos\theta$ model with $B_0 = 6.6$ Tesla and $a = 3$ cm. This storage ring is designed to store a beam of 20 TeV protons.

(a) Find the magnetic field energy per unit length $\mathcal{E}$ by these magnets.

(b) Find the total stored magnetic field energy in all magnets in the SSC. Use Homework 1.18 for more SSC parameters.

(c) Find the Lorentz force experienced by the current coils. Express the result as the pressure (force per unit area) at the two poles and the left- and right-equator of the coil package. Indicate the direction of the pressure.

Figure 1.11: Field patterns of thin $\cos\theta$ and $\cos 2\theta$ magnets.

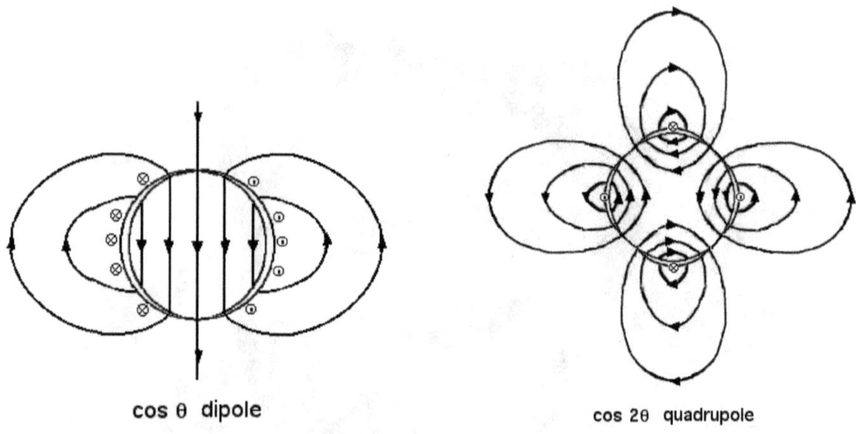

Solution

(a) $\mathcal{E} = 1.0 \times 10^5$ J/m.

(b) Total field energy = 6.2 GJ. The magnets store a great amount of energy. This is the kinetic energy carried by a 1000-kg mass at a velocity of 3.5 km/s.

Homework 1.27 Continue the previous Homework for the SSC. The energy stored in the magnets is provided through a power line during the energy ramping process. Assume the maximum power from the power line is 50 MW, what is the minimum required ramping time?

Solution If the ramping is done linearly, i.e. if the magnetic field is ramped linearly with time, then you might readily conclude that the required ramping time is given by the total stored energy divided by 50 MW, giving a result of 125 s. However, this is incorrect. Since the stored energy is quadratic in the magnetic field, a linear ramping demands higher power towards the end of the ramping process, and 50 MW will not be sufficient if you ask for a ramping time of 125 s. You may want to reprogram your ramping scenario and design an optimal ramping scenario for the given power line limit.

Homework 1.28

(a) Derive Eq. (1.15). It should not be as difficult as it might seem.

(b) Derive Eq. (1.16).

(c) For a thick-coil dipole with $b \approx a_1 \approx a_2$, it was mentioned in the text that the presence of the iron doubles the dipole magnetic field if we ignore iron saturation. Use Eq. (1.16) to find how much does the iron contribute for the case of a thick-coil quadrupole under similar conditions.

1.3.3 Iron-free kicker

To inject a beam into a storage ring, we need to provide a kicker that ideally kicks the injected beam from an injector accelerator towards the stored beam in the storage ring, while keeping the stored beam intact. Pulsed iron magnets are often used for the purpose. However, in case the kicker pulse needs to be very short and hysteresis becomes an issue, iron-free designs can be considered.

A simple design might look like that sketched in Fig. 1.12. Two thin current plates with width L provide the magnetic field that kicks the injected beam, and the direction of the kick is towards the stored beam. We assume the current plates are infinitely thin.

Figure 1.12: A fast kicker consisting of two thin current plates. The plates generate the pulsed magnetic field $\vec{B}$ that kicks the injected beam towards the stored beam.

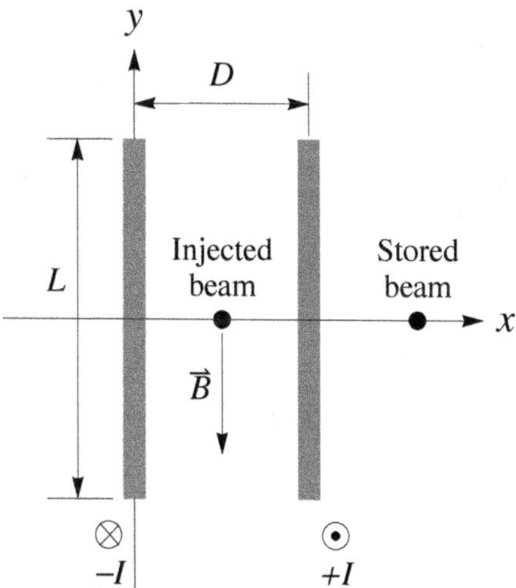

If the plates have infinite width ($L = \infty$), the stored beam will not see a magnetic field. But with finite L, we need to calculate the perturbing field seen by the stored beam and make sure it is tolerable. We already know from Eq. (1.12) or (1.13) that a current I flowing in the $\hat{z}$ direction in a straight line wire located at ($x = y = 0$) gives (this time in vector format)

$$\vec{B}(x,y) = \frac{\mu_0 I}{2\pi(x^2+y^2)}(-y\hat{x}+x\hat{y})\,. \tag{1.17}$$

For a current I flowing in the $-\hat{z}$ direction along a conducting plate on the left located from ($x = 0, y = -\frac{L}{2}$) to ($x = 0, y = \frac{L}{2}$), one can integrate Eq. (1.17) to obtain the magnetic field. Let us consider only the case when the observation point has $y = 0$. In this case, $B_x = 0$ by symmetry, and we find

$$B_y(x, y = 0) \;=\; -\frac{\mu_0 I}{\pi L} \tan^{-1} \frac{L}{2x} \,.$$

We now add a second conducting plate at location $x = D$. Let the width of this second plate be the same as the first plate, and let it be located from $(x = D, y = -\frac{L}{2})$ to $(x = D, y = \frac{L}{2})$. Let a current I flow on this second plate in the $\hat{z}$ direction. The field observed at location $(x, y = 0)$ will be

$$B_y(x, y = 0) \;=\; \frac{\mu_0 I}{\pi L} \tan^{-1} \frac{L}{2(x - D)} \,.$$

Adding the contributions from the two plates, we obtain the magnetic field

$$B_y(x, y = 0) \;=\; \frac{\mu_0 J}{\pi} \left[-\tan^{-1} \frac{L}{2x} + \tan^{-1} \frac{L}{2(x - D)} \right] ,$$

where the current density on the plates is given by $J = \frac{I}{L}$.

Figure 1.13 shows the magnetic field along the x axis of this kicker configuration for three values of L. As expected, when $L \to \infty$, the magnetic field is confined inside the region between the two plates and will be seen only by the injected beam and not the stored beam. However, when L is finite, the stored beam will see some disturbing magnetic field.

Figure 1.13: Magnetic field $\frac{\pi}{\mu_0 J} B_y(x, y = 0)$ for the two-plate kicker in Fig. 1.12 as a function of $u = \frac{x}{D}$ for various values of $v = \frac{L}{D}$. The red, green and blue curves correspond to $v = 1, 5, 100$, respectively.

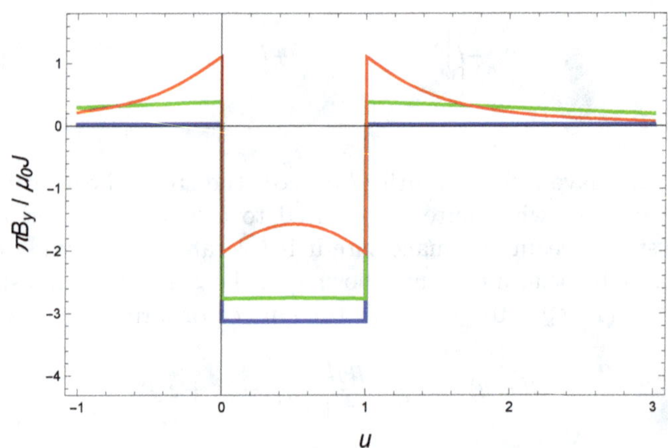

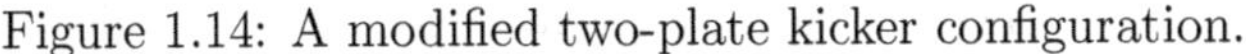
Figure 1.14: A modified two-plate kicker configuration.

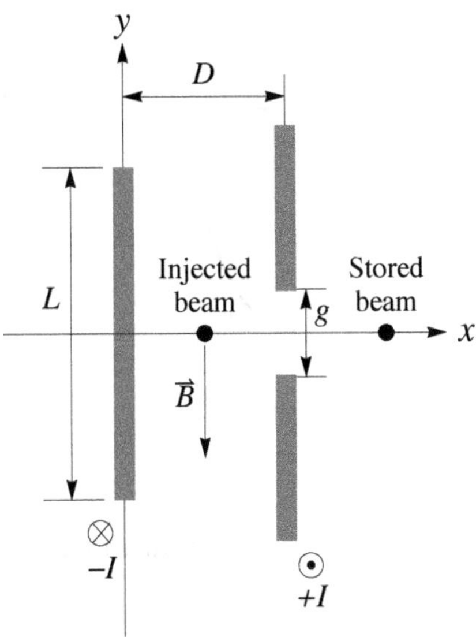

Homework 1.29 As mentioned in the text, a two-plate configuration of iron-free kicker will disturb the motion of the stored beam if the plates have finite width. One way to help the situation is to open up a gap on the second plate as sketched in the modified configuration of Fig. 1.14. This gap also helps the injected beam not to hit the plate during the transient process as it merges with the stored beam. You are asked to find the condition of the kicker parameters such that the stored beam does not experience a disturbing magnetic field.

Solution With the current density $J = \frac{I}{L}$ the same on the plates, the field can be obtained by a superposition of a plate of current density $-J$ and width L on the left, a wider plate of current density $+J$ and width $L+g$ on the right, and a third plate with current density $-J$ and width g also on the right. The result is

$$B_y(x, y=0) = \frac{\mu_0 J}{\pi}\left[-\tan^{-1}\frac{L}{2x} + \tan^{-1}\frac{L+g}{2(x-D)} - \tan^{-1}\frac{g}{2(x-D)}\right].$$

Defining $u = \frac{x}{D}$ and $v = \frac{L}{D}, w = \frac{g}{D}$, and take for example, $v = 4, w = 0.4$, we find a plot of $\frac{\pi}{\mu_0 J} B_y(x, y=0)$ versus u as shown in Fig. 1.15.

In this kicker configuration, we can probably inject the beam at $u = 0.1 - 0.3$ and locate the stored beam at $u = 1.44$. The stored beam sees a nearly zero field. The injected beam sees a field $B_y \approx 2.5 \times \frac{\mu_0 J}{\pi}$. Assuming $J = 100$ A/mm, then the injected beam sees a field ~ 0.10 T.

Figure 1.15: On-axis magnetic field for the modified two-plate kicker of Fig. 1.14.

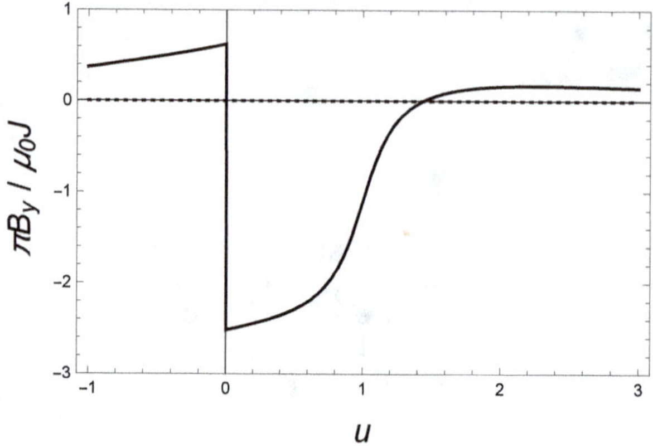

1.3.4 Field from elliptical current geometry

If a uniform current density J flows in a cable with an elliptical cross-section of semi-axes b and c, a formula that gives a magnetic field inside the elliptical region is

$$B_y + iB_x = \frac{\mu_0 J}{b+c}(cx - iby)\,. \tag{1.18}$$

Equation (1.18) is exact. Its field pattern inside the elliptical region is sketched in Fig. 1.16. The field components are linear in x and y, and vanishes at the origin $x = y = 0$. The magnitudes of the fields at the four extrema points along the long and the short axes are all equal. The field direction observes the right-hand rule of the Ampere's law.

Figure 1.16: Configuration of magnetic field applicable to Eq. (1.18). Red arrows are magnetic fields. The current flows in the $\hat{z}$ direction out of the board.

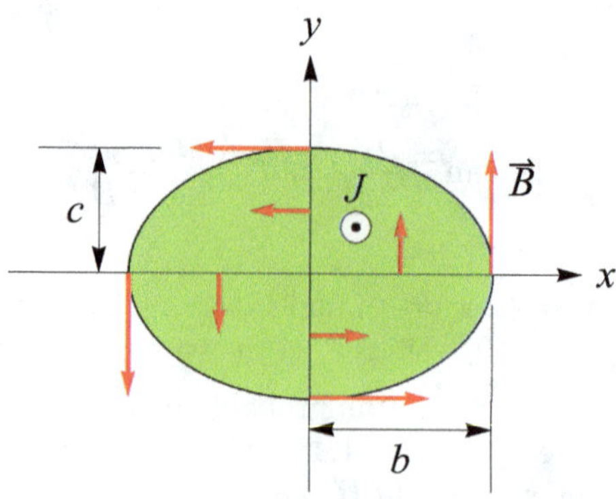

Figure 1.17: Uniform currents flowing in two elliptical areas as shown in these arrangements give a pure dipole (left) and a pure quadrupole (right) fields in their intersection regions, respectively.

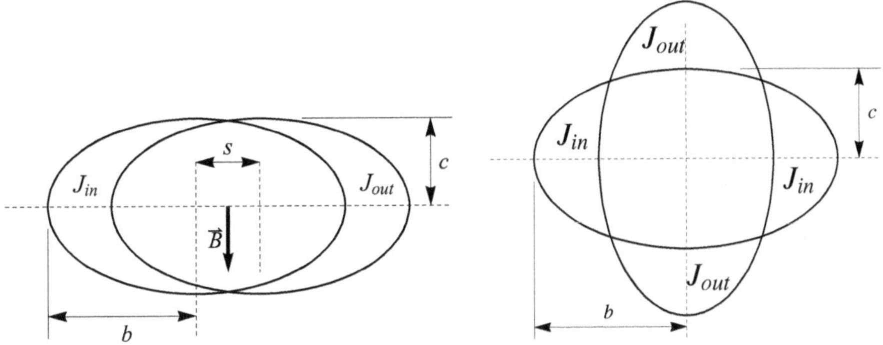

We will not prove the formula (1.18) here except to mention that for the special case when the current-carrying ellipse is circular with $b = c$, the formula gives

$$B_y + iB_x = \frac{\mu_0 J}{2}(x - iy).$$

This result can be cast in the vector form $\vec{B} = \frac{\mu_0 J r}{2}\hat{\theta}$, and we recognize can be obtained straight from Ampere's law. This result will be used in Sec. 2.2.5 when we discuss a device called lithium lens. Note that for a circular cable, this formula gives the field inside the cable. In contrast, Eq. (1.14) gives the field outside the filament.

Let us now make arrangements as shown in Fig. 1.17 with two intersecting ellipses. Apply the formula (1.18), we obtain by superposition pure dipole and pure quadrupole fields in the inner intersecting regions,

$$\begin{aligned} B_0 &= -\frac{\mu_0 Jcs}{b+c} \qquad \text{pure dipole}, \\ G &= \frac{\mu_0 J(b-c)}{b+c} \qquad \text{pure quadrupole}. \end{aligned} \tag{1.19}$$

In Fig. 1.17, J_{out} means current flowing out of the board, J_{in} flows into the board. The intersection regions in Fig. 1.17 have the currents from the two ellipses canceling each other and is therefore represented as empty free space where the beam is made to pass. The dipole case becomes a thin $\cos\theta$ magnet when the ellipses become circles and when $s \to 0$. Similarly, the quadrupole case becomes a thin $\cos 2\theta$ magnet when the ellipses have $b \to c$. The important point here however is the fact that the intersecting ellipses now allow the currents to be carried by wires with finite thickness and are thus much closer to practical designs than the thin-shell designs.

Figure 1.18: Magnets made from intersecting two ellipses that do not have the same aspect ratio, are not upright, but have the same area.

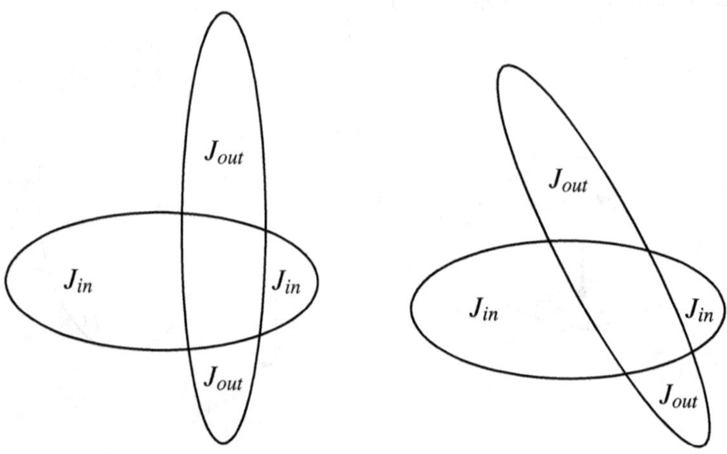

Homework 1.30

(a) Derive Eq. (1.19) using the formula (1.18).

(b) When $b = c$, show that the B_0 case of Eq. (1.19) reduces to the thin-sheet $\cos\theta$ field obtained earlier.

(c) When $c \to b$ and $s = 0$, show that the G case reduces to that for a thin-sheet $\cos 2\theta$ magnet after you learn Eq. (1.20) later.

Homework 1.31 The text explained how to construct dipoles and quadrupoles using intersecting ellipses. Can you construct a combined-function (dipole + quadrupole) in a similar manner? How about a skew quadrupole?

Solution Consider two ellipses of equal area intersecting as sketched in Fig. 1.18. The two ellipses do not have to have the same aspect ratio, and do not have to be upright.

Why do the ellipses have to have the same area? The reason is so that the two ellipses must carry equal and opposite current densities so that their overlap region becomes free space for the beam to pass. Given this condition, and given that the total current into and out of the board must be equal because of the way magnet windings are made, it follows that the two ellipses must have the same cross-sectional area.

Homework 1.32 More of a mathematical exercise is to derive Eq. (1.18) starting with Eq. (1.14). It is actually not too difficult.

1.3.5 Current-dominated multipole

One can also extend the analysis of the thin $\cos\theta$ magnet to that for a thin $\cos m\theta$ magnet. The sheet current will then have a $\cos m\theta$ distribution, and one obtains an m-multipole magnetic field whose complex magnetic field is proportional to z^{m-1} in region $r < a$, and proportional to $\frac{1}{z^{m+1}}$ In the region $r > a$.

More specifically, one constructs a $\cos m\theta$ magnet using current density

$$J(q,\phi) = \frac{mI_0}{2a}\,\delta(q-a)\,\cos m\phi\,,$$

where I_0 is the current flowing on the sheet from $\phi = -\frac{\pi}{2m}$ to $\frac{\pi}{2m}$. The magnetic field is found to be (Homework 1.33)

$$B_y + iB_x = \frac{\mu_0 m I_0}{4a}\begin{cases} -(z/a)^{m-1} & \text{for } r < a, \\ (a/z)^{m+1} & \text{for } r > a. \end{cases} \tag{1.20}$$

More generally, with an arbitrary current density

$$J = \frac{I_0}{2a}\,\delta(q-a)\sum_{n=1}^{\infty}(B_n\cos n\phi - A_n\sin n\phi)\,,$$

the magnetic field is (for $r < a$)

$$B_y + iB_x = -\frac{\mu_0 I_0}{4a}\sum_{n=1}^{\infty}(B_n + iA_n)\left(\frac{z}{a}\right)^{n-1}.$$

The reason the current can be expressed as a Fourier series is because it is necessarily periodic in ϕ. Also, the current and field are expressed as summations over n starting from $n = 1$. The $n = 0$ component is excluded because the current must vanish when integrated over θ. The total current into the board must exactly equal to the total current out of the board because that is how a magnet winding is made — and because of charge conservation.

It is often useful to express a static 2-D magnetic field as a *multipole expansion*,

$$B_y + iB_x = B_{y0}\sum_{m=0}^{\infty}(b_m + ia_m)z^m\,, \tag{1.21}$$

where $z = x + iy$, B_{y0} is the ideal value of B_y at the origin $z = 0$. Then

$$b_m = \frac{1}{B_{y0}\,m!}\left.\frac{\partial^m B_y}{\partial x^m}\right|_{x=y=0} = -\frac{\mu_0 I_0}{4B_{y0}}\frac{B_{m+1}}{a^{m+1}}\,,$$

$$a_m = \frac{1}{B_{y0}\,m!}\left.\frac{\partial^m B_x}{\partial x^m}\right|_{x=y=0} = -\frac{\mu_0 I_0}{4B_{y0}}\frac{A_{m+1}}{a^{m+1}}\,,$$

where $m = 0, 1, 2, \ldots$

Table 1.5: Table of multipoles established using Eq. (1.21).

m	magnet type	$B_y + iB_x =$	$B_y =$	$B_x =$
0	upright dipole	K_0	K_0	0
	skew dipole	iK_0	0	K_0
1	upright quad.	$K_1 z$	$K_1 x$	$K_1 y$
	skew quadrupole	$iK_1 z$	$-K_1 y$	$K_1 x$
2	upright sext.	$K_2 z^2$	$K_2(x^2 - y^2)$	$2K_2 xy$
	skew sextupole	$iK_2 z^2$	$-2K_2 xy$	$K_2(x^2 - y^2)$
3	upright octupole	$K_3 z^3$	$K_3(x^3 - 3xy^2)$	$K_3(3x^2 y - y^3)$
	skew octupole	$iK_3 z^3$	$-K_3(3x^2 y - y^3)$	$K_3(x^3 - 3xy^2)$
4	upright decapole	$K_4 z^4$	$K_4(x^4 - 6x^2 y^2 + y^4)$	$4K_4(x^3 y - xy^3)$
	skew decapole	$iK_4 z^4$	$-4K_4(x^3 y - xy^3)$	$K_4(x^4 - 6x^2 y^2 + y^4)$

Quantities b_m, a_m are the $2(m+1)$-th multipole coefficients. If the reference field B_{y0} is chosen to be $B_y(x = y = 0)$ of this same magnet, then $a_0 = 0, b_0 = 1$. Equation (1.21) gives the magnetic field inside the magnet. The same multipole magnet outside the magnet region will behave as z^{-m-2} in its m-th term.

A word of caution is in order concerning the notation adopted in Eq. (1.21). Here one often encounters a confusion involving the index m caused by the different US and the European conventions. Indeed, here the US notation, which we adopt, associates b_m and a_m with B_{m+1} and A_{m+1}.

Table 1.5 gives the field inside a multipole magnet. Upright multipoles have only a b_m field component with $a_m = 0$. Skew multipoles have only an a_m field component with $b_m = 0$. In Table 1.5, multipole strength is specified by the quantity K_m, which is identified in Eq. (1.21) as $B_{y0} b_m$ or $B_{y0} a_m$ according to whether the magnet is upright or skewed.

Bore tube winding corrector Consider a situation when a large magnet over a certain length contains some field error. Let the error be decomposed into a summation over a few multipoles,

$$\Delta B_y + i\Delta B_x = B_{y0} \sum_k (b_k + i a_k) z^k ,$$

inside the vacuum chamber. It is possible to consider a bore tube winding corrector to compensate for this error if we install a current winding on the vacuum chamber bore tube at radius a. It is easy to show that the required bore tube winding requires a current density given by

$$J(r,\theta) = \frac{B_{y0}}{\mu_0}\,\delta(r-a)\sum_k a^k\left[b_k\cos(k+1)\theta - a_k\sin(k+1)\theta\right].$$

Homework 1.33

(a) Prove Eq. (1.20) for a thin-sheet $\cos m\theta$ magnet. In the process, also calculate the field outside of the magnet. Caution: do not simply plug into mathematica to perform the necessary integration; you can be misled if not done properly.

(b) Calculate the stored magnetic field energy per unit length of the magnet inside the magnet and outside the magnet. Give expressions for these two results. The total stored energy per unit length, $\mathcal{E}$, is given by the sum of these two results. Give expression of $\mathcal{E}$.

(c) Take $I_0 = 10$ kA, $a = 10$ cm, $m = 2$. Find $\mathcal{E}$ in units of Joule/m.

Solution 1

(a) There is an integral to be made,

$$B_y + iB_x = \frac{m\mu_0 I_0}{4\pi}\int_0^{2\pi} d\phi\,\frac{\cos m\phi}{z - ae^{i\phi}},$$

and the main question is how to do this integral. One way is as follows. For $a > |z|$ (inside the magnet) and $m \geq 1$,

$$\frac{1}{z-ae^{i\phi}} = \frac{-1}{ae^{i\phi}}\left(1-\frac{z}{a}e^{-i\phi}\right)^{-1} = \frac{-1}{ae^{i\phi}}\sum_{k=0}^{\infty}\left(\frac{z}{a}e^{-i\phi}\right)^k$$

$$\Longrightarrow \int_0^{2\pi} d\phi\frac{\cos m\phi}{z-ae^{i\phi}} = \frac{-1}{2a}\sum_{k=0}^{\infty}\int_0^{2\pi} d\phi e^{-i\phi}(e^{im\phi}+e^{-im\phi})\left(\frac{z}{a}e^{-i\phi}\right)^k = \frac{-\pi}{a}\left(\frac{z}{a}\right)^{m-1}$$

$$\Longrightarrow\; B_y + iB_x = -\frac{m\mu_0 I_0}{4u}\left(\frac{z}{a}\right)^{m-1}.$$

For $|z| > a$ (outside the magnet) and $m \geq 1$, we have

$$\frac{1}{z-ae^{i\phi}} = \frac{1}{z}\left(1-\frac{a}{z}e^{i\phi}\right)^{-1} = \frac{1}{z}\sum_{k=0}^{\infty}\left(\frac{a}{z}e^{i\phi}\right)^k$$

$$\Longrightarrow \int_0^{2\pi} d\phi\,\frac{\cos m\phi}{z-ae^{i\phi}} = \frac{1}{2z}\sum_{k=0}^{\infty}\int_0^{2\pi} d\phi\,(e^{im\phi}+e^{-im\phi})\left(\frac{a}{z}e^{i\phi}\right)^k = \frac{\pi}{z}\left(\frac{a}{z}\right)^m$$

$$\Longrightarrow\; B_y + iB_x = \frac{m\mu_0 I_0}{4z}\left(\frac{a}{z}\right)^m.$$

Solution 2 Another neat (more sophisticated) way to perform the integral is to apply Cauchy residue theorem using complex variable integration. Let

$r = e^{i\phi}, d\phi = \frac{dr}{ir}$, and proceed with

$$B_y + iB_x = \frac{m\mu_0 I_0}{4\pi} \oint_{|r|=1} \frac{dr}{ir} \frac{(r^m - r^{-m})/2}{z - ar} .$$

Homework 1.34 Verify Table 1.5 of current-dominated multipoles. Do this for at least a few of the entries.

Homework 1.35 Write a computer program to explore multipole fields and visualize their patterns. Figure 1.11 gave the examples for $m = 0$ and $m = 1$.

Homework 1.36 A magnet is designed to consist of two current-carrying shells. Each shell extends an arc half-angle of θ_1 as shown in Fig. 1.19.

(a) Show that the magnet has upright multipoles given by

$$\frac{b_m}{b_0} = \frac{\sin[(m+1)\theta_1]}{(m+1)a^m \sin\theta_1}, \qquad m = \text{even} .$$

(b) Show that its sextupole component b_2 vanishes if one chooses $\theta_1 = 60°$. This constitutes an approximation (a poor man's approximation) to a $\cos\theta$ magnet. Show that the lowest order multipole field error to its desired dipole b_0 is an upright decapole with $\frac{b_4}{b_0} = -\frac{1}{5a^4}$. This error multipole has a strength that, at the magnet pole tip $r = a$, it is $\frac{1}{5}$ of that of the main dipole.

Figure 1.19: A current-sheet dipole magnet consisting of two current sheets.

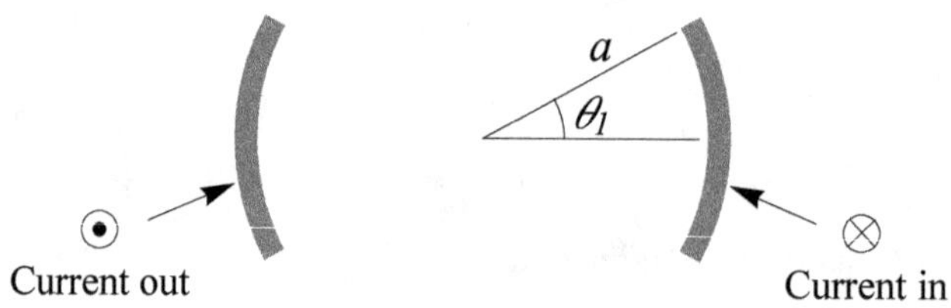

Solution

$$J(q, \phi) = \frac{I_0}{2a\theta_1}\delta(q - a) \sum_{n=1,\ \text{odd}}^{\infty} B_n \cos n\phi ,$$

where $B_n = \frac{4}{n\pi} \sin n\theta_1$.

Homework 1.37 The magnet of Fig. 1.19 is actually not good enough even if one chooses $\theta_1 = 60°$. To improve, use two current-sheet layers with arc angles

θ_1 and θ_2 as shown in Fig. 1.20. Assume the same current density in both coil packages, what should $\theta_{1,2}$ be chosen to eliminate b_2 and b_4 field errors? What is the lowest multipole error after this choice is made and how strong is it?

Figure 1.20: A current-sheet dipole magnet consisting of four current sheets.

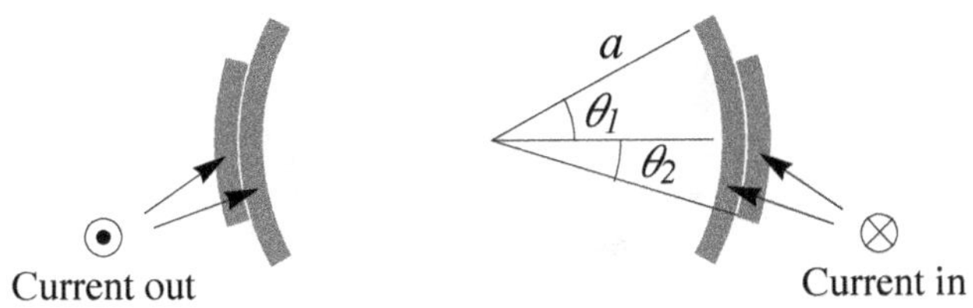

Solution

$$b_m \propto \sum_{k=1}^{2} \frac{\sin[(m+1)\theta_k]}{(m+1)a^m} \frac{I_k}{\theta_k},$$

with $\frac{I_1}{\theta_1} = \frac{I_2}{\theta_2}$. We demand the conditions

$$\sin 3\theta_1 + \sin 3\theta_2 = 0 \quad \text{to suppress } b_2\,,$$
$$\sin 5\theta_1 + \sin 5\theta_2 = 0 \quad \text{to suppress } b_4\,.$$

It is a bit surprising that these conditions lead to an explicit solution

$$\theta_1 = \frac{13\pi}{30}, \qquad \theta_2 = \frac{7\pi}{30}.$$

Homework 1.38 As a primitive approximation to a $\cos 2\theta$ quadrupole magnet, consider a magnet design modeled with four infinitely thin current sheets as shown in Fig. 1.21.

The sheets are located at radius a. Each sheet subtends an angle of $\frac{\pi}{2}$ and carries a total current of $\pm I_0$, where a positive (negative) current means the current comes out of (into) the board. The current density in each sheet is uniform. This magnet is installed into the accelerator with the x-y orientation as shown.

(a) In the region interior to the four sheets, obtain the magnetic field in terms of a multipole expansion. Show that this magnet has no skew multipole components; and among the normal components, it contains only multipoles of order $m = 2, 6, 10, 14, \cdots$. The $m = 2$ component is the quadrupole.

(b) Take $I_0 = 10$ kA, $a = 5$ cm, compute the quadrupole field gradient in units of T/m. Is this magnet going to be a focusing quadrupole or a defocusing quadrupole (consider horizontal motion for a positively charged particle)?

Figure 1.21: A current-sheet quadrupole magnet consisting of four current arcs.

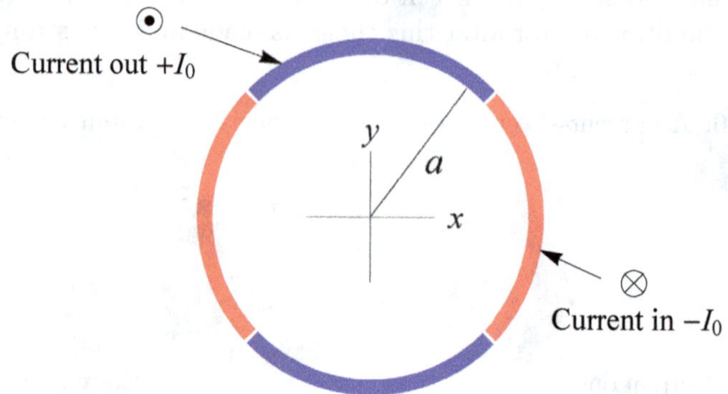

Figure 1.22: Current density as a function of θ for the magnet shown in Fig. 1.21.

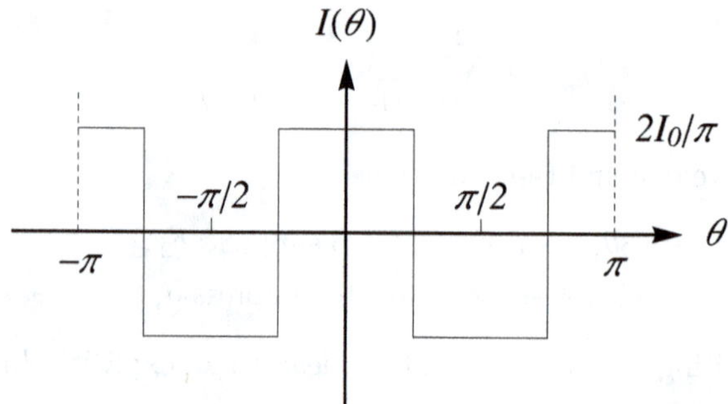

Solution The sheet current density as a function of θ is shown in Fig. 1.22,

$$\begin{aligned} I(\theta) &= \pm\frac{2}{\pi}I_0 \\ &= \frac{I_0}{2}\sum_{n=1}^{\infty}(B_n\cos n\theta - A_n\sin n\theta)\,, \end{aligned}$$

where I_0 is the total current flowing in each quadrant of the current sheet. The skew multipoles A_n all vanish. The upright components are determined by B_n with

$$\begin{aligned} \int_{-\pi}^{\pi} d\theta\; I(\theta)\cos n\theta &= \frac{\pi I_0}{2}B_n \\ \Longrightarrow \qquad B_n &= \frac{16}{\pi^2 n}\left(\sin\frac{n\pi}{4} - \sin\frac{3n\pi}{4}\right) \\ &= \frac{16}{\pi^2 n}\begin{cases} 2(-1)^m & \text{when } n = 2(2m+1), \\ 0 & \text{otherwise.} \end{cases} \end{aligned}$$

The upright components are nonvanishing only when $n = 2, 6, 10, \cdots$.

The multipole field expansion is given by

$$B_y + iB_x = -\frac{\mu_0 I_0}{4a} \sum_{n=1}^{\infty} B_n \left(\frac{x+iy}{a}\right)^{n-1} .$$

The field gradient is

$$\frac{\partial B_y}{\partial x} = -\frac{\mu_0 I_0}{4a^2} B_2 = -\frac{4\mu_0 I_0}{\pi^2 a^2} .$$

With $\mu_0 = 4\pi \times 10^{-7}$ Henry/m, $I_0 = 10$ kA, $a = 5$ cm, we obtain a gradient of -2.04 T/m. Since this gradient is negative, this is a defocusing quadrupole magnet (defocusing in x, focusing in y).

Homework 1.39

(a) The magnet configuration of Homework 1.38 can be stripped down to become four wires shown on the left panel of Fig. 1.23. Calculate the magnetic field near the magnet axis (near the origin of the figure). Calculate the quadrupole strength it produces.

(b) Once (a) is done, it is easy to give the skew-quadrupole strength for the configuration as shown on the right of Fig. 1.23.

Figure 1.23: Simplified 4-wire quadrupole and skew quadrupole.

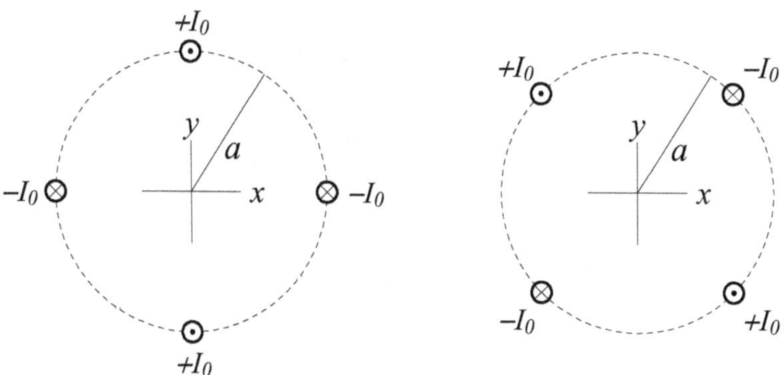

1.3.6 Iron-dominated multipole

The $\cos m\theta$ magnet discussed so far represents a current-dominated multipole magnet. For an iron-dominated magnet, pole face shaping is the key issue, their pole shape cross-sections are sketched in Fig. 1.24.

Figure 1.24: Cross-section of iron-dominated dipole, quadrupole and sextupole magnets.

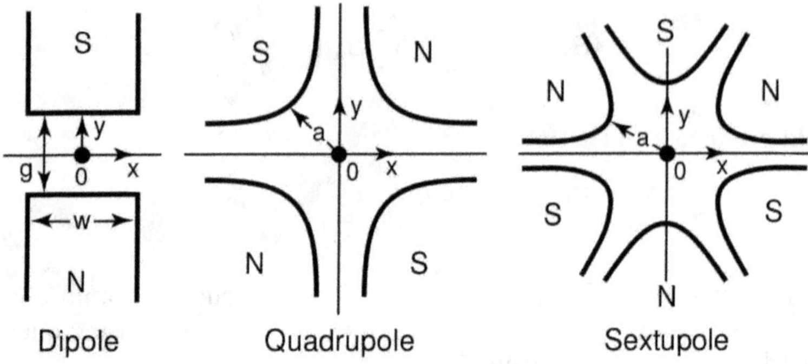

To determine the construction of the pole faces, the following formal treatment might be useful. Consider a complex potential function $\mathcal{F}(z)$ where $z = x + iy$. Since $\mathcal{F}$ depends on x and y only through z, it follows that

$$\frac{\partial \mathcal{F}}{\partial x} = \frac{\partial \mathcal{F}}{\partial z}, \qquad \frac{\partial \mathcal{F}}{\partial y} = i\frac{\partial \mathcal{F}}{\partial z},$$

and then

$$\left(\frac{\partial^2}{\partial x^2} + \frac{\partial^2}{\partial y^2}\right)\mathcal{F} = 0. \tag{1.22}$$

The unit of $\mathcal{F}$ is T-m. For each such a function $\mathcal{F}$, there is a corresponding magnetic field given by

$$\begin{aligned} B_y + iB_x &= \frac{\partial \mathcal{F}}{\partial z}, \\ B_y &= \frac{\partial \mathrm{Re}\mathcal{F}}{\partial x} = \frac{\partial \mathrm{Im}\mathcal{F}}{\partial y}, \\ B_x &= -\frac{\partial \mathrm{Re}\mathcal{F}}{\partial y} = \frac{\partial \mathrm{Im}\mathcal{F}}{\partial x}. \end{aligned} \tag{1.23}$$

For the special case of an m-multipole, Eq. (1.21), we have

$$\mathcal{F}(z) = \frac{B_{y0}}{m+1}(b_m + ia_m)z^{m+1}. \tag{1.24}$$

The magnetic field $\vec{B}$ is perpendicular to $\nabla\mathrm{Re}\mathcal{F}$ and parallel to $\nabla\mathrm{Im}\mathcal{F}$. This means the magnetic field lines lie along constant contours of $\mathrm{Re}\mathcal{F}$, and the magnet pole face is one of the contours of constant $\mathrm{Im}\mathcal{F}$. This formal treatment holds for any complex potential function $\mathcal{F}(x, y)$ that satisfies Eq. (1.23).

Of course as emphasized in Sec. 1.2.4, the magnet pole face must be shaped accurately to provide a useable magnet.

Homework 1.40 An iron-dominated magnet has a cross-section shown in Fig. 1.25. What type of multipole is this magnet? If $a = 5$ cm, and the pole face field strength at position A is 1 Tesla, how does one specify the strength of this magnet?

Figure 1.25: Cross-section of an iron-dominated magnet.

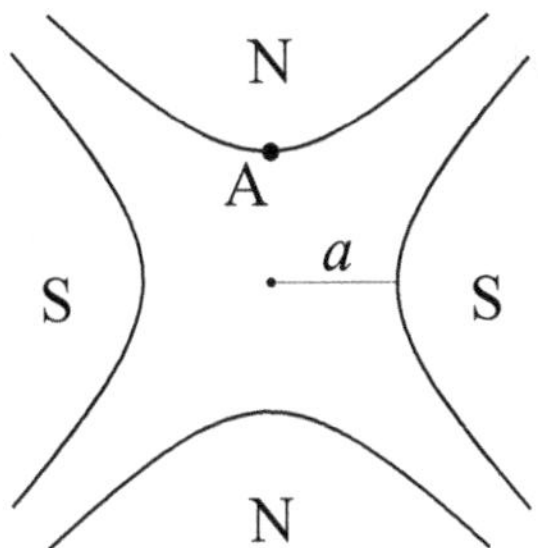

Homework 1.41

(a) Verify Eq. (1.22) for any analytical function $\mathcal{F}(z)$ with $z = x + iy$.

(b) Show that the field (1.23) obeys Maxwell equations $\nabla \times \vec{B} = 0$, $\nabla \cdot \vec{B} = 0$.

(c) Find the vector potential $\vec{A}$ that gives rise to field (1.23). How is $\vec{A}$ related to the quantity $\mathcal{F}$ in Eq. (1.24)?

Homework 1.42 A quadrupole cross-section is shown in the left panel of Fig. 1.26.[9] Apply Ampere's law to a circuit as shown on the right panel along the blue closed path. Assume the iron yoke has $\mu = \infty$. Show that the quadrupole strength is given by

$$\frac{\partial B_y}{\partial x} = \frac{2\mu_0 nI}{a^2},$$

where a is the magnet pole tip radius, nI is the current enclosed in the blue circuit.

Homework 1.43 A sextupole cross-section is shown in Fig. 1.27. Apply Ampere's law appropriately to show that the sextupole strength is given by

$$\frac{\partial^2 B_y}{\partial x^2} = \frac{6\mu_0 nI}{a^3}.$$

Assume the iron yoke has $\mu = \infty$.

[9] Figures 1.26 and 1.27 are referenced to K. Wille, The Physics of Particle Accelerators, Oxford (2001).

Figure 1.26: Cross-section of a quadrupole magnet, and calculation of its gradient.

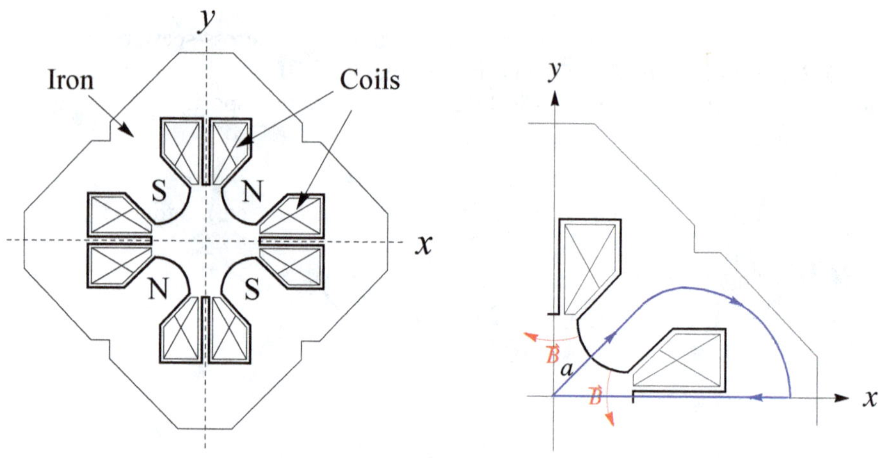

Figure 1.27: Cross-section of a sextupole magnet, and calculation of its gradient.

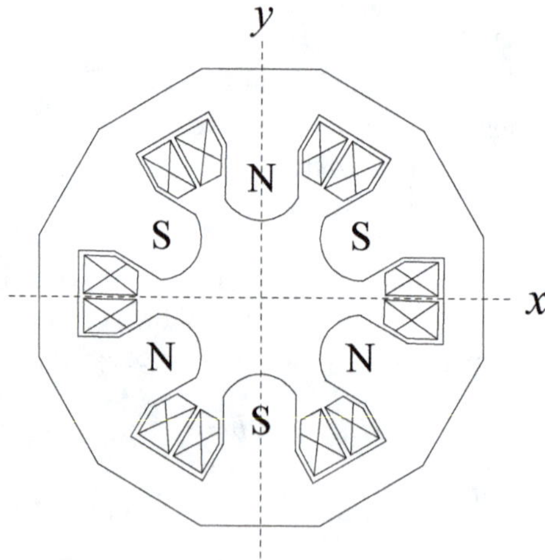

Homework 1.44 The text explained how to construct the pole faces of a pure m-multipole. Follow the line of reasoning of Sec. 1.3.6 in doing this homework.

(a) How to construct the pole faces of a combined-function (dipole and quadrupole) magnet? Show that these pole faces are equivalent to a simple quadrupole but is horizontally displaced. Express this displacement in terms of the dipole and the quadrupole strengths.

(b) Repeat the same for a (dipole and skew quadrupole), and show that these pole faces are equivalent to a simple quadrupole but is vertically displaced. Express this displacement in terms of the dipole and the skew quadrupole strengths.

(c) How to construct the pole faces of a combined-function (dipole, quadrupole, and sextupole)?

(d) To gain additional intuitive feel, give plots of pole face contours for a few example combined-function magnets, e,g, a dipole + quadrupole), a (dipole – quadrupole), a (quadrupole + sextupole), and a (quadrupole – sextupole).

(e) Analyze the feeddown effects of a misaligned sextupole magnet quantitatively by examining its corresponding $\mathcal{F}$ function and give the strengths of all the lower multipoles. Let the sextupole center be displaced by (x_0, y_0).

Solution If you need a hint first, the key is to first find an expression for the $\mathcal{F}(z)$ that gives the multipole you want.

(a) $\mathcal{F} = B_y z + \frac{G}{2} z^2$ is equivalent to $\mathcal{F} = \frac{G}{2}(z - z_0)^2$, where $z_0 = -\frac{B_y}{G}$. The magnet has $\vec{B} = Gy\hat{x} + (B_y + Gx)\hat{y}$. A more venturous reader should compare the pole face obtained here with that mentioned in Eq. (1.11).

(c) Find a constant-contour of Im$\mathcal{F}$, where $\mathcal{F} = B_y z + \frac{G}{2} z^2 + \frac{S}{3} z^3$. The magnet has $\vec{B} = (Gy + 2Sxy)\hat{x} + [B_y + Gx + S(x^2 - y^2)]\hat{y}$.

Homework 1.45 A wide dipole nominally has its two pole faces perfectly parallel to each other. What are the quadrupole and sextupole field errors if the two pole faces have

$$y_1(x) = \frac{g}{2}(1 + \epsilon_1 x + \epsilon_2 x^2), \quad y_2(x) = \frac{g}{2}(1 + \epsilon_3 x + \epsilon_4 x^2)?$$

Find approximate expressions for $\frac{\partial B_y}{\partial x}$ and $\frac{\partial^2 B_y}{\partial x^2}$ near $y = 0$ to first orders in the ϵ-coefficients.

Homework 1.46 A periodic structure (such as an undulator to be discussed in Chapter 7) can be obtained for example by considering

$$\mathcal{F}(z) = B_0 a \cos kz\,.$$

(a) Find analytic expressions of the magnetic field.

(b) Work out a numerical example of the pole faces and find its corresponding field lines.

Solution For this homework, you first have to note two things: (i) When you look for the pole faces, note that you need a pair of them, one determines the north pole face, the other south pole face. Field lines are provided between these two sets of pole faces. (ii) For this problem of an undulator magnet, the x in $z = x + iy$ is the *longitudinal* coordinate. Two examples of undulator pole face design are shown in Fig. 1.28. Note that if pole faces are determined this way, the desired pure sinusoidal undulator on-axis field will be exact.

Figure 1.28: Pole face shape for an undulator. Horizontal axis is the longitudinal coordinate. Vertical axis is the transverse coordinate. The red curves are the north pole faces, blue curves the south pole faces (or vice versa). Upper panel has the two pole faces symmetric. The lower panel shows a variation when they are not symmetric.

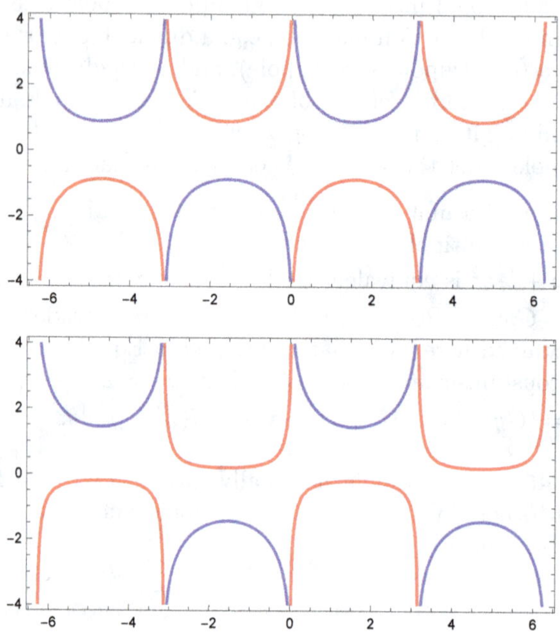

1.3.7 High-field superconducting magnet

Strong superconducting magnets are difficult to make. Naively, one imagines that all one has to do is to take a conventional magnet and replace the copper wires by superconducting ones, but this is far from the fact. Design of high-field superconducting magnets is a highly involved expertise, which evolved through a very long and treacherous learning process. Victims along the evolution process included a few very large science projects.

一把辛酸淚
誰解其中味? (紅樓夢)
Who truly appreciate these
Eyeful of tears? (Dream of Red Chamber)

一將功成萬骨枯
It takes ten thousand perished soldiers to bestow the glory of one general.

We are not in a position to discuss superconducting magnets in these lectures.[10] Perhaps we only comment on some of the main technical issues of the design as follows.

- The superconducting wire that replaces the copper wire is by itself an extremely complex hierarchy of objects,

  Coil package is constructed from cables;
  Cable is constructed from twisted strands;
  Strand is constructed of twisted superconducting *filaments* imbedded in a copper matrix.

  Clearly, this coil package is very far from that of an assembly of high-current wires! Very fine filaments (a few microns diameter) are needed to put eddy currents under control. Copper matrix is needed to stabilize quenches. To approximate $\cos\theta$ as much as possible, multiple wedges are inserted into the coil package. In addition, the evolution of the coil package had to take into consideration of issues including mechanical integrity, thermal expansion balance, field quality, thermal cycling, power cycling, manufacturing feasibility, and industrialization.

  A schematic illustration of the cable as conceived for the SSC is shown in Fig. 1.29.

- Field quality requirement is very high (better than 1×10^{-4}). This means coils must be placed accurately ($\lesssim 1/1000$ inch) to maintain the field integrity.

- Coils experience strong Lorentz forces — e.g. it was 5000 psi when powered for the case of the SSC with 6.6 T field,[11] and yet any small movement can release sufficient field energy to quench the magnet. Coil packages must be confined by strong collars heavily prestressed ($\sim$ 10000 psi) so that the coil packages remain motionless under repeated thermal cycles, power cycles, or quenches.[12]

- It is sometimes mistakenly considered the most difficult challenge is to reach a field as high as possible by stuffing currents into superconducting wires. This is not true. Pushing to achieve ever higher magnetic fields is far from what challenges a high-field high-quality superconducting accelerator magnet and is actually not too relevant. The real challenge is the mechanical design to strictly confine the superconducting wires, which

[10] See, for example, Martin N. Wilson, "Superconducting Magnets", Oxford Scientific Pub. (1983); K.-H. Mess, P. Schmüser, and S. Wolff, "Superconducting Accelerator Magnets", World Scientific (1996).

[11] This is like two compact cars piled up on a 1 in^2 area!

[12] The simplest cause of coil movement is due to thermal expansion even before the current begins to flow. Imagine how much the coil package expands and shrinks during thermal cycles between room temperature and cryogenic temperature.

Figure 1.29: A schematic illustration of a superconducting cable for a high-field and high-quality superconducting niobium-titanium (NbTi) magnet.

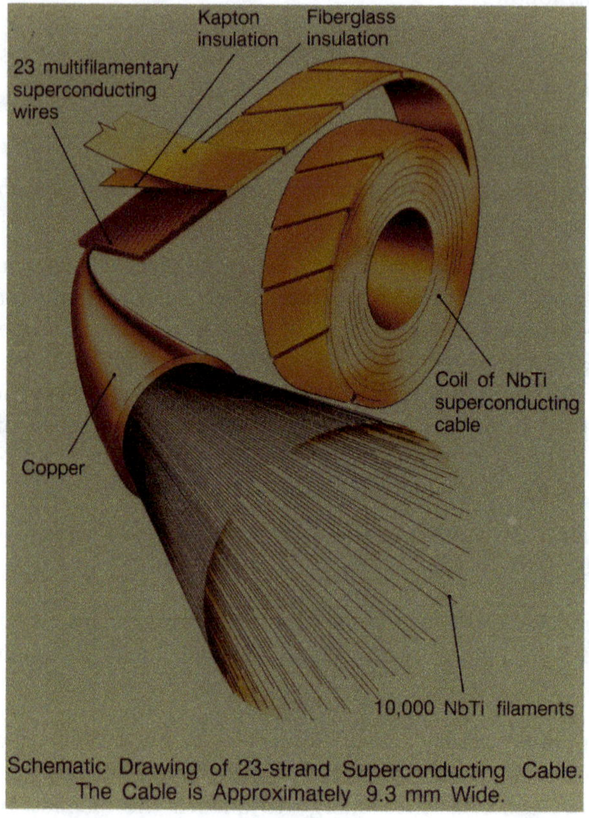

in turn forces a large compromise on the highest field reached. Furthermore, it is important to recognize that this mechanical issue becomes ever more critical as the required magnetic field increases because the Lorentz pressure grows quadratically with the field.

- The superconducting cables need to be connected through current leads to power supplies which are at room temperature. The current leads need to make the temperature transition as they go into and out of the warm and cold regions. They need to conduct currents well yet minimizing heat conduction to reduce the heat leak. This leads to the issues of properly choosing the conductor material, as well as its length and cross-section. A larger cross-section allows higher current, but also increases the heat leak. High temperature superconductor cables have been used as current leads for high-field superconducting magnets.

It goes without saying that a fundamental change of the basic magnet design, e.g. changing the superconducting material from NbTi to NbTn or high temperature superconductors, will necessarily require a complete overhaul starting from scratch of the knowledge acquired over several decades of magnet research focused on NbTi.

Homework 1.47 Cross-section of one of the Superconducting Super Collider dipole magnet designs is shown in Fig. 1.30. The superconducting cable carries a current of 6.5 kA. The dipole field is 6.6 T.[13]

(a) Approximate the dipole design as a $\cos\theta$ magnet (or as two displaced ellipses if you so prefer) using data estimated from Fig. 1.30, and convince yourself that it approximately produces 6.6 T field.

(b) The current-carrying cable immersed in a vertical magnetic field of 6.6 Tesla experiences a strong Lorentz force. What is the direction of this force, and how large is it per unit length of the cable?

(c) Use the result from (b), estimate the Lorentz pressure exerted on the magnet collars by the coil packages. Pay extra attention to the statement in the text that the pressure scales quadratically with the field strength.

1.4 Solenoid

We have so far discussed the design of multipole magnets. One magnet not of a multipole type is the undulator discussed in Homework 1.46. Another common non-multipole magnet is the solenoid.

1.4.1 Solenoid field

Body field To calculate the field of a solenoid, the 2-D complex Beth representation no longer applies. The currents no longer flows in infinitely-long straight filaments. We need to use a more general formula, the Biot-Savart law.

According to the Biot-Savart law, the static (no time dependence) magnetic field at position $\vec{r}$ due to a segment of current $Id\vec{\ell}$ at position $\vec{r}'$ is given by

$$\vec{B}(\vec{r}) = \frac{\mu_0}{4\pi}\oint\frac{Id\vec{\ell}\times\vec{R}}{R^3}, \tag{1.25}$$

where $\vec{R}=\vec{r}-r'$, and the integration over $d\vec{\ell}$ is performed over the entire path along which the current flows.

We will compute the magnetic field of a solenoid along its axis. To do so, we first compute the field due to a ring slice of current flow as shown in Fig. 1.31.

The field contribution from each ring slice of the solenoid is along the direction of the solenoid axis, i.e. along the $\hat{s}$-direction. Keeping only a component

[13]SSC Central Design Group, SSC-SR-2020, Conceptual Design of the Superconducting Super Collider (1986).

Figure 1.30: A sketch one of the cross-section designs of the SSC superconducting dipole magnet.

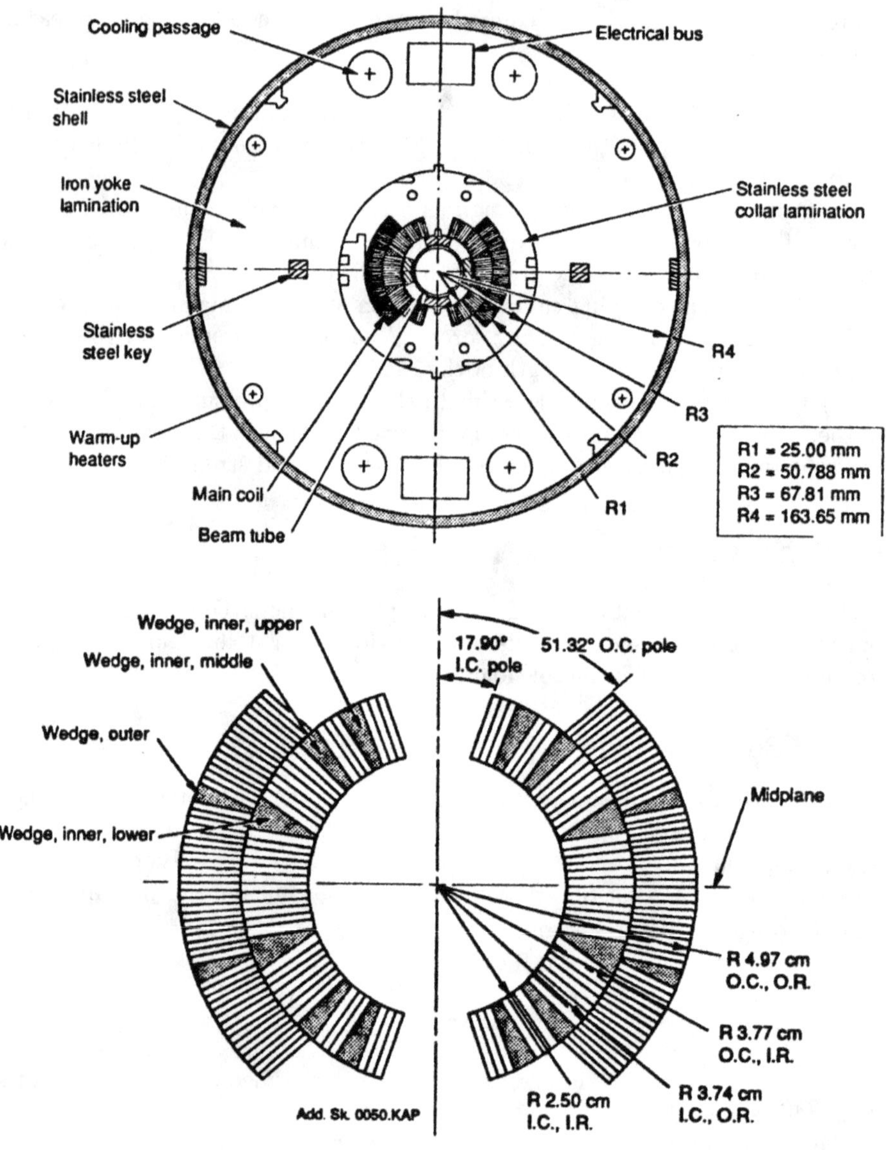

Figure 1.31: Calculation of magnetic field of a static solenoid.

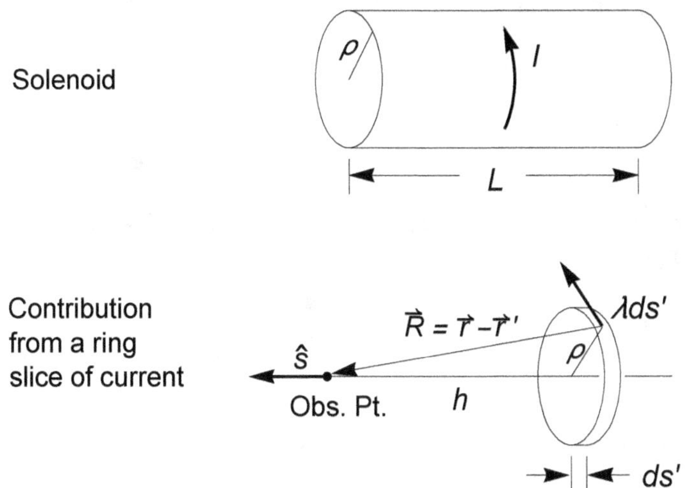

in the $\hat{s}$-direction, we have $d\vec{\ell} \times \vec{R} = \rho \hat{s} d\ell$ where ρ is the radius of the solenoid winding. We then set the ring slice current to be $I = \lambda ds'$ with $\lambda = \frac{NI_0}{L}$ the current density in Amp/m, where we have assumed that the solenoid has a total length of L and is made of N turns of winding of a wire carrying a current I_0. The ring slice contribution to the field, at a distance h from the ring center, is therefore

$$d\vec{B}(h) = \frac{\mu_0 N I_0 \rho^2}{2L(\rho^2 + h^2)^{3/2}} \, \hat{s} \, ds' \, .$$

Integrating over ds' of all slices then gives the solenoidal field along its axis (at a distance s from the center of the solenoid),

$$\begin{aligned} \vec{B}(s) &= \int_{-L/2}^{L/2} d\vec{B}(s - s') \\ &= \frac{\mu_0 N I_0}{2L} \hat{s} \left[\frac{\frac{L}{2} - s}{\sqrt{\rho^2 + (\frac{L}{2} - s)^2}} + \frac{\frac{L}{2} + s}{\sqrt{\rho^2 + (\frac{L}{2} + s)^2}} \right] . \end{aligned} \tag{1.26}$$

The solenoid field $\frac{LB_s}{\mu_0 N I_0}$ as a function of $\frac{s}{L}$ for the cases $\frac{\rho}{L} = \frac{1}{5}, \frac{1}{20}$ is plotted in Fig. 1.32.

End fields One observes from Fig. 1.32 that, for a long solenoid with $L \gg \rho$, the field as a function of s can be approximated as a step function, i.e. $B =$ constant B_0 inside the solenoid and $B = 0$ outside it. This is correct; however, there is a catch: Solenoids have *end fields* that must be considered. This is because of the fundamental rule that magnetic field lines do not simply stop in

Figure 1.32: Plot of solenoidal field along its axis. Solid curve is for $\frac{\rho}{L} = \frac{1}{20}$. Dashed curve is for $\frac{\rho}{L} = \frac{1}{5}$.

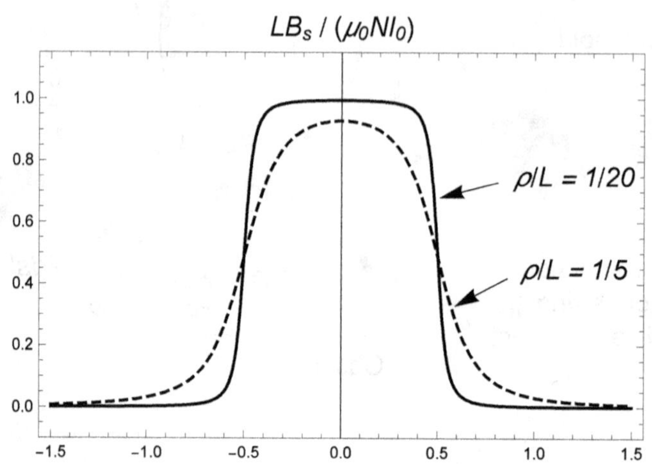

free space; they must form loops. Otherwise one violates the Maxwell equation $\nabla\cdot\vec{B} = 0$, which in turn will lead to wrong result of beam dynamics. At the ends of a solenoid, magnetic field has nowhere to go except to go transverse. This means that a correct representation of the solenoid field must involve transverse fields at its ends. Solenoids are tricky optical devices.

An alert reader might note the following: All multipole magnets have been approximated as 2-D magnets whose fields abruptly drops to zero at the magnet ends (the *hard-edge model*). Don't they violate the Maxwell equations? Strictly speaking, they do. The $\nabla\times\vec{B}$ equation is in fact put in jeopardy. However, this violation does not cause a fundamental crisis in beam dynamics.[14] In particular, the magnetic field lines, which all lie in the transverse plane, do not suddenly terminate in free space even in the hard edge model. So this violation of one of the Maxwell equations by hard-edge multipoles is permitted, while a hard-edge solenoid is not (unless we add the end fields) because it violates Maxwell equation in a more violent manner!

On the other hand, be reminded of our earlier observation when we discussed the magnetic field within the body of a quadrupole magnet, that B_y depending on x implied B_x depending on y was a result of $\nabla\times\vec{B} = \vec{0}$. The fact is that all of the four Maxwell equations are not equal. Some of them must not be violated, but some can be sometimes. One needs to choose carefully which ones to obey and which ones to compromise. For a hard-edge magnet, violating $\nabla\times\vec{B} = 0$

[14]That is assuming we stay at the level of linear optics. If we insist on hard-edge model and carry to higher orders, there are divergences waiting around the corner. We are not going in this area, however.

Figure 1.33: A hard-edge model of solenoid field lines with end fields added so that $\nabla \cdot \vec{B} = 0$ is obeyed and field lines do not stop in the air.

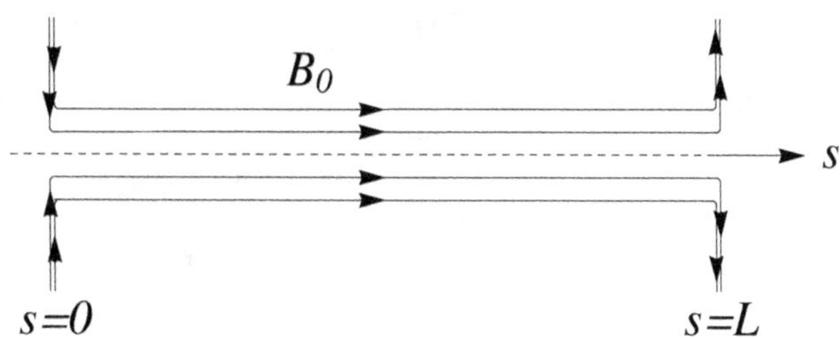

is OK while violating $\nabla \cdot \vec{B} = 0$ is not. Later we will see another example of rectangular dipole to see an end field effect on beam dynamics due to the consideration of the $\nabla \times \vec{B} = 0$ condition.

We now need to improve the hard-edge model for a solenoid in such a way that the field lines do not suddenly stop in the air. For a long solenoid, we can approximate the solenoid end by a δ-function transverse field as illustrated in Fig. 1.33.

At this point, we will change notation from s to z to denote the longitudinal coordinate along the solenoid axis. At the solenoid entrance end $z = 0$, field lines come in radially from infinity toward $r = 0$. These field lines enter the solenoid body and *become* the field lines in the solenoid body as they "leak out" longitudinally. Figure 1.33 in fact intends to illustrate the situation. A moment's reflection would indicate that at the entrance edge $z = 0$, B_r will be negative, and would be proportional to r,

$$B_r \propto -r\delta(z)\,.$$

At the exit end, similarly, we should have

$$B_r \propto r\delta(z - L)\,.$$

If we let (A is yet to be determined)

$$B_r = -Ar\delta(z) + Ar\delta(z - L)\,,$$

and let ($H(x)$ is the step function)

$$B_z = \begin{cases} B_0 & \text{if } L > z > 0 \\ 0 & \text{otherwise} \end{cases} = B_0 H(z) H(L - z)\,,$$

then the Maxwell equation $\nabla \cdot \vec{B} = 0$ leads to

$$\frac{1}{r}\frac{\partial}{\partial r}(rB_r) + \frac{\partial B_z}{\partial z} = 0 \qquad \Longrightarrow \qquad A = \frac{B_0}{2}\,.$$

These results combine to yield the complete solenoid field

$$\vec{B} = -\frac{B_0}{2}r\delta(z)\hat{r} + B_0 H(z)H(L-z)\hat{z} + \frac{B_0}{2}r\delta(z-L)\hat{r}\,, \qquad (1.27)$$

where the first term gives the solenoid entrance end field, the middle term give the solenoid body field, and the last term gives the exit end field. We will use this result later in Sec. 3.5.3 to derive a transfer map for a solenoid. We will see then that the end fields play an important role to beam dynamics.

Homework 1.48 Use the Biot-Savart law to verify Eq. (1.13),

$$B_y + iB_x = \frac{\mu_0 I}{2\pi z}, \quad z = x + iy\,,$$

for an infinitely long straight-line current wire along the s-axis.

Homework 1.49 Show that the solenoidal field at the solenoid center ($s = 0$) is given by

$$\vec{B}(s=0) = \frac{\mu_0 N I_0}{2L}\hat{s}\frac{L}{\sqrt{\rho^2 + (\frac{L}{2})^2}}\,.$$

When $L \gg \rho$, it becomes

$$\vec{B} = \frac{\mu_0 N I_0}{L}\hat{s}\,.$$

Show that the same result can be obtained by applying the Ampere's law for an infinitely long solenoid.

Homework 1.50 Using the general expression (1.26), show that

$$\int_{-\infty}^{\infty} \vec{B}(s)\cdot\hat{s}ds = \mu_0 N I_0\,,$$

and connect this result to a necessity required by Ampere's law.

Homework 1.51

(a) What is the magnetic field outside the solenoid near one of its ends? Let t be the distance to the solenoid end. Let $L \gg t \gg \rho$, the field along the axis $B_s \approx \frac{\mu_0}{4\pi L}\frac{(\pi\rho^2)(NI_0)}{t^2}$. The field near the solenoid end resembles that of a magnetic monopole. Find the equivalent monopole charge.

(b) Far away from the solenoid end with $t \gg L$ and $t \gg \rho$, one obtains $B_s \approx \frac{\mu_0}{2\pi}\frac{(\pi\rho^2)(NI_0)}{t^3}$. The field now resembles that of a magnetic dipole. Find the equivalent dipole moment.

Solution

(a) If we define a magnetic monopole by $B = \frac{M}{4\pi\epsilon_0 r^2}$ in its neighborhood, then $M = \frac{(\pi\rho^2)(NI_0)}{c^2 L}$.

(b) The equivalent magnetic dipole moment is $(\pi\rho^2)(NI_0)$. Refer to Homework 1.52 and Eq. (1.28).

Homework 1.52 Consider a current I flowing on a loop of radius a. It constitutes a magnetic dipole moment

$$\vec{\mu}_B = \frac{1}{2}\oint \vec{r}' \times I d\vec{\ell} = \pi a^2 I \hat{z}.$$

Consider a to be small and we are observing it from a far distance so that current loop can be considered a point dipole moment. Use the Biot-Savart law to calculate the magnetic field in the 3-D space around the current loop.

Solution In spherical coordinates with the observation point at polar radius R and polar angle θ,

$$\vec{B} = \frac{\mu_0}{4\pi}\frac{\mu_B}{R^3}(\hat{\theta}\sin\theta + 2\hat{R}\cos\theta). \tag{1.28}$$

The field pattern is shown in Fig. 1.34. Show that the parametric equation for the contours in Fig. 1.34 is $R^2 + z^2 = C|R|^{4/3}$.

We will continue this homework when we discuss the electron spin in Sec. 6.1.1.

Figure 1.34: Field pattern generated by a point magnetic dipole moment. The dipole moment is located at the origin.

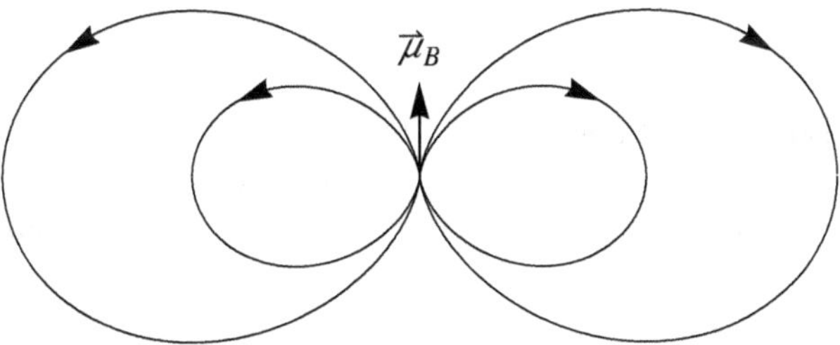

1.4.2 Paraxial ray approximation

The procedure we used to find the end field from the bulk body field of a solenoid magnet is a special case of an approximation called the paraxial ray approximation. More generally, one may be given the longitudinal solenoidal field B_s along the solenoid axis. The z-dependence of B_s implies a nonvanishing transverse field B_r by the condition $\nabla \cdot \vec{B} = 0$ (still assuming cylindrical symmetry, although we will stay with the Cartesian coordinates).

Given the z-dependent solenoidal field $B_s(z)$, it can be shown that the connection is given by (Homework 1.53)

$$\vec{B} = -\frac{1}{2}B_s'(z)(x\hat{x} + y\hat{y}) + B_s(z)\hat{z}, \tag{1.29}$$

to which the step-function hard-edge solenoid is a special case.

In passing, note that $\nabla\times\vec{B}=\frac{1}{2}B_s''(z)(y\hat{x}-x\hat{y})\neq 0$ — so Maxwell equations are not strictly fulfilled. This will have some (usually small) beam dynamics consequences. For example, for a step-function solenoid, this extra term involves $\delta'(z)$ — see Homework 1.54, which does not have a net beam dynamics consequence when integrated over.

The power of Eq. (1.29) lies in the fact that, with the input only of the field on axis, it also gives the magnetic field near the solenoid axis for small x and y. Furthermore, it can be extended to higher orders in x and y by iteration. Let

$$\begin{aligned}\vec{B} &= \left[-\frac{x}{2}B_s'(z)+x^3 f(z)\right]\hat{x}+\left[-\frac{y}{2}B_s'(z)+y^3 g(z)\right]\hat{y}\\ &\quad+\left[B_s(z)-(x^2+y^2)h(z)\right]\hat{z}\,,\end{aligned}\tag{1.30}$$

where $f(z),g(z),h(z)$ are connected by Maxwell equations. Condition $\nabla\cdot\vec{B}=0$ gives $f(z)=g(z)=\frac{1}{3}h'(z)$. Condition $\nabla\times\vec{B}=\vec{0}$ to second order in x and y gives $h(z)=\frac{1}{4}B''(s)$.

This algorithm is called the *paraxial ray approximation.* The longitudinal component of the field (1.26) along the solenoid axis can be used to calculate accurately the complete field inside the solenoid away from the axis using the paraxial ray approximation of sufficiently high order. Armed with the paraxial ray approximation, we sometimes quote only the on-axis fields without fear of losing information.

Homework 1.53

(a) Prove Eq. (1.29). Show that $\nabla\cdot\vec{B}=0$.

(b) The on-axis field of a long but finite solenoid is given by Eq. (1.26). Use Eq. (1.29) to show that away from but near its axis, there is a small transverse magnetic field

$$\vec{B}_\perp(z)\approx-\frac{\mu_0 N I_0\rho^2}{4L}\left[\frac{1}{[(\frac{L}{2}+z)^2+\rho^2]^{3/2}}-\frac{1}{[(\frac{L}{2}-z)^2+\rho^2]^{3/2}}\right](x\hat{x}+y\hat{y})\,.$$

This field, although linear in x and y, is neither a quadrupole nor a skew quadrupole (see also Fig. 3.19 later). It clusters around $z=\pm\frac{L}{2}$ and extends a range $\sim\rho$. Investigate how this paraxial ray result approaches the hard-edge result, Eq. (1.27).

Homework 1.54

(a) Prove the connections among $f(z),g(z),h(z)$ in Eq. (1.30) as stated in the text.

(b) Extend the transverse field in Homework 1.53(b) to second order in x,y. Although tedious, it gives a feel of the nonlinear effects due to solenoid end fields. Investigate how this second order end field approaches $\delta'(z)$ in the hard-edge limit as declared in the text.

1.4.3 Helical undulator-solenoid

Another example of current-dominated magnetic device is a helical undulator made from a helical wire. Consider a helical wire carrying a current I. Given the Biot-Savart law (1.25), let the geometry of the wire be

$$x = a\cos kz, \qquad y = a\sin kz\,,$$

where $\vec{r}\,' = (x, y, z)$ traces out the wire when z scans from $-\infty$ to ∞, k is the pitch angle. When z advances by $\frac{2\pi}{k}$, the wire completes one loop. We have

$$d\vec{\ell} = dz(-ak\hat{x}\sin kz + ak\hat{y}\cos kz + \hat{z})\,.$$

Consider an observation location $\vec{r} = (X, Y, Z)$. We have

$$\begin{aligned}
B_x(\vec{r}) &= \frac{\mu_0 I}{4\pi}\int_{-\infty}^{\infty} dz \frac{ak(Z-z)\cos kz - Y + a\sin kz}{[(X - a\cos kz)^2 + (Y - a\sin kz)^2 + (Z-z)^2]^{3/2}}\,,\\
B_y(\vec{r}) &= \frac{\mu_0 I}{4\pi}\int_{-\infty}^{\infty} dz \frac{ak(Z-z)\sin kz + X - a\cos kz}{[(X - a\cos kz)^2 + (Y - a\sin kz)^2 + (Z-z)^2]^{3/2}}\,,\\
B_z(\vec{r}) &= \frac{\mu_0 I}{4\pi}\int_{-\infty}^{\infty} dz \frac{-ak(X\cos kz + Y\sin kz) + a^2 k}{[(X - a\cos kz)^2 + (Y - a\sin kz)^2 + (Z-z)^2]^{3/2}}\,.
\end{aligned}$$

Consider on-axis field only, with $X = 0, Y = 0$, we have

$$\begin{aligned}
B_y(Z) + iB_x(Z) &= \frac{\mu_0 I}{4\pi}\int_{-\infty}^{\infty} dz \frac{-ae^{-ikz}[1 - ik(Z-z)]}{[a^2 + (Z-z)^2]^{3/2}}\\
&= \frac{\mu_0 I}{4\pi} e^{-ikZ}\int_{-\infty}^{\infty} dz \frac{-ae^{-ikz}(1 + ikz)}{(a^2 + z^2)^{3/2}}\\
&= -\frac{\mu_0 I k}{2\pi}\left[K_1(ak) + akK_0(ak)\right] e^{-ikZ}\,,\\
B_z(Z) &= \frac{\mu_0 I}{4\pi}\int_{\infty}^{\infty} dz \frac{a^2 k}{[a^2 + (Z-z)^2]^{3/2}}\\
&= \frac{\mu_0 I}{4\pi}\int_{-\infty}^{\infty} dz \frac{a^2 k}{(a^2 + z^2)^{3/2}}\\
&= \frac{\mu_0 I k}{2\pi}\,,
\end{aligned}$$

where $K_0(x), K_1(x)$ are Bessel functions. It is clear that this device contains a superposition of a transverse magnetic field that rotates with the pitch of the helical wire, and a longitudinal solenoidal field B_z. The transverse helical field results only when the helical pitch is significant [observing $K_{0,1}(ak) \to 0$ when $ak \to \infty$, i.e. when there is no pitching]. The expression for B_z should not come as a surprise.

Take for example $I = 100$ kA, $a = 6$ cm, $\frac{2\pi}{k} = 10$ cm, we have $B_\perp = 0.09$ T, $B_z = 1.26$ T. Such a superposition of fields might become useful when there is a need of inserting an undulator inside a solenoid. In that case, a single undulator-solenoid wire may serve the functions of both.

Homework 1.55 In case what one needs is a pure transverse helical field without a solenoidal component, what one can do is to consider a double-helix configuration with the two helical wires carry equal but opposite currents. Find the transverse field pattern of a double-helical configuration.

Homework 1.56 We found in the text that the helical undulator-solenoid has the on-axis field

$$\vec{B}(X=Y=0) \;=\; \frac{\mu_0 I}{2\pi a}\Big[ak\hat{z} - ak[K_1(ak) + akK_1(ak)](-\hat{x}\sin kZ + \hat{y}\cos kZ)\Big].$$

Apply the paraxial ray approximation, Sec. 1.4.2, to find the magnetic field near, but not exactly on, the axis.

Solution A plot of the function $x[K_1(x) + xK_0(x)]$ is given in Fig. 1.35. You may want to examine the field behavior when $k \to 0$ and $k \to \infty$.

Figure 1.35: Plot of $x[K_1(x) + xK_0(x)]$ as a function of x.

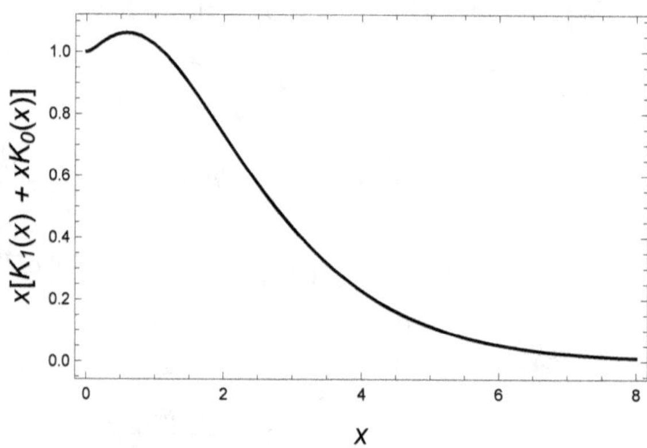

1.5 Space charge

So far we have been discussing the desired magnetic fields that a designer introduces into an accelerator. There are also undesired fields we need to understand, and we need to minimize their effects on the beam. One such case is the *space charge* effect.

Consider a cylindrically shaped beam with uniform distribution moving in the z-direction as shown in Fig. 1.36. Away from the ends of the beam, the electric field is in the r-direction. Drawing a cylinder of radius $r < a$ inside the

Figure 1.36: A cylindrically symmetric uniform beam with space charge.

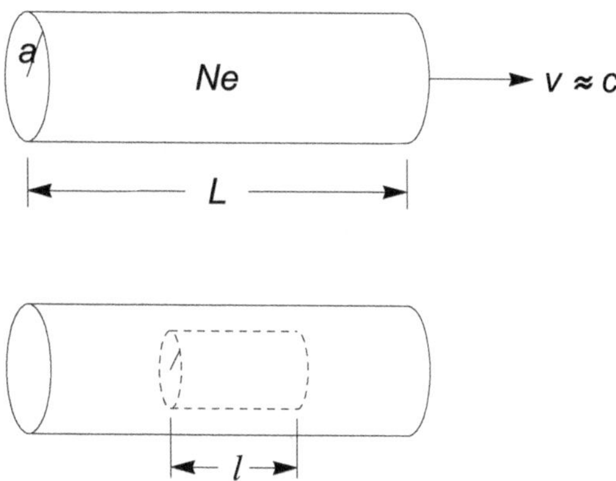

beam distribution (dotted cylinder in the lower panel of Fig. 1.36), and applying Gauss law, one obtains E_r at a radius r from the beam axis,

$$E_r = \frac{Ne}{2\pi\epsilon_0 L}\begin{cases} \frac{r}{a^2}, & r < a, \\ \frac{1}{r}, & r > a. \end{cases}$$

Incidentally, we said the space charge electric field is in the $\hat{r}$ direction. There are two reasons why it is so: (i) Lorentz contraction if the beam is relativistic, and (ii) the beam bunch is most likely long compared with its transverse size. Either one of these reasons will justify the statement.

Similarly, applying Ampere's law, one obtains the magnetic field inside the beam,

$$B_\theta = \frac{\mu_0 v Ne}{2\pi L}\begin{cases} \frac{r}{a^2}, & r < a, \\ \frac{1}{r}, & r > a. \end{cases}$$

The electric and magnetic fields are sketched in Fig. 1.37.

Gauss law We just mentioned the Gauss law. The Gauss law

$$\nabla \cdot \vec{E} = \frac{\rho}{\epsilon}, \qquad \text{or} \qquad \oint d\vec{S} \cdot \vec{E} = \frac{q}{\epsilon}, \tag{1.31}$$

is a truly remarkable law of physics. It states that integrating the electric field over the surface of a closed volume of arbitrary size and shape is equal to the total charge within the volume, divided by ϵ.

One of its remarkable features is that it applies even when the charges are moving, even relativistically, and in fact even if the charges are accelerated in some complex pattern. There is no need to worry about what causality could

Figure 1.37: The electric and magnetic fields of the space charge of a cylindrical uniform beam.

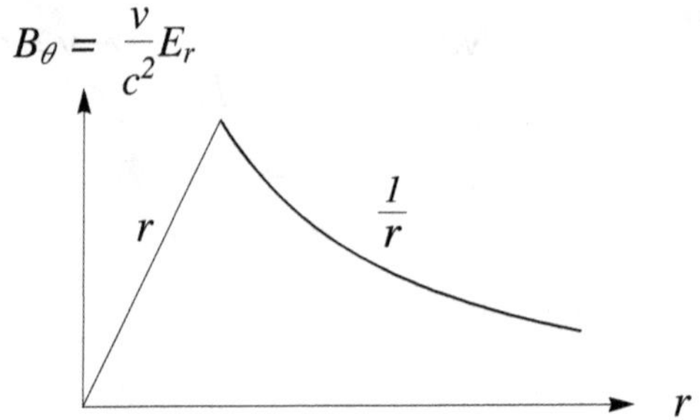

do here no matter how tricky the charge motion is made to be. Perhaps one may consider a point charge 1 nm inside a box, and let the charge move 2 nm to get out of the box. The field pattern has not changed much, but the surface integral of the electric field has just made a sudden discrete drop at the moment when the charge exits the box. Gauss law is an incredibly robust physical law. In comparison, Newton's law breaks down when relativity is introduced.

We will return to the subject of Gauss law in Chapter 8 again when we discuss wakefields.

Space charge force The next question to ask is what is the Lorentz force experienced by a particle in the beam due to the space charge fields it sees. To address this question, one needs E_r and B_θ inside the beam cylinder ($r < a$). The Lorentz force experienced by a particle in the beam is

$$\begin{aligned}\vec{F} &= e\vec{E} + e\vec{v}\times\vec{B} \\ &= \frac{Ne^2}{2\pi\epsilon_0 a^2 L}r\hat{r}\left(1-\frac{v^2}{c^2}\right) = \frac{Ne^2}{2\pi\epsilon_0 a^2 L\gamma^2}r\hat{r}\,,\end{aligned}$$

where $\gamma = (1-\frac{v^2}{c^2})^{-1/2}$ with v given by the same v that appeared in the expression of the magnetic field because the test particle is a particle in the beam and moves with the beam with the same velocity.

The $1-\frac{v^2}{c^2}$ cancellation The first term in the Lorentz force gives the electric component and the second term gives the magnetic component. These two components have opposite signs, i.e., they tend to cancel each other. The electric force is repulsive and defocusing. The magnetic force is attractive and focusing. This cancellation is almost complete when the beam motion is relativistic ($\gamma \to \infty$).[15] This almost-perfect cancellation between the electric and the mag-

[15]Actually, the cancellation is not pure luck. One can obtain the same conclusion by going to the moving frame of the beam, in which the only force is Coulomb repulsion (defocusing).

netic forces is very important, without which most accelerators will not work. Mother Nature can be kind sometimes, in spite of Murphy's law!

If the beam intensity is high, the space charge perturbation can be large. Space charge force, in spite of the cancellation between the electric and magnetic components, imposes a limit on the highest beam intensity for low-to-medium proton or heavy-ion accelerators.

At this point, a few remarks are in order.

- Can we invent a way to make the cancellation more complete? As it stands, unfortunately, the unfriendly electric force is slightly *larger* than the friendly magnetic force. How to make the magnetic force larger than the electric force? One possible answer: Consider the case when an electron beam traverses a background of positively charged ions. The stationary ions will neutralize the electric force of the electrons, but have no effect on the magnetic force.

- Are there situations when this fortunate cancellation is broken? If so, the instability threshold will be substantially lowered, and we will have to try very hard to avoid such situations. The answer to this question is yes. There are in fact many situations when this cancellation is broken and a lot of the research of accelerator physics has actually been devoted to recover it in such situations. We will revisit this subject in Chapter 8.

- The $1 - \frac{v^2}{c^2}$ cancellation was derived assuming a uniform cylindrically-symmetric beam. Would it still apply if the beam distribution is non-uniform, or non-cylindrically-symmetric? Answer: Fortunately yes. The only price to pay here is that the calculation of the fields and the Lorentz force becomes more involved but the cancellation remains.

- Would it still apply if the beam is not in free space, and there is a boundary wall that carries image charges and currents? The answer to this question depends on the type of boundary wall. In generally, the electric and magnetic forces do not cancel. This effect is then closely related to a phenomenon called *Laslett tune shift* in particular, and to *wakefield* effects in general. This consideration of wakefields leads to a significantly lowered limit for the beam intensity. There is however one important exception when the cancellation still applies, and that occurs when the boundary is a perfectly conducting pipe with a uniform transverse cross-section. For this special case, the $1 - \frac{v^2}{c^2}$ cancellation still applies. The challenge to the accelerator designers, therefore, becomes how to hold on to the cancellation as much as possible by constructing the accelerator with its vacuum chamber as conducting and as smooth as possible.

- Still another situation that fails the cancellation occurs in the *beam-beam interaction*. Beam-beam interaction occurs when two beams are made to

The extra factor of $\frac{1}{\gamma^2}$ is then a result of Lorentz transforming back to the laboratory frame.

collide with each other in a collider accelerator. This is the worst offender of the cancellation rule because in this case, the magnetic force reverses direction because the velocity of the test particle is now reversed as shown in Fig. 1.38. As a result, the electric and magnetic forces *add*. Beam-beam force therefore tends to be very strong, and is not helped by going to high beam energies. It is the leading limitation in the operation of collider accelerators. What saves us from the beam-beam force is the fact that this force acts only over the distance when the two beams cross each other, which is usually very short and occupies only a small fraction of the circumference, while space charge force permeates over the entire circumference. This large reduction, however, turns out insufficient. To further suppress the beam-beam force, clever and often very sophisticated beam optics are invented for storage ring colliders.

Figure 1.38: A test charge e^+ colliding with an on-coming beam in a collider arrangement.

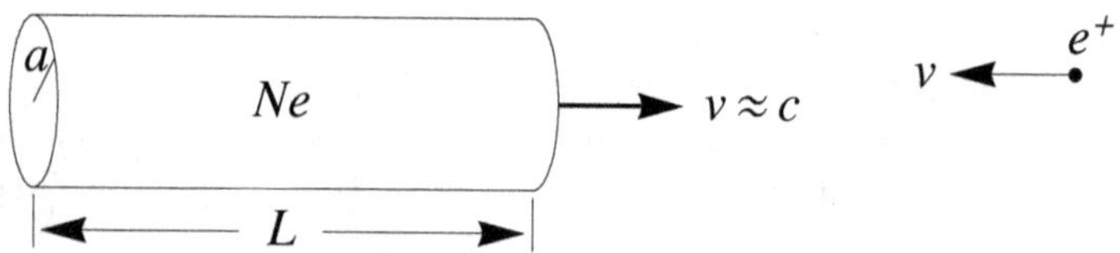

Homework 1.57

(a) Show that the electric and magnetic forces of the beam-beam interaction indeed add, as claimed in the text, and the suppression factor $1 - \frac{v^2}{c^2}$ becomes an enhancement factor $1 + \frac{v^2}{c^2}$, considering the arrangement in Fig. 1.38.

(b) What happens if we switch the test charge e^+ to e^-? What if we switch the beam from Ne to $-Ne$?

(c) If the beam moves with velocity v_1 to the right, and the test charge moves with velocity v_2 to the left, what does the enhancement factor $1 + \frac{v^2}{c^2}$ become?

Homework 1.58

(a) Consider a proton beam in Fig. 1.36 with $a = 2$ mm, $N = 10^{10}$, $L = 1$ m, $\gamma = 2$ (particle energy is 1.8 GeV). Calculate the radial electric field, the azimuthal magnetic field, and the radial Lorentz force experienced by a particle at the beam edge ($r = a$).

(b) Compute the force experienced by the same particle in a quadrupole magnet whose pole tip is at $r = 5$ cm with a pole tip field of 1 Tesla. Compare the result with the space charge force obtained in (a). Try to explain why the space charge force, although much weaker than the quadrupole force, remains a significant perturbation for beam motion.

(c) Still another force experienced by the proton is the Stern-Gerlach force because proton has spin. Calculate the Stern-Gerlach force $F = \frac{e\hbar g_p}{4m_p}G$ in this quadrupole magnet, where $g_p = 5.5856$ is the proton gyromagnetic ratio, G is the quadrupole gradient.

Solution

(a) $E_r = 1.4 \times 10^4$ V/m, $B_\theta = 4.2 \times 10^{-5}$ Tesla, $F_r = 5.7 \times 10^{-16}$ Newton.

(b) $F = 1.7 \times 10^{-12}$ Newton. This quadrupole force applies only over the magnet. Space charge force applies everywhere.

(c) 2.8×10^{-25} Newton.

1.6 Beam position monitor

Image fields Consider a beam moving in the z-direction inside a perfectly conducting circular metal pipe with inner radius b. Let the beam be represented as an infinitely long moving line charge with linear density λ. The beam is displaced transversely by $\vec{a} = (a\cos\phi, a\sin\phi)$ relative to the axis of the pipe. Our job is to calculate the beam-induced electric and magnetic fields. This can be done using the *method of images.*

Seen inside the pipe, the electric potential contains two contributions, one directly from the line charge of the beam, the other from a line image. The image has a line charge of linear density $-\lambda$, and is located at a position $\vec{a}\,' = (a'\cos\phi, a'\sin\phi)$ where $a' = \frac{b^2}{a}$. See Fig. 1.39.

Figure 1.39: A beam and its image in a circular pipe.

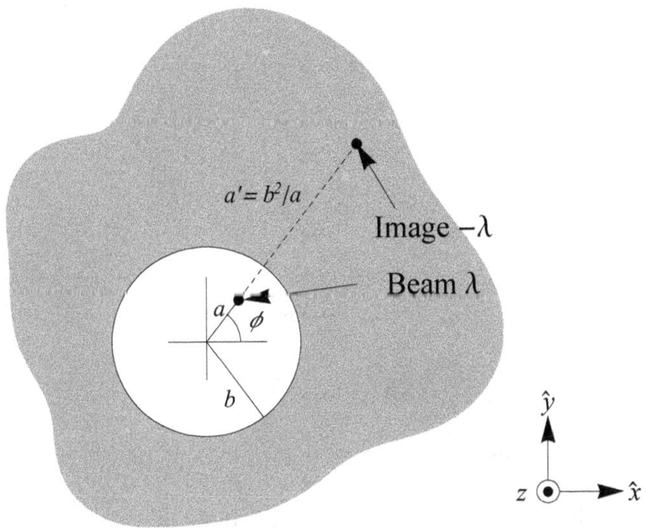

The electric potential at position $\vec{r}$ inside the pipe is given by (Homework 1.59)

$$\Phi(\vec{r}) = -\frac{\lambda}{2\pi\epsilon_0}\ln\left(\frac{|\vec{r}-\vec{a}|}{|\vec{r}-\vec{a}\,'|}\right). \tag{1.32}$$

Electric and magnetic fields inside the pipe are then given by

$$\begin{aligned}\vec{E} &= -\nabla\Phi\,,\\ \vec{B} &= \hat{z}\times\frac{v}{c^2}\vec{E}\,.\end{aligned}$$

BPM signal We are now ready to learn how a *beam position monitor* works. The job of a BPM is to detect the beam displacement in the vacuum chamber. A BPM can be simply modeled as a set of striplines attached to the side of the pipe wall and run longitudinally along the pipe. Each stripline is equipped to measure the total wall current carried on the stripline. By measuring the wall currents on a few strategically arranged striplines, the beam displacement $\vec{a}$ can be deduced.

In practice, the BPMs are made of buttons or striplines. Geometry of these buttons or striplines are not too important for our discussion provided that they are identical, and that they give sufficient signal-to-noise ratio. Different geometries have different frequency band widths and therefore give different beam signal responses as functions of time including transients, but beam displacement information can be extracted independently of the details of these time responses. If one knows the calibration in addition, then not only the beam position but also the beam intensity information can be extracted, and in this case, the BPM acts also as a beam intensity monitor.

To describe the function of a BPM, we first calculate the image charge and current of a displaced beam on the pipe wall. The surface charge Σ on the conducting pipe wall can be found as follows,

$$\begin{aligned}\Sigma(\theta) &= -\epsilon_0\,E_r\big|_{r=a} = \epsilon_0\left.\frac{\partial\Phi}{\partial r}\right|_{r=a}\\
&= -\frac{\lambda}{2\pi}\frac{\partial}{\partial r}\ln\left.\left(\frac{|\vec{r}-\vec{a}|}{|\vec{r}-\vec{a}\,'|}\right)\right|_{r=a}\\
&= -\frac{\lambda}{4\pi}\frac{\partial}{\partial r}\ln\left.\left[\frac{(r\cos\theta-a\cos\phi)^2+(r\sin\theta-a\sin\phi)^2}{(r\cos\theta-\frac{b^2}{a}\cos\phi)^2+(r\sin\theta-\frac{b^2}{a}\sin\phi)^2}\right]\right|_{r=a}\\
&= -\frac{\lambda}{2\pi b}\frac{b^2-a^2}{a^2+b^2-2ab\cos(\phi-\theta)}\,.\end{aligned}$$

As a simple check of this expression, one notes that $\Sigma(\theta) = -\frac{\lambda}{2\pi b}$ when $a=0$, as one would expect.

As an example, consider a BPM made of two stripline- or button-electrodes located symmetrically on the left and right side of a circular beam pipe. Each

Figure 1.40: A two-button (or two-stripline) BPM.

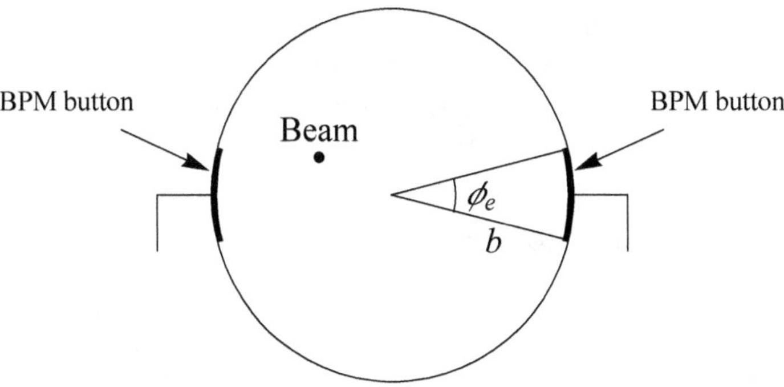

electrode subtends an angle φ_e as shown in Fig. 1.40. Homework 1.61 and 1.62 consider four-button BPMs to measure both x and y beam displacements. Homework 1.63 is for the possibility of a three-button BPM.

The signal seen by, for example, the button on the right side is obtained by integrating the wall current it carries, i.e.

$$\begin{aligned} R &= \int_{-\varphi_e/2}^{\varphi_e/2} d\theta \, \frac{b^2 - a^2}{a^2 + b^2 - 2ab\cos(\phi - \theta)} \\ &\approx \int_{-\varphi_e/2}^{\varphi_e/2} d\theta \left[1 + \frac{2a}{b}\cos(\phi - \theta) \right] \\ &= \varphi_e + \frac{4a}{b}\sin\frac{\varphi_e}{2}\cos\phi \;=\; \varphi_e + \frac{4x}{b}\sin\frac{\varphi_e}{2}, \end{aligned}$$

where $x = a\cos\phi$ ($x \ll b$) is the horizontal displacement of the beam, and we have assumed $a \ll b$, i.e. the beam is close to the center of the BPM.

Similarly, the left button measures the signal

$$\begin{aligned} L &= \int_{\pi-\varphi_e/2}^{\pi+\varphi_e/2} d\theta \, \frac{b^2 - a^2}{a^2 + b^2 - 2ab\cos(\phi - \theta)} \\ &\approx \varphi_e - \frac{4x}{b}\sin\frac{\varphi_e}{2}. \end{aligned}$$

One then combines the signals R and L to extract the horizontal beam position,

$$\frac{R - L}{R + L} = \frac{2x}{b}\,\frac{\sin(\varphi_e/2)}{(\varphi_e/2)}.$$

This measurement result is independent of the beam intensity. As mentioned, it is also insensitive to details of the button's engineering design. On the other hand, for a good BPM sensitivity, it is preferred to have a tight BPM (small b)

and a substantial button angle φ_e. To measure the vertical beam position, the two buttons will be located at the top and bottom of the beam pipe.

BPMs can in principle be built to measure not only the transverse dipole moments (i.e. x and y displacements) of the beam, but also its higher (quadrupole, sextupole $\cdots$) moments. For that purpose, it will be more convenient to consider the wall current as a multipole expansion ($b > a > 0$),

$$\frac{b^2 - a^2}{a^2 + b^2 - 2ab\cos(\phi - \theta)} = 1 + 2\sum_{m=1}^{\infty}\left(\frac{a}{b}\right)^m \cos m(\phi - \theta)\,.$$

The $m = 1$ term is the component being used to measure the beam dipole moment. Other terms will similarly allow measurements of higher beam moments using appropriate button configurations. However, higher multipole moments are increasingly difficult to measure because of the geometric suppression factor $(\frac{a}{b})^m$.

So far we have suggested a resistive BPM where the buttons or striplines detect the image currents flowing on the vacuum chamber wall. One may also consider an inductive BPM which detects the magnetic field carried by the beam as sketched in Fig. 1.41.[16] This is called a Rogowski BPM. One advantage of a Rogowski BPM is that it occupies much less space. The core of the BPM is made of a dielectric material. A ferromagnetic material enhances the signal but is not applicable because it couples the four coil readings. By connecting all four coils, one also have a beam current monitor. One disadvantage of the Rogowski BPM is that it must be installed inside the vacuum chamber because the vacuum chamber wall will disrupt the magnetic field.

Homework 1.59 Although Eq. (1.32) was given without proof, as a check of this potential, we may calculate the potential at the pipe wall. On the wall surface, we have $\vec{r} = (b\cos\theta, b\sin\theta)$ for some θ.

(a) Show that on the wall surface,

$$\frac{|\vec{r} - \vec{a}|}{|\vec{r} - \vec{a}\,'|} = \frac{a^2}{b^2}\,,$$

independently of θ and ϕ. We conclude that our expression gives a constant potential at all points on the wall surface.

(b) As to the magnetic field, show that it is tangential to the pipe wall surface.

Homework 1.60 The image charge and current induced by a displaced beam generates electric and magnetic fields that will affect the motion of particles in the beam. Derive an expression of the Lorentz force induced by the image charge and current as seen by a beam particle. This force is also the cause of Laslett tune shift mentioned in Sec. 1.5. Do you observe a $1 - \frac{v^2}{c^2}$ cancellation?

[16] W. Rogowski and W. Steinhaus, Archiv für Elektrotechnik 1, 141 (1912).

Figure 1.41: Schematic of a Rogowski BPM.

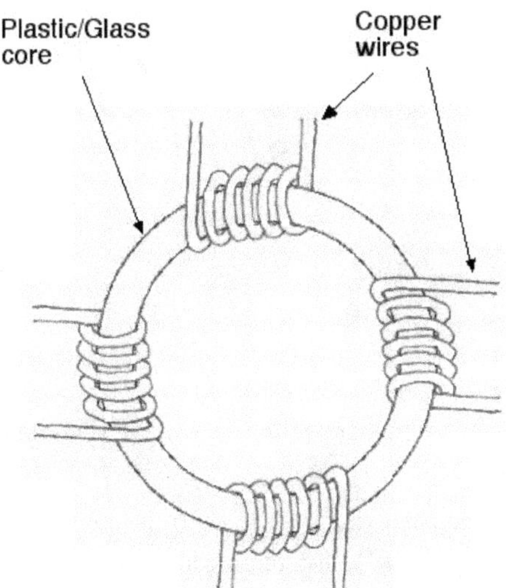

Homework 1.61 One can use a 4-button (left, right, top, bottom) BPM to measure the beam's quadrupole moments. Consider a beam with dipole moments $\langle x\rangle$, $\langle y\rangle$, and quadrupole moments $\langle x^2\rangle$, $\langle xy\rangle$, and $\langle y^2\rangle$.

(a) Find the appropriate combination of the four button signals that will allow the measurements of the two dipole moments without contaminations from the quadrupole moments.

(b) Can you measure the three second moments by some combinations of the four signals?

(c) If the answer to (b) is negative, how could you reconfigure your buttons to measure all three second moments?

Solution

$$\begin{aligned}
R &= \varphi_e + \frac{4}{b}\langle x\rangle \sin\frac{\varphi_e}{2} + \frac{2}{b^2}(\langle x^2\rangle - \langle y^2\rangle)\sin\varphi_e\,,\\
L &= \varphi_e - \frac{4}{b}\langle x\rangle \sin\frac{\varphi_e}{2} + \frac{2}{b^2}(\langle x^2\rangle - \langle y^2\rangle)\sin\varphi_e\,,\\
T &= \varphi_e + \frac{4}{b}\langle y\rangle \sin\frac{\varphi_e}{2} - \frac{2}{b^2}(\langle x^2\rangle - \langle y^2\rangle)\sin\varphi_e\,,\\
B &= \varphi_e - \frac{4}{b}\langle y\rangle \sin\frac{\varphi_e}{2} - \frac{2}{b^2}(\langle x^2\rangle - \langle y^2\rangle)\sin\varphi_e\,.
\end{aligned}$$

The four signals T, R, B, L can be used to extract the data of four quantities, namely $\langle x\rangle, \langle y\rangle, \langle x^2\rangle - \langle y^2\rangle$ and the overall normalization.

Figure 1.42: Configurations of the striplines for an upright and a tilted BPM.

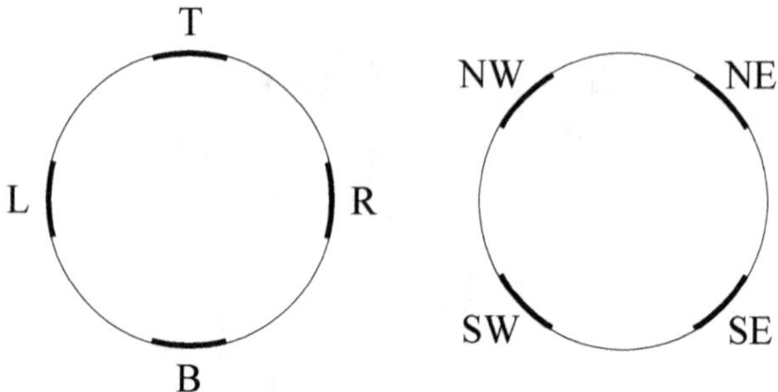

(a) Don't use $\frac{R-L}{R+L}$ to measure $\langle x\rangle$ if your BPM has four buttons. Use $\frac{2(R-L)}{R+L+T+B}$. Similarly measure $\langle y\rangle$ using $\frac{2(T-B)}{R+L+T+B}$.

(b) You can only measure $\langle x^2\rangle - \langle y^2\rangle$. It is contained in $\frac{R+L-T-B}{R+L+T+B}$. For a round beam with $\langle x^2\rangle = \langle y^2\rangle$, for example, it is easy to envision that these BPM signals will yield no information on the beam's second moments.

Homework 1.62 In an electron storage ring, sometimes the four BPM stiplines or buttons are located with 45° tilt relative to the upright configuration in order to avoid the synchrotron radiation. See Fig. 1.42.
What combinations of the four stripline/button signals are used to extract the first and second moments of the beam distribution in this case?

Solution

$$\begin{aligned}
NE &= \varphi_e + \frac{2\sqrt{2}}{b}(\langle x\rangle + \langle y\rangle)\sin\frac{\varphi_e}{2} + \frac{4}{b^2}\langle xy\rangle\sin\varphi_e\,,\\
NW &= \varphi_e - \frac{2\sqrt{2}}{b}(\langle x\rangle - \langle y\rangle)\sin\frac{\varphi_e}{2} - \frac{4}{b^2}\langle xy\rangle\sin\varphi_e\,,\\
SW &= \varphi_e - \frac{2\sqrt{2}}{b}(\langle x\rangle + \langle y\rangle)\sin\frac{\varphi_e}{2} + \frac{4}{b^2}\langle xy\rangle\sin\varphi_e\,,\\
SE &= \varphi_e + \frac{2\sqrt{2}}{b}(\langle x\rangle - \langle y\rangle)\sin\frac{\varphi_e}{2} - \frac{4}{b^2}\langle xy\rangle\sin\varphi_e\,.
\end{aligned}$$

The four signals can be used to extract the data of four quantities, namely $\langle x\rangle, \langle y\rangle, \langle xy\rangle$ and the overall normalization,

$$\begin{aligned}
\frac{NE+SE-NW-SW}{NE+SE+NW+SW} &\to \langle x\rangle\,,\\
\frac{NE+NW-SE-SW}{NE+SE+NW+SW} &\to \langle y\rangle\,,\\
\frac{NE+SW-NW-SE}{NE+SE+NW+SW} &\to \langle xy\rangle\,.
\end{aligned}$$

Homework 1.63 Homework 1.61 describes using four buttons to measure simultaneously $\langle x \rangle$ and $\langle y \rangle$ of a beam. Actually information obtained from any three of the buttons are sufficient to do the job. Find explicit expressions of $\langle x \rangle$ and $\langle y \rangle$ in terms of any three signals from either R, L, T, B or NE, NW, SW, SE. If one has four-button BPMs, the extra fourth value can be used as self-consistency check to detect flawed monitors. It is not clear why this application is not extensively used by the accelerator operators.

Homework 1.64 Consider a two-button $\langle x \rangle$ BPM with a construction error in that its left and right buttons have slightly different subtended angles φ_e and φ_e' respectively. As a result, beam position measurements are misinterpreted. Calculate the error of beam position $\langle x \rangle$ due to this construction error. What do you think should this construction error tolerance be?

Chapter 2

Transverse Motion

In Chapter 1, we mentioned that the 3-D motion of a charged particle in an accelerator is described in a 6-D phase space. The motion can be decomposed in terms of a horizontal, a vertical, and a longitudinal dimension, each contributing two degrees of freedom of phase space.

It turns out that in most accelerators, these three dimensions are decoupled, or at least nearly decoupled, by the accelerator design. As a result, the discussion of accelerator design and its accelerator physics can first be approximately described as three independent, uncoupled dimensions. When couplings occur, we treat them as perturbations to the nominal uncoupled case. Some of these perturbations causing coupling will be discussed in Chapter 3.

To zeroth order, therefore, our discussion of accelerator physics proceeds by first treating each of the three dimensions independently. Indeed, each of these three dimensions have their special considerations. For convenience, however, the horizontal and the vertical dimensions are sufficiently similar that we combine their treatments together in this chapter and Chapter 3, and we call that the transverse motion. The remaining dimension, the longitudinal, will be postponed till Chapters 4 and 5.

One might ask what is the reason that conventional accelerators always start with a decoupled design. In a sense, this is driven at least partly by a preference towards simplicity (or even a lack of imagination) and not necessarily based on a knowledge of the eventual optimized accelerator design. There are more venturous and ingenious cases when intrinsically coupled systems become advantageous. Those are, however, considered outside the scope of our discussions.

2.1 Equation of motion

2.1.1 Two treatments

In general, there are two treatments we can take to derive the equation of motion. One is a formal treatment starting with the *Hamiltonian* of the system.

Table 2.1: Two treatments to derive the equation of motion.

Starting point	Features
Hamiltonian	• formal • less intuitive • robust (small terms within the Hamiltonian can be dropped more readily)
Lorentz force equation	• less formal • more intuitive • less robust (dropping small terms in the force equation casually can violate fundamental principles of dynamics)

This is a lengthier but more rigorous treatment. The other treatment starting with the *Lorentz force equation* is briefer, but is more susceptible to mistakes when approximations are made by dropping some terms which may seem legitimately small but actually cannot be dropped without violating some deeper dynamical principles. When making approximations, it is safer to make them with the Hamiltonian than with the equations of motion.

We will adopt the second, less rigorous but more intuitive, treatment starting with the Lorentz equation.[1] The reader should be aware of this conscious choice made at the start. See Table 2.1 for a brief comparison between the two treatments.

2.1.2 Derivation

In the following, we shall develop the equation of motion with a goal. The goal is to analyze the particle motion in an accelerator whose design trajectory consists of a straight line or a planar curve, or a combination. By implication, we are developing the equation of motion for a linac, a synchrotron, or a storage ring. An accelerator whose design trajectory is not a piecewise combination of straight lines or planar curves will require extensions of our analysis. On the other hand, our analysis allows a mixed use of horizontal and/or vertical curvatures in these accelerators as long as each of the piecewise sections is planar.

Coordinate system Let us describe particle motion with a coordinate system shown in Fig. 2.1. The design trajectory has a local curvature ρ. The path length along the design trajectory is designated by s. Ultimately, s will be the independent "time" variable. At any point along the design trajectory, we define

[1] As a last step, the equations of motion obtained should be checked to see if they still obey Hamiltonian principles. This step will also be omitted in later discussions.

Figure 2.1: Coordinate system used to describe particle motion in a curved section of an accelerator. To apply to a straight line section, set $\rho = \infty$.

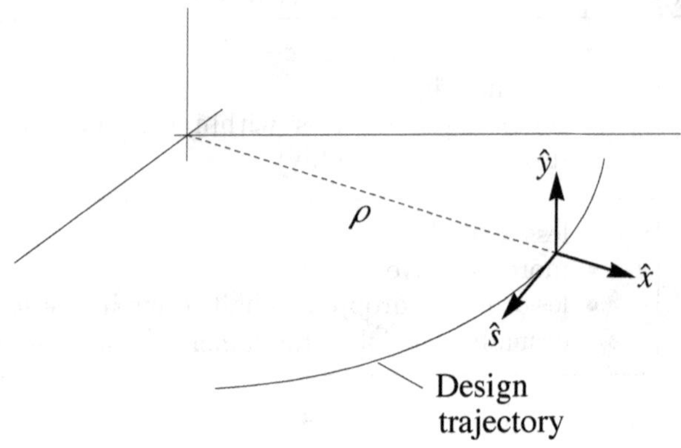

three right-handed unit vectors $\hat{x}$, $\hat{y}$ and $\hat{s}$. The position of a particle in this coordinate system is

$$\vec{R} = r\hat{x} + y\hat{y}$$

where $r = \rho + x$.

We are interested in knowing how the deviations x and y from the design trajectory behave as functions of s. The particle's transverse location is described by the quantities $x(s)$ and $y(s)$, both as functions of s. To describe the transverse motion of the particle, we will need two more quantities $x'(s) = \frac{dx}{ds}(s)$ and $y'(s) = \frac{dy}{ds}(s)$. The 4-D space (x, x', y, y') constitutes the transverse phase space.

At this point, a clear depiction of the coordinate system is advised. Let us consider a certain moment of time when a particle's location is known, what are exactly its x and y coordinates? To answer this question, one has to first identify what is the position s for this particle at this moment of time. The correspondence between the (arbitrary) particle position and accelerator coordinate s is unique. Given the particle position, there is a unique coordinate s. Given s, the unit vectors $\hat{x}(s)$ and $\hat{y}(s)$ are determined. The coordinates $x(s)$ and $y(s)$ of the particle are then determined. As the particle begins to move, definitions of $x'(s)$ and $y'(s)$ follow. The analysis that follows in the rest of this section is based on this physical picture of the coordinate system.

One might ask why we address x and y as functions of s, and not as the more conventional functions of time t. The answer is rather obvious. Accelerator physicists want to describe the particle motion relative to the position around the accelerator. All the accelerator beamline elements, such as magnets or beam position monitors, are located with fixed s, and their actions on a charged particle are executed as the particle passes by at the fixed s where the element is located. In contrast, the exact behavior in t is associated with times of arrivals of a particle at a specific location. The spacing between the arrival times of successive revolutions is approximately, but not exactly, equal to the revolution period T_0. The small variations of arrival times from revolution to revolution is sometimes important to the beam dynamics. This makes time t a *dynamical* quantity. It is not to be used as the independent time variable.

Kinematics Before we actually introduce the various forces exerting on the particles, let's first examine the *kinematics* in this coordinate system. The velocity is given by

$$\begin{aligned} \vec{v} &= \dot{\vec{R}} \\ &= \dot{r}\hat{x} + r\dot{\hat{x}} + \dot{y}\hat{y} \\ &= \dot{r}\hat{x} + r\dot{\theta}\hat{s} + \dot{y}\hat{y} \\ \text{or} \qquad & v_x = \dot{r}, \quad v_y = \dot{y}, \quad v_s = r\dot{\theta}, \end{aligned}$$

where v_s is the tangential component of the velocity. We have used the fact that $\hat{y}$ is a constant unit vector (independent of s) within this piecewise-curved section, while $\hat{x}$ changes according to the geometry of the coordinate system,

$$\dot{\hat{x}} = \dot{\theta}\hat{s}.$$

One also can calculate the acceleration vector

$$\begin{aligned} \ddot{\vec{R}} &= \ddot{r}\hat{x} + \dot{r}\dot{\hat{x}} + \dot{r}\dot{\theta}\hat{s} + r\ddot{\theta}\hat{s} + r\dot{\theta}\dot{\hat{s}} + \ddot{y}\hat{y} \\ &= \ddot{r}\hat{x} + 2\dot{r}\dot{\theta}\hat{s} + r\ddot{\theta}\hat{s} - r\dot{\theta}^2\hat{x} + \ddot{y}\hat{y}, \end{aligned}$$

where we have used

$$\dot{\hat{s}} = -\dot{\theta}\hat{x}.$$

The rules $\dot{\hat{x}} = \dot{\theta}\hat{s}$ and $\dot{\hat{s}} = -\dot{\theta}\hat{x}$ can be obtained by considering the geometry shown in Fig. 2.2.

Before we leave the subject of the kinematics, there is one more key quantity to calculate, i.e. the calculation of $\frac{ds}{dt}$. We will use this quantity momentarily when we need to convert the time variable from t to s. But first, one must take a note that it would be wrong to casually write $\frac{ds}{dt} = v_s$. To illustrate the problem, consider Fig. 2.3.

From Fig. 2.3, along its trajectory, a particle moves a distance $v_s\Delta t$ in time Δt. Note that v_s refers to the particle velocity along its trajectory, and not the velocity of the ideal design particle along the design orbit. Since we know $v_s = r\dot{\theta}$, this gives $\Delta\theta = \frac{v_s\Delta t}{r}$. We then refer to the design orbit to obtain the corresponding change in Δs, which is given by $\Delta s = \rho\Delta\theta$. Combining these observations then gives

$$\frac{ds}{dt} = v_s\frac{\rho}{r} = v_s\frac{\rho}{\rho + x}. \tag{2.1}$$

Figure 2.2: The rates of changing unit vectors $\hat{x}$ and $\hat{s}$.

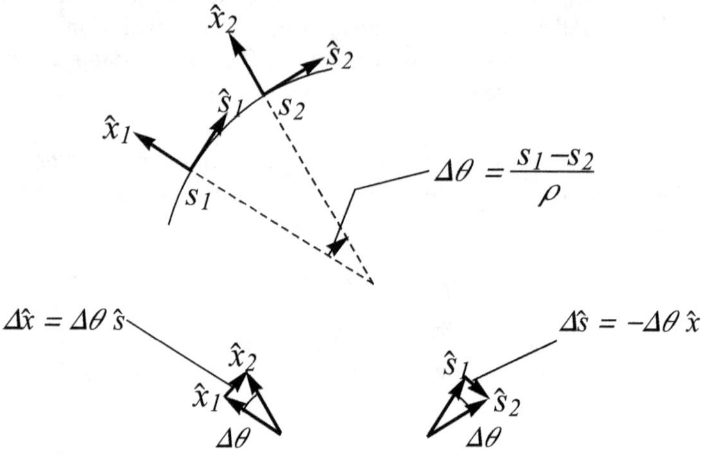

Figure 2.3: How the longitudinal coordinate s vary in time.

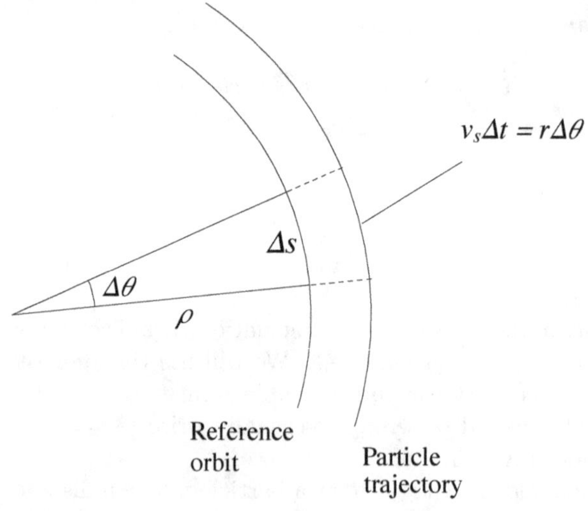

Dynamics We now add the *dynamics* by including the Lorentz force. For now, we assume that there are no electric field (no RF cavities), and that the magnetic field is purely transverse (only hard-edge multipoles, no solenoids), i.e. it has only x- and y-components. The equation of motion is

$$\begin{aligned}\frac{d\vec{p}}{dt} &= e\vec{v}\times\vec{B}\\ &= \text{Det}\begin{vmatrix}\hat{x} & \hat{y} & \hat{s}\\ ev_x & ev_y & ev_s\\ B_x & B_y & 0\end{vmatrix} = -ev_sB_y\hat{x} + ev_sB_x\hat{y} + e(v_xB_y - v_yB_x)\hat{s}\,,\end{aligned}$$

where the last term is very small because $v_{x,y} \ll v_s$.

Be reminded that for any particle's position as it propagates along the accelerator, one first needs to identify its unique corresponding coordinate s. In writing down the equation of motion using s as the independent time variable, all magnetic fields are expressed as functions strictly of s. This means the only meaningful beamline elements will have to be the hard edge devices with their fields changing sharply and cleanly according to s. Implicitly, therefore, all magnets are hard-edge magnets with their entrance and exit faces all perpendicular to $\hat{s}$, i.e. all perpendicular to the design beam trajectory. In case this is not the case, such as when a magnet has its entrance or exit faces not perpendicular to $\hat{s}$, an additional special treatment is required. One example of such is the case of edge focusing to be discussed in Sec. 2.2.8.

In a magnetic field, the energy of a particle does not change. This means γ is a constant of the motion, and therefore

$$\begin{aligned}\frac{d\vec{p}}{dt} &= \frac{d}{dt}m\gamma\dot{\vec{R}} = m\gamma\ddot{\vec{R}}\\ &= m\gamma\left[\ddot{r}\hat{x} + 2\dot{r}\dot{\theta}\hat{s} + r\ddot{\theta}\hat{s} - r\dot{\theta}^2\hat{x} + \ddot{y}\hat{y}\right].\end{aligned}$$

The transverse equations of motion then read

$$\begin{aligned}x\text{-component:}\qquad \ddot{r} - r\dot{\theta}^2 &= -\frac{eB_y}{m\gamma}v_s\,,\\ y\text{-component:}\qquad \ddot{y} &= \frac{eB_x}{m\gamma}v_s\,.\end{aligned}$$

There is also a longitudinal s-component of the equation of motion, but we are not interested in it here.[2]

We will next transform the independent variable of these equations from time t to path length s along the design orbit. To make this transformation, we need the quantity $\frac{ds}{dt}$ calculated in Eq. (2.1).

So far our analysis is exact for the x- and y-components of the particle motion. We now make approximations by assuming that the motion is relativistic, and that the bending curvature of the design trajectory is large, with $\rho \gg |x|$. Note that $\frac{ds}{dt}$ differs from v_s by a factor $\frac{\rho}{\rho+x} \approx 1$.

[2]Warning once more: we are taking a step of dropping terms – even though they are shown to be small – while adopting the equation of motion treatment. As mentioned, this is a step potentially dangerous.

For a relativistic particle, v_s is almost a constant of the motion because $v_s = \sqrt{v^2 - v_x^2 - v_y^2}$, where v is strictly a constant of the motion while $v_{x,y} \ll v$ contribute only second order corrections to v_s. In what follows, we will consider the kinematics part of the equations of motion only to first order in x and y, i.e. all nonlinear terms from kinematics will be dropped. We shall discuss nonlinear terms later, but all those nonlinearities will be assumed to come from the magnetic fields $B_{x,y}$ and not from the kinematics.

Dropping all nonlinear terms from kinematics is a good approximation in most (although not all) practical high energy accelerators. For low-energy storage rings, on the other hand, they are the source of much subtleties and headaches when applying various existing computer codes. Depending on the approximations made, different codes often give different answers. Users beware.

Back to our derivation of equation of motion, so far we have made three approximations,

- $\gamma \gg 1$,
- $\rho \gg |x|$,
- dropping all nonlinear terms from kinematics.

To simplify the analysis, we also assume ρ is constant over the range of s for the accelerator section under consideration. Strictly speaking, the equations of motion we obtain, therefore, apply only when ρ is piecewise constant in the accelerator, satisfying

$$\frac{1}{\rho} = \frac{eB_{y0}}{P},$$

where B_{y0} is the constant vertical magnetic field along the design orbit, and $P = m\gamma v$ is the design momentum of the particle. In general, however, our results will still apply for a piecewise-varying ρ by properly defining the design orbit.

With these approximations, we have

$$\begin{aligned}
\dot{r} &= \frac{v_s\rho}{\rho+x}\frac{dr}{ds} = \frac{v_s\rho}{\rho+x}\frac{dx}{ds}, \\
\ddot{r} &= -\frac{v_s\rho}{(\rho+x)^2}\left(\frac{dx}{ds}\right)^2\frac{ds}{dt} + \left(\frac{v_s\rho}{\rho+x}\right)^2\frac{d^2x}{ds^2} \\
&= \left(\frac{v_s\rho}{\rho+x}\right)^2\left[-\frac{1}{\rho+x}\left(\frac{dx}{ds}\right)^2 + \frac{d^2x}{ds^2}\right] \\
&\approx \left(\frac{v_s\rho}{\rho+x}\right)^2\frac{d^2x}{ds^2}, \\
\ddot{y} &\approx \left(\frac{v_s\rho}{\rho+x}\right)^2\frac{d^2y}{ds^2}.
\end{aligned}$$

The r-component of the equation of motion, for example, now reads

$$\left(\frac{\rho}{\rho+x}\right)^2 \frac{d^2x}{ds^2} - \frac{1}{\rho+x} = -\frac{eB_y}{m\gamma v_s}, \tag{2.2}$$

where use has been made of $r\dot{\theta} = v_s$.

We keep only to first order in x and y. The equations of motion then become

$$\begin{aligned} x\text{-component:} \quad & \frac{d^2x}{ds^2} + \frac{x}{\rho^2} \approx -\frac{e(B_y - B_{y0})}{P}, \\ y\text{-component:} \quad & \frac{d^2y}{ds^2} \approx \frac{eB_x}{P}, \end{aligned} \tag{2.3}$$

where use has also been made of $P = m\gamma v \approx m\gamma v_s$. The quantity $\frac{e}{P}$ is just the magnetic rigidity defined in Chapter 1,

$$\frac{e}{P} = \frac{1}{(B\rho)}.$$

Equations (2.3) are our transverse equations of motion. The x- and y-motions, at least the oscillating parts of the motions, obtained by solving them describe the *transverse betatron oscillations.* The nonoscillatory part is called the *closed orbit* or closed-orbit deviation. It is perhaps worth a remind that we have dropped in Eq. (2.3) all nonlinear terms in x and y of the kinematic origin. All nonlinearities will come from the magnetic fields B_x and B_y.

Note the asymmetry between the x- and the y-equations in (2.3). It is intuitively clear why the y-motion is driven by B_x. The x-motion, however, is not driven by B_y, but by $B_y - B_{y0}$. There is also an "extra" term $\frac{x}{\rho^2}$ on the left-hand-side due to the natural focusing of dipole bending as mentioned in Sec. 1.2.4. It may be instructive for the reader to think over the origin of this x-y asymmetry.

Homework 2.1 Prove, by using the Lorentz force equation, that magnetic fields bend the trajectory of a moving particle, but do not do any work on it. This is true for arbitrary magnetic fields. It does not change the particle's energy, i.e. in a magnetic field, $\gamma =$ constant. Magnetic force is an effortless force that deflects 1000 pounds with 4 ounces 四兩撥千斤. A lot of Chinese kung-fu must have extensively used the magnetic force.

Homework 2.2

(a) In the text, the Lorentz force equation was developed in the Cartesian $(\hat{x}, \hat{y}, \hat{s})$ system. Repeat the analysis in the cylindrical system $(\hat{r}, \hat{\theta}, \hat{z})$ given $\vec{E}$ and $\vec{B}$.

(b) Apply the result to the special case $\vec{E} = \vec{0}, \vec{B} = B_0\hat{z}$.

(c) Apply to the special case $\vec{B} = \vec{0}, \vec{E} = \frac{\lambda}{2\pi\epsilon_0 r}\hat{r}$.

Figure 2.4: Cartesian and cylindrical systems.

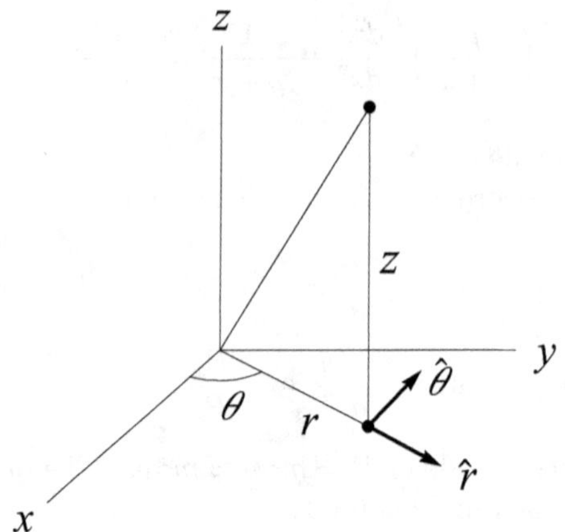

Solution

(a) See Fig. 2.4.

$$\begin{aligned}
\vec{R} &= r\hat{r} + z\hat{z}\,, \\
\dot{\hat{r}} &= \dot{\theta}\hat{\theta}\,, \\
\dot{\hat{\theta}} &= -\dot{\theta}\hat{r}\,, \\
\dot{\vec{R}} &= \dot{r}\hat{r} + r\dot{\hat{r}} + \dot{z}\hat{z} \;=\; \dot{r}\hat{r} + r\dot{\theta}\hat{\theta} + \dot{z}\hat{z}\,, \\
\ddot{\vec{R}} &= \ddot{r}\hat{r} + 2\dot{r}\dot{\theta}\hat{\theta} + r\ddot{\theta}\hat{\theta} - r\dot{\theta}^2\hat{r} + \ddot{z}\hat{z}\,.
\end{aligned}$$

(b)

$$m\gamma\ddot{\vec{R}} \;=\; e\dot{\vec{R}}\times\vec{B} \;=\; eB_0(r\dot{\theta}\hat{r} - \dot{r}\hat{\theta})$$

$$\Longrightarrow \begin{cases} \ddot{r} - r\dot{\theta}^2 = \frac{eB_0}{m\gamma}r\dot{\theta}, \\ 2\dot{r}\dot{\theta} + r\ddot{\theta} = -\frac{eB_0}{m\gamma}\dot{r}, \\ \ddot{z} = 0. \end{cases}$$

2.1.3 Combined-function magnet

A particularly important application of the transverse equations of motion occurs when the magnetic field contains a dipole component plus a quadrupole component as

$$\vec{B} \;=\; B_{y0}\hat{y} + G(y\hat{x} + x\hat{y})\,,$$

where $G = \frac{\partial B_y}{\partial x} = \frac{\partial B_x}{\partial y}$ is the quadrupole field gradient. This is a *combined-function magnet* as discussed in Chapter 1. The equations of motion become

$$\frac{d^2x}{ds^2} + \left(\frac{G}{(B\rho)} + \frac{1}{\rho^2}\right)x = 0,$$
$$\frac{d^2y}{ds^2} - \frac{G}{(B\rho)}y = 0, \tag{2.4}$$

where $(B\rho) = \frac{P}{e}$ is the rigidity. We will drop the parentheses from $(B\rho)$ from now on.

Consider a circular accelerator consisting only of (hard-edge) dipoles, quadrupoles, and drift spaces. In this accelerator, B_{y0} and G are considered functions of s, and they are necessarily periodic functions of s. The period is given by the accelerator circumference C. In case the accelerator circumference strings together P identical sections, each of length $L = \frac{C}{P}$, then the period becomes L.

The transverse equations of motion then assumes the form

$$u'' + K(s)u = 0, \tag{2.5}$$

where u represents either x or y, a prime indicates taking derivative with respect to s, and $K(s)$ is a periodic function of s with period L,

$$K_x = \frac{1}{B\rho}\frac{\partial B_y}{\partial x} + \frac{1}{\rho^2},$$
$$K_y = -\frac{1}{B\rho}\frac{\partial B_y}{\partial x}. \tag{2.6}$$

In the above focusing strengths, the $\frac{1}{\rho^2}$ term comes from dipoles, terms involving $\frac{\partial B_y}{\partial x}$ come from quadrupoles. The $\frac{1}{\rho^2}$ term testifies that there exists a *natural* x-focusing from dipole, but focusing in y requires quadrupoles (with a proper sign). This we learned before in Chapter 1.

Equation of the form (2.5) is a Hill's equation. Our challenge is to solve it. Note that the focusing function $K(s)$ must be periodic in s, but is otherwise arbitrary at this point. This periodicity condition on $K(s)$ is very critical and is not to be taken lightly. It imposes a strong constraint to the dynamical system, and it is because of this seemingly too-general-to-be-useful condition that permits a lot of the developments to be covered in this chapter.

2.1.4 Why not electric bends?

An alert reader may ask why do we use magnetic field to bend particle trajectories, and why not electric field. To illustrate this, consider the two arrangements, a magnetostatic magnet and an electrostatic high voltage device, as shown in Fig. 2.5.

Whether a magnetic device or an electric device is more effective depends on whether ρ_B or ρ_E is smaller. Let us take their ratio,

$$\frac{\rho_E}{\rho_B} = \frac{Bv}{E} = 300\frac{B\,[\mathrm{T}]}{E\,[\mathrm{MV/m}]} \times \frac{v}{c}. \tag{2.7}$$

Figure 2.5: Comparing a magnetic device (left) and an electric device (right) for the purpose of bending the trajectory of a beam.

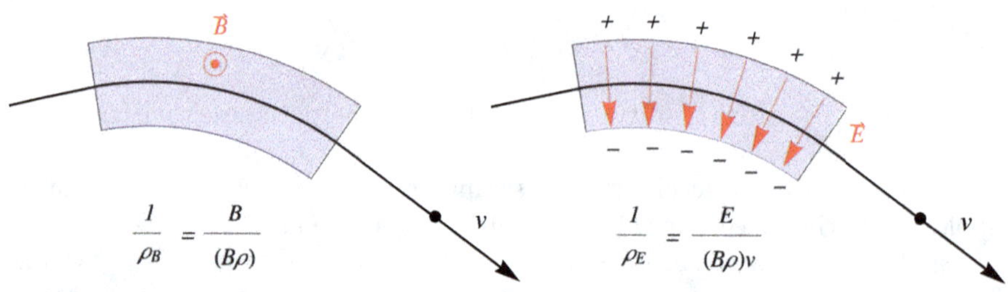

The electric device will be more effective if $E/B > v$.

Earlier in Chapter 1, we had a Homework 1.11 to calculate $\frac{E}{B}$ when their field energy densities are equal, and that occurs when $\frac{E}{B} = c$. Thus a 1-Tesla magnetic field has the same energy density as a 300-MV/m electric field. Here we have $\frac{E}{B} = v$ when their bending powers are equal. These two comparisons differ by a factor of $\frac{v}{c}$ because here the Lorentz force of magnetic field contains an additional factor of v. A magnet will not be effective if the beam is very nonrelativistic.

If we take a typical strong magnetic field of 1 Tesla and a strong electric field of 1 MV/m, then we see that a magnetic device is more effective when $\frac{v}{c} > \frac{1}{300}$ and an electric device is more effective when $\frac{v}{c} < \frac{1}{300}$. For most applications, therefore, magnetic devices are used for beam bending except for very close to the beam sources when the beam is still very nonrelativistic.

However, this strong preference to magnetic element applies only to the transverse beam dynamics because magnetic force is necessarily transverse. Longitudinally, parallel to the design orbit, magnetic force is ineffective. One has to resort to electric force for the control of the longitudinal beam dynamics. As a result, longitudinal motion of a particle in an accelerator is weak and slow compared with its transverse counterparts. Later this will show up, for example, in the fact that the synchrotron tune ν_s is typically much smaller than the betatron tunes ν_x and ν_y by a very large factor, which we crudely estimate here to be ~ 300.

One exception when transverse electric bends become required occurs in a storage ring collider that stores simultaneously electrons and positrons (or protons and antiprotons). When stored simultaneously, the two beams are made to circulate the ring in opposite directions and collide at some intended interaction points, while separated at other unintended positions. If we use exclusively magnetic bends, then the two beams will have exactly the same trajectories and will collide even at the unintended positions. In these cases, we need to separate the trajectories of the two beams using electric bends. Fortunately, the required separation is small enough that electric bends suffice for this purpose.

Homework 2.3

(a) Consider a beam of continuous stream of protons with momentum of 1 GeV/c being bent by a dipole of 1 Tesla and 1 m length. Let the proton beam current be 1 A. Find the magnitude and direction of the force experienced by the magnet due to the push by the proton beam.

(b) The same proton beam is being accelerated by an RF cavity. Let each proton gain an energy of 1 MeV. Find the magnitude and direction of the force experienced by the cavity.

(c) Compare the two forces obtained. What determines the ratio between their magnitudes?

Solution

(a) $F = IB\ell = 1$ Newton.

(b) $F = \frac{WI}{ev} = 0.0045$ Newton. Cavity length does not matter.

2.1.5 Weak focusing ring

In Chapter 1, we discussed the case in which the entire circular accelerator (or storage ring) is made of a single uniform combined-function magnet. In the present language, this corresponds to the case

$$\frac{B_{y0}}{B\rho} = \frac{1}{\rho}, \qquad \frac{G}{B\rho} = -\frac{n}{\rho^2},$$

where ρ is the uniform bending curvature and n is the uniform *field index* defined by

$$\frac{1}{B\rho}\frac{\partial B_y}{\partial x} = -\frac{n}{\rho^2},$$

which in turn is related to the tilt angle on the magnet pole face as discussed in Eq. (1.11).

The equations of motion (2.4) then become

$$x'' + \frac{1-n}{\rho^2}x = 0, \qquad y'' + \frac{n}{\rho^2}y = 0. \tag{2.8}$$

These are particularly simple special cases of Hill's equation. In this case, the focusing functions $K(s)$ are constant in s, and the Hill's equation reduces to that of a simple harmonic oscillator.

It is immediately obvious from (2.8) that for both x- and y-motions to be stable, one must have

$$1 > n > 0.$$

The part $1 > n$ is required for stable x motion. The part $n > 0$ is for a stable y motion. In the stable region, the x- and y-motions are sinusoidal,

$$x \sim \begin{matrix}\sin\\ \cos\end{matrix}\left(\frac{\sqrt{1-n}}{\rho}s\right),$$

$$y \sim \begin{matrix} \sin \\ \cos \end{matrix} \left(\frac{\sqrt{n}}{\rho} s \right),$$

with oscillation wavelengths $\frac{2\pi\rho}{\sqrt{1-n}}$ and $\frac{2\pi\rho}{\sqrt{n}}$, respectively.

In Sec. 1.2, we discussed an earliest ring design with uniform bending with no field gradient. This case corresponds to $n = 0$. We concluded then that in this ring, the y-motion was unstable, but the x-motion was stable, thanks to a natural focusing from the bending field. The situation was shown in Fig. 1.5.

By inspection of the horizontal motion in Fig. 1.5, we also concluded that the x-deviation oscillates with a period $2\pi\rho$, the ring circumference. Indeed, by setting $n = 0$ in Eq. (2.8), we obtain the solution

$$\begin{aligned} x &= x_0 \cos \frac{s}{\rho} + x_0' \rho \sin \frac{s}{\rho}, \\ y &= y_0 + y_0' s. \end{aligned} \tag{2.9}$$

There are four terms in these two expressions, each corresponds to one of the panels in Fig. 1.5.

The introduction of the field gradient ($n \neq 0$) now is to force the x-motion to share its natural focusing with the y-motion. The net result is that both x and y are stable, but both only weakly focused. Their oscillation periods are both longer than the ring circumference $2\pi\rho$. This is a *weak focusing* accelerator.

Most modern accelerators are *strong focusing*. Notable exceptions include the classical weak-focusing cyclotrons that aim for high continuous beam currents. In the strong focusing accelerators, x- and y-motions are tightly focused. Their corresponding betatron oscillation periods are both much shorter than the accelerator circumference. For those accelerators, we need to consider s-dependent focusing, i.e. $K_{x,y}$ depend on s. The art and science of arranging bending and quadrupole (and other) magnets into an accelerator that is not only stable, and not only strongly focusing, but also serves many special requirements optimized for various applications, is called *lattice design*.

Homework 2.4 In Chapter 1, we derived an expression for the field index n,

$$n \approx \frac{\rho g'}{g}, \qquad \frac{G}{B\rho} = -\frac{g'}{g\rho},$$

where g is the magnet pole face gap size, and g' is the pole face opening angle (opening toward outside if positive).

(a) Take $\frac{g}{\rho} = 0.5\%$ for example. If we choose $n = 0.3 \pm 0.01$, where ± 0.01 is the error tolerance, what should the pole gap opening angle be, and what is its tolerance? If the horizontal width of the magnet is 50 cm, the tolerance on Δg = (outer gap) - (inner gap) imposes a tight engineering demand.

(b) If a strong quadrupole is made by using the gap opening angle, and if G is required to have an accuracy of 10^{-3}, find the corresponding requirement on the accuracy of g'.

2.2 Matrix formalism

2.2.1 Transfer matrix

With the advent of separated-function magnets, a most commonly used focusing function $K(s)$ consists of piecewise constants as s varies around the storage ring. In each segment within which K is constant, its constant value does not have to observe the condition $1 > n > 0$. Although a weak focusing storage ring (the whole ring having the same value of n) requires the condition $1 > n > 0$, a strong focusing storage ring does not require it in each segment.

We now consider the equation of motion (2.5) in a region where K = constant, i.e.

$$u''(s) + Ku(s) = 0, \qquad \text{with} \qquad u(0) = u_0,\ u'(0) = u'_0 .$$

We know the solution of this equation is that u as a function of s is sinusoidal if $K > 0$ and hyperbolical if $K < 0$. But our job now is to introduce a matrix formalism to incorporate these solutions. To do so, we first introduce a vector

$$U(s) = \begin{bmatrix} u(s) \\ u'(s) \end{bmatrix} .$$

Then it is easy to show that the solution can be written as

$$U(s) = M(s|0)U(0) , \tag{2.10}$$

where

$$M(s|0) = \begin{cases} \begin{bmatrix} \cos\sqrt{K}s & \frac{\sin\sqrt{K}s}{\sqrt{K}} \\ -\sqrt{K}\sin\sqrt{K}s & \cos\sqrt{K}s \end{bmatrix}, & K>0, \text{ focusing} \\ \begin{bmatrix} 1 & s \\ 0 & 1 \end{bmatrix}, & K=0, \text{ drift space} \\ \begin{bmatrix} \cosh\sqrt{-K}s & \frac{\sinh\sqrt{-K}s}{\sqrt{-K}} \\ \sqrt{-K}\sinh\sqrt{-K}s & \cosh\sqrt{-K}s \end{bmatrix}, & K<0, \text{ defocusing} \end{cases} . \tag{2.11}$$

The matrix $M(s|0)$ is a 2×2 matrix called the *transfer matrix*. It transfers the initial conditions $U(0)$ of a particle motion at position $s = 0$ to $U(s)$ at position s provided that K is constant from $s = 0$ to s. Needless to say, (u, u') constitutes the phase space.

Factorization The reader will recall that the matrix formalism "factorizes" accelerator and beam properties. As mention in Sec. 1.1.7, the factorization property is why the matrix formalism is so powerful and elegant. Here, the factorization lies in the fact that the right-hand-side of Eq. (2.10) factorizes into $M(s|0)$ that depends only on accelerator properties and $U(0)$ that depends only on beam properties. The factorization says that the *same* transfer matrix $M(s|0)$ applies to all particles with different initial conditions.

To be more specific, consider the case if we had not introduced the matrix formalism. In this case, when we study the system, we might start with each particle with a specific initial condition, and follow its motion from $s = 0$ to s. We then repeat the same for all particles in the beam in order to find out the behavior of the accelerator-beam system. If we change the accelerator design, we have to repeat the cumbersome process to find out the effect of the redesign. With the matrix formalism, the information of the system is all contained in the transfer matrix. The process of repeating for all particles becomes unnecessary. Similarly, the cumbersome process is also avoided when the accelerator is unchanged but the beam property (e.g. the beam emittance) is changed.

Application of matrices to accelerator physics is a marvelous invention. More on this subject later.

Homework 2.5 Derive Eqs. (2.10) and (2.11).

Homework 2.6 So far we applied Eq. (2.11) to magnets with piecewise constant strengths. For an academic soluble exercise, consider a special magnet that lies between $s = s_1$ to $s = s_2$ with a focusing function $K(s) = \frac{G}{(s-s_0)^2}$, where $\frac{1}{4} > G$ and $s_2 > s_1 > s_0$. Ignoring the effects due to the solenoidal field within the magnet (dictated by Maxwell equations), show that the transfer map through the magnet is

$$M(s_2|s_1) = \frac{(t_1 t_2)^{\frac{1-\kappa}{2}}}{2\kappa} \begin{bmatrix} \frac{(1+\kappa)t_1^\kappa - (1-\kappa)t_2^\kappa}{t_1} & -2(t_1^\kappa - t_2^\kappa) \\ \frac{(1-\kappa^2)(t_1^\kappa - t_2^\kappa)}{2t_1 t_2} & -\frac{(1-\kappa)t_1^\kappa - (1+\kappa)t_2^\kappa}{t_2} \end{bmatrix},$$

where $t_1 = s_1 - s_0, t_2 = s_2 - s_0, \kappa = \sqrt{1-4G}$. Show that $\det M = 1$. It is conceivable that the fringe fields of a quadrupole or a bending magnet can be modeled following this soluble problem.

Solution An example case is shown in Fig. 2.6.

2.2.2 Pure quadrupole

As the first application of Eq. (2.11), we will be interested in knowing how U gets transferred when the particle passes through (entrance to exit) a pure quadrupole magnet in a hard-edge model. The answer is then given by Eq. (2.11) with appropriate value of K, and $\frac{1}{\rho} = 0$, and replacing s by the length of the quadrupole ℓ.

Usually the quadrupole magnet is not very long. This allows us to make a useful *thin-lens approximation*. We let $\ell \to 0, |K| \to \infty$, then

$$M = \begin{bmatrix} 1 & 0 \\ -\frac{1}{f} & 1 \end{bmatrix}, \quad \text{where} \quad \frac{1}{f} = K\ell. \tag{2.12}$$

Figure 2.6: The focusing strength $K(s)$ of the idealized magnet of Homework 2.6 is shown in blue. The cosine-like and the sine-like particle trajectories traversing the magnet are shown in red and green, respectively.

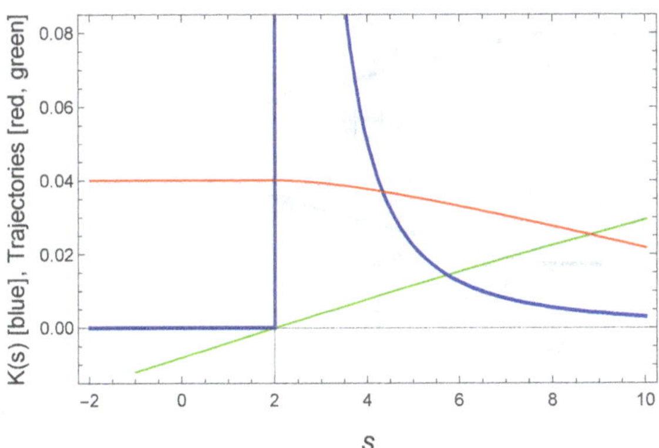

This expression holds for both signs of K. Focusing thin-lens quadrupole has $f > 0$, defocusing quadrupole has $f < 0$. See Fig. 2.7.

What is the validity criterion for a quadrupole to be considered a thin-lens? Generally it requires the argument in sin and cos terms be small, i.e. when $\sqrt{|K|}\ell \ll 1$, which translates to $\ell \ll |f|$, i.e. the magnet must be much shorter than its focal length. (Homework 2.7.) How much shorter? The answer depends on the purpose of the quadrupole. Some sensitive quadrupoles, especially those located at large β-functions (see later), will need to be accounted for very accurately. Thin-lens approximation will not work too well for them. However, for most other quadrupoles, thin-lens approximation can be very useful.

Another comment about the thin-lens matrix (2.12) can be found later in Sec. 2.3.2. One must not consider (2.12) simply as an approximation that is numerically approximately accurate. Fundamental physics must be obeyed when approximations are made, as already alluded to in Sec. 2.1.1. The thin-lens approximation is surprisingly beautiful not only because it is accurate numerically but also because it does the approximation while obeying the fundamental dynamical principles. We will discuss this when we come to Sec. 2.3.2.

From Eq. (2.6), we see that a thin-lens quadrupole with $\frac{\partial B_y}{\partial x} > 0$ is focusing in x and defocusing in y with the same focal length f. It should be noted that this statement holds if the magnet field does not contain a dipole component, or if the dipole focusing term $\frac{1}{\rho^2}$ is weak and can be ignored. Recall that in a weak focusing ring, the small $\frac{1}{\rho^2}$ term provides the only source of focusing. In a strong focusing ring here, in contrast, this $\frac{1}{\rho^2}$ term is so small that it is sometimes ignored.

Figure 2.7: A thin-lens focusing quadrupole (upper) and a thin-lens defocusing quadrupole (lower).

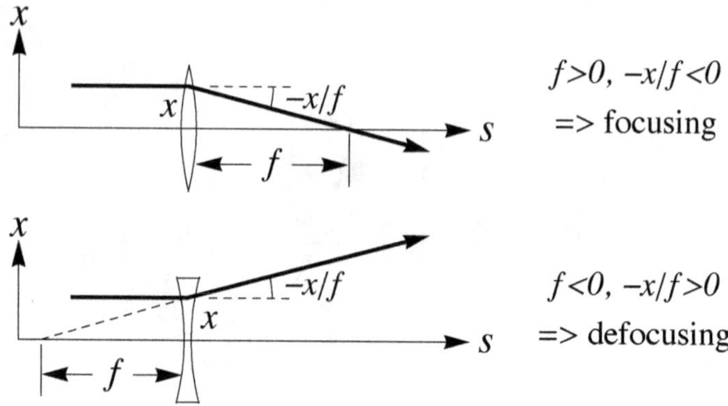

Similarly, a thin-lens quadrupole with $\frac{\partial B_y}{\partial x} < 0$ is defocusing in x and focusing in y with the same focal length f, i.e.

	x-motion	y-motion
$\partial B_y/\partial x > 0$	focusing	defocusing
$\partial B_y/\partial x < 0$	defocusing	focusing

In this situation, it is not justified to label a quadrupole of either sign to be focusing or a defocusing. However, by convention, we will call a quadrupole focusing or defocusing according to whether it is focusing or defocusing in the x-motion.

In light optics, we can focus or defocus the light simultaneously in x and y, independently of each other, by appropriate shaping of the optical lens. This is not doable in particle optics. Our quadrupoles focus in one dimension and defocus in the other. Since particle motion must be stable in *both* x- and y-motions, lattice design in accelerators requires correspondingly more care.

One can extend the matrix formalism to 4×4 by considering the transformation of $\begin{bmatrix} x \\ x' \\ y \\ y' \end{bmatrix}$. The transfer matrices for various cases are then

$$\begin{bmatrix} \cos & \frac{\sin}{\sqrt{K}} & 0 & 0 \\ -\sqrt{K}\sin & \cos & 0 & 0 \\ 0 & 0 & \cosh & \frac{\sinh}{\sqrt{K}} \\ 0 & 0 & \sqrt{K}\sinh & \cosh \end{bmatrix} \qquad \begin{bmatrix} \cosh & \frac{\sinh}{\sqrt{K}} & 0 & 0 \\ \sqrt{K}\sinh & \cosh & 0 & 0 \\ 0 & 0 & \cos & \frac{\sin}{\sqrt{K}} \\ 0 & 0 & -\sqrt{K}\sin & \cos \end{bmatrix}$$

thick-lens focusing quad. thick-lens defocusing quad.

$$\begin{bmatrix} 1 & 0 & 0 & 0 \\ -\frac{1}{f} & 1 & 0 & 0 \\ 0 & 0 & 1 & 0 \\ 0 & 0 & \frac{1}{f} & 1 \end{bmatrix} \qquad \begin{bmatrix} 1 & 0 & 0 & 0 \\ \frac{1}{f} & 1 & 0 & 0 \\ 0 & 0 & 1 & 0 \\ 0 & 0 & -\frac{1}{f} & 1 \end{bmatrix}$$

thin-lens focusing quad. thin-lens defocusing quad.

where cos, sin, cosh, sinh have the same argument $\sqrt{K}s$, and $K > 0, f > 0$.

Homework 2.7 Go through an analysis to show that for the thin-lens approximation to hold, the quadrupole length ℓ must be much shorter than the focal length f.

Solution When a particle enters a quadrupole with (x_0, x_0'), it exits with a slope change $\Delta x'_{\text{thick}} = -(\sqrt{K}\sin\sqrt{K}\ell)x_0 + (\cos\sqrt{K}\ell - 1)x_0'$. In the thin-lens approximation, it gives $\Delta x'_{\text{thin}} = -K\ell x_0$. When $\sqrt{K}\ell$ is not too large, their difference is $\frac{\Delta x'_{\text{thick}} - \Delta x'_{\text{thin}}}{x_0'} \approx -\frac{\ell}{2f}$. For the thin-lens approximation to be valid, this must be $\ll 1$.

Homework 2.8 Consider a relativistic particle going through a thin quadrupole with initial slope $x_0' = 0$ but a nonzero x_0. As the particle exits the quadrupole, it acquires a slope $x' = -\frac{x_0}{f}$. Ignore y-motion.

(a) This is a magnetic device, so the speed of the particle does not change and will remain v_0, but its velocity changed. Give an exact expression of the velocity of the particle as it exits the quadrupole.

(b) Explain how could it happen that, although there is no longitudinal force, the particle's longitudinal velocity has changed in your previous result in (a).

2.2.3 Focal length

Thin-lens magnet In the thin-lens approximation, one can define three quantities — the focal length f, the object distance s_0 and the image distance s_i — as sketched in Fig. 2.8.

The physical meanings of f, s_0 and s_i may be clearer if one recognizes that each particle ray (heavy arrows in Fig. 2.8) represents not the trajectory of *one* particle, but a whole family of trajectories for a *fan* of particles, as illustrated in Fig. 2.9. When drawing one particle ray such as a black arrowed line in Fig. 2.9, it also represents all the other green arrowed lines obtained by linearly stretching or compressing the x-deviation of the black arrowed particle trajectory. Quantities f, s_0 and s_i are clearly seen in Fig. 2.9. An object located at s_0 will be imaged at location s_i because the fan of rays emitted by the object is focused to form an image at s_i, and an observer will (mistakenly) think the object is located at s_i. We have assumed the thin-lens is a linear element whose transfer map can be represented by a matrix.

Figure 2.8: Illustrations of focal lengths for the case of thin-lens focusing quadrupole (upper) and defocusing quadrupole (lower).

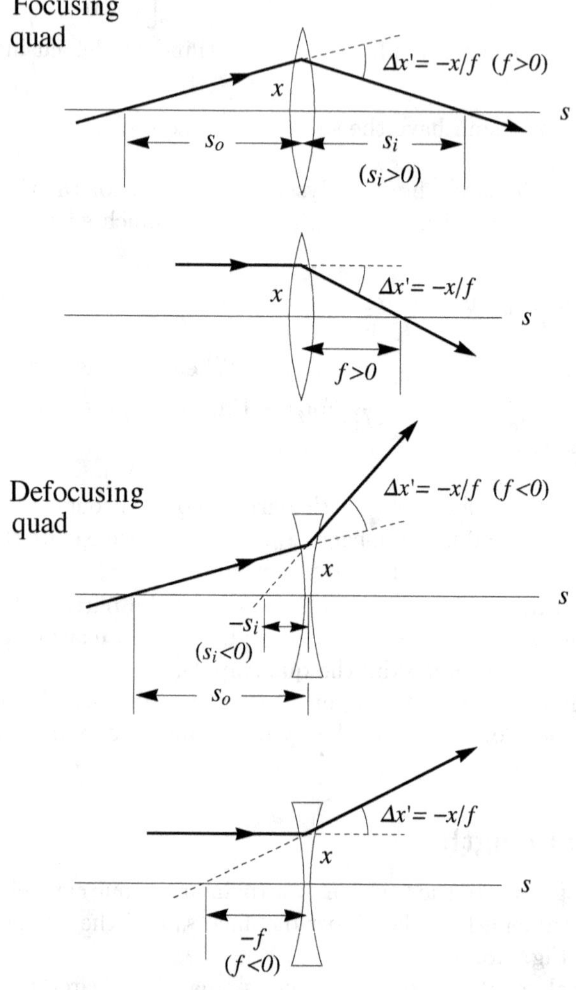

Thick magnet Definition of focal length f is straight forward in the thin-lens approximation. In a matrix formalism, $-\frac{1}{f}$ is the (2,1)-element of the 2×2 transfer matrix of the thin-lens. What should we define as the focal length in general when a thin-lens approximation is not applicable?

In the general case, consider a beamline section (it may consist of a single magnet or a string of magnets as long as they are linear elements) represented as the shaded element of Fig. 2.10. Consider an incident particle with displacement x_0 and zero slope as shown by the heavy arrow on the left. As this particle exits

Figure 2.9: Illustration of the meaning of focus and focal length for the case of a thin-lens focusing quadrupole (left) and a defocusing quadrupole (right).

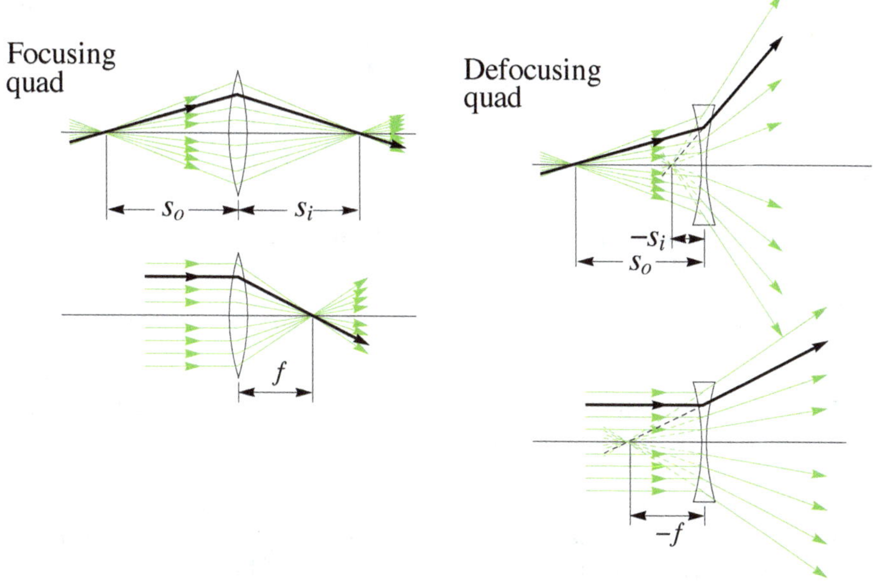

Figure 2.10: Illustration of focal length for a general beamline section of a finite length.

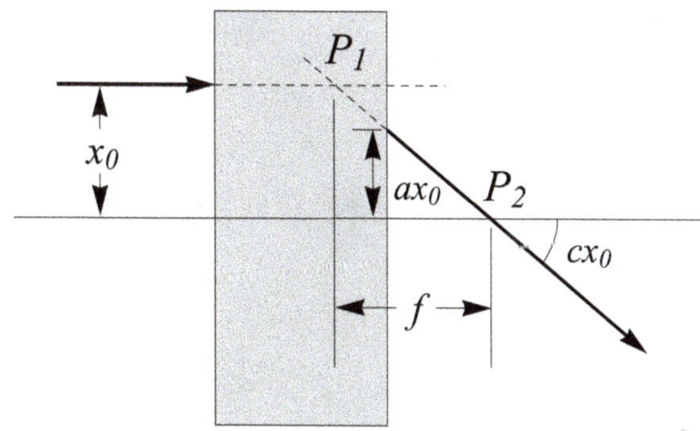

the beamline section, it has a displacement and slope shown as the other heavy arrow on the right. This exiting particle crosses the axis at location P_2. One then extrapolates these two heavy arrows to locate their intersection point P_1 (ignoring vertical y-motion). Position P_1 defines a principal focusing plane and the longitudinal separation between P_1 and P_2 defines the focal length f.

Let the transfer matrix through this beamline element be

$$M = \begin{bmatrix} a & b \\ c & d \end{bmatrix}.$$

The incident trajectory has $\begin{bmatrix} x_0 \\ 0 \end{bmatrix}$, while the exit vector will have $M \begin{bmatrix} x_0 \\ 0 \end{bmatrix} = \begin{bmatrix} ax_0 \\ cx_0 \end{bmatrix}$. In particular, the slope of the exiting trajectory is cx_0. From the geometry, we see that $x_0 = -fcx_0$, where a minus sign is introduced because the sign of c in the figure is negative.[3] It follows that the focal length of a general beamline element is given by

$$f = -\frac{1}{c},$$

i.e. it is solely determined by the (2,1)-element of the transfer matrix even if the beamline is not a simple thin-lens quadrupole.

Homework 2.9

(a) Draw the optical images for the four cases shown in Fig. 2.11, when an object is placed at a distance s_0 to the left of a lens of focal length f, and the observer to the right of the lens looks through the lens. How are the images oriented? Do they appear larger or smaller compared with the real object? Find the magnification factor.

Figure 2.11: Four cases to be addressed in Homework 2.9.

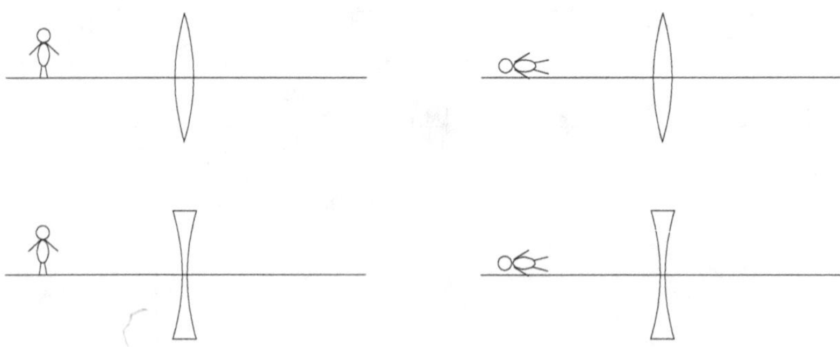

(b) Did you assume $s_0 > |f|$? Repeat the problem for the case $s_0 < |f|$, i.e. the object is very close to the lens.

[3] In this figure a is positive. Whether the system is focusing or defocusing depends also on the sign of a. Focal length is actually $f = -\frac{\text{sign}(a)}{c}$. It is focusing if $f > 0$ and defocusing if $f < 0$.

Homework 2.10 Establish the formula

$$\frac{1}{s_o} + \frac{1}{s_i} = \frac{1}{f}.$$

Amazingly, this formula holds for both focusing and defocusing lenses, provided that we follow the sign convention that f and s_i are positive for focusing lens and negative for defocusing lens.

Solution First consider the case of a focusing lens. Consider a particle trajectory that starts at the object distance with an initial slope x_0'. This particle has a displacement $x = s_o x_0'$ at the lens. The lens then bends the trajectory by an angle $\Delta x' = -\frac{x}{f} = -\frac{s_o x_0'}{f}$. The slope of the trajectory after the lens is then $\Delta x' + x_0' = (1 - \frac{s_o}{f})x_0'$. The trajectory then intersects the axis at a distance $s_i = -\frac{x}{\Delta x' + x_0'}$, which after simplifying gives the required formula. The case of a defocusing lens can be shown similarly. The same formula applies to both focusing and defocusing cases.

Homework 2.11

(a) Let m be the magnification factor, i.e. the (image size)/(object size) ratio. Establish the formula

$$m = -\frac{s_i}{s_o} = \frac{f}{f - s_o}.$$

Explain the meaning of the sign of m, e.g. what does it mean when $m < 0$? Does the formula apply to both $f > 0$ and $f < 0$? Also give a physical explanation of the divergence ($m \to \infty$) when $s_o = f$.

(b) The magnification factor considered in (a) refers to the transverse dimension. Define a longitudinal magnification factor and give an analytical expression for it in terms of s_o, s_i and f.

2.2.4 Alternating gradient

Consider a *doublet*, i.e. two thin quadrupoles of focal lengths f_1 and f_2 separated by a distance d. It is easy to show that its net focal length f is determined by

$$\frac{1}{f} = \frac{1}{f_1} + \frac{1}{f_2} - \frac{d}{f_1 f_2}. \tag{2.13}$$

There are two focal lengths, one for x and one for y. The signs of $f_{1,2}$ are reversed between x- and y-motions. A moment's reflection shows that there is no way to focus both of them if f_1 and f_2 have the same sign (Homework 2.12). We have to choose f_1 and f_2 to have opposite signs if we want the doublet to have a net focusing effect. In particular, if we choose $f_2 = -f_1$, then the net effect is focusing with $f = \frac{f_1^2}{d}$ for both x and y, regardless of the signs of $f_{1,2}$.

This analysis immediately yields two pleasant surprises for the accelerator designers.

- The two quadrupoles are such that one is focusing and the other is defocusing (either considering x-motion or y-motion), and the focusing and defocusing strengths are equal and opposite if $f_1 = -f_2$. One expects – correctly – that their leading contributions cancel, and after cancellation, the next order term gives the residual effect of the combined doublet system. The first pleasant surprise is that this residual effect is focusing, and is focusing for both planes.
- Then comes a second pleasant surprise. It is easy to choose the design parameters f_1, f_2 and d in such a way that this next-order residual term can in fact be pretty large. The net focusing can be rather strong in spite of the fact that their leading contributions in fact cancel and it is a second order effect that is contributing.

These surprising observations led to the invention of the *strong focusing principle*, also called the *alternating gradient principle*.[4] To have a strong focusing doublet, one could choose the quadrupole focal lengths as short as $f_{1,2} \sim d$. The net focal length of the doublet system is then $f \sim (f_{1,2})^2/d \sim d$, comparable to the separation of the two quadrupoles. Again, this new situation is to be compared with the case of weak focusing when the focal lengths are of the order of the accelerator circumference.

Homework 2.12

(a) Verify Eq. (2.13).

(b) Prove mathematically the statement that there is no way to focus both x and y motions if f_1 and f_2 are both focusing or both defocusing.

(c) Show that when f_1 and f_2 have opposite signs, the condition for a net focusing in both x and y is $d > |f_1 + f_2|$.

Solution (a) It would actually be easier (but more abstract) to derive Eq. (2.13) after you learn the fact that matrices of successive beamline elements can be multiplied together using matrix multiplication — a nontrivial thing to learn. We will discuss matrix multiplication in Sec. 2.2.10. But in this homework, it is intended for you to derive it by drawing the optics geometry in the spirit of this section. For the case $f_1 > 0, f_2 > 0$, for example, see Fig. 2.12,

$$\begin{cases} \Delta x' &= \Delta x_1' + \Delta x_2' &= -\frac{x_1}{f_1} - \frac{x_2}{f_2} &= -\frac{x_1}{f}, \\ x_2 &= x_1 - d\frac{x_1}{f}, \end{cases}$$

$$\Longrightarrow \qquad \frac{1}{f} = \frac{1}{f_1} + \frac{1}{f_2} - \frac{d}{f_1 f_2}.$$

[4] E.D. Courant, H.S. Snyder, and M.S. Livingston, Phys. Rev. 88, 1188 (1952); J. Blewett, Phys. Rev. 88, 1197 (1952).

Figure 2.12: The working of a thin-lens doublet.

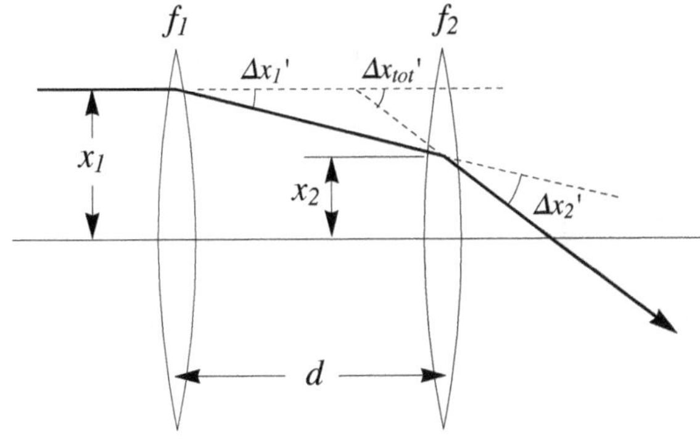

Homework 2.13 Consider a thin-lens quadrupole with focal length f. Do we make its focusing strength stronger or weaker if we split it in half and separate the two halves by a small distance? Does your answer depend on whether the original quadrupole is focusing or defocusing? Give your reason based on geometry of the optics.

Solution See Fig. 2.13. Alternatively, you can consider applying Eq. (2.13).

Figure 2.13: Strength of a split quadrupole.

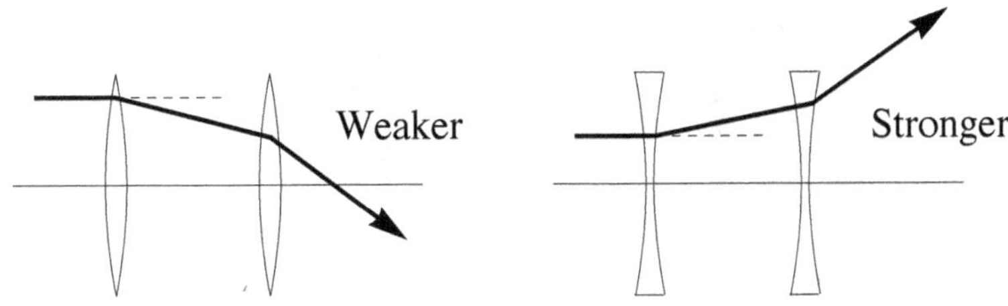

Homework 2.14 We discussed the case of a doublet in the text. Consider a *triplet* consisting of

$$\text{QF}/2 \quad \text{L} \quad \text{QD} \quad \text{L} \quad \text{QF}/2$$

where QF and QD are focusing and defocusing quadrupoles having the same focal length magnitude f, and L is a drift space of length L. Show that this triplet focuses both x and y if $f > \frac{L}{2}$, with net focal lengths

$$\frac{1}{f_x} = \frac{L}{2f^2}\left(1 - \frac{L}{2f}\right).$$

$$\frac{1}{f_y} = \frac{L}{2f^2}\left(1+\frac{L}{2f}\right).$$

For this polarity choice of the triplet, which dimension has a stronger focusing? Why do you think that is the case? It is OK to skip this homework in the present context, but return to it after you learn the matrix formalism of a magnet string.

2.2.5 Lithium lens

A lithium lens[5] is a special focusing element. Consider a very large current $I = 1$ MA flowing down a cylindrical lithium conductor of cross-sectional radius $a = 1$ cm and length $\ell = 10$ cm. An azimuthal magnetic field is established in the lithium conductor, by Ampere's law,

$$B = \frac{\mu_0 J r}{2} = Gr, \qquad \text{with} \quad J = \frac{I}{\pi a^2}.$$

Then consider a beam of antiprotons of momentum $P = 10$ GeV/c being sent down the conductor in the opposite direction as the beam current as shown in Fig. 2.14.

Figure 2.14: A lithium lens.

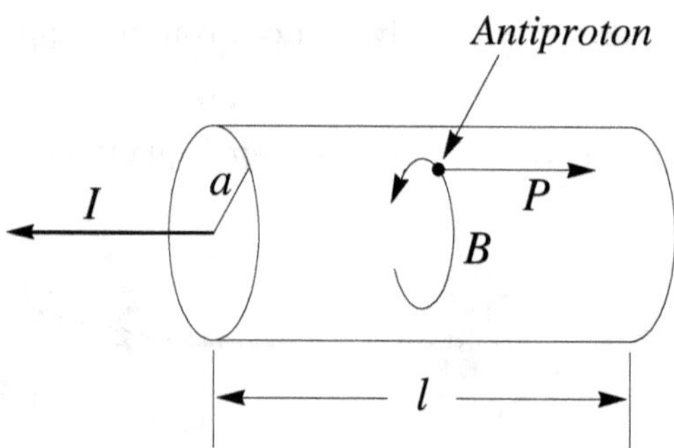

If we ignore the scattering of the antiprotons by the lithium atoms, the antiproton beam will be focused by the azimuthal magnetic field. The focal length of this lithium lens, in a thin-lens approximation, is given by

$$f[\mathrm{m}] = \frac{P[\mathrm{GeV}/c]}{0.3\,G[\mathrm{T/m}]\,\ell[\mathrm{m}]}. \tag{2.14}$$

where

$$G[\mathrm{T/m}] = 2000\,\frac{I[\mathrm{MA}]}{a[\mathrm{cm}]^2}.$$

[5]B. F. Bayanov and G. I. Silvestrov, Sov. Phys. Tech. Phys., Vol. 23, 1 (1978), p. 94; B.F. Bayanov, et al., Nucl. Instr. Meth. 190, 9 (1981); G. Dugan et al., Part. Accel. Conf. 3660 (1983).

The reason lithium is used in this application is that it is a metal, which conducts currents, and it has a low atomic number, which minimizes the scattering of the antiprotons. The lens is operated in pulsed mode because it cannot handle the heat generated in a continuous operation. Even in a pulsed mode, the lithium is likely to be liquid (melt). Note that this lens, unlike a quadrupole, focuses *both* the x- and y-motions. Furthermore, the focusing strength is typically very strong, as can be seen by its small value for the focal length.

The reason we can focus in both planes when it was not possible in a quadrupole magnet lies in the condition $\nabla \times \vec{B} = \vec{0}$ in the quadrupole, while it is violated in a lithium lens.[6]

Homework 2.15 Equation (2.14) is an expression obtained using thin-lens approximation. In this approximation, the linear dependence of the azimuthal magnetic field with respect to r allows a well-defined focal point. For a beam of monoenergetic antiprotons moving parallel to the lithium lens' axis, all antiprotons will be focused to the focal point regardless of the impact radius of the antiproton.

(a) Still consider the antiproton beam moving parallel to the lens' axis, what do you estimate to be the spread of the beam's focal point when the finite length of the lithium lens is considered?

(b) What if the beam has a small momentum spread $\frac{\Delta P}{P_0}$? Estimate the spread of the focal point due to $\frac{\Delta P}{P_0}$.

2.2.6 Sector dipole

We discussed a first application of Eq. (2.11) to pure quadrupoles in Sec. 2.2.2. In this section, we will discuss another application for a pure dipole for which $\frac{\partial B_y}{\partial x} = 0$ and $\frac{1}{\rho} \neq 0$. A hard-edge dipole is also called a sector dipole; its entrance and exit magnet faces are perpendicular to the design beam trajectory as sketched in the left panel of Fig. 2.15.

The vertical transfer matrix over a dipole sector of length ℓ and bending curvature ρ will be the same as a drift space of length ℓ because $K_y = 0$. (Note that ℓ is the length of the magnet measured along the arc, not along a straight line connecting the entrance and the exit points. This is the case even for the nonbending vertical plane.) The horizontal transfer matrix is given by, using Eq. (2.11),

$$M_x(s|0) = \begin{bmatrix} \cos\theta & \rho\sin\theta \\ -\frac{1}{\rho}\sin\theta & \cos\theta \end{bmatrix}, \tag{2.15}$$

where $\theta = \frac{\ell}{\rho}$ is the bend angle of the sector dipole.

---

[6]Incidentally, don't write $\nabla \times \vec{B} = 0$. The left hand side of your equation is a vector; the right hand side is a scalar; they cannot be equal. Write $\nabla \times \vec{B} = \vec{0}$.

Combining the x- and the y-motions, the general 4×4 matrix is

$$\begin{bmatrix} \cos\theta & \rho\sin\theta & 0 & 0 \\ -\frac{1}{\rho}\sin\theta & \cos\theta & 0 & 0 \\ 0 & 0 & 1 & \rho\theta \\ 0 & 0 & 0 & 1 \end{bmatrix}.$$

Homework 2.16 If $\theta = \pi$, we note that

$$M_x = \begin{bmatrix} -1 & 0 \\ 0 & -1 \end{bmatrix}.$$

If $\theta = 2\pi$,

$$M_x = \begin{bmatrix} 1 & 0 \\ 0 & 1 \end{bmatrix}.$$

Convince yourself that results for these special cases make physical sense.

Solution See Eq. (2.9), but you have learned this already in Chapter 1.

Homework 2.17 Consider an effort to design possibly a simplest storage ring. In this design, let the storage ring consists of only two back-to-back sector bending magnets, each bending an angle of π. At both of the locations where the two dipoles meet, insert a thin defocusing quadrupole to provide the betatron focusing.

(a) Is this storage ring stable? For stability, how is the quadrupole strength to be chosen?

(b) What if we replace the sector dipoles by rectangular ones (see next section)?

You may return to this homework after learning matrix map multiplication. It can be a helpful homework to practice the interplay between strong and weak focusing dynamics.

2.2.7 Rectangular dipole

The dipole we just discussed is called a *sector dipole* — it is necessarily a sector dipole because of the way we defined s as the independent variable and we have insisted that $\rho(s)$ is a function of s. Often a dipole can also be of the *rectangular* type. A comparison is shown in Fig. 2.15. The two types differ in their entrance and exit ends. Note that the reference trajectory in the rectangular magnet is curved, so it is not the same as the magnet axis.

It turns out that a rectangular dipole contains an extra focusing/defocusing effect at each of the two ends. As will be elaborated in Sec. 2.2.8, the transfer

Figure 2.15: A comparison of sector and rectangular dipoles.

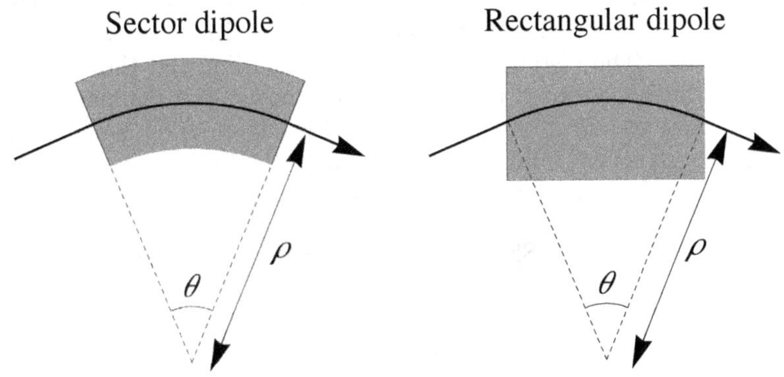

matrix for a rectangular dipole, including its entrance and exit ends, is given by

$$\begin{bmatrix} 1 & \rho\sin\theta & 0 & 0 \\ 0 & 1 & 0 & 0 \\ 0 & 0 & 1-\theta\tan\frac{\theta}{2} & \rho\theta \\ 0 & 0 & \frac{1}{\rho}\tan\frac{\theta}{2}(\theta\tan\frac{\theta}{2}-2) & 1-\theta\tan\frac{\theta}{2} \end{bmatrix}. \tag{2.16}$$

In Eq. (2.16), the horizontal map (the upper-left 2×2 corner) acts like a pure drift space. Interestingly, the natural dipole focusing coming from $\frac{1}{\rho}$ is exactly cancelled by the defocusing effects coming from the two ends. Unlike sector dipoles, there is no net focusing in a rectangular dipole in the bending plane. Furthermore, the effective drift length is $\rho\sin\theta$, which does not seem to have a simple geometric connection.[7]

The vertical matrix is more complicated. For $\theta \ll 1$, it is accurately equal to the horizontal map (2.15) of the sector magnet,[8] i.e. it seems that, in a rectangular dipole, the natural focusing of a sector dipole in the bending plane has been eliminated but is now fully transferred to the nonbending plane.

So the sector dipoles contribute slightly to horizontal focusing, while rectangular dipoles contribute to vertical focusing. For this reason, in a small circular accelerator with alternating, equal-strength QF and QD quadrupoles, and if all quadrupoles are powered by the same single coil in one long series, then there is a tendency for the QDs to be slightly longer than QFs with sector dipoles, and the opposite holds with rectangular dipoles.

[7]It is curious that the effective drift length should be $\rho\sin\theta$. The curved arc length is $\rho\theta$. The straight line distance between the entrance and exit points is $2\rho\sin\frac{\theta}{2}$. However, the reader is encouraged to find a geometric interpretation of his/her own.

[8]They are very close but not identical. The difference between these two matrices is of the order θ^6.

Homework 2.18 The text mentioned two dipole magnet configurations, sector and rectangular. A third variant is as illustrated in Fig. 2.16. The magnet is composed of rectangular slices but stacked together with a curvature. This is a laminated dipole. The distance between the curved line and the straight lines is called *sagitta* named after Latin word for a bow. The lamination is often used to minimize the effect of eddy currents. Each lamination is made identical to simplify construction, and the magnet sagitta is taken into account by a curvatured displacements of the laminations.

Figure 2.16: Sketch of a sagitta laminated dipole.

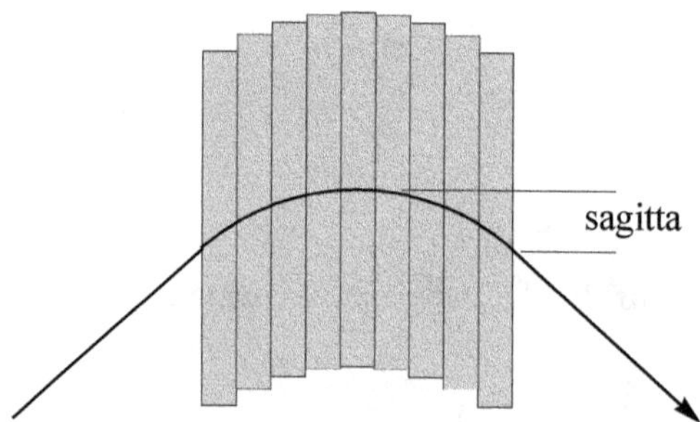

(a) In terms of beam dynamics, does this variant count as a sector dipole or a rectangular dipole?

(b) Assume the magnet has a bending curvature of ρ and total linear length of ℓ, find an expression for the sagitta distance.

2.2.8 Edge focusing

We learned that sector dipoles have natural horizontal focusing and conform to the curved coordinate s. For a rectangular magnet, although the design trajectory is curved inside the magnet, the magnetic field conforms to z, not s. It is conceivable that the horizontal weak focusing of a bend is therefore lost, and as a result the horizontal map acts like a drift space. But we claimed in the previous section that there is now an extra weak focusing in the vertical dimension. In a sense, the weak horizontal focusing in the magnet body is now transferred to vertical focusing at the magnet ends. Let us now examine closer where does this extra vertical focusing come from.

We mentioned before the importance of end fields when we discussed solenoids in Sec. 1.4.1. We pointed out that the solenoid end fields are dictated by the condition $\nabla \cdot \vec{B} = 0$. Here we are now discussing the end fields of a bending magnet, and it is when $\nabla \times \vec{B} = \vec{0}$ plays a role. At the entrance of a rectangular magnet, $B_y(z) = B_0 H(z)$ where $H(z)$ is the Heaviside step function. This gives

$$\begin{aligned} \nabla \times \vec{B} = \vec{0} \quad &\Longrightarrow \quad \frac{\partial B_y}{\partial z} = \frac{\partial B_z}{\partial y} \\ &\Longrightarrow \quad B_z = B_0 \delta(z) y\,, \end{aligned}$$

Figure 2.17: The geometry illustrating the vertical edge focusing of a rectangular dipole magnet. The geometry is shown in the x-z plane. The two arrows indicate the direction of motion of the design beam trajectory. The focusing kicks (not shown) at the entrance and the exit are in the y-direction.

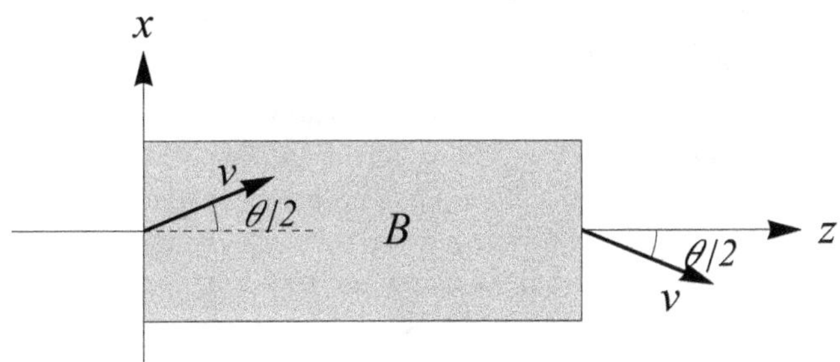

where $\delta(z)$ is the δ-function. Note this end field is in the z-direction and is proportional to y.

If a particle enters the dipole with normal incidence, it will not see any transverse force from the end field because the particle motion and the magnetic field are in the same direction. However, if the particle enters with an trajectory angle like the design trajectory does in the case of a rectangular magnet, its motion gets a kick. In this case, the entrance angle is $\frac{\theta}{2}$, and the Lorentz force $F_y = -ev_x B_z$ yields, for a small bending angle θ, a vertical kick $\Delta y' = -\frac{eB_0\theta y}{2P_0} = -\frac{\theta y}{2\rho}$. The sign of the kick is a focusing effect in the vertical plane. A similar analysis also yields the kick at the exit of the rectangular magnet. The geometry of the layout is illustrated in Fig. 2.17.

The vertical 2×2 map through the magnet is

$$\begin{bmatrix} 1 & 0 \\ -\frac{\theta}{2\rho} & 1 \end{bmatrix} \begin{bmatrix} 1 & \rho\theta \\ 0 & 1 \end{bmatrix} \begin{bmatrix} 1 & 0 \\ -\frac{\theta}{2\rho} & 1 \end{bmatrix} = \begin{bmatrix} 1 - \frac{\theta^2}{2} & \rho\theta \\ -\frac{\theta}{\rho} + \frac{\theta^3}{4\rho} & 1 - \frac{\theta^2}{2} \end{bmatrix}, \tag{2.17}$$

which agrees with (2.16) when $\theta \ll 1$ as it should.

The case when θ is not $\ll 1$ requires more thinking. In that case, Eq. (2.16) is a result of sandwiching the sector dipole matrix with the two edge field focusing maps, as follows.

Horizontal:

$$\begin{bmatrix} 1 & 0 \\ \frac{\tan\frac{\theta}{2}}{\rho} & 1 \end{bmatrix} \begin{bmatrix} \cos\theta & \rho\sin\theta \\ -\frac{1}{\rho}\sin\theta & \cos\theta \end{bmatrix} \begin{bmatrix} 1 & 0 \\ \frac{\tan\frac{\theta}{2}}{\rho} & 1 \end{bmatrix} = \begin{bmatrix} 1 & \rho\sin\theta \\ 0 & 1 \end{bmatrix},$$

Vertical:

$$\begin{bmatrix} 1 & 0 \\ -\frac{\tan\frac{\theta}{2}}{\rho} & 1 \end{bmatrix} \begin{bmatrix} 1 & \rho\theta \\ 0 & 1 \end{bmatrix} \begin{bmatrix} 1 & 0 \\ -\frac{\tan\frac{\theta}{2}}{\rho} & 1 \end{bmatrix} = \begin{bmatrix} 1 - \theta\tan\frac{\theta}{2} & \rho\theta \\ \frac{1}{\rho}\tan\frac{\theta}{2}(\theta\tan\frac{\theta}{2} - 2) & 1 - \theta\tan\frac{\theta}{2} \end{bmatrix}.$$

This result has been given in Eq. (2.16).

The above calculation of the horizontal map exhibits a product of the body bending map with two edge defocusing maps. The edge defocusing comes simply from the removal of the excess horizontal bending accounted for in the sector dipole but absent in the rectangular dipole. The product of the three matrices explicitly demonstrates the neat cancellation between the body focusing and the two edge defocusing effects.

A curiosity might be expressed as to why the transfer of focusing from x to y is not such that the vertical 2×2 map is simply the same as the x map of the sector magnet. Why are they different? The answer is they really should not be identical. The vertical edge focusing for rectangular magnet occurs strictly at the two ends. The horizontal focusing for sector magnet occurs uniformly through the magnet. Perhaps more curious is why they agree to such high power in θ in their Taylor expansions.

We will discuss more end field effects later in these lectures. In general, there are four types of end fields to be discussed, each corresponding to recovering one of the free-space Maxwell equations $\nabla \cdot \vec{B} = 0, \nabla \times \vec{B} = \vec{0}, \nabla \cdot \vec{E} = 0, \nabla \times \vec{E} = \vec{0}$, respectively. We have now discussed the first two types, one with solenoids and one with rectangular dipole magnets — solenoids for the $\nabla \cdot \vec{B} = 0$ type and rectangular dipoles for the $\nabla \times \vec{B} = \vec{0}$ type. We will discuss the two other types later. A summary of these four types of end fields is given in Table 4.1.

Homework 2.19 We will learn in Chapter 3 that, as a particle enters a solenoid from a free space, the end field of a solenoid transforms the coordinates (x, x', y, y') from the free space to the canonical coordinates $(x, x' - \frac{A_x}{B\rho}, y, y' - \frac{A_y}{B\rho})$ inside the solenoid. Is the same thing happening in a dipole magnet? In other words, the effect of dipole end field on the particle motion simply transforms (x, x', y, y') in the free space to the corresponding canonical coordinates inside the dipole?

Homework 2.20

(a) Extend the discussion of this section to a case when the dipole magnet is neither sector nor a rectangular but with slant entrance and exit angles θ_1 and θ_2 respectively.

(b) Can the analysis be extended to the case when the slant angles are in the y-z plane instead of in the x-z plane?

Homework 2.21 Extend the discussion to a case of a pure quadrupole magnet with slant entrance and exit angles θ_1 and θ_2 respectively. Find the edge maps. They may not be expressed as matrices because they may be nonlinear. However, this homework still offers a look at the nonlinear effects induced at the ends of hard-edge quadrupoles.

2.2.9 Mirror image system

If a beamline has a linear transfer map in matrix representation of

$$M = \begin{bmatrix} m_{11} & m_{12} \\ m_{21} & m_{22} \end{bmatrix},$$

what is the map for a beamline that is its *mirror image* as illustrated in Fig. 2.18? (or equivalently, what is the map for a beam of the same charge propagating in the opposite direction?)

Figure 2.18: The map for a mirror system.

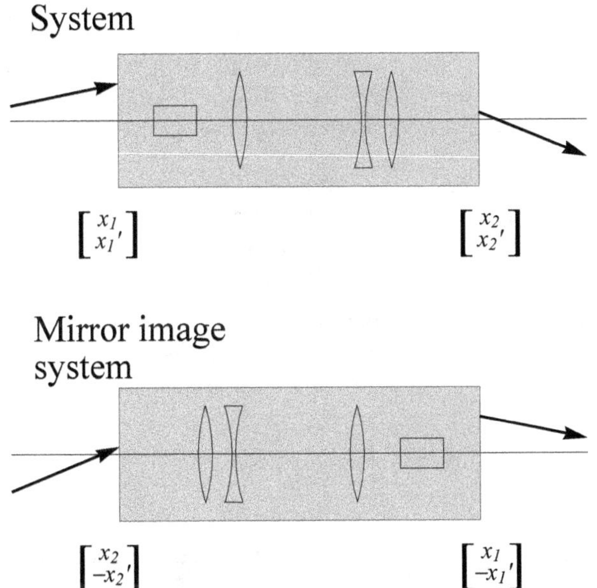

If your answer is M^{-1}, it is close but you would have to think again. The solution is more subtle and requires careful thinking. Consider a particle entering the original beamline with $x_0 = 1, x_0' = 0$. It will have $x = m_{11}, x' = m_{21}$ at the exit. This means a particle entering the mirror image beamline with $x_0 = m_{11}, x_0' = -m_{21}$ (the minus sign in x_0' is due to the reversal of the s direction) will exit with $x = 1, x' = 0$. Similarly, a particle entering the original beamline with $x_0 = 0, x_0' = 1$ will exit with $x = m_{12}, x' = m_{22}$. This means that a particle entering the image beamline with $x_0 = m_{12}, x_0' = -m_{22}$ will exit with $x = 0, x' = -1$. Writing these conditions in a matrix form, we have

$$M_{\text{mirror}} \begin{bmatrix} m_{11} & m_{12} \\ -m_{21} & -m_{22} \end{bmatrix} = \begin{bmatrix} 1 & 0 \\ 0 & -1 \end{bmatrix},$$

which can be solved to give

$$M_{\text{mirror}} = \begin{bmatrix} m_{22} & m_{12} \\ m_{21} & m_{11} \end{bmatrix} .$$

We have used $\det M = m_{11}m_{22} - m_{12}m_{21} = 1$. The mirror imaging switches the m_{11} and m_{22} elements while leaving the m_{12} and m_{21} elements unaffected. In particular, no terms change signs.

It follows from this exercise a theorem: The transfer map of a system that is mirror symmetric about its midpoint must have $m_{11} = m_{22}$.

Here are two notable examples of a system with mirror symmetry,

- Transfer map of a single pure element of any kind (e.g. a thick dipole, a drift, a thick quadrupole, or a combined dipole-quadrupole) of length ℓ is mirror symmetric about its middle point. As a result, all single elements, thin-lens or thick-lens, must have $m_{11} = m_{22}$. (Have you noticed this property before? See also Homework 2.23)

- As we will calculate in detail later, a FODO cell starting at the middle of the QF quadrupole has a mirror symmetry about the middle of the QD quadrupole. In this case, we also have $m_{11} = m_{22}$.

2-D and 3-D mirrors The derivation of the matrix for a mirror system can also be done in a more formal manner by thinking through the connection between M_{mirror} and M^{-1}. One in fact obtains the connection as

$$M_{\text{mirror}} = \begin{bmatrix} 1 & 0 \\ 0 & -1 \end{bmatrix} M^{-1} \begin{bmatrix} 1 & 0 \\ 0 & -1 \end{bmatrix} .$$

Physically, the matrix $\begin{bmatrix} 1 & 0 \\ 0 & -1 \end{bmatrix}$ represents the change of coordinates $x \to x$ and $x' \to -x'$ due to the reversal of the s-direction. One can easily check that this connection gives the correct answer of M_{mirror} for the 1-D case, but in addition, it gives the mirror map for 2-D or 3-D maps in general, if one replaces

$$\begin{bmatrix} 1 & 0 \\ 0 & -1 \end{bmatrix} \to T = \begin{cases} \begin{bmatrix} 1 & 0 & 0 & 0 \\ 0 & -1 & 0 & 0 \\ 0 & 0 & 1 & 0 \\ 0 & 0 & 0 & -1 \end{bmatrix}, & \text{2-D } x\text{-}y \text{ case}, \\ \begin{bmatrix} 1 & 0 & 0 & 0 & 0 & 0 \\ 0 & -1 & 0 & 0 & 0 & 0 \\ 0 & 0 & 1 & 0 & 0 & 0 \\ 0 & 0 & 0 & -1 & 0 & 0 \\ 0 & 0 & 0 & 0 & -1 & 0 \\ 0 & 0 & 0 & 0 & 0 & 1 \end{bmatrix}, & \text{3-D } x\text{-}y\text{-}z \text{ case} . \end{cases}$$

Note the flip of signs for the longitudinal dimension. This is because $z \to -z$ while $\delta \to \delta$ when the beam moves in the opposite direction. In the mirror reflection, for example, an accelerating RF becomes decelerating.

Since M is symplectic, we will show later in Sec. 2.3.1 that its inverse can be written as

$$M^{-1} = -S\tilde{M}S,$$

where S is the symplectic form matrix to be mentioned in Eq. (2.30). As a result the map of a mirror system can be written as

$$M_{\text{mirror}} = -TS\tilde{M}ST.$$

It follows by a simple calculation that the condition for a system to be mirror symmetric is, for a 2-D x-y case,

$$m_{11} = m_{22},\ m_{13} = m_{42},\ m_{14} = m_{32},\ m_{23} = m_{41},\ m_{24} = m_{31},\ m_{33} = m_{44}. \tag{2.18}$$

For a 3-D x-y-z case, in addition to the six conditions in Eq. (2.18), it requires

$$\begin{aligned} m_{15} = -m_{62},\ m_{16} = -m_{52},\ m_{25} = -m_{61},\ m_{26} = -m_{51}, \\ m_{35} = -m_{64},\ m_{36} = -m_{54},\ m_{45} = -m_{63},\ m_{46} = -m_{53},\ m_{55} = m_{66}. \end{aligned} \tag{2.19}$$

The number of constraints for a mirror symmetric system is equal to 1, 6, 15 for 1-D, 2-D and 3-D respectively. The degree of freedom remains 3, 10, and 21 coming from the symplecticity condition.

Homework 2.22 This is a simple homework. Let us say you have a storage ring lattice consisting of drifts, quadrupoles, and dipoles. You know there is a mirror symmetry around a position P. Suppose you know the one-turn 4×4 map for the betatron phase space around P is given by

$$\begin{bmatrix} 0.5 & 1\text{ m} & - & - \\ - & - & - & - \\ - & - & 1.2 & 2\text{ m} \\ - & - & - & - \end{bmatrix},$$

where four matrix elements are shown. Fill in the remaining elements.

Homework 2.23 Consider a 1-D case. Designate the 2×2 transfer map for a single element of length ℓ as $M(\ell)$. Consider ℓ as a variable.

(a) Convince yourself that it must have the form

$$\begin{bmatrix} a(\ell) & \int_0^\ell a(\ell')d\ell' \\ a'(\ell) & a(\ell) \end{bmatrix}.$$

Assign a physical meaning to this expression.

(b) Confirm that this form is fulfilled for a drift space, a dipole, a quadrupole, and a combined dipole-quadrupole. Explain why it is not satisfied by the thin-lens quadrupole. Does a solenoid fit this picture?

Homework 2.24 Show that the mirror image of a 1-D beamline has the same focal length as the original beamline. This theorem applies to beamlines in general, and not only to a thin-lens focusing or defocusing element.

Homework 2.25 If a storage ring lattice consists of a beamline M followed by its mirror image beamline $M_{\rm mirror}$, what is the stability condition of this storage ring? You will need the stability criterion $|{\rm Tr}(M)| \leq 2$ to be established in Sec. 2.3.4.

Solution $0 \geq m_{12}m_{21} \geq -1$. It does not matter what m_{11} or m_{22} or TrM are.

Homework 2.26 Derive Eqs. (2.18) and (2.19).

2.2.10 Chain of transfer matrices

Equation (2.11) gives transfer matrices of each and every beamline element that is either a drift, a quadrupole, a dipole, or a combined quadrupole-dipole. Once these component transfer matrices are obtained, how does one obtain the transfer matrix from an arbitrary position A in the accelerator to another arbitrary position B?

To do this, we will first need to establish the following rule. Let $M(s_1|s_0)$ and $M(s_2|s_1)$ be the transfer matrices from s_0 to s_1 and from s_1 to s_2, respectively. Then the transfer matrix from s_0 to s_2 is given by

$$M(s_2|s_0) \;=\; M(s_2|s_1)M(s_1|s_0)\,. \tag{2.20}$$

A word of caution is in order here immediately after writing down Eq. (2.20). Do not take this equation for granted! See Homework 2.27 for a more detailed discussion. In particular, Homework 2.27 also discusses the ordering of the matrix multiplication in Eq. (2.20).

Given this rule (2.20), it follows that the transfer matrix from A to B is then obtained by sequentially multiplying together the matrices of the component beamline elements. The ordering of the matrices is such that *earlier* elements (closer to the starting position A) will appear *later* in the multiplication, as shown in Fig. 2.19.

Associative law and lattice insertion Because matrix multiplication is *associative*, in the multiplication process, it is permitted to section the sequence of elements into any subgroups before performing the multiplication. For example, in a 3-element beamline with $M = M_3M_2M_1$, which is equal to $(M_3M_2)M_1$ and $M_3(M_2M_1)$ according to the associative law in matrix multiplication, we

Figure 2.19: Ordering of the beamline elements and the ordering of their matrices.

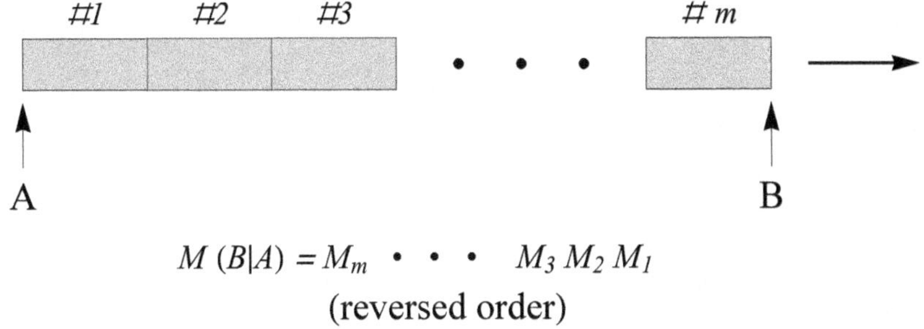

could consider elements 1 and 2 as one section $M_{12} = M_2 M_1$ and obtain M from $M = M_3 M_{12}$. We can also consider elements 2 and 3 as one section with $M_{23} = M_3 M_2$ and then obtain M from $M = M_{23} M_1$. Subgrouping can be made at will provided that the ordering is maintained. This is the associative law of matrix multiplication.

Physically this associative, or subgrouping, property has the important consequence that we are allowed to consider any sequence of elements as one single unit and therefore multiply their matrices first, waiting to be used later. It allows the design of "insertion sections", "matching sections", or "cells" as units of lattice segments and building blocks. The fact that "insertion sections" can make sense is a direct consequence of the associative law of matrix multiplication. One type of cells called FODO is to be introduced in Sec. 2.2.11.

A useful insertion section, for example, can be an "I-section", which gives a unit transfer matrix in both x- and y-motions. An I-section is obviously transparent for linear optics. Similarly, one may design a "$-I$-section" for special purposes. It should be noted, however, here we are speaking of $\pm I$-sections only to the linear order in x and y coordinates. Such sections can be represented by $\pm I$ matrix maps and can be inserted anywhere in the beam line without affecting the subsequent linear optics. Sometimes a more accurate $\pm I$-sections are needed for specialized insertion purposes. It is not a trivial matter to generalize this concept to higher orders and to 3-D (6-D phase space). Designs of high order $\pm I$-sections is an important specialized area of research.

In contrast to the associative law, the matrix algebra also has a noncommutative law, i.e. it says in general $M_1 M_2 \neq M_2 M_1$. This translates to the fact that you may not exchange the positions of beamline elements around arbitrarily.

One-turn map For circular accelerators, one particularly useful map is the one that, starting from a position s, multiplies all element maps together one turn around. This is called the *one-turn map* around s and is designated as $M(s+C|s)$, where C is the ring circumference.

One is often also interested in knowing what happens to a particle as it circulates around the accelerator for many (perhaps as many as 10^{10}) revolutions. The transfer map for N revolutions is given by $M^N(s+C|s)$, according to the rule (2.20).

Sometimes, the ring is divided into P identical sections, each of length $L=\frac{C}{P}$, then one would be interested in the map $M(s+L|s)$. In this case,

$$M(s+C|s) = M^P(s+L|s).$$

We next ask the following question: Suppose we are given the one-turn map $M(s+C|s)$ around position s, what is the one-turn map around another position s'? The answer is

$$M(s'+C|s') = M(s'|s)M(s+C|s)M^{-1}(s'|s), \tag{2.21}$$

where $M(s'|s)$ is the transfer map from s to s'.

The proof of Eq. (2.21) deserves some attention. It is valid only when s' is downstream of s. To prove it, consider a map from s to $s'+C$ — this is a map for a distance longer than the circumference C. We can calculate this map two ways. The first way is to go from s to s' and then from s' to $s'+C$. The total map is given by

$$M(s'+C|s')M(s'|s).$$

The second way is to go from s to $s+C$, then from $s+C$ to $s'+C$. The total map is given by

$$M(s'|s)M(s+C|s),$$

where we have used the fact that $M(s'+C|s+C)=M(s'|s)$. Equating these two expressions then yields Eq. (2.21).

But what if s' is upstream of s? At first glance, Eq. (2.21) should still hold. It might be expected that the map $M^{-1}(s'|s)$ first brings s to s' "backwards in time"; then $M(s+C|s)$ completes one turn and $M(s'+C|s+C)=M(s'|s)$ brings the needed answer. This is wrong. It is generally dangerous to invoke maps backward in time, and even worse to connect M^{-1} maps to that effect. In the proof of Eq. (2.21) just performed for $s'>s$, we invoked four maps, none of which goes backward in time. The appearance of M^{-1} at the end was only a mathematical step.

To find the transformation for the case $s>s'$, we proceed as follows. Consider the map from s to $s'+2C$ — this is a map going a distance between C and $2C$. It can be calculated two ways. The first way is to go from s to $s'+C$, then from $s'+C$ to $s'+2C$,

$$M(s'+C|s')M(s'+C|s),$$

where we have used $M(s'+2C|s'+C) = M(s'+C|s')$. The second way goes from s to $s+C$, then to $s'+2C$,

$$M(s'+C|s)M(s+C|s)\,.$$

Equating these two expressions gives

$$M(s'+C|s') \;=\; M(s'+C|s)M(s+C|s)M^{-1}(s'+C|s)\,. \tag{2.22}$$

Again, all four maps considered here go forward in time. Equations (2.21) and (2.22) give the connections among one-turn transformations, one for $s > s'$ and one for $s' > s$. We shall need these results in Sec. 3.3.2 and Homework 3.22.

Similarity transformation Equations (2.21) and (2.22) say that one-turn maps around different positions are related with one another by *similarity transformations.* One learns from matrix theory that matrices related by similarity transformations have identical determinants, identical eigenvalues, and identical traces. (They will have different eigenvectors, however.) As we will soon show, stability of our linear dynamical system is determined by the determinant, the eigenvalues and the traces of the one-turn maps. What we have just shown then implies that the stability of our system is the same regardless of where is the position around which the one-turn map is obtained. One moment's reflection will indicate that this is what one would expect. Indeed, observing a physical system at different locations in a circular accelerator should be a passive act. Stability of the system should not change simply because you have chosen a different observation point. This change of observation point is connected to a similarity transformation in matrix theory. It is not called *similarity* without reason.

As mentioned, the map for N turns around position s is given by $M^N(s+C|s)$. Equation (2.21) also says that the map for N turns around position s' is given by

$$\begin{aligned} M^N(s'+C|s') &= \left[M(s'|s)M(s+C|s)M^{-1}(s'|s)\right]^N \\ &= M(s'|s)M^N(s+C|s)M^{-1}(s'|s)\,. \end{aligned}$$

The same similarity transformation therefore applies to N-turn maps.

Matrix representation of lattice The matrix formalism we have introduced is referred to as a matrix *representation* of the particle dynamics. It is called a representation because it does not introduce any new physics. It only is intended to represent the already-established physics by an artificial formalism in a concise and elegant manner. By doing so, the wonderful thing is that our physical system – the accelerator lattice – now has an exact, intimate, and one-to-one correspondence with the matrix algebra. The entire arsenal of mathematical tools developed for the matrix theory — eigenvalues, eigenvectors, traces, determinants, similarity transformation and all the matrix theorems — are now transferrable readily for our use. Table 2.2 is attempted a summary of matrix formalism as applied to lattice design in accelerator physics.

The one-to-one correspondence between the physics of lattice design and the mathematics of matrix theory is quite remarkable. As Table 2.2 shows, the connection between them is rather intimate. In fact, there is always a one-to-one correspondence between mathematics and physics. For any piece of mathematics, there is a corresponding piece of physics, and vice versa. Their interplay is described best by "infinitely intertwined" 絲絲入扣.

Map We have quietly introduced in passing the important concept of *maps.* The matrix M is a map that brings initial state vector to its final values,

$$X(s) = M(s|s_0)X(s_0)\,.$$

In a linear case, the map M can be represented as a matrix. In other words, with the establishment of Table 2.2, what we did so far is to find a *matrix representation* of the map. Such a matrix representation applies when and only when the beam dynamical system is linear in u (u is either x or y as the case may be),

$$\text{(linear system)} \quad \Longleftrightarrow \quad \text{(matrices)}\,.$$

In a nonlinear system, matrices do not work any more. We will then have to resort to other map representations such as *Taylor map* (Taylor representation), or *Lie map* (Lie representation),

$$\text{(nonlinear system)} \quad \Longleftrightarrow \quad \begin{cases} \text{Taylor maps} \\ \text{Lie maps} \end{cases}.$$

In this chapter, we consider linear case only.

Homework 2.27 One should not take the rule (2.20) for granted. For example, why the rule has this form? Why are the component matrices multiplied together and not, for example, added together? Also, as we know, matrix multiplications don't commute, so why the component matrices are multiplied in the "reversed" order and not the "normal" order? If you question why this "obvious" rule requires any proof, it is likely that you have not appreciated the matrix formalism.

(a) Prove Eq. (2.20) by writing down explicitly two successive linear transformations, and show that the combined effect can be written in a matrix multiplication language if the definition of matrix multiplication is as defined by the convention,

$$\begin{bmatrix} a_1 & b_1 \\ c_1 & d_1 \end{bmatrix} \begin{bmatrix} a_2 & b_2 \\ c_2 & d_2 \end{bmatrix} = \begin{bmatrix} a_1a_2 + b_1c_2 & a_1b_2 + b_1d_2 \\ c_1a_2 + d_1c_2 & c_1b_2 + d_1d_2 \end{bmatrix}. \tag{2.23}$$

Table 2.2: A one-to-one correspondence comparison of the matrix formalism in mathematics and lattice design in accelerator physics.

Mathematics	Accelerator physics
• linear system • matrix	• phase space • vectors (e.g. $\begin{bmatrix} x \\ x' \end{bmatrix}$) and transfer matrices in phase space • factorization of beam properties (vector) and accelerator properties (transfer matrix)
matrix multiplication is defined in a particular and peculiar way (Homework 2.27)	chain of elements, beamlines
noncommutative law	cannot switch magnets around
associative law	lattice insertions, matching sections, cells
reversed ordering in multiplication (this is not mathematics; it is just a convention)	artifact due to convention, no physical meaning
maps, Poincaré section	transfer matrices (if linear) transfer maps (if nonlinear)
similarity transformation	observation of beam dynamics at different locations
eigenvalues	tunes (i.e. natural frequencies)
eigenvalues and trace are invariant under similarity transformations	• tunes don't change with observation location • stability/instability of beam dynamics does not change with observation location
symplecticity	• Hamiltonian dynamics • conservation of phase space, Liouville theorem
normal form	• Courant-Snyder analysis • β-function

In other words, Eq. (2.20) is intimately connected to this particular definition of matrix multiplication and is valid only if this definition is adopted.

Perhaps the way matrix multiplication as defined in (2.23) can be considered a bit peculiar. For example, why not "simply" multiply by a rule like this,

$$\begin{bmatrix} a_1 & b_1 \\ c_1 & d_1 \end{bmatrix} \begin{bmatrix} a_2 & b_2 \\ c_2 & d_2 \end{bmatrix} = \begin{bmatrix} a_1 a_2 & b_1 b_2 \\ c_1 c_2 & d_1 d_2 \end{bmatrix} ?$$

If so, Eq. (2.20) would not apply.

(b) Matrix multiplication $M = M_1 M_2$ is defined by multiplying the rows of M_1 into the columns of M_2. Had we chosen the definition to multiply the columns of M_1 into the rows of M_2 like this,

$$\begin{bmatrix} a_1 & b_1 \\ c_1 & d_1 \end{bmatrix} \begin{bmatrix} a_2 & b_2 \\ c_2 & d_2 \end{bmatrix} = \begin{bmatrix} a_1 a_2 + c_1 b_2 & b_1 a_2 + d_1 b_2 \\ a_1 c_2 + c_1 d_2 & b_1 c_2 + d_1 d_2 \end{bmatrix},$$

would it work, and if yes, what would the ordering of $M(s_2|s_1)$ and $M(s_1|s_0)$ be in Eq. (2.20)?

Readers who have now become familiar with and used to this counter-intuitive reversed ordering should try not to forget the puzzle when he/she first learned it. Otherwise, he/she will likely be puzzled again when learning the Lie algebra technique. In Lie algebra, the ordering of Lie operators are back to the "normal" ordering when multiplying Lie maps together.

(c) How did you prove Eq. (2.20) in (a)? If you did not explicitly use the matrix multiplication convention (2.23), your proof is wrong and you probably have not understood the matrix formalism.

Homework 2.28 Suppose it is found that the one-turn map of a storage ring has a determinant slightly larger than 1. First of all, this must have been a very nonconventional storage ring. Secondly, prove mathematically that motion in this ring is unstable. Give an estimate of how many revolutions a particle motion will last based on the value of the determinant.

Homework 2.29

(a) Show that the lattice shown in Fig. 2.20 gives a $-I$ transformation.

Figure 2.20: This is a $-I$ map.

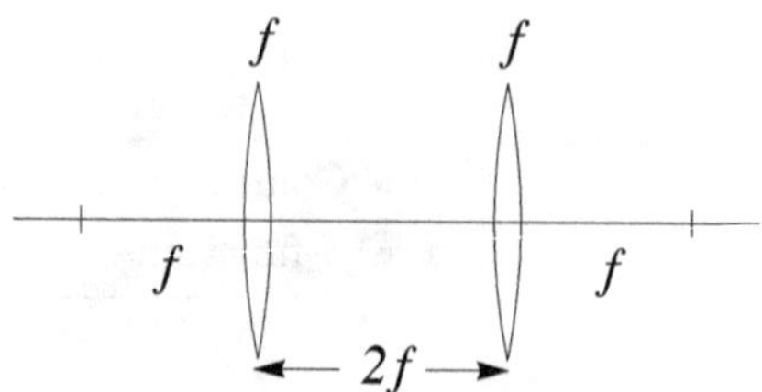

Figure 2.21: This is a $-M$ map.

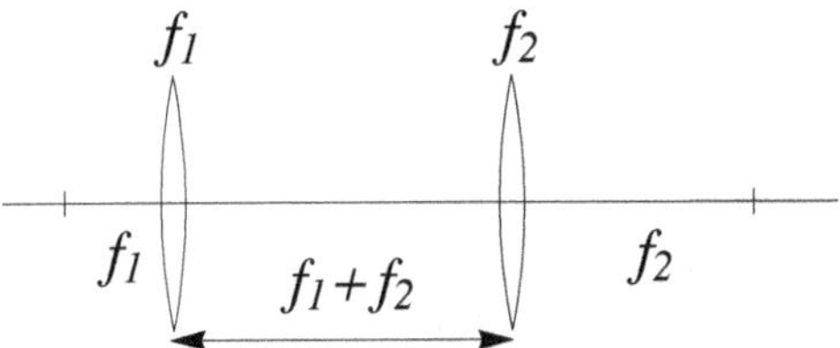

(b) Show that the lattice shown in Fig. 2.21 gives a $-M$ transformation. Show that the magnification factor $M = \frac{f_2}{f_1}$. Note that this is a 1-D configuration valid only in the focusing plane.

Homework 2.30 In Homework 2.29, an object at the beginning of the beamline would have its transverse size X inverted and magnified at the end of the beamline by a factor of 1 and $\frac{f_2}{f_1}$ for cases (a) and (b) respectively. By an analogous definition, what would be the longitudinal magnification factors M' for these two beamlines?

Solution $M' = -1$ and $M' = -(\frac{f_2}{f_1})^2$ respectively.

2.2.11 FODO cell

In the strong focusing system, one uses strong *separated function* quadrupoles in between bending dipoles in a periodic array of *cells*. An often used cell structure is that of a FODO cell, shown in Fig. 2.22. FODO cell analysis is a basic element in accelerator lattice design. As we develop various increasingly sophisticated physics of accelerator lattices, FODO cells will be used repeatedly as our benchmark example. We will return to the FODO cell topic, particularly Secs. 2.4.11, 4.7.1 and 4.7.2.

We assume the two quadrupoles are of the same strength. The convention is to call a quadrupole focusing when $\frac{\partial B_y}{\partial x} > 0$ and defocusing when $\frac{\partial B_y}{\partial x} < 0$. In other words, by convention, focusing or defocusing refer to the x-motion. For the unfortunate y-motion, a focusing quadrupole would be defocusing, and a defocusing quadrupole would be focusing.

The next thing to do is to calculate the one-cell map. To do so, we have to start with some position in the cell and multiply the matrices sequentially of the beamline elements according to the rule established in Sec. 2.2.10. We consider the quadrupoles to be thin. Starting from the middle of the QF quadrupole, the map for the x-motion, for one FODO cell, is given by

$$\begin{aligned} M_x &= \begin{bmatrix} 1 & 0 \\ -\frac{1}{2f} & 1 \end{bmatrix} \begin{bmatrix} 1 & L \\ 0 & 1 \end{bmatrix} \begin{bmatrix} 1 & 0 \\ \frac{1}{f} & 1 \end{bmatrix} \begin{bmatrix} 1 & L \\ 0 & 1 \end{bmatrix} \begin{bmatrix} 1 & 0 \\ -\frac{1}{2f} & 1 \end{bmatrix} \\ &= \begin{bmatrix} 1 - \frac{L^2}{2f^2} & 2L(1 + \frac{L}{2f}) \\ -\frac{L}{2f^2}(1 - \frac{L}{2f}) & 1 - \frac{L^2}{2f^2} \end{bmatrix}, \end{aligned}$$

Figure 2.22: Layout of a FODO cell.

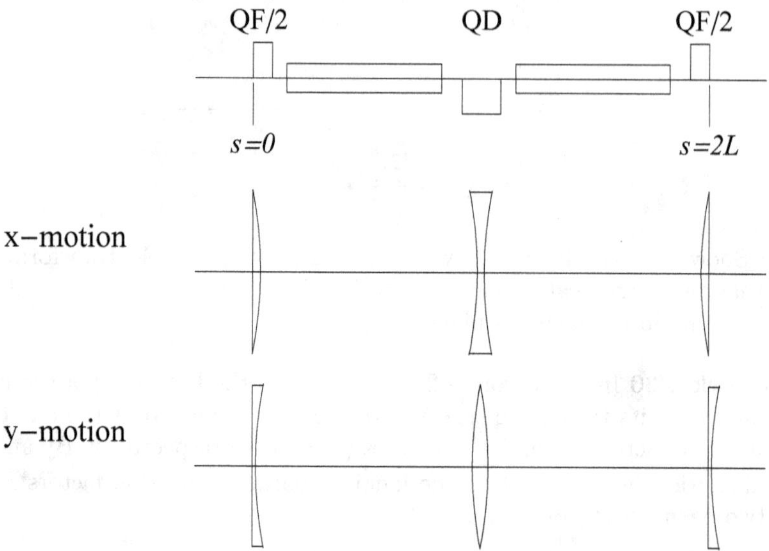

where L is the drift space length, f is the focal length of both the QF and QD. The total length of the cell is $2L$. The focusing by bending magnet has been ignored.

The one-cell map M_y around QF for y-motion can be obtained by replacing f by $-f$ in the expression of M_x. A moment's reflection indicates

$$\begin{aligned} & M_x(\text{middle of QF}) && = && M_y(\text{middle of QD}) \\ = \; & M_x(\text{middle of QD})|_{f\to -f} && = && M_y(\text{middle of QF})|_{f\to -f} \, . \end{aligned}$$

Symplecticity condition (Sec. 2.3) is satisfied, as can be checked from detM_x= detM_y =1. Note in passing that both M_x and M_y satisfy the condition $m_{11} = m_{22}$ dictated by the mirror symmetry of the system.

Consider a storage ring consisting of a string of this FODO cells. Stability of the ring is then determined by the stability of the cell lattice. Stability of FODO cell lattice requires stability of both x- and y-motions, which in turn requires (Sec. 2.3.4)

$$|\mathrm{Tr}\,(M_{x,y})| \le 2, \qquad \text{or} \quad |f| \ge \frac{L}{2} \, . \tag{2.24}$$

Since condition (2.24) applies for both M_x and M_y, and the same condition will be obtained if we calculate $M_{x,y}$ starting from any position in the FODO cell, we conclude that particle motion in a ring consisting of repeated identical FODO cells is stable if and only if condition (2.24) holds.

Figure 2.23: The case of an over-focus.

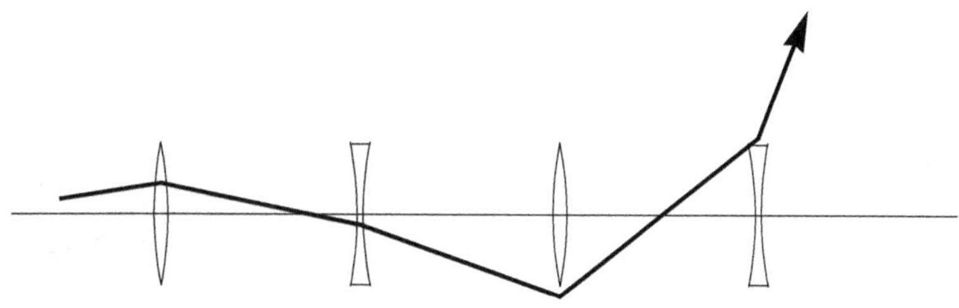

欲速則不達

Trying too hard to speed up only slows you down.

Equation (2.24), however, may look counter intuitive on the first sight. It says that the motion is stable when the quadrupoles are *weak*, and in fact, it is stable even when we let the quadrupoles have infinitesimal strength, $|f| \to \infty$. The system is unstable if quadrupoles are strong (small $|f|$), and in fact, the stronger the quadrupoles are, the more unstable the system becomes. The cause of this instability for overly strong quadrupoles is because the system is then *over-focused*, as illustrated in Fig. 2.23. The strong-focusing FODO system, by the way it is structured, is unstable not because of lack of focusing, but because of over-focusing.

Even though over-focusing must be avoided, within the stability region (2.24), it is indeed often favorable to have relatively strong quadrupoles. This means that f is often chosen to be of the order of L.

We make a side comment here about the general layout of an accelerator lattice. Next time you visit an accelerator tunnel and see some quadrupole magnets arranged along a beamline, bear in mind the following practical rule of thumb,

$$\begin{pmatrix}\text{focal length of}\\ \text{a quadrupole}\end{pmatrix} \overset{\substack{\text{very}\\ \text{roughly}}}{\sim} \begin{pmatrix}\text{spacing between}\\ \text{neighboring quadrupoles}\end{pmatrix} \sim \begin{pmatrix}\beta\text{-function in the}\\ \text{neighborhood}\end{pmatrix}. \tag{2.25}$$

You then get a rough estimate of the strengths of the quadrupoles and the local β-function (see later for the definition of β-function) just by a quick glance of the layout of the magnets around you.

One should again note that by combining QF's and QD's, remarkably, one obtains a net focusing for both x and y dimensions. Analysis of a FODO cell is an explicit demonstration of the strong focusing principle. The mathematical

Figure 2.24: Illustration of the mechanism of the focusing effect of a doublet.

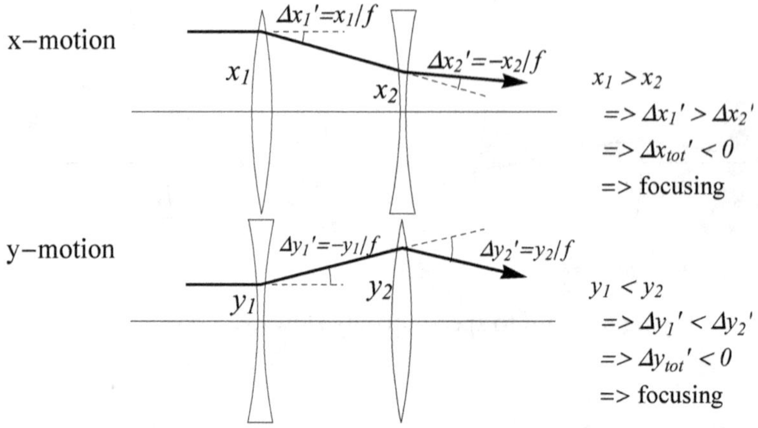

proof of this net focusing is given above by demonstrating the existence of a stability regime (2.24).

A more intuitive demonstration of the overall focusing effect is shown in Fig. 2.24. For the x-motion, the tendency is such that a particle has a larger x-displacement in QFs than in QDs. This means the focusing kicks are bigger than the defocusing kicks, leading to a net focusing force. For y-motion, a particle tends to have a larger y-displacement in QDs than in QFs. Again, this leads to a net focusing effect.

A moment's reflection also yields in passing a rule of thumb,

$$\begin{aligned} \sigma_x \text{ in QF's} &> \sigma_x \text{ in QD's}, \\ \sigma_y \text{ in QD's} &> \sigma_y \text{ in QF's}. \end{aligned} \tag{2.26}$$

where $\sigma_{x,y}$ are the x- and y-sizes of a beam of particles in the storage ring.

A mechanical model of the alternating gradient focusing mechanism is shown in Fig. 2.25.[9] If the setup is slanted, one also demonstrates the *adiabatic damping* mechanism, a topic to be discussed in Chapter 3.

In a storage ring, the O's in a FODO cell are most likely filled by dipoles. In a transport line, O is mostly a drift space. In a linac, it is most likely an RF acceleration section. See Homework 2.32. Beam optics is very similar (although with differences) in all these cases, mostly because the quadrupoles dominate the dynamics of the particle motion.

Recall that in a weak focusing ring, it will take more than one revolution to complete the focusing. In a FODO lattice, the focusing is done in a few cells, which typically occupies only a fraction of the ring circumference. A ring

[9] Jianming Xu, 徐建銘, 加速器原理, 科學出版社, 1974.

Figure 2.25: A mechanical model of strong focusing principle.

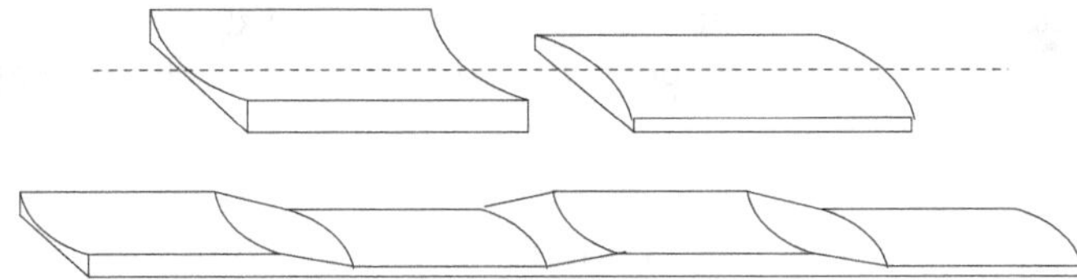

made of FODO cells constitutes a strong focusing lattice for a storage ring or a beamline.

Homework 2.31 The FODO cell map was derived at the observation point at the middle of QF. Repeat the calculation for a one-cell map starting at the middle of QD, and at the middle of one of the drift spaces. Demonstrate explicitly that the stability condition in all cases is given by Eq. (2.24).

Homework 2.32 It may be useful for your later use to prepare a more accurate transfer matrix of a FODO cell by replacing the map of drift space by the map of sector (or rectangular as the case may be) dipole. This will be used when chromatic effects are considered in Chapter 4. You may also repeat this exercise in preparation for application to a FODO cell in an accelerating linac.

Homework 2.33 Our analysis also applies to geometric light optics. Consider an infinite array of optical focusing lenses with focal length f, spaced by distance ℓ as shown in Fig. 2.26. Find the stability condition. One might want to focus the light at the midpoint of the drift spaces and create a waist there. With the temptation to focus the light as tightly as possible, one might then be led to choose $f = \frac{\ell}{2}$, which allows a light trajectory as depicted by the heavy lines in Fig. 2.26. Show that, unfortunately, this configuration is unstable.

Figure 2.26: An infinite array of optical lenses.

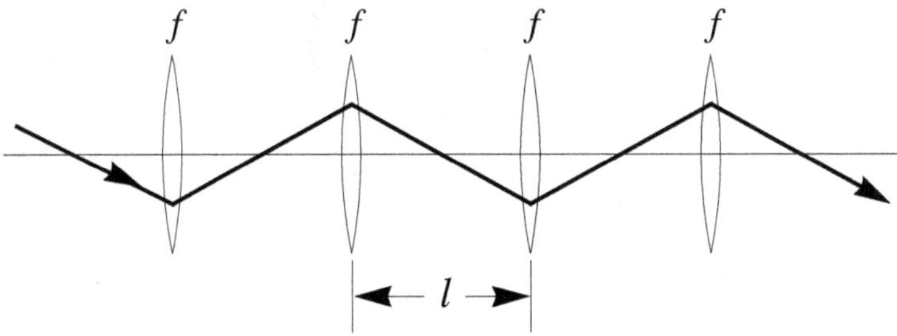

2.2.12 Triplet imaging

A triplet consisting of three alternating gradient quadrupoles is often used as an imaging system focusing in both x- and y-motions. The configuration is depicted in Fig. 2.27.

Figure 2.27: A triplet imaging system.

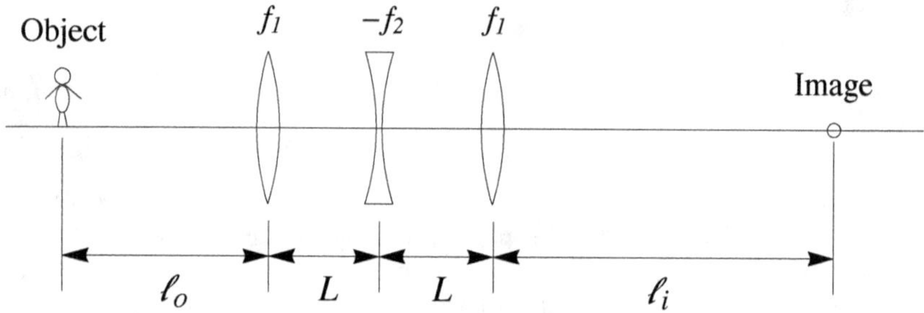

Consider x-motion. The object is located at a distance ℓ_o before the first quadrupole, and an image is to be formed at ℓ_i after the last quadrupole. To perform an imaging function, the (1,2) element of the 2×2 map has to vanish. This determines the imaging distance

$$\ell_{ix} = \frac{-f_1(2f_1f_2L + f_1L^2 + f_1f_2\ell_o + f_1L\ell_o - 2f_2L\ell_o - L^2\ell_o)}{f_1^2f_2 + f_1^2L - 2f_1f_2L - f_1L^2 + f_1^2\ell_o - 2f_1f_2\ell_o - 2f_1L\ell_o + 2f_2L\ell_o + L^2\ell_o}.$$

Similarly, considering the y-motion, its image occurs at

$$\ell_{iy} = \frac{f_1(2f_1f_2L - f_1L^2 + f_1f_2\ell_o - f_1L\ell_o + 2f_2L\ell_o - L^2\ell_o)}{-f_1^2f_2 + f_1^2L - 2f_1f_2L + f_1L^2 + f_1^2\ell_o - 2f_1f_2\ell_o + 2f_1L\ell_o - 2f_2L\ell_o + L^2\ell_o}.$$

To form an image, we demand $\ell_{ix} = \ell_{iy}$. A solution is given by choosing

$$f_1 = \frac{L\ell_o}{\sqrt{L^2 + L\ell_o + \frac{\ell_o^2}{2}}},$$

$$f_2 = \frac{L(2L + \ell_o)\sqrt{L^2 + L\ell_o + \frac{\ell_o^2}{2}}}{4L^2 + 2L\ell_o + \ell_o^2}.$$

In this case, the image is located at

$$\ell_{ix} = \ell_{iy} = \frac{2L^2\ell_o}{(2L^2 + \ell_o^2)}. \tag{2.27}$$

The equivalent focal length of the triplet is given by

$$\frac{1}{f} = \frac{1}{\ell_o} + \frac{1}{\ell_i} = \frac{4L^2 + \ell_o^2}{2\ell_o L^2}.$$

The magnification factors of the image size at the image position is given by the $(1, 1)$ map elements. Including their appropriate signs, they are found to be

$$
\begin{aligned}
M_x &= \frac{(2L^2 + \ell_o^2)(\ell_o - 2\sqrt{L^2 + L\ell_o + \frac{\ell_o^2}{2}})}{2L(2L + \ell_o)\sqrt{L^2 + L\ell_o + \frac{\ell_o^2}{2}}}, \\
M_y &= \frac{(2L^2 + \ell_o^2)(-\ell_o - 2\sqrt{L^2 + L\ell_o + \frac{\ell_o^2}{2}})}{2L(2L + \ell_o)\sqrt{L^2 + L\ell_o + \frac{\ell_o^2}{2}}}.
\end{aligned}
\tag{2.28}
$$

The fact that $M_x \neq M_y$ means the image, although cleanly formed at the focal point, is distorted with different magnification factors horizontally and vertically.

Homework 2.34 Derive Eqs. (2.27) and (2.28).

2.2.13 Negative drift

One can design a lattice insertion, consisting of quadrupoles and drifts, that together acts like a drift space of *negative* length. Such an insertion can be useful when dealing with path length dynamics. For example, as we will see later, one problem with dog-leg and chicane insertions is that they naturally introduce the chromatic path length change $R_{56} > 0$. It is difficult to switch the sign of R_{56} in a transport beamline. A negative drift section can resolve this problem. Let us try to find a few examples of negative drifts in this section.

As a first example, consider x-motion only. It is easy to show that the arrangement in Fig. 2.28 behaves as a negative drift $-L$ in x-motion. In the y-dimension, however, it is defocused,

$$
\begin{bmatrix} 17 & 3L \\ \frac{96}{L} & 17 \end{bmatrix}.
$$

Figure 2.28: This is a negative-drift map in the x-dimension.

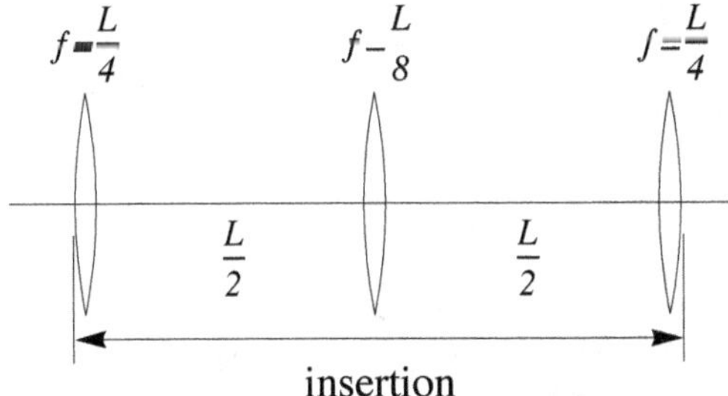

A more sophisticated design is shown in Fig. 2.29. It behaves as effective drifts

$$L_x = 2\left(f - L + \frac{L^2}{f}\right), \qquad L_y = -2\left(f + L + \frac{L^2}{f}\right).$$

In this design, it can be shown that L_x and L_y must have opposite signs and $L_x L_y \leq -12\,L^2$. For example, if we take $f = L$, then $L_x = 2L, L_y = -6L$.

Figure 2.29: This is a negative-drift map in one of the x- and y-dimensions.

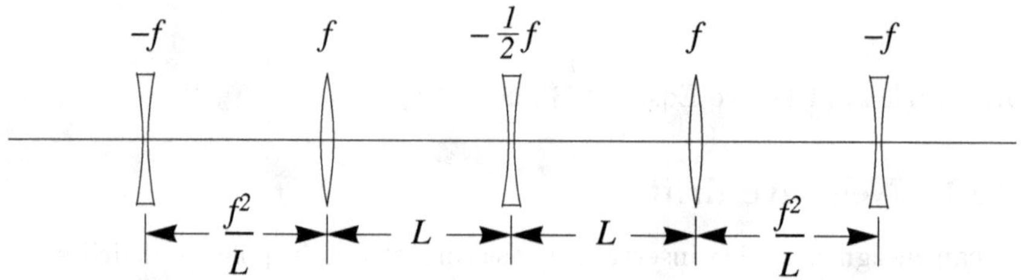

These designs have $L_x \neq L_y$. If $L_x = L_y$ is a requirement, one can join up a design as Fig. 2.29 with its reversed image where focusing and defocusing quadrupoles are reversed. The total arrangement is then as shown in Fig. 2.30. The total effective lengths are then $L_x = L_y = -4L$. The total physical length is $4(L + \frac{f^2}{L})$.

Figure 2.30: This is a negative-drift map in both x- and y-dimensions and with $L_x = L_y = -4L$.

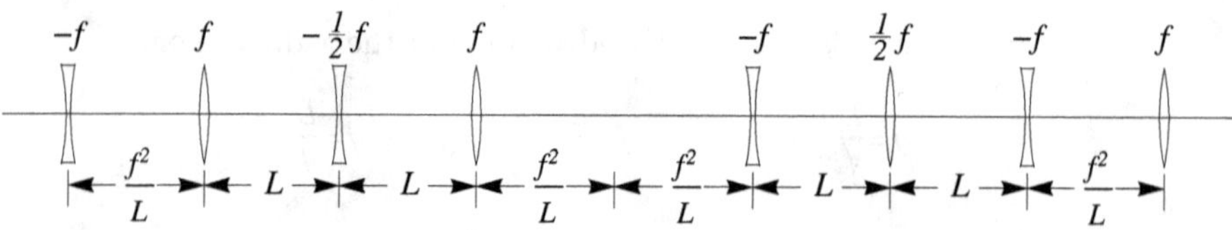

Homework 2.35 Let us try another negative-drift design. Show that the arrangement in Fig. 2.31 gives a drift space of length $-7f$ in the horizontal plane and a $-I$ transformation in the vertical plane.

Figure 2.31: Another quadrupole array that combines to provide the function of a negative drift.

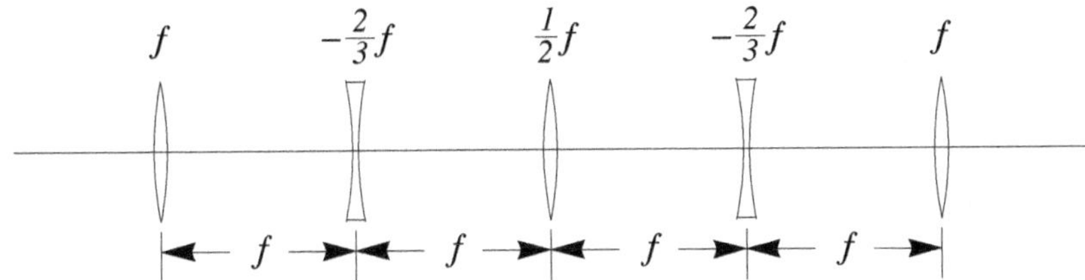

2.3 Symplecticity

2.3.1 Symplecticity condition

This is a very important and remarkable property of all Hamiltonian systems. It imposes a strong constraint on how our system must behave and on how we might formulate it. It follows from the mere fact that *there exists* a Hamiltonian that describes the particle motion. It does not matter what the Hamiltonian actually is.

It should cause the reader a pause to contemplate how such a condition (that there "exists" some unspecified Hamiltonian), seemingly too abstract to do anything, can impose any strong constraint on our system.

Here is another possible viewpoint: You might ignore Hamiltonian as an all-too-abstract construct and choose to ignore it, preferring more "tangible" tools like differential equations of motion, but regardless of your preference, behind the scene, Hamiltonian still strongly controls what we are allowed to do. Hamiltonian is the invisible hand 如來佛的手掌心.

We introduced the phase space in Sec. 1.2.3. At the time, we briefly mentioned symplecticity and emphasized its connection to the phase space. For a linear 1-D motion (2-D phase space), this important condition on symplecticity reduces to something very simple, i.e. the determinant of the 2×2 transfer matrix M must be equal to 1. The reason the map can be represented as a matrix is because we are considering a linear system. Indeed,

$$\text{1-D case} \qquad \text{symplecticity} \quad \Longleftrightarrow \quad \det(M) = 1\,.$$

The proof is as follows.

Consider two independent solutions $y_1(s)$ and $y_2(s)$ of a system described by the differential equation $y'' + K(s)y = 0$ [see Eq. (2.5)] of a 1-D linear Hamiltonian system. Form the *Wronskian*

$$W(s) \;\equiv\; y_1 y_2' - y_2 y_1'\,. \tag{2.29}$$

It is easy to show that W is a constant of the motion, i.e. $W' = 0$. One also observes that, given $M(s|s_0)$ as the transfer map,

$$\begin{bmatrix} y_1 & y_2 \\ y_1' & y_2' \end{bmatrix}_s = M(s|s_0) \begin{bmatrix} y_1 & y_2 \\ y_1' & y_2' \end{bmatrix}_{s_0} .$$

By taking the determinant on both sides, one sees $W(s) = \det(M)\, W(s_0)$, which leads to $\det(M) = 1$ because $W(s)$ is a constant of the motion. Q.E.D. What a neat proof!

For the n-D case, a linear motion can be described by a $2n \times 2n$ matrix. Symplecticity condition reads

$$\tilde{M} S M = S , \tag{2.30}$$

where a tilde means taking the transpose of a matrix, and S is the matrix, sometimes called the *symplectic form*,

$$S = \begin{bmatrix} S & 0 & & \\ 0 & S & & \\ & & \cdots & \\ & & \cdots & \\ & & & S \end{bmatrix} ,$$

consisting of a diagonal array of 2×2 matrix

$$S = \begin{bmatrix} 0 & 1 \\ -1 & 0 \end{bmatrix} .$$

The general proof of symplecticity condition for n-D linear or nonlinear systems can be found elsewhere and, although it can be instructive, is omitted here.

Symplecticity-imposed constraint The symplecticity condition imposes a very strong constraint on the n-D beam dynamical system through constraining its transfer maps. As we see in Homework 2.39, one might say that, due to the constraint of symplecticity, the 6-D phase space we live in is only $\frac{21}{36}$ filled in its available degrees of freedom.

Perhaps one would wonder where does the very strong symplecticity constraint originate from? To appreciate symplecticity, one should first appreciate the phase space, as we discussed already in Sec. 1.2.3. As an artifact and an abstract extension from 3-D space to 6-D, the phase space is necessarily largely empty, and the motion of a particle's phase space coordinates must be heavily constrained. After all, all its momentum coordinates $p_1, p_2, \cdots$ are directly related and dictated to be the time derivatives of the spatial coordinates $q_1, q_2 \cdots$. The symplecticity condition is a reflection of this constraint and in a sense is a measure of the degree of emptiness in phase space when it is generalized to 6-D from the 3-D physical space. What is most amazing is that this property of unfilled degrees of freedom in phase space, as abstract as it may appear, has such an elegant, simple, and necessary-and-sufficient expression, the symplecticity condition, to describe it.

What is also remarkable is that the symplecticity condition (2.30) applies also to a nonlinear system if we identify M to be the *Jacobian matrix* of the nonlinear map, whose elements are defined as

$$M_{\alpha\beta} = \frac{\partial X_\alpha}{\partial (X_0)_\beta},$$

where $(X_0)_\beta$ is the β-th component of the initial state of a particle at $s = 0$, X_α is the α-th component of the final state X of the particle at an arbitrary position s. In a linear system, the Jacobian matrix is just the transfer matrix, and is independent of the particle coordinates. In a nonlinear system, the Jacobian matrix M depends on the components of X_0, and the condition (2.30) must be satisfied for all X_0, thus imposing an even stronger constraint on the description of the nonlinear system.

A side remark is to be made here. A sharp observer might note that here is an exception to the factorization rule that a useful physical quantity is either purely accelerator-based or purely beam-based. The Jacobian for a nonlinear system has a mixed dependence and is therefore not factorized. So, somehow we need to factorize the Jacobian matrices. This observer deserves a congratulation. He/She has taken the first step towards the learning of the Lie algebra technique. In the mean time, Jacobian, a semi-cooked quantity, suffices for our present discussions.

Properties The symplecticity condition resembles a "unitarity" condition due to the fact that the left hand side of (2.30) is *quadratic* in M, while the right hand side is almost a unit matrix. Such a condition obviously imposes a very strong constraint on the matrix M. For example, had the symplectic condition read something such as $MS = S\tilde{M}$ with both sides of the equation linear in M, then M can be expanded or shrunk by some factor without consequence. If that were the case, beam dynamics would become very different – most likely much less interesting. One expects that the unitarity nature tends to force the elements of transfer matrices to take on trigonometric or hyperbolic functions, noting that $\sin^2 + \cos^2 = 1$ and $\cosh^2 - \sinh^2 = 1$.

The reason sines and cosines appear so often in the matrix elements is a consequence of the symplecticity condition. And that is also why particles in phase space naturally tend to *rotate*, although they prefer to move in straight lines in real space.

Some of the consequences of the symplecticity condition follow simply from Eq. (2.30). In particular, one observes that

- the matrix S is symplectic; so is the identity matrix I.
- if M is symplectic, then $\det M = 1$.
- if M is symplectic, so are $\tilde{M}$ and M^{-1}.

- if both M are N are symplectic, then MN is symplectic.
- Since M is real, if λ is an eigenvalue with eigenvector v, then λ^* is also an eigenvalue with eigenvector v^*.
- if λ is an eigenvalue of a symplectic matrix M with eigenvector v, then $\frac{1}{\lambda}$ is an eigenvalue of $\tilde{M}$ with eigenvector Sv.
- if λ is an eigenvalue of a symplectic matrix M, then so is $\frac{1}{\lambda}$.

(2.31)

Liouville theorem Another consequence of a symplectic map is that it obeys the *Liouville theorem*, i.e. the phase space volume is conserved as the system evolves according to the map. The phase space volume, together with all the particles contained in the volume, will be preserved. As a result the phase space density of particles is preserved.

Note that this density is preserved *only if* you move with the volume in the phase space. If you stand at a fixed point in the phase space and observe the moving phase space volumes flowing by, then particle density most likely will change.

Symplectic maps therefore are area-preserving maps. Liouville theorem follows because the Jacobian matrix, being symplectic, has unit determinant $\det M = 1$, which in turn assures that a volume element in phase space maintains its volume as it evolves in time.

Note that the Liouville theorem is a consequence of the symplecticity condition, but the reverse is not necessarily true — It is true only in a 1-D system. This is because the symplecticity condition implies $\det M = 1$, but the reverse is not true except for a 1-D system.

It should also be emphasized that Liouville theorem provides the backbone of the single-particle dynamics in accelerator physics. Particle motion — motion of single particles — and phase space dynamics are now intimately connected in one-to-one correspondence. Results predicted by the single-particle dynamics, including the most intricate effects down to the finest details, are rigorously preserved in phase space, and therefore last indefinitely,

$$\text{single particle dynamics} \quad \overset{\text{Liouville theorem}}{\Longleftrightarrow} \quad \text{phase space dynamics}\,.$$

Note that all these beam dynamical intricacies occur and are preserved in the 6-D phase space, while they are clueless in the 3-D real space.

As accelerators are required in more advanced applications, they are increasingly asked to perform intricate manipulations in phase space. A promising recent wave of applications has involved the generation and manipulation of *microbunches* in the accelerator beam. Examples include the free electron lasers, echoes, and steady-state microbunching storage rings. The fact that microbunches can be considered in the first place, with such detailed phase space features kept and maintained for applications, has to be attributed to the Liouville theorem.

Symplectic building blocks for 2-D systems Consider a linear 2-D system. Particle motion in this system is described by 4×4 symplectic matrices. We know these matrices have ten degrees of freedom to characterize them. This means there are ten basic matrices, with all 4×4 symplectic matrices expressible as products of these basic types, and one way to list these ten degrees of freedom is to give a list of these basic matrices. Exactly which ten get chosen is arbitrary; one such choice can be as follows,

$$\begin{bmatrix} 1 & 0 & 0 & 0 \\ 0 & 1 & t_1 & 0 \\ 0 & 0 & 1 & 0 \\ t_1 & 0 & 0 & 1 \end{bmatrix}, \begin{bmatrix} 1 & 0 & 0 & t_2 \\ 0 & 1 & 0 & 0 \\ 0 & t_2 & 1 & 0 \\ 0 & 0 & 0 & 1 \end{bmatrix}, \begin{bmatrix} 1 & 0 & 0 & 0 \\ 0 & 1 & 0 & t_3 \\ -t_3 & 0 & 1 & 0 \\ 0 & 0 & 0 & 1 \end{bmatrix},$$
$$\begin{bmatrix} 1 & 0 & t_4 & 0 \\ 0 & 1 & 0 & 0 \\ 0 & 0 & 1 & 0 \\ 0 & -t_4 & 0 & 1 \end{bmatrix}, \begin{bmatrix} 1 & t_5 & 0 & 0 \\ 0 & 1 & 0 & 0 \\ 0 & 0 & 1 & 0 \\ 0 & 0 & 0 & 1 \end{bmatrix}, \begin{bmatrix} t_6 & 0 & 0 & 0 \\ 0 & \frac{1}{t_6} & 0 & 0 \\ 0 & 0 & 1 & 0 \\ 0 & 0 & 0 & 1 \end{bmatrix},$$
$$\begin{bmatrix} 1 & 0 & 0 & 0 \\ t_7 & 1 & 0 & 0 \\ 0 & 0 & 1 & 0 \\ 0 & 0 & 0 & 1 \end{bmatrix}, \begin{bmatrix} 1 & 0 & 0 & 0 \\ 0 & 1 & 0 & 0 \\ 0 & 0 & 1 & t_8 \\ 0 & 0 & 0 & 1 \end{bmatrix}, \begin{bmatrix} 1 & 0 & 0 & 0 \\ 0 & 1 & 0 & 0 \\ 0 & 0 & t_9 & 0 \\ 0 & 0 & 0 & \frac{1}{t_9} \end{bmatrix}, \begin{bmatrix} 1 & 0 & 0 & 0 \\ 0 & 1 & 0 & 0 \\ 0 & 0 & 1 & 0 \\ 0 & 0 & t_{10} & 1 \end{bmatrix}. \tag{2.32}$$

Note that only the first four types involve cross-plane coupling. The remaining six involved only in-plane operations.

It might be instructive to identify the required accelerator hardware component for each of these basic operations. This means that, for each of the x-y, x-z and y-z coupled systems, there are four and only four different basic types of hardware components that provide cross-plane couplings in accelerators. Have all of them been invented? We will see later in Eq. (5.55) that, two of them, those involving t_2 and t_3, correspond to the betatron coordinate transformation for dispersion functions D and D' respectively, for the case of x-z cross-plane coupling. A t_1 type is the skew quadrupole discussed in Sec. 3.5.1. Another t_1 type is discussed next. Can all basic building blocks of (2.32) be constructed using combinations of known coupling elements together with quadrupoles and drift spaces?

Transverse deflecting RF cavity One of the 2-D symplectic building block matrices for the x-z cross-plane is that of a transverse radio frequency cavity, also known as the *crab cavity*. We will use this device to illustrate the symplecticity condition in this section.

Radio frequency cavities will be discussed in Chapter 5. But here, we will arrange an RF cavity in such a way that there is a time varying magnetic field in the y-direction, i.e. a time oscillating $B_y(t)$. The oscillating nature of B_y then kicks the passing particles with horizontal kicks $\Delta x'$ that depends on z, the longitudinal position of the particle as it passes through the deflecting cavity. If z is small, we can linearize so that $\Delta x' = kz$ with some parameter k. If we describe the transverse cavity by a linear map on the dynamical variables (x, x', z, δ) where $\delta = \frac{\Delta P}{P_0}$, then the transfer matrix would look like

$$\begin{bmatrix} 1 & 0 & 0 & 0 \\ 0 & 1 & k & 0 \\ 0 & 0 & 1 & 0 \\ 0 & 0 & 0 & 1 \end{bmatrix}.$$

However, a sharp-eyed reader would catch that this matrix is not symplectic. To make it symplectic, the matrix has better be like this,

$$\begin{bmatrix} 1 & 0 & 0 & 0 \\ 0 & 1 & k & 0 \\ 0 & 0 & 1 & 0 \\ k & 0 & 0 & 1 \end{bmatrix},$$

which is of course one of the building block map mentioned in Eq. (2.32). But one then needs to find the reason of the extra k element at the 41-position of the matrix.

The answer is readily given by Faraday's law, which states that $\frac{\partial B_y}{\partial t} = \frac{\partial E_z}{\partial x}$. The time-oscillating B_y dictates that there must be an E_z that depends on x. Therefore, δ must change according to x. The dependence is linear, and the proportionality constant is indeed k.[10] So symplecticity is assured because of Faraday's law!

It is in a sense rather miraculous that the symplecticity condition that results from the fact that the dynamical system is Hamiltonian has to be assured by a seemingly separate physical law called Faraday's law. Apparently, Maxwell equations hide underneath them the underlying reason the electromagnetic systems are Hamiltonian and it is not clear when and how they talked to each other.

We will give a more explicit calculation of the transverse cavity fields later with Eqs. (5.5) and (5.6).

Homework 2.36 Prove that the symplecticity condition (2.30) is satisfied if and only if det(M) =1 for the 1-D case.

Homework 2.37 Prove the properties listed in (2.31). A quick question: If M is symplectic, is aM also symplectic (a is a numerical number)?

Solution The second entry of Eq. (2.31) requires some thinking. It is easy to show that det$M = \pm 1$, but we want to show det$M = 1$.

If need help, make use of the fact that, for any matrix M, M and $\tilde{M}$ have the same set of eigenvalues. The condition det$M = 1$ then follows because the determinant is equal to the product of all eigenvalues.

[10] The reader is suggested to verify this by a simple analysis.

Homework 2.38

(a) Equation (2.30) lists the symplecticity condition as $\tilde{M}SM = S$. Show that it could have been listed as $MS\tilde{M} = S$ equally well.

(b) Show that, if M is symplectic, so is

$$\bar{M} \equiv -S\tilde{M}S\,.$$

The last matrix $\bar{M}$ is called the *symplectic conjugate* of M.

(c) In fact, if M is symplectic, then $\bar{M} = M^{-1}$. When a symplectic matrix is given, don't compute its inverse by inverting the matrix. Take its symplectic conjugate instead. This is a very useful expression because it would generally be time consuming to actually calculate M^{-1} by brute force, especially for the 2-D or 3-D cases. One surprising consequence is that M^{-1} will have the same matrix elements as M, just switched around positions and some flipped signs.

Homework 2.39

(a) Show that Eq. (2.30) imposes $n(2n-1)$ conditions on the matrix elements of M. The transfer matrix M therefore has only $n(2n+1)$ degrees of freedom in spite of the fact that it has $4n^2$ elements. For a 1-D system, there is one condition on the 4 matrix elements. The one condition is of course just $\det M = 1$.

(b) For an n-D system with $n > 1$, convince yourself that the symplecticity condition implies det(M) =1, but det(M) =1 does not imply symplecticity.

Solution

(a)

	$n=1$	$n=2$	$n=3$	$n=4$
# matrix elements	4	16	36	64
# conditions	1	6	15	28
degree of freedom	3	10	21	36

Homework 2.40 Let M be a symplectic matrix, and another matrix N is related to M by $N = AM\tilde{A}$ with some matrix A. What is the condition on the matrix A for N to be also symplectic?

Solution It is OK if the answer is A should be symplectic, i.e. $\tilde{A}SA = S$. In this case det$A$ =1. It is even better if the answer takes a note that $\tilde{A}SA = -S$ also works. In this case det$A$ = -1.

Homework 2.41 Same problem as Homework 2.40, but replace the connection between M and N,

(a) for the case $N = AM$.

(b) for the case $N = MA$.

(c) for the case $N = AMA^{-1}$.

Homework 2.42 Consider a nonsymplectic M, and you want to symplectify it by forming a new symplectic map $N = MA^{-1}$. What is the condition on A?

Solution $\tilde{A}SA = \tilde{M}SM$. In the 1-D case, this only requires $\det A = \det M$.

Homework 2.43

(a) A nonlinear 1-D map is given by

$$x_1 = f(x_0, p_0), \qquad p_1 = g(x_0, p_0).$$

What do the functions $f(x_0, p_0)$ and $g(x_0, p_0)$ have to obey for this map to be symplectic?

(b) Consider the map

$$x_1 = ax_0 + (bx_0 + cp_0)^2, \qquad p_1 = dp_0 + (ex_0 + fp_0)^2$$

Find the conditions on a, b, c, d, e, f for this map to be symplectic. There are several possible solutions, find at least two.

Solution

(a) Do not forget to say that the condition you find will have to be obeyed *for all* x_0 *and* p_0. For an alert reader, Poisson bracket is entering the scene.

(b) One nontrivial solution demands the conditions $ad = 1, c = -\frac{af^2}{bd}, e = -\frac{b^2 d}{af}$. Investigate this solution a bit and find out what makes this system symplectic.

Homework 2.44 Calculate the exact map (not in a matrix form because it is not to be linearized) for a field-free drift space of length L in the 6-D $(x, x', y, y', z, \delta)$ phase space. Prove the symplecticity of your map. This is not as simple a problem as it might seem. Define your dynamical quantities carefully.

Homework 2.45 Here is a homework concerning *kick maps* where the coordinates do not change and the momenta are changed as functions of the coordinates only. The map can be nonlinear otherwise.

(a) Suppose a map

$$\begin{aligned} x_1 &= x_0\,, \\ p_{x1} &= p_{x0} + U(x_0, y_0)\,, \\ y_1 &= y_0\,, \\ p_{y1} &= p_{y0} + V(x_0, y_0) \end{aligned}$$

is known to be symplectic. Find the condition the functions $U(x, y)$ and $V(x, y)$ need to obey.

(b) A map

$$\begin{aligned} x_1 &= x_0\,, \\ p_{x1} &= p_{x0} + U(x_0, y_0, z_0)\,, \\ y_1 &= y_0\,, \\ p_{y1} &= p_{y0} + V(x_0, y_0, z_0)\,, \\ z_1 &= z_0 \\ p_{z1} &= p_{z0} + W(x_0, y_0, z_0) \end{aligned}$$

is known to be symplectic. Find the conditions for $U(x, y, z), V(x, y, z)$ and $W(x, y, z)$.

Homework 2.46 Consider a 3-D dynamical system (6-D phase space). Let the equation of motion of the system have the form

$$\ddot{\vec{r}} = \vec{G}(\vec{r})\,,$$

for some vector function $\vec{G}(\vec{r})$.

(a) Find the Hamiltonian of the system.

(b) Show that this system is symplectic provided that $\nabla \times \vec{G} = \vec{0}$. This means if the force is derivable as a gradient of a potential, i.e. if $\vec{G} = -\nabla V$ for some potential function V, then the system is symplectic.

(c) If we replace the $\vec{G}(\vec{r})$ function by a more general case when it depends also on $\dot{\vec{r}}$, show that the system becomes nonsymplectic. For this system to be symplectic, the vector $\vec{G}$ must not have any dependence on momentum $\dot{\vec{r}}$.

Homework 2.47 Consider a linear system with Hamiltonian of a quadratic form

$$H(\vec{q}, \vec{p}, t) = \frac{1}{2}\tilde{X}HX\,,$$

where $X = (\vec{q}, \vec{p})$ and H is a symmetric matrix independent of $\vec{q}, \vec{p}$.

(a) Show that the Hamilton equation of the system can be written as

$$\dot{X} = SHX\,.$$

(b) The solution of $X(t)$ with initial condition $X(0)$ is

$$X(t) = e^{SHt}X(0)\,.$$

(c) Show that the map e^{SHt} is symplectic for all time t.

Homework 2.48 Express a 4×4 symplectic matrix M in 2×2 block form as

$$M = \begin{bmatrix} A & B \\ C & D \end{bmatrix}.$$

(a) Show that

$$\bar{M} = \begin{bmatrix} \bar{A} & \bar{C} \\ \bar{B} & \bar{D} \end{bmatrix} = \begin{bmatrix} (\det A)A^{-1} & (\det C)C^{-1} \\ (\det B)B^{-1} & (\det D)D^{-1} \end{bmatrix}.$$

(b) Show that the symplecticity condition gives six conditions

$$\det A + \det C = 1, \quad \det B + \det D = 1, \quad \bar{A}B + \bar{C}D = 0\,. \tag{2.33}$$

Check these conditions against the building block matrices (2.32).

(c) Since $\tilde{M}$ is also symplectic, applying the conditions (2.33) to $\tilde{M}$ gives

$$\det A + \det B = 1, \quad \det C + \det D = 1, \quad A\bar{C} + B\bar{D} = 0\,. \tag{2.34}$$

The six conditions in Eq. (2.33) and the six conditions in Eq. (2.34) are algebraically equivalent. They are also necessary and sufficient conditions for M to be symplectic because there should be six conditions in total to assure the sympecticity of M.

(d) Combining the first two members of Eq. (2.33) and the first two members of Eq. (2.34) gives the necessary properties

$$\det A = \det D, \qquad \det B = \det C\,.$$

These two conditions are necessary but not sufficient for M to be symplectic. Results in this homework are useful when dealing with linearly coupled motions in an accelerator.

Homework 2.49 Homework 2.48 can be extended to the 6×6 case. Show that if

$$M = \begin{bmatrix} M_1 & M_2 & M_3 \\ M_4 & M_5 & M_6 \\ M_7 & M_8 & M_9 \end{bmatrix}$$

is symplectic, then

$$\begin{aligned} & \det M_1 + \det M_2 + \det M_3 = \det M_4 + \det M_5 + \det M_6 \\ = & \det M_7 + \det M_8 + \det M_9 = \det M_1 + \det M_4 + \det M_7 \\ = & \det M_2 + \det M_5 + \det M_8 = \det M_3 + \det M_6 + \det M_9 = 1\,. \end{aligned}$$

The reader is encouraged to test these interesting conditions on the 6-D symplectic maps he/she learned in the past. Are these conditions necessary and/or sufficient for M to be symplectic?

2.3.2 Symplectification of maps

The transverse deflecting cavity discussed in the previous section offers one example to appreciate the intricacy of the symplecticity condition. For another glimpse of this intricacy, consider the 2×2 transfer matrix over a constant-gradient focusing quadrupole in Eq. (2.11),

$$M = \begin{bmatrix} \cos\sqrt{K}\ell & \frac{\sin\sqrt{K}\ell}{\sqrt{K}} \\ -\sqrt{K}\sin\sqrt{K}\ell & \cos\sqrt{K}\ell \end{bmatrix}.$$

We know that for a short magnet, we can make an approximation by letting $\ell \to 0$ while holding $K\ell = q$ fixed. The result is the thin-lens approximation, Eq. (2.12), valid to zeroth order in ℓ, and has been extensively used so far in our discussions,

$$M_{\text{0th order}} \approx \begin{bmatrix} 1 & 0 \\ -q & 1 \end{bmatrix}.$$

It is perhaps tempting to try to improve upon the thin-lens approximation by going to the next order in ℓ. Perhaps we can proceed by expanding the exact expression of M, again holding $K\ell = q$ fixed, in Taylor series in ℓ. This leads to a map to first order in ℓ,

$$M_{\text{1st order}} \approx \begin{bmatrix} 1 - \frac{q\ell}{2} & \ell \\ -q(1 - \frac{1}{6}q\ell) & 1 - \frac{q\ell}{2} \end{bmatrix}.$$

This expression is surely more accurate than the thin-lens approximation. But why is it not used? The answer lies in the fact that this more accurate map is not symplectic. We easily note that $\det M_{\text{0th order}} = 1$ while $\det M_{\text{1st order}} \neq 1$. We do not use it because higher accuracy loses its priority when integrity is in question.

That the symplecticity must be observed and must be checked at all times is an important lesson. It leads also to a branch of map physics called *symplectification.*[11] When higher accuracy is needed, and in doing so, symplecticity is compromised, one needs to find a way to maintain the desired higher accuracy while forcing the map to obey symplecticity.

To symplectify the map $M_{\text{1st order}}$, we first note that its determinant differs from 1 in second order in ℓ, i.e. the error is one order higher than the intended accuracy of the map. This is of course not unexpected; the error of this map is of the order $\mathcal{O}(\ell^2)$. To symplectify, the idea is in fact to artificially add some higher order terms in its expression to force its determinant to become 1. The accuracy is not compromised (in principle) because the artificially added terms are of higher orders.

For example, $M_{\text{1st order}}$ can be symplectified if we replace the two diagonal elements by $\sqrt{1 - q\ell + \frac{1}{6}q^2\ell^2}$. But such a step hardly has any physical meaning. Although there is nothing wrong with this way to symplectify $M_{\text{1st order}}$, a better way is to replace the map by

$$\begin{bmatrix} 1 & \frac{\ell}{2} \\ 0 & 1 \end{bmatrix} \begin{bmatrix} 1 & 0 \\ -q & 1 \end{bmatrix} \begin{bmatrix} 1 & \frac{\ell}{2} \\ 0 & 1 \end{bmatrix} = \begin{bmatrix} 1 - \frac{q\ell}{2} & \ell - \frac{q\ell^2}{4} \\ -q & 1 - \frac{q\ell}{2} \end{bmatrix}.$$

[11]See, for example, H. Yoshida, Sec. 2.3.8, Handbook Accel. Phys. & Eng., 2nd Ed., World Scientific (2013).

This is a symplectic map because all three component matrices are symplectic. It also has the same accuracy as $M_{\rm 1st\ order}$ because its error terms are of the order $\mathcal{O}(\ell^2)$. More importantly, this second way to symplectify $M_{\rm 1st\ order}$ has physical meaning and is in line with the idea of *kick maps*, which represent a complex dynamics by a series of δ-function kicks, interlaced by drift spaces. Both the δ-function kicks and the drifts are symplectic, so the total map will keep its symplecticity.

There are several ways to symplectify a map to high orders. We shall not discuss these techniques here. But the most straightforward candidate is as described above, i.e. to represent the map as a series of kick maps interlaced by drift spaces.

Consider for example a 1-D problem satisfying an equation of motion

$$x'' = G(x)\,.$$

This is a rather general potentially nonlinear equation, assuming the mechanism of dynamics is constant in s, i.e. we have an s-independent Hamiltonian and $G(x)$ does not depend on s. One section of this map from s to $s + \Delta s$, for example, can be approximated by a kick map

$$\begin{aligned} x_f &= x_i\,, \\ x'_f &= x'_i + G(x_i)\Delta s\,. \end{aligned}$$

Note two things: (a) the displacement x does not change in a kick map; only the slope x' gets changed. (b) This map is symplectic; its Jacobian matrix has unit determinant no matter what $G(x)$ is.

We then can split the whole section evenly into segments, each of length Δs. The map of each segment is given by the multiplication a map of drift space of $\frac{\Delta s}{2}$, followed by the kick map, followed by another drift of $\frac{\Delta s}{2}$. We then multiply together — or as the jargon goes, concatenate — all the maps of the segments to obtain the total map. This total map is symplectic because all component maps are symplectic.

This approach of kick maps interlaced by drift spaces, in fact, is what is most used in tracking simulation codes. When done this way, these simulation codes are symplectic.

Homework 2.50 The text took the quadrupole matrix to introduce the idea of symplectification. We can also use a sector dipole map (2.15) for the purpose. If the bend angle is small, then the horizontal transfer matrix is first given approximately by the drift space matrix, i.e.

$$M_x(s|0) \approx \begin{bmatrix} 1 & \rho\theta \\ 0 & 1 \end{bmatrix} = M_y(s|0)\,.$$

But doing so drops all effects coming from the magnet.

(a) So it is necessary to go for higher order terms. Proceeding to order $\mathcal{O}(\theta^2)$, for example, we have

$$M_x(s|0) \approx \begin{bmatrix} 1-\frac{\theta^2}{2} & \rho\theta \\ -\frac{\theta}{\rho} & 1-\frac{\theta^2}{2} \end{bmatrix}.$$

But now $\det M_x(s|0) = 1 + \frac{\theta^4}{4} \neq 1$.

(b) Find a way to symplectify the map.

Solution (b) Look up Eq. (2.17) for a hint.

Homework 2.51 It might be instructive to write a small simulation code to compare the effects of different approximations discussed in this section. Consider the 1-D motion of a particle in a focusing quadrupole of a long length L. Simulate the motion by dividing the long quadrupole into many steps, each with step size ℓ. Use your simulation code to track the motion of the particle using (a) the exact map, (b) the thin-lens approximation, (c) the $M_{\text{1st order}}$ as in the text, (d) the symplectified 1st order map. Present your results by graphs in phase space. Case (b) contains slight distortion of the particle motion compared with the exact case (a). Case (d) is similar to case (b), but the distortion becomes smaller. Observe the qualitative difference of case (c), whose deviation from the exact case (a) becomes bigger with the number of steps simulated.

2.3.3 Choice of dynamical variables

So far we have considered the dynamical variables (x, x', y, y') to describe our linear system. They are also considered the variables constituting the phase space. A question arises as to how flexible is the choice of dynamical variables, and therefore the phase space?

We shall limit our discussion to 2-D linear systems (4-D phase space). For a system to remain linear, meaning the transfer maps are linear in the dynamical variables, it is clear that any other choice of dynamical variables must be a linear combination of (x, x', y, y'). Let the new variables be related to (x, x', y, y') by a matrix T, i.e. let

$$V = TU, \qquad U = \begin{bmatrix} x \\ x' \\ y \\ y' \end{bmatrix}$$

be the new variables. What condition does T have to obey so that the new variables are legitimate dynamical variables? The jargon is that the transformation from U to V is *canonical.* We consider T to be independent of position s.

We know that the transfer maps for U (e.g. a map that brings U from position s_1 to s_2) are necessarily symplectic. In order to be legitimate, the transfer maps for V must also be symplectic. Consider a transfer map M for vector U from s_1 to s_2,

$$U_2 = MU_1, \quad \text{with} \quad \tilde{M}SM = S\,.$$

The same transfer map can be described in terms of the new vector V according to

$$V_2 = TU_2 = TMU_1 = TMT^{-1}V_1\,.$$

In other words, the transfer map now is TMT^{-1}. To be symplectic, we require

$$(\widetilde{TMT^{-1}})S(TMT^{-1}) = S\,,$$

or

$$\tilde{M}(\tilde{T}ST)M = (\tilde{T}ST)\,. \tag{2.35}$$

It is obvious that condition (2.35) is fulfilled if $\tilde{T}ST = S$, i.e. if the coordinate transformation is carried out by a matrix T that is symplectic. A change of dynamical variables by a symplectic matrix gives another legitimate set of dynamical variables. Canonical transformations, therefore, must be symplectic.

However, an alert reader will immediately notice that $\tilde{T}ST = S$ is not the only solution to allow (2.35). Any transformation with

$$\tilde{T}ST = \alpha S\,,$$

with any constant α can be legitimate. In particular, the determinant of T does not have to be unity. This property allows us extra freedom in choosing dynamical variables, and in fact is often used when we want to scale the dynamical variables by some factors of our convenience. Obviously, for example, a choice of $(2x, 2x', 2y, 2y')$ would be legitimate even though it is not strictly canonically related to the original coordinates. In fact, the system we use most often, the system of (x, x', y, y'), are not canonically connected to the more legitimate choice of (x, p_x, y, p_y). An application of this flexibility has already been routinely applied. See Homework 2.52.

Homework 2.52 Consider a Hamiltonian $H(q,p;t)$ with its canonical variables q, p and time variable t. Show that it is possible to linearly scale the coordinates by $Q = \lambda_1 q, P = \lambda_2 p$ with time variable $T = \lambda_3 t$, and the system remains Hamiltonian. Find the new Hamiltonian of the system. This transformation holds even if $\lambda_{1,2}$ might be negative.

Solution $K(Q,P;T) = \frac{\lambda_1\lambda_2}{\lambda_3} H(\frac{Q}{\lambda_1}, \frac{O}{\lambda_2}; \frac{T}{\lambda_3})$.

Homework 2.53

(a) Consider a canonical coordinate transformation by a symplectic matrix T. Establish the Hamilton's equation before the transformation. Then reestablish it using the new coordinate system. This homework assures the form of the Hamilton's equation stays invariant under canonical transformations.

(b) The text mentioned the possibility of choosing a coordinate transformation matrix T that is not strictly symplectic, i.e. a transformation with $\tilde{T}ST = \alpha S$ with $\alpha \neq 1$. What happens to the Hamilton's equation in this situation? In particular, what factors have been compromised when we use (x, x', y, y') instead of (x, p_x, y, p_y) as our coordinates?

Homework 2.54 If (x, x', y, y') constitutes a legitimate phase space, test each of the following coordinate systems to see if it is also legitimate (even when the transformation is not strictly canonical),

$$\begin{bmatrix} 2x \\ 2x' \\ 2y \\ 2y' \end{bmatrix}, \quad \begin{bmatrix} x \\ x' \\ 2y \\ 2y' \end{bmatrix}, \quad \begin{bmatrix} x \\ x' \\ 2y \\ \frac{1}{2}y' \end{bmatrix}, \quad \begin{bmatrix} x \\ x' \\ 2y \\ -\frac{1}{2}y' \end{bmatrix}, \quad \begin{bmatrix} x+y \\ x'+y' \\ x-y \\ x'-y' \end{bmatrix}.$$

Homework 2.55 Convince yourself that, no matter what set of dynamical variables you choose, the corresponding transfer matrices must be such that all its diagonal elements are dimensionless. Also, its determinate must be dimensionless.

Solution Hopefully, you have long since noticed this property. But please formulate your reasoning explicitly. This property had better be so, otherwise stability criterion like Tr$M < 2$ (see next section) and symplecticity condition like det$M = 1$ will not make sense.

2.3.4 Stability criterion

Consider a linear dynamical system that has a one-turn transfer map $M(s+C|s)$ around position s. One question to be addressed is whether particle motion in this dynamical system is stable or unstable. The answer must be fully contained in the matrix $M(s+C|s)$ and we should not need any more information than $M(s+C|s)$, but we have yet to find the concrete stability criterion.

A moment's reflection yields the following: Stability of particle motion in this dynamical system means that when a particle with initial condition (x_0, x_0') is mapped repeatedly by the one-turn map $M(s+C|s)$ for m turns, the resulting particle coordinates x and x' remains confined as $m \to \infty$. In other words,

$$\text{Stability} \quad \Longleftrightarrow \quad M^m \begin{bmatrix} x_0 \\ x_0' \end{bmatrix} = \text{finite as } m \to \infty\,,$$

where we have abbreviated $M = M(s+C|s)$.

To proceed, we need first to perform an eigenanalysis of the matrix M. Let $v_{1,2}$ be the eigenvectors of M with eigenvalues $\lambda_{1,2}$, i.e. (first consider a 1-D system with a 2-D phase space)

$$Mv_1 = \lambda_1 v_1, \qquad Mv_2 = \lambda_2 v_2\,.$$

In general, v's and λ's are complex.

If we form matrices

$$V = [v_1 \quad v_2], \qquad \Lambda = \begin{bmatrix} \lambda_1 & 0 \\ 0 & \lambda_2 \end{bmatrix},$$

where in matrix V, the elements v_1 and v_2 each represents a column vector, then we have

$$MV = V\Lambda .$$

(A quick comment here: Don't carelessly write $MV = \Lambda V$. It is not true!)

We then have

$$M = V\Lambda V^{-1},$$

which can be used to obtain

$$M^m = V\Lambda^m V^{-1} = V \begin{bmatrix} \lambda_1^m & 0 \\ 0 & \lambda_2^m \end{bmatrix} V^{-1} .$$

More specifically, we have, for any function $f(x)$,

$$f(M) = V \begin{bmatrix} f(\lambda_1) & 0 \\ 0 & f(\lambda_2) \end{bmatrix} V^{-1} . \tag{2.36}$$

Equation (2.36) is very useful. If you have any function $f(M)$ of the matrix M and you want to find the matrix expression of $f(M)$, this equation will come handy. For example, do you know how to calculate e^M or $\sqrt{M}$ given a matrix M? Keep this expression in mind.

It follows from Eq. (2.36) that, when M is applied repeatedly to any initial vector $\begin{bmatrix} x_0 \\ x_0' \end{bmatrix}$ $(m \to \infty)$, the resulting vector is confined if and only if (Remember that λ's are complex quantities.)

$$|\lambda_1| \le 1 \quad \text{and} \quad |\lambda_2| \le 1 .$$

But we are not done yet. Since the system is necessarily symplectic, we must have $\det M = 1$, which demands $\lambda_1\lambda_2 = 1$. This means $|\lambda_1|\,|\lambda_2| = 1$. The only way for the system to be stable, therefore, is

$$|\lambda_1| = |\lambda_2| = 1 .$$

Here one sees another manifestation of the power of the symplecticity condition. The condition $|\lambda_{1,2}| = 1$ is surely much more stringent than the condition $|\lambda_{1,2}| \le 1$.

The above analysis also holds for n-D linear systems. Stability of an n-D linear system is that all eigenvalues of the $2n \times 2n$ one-turn map have absolute values of 1. For a 1-D system, however, the stability condition can be written in yet another handier form. The eigenvalues are determined by

$$\det(M - \lambda I) = \begin{bmatrix} m_{11} - \lambda & m_{12} \\ m_{21} & m_{22} - \lambda \end{bmatrix} = \lambda^2 - (\mathrm{Tr}M)\lambda + 1 = 0\,,$$

where $\mathrm{Tr}M = m_{11} + m_{22}$ is the trace of M. It can be shown that $|\lambda| = 1$ for its two solutions if and only if

$$-2 \ \le \ \mathrm{Tr}M \ \le \ 2\,.$$

For a 1-D system, therefore, stability of the system can be simply found by calculating the trace of the one-turn matrix. One must keep in mind that the trace form of stability condition holds only for 1-D systems, although the form $|\lambda| = 1$ for all eigenvalues still holds for n-D systems. As noted in Homework 2.55, the trace of the transfer matrix must be dimensionless, and it is only then it makes sense when it is to be compared with a pure number 2.

Note that in passing we have shown that the stability condition is independent of the initial conditions. Although we did mention the initial condition $\begin{bmatrix} x_0 \\ x_0' \end{bmatrix}$, it never entered the stability analysis. This is an important property of a linear system. For a linear system, therefore, the stability of the system is completely specified by the one-turn map, independent of the initial conditions. In contrast, the stability of a nonlinear system depends not only on the map, but also on the initial launching conditions of the particle in question. Often it occurs that the motion is stable for a particle launched near the phase space origin, but becomes unstable when the coordinates get larger than some threshold values. This phenomenon is called *dynamic aperture* — see Sec. 4.7.4, and is an important subject of study in accelerator nonlinear dynamics. Keep in mind that the concept of dynamic aperture applies only to nonlinear systems. A linear system is either stable or unstable for the entire phase space, and there is no aperture limit to speak of.

The stability of a linear system is not only independent of the initial conditions of the particle motion, but also independent of the location s around which the one-turn map is obtained. It was pointed out before in Eq. (2.21) that one-turn maps around two different locations are related to each other by a similarity transformation, and that eigenvalues and trace of a matrix are invariant under similarity transformations. Indeed, it is also expected that the stability of a system must not change simply because one observes the particle motion from a different position.

It is perhaps instructive at this moment to reflect again on why a matrix formalism is so useful for us. Indeed, it seems all we have done was to make a simple notation change to write our results in a matrix form. Strictly speaking, no physics content has been added by adopting the matrix formalism. But that simple notation change has allowed us to introduce all the matrix techniques.

The mere change of notation therefore much enriched the physics contents for us. Through the matrix technique, our matrix formalism has become extremely powerful. And yet there are much more to come. One such beautiful accomplishment allowed by a matrix formalism is the Courant-Snyder transformation to be discussed momentarily.

Homework 2.56 For a stable 1-D linear system, the two eigenvalues satisfy $|\lambda_1| = |\lambda_2| = 1$. Show further that they are related by $\lambda_1 = \lambda_2^*$, i.e. they form complex conjugates of each other and they lie on the unit circle in the complex plane.

Homework 2.57
(a) Consider a storage ring that has a one-turn map

$$M = \begin{bmatrix} \cos\psi & \beta\sin\psi \\ -\frac{1}{\beta}\sin\psi & \cos\psi \end{bmatrix},$$

where β and ψ are some lattice parameters. Find its determinant, eigenvalues and eigenvectors.
(b) Repeat the problem for the case when

$$M = \begin{bmatrix} \cosh\psi & \beta\sinh\psi \\ \frac{1}{\beta}\sinh\psi & \cosh\psi \end{bmatrix}.$$

You of course know case (a) is stable, case (b) is unstable.

Homework 2.58 Suppose you have obtained a one-turn map and it is given by $M = \begin{bmatrix} a & b \\ c & d \end{bmatrix}$. You already know that your matrix satisfies $\det(M) = ad - bc = 1$ for symplecticity and that, for stability, it also satisfies $|\mathrm{Tr}(M)| = |a + d| \le 2$. Convince yourself of two more properties for a stable system.
(a) b and c must not be of the same sign.
(b) The product ad must not be larger than 1.
Have you noted these properties before? If not, make a note of it. If violated by the matrix M of your stable lattice design, look for an error.

Homework 2.59 Compute the following matrices. Express your result in 2×2 matrices.

$$\sin\begin{bmatrix} 1 & 2 \\ 0 & 1 \end{bmatrix}, \quad \begin{bmatrix} 2 & 1 \\ 0 & 1 \end{bmatrix}^{12}, \quad \exp\begin{bmatrix} a & a \\ a & a \end{bmatrix}.$$

Solution

$$\exp\begin{bmatrix} a & a \\ a & a \end{bmatrix} = e^a \begin{bmatrix} \cosh a & \sinh a \\ \sinh a & \cosh a \end{bmatrix},$$

$$\exp\begin{bmatrix} ia & ia \\ ia & ia \end{bmatrix} = e^{ia} \begin{bmatrix} \cos a & i\sin a \\ i\sin a & \cos a \end{bmatrix}.$$

Homework 2.60 Consider a nonsymplectic 2×2 map M with $\det(M) \neq 1$. Start with $X_0 = \begin{bmatrix} x_0 \\ x'_0 \end{bmatrix}$, and repeatedly apply the map M to obtain $X_k = M^k X_0$. Show analytically that X_k will grow with k if $\det(M) > 1$ and damp if $\det(M) < 1$. Find the growth/damping rate.

Homework 2.61 This homework is to practice your matrix manipulation skills. Consider a matrix

$$\Sigma = \begin{bmatrix} x^2 & xp \\ xp & p^2 \end{bmatrix}.$$

(a) Use Eq. (2.36) to show that $\Sigma^2 = (x^2 + p^2)\Sigma$.
(b) Give a matrix expression for $\sqrt{\Sigma}$.
(c) Find Σ^α for arbitrary α.

Solution (b) $\sqrt{\Sigma} = \frac{1}{\sqrt{x^2+p^2}}\Sigma$.

2.4 Courant-Snyder formalism

2.4.1 Courant-Snyder functions

The lattice is not always made up of simple FODO cells. To describe a general lattice, let us return to the Hill's equation (2.5),

$$u'' + K(s)u = 0.$$

The focusing function $K(s)$ is arbitrary, except that it is periodic in s with period L (L is either the ring circumference, or when the ring is composed of unit cells, then it is the length of each cell). As we shall see, after we learn the Courant-Snyder formalism,[12] we will almost always use transfer matrices, and will not refer back to the Hill's equation too much any more.

In passing, we note that here we have another example when an abstract mathematical condition makes a profound impact on the analysis of our system. As we will see, the seemingly innocent condition that $K(s)$ is periodic, but otherwise arbitrary, releases the mighty Courant-Snyder formalism.

There are several ways to introduce the Courant-Snyder formalism. We will discuss a couple of them in these lectures. We discuss one in this section. The others will be introduced as these lectures develop. As its first discussion, we introduce two functions $\beta(s)$ and $\psi(s)$ of s, with the ansatz (assuming stable motion)

$$u(s) = \sqrt{\epsilon\beta(s)}\cos(\psi(s) + \psi_0), \tag{2.37}$$

where $\beta(s)$ has the physical meaning related to the amplitude of u at position s, while $\psi(s)$ is a phase function. We want Eq. (2.37) to satisfy the Hill's equation (2.5).

[12]E. D. Courant and H. S. Snyder, Ann. Phys. 3, 1 (1958).

Figure 2.32: Ernest Courant (1920–2020). [Courtesy Jens Zorn and Paul Courant, 2020]

The quantity ϵ is a normalization constant (independent of s), which does not enter the Hill's equation because Hill's equation is linear in u and the factor $\sqrt{\epsilon}$ cancels out. As we shall see later, however, ϵ, called the *emittance*, has its own important physical meaning. The quantity ψ_0 is the initial phase at $s = 0$ and we shall impose the condition $\psi(0) = 0$. There is no loss of information doing so. The two initial conditions ϵ, ψ_0 are equivalent to, and are replacing, the two conditions u_0, u_0'.

Equation (2.37) is based on a technique often used in other branches of physics. When you know your dynamical variable tends to oscillate in time but in a somewhat complicated manner, such as our $u(s)$, it is often useful to decompose it into an amplitude and a phase, when both depend on time, such as in Eq. (2.37). The equation of motion for your dynamical variable (one equation) then decomposes to become coupled equations of motion for the amplitude and phase (two equations). In the case of simple harmonic oscillation, of course, the amplitude is a constant in time while the phase advances linearly with time. This technique is what we will adopt here.

We now discuss the two remaining quantities $\beta(s)$ and $\psi(s)$ of Eq. (2.37). The function $\beta(s)$ has two immediate properties,

- it is always positive ($\beta(s) > 0$);
- it is to be periodic in s with the period L.

$\beta(s)$ is a key quantity in the Courant-Snyder formalism, and its prominence is reflected by the fact that it has acquired a number of names: the amplitude function, the beam envelope function, the Floquet function, the betatron function, the Courant-Snyder β-function, the often but somewhat stretched use of the name Twiss parameters, or simply the β-function. The fact that we can write the solution $u(s)$ of the Hill's equation in terms of a periodic $\beta(s)$ is a result of the *Floquet theorem*, which we shall not go into.

The insistence of $\beta > 0$ is a convention. We can equally insist on $\beta < 0$. There is nothing to gain by doing so, however.

As one example of the physical significance of the β-function, note that a tightly focused lattice will mean a small beam size, which in turn means

$$\begin{aligned}(\text{tightly focused lattice}) \quad &\Longleftrightarrow \quad (\text{small beam size}) \\ &\Longleftrightarrow \quad (\text{small } \beta\text{-function}) \,.\end{aligned}$$

From Eq. (2.26), we also expect that there is a tendency that

$$\begin{aligned}\beta_x \text{ in QF's} \quad &> \quad \beta_x \text{ in QD's}\,, \\ \beta_y \text{ in QD's} \quad &> \quad \beta_y \text{ in QF's}\,.\end{aligned}$$

Substituting Eq. (2.37) into the left hand side of the Hill's equation, and grouping the many terms into two terms, one proportional to $\sin(\psi + \psi_0)$ and another proportional to $\cos(\psi + \psi_0)$. For Eq. (2.37) to hold for all s and for all ψ_0, these two terms must vanish separately. We then obtain, after some algebra,

$$\begin{aligned}\frac{1}{2}\left(\beta\beta'' - \frac{1}{2}\beta'^2\right) - \beta^2\psi'^2 + \beta^2 K \quad &= \quad 0\,, \\ \beta'\psi' + \beta\psi'' \quad &= \quad 0\,,\end{aligned} \tag{2.38}$$

where a prime means taking derivative with respect to s. A single Hill's differential equation has now decomposed into two coupled differential equations.

The second equation in (2.38) can be integrated immediately since $\beta'\psi' + \beta\psi'' = (\beta\psi')'$, giving

$$\beta\psi' = \text{constant} \; = \; 1\,, \tag{2.39}$$

where a specific normalization of $\psi(s)$ has been chosen by selecting the integration constant to be 1 (Homework 2.62). We then have

$$\psi(s) \; = \; \int_0^s \frac{ds'}{\beta(s')}\,.$$

The phase function is therefore simply related to the β-function by an integration over s, and knowing $\beta(s)$ along the beamline allows us to determine how the phase advances along the beamline. Since $\beta > 0$, it follows that $\psi(s)$ is a monotonic function of s, always increasing as s increases. It will be referred to as the *betatron phase advance*. By assigning $\psi(s)$ as the betatron phase, $\frac{1}{\beta(s)}$ now acquires the meaning of the local betatron oscillation wave number and $2\pi\beta(s)$ is the local betatron oscillation wavelength.

The first equation in (2.38) then gives a differential equation that contains only β and no ψ,

$$\frac{1}{2}\beta\beta'' - \frac{1}{4}\beta'^2 + \beta^2 K = 1. \tag{2.40}$$

In principle, knowing $K(s)$, Eq. (2.40) can be used to calculate the β-function — together with the periodicity condition of course. However, what one uses in practice is not Eq. (2.40). We will discuss the calculation of the β-function momentarily.

We now define two more functions,

$$\begin{aligned}
\alpha(s) &= -\frac{1}{2}\beta'(s), \\
\gamma(s) &= \frac{1+\alpha^2(s)}{\beta(s)}.
\end{aligned}$$

Like $\psi(s)$, functions $\alpha(s)$ and $\gamma(s)$ are determined completely once $\beta(s)$ is known. We will call the functions $\beta(s)$, $\alpha(s)$, $\gamma(s)$, and $\psi(s)$ the Courant-Snyder functions — we do not call them the Twiss parameters. The usefulness of $\alpha(s)$ and $\gamma(s)$ will become clear soon.

The quantities β, γ, ϵ are always positive, and as mentioned, ψ always increases with s monotonically. The sign of α can be positive or negative. Dimensionality of the various quantities are

$$\begin{aligned}
[\beta] &= \text{m}, \\
[\epsilon] &= \text{m-rad}, \\
[\alpha] &= \text{dimensionless}, \\
[\gamma] &= 1/\text{m}, \\
[\psi] &= \text{rad}.
\end{aligned}$$

Homework 2.62 In Eq. (2.39), the constant was chosen to be 1 by convention. Show that there is no loss of physical content by this specific choice.

Solution If we choose $\beta\psi' = \text{constant} = \kappa$, the new β-function is simply a factor κ larger than the old β-function, while ψ remains unchanged.

Homework 2.63 Apply Eq. (2.40) to a case of a thin quadrupole located at $s = 0$ with $K(s) = q\delta(s)$. Calculate the quantity $\alpha(0^+) - \alpha(0^-)$. Show that the β-function is bent upward or downward according to the sign of q.

Homework 2.64 Equation (2.40) usually is not too useful to calculate the β-function $\beta(s)$ knowing the focusing function $K(s)$. However, it can be useful to do the reverse. Sometimes we have a desired $\beta(s)$ — for example, for a matching section between two beamlines or plasma channels with known lattices — and there is a need to find the appropriate $K(s)$ to make the transition. As an example of this application, consider a need to match $\beta(s)$ from β_1 to β_2 with (Fig. 2.33)

$$\beta(s) = \beta_1 + \frac{\beta_2 - \beta_1}{2}\left(1 + \tanh\frac{s}{a}\right).$$

Find the required focusing function $K(s)$ and plot it as a function of s.

Figure 2.33: A possible transition region for matching β-function.

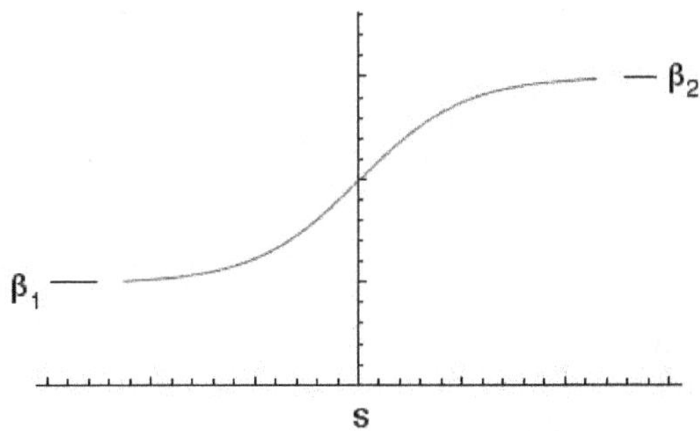

2.4.2 Phase space ellipse

Once the β-function, and therefore the Courant-Snyder functions, is known, the motion of a particle is completely specified by two constants, the emittance ϵ and the initial phase ψ_0. It is important to note that the same Courant-Snyder functions apply to *all* particles in the accelerator regardless of their initial phases or emittances. Courant-Snyder functions are strictly properties of the accelerator lattice design, and do not depend on any particle or beam properties.

Earlier when we introduced the Courant-Snyder transformation, we traded a single variable $u(s)$ with four variables $\epsilon, \beta(s), \psi(s), \psi_0$. Question may be raised if this is an unfavorable trade. Here let us repeat the reason why the trade is a good one. If we calculate $u(s)$ from the Hill's equation, we have to repeat the calculation for every particle, so the Hill's equation needs to be solved repeatedly. Using the Courant-Snyder description, we only have to calculate the β-function once, and solution to all particles follows. As mentioned in Sec. 1.1.7, here we see the power of factorization. Courant-Snyder formalism can simply be viewed as a way to perform this factorization for linear systems. For nonlinear systems, we implement Lie algebra.

We now have

$$\begin{aligned} u &= \sqrt{\epsilon\beta}\cos(\psi + \psi_0), \\ u' &= -\sqrt{\epsilon}\frac{\alpha}{\sqrt{\beta}}\cos(\psi + \psi_0) - \sqrt{\frac{\epsilon}{\beta}}\sin(\psi + \psi_0). \end{aligned} \tag{2.41}$$

These two equations can be combined to give

$$\beta u' + \alpha u = -\sqrt{\epsilon\beta}\sin(\psi + \psi_0) .$$

One sees that the quantity $\beta u' + \alpha u$ can be considered to be the momentum canonical conjugate to u, u being proportional to $\cos(\psi+\psi_0)$ and $\beta u+\alpha u'$ being proportional to $-\sin(\psi + \psi_0)$. This is a quantity that will repeatedly occur in later analysis, and is related to the *normal form* transformation (Sec. 2.5). Note the minus sign on the right hand side; it is the same minus sign that occurs in $\frac{d}{dx}\cos x = -\sin x$, and is very natural to have occurred.

We now have an important result that

$$u^2 + (\beta u' + \alpha u)^2 = \epsilon\beta ,$$

or equivalently

$$\gamma u^2 + 2\alpha u u' + \beta u'^2 = \epsilon . \tag{2.42}$$

In other words, as a particle moves along an accelerator with the motion described by $u(s)$ and $u'(s)$, if we form the quantity $\gamma u^2 + 2\alpha u u' + \beta u'^2$ as a function of s, we will find that its value is a constant even though u and u' vary. This constant of the motion, ϵ, is called the *emittance of the particle.* Equation (2.42) describes an ellipse of area $\pi\epsilon$ in the (u, u') space.

Consider a particle whose trajectory is described by $u(s)$ and $u'(s)$ as it moves along the accelerator. If we form a phase space (u, u'), then the particle's trajectory can be represented as a point in this space. As the particle moves along the accelerator, its representative point moves in the phase space. The Courant-Snyder formalism says that to describe the motion of the particle, one first construct the ellipse (2.42) at all locations s. Ellipses at different locations will look different, but they will all have the same area $\pi\epsilon$. The particle's representative point in the (u, u') space will then lie on these ellipses as it moves along the accelerator.

Not only the phase space ellipse evolves, but also the particle slides on the evolving ellipse with a rotational motion. As the particle makes one complete revolution (or one complete period), the ellipse will return to its original ellipse, but the particle's representative point would have slid a certain phase angle along the ellipse. The reader is suggested to imagine a combined motion of the ellipse evolution and a sliding motion on the ellipse. Hopefully Fig. 2.34 helps.

As mentioned, the particle will slide along the ellipse (which is itself evolving or breathing with period L). It would be instructive at this point to ask whether this sliding is clockwise or counterclockwise. The answer is clockwise. All phase space rotations are clockwise.[13] One way to tell this rule is to observe that in the upper half plane of the (u, u') space, particles have $u' > 0$, which means u will be increasing with s, and in turn means the representative point moves to the right. Similarly, particles in the lower half plane will move to the left. This

[13]Unless mass is negative. We will discuss the negative mass in Chapter 5.

Figure 2.34: Evolution of the phase space ellipse.

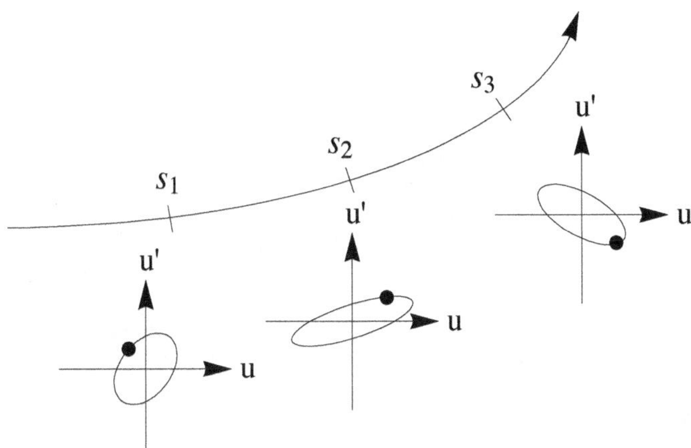

All ellipses have the same area $\pi\epsilon$.

sense of rotation can only be clockwise. This clockwise rotation is reflected in Fig. 2.34.

One may choose not to follow the evolving ellipse, but to observe a particle turn after turn at a fixed location in the ring, and collect the particle's motion turn after turn,

$$(u_1, u_1'), (u_2, u_2'), (u_3, u_3') \cdots$$

These points will all lie on an ellipse given by Eq. (2.42), where α, β, γ are the Courant-Snyder functions evaluated at the point of observation. Note that the locus of the particle does not trace out the ellipse in a continuous manner. Rather, it jumps from one point to next with a certain fixed clockwise sliding, and only after collecting data for sufficient number of turns does the shape of the ellipse emerge.

One might also ask if the sliding angle on the ellipse at the observation point is a constant at each revolution. The answer is no. A sliding angle as measured in this ellipse has no meaning, because the two coordinates u and u' do not even have the same dimensionality. On the other hand, when we transform the coordinates to normal form, as we will do later, the sliding angle does have physical meaning, and only in that case, the sliding angle is a constant from revolution to revolution.

The ellipse (2.42) is shown geometrically in Fig. 2.35. One can relate the Courant-Snyder functions to the various geometrical quantities on this phase space ellipse as shown.

The ellipse in Fig. 2.35 is drawn to describe the trajectory of a particle with emittance ϵ and phase ψ_0. What happens if we describe another particle with

Figure 2.35: The Courant-Snyder phase space ellipse.

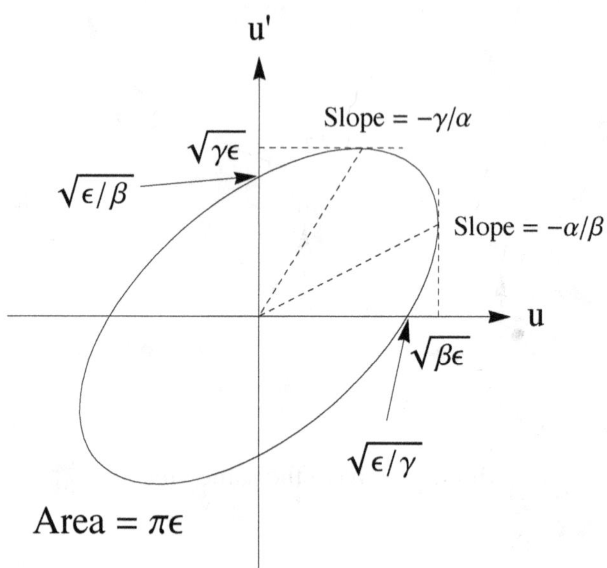

a different emittance and phase? The answer is that a particle with a different emittance will lie on a different ellipse of identical shape but a different size, while a particle with the same emittance but different phase will lie on the same ellipse but with a different phase angle as the particle circulates around the accelerator.

Signs of β and γ are always positive. How can one tell the sign of α from the shape and orientation of the phase space ellipse? One should first recognize from Eq. (2.42) that the ellipse will be upright if $\alpha = 0$. This means α is related to the tilting of the ellipse. A northeast-southwest tilt as shown in Fig. 2.35 will mean $\alpha < 0$, while a southeast-northwest tilt will mean $\alpha > 0$. Later we will calculate, for a distribution of many particles, the correlation distribution moment $\langle uu' \rangle = -\alpha\epsilon_{\rm rms}$. We see that $\langle uu' \rangle$ has the same sign as $-\alpha$. When $\alpha > 0$, for example, the beam distribution will have $\langle uu' \rangle < 0$, which indeed means it has a southeast-northwest orientation.

The sign of α also gives us some information on how the phase space ellipse will evolve in the next moment of time. This is because $\alpha = -\frac{1}{2}\beta'(s)$. A positive α, for example, means $\beta' < 0$, i.e. β is decreasing with s. Now from Fig. 2.35, we know that the u and u' extents of the ellipse scale like $\pm\sqrt{\beta}$ and $\pm\frac{1}{\sqrt{\beta}}$, respectively. A decreasing β therefore means that the ellipse will become narrower horizontally and taller vertically in the next moment. Similarly, when $\alpha < 0$, the ellipse will become wider and shorter in the next moment. It is therefore possible to more or less predict how the ellipse will evolve just by looking at its

Figure 2.36: Illustration of how the phase space ellipse evolves by examining the sign of α.

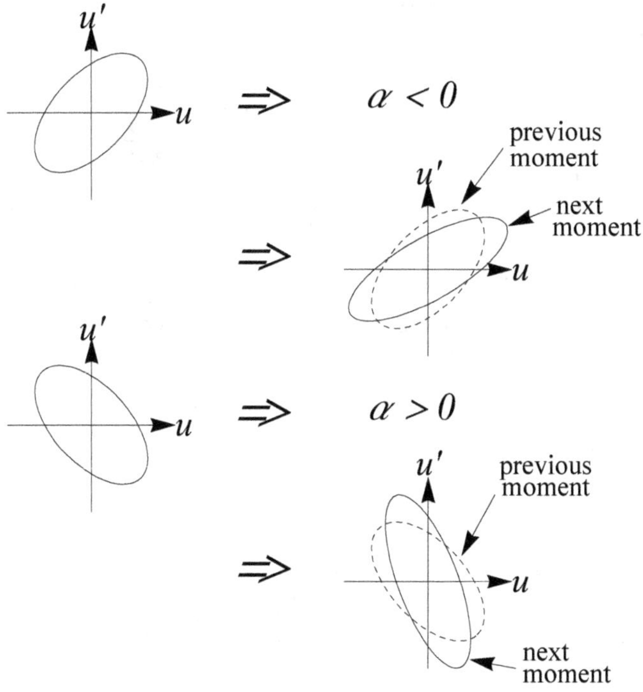

instantaneous shape and orientation. These observations are consistent with a clockwise phase space rotation. This conclusion is illustrated in Fig. 2.36.

Homework 2.65 One of the panels in Fig. 2.37 represents a converging beam, the other a diverging beam. Can you tell which is which? Give your reason.

Figure 2.37: Which case represents a converging beam, and which represents a diverging beam?

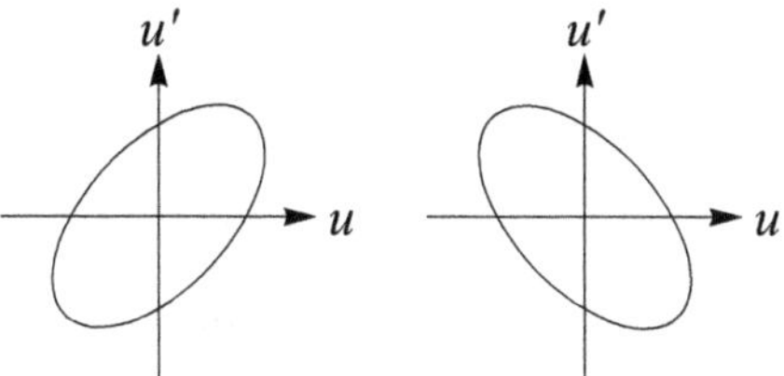

Homework 2.66 Develop the Courant-Snyder formalism for the simple harmonic oscillator that has the phase space (q,p) and Hamiltonian $H(q,p;t) = \frac{q^2}{2} + \frac{\omega^2 p^2}{2}$. What is your β-function? Follow the steps developed in the text for the Courant-Snyder formalism.

Solution $\beta = \frac{1}{\omega}$. In this case β-function is in units of seconds, not meters. High oscillation frequency means slow β-function.

Homework 2.67

(a) Prove that Eq. (2.42) gives an ellipse of area $\pi\epsilon$. This is valid regardless of the values of α, β, γ as long as $\gamma = \frac{1+\alpha^2}{\beta}$.

(b) Show that the geometry shown in Fig. 2.35 comes from Eq. (2.42).

Solution (b) One way to appreciate the ellipse is to connect it with the three Courant-Snyder functions as follows,

- β is related to the horizontal width of the ellipse.
- γ is related to its vertical height.
- α is related to its tilt.

Homework 2.68 Use Eq. (2.41) to show that

$$u'(s) = -\sqrt{\gamma} \sin\left(\int^s ds' \frac{K(s')}{\gamma(s')}\right).$$

Find also the phase of this sinusoidal expression in terms of the ψ_0 in Eq. (2.41). This important equation says that $\gamma(s)$ is the envelope function for u' just as much as $\beta(s)$ is the envelope function for u. There will be more discussion along this line later in Sec. 2.5.2.

Homework 2.69

(a) Show that

$$\frac{d}{ds}\begin{bmatrix}\alpha(s)\\ \beta(s)\\ \gamma(s)\end{bmatrix} = \begin{bmatrix}0 & K(s) & -1\\ -2 & 0 & 0\\ 2K(s) & 0 & 0\end{bmatrix}\begin{bmatrix}\alpha(s)\\ \beta(s)\\ \gamma(s)\end{bmatrix}.$$

(b) Use (a) to show that in a free drift space,

$$\begin{aligned}\gamma(s) &= \text{constant}\,,\\ \alpha(s) &= \text{linear in } s\,,\\ \beta(s) &= \text{quadratic in } s\,.\end{aligned}$$

(c) As a self-consistency check, use your result in (a) to show that $\gamma = \frac{1+\alpha^2}{\beta}$ is assured for all position s.

Solution (a) Very loosely speaking, one might develop a feel by observing the following:

$$\psi \sim \int \frac{1}{\beta}, \quad \alpha \sim \beta', \quad \gamma \sim \beta''.$$

Again, once β is determined, all the other Courant-Snyder functions assume the physical meanings of its integration or derivatives.

Homework 2.70 Continuing Homework 2.69, consider the case when $K(s) = K_0 = \text{constant}$, and let the matrix

$$T_0 = \begin{bmatrix} 0 & K_0 & -1 \\ -2 & 0 & 0 \\ 2K_0 & 0 & 0 \end{bmatrix},$$

then it is easy to see that

$$\begin{bmatrix} \alpha \\ \beta \\ \gamma \end{bmatrix} (s) = e^{T_0 s} \begin{bmatrix} \alpha \\ \beta \\ \gamma \end{bmatrix} (0).$$

Find a closed-form expression for $\beta(s)$ by expanding the matrix $e^{T_0 s}$. Compare your result with Eq. (2.55) later.

Solution This is a technical algebraic homework. Although not emphasized in our lectures, algebra is a basic required skill.

$$T_0^{2m} = (-4K_0)^{m-1} \begin{bmatrix} -4K_0 & 0 & 0 \\ 0 & -2K_0 & 2 \\ 0 & 2K_0^2 & -2K_0 \end{bmatrix},$$

$$T_0^{2m+1} = (-4K_0)^m T_0,$$

$\Longrightarrow$

$$\begin{aligned} e^{T_0 s} &= \sum_{n=0}^{\infty} \frac{1}{n!} (T_0 s)^n \\ &= \begin{bmatrix} \cos\xi & \frac{\sqrt{K_0}}{2} \sin\xi & -\frac{1}{2\sqrt{K_0}} \sin\xi \\ -\frac{1}{\sqrt{K_0}} \sin\xi & \frac{1+\cos\xi}{2} & \frac{1}{2K_0}(\cos\xi - 1) \\ \sqrt{K_0}\sin\xi & -\frac{K_0}{2}(\cos\xi - 1) & \frac{1+\cos\xi}{2} \end{bmatrix}, \end{aligned}$$

where $\xi = \sqrt{4K_0}\, s$ (assuming $K_0 > 0$)

$\Longrightarrow$

$$\beta(s) = -\frac{\alpha_0}{\sqrt{K_0}} \sin\xi + \beta_0 \left(\frac{1+\cos\xi}{2} \right) + \frac{1-\cos\xi}{2K_0} \frac{1+\alpha_0^2}{\beta_0}.$$

You may also calculate $\alpha(s), \gamma(s)$, and then confirm $\gamma(s) = \frac{1+\alpha^2(s)}{\beta(s)}$.

Homework 2.71

(a) Knowing the Courant-Snyder functions, the phase space ellipse of a particle with emittance ϵ looks like Fig. 2.38. Find the (u, u') coordinates for locations A, B, C, D, E, F, G, H in the phase space. Figure 2.35 gives most but not all the answers.

Figure 2.38: An exercise on the Courant-Snyder ellipse.

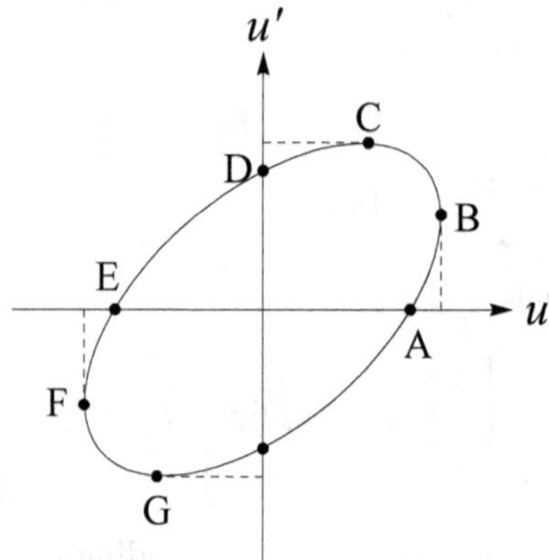

(b) We will learn later that if a one-period map has 90° phase advance, the map reads $M = \begin{bmatrix} \alpha & \beta \\ -\gamma & -\alpha \end{bmatrix}$. If a particle starts at position A, where does it end up after one period? Repeat the question for the other positions B through H.

Solution

(a) $A(\sqrt{\epsilon/\gamma}, 0)$, $B(\sqrt{\epsilon\beta}, \ -\alpha\sqrt{\epsilon\beta})$, $C(-\alpha\sqrt{\epsilon/\gamma}, \ \sqrt{\epsilon\gamma})$, $D(0, \sqrt{\epsilon/\beta})$, $E(-\sqrt{\epsilon/\gamma}, 0)$, $F(-\sqrt{\epsilon\beta}, \alpha\sqrt{\epsilon\beta})$, $G(\alpha\sqrt{\epsilon/\gamma}, -\sqrt{\epsilon\gamma})$, $H(0, -\sqrt{\epsilon/\beta})$.

Homework 2.72 As we learned, when observed at a fixed location in a storage ring, the trajectory of a particle in phase space is described by (u, u'). The turn-by-turn values of $(u_1, u_1'), (u_2, u_2'), \cdots$ trace out an ellipse $\gamma u^2 + 2\alpha u u' + \beta u'^2 = \epsilon$. In practice, however, u' is not easy to measure, and sometimes it is more convenient to represent the particle motion by tracing the particle's displacements u at two nearby locations u_a and u_b instead of u and u' at the same location. In a compromise, the particle motion is then described by the turn-by-turn values $(u_{a1}, u_{b1}), (u_{a2}, u_{b2}), \cdots$. The space (u_a, u_b) of course does not qualify to be a phase space.

(a) Show that the particle trajectory in the (u_a, u_b) space is also given by an ellipse,

$$\epsilon = \frac{1}{\sin^2\psi}\left(\frac{u_b^2}{\beta_b} - \frac{2u_a u_b}{\sqrt{\beta_a\beta_b}}\cos\psi + \frac{u_a^2}{\beta_a}\right),$$

where ψ is the betatron phase advance from position a to position b. This expression gives the connection between the single-particle emittance ϵ with the reading of its orbits at two nearby BPMs. This ellipse expression does not require knowing α_a and α_b.

(b) Find the area of the above ellipse in the (u_a, u_b) space. This area shrinks to zero if ψ is an integral multiple of π. Physical reason of this should be evident. State your understanding of this physical reason.

(c) There is no reason not to choose (u'_a, u'_b) to characterize the ellipse. Show that it is given by

$$\epsilon = \frac{\beta_a\beta_b\gamma_b u_a'^2 + \beta_a\beta_b\gamma_a u_b'^2 - 2\sqrt{\beta_a\beta_b}\,u'_a u'_b[(1+\alpha_a\alpha_b)\cos\psi + (\alpha_a-\alpha_b)\sin\psi]}{[(-\alpha_a+\alpha_b)\cos\psi + (1+\alpha_a\alpha_b)\sin\psi]^2}.$$

The expression is not as neat, but that is because the α-function is not defined tailored to the purpose. The symmetry to the (u_a, u_b) space is recovered only when $\alpha_a = \alpha_b = 0$.

Solution

(b) When $\psi = k\pi$, u_b is always proportional to u_a with $u_b = \sqrt{\frac{\beta_b}{\beta_a}}(-1)^k u_a$. Particle trajectory in the (u_a, u_b) space lies on a straight line.

2.4.3 Beam emittance

So far, we have defined an emittance of a particle, which we designate as ϵ. Given the accelerator lattice, the 1-D motion of a particle is completely determined by its emittance ϵ and its initial phase ψ_0. Different particles will have different values of ϵ and ψ_0 — although as emphasized, they all have the same Courant-Snyder functions. In particular, a particle that has $u = 0$ and $u' = 0$ initially will have a zero emittance $\epsilon = 0$, and will have $u = 0$ and $u' = 0$ at all times. This particle is of course just the ideal reference particle.

Beam distribution in phase space If we have a beam of particles, these particles will have a distribution of values of ϵ and ψ_0. For example, they may have a uniform distribution in ψ_0, and a "Gaussian distribution" in ϵ.[14] Since $\epsilon = \gamma u^2 + 2\alpha u u' + \beta u'^2 = \frac{1}{\beta}[u^2 + (\alpha u + \beta u')^2]$, the beam distribution in the (u, u') space is

$$\begin{aligned}\Psi(u,u') &= \frac{1}{2\pi\epsilon_{\rm rms}}\exp\left(-\frac{\epsilon}{2\epsilon_{\rm rms}}\right)\\ &= \frac{1}{2\pi\epsilon_{\rm rms}}\exp\left(-\frac{\gamma u^2 + 2\alpha u u' + \beta u'^2}{2\epsilon_{\rm rms}}\right)\\ &= \frac{1}{2\pi\epsilon_{\rm rms}}\exp\left[-\frac{u^2 + (\alpha u + \beta u')^2}{2\beta\epsilon_{\rm rms}}\right], \qquad (2.43)\end{aligned}$$

[14] Actually it is Gaussian distribution in u and u', not in ϵ. Here it means the distribution is exponential in ϵ but we still refer to it as "Gaussian".

where $\epsilon_{\rm rms}$ is called the *rms beam emittance.* In these expressions, α, β, γ are functions of s but $\epsilon_{\rm rms}$ is not. This Gaussian distribution depends on ϵ but is independent of ψ_0.

We have normalized $\Psi(u, u')$ according to

$$\int_{-\infty}^{\infty} du \int_{-\infty}^{\infty} du' \, \Psi(u, u') = 1 \, .$$

Equation (2.43) gives the expression of $\Psi(u, u')$ in terms of ϵ with this normalization condition. In case Ψ depends only on ϵ, we may want it to be expressed as $\Psi(\epsilon)$ with a normalization

$$\int_{0}^{\infty} d\epsilon \, \Psi(\epsilon) = 1 \, .$$

In that case, we use $dudu' = \pi d\epsilon$ to obtain

$$\Psi(\epsilon) = \frac{1}{2\epsilon_{\rm rms}} \exp\left(-\frac{\epsilon}{2\epsilon_{\rm rms}}\right) .$$

The independence of a distribution on ψ_0 is significant because that means it is an *equilibrium* distribution. Although particles constantly flow in the phase space, they flow in such a way that the overall beam distribution does not change turn after turn when observed at a fixed location in the accelerator. Distribution (2.43) is a special case of an equilibrium distribution, but any reasonably behaving function of ϵ can serve as an equilibrium beam distribution as long as it does not depend on ψ_0.

Beam distribution moments We can calculate the second moments of this Gaussian beam distribution (2.43) and obtain

$$\begin{aligned}
\langle u^2 \rangle &= \int_{-\infty}^{\infty} du \int_{-\infty}^{\infty} du' \, u^2 \Psi(u, u') = \beta \epsilon_{\rm rms} \, , \\
\langle uu' \rangle &= \int_{-\infty}^{\infty} du \int_{-\infty}^{\infty} du' \, uu' \Psi(u, u') = -\alpha \epsilon_{\rm rms} \, , \\
\langle u'^2 \rangle &= \int_{-\infty}^{\infty} du \int_{-\infty}^{\infty} du' \, u'^2 \Psi(u, u') = \gamma \epsilon_{\rm rms} \, .
\end{aligned}$$

These equations combine to yield

$$\epsilon_{\rm rms}^2 = \langle u^2 \rangle \langle u'^2 \rangle - \langle uu' \rangle^2 \, .$$

They also yield

$$\frac{\langle uu' \rangle}{\sqrt{\langle u^2 \rangle \langle u'^2 \rangle}} = \frac{-\alpha}{\sqrt{1+\alpha^2}} \, . \tag{2.44}$$

We see that each of the three Courant-Snyder functions α, β and γ, relates to one of the second moments, and that α is related to the second moment $\langle uu' \rangle$ responsible for the beam tilt in the phase space. Relating to the three second moments also provides a physical meaning of these three functions. Particularly important is the connection between the rms beam size and the β-function,

$$\text{rms beam size } (s) \;=\; \sqrt{\langle u^2(s) \rangle} \;=\; \sqrt{\beta(s)\epsilon_{\text{rms}}}\,. \tag{2.45}$$

The three moments can be combined to yield

$$\begin{aligned} \langle \epsilon \rangle &= \int_{-\infty}^{\infty} du \int_{-\infty}^{\infty} du'\, \epsilon \Psi(u, u') \\ &= \gamma \langle u^2 \rangle + 2\alpha \langle uu' \rangle + \beta \langle u'^2 \rangle \\ &= 2\epsilon_{\text{rms}}\,. \end{aligned} \tag{2.46}$$

We see that ϵ_{rms} is half the average value of ϵ of the beam. The annoying factor of 2 is a source of much confusion.

One more comment in passing: The important equation (2.45) has a counterpart given by

$$\text{rms beam slope } (s) \;=\; \sqrt{\langle u'^2(s) \rangle} \;=\; \sqrt{\gamma(s)\epsilon_{\text{rms}}}\,. \tag{2.47}$$

In terms of beam dynamics, this equation carries equal significance as Eq. (2.45). However, since beam slope is less convenient for observations, it has largely been neglected. We shall try to remedy this bias, e.g. by including a discussion of an alternative Courant-Snyder formalism in Sec. 2.5.2.

Definition of beam emittance For a beam distribution other than Gaussian, we may take the following as a definition of the rms beam emittance,

$$\epsilon_{\text{rms}} \;=\; \frac{1}{2}\langle \epsilon \rangle \;=\; \frac{1}{2}\int_{-\infty}^{\infty} du \int_{-\infty}^{\infty} du'\, \epsilon \Psi(u, u')\,.$$

In case Ψ is an equilibrium distribution, with $\Psi(u, u')$ depending only on the combined quantity $\gamma u^2 + 2\alpha uu' + \beta u'^2$,

$$\epsilon_{\text{rms}} \;=\; \frac{\pi}{2}\int_0^{\infty} \epsilon d\epsilon\, \Psi \Big|_{\gamma u^2 + 2\alpha uu' + \beta u'^2 \to \epsilon}\,.$$

We may ask what is the percentage of particles whose emittance is less than the rms emittance ($\epsilon < \epsilon_{\text{rms}}$). The answer is

$$P(\epsilon < \epsilon_{\text{rms}}) \;=\; \pi \int_0^{\epsilon_{\text{rms}}} d\epsilon\, \Psi \Big|_{\gamma u^2 + 2\alpha uu' + \beta u'^2 \to \epsilon}\,.$$

Take Gaussian distribution as example, we find

$$\begin{aligned} P(\epsilon < \epsilon_{\rm rms}) &= \frac{1}{2\epsilon_{\rm rms}} \int_0^{\epsilon_{\rm rms}} d\epsilon \, \exp\left(-\frac{\epsilon}{2\epsilon_{\rm rms}}\right) \\ &= 1 - e^{-1/2} = 0.39 \,. \end{aligned}$$

We find that for a Gaussian beam, 39% of its particles have emittances $\epsilon < \epsilon_{\rm rms}$, while 61% have $\epsilon > \epsilon_{\rm rms}$.

It is also easy to find the percentage of particles whose emittance is less than κ times the rms emittance. The answer is (Fig. 2.39)

$$P(\epsilon < \kappa\epsilon_{\rm rms}) = 1 - e^{-\kappa/2} \,. \tag{2.48}$$

Figure 2.39: Percentage of particles with emittance less than $\kappa\epsilon_{\rm rms}$ for a Gaussian beam.

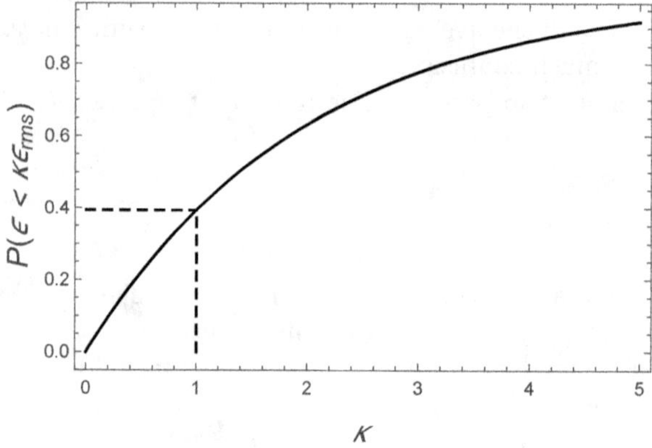

There is an endless ambiguity in the definition of the beam emittance. For us, we have defined an rms emittance $\epsilon_{\rm rms}$ as above — and we pointed out an annoying factor of 2 in Eq. (2.46). For some users, however, a definition that encompasses only 39% of the particles within it is not satisfactory. One then chooses an appropriate $\kappa > 1$ and defines the beam emittance to be $\kappa\epsilon_{\rm rms}$. Some of these users choose $\kappa = 4$, in which case 87% of the particles are within the beam emittance. Some others choose $\kappa = 6$ to encompass 95%.

As if these are not confusing enough, there is another source of confusion coming from the fact that some users try to identify the emittance with the phase space area. A beam emittance of $\kappa\epsilon_{\rm rms}$ then has an elliptical area of $\pi\kappa\epsilon_{\rm rms}$. Sometimes we add it to the units, therefore emittances are written, for example, not in mm-mrad but in π-mm-mrad. Although making a valid conceptual point, this turns out to be yet another source of confusion.[15]

[15]What goes beyond confusion, and is unforgivable, is to take this artificial factor of π from the unit of emittance and further multiply it numerically into the numerical value of ϵ.

For us, we shall stick to the definition with $\kappa = 1$, and there are 39% of particles encompassed within the rms emittance ellipse, and we shall use the unit of mm-mrad. As to the replacements of mm-mrad by μm-rad or simply μm, we shall avoid them too.

Envelope equation With emittance given by $\epsilon_{\rm rms} = \sqrt{\langle u^2\rangle\langle u'^2\rangle - \langle uu'\rangle^2}$, and let $\sigma = \sqrt{\langle u^2\rangle}$, it is easy to prove the *envelope equation*,

$$\frac{d^2\sigma(s)}{ds^2} + K(s)\sigma(s) = \frac{\epsilon_{\rm rms}^2}{\sigma^3(s)} . \qquad (2.49)$$

The rms beam size more or less oscillates with the focusing function, except for the centrifugal term on the right hand side, which prevents it from approaching zero. How close can σ approach zero, or equivalently how sharply the beam can come to focus, depends on how small the beam emittance $\epsilon_{\rm rms}$ is.

Given appropriate initial values, the envelope equation can be solved explicitly or numerically for $\sigma_x(s)$ and $\sigma_y(s)$ inside a drift space (Homework 2.75), a quadrupole, or a dipole magnet. However, the envelope equation is more often used when the space charge effect is considered by adding a driving term on the right hand side of Eq. (2.49).[16]

Σ matrix A Gaussian beam distribution (for n-D motion) can be generally written as

$$\exp\left(-\frac{1}{2}\tilde{U}AU\right),$$

where U is the $2n$-D state vector, A is a $2n \times 2n$ symmetric, positive definite matrix, and a normalization coefficient has been omitted.

Given the beam distribution, we want to compute the distribution moments of this distribution. For a Gaussian distribution, all we have to do is to compute the 2nd moments – all higher moments can be expressed in terms of the 2nd moments. Let Σ be the symmetric matrix formed by the 2nd moments of the beam distribution, i.e.

$$\Sigma_{ij} = \langle u_i u_j\rangle .$$

There is a neat theorem: The second moment Σ matrix of the Gaussian distribution $\exp\left(-\frac{1}{2}\tilde{U}AU\right)$ is given by

$$\Sigma = A^{-1}, \qquad \text{or} \qquad A = \Sigma^{-1} .$$

We will not prove the theorem here. We find content to check for the 1-D special case where the distribution is $e^{-au^2/2}$, then the second moment is given by $\langle u^2\rangle = \frac{1}{a}$.

[16] F. Sacherer, IEEE Trans. Nucl. Sci. NS-18, 1105 (1971).

Further content can perhaps be found for the distribution (2.43). We take $U = \begin{bmatrix} u \\ u' \end{bmatrix}$ and apply the theorem to obtain

$$\Sigma^{-1} = \frac{1}{\epsilon_{\rm rms}} \begin{bmatrix} \gamma & \alpha \\ \alpha & \beta \end{bmatrix} \quad \Longrightarrow \Sigma = \epsilon_{\rm rms} \begin{bmatrix} \beta & -\alpha \\ -\alpha & \gamma \end{bmatrix},$$

or

$$\langle u^2 \rangle = \beta\epsilon_{\rm rms}, \quad \langle uu' \rangle = -\alpha\epsilon_{\rm rms}, \quad \langle u'^2 \rangle = \gamma\epsilon_{\rm rms}\,,$$

It follows that

$$\langle \epsilon \rangle = \gamma\langle u^2 \rangle + 2\alpha\langle uu' \rangle + \beta\langle u'^2 \rangle = 2\epsilon_{\rm rms} = \frac{2\sigma^2}{\beta}\,.$$

These results have been obtained before.

As mentioned, the rms beam ellipse that encompasses 39% of particles in the Gaussian beam is given by

$$\tilde{U}\Sigma^{-1}U \;=\; \epsilon_{\rm rms}\,.$$

As the beam propagates in an accelerator, the beam ellipse evolves, and so does the second moments of the Gaussian beam. As the beam propagates from position s_1 to position s_2, with a linear transfer map described by $U_2 = M(s_2|s_1)U_1$, the evolution of its 2nd moments can be obtained by

$$\begin{aligned} \tilde{U}_2\Sigma_2^{-1}U_2 = \tilde{U}_1\Sigma_1^{-1}U_1 \quad &\Longrightarrow \quad \tilde{M}(s_2|s_1)\Sigma_2^{-1}M(s_2|s_1) = \Sigma_1^{-1} \\ &\Longrightarrow \quad \Sigma_2 = M(s_2|s_1)\Sigma_1\tilde{M}(s_2|s_1)\,. \end{aligned} \tag{2.50}$$

Knowing Σ at position s_1 and the linear map from position s_1 to position s_2, the Σ matrix at s_2 can be obtained. This relationship (2.50) is another way to derive how α, β, γ functions evolve from position s_1 to position s_2 (Homework 2.78), which can be checked against Eq. (2.55) later. Note that (2.50) is not a similarity transformation although it might look like one.

We know that $\det M = \det\tilde{M} = 1$. This then gives

$$\det\Sigma_2 \;=\; \det\Sigma_1\,,$$

or in other words, determinant of the Σ matrix is an invariant. Indeed,

$$\begin{aligned} \det\Sigma \;&=\; \langle u^2 \rangle\langle u'^2 \rangle - \langle uu' \rangle^2 \\ &=\; \epsilon_{\rm rms}^2(\beta\gamma - \alpha^2) \;=\; \epsilon_{\rm rms}^2\,. \end{aligned}$$

In passing we have also obtained a useful expression

$$\epsilon_{\rm rms} \;=\; \sqrt{\det\Sigma} \;=\; \sqrt{\langle u^2 \rangle\langle u'^2 \rangle - \langle uu' \rangle^2}$$

Incidentally, it is basic algebra to note that the quantity that appears in the square root, $\langle u^2\rangle\langle u'^2\rangle - \langle uu'\rangle^2$, is always positive for arbitrary beam distribution.

If a beam has an upright Gaussian distribution in the (u, u') phase space, then $\langle uu'\rangle = 0$, and we obtain

$$\epsilon_{\rm rms} \;=\; \sqrt{\langle u^2\rangle\langle u'^2\rangle} \;=\; \sigma_u\sigma_{u'}\,,$$

i.e. the beam rms emittance is given by the product of the rms in u and the rms in u'. However, it is important to remember the condition that the beam distribution must be upright. Otherwise the product $\sigma_u\sigma_{u'}$ can give a very large overestimate (never an underestimate) of the true beam rms emittance.

Homework 2.73 Prove Eq. (2.48).

Homework 2.74 Consider an uncoupled, bi-Gaussian beam distribution in the 4-D x and y phase space. The emittance distributions is given by

$$\Psi(\epsilon_x, \epsilon_y) \;=\; \frac{1}{4\epsilon_{x,\rm rms}\epsilon_{y,\rm rms}}\exp\left(-\frac{\epsilon_x}{2\epsilon_{x,\rm rms}} - \frac{\epsilon_y}{2\epsilon_{y,\rm rms}}\right).$$

Calculate the probabilities of the following cases,

(a) particles with $\epsilon_x < \kappa\epsilon_{x,\rm rms}$ regardless of $\epsilon_{y,\rm rms}$.
(b) particles with $\epsilon_y < \kappa\epsilon_{y,\rm rms}$ regardless of $\epsilon_{x,\rm rms}$.
(c) particles with $\epsilon_x < \kappa\epsilon_{x,\rm rms}$ and $\epsilon_y < \kappa\epsilon_{y,\rm rms}$.
(d) particles with $\frac{\epsilon_x}{\epsilon_{x,\rm rms}} + \frac{\epsilon_y}{\epsilon_{y,\rm rms}} < \kappa$.
(e) particles with $\epsilon_x + \epsilon_y < \kappa(\epsilon_{x,\rm rms} + \epsilon_{y,\rm rms})$.

The difference among the cases (a)-(d) is illustrated in Fig. 2.40. Case (e) might be a consideration near the resonance $\nu_x = \nu_y$.

Solution

$$P(\kappa) \;=\; \begin{cases} 1 - e^{-\kappa/2}, & \text{(a) and (b)} \\ (1 - e^{-\kappa/2})^2, & \text{(c)} \\ 1 - e^{-\kappa/2} - \frac{\kappa}{2}e^{-\kappa/2}, & \text{(d)} \\ 1 - \frac{1}{1-a}\left(e^{-\kappa(1+a)/2} - ae^{-\kappa(1+a)/(2a)}\right), & \text{(e), where } a = \frac{\epsilon_{y,\rm rms}}{\epsilon_{x,\rm rms}} \end{cases}.$$

The probabilities as functions of κ is shown in Fig. 2.41 for cases (a)-(d).

Homework 2.75

(a) Prove the envelope equation (2.49).
(b) Show that in a drift space, a solution to the envelope equation is

$$\sigma(s) \;=\; \sqrt{\sigma_0^2 + \frac{\epsilon_{\rm rms}^2}{\sigma_0^2}(s - s_0)^2}\,,$$

with arbitrary σ_0 and s_0 (to be determined by initial launching conditions).

Figure 2.40: Four cases of emittance probabilities. Here we choose $\kappa = 3$.

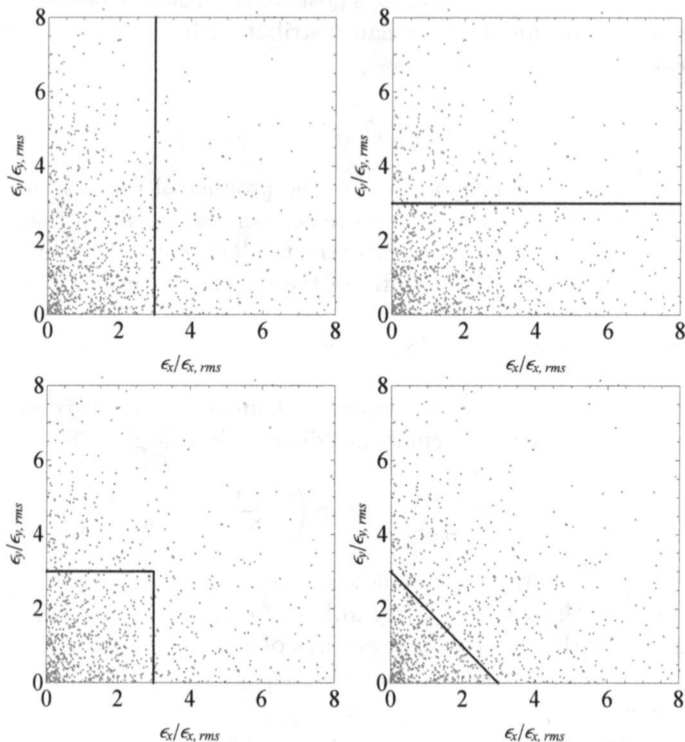

Homework 2.76 The Σ matrix contains the second moments of the beam distribution in phase space. It represents quantities that have direct physical meanings. One can also relate the Σ matrix to the Courant-Snyder functions, which are the more abstract quantities. Calculate α, β, γ and ϵ_{rms} if your beam has the following Σ matrix,

$$\begin{bmatrix} 10\ \text{mm}^2 & 10\ \text{mm-mrad} \\ 10\ \text{mm-mrad} & 20\ \text{mrad}^2 \end{bmatrix}.$$

Sketch the beam rms ellipse in phase space.

Homework 2.77

(a) A beam has a 1-D Gaussian distribution with first moments $\langle x \rangle = 0$, $\langle x' \rangle = 0$. It also has distribution second moments $\langle x^2 \rangle, \langle xx' \rangle, \langle x'^2 \rangle$. Write down the explicit expression of its distribution $\Psi(x, x')$, including the normalization coefficient, in terms of the second moments.

(b) Calculate the beam's higher moments $\langle x^4 \rangle, \langle x^3 x' \rangle, \langle x^2 x'^2 \rangle, \langle x x'^3 \rangle, \langle x'^4 \rangle$.

(c) Calculate $\langle x^{2n} \rangle$.

Figure 2.41: A few cases of emittance probabilities as functions of κ. Solid, dashed, dotted curves are for case (a) and (b), (c), and (d) respectively.

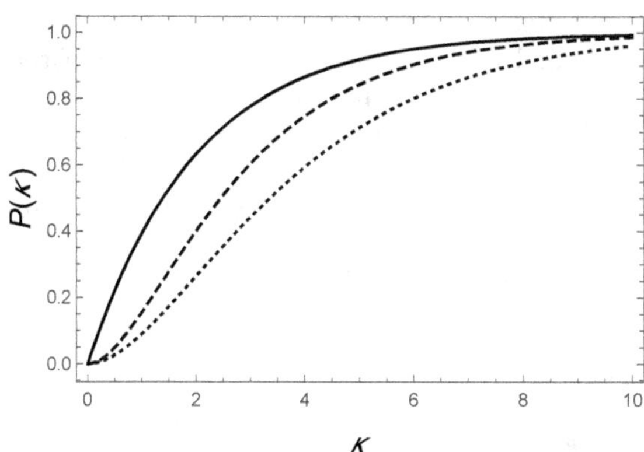

Solution

(a)

$$\frac{1}{2\pi\sqrt{\langle x^2\rangle\langle x'^2\rangle-\langle xx'\rangle^2}}\exp\left[-\frac{1}{2}\,\frac{x^2\langle x'^2\rangle-2xx'\langle xx'\rangle+x'^2\langle x^2\rangle}{\langle x^2\rangle\langle x'^2\rangle-\langle xx'\rangle^2}\right].$$

(b)

$$\begin{aligned}\langle x^4\rangle &= 3\langle x^2\rangle^2\,,\\ \langle x^3x'\rangle &= 3\langle x^2\rangle\langle xx'\rangle\,,\\ \langle x^2x'^2\rangle &= \langle x^2\rangle\langle x'^2\rangle+2\langle xx'\rangle^2\,,\\ \langle xx'^3\rangle &= 3\langle xx'\rangle\langle x'^2\rangle\,,\\ \langle x'^4\rangle &= 3\langle x'^2\rangle^2\,.\end{aligned}$$

(c) $(2n-1)!!\,\langle x^2\rangle^n$.

Homework 2.78

(a) If we know the values of the Courant-Snyder functions α,β,γ at a position $s=s_1$, we can use the transformation property of the Σ matrix, Eq. (2.50), to obtain their values at another position $s=s_2$ from the 2×2 map $M(s_2|s_1)$. Show that their connection can be written in a matrix form

$$\begin{bmatrix}\alpha_2\\ \beta_2\\ \gamma_2\end{bmatrix}=\begin{bmatrix}m_{11}m_{22}+m_{12}m_{21} & -m_{11}m_{21} & -m_{12}m_{22}\\ -2m_{11}m_{12} & m_{11}^2 & m_{12}^2\\ -2m_{21}m_{22} & m_{21}^2 & m_{22}^2\end{bmatrix}\begin{bmatrix}\alpha_1\\ \beta_1\\ \gamma_1\end{bmatrix}.$$

We will return to this later in Eq. (2.55).

(b) Show that $\gamma_2=\frac{1+\alpha_2^2}{\beta_2}$ is assured by this transformation.

2.4.4 Beam temperature

Consider a proton beam with design momentum P_0. The protons in the beam have rms angular spreads $\sigma'_x = \sigma'_y$ in the x- and y-directions, and they have an rms relative longitudinal momentum spread δ. Because of these spreads, the beam has a finite temperature. There are three temperatures corresponding to three dimensions. To find these temperatures, we need to go to the rest frame of the beam.

In the laboratory frame, consider a particle with momentum

$$p_x = P_0 x', \quad p_y = P_0 y', \quad p_z = P_0(1+\delta)\,.$$

Make a Lorentz transformation to the beam's rest frame. This particle has momentum components

$$\begin{aligned} p'_x &= p_x = P_0 x'\,, \\ p'_y &= p_y = P_0 y'\,, \\ p'_z &= \gamma\left(p_z - \frac{\beta E}{c}\right)\,. \end{aligned}$$

We also have in the laboratory frame,

$$\begin{aligned} E &= c\sqrt{p_x^2 + p_y^2 + p_z^2 + m^2c^2} = c\sqrt{P_0^2\left[x'^2 + y'^2 + (1+\delta)^2\right] + m^2c^2}\,, \\ \beta &= \frac{P_0}{\sqrt{P_0^2 + m^2c^2}}\,. \end{aligned}$$

If $1 \gg |\delta| \gg x'^2$ and y'^2, it follows that

$$p'_z = \frac{P_0}{\gamma}\delta\,.$$

In the rest frame of the beam, let us assume the individual particles move nonrelativistically. Identifying the kinetic energies in the three dimensions with their respective temperatures, then we have

$$\begin{aligned} \frac{k_B T_x}{2} &= \frac{{p'_x}^2}{2m} = \frac{P_0^2}{2m}{\sigma'_x}^2 = \frac{\beta^2\gamma^2}{2}{\sigma'_x}^2 mc^2\,, \\ \frac{k_B T_y}{2} &= \frac{{p'_y}^2}{2m} = \frac{P_0^2}{2m}{\sigma'_y}^2 = \frac{\beta^2\gamma^2}{2}{\sigma'_y}^2 mc^2\,, \\ \frac{k_B T_z}{2} &= \frac{{p'_z}^2}{2m} = \frac{P_0^2}{2m\gamma^2}{\sigma_\delta}^2 = \frac{\beta^2}{2}{\sigma_\delta}^2 mc^2\,, \end{aligned}$$

where k_B is the Boltzmann constant.

Taking for example $\sigma'_x = 10^{-3}, \sigma'_y = 10^{-3}, \sigma_\delta = 10^{-3}$, and $\gamma = \frac{1}{\sqrt{1-\beta^2}} = 10$ (for a 10 GeV proton beam), we find $k_B T_x = k_B T_y = 10^{-4}\,mc^2$, $k_B T_z =$

$10^{-6}\, mc^2$. Since they all are $\ll mc^2$, the motion is indeed nonrelativistic in all three dimensions. In terms of temperatures, we have $T_x = T_y = 1.1 \times 10^9$ K, $T_z = 1.1 \times 10^7$ K.

We note that the longitudinal temperature typically is much colder than the transverse temperatures. This means in the rest frame of the beam, the Coulomb scatterings typically take place in the transverse dimensions. Two particles scatter off each other in the transverse plane, and they scatter into longitudinal direction with much hotter longitudinal temperature than the rest of the particles in the beam. Often they then get lost from the beam when the accelerator cannot accept particles with such a hot longitudinal temperature. This is called the *Touschek scattering*.[17] Two particles are lost per each collision event. The Touschek lifetime is one of the important lifetime limitations of intense beams at low energies.

The beam temperatures we found here are related to $\sigma'_x, \sigma'_y, \sigma_\delta$. They do not relate to the beam emittances directly. However, $\sigma'_x, \sigma'_y, \sigma_\delta$ change from position to position. As the beam moves along the accelerator, its temperatures relate to the emittances according to ${\sigma'_{x,y}}^2 = \frac{\epsilon_{x,y}}{\beta_{x,y}}$. As the beam is accelerated, and when taking into consideration of adiabatic damping (Sec. 3.4), its transverse temperatures increase while its longitudinal temperature decreases.

Homework 2.79 We calculated the temperatures of a proton beam in the text. If we now have an electron beam, assuming all other parameters are the same as the proton example, what will happen to the three temperatures?

Solution The electron beam will have a much hotter transverse temperature because of the γ^2 dependence of T_x, T_y. This may render the particle motion to become relativistic in the beam's rest frame.

2.4.5 Scaling lattice design

Suppose one has completed the lattice design for a beamline consisting of some dipoles, quadrupoles, and drift spaces. The same design can sometimes be adopted to produce other designs by a scaling operation. This scaling property can be handy to obtain new beamline designs from existing ones. Parameters that can be considered by scaling include varying the design beam momentum, the magnet strengths, the emittances of the injected beam, and the overall length scale.

For the case of scaling the overall length, for example, it follows from the equation of motion $\frac{d^2u}{ds^2} + Ku = 0$ that once one lattice design has been made, the same beam motion applies also to a scaled beamline with the following scalings,

- change all drift spaces by a factor λ,
- change the lengths of all magnets by a factor λ,

[17]C. Bernardini et al., Phys. Rev. Lett. 10, 407 (1963); A. Piwinski, DESY Report DESY-98-179 (1998).

- change the bending curvature ρ of all dipoles by a factor λ,
- change all quadrupole strengths $\frac{G}{(B\rho)}$ by a factor of $\frac{1}{\lambda^2}$.

After the scaling, the scaled similar beam motion can be obtained for a beam with properly matched emittances. For example, the scaled beamline can accept beams with the original unscaled transverse beam sizes but with the transverse slopes x' and y' scaled by $\frac{1}{\lambda}$ or a beam with beam sizes scaled by λ but the slopes unscaled.

Homework 2.80 How does the Courant-Snyder functions α, β, γ scale in the overall length scaled design? How does the beam emittance scale?

Homework 2.81

(a) The text emphasized a length scaling. How do we introduce the scaling if we wish to scale the particle momentum by a factor λ'?

(b) Would it be possible to combine a double scaling with an overall length scaling by λ and a momentum scaling by λ'?

2.4.6 Transfer map in terms of Courant-Snyder functions

Linear Courant-Snyder map So far, we have been calculating transfer matrices by multiplying the element matrices along the accelerator beamline. After the calculation, we obtain an expression of the map for this beamline. In this section, we will give another way to express the transfer matrix in terms of the Courant-Snyder functions. These two expressions are of course supposed to be equal. By equating these two results, in the next section, we will in fact obtain a systematic way to calculate the Courant-Snyder functions along the accelerator.

Let us return to Eq. (2.41), which gives $u(s)$ and $u'(s)$ in terms of two constants ϵ and ψ_0. One can equally express $u(s)$ and $u'(s)$ in terms of another two constants, i.e. the initial conditions u_0 and u'_0 at $s = 0$. The reason of doing so is in order to apply the matrix formalism developed earlier (which requires u_0 and u'_0, not ϵ and ψ_0). To do so, we write

$$u(s) = a\sqrt{\beta(s)}\cos\psi(s) + b\sqrt{\beta(s)}\sin\psi(s),$$

where a and b are yet to be determined. Denoting $\beta(0) = \beta_0$ and $\alpha(0) = \alpha_0$ and noting $\psi(0) = 0$, the initial conditions give

$$\begin{aligned} a &= \frac{u_0}{\sqrt{\beta_0}}, \\ b &= \sqrt{\beta_0}\,u'_0 + \frac{\alpha_0}{\sqrt{\beta_0}}\,u_0, \end{aligned}$$

which are then substituted into $u(s)$ to give

$$u(s) = \sqrt{\frac{\beta(s)}{\beta_0}}\Big[\cos\psi(s) + \alpha_0\sin\psi(s)\Big]\,u_0 + \sqrt{\beta_0\beta(s)}\,u'_0\,\sin\psi(s),$$

$$u'(s) = \frac{1}{\sqrt{\beta_0\beta(s)}}\Big[(\alpha_0 - \alpha(s))\cos\psi(s) - (1+\alpha_0\alpha(s))\sin\psi(s)\Big]u_0 + \sqrt{\frac{\beta_0}{\beta(s)}}\Big[\cos\psi(s) - \alpha(s)\sin\psi(s)\Big]u_0'.$$

These expressions can be cast in a convenient matrix form (by identifying $\beta(s) \to \beta_2$ and $\beta_0 \to \beta_1$),

$$\begin{bmatrix} u(s_2) \\ u'(s_2) \end{bmatrix} = M(s_2|s_1)\begin{bmatrix} u(s_1) \\ u'(s_1) \end{bmatrix},$$

with

$$M(s_2|s_1) = \begin{bmatrix} \sqrt{\frac{\beta_2}{\beta_1}}(\cos\psi + \alpha_1\sin\psi) & \sqrt{\beta_1\beta_2}\sin\psi \\ \frac{\alpha_1-\alpha_2}{\sqrt{\beta_1\beta_2}}\cos\psi - \frac{1+\alpha_1\alpha_2}{\sqrt{\beta_1\beta_2}}\sin\psi & \sqrt{\frac{\beta_1}{\beta_2}}(\cos\psi - \alpha_2\sin\psi) \end{bmatrix}, \quad (2.51)$$

where subscripts 1 and 2 refer to the beginning and the end points of the transfer map, and $\psi = \psi(s_2) - \psi(s_1)$ is the betatron phase advance from position s_1 to position s_2. Equation (2.51) is our result for the transfer map expressed in terms of the Courant-Snyder functions.

Nonlinear map of a quadrupole channel Having obtained the Courant-Snyder expressions for the linear map, we can extend the application to find expressions of the nonlinear map elements. As an illustration, let us make a detour and consider the third-order Taylor map elements $T_{511}, T_{512}, T_{521}, T_{522}$.[18]

Assuming the beamline channel consists only of quadrupoles and drifts. Consider only the x-motion. A particle with x_0, x_0' at the beginning of the channel will acquire a path length shift at the end of the channel by an amount

$$\Delta z = -\frac{1}{2}\int_0^L ds\, x'^2(s).$$

According to Eq. (2.51), the particle executes a betatron oscillation,

$$x'(s) = A(s)x_0 + B(s)x_0',$$

where

$$A(s) = \frac{\alpha_0 - \alpha(s)}{\sqrt{\beta_0\beta(s)}}\cos\psi(s) - \frac{1+\alpha_0\alpha(s)}{\sqrt{\beta_0\beta(s)}}\sin\psi(s),$$

$$B(s) = \sqrt{\frac{\beta_0}{\beta(s)}}\Big[\cos\psi(s) - \alpha(s)\sin\psi(s)\Big],$$

[18] We assume the reader has some knowledge concerning Taylor map notations.

and $\psi(s)$ is the betatron phase advance from $s = 0$ to s.

It follows that

$$\begin{aligned} T_{511} &= -\frac{1}{2}\int_0^L ds\, A^2(s)\,, \\ T_{512} = T_{521} &= -\frac{1}{2}\int_0^L ds\, A(s)B(s)\,, \\ T_{522} &= -\frac{1}{2}\int_0^L ds\, B^2(s)\,. \end{aligned}$$

If we assume the channel is long, so that the total phase advance from $s = 0$ to $s = L$ is large. The interference terms involving $\sin\psi(s)\cos\psi(s)$ in the above integrals averages to approximately zero. Keeping only the $\sin^2\phi(s)$ and the $\cos^2\phi(s)$ terms, and after some algebra, we obtain

$$\begin{aligned} T_{511} &\approx -\frac{\gamma_0}{4}\int_0^L ds\,\gamma(s)\,, \\ T_{512} = T_{521} &\approx -\frac{\alpha_0}{4}\int_0^L ds\,\gamma(s)\,, \\ T_{522} &\approx -\frac{\beta_0}{4}\int_0^L ds\,\gamma(s)\,. \end{aligned}$$

As these nonlinear path length contributions originates from $x'(s)$, not $x(s)$, one expects a prominent role of the γ-function, instead of the β-function in the analysis. See comment following (2.47).

The total Δz is given by

$$\Delta z \;=\; T_{511}x_0^2 + 2T_{512}x_0x_0' + T_{522}x'^2_0 \;\approx\; -\frac{\epsilon_0}{4}\int_0^L ds\,\gamma(s)\,,$$

where $\epsilon_0 = \gamma_0{x_0}^2 + 2\alpha_0x_0x_0' + \beta_0{x_0'}^2$.

A particle executing betatron oscillation falls behind longitudinally relative to the ideal synchronous particle by a distance proportional to its emittance after traveling some distance. It is easy to show that $\Delta z < 0$ and

$$|\Delta z| \;>\; \frac{\epsilon_0\Phi}{4}\,,$$

where Φ is the total betatron phase advance along the beamline.

For the case of a planar storage ring, the contribution of the bending magnets, after being added to the quadrupole channel, is mostly included if the horizontal Courant-Snyder functions take into account of their extra focusing. It follows that the longitudinal shift Δz per revolution roughly obeys

$$|\Delta z| \text{ per revolution} > \frac{\pi}{2}(\nu_x\epsilon_x + \nu_y\epsilon_y)\,.$$

For example, if $\epsilon_x = 1 \times 10^{-9}$ m-rad, $\nu_x = 40$, and ignoring the ϵ_y contribution, we have $|\Delta z| > 60$ nm per revolution. This term contributes to a nonlinear synchrobetatron coupling effect.

Homework 2.82 As alluded to in the text, the two constants ϵ and ψ_0 are exchangeable with the two constants u_0 and u_0'. It might be instructive to find the explicit expressions that relate ϵ, ψ_0 to u_0, u_0'.

Homework 2.83 Invert the matrix (2.51) to obtain

$$M^{-1}(s_2|s_1) = \begin{bmatrix} \sqrt{\frac{\beta_1}{\beta_2}}(\cos\psi - \alpha_2 \sin\psi) & -\sqrt{\beta_1\beta_2}\sin\psi \\ \frac{-\alpha_1+\alpha_2}{\sqrt{\beta_1\beta_2}}\cos\psi + \frac{1+\alpha_1\alpha_2}{\sqrt{\beta_1\beta_2}}\sin\psi & \sqrt{\frac{\beta_2}{\beta_1}}(\cos\psi + \alpha_1\sin\psi) \end{bmatrix}. \tag{2.52}$$

Convince yourself that this result is reasonable by considering the exchange between position s_1 and s_2 and compare your result with $M(s_2|s_1)$.

2.4.7 Calculation of the Courant-Snyder functions

Equation (2.51) gives the map from s_1 to s_2 once the Courant-Snyder functions are known at its two end points. In practice, however, it is not the most common way it is used. To compute the transfer map from s_1 to s_2, the way is simply to multiply together, in appropriate order, the matrices of all the component elements contained in this section of beamline. Equation (2.51), instead, is used more often to compute the Courant-Snyder functions once the transfer maps are given. See also discussion following Eq. (2.40).

One-period map To do so, we first consider the one-period map from position s to $s + L$. The Courant-Snyder functions assume the values $\beta_1 = \beta_2 = \beta$, $\alpha_1 = \alpha_2 = \alpha$, and we designate the *phase advance per period* as $\psi_2 - \psi_1 = \Phi$. Then the one-period map reads

$$M(s+L|s) = \begin{bmatrix} \cos\Phi + \alpha\sin\Phi & \beta\sin\Phi \\ -\gamma\sin\Phi & \cos\Phi - \alpha\sin\Phi \end{bmatrix}, \tag{2.53}$$

where $\gamma = \frac{1+\alpha^2}{\beta}$.

The map (2.53) is used in practice to calculate α, β, γ-functions at position s. What one does is to first multiply the matrices of all accelerator elements to form a total matrix M of one period around position s. Once this matrix M is obtained, Eq. (2.53) is then used to extract α, β, γ and Φ. For example, the quantity $\cos\Phi$ is first obtained from the trace of M. Knowing Φ, one obtains β from the matrix element m_{12}. (Φ is chosen with the convention that β must be positive.) α and γ are then obtained from other matrix elements of M. More explicitly,

$$\Phi = \cos^{-1}\left(\frac{m_{11}+m_{22}}{2}\right) \quad \text{with } \sin\Phi \text{ the same sign as } m_{12}\,,$$

and

$$\begin{aligned} \beta &= \frac{m_{12}}{\sin\Phi}\,, \\ \alpha &= \frac{m_{11}-m_{22}}{2\sin\Phi}\,, \\ \gamma &= -\frac{m_{21}}{\sin\Phi}\,. \end{aligned} \tag{2.54}$$

Putting this in more vague terms, we might say

Φ is determined by $m_{11}+m_{22}$,
α is determined by $m_{11}-m_{22}$,
β is determined by m_{12},
γ is determined by m_{21}.

This calculation of Φ determines its value modulus integral multiples of 2π. An integral multiple of 2π will be missing. This information is not contained in the one-turn map, and has to be recovered only if one follows the map along as a function of s.

For stability, the accelerator lattice must be such that M, its map per period, satisfies $|\mathrm{Tr}M| \le 2$. Otherwise Φ does not have a real value solution.

Inter-position map We have now used the one-period matrix $M(s+L|s)$ to compute α, β, γ at one location s. Once they are obtained at one location s_1, their values at other locations can be obtained using the transfer map from s_1 to s_2. This is obtained using Eq. (2.21), i.e.

$$M(s_2+L|s_2) = M(s_2|s_1)M(s_1+L|s_1)M^{-1}(s_2|s_1)\,.$$

Designating the elements of $M(s_2|s_1)$ by m_{ij}, we obtain after some algebra — an alternative derivation was given in Homework 2.78, and yet another derivation is in Homework 2.88 — a relationship between $\alpha_2, \beta_2, \gamma_2$ and $\alpha_1, \beta_1, \gamma_1$. Writing in a matrix form, it reads

$$\begin{bmatrix} \alpha_2 \\ \beta_2 \\ \gamma_2 \end{bmatrix} = \begin{bmatrix} m_{11}m_{22}+m_{12}m_{21} & -m_{11}m_{21} & -m_{12}m_{22} \\ -2m_{11}m_{12} & m_{11}^2 & m_{12}^2 \\ -2m_{21}m_{22} & m_{21}^2 & m_{22}^2 \end{bmatrix} \begin{bmatrix} \alpha_1 \\ \beta_1 \\ \gamma_1 \end{bmatrix}. \tag{2.55}$$

Equation (2.55) gives the transformation of α, β, γ functions. But how does the 4th Courant-Snyder function Φ transform from s_1 to s_2? The answer is that $\Phi_2 = \Phi_1$. Indeed, the similarity transformation (2.21) gives the condition

$$\mathrm{Tr}M(s_2+L|s_2) = \mathrm{Tr}M(s_1+L|s_1)\,,$$

or

$$2\cos\Phi_2 = 2\cos\Phi_1, \qquad \text{Q.E.D.}$$

The phase advance per period Φ is an invariant, i.e. it is the same regardless of where the one-period matrix is evaluated. This is an important property. Unlike Φ, the other Courant-Snyder functions, α, β, γ, do not have this distinction.

The quantity Φ is related to another important quantity *betatron tune* per period,

$$\nu = \frac{\Phi}{2\pi} = \frac{1}{2\pi}\int_s^{s+L} \frac{ds'}{\beta(s')} = \frac{1}{2\pi}\oint \frac{ds'}{\beta(s')}.$$

The tune actually does not depend on s because $\beta(s)$ is periodic in s with period L. So ν can be evaluated at any position s around a storage ring.

Homework 2.84 Show explicitly that the matrices $M(s_2|s_1)$ of Eq. (2.51), $M^{-1}(s_2|s_1)$ of Eq. (2.52), and $M(s+L|s)$ of Eq. (2.53) have determinants equal to 1, and are thus symplectic.

Homework 2.85 Using expressions in Eq. (2.54), show that the condition $\det\begin{bmatrix} m_{11} & m_{12} \\ m_{21} & m_{22} \end{bmatrix}=1$ holds if $\gamma = \frac{1+\alpha^2}{\beta}$, and vice versa.

Solution One way to appreciate this homework is to note that M has four elements, but with $\det M = 1$, only three of them are independent. If we characterize M by four variables $\Phi, \alpha, \beta, \gamma$, then there must be one condition that relates them together, and that condition is $\gamma = \frac{1+\alpha^2}{\beta}$. The condition $\gamma = \frac{1+\alpha^2}{\beta}$ is what is required to assure of the symplecticity of the map.

Homework 2.86 Find the eigenvalues, the eigenvectors, and the eigentunes for the Courant-Snyder one-turn matrix (2.53). How do you transform these quantities when you move to another position s_2?

Homework 2.87 The one period map $M(s+L|s)$ can be written as

$$M(s+L|s) = I\cos\Phi + J(s)\sin\Phi,$$

where I is unit matrix, and

$$J(s) = \begin{bmatrix} \alpha(s) & \beta(s) \\ -\gamma(s) & -\alpha(s) \end{bmatrix}.$$

The first thing to note is that the s-dependence is all contained in $J(s)$. The quantities I and Φ are independent of s.

(a) It is interesting to observe that $\mathrm{Tr}J(s) = 0$ and $J^2(s) = -I$, not dissimilar to the imaginary unit i with $i^2 = -1$.

(b) Show that

$$M^{-1}(s+L|s) = I\cos\Phi - J(s)\sin\Phi.$$

(c) Show that

$$M(s+L|s) = e^{\Phi J(s)}.$$

(d) Show by mathematical induction that the map for n periods is given by

$$[M(s+L|s)]^n = I\cos n\Phi + J(s)\sin n\Phi.$$

Homework 2.88 In Homework 2.78, we derived Eq. (2.55) using the properties of the Σ-matrix. Here let us offer another perhaps shorter derivation. Note the invariant condition

$$\gamma_1 u_1^2 + 2\alpha_1 u_1 u_1' + \beta_1 {u_1'}^2 = \gamma_2 u_2^2 + 2\alpha_2 u_2 u_2' + \beta_1 {u_2'}^2.$$

Then substitute u_2 and u_2' by $\begin{bmatrix} u_2 \\ u_2' \end{bmatrix} = \begin{bmatrix} m_{11} & m_{12} \\ m_{21} & m_{22} \end{bmatrix} \begin{bmatrix} u_1 \\ u_1' \end{bmatrix}$ and demand the result to be valid for arbitrary u_1 and u_1'. You should obtain three conditions that lead to Eq. (2.55).

Homework 2.89 Suppose you have designed the linear lattice of a circular accelerator, and the one-period map happens to have the form

$$\begin{bmatrix} 0 & X \\ Y & 1 \end{bmatrix}.$$

By symplecticity, X and Y are related. Knowing X, you know Y, and vice versa.

(a) Show that $M^3 = -I$ and $M^6 = I$.

(b) Give a physical reason why the powers of M have such a simple result.

2.4.8 β-function in drift space and quadrupole

We now illustrate some applications of Eq. (2.55).

Drift space First consider a drift space. Given $\alpha_1, \beta_1, \gamma_1$ at position s_1, we can find $\alpha_2, \beta_2, \gamma_2$ at another position $s_2 = s_1 + s$, assuming the space between s_1 and s_2 is free drift space. Note that s can be negative, meaning transferring backwards in the drift space. The result is

$$\begin{aligned} \alpha_2 &= \alpha_1 - \gamma_1 s, \\ \beta_2 &= \beta_1 - 2\alpha_1 s + \gamma_1 s^2, \\ \gamma_2 &= \gamma_1. \end{aligned} \tag{2.56}$$

One sees that as one extends the drift space length s, and treat β_2 as a function of s, then $\beta(s)$ is quadratic in s. We see $\alpha(s) = -\frac{1}{2}\frac{d\beta(s)}{ds}$ as it should, and that

$\gamma(s)$ is invariant in drift space. One can also show that $\gamma = \frac{1+\alpha^2}{\beta}$ is obeyed. A quick way to remember, in a drift space, is

$$\begin{aligned}
\beta(s) &= \text{quadratic in } s\,, \\
\alpha(s) &= \text{linear in } s\,, \\
\gamma(s) &= \text{constant in } s\,, \\
\psi(s) &= \tan^{-1} \text{in } s\,.
\end{aligned}$$

Behavior around a focus Being a parabolic function of s, one can shift the coordinate origin to the location where β is minimum. This minimum location is called a *waist* of the β-function. It is located at $s^* = \frac{\alpha_1}{\gamma_1}$, and we rewrite Eq. (2.56) as

$$\begin{aligned}
\alpha(s) &= -\frac{s-s^*}{\beta^*}\,, \\
\beta(s) &= \beta^* + \frac{(s-s^*)^2}{\beta^*}\,, \\
\gamma(s) &= \frac{1}{\beta^*}\,,
\end{aligned} \tag{2.57}$$

where $\beta^* = \frac{1}{\gamma_1}$ is the minimal value of the β-function at the waist position $s = s^*$.

Because the beam size is proportional to $\sqrt{\beta}$, it reaches a minimum at the waist $s = s^*$. The smaller β^* is, the smaller the beam size is at the waist, i.e. the beam is more strongly focused at the focal point $s = s^*$. On the other hand, an inspection of the expression of $\beta(s)$ in Eq. (2.57) says that the smaller β^* is, also the quicker the beam gets defocused as one moves away from the focal point. The situation is illustrated in Fig. 2.42, which shows both $\beta(s)$ and $\sqrt{\beta(s)}$. The solid curves are for $\beta^* = 0.25$ m and the dashed curves are for $\beta^* = 1$ m. The figure for $\sqrt{\beta(s)}$ illustrates how the beam size varies as the beam propagates in a drift space. As $s \to \pm\infty$, the beam becomes a diverging beam with its size $\propto |s|$, as one expects.

The behavior of a weakly focused beam and a strongly focused beam in the phase space is illustrated in Fig. 2.43. For a strongly focused beam, β^* is small, and with given beam emittance, the phase space ellipse at the waist has to become tall, meaning u' is large, which means u changes quickly with s, and then in turn means the beam spreads out quickly.

The parametrization (2.57) in terms of a waist holds over any segment of drift space. It is not necessary that a waist actually occurs within the segment. In fact, in practice, an actual waist is realized only in very special circumstances (e.g. at the collision point in a collider, or at the photon radiation source point, etc.). In general, a waist is even to be avoided other than those few special circumstances because of the potentially rapidly diverging beam around it.

Figure 2.42: Behavior of $\beta(s)$ and $\sqrt{\beta(s)}$ around a focal point for two cases, a strongly focused case (solid) and a weakly focused case (dashed).

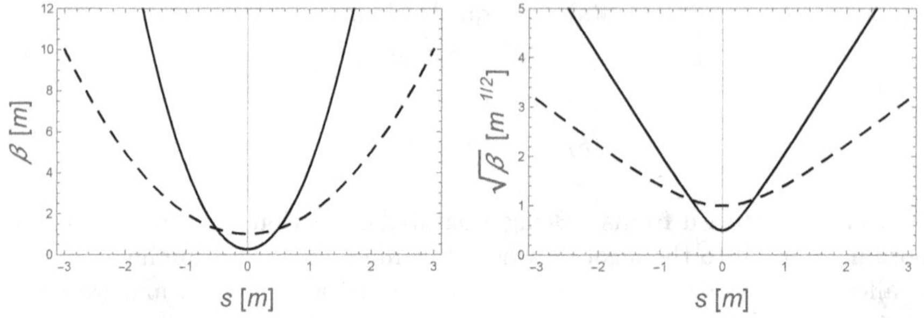

Figure 2.43: Behavior of a beam's phase space ellipse for two cases, a strongly focused case (left) and a weakly focused case (right). The two cases are assumed to have the same beam emittance (phase space area).

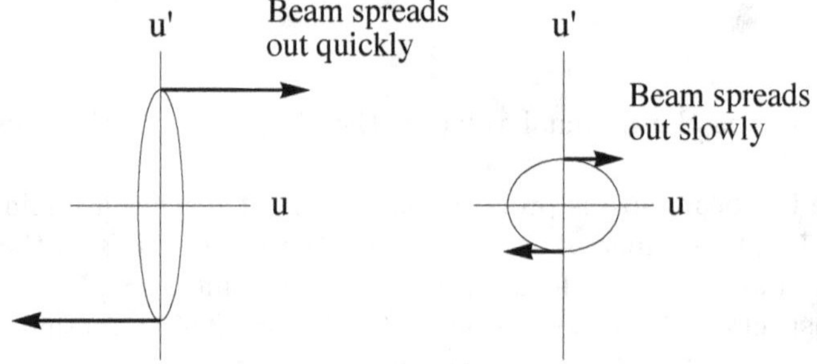

In laser physics, a laser is often focused to a focal point at the point of application. A similar situation occurs there and the laser spot will diverge in a distance called the *Rayleigh length.* Translated to our language here, the Rayleigh length is just β^*.

Quadrupole Applying Eq. (2.55) to the case of a quadrupole is straight forward. For example, as a beam passes through a thin-lens quadrupole with focal length f, the functions change by

$$\begin{aligned}\alpha_2 &= \alpha_1 + \frac{\beta_1}{f}, \\ \beta_2 &= \beta_1, \\ \gamma_2 &= \gamma_1 + \frac{2\alpha_1}{f} + \frac{\beta_1}{f^2}.\end{aligned} \tag{2.58}$$

Going across a thin-lens quadrupole, the β-function is continuous, but its slope has a discontinuity by the amount $-2\frac{\beta_1}{f}$. A focusing quadrupole ($f > 0$) will bend β-function downward toward the axis, while a defocusing quadrupole ($f < 0$) bends the β-function upwards away from the axis.

The case of a thick quadrupole is somewhat more involved (Homework 2.98). Let $\alpha_0, \beta_0, \gamma_0$ be the functions at the entrance of a thick-lens quadrupole with field strength K. The β-function at a distance s inside the quadrupole is given by

$$\beta = \begin{cases} \left(\frac{\beta_0}{2} - \frac{\gamma_0}{2K}\right)\cos 2\sqrt{K}s - \frac{\alpha_0}{\sqrt{K}}\sin 2\sqrt{K}s + \left(\frac{\beta_0}{2} + \frac{\gamma_0}{2K}\right), & \text{F quad.} \\ \left(\frac{\beta_0}{2} + \frac{\gamma_0}{2|K|}\right)\cosh 2\sqrt{|K|}s - \frac{\alpha_0}{\sqrt{|K|}}\sinh 2\sqrt{|K|}s + \left(\frac{\beta_0}{2} - \frac{\gamma_0}{2|K|}\right), & \text{D quad.} \end{cases} \tag{2.59}$$

Homework 2.90 To get more familiarity with phase space, try this simple homework. Consider a beam with a tilted trajectory traversing a drift space. A waist has been designed so that the beam in real space (not phase space) looks like Fig. 2.44 (upper).

Figure 2.44: A configuration of a drifting beam in real space (not phase space).

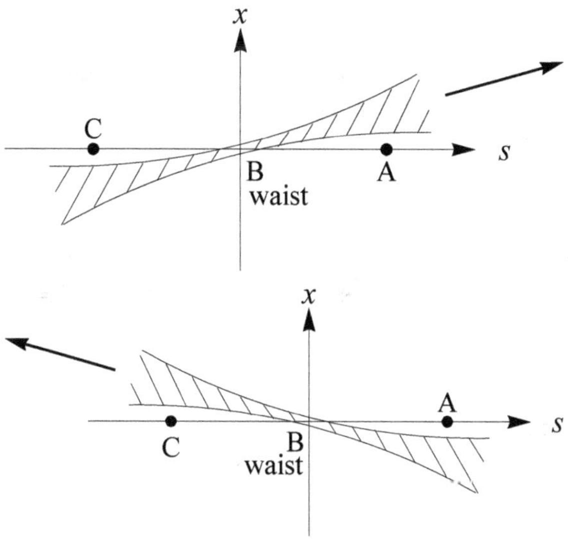

(a) Sketch the phase space distribution of the beam at positions A, B, C.

Now consider an identical beam going in the opposite direction with an opposite tilt angle like Fig. 2.44 (lower). The two beams are brought together and made to collide at point B. Note that the two beams overlap in real space, but they do not overlap in the phase space.

(b) Sketch the phase space distributions for the second beam. Note that $\hat{s}$ is defined in the same direction in both cases, so there is a flip of sign of x'.

Solution See Fig. 2.45. In this plot, all ellipses have the same area.

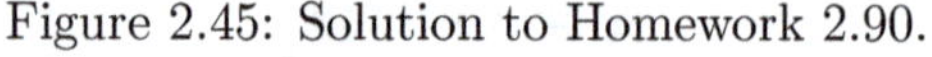

Figure 2.45: Solution to Homework 2.90.

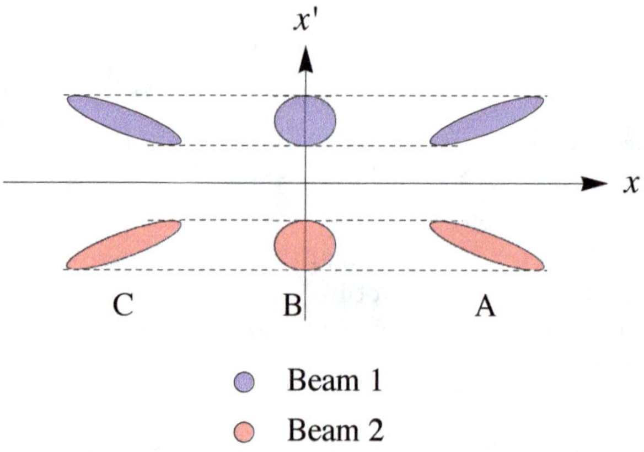

Homework 2.91 Consider the case of a drift space from $s = -\ell$ to $s = \ell$. A beam has a waist at $s = 0$. Work through the parametrization to convince yourself that the beam phase space distribution are as shown in Fig. 2.46 at the three locations $s = -\ell, 0, \ell$.

Figure 2.46: Behavior of a beam's phase space distribution as it propagates through the focal point in a drift space. We have defined $\ell^* = \sqrt{\beta^{*2} + \ell^2}$.

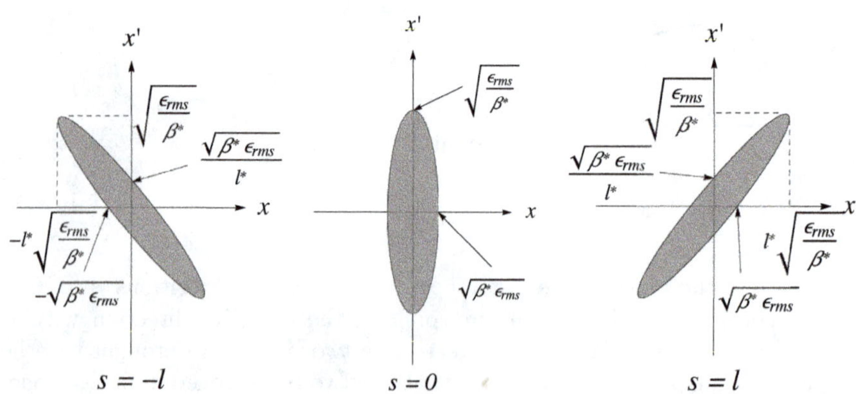

Homework 2.92 Show that in a drift space,

(a) $\alpha(s)$ as a function of s always *decreases* with s.

(b) $\beta(s)$ as a function of s is always concave *upwards*.

Homework 2.93 The transverse beam size $\sigma(s)$ is related to the β-function $\beta(s)$. Show that the evolution of $\sigma(s)$ in a drift space can be written as

$$\sigma^2(s) = \sigma^2(0) + \frac{\epsilon^2 s^2}{\sigma^2(0)},$$

where ϵ is the beam emittance. This result has been obtained earlier using the envelope equation in Sec. 2.4.3.

Homework 2.94 Suppose, in an analysis of a linear 1-D motion, we find at a certain position $s = 0$ in a drift space that $\beta = 10$ m, $\alpha = 1.0$, and an rms beam size of 1 mm.

(a) Find the rms beam emittance, and the rms beam divergence at $s = 0$. ("Divergence" means the spread in u'.) Plot the rms ellipse of beam distribution in the (u, u') phase space.

(b) Where is the beam waist located relative to $s = 0$? Is it upstream or downstream from $s = 0$? Find the value of the β-function, the rms beam size and rms beam divergence at the waist. Note that the beam divergences at $s = 0$ and at the waist are the same.

Homework 2.95 We have now learned two ways to describe the trajectory of a particle in a drift space. The first way is

$$x(s) = x_0 + x_0' s, \qquad x'(s) = x_0',$$

and another way is given in terms of the Courant-Snyder functions, Eq. (2.41) together with Eq. (2.56) or (2.57). Check the self-consistency by showing these two descriptions are identical.

Homework 2.96 Two thin quadrupoles are separated by a drift space of length L. You have managed to measure the β-functions at these quadrupoles to be β_1 and β_2 respectively.

(a) Give an expression of the β-function between the two quadrupoles in terms of $\beta_{1,2}$.

In a collider interaction region, the two quadrupoles could be the two strong quadrupoles adjacent to and symmetrically located relative to the interaction point at the middle of the drift space.

(b) Given the measured results of β_1 and β_2, what is the β-function β^* at exactly the interaction point?

(c) By design, the interaction point should be where the β-function waist occurs. But this is not happening because you found $\beta_1 \neq \beta_2$. Find the location of the waist to see how much the optics has missed your design.

Solution Parametrize the β-function by $\beta(s) = \beta^* + \frac{(s-s^*)^2}{\beta^*}$. We have $\beta(0) = \beta_1$ and $\beta(L) = \beta_2$. Solve for β^* and s^*.

Homework 2.97

(a) The betatron phase advance from s_1 to s_2 is, in general,

$$\Delta\psi = \int_{s_1}^{s_2} \frac{ds}{\beta(s)} .$$

Use this expression to show that, in a drift space, the betatron phase advance from a beam waist to a distance L away is

$$\Delta\psi = \tan^{-1} \frac{L}{\beta^*} .$$

As $L \to \infty$, we have $\Delta\psi \to \frac{\pi}{2}$.

(b) Following (a), show that a long field-free drift space containing a waist will contribute a $\Delta\psi$ slightly less than π, and can never exceed π, with the phase advance on each side of the waist not exceeding $\frac{\pi}{2}$. A beamline consisting of one quadrupole in a long drift cannot contribute a phase advance larger than 2π. What is the maximum phase advance with a string of N quadrupoles?

Homework 2.98

(a) Prove Eq. (2.58).

(b) Prove Eq. (2.59).

2.4.9 Transfer matrices must obey some constraints

In the previous section when we applied Eq. (2.56) to propagate the Courant-Snyder functions across a drift space, we introduced the viewpoint of treating s as the variable, and considering α, β, γ as functions of s. This consideration requires considering the matrix M as a function of s. Considering the map from position $s = 0$ to an arbitrary position s, then we write

$$\begin{bmatrix} u \\ u' \end{bmatrix}_s = \begin{bmatrix} m_{11}(s) & m_{12}(s) \\ m_{21}(s) & m_{22}(s) \end{bmatrix} \begin{bmatrix} u \\ u' \end{bmatrix}_0 .$$

One should appreciate the fact that values of transfer matrix elements are far from being arbitrary; there are multiple internal constraints among them. One notable constraint is of course the fact that

$$\det M = 1 \quad \text{for all } s ,$$

but there are other constraints. There is not much too profound here, but we list some of them below.

- Consider a particle with initial condition $\begin{bmatrix} 1 \\ 0 \end{bmatrix}$. Its trajectory is given by $u(s) = m_{11}(s), u'(s) = m_{21}(s)$. It follows that, considered as functions of s, we must have the condition

$$m_{21}(s) \ = \ m'_{11}(s) \, .$$

Similarly, considering a particle with initial condition $\begin{bmatrix} 0 \\ 1 \end{bmatrix}$, we must have

$$m_{22}(s) \ = \ m'_{12}(s) \, .$$

One can check these constraint conditions against known expressions for drift space, dipoles, and thick-lens quadrupoles.

Perhaps we bring this point closer to home this way: We know the map for a drift space is $\begin{bmatrix} 1 & s \\ 0 & 1 \end{bmatrix}$. Actually once we know $m_{11} = 1$, we can predict all the other three matrix elements. First, all single-element maps have mirror symmetry, which means $m_{22} = m_{11} = 1$. Then since $m'_{12} = m_{22}$, we know $m_{12} = s$. Finally, since $m_{21} = m'_{11}$, we know $m_{21} = 0$. Q.E.D. Similar exercises can be performed with the thick quadrupole and thick dipole elements.

On the other hand, this does not apply to the case of the thin-lens quadrupole matrix $\begin{bmatrix} 1 & 0 \\ -1/f & 1 \end{bmatrix}$ because this matrix has already integrated from the entrance to the exit of the quadrupole, and information on its dependence on s has been lost.

- Writing out the condition of unit determinant, we have

$$m_{11}(s)m'_{12}(s) - m_{12}(s)m'_{11}(s) \ = \ 1 \, .$$

One should also notice that this expression is simply the Wronskian (2.29) formed by the two independent trajectories with initial conditions $\begin{bmatrix} 1 \\ 0 \end{bmatrix}$ and $\begin{bmatrix} 0 \\ 1 \end{bmatrix}$. We did show that Wronskians are constants of the motion, but now we know it is related to the determinant of the transfer matrix being always equal to 1.

- We can also consider Eq. (2.55) as giving $\alpha(s)$, $\beta(s)$, $\gamma(s)$ in terms of the s-dependent matrix elements $m_{ij}(s)$. The above constraints on $m_{ij}(s)$ can be used to show that Eq. (2.55) satisfies the relations $\alpha(s) = -\beta'(s)/2$ and $\gamma(s) = \frac{1+\alpha(s)^2}{\beta(s)}$. This serves as a check of internal self-consistency of the formalism.

- The reader is to be reminded of a theorem we developed earlier that, for a beam line with mirror symmetry around its midpoint — a beam line consisting of a single-element is a simplest example,

$$m_{11}(s) = m_{22}(s) \quad \text{for mirror-symmetric beam lines.}$$

Homework 2.99 It is easy to see that the matrix elements for an element of length ℓ must satisfy

$$m_{11}(\ell) = 1 + \int_0^\ell d\ell' m_{21}(\ell'),$$
$$m_{12}(\ell) = \int_0^\ell d\ell' m_{22}(\ell').$$

Check these conditions are satisfied by the expressions of drift spaces, thick quadrupoles, and thick dipoles (note: $\theta = \frac{\ell}{\rho}$). See also Homework 2.23.

2.4.10 β-function in lattice

We now apply Eq. (2.55) to a lattice that consists of drift spaces, quadrupoles, and bends.

As mentioned, by convention, a focusing quadrupole focuses in x and defocuses in y. We learned that this quadrupole will contribute to the β_x-function a downward kick and to β_y an upward kick. Similarly a defocusing quadrupole will kick β_x upward and β_y downward. We also learned that, in between quadrupoles, either in a drift space or in a dipole magnet (ignoring weak horizontal focusing effect), both β_x and β_y are quadratic in s. With these information, together with the demand that the β-function be periodic around the ring circumference, we can now calculate the $\beta_{x,y}$-functions through any lattice consisting of quadrupole, dipoles, and drift spaces. A typical plot of functions β_x and β_y is sketched in Fig. 2.47.

Figure 2.47 is a plot of the two β-functions over one period of the lattice. Look closely and make the following observations,

- The β-functions are necessarily positive, and they are periodic with the lattice period, as evidenced by the fact that both their values and their slopes are equal at the two end-points.

- Note the appropriate upward and downward bending of the β_x- and β_y-functions at the quadrupoles.

- Note the quadratic behavior of $\beta_{x,y}$ in the drift space regions between quadrupoles. They are both concave upwards in the drift spaces. However, nowhere a waist is actually realized in this cell.

Figure 2.47: A typical plot of the β-functions over a periodic cell lattice. Underneath the plot shows the layout of the lattice. A rectangle straddling the center line is a dipole. A rectangle standing above and below the center line is a focusing and defocusing quadrupole, respectively [Courtesy Tenghui Rui, 2019].

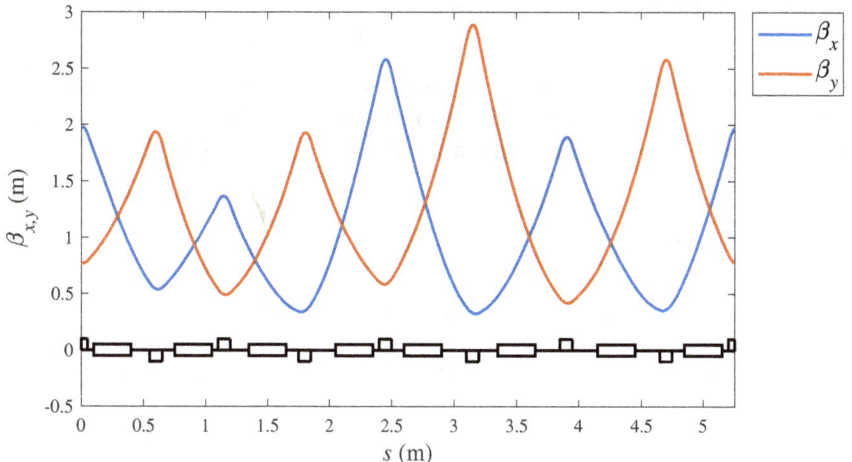

- The slight horizontal focusing action over the dipoles are included in the plot. However, the effect is not too noticeable.
- Observe that β_x tends to be larger in the focusing quadrupoles than in the defocusing quadrupoles, while the opposite holds for β_y.
- The magnitude of the β-functions is comparable to the spacing between adjacent quadrupoles [see Eq. (2.25)].
- Although not obvious, one should be able to estimate the quadrupole strengths from the lattice plot. It then follows that the magnitude of the β-functions is also comparable to the focal lengths f of the quadrupoles. See Homework 2.100.
- A strongly focused lattice will have small f, and therefore small $\beta_{x,y}$, and it follows that it will have large phase advances per period $\Phi_{x,y}$ and therefore large betatron tunes $\nu_{x,y}$.

Homework 2.100 Estimate the focal lengths for a few of the eight quadrupoles (the two at the ends are the same quadrupole in a periodic arrangement) in Fig. 2.47.

Solution Estimate the discontinuities of the slopes of $\beta_{x,y}$ across the quadrupoles, and apply $\Delta\beta' = -\frac{2\beta}{f}$. For each quadrupole, do this first for x-dimension, then repeat for the y-dimension. The focal lengths of each quadrupole should be the same (except for the signs). Indeed, one should observe directly from Fig. 2.47 that the kink angles induced at all the quadrupoles obey $\frac{\Delta\beta_x}{\Delta\beta_y} = -\frac{\beta_x}{\beta_y}$.

2.4.11 FODO cell - Revisit

Consider an accelerator composed of identical FODO cells. The one-period map of the FODO cell was discussed before in Sec. 2.2.11 without the help of the Courant-Snyder formalism. Starting from the middle of QF, we know that the one-period map for x-motion is given by, in the thin-lens approximation,

$$M = \begin{bmatrix} 1-\frac{L^2}{2f^2} & 2L(1+\frac{L}{2f}) \\ -\frac{L}{2f^2}(1-\frac{L}{2f}) & 1-\frac{L^2}{2f^2} \end{bmatrix},$$

where L is the drift space length between QF and QD, f is the focal length of QF or QD. For stability, we require $|f| \geq \frac{L}{2}$.

This map can now be used to find the betatron phase advance per cell Φ, and the α, β, γ-functions at the middle of QF. The result is

$$\begin{aligned} \cos\Phi &= \frac{1}{2}\text{Tr}(M) = 1-\frac{L^2}{2f^2} \qquad \Longrightarrow \qquad \sin\frac{\Phi}{2} = \frac{L}{2|f|}, \\ \beta_F &= \frac{2L(1+\sin\frac{\Phi}{2})}{\sin\Phi}, \\ \alpha_F &= 0, \\ \gamma_F &= \frac{1}{\beta_F}. \end{aligned} \tag{2.60}$$

We can safely choose $0 < \Phi < \pi$ for a FODO cell due to the conclusion made in Homework 2.97.

The α, β, γ-functions at QD can be obtained by reversing the sign of f. One first note that this sign reversal does not change Φ, as is expected because Φ is independent of the position of observation. The other functions at middle of QD are

$$\begin{aligned} \beta_D &= \frac{2L(1-\sin\frac{\Phi}{2})}{\sin\Phi}, \\ \alpha_D &= 0, \\ \gamma_D &= \frac{1}{\beta_D}. \end{aligned}$$

One can also obtain the α, β, γ-functions at the middle of the drift in a similar manner. The result is

$$\beta_{\text{midpoint}} = \frac{L}{\sin\Phi}\left(2-\sin^2\frac{\Phi}{2}\right),$$

$$\alpha_{\text{midpoint}} = \pm\frac{1}{\cos\frac{\Phi}{2}},$$
$$\gamma_{\text{midpoint}} = \frac{2}{L}\tan\frac{\Phi}{2}.$$

The $\pm$ sign refers to the two such midpoints.

A quick glance of these expressions should yield a very rough rule of thumb, i.e. the β-functions $\beta_{x,y}$, the focal length f, and the magnet spacing L are all of the same order of magnitude for some moderately chosen Φ. This point was mentioned in Eq. (2.25). By knowing the rough value of f and the intended beam energy, you also have an idea how strong are the magnets. The wide range of sizes of the various magnets then should come as an expected result. For example, remember to express an intelligent doubt when you see an excessively big quadrupole magnet installed neighboring a long drift space in an accelerator tunnel.

For the y-motion, all the above results apply if we change the sign of f. In particular, the β_y-function at QF is equal to β_D defined above, while the β_y-function at QD is equal to β_F defined above.

One should also calculate the β-function as a function of s. Consider a FODO cell with a half-QF at $s = 0$, a full QD at $s = L$ and another half-QF at $s = 2L$. Let the phase advance be Φ for the FODO cell. Let us parametrize the β-function as

$$\beta(s) = \beta^* + \frac{(s-s^*)^2}{\beta^*}.$$

It can be shown that in this FODO cell,

$$\begin{aligned}\beta^* &= \frac{L}{2}\cot\frac{\Phi}{2},\\ s^* &= \begin{cases}\frac{L}{2}(1+\csc\frac{\Phi}{2}), & \text{if } 0<s<L\\ \frac{L}{2}(3-\csc\frac{\Phi}{2}), & \text{if } L<s<2L\end{cases}.\end{aligned} \tag{2.61}$$

From the expression of s^*, which indicates the location of a waist, it follows that the waist occurs always outside of the drift spaces of the FODO. In a periodic FODO cell lattice, you will not find any β-function waist.[19]

Figure 2.48 (red curves) shows the Courant-Snyder functions as functions of s for the case when $\Phi = 90°$ and cell length of 20 m. The curve in blue will be mentioned later in Sec. 2.5.2.

It follows from the β-function (quadratic in s) that

$$\alpha(s) = -\frac{s-s^*}{\beta^*},$$
$$\gamma = \frac{1}{\beta^*}.$$

[19]You may want to review your answer to Homework 2.97(b) at this point.

Figure 2.48: Courant-Snyder functions of a FODO cell as a function of s. The blue curve applies for the alternative Courant-Snyder coordinates to be discussed in Sec. 2.5.2.

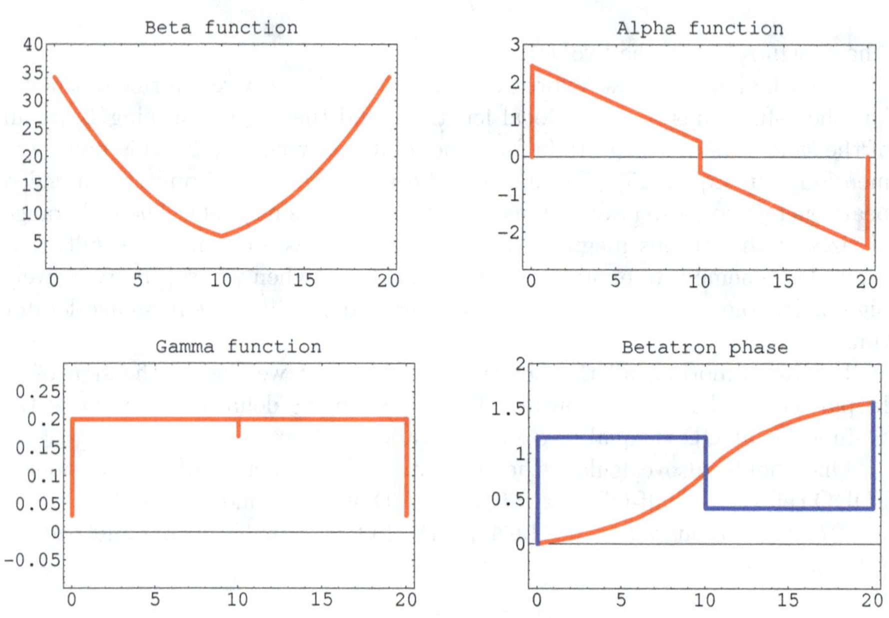

As mentioned earlier, α decreases monotonically in drift spaces.

One point should be made about the γ-function here. From the above expression, it looks like that $\gamma(s)$ = constant inside the entire FODO cell. This is not true. Note that $\gamma = \frac{1+\alpha^2}{\beta}$. Note from Fig. 2.48 that α changes sign across the quadrupoles. Since β does not change across the quadrupole, γ must change inside the quadrupole when α changes from a positive value through zero to a negative value, and it must assume a minimum value at the middle of the quadrupole. In fact, what happens is that $\gamma = \frac{1}{\beta^*} = \frac{2}{L} \tan \frac{\Phi}{2}$ = constant throughout the FODO, except that inside the quadrupoles, it dips to a lower value. It dips to $\frac{1}{\beta_F}$ in QF and $\frac{1}{\beta_D}$ in QD. These features are reflected in Fig. 2.48.

Homework 2.101 In the text, a nominal FODO cell was introduced with QF and QD having focal length $\pm f$ with the same f. Sometimes, however, a modified FODO cell whose QF and QD have different focal lengths is used. This modified FODO cell will have phase advances $\Phi_x \neq \Phi_y$. Analyze this modified FODO cell and derive expressions for $\Phi_{x,y}$. Also find expressions for α, β, γ-functions at QF and QD for both x- and y-motions.

Solution Let QF have focal length f_1 and QD have $-f_2$.

$$\begin{aligned}
\cos\Phi_x &= 1-\frac{L}{f_1}+\frac{L}{f_2}-\frac{L^2}{2f_1f_2}\,,\\
\alpha_{xF} &= 0\,,\\
\beta_{xF} &= \frac{1}{\sin\Phi_x}\left(2L+\frac{L^2}{f_2}\right)\,,\\
\cos\Phi_y &= 1-\frac{L}{f_2}+\frac{L}{f_1}-\frac{L^2}{2f_1f_2}\,,\\
\alpha_{yF} &= 0\,,\\
\beta_{yF} &= \frac{1}{\sin\Phi_y}\left(2L-\frac{L^2}{f_2}\right)\,.
\end{aligned}$$

Homework 2.102 Consider a beam transport line that is a periodic array of quadrupole doublets. Each doublet consists of a thin-lens QF with focal length f and a thin-lens QD with focal length $-f$. The separations between adjacent quadrupoles are L_1 and L_2 as shown in Fig. 2.49.

Figure 2.49: A doublet array.

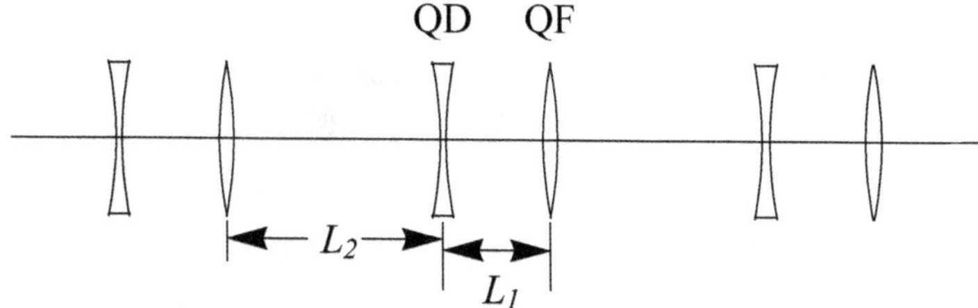

(a) Show that the betatron phase advance per cell is

$$\Phi_{x,y} = 2\sin^{-1}\left(\frac{\sqrt{L_1L_2}}{2f}\right)\,.$$

Stability requires $|f| \ge \frac{\sqrt{L_1L_2}}{2}$. Otherwise the lattice is over-focused.

(b) Show that the maximum horizontal β-function occurs at QF and is

$$\beta_{x,\max} = \frac{L_1+L_2+\frac{L_1L_2}{f}}{\sin\Phi_x}\,.$$

(c) Show that for the case when $f>\frac{L_1}{2}$ and $f>\frac{L_2}{2}$ the minimum horizontal β-function occurs at QD and is

$$\beta_{x,\min} = \frac{L_1+L_2-\frac{L_1L_2}{f}}{\sin\Phi_x}\,.$$

Anticipate how the minimum β-function should change when the above inequalities are not satisfied.

(d) Investigate the condition when there is a β-function waist generated within the cell lattice.

(e) Sketch the β-function over one cell assuming $L_2 \gg L_1$. Show that β_x and β_y are almost equal along the transport line.

Solution (d) You would have the intuition that a waist requires sufficiently long drift spaces and/or sufficiently short focal length. The criterion will require $|f|$ to be less than either $\frac{L_1}{2}$ or $\frac{L_2}{2}$ or some combination thereof. Work out the analysis to confirm this. Stay within the stable region $|f| \geq \sqrt{\frac{L_1}{2}}\sqrt{\frac{L_2}{2}}$.

Homework 2.103 Derive Eq. (2.61).

Homework 2.104 Design a straight section using two back-to-back FODO cells for a map with transfer matrix

$$-I = \begin{bmatrix} -1 & 0 & 0 & 0 \\ 0 & -1 & 0 & 0 \\ 0 & 0 & -1 & 0 \\ 0 & 0 & 0 & -1 \end{bmatrix}.$$

Solution Try two cells with 90° phase advances. In addition to the phase advances, are there other conditions required of your FODO cells? Two of your designs back-to-back then naturally gives a I map.

Homework 2.105 Write a computer program to study particle motion numerically in a lattice consisting of FODO cells. You may skip this homework now, but then find later that you need it anyway.

Homework 2.106 We have so far ignored the weak focusing in a dipole magnet, treating a dipole as a drift space. To demonstrate this focusing effect, admittedly weak, consider a sector dipole magnet with uniform bending magnetic field and with normal entrance and exit pole faces relative to the designed beam motion. Let the object beam source be located a distance s_o from the magnet entrance. Due to the focusing, a beam image is formed at a distance s_i downstream from the magnet exit, as shown in Fig. 2.50.

(a) Show the *Barber's rule* that the object point, the image point, and the center of the sector bend magnet lie on one straight line (as shown in Fig. 2.50), i.e. show that

$$\theta_o + \theta + \theta_i = \pi .$$

(b) As we slide the object position along the trajectory axis, the image point will also slide, and will slide according to the Barber's rule. One special case

Figure 2.50: Barber's rule.

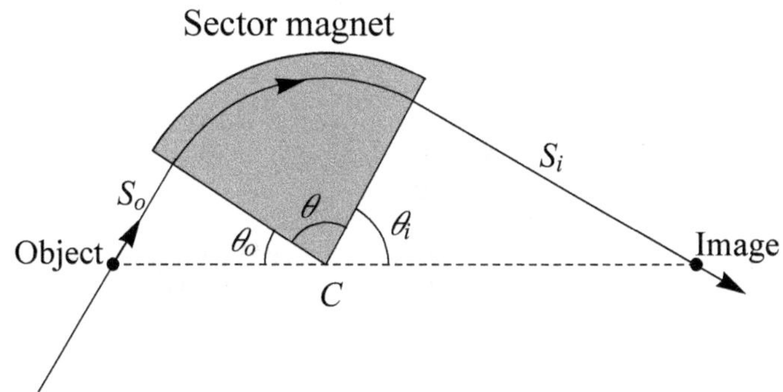

occurs when $\theta = \pi, \theta_i = \theta_o = 0$. Convince yourself that this is a focus-to-focus arrangement.

(c) Find the magnification factor of the image-to-object ratio.

Solution You first need to review what is meant by an image forming process and the condition for image forming in terms of transfer maps.

(a) The transfer map from the object to the image is

$$M = \begin{bmatrix} 1 & \rho\tan\theta_i \\ 0 & 1 \end{bmatrix} \begin{bmatrix} \cos\theta & \rho\sin\theta \\ -\frac{\sin\theta}{\rho} & \cos\theta \end{bmatrix} \begin{bmatrix} 1 & \rho\tan\theta_o \\ 0 & 1 \end{bmatrix},$$

where ρ is the bending radius of the dipole. In particular, the (1,2)-element of M is

$$m_{12} = \frac{\rho\sin(\theta_o + \theta + \theta_i)}{\cos\theta_o\cos\theta_i}.$$

For the image to form, the condition is

$$m_{12} = 0,$$

which implies that $\theta_o + \theta + \theta_i = \pi$.

Homework 2.107 A storage ring designer requests to produce perfectly round beam waists at two locations s_1 and s_2. We assume the beam has equal emittances in x and y. Consider the map from s_1 to s_2.

(a) Write down the requested conditions in the language of Courant-Snyder parameters. How many free parameters do you have left in your lattice design?

(b) Write down the transport matrix from s_1 to s_2.

Solution

(a) We note that this problem has to deal with 4×4 matrices. The lattice is required to bring a beam from a round waist to another round waist. The conditions are $\alpha_x(s_1) = \alpha_y(s_1) = \alpha_x(s_2) = \alpha_y(s_2) = 0$, $\beta_x(s_1) = \beta_y(s_1) = \beta_1$ with β_1 the β-function at s_1, and $\beta_x(s_2) = \beta_y(s_2) = \beta_2$ with β_2 the β-function at s_2.

(b)

$$M(s_2|s_1) = \begin{bmatrix} \cos\Delta\psi_x & \sqrt{\beta_1\beta_2}\sin\Delta\psi_x & 0 & 0 \\ \frac{-1}{\sqrt{\beta_1\beta_2}}\sin\Delta\psi_x & \cos\Delta\psi_x & 0 & 0 \\ 0 & 0 & \cos\Delta\psi_y & \sqrt{\beta_1\beta_2}\sin\Delta\psi_y \\ 0 & 0 & \frac{-1}{\sqrt{\beta_1\beta_2}}\sin\Delta\psi_y & \cos\Delta\psi_y \end{bmatrix}$$

2.4.12 Courant-Snyder transformation on the Hill's equation

Courant-Snyder formalism can be introduced in different ways We have not yet completed our discussion of Courant-Snyder formalism. Let us review what we have gone through so far. We first introduced it with Eq. (2.37) following the Hill's equation. This was a traditional way to introduce it. The second time we discussed it was actually through the normal form representation Eq. (2.42) but it was a discussion at best at a conceptual level and will be elaborated more in Sec. 2.5.

We now want to discuss in this section the Courant-Snyder formalism the third time. This approach is particularly useful when the Hill's equation contains additional nonlinear driving terms and one wishes to develop a perturbation analysis to the system. This approach will be repeated in Eq. (3.34).

We will discuss the fourth time in Sec. 3.7.2 when we depart from the equation of motion, the Hill's equation, and elevate to work on the Hamiltonian — after postponing it as much as possible. We will introduce a generating function to perform a canonical transformation on the Hamiltonian. The generating function has built in the Courant-Snyder transformation.

There is still a fifth step in discussing Courant-Snyder formalism which will apply Lie algebra. The subject is however beyond the scope of these lectures.

Hill's equation The Courant-Snyder formalism allows us to transform the Hill's equation into a simple harmonic motion. This should not be surprising after you learned normal form transformation, in particular Eq. (2.64), where the middle map R represents the simple harmonic aspect. But here we need to demonstrate this more explicitly starting with the Hill's equation (2.5).

If we apply the *Courant-Snyder transformation* and define

$$\eta = \frac{u}{\sqrt{\beta}}, \qquad \phi = \frac{1}{\nu}\int_0^s \frac{ds'}{\beta(s')}, \tag{2.62}$$

Figure 2.51: A schematic sketch of betatron phase advance as a function of s.

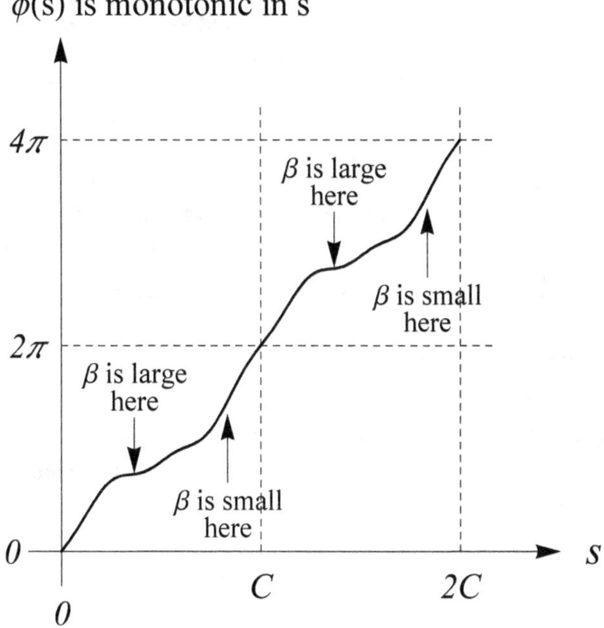

where $\nu = \frac{1}{2\pi}\int_0^C \frac{ds}{\beta(s)}$ is the betatron tune, it is straightforward to show that

$$\frac{d^2\eta}{d\phi^2} + \nu^2\eta = 0 . \tag{2.63}$$

The quantity ϕ is now the independent "time" variable, replacing the role of s, and η replaces the dynamical variable u. As s increases for one period of storage ring circumference, ϕ increases by 2π. See Fig. 2.51 of a schematic sketch of ϕ as a function of s. The slope of the curve at position s is related to the β-function at s, so one can tell the magnitude of β from the slope. Note that s and ϕ have a one-to-one correspondence because $\phi(s)$ *monotonically* increases with s. Other than incrementing by 2π's, the curve repeats itself every turn around the circumference.

Equations (2.62) and (2.63) constitute another way to summarize the Courant-Snyder formalism. With focusing strength $K(s)$ an arbitrary periodic function of s, Hill's equation is generally very complicated. The Courant-Snyder transformation allows us to drastically simplify it. The transformed equation is simple harmonic, whose focusing strength is now constant in ϕ.

Sometimes this step that allows replacing the s-dependent $K(s)$ by a ϕ-independent ν is referred to as the *smooth approximation.* This unfortunate use of the term is misleading. The Courant-Snyder transformation is not an approximation; it is exact. The smooth approximation is an accurate algorithm and is not to be interpreted as if it is a crude approximation of just replacing the storage ring lattice quantities by some kind of averages around the ring. Instead, it has an accurate mathematical meaning based on Courant-Snyder transformation and the approximation is made only when a small perturbation (e.g. a nonlinear magnet such as a sextupole) is introduced in an accelerator and then the subsequent analysis is carried out in a rigorous perturbation theory. Approximations then set in when and if we truncate the perturbation series. We will continue the discussion of smooth approximation in Sec. 3.7. Here we stay with the exact Eq. (2.62) and the unperturbed Hill's equation.

With $\eta \sim \begin{pmatrix} \cos\nu\phi \\ \sin\nu\phi \end{pmatrix}$ as the solution to Eq. (2.63), we easily see that $u \sim \sqrt{\beta}\begin{pmatrix} \cos\nu\phi \\ \sin\nu\phi \end{pmatrix}$ as described by Eq. (2.37).

The key to the Courant-Snyder transformation is the calculation of the periodic β-function. The betatron tune ν has the physical meaning that it is the number of oscillations a particle executes in one revolution around the ring. From the expression for ν, we also obtain a rough expression

$$\nu \approx \frac{R}{\langle\beta\rangle} \qquad \text{or} \qquad \langle\beta\rangle \approx \frac{R}{\nu},$$

where $\langle\beta\rangle$ is a rough average value of the β-function around the ring, $R = \frac{C}{2\pi}$ is the average radius of the ring. If $\nu \gg 1$, one expects $\langle\beta\rangle \ll R$. Note that $\langle\frac{1}{\beta}\rangle = \frac{\nu}{R}$ is valid rigorously, while the expression $\langle\beta\rangle \approx \frac{R}{\nu}$ is a rough approximate.

Homework 2.108 Derive Eq. (2.63). It might be a bit lengthy, but worthwhile. You will need to use the condition (2.40) in your proof.

2.5 Normal form

2.5.1 Normalized coordinates

As mentioned, the Courant-Snyder parametrization allows us to extract an invariant quantity Φ, phase advance per period, from the transfer map. To appreciate its physical meaning a bit further, let us return to the map (2.51) and rewrite it in the following instructive form,

$$\begin{aligned} M(s_2|s_1) &= A^{-1}(s_2)RA(s_1)\,, \qquad (2.64)\\ A(s) &= \begin{bmatrix} \frac{1}{\sqrt{\beta(s)}} & 0 \\ \frac{\alpha(s)}{\sqrt{\beta(s)}} & \sqrt{\beta(s)} \end{bmatrix} \quad \Longrightarrow \quad A^{-1}(s) = \begin{bmatrix} \sqrt{\beta(s)} & 0 \\ -\frac{\alpha(s)}{\sqrt{\beta(s)}} & \frac{1}{\sqrt{\beta(s)}} \end{bmatrix}, \\ R &= \begin{bmatrix} \cos\psi & \sin\psi \\ -\sin\psi & \cos\psi \end{bmatrix}. \end{aligned}$$

Equation (2.64) can be considered a summary of Courant-Snyder transformation. Let us observe it closely. The matrix $A(s)$ is the transformation from the physical coordinates u, u' to the *normalized* coordinates

$$\begin{bmatrix} v \\ p_v \end{bmatrix} = A \begin{bmatrix} u \\ u' \end{bmatrix} = \begin{bmatrix} \frac{u}{\sqrt{\beta}} \\ \frac{\beta u' + \alpha u}{\sqrt{\beta}} \end{bmatrix} .$$

These normalized coordinates were discussed earlier — see Eqs. (2.41) and (2.42). The transformation from (u, u') to (v, p_v) is a canonical transformation, which is assured because $\det A = 1$.

In terms of these normalized coordinates, the emittance can be written as

$$\epsilon = v^2 + {p_v}^2 . \tag{2.65}$$

If we describe particle motion in the normalized phase space (v, p_v), particle trajectories will trace out perfect circles (of radius $\sqrt{\epsilon}$) instead of tilted ellipses of Fig. 2.35. The circle covers a phase space area of $\pi\epsilon$.

The expression (2.65), as simple as it is, lies the secret of the Courant-Snyder formalism. Indeed, the first key step of the secret is to find an invariant and express it in terms of sum of perfect squares of your canonical dynamical variables like what is done in Eq. (2.65).

The middle matrix R of Eq. (2.64) describes a rotation by an angle ψ. We thus interpret Eq. (2.64) to mean the following: Start with $\begin{bmatrix} u \\ u' \end{bmatrix}$ at position s_1; we first transform it to the normalized coordinates at position s_1 by $A(s_1)$. In this normalized coordinate space, we then rotate by R with a clockwise rotation angle ψ given by the phase advance from s_1 to s_2. The result gives the normalized coordinates of the particle at position s_2. After the rotation, we then transform the normalized coordinates back to the physical space by $A^{-1}(s_2)$. The final result gives $\begin{bmatrix} u \\ u' \end{bmatrix}$ at position s_2.

Note the important observation that α, β, γ appear only in A, namely only in the transformation between physical coordinates and the normalized coordinates, while the remaining Courant-Snyder parameter ψ appears only in R. The four Courant-Snyder parameters $\alpha, \beta, \gamma, \psi$ behave distinctly differently in two camps. Note also that Eq. (2.64) looks like but is not a similarity transformation.

When discussing Fig. 2.34 earlier, we pointed out that it does not make sense to extract a "rotation angle" in the Courant-Snyder ellipse because the two axes have different dimensionalities. Here in the normalized coordinate system, however, a rotation is indeed legitimate, and the rotation angle is indeed ψ as expected. Incidentally, the rotation is clockwise, also as expected.

Equation (2.64) is called a *normal form representation* of the transfer matrix $M(s_2|s_1)$. If we have two beam-line sections connected together, we will then perform a multiplication as

$$\begin{aligned} M(s_3|s_2)M(s_2|s_1) &= A^{-1}(s_3)R(s_3|s_2)A(s_2)\; A^{-1}(s_2)R(s_2|s_1)A(s_1) \\ &= A^{-1}(s_3)R(s_3|s_2)R(s_2|s_1)A(s_1) \\ &= A^{-1}(s_3)R(s_3|s_1)A(s_1)\,. \end{aligned}$$

The middle two coordinate transformations at position s_2 cancel each other, and the new rotation has a rotation angle equal to the sum of the two original rotation angles. This is an important observation. What one learns is that the transformation A describes only a *static* — canonical — deformation of phase space. It is applied only at the beginning of particle motion. The entire particle motion can then be described much simpler, a simple rotation, in the normalized coordinates. After the motion is completed by the rotation, one then transform the normalized coordinates back to physical coordinates for observation by another static transformation A^{-1}. The beam *dynamics* is all contained in ψ. The other functions α, β, γ are not involved in beam dynamics. The transformations are illustrated schematically in Fig. 2.52.

Figure 2.52: Schematic illustration of the normal form transformation in the Courant-Snyder formalism. All three ellipses have the same area $\pi\epsilon$.

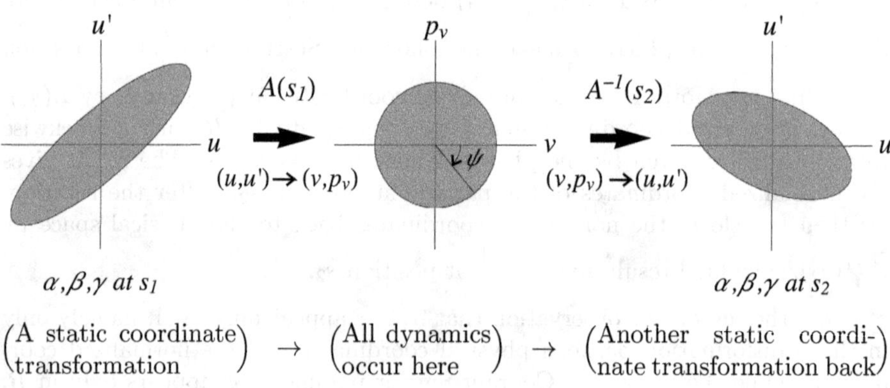

For the special case of a one-turn map, we have

$$M(s+C|s) = A^{-1}(s)R(s+C|s)A(s), \quad R(s+C|s) = \begin{bmatrix} \cos\Phi & \sin\Phi \\ -\sin_{\Phi} & \cos\Phi \end{bmatrix},$$

where $\Phi = 2\pi\nu$ is the total phase advance for one turn. This time it does become a true similarity transformation. As a result, Φ acquires an invariant property, i.e. its value does not change if calculated by using the one-turn map $M(s'+C|s')$ around a different location s'.

Homework 2.109 Although somewhat mundane, it is necessary that you go through the task of deriving Eqs. (2.64) and (2.65).

2.5.2 Courant-Snyder representation is not unique

Alternative Courant-Snyder formalism The Courant-Snyder representation as summarized in the form (2.64) is not unique. To appreciate this feature, we start by looking for other ways to represent the emittance ϵ. As mentioned, the key to a normal form is to express ϵ as a sum of two perfect squares, i.e. to write $\epsilon = v^2 + {p_v}^2$ for some conveniently chosen v and p_v. The Courant-Snyder representation (2.64) with the chosen coordinate transformation through

$$A = \begin{bmatrix} \frac{1}{\sqrt{\beta}} & 0 \\ \frac{\alpha}{\sqrt{\beta}} & \sqrt{\beta} \end{bmatrix}$$

is a good choice, but one may make infinitely many other choices — see Homework 2.113. Courant-Snyder has then chosen a particularly elegant one among the choices. But, for that matter, it is also not the only elegant one.[20] Another equally elegant choice, an alternative Courant-Snyder representation, is

$$\bar{A} = \begin{bmatrix} \sqrt{\gamma} & \frac{\alpha}{\sqrt{\gamma}} \\ 0 & \frac{1}{\sqrt{\gamma}} \end{bmatrix},$$

where α and γ are still the same functions we have defined and been using so far.

If we make this alternative choice, then we would have introduced the coordinates

$$\bar{v} = \frac{\gamma u + \alpha u'}{\sqrt{\gamma}}, \qquad \bar{p}_v = \frac{u'}{\sqrt{\gamma}},$$

and indeed we still maintain

$$\epsilon = \bar{v}^2 + \bar{p}_v^2 . \tag{2.66}$$

It can be shown that the map $M(s_2|s_1)$ can still be decomposed as (2.64) using the new choice of $\bar{A}(s)$, except that the rotation angle in matrix R becomes

$$\bar{\psi}(s) = \psi(s) + \tan^{-1}\alpha(s), \tag{2.67}$$

where ψ is the same betatron phase advance as we have defined always, and $\bar{\psi}$ is the new betatron phase advance if we choose the alternative Courant-Snyder parametrization.

[20]Admittedly, it may come as a surprise that there is more than one way to express the same quantity as sum of two perfect squares, and at least two of them are equally elegant.

In the Courant-Snyder representation, we have chosen the parametrization so that

$$u = \sqrt{\beta}\cos(\psi + \phi)\,. \tag{2.68}$$

The expression in parallel in the alternative Courant-Snyder representation is

$$u' = \sqrt{\gamma}\cos(\bar{\psi} + \bar{\phi})\,. \tag{2.69}$$

In other words, Courant and Snyder chose a particularly neat expression for u, as in Eq. (2.68), while leaving u' expression complicated. Equally well justified, we can choose u' to have a neat expression, as in Eq. (2.69), and in that case, u becomes complicated.

If we take the case for one period, with $s_2 = s_1 + C$, it follows from Eq. (2.67) that

$$\bar{\psi}(s_2) - \bar{\psi}(s_1) = \psi(s_2) - \psi(s_1) = \Phi = 2\pi\nu\,.$$

In other words, the phase advance per period, and thus the tune, is not only an invariant independent of the choice of the point of observation — as we emphasized before — but also a "super" invariant independent of the choice of Courant-Snyder parametrization. Although there are an infinite number of normal form representations, all of them give the same phase advance per period, and they are all equal to Φ. This is an important observation, and a moment's reflection indicates that it is also expected because, unlike the other Courant-Snyder parameters α, β, γ, the parameter Φ is related to the *physical* quantity of the tune; it must not depend on the choice of representation.

If we take an infinitesimal step, with $s_2 = s_1 + \Delta s$, and let $\Delta s \to 0$, we obtain

$$\begin{aligned}\frac{d\bar{\psi}(s)}{ds} &= \frac{d\psi(s)}{ds} + \frac{d}{ds}\tan^{-1}\alpha(s) \\ &= \frac{1}{\beta} + \frac{\alpha'}{1+\alpha^2} = \frac{K(s)}{\gamma(s)}\,.\end{aligned}$$

In other words, the new betatron phase is given by

$$\bar{\psi}(s) = \int_0^s ds'\,\frac{K(s')}{\gamma(s')}\,. \tag{2.70}$$

The fundamental role of the β-function in the Courant-Snyder formalism is taken over by the γ-function in the alternative formalism.

According to Eq. (2.70), in contrast to the monotonically progressing phase advance $\psi(s)$ in the normal Courant-Snyder representation, $\bar{\psi}(s)$ progresses forward in focusing quadrupoles and backward in defocusing quadrupoles, and stays constant over drift spaces. Figure 2.53 illustrates a comparison between the normalized coordinates in the familiar Courant-Snyder (left) and the alternative (right) conventions. Particle trajectory lies on a circle for both cases. Betatron phase advance in the normal Courant-Snyder case has a monotonic clockwise rotation on the circle, while in the alternative case, the phase advance is not monotonic and can go backwards.

Figure 2.53: Betatron phase advance for the Courant-Snyder and the alternative Courant-Snyder transformations.

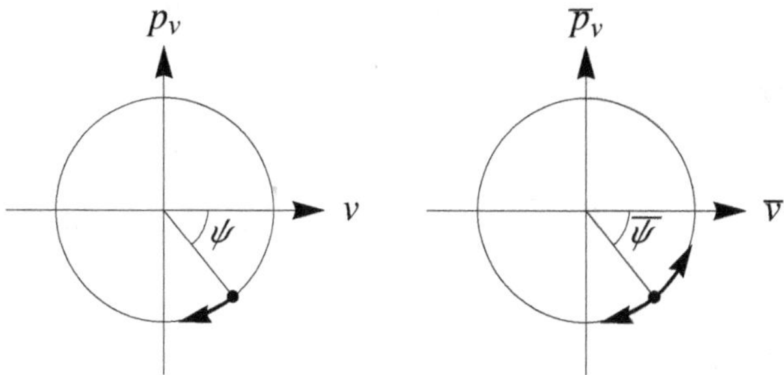

β-function is an artifact It is clear that one could develop a formalism with the alternative choice which is equally elegant as the Courant-Snyder parametrization. All physical results we have obtained will remain identical, however. The point to be made here is that any result that depends on the representation chosen to describe the problem can only be an artifact and cannot be physical. The Courant-Snyder transformation and the Courant-Snyder functions $\alpha, \beta, \gamma, \psi$, as useful as they are, are artifacts. Among the Courant-Snyder functions and parameters, only the total tune $\nu = \frac{\Phi}{2\pi}$ is a true physical quantity. A brief comparison between the Courant-Snyder and the alternative formalisms might be summarized in Table 2.3.

Table 2.3: A brief comparison between the Courant-Snyder and the alternative formalisms.

	envelope function	betatron phase	action-angle form
Courant-Snyder	$\beta(s)$	$\psi(s) = \int^s \frac{ds'}{\beta(s')}$	$u = \sqrt{\epsilon\beta}\cos\psi$
Alternative	$\gamma(s)$	$\bar{\psi}(s) = \int^s \frac{K(s')ds'}{\gamma(s')}$	$u' = \sqrt{\epsilon\gamma}\cos\bar{\psi}$

Had Courant-Snyder chosen the alternative representation for their classic paper in 1958, today we would be using a different accelerator physics language. We would probably not recognize any of the equations familiar in our present-day textbooks. Specifically, even the expressions of u and u' in terms of the β-functions will look different. And yet, both representations would give identical results for all physical quantities. What Courant-Snyder formalism provided is an artifact without direct physical meanings. However, that is what a spoken language does, and its importance is not to be underestimated. Languages are clearly artifacts, but we can't live without them.

Homework 2.110
(a) Prove Eq. (2.66). As mentioned in the text, it is curious that the same quantity ϵ has different ways to express it as sum of two perfect squares.

(b) Prove Eq. (2.67).

(c) Show that the $(\bar{v}, \bar{p}_v)$ coordinates are related to the (v, p_v) coordinates by

$$\begin{bmatrix} \bar{v} \\ \bar{p}_v \end{bmatrix} = \begin{bmatrix} \cos\theta_0 & \sin\theta_0 \\ -\sin\theta_0 & \cos\theta_0 \end{bmatrix} \begin{bmatrix} v \\ p_v \end{bmatrix}, \qquad \text{where} \quad \theta_0 = \tan^{-1}\alpha\,. \tag{2.71}$$

(d) Prove Eq. (2.70).

Homework 2.111 Prove Eq. (2.69) by substituting Eq. (2.67) into Eq. (2.68) and taking a derivative.

Homework 2.112 By direct differentiation, show that for the Courant-Snyder parametrization, the normalized coordinates satisfy

$$\frac{dv}{d\psi} = p_v, \qquad \frac{dp_v}{d\psi} = -v\,,$$

and for the alternative Courant-Snyder parametrization, we have

$$\frac{d\bar{v}}{d\bar{\psi}} = \bar{p}_v, \qquad \frac{d\bar{p}_v}{d\bar{\psi}} = -\bar{v}\,.$$

Homework 2.113 There are infinite number of ways to express the emittance ϵ as the sum of two perfect squares. The text gave two examples. See if you can find another one. It is not going to be as elegant. But it will work just the same.

Solution A general transformation can be written as

$$\begin{bmatrix} v \\ p_v \end{bmatrix} = \begin{bmatrix} \cos\Phi & \sin\Phi \\ -\sin\Phi & \cos\Phi \end{bmatrix} \begin{bmatrix} \frac{1}{\sqrt{\beta}} & 0 \\ \frac{\alpha}{\sqrt{\beta}} & \sqrt{\beta} \end{bmatrix} \begin{bmatrix} u \\ u' \end{bmatrix},$$

where $\Phi = \frac{2\theta}{\pi}\tan^{-1}\alpha$ for some parameter θ. When $\theta = 0$, this gives the normal Courant-Snyder. When $\theta = \frac{\pi}{2}$, it gives the alternative Courant-Snyder — see Eq. (2.71). In principle, any choice of θ would work other than the sacrifice of elegance — but there is nothing to gain either.

For example, one might choose $\theta = \pi$. Then Eq. (2.71) becomes

$$\begin{aligned} \begin{bmatrix} \bar{\bar{v}} \\ \bar{\bar{p}}_v \end{bmatrix} &= \begin{bmatrix} \cos 2\theta_0 & \sin 2\theta_0 \\ -\sin 2\theta_0 & \cos 2\theta_0 \end{bmatrix} \begin{bmatrix} v \\ p_v \end{bmatrix} \\ &= \begin{bmatrix} \frac{-1+\alpha^2}{1+\alpha^2} & \frac{2\alpha}{1+\alpha^2} \\ -\frac{2\alpha}{1+\alpha^2} & \frac{-1+\alpha^2}{1+\alpha^2} \end{bmatrix} \begin{bmatrix} v \\ p_v \end{bmatrix} \end{aligned}$$

$$= \begin{bmatrix} \frac{-1+3\alpha^2}{(1+\alpha^2)\sqrt{\beta}} & \frac{2\alpha\sqrt{\beta}}{1+\alpha^2} \\ \frac{\alpha(-3+\alpha^2)}{(1+\alpha^2)\sqrt{\beta}} & \frac{(-1+\alpha^2)\sqrt{\beta}}{1+\alpha^2} \end{bmatrix} \begin{bmatrix} u \\ u' \end{bmatrix}.$$

It still follows that

$$\bar{\bar{v}}^2 + \bar{\bar{p}}_v^2 = \epsilon .$$

Homework 2.114 As mentioned in the text, the alternative representation has a different definition of betatron phase given by $\bar{\psi}(s)$, Eq. (2.70). Calculate $\bar{\psi}(s)$ through a FODO cell. Compare it with the betatron phase $\psi(s)$ in the normal Courant-Snyder representation.

Solution The Courant-Snyder functions $\beta(s), \alpha(s), \gamma(s)$ and $\psi(s)$ are shown as red curves in Fig. 2.48. The blue curve gives $\bar{\psi}(s)$. One sees that ψ and $\bar{\psi}$ look very different. For example, ψ advances when the particle goes through a drift space, while $\bar{\psi}$ stands still in a drift space and advances only when going through a quadrupole. Also, ψ always advances monotonically, while $\bar{\psi}$ advances going through a focusing quadrupole and goes backwards going through a defocusing quadrupole. The expressions of u and u' in terms of the β-functions are different. We however emphasize that either representation gives identical physical results. In particular, the total phase advance over the cell in the two representations are equal, as shown by comparing the red and the blue curves.

2.6 3-D radio-frequency particle trap

2.6.1 RF particle trap

Accelerator physics studies the motion of charged particles in an accelerator. Most likely the particles collectively takes the form a *beam*, i.e. all particles move with more or less the same average momentum so that they stay together in a confined space as well as phase space as they move ahead in time. One exception of this description of accelerator physics is the case of a particle trap.

In a particle trap, charged particles do not move collectively. They move, but only in such a way confined in a physical space and have no net momentum moving in any direction. These particles are therefore said to be trapped. A mechanism is needed to do the trapping, thus the subject of *particle trap*. Once trapped, the particles form a distribution around a stationary trap center.

Since particle traps mostly use relatively weak mechanisms to confine the particles, there is a requirement that the particles must not move too violently. This then leads to a requirement that the trapped particles must be cold enough. Particles too energetic, or too hot, may overcome the trapping mechanism and escape the trap.

Particle traps are used in various applications, e.g. in the research of atomic physics, condensed matter physics, or low temperature quantum mechanics. The design of particle traps is a large subject involving clever inventions of various trapping and cooling techniques. To appreciate why designing a particle trap requires such a special effort, let us first consider the Maxwell equations in free space. In the absence of material other than the charged particles themselves, any electromagnetic fields used to provide the trapping mechanism must obey the free-space Maxwell equations near the neighborhood around the trapped beam.

To constitute a particle trap, the particle motion must be stable about the trap center. In other words, as the particles move about in space, their motion must be focused towards the trap center. The trapping mechanism is to be configured in such a way that the ensuing free-space Maxwell equations focus particle motion in all x-, y- and z-dimensions.

Maxwell equations in free space, however, says there is no way to arrange the electric and magnetic fields so that the motions in all three dimensions be focusing simultaneously.[21] If two dimensions are focusing at a given moment of time, then at that moment, the third dimension can only be defocusing. (Similarly, one cannot have all three dimensions be simultaneously defocusing.) There is no way to simply focus the particle motion at all times in all three dimensions. Existing particle traps have been designed in which two of the dimensions are made focusing (and that is easy to do), and then the third dimension is configured in such a way that makes particles extremely difficult to escape. Particles do still escape, though, so there is associated a certain lifetime of the beam in these types of traps.

The above free-space condition mentioned above requires that there are no material immediately at the location of the charge particles, i.e. the relevant fields describing particle motion satisfy Maxwell equations in free space. It does not forbid any material at a distance from the charges, e.g. a metal wall that provides appropriate boundary conditions. Away from the metal walls, where the trapped particles reside, the electromagnetic fields still satisfy free-space Maxwell equations, and the above statement then applies.

In this section, we describe a simple particle trap of cold charged particles. The trap simply uses a closed radio-frequency cavity and we call this a radio-frequency trap.[22] The device is simple, and it has the advantage that motions in all three dimensions are focused and therefore, in principle, there is no particle loss. The trick is to use time-varying fields for the trapping mechanism. If not all three dimensions are focusing at any given time, we can let the three dimensions take turns to focus. The idea makes use of the principle of alternate focusing we learned in this chapter.

In principle, a similar RF cavity trap can be used as a tokomak. However, it does not work in this case for two reasons. One is that the trapped particles are too hot. The other is that they are also too intense so that their self-fields overwhelm the RF field.

---

[21]This is because $\nabla \cdot \vec{E} = \vec{0}$ and $\nabla \cdot \vec{B} = 0$. More on this later in this section.

[22]Alex Chao, SLAC Report SLAC-PUB-7072 (1993).

2.6.2 Field in the cavity

Consider a perfectly conducting RF cavity. For simplicity, let this cavity be a closed rectangular box with three dimensions ℓ_x, ℓ_y and ℓ_z. Let there be a mode excited in this cavity,

$$\begin{aligned} E_x &= E_{x0} \sin\frac{\pi x}{\ell_x} \cos\frac{\pi y}{\ell_y} \cos\frac{\pi z}{\ell_z} e^{-i\omega t}\,, \\ E_y &= E_{y0} \cos\frac{\pi x}{\ell_x} \sin\frac{\pi y}{\ell_y} \cos\frac{\pi z}{\ell_z} e^{-i\omega t}\,, \\ E_z &= E_{z0} \cos\frac{\pi x}{\ell_x} \cos\frac{\pi y}{\ell_y} \sin\frac{\pi z}{\ell_z} e^{-i\omega t}\,, \end{aligned} \tag{2.72}$$

where ω is the mode frequency.

The field (2.72) satisfies the boundary conditions

$$\begin{aligned} E_x = 0 \quad &\text{when} \quad y = \pm\frac{\ell_y}{2} \quad \text{or} \quad z = \pm\frac{\ell_z}{2}\,, \\ E_y = 0 \quad &\text{when} \quad x = \pm\frac{\ell_x}{2} \quad \text{or} \quad z = \pm\frac{\ell_z}{2}\,, \\ E_z = 0 \quad &\text{when} \quad x = \pm\frac{\ell_x}{2} \quad \text{or} \quad y = \pm\frac{\ell_y}{2}\,. \end{aligned}$$

To satisfy $\nabla \cdot \vec{E} = 0$, we require

$$\frac{E_{x0}}{\ell_x} + \frac{E_{y0}}{\ell_y} + \frac{E_{z0}}{\ell_z} = 0\,. \tag{2.73}$$

It is obvious that not all three components E_{x0}, E_{y0}, E_{z0} can be of the same sign, which in turn leads to the fact that not all three dimensions can be focused at any given moment of time.

To find the magnetic field, we use

$$\vec{B} = -\frac{i}{\omega} \nabla \times \vec{E}\,,$$

which leads to

$$\begin{aligned} B_x &= -\frac{i}{\omega}\left(\frac{\partial E_z}{\partial y} - \frac{\partial E_y}{\partial z}\right) = \frac{i\pi}{\omega}\left(\frac{E_{z0}}{\ell_y} - \frac{E_{y0}}{\ell_z}\right) \cos\frac{\pi x}{\ell_x} \sin\frac{\pi y}{\ell_y} \sin\frac{\pi z}{\ell_z} e^{-i\omega t}\,, \\ B_y &= -\frac{i}{\omega}\left(\frac{\partial E_x}{\partial z} - \frac{\partial E_z}{\partial x}\right) = \frac{i\pi}{\omega}\left(\frac{E_{x0}}{\ell_z} - \frac{E_{z0}}{\ell_x}\right) \sin\frac{\pi x}{\ell_x} \cos\frac{\pi y}{\ell_y} \sin\frac{\pi z}{\ell_z} e^{-i\omega t}\,, \\ B_z &= -\frac{i}{\omega}\left(\frac{\partial E_y}{\partial x} - \frac{\partial E_x}{\partial y}\right) = \frac{i\pi}{\omega}\left(\frac{E_{y0}}{\ell_x} - \frac{E_{x0}}{\ell_y}\right) \sin\frac{\pi x}{\ell_x} \sin\frac{\pi y}{\ell_y} \cos\frac{\pi z}{\ell_z} e^{-i\omega t}\,. \end{aligned} \tag{2.74}$$

Equation (2.74) satisfies automatically $\nabla \cdot \vec{B} = 0$. It also automatically satisfies the boundary conditions

$$\begin{aligned} B_x &= 0 \quad \text{when} \quad x = \pm\frac{\ell_x}{2}\,, \\ B_y &= 0 \quad \text{when} \quad y = \pm\frac{\ell_y}{2}\,, \\ B_z &= 0 \quad \text{when} \quad z = \pm\frac{\ell_z}{2}\,. \end{aligned}$$

We are still left with the condition, the last of the Maxwell equations,

$$\vec{E} = \frac{ic^2}{\omega}\nabla \times \vec{B}\,.$$

With fields given by Eqs. (2.72) and (2.74), we find that this remaining condition is satisfied provided

$$\omega = \pi c\sqrt{\frac{1}{\ell_x^2} + \frac{1}{\ell_y^2} + \frac{1}{\ell_z^2}}\,. \tag{2.75}$$

Equation (2.75) determines the mode frequency ω.

Homework 2.115 Derive Eq. (2.75). You will need to use Eq. (2.73).

2.6.3 Ion motion in the cavity

Consider the motion of a cold ion near the center of the cavity $|x| \ll \ell_x, |y| \ll \ell_y, |z| \ll \ell_z$. The fields the ion sees are approximately given by

$$\begin{aligned} \vec{E} &\approx \pi e^{-i\omega t}\left[\frac{E_{x0}}{\ell_x}x\hat{x} + \frac{E_{y0}}{\ell_y}y\hat{y} + \frac{E_{z0}}{\ell_z}z\hat{z}\right], \\ \vec{B} &\approx 0\,. \end{aligned}$$

The magnetic field is negligible not only because it is second order in x, y, z, but also because the motion of nonrelativistic ions is insensitive to magnetic fields. For what follows, we replace $e^{-i\omega t}$ by $\cos\omega t$.

The equation of motion of a nonrelativistic ion with rest mass M and charge Q is

$$M\ddot{x} = QE_x \approx Q\frac{\pi E_{x0}}{\ell_x}x\cos\omega t\,,$$

and similarly for the y- and z-equations. The three dimensions are decoupled, and they are all of the form

$$\ddot{u} + (K\cos\omega t)u = 0\,. \tag{2.76}$$

where K is a constant, which may either be positive or negative. An inspection of Eq. (2.73) shows that the values of K for the three dimensions must not all have the same sign. For the x-motion,

$$K_x = -\frac{\pi Q E_{x0}}{M\ell_x} .$$

As mentioned, the present idea of RF trapping is based on the alternate focusing principle for Eq. (2.76). In Eq. (2.76), the focusing gradient is given by $K\cos\omega t$. The phase ωt determines whether the field is focusing or defocusing at a given time. The alternating sign of this gradient provides an opportunity for the u-motion to be focused regardless of the sign of K. Although the field cannot focus all three dimensions at a given instant of time, the RF field may focus all three dimensions as a net result provided some stability conditions are met.

2.6.4 Stability of ion motion — crude model

Equation (2.76) is in fact the Mathieu's equation. In order for the u-motion to be stable, K and ω must satisfy some condition. In this and the next sections, we look for this stability condition.

Before we solve the Mathieu's equation, however, it may be instructive to examine the qualitative behavior of the ion motion using a simplified model. To do so, consider the replacement

$$K\cos\omega t \rightarrow \begin{cases} |K| & \text{when } K\cos\omega t > 0 \\ -|K| & \text{when } K\cos\omega t < 0 \end{cases} . \tag{2.77}$$

In other words, replace the $K\cos\omega t$ focusing by a piecewise-constant alternating focusing. We expect this simplified model to be easier to solve and yet it describes faithfully the important qualitative behavior of the ion motion.

With the replacement (2.77), the ion motion over one period of time (specifically from $\omega t = -\frac{\pi}{2}$ to $\omega t = \frac{3\pi}{2}$ when $K > 0$) is described by a matrix map

$$\begin{bmatrix} u \\ \frac{\dot{u}}{\sqrt{|K|}} \end{bmatrix}_{\text{final}} = T \begin{bmatrix} u \\ \frac{\dot{u}}{\sqrt{|K|}} \end{bmatrix}_{\text{initial}} ,$$

where

$$T = \begin{bmatrix} \cosh\xi & \sinh\xi \\ \sinh\xi & \cosh\xi \end{bmatrix} \begin{bmatrix} \cos\xi & \sin\xi \\ -\sin\xi & \cos\xi \end{bmatrix} ,$$

and we have defined

$$\xi = \sqrt{|K|}\,\frac{\pi}{\omega} .$$

Stability of ion motion is determined by the condition

$$|\text{trace}(T)| \leq 2 ,$$

or

$$|\cos\xi\cosh\xi| \;\le\; 1\,.$$

Figure 2.54 shows the function $|\cos\xi\cosh\xi|$ vs ξ. There are infinite number of bands of values of ξ when the RF trap gives stable ion motion.

Figure 2.54: Stability criterion of an RF particle trap in the crude model. Stability requires the curve to stay below the dashed line.

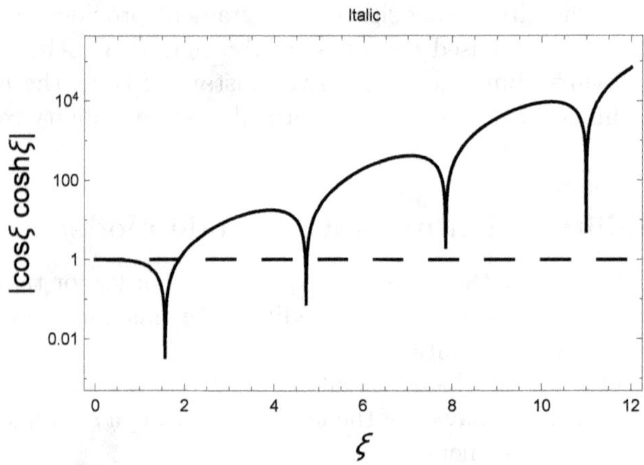

The stable band near $\xi = 0$ has the largest width, and it spans the region

$$0 \;\le\; \xi \;<\; 1.875\,. \tag{2.78}$$

Higher values of ξ becomes unstable due to over-focusing. The stability condition therefore reads

$$\frac{|K|}{\omega^2} \;<\; 0.356\,. \tag{2.79}$$

There are also higher order stability regions, but their widths becomes rapidly much narrower. For example, the next stable region has $4.694 < \xi < 4.73$. We will not consider the use of these narrow stable regions for our application.

For $\xi \lesssim 1$, we have

$$\cos\xi\cosh\xi \;\approx\; 1 - \frac{\xi^4}{6}\,. \tag{2.80}$$

If we ignore all higher order terms, the stability region near $\xi = 0$ would be, according to (2.80), $0 \le \xi < 12^{1/4} = 1.861$, which agrees reasonably well with (2.78).

In the stable region, the ions move with an oscillation frequency

$$f \;=\; \frac{\omega}{2\pi}\frac{1}{2\pi}\cos^{-1}(\cos\xi\cosh\xi)\,.$$

If $\xi \lesssim 1$, we have

$$f \approx \frac{\omega\xi^2}{4\pi^2\sqrt{3}} = \frac{|K|}{4\sqrt{3}\,\omega}. \tag{2.81}$$

The three oscillation frequencies are therefore

$$f_x \approx \frac{\pi Q E_{x0}}{4\sqrt{3}M\ell_x\omega}, \qquad \text{similarly for } f_y, f_z\,. \tag{2.82}$$

Homework 2.116 According to the text, the motion of an ion in x, y and z are decoupled, as given by Eqs. (2.76) and (2.77) for the crude model. The stability conditions and the resulting net focusing frequencies were derived.

(a) Continue the analysis to find the Courant-Snyder invariant in terms of u and $\frac{\dot{u}}{\sqrt{|K|}}$ at the observation time for the one-period map T.

(b) Find the expressions of these invariants when the focusing strength is weak with $\xi \ll 1$.

(c) Show that the equivalent temperatures of the trapped particles are given by

$$k_B T_i \approx \frac{\pi^2}{3} \frac{Q^2 E_{i0}^2}{Mc^2\left(\frac{1}{\ell_x^2}+\frac{1}{\ell_y^2}+\frac{1}{\ell_z^2}\right)} \frac{\sigma_i^2}{\ell_i^2}, \qquad i = x, y, z\,.$$

We have assumed $\sigma_i \ll \ell_i$. These invariants can be used to calculate a particle's x-, y-, z-oscillation amplitudes given the initial launching conditions of the particle.

Solution Observed at the time where T is evaluated, the invariant is

$$\begin{aligned} J &= u^2(\cosh\xi\sin\xi - \sinh\xi\cos\xi) - 2u\frac{\dot{u}}{\sqrt{K}}\sinh\xi\sin\xi \\ &\quad + \frac{\dot{u}^2}{K}(\cosh\xi\sin\xi + \sinh\xi\cos\xi)\,. \end{aligned}$$

When $\xi \ll 1$, it reads

$$J \approx \frac{2\xi}{3}\left(\xi^2 u^2 - 3\xi u\frac{\dot{u}}{\sqrt{K}} + 3\frac{\dot{u}^2}{K}\right).$$

2.6.5 Mathieu equation analysis

We now analyze Eq. (2.76) more carefully as the Mathieu equation. We expect to obtain a stability condition similar to Eq. (2.78) and oscillation frequencies similar to Eq. (2.82) but with somewhat different numerical coefficients.

First change variable from t to $x = \frac{1}{2}\omega t$. Equation (2.76) reads

$$\frac{d^2u}{dx^2} + (\beta\cos 2x)u = 0\,, \tag{2.83}$$

where

$$\beta = \frac{4K}{\omega^2}.$$

Equation (2.83) is a special case of the Mathieu equation.[23] The boundary of the stability regions is determined by the existence of periodic solutions. In particular, the boundary of the stable region near $\beta = 0$ is determined by the existence of two periodic solutions which can be written as Fourier series,

$$\begin{aligned} C(x) &= \sum_{m=0}^{\infty} A_m \cos(2m+1)x\,, \\ S(x) &= \sum_{m=0}^{\infty} A_m \sin(2m+1)x\,. \end{aligned} \tag{2.84}$$

Let us examine the $C(x)$ solution first. We will look for the value of β for which the periodic solution $C(x)$ exists. Substituting (2.84) into (2.83) gives

$$-\sum_{m=0}^{\infty}(2m+1)^2 A_m \cos(2m+1)x + \frac{\beta}{2}\sum_{m=0}^{\infty} A_m[\cos(2m+3)x + \cos(2m-1)x] = 0\,,$$

or

$$-A_0\cos x - \sum_{m=1}^{\infty}(2m+1)^2 A_m \cos(2m+1)x + \frac{\beta}{2}\sum_{m=1}^{\infty} A_{m-1}\cos(2m+1)x$$
$$+\frac{\beta}{2}A_0\cos x + \frac{\beta}{2}A_1\cos x + \frac{\beta}{2}\sum_{m=1}^{\infty} A_{m+1}\cos(2m+1)x = 0\,.$$

Equating different $\cos(2m+1)x$ terms to zero, we obtain the recurrence equations

$$\begin{aligned} -A_0 + \frac{\beta}{2}A_0 + \frac{\beta}{2}A_1 &= 0\,, \\ -(2m+1)^2 A_m + \frac{\beta}{2}A_{m-1} + \frac{\beta}{2}A_{m+1} &= 0\,, \end{aligned} \tag{2.85}$$

where the second equation applies for $m \geq 1$.

The first of Eq. (2.85) gives

$$\frac{A_1}{A_0} = \frac{2}{\beta} - 1\,. \tag{2.86}$$

---

[23]Mathieu equation reads

$$\frac{d^2u}{dx^2} + (\alpha + \beta\cos 2x)u = 0\,.$$

Here we have the special case when $\alpha = 0$.

On the other hand, the second of (2.85) gives

$$\frac{A_m}{A_{m-1}} = \frac{\beta}{2(2m+1)^2 - \beta \frac{A_{m+1}}{A_m}},$$

which gives a continued fraction expression

$$\frac{A_1}{A_0} = \frac{\beta}{2 \cdot 3^2 - \frac{\beta^2}{2 \cdot 5^2 - \frac{\beta^2}{2 \cdot 7^2 - \cdots}}}. \tag{2.87}$$

Equating Eqs. (2.86) and (2.87), for the periodic solution $C(x)$ to exist, the value of β must satisfy

$$2 - \beta = \frac{\beta^2}{2 \cdot 3^2 - \frac{\beta^2}{2 \cdot 5^2 - \frac{\beta^2}{2 \cdot 7^2 - \cdots}}}.$$

The successive approximate values of β are found to be

$$\beta \approx 2, \frac{16}{9}, 1.824, 1.8145, \cdots$$

A numerical calculation gives the result to be $\beta = 1.816$.

Similar analysis can be applied to the periodic solution $S(x)$ of Eq. (2.84). For $S(x)$ to exist, we need

$$-\sum_{m=0}^{\infty}(2m+1)^2 A_m \sin(2m+1)x + \frac{\beta}{2}\sum_{m=0}^{\infty} A_m[\sin(2m+3)x + \sin(2m-1)x] = 0,$$

or

$$-A_0 \sin x - \sum_{m=1}^{\infty}(2m+1)^2 A_m \sin(2m+1)x + \frac{\beta}{2}\sum_{m=1}^{\infty} A_{m-1}\sin(2m+1)x$$
$$-\frac{\beta}{2}A_0 \sin x + \frac{\beta}{2}A_1 \sin x + \frac{\beta}{2}\sum_{m=1}^{\infty} A_{m+1}\sin(2m+1)x = 0.$$

Equating different $\sin(2m+1)x$ terms to zero, we obtain the recurrence equations

$$\begin{aligned} -A_0 - \frac{\beta}{2}A_0 + \frac{\beta}{2}A_1 &= 0, \\ -(2m+1)^2 A_m + \frac{\beta}{2}A_{m-1} + \frac{\beta}{2}A_{m+1} &= 0. \end{aligned} \tag{2.88}$$

The first of Eq. (2.88) gives

$$\frac{A_1}{A_0} = \frac{2}{\beta} + 1. \tag{2.89}$$

The second of Eq. (2.88) gives the same expressions as (2.87). Equating Eqs. (2.89) and (2.87) yields

$$2+\beta = \frac{\beta^2}{2\cdot 3^2 - \frac{\beta^2}{2\cdot 5^2 - \frac{\beta^2}{2\cdot 7^2 - \cdots}}}.$$

The result is $\beta = -1.816$.

Combining the results so far, we find that the stability condition is

$$|\beta| < 1.816, \qquad \text{or} \qquad \frac{|K|}{\omega^2} < 0.454\,. \tag{2.90}$$

which is to be compared with the crude estimate (2.79).

To appreciate the physical meaning of the condition (2.90), let us rewrite its scaling as

$$QE_0\ell \lesssim Mc^2\,,$$

where E_0 is the cavity field strength, ℓ is the size of the cavity. This condition is equivalent of saying the energy gained from the cavity electric field over the distance of the cavity dimension must not render the ion relativistic. Such a condition can easily be satisfied. In fact, it is difficult to violate it.

2.6.6 Tracking results

We can track the motion of a particle using Eq. (2.83). The upper panel of Fig. 2.55 shows the results for tracking a particle with initial conditions $u(0) = 1$ and $\frac{du}{dx}(0) = 0$ for 4 values of $\beta = 0$ (solid curve), 0.5 (dashed curve), 1 (dotted curve), and 1.5 (solid curve), for the range of $x = \frac{1}{2}\omega t$ from 0 to 100. The motion is stable for these values of β. The lower panel is the same with the initial conditions $u(0) = 0$ and $\frac{du}{dx}(0) = 1$. A strong focusing occurs when $\beta \sim 1$, i.e. away from 0 and 1.816.

Figure 2.56 shows the particle motion near the threshold $\beta = 1.816$. The two values of β are 1.814 (solid curve) and 1.818 (dotted curve). Again the upper panel is for the cosine-like and lower panel is for the sine-like trajectories.

The period of oscillation can be read off from these two figures. This gives $X = \frac{1}{2}\omega T \approx 35,\ 17,\ 10,\ 6$ for $\beta = 0.5$, 1, 1.5 and 1.814, respectively. These periods can be compared with the approximate expression (2.81), which gives

$$X \approx \frac{8\sqrt{3}}{\beta}\,. \tag{2.91}$$

For $\beta = 0.5$, 1, 1.5 and 1.814, Eq. (2.91) gives $X = 28$, 14, 9, and 7.6 in reasonable agreement.

Figure 2.55: Particle motion in the RF trap is studied by computer tracking.

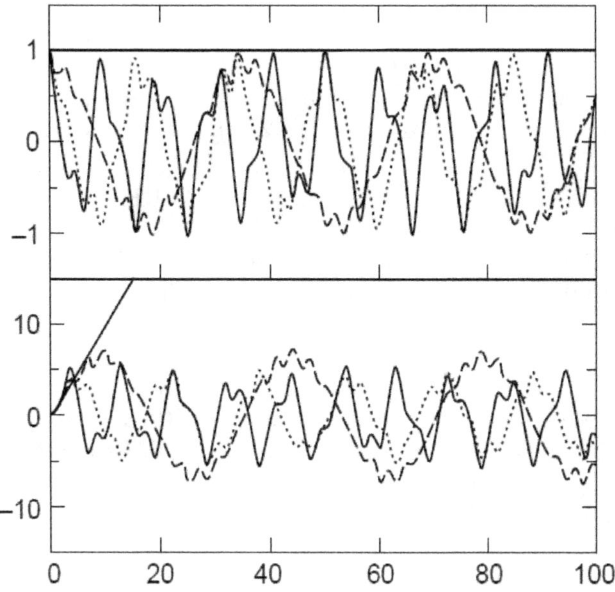

Figure 2.56: Same as Fig. 2.55 but with β value close to the threshold.

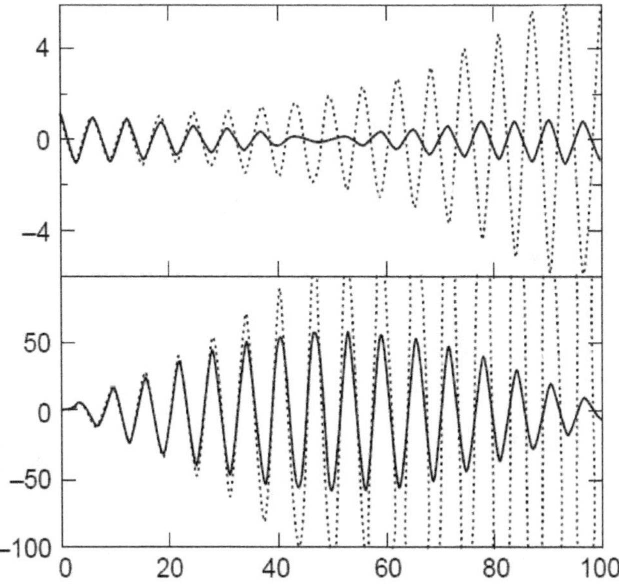

2.6.7 Design example

Take $\ell_x = \ell_y = \ell_z = 30$ cm, Eq. (2.75) gives $\omega = 0.54 \times 10^{10}$ s^{-1} (0.86 GHz).[24] Consider antiprotons, we have $Q = e$, $Mc^2 = 938$ MeV. Let the cavity field be such that $E_{x0} = E_{y0} = 100$ KV/cm and, from Eq. (2.73), $E_{z0} = -200$ KV/cm. We have $|K_x| = |K_y| = 1.0 \times 10^{16}\,\mathrm{s}^{-2}$ and $|K_z| = 2.0 \times 10^{16}\,\mathrm{s}^{-2}$. The stability condition, according to Eq. (2.90), is therefore easily satisfied with $\frac{|K_x|}{\omega^2} = \frac{|K_y|}{\omega^2} = \frac{1}{2}\frac{|K_z|}{\omega^2} = 3.4 \times 10^{-4}$. Using Eq. (2.82), we estimate the oscillation frequencies to be $f_x = f_y = 270$ KHz and $f_z = 540$ KHz.

Homework 2.117 Use the result of Homework 2.116(c) to estimate the equivalent temperature of the trapped antiprotons assuming the trapped particles has an rms beam size $\sigma_x = \sigma_y = \sigma_z = 3$ cm.

Homework 2.118

(a) Compose a set of design parameters of an RF cavity of particle trap for a heavy ion Cs^+ beam with a thermal energy of 1 eV. Estimate the size of your trapped beam.

(b) It is a different challenge to design a trap for electrons.

[24]One may want to avoid this highly degenerate mode by having $\ell_{x,y,z}$ slightly different from one another, but this does not affect a crude numerical example being considered here.

Chapter 3

Perturbation

We have now completed the basics of the 1-D Courant-Snyder analysis. There are a few considerations at this point,

1. How to extend the successful Courant-Snyder formalism to coupled 2-D (and 3-D) motion?
2. How to extend it to nonlinear motion, even 1-D?
3. How to incorporate error effects in the Courant-Snyder formalism?

There are several approaches to address these questions. Our lectures can therefore take a number of directions at this junction.

To address question 1, for example, various linear coupling formalisms can be introduced. Based on our analysis of Sec. 2.5.2 when we came to the conclusion that the nominal Courant-Snyder formalism is in fact an artifact designed exclusively for the 1-D decoupled case, however, we shall not adopt this direction of searching for extensions of the Courant-Snyder formalism to the 2-D coupled cases. Instead, we shall postpone the subject and introduce later in Sec. 6.5 one approach that skips the attempt for formalisms and proceed directly to a 6-D eigenanalysis.

The remaining two questions are the subject of this and the next chapters. Deviations of the accelerator from the perfect linear design are considered errors. Questions 2 and 3 concern the situation when the accelerator environment contains errors, which can be linear or nonlinear. When considering these error effects, we assume they are weak compared with the linear uncoupled system we have designed the accelerator to be. As such, they are considered perturbations and a perturbative approach can be applied to treat them.

There are two kinds of error effects we need to address. First, the lattice so far is considered to be perfect, consisting of perfect beamline elements and are constructed according to the design. Second, all particles so far are assumed to have the design beam momentum. Neither is true in reality, of course. The magnets in a real accelerator will inevitably have field errors, and they might not be installed perfectly. Particles will inevitably have momentum deviations. In this chapter, we will discuss the effects associated with these imperfect beamline elements (mainly magnets) with field errors or installation errors. In the next chapter, we will discuss effects associated with momentum errors of the particles.

3.1 Field error

We first treat the case of magnetic field errors. These field errors are considered perturbations to the otherwise perfect accelerator by design.

The magnetic field with errors can be written as

$$\begin{aligned} B_x &= Gy + \Delta B_x \,, \\ B_y &= B_{y0} + Gx + \Delta B_y \,, \end{aligned}$$

where $G, B_{y0}, \Delta B_x, \Delta B_y$ are periodic functions of s. The equations of motion (2.3) in Chapter 2 are reproduced here,

$$\begin{aligned} x'' + \frac{x}{\rho^2} &= -\frac{e(B_y - B_{y0})}{P} \,, \\ y'' &= \frac{eB_x}{P} \,. \end{aligned} \tag{3.1}$$

As mentioned in Chapter 2, Eq. (3.1) has dropped all the nonlinear *kinematic* terms. Nonlinearities are assumed to be dominated by the nonlinear magnetic fields in ΔB_x and ΔB_y.

In this chapter, we consider only on-momentum particles. For on-momentum particles, $\frac{e}{P} = \frac{1}{B\rho}$ where $B\rho$ is the magnetic rigidity. The equations of motion become

$$\begin{aligned} x'' + K_x x &= -\frac{\Delta B_y}{B\rho} \,, \\ y'' + K_y y &= \frac{\Delta B_x}{B\rho} \,, \end{aligned} \tag{3.2}$$

where $K_x = \frac{1}{\rho^2} + \frac{G}{B\rho}, K_y = -\frac{G}{B\rho}$. We have also assumed there are no electric field devices in the accelerator so that momentum P, and therefore the magnetic rigidity $B\rho$, does not change.

One particular form of magnet errors is expressed in multipole expansion, Eq. (1.21),

$$\Delta B_y + i\Delta B_x = B_{y0} \sum_{m=0}^{\infty} (b_m + ia_m)(x + iy)^m \,,$$

where B_{y0} is some convenient reference field. We shall take it as the nominal design field of the bending magnet in the lattice design, even when considering field error of some magnet other than dipole magnets.

Table 3.1: Dipole and quadrupole field errors, their effects, and their sources.

	multipole coefficient	effects	sources
$m = 0$	a_0	vertical orbit distortion	• bending magnet roll • quadrupole vertical misalignment
	b_0	horizontal orbit distortion	• bending magnet field or length error • quadrupole horizontal misalignment
$m = 1$	a_1	linear x-y coupling	• quadrupole magnet roll • feeddown of sextupole with vertical orbit • sextupole vertical misalignment
	b_1	• tune shift • β-function beat	• quadrupole field or length error • feeddown of sextupole with horizontal orbit • sextupole horizontal misalignment

Different multipole errors produce different beam dynamical effects. Dipole ($m = 0$) errors produce orbit distortions. Quadrupole ($m = 1$) errors produce betatron tune shifts and distortions in the β-functions. See Table 3.1. For the dipole and quadrupole perturbations, exact treatments are possible. For higher multipoles ($m \geq 2$), we enter the regime of nonlinear dynamics. Exact treatment is in general not available. In this chapter, we shall first discuss the dipole and quadrupole perturbations. A brief discussion of nonlinear perturbations will also be included.

Take the dipole errors for example. The error fields are

$$\Delta B_x = B_{y0} a_0, \qquad \Delta B_y = B_{y0} b_0 .$$

The quantities a_0 and b_0 are the field multipole coefficients. For dipole errors, they are dimensionless. In a dipole magnet, they are the relative strengths of the error fields to the nominal bending field and are typically of the order of

10^{-4} for precision magnets. Sources of ΔB_y or b_0 come from errors in magnetic field strength or in the magnet length. Sources of ΔB_x or a_0 include a roll misalignment of the magnet's vertical axis.

For quadrupole errors,

$$\Delta B_x = B_{y0}(a_1 x + b_1 y), \qquad \Delta B_y = B_{y0}(b_1 x - a_1 y).$$

The a_1 terms correspond to a skew quadrupole, while the b_1 terms correspond to a quadrupole. Coefficients a_1, b_1 have dimensionality of m^{-1}. Sources of a_1 include a roll misalignment of the quadrupole magnet under consideration. Sources of b_1 include errors in the quadrupole field strength or in the magnet length.

Table 3.1 is incomplete in that it has not included the nonlinear ($m \geq 2$) perturbations. It has also not included one additional column on the right to discuss the cures for the adverse effects each perturbation causes. A wealth of correction schemes and algorithms have been invented to compensate for these effects under various conditions. Error compensation is an active area of accelerator physics research.

3.2 Orbit distortion

3.2.1 Single dipole kick

When there are dipole field errors in a storage ring, what happens is that the closed orbit of the stored beam will no longer coincide with the design trajectory. The deviation of the closed orbit from the design trajectory is referred to as a *closed orbit distortion* (COD). When $a_0 \neq 0$, there is a horizontal dipole field error, causing a vertical COD. When $b_0 \neq 0$, there is a vertical dipole field error, causing a horizontal COD.

When there is a COD, the transverse particle motion can be decomposed into a COD component and a betatron oscillation component as

$$x = x_{\mathrm{COD}} + x_{\mathrm{betatron}}, \tag{3.3}$$

and a corresponding expression for y. What we discussed in Chapter 2 addresses the betatron components. With dipole field errors, we have the additional contribution of the horizontal and vertical CODs.

Consider a dipole field error located at position $s = s_0$ in the storage ring. For simplicity, we represent this dipole perturbation by a thin-lens dipole with integrated field of $\Delta B\ell$. This dipole will produce an angular kick to the passing beam,

$$\theta = \begin{cases} -\frac{\Delta B_y \ell}{B\rho}, & \text{horizontal kick,} \\ \frac{\Delta B_x \ell}{B\rho}, & \text{vertical kick.} \end{cases} \tag{3.4}$$

All particles in the beam receive the same kick. In the rest of this section, assuming no x-y coupling, we will consider either the x- or the y-motion, and use the generic u-coordinate to designate either x or y.

The COD induced by this thin-lens dipole error will be a function of s along the storage ring, and will be periodic in s with period C. When thinking about the COD, we will not only consider the displacement u, but also the slope u', i.e. we have its phase space behavior in mind. To find this COD, we will first calculate the COD at the position of the error dipole. Let the distorted orbit be such that at the exit of the error dipole, we have a COD given by

$$\begin{bmatrix} u_0 \\ u_0' \end{bmatrix} .$$

Then at the entrance of the error dipole, we must have a COD given by

$$\begin{bmatrix} u_0 \\ u_0' - \theta \end{bmatrix} .$$

The COD displacement is continuous at s_0, but the COD slope has a kink by an amount θ.

In order for u_0 and u_0' to describe the COD, they must satisfy the condition

$$\begin{bmatrix} u_0 \\ u_0' - \theta \end{bmatrix} = M(s_0 + C|s_0) \begin{bmatrix} u_0 \\ u_0' \end{bmatrix} , \tag{3.5}$$

where $M(s_0 + C|s_0)$ is the one-turn map around the position of the error dipole.

Written in terms of the Courant-Snyder parameters at the position s_0, we already know that

$$M(s_0 + C|s_0) = \begin{bmatrix} \cos 2\pi\nu + \alpha_0 \sin 2\pi\nu & \beta_0 \sin 2\pi\nu \\ -\gamma_0 \sin 2\pi\nu & \cos 2\pi\nu - \alpha_0 \sin 2\pi\nu \end{bmatrix} ,$$

where $\alpha_0, \beta_0, \gamma_0$ at the kick position is continuous, i.e. they are the same before and after the kick. We then solve the COD condition and obtain

$$\begin{aligned} u_0 &= \frac{\beta_0 \theta}{2} \cot \pi\nu , \\ u_0' &= -\frac{\theta}{2}(\alpha_0 \cot \pi\nu - 1) \qquad \text{(COD slope at exit)} , \\ u_0' - \theta &= -\frac{\theta}{2}(\alpha_0 \cot \pi\nu + 1) \qquad \text{(COD slope at entrance)} . \end{aligned} \tag{3.6}$$

Once the COD at s_0 is obtained, the COD at other positions s can be found by the transfer map between s_0 and s,

$$\begin{bmatrix} u(s) \\ u'(s) \end{bmatrix} = M(s|s_0) \begin{bmatrix} u_0 \\ u_0' \end{bmatrix} .$$

Again in terms of Courant-Snyder parametrization,

$$M(s|s_0) = \begin{bmatrix} \sqrt{\frac{\beta}{\beta_0}}(\cos\Delta\psi + \alpha_0 \sin\Delta\psi) & \sqrt{\beta_0\beta}\sin\Delta\psi \\ \frac{\alpha_0-\alpha}{\sqrt{\beta_0\beta}}\cos\Delta\psi - \frac{1+\alpha_0\alpha}{\sqrt{\beta_0\beta}}\sin\Delta\psi & \sqrt{\frac{\beta_0}{\beta}}(\cos\Delta\psi - \alpha\sin\Delta\psi) \end{bmatrix},$$

where $\Delta\psi = \psi(s) - \psi(s_0)$. We thus obtain the formula for the COD (Homework 3.1),

$$u(s) = \theta\frac{\sqrt{\beta_0\beta(s)}}{2\sin\pi\nu}\cos(\pi\nu - |\psi(s) - \psi(s_0)|) . \tag{3.7}$$

For the accompanying $u'(s)$, see Homework 3.5.

Nominally the way we derived Eq. (3.7) assumes s is downstream of s_0, i.e. $\psi(s) > \psi(s_0)$. However, it is curiously interesting that, by taking absolute value on $\Delta\psi$ in this expression, one can extend it to the case when s is upstream of s_0. So that is what we did to the expression (3.7). In particular, one might use the expression with absolute sign to show that the kink in COD slope between $s = s_0^+$ and $s = s_0^-$ is indeed equal to θ. Note however that Eq. (3.7) is valid for one turn, either forward or backward, but not valid for multiple turns.

It should be mentioned here that Eq. (3.7) is not the complete story when a horizontal kick is applied to a stored beam. Whether due to an error field or an orbit corrector, Eq. (3.7) would be fine if there is no RF system, but it needs to be added another term if the accelerator has an RF system. This will be discussed in Sec. 4.5.3.

Closed orbits are measured by beam position monitors (Sec. 1.6). We make a side comment here on what a beam position monitor really measures. In actual operation of a storage ring, the COD refers to the center of charge $\langle x\rangle, \langle y\rangle$ of a *beam of particles.* What we have calculated above gives the COD of the motion of a *single particle.* One might therefore ask if that correctly gives the COD of a beam of particles as seen by the BPMs. The answer is that Eq. (3.7) gives the COD of the beam correctly if the accelerator optics is linear (i.e. the storage ring contains only dipoles and quadrupoles) because then all particles will have the same COD (although each particle still has its own betatron oscillation on top of the COD). Strictly speaking, when there are nonlinear elements such as sextupoles, or when the beam has an energy spread, different particles will have different CODs, and what the BPM measures will be different from Eq. (3.7). The difference depends on the emittance and energy spread of the beam, and is yet to be added to Eq. (3.7).

Homework 3.1

(a) Derive Eq. (3.6).

(b) Derive Eq. (3.7) including the use of the absolute value of $\psi(s) - \psi(s_0)$ for the COD upstream of the orbit kicker.

Solution

(b) A direct solution might miss the absolute value sign. That is because the derivation assumed the observation point s is downstream of the dipole error. The COD at an upstream point ($\psi(s) < \psi(s_0)$) can only be reached in the downstream direction, i.e. you should use the downstream phase $\psi(s) + 2\pi\nu$ in your analysis.

Homework 3.2 Consider a thin-lens quadrupole with focal length f being misaligned with a horizontal displacement Δx_0. The misaligned quadrupole contribute an angular kick $\theta = -\frac{\Delta x_0}{f}$ to the beam trajectory. To find the COD, we substitute this θ into expression (3.7). But, is this calculation correct? After we include the COD, the beam's position in the magnet is no longer Δx_0; it is now $\Delta x_0 + x_{\rm COD}$. But we assumed the position to be Δx_0 when we calculated the COD. So what should we do in this case? Why?

Solution This is an important homework, deserving some thinking.

3.2.2 COD and integer resonance

Observations on the COD Now let us inspect the COD expression (3.7) closely.

- The COD is periodic around the ring. It oscillates around $x = 0$ with the betatron oscillation except at the kick position. If the COD is generated by a single dipole kick, the COD will have a prominent oscillatory look, and the total number of oscillations will be an integer very close to the betatron tune. This same behavior will be preserved for the case of multiple dipole kicks. But will be lost when COD corrections (Sec. 3.2.6) are applied.
- The COD has a slope $u_0' \neq 0$ at s_0. This is of course expected because, after all, we gave a kick angle to the particle.
- But one notes that the COD displacement u_0 is also $\neq 0$. The beam has acquired an orbit displacement although the error dipole did not ask for it. Furthermore, the displacement has to be of a specific value given by u_0. The reason of this is that after the kick, the beam will have to adjust its COD to satisfy a self-consistency condition (3.5) to close the one-turn periodicity. Had we suddenly turned on the dipole kick, then the beam will have to make large transients in its subsequent motion after the kick conforming to Eq. (3.3), and in that case, the COD gives only the *average* of beam motion. The beam will quiet down to follow the COD only after this transient betatron motion is damped out, e.g. by a feedback system or some other damping mechanisms. But after it is damped down, it necessarily follows the COD.
- The COD is proportional to θ. This is expected.

- We also observe that it is proportional to $\frac{1}{\sin \pi\nu}$. This means the COD diverges when ν approaches an integer.

Integer resonance We will continue to make more observations on the COD formula (3.7), but here let us make a detour to discuss the integer divergence. The last observation is of particular significance to us. Here we are encountering for the first time a phenomenon called *resonances* in a storage ring. Near a resonance, perturbations due to small errors accumulate turn after turn systematically to build up into a large perturbation to particle motion and potentially render the particle motion unstable.

Note the notion of slow, systematic, turn-by-turn build-ups; it is the essence of all resonance mechanisms. Because of its nature of building up turn-by-turn, you expect a general nature of all resonances, namely their effects grow linearly in time, in contrast to other instability effects. We will return to this comment momentarily in Fig. 3.3.

The resonance of COD occurs when ν is close to an integer. To visualize the resonance effect, let us start with the COD in the (u, u') phase space away from resonance. In the phase space, a COD is represented by a closed contour. As a particle circulates around the accelerator, its representative point in phase space traces out a closed contour as shown in Fig. 3.1. The particle that has just received the dipole kick would start at position A in phase space; it then traverses along the Courant-Snyder ellipse for multiple number of turns (the number of turns is determined by ν) and ends up at position B right before it reaches the kick dipole again. At the dipole, it then receives a kick θ in the u' direction, and as a closed orbit, the particle will return exactly to position A. The only way to make this possible is to choose the positions A and B according to Eq. (3.6).

Figure 3.1: The COD of a particle shown in phase space. The phase space ellipse is shown as being upright for simplicity. Recall: clockwise rotation.

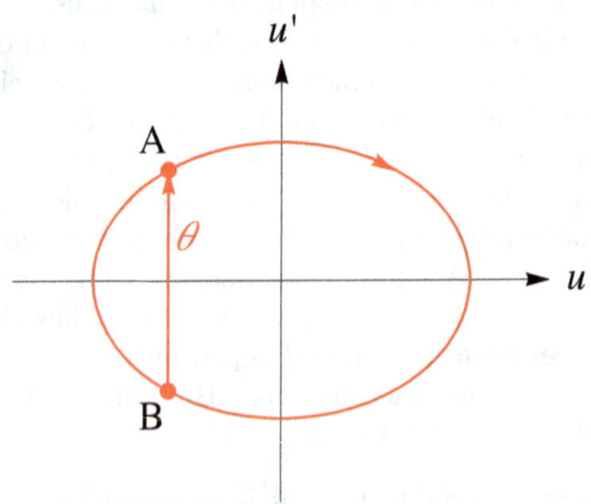

As the tune approaches an integer, the COD becomes larger. The reason is illustrated in Fig. 3.2. The two ellipses have the same shape because they are determined by the same set of Courant-Snyder functions. But they have different sizes. The point is that the two vertical arrows are the same for a given kicking angle θ, while the trajectory in the rest of the accelerator will have to trace out a bigger ellipse if tune ν gets closer to an integer.

Figure 3.2: Illustration of integer resonance of COD in phase space.

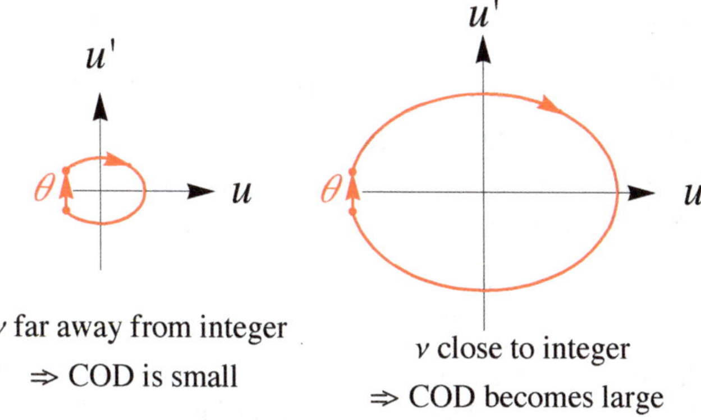

When ν becomes exactly an integer, a COD can no longer be found. Figure 3.3 shows the mechanism of the COD resonance when $\nu =$ integer.

Figure 3.3: Orbit distortion when ν is exactly an integer. No closed orbit can be found.

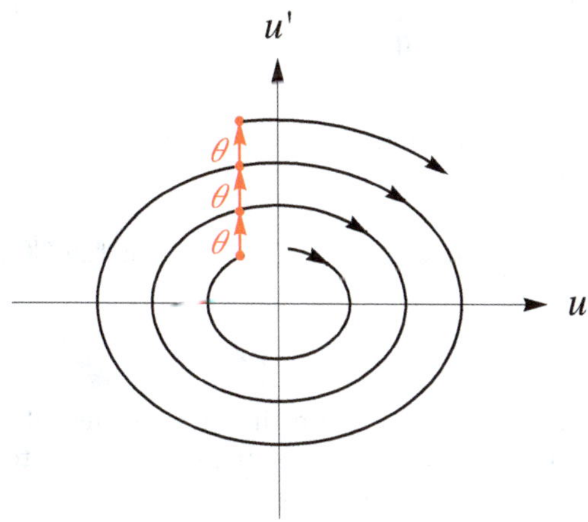

At the resonance, the particle returns to the same phase space point when completing one revolution (points A and B coincide). The little red vertical arrows due to the error dipole kicks, each of magnitude θ, pile up turn after turn, accumulating to a resonance.

Let us make a few remarks specifically for $\nu =$ integer.

- As just mentioned, the particle's amplitude increases *linearly* in time. This linear growth is a characteristic of resonances for a linear system. In contrast, an antidamping effect or a collective instability effect in a linear system would have an *exponential* growth. This is a fundamental difference between a resonance instability and a "true" instability.

- In general, there are three types of instabilities in physical mechanisms,

    1. Diffusion instability $\Longrightarrow$ the growth is $\propto \sqrt{t}$;
    2. Resonance instability $\Longrightarrow$ the growth is $\propto t$;
    3. "True" instability $\Longrightarrow$ the growth is $\propto e^t$.

    The first two are typically external driven, e.g. due to noise or resonance driving. The third is often self-driven, e.g. in a collective instability.

- It might also be interesting to note that at the location of the dipole error, it is the orbital slope that keeps on increasing linearly with time. The orbital displacement at the kicker keeps its initial value turn after turn, as shown in Fig. 3.3.

- What happen to the small red arrows when ν is away from an integer is that they will occur at different locations in phase space, slipping by some angle each turn. Although the arrows are always pointing vertically (upward if $\theta > 0$), they can sometimes increase the beam oscillation amplitude and sometimes decrease it. The net result is that the oscillation amplitude does not go to infinity. Instead, it will make a transient oscillation around the COD.

More observations on the COD We now return to more observations on the COD formula (3.7).

- Away from the integer resonance, one sees that the COD has an order of magnitude

$$u \sim \frac{\theta}{2}\sqrt{\beta_0 \beta(s)}\,. \tag{3.8}$$

  For a rough order of magnitude estimation, let us take $\beta \sim$ spacing between quadrupole magnets ~ 10 m. If we would like to have $u \lesssim 1$ mm, and assuming the COD comes solely from one dipole error, then the dipole error must be such that $\theta \lesssim 2 \times 10^{-4}$ rad.

- One also sees that the COD is proportional to the β-function. It is a general rule that large β-functions make the beam dynamics sensitive to errors, i.e. the β-function at position s serves as a magnification factor of the beam dynamics effect due to a perturbation at the same position.

  When possible, it is generally better to minimize the β-functions, and the way to do so is to tighten up the focusing of the lattice design. We learned before that in a long straight section with a tight beam waist the β-function will become large at the two ends of the straight section. Indeed, magnet errors at these end locations with large β-functions will have particularly strong effect on particle motion and must be handled with extra care. This situation occurs for example around the interaction region in a collider storage ring.

Multiturn closed orbit The closed orbit discussed so far has the imposed condition that it closes onto itself after completion of one revolution. The COD, however, is not the only circumstance when the trajectory of a particle motion repeats itself.

Consider the free betatron oscillation of a particle in a perfect accelerator without closed orbit distortion. Consider the case when the betatron tune is a fractional number, say $\nu = \frac{n}{m}$ with integral n and m. It is clear that the betatron oscillation will repeat itself every m revolutions. Presented in phase space as observed at a fixed location in the accelerator, the particle trajectory consists of m distinct points distributed around a Courant-Snyder ellipse. The trajectory simply hops around these m points and return to the original position after m turns. This may be considered a case of an "m-turn closed orbit".

A difference between this multiturn closed orbit with the single-turn COD is that this closed orbit results not from field errors but simply from the fractional tune. Another difference is that this multiturn orbit is not unique; any particle with any initial value (x_0, x_0') will generate a new multiturn closed orbit.

In case the accelerator has a dipole error kick, thus a nonzero COD, and the betatron tune is fractional, then the net result is that the trajectory of a particle can be decomposed into a simple superposition of the COD and the multiturn closed orbit.

Figures 3.4 shows an example when $\nu = \frac{1}{3}$ and with a dipole error field. The particle trajectory is observed continuously, so the trajectory fills out continuous curves. If observation is made at a fixed location in the accelerator, then only three distinct dots will be seen. The panel on the left shows the COD and the betatron oscillation separately. The COD is uniquely determined once the perturbation kick is given, while the betatron circle is for a particle with an arbitrary amplitude. The right panel shows the combined COD and betatron oscillation. For completeness, we also show another example when $\nu = \frac{2}{3}$ in Fig. 3.5.

The multiturn closed orbits play a role when nonlinearities are included, and in fact are the root cause of nonlinear resonance effects. In particular, they

Figure 3.4: Phase space plot of a 3-turn COD when $\nu = \frac{1}{3}$.

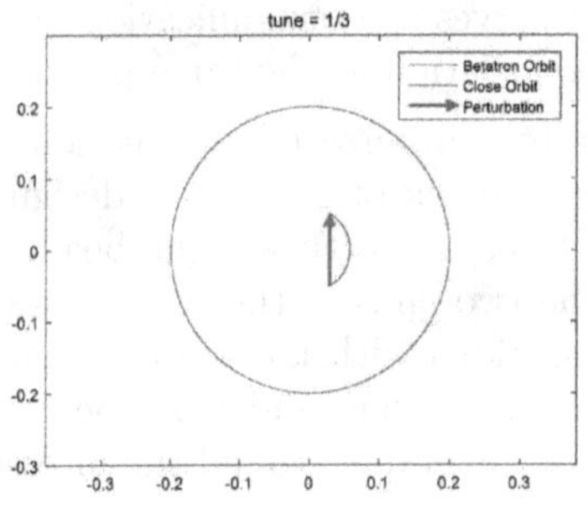

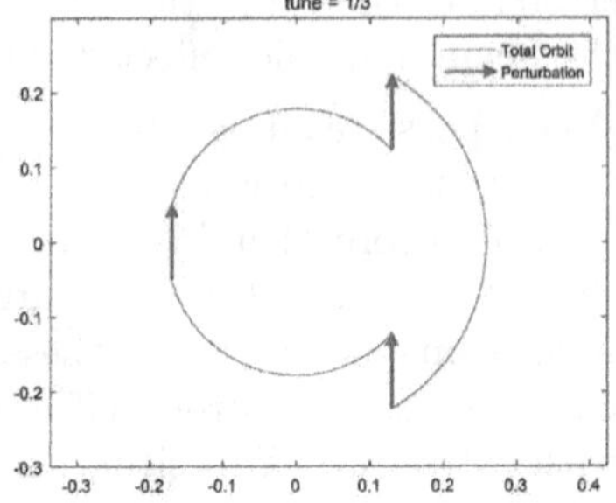

Figure 3.5: Phase space plot of a 3-turn COD when $\nu = \frac{2}{3}$.

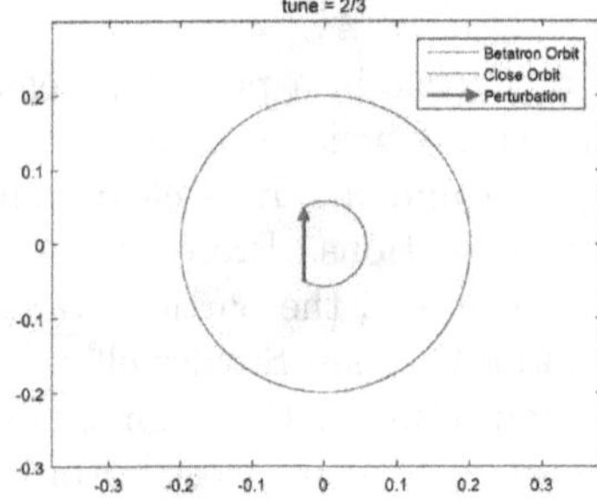

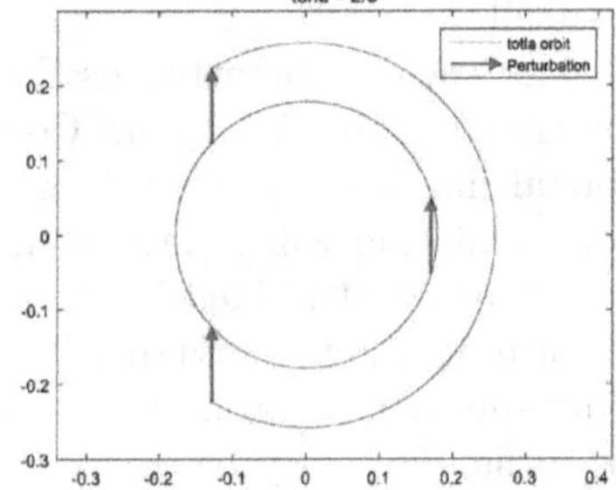

generate a possibility of an island structure with particles hopping around among islands in the phase space. We will return to this subject later in Sec. 3.6.2.

Homework 3.3 In the text, we showed a plot of phase space trajectory of a COD when the tune ν approaches an integer, thus illustrating the mechanism of integer resonances.

(a) What happens to the COD phase space plot if $\nu = \frac{1}{2}$?

(b) Try to derive an expression of the COD when $\nu = \frac{1}{2}$ geometrically using your plot obtained in (a). Compare your result with the expression if you directly apply Eq. (3.7).

Solution (a) See Fig. 3.6.

Homework 3.4 A theorem was proven in Courant and Snyder's classic paper that a COD always exists as long as all magnetic fields that influence the particle motion are bounded in space. On the other hand, we showed that when $\nu =$ integer, the COD extends to infinity. Is the theorem violated here?

Figure 3.6: Phase space plot of the COD when $\nu = \frac{1}{2}$.

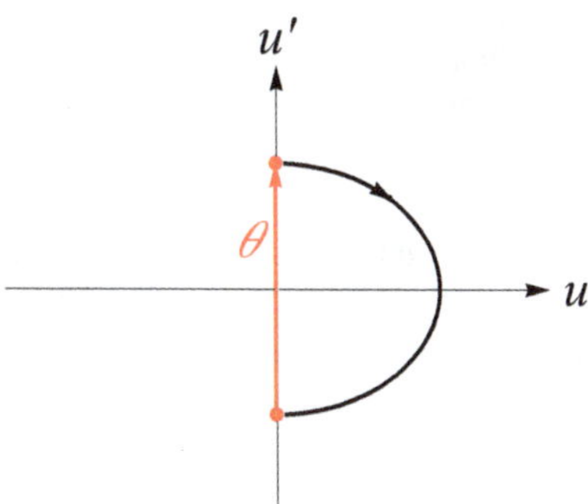

Solution The theorem is not violated. When ν = integer, COD extends to infinity; but that assumes that all magnets of the ring (not only the error dipole, but also all the quadrupoles in the ring) have fields extending transversely to infinity. The condition that the fields are bounded is therefore violated. The theorem is OK.

Homework 3.5

(a) Sometimes one needs an expression of the slope of the COD. Derive an expression of $u'(s)$ using Eq. (3.7).

(b) Show that the expression you obtained for u' does give a kicking angle of θ at position $s = s_0$.

(c) What is the order of magnitude of u' away from resonances? Do this to pair up with Eq. (3.8).

Solution

(a)

$$u'(s) = -\frac{\theta}{2\sin\pi\nu}\sqrt{\frac{\beta_0}{\beta}}\left[\alpha\cos(\pi\nu - |\Delta\psi|) - \text{sgn}(\Delta\psi)\sin(\pi\nu - |\Delta\psi|)\right],$$

where $\Delta\psi = \psi(s) - \psi(s_0)$.

(c) You should have expected $\gamma(s)$ to play the role in u' similar to the role $\beta(s)$ plays in u.

Homework 3.6 Consider the case of a single error dipole as in the text. We now consider adding a feedback system, whose net function is to introduce a change in the particle's displacement proportionally to the instantaneous value of the displacement, i.e. it provides an additional kick $\Delta x = -\alpha x$, where α is a positive quantity depending on the strength of the feedback. For simplicity, consider the feedback action be applied at the entrance of the dipole error. The feedback system aims to damp dynamical motions of the beam on turn-by-turn basis, but here it acts as a static device because the motion of the beam along the COD is static.

(a) Calculate the COD (u_0, u_0') immediately after the error dipole.

(b) Does the application of a feedback avoid the integer divergence of the COD?

Homework 3.7 Consider a linear accelerator. Let the accelerator be perfectly aligned using a liquid-level alignment system, so that it is not really linear but follows perfectly the earth curved surface. Let the accelerator contain quadrupoles along its path in such a way that the focusing effect can be approximated as being uniform, with a constant β-function of β_0.

(a) First switch off the quadrupoles so that there is no focusing in this accelerator. Launch a beam perfectly at the injection end and let the beam go in a straight line towards the end of the linac. Let the earth radius be R_e. What is the beam misalignment relative to the accelerator axis at the end of linac?

(b) Repeat the calculation now with the quadrupoles turned on. Discuss your result. Appreciate the fact that the beam is *guided* by the quadrupoles to follow the curved earth surface and this is done automatically without using dipole magnets.

(c) Ignore the earth's curvature. Now calculate the COD at the end of the linac with and without quadrupoles when the earth's gravity is included. The effect is expected to be small. But how small? For a linear collider, after all, the beam size at the end of the linac is only nm size.

(d) Ignore both the earth's curvature and gravity, repeat the question for the case considering the earth's magnetic field. Take the field to be 0.5 Gauss pointing in the south direction and let the linac to go from west to east (like the SLAC linac).

Homework 3.8 Homework 3.7(b) also triggers the consideration of this homework [Courtesy Gennady Stupakov].

Consider a peculiar storage ring made purely of quadrupoles without any bending magnets. The ring consists of N_{cell} thin-lens FODO cells whose quadrupoles are aligned uniformly along a circle. A case with $N_{\text{cell}} = 12$ is illustrated in Fig. 3.7. A motivation of this problem is to answer the following question: What if one builds a linear accelerator (such as in a linear collider), which contains no dipoles, and is forced to follow the curvature of the earth? This example pretends this linac makes one complete turn around the earth.

Assume the focal lengths of QF and QD be f_1 and $-f_2$ respectively. Find the COD (in this case, the COD is the design orbit relative to the perfect circle) of this storage ring. Use the numerical parameters: ring circumference = 40,000 km, FODO cell length = 30 m, and phase advance per cell = 100°.

Solution The COD has only a horizontal component in the storage ring plane. Let the COD at the entrance to QF be $\begin{bmatrix} x_0 \\ x_0' \end{bmatrix}$. Relative to the entrance coordi-

Figure 3.7: A storage ring consisting only of quadrupoles and no bending magnets.

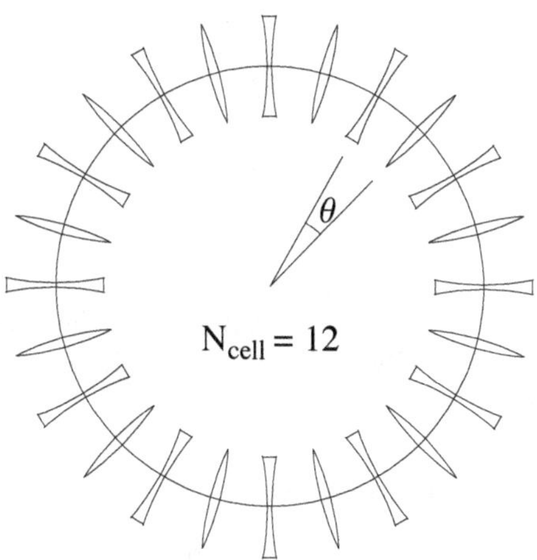

nate system, QF is slanted by an angle θ. First consider the case $f_1 = f_2$. To first order in θ, the COD at the exit of QF, again relative to the local coordinate system is

$$\begin{bmatrix} x_0 \\ -\frac{x_0}{f} + x_0' + \theta \end{bmatrix} .$$

This map can be written in terms of a 3×3 matrix

$$\begin{bmatrix} 1 & 0 & 0 \\ -\frac{1}{f} & 1 & \theta \\ 0 & 0 & 1 \end{bmatrix} .$$

Similarly we find the 3×3 matrices for the drift spaces and the QD. The self-consistency condition for the COD then requires

$$\begin{bmatrix} 1 & L & 0 \\ 0 & 1 & 0 \\ 0 & 0 & 1 \end{bmatrix} \begin{bmatrix} 1 & 0 & 0 \\ \frac{1}{f} & 1 & \theta \\ 0 & 0 & 1 \end{bmatrix} \begin{bmatrix} 1 & L & 0 \\ 0 & 1 & 0 \\ 0 & 0 & 1 \end{bmatrix} \begin{bmatrix} 1 & 0 & 0 \\ -\frac{1}{f} & 1 & \theta \\ 0 & 0 & 1 \end{bmatrix} \begin{bmatrix} x_0 \\ x_0' \\ 1 \end{bmatrix} = \begin{bmatrix} x_0 \\ x_0' \\ 1 \end{bmatrix} ,$$

which can be solved to yield

$$\begin{bmatrix} x_0 \\ x_0' \end{bmatrix}_{\mathrm{QF}} = \frac{f\theta}{L} \begin{bmatrix} 4f + L \\ 2 \end{bmatrix} .$$

Similarly one can calculate the COD at other locations around the ring.

A few comments:

- Extending to the case $f_1 \neq f_2$ gives

$$x_0 = \theta \begin{cases} \frac{f_1(L+4f_2)}{L-2f_1+2f_2}, & \text{at QF} \\ -\frac{f_2(L-4f_1)}{L-2f_1+2f_2}, & \text{at QD} \end{cases} .$$

- The net bending over one cell is $-\frac{x_F}{f_1} + \frac{x_D}{f_2} = -2\theta$ as it should.
- If the linac follows the earth's curvature, but is only 10 km long, what is the COD at the quadrupoles? (Even though the orbit does not actually close so it does not justify the name COD, it assumes the meaning of a design trajectory.)

3.2.3 Distributed dipole error

So far we considered the case with one single thin-lens dipole field error. In reality, dipole field errors are distributed and can be described as $\Delta B(s)$ as a function of s. The COD due to a distributed dipole error is obtained by superposition,

$$u(s) = \int_s^{s+L} ds' \, \frac{\Delta B(s')}{B\rho} \frac{\sqrt{\beta(s)\beta(s')}}{2\sin\pi\nu} \cos(\pi\nu - |\psi(s') - \psi(s)|) . \qquad (3.9)$$

In this expression, $\Delta B = -\Delta B_y$ for x-orbit and $\Delta B = \Delta B_x$ for y-orbit, as similarly seen in Eq. (3.4).

In case there are N thin-lens error dipoles, the resulting COD is the linear sum of contributions from each error dipole. If these error dipoles are located randomly in their betatron phases, it can be shown that a statistical rms value of the COD is given by

$$\frac{\langle u^2(s)\rangle}{\beta(s)} \sim \frac{N\langle\beta_i\theta_i^2\rangle}{8\sin^2\pi\nu} , \qquad (3.10)$$

where $\langle\beta_i\theta_i^2\rangle = \frac{1}{N}\sum_{i=1}^N \beta_i\theta_i^2$. The magnitude of COD therefore scales like $\sqrt{N}$. For large storage rings with many magnet elements, and therefore many sources of error dipoles, the COD becomes large. For example, if we take $\langle\beta_i\theta_i^2\rangle = 10^{-6}$ m, $N = 500$, and $\sin\pi\nu = 1$, then the expected rms COD at a position with $\beta = 20$ m would be 3.5 cm. For those cases, one must install correctors to control this COD.

Orbit correctors are small dipole magnets, each individually wired to have its own field strength. There are two sets of such orbit correctors, one for x- and one for y-COD. There are a number of algorithms to correct the COD's effectively. In the following, we shall mention only two of the more commonly used schemes, a closed orbit bump scheme (Sec. 3.2.4) and an rms correction scheme (Sec. 3.2.6).

We should mention here that orbit correction is only a special case of error corrections. Error corrections are an important activity in accelerator design and accelerator physics. In addition to the corrections of COD, there are corrections of the betatron functions and tunes, chromatic and coupling effects, and beyond these more straightforward applications, one enters more advanced research areas such as *beam-based alignment* techniques, *singular value decomposition* techniques, *model-independent analysis*, *genetic algorithms*, turn-by-turn beam diagnostics, and machine learning. Discussion of these topics in detail is beyond our scope here.[1]

Homework 3.9 Derive the statistical COD expression Eq. (3.10).

Homework 3.10 The text gave a COD expression (3.7) for a thin-lens dipole kick. Try to extend the expression to the case of a single thick dipole error.

3.2.4 Closed orbit bump

A closed orbit bump is an arrangement of COD correctors and their strengths in such a way that the COD they generate strictly vanishes outside the region they occupy, thus forming a localized COD "bump".

A most common COD bump consists of three correctors. In general, the COD caused by a 3-bump system is given by

$$u(s) = \frac{\sqrt{\beta(s)}}{2\sin\pi\nu}\sum_{i=1}^{3}\theta_i\sqrt{\beta_i}\cos(\pi\nu - |\psi(s)-\psi_i|)\,.$$

A perfect orbit bump is created if the COD is zero outside the region of the three correctors. This can be achieved by demanding that

$$u(s_3) = 0 \qquad \text{and} \qquad u'(s_3) = 0\,,$$

which yields

$$\begin{aligned}
\theta_1\sqrt{\beta_1}\cos(\pi\nu-\psi_3+\psi_1)+\theta_2\sqrt{\beta_2}\cos(\pi\nu-\psi_3+\psi_2)+\theta_3\sqrt{\beta_3}\cos\pi\nu &= 0\,,\\
\theta_1\sqrt{\beta_1}\sin(\pi\nu-\psi_3+\psi_1)+\theta_2\sqrt{\beta_2}\sin(\pi\nu-\psi_3+\psi_2)+\theta_3\sqrt{\beta_3}\sin\pi\nu &= 0\,.
\end{aligned}$$

The solution is

$$\begin{aligned}
\theta_2 &= -\theta_1\sqrt{\frac{\beta_1}{\beta_2}\frac{\sin(\psi_3-\psi_1)}{\sin(\psi_3-\psi_2)}}\,,\\
\theta_3 &= \theta_1\sqrt{\frac{\beta_1}{\beta_3}\frac{\sin(\psi_2-\psi_1)}{\sin(\psi_3-\psi_2)}}\,.
\end{aligned} \tag{3.11}$$

[1]W.H. Press et al., Numerical Recipes, Cambridge Univ. Press (1988); B. Autin and Y. Marti, CERN report ISR MA/73-17 (1973); D. Rice, Sec. 4.7.2, J. Wenninger, Sec. 4.7.3, D. Rubin, Sec. 4.7.4, J. Safranek, Sec. 4.7.5, R. Jones and R. Steinhagen, Sec. 4.7.7, J. Irwin, Sec. 4.8.1, Handbook Accel. Phys. & Eng., 2nd Ed., World Scientific (2013).

This solution gives θ_2 and θ_3 in terms of θ_1 for a perfect 3-corrector COD bump. The magnitude of θ_1 is to be determined by how large the COD is to be inside the bump.

It follows from Eq. (3.11) that in general it is preferred to locate the orbit correctors at positions with high β-functions. This in turn leads to the preference to locate the horizontal correctors near QFs and the vertical correctors near the QDs.

Homework 3.11 Derive the condition for a 3-bump orbit correction, Eq. (3.11). Under what condition can there be a 2-corrector bump? Find the expression of the 2-bump COD between the two correctors.

Solution $\Delta\psi = \pi, \theta_1\sqrt{\beta_1} = \theta_2\sqrt{\beta_2}$.

Homework 3.12 One may consider a 4-corrector bump as well. Show that the 4-bump is specified by

$$\begin{aligned}
\theta_3 &= -\frac{\theta_1\sqrt{\beta_1}\sin(\psi_4-\psi_1)+\theta_2\sqrt{\beta_2}\sin(\psi_4-\psi_2)}{\sqrt{\beta_3}\,\sin(\psi_4-\psi_3)}, \\
\theta_4 &= \frac{\theta_1\sqrt{\beta_1}\sin(\psi_3-\psi_1)+\theta_2\sqrt{\beta_2}\sin(\psi_3-\psi_2)}{\sqrt{\beta_4}\,\sin(\psi_4-\psi_3)}.
\end{aligned}$$

The two free parameters θ_1 and θ_2 will allow this COD not only to reach a specific needed orbit displacement but also a needed orbit slope at some desired location within the bump region, for example, for the application of an injection beamline. In particular, such a versatile injection beamline may be needed in a *painting scheme* to inject H^- ions into a proton synchrotron.

3.2.5 Orbit correction in a transport beamline

Consider a transport beamline (not necessarily in a storage ring or synchrotron) that consists of a periodic array of cells. Let the beam position monitors and the orbit correctors be located with the same period as the cells and let the BPMs and the correctors interlace each other as shown in Fig. 3.8. Let us consider only one of the transverse dimensions, say the x-dimension.

The arrangement, although looks reasonable from first sight and just might be the first thing a designer would try, may not work. To see this, consider the n-th cell in the transport line. Let errors in this cell be such that they produce an orbit displacement e_n and angle e'_n at the end of the cell where the n-th BPM is located. The orbit at the entrance to the n-th cell has been corrected so that it has a zero displacement but has an angle of x'_{n-1} because the orbit slope is not corrected at the BPMs.

Let T be the one-cell transport matrix from BPM to BPM. Let the kick angle of the n-th corrector be θ_n. Let the displacement and angle at the BPM

Figure 3.8: A possible trajectory correction scheme for a transport line consisting of a periodic array of cells. Stability of its trajectory correction function needs to be analyzed.

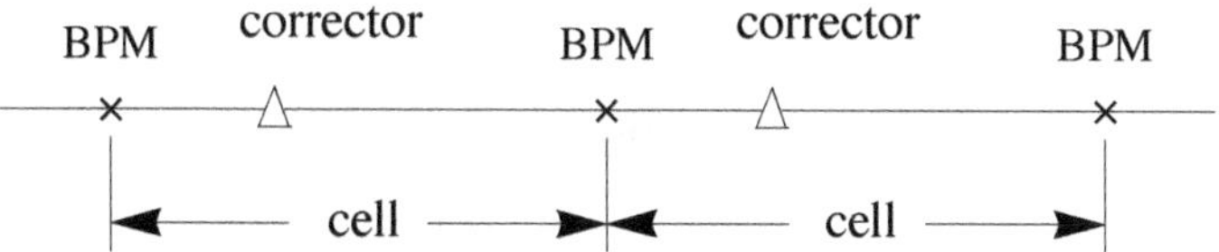

at the end of the n-th cell due to the n-th corrector be given by $\theta_n \begin{bmatrix} \eta \\ \eta' \end{bmatrix}$. We then have

$$\begin{bmatrix} x_n \\ x'_n \end{bmatrix} = T \begin{bmatrix} 0 \\ x'_{n-1} \end{bmatrix} + \theta_n \begin{bmatrix} \eta \\ \eta' \end{bmatrix} + \begin{bmatrix} e_n \\ e'_n \end{bmatrix}.$$

The corrector strength θ_n is chosen to make $x_n = 0$. This requires

$$\theta_n = -\frac{T_{12}x'_{n-1} + e_n}{\eta}.$$

Substituting into the expression for x'_n then gives

$$x'_n = \left(T_{22} - \frac{\eta'}{\eta}T_{12}\right)x'_{n-1} + \left(e'_n - \frac{\eta'}{\eta}e_n\right).$$

The angular divergence of the COD at the BPMs after correction contains two terms. The first term comes from the propagation of the residual angular divergence upstream of the cell under consideration. The second term is the residual noise contribution from errors in the cell. The first term is multiplied by a factor $A = T_{22} - \frac{\eta'}{\eta}T_{12}$. It is clear that this orbit correction scheme is stable only when $|A| < 1$. The COD grows exponentially if $|A| > 1$. This COD instability is sketched in Fig. 3.9.

In terms of Courant-Snyder parameters, we may write $\eta = \sqrt{\beta_c\beta_m}\sin\psi, \eta' = \sqrt{\frac{\beta_c}{\beta_m}}(\cos\psi - \alpha_m\sin\psi)$, where the subscripts c and m refer to corrector and PBM respectively, ψ is the phase advance from corrector to its downstream BPM. In addition, we also have $T_{12} = \beta_m\sin\psi_{cell}, T_{22} = \cos\psi_{cell} - \alpha_m\sin\psi_{cell}$. It then follows that

$$A = -\frac{\sin(\psi_{cell} - \psi)}{\sin\psi}.$$

The stability criterion $|A| < 1$ does not depend on the β-function at the correctors or monitors. A strong damping of COD propagation occurs when $|A| \ll 1$, which can be achieved if the phase advance from monitor to corrector $\psi_{cell} - \psi$ is zero or close to a multiple of π. An optimum arrangement is to

Figure 3.9: An instability of trajectory correction in a transport line. The dotted curve indicates the beam trajectory after correction.

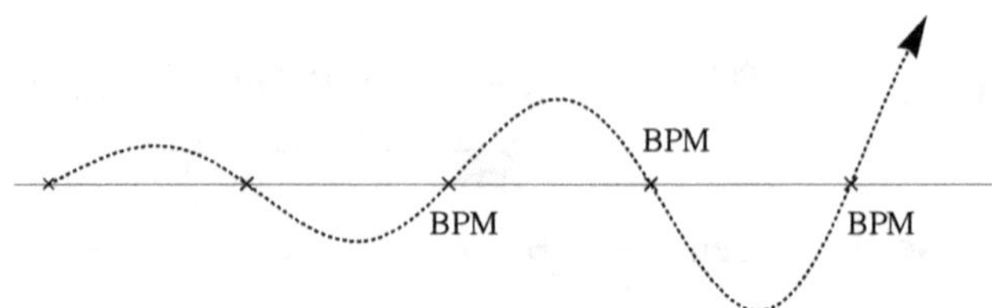

have 90° cells and to have the correctors immediately after the monitors, which is not a surprising result. In contrast, if we simply switch the positions so that the corrector becomes immediately before the monitor, we would have a worst possible unstable design.

This example indicates that COD correction may not be as straightforward or mundane as one might think. COD correction schemes can be effective or ineffective, and there can even be schemes that do more harm than good if not designed properly.

Homework 3.13 Consider a long transport line as discussed in the text. Assuming the error contributions in each cell is uncorrelated to other cells, one may define an rms value of the error parameter σ_e by $\sigma_e^2 = \langle (e_n' - \frac{\eta'}{\eta} e_n)^2 \rangle$. Show that, for the case of stable system ($|A| < 1$), the statistical rms COD slope at the BPMs is given by

$$\sigma_x' = \frac{\sigma_e}{\sqrt{1 - A^2}} .$$

Homework 3.14 The text gave a design example that can do more harm than good for a long transport beamline. If a storage ring is designed to contain such a long section of cells, what do you think will happen? What role does the parameter A play in this situation? Should we try to design the lattice and the correction scheme so that $|A| \ll 1$ in a storage ring?

3.2.6 RMS orbit correction

We now return to the discussion of COD correction in storage rings. We mentioned an orbit bump scheme earlier. Another approach, to be discussed now, is based on minimizing the rms of the COD. This is called an rms orbit correction scheme. In contrast to the orbit bump scheme, this scheme corrects the orbit globally instead of locally.

Response matrix Consider a circular accelerator with M BPMs and N orbit correctors around its circumference. To correct the COD, we want to power the

N correctors to make the orbit readings at the M monitors as small as possible. If the design has $N = M$, we can power the M correctors to make the M monitors all read zero, i.e. a "perfect" orbit correction — even though the orbit distortion between the BPMs can still be nonzero, which we don't have a way to observe and have no way to control. This means there is no gain to be made by having $N > M$. As a result, we only consider the case with $N \leq M$.

Let us define a *response matrix* R, whose ij-element is[2]

$$R_{ij} = \frac{\sqrt{\beta_i \beta_j}}{2 \sin \pi\nu} \cos(\pi\nu - |\psi_j - \psi_i|)\,.$$

The COD contribution due to the i-th corrector at the j-th monitor is $\Delta x_j = R_{ij}\theta_i$. The matrix R is determined by the accelerator lattice design. With $N \leq M$, the matrix R is taller than wide. In a matrix form, we write

$$\Delta X = RC\,,$$

where

$$\Delta X = \begin{bmatrix} \Delta x_1 \\ \Delta x_2 \\ \cdot \\ \cdot \\ \Delta x_M \end{bmatrix}, \qquad C = \begin{bmatrix} \theta_1 \\ \theta_2 \\ \cdot \\ \cdot \\ \theta_N \end{bmatrix}.$$

RMS correction Now let

$$\Delta X_0 = \begin{bmatrix} \Delta x_{01} \\ \Delta x_{02} \\ \cdot \\ \cdot \\ \Delta x_{0M} \end{bmatrix}$$

be the COD to be corrected. After powering the correctors to make the COD correction, the COD at the i-th monitor will have a reading $\Delta x_{i0} + \sum_{j=1}^{N} R_{ij}\theta_j$.

Define the error parameter after COD correction,

$$\epsilon \equiv \sum_{i=1}^{M} \left(\Delta x_{i0} + \sum_{j=1}^{N} R_{ij}\theta_j \right)^2 .$$

Physically,

$$\epsilon = M\sigma^2\,,$$

where σ is the rms COD at all the BPMs after correction. To minimize the COD in the rms correction scheme, we demand a minimal possible value of ϵ.

[2]Be reminded that this formula applies whether ψ_j is upstream or downstream of ψ_i, but does not apply if they separate by more than one whole revolution.

This requires

$$\frac{\partial \epsilon}{\partial \theta_k} = 2\sum_{i=1}^{M} R_{ik}\left(\Delta x_{i0} + \sum_{j=1}^{N} R_{ij}\theta_j\right) = 0, \quad \text{for all } k = 1, 2, ..., N\,.$$

It follows that

$$\begin{aligned} \sum_{i=1}^{M} \Delta x_{0i} R_{ik} &= -\sum_{i=1}^{M}\sum_{j=1}^{N} R_{ij}\theta_j R_{ik} \\ \Longrightarrow \quad \tilde{R}\Delta X_0 &= -(\tilde{R}R)C \\ \Longrightarrow \quad C &= -(\tilde{R}R)^{-1}\tilde{R}\,\Delta X_0\,. \end{aligned} \tag{3.12}$$

Equation (3.12) is an explicit expression of the required corrector settings to minimize the rms COD as requested.

Note that although R is not a square matrix, the matrix

$$K = \tilde{R}R$$

is square, so it possesses an inverse as required by the formula (3.12). Matrix K is symmetric.

The COD after correction is

$$\Delta X_1 = (1 - RK^{-1}\tilde{R})\Delta X_0\,.$$

The rms COD after correction, having its minimum value by the optimum choice of C, is given by, after a simple algebra,

$$\epsilon = M\sigma^2 = \tilde{\Delta X_1}\Delta X_1 = \tilde{\Delta X_0}(1 - RK^{-1}\tilde{R})\Delta X_0\,.$$

When $M = N$, the number of correctors is equal to the number of BPMs. In that special case, R is a square matrix (so that we can define the matrix R^{-1}), and we have $K = R^{-1}(\tilde{R})^{-1}$. It follows that the optimum choice of C is

$$C = -(\tilde{R}R)^{-1}\tilde{R}\Delta X_0 = -R^{-1}\Delta X_0\,,$$

and that the final rms COD after correction is $\epsilon = 0$. This is expected as the COD at all BPMs are corrected to zero.

Weight function In some cases, we may want to emphasize some of the BPMs because the COD at those locations are of more importance and need to be controlled tighter than the other locations. In those cases, we may want to introduce weighting functions $w_i, i = 1, 2, \cdots M$, where $w_i \geq 0$. The rms distortion is then defined as

$$\epsilon = \sum_{i=1}^{M} w_i(\Delta X_0 + RC)_i^2\,.$$

Following the same procedure as before, the optimal corrector setting can be found to be (Homework 3.15)

$$C = -(\tilde{R}WR)^{-1}\tilde{R}W\Delta X_0 , \tag{3.13}$$

where W is a diagonal matrix whose diagonal elements are $w_1, w_2, \cdots w_M$. Normalization of the weighting functions is not important. Only the relative weighting has an effect on the outcome.

The rms method discussed here is not the only global COD correction scheme available. In this scheme, all M correctors are applied to make the COD correction. However, not all correctors are equal. Some of them are more effective than others in the effort to minimize ϵ and which ones are more effective depends not only on the response matrix R but also the initial COD ΔX_0 to be corrected. It may be very useful if we can order the correctors according to their effectiveness, and after doing so, we have the freedom to correct only using a few correctors instead of invoking all of them obligatorily. The technique for this purpose is called *singular value decomposition.* SVD is a powerful technique, but we shall not discuss it considering it outside of the present scope. Suffice it to mention that, as one might expect, if we choose to invoke all correctors to do the COD correction using a SVD scheme, the result will be the same as the rms scheme.

It also should be mentioned that these correction schemes do not apply only to COD corrections. Corrections of any other error effects can be applied similarly as long as there is a linear correspondence between the correctors and the intended corrections. For example, correction of dispersion functions, or beta-beating effects, or any combination of these effects together with the COD, can also be considered in a grand correction package.

Homework 3.15 Prove the COD correction scheme with weighting functions as given by Eq. (3.13).

Homework 3.16 Suppose you have carefully designed a system (it does not have to be an accelerator). But now it is being compromised by two error effects that yield errors ϵ_1 and ϵ_2. You are given a single corrector knob to rescue your system. The errors can be changed by your corrector by the relationship $\Delta\epsilon_1 = t_1\theta$ and $\Delta\epsilon_2 = t_2\theta$ for the two errors, where θ is the corrector setting.

(a) Find the optimum choice of your corrector setting according to the rms method suggested in the text.

(b) Find the values of the two errors after correction.

(c) Find the expression for σ^2 after correction.

(d) Extend the above results to general expressions for the case when there are $N > 1$ monitors and only one corrector.

Solution

(a) In this system, we have $M = 2, N = 1$.

$$R = \begin{bmatrix} t_1 \\ t_2 \end{bmatrix},$$

$$\tilde{R}R = \begin{bmatrix} t_1 & t_2 \end{bmatrix} \begin{bmatrix} t_1 \\ t_2 \end{bmatrix} = t_1^2 + t_2^2,$$

$$K^{-1} = \frac{1}{t_1^2 + t_2^2},$$

$$C = -K^{-1}\tilde{R}\Delta X_0 = -\frac{t_1\epsilon_1 + t_2\epsilon_2}{t_1^2 + t_2^2}.$$

A straightforward algebraic calculation without using our formalism can of course be done for this simple example, yielding the same result.

(b)

$$\begin{bmatrix} \Delta x_1 \\ \Delta x_2 \end{bmatrix} = \frac{t_2\epsilon_1 - t_1\epsilon_2}{t_1^2 + t_2^2} \begin{bmatrix} t_2 \\ -t_1 \end{bmatrix}.$$

(c)

$$\sigma^2 = \frac{(t_2\epsilon_1 - t_1\epsilon_2)^2}{2(t_1^2 + t_2^2)}.$$

Two special cases: (i) The correction is perfect if $\frac{t_1}{t_2} = \frac{\epsilon_1}{\epsilon_2}$, as one could expect. (ii) The corrector makes no improvement on the rms COD if $\frac{t_1}{t_2} = -\frac{\epsilon_2}{\epsilon_1}$.

(d) It should be easy to see that for the case of multiple errors being corrected by a single corrector, the optimal choice is

$$\theta = -\frac{\sum_{i=1}^{M} t_i\epsilon_i}{\sum_{i=1}^{M} t_i^2}.$$

Homework 3.17 In Homework 3.16, we have a system with one corrector and two errors. What happens when we have two correctors and one error? What happens to the matrix $(\tilde{R}R)^{-1}$ in this case? Explain what has happened.

Homework 3.18 Redo Homework 3.16 with weighting functions w_1 and w_2. Show that the optimal corrector strength is

$$\theta = -\frac{w_1t_1\epsilon_1 + w_2t_2\epsilon_2}{w_1t_1^2 + w_2t_2^2}.$$

The result recovers Homework 3.16 when $w_1 = w_2$, as it should. In case $w_1 = 0$, then the optimal corrector is $\theta = -\frac{\epsilon_2}{t_2}$.

Homework 3.19 In this homework, we intend to illustrate the fact that errors do not have strictly to be the BPM readings. Consider the case when we do not really care what the COD readings are at two neighboring BPMs, but we do care that their readings must be as equal as possible. How do we proceed in this case?

Solution Let the COD at the two BPMs be Δx_{0a} and Δx_{0b}. The two errors can be considered to be $\epsilon_1 = \Delta x_{0a} + \Delta x_{0b}$ and $\epsilon_2 = \Delta x_{0a} - \Delta x_{0b}$. We correct the two errors with different weighting functions. For example, we may choose $w_1 = 0$ and $w_2 = 1$ for the present homework. After correction of the two errors, remember to translate the final values of the corrected errors to the values of Δx_a and Δx_b.

3.2.7 Alternating current dipole

The beam's dynamics can often be explored by intentionally kicking the beam, say using a small dipole magnet, and study its response to the kick. There are basically two ways to kick the beam in such studies. One is to kick the beam sharply in less than one turn. The beam responds to this sharp kick as one measures its Green's function response. This is a time-domain approach. The other way is to kick the beam continuously turn by turn, but with the dipole field oscillating with some specific frequency. The beam then responds to the alternating-current dipole field.[3] This is a frequency domain approach. The beam's response in this case is called a *beam transfer function*. Both approaches in principle contain the same physical information, but they are different in terms of which physical information is easier to extract in practice. Both the time-domain Greens' function approach and the frequency-domain beam transfer function approach can yield a wealth of beam dynamics information.[4]

Consider a 1-D case in the horizontal dimension. Let us first consider a static kick $\Delta x' = \theta$ every turn. The orbit one turn after the kick is

$$TX_0, \qquad \text{with} \quad X_0 = \begin{bmatrix} 0 \\ \theta \end{bmatrix},$$

where T is the one-turn matrix map around the position of the kicker.

Many turns later, the beam has (still observing at the kicker location)

$$\begin{aligned} X &= X_0 + TX_0 + T^2X_0 + \cdots \\ &= (1 + T + T^2 + \cdots)X_0 \\ &= (1 - T)^{-1}X_0 \, . \end{aligned}$$

[3] M. Bai et al., Phys. Rev. E56, 6002 (1997); M. Bai et al., Proc. IEEE Part. Accel. Conf., New York (1999).

[4] See for example, O. Boine-Frankenheim, V. Kornilov, S. Paret, Sec. 7.4.17.1, E. Shaposhnikova, Sec. 7.4.17.2, Handbook Accel. Phys. & Eng., 2nd Ed., World Scientific (2013).

So we need to calculate $(1-T)^{-1}$. The assumed convergence of the infinite series is equivalent to assuming we are calculating the steady-state response of the beam to this kicker. The fact that the infinite series starts with 1 is because we are calculating the trajectory immediately after the kick location.

Let the one-turn transport matrix be

$$T = \begin{bmatrix} \cos\mu + \alpha\sin\mu & \beta\sin\mu \\ -\frac{1+\alpha^2}{\beta}\sin\mu & \cos\mu - \alpha\sin\mu \end{bmatrix}.$$

The two eigenvectors of T are

$$\begin{bmatrix} -\beta \\ i+\alpha \end{bmatrix} \quad \text{and} \quad \begin{bmatrix} -\beta \\ -i+\alpha \end{bmatrix},$$

with eigenvalues $e^{-i\mu}$ and $e^{i\mu}$ respectively. This means T can be expressed as

$$T = E\begin{bmatrix} e^{-i\mu} & 0 \\ 0 & e^{i\mu} \end{bmatrix}E^{-1}, \qquad \text{where} \qquad E = \begin{bmatrix} -\beta & -\beta \\ i+\alpha & -i+\alpha \end{bmatrix}.$$

Using this property, we obtain, with some algebra,

$$\begin{aligned} (1-T)^{-1} &= E\begin{bmatrix} \frac{1}{1-e^{-i\mu}} & 0 \\ 0 & \frac{1}{1-e^{i\mu}} \end{bmatrix}E^{-1} \\ &= \begin{bmatrix} \frac{1-\cos\mu+\alpha\sin\mu}{2(1-\cos\mu)} & \frac{1}{2}\beta\cot\frac{\mu}{2} \\ -\frac{(1+\alpha^2)\cot\frac{\mu}{2}}{2\beta} & \frac{1-\cos\mu-\alpha\sin\mu}{2(1-\cos\mu)} \end{bmatrix}. \end{aligned}$$

This matrix is real although the eigenvalues and eigenvectors of T are complex. By multiplying this matrix to X_0, this then reconfirms the COD at the kicker, Eq. (3.6).

Now we consider the case of an alternating current dipole. The AC dipole gives a beam trajectory

$$(e^{im\phi} + e^{i(m-1)\phi}T + e^{i(m-2)\phi}T^2 + \cdots)X_0 = e^{im\phi}(1-e^{-i\phi}T)^{-1}X_0\,,$$

where we have assumed the AC dipole delivers a kick θe^{imx} for the m-th turn, with ϕ the kicker phase advance per turn, understanding that we will keep only the real part at the end. We have

$$\begin{aligned} (1-e^{-i\phi}T)^{-1} &= E\begin{bmatrix} \frac{1}{1-e^{-i\phi-i\mu}} & 0 \\ 0 & \frac{1}{1-e^{-i\phi+i\mu}} \end{bmatrix}E^{-1} \\ &= \begin{bmatrix} e^{i\phi}\,\frac{e^{i\phi}-\cos\mu+\alpha\sin\mu}{1+e^{2i\phi}-2e^{i\phi}\cos\mu} & \frac{\beta e^{i\phi}\sin\mu}{1+e^{2i\phi}-2e^{i\phi}\cos\mu} \\ -\frac{(1+\alpha^2)}{\beta}\,\frac{e^{i\phi}\sin\mu}{1+e^{2i\phi}-2e^{i\phi}\cos\mu} & e^{i\phi}\,\frac{e^{i\phi}-\cos\mu-\alpha\sin\mu}{1+e^{2i\phi}-2e^{i\phi}\cos\mu} \end{bmatrix}. \end{aligned}$$

This result means that immediately after the kick $\theta e^{im\phi}$ at the m-th turn, the steady-state AC orbit displacement of the beam is given by the real part of

$$\theta e^{im\phi}\,\frac{\beta\sin\mu}{2(\cos\phi-\cos\mu)}.$$

The steady-state response, observed at the kicker location, oscillates in phase with the oscillating kicker strength. When $\phi = 0$, the kick is static, we recover the static COD expression as before. Also, the presence of a resonance when $\phi = \pm\mu$ should not be a surprise. A resonance occurs when $\phi \mp \mu = 2\pi\times$ integer.

Advantages of considering an AC dipole for beam diagnostics purposes include: (i) an oscillating signal from a BPM is easier to detect with higher accuracy than a static signal which is prone to systematic errors, and (ii) a possible enhancement factor when the AC dipole frequency approaches the betatron frequency.

3.3 Tune shift

3.3.1 Quadrupole magnet field errors

Consider a thin-lens quadrupole field error located at position s_0 and with integrated strength of $\int ds\, k(s) = q$. The focal length due to the error quadrupole is $\frac{1}{q}$. The equation of motion is

$$u'' + [K(s) + k(s)]u \;=\; 0\,,$$

where $K(s)$ is the unperturbed focusing strength, periodic in s with period C. When a particle passes through the thin-lens error quadrupole, its displacement is unchanged. Its trajectory slope changes by

$$\begin{aligned}\Delta u' &= \int_{s_0^-}^{s_0^+} ds' u''(s') \;=\; \int_{s_0^-}^{s_0^+} ds'[-k(s')u(s')] \\ &= -u(s_0)\int_{s_0^-}^{s_0^+} ds' k(s') \;=\; -qu(s_0)\,.\end{aligned}$$

The transfer matrix going through the error quadrupole is therefore

$$\begin{bmatrix} 1 & 0 \\ -q & 1 \end{bmatrix}.$$

Starting with the exit of the error quadrupole, the one-turn map is then given by

$$M(s_0+C|s_0) \;=\; \begin{bmatrix} 1 & 0 \\ -q & 1 \end{bmatrix}\begin{bmatrix} \cos 2\pi\nu + \alpha_0 \sin 2\pi\nu & \beta_0 \sin 2\pi\nu \\ -\gamma_0 \sin 2\pi\nu & \cos 2\pi\nu - \alpha_0 \sin 2\pi\nu \end{bmatrix}$$

$$= \begin{bmatrix} \cos 2\pi\nu + \alpha_0 \sin 2\pi\nu & \beta_0 \sin 2\pi\nu \\ -\gamma_0 \sin 2\pi\nu - q(\cos 2\pi\nu + \alpha_0 \sin 2\pi\nu) & \cos 2\pi\nu - \alpha_0 \sin 2\pi\nu - q\beta_0 \sin 2\pi\nu \end{bmatrix}, \tag{3.14}$$

where $\alpha_0, \beta_0, \gamma_0$ are the unperturbed Courant-Snyder functions at s_0, ν is the unperturbed betatron tune.

The matrix $M(s_0 + C|s_0)$ is the *perturbed* one-turn map around the exit point of the error quadrupole. If we designate the perturbed tunes and Courant-Snyder functions as $\bar{\nu}, \bar{\alpha}(s), \bar{\beta}(s), \bar{\gamma}(s)$, then it can also be written as

$$M(s_0+C|s_0) = \begin{bmatrix} \cos 2\pi\bar{\nu} + \bar{\alpha}_0 \sin 2\pi\bar{\nu} & \bar{\beta}_0 \sin 2\pi\bar{\nu} \\ -\bar{\gamma}_0 \sin 2\pi\bar{\nu} & \cos 2\pi\bar{\nu} - \bar{\alpha}_0 \sin 2\pi\bar{\nu} \end{bmatrix} .$$

Tune shift By equating the traces of the two expressions of $M(s_0+C|s_0)$, we obtain

$$2\cos 2\pi\bar{\nu} = 2\cos 2\pi\nu - q\beta_0 \sin 2\pi\nu \,. \tag{3.15}$$

This is an important expression that relates the perturbed tune in terms of the unperturbed quantities.

When the error quadrupole is weak, the perturbed tune does not change much from the unperturbed tune, and the *tune shift* is approximately given by

$$\Delta\nu = \bar{\nu} - \nu \approx \frac{q\beta_0}{4\pi} \,. \tag{3.16}$$

The betatron tune shift is proportional to the β-function at the error quadrupole. We see here again that the β-function gives a magnification factor of the effect on particle motion due to errors. Quadrupole errors at locations of large β-function are particularly damaging for beam stability.

The tune shift is also proportional to q. In particular, when $q > 0$, meaning the error field is focusing, the betatron tune is shifted up. Similarly, a defocusing quadrupole error shifts the tune down. These are as one would expect.

The reader is reminded of the sign convention that a "focusing" quadrupole whose strength is slightly too large will be a focusing field error in x but a defocusing error in y. As a result, in most circumstances, the error-induced x- and y-tune shifts are of opposite signs.

Recall that the opposite focusing signs in x and y was dictated by the Maxwell's equation (particularly $\frac{\partial B_x}{\partial y} = \frac{\partial B_y}{\partial x}$), but it was the Maxwell's equation in free space. Exceptions to the rule of $\Delta\nu_x$ and $\Delta\nu_y$ having opposite signs therefore can often be traced to cases when there are perturbing charges or currents *within* the beam distribution, i.e. not in free space. Here are a few examples,

- ions from residual gas,
- space charge,
- beam-beam,
- lithium lens,
- electron cloud,
- a special example is given in Homework 3.20.

When there is a distribution of quadrupole error $k(s)$ around the storage ring, the approximate tune shift can be found by superposition, to first order in quadrupole error,

$$\Delta\nu \approx \frac{1}{4\pi}\oint ds\,\beta(s)k(s)\,. \tag{3.17}$$

β-function measurement The tune shift formula (3.16) can be used to measure the β-functions at locations where there are quadrupoles. If the quadrupole strength is varied by an amount $\Delta k\ell$, and the tunes are measured to change by $\Delta\nu_x$ and $\Delta\nu_y$, then the β-functions are found by

$$\beta_x = 4\pi\frac{\Delta\nu_x}{\Delta k\ell}, \qquad \beta_y = -4\pi\frac{\Delta\nu_y}{\Delta k\ell}\,.$$

A positive $\Delta k\ell$ will cause a positive tune shift in x and negative tune shift in y.

This method is commonly used to measure the β-functions in storage rings. On the other hand, it allows measurements only at quadrupole locations. Sometimes a string of quadrupoles have their power supplies strung in series, in which case we measure only their average values. Also, if the tune is measured by a beam position monitor, the tune signal is measured modulus the revolution frequency. This can complicate the measurement procedure, especially concerning the signs of the tune shifts.

Homework 3.20 Consider a lattice around a position s with known Courant-Snyder parameters α, β, γ. Now the accelerator designer has asked that a small drift space $\Delta\ell$ be inserted at position s. Show that, to first order in $\Delta\ell$, this additional drift will cause a tune shift

$$\Delta\nu = \frac{\gamma(s)}{4\pi}\Delta\ell\,.$$

Note that this tune shift is always positive (assume $\Delta\ell > 0$) and has the same sign in ν_x and ν_y, and it occurs in free space!

Solution Did you try to guess the answer using $\Delta\psi = \frac{\Delta\ell}{\beta(s)}$? If so, you would get a wrong answer.

Homework 3.21 Consider a weak, linear symplectic perturbation occurring at location $s = 0$ in a storage ring. Let the perturbation be described by a symplectic matrix $T = I + \epsilon$ where I is the unit matrix and ϵ is a small perturbation matrix.

(a) Show that the fact that T is symplectic implies that, to first order in ϵ, the matrix ϵ must satisfy the condition

$$\epsilon = S\tilde{\epsilon}S\,.$$

(b) Applied to the case of 1-D motion, so that T and ϵ are 2×2 matrices, show that the condition on ϵ becomes

$$\mathrm{tr}(\epsilon) \;=\; 0\,.$$

(c) Let the symplectic perturbation be applied at location $s = 0$, show that to first order in ϵ, the betatron tune shift is given by

$$\Delta\nu \;=\; \frac{1}{4\pi}(\gamma_0\epsilon_{12} - 2\alpha_0\epsilon_{11} - \beta_0\epsilon_{21})\,.$$

The usual tune shift formula is recovered when $\epsilon = \begin{bmatrix} 0 & 0 \\ -q & 0 \end{bmatrix}$. Homework 3.20 can also be recovered as another special case.

3.3.2 β-beat and half-integer resonances

Beta beat In addition to tune shifts, quadrupole errors also cause distortion of the β-function, designated as $\Delta\beta(s)$. Consider again a thin-lens error quadrupole with integrated gradient error q. The one-period map around position s is related to the one-turn map $M(s_0 + C|s_0)$ by Eq. (2.21),

$$M(s + C|s) = M(s|s_0)M(s_0 + C|s_0)M^{-1}(s|s_0)\,,$$

where $M(s|s_0)$ is the transfer map from s_0 to s. Since the perturbation occurs only at s_0, $M(s|s_0)$ is given by the unperturbed map

$$M(s|s_0) \;=\; \begin{bmatrix} \sqrt{\frac{\beta}{\beta_0}}(\cos\Delta\psi + \alpha_0\sin\Delta\psi) & \sqrt{\beta_0\beta}\sin\Delta\psi \\ \frac{\alpha_0-\alpha}{\sqrt{\beta_0\beta}}\cos\Delta\psi - \frac{1+\alpha_0\alpha}{\sqrt{\beta_0\beta}}\sin\Delta\psi & \sqrt{\frac{\beta_0}{\beta}}(\cos\Delta\psi - \alpha\sin\Delta\psi) \end{bmatrix}\,,$$

with $\Delta\psi = \psi(s) - \psi(s_0)$ the betatron phase advance from s_0 to s.

Substituting this expression of $M(s|s_0)$ and the expression of $M(s_0 + C|s_0)$ from Eq. (3.14), we obtain the map $M(s + C|s)$ whose (1,2)-element is

$$\beta\sin 2\pi\nu + q\frac{\beta_0\beta}{2}[\cos 2\pi\nu - \cos(2\pi\nu - 2\Delta\psi)]\,.$$

In terms of the perturbed Courant-Snyder parameters, this element can also be written as $\bar\beta\sin 2\pi\bar\nu$. Therefore,

$$\bar\beta\sin 2\pi\bar\nu \;=\; \beta\sin 2\pi\nu + q\frac{\beta_0\beta}{2}[\cos 2\pi\nu - \cos(2\pi\nu - 2\Delta\psi)]\,.$$

When the quadrupole perturbation is weak, we linearize the expression with respect to q and, after some algebra, obtain

$$\frac{\Delta\beta}{\beta} \;=\; \frac{\bar\beta - \beta}{\beta} \;\approx\; -\frac{q\beta_0}{2\sin 2\pi\nu}\cos[2\pi\nu - 2\psi(s) + 2\psi(s_0)]\,.$$

As mentioned after Eq. (2.21), this result is applicable only when s is downstream of the error quadrupole. When s is upstream, it is necessary to apply Eq. (2.22). When the quadrupole perturbation is weak, we combine the results (Homework 3.22) to obtain

$$\frac{\Delta\beta(s)}{\beta(s)} \approx -\frac{q\beta_0}{2\sin 2\pi\nu}\cos(2\pi\nu - 2|\psi(s) - \psi(s_0)|)\,. \tag{3.18}$$

An absolute value operation is introduced so that the formula applies also when s is upstream of s_0.

As in the case of COD expression (3.7), Eq. (3.18) applies only when s and s_0 are separated either upstream or downstream of each other less than the one-turn circumference C.

The β-function therefore is distorted when there are quadrupole errors. The distortion oscillates with *twice* the betatron frequency and with an amplitude that is proportional to q and β_0. This phenomenon is called a β-beat, or β-wave.

For a distributed quadrupole error $k(s)$, the β-beat is given by superposition, to first order in k,

$$\frac{\Delta\beta(s)}{\beta(s)} \approx -\frac{1}{2\sin 2\pi\nu}\oint ds'\,\beta(s')k(s')\cos(2\pi\nu - 2|\psi(s) - \psi(s')|)\,. \tag{3.19}$$

Half-integer resonance There is a factor $\frac{1}{\sin 2\pi\nu}$ in the β-beat. This means $\Delta\beta \to \infty$ when $\nu =$ half integer. The origin of this divergence is a resonance effect similar to the resonance in COD when $\nu =$ integer. The dynamics of this half-integer resonance is illustrated in phase space in Fig. 3.10. The vertical red arrows add up to a *linear* (not exponential) growth of betatron amplitude of the particle.

The orbital displacement of the particle, when observed at the error quadrupole, does not grow in time, but the orbital slope increases linearly in time. Each arrow has a magnitude of $q|u_0|$ where $\pm u_0$ is the particle's displacement at the error quadrupole with alternating signs in alternating turns.

In Fig. 3.10, recall that the rotation in phase space is clockwise. The phase space ellipse is shown as being upright for simplicity. This figure applies when ν is half-integer. For integer ν, it changes but linear growth remains the same — quadrupole errors also drive integer resonances. The reader is suggested to modify Fig. 3.10 for the case when $\nu =$ integer.

Distributed quadrupole error In case there are many uncorrelated sources of quadrupole field errors, one may perform a statistical estimate. The expected rms tune shift is given by

$$\langle\Delta\nu^2\rangle = \frac{N}{32\pi^2}\langle\beta_i^2 k_i^2 \ell_i^2\rangle\,,$$

Figure 3.10: Illustration of half-integer resonance mechanism due to a single error quadrupole.

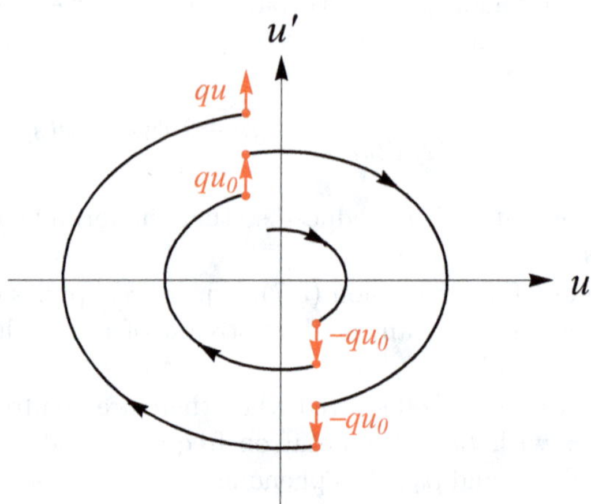

where $\langle \beta_i^2 k_i^2 \ell_i^2 \rangle$ is the average value over all the error field sources. The same quadrupole field errors also give rise to a β-beat, whose statistical estimate is given by

$$\left\langle \left(\frac{\Delta\beta}{\beta} \right)^2 \right\rangle = \frac{N}{8 \sin^2 2\pi\nu} \langle \beta_i^2 k_i^2 \ell_i^2 \rangle \, .$$

These estimates are obviously related,

$$\left\langle \left(\frac{\Delta\beta}{\beta} \right)^2 \right\rangle^{1/2} = \frac{2\pi}{\sin 2\pi\nu} \langle \Delta\nu^2 \rangle^{1/2} \, . \tag{3.20}$$

Take for example $N = 500, \sin 2\pi\nu = 1$ and $\langle \beta_i^2 k_i^2 \ell_i^2 \rangle^{1/2} = 0.01$, we obtain $\langle \Delta\nu^2 \rangle^{1/2} = 0.013$ and $\left\langle \left(\frac{\Delta\beta}{\beta} \right)^2 \right\rangle^{1/2} = 0.08$.

Away from resonances, a rough connection can be written,

$$\frac{\Delta\beta}{\beta} \sim 2\pi\Delta\nu \, .$$

For example, an error quadrupole that gives rise to a tune shift of ~0.01 is expected to give a β-beat of ~6%. But there is also a resonance factor $\sin 2\pi\nu$ in Eq. (3.20). We shall focus on this factor in a minute.

Homework 3.22 The text derived $\Delta\beta(s)$ when s is downstream of s_0, where the error quadrupole is located, using Eq. (2.21). Repeat the derivation when s is upstream of s_0 using Eq. (2.22). Justify the introduction of absolute value in Eq. (3.18).

Homework 3.23 Show that, in terms of the Courant-Snyder variables,

$$\frac{d^2}{d\psi^2}\left(\frac{\Delta\beta}{\beta}\right)+4\left(\frac{\Delta\beta}{\beta}\right) = -2\beta^2 k\,.$$

Equation (3.19) can be considered the periodic solution of this differential equation. The fact that $\frac{\Delta\beta}{\beta}$ naturally beats at twice the betatron frequency becomes obvious from this equation.

Homework 3.24 The text calculated the perturbations on ν and $\beta(s)$. There is also perturbation on $\alpha(s)$ and $\gamma(s)$. To first order in quadrupole perturbation $k(s)$, show that

$$\begin{aligned}\Delta\alpha(s) &= \alpha\frac{\Delta\beta}{\beta}\\ &+\frac{1}{2\sin 2\pi\nu}\oint ds'\beta(s')k(s')\,\mathrm{sgn}[\psi(s)-\psi(s')]\,\sin(2\pi\nu-2|\psi(s)-\psi(s')|)\,.\end{aligned}$$

For completeness, the change in $\Delta\gamma$ is given by

$$\frac{\Delta\gamma}{\gamma} = \frac{2\alpha\Delta\alpha}{1+\alpha^2}-\frac{\Delta\beta}{\beta}\,.$$

Homework 3.25 We derived an expression of β-beat and an expression of tune shift in the text. These two expressions must be related because the tune shift originates from the β-beat,

$$\Delta\nu = \frac{1}{2\pi}\oint ds\Delta\left(\frac{1}{\beta(s)}\right) = \frac{1}{2\pi}\oint ds\left(\frac{-\Delta\beta(s)}{\beta^2(s)}\right)\,.$$

Verify the validity of this expression explicitly for the case of a single weak thin-lens quadrupole of strength q to first order in q.

Homework 3.26 Suppose you are given accurately the β-function distortion $\Delta\beta(s)$ around the circular accelerator. Find an expression of the tune shift $\Delta\nu$ valid to second order in $\Delta\beta(s)$. Refer to Homework 3.25.

Homework 3.27 In analogy to a closed-orbit bump, develop a way to construct a β-function bump using (a) two quadrupoles, and (b) three quadrupoles.

Homework 3.28 Continuing Homework 3.27, consider two thin quadrupoles with β-functions β_1 and β_2, and a phase advance $\Delta\psi$ between them. You are asked to generate a perfect β-bump between them without generating a tune shift. Find the conditions for you to do so. Consider 1-D x-motion only.

Homework 3.29 Consider a lattice arranged with mirror symmetry about a waist position $s = 0$. Suppose you have two weak quadrupoles available symmetrically located around $s = 0$ and with betatron phase Φ upstream and downstream from the waist. Your two quadrupoles are independently powered.

(a) How do you power your quadrupoles to produce a net change of β-function at the waist? You are not supposed to move the position of the waist from $s = 0$. How do you choose optimally the value of Φ if it is under your control?

(b) How do you power them if you want to shift the waist position by a small distance Δs^* without changing the value of β-function at $s = 0$?

Homework 3.30 This homework is related but not the same as the previous one. A storage ring is found to need to control its β_x-function locally at one location in the ring with minimum disturbance to the β_x-function elsewhere. This is a job of a local β_x-bump. To do this job, you are given two thin-lens quadrupoles with strengths $q_{1,2} = \left(\frac{\ell}{B\rho}\frac{\partial B_y}{\partial x}\right)_{1,2}$ at positions s_1 and s_2 (s_2 is downstream of s_1). Let the designed β_x-functions at these two locations be β_{x1} and β_{x2} respectively. The betatron phase advance between the two locations is designed to be $\Delta\psi_x = \frac{\pi}{2}$. The quadrupole strengths are under your control.

(a) You are asked to provide a β_x-bump within the region between s_1 and s_2. Give the condition on q_1 and q_2 so that they produce a β_x-bump.

(b) The two quadrupoles will also introduce tune shifts $\Delta\nu_x$ and $\Delta\nu_y$. Find their expressions in terms of quantities introduced above.

We assume β_y-function is somewhat flexible for this storage ring, so that the effect on β_y-function by the two extra quadrupoles is not a concern.

3.3.3 Stopband

One curious thing in our expressions of $\Delta\nu$ and $\Delta\beta$, both caused by the same error quadrupole, is that the resonance denominator occurs only to $\Delta\beta$, not to the tune shift $\Delta\nu$, even though they are necessarily connected as established in Homework 3.25. In fact, $\Delta\nu$ is independent of ν! In particular, the tune shift remains finite even when ν is close to a half integer, while the effect on $\Delta\beta$ is drastic. This effect has been evident already in Eq. (3.20).

This observation, although basically correct, is somewhat misleading because when ν is close to a half integer, the first order expression (3.16) for $\Delta\nu$ breaks down and we should use the exact expression (3.15). What one finds then is that the resonance does affect the tune. In particular, when ν is close to a half-integer $\frac{n}{2}$, there is even a *stopband* around $\nu = \frac{n}{2}$ inside of which there is no real solution for $\bar{\nu}$, meaning the particle motion has been rendered unstable by the error quadrupole.

Figure 3.11 shows an example. Results using the exact (3.15) are shown by the solid red curves. The two panels on the left show the case $q\beta_0 = 0.2$; the two on the right show the case $q\beta_0 = -0.2$. The two upper panels plot the perturbed tune $\bar{\nu}$ versus the nominal tune ν. The two lower ones plot the tune shift $\bar{\nu} - \nu$ versus ν. The grey shaded regions indicate the stopbands. The

Figure 3.11: Stopband structure as seen in the behavior of the perturbed tune $\bar{\nu}$ versus the nominal tune ν.

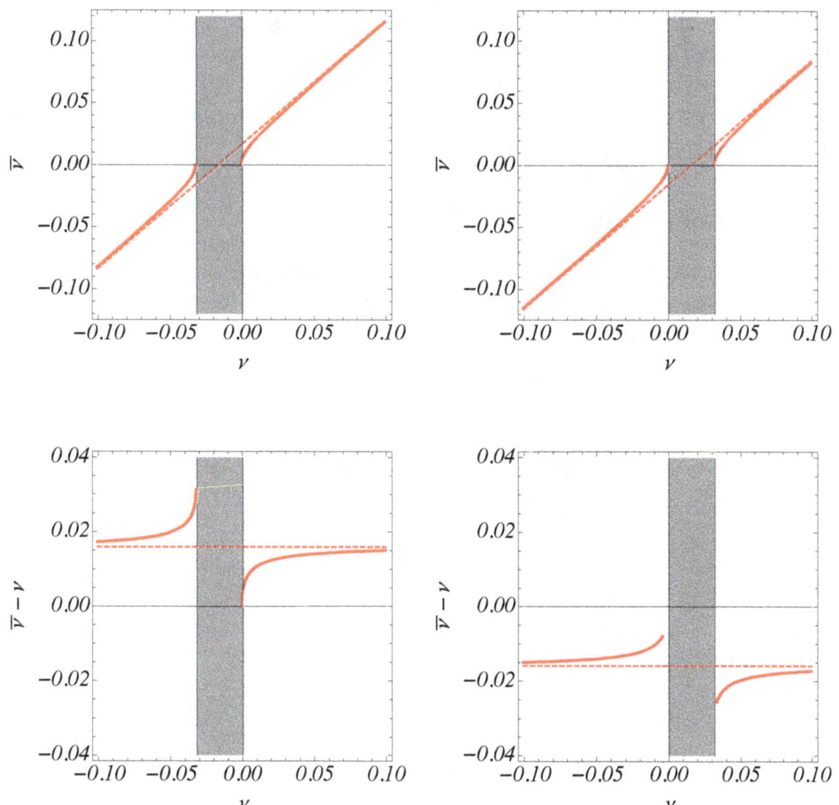

dashed red straight lines are first order result (3.16), which reflects a tune shift $\bar{\nu} - \nu$ independent of ν.

One observes that

- The stopband occurs near the resonance $\nu = \frac{n}{2}$.
- The first order expression (3.16) is good if one stays sufficiently away from the stopband (it is not enough to be outside of the stopband).
- The stopband covers only one side of the resonance, not straddling over both sides. Whether it covers the left-side or the right-side depends on the sign of q.
- One could provide a quantitative value of the stopband by identifying the width of the stopband in tune units. The stopband gap width is

$$\Delta\nu_{\text{stopband}} = \frac{1}{\pi} \tan^{-1} \frac{q\beta_0}{2} . \tag{3.21}$$

Homework 3.31 Prove Eq. (3.21).

Solution Hidden in Eq. (3.21) is the definition of the stopband. Think through the definition. It is defined as the range in the unperturbed tune ν when the perturbed tune becomes undefined. So when you hear about a stopband mentioned, it refers to a forbidden region on the *unperturbed* tune. The actual tune, the perturbed one, is undefined within the stopband. Outside of the stopband, it in fact covers all values; it itself does not have a stopband.

3.3.4 Space charge tune shift

In Sec. 1.5, we calculated the space charge force. We can use that result now to calculate the space charge tune shift. The space charge force seen by a particle in a uniform cylindrical beam (radius a, length ℓ) is given by

$$\vec{F} = \frac{2Nr_0mc^2}{a^2\ell\gamma^2}\, r\hat{r}\,,$$

where r_0 is the classical radius of the particle under consideration, N is the number of particles in the beam bunch. We elaborated on the factor $\frac{1}{\gamma^2}$ before in Sec. 1.5.

The fact that the force is directed radially outward indicates that this is a defocusing force in *both* x and y. We expect tune shifts in both x and y to be negative. The space charge force contributes to an additional term to the equation of motion

$$m\gamma\ddot{u} = \frac{2Nr_0mc^2}{a^2\ell\gamma^2}\, u\,.$$

By relating $\ddot{u} = \bar{\beta}^2c^2u''$,[5] we obtain the space-charge induced focusing error

$$k(s) = -\frac{2Nr_0}{a^2(s)\ell\bar{\beta}^2\gamma^3}\,,$$

where we have allowed the beam radius a to depend on s. We then apply the tune shift formula to obtain

$$\Delta\nu = -\frac{Nr_0}{2\pi\ell\bar{\beta}^2\gamma^3}\oint ds\frac{\beta(s)}{a^2(s)}\,.$$

The quantity $\frac{a^2}{\beta}$ is basically what we introduced earlier as the beam emittance. For a uniform cylinder beam with radius a, we identify it to be

$$\frac{a^2(s)}{\beta(s)} = 4\epsilon_{\rm rms}\,,$$

where $\epsilon_{\rm rms}$ is the rms beam emittance. In particular, we note that this quantity is an invariant and is independent of s. In this discussion, we have assumed

[5] $\bar{\beta} = \frac{v}{c}$ is used here with a bar on top to avoid confusion with the β-function. But we do not take the same precaution for γ.

Figure 3.12: Plots of $\frac{1}{\bar{\beta}\gamma^2}$ versus $\bar{\beta}$ (left) and γ (right) explain the sensitivity of space charge effect to beam energy. The left panel emphasizes nonrelativistic beams. The right panel emphasizes relativistic beams.

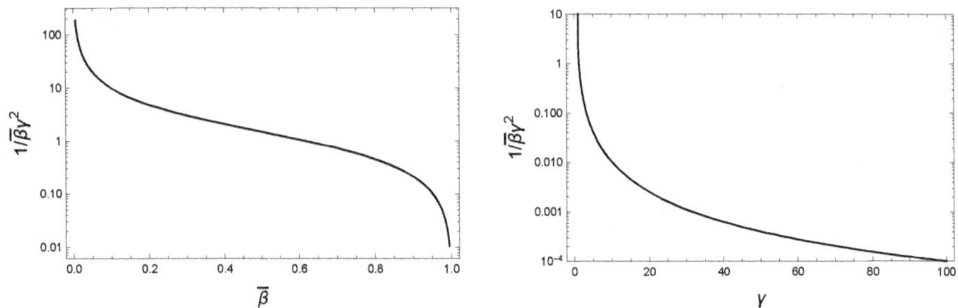

that the beam size comes solely from a betatron emittance, and have ignored, for example, any contribution from an energy spread of the beam.

We then obtain

$$\Delta\nu = -\frac{Nr_0}{8\pi\bar{\beta}^2\gamma^3\epsilon_{\rm rms}}\frac{2\pi R}{\ell},$$

where $2\pi R$ is the storage ring circumference, and $\frac{\ell}{2\pi R}$ is the bunch filling factor.

In a moment, we will mention a phenomenon called adiabatic damping, which will show that, as the beam momentum is varied due to acceleration or deceleration, the beam emittance will vary in such a way that $\epsilon_{\rm rmsN} \equiv \bar{\beta}\gamma\epsilon_{\rm rms}$ is constant. In terms of this *normalized* emittance, we have

$$\Delta\nu = -\frac{Nr_0}{8\pi\bar{\beta}\gamma^2\epsilon_{\rm rmsN}}\frac{2\pi R}{\ell}. \tag{3.22}$$

We see that $\Delta\nu$ is proportional to the line charge density $\frac{N}{\ell}$ and the storage ring radius R. The physics reason for this should be obvious. We see that $\Delta\nu \propto \frac{1}{\epsilon_{\rm rmsN}}$. This can be expected as well.

We also see $\Delta\nu \propto \frac{1}{\bar{\beta}\gamma^2}$. The $\frac{1}{\gamma^2}$ suppression factor of the space charge force is explicitly seen here, but there is an additional factor of $\frac{1}{\bar{\beta}}$. The space charge effect is mainly a nonrelativistic effect because of the factor γ^2, but it particularly becomes sensitive at the very low energies due to the factor $\bar{\beta}$. The extreme sensitivity at low energies reiterates the fact that an intense beam will require rapid acceleration as soon as possible when it is generated from a beam source. Figure 3.12 shows the factor $\frac{1}{\bar{\beta}\gamma^2}$ as a function of β and γ. The left figure exhibits the extreme sensitivity of $\frac{1}{\bar{\beta}}$ at nonrelativistic energies while the right figure highlights mainly the relativistic suppression of $\frac{1}{\gamma^2}$.

This tune shift (3.22) is called the *space charge tune shift.* In our derivation, we assumed a uniform cylindrical beam distribution, and what we obtained is a true tune shift, meaning that all particles in the beam suffer the same tune shift. In reality the beam distribution is not as simple, and perhaps more like a 3-D Gaussian. In those cases, what we have obtained will be a tune *spread.* There are two reasons for this,

- The space charge force seen by a particle is proportional to the *local* line density of the charge distribution. The tune shift therefore will be small towards the bunch head and the bunch tail and reaches a maximum at the bunch center. Different particles therefore will have different tune shifts. Equation (3.22) then gives a space charge tune spread if we let

$$\frac{N}{\ell} \to \frac{N}{\sqrt{2\pi}\,\sigma_z},$$

  with σ_z the rms bunch length for a Gaussian distribution.

- Another reason that this expression should be interpreted as a tune spread is that it gives the tune shift only for particles with small betatron oscillations. Particles with large betatron oscillations will have diminishing tune shifts because they spend some of its time outside of the charge distribution.

As mentioned, space charge effect dominates the beam dynamics near the beam source at very low energies. At higher energies, the space charge tune shift (or tune spread) also imposes an important limit on the intensity of low-to-medium energy synchrotrons or storage rings. As the beam energy is increased, the factor $\bar{\beta}\gamma^2$ in the denominator quickly suppresses the space charge effect. On the other hand, the tune shift also increases with the circumference, so its effect also increases for large accelerators even when its energy is higher.

Homework 3.32 The text assumed a uniform cylindrical beam. Now consider a relativistic beam with a transverse distribution of a round Gaussian and uniform longitudinal density traveling in a circular accelerator with speed v. Let N be the number of charged particles in the beam, each particle with charge e. Let L be the beam's longitudinal length. The transverse distribution of the beam is

$$\rho(r) = \frac{Ne}{2\pi\sigma^2 L} e^{-r^2/2\sigma^2},$$

with normalization $\int_0^\infty 2\pi r dr \rho(r) = \frac{Ne}{L}$.

(a) Find the electric and magnetic fields carried by the beam as functions of r.

(b) Find the Lorentz force, including its direction, seen by a test particle in the beam and moving with the beam.

(c) If the test particle has a small $r \ll \sigma$, find expressions for the space charge tune shifts $\Delta\nu_{x,y}$. Make sure you have the correct signs of these tune shifts.

(d) Try to understand the mechanism that, for a particle executing betatron oscillation with an amplitude $\gg \sigma$, its space charge tune shift $\Delta\nu \to -0$.

3.3.5 Characterizing β-function mismatch

Consider a section of a linear accelerator or beam transport line. Suppose its lattice design has been made and it is described by the nominal $\alpha_0(s), \beta_0(s), \gamma_0(s)$ functions. Now unfortunately the lattice upstream of the lattice section has an error, so that there is a mismatch of the Courant-Snyder functions as the beam enters your otherwise-perfect section. As a result of this mismatch, when you measured these functions in your section, e.g. by measuring the rms beam sizes $\langle u^2\rangle, \langle uu'\rangle, \langle u'^2\rangle$, we found $\alpha_1(s), \beta_1(s), \gamma_1(s)$ instead, where α_1, β_1 are obtained by measurements and $\gamma_1 = \frac{1+\alpha_1^2}{\beta_1}$. Question then arises how to quantify the lattice mismatch by a single convenient error parameter $\epsilon_{\rm err}$.

The solution is not unique, of course, but to be useful, the parameter should at least have these properties:

1. $\epsilon_{\rm err}$ should vanish when $\alpha_1 = \alpha_0, \beta_1 = \beta_0, \gamma_1 = \gamma_0$.
2. $\epsilon_{\rm err}$ should be an invariant along your section where there are no errors, so that once $\epsilon_{\rm err}$ is corrected at one position in your section (e.g. at the entrance point), it will vanish over the entire section of yours.
3. Definition of $\epsilon_{\rm err}$ should not depend on anything specific to your lattice design details such as the focusing function $K(s)$, so that it has a wider application.

A clever combination was suggested for the SLAC Linear Collider linac,[6]

$$\epsilon_{\rm err} = \gamma_0\beta_1 + \gamma_1\beta_0 - 2\alpha_1\alpha_0 - 2\,. \tag{3.23}$$

It is obvious that $\epsilon_{\rm err} = 0$ when $\alpha_1 = \alpha_0, \beta_1 = \beta_0, \gamma_1 = \gamma_0$. More discussions of $\epsilon_{\rm rms}$ are included in the Homework problems 3.33 and 3.34.

Homework 3.33

(a) Show that $\epsilon_{\rm err}{}' = 0$, i.e. $\epsilon_{\rm err}$ is an invariant in the section without errors.

(b) Show that another expression of $\epsilon_{\rm err}$ is

$$\epsilon_{\rm err} = \left(\sqrt{\frac{\beta_0}{\beta_1}} - \sqrt{\frac{\beta_1}{\beta_0}}\right)^2 + \left(\alpha_1\sqrt{\frac{\beta_0}{\beta_1}} - \alpha_0\sqrt{\frac{\beta_1}{\beta_0}}\right)^2 .$$

[6] F.J. Decker et al., Proc. Part. Accel. Conf., 11 (1991).

(c) Show that still another expression of $\epsilon_{\rm err}$ is

$$\epsilon_{\rm err} = -\det[J_0 - J_1],$$

where $J_i = \begin{bmatrix} \alpha_i & \beta_i \\ -\gamma_i & -\alpha_i \end{bmatrix}$, $i = 1, 2$, was introduced in Homework 2.86.

Solution (a) Use Homework 2.68.

Homework 3.34 When the beam enters the beamline section with the mismatch, its emittance will filament conforming to the local lattice functions. Show that the filamented beam emittance will dilute by the factor

$$\frac{\epsilon_{\rm rms,1}}{\epsilon_{\rm rms,0}} = 1 + \frac{\epsilon_{\rm err}}{2}.$$

Note $\epsilon_{\rm err}$ is always positive (or zero).

3.4 Adiabatic damping

So far we have not considered electric fields in the accelerator. To accelerate the beam, we will have to add a longitudinal electric field to the accelerator design. In addition to *acceleration*, another equally important but more subtle reason to add a longitudinal electric field is to provide *longitudinal focusing* to particle motion. That will be the subject of Chapter 5.

3.4.1 Equation of motion

To proceed, let us consider a beam moving in the environment with fields

$$\vec{B} = (B_x, B_y, 0), \qquad \vec{E} = (0, 0, E_z).$$

The particle energy changes because of E_z.

Take y-motion for example. The equation of motion is (we return to the notation $\beta = \frac{v}{c}$ here because β-function is not used in this section)

$$\frac{d}{dt}(m\gamma\dot{y}) = e\beta c B_x,$$

where a dot means taking $\frac{d}{dt}$ and we write the longitudinal velocity as βc. With $\frac{d}{dt} = \beta c \frac{d}{ds}$, we have

$$\begin{aligned} \beta c(m\beta c\gamma y')' &= e\beta c B_x \\ \Longrightarrow \quad (\beta\gamma y')' &= \frac{eB_x}{mc} \\ \Longrightarrow \quad y'' + \frac{(\beta\gamma)'}{(\beta\gamma)} y' + K_y y &= 0, \end{aligned}$$

where a prime means taking $\frac{d}{ds}$.

We have written $B_x = -\frac{P}{e}K_y y = -\frac{mc\beta\gamma}{e}K_y y$. This means we have assumed that the transverse focusing and defocusing magnetic fields are made to scale with the beam momentum at all times as the beam is accelerated. This is a reasonable assumption because it is the way we will naturally design our accelerator. Incidentally, we scale the magnet strengths with the beam momentum, not the beam energy.

If the acceleration comes from a constant electric field E_z, then $\gamma' = \frac{eE_z}{mc^2} = g$, and the equation of motion becomes

$$y'' + \frac{\gamma g}{\gamma^2 - 1}\, y' + K_y y \;=\; 0\,.$$

When $\gamma \gg 1$, the two independent solutions of this equation of motion are

$$J_0\left(\frac{k}{g}\gamma(s)\right), \qquad \text{and} \qquad N_0\left(\frac{k}{g}\gamma(s)\right),$$

given by the Bessel and Neumann functions.

We now define a quantity $\tilde{y}$ by

$$y \;=\; \frac{\tilde{y}}{\sqrt{\beta\gamma}}\,.$$

Substituting into the equation of motion, we obtain an equation for $\tilde{y}$,

$$\tilde{y}'' - \frac{1}{2}\left[\frac{(\beta\gamma)''}{(\beta\gamma)} - \frac{((\beta\gamma)')^2}{2(\beta\gamma)^2}\right]\tilde{y} + K_y\tilde{y} \;=\; 0\,.$$

The $\tilde{y}'$ term has been eliminated, as intended by the transformation from y to $\tilde{y}$.

Let us perhaps make an observation in passing. When you see an equation like $Ay'' + By' + Cy = 0$, your first instinct should be to identify the C term as focusing/defocusing, and the B term as "damping/antidamping". Your second instinct should be to transform the B term away. Here we see there is damping in y and no damping in $\tilde{y}$. The damping effect is absorbed into the transformation from y to $\tilde{y}$. This will be seen explicitly momentarily.

In most applications, the rate of acceleration is slow compared with the rate the particle motion is being transversely focused. This is because acceleration is done by electric fields, and transverse focusing by magnetic fields. As we emphasized in Sec. 2.1.4, electric fields are generally less effective than magnetic fields in terms of affecting particle motion. As a result, we expect the rate of acceleration to be slower than the rate of transverse oscillation, at least at moderate-to-high energies.

This is called the adiabatic approximation. It leads to an adiabatic damping, subject of this section. Strictly speaking, our analysis breaks down at very low energies. However, even at very low energies, adiabatic damping still occurs; it is only that the quantitative agreement with our analysis becomes compromised.

In the adiabatic case, intended for medium-to-high energy linacs, we require

$$K_y \gg \frac{(\beta\gamma)''}{(\beta\gamma)} \quad \text{and} \quad \frac{((\beta\gamma)')^2}{(\beta\gamma)^2} ,$$

and the equation of motion becomes

$$\tilde{y}'' + K_y \tilde{y} \approx 0 .$$

The quantity $\tilde{y}$ therefore is approximately sinusoidal, just like an ordinary betatron oscillation without acceleration with constant betatron amplitude. It follows that

$$y \approx \frac{1}{\sqrt{\beta\gamma}} \times (\text{sinusoidal betatron oscillation}) .$$

As the beam is accelerated, the betatron oscillations of all the beam particles therefore damp inversely proportionally to square root of the beam momentum (not beam energy) $\beta\gamma$. This is called the *adiabatic damping* effect.

Normalized emittance With adiabatic damping, the beam emittance scales like

$$\epsilon \propto \frac{1}{\beta\gamma} .$$

One can define a *normalized emittance*

$$\epsilon_{\mathrm{N}} = \beta\gamma\epsilon .$$

The normalized emittance is invariant under acceleration (provided the acceleration is sufficiently adiabatic), and is a very useful generic quantity when a beam is accelerated over a large range in energy. It is also useful to compare the beam qualities of accelerators of different energies, or even different types of accelerators by comparing their normalized emittances.

Adiabatic damping mechanism Physically, adiabatic damping comes from the following. When a particle with momentum P and with phase space coordinates (y, y') is accelerated to gain a longitudinal momentum ΔP, its y-displacement is unchanged, but its slope y' is reduced a little to become

$$y' \to \frac{P_y}{P_z} = \frac{P}{P + \Delta P} y' ,$$

as illustrated in Fig. 3.13.

As a particle executes betatron oscillation while being accelerated, its y' coordinate is constantly being damped proportionally to $\frac{1}{P}$, while y is left unperturbed. As the particle's coordinates rotate in the phase space, the damping action in y' gets transferred into damping in y as well, the net result is that both

Figure 3.13: Mechanism of adiabatic damping.

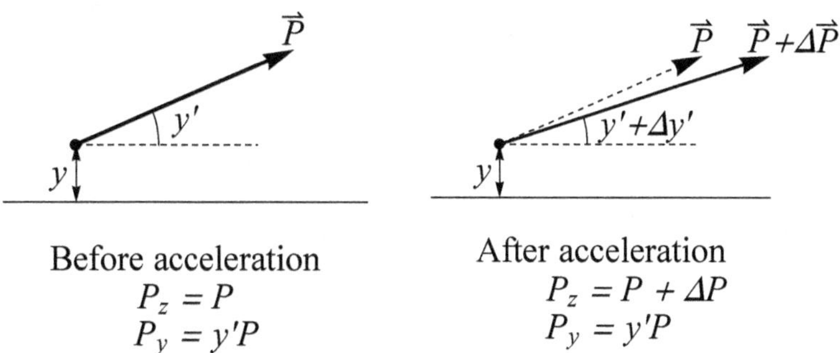

y and y' are damped proportionally to $\frac{1}{\sqrt{P}}$, which is the adiabatic damping. A mechanical model of adiabatic damping was mentioned in Fig. 2.25.

Adiabatic damping is not a true dynamical damping, but results from pure kinematics. The internal temperature of the beam has not been removed by the acceleration. The decrease of the emittance of a beam as it is accelerated will reappear if the beam is decelerated. This decrease of apparent emittance is not true damping. Also see Homework 3.35(a).

In Chapter 7, we will discuss in detail another effect called *radiation damping*, which is closely related to the adiabatic damping discussed here, but in that case, it is true damping. It is perhaps arguably legitimate to call the adiabatic damping a damping effect, but it is not to be referred to as a cooling effect. In contrast, radiation damping can justifiably be called radiation cooling.

Rain drops falling from the sky are adiabatically damped as they are accelerated by gravity. See Fig. 3.14. Real damping occurs if air friction is taken into account.

Homework 3.35

(a) Acceleration is a symplectic process induced by a longitudinal electric field. Here we seem to be discussing the fact that it induces a damping mechanism, which necessarily violates symplecticity. It is suggested to reconcile these two observations.

(b) A proton beam is accelerated by a chain of linacs and synchrotrons from 1 MeV/c to 1 TeV/c. Its emittance is 0.5 m-rad initially and is found to be 1 mm-mrad upon reaching 1 TeV. Find the emittance dilution factor over the acceleration process.

Homework 3.36

(a) Convince yourself that in general the equation of motion with acceleration has the following form in the adiabatic approximation when $\beta \approx 1$,

$$(\gamma y')' + \gamma K_y y = 0\,, \tag{3.24}$$

Figure 3.14: Emittance of rain drops are adiabatically damped as they fall.

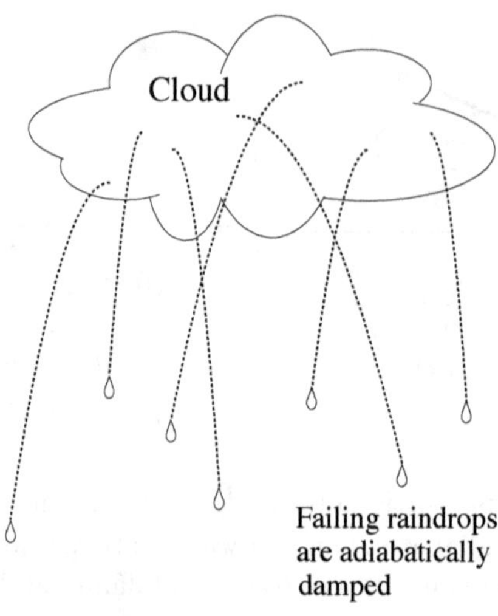

(b) Define $\epsilon^2 = \langle y^2\rangle\langle y'^2\rangle - \langle yy'\rangle^2$. Show directly by Eq. (3.24) that $\epsilon_{\rm N} = \gamma\epsilon$ is a constant of the motion.

Solution (b) Just show that $\frac{d}{dt}(\gamma\epsilon) = 0$. This conclusion holds for arbitrary $K_y(s)$ as long as $\beta \approx 1$.

Homework 3.37 The adiabatic criterion was derived as $K_y \gg \frac{(\beta\gamma)''}{(\beta\gamma)}$ and $\frac{((\beta\gamma)')^2}{(\beta\gamma)^2}$. Give some typical values of K_y and E_z. Does your typical example satisfy the adiabatic criterion?

Homework 3.38 What if the acceleration is not done continuously, but done discretely? In particular, what happens to the adiabatic damping of the motion of a single particle if acceleration is done only every π phase shift in its betatron oscillation? What happens to the emittance of a beam of particles?

Solution As alluded to in the text, radiation damping that predicts $y, y' \propto \frac{1}{\sqrt{\beta\gamma}}$ requires rapid mixing between y and y' by betatron oscillation. The reason of asking this homework is illustrated in Fig. 3.15. Consider the case when acceleration occurs every π phase shift. The upper panel shows what happens to a particle whose betatron motion is such that acceleration occurs at the nodes of its betatron oscillation. The reader should be able to see that this particle has an adiabatic damping with $y \propto \frac{1}{P}$. The lower panel shows what happens if acceleration occurs when the particle's betatron oscillation is at its crests. In this case, adiabatic damping is ineffective, i.e. y is independent of P.

Figure 3.15: What happens to adiabatic damping if the acceleration is done sporadically and not continuously? This figure shows the case when acceleration cavities are located with π betatron phase apart. A particle in the upper panel sees full damping; a particle in the lower panel sees none.

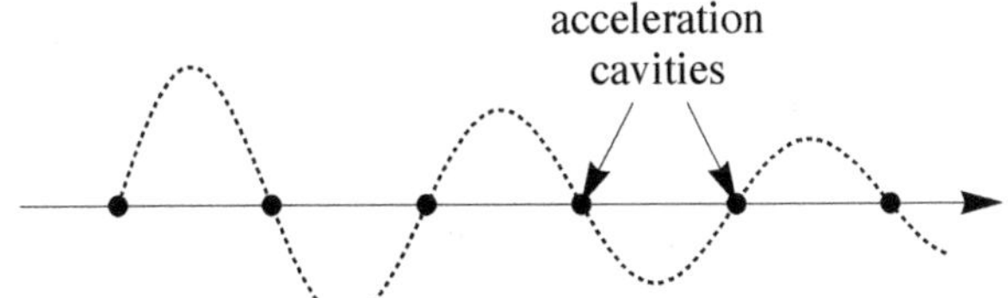

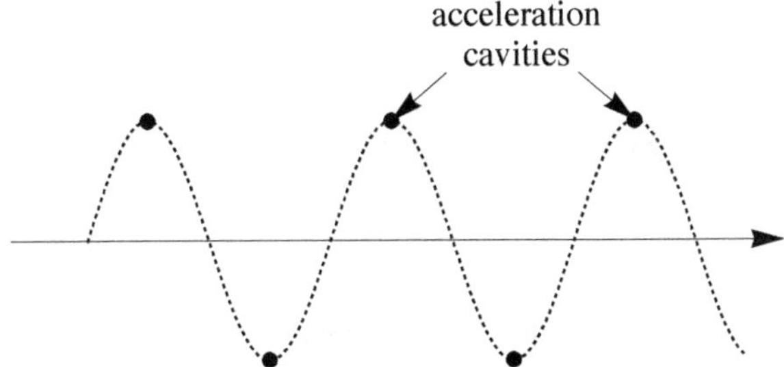

For a beam of particles, the reader should show that the beam's phase space distributions right after each step of acceleration are sketched in the upper panel of Fig. 3.16. The u' dimension of the beam distribution damps according to $\frac{1}{P}$ while the u dimension is invariant. If acceleration is continuous, the beam distributions are sketched in the lower panel. In this case, both the u' and the u dimensions damp according to $\frac{1}{\sqrt{P}}$. In both cases, the damping in phase space area is the same, i.e. area $\propto \frac{1}{P}$. The same statements can be cast in the language of β-function. In terms of β-function, we would say $\beta \propto P$ for the case of discrete acceleration and $\beta \propto P^0$ for the case of continuous acceleration.

Homework 3.39 The raindrop picture Fig. 3.14 would look similar if we replace the raindrops by charged particles and replace gravity by electric field in between two parallel plates. However, consider the physical system consisting of water coming out of a faucet. The water flow will have a shape that looks like some damping curve, like Fig. 3.17 (left). Another physical system one might consider is a water spouting vertically up. The water flow also has a shape that looks like some damping curve (antidamping here), like that sketched in Fig. 3.17 (right).

These curves look like adiabatic damping but are not. They have a very different physical origin. Find the curves $r(z)$ of water faucet and water jet. What is the difference from adiabatic damping mechanism?

Figure 3.16: The upper panel shows the behavior of beam emittance if acceleration is done sporadically. The lower panel is when acceleration is continuous.

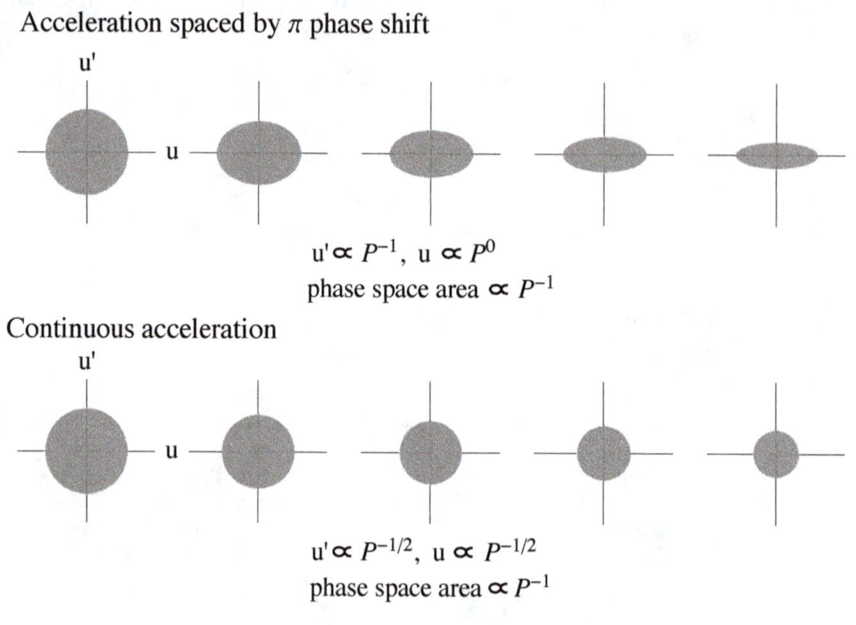

Figure 3.17: An apparent adiabatic damping of a downward water faucet (left) and an apparent adiabatic antidamping of an upward water jet (right).

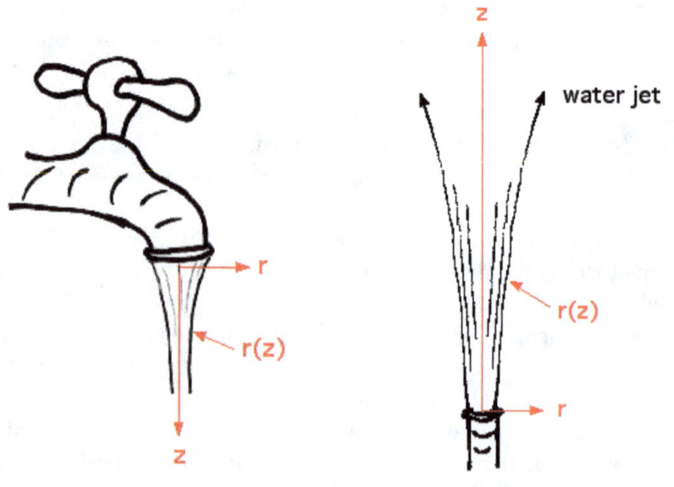

Solution The water flows as an incompressible fluid. This fluid is incompressible in the physical 3-D space. Charged particle beam flows also as an incompressible fluid, but it is incompressible in the phase space. These two systems both behave as incompressible fluid, but in different spaces.

3.5 Linear coupling

So far, we assumed the x- and y-motions are decoupled. A kick in x', for example, will induce a pure x-oscillation in the subsequent motion of the beam. In reality, beamlines contain error fields that cause x- and y-motions to be coupled. Some of these coupled motions are linear, some nonlinear. We discuss linear motion in this section and postpone the nonlinear coupled motion to Sec. 3.6.3.[7] Linear x-y coupled motion is mainly due to skew quadrupolar and solenoidal fields. A rolled normal quadrupole gives a skew quadrupole field component. We will also discuss a linear x-z or y-z coupling due to an acceleration section.

3.5.1 x-y coupling due to skew quadrupole

Consider a single thin-lens skew quadrupole in an otherwise ideal unperturbed storage ring. The skew quadrupole perturbation to particle motion is given by

$$\Delta x' = -\frac{B_y \ell}{B\rho} = -ky, \qquad \Delta y' = \frac{B_x \ell}{B\rho} = -kx\,,$$

where k specifies the skew quadrupole strength.

To describe the linear coupled motion, we need 4×4 matrices. The vector being transformed is chosen to be

$$\begin{bmatrix} x \\ x' \\ y \\ y' \end{bmatrix}.$$

The matrix that describes the skew quadrupole action is

$$M_{\text{sq}} = \begin{bmatrix} 1 & 0 & 0 & 0 \\ 0 & 1 & -k & 0 \\ 0 & 0 & 1 & 0 \\ -k & 0 & 0 & 1 \end{bmatrix}.$$

[7]For linear coupling analysis, see, for example, V. Lebedev, Sec. 2.2.5 and D. Rubin, Sec. 4.7.4, Handbook Accel. Phys. & Eng. 2nd Ed., World Scientific (2013); D.A. Edwards and L.C. Teng, IEEE Trans. Nucl. Sci. 20, 3 (1973).

The map carrying one turn around the skew quadrupole is $M_0 =$

$$\begin{bmatrix} \cos 2\pi\nu_x + \alpha_x \sin 2\pi\nu_x & \beta_x \sin 2\pi\nu_x & 0 & 0 \\ -\gamma_x \sin 2\pi\nu_x & \cos 2\pi\nu_x - \alpha_x \sin 2\pi\nu_x & 0 & 0 \\ 0 & 0 & \cos 2\pi\nu_y + \alpha_y \sin 2\pi\nu_y & \beta_y \sin 2\pi\nu_y \\ 0 & 0 & -\gamma_y \sin 2\pi\nu_y & \cos 2\pi\nu_y - \alpha_y \sin 2\pi\nu_y \end{bmatrix},$$

where all the Courant-Snyder functions for both x and y are unperturbed values evaluated at the location of the skew quadrupole.

The one-turn map around the exit of the skew quadrupole is given by

$$M_{\rm tot} = M_{\rm sq} M_0 .$$

This map describes the coupled motion of the particle. In particular, one calculates the eigenvalues of $M_{\rm tot}$ to obtain the eigentunes of the coupled motion. The matrix $M_{\rm tot}$ has four eigenvalues $e^{\pm i2\pi\nu_+}$ and $e^{\pm i2\pi\nu_-}$ with two eigentunes $\nu_\pm$. After some algebra, $\nu_\pm$ are found to be given by

$$(\cos 2\pi\nu_\pm - \cos 2\pi\nu_x)(\cos 2\pi\nu_\pm - \cos 2\pi\nu_y) = \frac{k^2\beta_x\beta_y}{4} \sin 2\pi\nu_x \sin 2\pi\nu_y ,$$

or

$$\begin{aligned} 2\cos 2\pi\nu_\pm &= \cos 2\pi\nu_x + \cos 2\pi\nu_y \\ &\pm \sqrt{(\cos 2\pi\nu_x - \cos 2\pi\nu_y)^2 + k^2\beta_x\beta_y \sin 2\pi\nu_x \sin 2\pi\nu_y} . \quad (3.25) \end{aligned}$$

When $k = 0$, $\nu_\pm$ become ν_x and ν_y.

Homework 3.40 Rotate a quadrupole magnet by an angle θ around its axis. The quadrupole now contains a skew quadrupolar component. Calculate the 4×4 transfer matrix for this rotated magnet. For simplicity, assume a thin-lens quadrupole. Obtain the special cases of $\theta = 45^\circ$, and $\theta = 90^\circ$ to make sure your result reduces to answers that you expect.

Solution There is an easier way to get to the solution without solving the equation of motion. If the quadrupole is rotated by an angle θ, the same effect can be obtained by first rotating the beam by an angle $-\theta$, letting the beam go through an upright quadrupole, and then rotate the beam back by an angle $+\theta$. The total map is

$$R^{-1} \begin{bmatrix} 1 & 0 & 0 & 0 \\ k & 1 & 0 & 0 \\ 0 & 0 & 1 & 0 \\ 0 & 0 & -k & 1 \end{bmatrix} R = \begin{bmatrix} 1 & 0 & 0 & 0 \\ k\cos 2\theta & 1 & -k\sin 2\theta & 0 \\ 0 & 0 & 1 & 0 \\ -k\sin 2\theta & 0 & -k\cos 2\theta & 1 \end{bmatrix},$$

where

$$R = \begin{bmatrix} \cos\theta & 0 & -\sin\theta & 0 \\ 0 & \cos\theta & 0 & -\sin\theta \\ \sin\theta & 0 & \cos\theta & 0 \\ 0 & \sin\theta & 0 & \cos\theta \end{bmatrix}.$$

Homework 3.41

(a) Verify Eq. (3.25).

(b) Plot $\nu_\pm$ versus ν_x as ν_y is held fixed. Observe how $\nu_\pm$ approach each other when ν_x approaches ν_y in the neighborhood $\nu_x \approx \nu_y$.

(c) Repeat (b) but this time not when $\nu_x \approx \nu_y$ but when $\nu_x + \nu_y \approx$ integer. You will now see a stopband when there are no solution for $\nu_\pm$. Examine the stopband and observe how the eigentunes behave when the stopband is approached.

3.5.2 Coupling resonance stopband

Stopband Homework 3.41 already introduced stopbands. Figure 3.18 shows the contour plots of $\nu_\pm$ in the (ν_x, ν_y) space. The contour lines correspond to $\nu_\pm = 0.1, 0.2, 0.3, 0.4, 0.5$ respectively. The white regions are unstable with no solutions for $\nu_\pm$. The coupling strength is chosen to be an exaggerated value of $k^2\beta_x\beta_y = 0.1$ to show the stopbands more clearly.

Figure 3.18: Perturbation by a single skew quadrupole viewed as contour plots of ν_+ (left) and ν_- (right) in the (ν_x, ν_y) plane. White regions are unstable.

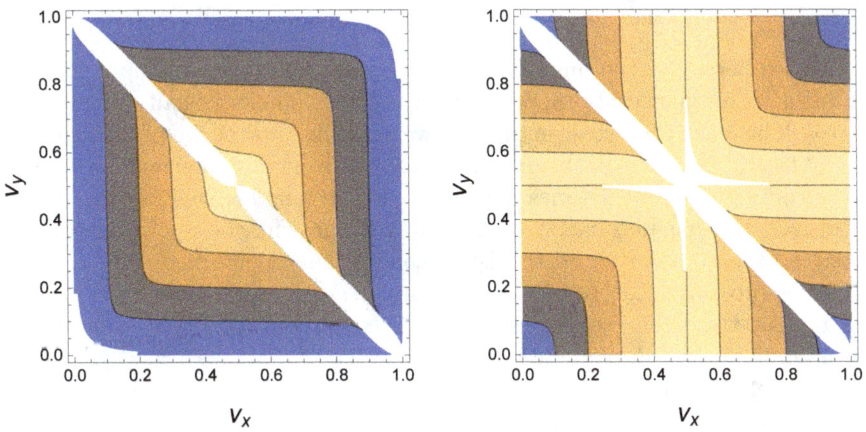

A skew quadrupole drives resonances at

$$\begin{aligned} \nu_x + \nu_y &= n, && \text{sum resonance}, \\ \nu_x - \nu_y &= n, && \text{difference resonance}. \end{aligned}$$

For either case, we have $\cos 2\pi\nu_x = \cos 2\pi\nu_y$, but near the sum resonance, $\sin 2\pi\nu_x$ and $\sin 2\pi\nu_y$ have opposite signs and the quantity

$$k^2\beta_x\beta_y \sin 2\pi\nu_x \sin 2\pi\nu_y$$

in Eq. (3.25) is negative. The motion is therefore unstable near a *sum resonance* (the diagonal white regions in Fig. 3.18). Near a *difference resonance*, on the other hand, the motion is stable.

Let us examine Eq. (3.25) closer. We already said that exactly on the sum resonance, the motion is unstable. How far do we have to stay away from the exact sum resonance to avoid this instability? Using Eq. (3.25), it can be shown that the edge of instability occurs when $\nu_x + \nu_y = n \pm 2\Delta\nu_{\rm SB}$, where

$$\Delta\nu_{\rm SB} \approx \frac{|k|}{4\pi}\sqrt{\beta_x\beta_y}\,. \tag{3.26}$$

Condition (3.26) defines a stopband. Note that unlike the stopband of a quadrupole error, this stopband occurs on both sides of the resonance. The quantity $\Delta\nu_{\rm SB}$ specifies the stopband half-width. See Homework 3.43 and Sec. 3.7.8 later.

Coupling coefficient Equation (3.25) also yields more information about the difference resonance. The skew quadrupole drives both the sum and the difference resonances — only their stabilities are different. It can be shown (Homework 3.43) that near a difference resonance, the eigentunes are given by

$$\nu_\pm = \frac{\nu_x + \nu_y}{2} \pm \sqrt{\frac{(\nu_x - \nu_y - n)^2}{4} + \Delta\nu_{\rm SB}^2}\,. \tag{3.27}$$

We see here that the quantity $\Delta\nu_{\rm SB}$ that appeared as the stopband width for the sum resonance also appears here. It now represents half of the closest approach between the two eigentunes as we vary the unperturbed betatron tunes and let them pass through the difference resonance. It characterizes the strength of the coupling, and sometimes referred to as the *coupling coefficient.*

In fact, this has become one reliable way to measure the x-y linear coupling strength of a storage ring optics. By sweeping the unperturbed tune ν_x across the other tune ν_y, measure two tunes using a tune monitor (what a tune monitor measures are the eigentunes) as the sweeping is taking place, and record the minimal approach of these two tunes. In effect, by measuring the nearest approach of the two tunes near a difference resonance, we obtain the stopband width around the sum resonance because they are really the same driving term showing up on two different occasions.

When a normal quadrupole with focal length f is misaligned by a tilt angle ϕ, then $k = \frac{2\phi}{f}$ (note the extra factor of 2 in this expression, see Homework 3.40). For example, if $f = 10$ m, $\beta_x = 35$ m, $\beta_y = 12$ m (these β-functions are for a QF quadrupole in a FODO cell with phase advance per cell of 60°), and $\phi = 1$ mrad, then we estimate the coupling coefficient of $\Delta\nu_{\rm SB} = 0.3 \times 10^{-3}$.

Emittance beating The coupled x-y motion allows the two beam emittances ϵ_x and ϵ_y to exchange. The unperturbed emittances ϵ_x and ϵ_y are no longer constants of the motion. Instead they will slowly exchange near a difference resonance. The beating frequency at which they exchange and the percentage depth of the unperturbed emittances to be exchanged depend on the distance to the resonance as well as the coupling coefficient. For a linear coupling resonance, the beating frequency is expected to be given by the split between the two eigentunes,

$$\nu_+ - \nu_- = 2\sqrt{\frac{(\nu_x - \nu_y - n)^2}{4} + \Delta\nu_{\rm SB}^2} \tag{3.28}$$

This subject will be discussed further in Secs. 3.7.6 and 3.7.7.

As will be shown soon, it is a general result in nonlinear dynamics that near a nonlinear resonance

$$m_x\nu_x + m_y\nu_y = n\,,$$

particle motion is unstable for sum resonances (i.e. when m_x and m_y have the same sign) and stable for difference resonances (when m_x and m_y have opposite signs). The exception occurs when synchrotron motion is replacing one of the betatron dimensions. In that case, we have a nonlinear *synchrobetatron resonance.*[8] A *negative mass* phenomenon, when it occurs, specific to synchrotron motion, will switch the stability criterion between sum and difference resonances.

Homework 3.42 Extend your 2×2 simulation code to 4×4 to track particle motion in a FODO lattice. Add a weak skew quadrupole to your FODO cell. Observe how the x- and y-emittances transfer back and forth slowly near a difference resonance. Repeat near a sum resonance. There are analytic expressions governing these behaviors. See Sec. 3.7.7.

Homework 3.43

(a) Use Eq. (3.25) to find expressions for $\nu_\pm$ near the sum resonance $\nu_x+\nu_y = n$ and verify Eq. (3.26). Show that

$$\nu_+ - \nu_- = 2\sqrt{\frac{(\nu_x + \nu_y - n)^2}{4} - \Delta\nu_{\rm SB}^2}\,. \tag{3.29}$$

Compared with Eq. (3.28). This result will reappear in Sec. 3.7.8.

(b) Verify Eq. (3.27) near the difference resonance $\nu_x - \nu_y = n$.

Homework 3.44 An alert reader would have noticed there are white unstable regions in Fig. 3.18 in addition to the linear sum resonance stopband. Those are in fact the skew quadrupole driving integer and half-integer resonances. Strictly, they are not the linear coupling resonances discussed in this section, but they are nevertheless driven by the single skew quadrupole. Derive the expressions for their locations and their stopband widths.

[8] A. Piwinski, Sec. 2.3.4, Handbook Accel. Phys. & Eng., 2nd Ed., World Scientific (2013); A. Piwinski and A. Wrulich, DESY Report 76/07 (1976); T. Suzuki, Part. Accel. 27, 157 (1990).

Solution This time, in Eq. (3.25), the quantity inside the square root is positive. But the stability is violated anyway because the right-hand-side becomes > 2 or < -2. The boundaries of these stopbands are determined by

$$\tan \pi\nu_x \tan \pi\nu_y = \frac{1}{4}k^2\beta_x\beta_y, \qquad \cot \pi\nu_x \cot \pi\nu_y = \frac{1}{4}k^2\beta_x\beta_y,$$

respectively. Derive these expressions and confirm these stopband widths in Fig. 3.18.

Unlike the linear coupling resonances, the stopband widths of these resonances are quadratic in k. In a case with a more realistic value of k, these stopbands will not be easily visible in Fig. 3.18.

3.5.3 Linear x-y coupling due to solenoid

Body field Solenoids are another important element that couples x and y. Equation of motion in a uniform solenoidal field $B_0\hat{z}$ is given by ($\gamma =$ constant)

$$\begin{aligned}
& m\gamma\dot{\vec{v}} = e\vec{v}\times B_0\hat{z} \\
\Longrightarrow\quad & \dot{v}_x = \frac{eB_0}{m\gamma}v_y, \quad \dot{v}_y = -\frac{eB_0}{m\gamma}v_x \\
\Longrightarrow\quad & v_x' = Kv_y, \quad v_y' = -Kv_x, \quad \text{with} \quad K = \frac{B_0}{B\rho} \\
\Longrightarrow\quad & x'' = Ky', \quad y'' = -Kx'.
\end{aligned}$$

Let $u = x + iy$, then

$$\begin{aligned}
& u'' = -iKu' \\
\Longrightarrow\quad & u'(s) = u'(0)e^{-iKs} \\
\Longrightarrow\quad & u(s) = u(0) + i\frac{u'(0)}{K}(e^{-iKs} - 1) \\
\Longrightarrow\quad & M(L|0) = \begin{bmatrix} 1 & \frac{\sin KL}{K} & 0 & \frac{1-\cos KL}{K} \\ 0 & \cos KL & 0 & \sin KL \\ 0 & -\frac{1-\cos KL}{K} & 1 & \frac{\sin KL}{K} \\ 0 & -\sin KL & 0 & \cos KL \end{bmatrix}.
\end{aligned}$$

It should be immediately pointed out that this map is not symplectic — even though its determinant is 1!

End fields As emphasized in Sec. 1.4, we must not forget the end fields. In this section below, we will assume a hard edge model for our solenoid — see Fig. 1.33. It should be mentioned that this model is acceptable only to first order linear optics. It breaks down, and in fact diverges, when one pushes it to higher orders. From that point of view, the hard edge model is unphysical. But it is fine for our purpose.

At the solenoid entrance, in the hard edge model, from Eq. (1.27),

$$\begin{aligned}
\vec{B} &= -\frac{1}{2}B_0(x\hat{x} + y\hat{y})\delta(z) \\
\Longrightarrow \quad m\gamma\dot{\vec{v}} &= -\frac{eB_0}{2}v\hat{z} \times (x\hat{x} + y\hat{y})\delta(z) = -\frac{eB_0}{2}v(x\hat{y} - y\hat{x})\delta(z) \\
\Longrightarrow \quad \Delta x' = \frac{K}{2}y, \quad \Delta y' &= -\frac{K}{2}x \\
\Longrightarrow \quad M_{\text{entrance}} &= \begin{bmatrix} 1 & 0 & 0 & 0 \\ 0 & 1 & \frac{K}{2} & 0 \\ 0 & 0 & 1 & 0 \\ -\frac{K}{2} & 0 & 0 & 1 \end{bmatrix}.
\end{aligned}$$

Similarly

$$M_{\text{exit}} = \begin{bmatrix} 1 & 0 & 0 & 0 \\ 0 & 1 & -\frac{K}{2} & 0 \\ 0 & 0 & 1 & 0 \\ \frac{K}{2} & 0 & 0 & 1 \end{bmatrix}.$$

The total solenoid matrix is therefore

$$\begin{aligned}
M_{\text{solenoid}} &= M_{\text{exit}} M(L|0) M_{\text{entrance}} \\
&= \begin{bmatrix} \frac{1+C}{2} & \frac{S}{K} & \frac{S}{2} & \frac{1-C}{K} \\ -\frac{KS}{4} & \frac{1+C}{2} & -\frac{K(1-C)}{4} & \frac{S}{2} \\ -\frac{S}{2} & -\frac{1-C}{K} & \frac{1+C}{2} & \frac{S}{K} \\ \frac{K(1-C)}{4} & -\frac{S}{2} & -\frac{KS}{4} & \frac{1+C}{2} \end{bmatrix}, \qquad (3.30)
\end{aligned}$$

with $C = \cos KL, S = \sin KL$. We note that all 16 elements are filled, and that the final matrix looks very different if the solenoid ends are ignored. Homework 3.47 shows this map is symplectic, and it correctly represents the map of the hard-edge solenoid.

Another equivalent, perhaps more abstract, way to describe the dynamics is as follows. The map M_{entrance} is what transforms (x, x', y, y') to the canonical coordinates

$$\begin{bmatrix} x \\ p_x \\ y \\ p_y \end{bmatrix} = \begin{bmatrix} x \\ x' - \frac{A_x}{B\rho} \\ y \\ y' - \frac{A_y}{B\rho} \end{bmatrix},$$

where $A_x = -\frac{B_0}{2}y, A_y = \frac{B_0}{2}x$ are the components of the vector potential. The map M_{exit} is what transforms (x, p_x, y, p_y) back to (x, x', y, y'). Inside the solenoid proper, and when the two edge maps are included, the matrix $M(L|0)$ acts on the canonical coordinates only. The twists at the solenoid edges come about only because we used our beloved (x, x', y, y') for the coordinate system. Had we been using the more formal choice of (x, p_x, y, p_y), then there is no need to introduce the edge maps. The two descriptions are completely equivalent, however.

Figure 3.19: The red arrows indicate the kicks a particle receives going through various coupling elements.

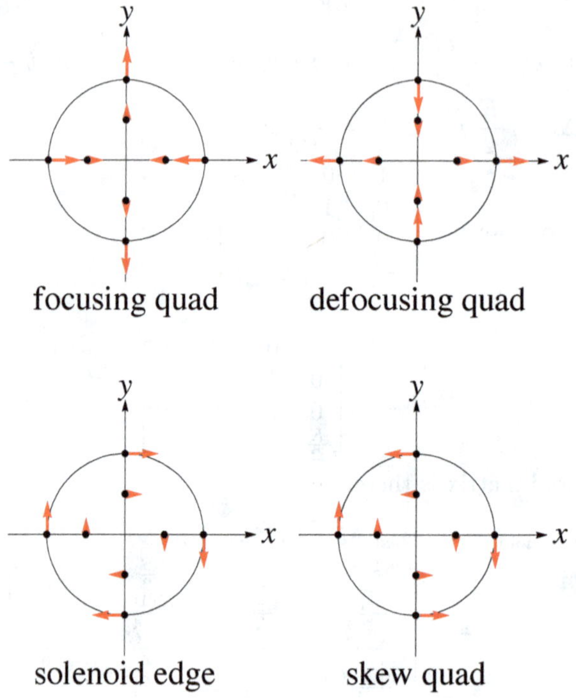

How strong are the solenoid end fields? The strength of the field at the solenoid edges can be compared with that of a typical quadrupole as follows. The kick angle delivered to a particle by a quadrupole is given by $\frac{G\ell}{B\rho}x$. In comparison, the angle delivered by the solenoid edge field is $\frac{1}{2}Kx$. If we take $G \sim \frac{B_{\text{poletip}}}{a}$ with a the magnet pole tip radius, then their strength ratio is

$$\sim \left(\frac{1}{2}\frac{B_0}{B\rho}\right) \Big/ \left(\frac{B_{\text{poletip}}\ell/a}{B\rho}\right)$$

If we take $B_{\text{poletip}} \sim B_0$, then it becomes $\sim \frac{a}{2\ell}$. Typically, $\ell \gg a$, so the solenoid edge field are relatively weak compared with a typical quadrupole magnet.

Figure 3.19 illustrates the patterns in the physical x-y space of kicks given to a beam as it passes through various coupling elements. The quadrupole and skew quadrupole produce a squeezing pattern, while a solenoid edge produces a twisting pattern.

Immersed gun in solenoid This unique twisting feature of a solenoid edge can be made to use in elegant manners under the generic subject of *lattice* *adaptors.*[9] These are gadgets in the magic bag developed by the lattice designers to be used to make transitions between lattices for various special applications. We shall not indulge in this interesting subject in our lectures except to mention one notable example of a lattice adaptor is when an electron gun is installed inside a solenoid.

This is called "immersed gun" technique. The beam is produced inside the solenoid. As it exits the solenoid, it is made to twist by the solenoid end field, and this additional twist does the magic. In contrast, if the beam is produced outside and made to go through both the entrance and exit of a solenoid, the two twists would cancel each other.

The solenoid-edge twist map is nonsymplectic. The immersed gun technique is an elegant way to manipulate the x- and y-emittances of a beam. A beam produced round at the source, for example, can be made to exit the solenoid as a flat beam. The aspect ratio of the flat beam can be varied by the solenoid strength over a wide range.

Later in Sec. 4.3, we shall mention another example of immersed gun technique, i.e. if we immerse the gun inside another device called Wien filter, it in principle can manipulate the emittances between one of the transverse dimensions and the longitudinal dimension similar to the solenoid-edge used to manipulate between the two transverse dimensions.

Homework 3.45 When $KL = \pi$ for a solenoid, calculate the solenoid body map $M(L|0)$ (not including the solenoid edges). Give physical picture of what has happened. Repeat the problem for the case $KL = 2\pi$.

Homework 3.46 Show that the angular momentum $L_z = xy' - yx'$ is preserved when a particle goes through a solenoid. Show explicitly that L_z is preserved only if the entrance and exit fringe fields are taken into account.

Homework 3.47 Matrices $M(L|0), M_{\text{entrance}}, M_{\text{exit}}$ are not symplectic. But show that the matrix M_{solenoid} is symplectic.

Homework 3.48 A solenoid is inserted in a straight beamline as in Fig. 3.20. It is desired to make a point-to-point focusing imaging (both x and y) from position A to position B. Design your system accordingly by choosing the parameters L_1, L_2, K, L. Can A and B be switched as object and image?

Solution $KL_2 = \frac{2KL_1 \cos\frac{KL}{2} + 4\sin\frac{KL}{2}}{KL_1 \sin\frac{KL}{2} - 2\cos\frac{KL}{2}}$. For example, we might take $KL = \pi$. Then the imaging condition requires $K^2 L_1 L_2 = 4$. This means A and B can be traded off against each other, including exchanging positions.

Homework 3.49 Solenoids sometimes make large perturbation to the x- and y optics in an accelerator. It leads to the question of whether two solenoids

[9] This is a large subject first initiated by Y. Derbenev, Michigan Univ. Report No. 91-2 (1991); UM HE 93-20 (1993); UM HE 98-04 (1998).

Figure 3.20: Point-to-point focusing of a solenoid. A and B are the focal points.

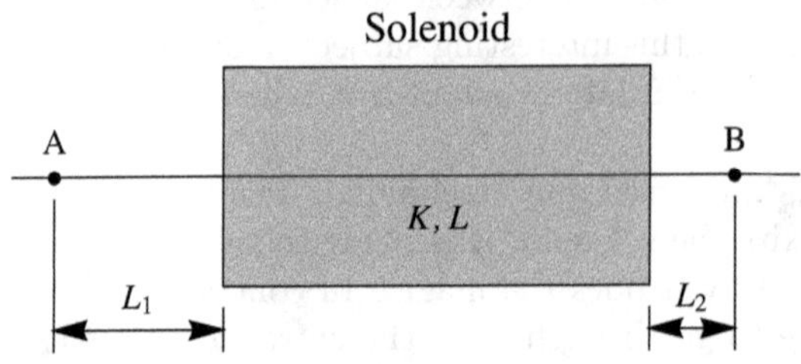

can be installed in such a way that their net perturbation effects are minimized. For example, one might consider two solenoids of equal strength but opposite polarity be installed back-to-back. Or one may consider two solenoids of the same strength and polarity installed with a $-I$ straight section in between. Find the net 4×4 map for these two solenoid systems. Such cancellation tricks have actually been considered.

Solution Because the solenoid is necessarily thick element, be careful with the $-I$ map you introduce. It should be the map between the two solenoid entrances or between their exits, but not from one exit to the next entrance.

Homework 3.50 It is well known that the vector potential $\vec{A}$ is not unique. One can add to it any quantity ∇G without affecting the magnetic field, and thus without affecting any physical results. In the text, we demonstrated the fact that the solenoid edge field simply transforms the coordinates from (x, x', y, y') to the appropriate canonical momentum as a particle enters the solenoid. However, that statement was made assuming a particular choice of $\vec{A}$. What happens when we choose a different $\vec{A}$? Will this statement still hold?

Homework 3.51 Just like a skew quadrupole, a solenoid can also produce stopbands. Consider an otherwise-perfect storage ring with a solenoid inserted. Calculate the stopbands in the (ν_x, ν_y) space. If needed, assume the solenoid is weak and find the stopband to first order in the solenoid strength.

Solution Be careful when you insert a solenoid. As a thick-lens object, when you insert it into your lattice, you must remove a drift space length to make room for it, i.e. the matrix you insert should be $M_{\text{solenoid}} M_{\text{negative drift}}$.

Homework 3.52 This homework is a specific example of Homework 3.51. Sometimes a collider storage ring is required to manipulate the spin polarization of the circulating beam. An ingenious device called the "Siberian snake" was invented by Derbenev and Kondratenko[10] to manipulate the spin dynamics. We

[10] Ya.S. Derbenev and A.M. Kondratenko, Sov. Phys. Doklady 20, 562 (1976); Ya.S. Derbenev et al., Part. Accel. 8, 115 (1978). The name "Siberian snake" was coined by Ernest Courant.

shall not discuss such devices in these lectures, but one of the snake designs, the simplest version, consists of simply a solenoid.

A drift space of length L is replaced by a solenoid of the same length in a collider storage ring. The purpose of the solenoid is to rotate the spin of beam particles by π around the z-axis. The required solenoid strength is $KL = \frac{B_0 L}{B\rho} = \frac{2\pi}{1+a}$, where a is the anomalous gyromagnetic ratio of the beam particle, $a = 0.001160$ for electrons and 1.7928 for protons. Consider the case of 10 GeV/c protons, the required solenoid strength is $B_0 L = 75$ T-m. This strong solenoid may take $L = 15$ m and $B_0 = 5$ T.

Consider now a drift space in a collider where the one-turn map around its center is

$$M_0 = \begin{bmatrix} \cos\mu_x & \beta_x \sin\mu_x & 0 & 0 \\ -\frac{1}{\beta_x}\sin\mu_x & \cos\mu_x & 0 & 0 \\ 0 & 0 & \cos\mu_y & \beta_y \sin\mu_y \\ 0 & 0 & -\frac{1}{\beta_y}\sin\mu_y & \cos\mu_y \end{bmatrix},$$

where $\mu_{x,y} = 2\pi\nu_{x,y}$. When the solenoid is installed, the one-turn map becomes

$$M_1 = M_{\text{solenoid}} M_{-L/2} M_0 M_{-L/2}\,,$$

where M_{solenoid} is given by Eq. (3.30) and $M_{-L/2}$ is a drift space of length $-\frac{L}{2}$.

Establish the one-turn map with the Siberian snake solenoid. Study the stability region in the (ν_x, ν_y) space. Because the solenoid is very strong, you will find the stable regions are rather limited, and in the unstable region, the growth rate is very fast.

Solution If you are curious, you can work it out yourself, or here is the result. The shaded regions in the (ν_x, ν_y) space of Fig. 3.21 are stable. White regions are unstable. There are two modes. Beam stability requires *both* modes, shown in two contour plots respectively in Fig. 3.21, to be stable. Not much tune space is available for stable operation.

3.5.4 Linear x-z and y-z coupling due to electrostatic acceleration section

The end field of a solenoid edge comes from the fact that the magnetic field cannot simply suddenly stop at the ends of a solenoid and it has to bend into transverse direction to terminate somewhere outside the accelerator pipe. A similar situation occurs to an electrostatic accelerating section. The longitudinal electric field must also go transverse to be terminated at both ends of the acceleration section, as sketched in Fig. 3.22.

Perhaps one can make a detour observation on Fig. 3.22. The realistic system on the left is to be replaced by the hard-edge model on the right. It is instructive to see that the field pattern has been altered significantly in the hard-edge

Figure 3.21: Stable regions in the tune space when a Siberian snake solenoid is installed. The shaded regions are stable (|eigenvalue| = 1). Parameters assumed are $B_0 = 5$ T, $L = 15$ m, $\beta_x = \beta_y = 1$ m, $P = 10$ GeV/c, proton beam. Stability of two modes are shown. Beam stability require both modes to be stable. Reminiscence of sum resonance stopbands are apparent.

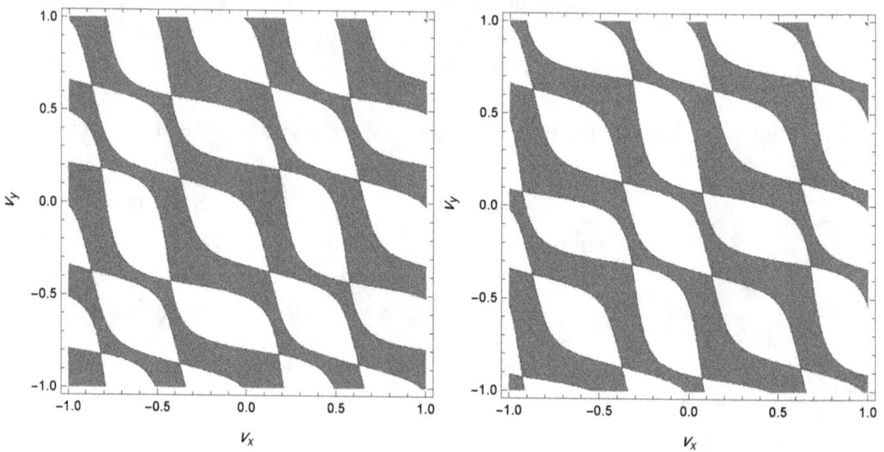

approximation in a very compromising manner. However, the key point, the point that the field lines do not terminate midair, is the one feature that must not be compromised. Physics is an art after all, an art of how to make valid approximations.

In addition to solenoids and acceleration sections, there is one more example when end fields play a principal role, namely the case of a Wien filter. We will discuss that case in Sec. 4.3.

Body field In this case, since there is an electric field involved, the simplifying condition $\gamma =$ constant no longer holds. We need to return to the equation of motion. First, let us look at the motion inside the bulk of the acceleration section — no edge fields yet. We shall see that as the beam energy varies in this case, the algebra becomes much more involved.

Considering a particle with initial conditions $x_0, p_{x0}, y_0, p_{y0}, z_0, E_0$. Inside the body of the acceleration section, we have

$$\frac{dp_x}{dt} = 0, \quad \frac{dp_y}{dt} = 0, \quad \frac{dp_z}{dt} = eE_z \,.$$

We first immediately have

$$p_x = p_{x0}, \qquad p_y = p_{y0} \,.$$

Figure 3.22: Field pattern of an electrostatic acceleration section. A more realistic layout is on the left, and a hard edge model is shown on the right.

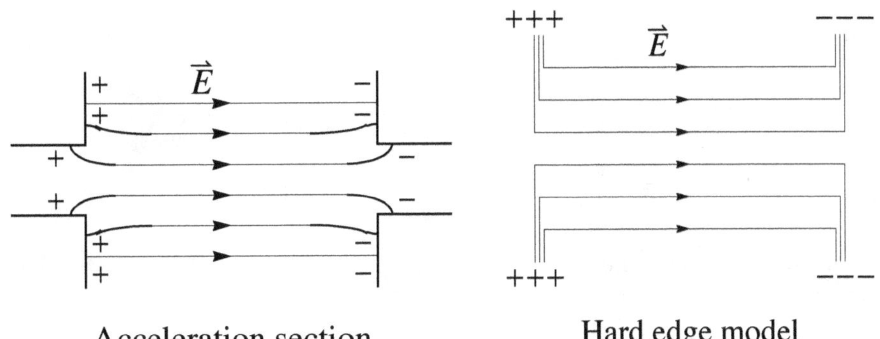

We then note that the particle energy gain satisfies the condition

$$\begin{aligned} E^2(t) &= p_{x0}^2c^2 + p_{y0}^2c^2 + p_z^2(t)c^2 + m^2c^4\,, \\ \Longrightarrow \quad 2E\frac{dE}{dt} &= 2p_zc^2eE_z \\ \Longrightarrow \quad \frac{dE}{dt} &= eE_z\frac{p_zc^2}{E} = eE_zv_z \\ \Longrightarrow \quad \frac{dE}{dz} &= \frac{1}{v_z}\frac{dE}{dt} = eE_z\,, \end{aligned}$$

from which we obtain

$$E(z) = E_0 + eE_zz\,. \tag{3.31}$$

Equation (3.31) just derived the well-known fact that energy gain is equal to the potential drop. But this well-known fact is a neat consequence of the Lorentz force equation, plus the kinematics in relativity. It is not meant to be a triviality.

Once we have $E(z)$, we have

$$\begin{aligned} (p_z^2(z) - p_{z0}^2)c^2 &= E(z)^2 - E_0^2 \\ \Longrightarrow \quad p_z(z) &= \sqrt{\frac{1}{c^2}(E(z)^2 - E_0^2) + p_{z0}^2} = \frac{1}{c}\sqrt{2E_0eE_zz + e^2E_z^2z^2 + c^2p_{z0}^2}\,. \end{aligned}$$

and

$$v_z(z) = \frac{p_z(z)c^2}{E(z)} = \frac{c\sqrt{2E_0eE_zz + e^2E_z^2z^2 + c^2p_{z0}^2}}{E_0 + eE_zz}\,.$$

We also have $v_{x,y}$,

$$v_x(z) = \frac{p_x(z)c^2}{E(z)} = \frac{p_{x0}c^2}{E_0 + eE_zz}\,,$$

$$v_y(z) = \frac{p_y(z)c^2}{E(z)} = \frac{p_{y0}c^2}{E_0 + eE_z z}.$$

and x', y',

$$x'(z) = \frac{v_x(z)}{v_z(z)} = \frac{p_{x0}c}{\sqrt{2E_0 eE_z z + e^2E_z^2 z^2 + c^2p_{z0}^2}},$$
$$y'(z) = \frac{v_y(z)}{v_z(z)} = \frac{p_{y0}c}{\sqrt{2E_0 eE_z z + e^2E_z^2 z^2 + c^2p_{z0}^2}}.$$

We now calculate the more difficult quantities x and y. We will calculate their values at $z = L$, the end of the accelerator section,

$$\begin{aligned} x(L) &= x_0 + \int_0^L x'(z)dz \\ &= x_0 + p_{x0}c\int_0^L \frac{dz}{\sqrt{2E_0 eE_z z + e^2E_z^2 z^2 + c^2p_{z0}^2}} \\ &= x_0 + \frac{p_{x0}c}{eE_z}F, \end{aligned}$$

where[11]

$$F = \ln\left[\frac{E_0 + eE_zL + \sqrt{c^2p_{z0}^2 + eE_zL(2E_0 + eE_zL)}}{cp_{z0} + E_0}\right].$$

Similarly,

$$y(L) = y_0 + \frac{p_{y0}c}{eE_z}F.$$

When $g = \frac{eE_zL}{cp_{z0}} \ll 1$, we have

$$F \approx g\left(1 - \frac{E_0}{2cp_{z0}}g - \frac{c^2p_{z0}^2 - 3E_0^2}{6c^2p_{z0}^2}g^2\right),$$

and therefore

$$x(L) \approx x_0 + L\frac{p_{x0}}{p_{z0}}\left(1 - \frac{E_0}{2cp_{z0}}g - \frac{c^2p_{z0}^2 - 3E_0^2}{6c^2p_{z0}^2}g^2\right).$$

The linear term in g, with a minus sign, indicates an adiabatic damping effect due to the acceleration (Homework 3.53).

[11] We have used

$$\int dx\frac{1}{\sqrt{1 + ax + bx^2}} = \frac{1}{\sqrt{b}}\ln\left[a + 2bx + 2\sqrt{b}\sqrt{1 + ax + bx^2}\right].$$

End field We now consider the end fields. At the entrance of the acceleration section, in the hard edge model, a similar expression like Eq. (1.27) holds, and we have an end field

$$\vec{E} = -\frac{1}{2}E_z(x\hat{x} + y\hat{y})\delta(z)\,.$$

For an acceleration field $E_z > 0$, the entrance end field points inward toward the axis, as shown in Fig. 3.22 (right). This end field contributes to an inward transverse momentum kick

$$\begin{aligned}\Delta\vec{p} &= \int dt\, e\vec{E} = \int \frac{dz}{v_z}\, e\vec{E} \\ &= -\frac{E_0}{2c^2 p_{z0}}\, eE_z\,(x\hat{x} + y\hat{y})\,.\end{aligned}$$

The corresponding kicking angles are

$$\Delta x' = -\frac{E_0}{2c^2p_{z0}^2}\, eE_z x_0, \qquad \Delta y' = -\frac{E_0}{2c^2p_{z0}^2}\, eE_z y_0\,.$$

Similarly, the exit end field contributes to kicking angles

$$\begin{aligned}\Delta x' &= \frac{E(L)}{2c^2p_z^2(L)}\, eE_z x(L) \\ &= \frac{eE_z}{2}\,\frac{E_0 + eE_zL}{2E_0eE_zL + e^2E_z^2L^2 + c^2p_{z0}^2}\left(x_0 + \frac{p_{x0}c}{eE_z}F\right), \\ \Delta y' &= \frac{eE_z}{2}\,\frac{E_0 + eE_zL}{2E_0eE_zL + e^2E_z^2L^2 + c^2p_{z0}^2}\left(y_0 + \frac{p_{y0}c}{eE_z}F\right).\end{aligned}$$

Net focusing The inward kick at the entrance provides focuses the particle motion both in the x- and the y planes. The outward kick at the exit is defocusing. To see if the net effects is focusing or defocusing, on top and beyond the adiabatic damping due to the acceleration, let us consider the total x kick angle for the case $p_{x0} = 0$,

$$\Delta x' = \frac{eE_z}{2}\left[-\frac{E_0}{c^2p_{z0}^2} + \frac{E_0 + eE_zL}{2E_0eE_zL + e^2E_z^2L^2 + c^2p_{z0}^2}\right] x_0\,.$$

In case $g = \frac{eE_zL}{cp_{z0}} \ll 1$, we have

$$\Delta x' \approx -\frac{2E_0^2 - c^2p_{z0}^2}{2c^2p_{z0}^2}\,\frac{g^2}{L}x_0\,. \tag{3.32}$$

The net effect is therefore focusing because the kick is proportional to $-x_0$. Similar expression holds for y. The focal length for both planes is

$$f = \frac{2c^2p_{z0}^2}{2E_0^2 - c^2p_{z0}^2}\,\frac{L}{g^2}\,.$$

Take for example, a case of a proton with kinetic energy of 1 MeV (i.e. $E_0 = mc^2 + 1$ MeV), and an acceleration section with $E_z = 100$ kV/m, $L = 1$ m, we have $g = 2.3 \times 10^{-3}$, and $f = 400$ m. This focusing is weak and is mainly a low-energy effect.

The focusing effect occurs at the entrance and the defocusing effect occurs at the exit. Since the particle energy is higher at the exit than at the entrance (for the case when $E_z > 0$), the defocusing kicking angle at the exit is smaller than the focusing kicking angle at the entrance. The net result is focusing.

For the case of a decelerating section, the particle energy is lower at the exit with a larger kicking angle. But this time, with $E_z < 0$, this larger kicking angle is focusing this time. The net result is that the decelerating section is also net focusing. This effect of a net focusing for both accelerating and decelerating sections is reflected by the fact that, in Eq. (3.32), the net kicking angle is quadratic in g.

Paraxial ray approximation So far we have assumed a hard-edge model with

$$\vec{E} \;=\; -\frac{1}{2}E_z r\delta(z)\hat{r} + E_z H(z)H(L-z) + \frac{1}{2}E_z r\delta(z-L)\hat{r}\,,$$

similarly to Eq. (1.27) for a hard-edge solenoid. In case E_z is not a hard-edge but is represented by $E_z(z)$, then similarly to Eq. (1.29), to first order in x, y, we have

$$\vec{E} \;\approx\; E_z(z)\hat{z} - \frac{1}{2}E_z'(z)r\hat{r}\,.$$

Higher order expressions can be obtained like Eq. (1.30).

Homework 3.53 The text calculated in detail the particle trajectory going through the body of the acceleration section. As the particle is accelerated, one expects to see adiabatic damping. Consider the case of a small acceleration step with small $g = \frac{eE_zL}{cp_{z0}}$. Identify where the adiabatic damping effect is to be observed. The map in the body field is nonsymplectic.

Solution We first calculate $x'(L)$ for small g. Show that it is given by

$$x'(L) \;\approx\; x_0'\left(1 - \frac{E_0}{cp_{z0}}g\right).$$

This means an adiabatic damping decrement of $\Delta x' \approx -x_0'\frac{E_0}{cp_{z0}}g$, which must be connected to $-x_0'\frac{\Delta p_z}{p_{z0}}$ to establish the adiabatic damping mechanism. Indeed, this is established because $\Delta p_z = eE_z\Delta t = eE_z\frac{L}{v_z} = eE_z\frac{LE_0}{c^2p_{z0}} = \frac{E_0}{cp_{z0}}gp_{z0}$.

Homework 3.54 The text calculated $x(z), p_x(z), y(z), p_y(z), E(z), p_z(z)$ over the body of the acceleration section. There is one notable missing quantity, namely the time of arrival $t(L)$ of the particle when it reaches $z = L$. Calculate $t(L)$ for a particle with appropriate initial conditions.

Solution First note that in the body of the acceleration section $t(L)$ depends on the initial values of p_{z0}, t_0, E_0, but not on x_0, p_{x0}, y_0, p_{y0},

$$\begin{aligned} t(L) - t_0 &= \int_0^L \frac{dz}{v_z(z)} = \int_0^L \frac{dz}{c} \frac{E_0 + eE_z z}{\sqrt{2E_0 eE_z z + e^2 E_z^2 z^2 + c^2 p_{z0}^2}} \\ &= \frac{L}{c} \frac{\sqrt{2E_0 eE_z L + e^2 E_z^2 L^2 + c^2 p_{z0}^2} - cp_{z0}}{eE_z L} . \end{aligned}$$

In the limit $g = \frac{eE_z L}{cp_{z0}} \ll 1$,

$$t(L) - t_0 \approx \frac{L}{c}\left(\frac{E_0}{cp_{z0}} - \frac{E_0^2 - c^2 p_{z0}^2}{2c^2 p_{z0}^2} g\right) .$$

When $g = 0$, it reduces to $t(L) = \frac{LE_0}{c^2 p_{z0}} = \frac{L}{v_{z0}}$ as it should.

3.5.5 Misaligned element

Rotation Consider a beamline element (or a beamline section altogether) whose x-y coupled 4×4 linear map is T. What is the map for the same beamline but now rotated by a clockwise angle θ around its axis? The answer can be provided by first rotating the coordinate system by a counter-clockwise angle $-\theta$, go through the map T, and then rotate the coordinate system back by a clockwise angle θ. The result is a map given by

$$T_{\text{rotated}} = R^{-1} T R, \quad \text{where} \quad R = \begin{bmatrix} \cos\theta & 0 & -\sin\theta & 0 \\ 0 & \cos\theta & 0 & -\sin\theta \\ \sin\theta & 0 & \cos\theta & 0 \\ 0 & \sin\theta & 0 & \cos\theta \end{bmatrix} .$$

A simple example occurs when T is uncoupled, with

$$T = \begin{bmatrix} T_x & 0 \\ 0 & T_y \end{bmatrix} ,$$

where T_x and T_y are 2×2 maps. Then

$$T_{\text{rotated}} = \begin{bmatrix} T_x \cos^2\theta + T_y \sin^2\theta & -(T_x - T_y)\sin\theta\cos\theta \\ -(T_x - T_y)\sin\theta\cos\theta & T_x \sin^2\theta + T_y \cos^2\theta \end{bmatrix} .$$

A special example is when $T_x = T_y$, then the rotation does not change the map, as one would expect.

Another example is thin-lens quadrupole,

$$T_x = \begin{bmatrix} 1 & 0 \\ -\frac{1}{f} & 1 \end{bmatrix}, \quad T_y = \begin{bmatrix} 1 & 0 \\ \frac{1}{f} & 1 \end{bmatrix},$$

$$\Longrightarrow \quad T_{\text{rotated}} = \begin{bmatrix} 1 & 0 & 0 & 0 \\ -\frac{\cos 2\theta}{f} & 1 & \frac{\sin 2\theta}{f} & 0 \\ 0 & 0 & 1 & 0 \\ \frac{\sin 2\theta}{f} & 0 & \frac{\cos 2\theta}{f} & 1 \end{bmatrix} .$$

Displacement A similar technique also applies to the case when a beamline or a beamline element is misaligned by a transverse displacement $(\Delta x, \Delta y)$. Its map can be obtained by first introducing a displacement $(-\Delta x, -\Delta y)$ to the particle coordinates, followed by the beamline map, and then a displacement map $(\Delta x, \Delta y)$. The only slight difference from the case of a beamline rotation is that in this case the map cannot be described by 4×4 matrices, although it can still be described if we apply 5×5 matrices (Homework 3.56).

Homework 3.55 Show that the map of a solenoid, Eq. (3.30), is invariant under axis rotation.

Homework 3.56 The text gives the map for a rotated thin-lens quadrupole. Extend the map to describe a map for a thin-lens rotated quadrupole with a transverse displacement $(\Delta x, \Delta y)$.

Solution Consider the vector $(x, x', y, y', 1)$. The map can be cast as a 5×5 matrix,

$$\begin{bmatrix} 1 & 0 & 0 & 0 & \Delta x \\ 0 & 1 & 0 & 0 & 0 \\ 0 & 0 & 1 & 0 & \Delta y \\ 0 & 0 & 0 & 1 & 0 \\ 0 & 0 & 0 & 0 & 1 \end{bmatrix} \begin{bmatrix} 1 & 0 & 0 & 0 & 0 \\ -\frac{\cos 2\theta}{f} & 1 & \frac{\sin 2\theta}{f} & 0 & 0 \\ 0 & 0 & 1 & 0 & 0 \\ \frac{\sin 2\theta}{f} & 0 & \frac{\cos 2\theta}{f} & 1 & 0 \\ 0 & 0 & 0 & 0 & 1 \end{bmatrix} \begin{bmatrix} 1 & 0 & 0 & 0 & -\Delta x \\ 0 & 1 & 0 & 0 & 0 \\ 0 & 0 & 1 & 0 & -\Delta y \\ 0 & 0 & 0 & 1 & 0 \\ 0 & 0 & 0 & 0 & 1 \end{bmatrix}$$

$$= \begin{bmatrix} 1 & 0 & 0 & 0 & 0 \\ -\frac{\cos 2\theta}{f} & 1 & \frac{\sin 2\theta}{f} & 0 & \frac{1}{f}(\Delta x \cos 2\theta - \Delta y \sin 2\theta) \\ 0 & 0 & 1 & 0 & 0 \\ \frac{\sin 2\theta}{f} & 0 & \frac{\cos 2\theta}{f} & 1 & -\frac{1}{f}(\Delta y \cos 2\theta + \Delta x \sin 2\theta) \\ 0 & 0 & 0 & 0 & 1 \end{bmatrix}.$$

Physical meaning should be clear.

3.6 Nonlinear resonance

So far we have discussed linear resonances due to dipole and quadrupole perturbations driving integer and half-integer resonances, respectively, and linear coupling resonances driven by skew quadrupoles and solenoids. All these were linear resonances for which exact analytical results could be obtained using matrix algebra. Below we briefly discuss the nonlinear resonances. Consider a thin-lens nonlinear perturbation at location $s = 0$. As we will see, exact solutions are now not possible in general, and we shall depend on a perturbation calculation assuming the nonlinear perturbation is weak.

3.6.1 Nonlinear 1-D resonance

Let us first give a general, qualitative discussion of the effect of these nonlinear perturbations before we enter the more detailed systematic treatment in a later section 3.7 under the subject of smooth approximation. For this purpose, let us consider a 1-D case and let the nonlinearity give $\Delta x' \propto x^m$. We first perform a Courant-Snyder transformation on the system, and borrow a result, Eq. (3.35), from Sec. 3.7, the equation of motion will look like

$$\frac{d^2\eta}{d\theta^2} + \nu^2\eta = \epsilon\eta^m\delta(\theta),$$

where ϵ specifies the nonlinearity strength; $\delta(\theta)$ is a periodic δ-function with period 2π in θ; ν is the unperturbed betatron tune. The perturbation is nonlinear when $m \geq 2$ ($m = 2$ for sextupole, $m = 3$ for octupole, etc.).

The left-hand-side of the equation of motion describes a simple harmonic oscillation with a natural frequency ν. The unperturbed solution is

$$\eta \sim \begin{cases} \sin\nu\theta \\ \cos\nu\theta \end{cases}.$$

Substituting this solution into the right-hand-side, we see that the simple harmonic oscillator is now driven by a force, represented by the right-hand-side of the equation of motion, that behaves like

$$\sim \delta(\theta)\begin{cases} \sin^m \nu\theta \\ \cos^m \nu\theta \end{cases}.$$

The frequency content of this driving force contains frequencies $K \pm m\nu, K \pm (m-2)\nu, K \pm (m-4)\nu, \cdots$, where the extra integer frequency K comes from the δ-function $\delta(\theta)$.

A resonance occurs when the driving force on the right hand side of the equation has a frequency content that coincides with the natural frequency ν on the left hand side of the equation, i.e. when

$$\nu = K \pm \begin{cases} m\nu \\ (m-2)\nu \\ (m-4)\nu \\ \cdot \\ \cdot \\ \cdot \end{cases}.$$

Solving for ν gives the resonance condition to first order of the driving strength.

Take octupole ($m = 4$) for example. Resonances occur when $\nu = K \pm 3\nu$ and $\nu = K \pm \nu$. This means either $\nu = \frac{K}{4}$ or $\nu = \frac{K}{2}$ for some integer K.

More explicitly, we have resonances occurring as shown in Table 3.2. A few comments are in order.

- An $2m$-th order multipole drives m-th order resonance and resonances of order $m-2, m-4, \cdots$. However, for a given resonance order, most likely it is the lowest multipole order that dominate. Dominant drivers are marked by XX's in Table 3.2.

Table 3.2: Nonlinear 1-D resonances driven by nonlinearities of various orders. Table lists the 1-D resonances to first order in the strength of the nonlinearity. An X means the resonance is driven by the nonlinearity. Strongest resonances are driven by nonlinearity of the lowest order, as indicated by XX.

Nonlinearity	m	$\nu = K$	$\nu = \frac{K}{2}$	$\nu = \frac{K}{3}$	$\nu = \frac{K}{4}$	$\nu = \frac{K}{5}$
dipole	1	XX				
quadrupole	2		XX			
sextupole	3	X		XX		
octupole	4		X		XX	
decapole	5	X		X		XX

- Our analysis above is based on a first-order perturbative calculation. It applies only to first order in the driving force strength ϵ. Table 3.2 therefore is valid to the same order. In case of very strong nonlinearities, higher order effects in ϵ may become important. This is the case for some of the synchrotron radiation facilities or collider storage rings where very strong sextupoles are implemented. In those cases, their sextupoles easily drive resonances substantially higher than third order.

- So far, the nonlinearity considered is localized as a δ-function in θ. When the nonlinearity has a distribution, the effective strength ϵ will be given by a specific k-th Fourier harmonic of the driving force related to the order of the resonance — details will be spelled out more explicitly in Sec. 3.7. To mitigate the harm by the nonlinearity, for example, one may try to control the specific resonance-driving Fourier harmonic of its distribution. In doing so, it is not required to eliminate the nonlinearity altogether, and in fact thus allowing to utilize its other harmonics for good purposes. For example, in synchrotron radiation facilities, strong sextupoles are used in such a way that their zeroth harmonic is used to compensate for the chromaticities, while their other harmonics are controlled to minimize the resonance driving effects.

Homework 3.57 Consider the nonlinear 1-D horizontal betatron motion of a particle in a storage ring of circumference C. What linear and nonlinear resonances do you qualitatively expect would be driven under the following circumstances,

(a) resonances driven by a single thin-lens sextupole to first order in the sextupole strength?

(b) resonances driven by a single thick-lens sextupole of length $\ell = \frac{C}{100}$ to first order in the sextupole strength?

(c) resonances driven by a single thin-lens sextupole to first order in the sextupole strength, but instead of a storage ring, we have a synchrotron that stores the beam only 1000 revolutions?

(d) resonance driven by a single thin-lens sextupole to second order sextupole strength?

(e) resonances driven by a thin-lens beam-beam head-on collision to first order in beam-beam strength?

Solution

(a) $\nu = \frac{K}{3}, \nu = K$.

(b) Same as (a) but $-100 \lesssim K \lesssim +100$.

(c) This is a difficult question. Needs thinking.

Homework 3.58 Extend the 1-D analysis in the text, in particular Table 3.2, to second order in ϵ. The case of a strong sextupole has been dealt with in Homework 3.57.

3.6.2 Multiturn closed orbit

We have been considering nonlinear dynamics with a view of particle motion on a turn-by-turn basis. It is possible to adopt another point of view in terms of multiturn dynamics. In this case, we first look for the "multiturn closed orbit" in the presence of a nonlinearity. The two viewpoints are of course intimately connected. We will adopt the turn-by-turn view in our lectures, mentioning the multiturn point of view only very briefly in this section.

We discussed the multiturn closed orbit effect for linear systems on page 223. We pointed out that, in a linear system with a one-turn matrix map M, an m-turn closed orbit requires $M^m = I$. However, those multiturn orbits occur only when the tune $\nu = \frac{K}{m}$ is exactly a rational number, and they repeat after m turns regardless of the particle's initial conditions. Every particle executes its own multiturn closed orbit. These are not true closed orbits because they do not constitute closed trajectories around which other particles oscillate. See Homework 3.59.

The situation changes when nonlinearities are introduced. True multiturn closed orbits become possible. This is because now the one-turn map M depends on the particle's initial conditions. As a result, the condition $M^m = I$ can be found for a set of discrete locations in phase space for each $m \neq 1$. These locations then constitute the phase space fixed points for the map M^m and they constitute the m-turn closed orbits.

When that happens, a particle starting exactly on the closed-orbit will return to this point after m turns. One then asks the question how do particles near the multiturn closed-orbit behave. A stable closed-orbit means neighboring particles oscillate around it; unstable closed-orbit means neighboring particles move exponentially away from it.

Extending along this line of discussion based on multiturn closed orbits, we will then enter topics such as phase space islands, chaos, and dynamic aperture. The centers of islands constitute the multiturn COD. As mentioned, these discussions parallel those based on the turn-by-turn view, and, although providing additional insights, we shall not elaborate on them further.

Homework 3.59 Investigate the possibility of multiturn closed orbit in a linear system. Show that they all degenerate to the same 1-turn closed orbit unless $M^m = I$.

Solution To show there is no multiturn closed orbit in a linear system mathematically, consider a case of calculating the closed-orbit distortion after a kick,

$$MX_1 + \Delta X = X_1 \quad \Longrightarrow \quad X_1 = (I - M)^{-1}\Delta X\,,$$

where M is the one-turn 2×2 matrix map, $X_1 = \begin{bmatrix} x_1 \\ x_1' \end{bmatrix}$ is the 1-turn closed orbit after the kick ΔX.

To find a 2-turn closed orbit X_2, we need

$$M(MX_2 + \Delta X) + \Delta X = X_2 \quad \Longrightarrow \quad X_2 = (I - M^2)^{-1}(I + M)\Delta X\,.$$

It is not difficult to prove $X_2 = X_1$ unless $M^2 = I$.

Homework 3.60 The discussion in the text might be a bit abstract. This homework intends to offer a concrete example. Consider a thin sextupole inserted at a location with one-turn map

$$M = \begin{bmatrix} \cos\mu & \beta\sin\mu \\ -\frac{\sin\mu}{\beta} & \cos\mu \end{bmatrix}.$$

Let the map at the sextupole be

$$\Delta x = 0, \qquad \Delta x' = \epsilon x^2\,.$$

(a)Find the 1-turn closed-orbit at the exit of the sextupole. The origin $(0,0)$ is one of them. There is another one.

(b) Find the 3-turn closed-orbit at the exit of the sextupole numerically.

Solution

(a) In addition to the stable $(0,0)$, the other unstable 1-turn closed orbit is located at

$$x_0 = 2\,\frac{\tan\frac{\mu}{2}}{\beta\epsilon}, \qquad x_0' = 2\,\frac{\tan^2\frac{\mu}{2}}{\beta^2\epsilon}\,.$$

Figure 3.23: Unstable fixed point for the case of a sextupole in an otherwise linear optics. Plotted is $\beta\epsilon x_0$ (solid) and $\beta^2\epsilon x_0'$ (dashed) at the exit of the sextupole as function of μ.

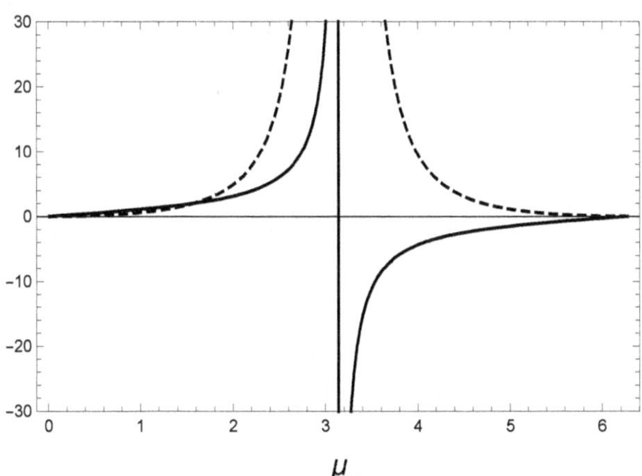

Figure 3.23 shows the unstable 1-turn closed-orbit. Note that when x_0 and x_0' are large, it mean they are far from the region of interest, and become not relevant.

(b) You will find six 3-turn closed orbit, three stable and three unstable ones. You will not find 2-turn closed orbits.
As mentioned in the text, pursuing further along this line, we will soon enter the topics of phase space islands and chaos.

3.6.3 Nonlinear coupling resonance

The discussion so far is for a 1-D system. In a 2-D case involving x and y motions, a resonance occurs when

$$m_x\nu_x + m_y\nu_y = n\,,$$

where m_x, m_y, n are integers (positive and negative). For the 2-D case, we have seen in Sec. 3.5.2 that a single skew quadrupole excites $\nu_x \pm \nu_y = n$, $\nu_{x,y} = n$, $2\nu_{x,y} = n$ resonances.

The *order* of a resonance is given by the integer $|m_x|+|m_y|$. As resonance of higher and higher orders are taken into account, the resonance population becomes increasingly dense in the (ν_x, ν_y) space. Figure 3.24 shows the resonance lines up to the 8th order. The storage ring's *working point* in the (ν_x, ν_y) space will have to be chosen to avoid these resonance lines. It is clear that as higher orders are included, the available tune space for the choice of a good working point quickly diminishes.

Figure 3.24: Resonance lines in the (ν_x, ν_y) space up to the 8th order. Resonances are represented as straight lines. Different resonance orders are indicated by different colors. Lower order resonances have thicker lines. The pattern repeats for each integer block of the (ν_x, ν_y) space. Sum resonances are depicted as solid lines, difference resonances by dashed lines.

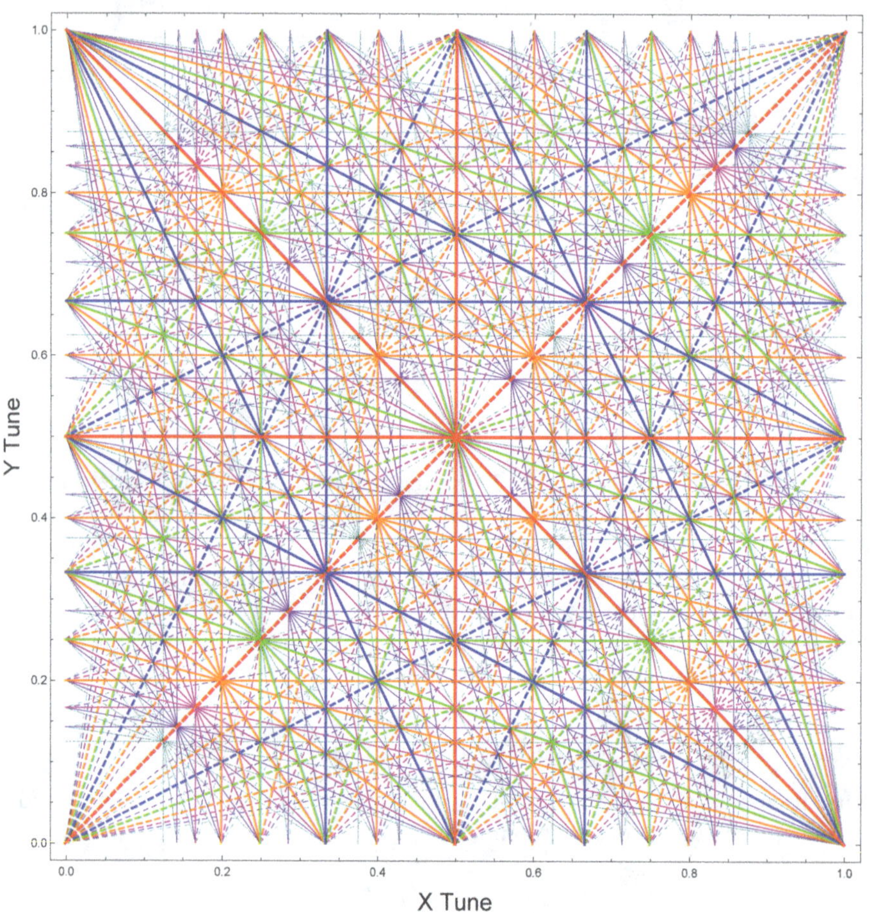

Homework 3.61 Consider a 2-D motion in a storage ring with a localized coupling nonlinearity. Let the equations of motion be

$$\begin{aligned}
\frac{d^2\eta_x}{d\theta^2} + \nu_x^2\eta_x &= -\epsilon m_x \nu_x^2 \eta_x^{m_x-1}\eta_y^{m_y}\delta(\theta)\,, \\
\frac{d^2\eta_y}{d\theta^2} + \nu_y^2\eta_y &= -\epsilon m_y \nu_y^2 \eta_x^{m_x}\eta_y^{m_y-1}\delta(\theta)\,.
\end{aligned}$$

where θ is the angular coordinate around the storage ring, ϵ specifies the strength of the nonlinear perturbation, and $m_x \geq 1, m_y \geq 1$.

(a) First establish the fact that this system is Hamiltonian.

(b) Follow the steps similar to the 1-D case, derive the resonance conditions for the system to first order in ϵ.

(c) A somewhat tedious but still instructive exercise is to identify the resonance conditions to second order in ϵ. Do it for the special case of a sextupole resonance, $m_x = 1, m_y = 2$.

Solution (a)

$$H(\eta_x, p_x, \eta_y, p_y, \theta) = \frac{1}{2}(\eta_x^2 + \nu_x p_x^2 + \eta_y^2 + \nu_y^2 p_y^2) + \epsilon \eta_x^{m_x} \eta_y^{m_y} \delta(\theta)\,.$$

3.6.4 Tune and irrational numbers

The dilemma We detour for an academic discussion in this section. It follows from the previous discussion that, for a 1-D system, whenever $\nu =$ any rational number, there will be potentially some small nonlinear perturbation that will excite a resonance and will drive the particle motion unstable. Since rational and irrational numbers are both infinitely densely populated on the real-number axis and they are infinitely mutually intertwined, it follows that no matter how we choose ν, there will always be some rational number infinitely close to it, and when that happens, an infinitely weak nonlinear perturbation can potentially drive the particle motion unstable. Here we are coming very close to making an (incorrect) conclusion that there is no way that a storage ring can be stable against nonlinear resonances. In a 2-D case, the same question can be raised by pointing out that as we increase the order of resonances, the (ν_x, ν_y) diagram Fig. 3.24 will be filled with resonance lines and there will be no space left for locating the working point.

The proper addressing of this question lies in the realm of nonlinear dynamics, and is beyond the scope of these notes. We shall only mention that the answer to this question is by no means obvious, and indeed this had been a fundamental uncertainty for storage ring beam dynamists in the early days until the first storage rings were built. Even from the hindsight, however, the truth is that this question has never been really settled.

Even today, mathematicians still question (correctly) whether storage rings are really stable. After all, a typical storage ring stores the beam for 10^{11-12} turns, while the earth has evolved around the sun merely a few times 10^9 turns. Stability of a nonlinear dynamical system remains an intricate and challenging subject of research, for which storage rings are perfect experimental devices, at least better than the earth.

There is a theorem in nonlinear dynamics called the Kolmogorov-Arnold-Moser (KAM) theorem.[12] It states that if the nonlinear perturbation is suffi-

[12] A. N. Kolmogorov, Dokl. Akad. Nauk SSR 98 (1954); J. Moser, Nachr. Akad. Wiss. Göttingen Math.-Phys. Kl. II 1962 (1962), 1–20; V. I. Arnold, Uspekhi Mat. Nauk 18 (1963).

ciently weak, parts of the phase space will maintain their integrability. If we locate the working point at those values, then in principle the particle motion will be stable. However, when putting in realistic values according to the KAM algorithm, it is clear that these integrable regions can be finite but extremely small, not even close to what is needed for a realistic beam in a storage ring.

The catch of this dilemma lies in the fact that the KAM theorem addresses the stability for truly infinite number of turns. For storage rings, the beam has to survive for a mere 10^{11-12} turns. These two criteria of stability are distinctly different. For one thing, to store a beam for infinite number of turns in a storage ring, we have to worry about resonances of infinitely high orders. To store 10^{12} turns, we need to worry only about up to the 10^{12}-th order resonances.

Irrational numbers So we learned that the betatron tune ν must be chosen to avoid rational numbers. In fact, it is best to choose ν as far away as possible from all rational numbers. This means the best choice of ν should be an irrational number that is as irrational as possible.

One should then note that not all irrational numbers are equally irrational. The most famous irrational number π, for example, is not very irrational after all. From advanced calculus, one knows that the number of irrational numbers is infinitely greater than the number of rational numbers (although both numbers are infinite). If one arbitrarily chooses ten numbers, then chances are that all ten of them will be irrational. It turns out that probably 8 or 9 of them will be more irrational than π. This is reflected by the fact that the three successive rational approximations of π are given by

$$\begin{aligned}
\pi &\approx 3\,, \\
\pi &\approx 3+\frac{1}{7} = \frac{22}{7}\,, \\
\pi &\approx 3+\frac{1}{7+\frac{1}{16}} = \frac{355}{113}\,,
\end{aligned}$$

and the fact that they converge to the value of $\pi \approx 3.1415926535897932\cdots$ surprisingly rapidly.

The golden mean The one irrational number that keeps its rational neighbors (there are an infinite number of them, and infinitely close by) the farthest distances is the golden mean

$$\Gamma = \frac{1+\sqrt{5}}{2} \approx 1.618\cdots\,.$$

The underlying reason why Γ is the most irrational number traces to the fact that its *continued fraction* representation is

$$\Gamma = 1+\frac{1}{1+\frac{1}{1+\frac{1}{1+\cdots}}}\,.$$

It is said that the most beautiful rectangle has a height-to-width ratio equal to the golden mean. This principle is amply applied in architecture designs and in Japanese flower arrangements. The golden mean rectangle has the property that after cutting off a perfect square from the bottom of the rectangle, the remaining rectangle still is a golden mean rectangle.[13] Let the height-to-width ratio of the original rectangle be x. After cutting off the perfect square, the width-to-height ratio is $\frac{1}{x-1}$. Thus the golden mean rectangle condition is given by

$$x = \frac{1}{x-1},$$

which has the solution $x = \Gamma$.

It might be of academic interest to contemplate that the optimal choice of a tune value is $\nu = \Gamma$. Failing to fulfill that, the next optimal thing to do, it turns out, is to choose[14]

$$\nu = \frac{m + n\sqrt{5}}{\ell},$$

with m, n, ℓ integers, because these irrational numbers are more irrational than others in their respective neighborhoods.[15]

The 2-D case In a 2-D system, resonances occur when $m_x\nu_x + m_y\nu_y = n$. The working point will have to avoid getting too close to these resonance lines. A question arises as to how can this be done. Clearly, as the resonance order is increased, the room left for tune choice becomes increasingly restricted, and pretty soon there will be no room left.

But then one might say, "Yes, but the higher order resonances tend to be weaker..." That is true, but they are also closer to the desired working point. So at some point, one has to address the question on stability when

> an infinite number of infinitely weak resonances drive the motion infinitely close by. (3.33)

We shall not explore this subtlety, except for a short return to the topic in Sec. 5.4 when we briefly talk about chaos. Pushed too hard, we will discuss the intricate dynamics of *chaotic* motion, not to mention chaotic discussions.

Leap year An average year (so-called a "tropical year") has 365.2422 days. This can be expressed approximately as 365 days, 5 hours, 48 minutes, 45 seconds, or $\frac{46751}{128}$ days. Expressed as continued fraction,

$$\frac{46751}{128} = 365 + \frac{1}{4 + \frac{1}{7 + \frac{1}{1 + \frac{1}{3}}}}.$$

---

[13]The fact that this constitutes a most beautiful rectangle aesthetically makes a lot of sense as the square cutting process can be continued indefinitely.

[14]A. W. Chao, Proc. Euro. Part. Accel. Conf. Rome, 228 (1988).

[15]Someone should think about why the most irrational numbers would involve $\sqrt{5}$, and not, say $\sqrt{2}$. What's so special about $\sqrt{5}$?

This means 365.2422 has the successively accurate best rational approximations as follows,

$$365, \ \frac{1461}{4}, \ \frac{10592}{29}, \ \frac{12053}{33}, \ \frac{46751}{128},$$

or numerically,

$$365, \ 365.25, \ 365.2414, \ 365.2424, \ 365.2422\,.$$

It then seems that one way — the mathematically optimal – of choosing the leap years is to follow the following pattern,

1. A normal year should have 365 days. Year one has 365 days.

2. Every 4th year, have one leap year, i.e. that year will have 366 days. At the end of this 4-year cycle, the average year has $\frac{1461}{4}$ days.

3. Complete seven of the 4-year cycles. The 29th year is normal year. After the completion of that year, the average year has $\frac{10592}{29}$ days. A new cycle of leap-year counting now starts with each cycle having 29 years.

4. In this new cycle, the 33rd year is leap-year. After the completion of that year, the average year has $\frac{12053}{33}$ days, and another new cycle, each with 33 years, starts.

5. The 33-year cycle goes on for three cycles. After that, proceed with one more 33-year cycle but stop at the 128-th year. That year is a normal year. At its completion, the average year has $\frac{46751}{128}$ days.

6. The complete 128-year then repeats.

Starting with the year 2021, the leap year sequence for one 128-year cycle would look like Table 3.3. In reality, unfortunately, this does not seem to be what we do.

Homework 3.62 The text gives three successive rational approximations of $\pi \approx 3, \frac{22}{7}, \frac{355}{113}$. Find the next in the series.

Homework 3.63

(a) Find the series of successive rational approximations of the Golden Mean Γ using the continued fraction representation.

(b) Consider why the continued fractions form for Γ predicts it to be the most irrational of all numbers.

Solution The convergence is very slow: $2, \frac{3}{2}, \frac{5}{3}, \frac{8}{5}, \frac{11}{8}$...., as shown in Fig. 3.25.

Table 3.3: A continued-fraction suggestion of the leap-year pattern. Suggested leap years are in bold face.

2021	2022	2023	**2024**	2025	2026	2027	**2028**
2029	2030	2031	**2032**	2033	2034	2035	**2036**
2037	2038	2039	**2040**	2041	2042	2043	**2044**
2045	2046	2047	**2048**	2049	2050	2051	2052
2053	2054	2055	2056	**2057**	2058	2059	2060
2061	2062	2063	2064	**2065**	2066	2067	2068
2069	2070	2071	2072	**2073**	2074	2075	2076
2077	2078	2079	2080	**2081**	2082	2083	2084
2085	2086	2087	2088	2089	**2090**	2091	2092
2093	**2094**	2095	2096	2097	**2098**	2099	2100
2101	**2102**	2103	2104	2105	**2106**	2107	2108
2109	**2110**	2111	2112	2113	**2114**	2115	2116
2117	**2118**	2119	2120	2121	2122	**2123**	2124
2125	2126	**2127**	2128	2129	2130	**2131**	2132
2133	2134	**2135**	2136	2137	2138	**2139**	2140
2141	2142	**2143**	2144	2145	2146	**2147**	2148

Figure 3.25: Successive rational approximations of the Golden Mean.

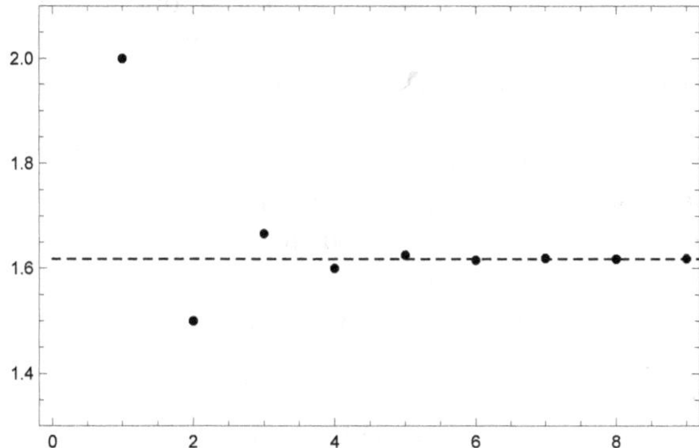

Homework 3.64

(a) Show that

$$\lambda_{n,m} = n + \cfrac{m}{n + \cfrac{m}{n + \cfrac{m}{n + \cdots}}} = \frac{n + \sqrt{n^2 + 4m}}{2}.$$

Golden mean is $\Gamma = \lambda_{1,1}$.

(b) Not all of them are irrational, however. For example, $\lambda_{1,2} = 2$. In fact, $\lambda_{n,\pm n+1} = n \pm 1$ is rational. Also show that

$$\lambda_{kn,k^2m} = k\lambda_{n,m}, \qquad \lambda_{2n,4m} = 2\lambda_{n,m}$$
$$\lambda_{n,2n^2} = 2n. \qquad \lambda_{n,n^3(n+1)} = n(n+1)\,.$$

3.7 Smooth approximation

In Sec. 3.6, we briefly discussed nonlinear resonances, but only qualitatively. In the remaining of this chapter, we introduce and apply a quantitative analysis of nonlinear dynamics, called *smooth approximation.*[16] This is not the only approach, but it has been one of the more basic and classic approaches to treat nonlinear dynamics in accelerator physics.

We will develop the smooth approximation in steps. In spite of all the subtleties mentioned earlier, here are the three steps we will adopt:

1. the system is nonlinear, but we are away from all resonances;
2. the system is nonlinear, and there are resonances, but there is one and only one resonance of any importance;
3. the system is nonlinear, and there are multiple resonances playing the role.

Smooth approximation applies to steps 1 and 2. Towards the end of this section, we will apply the smooth approximation to a few examples. As to the step 3, this is a system vulnerable to chaotic motion, and the smooth approximation breaks down. Efforts were made to study the case when two and only two resonances are playing the role, but even in that simplified case, conclusive insight has yet to be obtained.

Incidentally, as warned in a discussion following Eq. (2.63), the smooth approximation does not mean taking an eye-ball average of the accelerator and make approximate estimates of the beam quantities accordingly. Smooth approximation is a precise mathematical treatment of the motion in an accelerator, based on well-defined perturbative operations on the Hamiltonian dynamics of the system.

The well-defined mathematical procedure involves removing fast oscillating terms in the Hamiltonian — the smoothing process. The resulting Hamiltonian of the system then becomes quasistatic, or slowly varying in time. The Hamilton's equations describing the evolution of the system can then be integrated. One important feature of this approach is that *differential equations* now replace the maps or matrices we have been using extensively so far. Time evolves continuously rather than in discrete steps. Later we will encounter another situation when maps are replaced by differential equations in a slowly varying system when we discuss the synchrotron motion in a circular accelerator in Chapter 5.

[16] A. Schoch, CERN report 57-23 (1958); P.A. Sturrock, Ann. Phys. 3, 113 (1958); G. Guignard, CERN report 76-06 (1976).

3.7.1 Smooth approximation away from resonances

We mentioned a three-step process of introducing the smooth approximation. As the first step, given in this section, we consider the ideal case when the tune is free from any nearby resonances. The quantity we will compute is called *nonlinear tune shift*. The effect is called *detuning*. Resonance effects will be considered later.

Let us start with a 1-D motion of a particle described by the equation

$$x'' + K(s)x = f(x,s),$$

where $K(s)$ is the focusing function of the linear, unperturbed accelerator environment, $f(x,s)$ is a perturbation that depends on the instantaneous displacement of the particle x and the position coordinate s. The perturbation $f(x,s)$ is considered to be periodic in s with period $2\pi R$, the circumference of the accelerator. Symplecticity demands $f(x,s)$ be independent of x'.

We learned from the Courant-Snyder treatment that the problem simplifies if we introduce the transformation [see Eqs. (2.62, 2.63)]

$$\begin{aligned} \eta &= \frac{x}{\sqrt{\beta(s)}}, \\ \theta &= \frac{1}{\nu}\int_0^s \frac{ds'}{\beta(s')}, \end{aligned} \tag{3.34}$$

where $\beta(s)$ is the unperturbed β-function found in the absence of $f(x,s)$, and ν is the unperturbed tune. With Eq. (3.34), the dynamic variable x is replaced by η, and the independent time-variable s becomes θ. The functions $\beta(s), K(s)$ and $f(x,s)$ become $\beta(\theta), K(\theta)$ and $f(\eta,\theta)$, and they are periodic in θ with period 2π.

Equation of motion in the new variables reads (Homework 3.65)

$$\frac{d^2\eta}{d\theta^2} + \nu^2\eta = \nu^2\beta^{3/2}(\theta)f(\eta,\theta) \equiv F(\eta,\theta). \tag{3.35}$$

In passing, note the factor $\beta^{3/2}(\theta)$ in the definition of $F(\eta,\theta)$. It reiterates the earlier statement that β-function serves as a magnification factor to the effects of perturbations.

In the absence of perturbation, i.e. when $F(\eta,\theta) = 0$, η is described by a simple harmonic motion in θ. This means we can write

$$\eta = \sqrt{2J}\cos\phi \tag{3.36}$$

and

$$\frac{d\eta}{d\theta} = -\nu\sqrt{2J}\sin\phi, \tag{3.37}$$

where J and $\frac{d\phi}{d\theta} = \nu$ are constants. Equations (3.36-3.37) are of course just the action-angle transformation[17] of the dynamical variables from $(\eta, \frac{d\eta}{d\theta})$ to (ϕ, J).

With a perturbation, we could insist on the action-angle transformation of the form (3.36-3.37) except that J and $\frac{d\phi}{d\theta}$ now depend on θ and we need to solve for them. The only thing we may expect is that J and $\frac{d\phi}{d\theta}$ are still approximately constants so that they will change slowly with θ because the perturbation is assumed to be weak.

To proceed, first note that by substituting (3.36) into (3.37), one obtains a self-consistency condition

$$\frac{1}{2}\frac{dJ}{d\theta}\cos\phi - \left(\frac{d\phi}{d\theta} - \nu\right) J\sin\phi \;=\; 0\,. \tag{3.38}$$

Secondly, substituting Eqs. (3.36-3.37) into Eq. (3.35) gives another condition

$$-\frac{1}{2}\frac{dJ}{d\theta}\sin\phi - \left(\frac{d\phi}{d\theta} - \nu\right) J\cos\phi \;=\; \frac{\sqrt{J}}{\sqrt{2}\,\nu} F(\sqrt{2J}\cos\phi, \theta)\,. \tag{3.39}$$

Equations (3.38-3.39) can be combined to give

$$\begin{aligned} \frac{dJ}{d\theta} &= -\frac{\sqrt{2J}}{\nu}\sin\phi\; F(\sqrt{2J}\cos\phi, \theta)\,, \\ \frac{d\phi}{d\theta} &= \nu - \frac{1}{\nu\sqrt{2J}}\cos\phi\; F(\sqrt{2J}\cos\phi, \theta)\,. \end{aligned} \tag{3.40}$$

Effectively we have decomposed a second order differential equation (3.35) of one variable η into two first order differential equations (3.40) of two variables J and ϕ. So far no approximations have been made; Eq. (3.40) is exact.

An alert reader may notice that what we have been doing, decomposing one equation (3.35) into two equations (3.40), is just reversing the process of combining the two Hamilton equations into an equation of motion. Hidden behind all this is the Hamiltonian. We shall discuss the Hamiltonian momentarily. After all, it is the real key.

As mentioned, the quantities J and $\frac{d\phi}{d\theta}$ are expected to change slowly with time variable θ. We can approximate the expressions (3.40) by keeping only the slowly varying terms and dropping all fast oscillating terms on the right hand sides. In other words, we make the "smooth approximation",

$$\begin{aligned} \frac{dJ}{d\theta} &\approx -\frac{\sqrt{2J}}{\nu}\langle\sin\phi\; F(\sqrt{2J}\cos\phi, \theta)\rangle\,, \\ \frac{d\phi}{d\theta} &\approx \nu - \frac{1}{\nu\sqrt{2J}}\langle\cos\phi\; F(\sqrt{2J}\cos\phi, \theta)\rangle\,, \end{aligned} \tag{3.41}$$

[17]It is unfortunate that we are trained to use the term "action-angle". It really should be "angle-action". In the Hamiltonian formalism, it is necessary to define exactly which is coordinate and which is momentum. Here, ϕ is the coordinate, J is the momentum, not the other way around.

where the angular brackets mean the smoothing procedure. Fast and slow refer to a comparison with the revolutions. The angular variables θ and ϕ are considered to be fast variables, while the changes in J and $\frac{d\phi}{d\theta}$ per revolution are considered to be slow for sufficiently weak perturbations.

Consider the case when the tune is away from resonances. What that implies is that there is no correlation in how θ and ϕ progresses in time. The smoothing in this case is obtained by a straightforward averaging over the angular variables θ and ϕ separately and independently, i.e.

$$\langle\,\rangle = \frac{1}{4\pi^2}\int_0^{2\pi} d\phi \int_0^{2\pi} d\theta\,. \tag{3.42}$$

Equation (3.42) is a well-defined operation, and is the essence of the smooth approximation, at east away from resonances, bearing in mind that the perturbation is assumed to be weak.

If there were any resonances nearby and playing a role, there will be correlations between θ and ϕ, Eq. (3.42) then breaks down, or at least is incomplete, as will be elaborated in Sec. 3.7.2.

After averaging, Eq. (3.41) becomes

$$\begin{aligned} \frac{dJ}{d\theta} &\approx 0\,, \\ \frac{d\phi}{d\theta} &\approx \nu - \frac{1}{4\pi^2\nu\sqrt{2J}}\int_0^{2\pi} d\phi \int_0^{2\pi} d\theta \cos\phi\; F(\sqrt{2J}\cos\phi,\theta)\,. \end{aligned} \tag{3.43}$$

The reason $\frac{dJ}{d\theta}$ vanishes is that $F(\sqrt{2J}\cos\phi,\theta)$ contains terms with integer powers of $\cos\phi$; when multiplied by $\sin\phi$, its average over ϕ always vanishes.

Equation (3.43) says that the action variable J is approximately a constant of the motion even in the presence of (weak) perturbations provided the unperturbed tune is away from resonances. This is of course a critically important conclusion.

The quantity $\frac{d\phi}{d\theta}$, which assumes the physical meaning of the perturbed tune, depends on J. The motion of particles in the polar $(\sqrt{2J},\phi)$ phase space is such that particles rotate around the origin along circles of $\sqrt{2J}$ = constant with constant angular speeds. This constant angular speed, however, depends on the particle's amplitude J and therefore is different for different particles, creating a sheering effect on the distribution of particles as they rotate in the phase space. The dependence of the perturbed tune on J specifies a *detuning* effect. The *tune shift* is given by

$$\Delta\nu(J) = -\frac{1}{4\pi^2\nu\sqrt{2J}}\int_0^{2\pi} d\phi \int_0^{2\pi} d\theta \cos\phi\; F(\sqrt{2J}\cos\phi,\theta)\,. \tag{3.44}$$

Detuning, i.e. a tune shift that depends on particle's betatron oscillation amplitude, is a self-stabilizing mechanism. This is because if an unfortunate particle has its tune falling too close to a resonance, and as a result, its betatron oscillation grows as it proceeds toward instability, then its tune shifts and as a result gets off the resonance and stabilizes. Detuning in fact is what the KAM stability relies on.

From Eq. (3.44), one notes that if the perturbation $f(x, s)$ is an even function of x, then to first order of the perturbation, the tune shift vanishes. Since the perturbation due to sextupoles, decapoles, etc. are even in x, these multipoles do not contribute to tune shifts for on-momentum particles to first order of their strengths. Quadrupoles, octupoles, and the beam-beam force, on the other hand, do contribute to first order tune shifts.

Let us apply Eq. (3.44) to two examples below. More elaborate applications can be found in Secs. 3.7.3 and 3.7.4.

Thin-lens error quadrupole Consider first a thin-lens error quadrupole with focal length f located as the position $s = 0$ in the circular accelerator. In this case, we have

$$f(x,s) \;=\; -\frac{x}{f}\delta(s)\,,$$

where $\delta(s)$ is the periodic δ-function with period $2\pi R$. The kick a particle receives as it traverses the error quadrupole is $\Delta x' = -\frac{x}{f}$. We have

$$F(\eta,\theta) \;=\; \nu^2\beta^{3/2}(\theta)f(\eta\sqrt{\beta(\theta)},\theta) \;=\; -\nu\beta(0)\frac{\eta}{f}\delta(\theta)\,,$$

where we have used $\delta(s) = \frac{1}{\nu\beta(\theta)}\delta(\theta)$. Substitution into Eq. (3.44) yields

$$\Delta\nu \;\approx\; \frac{\beta(0)}{4\pi f}\,, \tag{3.45}$$

which is a familiar result. In particular, in this case, $\Delta\nu$ is independent of J. All particles suffer the same tune shift. Also note that this result requires absence of resonances — both in the context of the smooth approximation and in the context that, as we know for the case of a single error quadrupole, this result applies only if we are away from half-integer resonances.

Thin-lens error octupole Now consider a more complex problem, namely the case of a thin-lens error octupole with

$$f(x,s) \;=\; \epsilon x^3\delta(s)\,. \tag{3.46}$$

It follows that

$$F(\eta,\theta) \;=\; \epsilon\nu\beta^2(\theta)\eta^3\delta(\theta)\,.$$

Substituting into Eq. (3.44) gives

$$\Delta\nu \;\approx\; -\frac{1}{2\pi^2}\epsilon\beta^2(0)J\int_0^{2\pi}d\phi\cos^4\phi \;=\; -\frac{3}{8\pi}\epsilon\beta^2(0)J\,. \tag{3.47}$$

This tune shift is proportional to the action J.

Hamiltonian is for real It may be instructive at this point to make the following observation. If one is given a perturbation such as that given by Eq. (3.46) and is asked to calculate the tune shift, and had he (or she) not known the smooth approximation, what would he do? First, he might note that the focusing effect changes from turn to turn, so some kind of averaging would be unavoidable in order to calculate the tune shift, and he would have to first introduce an "instantaneous" tune and then try to average it over many turns. Naively, the first quantity that comes to mind to be averaged is the force, because the force is what is "physically" applied to the particle. The instantaneous force is given by $f(x, s)$, which is proportional to x^3. Now since x is basically sinusoidal in time, the average of the force over many turns is zero, which of course gives a wrong answer.

He might then consider perhaps the averaging should be performed over the "gradient" of the force. Although the gradient is not what the particle experience directly, the intuition of averaging over the gradient may be supported by the direct link between tune shift and gradient, as Eq. (3.45) suggests. He then computes the gradient which is connected to an instantaneous focal length f with $\frac{1}{f} = -3\epsilon x^2$. Substituting into Eq. (3.45) and averaging over x give

$$\Delta\nu \approx \frac{\beta(0)}{4\pi}\langle -3\epsilon x^2 \rangle = -\frac{3}{4\pi}\epsilon\beta^2(0)J \,.$$

This result gets closer to the correct answer (3.47) but still is wrong.

The right answer, surprisingly, is to average over the *potential* of the force, i.e. the average should be performed over not the force, not the gradient (i.e. the derivative of the force), but the potential (i.e. the *integral* of the force),

$$V(x, s) = -\int_0^x dx' f(x', s) \,.$$

Noting the fact that the potential is a term contained in the Hamiltonian, he then recognizes that what is being smoothed over is the Hamiltonian![18]

In fact, that was what we observed in the smooth approximation developed all along. What we have done, although not reasoned or derived this way, was in fact a smoothing over the Hamiltonian. Indeed, the averaged potential is given by

$$\begin{aligned} \langle V \rangle_\phi &= \frac{1}{2\pi}\int_0^{2\pi} d\phi\, V(\sqrt{2J\beta}\cos\phi, s) \\ &= -\epsilon\beta^2(0)J^2\langle\cos^4\phi\rangle\delta(s) = -\frac{3}{8}\epsilon\beta^2(0)J^2\delta(s) \,. \end{aligned} \tag{3.48}$$

In terms of the potential V, the tune shift (3.44) can be written as

$$\Delta\nu = \frac{1}{2\pi}\oint ds \frac{\partial\langle V \rangle_\phi}{\partial J} \,. \tag{3.49}$$

[18]Who would have guessed it is the integral of the force that is supposed to be averaged over? And who said Hamiltonian was merely a mathematical construct?

Substituting Eq. (3.48) into Eq. (3.49), we recover the correct tune shift (3.47). As funny looking or nonintuitive as Eq. (3.49) is, it is the correct answer. Here, the reader would recall a discussion connected to Table 2.1, when we cautioned against a subtlety which is best avoided by implementing the Hamiltonian. In case approximations are to be made, such as the smoothing procedure in the smooth approximation, it is safest to do them on the Hamiltonian.

Homework 3.65 Derive Eq. (3.35).

Solution Algebra is lengthy but straightforward. This effort is worthwhile because Eq. (3.35) is an important and rich result. The neat cancellation of terms is indication that you are on the right track. You will need the expression $2\beta\beta'' - \beta'^2 + 4\beta^2 K = 4$.

Homework 3.66 Show that Eqs. (3.49) and (3.44) are equivalent.

3.7.2 Smooth approximation near a single resonance

We now repeat the smooth approximation, but this time for the case when the tune ν is close to a resonance condition

$$\nu \approx \frac{n}{m} .$$

In practice, we are mostly concerned with low order resonances, i.e. rational tune with relatively small m.

What happens to the smooth approximation when $\nu \approx \frac{n}{m}$? The smooth approximation is based on removing all fast oscillating terms, while keeping only the slowly varying terms. Away from resonances, the only slowly varying term is the perturbation averaged over accelerator circumference and averaged over betatron oscillation, i.e. the $\langle\ \rangle$ represented by Eq. (3.42). Near a resonance, we will have another new slowly varying quantity, namely $m\phi$, which is near $2\pi n$ because the tune is close to $\frac{n}{m}$. With the oscillating nature of the betatron motion, any quantity with $\sin m\phi$ or $\cos m\phi$ will effectively be slowly varying. Our job now is to clearly identify all the slowly varying terms in the present case.

This time around, having learned the lesson in the last section, we shall start with Hamiltonian instead of equation of motion (although there is nothing wrong to do so). The analysis assumes you are familiar with Hamiltonians and canonical transformations, so hold on tight!

Canonical transformation to slow variables Consider a 1-D Hamiltonian

$$H(x, p, s) = \frac{p^2}{2} + \frac{K(s)x^2}{2} + g(x, s) , \tag{3.50}$$

where $g(x, s)$ is a weak nonlinear perturbation periodic in s with period $2\pi R$. Symplecticity requires g not to depend on p.

Incidentally, do not write the Hamiltonian (3.50) with p replaced by x'. The fact that $p = x'$ is a *consequence* of the Hamiltonian (3.50). Replacing p by x' at the Hamiltonian level is a misuse of the Hamiltonian formalism.

In the presence of the resonance $\nu = \frac{n}{m}$, we make the canonical transformation from (x, p) to (ϕ, J) with generating function

$$\begin{aligned}
G(x,\phi) &= -\frac{x^2}{2\beta(s)}\left[\tan\phi_1 - \frac{\beta'(s)}{2}\right], \\
\text{with } \phi_1 &= \phi + \frac{n}{m}\frac{s}{R} + \int_0^s ds' \left(\frac{1}{\beta(s')} - \frac{\nu}{R}\right),
\end{aligned}$$

to obtain

$$\begin{aligned}
p &= \frac{\partial G}{\partial x} = -\frac{x}{\beta(s)}\left[\tan\phi_1 - \frac{\beta'(s)}{2}\right], \\
J &= -\frac{\partial G}{\partial \phi} = \frac{x^2}{2\beta(s)}\sec^2\phi_1,
\end{aligned}$$

or

$$\begin{aligned}
x &= \sqrt{2J\beta(s)}\cos\phi_1, \\
p &= -\sqrt{\frac{2J}{\beta(s)}}\left[\sin\phi_1 - \frac{\beta'(s)}{2}\cos\phi_1\right], \\
p - \frac{\beta'(s)}{2\beta(s)}x &= -\sqrt{\frac{2J}{\beta(s)}}\sin\phi_1,
\end{aligned}$$

or

$$\begin{aligned}
J &= \frac{x^2}{2\beta(s)} + \frac{\beta(s)}{2}\left[p - \frac{\beta'(s)}{2\beta(s)}x\right]^2, \\
\tan\phi_1 &= -\beta(s)\frac{p}{x} + \frac{\beta'(s)}{2}.
\end{aligned}$$

The canonical transformation we just performed is exact. It does not involve the perturbation $g(x, s)$. The function $\beta(s)$ refers to the unperturbed β-function. The action J is just the Courant-Snyder emittance of the particle under consideration. We shall sometimes also refer to it as the emittance of the particle.

Let us look closer into what we just did. From the above expressions, it is clear that ϕ_1 is the betatron phase contained in x and $p - \frac{\beta'}{2\beta}x$. This means that, near the resonance, ϕ_1 evolves in time very close to $\frac{n}{m}\frac{s}{R}$. We then note in the expression for ϕ_1, the integral $\int_0^s ds' \left(\frac{1}{\beta(s')} - \frac{\nu}{R}\right)$ does not accumulate from turn to turn because it vanishes each time s propagates for one revolution. A moment's reflection says that the new time variable ϕ must vary only slowly in time. It does not change much from turn to turn. The canonical transformation has transformed from the fast variables x, p to slow variables ϕ, J.

The new Hamiltonian is

$$\begin{aligned}
K(\phi, J, s) &= H + \frac{\partial G}{\partial s} \\
&= \frac{p^2}{2} + \frac{K(s)x^2}{2} + g(x,s) + \frac{x^2\beta'(s)}{2\beta^2(s)}\left[\tan\phi_1 - \frac{\beta'(s)}{2}\right] \\
&\quad -\frac{x^2}{2\beta(s)}\left[\sec^2\phi_1\left(\frac{n}{m}\frac{1}{R} + \frac{1}{\beta(s)} - \frac{\nu}{R}\right) - \frac{\beta''(s)}{2}\right] \\
&= \frac{J}{R}\left(\nu - \frac{n}{m}\right) + g\left(\sqrt{2J\beta(s)}\cos\phi_1, s\right), \qquad (3.51)
\end{aligned}$$

where one sees many terms cancel nicely in the last step. Use has been made of the property $2\beta\beta'' - \beta'^2 + 4\beta^2 K = 4$. This result is still exact.

Smoothing the Hamiltonian We now Fourier decompose g,

$$g(\sqrt{2J\beta(s)}\cos\phi_1, s) = \sum_{j,\ell} g_{j,\ell}(J) e^{ij(\phi + \frac{n}{m}\frac{s}{R}) + i\ell\frac{s}{R}},$$

which is possible because g is periodic in both $\frac{s}{R}$ and in $\phi + \frac{n}{m}\frac{s}{R}$, both with period of 2π,

$$\begin{aligned}
g_{j,\ell} &= \frac{1}{4\pi^2 R}\int_0^{2\pi} d\xi \int_0^{2\pi R} ds\, e^{-ij\xi - i\ell\frac{s}{R}} \\
&\times\; g\left(\sqrt{2J\beta(s)}\cos\left[\xi + \int_0^s ds'\left(\frac{1}{\beta(s')} - \frac{\nu}{R}\right)\right], s\right), \qquad (3.52)
\end{aligned}$$

with $g_{-j,-\ell} = g^*_{j,\ell}$.

As mentioned, the phase quantity $\int_0^s ds'\left(\frac{1}{\beta(s')} - \frac{\nu}{R}\right)$ in Eq. (3.52) contains two terms. The first term is the unperturbed betatron phase from $s = 0$ to s. The second term is a reference phase advancing uniformly with s. The two terms both advance in s but with different rates, the first term according to the unperturbed β-function, the second term uniformly. However, as s completes one revolution, the two phase terms cancel to zero. This phase term therefore evolves within one revolution, but does not accumulate from turn to turn, which in turn allowed the Fourier decomposition to yield $g_{j\ell}$.

The coefficient $g_{j,\ell}$ given in Eq. (3.52) defines the coupling to the nonlinearity as observed at position $s = 0$. If the observation point is changed to $s = s^*$, the phase $\int_0^s ds'\left(\frac{1}{\beta(s')} - \frac{\nu}{R}\right)$ becomes $\int_{s^*}^s ds'\left(\frac{1}{\beta(s')} - \frac{\nu}{R}\right)$. It can be shown

(Homework 3.68) that the coefficients are related by

$$g_{j,\ell}\Big|_{s=s^*} = g_{j,\ell}\Big|_{s=0} \exp\left[ij\int_0^{s^*} ds'\left(\frac{1}{\beta(s')}-\frac{\nu}{R}\right)\right]. \tag{3.53}$$

It follows that the magnitudes of the driving coefficients are invariant around the ring. However, its phase depends on the observation point. If $s^* = 2\pi R$, then the coefficients return to their value at $s = 0$.

We now apply the smooth approximation. We drop fast oscillating terms which means the Hamiltonian is approximated by replacing the g term in Eq. (3.51) by its average over s over the range of $(0, 2\pi mR)$,

$$\langle g(s)\rangle_s = \frac{1}{2\pi mR}\int_0^{2\pi mR} ds\, g(s)\,.$$

The new Hamiltonian reads

$$\begin{aligned} K(\phi, J, s) &\approx \frac{J}{R}\left(\nu-\frac{n}{m}\right)+\left\langle g\left(\sqrt{2J\beta(s)}\cos\phi_1, s\right)\right\rangle_s \\ &= \frac{J}{R}\left(\nu-\frac{n}{m}\right)+\sum_{k=-\infty}^{\infty} g_{mk,-nk}(J)\, e^{imk\phi}\,. \end{aligned} \tag{3.54}$$

The $e^{imk\phi}$ factor is slowly varying near the resonance. Away from resonances, only the $k = 0$ term, i.e. only $g_{0,0}$, remains in Eq. (3.54). Near the resonance $\nu = \frac{n}{m}$, terms with other coefficients $g_{mk,-nk}$ with $k \neq 0$ are also slowly varying and therefore also contribute to the slow Hamiltonian.

The fact that our result (3.54) contains a summation over k reflects the fact that the analysis has included not only the $\nu = \frac{n}{m}$ but also all the accompanying resonances $\nu = \pm\frac{n}{m}, \pm\frac{2n}{2m}, \cdots$, etc.

This smooth approximation is valid to first order in the nonlinear perturbation. Extensions of the analysis to higher orders is also possible. Suffice it to say here that the result will not look as clean as the first order result.

The new Hamiltonian $K(\phi, J, s)$ depends on ϕ and J but no longer depends on s. It is important to realize that something rather significant has just occurred. The smooth approximation has allowed the replacement of a time-dependent Hamiltonian by a time-independent one. The particle motion is described – approximately – by applying the Hamilton's equations to the new Hamiltonian. Since this new Hamiltonian is independent of s, it is a constant of the motion. The problem is solved — mathematical jargon is, "the problem is integrated." So the amazing thing is that, as long as there is one and only one resonance playing the role, a sufficiently weak nonlinear system is integrable by the application of the smooth approximation.

One may also note that the Courant-Snyder transformation for the unperturbed linear system is in fact a special case of smooth approximation in the sense that the s-dependent focusing function $K(s)$ has been replaced by the constant value of tune ν. The linear case, on the other hand, enjoys the distinction that the Courant-Snyder transformation is exact so there is no approximation needed.

As we have so far insisted, we have assumed that there is one and only one resonance near the tune, i.e. there is one and only one rational number near ν. This is in principle absurd because for any ν there will be an infinite number of rational numbers near it. This is a fundamental flaw of the smooth approximation. It is in fact a fundamental flaw of all perturbative analysis of nonlinear Hamiltonian dynamics. The unending series of nearby rational numbers and the possibility of interplay of multiple number of distinct resonances are the causes of *chaotic* behavior of Hamiltonian motion. For now, we shall sweep aside this subtlety and continue below with a few example applications of the smooth approximation.

Harmonic correction Let us make an important observation here. The coefficients $g_{mk,-nk}$ are the resonance coefficients. They are responsible for the resonance driving. In case the nonlinearities are distributed around the storage ring, it is the particular Fourier component of the distribution that enters these coefficients. In the smooth approximation, a localized nonlinearity and a distributed nonlinearity will have the same effect as long as they have the same corresponding Fourier component. See Homework 3.71.

This property also leads to the possibility of nonlinear harmonic correction schemes. In case a resonance is known to be harmful to the beam, it is possible to introduce a set of correction knobs distributed around the ring with the intended harmonics. Without knowing where the error nonlinearities come from, one can compensate for their nonlinear resonance effects by tweaking the set of correctors as long as we know which resonance is responsible for the problem. This is referred to as a harmonic correction scheme for the nonlinear resonance effects.

This is storage ring, not linac To appreciate the smooth approximation in its role to integrate the nonlinear dynamics system, one can perhaps take a step back and ask oneself the question what would he do to calculate the motion of a particle to first order in the nonlinear perturbation ϵ if not applying the smooth approximation? He might consider to start with some initial condition x_0, x_0' and follow its unperturbed betatron oscillation until it hits the nonlinearity. Then the particle gets a nonlinear kick. Following the motion turn after turn, then he concludes that to first order in ϵ, he just adds up all the kicks, and the magnitude of each kick is evaluated as if the motion is unperturbed up to the kick. It is conceivable that he develops an analysis following this line, and the calculation may even be performed analytically to some extent. Seemingly reasonable, however, this approach is a time-domain approach, and is applicable only to a transport line or a linac, not to a storage ring. The repetitive nature of a storage ring has not been taken into consideration. The system cannot be integrated this way. In particular, the smooth approximation yields a tune shift that is valid to first order in ϵ, that the time-domain approach does not give. Culprit of this mistake is traced again to the view based on the equation of motion versus the view based on the Hamiltonian.

A cartoonist's view might illustrate the point. In much of storage ring accelerator physics, we look for a frequency shift. Dynamics we study can be written in the form of $\sin[(\omega+\Delta\omega)t]$, where $\Delta\omega$ represents a frequency shift if it is real, and a growth rate if imaginary; the focus of attention is $\Delta\omega$. The difference between linac and storage ring dynamics can be viewed to be

$$\begin{aligned} \text{Storage ring view:} \quad & \sin[(\omega+\Delta\omega)t] \\ \text{Linac view:} \quad & \sin\omega t + \Delta\omega t\cos\omega t - \frac{(\Delta\omega t)^2}{2}\sin\omega t + \cdots . \end{aligned}$$

The reader is advised to take a note their distinct difference. A linac is a time-domain device. Its analysis applies an initial-value problem. A storage ring is a frequency-domain device. Its dynamics applies an eigenanalysis.

	The problem	Treatment	Analysis
Linac	initial-value problem	time-domain	• initial values • ray tracking
Storage ring	eigenvalue problem	frequency-domain	• eigenvalues & eigenvectors • one-turn map • normal form

Detuning and resonance The function $\langle g\rangle_s$ is periodic in ϕ with period $\frac{2\pi}{m}$. As ϕ varies, $\langle g\rangle_s$ contains a constant part given by

$$\begin{aligned} g_{0,0} = D_\nu(J) &= \frac{1}{4\pi^2 R}\int_0^{2\pi} d\xi \int_0^{2\pi R} ds \\ &\times\ g\left(\sqrt{2J\beta(s)}\cos\left[\xi + \int_0^s ds'\left(\frac{1}{\beta(s')} - \frac{\nu}{R}\right)\right], s\right), \end{aligned} \tag{3.55}$$

plus a remaining oscillating part

$$g_1(\phi, J) = \sum_{k\neq 0} g_{mk,-nk}(J)e^{imk\phi} = \sum_{k=1}^{\infty}\left[g_{mk,-nk}(J)e^{imk\phi} + \text{c.c}\right], \tag{3.56}$$

i.e. we decompose $\langle g\rangle_s$ into two terms,

$$\langle g\rangle_s = D_\nu(J) + g_1(\phi, J). \tag{3.57}$$

We now have the Hamiltonian

$$K(\phi, J, s) = \frac{J}{R}\left(\nu - \frac{n}{m}\right) + D_\nu(J) + g_1(\phi, J),$$

which, again, is a constant of the motion.

The two terms D_ν and g_1 have important physical meanings. The term D_ν was found in the previous section already. If we drop $g_1(\phi, J)$, the Hamilton equation gives

$$\frac{d\phi}{ds} = \frac{\partial K}{\partial J} = \frac{1}{R}\left(\nu - \frac{n}{m}\right) + D'_\nu(J).$$

Away from resonances, it gives a tune shift as a function of amplitude J according to

$$\nu(J) = \nu + RD'_\nu(J). \tag{3.58}$$

The function g_1 takes into account of the effect of the single resonance, and can be associated with the resonance strength. We will give explicit examples below.

Homework 3.67 Show that Eqs. (3.58) and (3.44) are equivalent.

Homework 3.68 Prove Eq. (3.53).

3.7.3 Multipole field error

We now apply the results obtained so far to a specific example of a thin-lens multipole error. Consider a perturbation of the form

$$g(x, s) = \epsilon x^m \delta(s), \tag{3.59}$$

where $\delta(s)$ is a periodic δ-function with period $2\pi R$.

This $g(x, s)$ represents a nonlinear perturbation located at $s = 0$ that gives δ-function kicks to the particle as it passes by each turn. When the perturbation is given by a magnet multipole, we take (Table 1.5)

$$\epsilon = \frac{\ell}{m!(B\rho)} \frac{\partial^{m-1} B_y}{\partial x^{m-1}},$$

where $(B\rho) = \frac{P}{e}$ is the magnetic rigidity of the particle, ℓ is the length of the multipole magnet. A quadrupole would have $m = 2$. A nonlinear multipole would have $m \geq 3$.

Following the first-order smooth approximation, we first obtain

$$\langle g \rangle_s = \frac{\epsilon}{2\pi m R} \sum_{k=1}^{m} \left[\sqrt{2J\beta} \cos\left(\phi + 2\pi \frac{kn}{m}\right)\right]^m,$$

and then decompose $\langle g \rangle_s$ into a detuning term plus an oscillating term according to Eq. (3.57). To first order in ϵ, using Eq. (3.55), we have

$$\begin{aligned} D_\nu(J) &= \frac{\epsilon}{4\pi^2 R} \int_0^{2\pi} d\xi \, (\sqrt{2J\beta} \cos \xi)^m \\ &= \frac{\epsilon}{\pi R} \left(\frac{J\beta}{2} \right)^{m/2} \begin{cases} \frac{m!}{2[(m/2)!]^2}, & \text{if } m = \text{even}, \\ 0, & \text{if } m = \text{odd}, \end{cases} \end{aligned} \quad (3.60)$$

where β is the β-function evaluated at the multipole.

To obtain $g_1(\phi, J)$ using Eq. (3.56), we first note that only the $k = \pm 1$ terms contribute. Then it follows that

$$g_1(\phi, J) = \frac{\epsilon}{\pi R} \left(\frac{J\beta}{2} \right)^{m/2} \cos m\phi \, .$$

We see that $D_\nu(J)$ is provided by the multipole if m is even. When m is odd, it vanishes. We also see that the resonance strength g_1 is proportional to $J^{m/2} \cos m\phi$. As mentioned, the $\cos m\phi$ factor assures it is slowly varying.

Once we obtain D_ν and g_1, the constant of the motion, the Hamiltonian $K(\phi, J)$ is obtained. We then plot the constant-K contours on polar coordinates $(\sqrt{J}, \phi)$. Each constant-K contour represents the motion of a particle. As the particle moves along, its trajectory in phase space will stay on the contour.

Away from resonances, the contours are simply circles (J = constant). Resonances will deform these circular contours due to the term g_1 in the Hamiltonian. Near a resonance $\nu \approx \frac{n}{m}$, one would observe *island* structures near some particular amplitude $J = J_{island}$. There will be m such islands.

Sometimes the contours surrounding the islands will extend to infinity beyond the islands. Sometimes they are bounded. In the former case, motion is unstable for particles following those contours. In the latter case, motion is stable. This behavior depends on the relative magnitudes of D_ν versus g_1. When D_ν dominates, the islands are closed and motion is stable. When g_1 dominates, the motion is unstable.

Sextupole and $m = 3$ resonance Take the case of a sextupole nonlinearity for example. In this case, $m = 3$, and we have $D_\nu = 0$ and

$$RK(\phi, J, s) = J\Delta\nu + \frac{\epsilon}{\pi} \left(\frac{\beta}{2} \right)^{3/2} J^{3/2} \cos 3\phi \, , \quad (3.61)$$

where $\Delta\nu = \nu - \frac{n}{m}$ is the distance of ν from the resonance.

Figure 3.26 shows some contour plots for the case with $\frac{\epsilon}{\pi} \left(\frac{\beta}{2} \right)^{3/2} = 0.05$. When the tune is far from the resonance, the phase space contours form circles. As the tune gets close to the resonance, the circles are deformed into triangular shape. The triangle is called a *separatrix* in the phase space plot. Particles

Figure 3.26: Phase space contours in polar coordinates $(\sqrt{J}, \phi)$ when the tune is close to a $\frac{1}{3}$ resonance and there is a sextupole field error present with $\frac{\epsilon}{\pi}\left(\frac{\beta}{2}\right)^{3/2} = 0.05$. The plots show what happens when $\Delta\nu$ is scanned through the resonance. All five panels have the same plot scales, so the size and orientation of the triangles, for example, can be meaningfully compared.

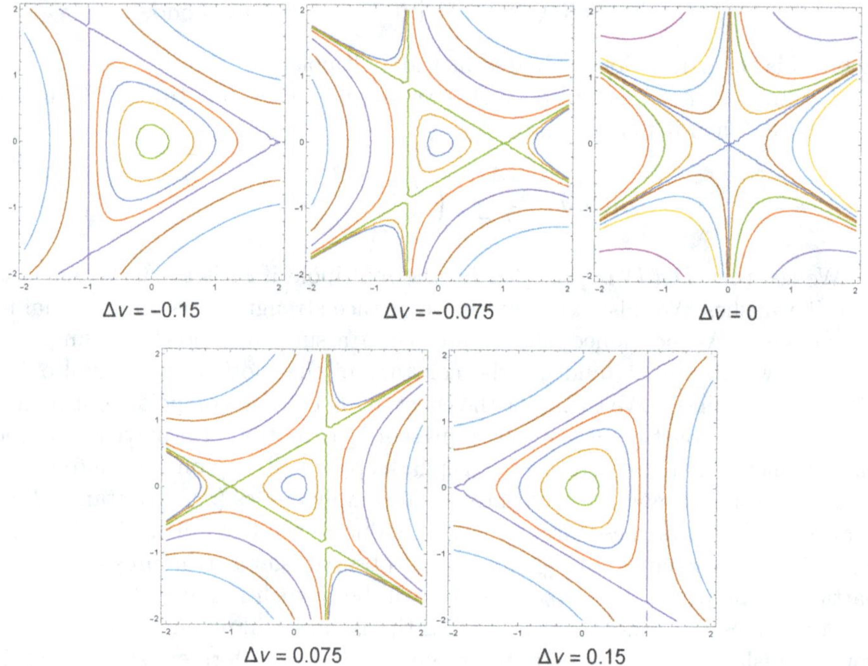

inside the triangles are bounded and are stable. Outside the triangle, particles move without bound and will eventually be lost from the beam. The stable triangular region shrinks as the tune approaches the resonance and vanishes exactly on resonance.

This resonance behavior has been put to use to slowly extract particles, in a controlled fashion, from a stored beam in a synchrotron for medical or research applications. A beam is first stored, and a sextupole nonlinearity is introduced, first with a tune away from a $\frac{1}{3}$ resonance, so the sextupole perturbation on particle motion is negligible. When a slow beam extraction is requested, the tune is brought to slowly approach the $\frac{1}{3}$ resonance either from above or from below. The stable region shrinks, squeezing particles slowly out of the synchrotron towards the desired application facility. The speed of resonance approach determines the rate of particle spill.[19]

[19] See, for example, P.J. Bryant, Sec. 4.12.1, Handbook Accel. Phys. & Eng., 2nd Ed., World Scientific (2013).

Octupole and $m = 4$ resonance The sextupole case has no detuning, i.e. $D_\nu(J) = 0$. This is not the case for octupoles. For the octupole case, we have

$$RK(\phi, J) = J\Delta\nu + 3\epsilon_1 J^2 + \epsilon_1 J^2 \cos 4\phi\,,$$

where $\epsilon_1 = \frac{\epsilon\beta^2}{4\pi} = \frac{1}{96\pi}\frac{\ell}{B\rho}\frac{\partial^3 B_y}{\partial x^3}$.

Phase space behavior is shown in Fig. 3.27. One sees here the generation of islands mentioned earlier. Formation of closed islands requires the existence of the detuning term $D_\nu(J)$. As shown, islands are formed only on one side of the resonance. With or without islands, particle motion in this 1-D octupole system is bounded and is therefore stable, at least to first order of the octupole strength.

Figure 3.27: Phase space contours in polar coordinates $(\sqrt{J}, \phi)$ when the tune is close to a $\frac{1}{4}$ resonance and there is an octupole field error present with $\frac{\epsilon\beta^2}{4\pi} = 0.05$.

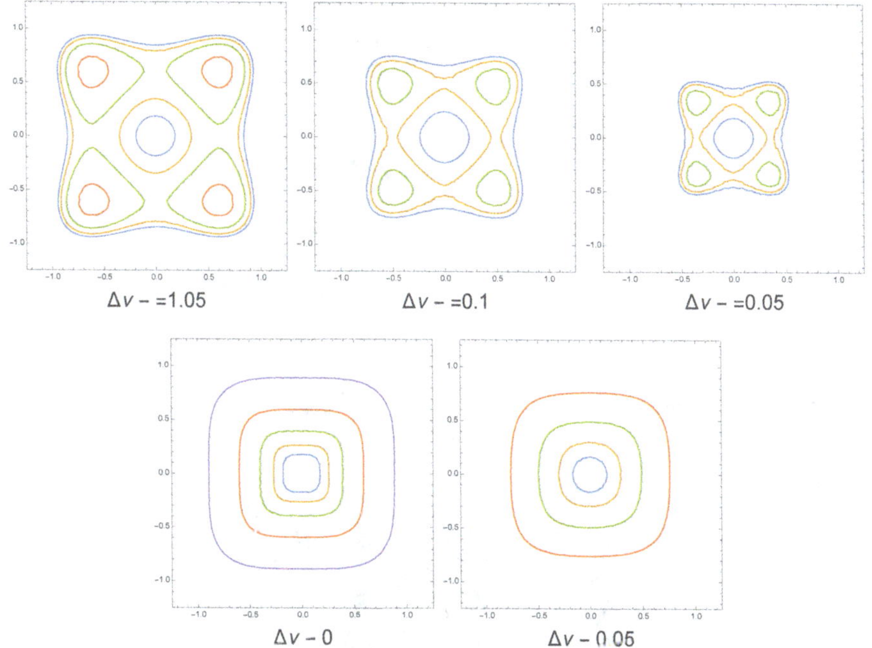

IUCF data Experiments of phase space contours were studied at the Indiana University Cyclotron Facility in the early 1990s. Some of the results are shown in Fig. 3.28.[20] In these experiments, the trajectories of a cooled beam after

[20] D.D. Caussyn, et al., Phys. Rev. A46, 7942 (1992).

an initial kick in the x plane are recorded turn-by-turn at two adjacent beam position monitors. Phase space records of the beam motion are then plotted. The upper panel is the case when $\nu_x \approx \frac{11}{3}$. The lower panel is when $\nu_x \approx \frac{15}{4}$. The data trace out the phase space beautifully, including the triangle and the islands, all as expected.

Figure 3.28: Experimental results exploring the phase space dynamics of a stored beam at the IUCF facility. [Courtesy S.Y. Lee, 2019]

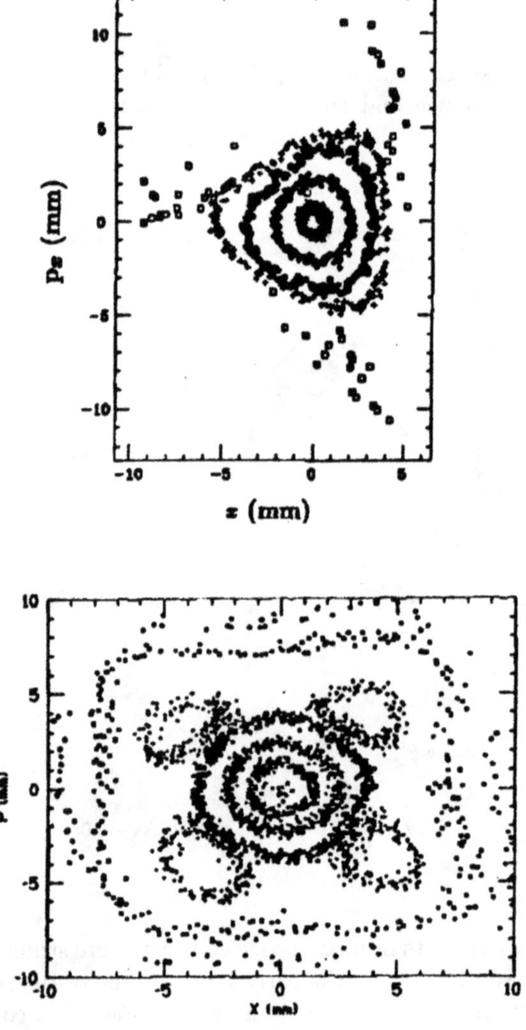

Incidentally, a strong signal for the existence of islands is most readily seen when the turn-by-turn data was Fourier transformed. In the Fourier spectrum, a sharp peak occurred at exactly the value of $\nu_x = \frac{15}{4}$ (not shown in Fig. 3.28). This is a clear sign of particles trapped in the islands, whose oscillation frequency is locked to be exactly $\frac{15}{4}$.

Homework 3.69 Smooth approximation can also be applied to a linear perturbation such as the case of a quadrupole with $m = 2$, even though the problem can be solved exactly using matrices. The text gives a way to calculate the Hamiltonian after smoothing. Show that it is given by

$$RK(\phi, J, s) = J\Delta\nu + \frac{\epsilon\beta}{2\pi}J(1 + \cos 2\phi),$$

where $\Delta\nu = \nu - \frac{n}{2}$ is the distance of the tune from the resonance. Show that the expression of $D_\nu(J)$ yields the expected expression of tune shift. Plot the constant contour of particle trajectories in the polar phase space $(\sqrt{J}, \phi)$.

Solution Some contour plots are shown in Fig. 3.29 for the case $\frac{\epsilon\beta}{2\pi} = 0.05$ for various values of $\Delta\nu$. Particle motion is unstable when $-0.1 < \Delta\nu < 0$. In other words, instability condition is $\frac{\epsilon\beta}{2\pi} > (\Delta\nu - \frac{\epsilon\beta}{2\pi}) > -\frac{\epsilon\beta}{2\pi}$. From this instability condition, one observes that the quantity $\frac{\epsilon\beta}{2\pi}$ assumes two physical meanings: it is both the tune shift and the stopband width. Note that $\epsilon = \frac{1}{2f}$ if one prefers to use the focal length f to specify the perturbation strength.

Homework 3.70

(a) The constant Hamiltonian for the $m = 3$ sextupole case was given in Eq. (3.61). Expressed it again in terms of the Cartesian coordinates $x = \sqrt{J}\cos\phi, y = \sqrt{J}\sin\phi$.

(b) Find the expressions of the straight lines that determine the boundaries of the triangular region.

(c) Find the value of the constant Hamiltonian RK on the triangle. Make sure the same value applies to all three segments of the triangle.

Solution

(a) Let $\epsilon_1 = \frac{\epsilon}{\pi}\left(\frac{\beta}{2}\right)^{3/2}$,

$$RK(x, y) = \Delta\nu(x^2 + y^2) + \epsilon_1(x^3 - 3xy^2).$$

(b)

$$\begin{aligned} x &= \frac{\Delta\nu}{3\epsilon_1}, \\ x + \sqrt{3}y &= -\frac{2\Delta\nu}{3\epsilon_1}, \\ x - \sqrt{3}y &= -\frac{2\Delta\nu}{3\epsilon_1}. \end{aligned}$$

(c) $RK = \frac{4\Delta\nu^3}{27\epsilon_1^2}$.

Figure 3.29: Phase space contours when the tune is close to a half integer and there is a quadrupole field error present with $\frac{\epsilon\beta}{2\pi} = 0.05$. The plots show how the phase space is distorted when $\Delta\nu$ is scanned through the resonance. Far away from the resonance, the phase space contours are circles.

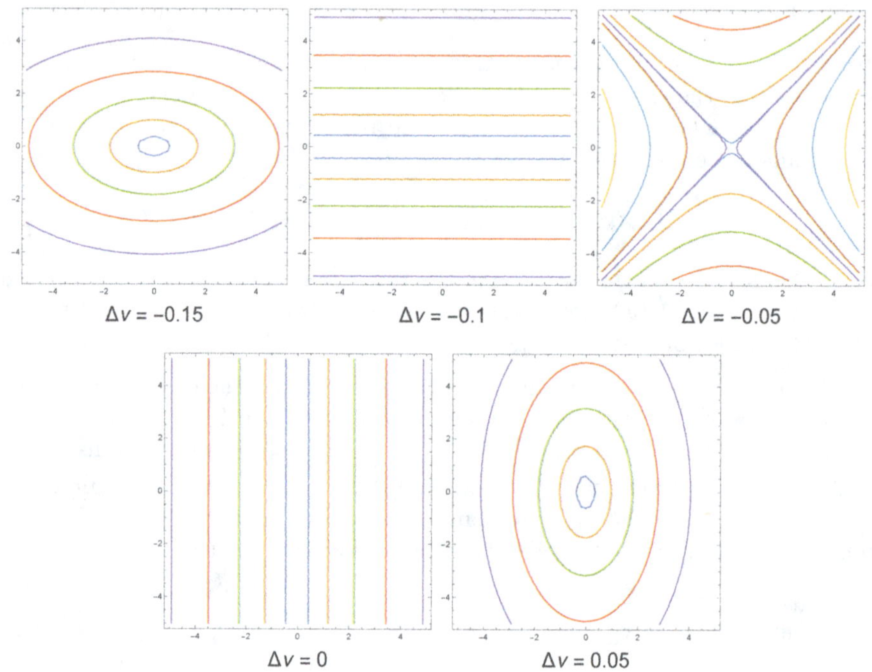

Homework 3.71 Equation (3.61) gives the smoothed Hamiltonian near the $\nu = \frac{1}{3}$ resonance for a single sextupole of strength ϵ at one lumped location at $s = 0$. Consider the case when there are N thin sextupoles evenly distributed around the storage ring. All distributed sextupoles are of strength $\frac{\epsilon}{N}$ and all located with the same β-function. Let the adjacent sextupoles be separated by a betatron phase χ. In this case, Eq. (3.61) still applies if the parameter ϵ is appropriately modified. How should ϵ be modified?

3.7.4 Beam-beam perturbation

Now we apply our results to another example of the beam-beam perturbation. In a colliding-beam storage ring, the head-on collisions of the colliding beams produce a perturbation of the form (3.50). Consider a particle circulating in a collider ring. Each time it passes by the collision point $s = 0$, it encounters and collides with an oncoming beam. Let the oncoming beam have N particles. To simplify our analysis, we assume the oncoming beam has a round bi-Gaussian distribution with rms transverse size σ. The general case with elliptical cross-section can also be performed although the analysis becomes lengthy.

Consider a particle executing x-motion without any y-motion. As it passes through the collision point, it gets a kick (Fig. 1.38)

$$\Delta x' = \frac{2Nr_0}{\gamma}\frac{1}{x}\left(1-e^{-x^2/2\sigma^2}\right),$$

where r_0 is the classical radius and γ is the relativistic energy factor for the particle under consideration — not for the particles in the oncoming beam. We have assumed $\gamma \gg 1$.

In this treatment of the beam-beam perturbation, we assume the on-coming beam has a continuum distribution. In reality the beam has a collection of discrete point charges. It remains to be justified that a smooth distribution can represent the collection of singular point charges faithfully. This point will be mentioned again in Fig. 8.29 later. Here we shall proceed with the continuum picture.

The beam-beam interaction can be described by the Hamiltonian (3.50) with

$$g(x,s) = -\frac{2Nr_0}{\gamma}\,\delta(s)\int_0^x \frac{dx'}{x'}\left(1-e^{-x'^2/2\sigma^2}\right).$$

With s evaluated only at $s=0$ each turn, the integral $\int_0^s ds'\left(\frac{1}{\beta(s')}-\frac{\nu}{R}\right)$ in Eq. (3.52) vanishes. We have

$$\begin{aligned}
g_{j,\ell}(J) &= -\frac{Nr_0}{2\pi^2 R\gamma}\int_0^{2\pi} d\xi\, e^{-ij\xi}\int_0^{\sqrt{2J\beta_0}\,\cos\xi}\frac{dx'}{x'}\left(1-e^{-x'^2/2\sigma^2}\right)\\
&= -\frac{Nr_0}{2\pi^2 R\gamma}\int_0^{2\pi} d\xi\, e^{-ij\xi}\int_0^{\sqrt{2J\beta_0}/\sigma}\frac{dy}{y}\left(1-e^{-\frac{1}{2}y^2\cos^2\xi}\right)\\
&= -\frac{Nr_0}{\pi R\gamma}\int_0^{\sqrt{2J\beta_0}/\sigma}\frac{dy}{y}A_j(y),
\end{aligned}$$

where β_0 is the β-function at the collision point $s=0$, and we have made a change of integration variable from x' to y, where $x'=\sigma y\cos\xi$. The Fourier harmonic is independent of the index ℓ, and we shall then drop it from our indices. We have also defined

$$\begin{aligned}
A_j(y) &= \frac{1}{2\pi}\int_0^{2\pi} d\xi\, e^{-ij\xi}\left(1-e^{-\frac{1}{2}y^2\cos^2\xi}\right)\\
&= \begin{cases} 1-e^{-y^2/4}I_0(y^2/4), & j=0\\ 0, & j=\text{odd}\\ -(-1)^{j/2}e^{-y^2/4}I_{j/2}(y^2/4), & j=\text{even}\neq 0\end{cases}.
\end{aligned}$$

Beam-beam detuning Away from resonances, we have a detuning given by

$$\nu(J) = \nu + R g_0'(J),$$

with

$$g_0(J) = -\frac{N r_0}{\pi R \gamma} \int_0^{\sqrt{2\beta_0 J}/\sigma} \frac{dy}{y} \left[1 - e^{-y^2/4} I_0\left(\frac{y^2}{4}\right) \right].$$

Although there is no simple closed form expression for $g_0(J)$, there is a simple expression for $g_0'(J)$,

$$\Delta\nu_{BB}(J) = R g_0'(J) = -\frac{N r_0 \beta_0}{4\pi\sigma^2\gamma} \left[\frac{1 - e^{-\chi} I_0(\chi)}{\chi} \right]_{\chi = \frac{\beta_0 J}{2\sigma^2}}.$$

The quantity

$$\Delta\nu_{BB}(0) = \frac{N r_0 \beta_0}{4\pi\sigma^2\gamma}$$

is of course the beam-beam tune shift parameter. The overall minus sign is because we have assumed equal sign of electric charges of the two colliding beams. If they have opposite signs, then the minus sign is to be removed.

Figure 3.30 shows $\frac{1-e^{-\chi} I_0(\chi)}{\chi}$ as a function of χ. For a particle with small betatron oscillation amplitude, $\chi \to 0$, the particle receives a maximum tune shift to the full extent of $\Delta_{BB}(0)$. As the oscillation amplitude increases, the tune shift drops. When $\chi \to \infty$, the particle spends most of the time outside the oncoming beam distribution; its tune shift drops to zero.

Figure 3.30: A plot of the beam-beam tune shift function $\frac{1-e^{-\chi} I_0(\chi)}{\chi}$.

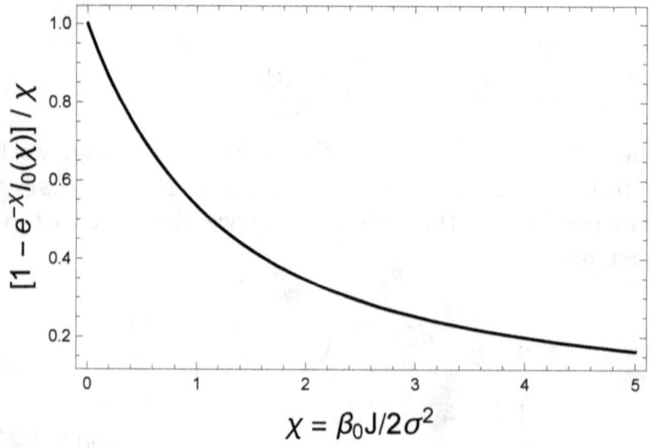

Beam-beam resonance Beam-beam interaction excites even-order resonances. Consider the case of a resonance near $\nu = \frac{1}{4}$, i.e. the case when $m = 4$. The Hamiltonian is now given by

$$RK(\phi, J) = J\Delta\nu + Rg_0(J) + 2R\sum_{k=1}^{\infty} g_{4k}(J)\cos(4k\phi),$$

where $\Delta\nu = \nu - \frac{1}{4}$ is the distance to the resonance. We also have

$$\begin{aligned} g_{4k}(J) &= -\frac{Nr_0}{\pi R\gamma}\int_0^{\sqrt{2\beta_0 J}/\sigma} \frac{dy}{y} A_{4k}(y) \\ &= \frac{Nr_0}{\pi R\gamma}\int_0^{\sqrt{2\beta_0 J}/\sigma} \frac{dy}{y} e^{-y^2/4} I_{2k}\left(\frac{y^2}{4}\right). \end{aligned}$$

Contour plots of the beam-beam dynamics near a $\frac{1}{4}$ resonance can be readily obtained. Figure 3.31 shows a plot of the functions g_0, g_4, g_8 as functions of $\chi = \frac{\beta_0 J}{2\sigma^2}$.

Figure 3.31: A plot of the beam-beam detuning and resonance functions $-\left(g_0/\frac{Nr_0}{\pi R\gamma}\right)$ (solid), $\left(g_4/\frac{Nr_0}{\pi R\gamma}\right)\times 10$ (dashed), $\left(g_8/\frac{Nr_0}{\pi R\gamma}\right)\times 100$ (dotted) as functions of $\chi = \frac{\beta_0 J}{2\sigma^2}$.

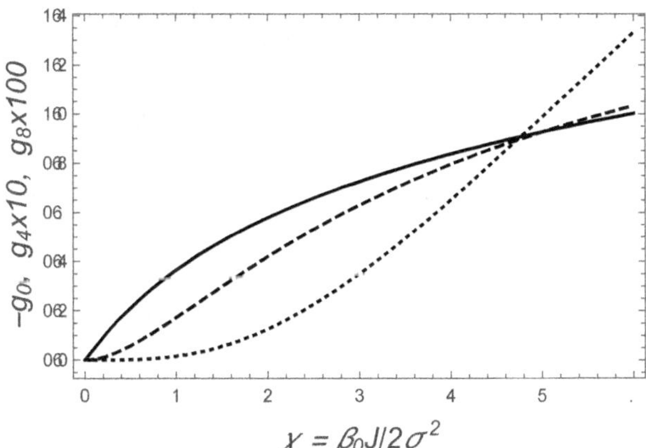

Homework 3.72 With g_0, g_4, g_8 given in the text, plot the phase space contour plot to show the effect of a $\frac{1}{4}$ resonance due to beam-beam interaction. Dropping higher orders, the Hamiltonian is $RK = J\Delta\nu + Rg_0(J) + 2Rg_4(J)\cos 4\phi + 2Rg_8(J)\cos 8\phi$. Look for formation of four distinct islands.

Solution A contour plot in $(\sqrt{J}, \phi)$ phase space is shown in Fig. 3.32 near a $\frac{1}{4}$ resonance. Islands are formed only on one side of the resonance.

Figure 3.32: Contour plot of phase space in the presence of beam-beam perturbation near a $\frac{1}{4}$ resonance. Parameters chosen are $\frac{Nr_0}{\pi R\gamma} = 0.1$, $\frac{\sqrt{2\beta_0}}{\sigma} = 20$. The three panels show what happens when the tune is scanned across the resonance.

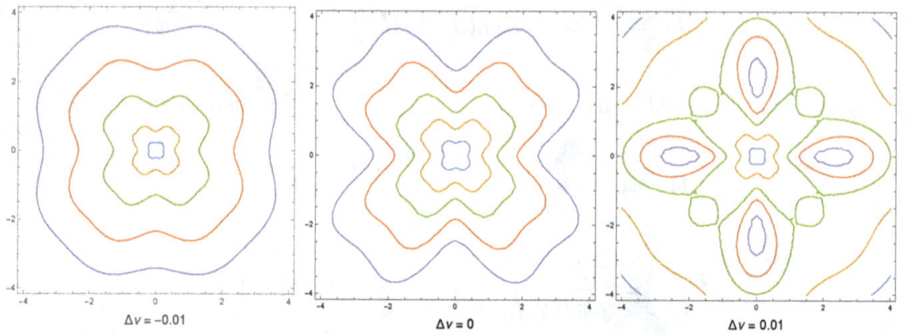

Homework 3.73 The Bessel functions are pretty awkward to work with. In case we are interested only in the part of phase space $\beta_0 J \lesssim \sigma^2$, show that

$$
\begin{aligned}
g_0(J) &\approx -\frac{Nr_0}{4\pi R\gamma}\frac{\beta_0 J}{\sigma^2}, \\
g_{4k}(J) &\approx \frac{Nr_0}{\pi R\gamma}\frac{1}{k\,(2k)!\,2^{4k+2}}\left(\frac{\beta_0 J}{\sigma^2}\right)^{2k}, \\
g_4(J) &\approx \frac{Nr_0}{128\,\pi R\gamma}\left(\frac{\beta_0 J}{\sigma^2}\right)^2, \\
g_8(J) &\approx \frac{Nr_0}{49152\,\pi R\gamma}\left(\frac{\beta_0 J}{\sigma^2}\right)^4.
\end{aligned}
$$

The dependences on J of these functions becomes clearer, reinforcing Fig. 3.31.

3.7.5 Coupled system

So far we have considered smooth approximation only to 1-D systems. Even for the beam-beam discussion, an intrinsically coupling mechanism, we treated only the case of a round beam cross-section, which reduces the problem to 1-D. Smooth approximation can also be extended to 2-D coupled systems, however, in a rather straight forward manner.

The original Hamiltonian of the system is

$$H(x, p_x, y, p_y, s) = \frac{p_x^2}{2} + \frac{K_x(s)x^2}{2} + \frac{p_y^2}{2} + \frac{K_y(s)y^2}{2} + g(x, y, s).$$

A canonical transformation similar to the 1-D case then leads to a new Hamiltonian

$$K(\phi_x, J_x, \phi_y, J_y, s) = \frac{\nu_x J_x}{R} + \frac{\nu_y J_y}{R} + g\left(\sqrt{2\beta_{x0}J_x}\cos\phi_x, \sqrt{2\beta_{y0}J_y}\cos\phi_y, s\right).$$

This is an exact result. We then make the smooth approximation to obtain a 2-D extension of Eqs. (3.52) and (3.54),

$$\begin{aligned} K(\phi_x, J_x, \phi_y, J_y, s) &\approx \frac{J_x}{R}(\nu_x - \nu_{x0}) + \frac{J_y}{R}(\nu_y - \nu_{y0}) \\ &+ \sum_{k=-\infty}^{\infty} g_{m_x k, m_y k, -nk}(J_x, J_y) e^{ik(m_x\phi_x + m_y\phi_y)}, \end{aligned}$$

where $m_x\nu_{x0} + m_y\nu_{y0} = n$ defines the nearby resonance, and

$$\begin{aligned} g_{j_x, j_y, \ell} &= \frac{1}{8\pi^3 R}\int_0^{2\pi} d\xi_x \int_0^{2\pi} d\xi_y \int_0^{2\pi R} ds e^{-ij_x\xi_x - ij_y\xi_y - i\ell\frac{s}{R}} \\ &\times g\Bigg(\sqrt{2J_x\beta_x(s)}\cos\left[\xi_x + \phi_{x0}(s) - \nu_x\frac{s}{R}\right] \\ &\qquad \times \sqrt{2J_y\beta_y(s)}\cos\left[\xi_y + \phi_{y0}(s) - \nu_y\frac{s}{R}\right], s\Bigg), \end{aligned}$$

with $\phi_{x0,y0} = \int_0^s \frac{ds'}{\beta_{x,y}(s')}$ the unperturbed betatron phase advances, not to be confused with the dynamical variables $\phi_{x,y}$. We also have $g_{-j_x,-j_y,-\ell} = g^*_{j_x,j_y,\ell}$.

Skew quadrupole distribution For the case of a linear coupling resonance $\nu_{x0} - \nu_{y0} = n$ driven by a distribution of skew quadrupoles, we have

$$g(x, y, s) = k(s)xy,$$

where $k(s) = \frac{1}{B\rho}\frac{\partial B_y}{\partial y}$. The smoothed Hamiltonian receives contributions only from two terms with $k = \pm 1$, i.e. $g_{1,-1,-n}$ and $g_{-1,1,n}$. All other coefficients vanish. We have

$$\begin{aligned} K(\phi_x, J_x, \phi_y, J_y, s) &\approx \frac{J_x}{R}(\nu_x - \nu_{x0}) + \frac{J_y}{R}(\nu_y - \nu_{y0}) \\ &+ \left[g_{1,-1,-n}e^{i(\phi_x - \phi_y)} + g_{-1,1,n}e^{-i(\phi_x - \phi_y)}\right], \end{aligned}$$

where

$$\begin{aligned} g_{1,-1,-n} &= g^*_{-1,1,n} \\ &= \frac{\sqrt{J_x J_y}}{4\pi R}\int_0^{2\pi R} ds\sqrt{\beta_x(s)\beta_y(s)}\,k(s)e^{i\phi_{x0}(s) - i\phi_{y0}(s) - i(\nu_x - \nu_y - n)\frac{s}{R}}. \end{aligned}$$

Denoting $g_{1,-1,-n}$ by $\frac{\sqrt{J_x J_y}}{R} g$, we have defined a dimensionless complex *linear coupling coefficient,*

$$g = \frac{1}{4\pi} \int_0^{2\pi R} ds \sqrt{\beta_x(s)\beta_y(s)}\, k(s) e^{i\phi_{x0}(s) - i\phi_{y0}(s) - i(\nu_x - \nu_y - n)\frac{s}{R}}. \tag{3.62}$$

We will return to this coupling coefficient later in Sec. 6.4.13, and will assign it a different notation ϵ.

It is instructive to compare the expression (3.62) with that of the linear tune shift (3.17). Including the factor of $\frac{1}{4\pi}$ and the weighting by the β-functions, they really have the same form. The coupling coefficient resembles a tune shift in the coupled dynamics. It has the dimension of tune shift, except that it is complex. The phase factor takes into account the phase space rotation in the coupled x and y dimensions.

In case there is only one thin skew quadrupole around the ring with $k(s) = \frac{\ell}{B\rho}\frac{\partial B_y}{\partial y}\delta(s)$, then

$$\begin{aligned} K(\phi_x, J_x, \phi_y, J_y, s) &\approx \frac{J_x}{R}(\nu_x - \nu_{x0}) + \frac{J_y}{R}(\nu_y - \nu_{y0}) \\ &+ \frac{\sqrt{J_x J_y}}{2\pi R} \sqrt{\beta_x(0)\beta_y(0)} \left(\frac{\ell}{B\rho} \frac{\partial B_y}{\partial y} \right) \cos(\phi_x - \phi_y). \end{aligned}$$

Single octupole detuning Consider for example a single thin octupole. We have the perturbation

$$g(x, y, s) = \epsilon(x^4 - 6x^2y^2 + y^4)\, \delta(s).$$

The parameter $\epsilon = \frac{1}{24}\frac{\ell}{B\rho}\frac{\partial^3 B_y}{\partial x^3}$ can be written as $\frac{1}{4}\frac{K_3 \ell}{B\rho}$ if it is to be related to Table 1.5.

We will consider the case away from resonances. Near a resonance, there will be an emittance sum rule and an emittance beating effect, which will be discussed in Secs. 3.7.6 and 3.7.7 momentarily. Away from resonances, smooth approximation is made by replacing g function by its time average, i.e. its zeroth harmonic,

$$\begin{aligned} \langle g \rangle &= \int_0^{2\pi} \frac{d\phi_x}{2\pi} \int_0^{2\pi} \frac{d\phi_y}{2\pi} \int_0^{2\pi R} \frac{ds}{2\pi R} g(\sqrt{2\beta_{x0} J_x} \cos\phi_x, \sqrt{2\beta_{y0} J_y} \cos\phi_y, s) \\ &= \frac{\epsilon}{2\pi R} \int_0^{2\pi} \frac{d\phi_x}{2\pi} \int_0^{2\pi} \frac{d\phi_y}{2\pi} \\ &\quad \times (4\beta_{x0}^2 J_x^2 \cos^4\phi_x - 24\beta_{x0}\beta_{y0} J_x J_y \cos^2\phi_x \cos^2\phi_y + 4\beta_{y0}^2 J_y^2 \cos^4\phi_y) \\ &= \frac{\epsilon}{2\pi R} \left(\frac{3}{2}\beta_{x0}^2 J_x^2 - 6\beta_{x0}\beta_{y0} J_x J_y + \frac{3}{2}\beta_{y0}^2 J_y^2 \right). \end{aligned}$$

The Hamilton equations yield J_x and J_y are constants of the motion, and

$$\frac{d\phi_x}{ds} = \frac{\partial \langle g \rangle}{\partial J_x}, \qquad \frac{d\phi_y}{ds} = \frac{\partial \langle g \rangle}{\partial J_y}.$$

Tune shifts are then given by

$$\begin{aligned} \Delta\nu_x(J_x, J_y) &= R\frac{\partial \langle g \rangle}{\partial J_x} = \frac{3\epsilon}{2\pi}\beta_{x0}(\beta_{x0}J_x - 2\beta_{y0}J_y), \\ \Delta\nu_y(J_x, J_y) &= R\frac{\partial \langle g \rangle}{\partial J_y} = \frac{3\epsilon}{2\pi}\beta_{y0}(-2\beta_{x0}J_x + \beta_{y0}J_y). \end{aligned} \tag{3.63}$$

One notes that the octupole shifts both the betatron tunes of a particle according to its betatron amplitudes. The shifts are linear in J_x and J_y. The shift of ν_x depending on J_x and the shift of ν_y depending on J_y have the same sign, and both having the sign of ϵ. The shift of ν_x depending on J_y and the shift of ν_y depending on J_x have the opposite sign as ϵ.

An alert reader would note that $\frac{\partial \Delta\nu_x}{\partial J_y} = \frac{\partial \Delta\nu_y}{\partial J_x}$. This in fact involves a theorem. We will discuss it next.

Hamiltonian and symplecticity An important observation on Eq. (3.63) is the theorem,

$$\frac{\partial \Delta\nu_x}{\partial J_y} = \frac{\partial \Delta\nu_y}{\partial J_x}.$$

The dependence of $\Delta\nu_x$ on J_y is in fact equal to the dependence of $\Delta\nu_y$ on J_x. An alert reader should not be surprised of this result. It follows simply from the fact that $\Delta\nu_x = \frac{\partial \langle g \rangle}{\partial J_x}$ and $\Delta\nu_y = \frac{\partial \langle g \rangle}{\partial J_y}$ with the same $\langle g \rangle$ contained in the Hamiltonian, and is a general result, not only for octupoles.

Here let us make the observation again that such a general rule follows from the Hamiltonian. After we make the smooth approximation on the Hamiltonian, this rule follows almost trivially. There are many examples of such general rules following from the Hamiltonian. Another general rule — another theorem — will be discussed in the next section 3.7.6. Still one more example will be mentioned in Eq. (4.10) concerning a link between a closed-orbit distortion and the dispersion function. See also Homework 3.76.

These general rules, when first appear, may sound miraculous, seemingly emerging from nowhere, and they all look neat — hard to imagine to derive them from equations of motion. Since Hamiltonian and symplecticity are intrinsically linked, it is often stated, also correctly, that such rules follow from the symplecticity condition.

Homework 3.74 Consider a storage ring with a large circumference. All its quadrupoles are perfectly aligned with no tilt errors. In principle its linear coupling coefficient (3.62) vanishes. However, the alignment is done relative to the local gravitation. This means all magnets are actually misaligned with a tilt angle

$$\theta_0 = \frac{\text{storage ring radius}}{\text{earth radius}} \,.$$

Each quadrupole then contributes to an effective skew quadrupole field gradient of $2\theta_0 \frac{\partial B_y}{\partial x}$. Estimate the magnitude of the linear coupling coefficient of this storage ring. It may become large if the storage ring is large.

Homework 3.75 Sextupoles and decapoles do not contribute to tune shifts in the first-order smooth approximation. The text calculated the tune shifts for the case of a localized octupole. Repeat the calculation of tune shifts of a localized dodecapole, for which

$$g(x, y, s) = \epsilon(x^6 - 15\,x^4 y^2 + 15\,x^2 y^4 - y^6)\,\delta(s)\,.$$

Check the theorem $\frac{\partial \Delta\nu_x}{\partial J_y} = \frac{\partial \Delta\nu_y}{\partial J_x}$.

Homework 3.76 As another revelation of the power of the Hamiltonian and sympecticity, we leap the time and consider a circular accelerator with a finite horizontal or vertical chromaticity (Chapter 4) and let there be an RF so that there is also a synchrotron oscillation (Chapter 5). Let us consider horizontal chromaticity; the unperturbed Hamiltonian of the accelerator can be written as

$$H_0(\phi_x, J_x, z, \delta, s) = \nu_x J_x + a z^2 + b\delta^2 \,,$$

where z, δ are the synchrotron dynamical variables, and we have let the synchrotron oscillation be a simple harmonic motion. The coefficients a, b will be found in Chapter 5 but are not too relevant here.

(a) When there is introduced a finite chromaticity ξ_x, it means $\nu_x = \nu_{x0} + \xi_x \delta$. The design of the accelerator becomes

$$H_1(\phi_x, J_x, z, \delta, s) = (\nu_{x0} + \xi_x \delta) J_x + a z^2 + b\delta^2 \,.$$

That is all we need to prove a powerful theorem. Show the theorem that when there is a finite chromaticity, then there must be associated a circumference path length dependence on the betatron oscillation amplitude. The connection is given by

$$\nu_x = \nu_{x0} + \xi_x \delta \qquad \Longleftrightarrow \qquad \Delta C = -\xi_x J_x \,.$$

Similar effect applies to the vertical case.

(b) Pursuing further, if you are given a Hamiltonian in Cartesian coordinates,

$$H_2(x, p_x, z, \delta, s) = H_0(x, p_x, z, \delta, s) + (C_{116} x^2 + C_{126} x p_x + C_{226} p_x^2)\delta \,,$$

you can work out the inner connection between each of the perturbation terms and its specific effect on the coupled dynamics. These connections can be explored to investigate the nonlinear perturbations term by term in a given accelerator lattice.

(c) Find the connection between ξ_x and $C_{116}, C_{126}, C_{226}$ in the smooth approximation.

3.7.6 Sum and difference resonances

We found earlier that a perturbation caused by a thin-lens skew quadrupole would make the accelerator unstable near a sum resonance, while near a difference resonance, the particle motion is strongly coupled but nevertheless remains stable. We will show in this section that this property of instability near sum resonances and coupling-with-stability near difference resonances is true in general, at least in the smooth approximation to first order in the strength of the nonlinearity.

The theorem In the application of smooth approximation to a 2-D system in the presence of a coupling resonance $m_x\nu_x + m_y\nu_y \approx n$, the important thing is to note that the smoothed Hamiltonian depends on $J_x, J_y, m_x\phi_x + m_y\phi_y$,

$$K(\phi_x, J_x, \phi_y, J_y, s) \approx K(J_x, J_y, m_x\phi_x + m_y\phi_y)\,. \tag{3.64}$$

Its dependences on ϕ_x and ϕ_y are such that, near the resonance, it depends only on the *combined* variable $\Phi = m_x\phi_x + m_y\phi_y$. This turns out to have a profound consequence on particle motion even without knowing any other details of the Hamiltonian. The only thing we need to know is that the system obeys a single-resonance smooth approximation.

With the smoothed Hamiltonian independent of s, it is a constant of the motion. But that is not what we are pursuing here.

Hamilton equations give

$$\begin{aligned}\frac{dJ_x}{ds} &= -\frac{\partial K}{\partial \phi_x} = -m_x\frac{\partial K}{\partial \Phi}\,,\\ \frac{dJ_y}{ds} &= -\frac{\partial K}{\partial \phi_y} = -m_y\frac{\partial K}{\partial \Phi}\,,\end{aligned}$$

which can be combined to yield

$$m_y\frac{dJ_x}{ds} - m_x\frac{dJ_y}{ds} = 0\,,$$

or

$$m_yJ_x - m_xJ_y = \text{constant}\,. \tag{3.65}$$

In case m_x and m_y have opposite signs, we have a situation that both J_x and J_y can only be bounded during the particle motion because both J_x and J_y must be positive. The particle's x- and y-amplitudes (measured by J_x and J_y respectively) can exchange among themselves, but neither will be able to move away indefinitely. This means particle motion is bounded and therefore stable near difference resonances.

The situation is different for the sum resonances. In that case, J_x and J_y are allowed to grow indefinitely. In fact, it dictates that if one of them grows, the other one must also do so in such a way that $m_y J_x - m_x J_y$ maintains its constant value.

To appreciate the sum rule (3.65), note that the quantities J_x and J_y are the two *unperturbed* Courant-Snyder invariants. The theorem says that in the smooth approximation, and to first order of the perturbation strength, these quantities, still defined the same way without perturbation, will change but their changes obey a sum rule that is not obvious from any other point of view.

The quantities J_x and J_y are also referred to as the *apparent* emittances, meaning that they are the emittance that appear in the uncoupled x-y space. They are not constants of the motion, and they beat in time obeying the sum rule. In symplectic systems, we will show later in Sec. 5.6 that the true emittances, the *eigenemittances*, are true constants of the motion. You cannot change the eigenemittances by any symplectic devices. But here we are talking about the sum rule of the two apparent emittances, not the eigenemittances.

We thus arrive at an important theorem, i.e. particle motion is stable near difference resonances, and unstable near sum resonances – and all we have assumed is the system is Hamiltonian in its 2-D dynamics. One should keep in mind however that this theorem holds only when the single-resonance smooth approximation holds.

3.7.7 Emittance beating near difference resonance

The slow Hamiltonian (3.64) near the resonance $\Delta\nu = m_x\nu_x + m_y\nu_y - n \approx 0$, to first order in the nonlinearity ϵ, has the more explicit form

$$\begin{aligned} K(\phi_x, J_x, \phi_y, J_y, s) \; \approx \; & \frac{J_x}{R}(\nu_x - \nu_{x0}) + \frac{J_y}{R}(\nu_y - \nu_{y0}) + D_\nu(J_x, J_y) \\ & + \epsilon J_x^{\frac{|m_x|}{2}} J_y^{\frac{|m_y|}{2}} \cos(m_x\phi_x + m_y\phi_y + \chi) \,, \end{aligned} \tag{3.66}$$

where ν_{x0}, ν_{y0} are the reference tune values with $m_x\nu_{x0} + m_y\nu_{y0} = n$, $\Delta\nu$ is the distance of the unperturbed tunes from the resonant values, and χ is the phase of the resonance driving harmonic.

The Hamilton equations have been used to derive the theorem (3.65). Near a difference resonance, while obeying the sum rule, J_x and J_y still evolve in time, exchanging with each other. This is the emittance beating effect. In this section, we study how these apparent emittances beat.

Simulation $\nu_x - \nu_y \approx n$ resonance Consider the case of a linear resonance near $\Delta\nu = \nu_x - \nu_y - n \approx 0$ driven by a localized skew quadrupole with strength $\epsilon = \frac{\sqrt{\beta_x\beta_y}}{2\pi R}\frac{\ell}{B\rho}\frac{\partial B_y}{\partial y}$. Figure 3.33 shows some of our results by a computer simulation. In this simulation, the motion of a particle is tracked turn

Figure 3.33: The $\nu_x - \nu_y = 0$ resonance driving by a skew quadrupole in a computer simulation. The behaviors of $J_x, J_y, J_x + J_y$ are shown for three times the emittance beating period $\frac{1}{\sqrt{\Delta\nu^2+\epsilon^2}}$. The initial launching conditions are $J_{x0} = 0.8, J_{y0} = 0.2, \phi_{x0} = \phi_{y0} = 0$. The two solid curves are J_x, J_y. The dotted curves are $J_x + J_y$. Skew quadrupole strength is $\epsilon = 0.01$. Simulations are done with normalized units with $\beta_x = \beta_y = 1$. The five plots are for $\Delta\nu = 0.04, 0.01, 0, -0.01, -0.04$.

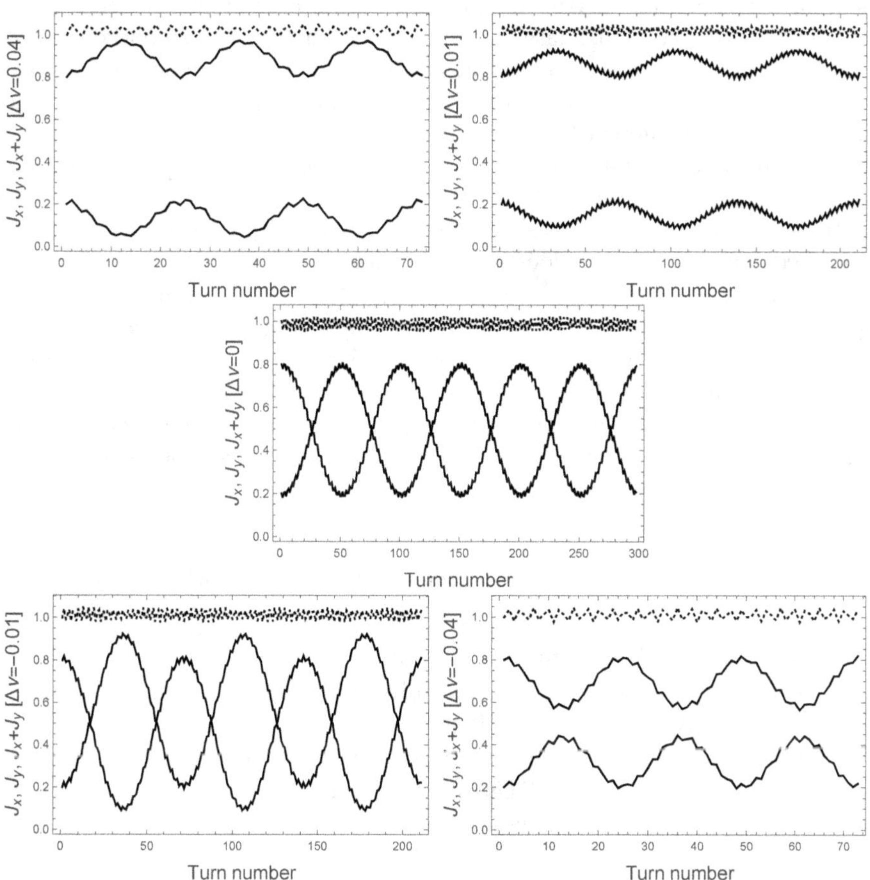

after turn with unperturbed tunes ν_x, ν_y, and at the completion of each turn, the particle is kicked by a localized skew quadrupole. The simulation does not assume knowledge of smooth approximation.

Figure 3.33 shows the tracking for a single particle with initial conditions $J_{x0} = 0.8, J_{y0} = 0.2, \phi_{x0} = \phi_{y0} = 0$. Observe the sum rule $J_x + J_y =$ constant being confirmed. Also observe the emittance beating effect. The emittance beating is visibly to be sinusoidal. The beating period is $\frac{1}{\sqrt{\Delta\nu^2+\epsilon^2}}$ turns. These properties will be derived analytically soon below. This beating frequency is equal to $\nu_+ - \nu_-$ as predicted by Eq. (3.28).

It is important to know that the emittance beating discussed here is only the emittance evolution of a single particle. It is not to be confused with the emittance evolution of a beam. If we change the initial conditions, for example changing ϕ_{x0}, the emittance evolution pattern will change significantly. As mentioned, the parameter χ in the smoothed Hamiltonian (3.66) depends on the particle's initial betatron phases. If the phases change, the smoothed Hamiltonian changes. A consequence of this is that if a beam is made to approach a resonance, forcing an emittance exchange for all its particles, the exchanges will be sensitively different for different particles, and there will be no clear systematic emittance evolution of the beam as a whole — the perturbation term vanishes if the Hamiltonian (3.66) is averaged over χ. The clean exchange of emittances observed here for a single particle is not going to occur for the whole beam because each particle does its own exchange in its own way.

On the other hand, if the entire beam is kicked with coupling turned on, the whole beam will behave as if it were a single particle if the kick is larger than the beam's natural emittance. In that case, there will be an emittance beating effect observed before the beam decoheres. In fact, such kicked beam measurements were behind the data discussed in Fig. 3.28. Furthermore, this difference between a single particle and a whole beam is also not to be confused with the emittance exchange mechanisms we will discuss in Sec. 5.6.

Simulation $\nu_x - 2\nu_y \approx n$ resonance Consider a case when the resonance $\nu_x - 2\nu_y = n$ is driven by a sextupole of strength $\epsilon = -\frac{\sqrt{2\beta_x}\beta_y}{8\pi R}\frac{\ell}{B\rho}\frac{\partial^2 B_y}{\partial x^2}$. Let $\Delta\nu = \nu_x - 2\nu_y - n$. A few simulation results are shown in Fig. 3.34.

The sum rule of the emittances dutifully applies here as well. The beating behavior is clearly seen, although it is seen not to be purely sinusoidal. The beating period in this nonlinear case is still approximately $\frac{1}{\sqrt{\Delta\nu^2+\epsilon^2}}$ as in the linear skew quadrupole case. Due to the nonlinearities, however, the beating period depends somewhat on the initial emittances J_{x0}, J_{y0}.

Analysis $\nu_x - \nu_y \approx n$ resonance The discussion of beating of single-particle emittances has been given in terms of simulations. One can also analyze the case of the resonance $\Delta\nu = \nu_x - \nu_y - n \approx 0$ driven by a single skew quadrupole following the smooth approximation. Not surprisingly, this problem, a linear case, is exactly solvable.

Near the resonance $\nu_x - \nu_y \approx n$, the smoothed Hamiltonian reads

$$K(\phi_x, J_x, \phi_y, J_y, s) = \frac{J_x}{R}(\nu_x - \nu_{x0}) + \frac{J_y}{R}(\nu_y - \nu_{y0}) + \epsilon\sqrt{J_x J_y}\cos(\phi_x - \phi_y)\,,$$

Figure 3.34: The $\nu_x - 2\nu_y = 0$ resonance driving by a sextupole in a computer simulation. The behavior of $J_x, J_y, 2J_x + J_y$ are shown for a time span of $\frac{3}{\sqrt{\Delta\nu^2+\epsilon^2}}$ turns. The initial launching conditions are $J_{x0} = 0.8, J_{y0} = 0.2, \phi_{x0} = \phi_{y0} = 0$. The two solid curves are J_x, J_y. The dotted curves are $2J_x + J_y$. Sextupole strength is $\epsilon = 0.005$. Simulations are done with normalized units with $\beta_x = \beta_y = 1$. The five plots are for $\Delta\nu = 0.006, 0.003, 0, -0.003, -0.006$.

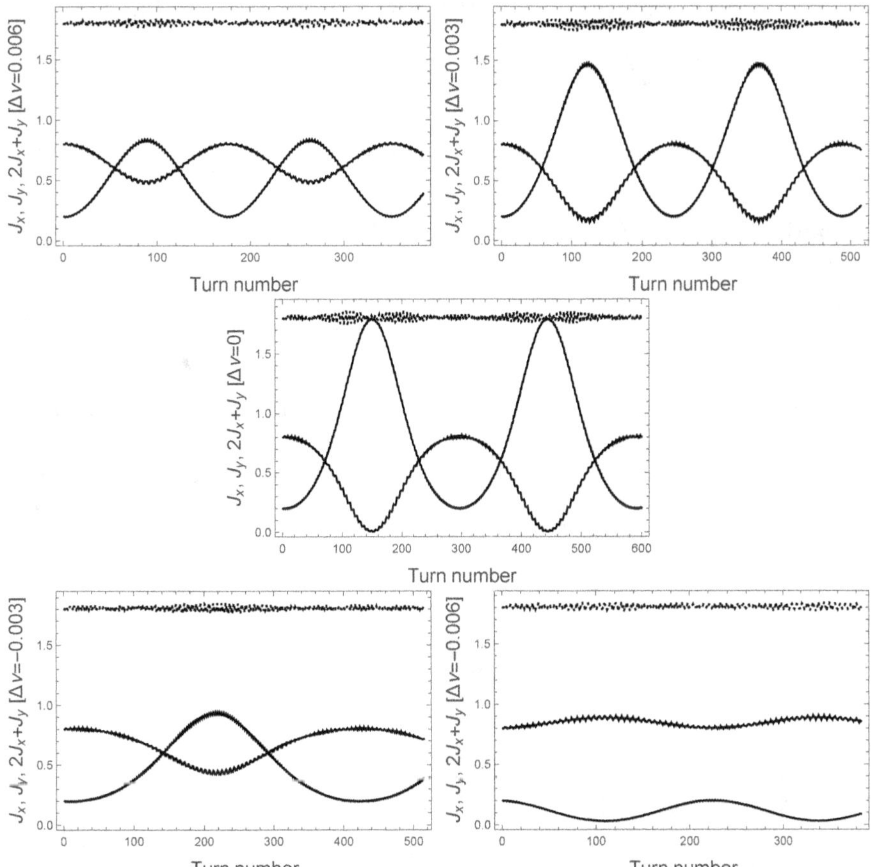

where $\epsilon = \frac{\sqrt{\beta_x\beta_y}}{2\pi R}\frac{\ell}{B\rho}\frac{\partial B_y}{\partial y}$. We then have

$$\begin{aligned}
\frac{d\phi_x}{ds} &= \frac{\partial K}{\partial J_x} = \frac{\nu_x - \nu_{x0}}{R} + \frac{\epsilon}{2}\sqrt{\frac{J_y}{J_x}}\cos(\phi_x - \phi_y)\,, \\
\frac{d\phi_y}{ds} &= \frac{\partial K}{\partial J_y} = \frac{\nu_y - \nu_{y0}}{R} + \frac{\epsilon}{2}\sqrt{\frac{J_x}{J_y}}\cos(\phi_x - \phi_y)\,,
\end{aligned}$$

$$\begin{aligned}\frac{dJ_x}{ds} &= -\frac{\partial K}{\partial \phi_x} = \epsilon\sqrt{J_x J_y}\sin(\phi_x-\phi_y)\,,\\ \frac{dJ_y}{ds} &= -\frac{\partial K}{\partial \phi_y} = -\epsilon\sqrt{J_x J_y}\sin(\phi_x-\phi_y)\,.\end{aligned}$$

Let $J_x = \frac{J_{x0}+J_{y0}}{2}(1+A), J_y = \frac{J_{x0}+J_{y0}}{2}(1-A), \delta\phi = \phi_x - \phi_y$. We have a set of coupled differential equations for the two quantities $A(\theta), \delta\phi(\theta)$,

$$\begin{aligned}\frac{d\delta\phi}{ds} &= \frac{\Delta\nu}{R} - \epsilon\frac{A}{\sqrt{1-A^2}}\cos\delta\phi\,,\\ \frac{dA}{ds} &= \epsilon\sqrt{1-A^2}\sin\delta\phi\,.\end{aligned} \tag{3.67}$$

The initial conditions are $A(0) = \frac{J_{x0}-J_{y0}}{J_{x0}+J_{y0}}, \delta\phi(0) = \phi_{x0} - \phi_{y0}$.

The two equations (3.67) are the Hamilton equations from a new Hamiltonian

$$K_1(\delta\phi, A, \theta) = \frac{\Delta\nu}{R}A + \epsilon\sqrt{1-A^2}\cos\delta\phi\,.$$

The fact that K_1 is explicitly independent of the time variable θ means it is a constant of the motion. Its value is $K_1 = \Delta\nu A(0) + \epsilon\sqrt{1-A^2(0)}\cos\delta\phi(0)$.

Equation (3.67) can be combined to derive an equation of motion for A,

$$\frac{d^2A}{ds^2} + \frac{\lambda^2}{R^2}A = \frac{\Delta\nu}{R}K_1\,, \tag{3.68}$$

where

$$\lambda = \sqrt{\Delta\nu^2 + R^2\epsilon^2}\,.$$

The beating $A(s)$ is sinusoidal with beating wavelength $\frac{2\pi R}{\lambda}$. It oscillates around the value $\frac{\Delta\nu R}{\lambda^2}K_1$. The general solution is

$$A(s) = \frac{\Delta\nu R}{\lambda^2}K_1 + \left(A(0) - \frac{\Delta\nu R}{\lambda^2}K_1\right)\cos\frac{\lambda s}{R} + \left(\frac{\epsilon R}{\lambda}\sqrt{1-A^2(0)}\sin\delta\phi(0)\right)\sin\frac{\lambda s}{R}.$$

The reader may explore the various special cases with this rather complex-looking expression. In particular, check that this result agrees well with the simulation Fig. 3.33. Pay special note to the fact that the behavior changes sensitively to the value of the initial phase $\delta\phi(0)$.

We have only solved A. The behavior of $\delta\phi$ is more complex. Fortunately we are not too interested in it. Fig. 3.35 illustrates the behavior of $\delta\phi$ obtained by numerically solving Eq. (3.67). One sees that in the middle of the resonance width, $\delta\phi$ oscillates like sinusoidal (but is not). Approaching the edges of the resonance bandwidth, it migrates away. Figure 3.35 plots $\delta\phi$ modulus π. With modulus of π, $\delta\phi$ also has period of $\frac{1}{\sqrt{\Delta\nu^2+\epsilon^2}}$, even though not sinusoidal.

Figure 3.35: Behavior of $\delta\phi$ near the coupling resonance $\nu_x - \nu_y = n$. The inputs are the same as in Fig. 3.33, i.e. $J_{x0} = 0.8, J_{y0} = 0.2, \epsilon = 0.01, \delta\phi(0) = 0$. The solid curve is for $\Delta\nu = 0$. Dashed is for $\Delta\nu = 0.019$. Dotted is for $\Delta\nu = -0.005$.

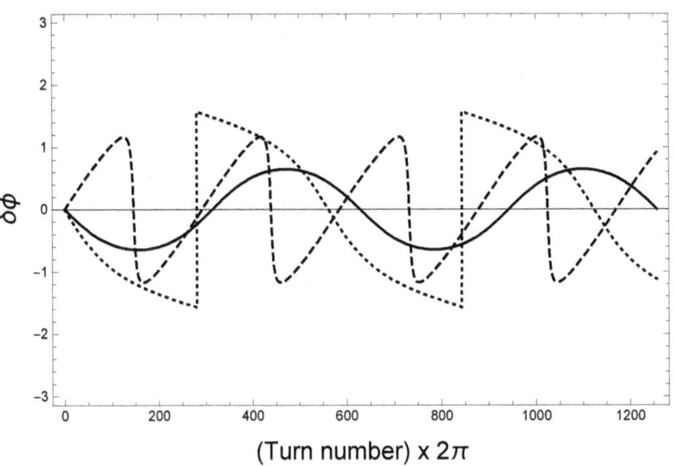

Analysis $\nu_x - 2\nu_y \approx n$ resonance The analysis provided for the $\nu_x - \nu_y$ resonance can be extended to nonlinear coupling resonances. Take the sextupole-driven $\nu_x - 2\nu_y$ resonance for example. The smoothed Hamiltonian is

$$K(\phi_x, J_x, \phi_y, J_y, s) \approx \frac{J_x}{R}(\nu_x - \nu_{x0}) + \frac{J_y}{R}(\nu_y - \nu_{y0}) + \epsilon\sqrt{J_x}J_y \cos(\phi_x - 2\phi_y),$$

where $\epsilon = -\frac{\sqrt{2\beta_x}\beta_y}{8\pi R}\frac{\ell}{B\rho}\frac{\partial^2 B_y}{\partial x^2}$.

Define $\delta\phi = \phi_x - 2\phi_y, J_x = \frac{2J_{x0}+J_{y0}}{3}(1 + A), J_y = \frac{2J_{x0}+J_{y0}}{3}(1 - 2A)$ and $\Delta\nu = \nu_x - 2\nu_y - n$, then the equations of motion are

$$\frac{d\delta\phi}{ds} = \frac{\Delta\nu}{R} - q\frac{3 + 6A}{2\sqrt{1 + A}}\cos\delta\phi,$$

$$\frac{dA}{ds} = q\sqrt{1 + A}\,(1 - 2A)\sin\delta\phi.$$

The initial conditions are $A(0) = \frac{J_{x0}-J_{y0}}{2J_{x0}+J_{y0}}$ and $\delta\phi(0) = \phi_{x0} - 2\phi_{y0}$. We have also defined $q = \epsilon\sqrt{\frac{2J_{x0}+J_{y0}}{3}}$.

A constant of the motion, the Hamiltonian, is

$$K_1 = \frac{\Delta\nu}{R}A + q\sqrt{1 + A}\,(1 - 2A)\cos\delta\phi.$$

It follows that A satisfies the equation

$$\frac{d^2A}{ds^2} + \frac{\Delta\nu^2}{R^2}A + q^2\frac{3(1 - 4A^2)}{2} = \frac{\Delta\nu}{R}K_1. \tag{3.69}$$

This analysis is similar to the linear $\nu_x - \nu_y$, but this time it is no longer a sinusoidal behavior, and there is no obvious appearance of the frequency $\sqrt{\Delta\nu^2 + \epsilon^2}$.

Homework 3.77 Solve Eq. (3.69) numerically for $A(s)$ and confirm the simulation results of Fig. 3.34. The agreement should be rather accurate.

Homework 3.78 The analysis of emittance beating was solved exactly for the linear $\nu_x - \nu_y$ resonance and numerically for the nonlinear $\nu_x - 2\nu_y$ resonance in the text via Eqs. (3.68) and (3.69), respectively. Follow similar steps to extend the analysis to general nonlinear coupling resonances with smoothed Hamiltonian (3.66). Include the detuning term D_ν as far as you can.

Solution Let the resonance be designated $m_x\nu_x - m_y\nu_y = n$, so that $m_{x,y}$ are both positive. Let $\delta\phi = m_x\phi_x - m_y\phi_y + \chi, \Delta\nu = m_x\nu_x - m_y\nu_y - n$. The sum rule says $m_yJ_x + m_xJ_y = J_0$ is constant. Define $J_x = \frac{J_0}{m_x+m_y}(1 + m_xA), J_y = \frac{J_0}{m_x+m_y}(1 - m_yA)$. Initial condition is $A(0) = \frac{J_{x0}-J_{y0}}{J_0}$.

The coupled equations are

$$\begin{aligned}\frac{d\delta\phi}{ds} &= \frac{\Delta\nu}{R} + \frac{m_x}{R}\frac{\partial D_\nu}{\partial J_x} - \frac{m_y}{R}\frac{\partial D_\nu}{\partial J_y} + \frac{q}{2}(1+m_xA)^{(m_x-2)/2}(1-m_yA)^{(m_y-2)/2} \\ &\quad\times(m_x + m_y)(m_x - m_y - m_xm_yA)\cos\delta\phi\,, \\ \frac{dA}{ds} &= q(1 + m_xA)^{m_x/2}(1 - m_yA)^{m_y/2}\sin\delta\phi\,,\end{aligned}$$

where $q = \epsilon(\frac{J_0}{m_x+m_y})^{(m_x+m_y-2)/2}$.

To proceed further, we will drop the detuning term D_ν. Then

$$K_1 = \frac{\Delta\nu}{R}A + q(1 + m_xA)^{m_x/2}(1 - m_yA)^{m_y/2}\cos\delta\phi$$

is a constant of the motion. The final emittance beating equation is then

$$\begin{aligned}\frac{d^2A}{ds^2} + \frac{\Delta\nu^2}{R^2}A - \frac{q^2}{2}(1 + m_xA)^{m_x-1}(1 - m_yA)^{m_y-1}& \\ \times(m_x + m_y)(m_x - m_y - m_xm_yA) &= \frac{\Delta\nu}{R}K_1\,.\end{aligned}$$

Equations (3.68) and (3.69) are its special cases.

3.7.8 Emittance growth near sum resonance

A similar analysis can also be applied to find out the growth behavior of the apparent emittances near a sum resonance. Although unstable, the beam's behavior still obeys the smooth approximation. The Hamiltonian (3.66) still applies in general. We shall however only analyze the linear case in this section. Extension to the general case is straight forward.

Near the $\nu_x + \nu_y \approx n$ resonance, we expect the problem to be exactly soluble. The smoothed Hamiltonian is

$$K(\phi_x, J_x, \phi_y, J_y, s) = \frac{J_x}{R}(\nu_x - \nu_{x0}) + \frac{J_y}{R}(\nu_y - \nu_{y0}) + \epsilon\sqrt{J_x J_y}\cos(\phi_x + \phi_y)\,.$$

This time $J_x - J_y$ is an invariant. We define $\Delta\nu = \nu_x + \nu_y - n$, $J_x = \frac{J_{x0}-J_{y0}}{2}(A+1)$, $J_y = \frac{J_{x0}-J_{y0}}{2}(A-1)$, $\delta\phi = \phi_x + \phi_y$. Initial condition is $A(0) = \frac{J_{x0}+J_{y0}}{J_{x0}-J_{y0}}$. A procedure similar to that used on the difference resonances then leads to

$$\begin{aligned} \frac{d\delta\phi}{ds} &= \frac{\Delta\nu}{R} + \epsilon\frac{A}{\sqrt{A^2-1}}\cos\delta\phi\,, \\ \frac{dA}{ds} &= \epsilon\sqrt{A^2-1}\,\sin\delta\phi\,. \end{aligned}$$

A constant of the motion is

$$K_1 = \frac{\Delta\nu}{R}A + \epsilon\sqrt{A^2-1}\cos\delta\phi\,.$$

Combining the equations, we obtain the final result

$$\frac{d^2A}{ds^2} + \frac{1}{R^2}(\Delta\nu^2 - R^2\epsilon^2)A = \frac{\Delta\nu}{R}K_1\,.$$

The final result should not be too surprisingly. It has already appeared before in Eq. (3.29). It means there is a stopband, the width of which is given by $\Delta\nu = \pm R\epsilon$. Outside the stopband, the motion is sinusoidal. Inside of it, the motion is hyperbolic. Immediately outside of the stopband, the emittance beating frequency is zero.

Chapter 4

Chromatic Effect

A special type of perturbations to particle motion comes from the fact that the particles in a beam inevitably have a spread of momentum errors. This chapter is dedicated to this special type of perturbation, and the beam dynamical effects caused by momentum errors are referred to as *chromatic effects*.

4.1 Dispersion function

4.1.1 Momentum error

In this treatment, we shall consider the momentum of a particle to be fixed. If it contains an error, and the errors are different for different particles, this momentum error for each particle is considered fixed in time. We are interested in knowing how this fixed momentum error affects particle's motion, particularly its transverse motion. In this sense, this momentum error may be considered an externally given parameter for each particle, and not considered to be a dynamical variable that changes with time.

In the next chapter, this momentum error will be allowed to change as would be the case when the accelerator contains an acceleration device such as an RF cavity, and in that case, momentum error becomes a dynamical variable. In this chapter, however, that complication is not considered.

Another way of saying the same thing is as follows. In Chapters 2 and 3, we discussed the transverse motion. The phase space of interest was 2-D if decoupled and 4-D if coupled. In Chapter 5 later, we shall introduce the longitudinal motion, and when added to the transverse motion, we shall have a 6-D phase space. This chapter is somewhere in between. We may insist on considering the momentum error as a dynamical variable in this chapter too, but then we would be considering a 5-D phase space in which the last element, the momentum error, does not change and does not really participate in any *dynamics*. The discussions of 6-D phase space in the next chapter will use many of the results we learn in this chapter on the 5-D phase space.

We dedicated Chapter 3 to the study of perturbations due to error fields. In this chapter. we assume that there are no error fields in the accelerator, and all perturbations are due to chromatic errors.

Chromatic effects originate from an effect of *momentum dispersion*, which in turn originates from the fact that for given strength of a magnet, its effect on particle motion depends on the particle's momentum. Take a dipole for example; it bends higher momentum particles slightly less than it bends the lower momentum ones.

Incidentally, there are two conventions to symbolize dipoles and quadrupoles in a sketch of a beamline, as illustrated in Fig. 4.1. We see both notations being used in the literature.

Figure 4.1: Two ways to symbolize dipoles and quadrupoles in a beamline.

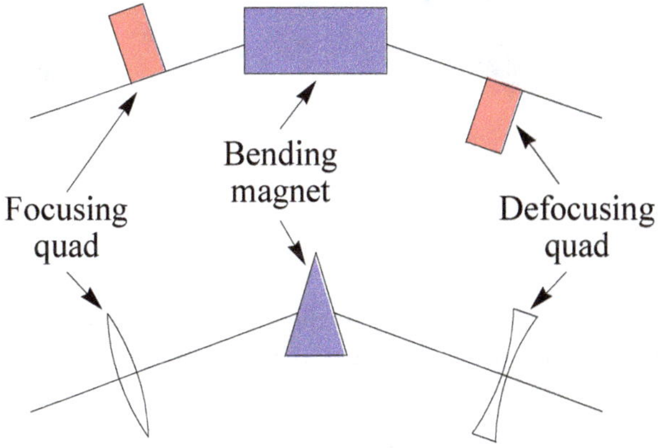

It should be noted that in the lower panel in Fig. 4.1, the dipole symbol is made by analogy to a prism in optics, just like the quadrupoles are represented as optical lenses. However, it should be noted the appropriate orientation of the prism. Drawing the orientation too casually can lead to an error, as illustrated in Fig. 4.2.

Perhaps there is one more detail to discuss. Note that the red light is bent less than the purple light by a prism. If we associate red light with low energy particles and blue light with the high energy particles, is the prism really a correct analogy to a magnetic dipole?

The small parameter So far we have been considering betatron motion of a particle that has the design momentum P_0, i.e. we have been considering only *on-momentum* particles. In general, a particle's momentum will be $P_0 + \Delta P =$

Figure 4.2: Correct and incorrect ways to draw a prism as symbol for dipole.

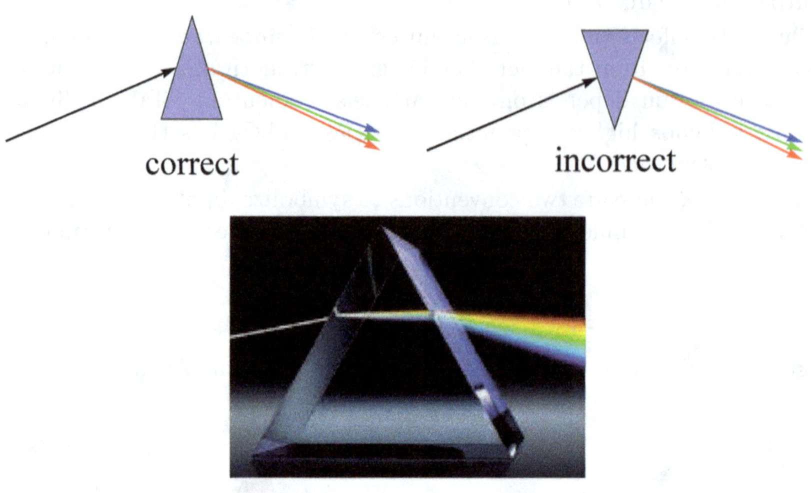

$P_0(1+\delta)$, where

$$\delta = \frac{\Delta P}{P_0},$$

is a small deviation. Note that δ specifies momentum deviation, not energy deviation. Energy deviation is related to δ according to

$$\frac{\Delta E}{E_0} = \beta_0^2 \frac{\Delta P}{P_0} = \beta_0^2 \delta .$$

They are approximately equal for relativistic particles.

We just introduced δ as the perturbation parameter. It may be raised a question what should be used as the small perturbation parameter for chromatic effects. We have several options,

$$\frac{\Delta E}{E_0}, \quad \frac{\Delta E}{E_0+\Delta E}, \quad \frac{\Delta P}{P_0}, \quad \frac{\Delta P}{P_0+\Delta P}.$$

Loosely speaking, the proton accelerator physicists tend to take $\frac{\Delta P}{P_0}$. Electron people are less critical about the choice (for obvious reason that all these choices are so nearly equal) and often settle with $\frac{\Delta E}{E_0}$. They can all work. We have chosen $\frac{\Delta P}{P_0}$ in this chapter. Strictly, however, the best one to use could be the unconventional $\frac{\Delta P}{P_0+\Delta P}$, because then a magnetic kick will have only a first order linear chromatic perturbation. This may simplify the kinematics. It is not obvious why this option is not more widely adopted.

4.1.2 Equation of motion

Equation for the dispersion function We now return to the equation of motion Eq. (2.2) of Chapter 2, which is rewritten as

$$x'' - \frac{\rho + x}{\rho^2} = -\frac{eB_y}{P_0(1+\delta)} \left(1 + \frac{x}{\rho}\right)^2 ,$$

where ρ is the design bending radius for on-momentum particles and we have set $m\gamma v_s = P_0(1+\delta)$ to be the momentum of the particle under consideration.

Let $B_y = B_{y0} + Gx$ with the bending magnetic field B_{y0} satisfying the condition $\frac{eB_{y0}}{P_0} = \frac{1}{\rho}$. To first order in x but keeping all order in δ, we obtain

$$x'' + \frac{1}{1+\delta}\left(\frac{1-\delta}{\rho^2} + \frac{G}{B_{y0}\rho}\right) x \approx \frac{\delta}{1+\delta} \frac{1}{\rho} . \tag{4.1}$$

It follows that, to first order in x and δ,

$$x'' + K_x(s)x \approx \frac{\delta}{\rho(s)} , \tag{4.2}$$

with $K_x = \frac{1}{\rho^2} + \frac{G}{B_{y0}\rho}$ as defined before. Both $\rho(s)$ and $K_x(s)$ are periodic in s with period L.

Compared with the betatron equation of motion for on-momentum particles, Eq. (4.2) contains an extra term $\frac{\delta}{\rho}$ on the right hand side. This extra term will drive the x-motion of an off-momentum particle which we call the *horizontal dispersion*, or simply the *dispersion*.

Solving for the dispersion From the general theory of ordinary differential equations, the solution to Eq. (4.2) for $x(s)$ can be written as a sum of two terms, one is a general solution to the differential equation with the right hand side replaced by zero, the other is a particular solution of Eq. (4.2) with a nonzero right hand side. There is an infinite number of general solutions, and there needs to be only one particular solution. The infinite number of general solutions are simply the betatron oscillations we discussed in Chapter 2. The one particular solution will be related to the dispersion function to be elaborated below.

Since we need only one candidate for the particular solution, we will look for one that is unique and well defined. The uniqueness will come from demanding a strong condition that it be periodic in s with the same period as the function $K(s)$. Once the particular solution is found, any solution to Eq. (4.2) can be written as this particular solution plus an infinite number of possible betatron oscillations.

We may also look for the equation of y-motion for an off-momentum particle. To first order in y and δ, we find that it is the same as the on-momentum equation. To the linear order, therefore, there is no vertical dispersion effect.

Here we see a loss of symmetry between the horizontal and vertical motions in a planar accelerator design.

Equation (4.2) resembles closely that for the closed orbit distortion with $\frac{\delta}{\rho}$ standing for the orbit dipole kick [see Eq. (3.2)]. In fact, dispersion is introduced exactly to characterize the difference between the closed orbit of an on-momentum particle and that of an off-momentum particle. For an off-momentum particle with $\delta > 0$, each dipole magnet effectively acts as an outward kick (inward if $\delta < 0$) with a kick angle of $\frac{\ell}{\rho}\delta$. A closed orbit solution of x for the off-momentum particle therefore can be written as, using the COD formula (3.9),

$$\begin{aligned} x(s) &= D(s)\delta\,, \\ D(s) &= \frac{\sqrt{\beta_x(s)}}{2\sin\pi\nu_x}\int_s^{s+L} ds'\,\frac{\sqrt{\beta_x(s')}}{\rho(s')}\cos(\pi\nu_x - |\psi_x(s') - \psi_x(s)|)\,, \end{aligned} \tag{4.3}$$

where $D(s)$ is called the *dispersion function.* In other words, we have defined

$$\begin{pmatrix}\text{dispersion}\\ \text{function}\end{pmatrix} = \frac{\begin{pmatrix}\text{closed orbit distortion for a particle}\\ \text{due to a fixed momentum error } \delta\end{pmatrix}}{\delta}\,.$$

The dispersion function satisfies

$$D'' + K_x(s)D = \frac{1}{\rho(s)}\,,$$

and, like $K_x(s)$ and $\rho(s)$, is periodic in s with period L. The general solution for x of an off-momentum particle is given by

$$x(s) = x_\beta(s) + D(s)\delta\,. \tag{4.4}$$

There are two ways to view Eq. (4.4). In one view, the horizontal displacement consists of two separate terms, an on-momentum betatron oscillation term x_β and an off-momentum dispersion term. Another viewpoint is to think of $D(s)\delta$ as a closed-orbit term, while $x_\beta(s)$ is a betatron oscillating term around $D(s)\delta$. The former viewpoint is somewhat more accurate when δ executes synchrotron oscillation in time (Chapter 5). When the synchrotron oscillation is slow, as usually is the case, the difference is often neglected. When the synchrotron motion is faster, or when near a phenomenon called *synchrobetatron resonances*, however, the second viewpoint breaks down.

Dispersion function is thus defined only when the momentum error δ is a constant around the accelerator. The justification of using dispersion function therefore depends on the fact that momentum changes in accelerators are typically difficult to occur, making δ at least approximately a constant in one revolution. This in turn, will appear as the condition that the synchrotron oscillation tune ν_s must be $\ll 1$. We thus conclude by saying that the dispersion function is meaningful only to an accuracy $\sim \mathcal{O}(\nu_s)$. When ν_s is larger, beware.

Figure 4.3: A simple model of a uniform-field storage ring with dispersion $D(s) = R$.

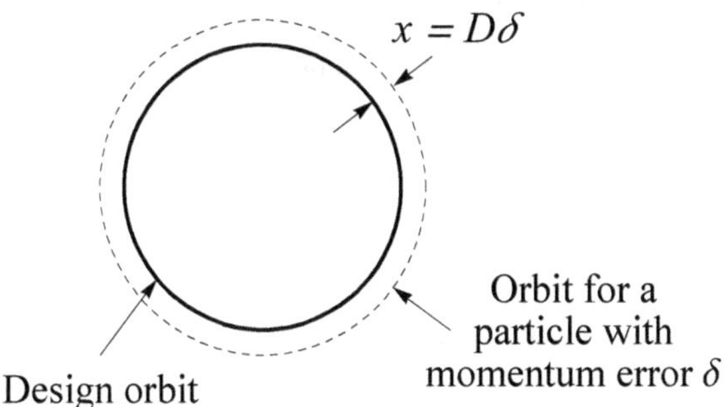

A simple case We just gave a solution (4.3) for the dispersion function. However, it is not the most efficient way to calculate $D(s)$. Before we enter the next section to calculate it, it is perhaps instructive here to first obtain a physical feel of the dispersion function. We do this by considering an over-simplified example as follows.

Consider a storage ring that consists simply of uniform bending magnet without gradient.[1] The design orbit for an on-momentum particle with momentum P_0 is a simple circle of radius R in the horizontal plane. Now consider a particle with momentum $P_0(1+\delta)$. What is its closed orbit? The answer is a circle with radius $R(1+\delta)$. This off-momentum particle, therefore, has a horizontal closed orbit (relative to the design orbit) of $R\delta$. This means the dispersion of this storage ring is $D = R$. See Fig. 4.3.

A question may be raised here. In Fig. 4.3, the off-momentum particle's circle is shown to be concentric with the design orbit circle. But an off-momentum particle can also make its circle displaced from this concentric circle. If we do that, then our dispersion function would be different. The answer to that question is that in that case, the trajectory of the off-momentum particle can be decomposed into a synchrotron component $D\delta$, plus a betatron component. In this case, however, the betatron component is not oscillatory because the horizontal tune is equal to 1, and the betatron oscillation shows up as if it is part of a COD.

Keep in mind that for this simplest storage ring, the dispersion function $D = R$. Later you will learn that in practice D is almost always very much smaller than R.

Higher order dispersion Equation (4.1) can be used to find the higher order dispersion function $D_\delta(s)$ when the storage ring contains only dipoles,

[1] This is allowed because we shall not consider vertical motion, which you recall, is unstable in this case.

quadrupoles and drifts — no sextupoles, for example. The function $x = D_\delta(s)$ satisfies Eq. (4.1) together with a periodicity condition $D_\delta(s+C) = D_\delta(s)$.

Let $D_\delta(s)$ be Taylor expanded,

$$D_\delta(s) = D_0(s) + D_1(s)\delta + D_2(s)\delta^2 + \cdots .$$

It is easy to see that the leading term $D_0(s)$ satisfies

$$D_0'' + \left(\frac{1}{\rho^2} + \frac{G}{B\rho}\right) D_0 = 0 .$$

This is of course just the equation for betatron oscillation and here with the periodicity condition, the solution is $D_0(s) = 0$.

The higher order dispersions satisfy a set of equations

$$D_1'' + \left(\frac{1}{\rho^2} + \frac{G}{B\rho}\right) D_1 = \frac{1}{\rho} ,$$
$$D_2'' + \left(\frac{1}{\rho^2} + \frac{G}{B\rho}\right) D_2 = -\frac{1}{\rho} + \left(\frac{2}{\rho^2} + \frac{G}{B\rho}\right) D_1 .$$

The quantity $D_1(s)$ is just the linear dispersion function treated before. The higher order dispersions can be obtained iteratively.

Homework 4.1 Consider an idealistic planar storage ring that consists of segments of pure sector bends and drift spaces (no quadrupoles). Let all bending sections have the same bending radius R, as sketched in Fig 4.4. What is the dispersion function for this storage ring?

Figure 4.4: Another simple model whose dispersion function is to be found.

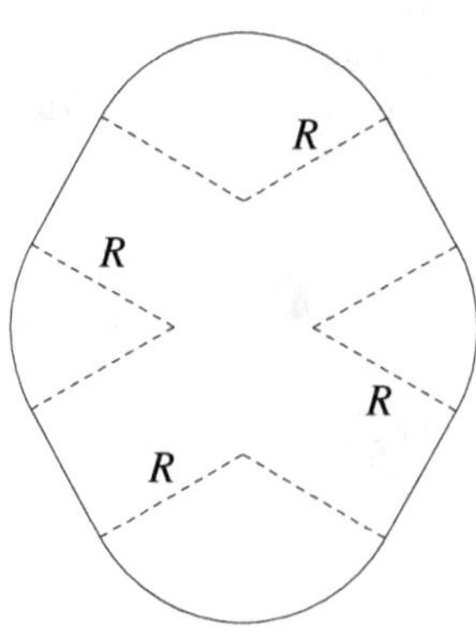

Solution $D = R$. The dispersion function is the same everywhere, including in the straight sections where there are not magnets. You should learn here that dispersion function is a global quantity determined by the ring considered as a whole, and is not solely determined by the local arrangements. In this example, the dispersion function in a straight section has nothing to do with any local properties of the straight section.

Homework 4.2 Some of the bends may have the opposite sign, like the figure-8 ring in Fig. 4.5. Find its dispersion function.

Figure 4.5: One more simple model whose dispersion function is to be found.

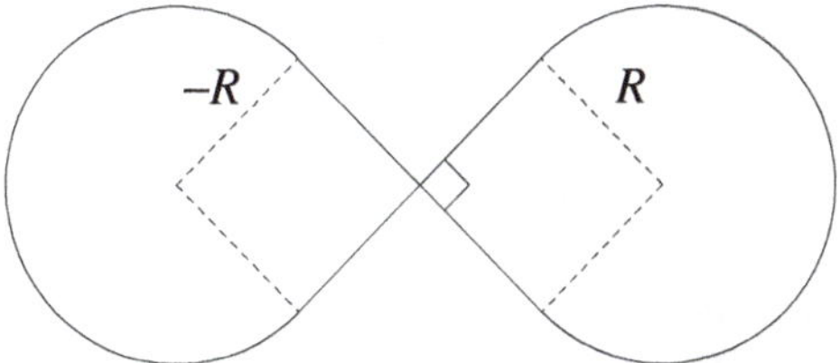

Solution You need to find the closed orbit for an off-momentum particle. It is given by the red orbit shown in Fig. 4.6. At position A, for example, the dispersion function has $D = \frac{R}{2}, D' = \frac{1}{2}$.

Figure 4.6: Solution to Homework 4.2.

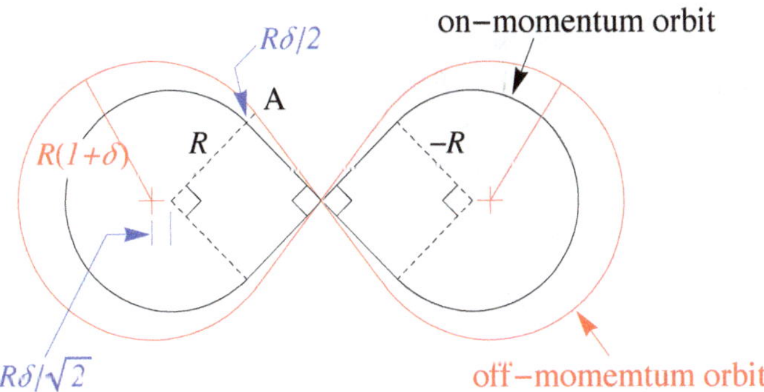

Homework 4.3 If some of the bends in this storage ring have a different bending radius, such as a ring sketched in Fig. 4.7, what is the dispersion function?

Solution It is doable, but if you have some difficulty now, you may delay it to the next section after you learn the matrix technique.

Figure 4.7: Still one more model whose dispersion function is to be found.

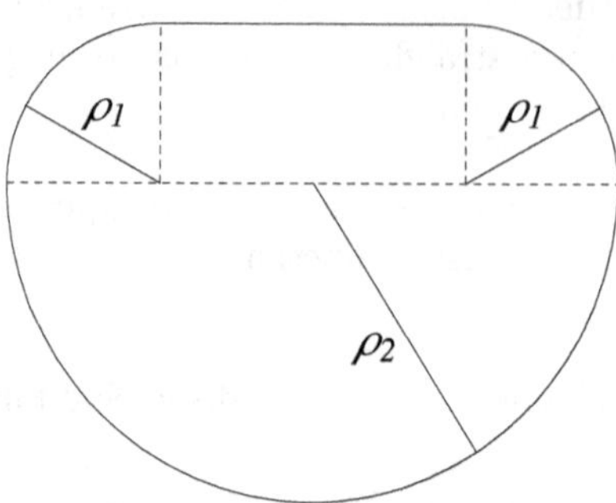

4.1.3 Dispersion-free storage ring

So far we have assumed that there are no transverse electric fields. It turns out that in principle, very interestingly, one can design a dispersion-free storage ring if one introduces a horizontal electric field E_x in addition to a vertical magnetic field B_y in all the bending magnets.

Note first that the transverse electric field E_x does not affect the particle energy. Equation (4.1) becomes

$$x'' + \frac{x}{\rho^2} = \frac{1}{\rho} + \frac{-ev_s B_y + eE_x}{m\gamma v_s^2} .$$

We then note that, when there is a momentum error δ, the velocity

$$v_s \approx v_{s0}\left(1 + \frac{\delta}{\gamma^2}\right) ,$$

with v_{s0} the design velocity. Let $B_y = B_{y0} + Gx$, and set the bending radius with

$$\frac{1}{\rho} = \frac{eB_{y0}}{P_0} - \frac{eE_x}{P_0 v_{s0}} .$$

Keeping to first order in x and δ, we obtain

$$x'' + K_x x = \left(\frac{1}{\rho} - \frac{eE_x}{P_0 v_{s0}\gamma^2}\right)\delta .$$

This means the dispersion function satisfies

$$D'' + K_x D = \frac{1}{\rho} - \frac{eE_x}{P_0 v_{s0}\gamma^2} .$$

If we now choose

$$\frac{eE_x}{P_0} = \frac{v_{s0}\gamma^2}{1+\gamma^2}\frac{eB_{y0}}{P_0} = \frac{v_{s0}\gamma^2}{\rho} ,$$

the storage ring can be made intrinsically dispersion free!

What one does here is to replace each dipole magnet by a combined dipole magnet plus a set of high voltage plates. In this arrangement, the electric field bends the beam *outward*, while the magnetic field overcomes it and keeps the particle moving inward. Reader of this section may also read Sec. 4.3 about a device called Wien filter.

Homework 4.4 Consider a dispersion-free storage ring for a beam of protons with kinetic energy of 1 GeV. Find the values of B_{y0} and E_x needed to replace a dipole magnet of 1 Tesla. Is this interesting idea practical in this case?

Solution $E_x = 1.12$ GV/m, $B_{y0} = 5.3$ T. It is more practical for nonrelativistic protons — not a surprising conclusion.

4.2 Calculation of dispersion function

4.2.1 The 3 × 3 matrix formalism

In practice, $D(s)$ is most often obtained not by the closed form integral (4.3), but by the technique of matrices.

Consider a separated-function storage ring. Consider a sector dipole with bending radius ρ. At the entrance to the dipole, $s = 0$, let us assume we have dispersion function $D(0)$ and $D'(0)$. We now calculate the dispersion $D(s)$ through the dipole magnet till its exit. Inside the dipole, the dispersion function satisfies

$$D''(s) + \frac{D(s)}{\rho^2} = \frac{1}{\rho}, \qquad \rho = \text{constant}.$$

The solution with the given initial conditions can easily be found. In a matrix form, it can be written as

$$\begin{bmatrix} D(s) \\ D'(s) \end{bmatrix} = \begin{bmatrix} \cos\frac{s}{\rho} & \rho\sin\frac{s}{\rho} \\ -\frac{1}{\rho}\sin\frac{s}{\rho} & \cos\frac{s}{\rho} \end{bmatrix} \begin{bmatrix} D(0) \\ D'(0) \end{bmatrix} + \begin{bmatrix} \rho(1-\cos\frac{s}{\rho}) \\ \sin\frac{s}{\rho} \end{bmatrix}.$$

The first term on the right hand side satisfies the homogeneous differential equation. The second term comes from the inhomogeneous driving term $\frac{1}{\rho}$.

At this point, we introduce a clever trick of matrices that allows us to combine the homogeneous and the inhomogeneous contributions into a single matrix, and treat them on equal footing. This is done by introducing a vector[2]

$$\begin{bmatrix} D \\ D' \\ 1 \end{bmatrix}.$$

[2]The same trick could be introduced to calculate COD earlier, but we did not do so then.

The transfer maps for this vector are now 3×3. For the sector dipole, the map is

$$\begin{bmatrix} D(s) \\ D'(s) \\ 1 \end{bmatrix} = \begin{bmatrix} \cos\frac{s}{\rho} & \rho\sin\frac{s}{\rho} & \rho(1-\cos\frac{s}{\rho}) \\ -\frac{1}{\rho}\sin\frac{s}{\rho} & \cos\frac{s}{\rho} & \sin\frac{s}{\rho} \\ 0 & 0 & 1 \end{bmatrix} \begin{bmatrix} D(0) \\ D'(0) \\ 1 \end{bmatrix} .$$

Let us here make a general side remark concerning the use of matrices in accelerator physics. Matrices are used extensively in accelerator optics. We have learned to use 2×2 in Courant-Snyder formalism, and 3×3 in the chromatic dynamics in this Chapter. We learned 4×4 when we discussed x-y coupled optics. When chromatic effects are to be discussed with x-y coupling, we will need 5×5 matrices. In Chapter 5, we shall introduce synchrotron oscillation by adding the third longitudinal dimension of beam dynamics, then we shall need 6×6 when we include all possible coupling effects among x, y, z. For a full 3-D coupled case, a COD will have to be found using 7×7 matrices. Finally, when we further include the fourth dimension of beam dynamics – the spin motion, then enter the 8×8 matrices. All these variations have been used in accelerator physics.

Only the even-dimensioned matrixes are describing faithful beam dynamics. For example, they must obey strict fundamental principles such as the symplecticity condition. The odd-dimensioned matrices here are only shorthand conveniences useful to perform accelerator lattice calculations.

We now return to the discussion of 3×3 matrices. For a short sector dipole with bending angle $\theta = \frac{\ell}{\rho} \ll 1$, the map from its entrance to its exit is

$$\begin{bmatrix} 1 & \ell & \frac{\ell\theta}{2} \\ 0 & 1 & \theta \\ 0 & 0 & 1 \end{bmatrix} .$$

The upper-left 2×2 corner represents just the drift space. The third column can be easily obtained by a thin-lens dipole kick as illustrated in Fig. 4.8. Consider a particle with momentum error δ passing through this short dipole. Relative to the design trajectory, the particle receives an extra kick $\theta\delta$ outward. The kick can be approximated by a δ-function kick at the middle of the dipole, as evidenced by the observation

$$\begin{bmatrix} 1 & \frac{\ell}{2} & 0 \\ 0 & 1 & 0 \\ 0 & 0 & 1 \end{bmatrix} \begin{bmatrix} 1 & 0 & 0 \\ 0 & 1 & \theta \\ 0 & 0 & 1 \end{bmatrix} \begin{bmatrix} 1 & \frac{\ell}{2} & 0 \\ 0 & 1 & 0 \\ 0 & 0 & 1 \end{bmatrix} = \begin{bmatrix} 1 & \ell & \frac{\ell\theta}{2} \\ 0 & 1 & \theta \\ 0 & 0 & 1 \end{bmatrix} .$$

At the end of the dipole, the particle's trajectory will be $x = \frac{1}{2}\ell\theta\delta$ and $x' = \theta\delta$.

For a quadrupole magnet, there is no driving term for dispersion. The 3×3 map is given by

$$\begin{bmatrix} m_{11} & m_{12} & 0 \\ m_{21} & m_{22} & 0 \\ 0 & 0 & 1 \end{bmatrix} ,$$

where the upper-left 2×2 corner is simply the usual betatron map.

Figure 4.8: The kick received by an off-momentum particle can be approximated by a kick of an angle $\theta\delta$ relative to the design orbit. For a short dipole, the kick is lumped into a single kick at the middle of the dipole. The bending of the design orbit is not shown. Only the kick given to the off-momentum particle relative to the design orbit is shown.

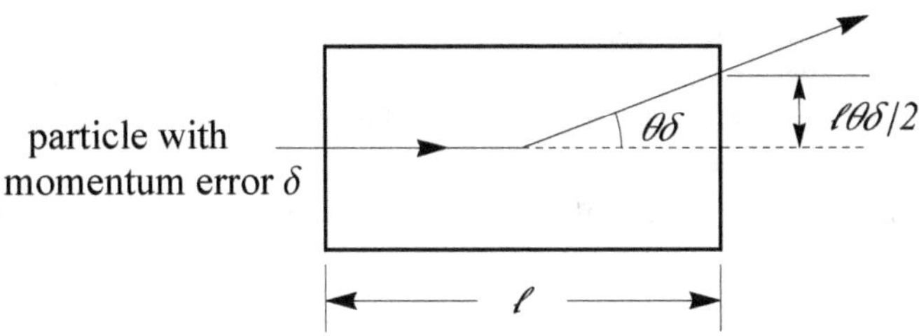

We now have obtained the 3×3 matrices of all single elements. Although introduced here to describe the dispersion function D in terms of the vector $(D, D', 1)$, these matrices also apply to the general case as the map to any particle with entrance coordinates $(x, x', \delta)_{\rm in}$ to find the exit coordinates $(x, x', \delta)_{\rm out}$.

For each element, the second row of the matrix is necessarily equal to the s-derivative of the first row. Also, the (1,3)- and the (2,3)-elements must vanish at the initial position $s = 0$ because they correspond to driven solutions.

Given the lattice design of a beamline, we then multiply the string of 3×3 matrices together to form the total transfer matrix of the beamline. All 3×3 transfer matrices we just introduced, whether they are for single elements or for the combined beamline, have the following properties,

- The last row of elements of these maps must be 0,0,1. This assures that the third element in the vector stays always to be 1 when manipulated by these maps.

- The upper-left 2×2 corner of the map is just the usual 2×2 transfer map of the betatron motion. Adding the third dimension does not affect the betatron part of the beam dynamics. In all subsequent matrix operations when beamline elements are combined, it is only allowed that the upper-left 2×2 corner to influence the third column, and never the other way around.

Given the lattice design of a storage ring, we multiply the 3×3 matrices around a position s_0 to obtain a one-period map $M(s_0 + L|s_0)$. The dispersion function D and its slope D' at s_0 are found using the fact that they must be periodic and therefore

$$\begin{bmatrix} D \\ D' \\ 1 \end{bmatrix} (s_0) \;=\; M(s_0 + L|s_0) \begin{bmatrix} D \\ D' \\ 1 \end{bmatrix} (s_0)\,,$$

i.e.

$$\begin{bmatrix} D \\ D' \\ 1 \end{bmatrix} (s_0) \;=\; \text{eigenvector of } M(s_0+L|s_0) \text{ with eigenvalue } 1\,.$$

The fact that the 3×3 one-period matrix always has an eigenvalue of 1 is assured by its form. This eigenmode with eigenvalue 1 is the closed-orbit static solution of the system, and does not affect the beam dynamics. The other two eigenmodes are the betatron modes. They are responsible for the beam dynamics around the closed orbit.

Once D and D' are found at position s_0, D and D' at other positions are found by

$$\begin{bmatrix} D \\ D' \\ 1 \end{bmatrix} (s) \;=\; M(s|s_0) \begin{bmatrix} D \\ D' \\ 1 \end{bmatrix} (s_0)\,.$$

Homework 4.5 What happens to the particle motion and the dispersion function when two dipoles of opposite polarities and same length ℓ are joined back to back in a beamline?

(a) Consider the case of two sector dipoles.

(b) Consider two rectangular dipoles.

Homework 4.6 Is there a way for a string of dipoles to have their chromatic effects self-cancel without the help of additional quadrupoles? Consider an insertion consisting of three long sector dipoles as sketched in Fig. 4.9. The net bending is $2\theta_1 - 2\theta_2$. the total length of the insertion is $2\rho(\theta_1+\theta_2)+2L_0$. This insertion can be used to guide the accelerator trajectory layout without affecting its dispersion.

(a) Use the 3×3 matrix of sector dipoles to show that the insertion is achromatic, i.e. it does not generate D and D' of its own, if

$$L_0 \;=\; \frac{\sin(\theta_1+\theta_2) - 2\sin\theta_2}{\sin\theta_1 \sin\theta_2}\,\rho\,.$$

Note that this condition necessarily requires the bending angles cannot be too small, otherwise the insertion becomes too long. Figure 4.9 shows the case $\theta_1 = \frac{\pi}{4}, \theta_2 = \frac{\pi}{8}$. If $\theta_{1,2} \ll 1$, we have $L_0 \approx (\frac{1}{\theta_2} - \frac{1}{\theta_1})\rho$.

(b) With L_0 given by the result in (a), show that this insertion affects the betatron motion by the map

$$\begin{bmatrix} 1 - 4\cos\theta_1 + \frac{2(2\cos\theta_1 - 1)\sin 2\theta_2}{\sin\theta_1} & \frac{\rho(2\cos\theta_1 - 1)}{\sin^2\theta_1}[\sin 2\theta_1 - \sin 2\theta_2(2\cos\theta_1 - 1)] \\ \frac{4}{\rho}(\sin\theta_1 - \sin 2\theta_2) & 1 - 4\cos\theta_1 + \frac{2(2\cos\theta_1 - 1)\sin 2\theta_2}{\sin\theta_1} \end{bmatrix}.$$

Figure 4.9: An achromatic insertion consisting of three sector dipoles.

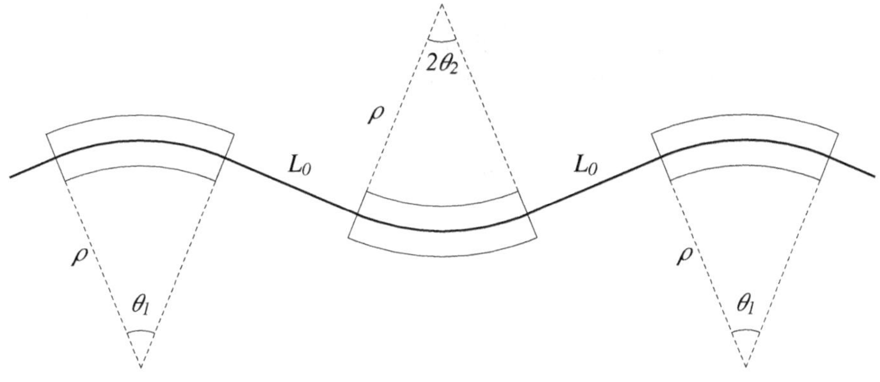

(c) If $\theta_{1,2} \ll 1$, this betatron map becomes

$$\begin{bmatrix} \frac{4\theta_2}{\theta_1} - 3 & \frac{2\rho}{\theta_1^2}(\theta_1 - \theta_2) \\ \frac{4}{\rho}(\theta_1 - 2\theta_2) & \frac{4\theta_2}{\theta_1} - 3 \end{bmatrix}.$$

But wait, is this approximation legitimate?

(d) For L_0 to be positive, it is clear that θ_2 cannot exceed 30° — for which case the only solution for θ_2 is 60° and $L_0 = 0$. However, if the middle bend is made to complete a full circle, i.e. allowing $\theta_2 > \pi$, show that the same configuration remains achromatic regardless of how many turns the beam is made to circulate.

Homework 4.7 For a combined-function magnet with simultaneously a constant K_x and a constant ρ, the 3×3 map is more involved. Find this map by solving the equation for $D(s)$. Refer to Eq. (2.11).

Solution

$$M(s|0) = \begin{cases} \begin{bmatrix} \cos\sqrt{K_x}s & \frac{\sin\sqrt{K_x}s}{\sqrt{K_x}} & \frac{1-\cos\sqrt{K_x}s}{\rho K_x} \\ -\sqrt{K}\sin\sqrt{K_x}s & \cos\sqrt{K_x}s & \frac{\sin\sqrt{K_x}s}{\rho\sqrt{K_x}} \\ 0 & 0 & 1 \end{bmatrix}, & K_x > 0, \\ \begin{bmatrix} \cosh\sqrt{-K_x}s & \frac{\sinh\sqrt{-K_x}s}{\sqrt{-K_x}} & \frac{1-\cosh\sqrt{-K_x}s}{\rho K_x} \\ \sqrt{-K_x}\sinh\sqrt{-K_x}s & \cosh\sqrt{-K_x}s & \frac{\sinh\sqrt{-K_x}s}{\rho\sqrt{-K_x}} \\ 0 & 0 & 1 \end{bmatrix}, & K_x < 0. \end{cases}$$

Homework 4.8 We performed the closed orbit calculation in Sec. 3.2.1 using 2×2 matrices. Recast that analysis using a 3×3 formalism.

Homework 4.9

(a) We are adding a *chicane* to dispersion-free straight section as shown in the upper panel of Fig. 4.10. Four thin-lens dipoles of kicking angles $\theta, -\theta, -\theta, \theta$ are inserted as shown. All kicking angles are in the horizontal plane. A beam enters the straight section from the left with dispersion function $D = 0, D' = 0$. Find the dispersion function $D(s)$ of the optics in the system, including inside the chicane and the drift space downstream of the chicane.

Figure 4.10: A chicane arrangement and a double dog-leg arrangement.

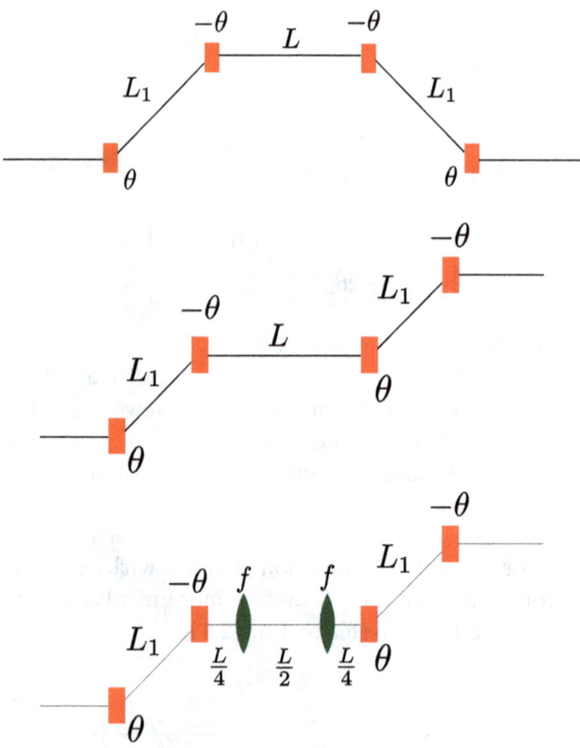

(b) Repeat the problem (a) for a *double dog-leg* beamline, as illustrated in the middle panel of Fig. 4.10. The four dipole kicks are now $\theta, -\theta, \theta, -\theta$.

(c) The middle drift space L of the double dog-leg is now supplemented by two thin-lens focusing quadrupoles at positions as indicated in the lower panel of Fig. 4.10. The two quadrupoles have the same focal length $f = \frac{L}{4}$. Repeat problem (b) with this new configuration. What is the function of the two quadrupoles? (See also Fig. 5.60.)

4.2.2 Dispersion function of FODO cell

FODO cell Consider a FODO cell with thin-lens quadrupoles. In the past, we often ignored the dipoles when we calculated the Courant-Snyder functions for a FODO cell as in Secs. 2.2.11 and 2.4.11. The two FODO dipoles were considered simply as two drift spaces. To calculate the dispersion function, however, we must not do so because after all dispersions are driven by the dipoles. On the other hand, for simplicity, let us consider thin-lens dipoles in this section.

We will first calculate the lattice functions at the middle of the QF quadrupole for a FODO cell,

$$\frac{\text{QF}}{2} \quad \text{B} \quad \text{QD} \quad \text{B} \quad \frac{\text{QF}}{2} .$$

We are interested only in the x-motion. The one-cell map is given by

$$\begin{aligned} M &= \begin{bmatrix} 1 & 0 & 0 \\ -\frac{1}{2f} & 1 & 0 \\ 0 & 0 & 1 \end{bmatrix} \begin{bmatrix} 1 & L & \frac{L\theta}{2} \\ 0 & 1 & \theta \\ 0 & 0 & 1 \end{bmatrix} \begin{bmatrix} 1 & 0 & 0 \\ \frac{1}{f} & 1 & 0 \\ 0 & 0 & 1 \end{bmatrix} \begin{bmatrix} 1 & L & \frac{L\theta}{2} \\ 0 & 1 & \theta \\ 0 & 0 & 1 \end{bmatrix} \begin{bmatrix} 1 & 0 & 0 \\ -\frac{1}{2f} & 1 & 0 \\ 0 & 0 & 1 \end{bmatrix} \\ &= \begin{bmatrix} 1 - \frac{L^2}{2f^2} & 2L(1 + \frac{L}{2f}) & 2L\theta(1 + \frac{L}{4f}) \\ -\frac{L}{2f^2}(1 - \frac{L}{2f}) & 1 - \frac{L^2}{2f^2} & 2\theta(1 - \frac{L}{4f} - \frac{L^2}{8f^2}) \\ 0 & 0 & 1 \end{bmatrix} , \end{aligned}$$

where L and $\theta = \frac{L}{\rho}$ are the length and bending angle of each dipole, f is the quadrupole focal length. The upper-left 2×2 corner is just what was obtained in Sec. 2.4.11. The new results are the (1,3)- and the (2,3)-elements of the map.

The dispersion has to satisfy the closed-orbit condition

$$\begin{bmatrix} D_F \\ D'_F \\ 1 \end{bmatrix} = M \begin{bmatrix} D_F \\ D'_F \\ 1 \end{bmatrix} ,$$

which can be solved to yield

$$D_F = \frac{L\theta(1 + \frac{1}{2}\sin\frac{\Phi_x}{2})}{\sin^2\frac{\Phi_x}{2}} , \qquad D'_F = 0 ,$$

where $\Phi_x = 2\sin^{-1}\frac{L}{2|f|}$ is the horizontal betatron phase advance per cell obtained in Eq. (2.60).

Since in the thin-lens dipole approximation the weak horizontal focusing in the dipoles are ignored, we have the situation when $\Phi_x = \Phi_y$. The dipole focusing, if included, will break this degeneracy (see Homework 4.11).

Dispersion at other locations in the FODO cell can be obtained by transforming the dispersion vector by the 3×3 maps. For region in the first dipole, $L > s > 0$, for example, we have

$$D(s) = \frac{s^2}{2\rho} + \frac{L}{\rho}(-s\sin\frac{\Phi_x}{2} + L)\left(\frac{1 + \frac{1}{2}\sin\frac{\Phi_x}{2}}{\sin^2\frac{\Phi_x}{2}}\right) , \tag{4.5}$$

$$D'(s) = \frac{s}{\rho} - \frac{L}{\rho}\left(\frac{1+\frac{1}{2}\sin\frac{\Phi_x}{2}}{\sin\frac{\Phi_x}{2}}\right).$$

In this calculation, although we assumed thin-lens dipoles, we have considered the dipole map to be a thick element. In other words, we have taken the map from $s = 0$ to any s interior of the dipole ($L > s > 0$) to be

$$\begin{bmatrix} 1 & s & \frac{s^2}{2\rho} \\ 0 & 1 & \frac{s}{\rho} \\ 0 & 0 & 1 \end{bmatrix}.$$

This is of course an approximation for small bending angle. Exact expression will have to contain sine and cosine terms instead of s and s^2, and the weak dipole focusing would have to be considered. In case a more accurate expression is desired, refer to Homework 4.11.

At the middle of the QD quadrupole, we have

$$D_D = \frac{L\theta(1-\frac{1}{2}\sin\frac{\Phi_x}{2})}{\sin^2\frac{\Phi_x}{2}}, \qquad D'_D = 0.$$

For a FODO cell, we calculated in Chapter 2 the Courant-Snyder $\alpha_{x,y}$, $\beta_{x,y}$, and $\gamma_{x,y}$ functions, and the phase advances per cell Φ_x, Φ_y. We just now calculated the dispersion functions D and D'. For completeness, let us write down, also for $L > s > 0$,

$$\begin{aligned}
\beta_x(s) &= \frac{2s^2}{L}\tan\frac{\Phi_x}{2} + \frac{2(1+\sin\frac{\Phi_x}{2})(L-2s\sin\frac{\Phi_x}{2})}{\sin\Phi_x}, \\
\alpha_x(s) &= -\frac{2s}{L}\tan\frac{\Phi_x}{2} + \frac{1+\sin\frac{\Phi_x}{2}}{\cos\frac{\Phi_x}{2}}, \\
\gamma_x(s) &= \frac{2}{L}\tan\frac{\Phi_x}{2}.
\end{aligned}$$

Figure 4.11 shows a plot of the dispersion function $D(s)$ together with the β-functions in a FODO cell for the case when the cell's betatron phase advance is $\Phi = 90°$, with $L_p = 2L =$ length of whole cell and $\varphi =$ bending angle of whole cell.

Later in this chapter we shall calculate the momentum compaction factor α_c per cell and chromaticities ξ_x, ξ_y per cell. The functions $\alpha_{x,y}$, $\beta_{x,y}$, $\gamma_{x,y}$, D, and D' are s-dependent functions. The quantities $\Phi_{x,y}, \alpha_c, \xi_{x,y}$ are global quantities and therefore are s-independent. In Chapter 5, we shall calculate some additional lattice functions involved in synchrotron radiation effects for a FODO cell. FODO cells are conceptually the simplest building blocks of a beamline.

Figure 4.11: Lattice functions β_x, β_y, D as functions of s for a FODO cell. [Courtesy Eberhard Keil]

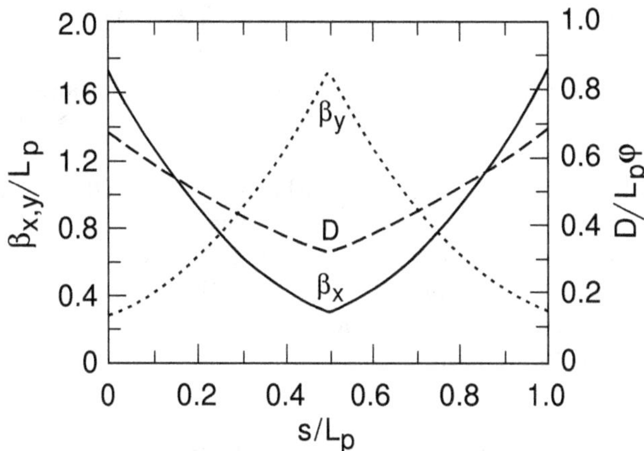

Approximate expressions of lattice functions The dispersion is maximum at QF and minimum at QD. For a 90° FODO cell, the ratio $\frac{D_F}{D_D}$ is ≈ 2. One may obtain an order of magnitude of the dispersion by considering the average of D_F and D_D, i.e.

$$\langle D \rangle \approx \frac{D_F + D_D}{2} = \frac{L\theta}{\sin^2 \frac{\Phi}{2}} .$$

We obtained earlier a rough estimate of the β-functions,

$$\langle \beta_x \rangle \approx \frac{R}{\nu_x}, \qquad \langle \beta_y \rangle \approx \frac{R}{\nu_y},$$

where $2\pi R$ is the storage ring circumference. This led to the conclusion that for a strong focusing lattice with $\nu_{x,y} \gg 1$, we have $\beta_{x,y} \ll R$. For cells with $\Phi \lesssim 90°$, we let $\sin \frac{\Phi}{2} \approx \frac{\Phi}{2}$, and therefore

$$\langle D \rangle \approx \frac{4L\theta}{\Phi^2} .$$

If the storage ring consists of N FODO cells, then $N\Phi = 2\pi\nu_x$, $2N\theta = 2\pi$, and $2NL = 2\pi R$. It follows that

$$\langle D \rangle \approx \frac{R}{\nu_x^2} . \tag{4.6}$$

What we learned here says that the dispersion is suppressed by a strong focusing lattice even more than the β-functions are. Strong focusing has significantly reduced the transverse beam sizes in the accelerator. Take for example a storage ring with $R = 100$ m and $\nu_x \approx 10$, we will have $\beta_x \approx 10$ m and

$D \approx 1$ m. A particle with momentum error of $\delta = 1\%$ has a dispersive orbit of $D\delta \approx 1$ cm. To appreciate the strong suppression effect on dispersion, one may consider a particle moving in a uniform magnetic field. Recalling that the dispersion function $D = R$ in that case (Fig. 4.3), a particle with 1% momentum error will have an orbit as large as 1 m, which is obviously not very acceptable.

In Chapter 7, we shall introduce a hierarchy of length scales in an electron storage ring. The three quantities R, β, and D (with $R \gg \beta \gg D$) will be part of that length scale hierarchy.

It should be mentioned here that, although suppressed by a large factor, the average dispersion generally remains positive. In other words, a particle with momentum higher than the design value ($\delta > 0$) will tend to have an average orbit horizontally displaced *outward* relative to the design orbit as one would expect intuitively. We will come back to this point when we discuss the momentum compaction factor.

Homework 4.10

(a) Derive the expressions D_D and D_F in the text.

(b) Derive the expressions $D(s)$ and $D'(s)$ in Eq. (4.5).

Solution A sharp-eyed reader might ask why D_D is not obtained by simply setting $f \to -f$ in the expression of D_F. It is considered part of this homework to understand why.

Homework 4.11

(a) In the text, we made the approximation of thin-lens dipole. It assumes the bending is weak, and in doing so, also ignored the weak focusing in the dipoles. Follow the text and repeat the calculation using the exact 3×3 dipole matrix, including the weak dipole focusing, to obtain the exact values of f, β_F, β_D, D_F and D_D, assuming Φ_x is given. Assume sector dipoles.

(b) It follows from your result in part (a) that the additional dipole focusing has altered the stability criterion slightly. Explain the original of this change.

Solution

(a) The problem asks for using the dipole map

$$\begin{bmatrix} \cos\frac{s}{\rho} & \rho\sin\frac{s}{\rho} & \rho(1-\cos\frac{s}{\rho}) \\ -\frac{1}{\rho}\sin\frac{s}{\rho} & \cos\frac{s}{\rho} & \sin\frac{s}{\rho} \\ 0 & 0 & 1 \end{bmatrix}.$$

Follow the text for the rest of the calculations,

$$\begin{aligned} \cos\Phi_x &= \cos 2\theta - \frac{\rho^2\sin^2\theta}{2f^2}, \\ \beta_F &= \frac{\rho\sin\theta}{\sin\Phi_x}\left(2\cos\theta + \sqrt{2}\sqrt{\cos 2\theta - \cos\Phi_x}\right), \\ \beta_D &= \frac{\rho\sin\theta}{\sin\Phi_x}\left(2\cos\theta - \sqrt{2}\sqrt{\cos 2\theta - \cos\Phi_x}\right), \end{aligned}$$

$$D_F = \frac{\rho \sin^2 \frac{\theta}{2}}{\sin^2 \frac{\Phi_x}{2}} \left(2 + 2\cos\theta + \sqrt{2}\sqrt{\cos 2\theta - \cos\Phi_x}\right),$$
$$D_D = \frac{\rho \sin^2 \frac{\theta}{2}}{\sin^2 \frac{\Phi_x}{2}} \left(2 + 2\cos\theta - \sqrt{2}\sqrt{\cos 2\theta - \cos\Phi_x}\right).$$

In this case, Φ_y is slightly different from Φ_x even though the two quadrupoles have the same strength because the dipoles focus in x and not in y.

(b) We have required an additional stability condition $\cos 2\theta - \cos\Phi_x > 0$ in (a). When dipole focusing is ignored, this condition reads $1 > \cos\Phi_x$ which is always obeyed. It should be observed that the additional focusing has made the system *less* stable, possibly contrary to intuition. Physically this comes from the fact that the FODO lattice instability originates from over-focusing, and that the additional focusing of the dipoles helps reaching the over-focusing instability limit earlier.

Homework 4.12 We wrote $\langle D\rangle \approx \frac{R}{\nu_x^2}$ instead of $\langle D\rangle \approx \frac{R}{\nu_y^2}$. Although this is expected intuitively, our example FODO cell has $\Phi_x = \Phi_y$, so there is an ambiguity between ν_x and ν_y. Resolve this ambiguity by analyzing a FODO variation that has $\Phi_x \neq \Phi_y$.

Homework 4.13 Consider a storage ring consisting of identical FODO cells in a plane. But now the design of the storage ring requires it to contain a fold geometry with n_1 cells in one plane and the remaining n_2 cells in another plane with a tilting angle, as sketched in Fig. 4.12. How should the unit cells be designed so that the folding does not generate interfering dispersive effects? For this exercise, ignore complications due to resonances when tunes assume rational numbers.

Figure 4.12: A ring with a fold geometry.

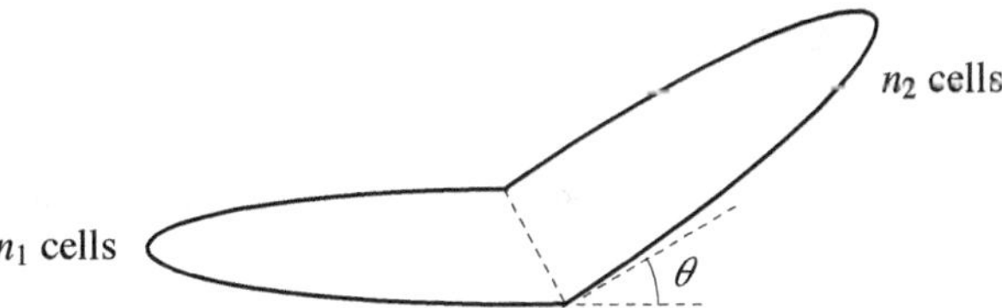

4.2.3 General expression of the 3 × 3 map

We just calculated the dispersion in a FODO cell. In general, if the 3 × 3 one-period map around position s is given by

$$\begin{bmatrix} m_{11} & m_{12} & m_{13} \\ m_{21} & m_{22} & m_{23} \\ 0 & 0 & 1 \end{bmatrix},$$

then it is easy to show that the closed-orbit condition gives the dispersion at position s,

$$\begin{aligned} D &= \frac{m_{13}(1-m_{22})+m_{12}m_{23}}{2-m_{11}-m_{22}}, \\ D' &= \frac{m_{21}m_{13}+m_{23}(1-m_{11})}{2-m_{11}-m_{22}}, \end{aligned}$$

where use has been made of $m_{11}m_{22} - m_{12}m_{21} = 1$.

This result can be used to calculate the dispersion function at any position s around the storage ring. What one does is to sequentially multiply the 3×3 maps element by element to obtain the one-period map around position s. The values of D and D' at s then follow.

Since the upper-left 2×2 corner is the same as the betatron map, the above expressions can be rewritten in terms of m_{13}, m_{23}, and the Courant-Snyder functions,

$$\begin{aligned} D &= \frac{1}{2}\left[m_{13} + (\alpha m_{13} + \beta m_{23}) \cot\frac{\Phi}{2}\right], \\ D' &= \frac{1}{2}\left[m_{23} - (\gamma m_{13} + \alpha m_{23}) \cot\frac{\Phi}{2}\right], \end{aligned}$$

where α, β, γ are the horizontal Courant-Snyder functions evaluated at s.

Alternatively, one may solve m_{13} and m_{23} in terms of D and D'. When substituted into the one-period map, one obtains a general expression of the 3×3 one-period map around position s in terms of the horizontal lattice functions at position s,

$$\begin{bmatrix} \cos\Phi + \alpha\sin\Phi & \beta\sin\Phi & (1-\cos\Phi-\alpha\sin\Phi)D - \beta D'\sin\Phi \\ -\gamma\sin\Phi & \cos\Phi - \alpha\sin\Phi & \gamma D\sin\Phi + (1-\cos\Phi+\alpha\sin\Phi)D' \\ 0 & 0 & 1 \end{bmatrix}.$$

Once D and D' are obtained at position s_1, dispersion at some other position s_2 can be obtained from

$$\begin{bmatrix} D \\ D' \\ 1 \end{bmatrix}_{s_2} = M(s_2|s_1) \begin{bmatrix} D \\ D' \\ 1 \end{bmatrix}_{s_1}.$$

Knowing the lattice functions at positions s_1 and s_2, and using Eq. (2.51), the 3×3 map from s_1 to s_2 can be written as

$$M(s_2|s_1) = \begin{bmatrix} m_{11} & m_{12} & D_2 - m_{11}D_1 - m_{12}D_1' \\ m_{21} & m_{22} & D_2' - m_{21}D_1 - m_{22}D_1' \\ 0 & 0 & 1 \end{bmatrix}, \qquad (4.7)$$

$$
\begin{aligned}
m_{11} &= \sqrt{\frac{\beta_2}{\beta_1}}(\cos\psi + \alpha_1 \sin\psi)\,, \\
m_{12} &= \sqrt{\beta_1\beta_2}\sin\psi\,, \\
m_{21} &= \frac{\alpha_1 - \alpha_2}{\sqrt{\beta_1\beta_2}}\cos\psi - \frac{1+\alpha_1\alpha_2}{\sqrt{\beta_1\beta_2}}\sin\psi\,, \\
m_{22} &= \sqrt{\frac{\beta_1}{\beta_2}}(\cos\psi - \alpha_2 \sin\psi)\,,
\end{aligned}
$$

where ψ is the horizontal betatron phase advance from s_1 to s_2.

Homework 4.14 Derive the expression for the 3×3 map $M(s_2|s_1)$ in Eq. (4.7). When $s_2 = s_1 + L$, show that the result reduces to that for one-period map.

4.3 Wien filter

There is a neat device that maintains a straight line motion of a beam without curvature, while can still provide a dispersion. It is called the Wien filter.[3] Strictly, what it filters is not particles momentum or energy, but its velocity. So it is mostly a low-energy device. In fact, we will analyze this device in a nonrelativistic limit for a change — the Wien filter was invented before the theory of relativity anyway.

Before starting, perhaps we ask how would you proceed to design a velocity filter? A long time-of-flight straight section with a fast shutter at its entrance and another fast shutter at its exit might serve the purpose? Or maybe a dipole magnet followed by a long drift space to disperse the particle motion according to its momentum, followed by an aperture slit to select the desired momentum that corresponds to the desired velocity? Wien filter is a much more clever invention. It basically makes use of the fact that the electric force is independent of a particle's velocity, while the magnetic force is proportional to it.

Body field Consider a region with combined uniform static electric field $E_0\hat{x}$ and uniform static magnetic field $B_0\hat{y}$. Let the beam move in the $\hat{z}$ direction. A particle in the beam has the equation of motion

$$
m(\ddot{x}\hat{x} + \ddot{y}\hat{y} + \ddot{z}\hat{z}) = qE_0\hat{x} + q(\dot{x}\hat{x} + \dot{y}\hat{y} + \dot{z}\hat{z}) \times B_0\hat{y}\,,
$$

or

$$
\begin{aligned}
m\ddot{x} &= qE_0 - qB_0\dot{z}\,, \\
m\ddot{y} &= 0\,, \\
m\ddot{z} &= qB_0\dot{x}\,.
\end{aligned}
$$

[3]W. Wien, Annalen der Physik, Vol. 301, Issue 6, 440 (1898).

It immediately follows that if we choose the longitudinal velocity of a particle $\dot{z} = \frac{E_0}{B_0}$, the electric and magnetic forces it sees cancel each other, and the particle will not experience any transverse deflections and will stay moving on the z-axis. A faster particle with $\dot{z} > \frac{E_0}{B_0}$ will be deflected toward $-\hat{x}$ direction and a slower particle in $\hat{x}$ direction — thus the velocity filtering action.

Define a small dynamical variable δ_v by

$$\dot{z} = \frac{E_0}{B_0} + \delta_v, \quad |\delta_v| \ll \frac{E_0}{B_0}.$$

The equations of motion can be rewritten,

$$\ddot{x} = -\omega\delta_v, \qquad \dot{\delta}_v = \omega\dot{x},$$

with $\omega = \frac{qB_0}{m}$, which can be combined to yield

$$\frac{d^2\dot{x}}{dt^2} + \omega^2\dot{x} = 0,$$

The dynamics in this Wien system is basically sinusoidal motions in x and z and a drifting motion in y.

The solution is, with appropriate initial conditions,

$$\begin{aligned} x(t) &= A\cos\omega t + B\sin\omega t + C, \\ \dot{x}(t) &= -\omega A\sin\omega t + \omega B\cos\omega t, \\ \ddot{x}(t) &= -\omega^2 A\cos\omega t - \omega^2 B\sin\omega t. \end{aligned}$$

Initial conditions are $x_0 = A + C$, $\dot{x}_0 = \omega B$, $\ddot{x}_0 = -\omega\delta_{v0} = -\omega^2 A$, which gives the solution $A = \frac{\delta_{v0}}{\omega}$, $B = \frac{\dot{x}_0}{\omega}$, $C = x_0 - \frac{\delta_{v0}}{\omega}$. We then have

$$\begin{aligned} x(t) &= \frac{\delta_{v0}}{\omega}\cos\omega t + \frac{\dot{x}_0}{\omega}\sin\omega t + x_0 - \frac{\delta_{v0}}{\omega}, \\ \dot{x}(t) &= -\delta_{v0}\sin\omega t + \dot{x}_0\cos\omega t. \end{aligned}$$

We also have

$$\delta_v(t) = -\frac{\ddot{x}(t)}{\omega} = \delta_{v0}\cos\omega t + \dot{x}_0\sin\omega t.$$

The quantity δ_v defines the longitudinal velocity of the particle relative to the ideal velocity of $\frac{E_0}{B_0}$. We can also define a longitudinal displacement variable δ_z with

$$z(t) = \frac{E_0}{B_0}t + \delta_z(t).$$

Integrating $\delta_v(t)$, we obtain

$$\delta_z(t) = \delta_{z0} + \frac{\delta_{v0}}{\omega}\sin\omega t + \frac{\dot{x}_0}{\omega}(1 - \cos\omega t).$$

The total passage of the device can be obtained by setting $t = \frac{L}{E_0/B_0}$, where L is the length of the device.

The particle motion can be described by a linear transfer map for the vector

$$V = \begin{bmatrix} x \\ x' \\ y \\ y' \\ \delta_z \\ \delta_z' \end{bmatrix},$$

where $x' = \frac{B_0}{E_0}\dot{x}$, $y' = \frac{B_0}{E_0}\dot{y}$, $\delta_z' = \frac{B_0}{E_0}\delta_v$. The transfer matrix over the body of the device can be written as

$$M = \begin{bmatrix} 1 & \rho\sin\frac{L}{\rho} & 0 & 0 & 0 & -\rho(1-\cos\frac{L}{\rho}) \\ 0 & \cos\frac{L}{\rho} & 0 & 0 & 0 & -\sin\frac{L}{\rho} \\ 0 & 0 & 1 & L & 0 & 0 \\ 0 & 0 & 0 & 1 & 0 & 0 \\ 0 & \rho(1-\cos\frac{L}{\rho}) & 0 & 0 & 1 & \rho\sin\frac{L}{\rho} \\ 0 & \sin\frac{L}{\rho} & 0 & 0 & 0 & \cos\frac{L}{\rho} \end{bmatrix},$$

where $\rho = \frac{mE_0}{qB_0^2}$ is the cyclotron curvature of particle motion due to the magnetic field for a particle with velocity $\frac{E_0}{B_0}$.

Canonical variables But the matrix M is not symplectic! Experience from solenoids and acceleration sections suggests that the problem is that we have not used the canonical variables, and the transformation from the conventional coordinates V to the canonical variables shall be imbedded in the effect due to end fields.

First, let us touch base and establish the fact that the system is Hamiltonian — and therefore it *should* be symplectic. The Hamiltonian is

$$\begin{aligned} H(x, p_x, y, p_y, \delta_z, p_{\delta v}, t) &= \frac{1}{2m}[p_x^2 + p_y^2 + (p_{\delta z} + qB_0 x)^2] - qE_0 x \\ \Longrightarrow \quad \dot{x} &= \frac{\partial H}{\partial p_x} = \frac{p_x}{m}, \\ \dot{p}_x &= -\frac{\partial H}{\partial x} = qE_0 - \frac{qB_0}{m}(p_{\delta z} + qB_0 x), \\ \dot{y} &= \frac{\partial H}{\partial p_y} = \frac{p_y}{m}, \\ \dot{p}_y &= -\frac{\partial H}{\partial y} = 0, \\ \dot{\delta}_z &= \frac{\partial H}{\partial p_{\delta z}} = \frac{1}{m}(p_{\delta z} + qB_0 x), \\ \dot{p}_{\delta z} &= -\frac{\partial H}{\partial \delta_z} = 0, \end{aligned} \tag{4.8}$$

which can be checked to reproduce our previous results. So the system is indeed Hamiltonian, but it is so only if we use the canonical coordinates x, p_x, y, p_y, δ_z, and $p_{\delta z}$.

This means that, to be symplectic, it is required to use the canonical variable $p_{\delta z} = m\dot{\delta}_z - qB_0 x$, i.e. to use the canonical momentum $\delta_z' - \frac{x}{\rho}$ instead of the mechanical momentum δ_z' in our matrix map. The symplectic map is therefore to be obtained for the matrix NMN^{-1}, where

$$N = \begin{bmatrix} 1 & 0 & 0 & 0 & 0 & 0 \\ 0 & 1 & 0 & 0 & 0 & 0 \\ 0 & 0 & 1 & 0 & 0 & 0 \\ 0 & 0 & 0 & 1 & 0 & 0 \\ 0 & 0 & 0 & 0 & 1 & 0 \\ -\frac{1}{\rho} & 0 & 0 & 0 & 0 & 1 \end{bmatrix}.$$

The transfer map for the vector V should be NMN^{-1},

$$NMN^{-1} = \begin{bmatrix} \cos\frac{L}{\rho} & \rho\sin\frac{L}{\rho} & 0 & 0 & 0 & -\rho(1-\cos\frac{L}{\rho}) \\ -\frac{\sin\frac{L}{\rho}}{\rho} & \cos\frac{L}{\rho} & 0 & 0 & 0 & -\sin\frac{L}{\rho} \\ 0 & 0 & 1 & L & 0 & 0 \\ 0 & 0 & 0 & 1 & 0 & 0 \\ \sin\frac{L}{\rho} & \rho(1-\cos\frac{L}{\rho}) & 0 & 0 & 1 & \rho\sin\frac{L}{\rho} \\ 0 & 0 & 0 & 0 & 0 & 1 \end{bmatrix}.$$

The maps N, M are not symplectic, but NMN^{-1} is. The extra maps N and N^{-1} must be caused by actual physical effects, not by an abstract change of variables. The physical effects are due to the end fields.

End fields Consider the entrance edge of the Wien filter. The electric field requires $\vec{E} = E_0\hat{x}H(z)$. It is an x-component that depends on z. The condition $\nabla\times\vec{E} = \vec{0}$ then requires adding a z-component that depends on x, i.e. $E_0 x\delta(z)\hat{z}$, to the field. This additional longitudinal electric field then yields an additional longitudinal acceleration $\Delta\delta_z' = \frac{x}{\rho}$. This then gives rise to the map N^{-1} at the entrance to the Wien filter. A similar consideration at the exit gives the additional map N.

It might be useful to compare the end field effect of a Wien filter with that of a dipole magnet. Recall that, for a dipole magnet, the magnetic field required to fulfill the condition $\nabla \times \vec{B} = \vec{0}$ at the magnet ends is also longitudinal. A longitudinal magnetic field normally does not affect the particle motion because the particle moves mainly in the longitudinal direction. However, if there is a design entrance angle, like the case of a rectangular dipole, then this end field can introduce an edge focusing effect. This has been discussed in Sec. 2.2.8.

Here in the Wien filter, there is no design entrance angle, so the condition $\nabla \times \vec{B} = \vec{0}$ does not produce an effect. In a Wien filter, on the other hand,

Table 4.1: A comparison of four types of end field effects.

Device	Condition	End field	Effect	Ref.
Dipole	$\nabla \times \vec{B} = \vec{0}$	$B_z \propto y$	focusing, design orbit angle	Sec. 2.2.8
Solenoid	$\nabla \cdot \vec{B} = 0$	$B_r \propto r$	twist in x-y plane	Sec. 1.4.1 Sec. 3.5.3
Accel. sec.	$\nabla \cdot \vec{E} = 0$	$E_r \propto r$	uncoupled kicks in x and y	Sec. 3.5.4
Wien filter	$\nabla \times \vec{E} = \vec{0}$	$E_z \propto x$	twist in x-z plane	Sec. 4.3

there is also an electric field. The condition $\nabla \times \vec{E} = \vec{0}$ generates a necessary longitudinal electric fields at the ends. A longitudinal electric field, unlike a longitudinal magnetic field, does affect directly the particle motion even when there is no design entrance angle. And it is this effect we are observing here.

Note that although we are discussing the Wien filter here. The same end field effect exists also for a device with only E_0 without B_0. The same end fields apply with just a pure electric bend.

Note also that, according to the expression of the matrix NMN^{-1}, the longitudinal momentum δ_z' does not change, while an inspection of the matrix M says that it does change inside the body of the Wien filter. The contributions to the changes in longitudinal momentum at the two ends exactly cancel the contribution from the body. The reason of this happening is simply because the canonical momentum obeys $\dot{p}_{\delta v} = 0$ as derived in Eq. (4.8).

Four types of end fields Table 4.1 is a comparison among the various end field effects we have studied. So far, we have studied all four types of end fields, imposed by four Maxwell equations $\nabla \times \vec{B} = \vec{0}, \nabla \cdot \vec{B} = 0, \nabla \cdot \vec{E} = 0, \nabla \times \vec{E} = \vec{0}$, respectively. Which ones of these effects are symplectic and which are not?

Velocity filtering As can be seen, while keeping the design trajectory in a straight line, the device allows the generation of a velocity dispersion downstream. An initial velocity spread $\Delta\delta_{v0}$ causes a spread in the transverse direction at the exit of the filter. To optimize the design, one may choose $\frac{L}{\rho} = \frac{\pi}{2}$. In that case, we have

$$NMN^{-1} = \begin{bmatrix} 0 & \rho & 0 & 0 & 0 & -\rho \\ -\frac{1}{\rho} & 0 & 0 & 0 & 0 & -1 \\ 0 & 0 & 1 & L & 0 & 0 \\ 0 & 0 & 0 & 1 & 0 & 0 \\ 1 & \rho & 0 & 0 & 1 & \rho \\ 0 & 0 & 0 & 0 & 0 & 1 \end{bmatrix}.$$

An initial spread $\Delta\delta_{v0}$ gives rise to these analyzing powers,

$$\begin{aligned}\Delta x &= \rho\Delta\delta_{v0}\,,\\ \Delta x' &= \Delta\delta_{v0}\,,\\ \Delta\delta_v &= 0\,.\end{aligned}$$

This may serve as a velocity filter if the initial beam is sufficiently collimated.

Take for example a device with $E_0 = 400$ kV/m, $B_0 = 0.5$ Tesla, and for a beam of gold ions with $q = 4\,e, m = 79\,m_p$. We then have $\rho = 0.33$ m. The condition $\frac{L}{\rho} = \frac{\pi}{2}$ requires $L = 0.52$ m. The gold ion has velocity 8×10^5 m/s, and carries a kinetic energy of 0.26 MeV.

Immersed gun in a Wien filter Let us immerse a particle source inside a Wien filter or an electrostatic bending device. The beam comes out the exit without having to enter it, then the map for the vector V from the source to a point immediately after the exit will be NM. It is nonsymplectic. In principle, the x- and the z-emittances of the beam can be manipulated similar to the case of an immersed gun in a solenoid, where the x- and the y-emittances are manipulated, as discussed in Sec. 3.5.3. However, the end field of a Wien filter most likely is very weak.

Homework 4.15 Show explicitly that the matrix NMN^{-1} is symplectic.

Homework 4.16 The Wien filter also serves as a filter according to the $\frac{q}{m}$ ratio of the particles in the beam. Consider a well-collimated beam with negligible initial transverse velocity spreads. The beam particles have a spread in $\frac{q}{m}$, but all particles have just been produced and accelerated by a common electrostatic voltage V_0. Derive the analyzing power of the Wien filter as a $\frac{q}{m}$ filter device.

4.4 Dispersion suppressor

Often in accelerator designs, there is a need to suppress the dispersion in a straight section. The lattice containing a string of FODO cells, for example, is to be connected to a straight section and then connected back to the FODO string. Depending on the purpose of the straight section, often it is necessary to make it dispersion free. The question then is how to arrange the lattice to fulfill this design requirement.

There are many ways to make a dispersion free straight section. We will discuss a particularly elegant one below.[4] Consider a string of FODO cells of the form

$$\frac{\text{QF}}{2} \quad \text{B} \quad \text{QD} \quad \text{B} \quad \frac{\text{QF}}{2}\,.$$

[4]R.H. Helm, SLAC-PUB-3278 (1984).

Figure 4.13: A dispersion suppressor design.

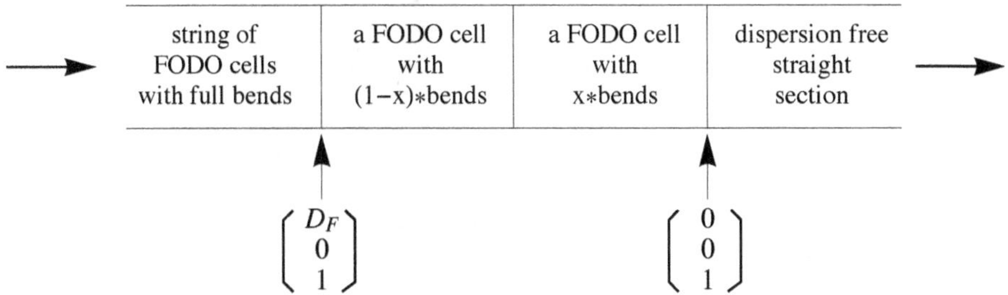

The 3×3 map for this cell is

$$M_0 = \begin{bmatrix} \cos\Phi & \beta_F \sin\Phi & (1-\cos\Phi)D_F \\ -\frac{1}{\beta_F}\sin\Phi & \cos\Phi & \frac{\sin\Phi}{\beta_F}D_F \\ 0 & 0 & 1 \end{bmatrix},$$

and the dispersion vector at the entrance and exit of this FODO cell

$$\vec{D}_F = \begin{bmatrix} D_F \\ 0 \\ 1 \end{bmatrix}$$

of course satisfies the condition $M_0\vec{D}_F = \vec{D}_F$, or another way of saying the same thing, $\vec{D}_F$ is an eigenvector of M_0 with eigenvalue 1.

We now insert, at the end of this string of FODO cells, two more FODO cells which have exactly the same quadrupoles and the same bending magnet length as the regular cells, but their bending magnet strengths are reduced by a factor of x and $1-x$ respectively, i.e.

$$\frac{\text{QF}}{2} \quad (1-x)\text{B} \quad \text{QD} \quad (1-x)\text{B} \quad \text{QF} \quad x\text{B} \quad \text{QD} \quad x\text{B} \quad \frac{\text{QF}}{2}.$$

The situation is sketched in Fig. 4.13.

The 3×3 maps for these two cells are

$$M_x = \begin{bmatrix} \cos\Phi & \beta_F \sin\Phi & (1-\cos\Phi)xD_F \\ -\frac{\sin\Phi}{\beta_F} & \cos\Phi & \frac{\sin\Phi}{\beta_F}xD_F \\ 0 & 0 & 1 \end{bmatrix},$$

$$M_{1-x} = \begin{bmatrix} \cos\Phi & \beta_F \sin\Phi & (1-\cos\Phi)(1-x)D_F \\ -\frac{\sin\Phi}{\beta_F} & \cos\Phi & \frac{\sin\Phi}{\beta_F}(1-x)D_F \\ 0 & 0 & 1 \end{bmatrix}.$$

At the end of these two added cells, we wish to obtain a dispersion free optics. This requires the condition

$$M_x M_{1-x} \vec{D}_F = \begin{bmatrix} 0 \\ 0 \\ 1 \end{bmatrix} .$$

We omit the straightforward algebra here, but amazingly, there is a solution for x, namely,

$$x = \frac{1}{2(1-\cos\Phi)} .$$

In other words, if at the end of a string of FODO cells, we insert two additional FODO cells prescribed here, then the downstream straight section will be dispersion free as long as it contains only quadrupoles and no dipoles. At the end of the straight section, to return to the repetitive FODO arrangement, all one has to do is to install a mirror image of the dispersion suppressor. This elegant design applies to FODO cells of arbitrary phase advance Φ.

Figure 4.14 shows x and $(1-x)$ as functions of $\frac{\Phi}{\pi}$ from 0 to 1. One special case occurs when $\Phi = 90°$. Then $x = \frac{1}{2}$, and the two added cells are identical with half bends. Another special case is when $\Phi = 60°$. One of the two FODOs is then a missing dipole cell and the other cell is a cell with the nominal dipole strength. Needless to say, if we have $x > 1$ or $x < 0$, we have reverse bends in the lattice.

Figure 4.14: The parameters x (solid) and $1-x$ (dashed) as functions of cell phase advance $\frac{\Phi}{\pi}$ in the dispersion suppressor design of Fig. 4.13.

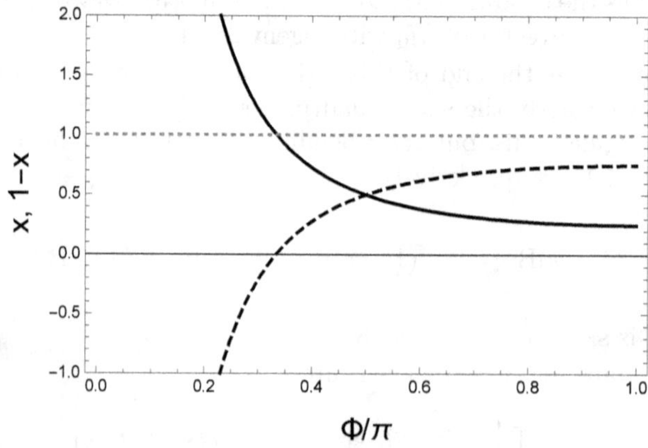

Homework 4.17 Show that the arrangement in Fig. 4.15 gives another dispersion suppressor. The case with $n = 2$ becomes the same as the dispersion suppressor discussed in the text when $\Phi = 90°$.

Figure 4.15: Another dispersion suppressor design.

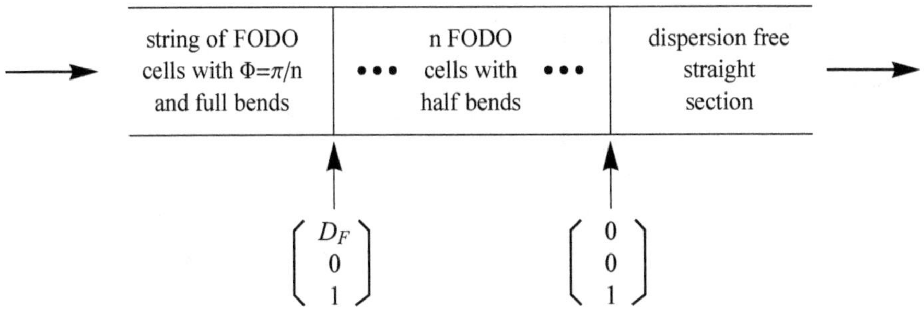

string of FODO cells with $\Phi=\pi/n$ and full bends	••• n FODO cells with half bends •••	dispersion free straight section

$\begin{pmatrix} D_F \\ 0 \\ 1 \end{pmatrix}$ $\begin{pmatrix} 0 \\ 0 \\ 1 \end{pmatrix}$

4.5 Momentum compaction

4.5.1 Path length

Dispersion function describes the horizontal closed orbit distortion of an off-momentum particle. In the ideal case, the momentum error does not cause any vertical orbit effect. However, it does cause a longitudinal effect because the total circumference of an off-momentum particle will no longer be given by the design circumference C. As an off-momentum particle circulates around the accelerator, its longitudinal position relative to the on-momentum reference particle has to change from turn to turn due to the difference in circumference. To discuss the dynamics of an off-momentum particle, therefore, it is possible to ignore the vertical motion, at least for the ideal uncoupled accelerator, but it is unavoidable to involve the longitudinal dynamics. Exactly how this circumference affects the longitudinal motion of the particle will be the subject of the next chapter, but we will here calculate the circumference error ΔC associated with a momentum error δ.

In general, path length change occurs when a particle passes through a bending magnet with a horizontal displacement. There are two conditions for the generation of a path length deviation: (a) a displacement, and (b) a trajectory bending in the same plane as the displacement. Both conditions must be fulfilled in order to cause a path length effect. A moment's reflection indicates that an x-displacement in a quadrupole will cause a path length contribution only to second order in x. By the same token, bending is not required if one considers higher order effects, as Homework 4.19 intends to illustrate.

A path length contribution is illustrated in Fig. 4.16(left) for a sector dipole and for a particle with a displacement in x. The particle experiences a path length deviation as it passes through the magnet. The deviation is given by

$$\Delta C = (\rho + x)\theta - \rho\theta = x\theta .$$

Figure 4.16: Momentum compaction factor due to a sector dipole (left) and rectangular dipole (right).

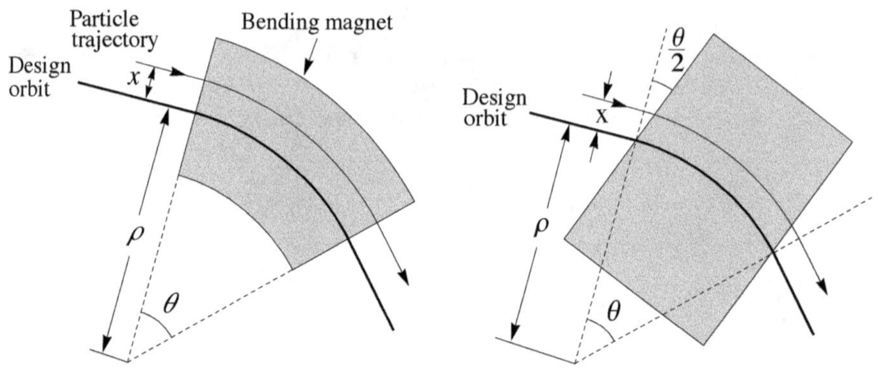

For a rectangular dipole, it is easy to show that the path length change is

$$\Delta C = 2x \tan\frac{\theta}{2},$$

which for small θ is approximately the same as the ΔC for a sector dipole.

The change in circumference due to a horizontal closed orbit distortion is obtained by integrating over the storage ring,

$$\Delta C = \oint \frac{x_{\rm COD}(s)}{\rho(s)} ds\,. \tag{4.9}$$

Homework 4.18 The text derived Eq. (4.9), the path length change in a storage ring. Strictly it is only if all bends are sector dipoles. What changes have to be made for the case of a rectangular dipole? Give an estimate of an error made if one uses Eq. (4.9) when the dipoles are all rectangular.

Homework 4.19 In a region without bending, the path length changes according to

$$\Delta C = \int_0^\ell ds \left[\sqrt{1 + x'^2(s) + y'^2(s)} - 1\right].$$

Give analytic expressions of ΔC as a particle traverses a drift space or a quadrupole in terms of a particles entrance coordinate deviations $x_0, x_0', y_0, y_0', z_0, \delta_0$ to second order in these deviations.

4.5.2 Momentum compaction factor

For the present discussion, we are interested in the horizontal displacement of a particle due to its momentum deviation δ. In the case of a COD due to a momentum error, this gives

$$\Delta C = \delta \oint \frac{D(s)}{\rho(s)} ds .$$

A *momentum compaction factor* α_c is defined by

$$\frac{\Delta C}{C} = \alpha_c \delta, \qquad \text{or} \qquad \alpha_c = \frac{1}{\delta}\frac{\Delta C}{C} .$$

We have then

$$\alpha_c = \frac{1}{C} \oint \frac{D(s)}{\rho(s)} ds .$$

For an order of magnitude estimate,

$$\alpha_c \approx \frac{\langle D \rangle}{R} \approx \frac{1}{\nu_x^2} ,$$

where we have let $C \approx 2\pi R, \rho \approx R$, and $\langle D \rangle \approx \frac{R}{\nu_x^2}$ using Eq. (4.6). As discussed earlier, in a uniform magnetic field, we have $D = R$, and then we have a large momentum compaction factor $\alpha_c = 1$. Strong focusing has very much suppressed the value of α_c as $\nu_x \gg 1$.

The momentum compaction factor is a fundamental parameter in a lattice design. A large α_c means the path length varies by a large amount with a small momentum error. In such a storage ring, a stored bunched beam will tend to have a long bunch length because a large α_c makes it easy for particles to spread out longitudinally. Therefore, large α_c means long bunch length, i.e. the beam bunch is less compact. Similarly, small α_c means the beam bunch is more compact. In a sense, α_c should have been called the momentum "dilution" factor instead of a momentum "compaction" factor, but the name of course stays.

As discussed earlier following Eq. (4.6), a natural lattice will have an average dispersion $\langle D \rangle > 0$. This leads to the result that $\alpha_c > 0$. Sometimes special lattices are designed to counter this natural tendency. One then designs an *isochronous* lattice that the average orbit is independent of δ, at least to first order in δ. Or one may even design a *negative momentum compaction* lattice with average orbit on the inward side (for $\delta > 0$) relative to the design orbit. Attention will be needed to study the beam dynamics of these special lattices.

Homework 4.20 Show that there is no way to use periodic FODO cells to make an isochronous storage ring.

Solution Show first that in the middle of dipole, dispersion is given by

$$D_{\text{mid-dipole}} = D_F \left(1 - \frac{L}{4f}\right) .$$

It might seem one can make the FODO cell isochronous by choosing some value of f less than $\frac{L}{4}$. But that FODO cell is unstable due to over-focusing.

Homework 4.21 Find an expression for the momentum compaction per cell α_c for a FODO cell. Assume thin-lens quadrupoles and dipoles.

Solution Use Eq. (4.5) to obtain

$$\alpha_c \text{ per cell} = \frac{L^3}{C\rho^2}\left(\frac{6-\frac{1}{2}\sin^2\frac{\Phi_x}{2}}{3\sin^2\frac{\Phi_x}{2}}\right).$$

4.5.3 Path length correction to COD

Consider a storage ring that contains a horizontal orbital kick with angle θ at position s_0. The kick generates a horizontal COD around the ring given by Eq. (3.7). This COD in turn gives rise to a change of orbital circumference by an amount

$$\Delta C = \oint ds \frac{x_{\text{COD}}(s)}{\rho(s)}.$$

Substituting explicit expression of the COD, Eq. (3.7), into ΔC and apply Eq. (4.3), one obtains a very neat expression

$$\Delta C = \theta D(s_0). \tag{4.10}$$

Thus a horizontal orbital kick automatically generates a path length deviation of the beam on its subsequent revolutions.

The fact that $\Delta C \propto \theta$ is not surprising. What is surprising is the fact that the proportionality constant is just the dispersion function. On the surface, there is really no reason that this has to be true. After all, ΔC plays a role in the longitudinal beam dynamics in terms of the z-coordinate, while dispersion is basically a transverse quantity and relates only to the δ-coordinate. Besides, why does the lengthy mathematics of substituting the complex expression (3.7) into the expression of ΔC yields such a neat result? It cannot be purely accidental. Indeed, the underlying reason of this neat expression can be traced to the symplecticity condition. We discussed this in Sec. 3.7.5.

One can prove this neat expression by another way as follows. Observe that

$$\begin{aligned} x''_{\text{COD}} + K x_{\text{COD}} &= \theta\delta(s-s_0), \\ D'' + KD &= \frac{1}{\rho}. \end{aligned}$$

These expressions can be combined to give

$$x_{\text{COD}} D'' - D x''_{\text{COD}} = \frac{x_{\text{COD}}}{\rho} - \theta D \delta(s-s_0).$$

Integrate both sides over s over one revolution. The left-hand-side vanishes, which then establishes the proof.

At the first impression, the miraculous result (4.10) is coincidental – it is wonderful and it just happened. At a second more technical level, one traces it back and proves it using its equation of motion as we just did. Much satisfaction can be drawn by this step. The third level then asks for the physical origin of the equation, and finally traces it to the principle of symplecticity.

As a rough order of magnitude estimate, the COD induced by the kick is of the order $\sim \theta\beta_x$, and that the path length change due to the kick is $\sim \theta D$. Typically β is of the order of $\frac{R}{\nu_x}$ and D is of the order of $\frac{R}{\nu_x^2}$, where R is the average radius of the ring. This means

$$\frac{\Delta C}{\Delta x} \sim \frac{1}{\nu_x}.$$

The path length change due to a kick is much less than the magnitude of the COD it generates.

What happens to the COD if we take into account that the dipole kick has produced this ΔC in addition to the COD given by Eq. (3.7)? In case the beam is bunched by an RF system as in the case of a synchrotron or an electron storage ring, the circumference is dictated by the RF frequency and is not to be changed. This additional ΔC has to be compensated by an adjustment of its average energy by an amount

$$\Delta\delta = -\frac{1}{\alpha_c}\frac{\Delta C}{C} = -\frac{1}{\alpha_c}\frac{\theta D(s_0)}{C}.$$

Taking this into account, for these accelerators, a more accurate COD formula for a horizontal dipole kick is not given by Eq. (3.7) but by

$$x_{\rm COD}(s) = \theta\frac{\sqrt{\beta(s_0)\beta(s)}}{2\sin\pi\nu}\cos(\pi\nu - |\psi(s)-\psi(s_0)|) - \theta\frac{D(s_0)D(s)}{\alpha_c C}.$$

4.6 Achromat cell

We discussed dispersion suppressors in Sec. 4.4. The idea was to cancel out the existing dispersion by a suppressor insertion. There are other types of insertions. One of them is the *achromatic* insertion, which has the property that, in spite of the fact that it contains bending magnets, it does not generate a new dispersion in an already-dispersion-free region. If we use achromat insertions as the bending arcs of an electron storage ring, for example, then the adjacent straight sections can be dispersion free — dispersion suppressing is automatic. That can be useful for compact designs of synchrotron radiation facilities.[5]

A *double-bend achromat* basically consists of

$$\text{B} \quad \text{L} \quad \text{QF} \quad \text{L} \quad \text{B} \, ,$$

[5] R. Chasman, K. Green, and E. Rowe, IEEE Trans. NS 22, 1765 (1975); C. Steier, Handbook Accel. Phys. & Eng., 2nd Ed., World Scientific (2013).

where L is a drift of length L, B is a bend with bend angle θ, and QF is a horizontal focusing quadrupole with focal length f. The idea of a DBA is that, in a dispersion-free straight section, one can insert this DBA cell at any location to provide bending without disturbing its dispersion-free nature.

The condition for a DBA is therefore that it has $D = 0$ and $D' = 0$ on both sides of the arrangement. The condition is fulfilled (in the thin-lens approximation) if

$$f = \frac{L}{2}, \tag{4.11}$$

and the dispersion at QF is $D_{\text{QF}} = L\theta$. When condition (4.11) is fulfilled, this DBA insertion is achromatic. Its layout and dispersion function are shown in Fig. 4.17.

Figure 4.17: Schematic illustration of the DBA principle, its layout and its dispersion function in the thin-lens approximation.

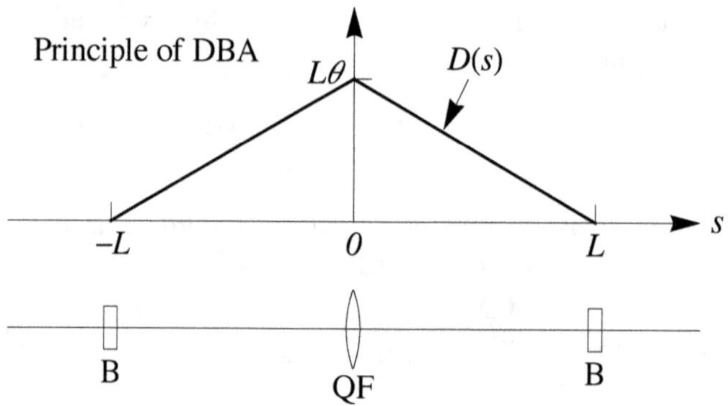

In practice a DBA cell is more complicated — for one thing, without any QD in it, the DBA as shown in Fig. 4.17 is vertically defocused. For example, on each side of the arrangement will likely be inserted some focusing triplets. A more practical design is shown in Fig. 4.18 (upper).

Also shown in Fig. 4.18(lower) is a *triple-bend achromat*, which is a variation of DBA. What happens here is that the middle QF of the DBA is first replaced by a triplet, and then the middle quadrupole of this triplet is split into two halves, with a dipole inserted in between. One purpose of the third inserted bend is to have an easier handle on chromatic corrections.

Homework 4.22 Use the 3×3 formalism to show that, in a thin-lens approximation, (a) the DBA arrangement of Fig. 4.17 fulfills the achromat requirement if $f = \frac{L}{2}$, and (b) the self-induced dispersion inside the arrangement is such that $D_{\text{QF}} = L\theta$.

Figure 4.18: Example designs of a DBA cell (upper) and a TBA cell (lower). The DBA example is being used at the NSLS-II storage ring at BNL. The TBA example was an original design for the ALS storage ring at LBNL. The TBA example forms a complete unit cell. The DBA example is to be used back-to-back with its mirror image to form a unit cell. [Courtesy C. Steier (2019)]

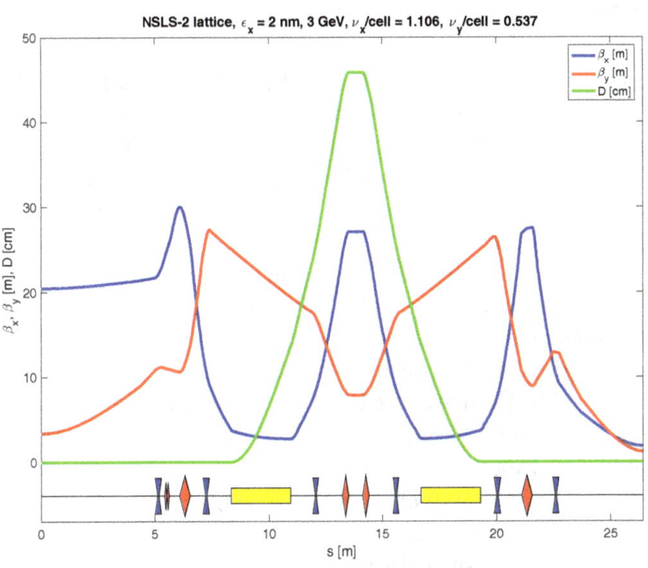

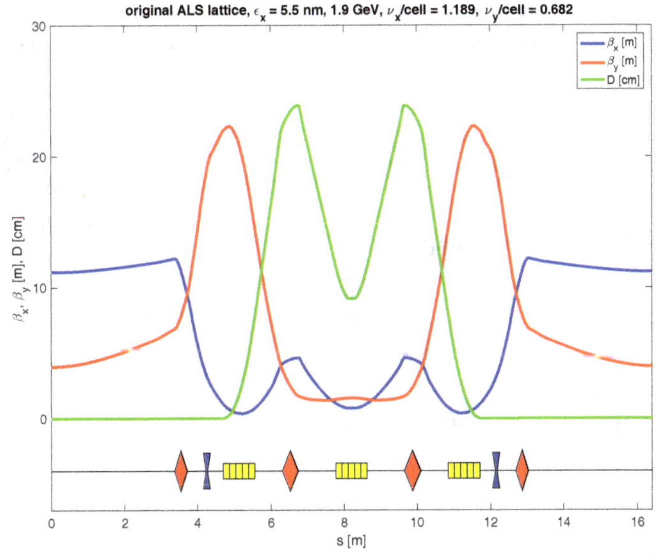

Homework 4.23 Examine the two lattice designs of Fig. 4.18 closer. Based on the plots of the beta- and the dispersion functions, estimate the required strengths of the following magnets,

(a) the dipole bend angle of the DBA lattice,

(b) the focal length of the two middle defocusing quadrupoles of the DBA lattice,

(c) the bending angle of the middle dipole of the TBA lattice.

Solution

(a) From the plot, dispersion at the mid-quadrupole is $D_{QF} \approx 0.5$ m, while the distance between the mid-quadrupole and the dipole is $L \approx 4$ m. The dipole bend angle must then be $\theta \approx \frac{D_{QF}}{L} \approx 8°$ for each of the two dipoles.

Another way to estimate the bend angle is to apply the equation $\Delta D' = \theta$ when dispersion goes across a thin dipole. From the plot, we estimate $\delta D' \approx 0.1$, which gives $\theta \approx 7°$.

(b) $f = \frac{L}{2} \approx 2$ m. But there are two quadrupoles in the middle, so each has $f \approx 4$ m. The additional QFs on the sides of the middle straight section are ignored in this estimate.

Another way to estimate the quadrupole strength is to use the condition $\Delta\beta' = -\frac{2\beta}{f}$, derived in Homework 2.100. Across each of the two middle QFs, we estimate $\Delta\beta' \approx 8$. We also estimate $\beta \approx 17$ m in the middle of the quadrupole. This also yields $f \approx 4$ m.

(c) Apply the expression $\Delta D' = \theta$. The result will have a small error because the middle dipole contains a slight defocusing quadrupole component, as evidenced by the behavior of β_y inside the dipole.

Homework 4.24 Consider a string of n identical cells, each cell containing quadrupoles, bends and drifts. It is of course an achromat if each cell is achromatic.

(a) Otherwise, show that another way for the string to be achromatic is when the betatron 2×2 map of the string is a unit map. Each cell therefore contributes a phase advance of $\frac{2\pi}{n}$.[6]

(b) Are there any additional ways?

Solution (a) Let the 3×3 map of each cell be

$$M_{3\times3} = \begin{bmatrix} M_{2\times2} & \vec{d} \\ 0 & 1 \end{bmatrix}.$$

The map of the string is

$$\begin{aligned} M_{3\times3}^n &= \begin{bmatrix} M_{2\times2}^n & (M_{2\times2}^{n-1} + M_{2\times2}^{n-2} + \cdots + 1)\vec{d} \\ 0 & 1 \end{bmatrix} \\ &= \begin{bmatrix} M_{2\times2}^n & (M_{2\times2}^n - I)(M_{2\times2} - I)^{-1}\vec{d} \\ 0 & 1 \end{bmatrix}. \end{aligned}$$

[6]Roger V. Servranckx and Karl L. Brown, AIP Con. Proc. No. 153, Fermilab 1984, SLAC 1985, p. 121.

Condition of achromaticity requires either $M_{2\times2}^n = I$ or $\vec{d} = \vec{0}$. The case $\vec{d} = \vec{0}$ is the trivial case when each cell is achromatic. The case $M_{2\times2}^n = I$ says the 2×2 part of the string is a unit map.

(b) Yes, there are other ways. For example, one may choose (* means some arbitrary value)

$$M_{2\times2}^n = \begin{bmatrix} 1 & 0 \\ * & 1 \end{bmatrix}, \quad \vec{d} = (M_{2\times2} - I) \begin{bmatrix} 0 \\ * \end{bmatrix},$$

or

$$M_{2\times2}^n = \begin{bmatrix} 1 & * \\ 0 & 1 \end{bmatrix}, \quad \vec{d} = (M_{2\times2} - I) \begin{bmatrix} * \\ 0 \end{bmatrix}.$$

4.7 Chromaticity

In Chapter 3, we said there were two types of error effects, one due to magnetic field errors of the accelerator, the other due to a momentum spread of the beam. Field errors were discussed in Chapter 3. In Chapter 3, we first analyzed a closed orbit distortion due to dipole field errors, then a tune shift and β-beat due to quadrupole errors, and finally we talked about effects due to nonlinear field errors. In this chapter for the chromatic effects, we follow a similar line of analysis.

So far, we have talked about the first chromatic effect equivalent to the closed orbit, i.e. the dispersion caused by dipoles. In this section we will discuss another important chromatic effect equivalent to tune shifts and are caused by the quadrupoles.

Basically what happens is the following. Higher momentum particles ($\delta > 0$) have higher rigidity, and therefore experience weaker effect due to magnetic fields. Dispersion discussed so far is a result of the weakened dipoles. The weakened quadrupoles will introduce chromaticities, i.e. the betatron tunes will depend on δ, and if we keep only the first order perturbation,

$$\nu_{x,y}(\delta) = \nu_{x,y}(0) + \xi_{x,y}\delta, \tag{4.12}$$

where the parameters $\xi_{x,y}$ are the *chromaticities.*

A comment on the notation is in order. The definition (4.12) is the US definition. Another definition commonly used in Europe is $\nu_{x,y}(\delta) = \nu_{x,y}(0)(1+\xi_{x,y}\delta)$, which does not make sense. What we care in beam dynamics and beam stability is the betatron tune. The US definition addresses the tune change with momentum. The European definition addresses the *percentage* change of tune with momentum, but who cares about the percentage tune?

4.7.1 Natural chromaticity

To analyze the problem, we return to Eq. (4.1). In the derivation for the dispersion function, we have kept terms linear in x and δ in Eq. (4.2). Terms proportional to $x\delta$ were considered nonlinear and have been dropped. Now for the chromaticity analysis, we will need to keep these terms. We obtain from Eq. (4.1),

$$x'' + K_x x \approx \frac{\delta}{\rho} + \left(\frac{2}{\rho^2} + \frac{G}{B\rho}\right) x\delta ,$$

where $K_x = \frac{1}{\rho^2} + \frac{G}{B\rho}$ with $G = \frac{\partial B_y}{\partial x}$. The term $\frac{\delta}{\rho}$ on the right hand side was fully analyzed before. The second term on the right hand side can be considered to come from an error quadrupole of strength

$$\Delta K_x = -\left(\frac{2}{\rho^2} + \frac{G}{B\rho}\right)\delta .$$

Similarly, a simpler analysis for the y-motion gives a chromatic perturbation to the vertical focusing strength to be

$$\Delta K_y = \frac{G}{B\rho}\delta ,$$

with the unperturbed focusing strength given by $K_y = -\frac{G}{B\rho}$.

Given these quadrupole perturbations, the tune shift with momentum deviation can be obtained. From those results, we extract the chromaticities,

$$\begin{aligned}
\xi_{x,\mathrm{nat}} &= -\frac{1}{4\pi}\oint ds\, \beta_x(s)\left[\frac{2}{\rho^2(s)} + \frac{G(s)}{B\rho}\right] \\
&\approx -\frac{1}{4\pi}\oint ds\, \beta_x(s)\frac{G(s)}{B\rho} , \\
\xi_{y,\mathrm{nat}} &= \frac{1}{4\pi}\oint ds\, \beta_y(s)\frac{G(s)}{B\rho} ,
\end{aligned} \tag{4.13}$$

where the subscripts "nat" indicate these are the *natural chromaticities* coming from the quadrupole magnets in the lattice.

In the expression for $\xi_{x,\mathrm{nat}}$ of Eq. (4.13), we often drop the $\frac{2}{\rho^2}$ term for strong focusing lattice because it is small. However, in Homework 4.27 for a weak focusing ring, this approximation should not be made.

The reader is reminded that these chromaticities are derived using Eq. (3.17), which holds away from half-integer resonances. Near these resonances, the numerical difference can be substantial, and one must then use Eq. (3.15).

Equation (4.13) tends to give negative values for the natural chromaticities, for both x and y dimensions. Physically this comes about because the magnetic fields have weaker effects for a particle with $\delta > 0$. The betatron tunes, which are provided by the focusing effects of the quadrupoles, therefore become lower with $\delta > 0$. Mathematically, the same conclusion comes about from Eq. (4.13) because β_x tends to be larger at the focusing quadrupoles ($G > 0$) than in the defocusing quadrupoles ($G < 0$), while β_y has the opposite tendency.

We mentioned before that a quadrupole field error causes tune shifts in x and y, and they have opposite signs. Here, in contrast, we have the tune shifts due to chromatic error δ, and we find that they both tend to have same negative sign.

FODO cell For a FODO cell, one can calculate the chromaticities readily,

$$\xi_{x,\text{nat}} \text{ per cell } = \xi_{y,\text{nat}} \text{ per cell } = -\frac{\beta_F - \beta_D}{4\pi f} \,,$$

where

$$\beta_F = \frac{2L(1+\sin\frac{\Phi}{2})}{\sin\Phi}, \quad \beta_D = \frac{2L(1-\sin\frac{\Phi}{2})}{\sin\Phi}, \quad \sin\frac{\Phi}{2} = \frac{L}{2f} \,.$$

We then have

$$\xi_{\text{nat}} \text{ per cell } \approx -\frac{1}{\pi}\tan\frac{\Phi}{2} \quad (\text{both } x \text{ and } y)\,. \tag{4.14}$$

For a 90° FODO lattice, for example, each cell will contribute $-\frac{1}{\pi}$ units to $\xi_{x,\text{nat}}$ and the same amount to $\xi_{y,\text{nat}}$.

For $\Phi \lesssim 90°$, we have

$$\xi_{\text{nat}} \approx -\frac{N_{\text{cell}}\Phi}{2\pi} \approx -\nu \,,$$

where $\nu = \frac{N_{\text{cell}}\Phi}{2\pi}$ is the total betatron tune of the storage ring. We thus conclude the natural chromaticities are roughly equal to the negative of the betatron tunes.

The fact that ξ_{nat} is approximately equal to $-\nu$ can be appreciated as follows. It is easy to understand qualitatively that the focusing effect of quadrupoles becomes weaker for a particle with larger momentum, and that this leads to a reduction of the betatron tunes. But what we now learn in addition is that quantitatively there is a strong tendency for the tune to behave almost exactly proportionally to the effective quadrupole strength, and as a result, the tune is reduced from ν almost exactly to $\frac{\nu}{1+\delta}$. This then leads to $\xi_{\text{nat}} \approx -\nu$.

On the other hand, the conclusion $\xi_{\text{nat}} \approx -\nu$ holds more for a FODO type lattice. If there is a special insertion in the regular FODO array, there is a tendency that ξ becomes even more negative than $-\nu$.

If you see $\xi_{\text{nat}} \approx -\nu$ in a storage ring lattice, you most likely can conclude that this is a regular FODO-like lattice. If ξ_{nat} is more negative than $-\nu$, you can probably guess there are significant special insertions. It is generally unlikely that ν_{nat} is substantially more positive than $-\nu$. In fact, there is a theorem,

$$\xi_{x,\text{nat}} \le -\frac{\nu_x}{2}, \qquad \xi_{y,\text{nat}} \le -\frac{\nu_y}{2} \,. \tag{4.15}$$

See Homework 4.26(b) for a proof of this theorem. If you see a storage ring design with chromaticities with much smaller magnitudes, it is because sextupoles have been added — as to be discussed in the next section, and we are no longer speaking of the natural chromaticities.

Homework 4.25 The text derived the natural chromaticities of a FODO cell by applying the formula (4.13). We may prefer to derive it using first principles. Consider a FODO lattice for an off-momentum particle. All that happens to betatron focusing is that the effective focal length f is to be replaced by $f(1+\delta)$ because the change in magnetic rigidity.

(a) Use this observation to derive Eq. (4.14).

(b) Consider now a FODO channel. Extend the calculation to find the second order chromaticity, i.e. tune shift quadratic in δ, per FODO cell. Ignore the dipole focusing in this calculation.

(c) Find a general expression for higher-order chromaticities to the n-th order.

Solution The phase advance per cell Φ satisfies $\sin\frac{\Phi}{2} = \frac{L}{2f}$. The perturbed phase advance satisfies $\sin\left(\frac{\Phi+\Delta\Phi}{2}\right) = \frac{L}{2f(1+\delta)}$. A simple algebra establishes Eq. (4.14).

Homework 4.26

(a) Use Eq. (2.40) to prove that the natural chromaticities of a storage ring with linear optics and without x-y coupling are necessarily negative for all stable lattices.

(b) Prove that, in fact, the natural chromaticities not only have to be negative, but also they must obey the inequalities in the theorem (4.15).

(c) What are the conditions in the lattice design in order for the equalities $\xi_{x,\text{nat}} = -\frac{\nu_x}{2}, \xi_{y,\text{nat}} = -\frac{\nu_y}{2}$ to hold? Under these conditions, in fact, the equalities can be derived directly using the original Hill's equation and noting that $K \to \frac{K}{1+\delta}$ for chromaticity consideration. Provide that direct derivation.

Solution (a) and (b)

$$\frac{1}{2}\beta\beta'' - \frac{1}{4}\beta'^2 + \beta^2 K = 1$$

$$\Longrightarrow \quad \oint ds\beta K = \oint ds\left(\frac{1+\frac{1}{4}\beta'^2}{\beta}\right) - \oint ds\frac{\beta''}{2}.$$

The second term vanishes.

(c) It occurs when $K(s) =$ constant and $\beta(s) = \frac{1}{\sqrt{K}} =$ constant. It is easy to show that $\xi_{\text{nat}} = -\frac{\nu}{2}$ in that case.

Homework 4.27 Calculate the Courant-Snyder functions, the dispersion function, and the natural chromaticities for a weak focusing ring with field index $1 > n > 0$. This homework will continue in Homework 6.50.

Solution With field index $1 > n > 0$, we have $K_x = \frac{1-n}{\rho^2}$, $K_y = \frac{n}{\rho^2}$, and

$$\begin{aligned}
M_x &= \begin{bmatrix} \cos(2\pi\sqrt{1-n}) & \frac{\rho}{\sqrt{1-n}}\sin(2\pi\sqrt{1-n}) \\ -\frac{\sqrt{1-n}}{\rho}\sin(2\pi\sqrt{1-n}) & \cos(2\pi\sqrt{1-n}) \end{bmatrix}, \\
M_y &= \begin{bmatrix} \cos(2\pi\sqrt{n}) & \frac{\rho}{\sqrt{n}}\sin(2\pi\sqrt{n}) \\ -\frac{\sqrt{n}}{\rho}\sin(2\pi\sqrt{n}) & \cos(2\pi\sqrt{n}) \end{bmatrix}, \\
\beta_x &= \frac{\rho}{\sqrt{1-n}}, \\
\beta_y &= \frac{\rho}{\sqrt{n}}, \\
\alpha_x &= \alpha_y = 0, \\
\nu_x &= \sqrt{1-n}, \\
\nu_y &= \sqrt{n}, \\
D &= \frac{\rho}{1-n}, \\
\alpha_c &= \frac{1}{1-n}, \\
\xi_{x,\text{nat}} &= -\frac{1}{2}\frac{2-n}{\sqrt{1-n}}, \\
\xi_{y,\text{nat}} &= -\frac{\sqrt{n}}{2}.
\end{aligned}$$

Homework 4.28 In the derivation of $\xi_{x,\text{nat}}$ per cell for the FODO cell, we made an approximation by dropping a term $-\frac{1}{2\pi}\oint_{\text{cell}} ds\,\frac{\beta_x(s)}{\rho^2}$, claiming it was a small term. Verify that using thin-lens dipoles in the FODO cell. When might this approximation break down?

4.7.2 Chromaticity correction

In many applications, these natural chromaticities will need to be corrected. A momentum spread in a beam will induce a tune spread if $\xi_{x,y} \neq 0$. For example, if $\xi_x = -30$, then a momentum spread of 1% will give rise to a tune spread of 0.3, which is way too large to avoid the dangerous nonlinear resonances.

Furthermore, a negative $\xi_{x,y}$ (such as those arising from natural chromaticities) also has another unexpected problem called *head-tail instability* (Sec. 8.7.4), which we will have to avoid when the beam intensity is sufficiently high. To mitigate the head-tail instability, we require the chromaticities to be small and slightly positive. For these reasons, chromaticities will need to be corrected — either to zero, or to some small and slightly positive values.

The way to correct chromaticities is to use sextupoles. Unfortunately, sextupoles, although effective to cure the chromaticity disease, have their strong side effects. This is because sextupoles are nonlinear. The troubles with nonlinear effects were discussed in Chapter 3, and for here, we will discuss only their helpful effect, namely the effect of correcting chromaticities. For chromaticity correction, we need two families of sextupoles.

The harmful effects of sextupoles can be partially mitigated by pairing them (Sec. 4.7.3). As the demand of nonlinear corrections becomes higher for high-performance storage rings, however, even the pairing idea becomes insufficient. More intricate arrangements with multiple sextupole families, aimed to cancel out specific higher-order resonance driving terms, become necessary. This subject will be briefly touched upon in Sec. 4.7.4. With not too much exaggeration, however, one might still say that sextupoles are one area where the otherwise clever accelerator physicists are running out of ideas. Beyond a certain point, the only idea seems just to try to live with the devil!

We now proceed to establish the two-family sextupole correction of the chromaticities. A sextupole magnet has fields

$$B_x = Sxy, \qquad B_y = \frac{S}{2}(x^2 - y^2),$$

where $S = \frac{\partial^2 B_y}{\partial x^2}$ specifies the sextupole strength. An off-momentum particle passing through the sextupole has displacements

$$x = x_\beta + D\delta, \qquad y = y_\beta,$$

where subscript β means the betatron component of the displacement, which is δ-independent. The sextupole fields seen by the particle, therefore, are

$$\begin{aligned} B_x &= Sx_\beta y_\beta + Sy_\beta D\delta, \\ B_y &= \frac{S}{2}(x_\beta^2 - y_\beta^2) + Sx_\beta D\delta + \frac{S}{2}D^2\delta^2. \end{aligned}$$

As mentioned, these sextupole fields contain harmful terms, but we shall keep here only the helpful ones, namely

$$B_x = Sy_\beta D\delta, \qquad B_y = Sx_\beta D\delta.$$

The other terms $Sx_\beta y_\beta, \frac{S}{2}(x_\beta^2 - y_\beta^2), \frac{S}{2}D^2\delta^2$ are simply dropped. One should note that not even written down are additional terms coming from closed orbits $\Delta x_{\rm COD}, \Delta y_{\rm COD}$ and misalignments $\Delta x_{\rm align}, \Delta y_{\rm align}$. Sextupoles are truly complicated objects.

To take advantage of these helpful terms, sextupoles must be located at dispersive locations, i.e. $D \neq 0$ at the sextupoles. The horizontal dispersive displacement $D\delta$ has rendered each sextupole to become an effective quadrupole whose strength is $G = \frac{\partial B_y}{\partial x} = SD\delta$. These additional quadrupoles at the sextupoles in turn contribute to tune shifts in both x and y according to

$$\Delta\nu_x = \frac{1}{4\pi}\oint ds \frac{S(s)}{B\rho}\beta_x(s)D(s)\delta.$$

$$\Delta\nu_y = -\frac{1}{4\pi}\oint ds \frac{S(s)}{B\rho}\beta_y(s)D(s)\delta .$$

We recognize these tune shifts with momentum are additional contributions to the two chromaticities. This time, the two contributions, unlike their natural counterparts, have opposite signs.

Combining the contributions of natural and sextupole chromaticities, we obtain the total chromaticities

$$\begin{aligned}\xi_x &= -\frac{1}{4\pi}\oint ds\,\beta_x(s)\left[\frac{2}{\rho^2(s)} + \frac{G(s)}{B\rho} - \frac{S(s)}{B\rho}D(s)\right] ,\\ \xi_y &= \frac{1}{4\pi}\oint ds\,\beta_y(s)\left[\frac{G(s)}{B\rho} - \frac{S(s)}{B\rho}D(s)\right] .\end{aligned}$$

FODO cell Take again a FODO lattice for example. Let us insert a thin SF sextupole next to the thin QF quadrupole, and an SD sextupole next to the QD quadrupole. Define for the thin sextupoles

$$T = \frac{S\ell}{B\rho},$$

then the chromaticities per cell are given by

$$\begin{aligned}\xi_x \text{ per cell} &= -\frac{1}{4\pi}\left(\frac{\beta_F-\beta_D}{f} - \beta_F T_F D_F - \beta_D T_D D_D\right) ,\\ \xi_y \text{ per cell} &= -\frac{1}{4\pi}\left(\frac{\beta_F-\beta_D}{f} + \beta_D T_F D_F + \beta_F T_D D_D\right) ,\end{aligned}$$

where we have again dropped the $\frac{2}{\rho^2}$ term in ξ_x.

In case our goal is to zero the chromaticities, then we set $\xi_x = \xi_y = 0$ and obtain the required sextupole strengths,

$$\begin{aligned}T_F &= \frac{1}{fD_F} = \frac{1}{2f^2\theta}\left(\frac{\sin\frac{\Phi}{2}}{1+\frac{1}{2}\sin\frac{\Phi}{2}}\right) ,\\ T_D &= -\frac{1}{fD_D} = -\frac{1}{2f^2\theta}\left(\frac{\sin\frac{\Phi}{2}}{1-\frac{1}{2}\sin\frac{\Phi}{2}}\right) .\end{aligned} \tag{4.16}$$

One notes that S_F is necessarily positive and S_D negative for a FODO cell, and that S_D is always stronger than S_F.

In this section, we have concentrated on the first-order chromatic shifts of the tunes, first the natural values due to the quadrupoles and then the corrections due to the sextupoles. Obviously the other lattice functions also have chromatic shifts. For example, there are chromatic distortions to the β-functions. They can be obtained from Eq. (3.19) with the perturbations coming from the same quadrupole and sextupole fields.

Homework 4.29 Give a physical explanation why T_F and T_D have the simple solution $T_F = \frac{1}{fD_F}$ and $T_D = -\frac{1}{fD_D}$ in Eq. (4.16).

Homework 4.30 For $\Phi \ll 1$, show that a rough estimate of the sextupole strengths is

$$T_{F,D} \sim \pm \frac{2\nu^3}{R^2},$$

where $2\pi R$ is the storage ring circumference and ν is the total betatron tune. Take $\nu = 10, R = 100$ m, we have $T_{F,D} \sim \pm 0.2$ m^{-2}. Give a physical argument why this scaling makes sense.

Homework 4.31 An electron synchrotron consists of 48 FODO cells with quadrupole gradient 12 T/m and 0.5 m in quadrupole length. The distance between the centers of adjacent quadrupoles is 5.5 m. The magnetic rigidity is 24 T-m and dipole length is 3 m.

(a) Evaluate phase advance per cell, betatron tunes of the ring, β-functions at quadrupole centers, dispersion functions at quadrupole centers, and natural chromaticities.

(b) Put thin-lens sextupoles at quadrupole centers to correct chromaticity to zero. Find the integrated sextupole field strengths.

(c) If the estimated working tune is not as good as expected, what would you do?

(d) Estimate the expected rms closed orbit errors due to 0.5 mm rms alignment errors in quadrupoles.

(e) Calculate the expected rms tune shift due to 1% rms quadrupole field errors.

Solution

(a) $\beta_{F,D} = 18.6$ m, 3.44 m; $\nu_{x,y} = 11.582$; $D_{F,D} = 1.02$ m, 0.50 m; $\xi_{x,y,\text{nat}} = -14.46$.

(b) $T_F = 0.24$ m^{-2}, $T_D = -0.50$ m^{-2}.

(d) $y_{\text{COD}} \approx \frac{\beta_y \Delta y}{2\sqrt{2}f|\sin \pi\nu|}\sqrt{N_{\text{quad}}} = 3.0$ mm if we take $\beta_y \approx \frac{R}{\nu}$.

(e) $\Delta\nu_{x,y} \approx 0.14$.

Homework 4.32 We derived the natural chromaticities for a nominal FODO cell, and $\xi_{x,\text{nat}}$ and $\xi_{y,\text{nat}}$ are equal.

(a) Derive the expressions of the natural chromaticities for a modified FODO cell whose QF has focal length f_1 and QD has focal length $-f_2$ (therefore $\Phi_x \neq \Phi_y$).

(b) Find the sextupole strengths T_F and T_D that compensate for the natural chromaticities.

Solution

$$\beta_{xD} = \frac{L}{\sin \Phi_x}\left(2 - \frac{L}{f_1}\right),$$

$$\begin{aligned}
\beta_{yD} &= \frac{L}{\sin\Phi_y}\left(2+\frac{L}{f_1}\right), \\
\beta_{xF} &= \frac{L}{\sin\Phi_x}\left(2+\frac{L}{f_2}\right), \\
\beta_{yF} &= \frac{L}{\sin\Phi_y}\left(2-\frac{L}{f_2}\right), \\
\xi_{x,\mathrm{nat}} &= -\frac{L(L-f_1+f_2)}{2\pi f_1 f_2\sin\Phi_x}, \\
\xi_{y,\mathrm{nat}} &= -\frac{L(L+f_1-f_2)}{2\pi f_1 f_2\sin\Phi_y}, \\
T_F &= \frac{1}{D_F f_1}, \\
T_D &= -\frac{1}{D_D f_2}.
\end{aligned}$$

4.7.3 Sextupole pair

Sextupoles at dispersive locations provide a way to control the chromaticities as described in the previous section. But also as mentioned, sextupoles are nonlinear devices, and they impose some necessary evils on the beam dynamics. One way to minimize their nonlinear effects on beam dynamics is to form pairs of sextupoles in such a way that their harmful effects cancel each other as much as possible while, hopefully, the helpful effects are maintained.[7]

Undesired nonlinearities cancel Consider two thin sextupoles of strengths S_a and S_b at locations s_a and s_b. A particle arriving at the first sextupole receives an angular kick, and its trajectory is changed according to

$$\begin{aligned}
x_2 &= x_1\,, \\
x_2' &= x_1' - \frac{T_a}{2}(x_1^2 - y_1^2)\,, \\
y_2 &= y_1\,, \\
y_2' &= y_1' + T_a x_1 y_1\,,
\end{aligned} \tag{4.17}$$

where subscripts 1 and 2 refer to before and after the sextupole kick.

We now design the lattice so that the transport map between the two sextupoles is $-I$ for both the horizontal and the vertical dimensions. Then the coordinates of the particle when it arrives at the second sextupole will be

$$\begin{bmatrix} x_3 \\ x_3' \\ y_3 \\ y_3' \end{bmatrix} = - \begin{bmatrix} x_2 \\ x_2' \\ y_2 \\ y_2' \end{bmatrix}.$$

[7]K.L. Brown, IEEE Trans. Nucl. Sci. Vol. NS-26, No.3 (1979).

As the particle then passes through the second sextupole, its trajectory changes according to

$$\begin{aligned} x_4 &= x_3 \,, \\ x_4' &= x_3' - \frac{T_b}{2}(x_3^2 - y_3^2) \,, \\ y_4 &= y_3 \,, \\ y_4' &= y_3' + T_b x_3 y_3 \,, \end{aligned}$$

which can be written in terms of the particle's initial coordinates as

$$\begin{aligned} x_4 &= -x_1 \,, \\ x_4' &= -x_1' - \frac{T_b - T_a}{2}(x_1^2 - y_1^2) \,, \\ y_4 &= -y_1 \,, \\ y_4' &= -y_1' + (T_b - T_a) x_1 y_1 \,. \end{aligned}$$

It is clear that if we choose the transport map between the two sextupoles to be $-I$ and let the two thin sextupoles to have the same strength $T_a = T_b$, then their nonlinear effects cancel each other perfectly. This then forms the basis of using sextupole *pairs* to control chromaticities with minimum side effects of their nonlinearities. What remains to be shown, as will be done momentarily, is the fact that the beneficial effects of this sextupole pair do not also vanish by the pairing.

Note that the above cancellation relies however on the following conditions:

1. The sextupoles are thin. When the required sextupole strength is too strong, so that the sextupoles need to be long, this cancellation will be compromised.

2. The transport map between the sextupoles is exactly $-I$. This requires of course the lattice design and engineering construction to be accurate. In addition, more importantly, is the fact that most likely this condition, even fulfilled, can only be fulfilled for on-momentum particles. Cancellation of sextupole nonlinearities for off-momentum particles therefore is not perfect, leading to important residual chromatic nonlinearities to be dealt with.

3. When more than one pair is needed, sextupole pairs must not be interlaced. If interlaced, high order nonlinearities will leak out of the interlaced sextupole-pairs region. In Fig. 4.19, the top panel has imperfect cancellation, while the lower two cases maintains cancellation, at least for on-momentum particles.

Figure 4.19: A few possible configurations of interlaced and noninterlaced sextupole pairs.

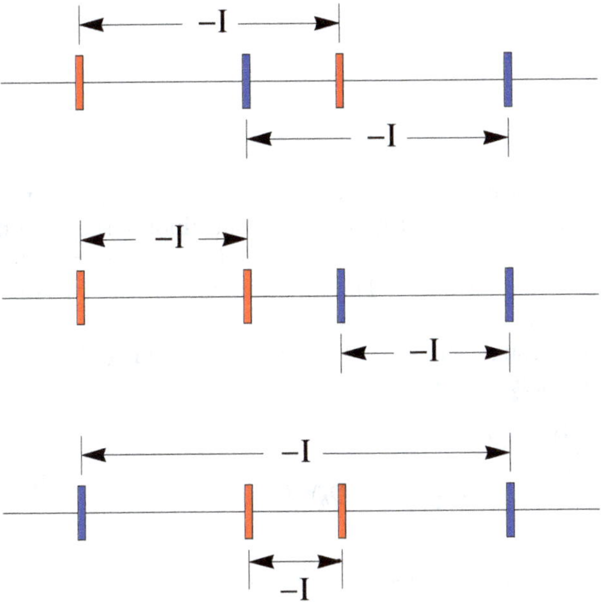

Desired nonlinearities add We want to use the sextupoles to control chromaticities. The above analysis applies to an on-momentum particle. For an off-momentum particle, it is to be modified. In particular, we want the two sextupoles to contribute to chromaticities, and their contributions must not also cancel like the paired nonlinearities.

For an off-momentum particle, its x, x' coordinates contains additional dispersion terms,

$$\begin{aligned} x_1 &= x_{1\beta} + D_a\delta\,, \\ x_1' &= x_{1\beta}' + D_a'\delta\,. \end{aligned}$$

The kick at the first sextupole is still given by Eq. (4.17). The transformation to the second sextupole requires some thought. The betatron part of the coordinates are transformed with a $-I$ transformation — here we ignore the energy dependence of the $-I$ map. But then there is also a dispersion component to be considered. This is done by the 3×3 map [see Eq. (4.7)]

$$\begin{bmatrix} -1 & 0 & D_a + D_b \\ 0 & -1 & D_a' + D_b' \\ 0 & 0 & 1 \end{bmatrix} .$$

This means

$$
\begin{aligned}
x_3 &= -x_2 + (D_a + D_b)\delta\,, \\
x'_3 &= -x'_2 + (D'_a + D'_b)\delta\,, \\
y_3 &= -y_2\,, \\
y'_3 &= -y'_2\,,
\end{aligned}
$$

where D_b, D'_b are the dispersions at the second sextupole. The fact that we have not imposed any conditions on them relative to D_a, D'_a implies that there must be dipoles between the two sextupoles. Without dipoles, they will necessarily be related by $D_b = -D_a, D'_b = -D'_a$ just like the betatron part of the coordinates, and that is not what we want. The dipoles between the two sextupoles are a necessary requirement for the scheme.

We then go through the second sextupole, receiving the second sextupole kick. The final result is

$$
\begin{aligned}
x_4 &= -x_1 + (D_a + D_b)\delta\,, \\
x'_4 &= -x'_1 + (D'_a + D'_b)\delta - \frac{T_b}{2}(D_a + D_b)^2\delta^2 + T_b(D_a + D_b)x_1\delta + \frac{T_a - T_b}{2}(x_1^2 - y_1^2)\,, \\
y_4 &= -y_1\,, \\
y'_4 &= -y'_1 - (D_a + D_b)T_b y_1\delta - (T_a - T_b)x_1 y_1\,. \qquad (4.18)
\end{aligned}
$$

There are many nonlinear terms contained in this sextupole pair, as evidenced by Eq. (4.18). As mentioned, some cancellation of the undesired nonlinearities require $T_b = T_a$. Equation (4.18) then becomes

$$
\begin{aligned}
x_4 &= -x_1 + (D_a + D_b)\delta\,, \\
x'_4 &= -x'_1 + (D'_a + D'_b)\delta - \frac{T_a}{2}(D_a + D_b)^2\delta^2 + T_a(D_a + D_b)x_1\delta\,, \\
y_4 &= -y_1\,, \\
y'_4 &= -y'_1 - (D_a + D_b)T_a y_1\delta\,. \qquad (4.19)
\end{aligned}
$$

It should be pointed out that the sextupole paring and the choice of $T_b = T_a$ are very helpful, but they do not eliminate all of the undesired nonlinearities. Two nonlinear terms still exist in the x'_4 expression. The nonlinear term proportional to $x_1\delta$ is the useful term for chromaticity correction. The other chromatic term proportional to δ^2 still can cause trouble. In addition, the effects mentioned earlier of the finite sextupole lengths and the chromatic dependence of the $-I$ map are not addressed here. The slope quantities D'_a, D'_b do not contribute to any net nonlinear effects, as one might have guessed.

For an off-momentum particle, its coordinate x_1 contains a term $D_a\delta$. Coordinate x'_2 therefore contains a term $-T_a x_{1\beta} D_a\delta$. This term acts like a quadrupole with strength $\frac{1}{B\rho}\frac{\partial B_y}{\partial x} = T_a D_a \delta$, and contributes chromaticities $\xi_x = \frac{\beta_x}{4\pi}T_a D_a$, $\xi_y = -\frac{\beta_y}{4\pi}T_a D_a$.

When the particle arrives the second sextupole, its x_3 acquires the extra term $D_b\delta$. An additional contribution to the chromaticities is to be added. Since the β-functions at the two sextupole locations are the same due to the $-I$ map between them, the total contributions of the sextupole pair to the chromaticities are

$$\begin{aligned} \xi_x &= \frac{\beta_x T_a}{4\pi}(D_a + D_b)\,, \\ \xi_y &= -\frac{\beta_y T_a}{4\pi}(D_a + D_b)\,. \end{aligned}$$

If the transport between the two sextupoles, although flipping the signs of the betatron coordinates, keeps their dispersion functions free, i.e. if $D_b \neq -D_a$ (as mentioned, this requires having dipoles in this transport section), then this sextupole pair can be used to control chromaticities.

Homework 4.33

(a) Show that the nonlinear map of a thin-lens sextupole, Eq. (4.17), is symplectic.

(b) Show that the nonlinear map of the combined sextupole pair, Eq. (4.18), is symplectic.

Homework 4.34

(a) Repeat the above thin-lens sextupole pairs to see if there is a perfect cancellation condition when we choose the transport map to be $+I$ between the two sextupoles. How do we choose their strengths for a perfect cancellation in this case?

(b) In case (a) above, how would these sextupoles be used for chromaticity compensation? Is there a sign requirement on the dispersion functions at the two sextupole locations?

(c) Would the choice $-I$ for x-dimension and $+I$ in the y-dimension still minimize the sextupole nonlinear effects?

4.7.4 Dynamic aperture

Consider a particle moving along the closed orbit with the design momentum, it will circulate around the storage ring turn after turn. We emphasized in Sec. 1.2.2 that a storage ring must not only provide a closed orbit, but also assure stable motion of all 6-D small phase-space deviations about the design closed orbit in order for it to be declared stable.

Question arises as to how small the small deviations have to be, and as these deviations increase, when does the motion become unstable. To address this issue, we look for the largest deviations allowed for stable motion in three dimensions. In the x-dimension, the largest deviation amplitude, designated by A_x, will be called the horizontal dynamic aperture. It is envisioned that the two

Figure 4.20: Various ways to present the result of dynamic aperture studies. Shaded regions are stable for particle motion.

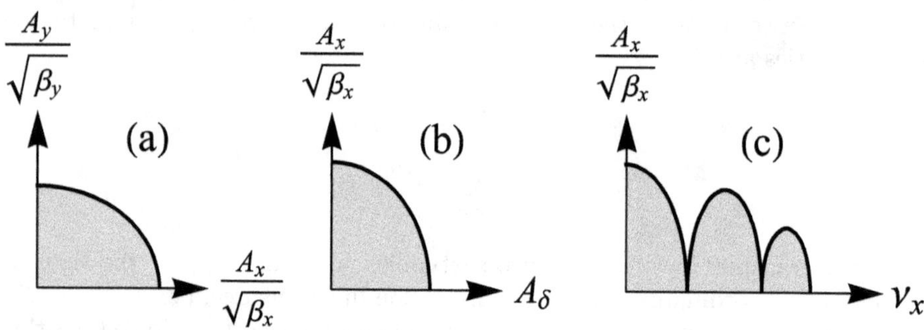

phase-space coordinates x and x' are intimately linked and transforming into each other rapidly by the betatron oscillation, so that a single limit A_x describes the horizontal dynamic aperture. The quantity A_x will depend on the location where the x-amplitude is observed. In a storage ring without x-y coupling, however, it is expected that A_x at position s simply scales with $\sqrt{\beta_x(s)}$, so it suffices to select a reference position to define A_x for the whole storage ring.

Similarly, we can define a vertical dynamic aperture A_y at the same reference position. There is also a maximum δ that allows stable motion in the storage ring. The maximum value, designated as A_δ, is called the momentum aperture. The three apertures A_x, A_y, A_δ can collectively be referred to as the dynamic apertures of the storage ring.

There are several ways to present the results of dynamic apertures. To emphasize the transverse motion, we can plot the stable region in the $\left(\frac{A_x}{\sqrt{\beta_x}}, \frac{A_y}{\sqrt{\beta_y}}\right)$ space. The result may schematically look like Fig. 4.20(a). Similarly, if we emphasize the momentum aperture, the result may look like Fig. 4.20(b). Figure 4.20(c) is a plot of 1-D dynamic aperture in the x-dimension, with emphasis on its dependence on the choice of tune working point. A couple of resonances appear as the dynamic aperture dips.

The study of dynamic apertures is a critical and specialized subject in accelerator physics.[8] We shall not discuss it in any detail, as it is considered to be beyond the scope of these lectures. In the remaining of this section, we only present an outline of the logic flow and the direction of efforts required. For this discussion, let us focus on the case illustrated in the case of Fig. 4.20(b). Evolution of the dynamic aperture studies can probably be presented as the following steps.

[8]This is a very extensive subject. See, for example, A. Wolski, Handbook Accel. Phys. & Eng., 2nd Ed., Sec. 2.3.9, World Scientific (2013).

Figure 4.21: Evolution of the dynamic aperture studies. Shaded regions are stable for particle motion.

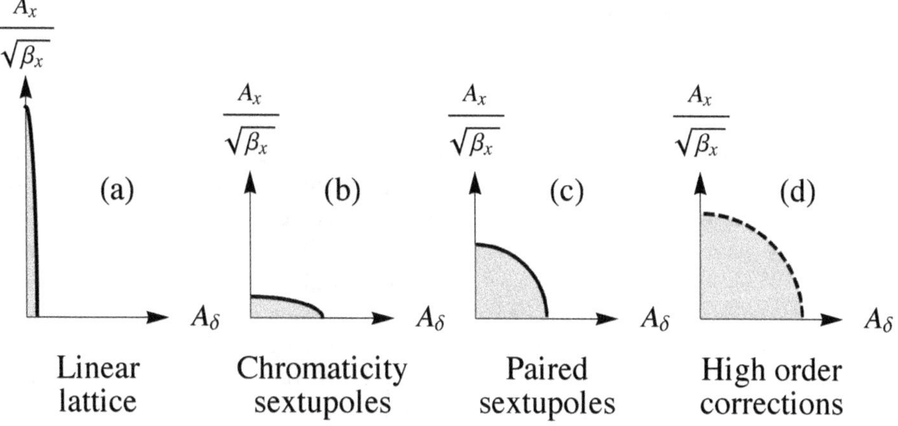

1. Linear lattice: When we get started, we had a linear lattice. The storage ring has only dipoles and quadrupoles. All motion are stable by the Courant-Snyder analysis. However, as mentioned in Sec. 4.7.2, this linear lattice will have a very small momentum aperture because of the large natural chromaticities. The dynamic aperture for this happy linear lattice will look like the sketch in Fig. 4.21(a), and is inappropriate for any real beams.

2. Chromaticity sextupoles: As mentioned in Sec. 4.7.2, in order to correct for the natural chromaticities, we add two families of chromaticity sextupoles to the lattice. As a result, the momentum aperture is enlarged. However, the nonlinear effects of these sextupoles make the betatron aperture shrink, as illustrated in Fig. 4.21(b).

3. Sextupole pairing: The clever idea of pairing the sextupoles then is applied as discussed in Sec. 4.7.3. The result is that the betatron dynamic aperture increases, as illustrated in Fig. 4.21(c). The momentum aperture is basically unaffected by the pairing. There remains two families of sextupoles in the storage ring at this stage. In addition to the paired sextupole, magnet field error and magnet misalignment effects are to be included here.

4. High-order nonlinearities: Up to this step, the lattice design is progressing in a relatively well-defined manner, each step has been with well-defined purpose and methodology. Even with pairing, however, sextupoles often still cause harm to particle motion by their residual nonlinear perturba-

tions. The reason sextupoles are so potentially harmful is because they are often very strong. Unlike most of the other nonlinear perturbations that come from construction or installation errors, these sextupoles are intentional installations and their strengths are designed to compensate for the large natural chromaticities. Their strengths are typically very much larger than the error fields.

To make the situation worse, many modern applications of accelerators require high brightness of the beam, which demands small transverse beam sizes. Small transverse beam sizes demand high betatron tunes because of the need of smaller β-function (beam size $\propto \sqrt{\beta}$, and $\beta \sim \frac{R}{\nu}$). But as we have learned, a higher tune reduces the dispersion function even more ($D \sim \frac{R}{\nu^2}$) than it reduces the beta-functions. Sextupole strengths to compensate for the natural chromaticities, as we have shown, are inversely proportional to the dispersion function, so they steeply become stronger in these modern applications. For these applications, the dynamic aperture obtained even after sextupole pairing, as shown in Fig. 4.21(c) is not sufficient.

Furthermore, in the discussion so far, we have implicitly assumed that δ, the off-momentum deviation, is a constant as a particle circulates around the storage ring. This is not necessarily the case. In synchrotrons and most storage rings, δ executes synchrotron oscillation (Chapter 5). For those cases, it is very important to include synchrotron oscillation in the evaluation of momentum aperture. An oscillating δ has much richer dynamical effects to worry about than a static δ. Momentum aperture tends to become further restricted when synchrotron oscillation is included.

To improve both the betatron and the momentum apertures, sophisticated schemes are invented to deal with the various higher order nonlinearities and resonances. Unlike the previous steps, here there is presently not a well-defined routine to follow, although there are many cases of success in actual operation. A price to pay is there typically requires a much more sophisticated scheme with several families of sextupoles, and sometimes also octupoles. The end result is enlarged dynamic aperture, as illustrated in Fig. 4.21(d). We shall not discuss this subject, considering it beyond our scope.

5. Oscillating perturbations: As just alluded to, oscillating coordinates much enrich the resonance dynamics, and can be sources of detrimental effects for particle stability. Synchrotron oscillation of δ is a prime example. Oscillations or vibrations of other external parameters can be equally detrimental; examples include ground vibration, power supply ripple, noise from RF system, noise from water heating. These effects may sometimes also affect the dynamic aperture.

Chapter 5

Longitudinal Motion

So far we have studied the 4-D transverse (x, x') and (y, y') motions in Chapters 2 and 3. For beam stability, we know we must study 6-D phase space, and we still have the remaining 2-D to study, namely the motion involving the coordinate z, the longitudinal displacement of a particle relative to the trajectory of the idealized reference particle. This motion is longitudinal, and is called *synchrotron* motion.

By drawing an analogy with the transverse analysis, we naturally expect the longitudinal dynamics will be described by the coordinates z and z'. Indeed this is what we will do, except that the quantity z', meaning the rate at which z changes as a particle moves along s, is not easy to visualize. This is not the case for x' and y' which are easily visualized as slopes of the direction of motion in the transverse planes. So for the longitudinal analysis, we look for a quantity closely related to z' but something easier to visualize, and what we found convenient is the momentum error $\delta = \frac{\Delta P}{P_0}$. The quantity δ has been used when we discussed the dispersion effects in Chapter 4. Exactly how z' is related to δ will become clear soon.

Again by drawing analogy to the transverse analysis, we expect to introduce the longitudinal Courant-Snyder functions, which presumably we designate as $\alpha_z(s), \beta_z(s), \gamma_z(s)$. This is indeed possible, but is also not what we normally do. One main difference between synchrotron and betatron motions is the fact that synchrotron motion tends to be very slow. From relativistic beam dynamics, it turns out that it is very difficult to exert longitudinal forces to particles. A magnetic field exerts force in the direction $\vec{v} \times \vec{B}$ and is therefore necessarily transverse. Whereas both electric and magnetic fields can be used to control the transverse motion of particles, electric field is the only choice we have to control longitudinal motion.

However, electric forces are generally not very effective in influencing particle's motion for relativistic beams (Sec. 2.1.4). As a result, particles do not respond quickly and their longitudinal motion tends to be very slow, not only slower than the transverse motion, but also slower than the particle's revolution around the storage ring. In other words, betatron motions controlled by magnetic fields are much faster than the beam's revolution around the circular accelerator, which is in turn much faster than the synchrotron motion. In terms of the betatron tunes $\nu_{x,y}$ and synchrotron tune ν_s, we have

$$\nu_{x,y} \gg 1 \gg \nu_s \,. \tag{5.1}$$

Incidentally, here one finds another cosmetic asymmetry between transverse and longitudinal languages: we customarily denote the longitudinal tune as ν_s instead of the more analogous ν_z.

The fact that $\nu_{x,y} \gg 1$ comes from the strong focusing principle. The fact that ν_s is small but allows focusing at all comes from the phase stability principle, which we will elaborate on in Sec. 5.3.1. These two principles constitute the prime cornerstones of all modern accelerators.

It might be pointed out here that, ironically, as will become clear at a later point (although we will not emphasize it again), the strong focusing that pushes up the values of the betatron tunes contributes partly to *lower* the value of ν_s. The higher is ν_x, the lower will be ν_s.

Later in Chapter 7, we shall introduce a hierarchy of time scales in an electron storage ring. Equation (5.1) will then be part of that hierarchy.

Because the synchrotron motion is slow, we can ignore the s-dependent effects around the storage ring. This allows us to avoid the Courant-Snyder analysis for synchrotron motion. The fact $\nu_s \ll 1$ allows us to consider one revolution as a single "small time step" when we analyze the synchrotron motion, while the same is not allowed for analyzing the transverse motions. Synchrotron motion is therefore simpler to analyze than the betatron motion in this regard. A longitudinal Courant-Snyder transformation can often be bypassed (see however a discussion in Sec. 5.3.6).

5.1 Radio-frequency cavity

Consider a storage ring with an accelerating radio-frequency (RF) cavity as sketched in Fig. 5.1. As a beam circulates around the storage ring, it receives an RF voltage each time it traverses the RF cavity.

The RF voltage applied to the particles as they traverse the RF cavity is sinusoidal in time t,

$$V(t) = V_0 \sin \omega_{\text{rf}} t \,.$$

The RF frequency ω_{rf} is chosen to be an integral multiple of the revolution frequency ω_0,

$$\omega_{\text{rf}} = h\omega_0 \,,$$

where the integer h is called the *harmonic number*. The reason of this integral relationship is so that the beam nominally receives exactly the same voltage at all passages turn after turn. It is clear that one would want to impose this

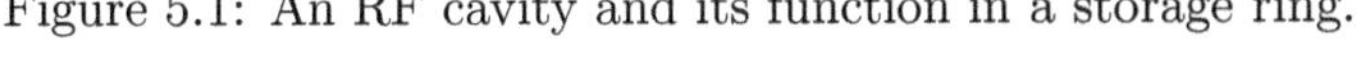

Figure 5.1: An RF cavity and its function in a storage ring.

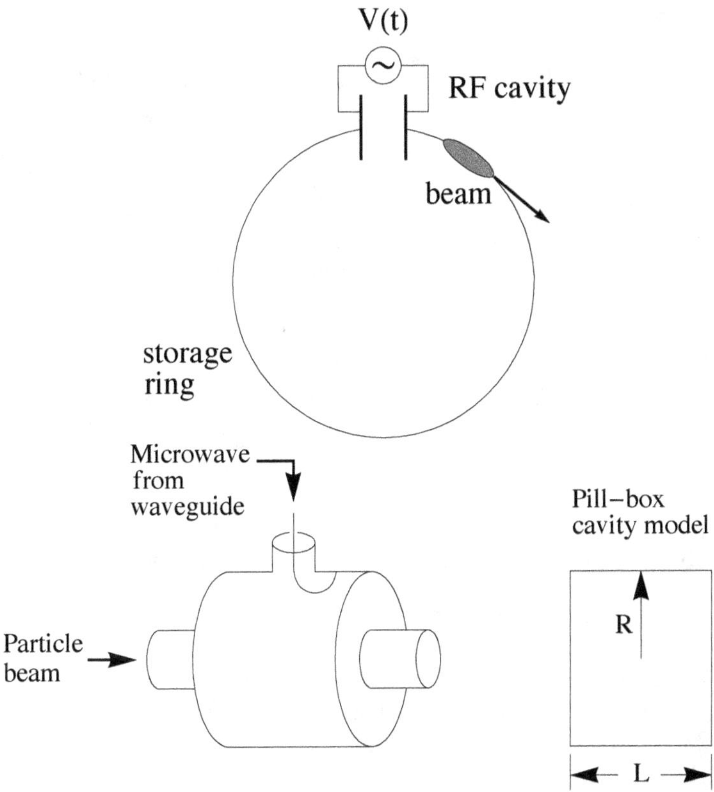

condition if one wants to continually accelerate the beam at each passage. We will return to the harmonic number when we address the phase stability principle later.

As will be elaborated in this chapter, it turns out that this synchronization condition is needed even if we do not accelerate the beam but just to store the beam in the storage ring, the latter in fact being the more common situation. In cases when the beam is not to be accelerated but only to be stored, the phase of the RF voltage will be such that the center of the beam arrives at the RF cavity at times when $\sin\omega_{\text{rf}}t = 0$.

5.1.1 Cavity mode

A simplified model of the RF cavity is a *pillbox cavity* with length L and radius R as illustrated in Fig. 5.1. We first assume the cavity wall is perfectly conducting. Inside the cavity, we will excite an electromagnetic mode with frequency ω_{rf}.

This excited mode is to be driven by sending microwave of the same frequency from a waveguide.

Field pattern In addition to the mode frequency, the mode in the cavity is also characterized by its electromagnetic field's spatial pattern. For a mode, the field in the cavity oscillates with the mode frequency while conforming to this spatial pattern at all times. For the mode we are interested in, the field pattern is particularly simple. Its electric field has only the z-component and its magnetic field has only a θ-component, i.e. $\vec{E} = E_z\hat{z}$ and $\vec{B} = B_\theta\hat{\theta}$. Furthermore, both E_z and B_θ depend on r and t, and do not depend on z and θ. Inside the cavity, Maxwell equations require

$$\nabla \times \vec{B} = \frac{1}{c^2}\frac{\partial \vec{E}}{\partial t} \quad \Longrightarrow \quad \frac{1}{r}\frac{\partial}{\partial r}(rB_\theta) = \frac{1}{c^2}\frac{\partial E_z}{\partial t},$$

$$\nabla \times \vec{E} = -\frac{\partial \vec{B}}{\partial t} \quad \Longrightarrow \quad \frac{\partial E_z}{\partial r} = \frac{\partial B_\theta}{\partial t}.$$

In a pure oscillating mode with frequency ω (we drop the subscript "rf" temporarily), we let all field components to behave in time like

$$E_z(r,t) = E_z(r)e^{-i\omega t}, \qquad B_\theta(r,t) = B_\theta(r)e^{-i\omega t}.$$

Both fields are the same at all z-positions, and both oscillate in time up and down with frequency ω, while rigidly holding the spatial pattern. All fields here are expressed in complex notation. Only the real parts are meaningful.

Note the difference here from the case of the wire formula Eq. (1.13), when we applied the complex numbers to represent the 2-D magnetic field. In Eq. (1.13), both the real and imaginary parts are physically meaningful.[1]

Maxwell equations then become

$$\begin{aligned} \frac{1}{r}\frac{\partial}{\partial r}(rB_\theta) &= -\frac{i\omega}{c^2}E_z, \\ \frac{\partial E_z}{\partial r} &= -i\omega B_\theta. \end{aligned}$$

These equations are to be solved to find the electromagnetic field mode pattern. It is a standard technique in waveguide and cavity analysis that after writing out the Maxwell equations, the thing to do is to first try to express the transverse field components ($E_r, E_\theta, B_r, B_\theta$ in cylindrical coordinates) in terms of the longitudinal components (E_z, B_z). Then one tries to find the wave

[1]Complex numbers are used in physics in many different ways. They are not all the same and cover a wide range from simple shorthand to deep mathematics, and from merely simplifying derivations to insight of theoretical physics. Further extension of the complex numbers involves the use of quaternions. Discussion of the use of complex numbers and quaternions in physics should deserve a separate discussion.

equations for E_z and B_z and solve them with appropriate boundary conditions. The transverse field components are then calculated using the E_z, B_z obtained.

This procedure is particularly handy for the pillbox cavity case at hand. By eliminating B_θ, we obtain the wave equation for E_z,

$$\frac{\partial^2 E_z}{\partial r^2} + \frac{1}{r}\frac{\partial E_z}{\partial r} + \frac{\omega^2}{c^2}E_z = 0 \, .$$

The solution for E_z is given by the Bessel function,

$$E_z(r) = E_0 J_0\left(\frac{\omega}{c}r\right) ,$$

where the electric field strength is specified by E_0. The electric field E_z is a function of r, and reaches its maximum along the cavity axis $r = 0$. Boundary condition requires $E_z = 0$ at $r = R$ at all times t. This determines the mode frequency,

$$\frac{\omega}{c}R = 2.405 \qquad \text{or} \qquad \omega = 2.405\,\frac{c}{R} , \tag{5.2}$$

where 2.405 is the first root of $J_0(x) = 0$.

The frequency of this simplest cavity mode (the *fundamental mode*) depends only on the cavity radius R, and is independent of cavity length L. For example, regardless of the length L, a pillbox cavity of radius $R = 30$ cm will have a fundamental mode frequency of $f_0 = \frac{\omega_0}{2\pi} \approx 380$ MHz. Cavity of smaller radius will have higher mode frequency.

Given $E_z(r)$, the magnetic field is found by

$$B_\theta(r) = \frac{i}{\omega}\frac{dE_z(r)}{dr} = -\frac{iE_0}{c}J_1\left(\frac{\omega}{c}r\right) ,$$

where we have used the property $J_0'(x) = -J_1(x)$.

Field oscillation phase The fact that B_θ contains an extra factor of $-i$ means that its oscillation in time is 90° out of phase with respect to E_z. Indeed by taking real parts of the expressions, we have

$$E_z(r) = E_0 J_0\left(\frac{\omega}{c}r\right)\cos\omega t, \qquad B_\theta(r) = -\frac{E_0}{c}J_1\left(\frac{\omega}{c}r\right)\sin\omega t \, .$$

Let us make a comment about the relative phase between electric and magnetic fields in two different cases. Here, in an RF cavity, $\vec{E}$ and $\vec{B}$ oscillate in an alternating way (90° out of phase). This is in sharp contrast to a laser in free space, when $\vec{E}$ and $\vec{B}$ oscillate together (in phase). This is an important difference, often not emphasized in textbooks. The reader should contemplate what causes this difference.

Let us extend this comment a bit further. In a closed cavity like in the present case, field energy is conserved. The sum of electric and magnetic field energies is a constant. If one of them increases, the other must decrease, and vice versa. So the fact that the fields oscillate out of phase is expected simply from energy conservation. But now how does energy conserve in a laser in free space when both electric and magnetic field energies go up and down together?

Returning to the discussion of RF cavity, the evolution of E_z and B_θ is sketched in Fig. 5.2. The upper panel shows the moment when $E_z = 0$ throughout the cavity. At this moment of time, B_θ reaches its maximum. Nowhere on the cavity wall has accumulated any surface charge σ. The instantaneous wall currents J that produce the magnetic field are sketched as red arrows.

Figure 5.2: Instantaneous snapshots of the magnetic field (upper) and electric field (lower) in the fundamental mode of pillbox RF cavity.

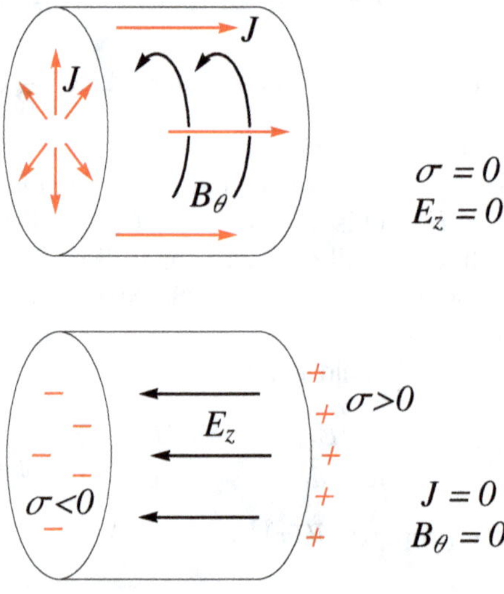

The wall currents shown in the upper panel of Fig. 5.2 lead to surface charges accumulated at the two end plates of the pillbox cavity, one end plate accumulating positive charges and the other one accumulating negative charges, which in turn induce a longitudinal electric field with a polarity shown in the lower panel of Fig. 5.2, taken 90° phase later. Magnetic field and wall current vanish at this moment. The oscillating cavity field then follows a repetition of the two patterns of Fig. 5.2 with alternating signs.

At no instance of time will the outer cylindrical walls carry any electric charges. Surface charges, if any, are accumulated only at the end plates. Convince yourself of this and understand its reason. Direct your reasons toward a conclusion that uniform currents never lead to charge accumulation — equation of continuity (1.4).

Stored field energy The field energy stored in the RF cavity can be obtained by integrating the field energy density at either time moments of Fig. 5.2. Take the moment when $B_\theta = 0$, for example, the stored field energy is found to be

$$\begin{aligned} U &= \frac{\epsilon_0}{2} \int dV \, E_z^2 \\ &= \frac{\epsilon_0}{2} E_0^2 \pi R^2 L J_1^2(2.405) , \end{aligned} \tag{5.3}$$

where $J_1(2.405) = 0.52$, and we have used the identity

$$\int_0^1 x dx \, J_0^2(\alpha x) = \frac{1}{2} J_0^2(\alpha) + \frac{1}{2} J_1^2(\alpha) .$$

Compared with a pillbox cavity uniformly filled with a field of E_0, our cavity has a stored energy a factor of $J_1^2(2.405) = 0.25$ less.

Homework 5.1 The field energy density in the cavity at an arbitrary time t is given by

$$u(r,t) = \frac{\epsilon_0 E_0^2}{2} \left[J_0^2(\frac{\omega}{c} r) \cos^2 \omega t + J_1^2(\frac{\omega}{c} r) \sin^2 \omega t \right] ,$$

where the first term is the electric field energy and the second term is magnetic.

(a) Compute the total field energy in the pillbox cavity at an arbitrary time t,

$$U(t) = L \int_0^R 2\pi r dr \, u(r,t) .$$

Show that $U(t)$ is constant in t, as it should.

(b) Plot $ru(r,t)$ as a function of r for a few different time instances. The area under the curve should be the same for all t. Obtain from this plot a feel of how the field energy is flowing inside the cavity. Compare what you learned with the Poynting vector $\vec{S} = \frac{1}{\mu_0} \vec{E} \times \vec{B}$.

Solution (b) See Fig. 5.3. The green, blue, red curves are for $\omega t = 0, \frac{\pi}{4}, \frac{\pi}{2}$, respectively. The electric field energy concentrates more toward the cavity center, while the magnetic field energy lies more toward the cavity's outer wall. There is a node of field energy pattern at $\frac{r}{R} \approx 0.6$, which occurs with the condition $J_0(2.405 \frac{r}{R}) = J_1(2.405 \frac{r}{R})$.

As the field energy oscillates between electric and magnetic fields, the Poynting vector describes its flow pattern. The Poynting vector vanishes at $\omega t = 0$ and $\frac{\pi}{2}$, and points in the $\hat{r}$-direction at other times.

Homework 5.2 Give an expression for the number of RF photons in the pillbox cavity. Give a numerical example. What is the order of magnitude of your result?

Figure 5.3: Three snapshots of the field energy density in an RF cavity storing a fundamental mode.

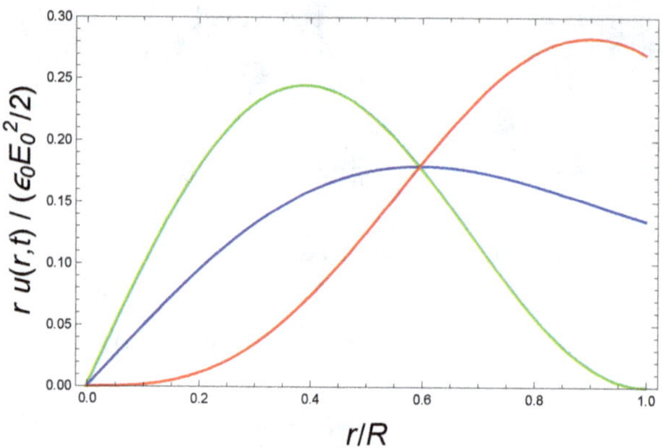

Homework 5.3

(a) This section considers the field pattern of the fundamental mode of a closed pillbox cavity in which the electric field E_z is independent of z. In general, for a more realistic cavity shape, one might have instead $E_z(z)\cos\omega t$ with E_z a function of z along the cavity axis $r = 0$. Such a deviation from the pillbox case can for example be due to small openings at the beam entrance and exit. Use paraxial ray approximation to find the transverse electric field E_r and the azimuthal magnetic field B_θ to first order in r.

(b) The pillbox fields are the case of a hard-edge model with an on-axis field $E_z(z) = E_0 H(z) H(L-z)$. Apply your result of (a) to find the transverse field components when $r \ll R$. Compare with the exact expressions.

5.1.2 Transit time factor

We now consider what happens to a particle when it passes through the RF cavity along the cavity axis ($r = 0$) in the z-direction. Let the RF voltage be sufficiently low and/or the energy of the particle be sufficiently high, so that its velocity v does not change much during the passage. Consider a particle with $z = vt$, corresponding to a particle that passes the center of the cavity at time $t = 0$ when its accelerating electric field reaches maximum. As the particle passes through the cavity, it is accelerated by E_z. During the passage, the electric field seen by the particle varies because, although the field is independent of z, it does oscillate in time. To find the net acceleration on the particle, we need to integrate E_z as follows,

$$\Delta P_z = e\int_{-\frac{L}{2v}}^{\frac{L}{2v}} dt\, E_z$$

$$= \frac{e}{v}\int_{-\frac{L}{2}}^{\frac{L}{2}} dz\, E_z$$

$$= \frac{eE_0}{v}\int_{-\frac{L}{2}}^{\frac{L}{2}} dz \cos\frac{\omega z}{v}$$

$$= \frac{2eE_0}{\omega}\sin\frac{\omega L}{2v}\,.$$

Compared with the case when a particle receiving a maximum E_0 throughout the cavity passage, the momentum gain by this particle is reduced by a factor

$$T = \frac{\sin u}{u}, \qquad \text{where} \qquad u = \frac{\omega L}{2v}\,.$$

This factor T is called the *transit time factor.* It is always less than 1. The momentum gain by the particle is $\Delta P_z = \frac{eE_0LT}{v}$.

The reader is asked then to use this result to show that its energy gain is $\Delta E = eE_0LT$, as one might have guessed but have not proven so.

An efficient design of an RF cavity will have a large T. For example, one might want $T = 0.9$. This means $u = 0.8$. For a relativistic beam with $v \approx c$, this means

$$\frac{\omega L}{2c} = 2.405\,\frac{L}{2R} \approx 0.8, \qquad \text{or} \qquad \frac{L}{R} \approx \frac{2}{3}\,.$$

If we further take $E_0 = 3$ MV/m and $L = 20$ cm, then $R = 30$ cm and the stored field energy is $U = 0.6$ Joule. If we send a beam of $N = 10^{10}$ protons through this cavity, the energy extracted by the beam from the stored energy is $NeE_0LT = 0.54 \times 10^{16}$ eV, or 0.86 mJ, which is a small fraction of the stored field energy. The energy extracted by the beam will of course have to be replenished by the RF source before the beam returns on its next passage.

With a slight exaggeration, one might state that all efficient RF cavities have a shape $\frac{L}{2R} = \frac{1}{3}$, where $2R$ is the diameter of the cavity pillbox. One then ends up with a rather shallow pillbox — a beer can, for example, will not make a very efficient RF cavity. Figure 5.4 sketches three RF cavity designs. The design on the left is efficient; the other two are not.

The middle cavity in Fig. 5.4 is inefficient because T is small. When $L \gtrsim R$, the field has over-oscillated during the passage of the particle, and the energy gain is reduced, or even becomes negative. As seen in Fig. 5.5, even a squarish box with $L = 2R$ would give a very small T. Why is the third cavity inefficient? The answer is that the transit time factor increases only a little compared with the first figure, while its ohmic heating (see next section) becomes excessive.

It should be noted that the transit time factor T depends only on the cavity aspect ratio $\frac{L}{R}$, and not on L or R independently. In other words, it depends only on the *shape*, and not on the *size*, of the cavity.

To extend the discussion a bit further, with an efficient RF cavity shape $\frac{L}{2R} = \frac{1}{3}$, one also notes that a practical cavity size might cover the range from

Figure 5.4: Efficient and inefficient cavity shapes.

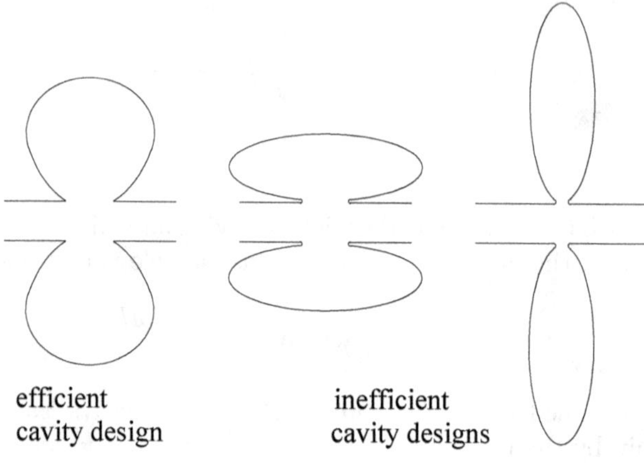

Figure 5.5: Transit time factor as a function of $\frac{c}{v}\frac{L}{R}$ for a pillbox cavity.

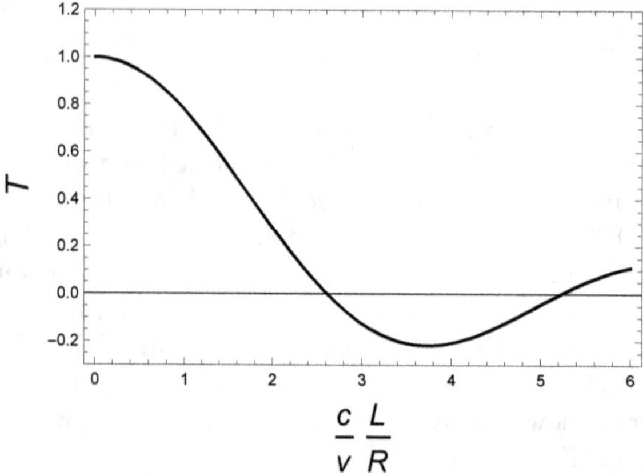

$R \sim 1$ cm to ~ 1 m. It follows that a practical range of microwave frequency $f = \frac{2.405}{2\pi}\frac{c}{R}$ covers a range from ~ 100 MHz to ~ 10 GHz.

Homework 5.4 In the text, T was obtained for a particle reaching the cavity center when the accelerating electric field reaches maximum. How does T depend on a time offset from this optimal condition?

Solution If a particle enters the cavity when the cavity voltage has a phase offset $\Delta\phi$ from its peak, the momentum gain becomes $\Delta P_z = \frac{eE_0LT}{v}\cos\Delta\phi$ with the same transit time factor T as defined in the text. Don't take for granted that the same transit time factor can be applied in this case. It is true, but needs to be proven first.

Homework 5.5 In the text, we derived ΔP_z and ΔE approximating $v =$ constant during the cavity passage. Derive the more general expressions without this approximation.

Solution ΔP_z becomes more complicated, but it would be instructive to show that ΔE remains the same – as it should.

Homework 5.6 We should not forget that as the particle traverses the pillbox cavity, its transverse motion is also affected by the oscillating magnetic field. Calculate the transverse kick given to (a) the optimal particle, i.e. a particle reaching the cavity center when the accelerating electric field reaches maximum, (b) a particle arriving at the same time as the optimal particle, but with a finite transverse displacement r, (c) a particle with $r = 0$ but arriving the cavity with a time lag τ relative to the optimal particle.

5.1.3 Transverse deflecting cavity

Transverse deflecting cavity is also called crab cavity.[2] We mentioned it when we discussed the symplecticity condition (Sec. 2.3.1) and the intriguing connection between the Maxwell equations and symplecticity. But we did not discuss what physically constitutes a transverse deflecting cavity.

We can basically consider the same pillbox cavity as we have been considering so far, except that we now excite a different, higher order mode in this cavity at a higher frequency than the fundamental mode. This time, the fields will have dependences on θ and z. There are two such modes available to be chosen for a transverse deflecting cavity. One is a transverse electric (TE) mode, the other is a transverse magnetic (TM) mode.

TM mode The lowest TM mode was the fundamental mode found earlier. The next TM mode has the field component $E_z \propto J_1(3.832\,\frac{r}{R})\cos\theta$, where 3.832 is the first zero of the function $J_1(x)$. The cavity fields are

$$\begin{aligned}
E_z &= E_0 J_1(k_r r)\cos\theta\cos k_z z\cos\omega t\,, \\
B_z &= 0\,, \\
B_r &= \frac{\omega}{c^2 k_r^2 r} E_0 J_1(k_r r)\sin\theta\cos k_z z\sin\omega t\,, \\
B_\theta &= \frac{\omega}{c^2 k_r} E_0 J_1'(k_r r)\cos\theta\cos k_z z\sin\omega t\,,
\end{aligned}$$

[2]See, for example, K. Akai and M. Tigner, Sec. 7.2.10, Handbook Accel. Phys. & Eng., 2nd Ed., World Scientific (2013). The illustrative name of crab cavity was coined by Robert Palmer.

$$
\begin{aligned}
E_\theta &= \frac{k_z}{k_r^2 r} E_0 J_1(k_r r) \sin\theta \sin k_z z \cos\omega t\,, \\
E_r &= -\frac{k_z}{k_r} E_0 J_1'(k_r r) \cos\theta \sin k_z z \cos\omega t\,,
\end{aligned} \tag{5.4}
$$

where $k_r = \frac{3.832}{R}$, $k_z = \frac{\pi}{L}$ and $\omega = c\sqrt{k_r^2 + k_z^2}$. Boundary conditions are satisfied with $E_z(r = R) = 0, E_r(z = \pm\frac{L}{2}) = E_\theta(z = \pm\frac{L}{2}) = 0$. The electric and magnetic fields oscillate out of phase by 90°.

Near-axis fields and symplecticity The transverse field pattern near the cavity axis looks like Fig. 5.6. For an on-axis particle, the electric and magnetic forces kick its transverse motion in the same direction.

Figure 5.6: Field pattern of a TM mode near the axis of a pillbox cavity. The two patterns however are 90° out of phase in time, so they occur alternatingly.

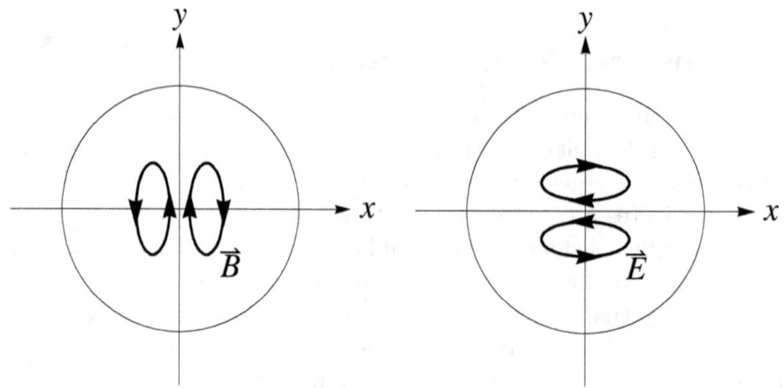

Using Eq. (5.4), one can calculate the near-axis fields to be

$$
\begin{aligned}
\vec{B} &\approx \hat{y}\,\frac{\omega}{2c^2 k_r} E_0 \cos k_z z \sin\omega t\,, \\
\vec{E} &\approx -\hat{x}\,\frac{k_z}{2k_r} E_0 \sin k_z z \cos\omega t + \hat{z}\,\frac{k_r x}{2} E_0 \cos k_z z \cos\omega t\,.
\end{aligned}
$$

As mentioned in Sec. 2.3.1, the time oscillating transverse magnetic field near-axis necessarily introduces a longitudinal electric field that contains a transverse gradient by the Maxwell equations. This in turn assures symplecticity.

More explicitly, let us calculate the momentum impulse received by a particle moving with velocity $v\hat{z}$ through a cavity. Consider the cavity to be short so that we can ignore the transit time. Let the nominal arrival of the beam occur when $\sin\omega t = 0$, but the particle has a displacement x horizontally and

Δz longitudinally as it enters the cavity. The impulse calculation requires an integration over z over the length of the cavity. The magnetic force is in the x-direction, and the x-impulse delivered to the particle, with $\omega t = -\omega\frac{\Delta z}{v}$, is

$$\Delta p_{x1} = \int_{-L/2}^{L/2} \frac{dz}{v}(-evB_y) = \frac{\omega^2 eE_0 L}{2c^2 k_r v}\Delta z\,.$$

For a particle with a longitudinal displacement Δz, the x-component of electric field contributes an impulse

$$\Delta p_{x2} = \int_{-L/2}^{L/2} \frac{dz}{v} eE_x = -\frac{k_z^2 eE_0 L}{2vk_r}\Delta z\,.$$

Both the magnetic and the electric fields contributes to the transverse kick. The total x-impulse is

$$\Delta p_x = \Delta p_{x1} + \Delta p_{x2} = \frac{eE_0 L}{v}\frac{k_r}{2}\Delta z\,. \tag{5.5}$$

The integrated longitudinal impulse is from the electric field,

$$\Delta p_z = \int_{-L/2}^{L/2} \frac{dz}{v}\, eE_z = \frac{eE_0 L}{v}\frac{k_r}{2}x\,. \tag{5.6}$$

Equations (5.5) and (5.6) establish the symplecticity of the map as discussed in Sec. 2.3.1. With the near-axis approximation made, the symplecticity is established only to first order in the matrix form.

Take for example the case of a relativistic electron beam passing through a TM transverse cavity. We have $v = c, k_r = \frac{3.832}{R}$. Then we have

$$\Delta x' = k\Delta z, \quad \Delta\delta = kx, \qquad \text{with } k = \frac{3.832}{2}\frac{eE_0}{\gamma mc^2}\frac{L}{R}\,.$$

This result will appear again in Eq. (5.46). If $E_0 = 10$ MV/m, $\gamma mc^2 = 1$ GeV, $\frac{L}{R} = \frac{2}{3}$, then $k = 0.013$/m.

As mentioned in Eq. (2.32), transverse cavity is one of the ten fundamental building blocks of coupled symplectic units. There are many applications of transverse deflecting cavities, and more applications might be expected in the future. Notable examples include beam separator, beam diagnostics, beam longitudinal slicing, crab collision, and emittance exchanges. We shall not elaborate on this advanced subject. A brief mention of emittance exchange will be made later in Sec. 5.6.

TE mode The lowest TE mode has the field component $B_z \propto J_1(1.841\,\frac{r}{R})\cos\theta$, where 1.841 is the first zero of the function $J_1'(x)$. For completeness, we give the cavity fields,

$$\begin{aligned}
B_z &= B_0 J_1(k_r r)\cos\theta \sin k_z z \cos\omega t\,,\\
E_z &= 0\,,\\
B_r &= \frac{k_z}{k_r} B_0 J_1'(k_r r)\cos\theta\cos k_z z\cos\omega t\,,\\
B_\theta &= -\frac{k_z}{k_r^2 r} B_0 J_1(k_r r)\sin\theta\cos k_z z\cos\omega t\,,\\
E_\theta &= -\frac{\omega}{k_r} B_0 J_1'(k_r r)\cos\theta\sin k_z z\sin\omega t\,,\\
E_r &= -\frac{\omega}{k_r^2 r} B_0 J_1(k_r r)\sin\theta\sin k_z z\sin\omega t\,,
\end{aligned} \tag{5.7}$$

where $\omega = c\sqrt{k_z^2 + k_r^2}$, $k_r = \frac{1.841}{R}$, $k_z = \frac{\pi}{L}$. The boundary conditions are satisfied. See Homework 5.9.

Homework 5.7 Although not much more insight to gain, a basic exercise is to confirm that Eqs. (5.4) and (5.7) do satisfy the Maxwell equations and the boundary conditions.

Homework 5.8 Another good practice is to establish the energy conservation. Take either the TM or the TE mode for example. When $\omega t = 0$, all field energy is electric. Calculate the total electric energy. When $\omega t = \frac{\pi}{2}$, all fields are magnetic. At any other times, calculate the total field energy. It should be a constant in time.

Solution For the TM case,

$$U_{\rm tot} = U_E + E_B = \frac{\pi\epsilon_0 L R^2 E_0^2}{8}\left(1 + \frac{k_z^2}{k_r^2}\right) J_0(3.832)^2\,.$$

You may set $J_0(3.832)^2 = 0.1622$, and $\frac{k_z}{k_r} = \frac{\pi}{3.832}\frac{R}{L}$. To derive these expressions, you need

$$\int_0^a x dx J_1^2(x) = \frac{a^2}{2} J_0^2(a) - a J_0(a) J_1(a) + \frac{a^2}{2} J_1^2(a)\,,$$

$$\int_0^a x dx \left[\frac{J_1^2(x)}{x^2} + J_1'^2(x)\right] = -J_1^2(a) + \frac{a^2}{2}[J_0^2(a) + J_1^2(a)]\,.$$

The stored energy is a factor $\frac{1}{4}\left(1 + \frac{k_z^2}{k_r^2}\right)$ times the stored energy of a uniform field E_0 filling the cavity volume. This factor is larger than 1 for a shallow cavity with $R > 2.11\,L$.

Homework 5.9 Follow the discussion of the TM mode, find the expressions of near-axis fields of the TE mode, sketch the near-axis TE field pattern similar to that of Fig. 5.7, and calculate the 4×4 matrix map for a thin cavity and establishes its symplecticity.

Solution Near axis,

$$\begin{aligned}\vec{E} &\approx -\hat{y}\frac{\omega}{2k_r}B_0 \sin k_z z \sin\omega t\,,\\ \vec{B} &\approx \hat{x}\frac{k_z}{2k_r}B_0 \cos k_z z \cos\omega t + \hat{z}\frac{k_r x}{2}B_0 \sin k_z z \cos\omega t\,.\end{aligned}$$

Figure 5.7 shows the field patterns at the moment $\omega t = 0$ when it is pure magnetic (left) and when the moment $\omega t = \frac{\pi}{2}$ when it is the pure electric (right).

Figure 5.7: Field pattern of a TE mode near the axis of a pillbox cavity.

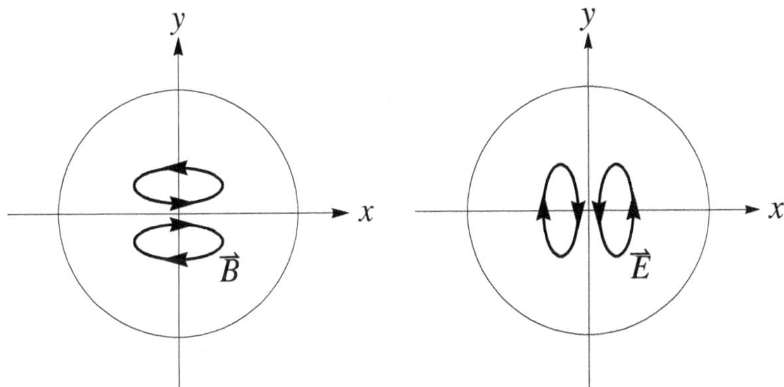

Homework 5.10 Equations (5.4) and (5.7) are given for two low order TM and TE modes. The hard work has been done. Given their forms, it is not difficult to write down all higher order TM and TE modes. For one thing, the higher order longitudinal modes have exactly the same forms, the only required change is to set $k_z = \frac{m\pi}{L}$ with other integers $m > 1$. How does the azimuthal higher modes and radial higher modes look like? You need a total of three integer indices to specify a mode.

5.1.4 Ohmic heating

Defining metal and insulator We now turn our attention to the cavity walls. Knowing the electromagnetic fields in the cavity, it would be useful to calculate the surface charge density σ and the surface current density $\vec{K}$ on the cavity walls. Assuming the walls are perfectly conducting, we apply Gauss' law to Fig. 5.8 to obtain

$$\sigma = \epsilon_0 \vec{E}\cdot\hat{n}\,,$$

where $\vec{E}$ is the electric field on the cavity surface at the location of interest, $\hat{n}$ is the unit vector perpendicular to the surface and pointing out of the metal. By Gauss law, surface charge density is simply related to the normal component of the electric field at the surface.

Figure 5.8: Gauss law on a metal surface.

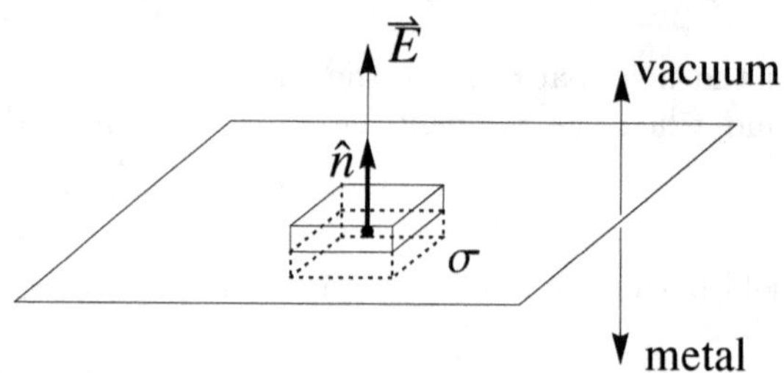

Similarly we apply Ampere law to Fig. 5.9 to obtain

$$\vec{K} = -\frac{1}{\mu_0}\vec{B}\times\hat{n},$$

where $\vec{B}$ is the magnetic field immediately on the vacuum side of the cavity wall at the point of interest. Surface current density is simply related to the tangential component of the magnetic field at the surface.

Figure 5.9: Ampere law on a metal surface.

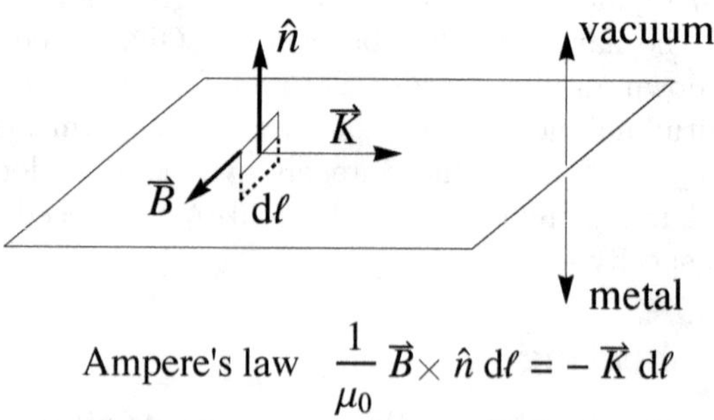

Note the symmetry between electric and magnetic fields, and between charge and current densities, as summarized in Fig. 5.10. A revelation of the elegance and internal self-consistency and completeness has been attempted by this brief summary.

Figure 5.10: A brief summary of the Maxwell system. A symmetry between the electric family (left) and the magnetic family (right) is illustrated. The two families are linked twice. Their sources are linked by the equation of continuity (1.4). Their fields are linked by Maxwell equations.

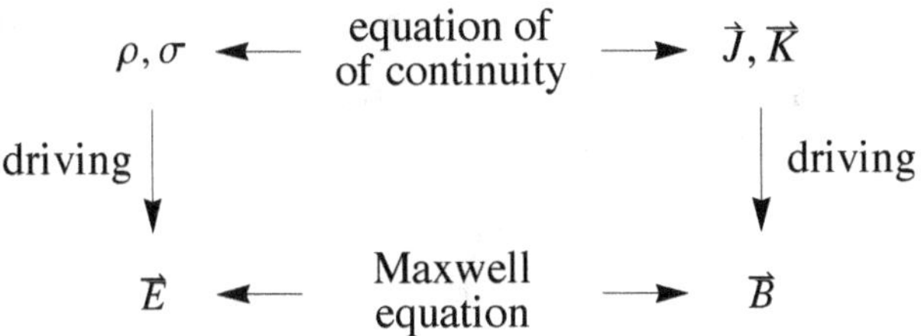

This symmetry, however, holds only in vacuum. It is lost when we consider a metal or an insulator. Metals break the symmetry by making a preference to the magnetic family $(\vec{B}, \vec{J})$, while insulators make a preference in favor of the electric family $(\vec{E}, \rho)$. No charges are allowed inside a metal while currents are allowed to penetrate. Inside a metal, therefore, there is more magnetic field than electric field. On the other hand, currents are forbidden inside an insulator, and there is more electric field than magnetic field. So the electric and the magnetic families transform into each other as equal partners only in vacuum. Conductors discriminates against electric family, while insulators discriminates against magnetic family.

To extend the discussion a bit further. One can define metals and insulators from the point of view of electrodynamics,[3]

$$\begin{aligned} &\text{Definition of metal/conductor:} \qquad \rho = 0, \ \vec{J} \propto \vec{E}\,, \\ &\text{Definition of insulator:} \qquad \vec{J} = \vec{0}, \ \rho \propto \nabla \cdot \vec{E}\,. \end{aligned} \tag{5.8}$$

Wall heating So far we have assumed the cavity wall is perfectly conducting. In reality, the wall will be slightly resistive. Field energy stored in the cavity will dissipate slowly due to ohmic loss on the wall. Consider the fundamental RF mode for a pillbox cavity. To calculate its ohmic loss, we first need to find the surface current on the cavity walls. On the cylindrical part of the surface, the surface current is in the $-\hat{z}$ direction and is constant, independent of z,

$$\vec{K} = -\frac{1}{\mu_0} B_\theta \hat{z} = \frac{iE_0}{c\mu_0} J_1(2.405)\, e^{-i\omega t}\, \hat{z}\,.$$

On the end plates, it is in the radial direction,

$$\vec{K} = \pm\frac{1}{\mu_0} B_\theta \hat{r} = \mp\frac{iE_0}{c\mu_0} J_1\left(2.405\,\frac{r}{R}\right) e^{-i\omega t}\, \hat{r}\,,$$

where one end plate has + sign, the other has − sign.

[3]Set aside what you learned from solid state physics and material science for a moment.

These surface currents do not stay strictly on the surface, but for non-perfectly conducting metals, they penetrate a depth given by the *skin depth* [see also Eq. (8.13)]

$$\delta_{\text{skin}} = \sqrt{\frac{2\rho_c}{\mu\omega}}, \tag{5.9}$$

where ρ_c is the resistivity of the metal. Take copper for example, we have $\mu = \mu_0 = 4\pi \times 10^{-7}$ Henry/m, $\rho_c = 1.7 \times 10^{-8}$ Ω-m, and

$$\delta_{\text{skin}} = \frac{66\,\mu\text{m}}{\sqrt{f\,[\text{MHz}]}} \qquad (\text{for Cu})\,.$$

For 380 MHz, copper would have a skin depth of 3.35 μm.

Ohmic heating originates from a surface current on the cavity walls, i.e. it comes from $\vec{K}$. The surface current $\vec{K}$, in turn, is driven by the magnetic field next to the wall surface. The culprit of ohmic heating is the magnetic fields next to the cavity walls.

If we draw a rectangular area $dxdy$ on the metal surface where dx is parallel to the surface current and dy is perpendicular to it, as shown in Fig. 5.11, then the total current flowing on the area element is $i = Kdy$. With the current being carried in a cross-sectional area of $dy\delta_{\text{skin}}$, the resistance of this area element is

$$R = \rho_c \frac{dx}{dy\delta_{\text{skin}}}\,.$$

The ohmic heating power produced over this area element is

$$P = \frac{1}{2} i^2 R = \frac{\rho_c K^2}{2\delta_{\text{skin}}}\, dxdy\,,$$

where the factor $\frac{1}{2}$ comes from time averaging over the sinusoidal oscillating currents.

Figure 5.11: A surface area of the cavity wall with a flowing surface current generating ohmic heating.

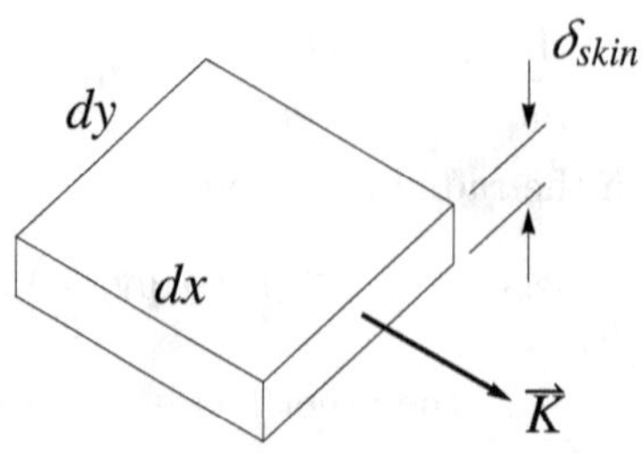

We are now ready to calculate the total ohmic heating power by integrating over the cavity wall surface. It is given by

$$\begin{aligned} P &= \frac{\rho_c}{2\delta_{\rm skin}} \oint dAK^2 \\ &= \frac{\rho_c}{2\delta_{\rm skin}} \frac{E_0^2}{c^2\mu_0^2} \left[2\pi RLJ_1^2(2.405) + 2\times 2\pi \int_0^R rdrJ_1^2\left(2.405\frac{r}{R}\right)\right] \\ &= \frac{\rho_c}{2\delta_{\rm skin}} \frac{E_0^2}{Z_0^2} 2\pi RLJ_1^2(2.405)\left(1+\frac{R}{L}\right), \end{aligned} \tag{5.10}$$

where we have used

$$\int J_1^2(u)udu = \frac{u^2}{2}[J_1^2(u) - J_0(u)J_2(u)],$$

and replaced $c\mu_0$ by

$$\begin{aligned} Z_0 &= c\mu_0 = \sqrt{\frac{\mu_0}{\epsilon_0}} = \frac{1}{c\epsilon_0} \\ &= 377\,\Omega, \end{aligned}$$

the *free space impedance.* There are two terms in the expression (5.10). The first term is due to heating on the cylindrical wall. The second term is due to heating on the two end plates. Their relative ratio is $\frac{R}{L}$.

Yes, free space has impedance! An oscillating charge and current in an antenna radiate electromagnetic radiation into free space. There is an energy loss by the antenna due to this radiation, and its reaction back to the radiating charge and current is described by Z_0, the impedance of free space.

Homework 5.11 Sea water has a finite conductivity $\rho_c = 0.25$ Ω-m, and a dielectric constant at radio frequencies $\frac{\epsilon}{\epsilon_0} = 80$. Calculate the skin depth of sea water at 400 MHz. Would a strong communication radar work for a submarine?

5.1.5 Lorentz deformation

Lorentz pressure The surface charges and currents on the cavity wall interact with the electromagnetic fields in the cavity next to the wall surface. This interaction generates a Lorentz force that in turn gives rise to field pressure on the wall surface. The pressure can be easily found from the Lorentz force equation,

$$\vec{P} = \frac{\text{Lorentz force}}{\text{unit area}} = \frac{1}{2}\left(\epsilon_0\vec{E}^2 - \frac{1}{\mu_0}\vec{B}^2\right)\hat{n},$$

where $\vec{E}, \vec{B}$ are the electromagnetic fields next to the wall surface, $\hat{n}$ is a unit vector pointing out of the metal surface wall. On the right-hand-side, the first term comes from electric field acting on the surface charge; the second term comes from magnetic field acting on the surface current; both are in the $\hat{n}$-direction. An overall factor of $\frac{1}{2}$ comes from the fact that the electromagnetic fields diminishes into the wall surface, and does not have the full value of $\vec{E}, \vec{B}$ throughout the skin depth. One notes that magnetic force would push the cavity walls outwards, while electric force would suck the walls into the cavity. Superimposed with the Lorentz force is of course the usual vacuum pressure force if the RF cavity is under vacuum.

It would be instructive for the reader to think through the physical reason why magnetic force is always pushing outwards while electric force is always sucking inwards. Figure 5.12 hopefully helps.

Figure 5.12: Configurations of electric field, magnetic field, surface charge and surface current at the cavity wall, illustrating the Lorentz pressure.

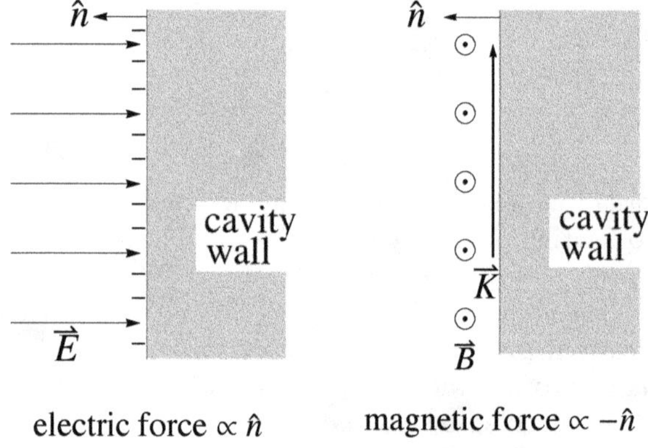

The instantaneous pressure changes rapidly at twice the RF frequency. The pressure then is obtained by time-averaging over this oscillation. For the pillbox cavity, we then obtain for the pressure on the two end-plates,

$$\vec{P} = \pm\frac{\epsilon_0 \vec{E}_0^2}{4}\hat{z}\left[J_0^2\left(2.405\frac{r}{R}\right) - J_1^2\left(2.405\frac{r}{R}\right)\right],$$

where the first term comes from electric force and the second term comes from the magnetic force. The $\pm$ sign refers to pressures on the two end-plates having opposite directions $\pm\hat{z}$. Figure 5.13 shows the function $J_0^2(x) - J_1^2(x)$.

The pressure on the cylindrical wall is purely magnetic (because electric field vanishes on the wall surface),

$$\vec{P} = \frac{\epsilon_0 \vec{E}_0^2}{4} J_1^2(2.405)\hat{r}.$$

Figure 5.13: The function $J_0^2(x) - J_1^2(x)$. Although plotted over a wider range, the plot applies to the pillbox cavity only up to $x = 2.405$, where only magnetic field is doing the pushing.

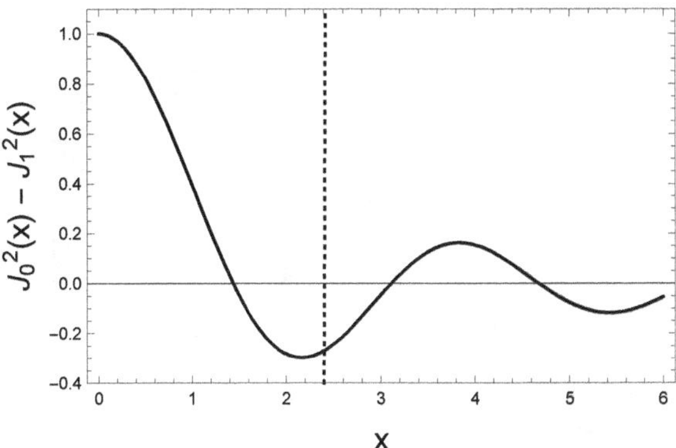

The pressure is always pushing the wall outwards. The Lorentz pressure on the cavity walls is as sketched in Fig. 5.14 (left).

Deformation Due to this pressure, the RF cavity will be deformed. Details of how it is deformed depends on the mechanical properties of the cavity material and construction. But an exaggerated deformation might look like the sketch of Fig. 5.14 (right). The electric pressure pulls the two end plates inwards, making the cavity length L shorter. The magnetic pressure pushes the cylindrical walls outwards, making the cavity radius R larger.

When the cavity shape is deformed, the RF mode frequency $\omega_{\rm rf}$ changes. This change is called *Lorentz detuning*, and is a significant consideration concerning the mechanical integrity of an RF cavity, particularly the superconducting ones.

For a pillbox cavity, since the deformation tends to make the cavity's effective radius larger, Lorentz detuning tends to shift the fundamental RF mode frequency *downwards* — see Eq. (5.2). Lorentz pressure also tends to shorten the effective cavity length, but the mode frequency is not sensitive to this deformation in cavity length. Transit time factor might be affected slightly, but that is not a serious concern.

Large Lorentz detuning often occurs for superconducting cavities because (a) niobium material is softer than copper, and (b) the stronger superconducting RF fields.

Homework 5.12 How do you expect the Lorentz deformation to behave for a transverse deflecting cavity for the TM and for the TE modes? How about the Lorentz detuning? Do the mode frequencies move up or down?

Figure 5.14: A sketch of the Lorentz pressure on the pillbox cavity walls (left) and its deformation (right).

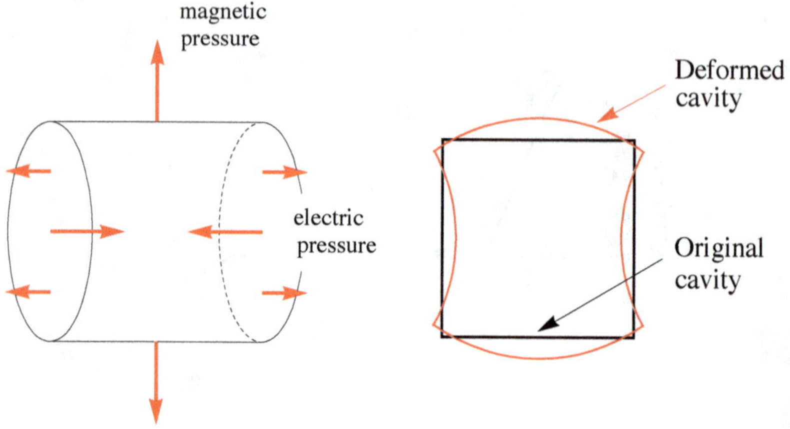

5.1.6 Quality factor

We now define an important parameter for the RF cavity, its *quality factor*, or Q value,

$$Q = \frac{\text{stored field energy}}{\text{ohmic loss per radian of RF oscillation}} = \frac{\omega U}{P}.$$

In other words, it takes $\frac{1}{2\pi} \times Q$ oscillations of the cavity fields for the ohmic heating to significantly dissipate the stored field energy. The stored energy in a cavity, if left alone, will dissipate according to

$$\dot{U} = -P = -\frac{\omega}{Q}U$$
$$\Longrightarrow \quad U(t) = U_0\, e^{-\omega t/Q}.$$

It follows, embedded in its definition, that for a given Q-value, higher frequency cavities dissipate faster. This is made even more so when taking into account the physical fact that $Q \propto \frac{1}{\sqrt{\omega}}$ (see below).

In terms of the number of microwave oscillations $n = \frac{\omega t}{2\pi}$, the stored energy reads

$$U(n) = U_0\, e^{-2\pi n/Q}.$$

Fundamental mode of pillbox cavity Substituting our results, ω from Eq. (5.2), U from Eq. (5.3), and P from Eq. (5.10), we find for the fundamental mode of a pillbox cavity,

$$Q = \frac{2.405\, Z_0 \delta_{\text{skin}}}{2\rho_c(1+\frac{R}{L})},$$

or equivalently, using Eq. (5.9),

$$Q = \frac{RL}{\delta_{\text{skin}}(R+L)}. \tag{5.11}$$

We see that Q is independent of the field strength filling the cavity, and is therefore an intrinsic parameter of the cavity alone. We see that Q depends linearly on its size, except for the skin depth, which is proportional to $\frac{1}{\sqrt{\omega}} \propto \sqrt{\text{cavity size}}$. Taking skin depth into consideration, we see that

$$Q \propto \frac{1}{\sqrt{\omega}} \propto \sqrt{\text{cavity size}}.$$

Larger cavities tend to have higher Q-values. Also, cavities made of higher conductivity material have higher Q-values. Superconducting cavities may have $Q \sim 10^9$ or 10^{10}. For room temperature cavities, consider for example a copper pillbox cavity with $R = 30$ cm and $L = 20$ cm. Its fundamental mode frequency is 380 MHz, and corresponding skin depth is 3.35 μm. This leads to a Q-value of $\sim 3.5 \times 10^4$.

The Q-value of a particular mode of a cavity has the physical meaning that it describes the decay rate of its stored field energy when the field filling the cavity is in that particular cavity mode. Different mode of the same cavity have different Q-values.

There is a second physical meaning of the Q-value as follows. If the cavity is driven not exactly at the mode frequency ω_0 of the specified mode, but at a frequency $\omega_0 + \Delta\omega$, then the cavity can only be driven efficiently when $\frac{\Delta\omega}{\omega_0} \lesssim \frac{1}{Q}$. So Q-value is also related to a *width* of the resonant frequency of the particular cavity mode under consideration.

Arbitrary cavity We now try to find a rough estimate of Q for an arbitrary cavity shape. Let the cavity be filled with microwave with frequency ω and peak electric field $\sim E_0$. The stored field energy is approximately

$$U \sim \frac{\epsilon_0}{2} E_0^2 \mathcal{V},$$

where $\mathcal{V}$ is the volume contained by the cavity. The magnetic field is going to be $B \sim \frac{E_0}{c}$. Surface current induced by the magnetic field $K \sim \frac{B}{\mu_0} \sim \frac{E_0}{\mu_0 c}$. Ohmic heating power

$$P \sim \frac{\rho_c}{2\delta_{\text{skin}}} \mathcal{A} K^2 \sim \frac{\rho_c}{2\delta_{\text{skin}}} \mathcal{A} \frac{E_0^2}{\mu_0^2 c^2},$$

where $\mathcal{A}$ is the surface area of the cavity. The Q-value is then found to be

$$Q = \frac{\omega U}{P} \sim \frac{2\mathcal{V}}{\delta_{\text{skin}} \mathcal{A}}. \tag{5.12}$$

In other words, Q is roughly equal to two times the ratio between the cavity volume to the cavity "surface volume" of cavity interior surface with a depth given by the skin depth. Note the following,

- For a pillbox cavity, the exact and the rough expressions of Q give the same answer.
- For cavities of similar shape, larger cavities tend to have larger values of $\frac{\mathcal{V}}{\mathcal{A}}$, there is a tendency that the larger the cavities, the larger their Q-values. This is true even taking into account of the fact that $\delta_{\rm skin} \propto \sqrt{\text{cavity size}}$.
- For cavities of similar size, round shaped cavities tend to have larger values of $\frac{\mathcal{V}}{\mathcal{A}}$, there is a tendency that the rounder the cavities, the larger their Q-values. Refer to Fig. 5.4.

Homework 5.13 Apply the rough formula (5.12) to find the Q-value for a pillbox cavity, and show that it predicts the same Q-value as the more precise formula (5.11).

Homework 5.14 The field patterns of TM and TE modes of a pillbox cavity were given by Eqs. (5.4) and (5.7) respectively.

(a) Find the expressions of ohmic heating power for these cavity modes.

(b) Find the expressions of Q-value for these cavity modes. Compare with the rough estimation (5.12).

5.1.7 Shunt impedance

There are three figures of merit of an RF cavity: its mode frequency $\omega_{\rm rf}$ that specifies the cavity size, the ratio $\frac{R}{L}$ that specifies its shape, and the skin depth $\delta_{\rm skin}$ that specifies its construction material. Alternatively and more conveniently for an accelerator designer, these three parameters can be translated into another set of three parameters $\omega_{\rm rf}$, the quality factor Q, and a third parameter called the *shunt impedance* designated by R_s.

The same three physical quantities can be viewed differently by different professions,

Pedestrian	Engineer	Accelerator designer
Size	L	$\omega_{\rm rf}$
Shape	R/L	Q
Material	ρ_c	R_s

The shunt impedance is defined as

$$R_s = \frac{(\text{voltage gain by a particle})^2}{\text{ohmic loss power}} = \frac{(E_0LT)^2}{P},$$

where E_0 is the peak electric field, T is the transit time factor, P is the ohmic heating power. Sometimes the definition includes an additional multiplicative factor of $\frac{1}{2}$ on the right-hand-side, and sometimes, it does not include the transit time factor T^2. So one has to be careful here. When the factor of $\frac{1}{2}$ is included, it is called the "circuit definition", while ours would be the "linac definition". Usually we want R_s to have a high value.

Pillbox cavity For a pillbox cavity, we have

$$R_s = \frac{Z_0^2 \delta_{\text{skin}}}{\pi \rho_c} \frac{T^2}{\frac{R}{L}(1+\frac{R}{L})J_1^2(2.405)} \, .$$

Shunt impedance does not depend on the field strength filling the cavity and therefore, like Q, is another intrinsic property of the cavity. In addition, it depends only on the cavity shape and not on its size except for the dependence through δ_{skin}, i.e.

$$R_s \propto \delta_{\text{skin}} \propto \frac{1}{\sqrt{\omega}} \propto \sqrt{\text{cavity size}} \, .$$

On the other hand, through the skin depth and resistivity, both Q and R_s depend on the material that the cavity is made of. It turns out that the quantity $\frac{R_s}{Q}$, however, is not only independent of the cavity size but also independent of the material. For instance, for a given cavity shape, one could construct one cavity made of superconducting material and another cavity of a much different size from copper. These two cavities will have very different Q's and R_s's, but they will have the same $\frac{R_s}{Q}$. An illustration to this point is in Fig. 5.15.

Figure 5.15: Illustration of $\frac{R_s}{Q}$ being independent of either the size or the material of the cavity. Same comment applies also to the transit time factor T.

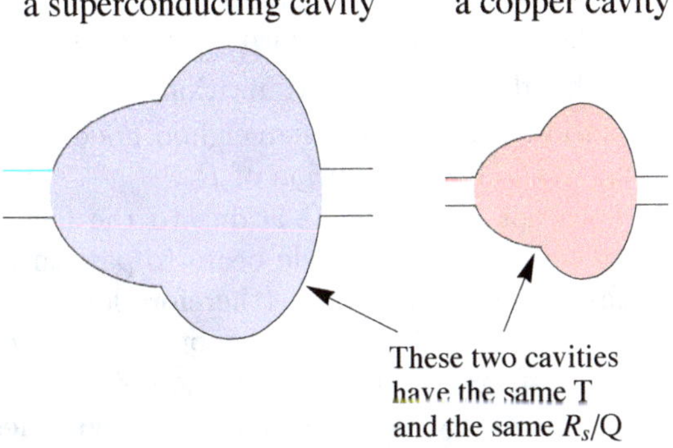

For a pillbox cavity, fundamental mode, we have

$$\frac{R_s}{Q} = 0.817 \, Z_0 \frac{\sin^2 u}{u}, \quad u = \frac{2.405 \, L}{2R} \, .$$

The exclusive dependence on cavity shape is evident.

Figure 5.16: Behavior of $\frac{R_s}{Q}$ (red) and T (green) as functions of $\frac{L}{R}$ for a pillbox cavity fundamental mode.

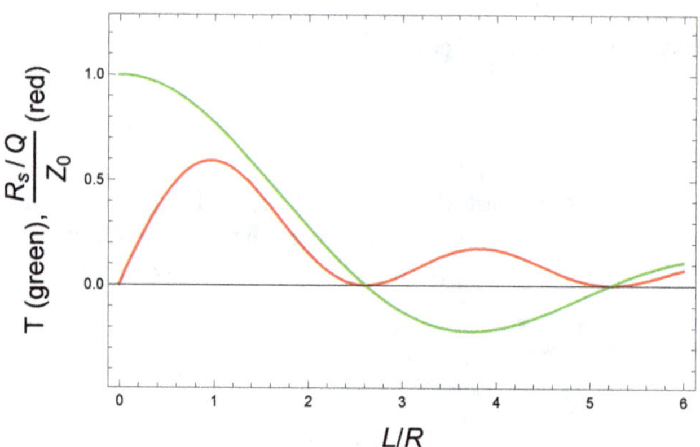

In Fig. 5.16, the green curve is the transit time factor replotted as a function of $\frac{L}{R}$ (setting $v = c$), and the red curve is $\frac{R_s}{Q}$. The order of magnitude of $\frac{R_s}{Q}$ is always a fraction of $Z_0 = 377\ \Omega$. A cavity structure with lots of empty space inside acts not as open as a free space, but is close to it. The oscillatory nature of $\frac{R_s}{Q}$ comes from the inclusion of the transit time factor in its definition.

Cavity design optimization In a cavity design, it is desirable to minimize the ohmic heating loss, therefore we maximize Q. It is also desirable to have an efficient acceleration, which means we maximize R_s. One sees from Fig. 5.16 that in terms of $\frac{R_s}{Q}$, there is an optimal with $\frac{R_s}{Q} \approx 200\,\Omega$ which occurs when $\frac{L}{R} \approx 1$. On the other hand, the other optimization considerations restrict the value of $\frac{R_s}{Q}$ to be not too far away from the neighborhood of 75-100 Ω or so — be reminded that ours is the linac definition of R_s.

One reason $\frac{R_s}{Q}$ does not reach 200 Ω is due to the fact that the pillbox needs to open the end holes for the particle beam to pass through. One might have also observed that a lower value of $\frac{L}{R}$ (therefore lower value of $\frac{R_s}{Q}$) helps increasing the transit time factor T, but that is not the main reason because we have already included the factor T^2 in our definition of $\frac{R_s}{Q}$.

Starting with a pillbox cavity, how to change the cavity design to optimize their R_s and Q? First of all, we will need two opening holes for the beam passage. For given stored field energy, the transit time factor, and therefore R_s, can be improved by adding nose cones to the cavity because nose cones help concentrating the electric field to have a shorter acceleration region. The next step is to improve the Q-value by increasing the volume to surface ratio of the cavity, and that makes the cavity more spherically shaped. These optimizations are sketched in Fig. 5.17.

Figure 5.17: Evolution of the RF cavity shape.

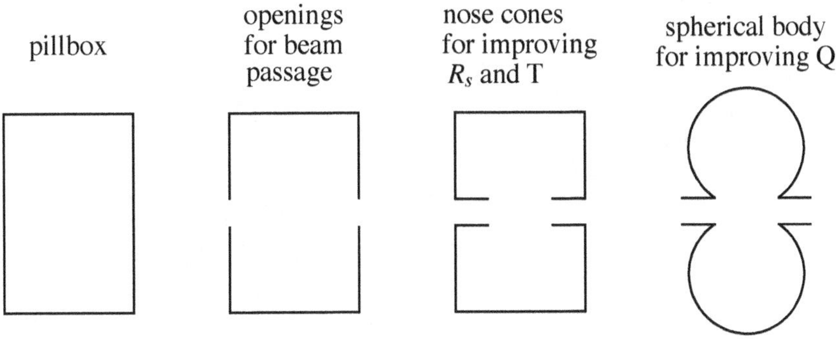

pillbox

openings for beam passage

nose cones for improving R_s and T

spherical body for improving Q

5.1.8 Breakdown

The peak radio-frequency field in a cavity in vacuum is limited by breakdowns. One often uses the *Kilpatrick limit*[4] to determine where the breakdown might occur. The Kilpatrick limit is an empirical relation derived from data taken before the era of ultra-high-vacuum technology. The maximum field E_k [MV/m] at an RF frequency f [MHz], according to this criterion, is determined by the equation

$$f = 1.64\, E_k^2 \exp\left(-8.5/E_k\right) . \tag{5.13}$$

At the S-band frequency 2856 MHz, for example, Kilpatrick predicts a breakdown limit at 46 MV/m.

The breakdown limit increases as the RF frequency is increased. This is one reason why linear colliders tend to push for higher frequency technologies. Obviously the Kilpatrick formula does not apply to DC breakdown, and neither did it intend to.

Today, with ultra-high-vacuum technology, much higher fields are often achieved. Indeed a more recent fit gives

$$E_k[\text{MV/m}] = 220(f[\text{GHz}])^{1/3} . \tag{5.14}$$

It should be added that more studies are being carried out in this active research area of breakdown limit.[5] Some researches pursue a possibility that the limit is in fact not imposed by a high peak electric field but by a high spontaneous heating due to surface currents. In addition, a black-and-white breakdown limit has yielded to a softer definition in terms of breakdown probability. Here we shall simply take the peak electric field as an indicator of breakdown limit without delving into the subject. As shown in Fig. 5.18, the new fit (5.14) gives

[4] W.D. Kilpatrick, Rev. Sci. Instru., 28 (10), 824 (1957).

[5] See for example G.A. Loew, S. Tantawi, and J.W. Wang, Handbook Accel. Phys. & Eng., 2nd Ed., World Scientific, p. 507.

Figure 5.18: Comparison of two break down formulae, dashed curve for (5.13) and solid curve for (5.14).

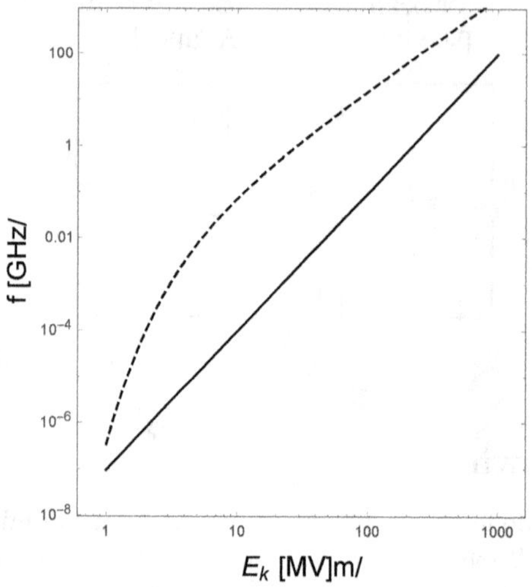

breakdown fields much higher than the Kilpatrick values (5.13). For historical reason, however, the Kilpatrick limit is still being used as the benchmark, and one often then speaks of fields several times that of this "limit".

The formula (5.14) simply expresses the attainable electric field as a function of frequency f and it says $E_k \propto f^{1/3}$. The Kilpatrick formula on the other hand has the peculiar form expressing f in terms of E_k. Asymptotically, it says $E_k \propto f^{1/2}$ when $E_k \gg 8.5$ MV/m, and E_k is suppressed exponentially at low frequencies when $E_k \ll 8,5$ MV/m. A serious examination of these differences, however, might be a moot point. Neither formulae apply to the DC case $f = 0$. Both break down formulae break down for DC.

On the other hand, the push for high fields (and therefore high acceleration gradients) is much aimed for linear accelerators. In these applications, the RF system is operated in a pulsed mode, i.e. the RF microwave is fed into the cavities only rarely, e.g. $\sim\mu$s RF pulses at 120 Hz. In contrast, in storage rings, RF system is operated in a CW mode, and one usually operates below the Kilpatrick limit to avoid too much heating power deposited on the cavity wall, not for the fear of breakdown (Homework 5.15).

Homework 5.15 Consider a copper pillbox cavity. We want it to have a transit time factor of $T = 0.9$, a mode frequency of $f = 380$ MHz, and a peak accelerating electric field equal to the Kilpatrick limit.

(a) Find the cavity radius R, length L, the stored field energy per cavity U.

(b) A beam of 10^{10} protons passes through the cavity at the phase optimized for maximum energy gain. Find the fraction of stored field energy extracted by the beam. How many cavities are needed to accelerate the beam by 20 MeV?

(c) Find the skin depth δ_{skin}, the heating power per cavity P, the quality factor Q, and the shunt impedance per cavity R_s of the cavity. Note the very high heating power.

(d) What is the field pressure exerted on the cavity's cylindrical wall and the center of the two end plates? Indicate the directions of the pressure.

Solution

(a) $R = 30$ cm, $L = 20$ cm, $U = 25$ J.

(b) $\frac{\Delta U}{U} = 2.4 \times 10^{-4}$.

(c) $P = 1.7$ MW, $Q = 3.5 \times 10^4$, $R_s = 7.2$ MΩ, $\frac{R_s}{Q} = 204$ Ω.

(d) 224 Ps on the cylindrical wall, 833 Ps at the center of the end plates.

5.2 Cavity array

In practice, cavities are often grouped into an array and driven using a single RF microwave source as sketched in Fig. 5.19.

Figure 5.19: A cavity array driven by a single microwave line.

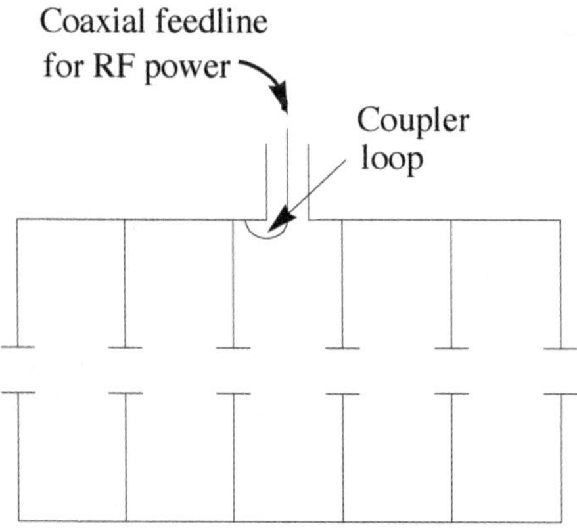

Figure 5.19 actually shows one way of operation called "standing wave" operation of cavity arrays. The RF power resonates and stays inside the array. Sketched in Fig. 5.19 is a coaxial cable feeding microwave power into the cavity array. The incoming microwave has a frequency that matches the chosen resonant frequency ω_{rf} of the RF cavity array. The drive frequency must be within the bandwidth $\frac{\omega_{\text{rf}}}{Q}$, otherwise the cavity field will be miserably weak. The microwave feed can be located in the middle of the cavity array, as shown, or sometimes at the one end of the array.

5.2.1 Equivalent circuit model

An array of pillbox cavities is a reasonable approximate model of an accelerating structure for a linac. All cavity cells in this array are identical, and if driven separately, all have the same natural mode frequency ω_0 — we consider the fundamental mode of each cavity only. However, when there are N cavities coupled into an array, because of the cavity-cavity coupling, there will be N perturbed eigenmode frequencies clustered around the unperturbed frequency ω_0. Each of the eigenmodes has a specific pattern in terms of the field oscillation phase between adjacent cavities. The frequency of the microwave source will then be carefully chosen to drive one and only one of these eigenmodes with a specific mode frequency and a specific mode pattern. The microwave drive frequency must match the frequency of that particular mode within its bandwidth in spite of the fact that there are $N-1$ other modes nearby.

LC circuit To analyze this cavity array problem, we model each cavity by an LC circuit, where the inductance L represents the cavity volume where the magnetic field has to fill, and the capacitance C represents the gap across which electric field fills and a passing beam receives the electric voltage across the capacitance C. In this circuit model, magnetic field behavior is described by L, electric field by C. The circuit equation is established considering the inductive voltage induced by the changing magnetic field and the capacitive voltage induced by the electric field as sketched in Fig. 5.20. Exchange between magnetic and electric field energies is represented by exchange between the inductive and capacitive energies in the circuit.

Figure 5.20: The electric and magnetic fields executing an LC oscillation in an RF cavity.

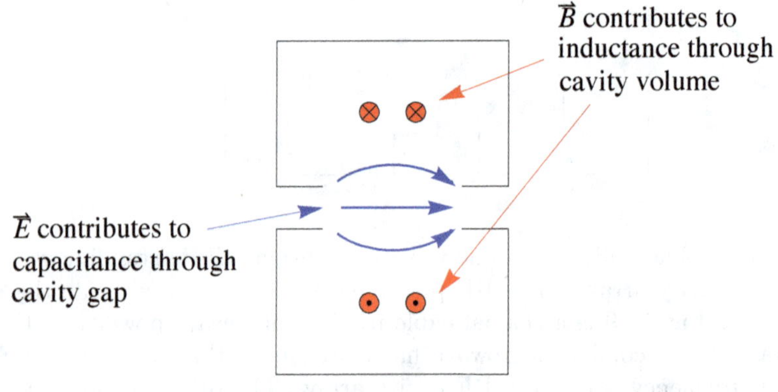

We then construct an equivalent circuit model of a magnetically-coupled cavity array as in Fig. 5.21. Coupling between adjacent cavities through the cavity

Figure 5.21: An equivalent LC circuit for an RF cavity array assuming inductive coupling between adjacent cavities.

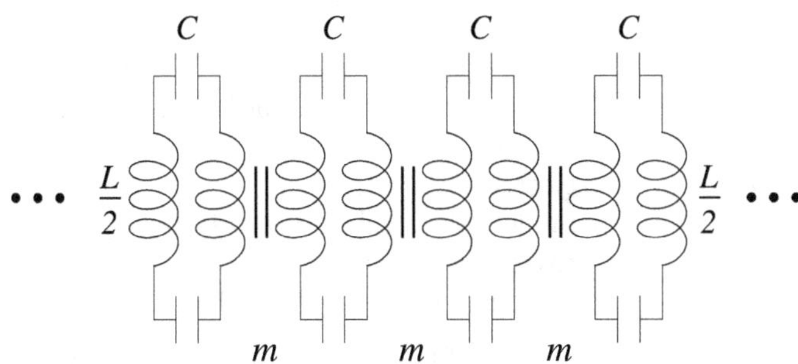

openings is represented by a small coupling coefficient m which is dimensionless and $|m| \ll 1$. The figure indicates a positive coupling coefficient $m > 0$. If one of the inductive windings is reversed (clockwise versus counter-clockwise), in principle, a negative m is obtained.

Modes In this circuit model, a single cavity has an unperturbed frequency $\omega_0 = \frac{1}{\sqrt{LC}}$. For simplicity, we ignore the end-effects and consider the n-th cavity only, with coupling to its neighbors on both sides. The circuit equation for the current I_n in the n-th cavity is

$$I_n \left(i\omega L + \frac{1}{i\omega C} \right) + m \frac{i\omega L}{2} (I_{n-1} + I_{n+1}) = 0 \,,$$

or

$$I_n \left(1 - \frac{\omega_0^2}{\omega^2} \right) + \frac{m}{2} (I_{n-1} + I_{n+1}) = 0, \quad n = 0, 1, 2, \cdots \, (N-1) \,,$$

where ω is the mode frequency yet to be calculated. With N cavities in the array, there are N eigenmodes of this coupled-cavity system. Ignoring end-effects, the solution for the q-th eigenmode is

$$\begin{aligned} \omega^{(q)} &= \frac{\omega_0}{\sqrt{1 + m \cos \frac{\pi q}{N}}} \,, \\ I_n^{(q)} &= \cos \frac{n\pi q}{N} \quad \text{or} \quad \sin \frac{n\pi q}{N} \,, \end{aligned} \tag{5.15}$$

as can be checked by back substitution. The accelerating electric field in the n-th cavity is proportional to $\frac{I_n}{i\omega C}$. The $\cos \frac{n\pi q}{N}$ and $\sin \frac{n\pi q}{N}$ solutions represent two degenerate standing wave modes.

Table 5.1 gives a summary for the equivalent circuit model. Note that I_n and $\frac{I_n}{i\omega C}$ are 90° out of phase, reflecting the wall current and the gap voltage are 90° out of phase (see Fig. 5.2). Note also that the voltage is proportional to

Table 5.1: A summary of the LC circuit model representing an RF cavity array.

circuit quantity	the corresponding cavity property
L	cavity volume
$\frac{1}{C}$	cavity gap
m	opening area between adjacent cavities
energy stored in L	magnetic field energy
energy stored in C	electric field energy
q	mode index and inter-cavity field pattern
$\omega^{(q)}$	cavity mode frequency
I_n	current flowing on the walls of the n-th cavity
$\frac{I_n}{i\omega C}$	accelerating voltage of the n-th cavity

$\frac{1}{C}$, which in turn is proportional to the cavity gap length, as one would expect. The one-to-one correspondence between the cavity fields and the LC circuit parameters is rather surprisingly close and accurate.

The mode index q assumes the values

$$q = 0, 1, 2, \cdots (N-1) .$$

Each mode is characterized by one mode frequency $\omega^{(q)}$ and a set of N values of $\{I_n^{(q)}, n = 0, 1, 2, ...N-1\}$ that gives the inter-cavity mode pattern. The $q = 0$ mode, for example, has a mode frequency $\frac{\omega_0}{\sqrt{1+m}}$ and a mode pattern in which all cavities oscillate in phase. This mode is referred to as the 0-mode.

A relationship linking the mode frequency to mode pattern, such as the relationship of $\omega^{(q)}$ to q for the case of our cavity array, is called a *dispersion relation*. When N is a large number, our dispersion relation is as shown in Fig. 5.22 for a case with positive m.

The quantity $\frac{\pi q}{N}$ is the relative phase between adjacent cavities. The mode with $\frac{\pi q}{N} = \frac{\pi}{2}$ is referred to as the $\frac{\pi}{2}$-mode, and the mode with $\frac{\pi q}{N} = \pi$ is referred to as the π-mode. Obviously there will not be an exact $\frac{\pi}{2}$-mode unless N is even, and the π-mode condition $\frac{\pi q}{N} = \pi$ is never fulfilled because the maximum extent of q is $N-1$; but we will not worry about that here when $N \gg 1$. The quantity $\frac{\pi q}{N}$ becomes a continuum when $N \to \infty$. Table 5.2 gives a summary of mode patterns.

When Table 5.2 refers to a "0" for the mode pattern of the $\frac{\pi}{2}$ case, it means the field is zero and vanishes all the time. This is not to be confused with the case when the field is oscillating and vanishes at some instants of time. Here the reader is suggested to contemplate how is it possible to have zero field in a cavity while still providing the needed coupling between its two adjacent cavities.

The "snap shot" cavity field patterns for the 0-, the $\frac{\pi}{2}$-, and the π-modes are illustrated in Fig 5.23 (with $N \gg 1$).

Figure 5.22: Dispersion relation for an RF cavity array in the LC circuit model for the case $m = 0.3$.

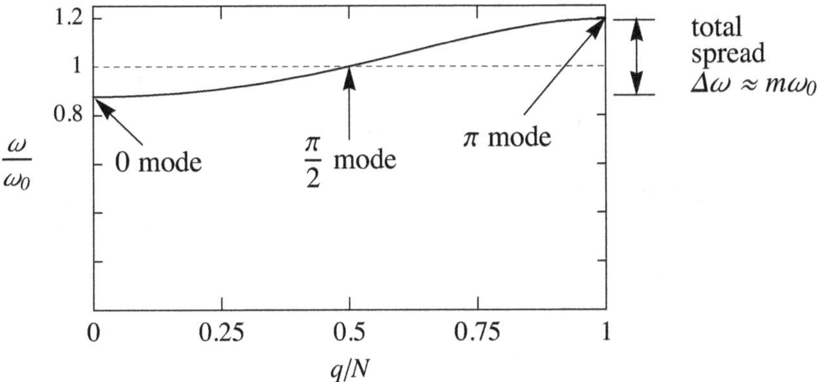

Table 5.2: A summary of the mode patterns of a LC circuit model.

Mode	time dependence	field pattern
General	$e^{i\omega^{(q)}t}$	$1 \quad \cos\frac{\pi q}{N} \quad \cos\frac{2\pi q}{N} \quad \cdots\cdot \quad \cos\frac{(N-1)\pi q}{N}$
0-mode	$e^{i\omega^{(0)}t}$	$1 \quad 1 \quad 1 \quad \cdots\cdot \quad 1$
$\frac{\pi}{2}$-mode	$e^{i\omega_0 t}$	$1 \quad 0 \quad -1 \quad 0 \quad 1 \quad \cdots\cdot \quad 0$
π-mode	$e^{i\omega^{(N)}t}$	$1 \quad -1 \quad 1 \quad -1 \quad \cdots\cdot \quad (-1)^N$

The 0-mode has a mode pattern that all cavity fields oscillate in phase. The little space near the opening holes between two adjacent cavities are available for both cavities when $m > 0$. As a result, the effective cavity size available for the field is slightly larger than the case of a single cavity, which in turn means the inductance is effectively increased and the mode frequency is slightly lowered. This is reflected by the mode frequency formula for $\omega^{(0)}$. Similarly, for the π-mode, the fields of adjacent cavities oppose each other, and the mode frequency is slightly higher than the single-cavity value. The $\frac{\pi}{2}$-mode has a special feature that adjacent cavities behave as if uncoupled; and that is why its mode frequency remains ω_0.

For each mode specified by the mode index q, it accommodates also the mode $-q$. The mode frequency and mode pattern are unchanged by replacing q by $-q$. The above dispersion curve therefore has its mirror symmetry, and in fact the dispersion curve extends with $\frac{\pi q}{N}$ ranging from $-\infty$ to ∞ with a period of 2π.

Figure 5.23: Snapshot pictures of the field patterns in a cavity array in the $0, \frac{\pi}{2}, \pi$ modes.

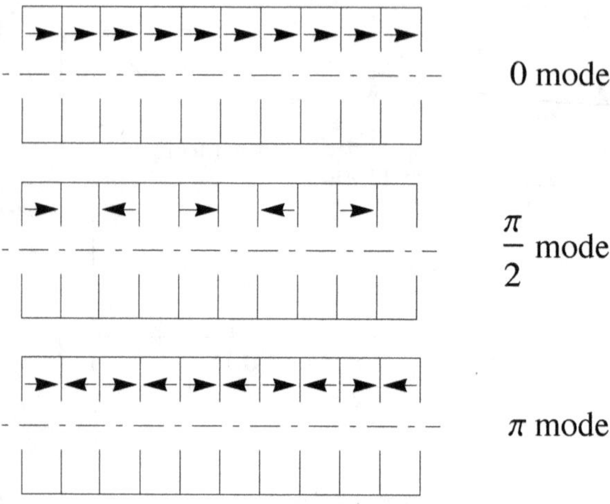

Associating $\kappa = \frac{\pi q}{N}$ with kL where L is the cavity cell length, the phase velocity $v_{\rm ph} = \frac{\omega}{k}$ can be read off the extended dispersion curve, and can propagate in the forward or backward directions. Each mode also has a group velocity given by $\frac{d\omega}{dk}$,

$$v_g = \frac{d\omega}{dk} = \frac{m\omega_0 L}{2} \frac{\sin\kappa}{(1 + m\cos\kappa)^{3/2}} . \tag{5.16}$$

One sees that the group velocity is positive when $\pi > \kappa > 0$ (as should be apparent from the dispersion curve) and negative when $0 > \kappa > -\pi$ for the case $m > 0$. It vanishes for the 0- and the π-modes. The group velocity must not be larger than c, which is assured because $|m| \ll 1$ and typically $\omega_0 L \sim 2\pi c$.

Homework 5.16 The text considered a long array of coupled cavity cells ignoring the end effects. Our result breaks down, for example, when $N = 1$. To get a feel of what end effects to expect, consider a small number of cells.

(a) With two coupled cells, show that the two modes are

$$\omega_\pm = \frac{\omega_0}{\sqrt{1 \pm \frac{m}{2}}}, \qquad I_0 = \pm I_1 .$$

(b) With three coupled cavities,

$$\omega_0 \text{ unperturbed}, \qquad \omega_\pm = \frac{\omega_0}{\sqrt{1 \pm \frac{m}{\sqrt{2}}}},$$

$$\{I_0, I_1, I_2\}_0 = \{1, 0, -1\}, \quad \{I_0, I_1, I_2\}_\pm = \{1, \pm\sqrt{2}, 1\} .$$

Homework 5.17 The circuit analysis in the text assumed the circuit was perfectly conducting with no resistance. The result applies to the case when the quality factor $Q = \infty$. Finite Q-value can be modeled by adding a resistance to the equivalent circuit, making it an LRC circuit. Perform this analysis and show explicitly the condition for overlapping modes when $Q \neq \infty$.

Homework 5.18 The text developed the circuit model assuming inductive coupling between adjacent cells. Follow the text to develop a circuit assuming capacitive coupling. The circuit might look like Fig. 5.24 for the case $m > 0$.

Figure 5.24: An equivalent LC circuit for an RF cavity array assuming capacitive coupling between adjacent cavities.

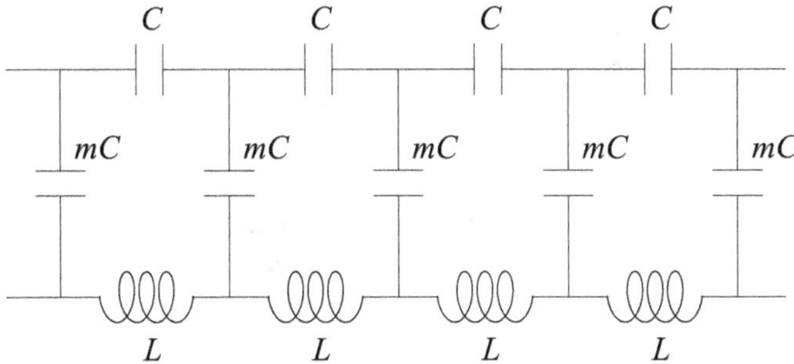

5.2.2 Mode driving

In Fig. 5.23, the arrows represent the standing wave patterns of the cavity array, including their relative phases. This pattern as a whole represents a mode of the cavity array. There are a total of N pairs of cavity array modes with each pair consisting of two degenerate modes – one cosine pattern and one sine pattern. The 0- and π-modes are exceptions in that each has only one mode. The total number of modes is $2N$.

Exactly which of the $2N$ modes is driven depends on the microwave driving frequency. For the two degenerate modes with same mode frequency, the mode pattern will be determined by the location of the microwave drive coupler within the cavity array.

Each cavity cell has an infinite number of modes of its own. The mode we are dealing with assumes that each cavity cell is filled with the lowest mode, i.e. its own fundamental mode. Since all cavities are assumed identical, their unperturbed fundamental modes all have the same ω_0. As a result, the cavity array modes will have their mode frequency clustered around ω_0, as shown by the expression (5.15) for $\omega^{(q)}$. Figure 5.25 illustrates the mode frequency splitting due to coupling assuming all cavities were constructed identical.

Figure 5.25: Splitting of mode frequency when 13 cavities are coupled into a cavity array.

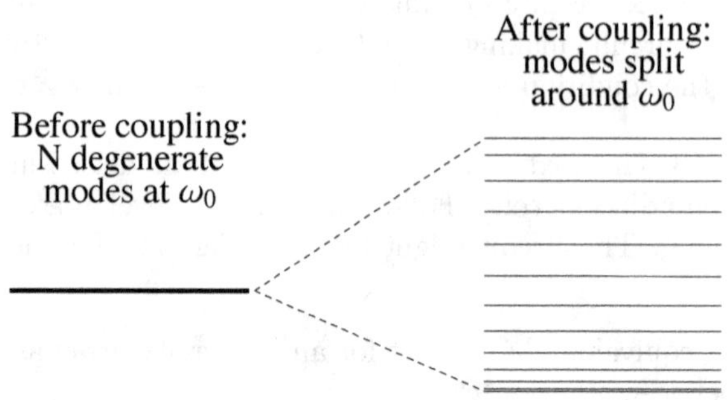

Having obtained the eigenmode frequencies, we then need to choose one of them to be driven by the microwave source. The optimal choice is near the $\frac{\pi}{2}$-mode. The reason is that near the 0- or the π-modes, the eigenmode frequencies are too crowded together — see Fig. 5.25. It is much easier to drive a single eigenmode when the neighboring eigenmode frequencies are far away. For $N \gg 1$, it is easy to show that near the 0- and the π-modes, the frequency spacing between adjacent modes is given by

$$\frac{\Delta\omega}{\omega_0} \approx \frac{m\pi^2}{4N^2}, \tag{5.17}$$

while the spacing near the $\frac{\pi}{2}$-mode is much larger,

$$\frac{\Delta\omega}{\omega_0} \approx \frac{m\pi}{2N}. \tag{5.18}$$

The $\frac{\pi}{2}$-mode is therefore much more cleanly driven by a microwave source.

Even for the $\frac{\pi}{2}$-mode, the mode spacing must be wide enough to avoid mixing between adjacent modes. This requires $\frac{\Delta\omega}{\omega_0} \gg \frac{1}{Q}$, or

$$\frac{|m|\pi}{2N} \gg \frac{1}{Q} \quad \Longrightarrow \quad |m| \gg \frac{2N}{\pi Q},$$

where Q is the quality factor of the cavities. So the condition for $|m|$ is such that it has to be much less than 1, but it also must be much larger than $\frac{N}{Q}$. For example, if $Q = 10^4, N = 100$ — this long cavity array is typical for an accelerator section of a linac, not for a storage ring — then we must have $1 \gg |m| \gg 0.01$. Since m is related to the size of the opening holes, this means the opening holes must not be too large or too small, and is far from being an arbitrary engineering parameter.

Homework 5.19 Verify Eqs. (5.17) and (5.18).

5.2.3 Structure optimization

There is, however, an important drawback for the $\frac{\pi}{2}$-mode, namely, the field pattern is such that every other cavity is not excited. Only cavities with $n =$ even (for the cosine pattern, or odd for the sine pattern) will ever have fields in them, and the cavities with $n =$ odd serve only as coupling agent and are always empty in fields. For the purpose of beam acceleration, this is a very wasteful thing to do. Accelerator physicists then invented several clever ideas to circumvent this problem, as illustrated in Fig. 5.26.

Figure 5.26: Evolution of accelerating structure.

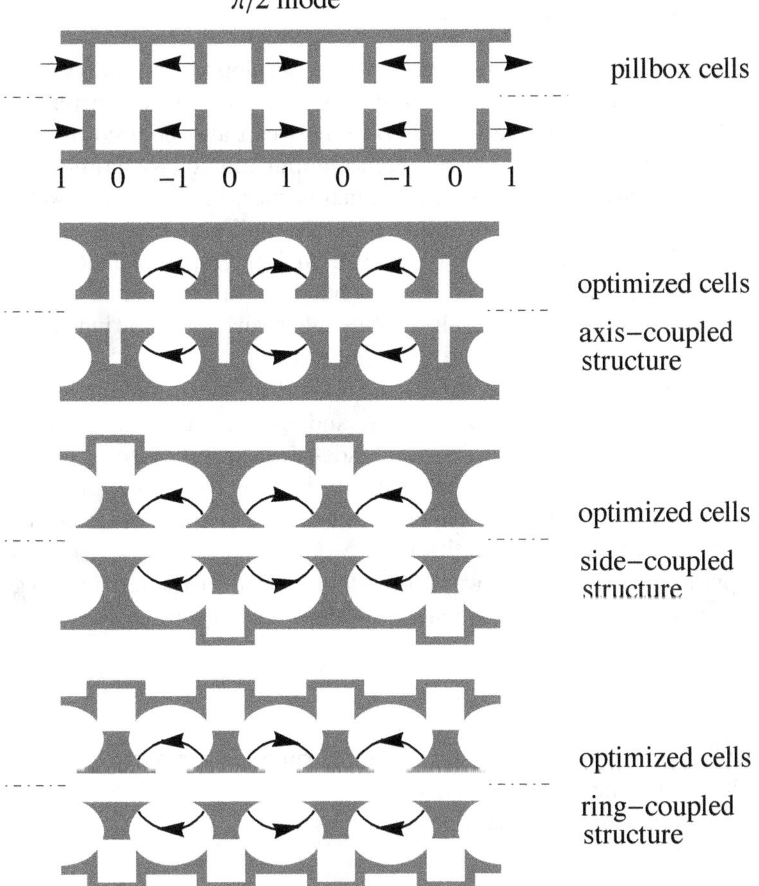

After optimization of the geometry of the individual cells as illustrated in Fig. 5.17, one proceeds to optimize the cavity array. As illustrated in Fig. 5.26, one way to optimize the $\frac{\pi}{2}$-mode array structure is to reduce the size of the zero-field cavities. To push that idea to an extreme, these zero-field cavities are pushed to the sides, as *side-coupled cavity structure*, so that they do not occupy any longitudinal space. When operated in the $\frac{\pi}{2}$-mode, a side-coupled structure behaves as a π-mode as seen by the beam.

In the side-coupled structure, fields in a cell are supposed to couple to the side cavities, and not to couple to the neighboring cavities. This is why the on-axis cylindrical channels connecting the adjacent cavities (beam passage) must be sufficiently long and narrow, so that the microwave cannot propagate through them (below cut-off) and is forced to go via the coupling cavities. The nose cones, as shown in Fig. 5.26, help in this regard.

5.2.4 Synchronism condition

Standing wave structure So far we have developed our analysis assuming the RF wave in the cavity array establishes a standing wave pattern. This is usually the case for a storage ring operation. It can also be used in an accelerating linac. However, in a linac, it becomes important to pay more attention to the efficiency of acceleration. As mentioned earlier, in a standing wave mode, some of the cavities are not efficiently providing full accelerating voltage. In a $\frac{\pi}{2}$-mode, for instance, every other cavity is in fact empty of fields. Even the other non-$\frac{\pi}{2}$-modes will have the cavities not filled up to their maximum field strength. The only modes with all cavities filled with accelerating fields to the maximum strength are the 0- and the π-modes.

As mentioned in Sec. 5.2.2, the 0- and π-modes tend to be unstable because the neighboring modes are too close by, and we tend to prefer $\frac{\pi}{2}$-mode, and consequently the inventions of axis- and side-coupled structures (Sec. 5.2.3). It turns out that even if the drive stability problem is overcome and it becomes possible to operate in the 0- or π-modes, they are still not useable because they do not fulfill a synchronism condition.

To operate in the standing wave 0-mode, in order for the particle to gain the maximum voltage every time it passes through a cavity, we have the condition that

$$\omega^{(0)}\frac{L}{v} = 2\pi\,,$$

where $\omega^{(0)}$ is the 0-mode frequency, $\frac{L}{v}$ is the time for the particle to pass from one cavity to the next. Similarly, to operate in a standing wave π-mode, we require a synchronism condition

$$\omega^{(N)}\frac{L}{v} = \pi\,.$$

Consider the case when the mode frequencies are near the unperturbed frequency ω_0 and when the particle motion is relativistic with $v \approx c$, the synchronism condition becomes

$$\omega_0 \frac{L}{c} \approx \begin{cases} 2\pi, & \text{0-mode} \\ \pi, & \pi\text{-mode} \end{cases} .$$

If we take a pillbox approximation $\omega_0 = 2.405\,\frac{c}{R}$, then we have

$$2.405\,\frac{L}{R} \approx \begin{cases} 2\pi, & \text{0-mode} \\ \pi, & \pi\text{-mode} \end{cases} .$$

It follows that the transit time factor $T = \sin(\frac{2.405\,L}{2R})/(\frac{2.405\,L}{2R})$ is given by

$$T \approx \begin{cases} 0, & \text{0-mode} \\ 0.64, & \pi\text{-mode} \end{cases} .$$

It is clear that the 0-mode is useless because the accelerating field has become equally decelerating due to its needed long cavity length. Even for the π-mode, its transit time factor is marginal at best.

There is another consideration of a π-mode standing wave operation, namely the group velocity, Eq. (5.16), of the microwave is zero for the π-mode. The microwave does not propagate from cell to cell were it not because of some manufacturing imperfections or when the cavity material is slightly lossy, and even then, the group velocity can only be slow. As a result, standing wave structures often do not have a large number (e.g. < 10) of cells per structure.

So standing wave structures have some issues when used for high gradient linacs. For the π-mode, there are three issues: (a) mode frequency too close to neighboring modes, (b) transit time factor is only ~ 0.64, and (c) group velocity is nearly zero, forcing the cavity array to be short. For the standing wave 0-mode, one problem is fatal: transit time factor nominally vanishes due to the synchronism condition. For the standing wave $\frac{\pi}{2}$-mode, transit time factor ≈ 0.9 (Homework 5.20), and its mode frequency is far from neighboring modes, but half of the cavities are ineffective.

Traveling wave structure We will describe another operational mode, i.e. a "traveling wave" mode. In contrast to standing wave operation, the number of cells for a traveling wave structure tends to be large (e.g. ~ 100). This is because in a traveling wave operation we do not have to operate in or near the π-mode to maximize the acceleration gradient, and furthermore, we would want to use up the microwave energy as it propagates down the structure until it is almost completely (e.g. $\sim 95\%$) consumed by the resistive wall of the structure or absorbed by the beam.

A comparison of standing wave and traveling wave structures is sketched in Fig. 5.27. The standing wave structures are typically used in storage rings and have relatively fewer cells. The coupling feed into a standing wave structure can be at some optimized location along the structure depending on the desired

Figure 5.27: A comparison of standing wave and traveling wave structures of a cavity array.

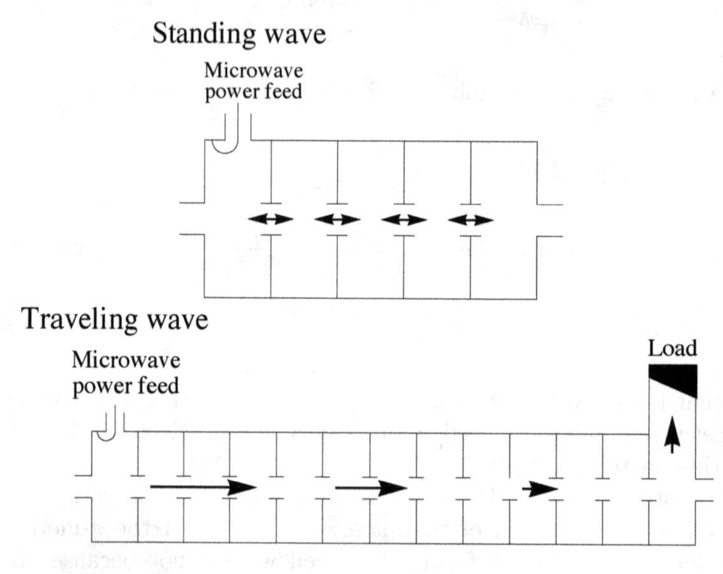

mode pattern – Fig. 5.27 shows a feed at the beginning of the structure. The traveling wave structures are often used in linacs and each structure may contain ∼100 cells. In the traveling wave case, the RF power is fed from the entrance point of the array and dumped at the other end.

Standing wave often (not always) is used for superconducting structures. This is especially the case with a π-mode traveling wave due to its small group velocity. Traveling wave often is used for normal conducting structures.

A long traveling wave structure can be modeled as a cavity array that has an infinite length. The equivalent circuit equation then reads

$$I_n\left(1-\frac{\omega_0^2}{\omega^2}\right)+\frac{m}{2}(I_{n-1}+I_{n+1}) = 0, \qquad n = -\infty,\cdots\infty.$$

Solution to this coupled equation, as can be shown by back substitution, is

$$\omega^{(\kappa)} = \frac{\omega_0}{\sqrt{1+m\cos\kappa}},$$

$$I_n^{(\kappa)} = e^{-i\omega^{(\kappa)}t+in\kappa},$$

where κ is a continuum of mode index. Physically, κ represents the relative phase of accelerating field from one cavity cell to its next neighbor. In a traveling wave mode, all cavities take turn to reach the same value of peak field. No cavities are ineffective.

To maintain synchronism for a traveling wave operation, we require

$$\omega^{(\kappa)}\frac{L}{v} = \kappa,$$

where L is the length of each cavity, v is the longitudinal velocity of the particle being accelerated.

For a relativistic beam with $v \approx c$, and a pillbox cavity with $\omega^{(\kappa)} \approx \omega_0 = 2.405\frac{c}{R}$, the synchronism condition yields

$$\frac{R}{L} \approx \frac{2.405}{\kappa}.$$

For a given phase advance between neighboring cells, synchronism condition is the same whether it is standing wave or traveling wave. Note that synchronism imposes a condition that the microwave's phase velocity matches the beam velocity, $v_{ph} = v$. It does not require a condition on the group velocity v_g.

Although mentioned before, it may be instructive to highlight the difference between the standing wave and traveling wave modes in terms of their accelerating fields. Let κ be the phase advance between adjacent cells, the acceleration field in the n-th cavity is

$$E_n = E_0 \times \begin{cases} \cos(n\kappa)\cos(\omega^{(\kappa)}t), & \text{standing wave} \\ \cos(\omega^{(\kappa)}t - n\kappa), & \text{traveling wave} \end{cases},$$

or

$$E_n = E_0 \times \begin{cases} \sin(n\kappa)\cos(\omega^{(\kappa)}t), & \text{standing wave} \\ \sin(\omega^{(\kappa)}t - n\kappa), & \text{traveling wave} \end{cases}.$$

It is clear that some cavities are not fully filled in the standing wave mode (except when $\kappa = 0$ or π), while all cavities are filled with full fields in the traveling wave mode.

Optimum cell-to-cell phase We also make a few remarks on optimal choice of phase advance κ,

- Take the $\frac{\pi}{2}$-mode with $\kappa = \frac{\pi}{2}$, the synchronism condition requires $\frac{L}{R} \approx 0.65$ which happens to be close to our numerical example which gave a transit time factor of 0.9 (Homework 5.20).
- For a π-mode, we would have $\frac{L}{R} \approx 1.3$. The transit time factor is reduced to 0.64 as discussed earlier.
- For a $\frac{\pi}{2}$-mode using a side-coupled structure, the synchronism condition is the same as that for the π-mode. In particular, the relatively low transit time factor is one disadvantage of the side-coupled structures.
- We mentioned earlier that the optimum for $\frac{R_s}{Q}$ occurs when $\frac{L}{R} \approx 1$. Synchronism for this cavity geometry occurs if one chooses $\kappa \approx 2.405 \approx \frac{3\pi}{4}$. This means the optimal coupled-cavity mode from the point of view of

shunt impedance is the $\frac{3\pi}{4}$-mode, while the $\frac{\pi}{2}$-mode is optimum from the point of view of avoiding coupling to adjacent modes. It is not inconceivable, therefore, that a broad optimum near the $\frac{2\pi}{3}$ is sometimes chosen.

Homework 5.20 Find the transit time factor T for a $\frac{\pi}{2}$-mode acceleration structure, whether standing wave or traveling wave, satisfying the synchronism condition for a relativistic beam.

5.3 Longitudinal dynamics

5.3.1 Principle of phase stability

So far, we have discussed the functions of an RF cavity or an RF cavity array, but have not yet discussed the longitudinal beam dynamics. To do so, we need to return to our longitudinal coordinates z and δ of the particle motion and establish their equations of motion.

In setting up the equation of motion, we shall consistently stay in the laboratory frame. This has been the obvious choice for the transverse motion — the coordinates x, x', y, y' are the displacements and slopes as seen in the laboratory frame. We now make the same choice for the two longitudinal coordinates z, δ. However, there is one difference between the transverse and the longitudinal coordinates, namely, z, δ are measured relative to a moving reference. The reference is a fictitious idealized particle called the *synchronous particle* that has exactly the design momentum and moves along the design orbit. Coordinates z, δ are therefore measured relative to a moving reference in the laboratory frame.

Incidentally, (measured relative to a moving reference in the laboratory frame) $\neq$ (measured in the rest frame of the moving reference). The quantity z refers to the relative longitudinal displacement as viewed in the laboratory frame. No Lorentz transformation is invoked or implied here.

We assume the longitudinal voltage across an RF cavity is

$$V(t) = V_0 \sin(\omega_{\mathrm{rf}} t + \phi_s)\,, \tag{5.19}$$

where ϕ_s is the RF phase of the reference synchronous particle. The RF frequency ω_{rf} is an integer multiple of the revolution frequency ω_0, i.e.

$$\omega_{\mathrm{rf}} = h\omega_0\,,$$

where h is called the harmonic number. The synchronous particle arriving the cavity at frequency ω_0 sees a fixed voltage of $V_0 \sin\phi_s$ turn after turn. A nonsynchronous particle will receive a voltage slightly different from the synchronous voltage and different from turn to turn depending on its exact time of arrival at the cavity. We have ignored the r-dependence of V here because we consider on-axis field with $r = 0$.

As mentioned, h has to be exactly an integer. Otherwise we will lose the synchronism and lose the ability to accelerate the particle. One might ask how exactly does this condition have to be fulfilled? What if, for example, $\frac{\omega_{\rm rf}}{\omega_0} = 200.000001$? If this were the case, then after $\frac{1}{2} \times 10^6$ turns, the RF voltage will get out of phase with the beam's arrival time, and we will be decelerating the beam. Since a beam is to be stored much longer than $\frac{1}{2} \times 10^6$ turns, apparently, any tiny mismatch of frequencies must not be allowed.

This difficulty was resolved by the important *phase stability principle* of McMillan and Veksler.[6] What happens is that under some condition of stability (which we shall provide later), the beam will settle this problem by itself. In particular, the phase stability principle makes the following two statements,

1. You first choose your $\omega_{\rm rf}$. Once $\omega_{\rm rf}$ is chosen, the beam — at least its synchronous particle — will adjust its revolution frequency ω_0 in such a way that it becomes *exactly* equal to $\frac{\omega_{\rm rf}}{200}$ even though its designed ω_0 is slightly off.

2. A particle with slight deviations in z, δ from the synchronous particle will oscillate around the synchronous particle, and these deviations will not grow indefinitely with time.

As mentioned, the beam will choose the synchronous particle once you choose your $\omega_{\rm rf}$ so that the exact synchronism condition is fulfilled. This is done by the beam slightly adjusting its beam momentum, and with this adjustment, the revolution circumference changes through the finite momentum compaction factor α_c by $\frac{\Delta C}{C} = \alpha_c \frac{\Delta P}{P_0}$. The change in circumference then changes the revolution frequency ω_0.

Incidentally, that is also how the beam copes with any construction error in the storage ring circumference. The phase stability principle says that the beam will adjust its momentum slightly to assure synchronism.

Measuring D_x, D_y, α_c The beam's ability to self-adjust its energy to comply with $\omega_{\rm rf}$ is put to use in electron storage rings as a way to measure the dispersion functions D_x and D_y at the beam position monitors. By measuring the x- and y-CODs as we vary $\omega_{\rm rf}$, it follows that

$$D_x = -\frac{\omega_{\rm rf}}{\alpha_c}\frac{d\Delta x_{\rm COD}}{d\omega_{\rm rf}}, \qquad D_y = -\frac{\omega_{\rm rf}}{\alpha_c}\frac{d\Delta y_{\rm COD}}{d\omega_{\rm rf}}. \tag{5.20}$$

Similarly, one can also measure the chromaticities ξ_x and ξ_y using

$$\xi_{x,y} = -\alpha_c \omega_{\rm rf}\frac{d\nu_{x,y}}{d\omega_{\rm rf}}. \tag{5.21}$$

If α_c is known, the chromaticities are obtained.

[6] V.I. Veksler, Comptes Rendus (Dokaldy) de l'Academie Sciences de l'URSS, 43, 8, 329 (1944); E.M. McMillan, Phys. Rev. 68, 143 (1945).

The phase stability is an important principle in accelerator physics. Together with the strong focusing principle, they provide the two foundations for all modern accelerators, phase stability addressing the longitudinal dynamics while strong focusing addressing the transverse dynamics. More on phase stability principle will continue in this chapter.

Homework 5.21 It seems we are sweeping something under the rug when we wrote Eq. (5.19). The reader is likely to note that the electric field is time-oscillating, that necessarily induces an oscillating magnetic field in the RF cavity. It then raises the question what does the magnetic field do to the passing particles. Under some circumstances, for example in our simplified case of a pillbox RF cavity, the magnetic field can cause a betatron tune shift. Find this tune shift. Do the horizontal and vertical tune shifts have the same sign or opposite signs? Does ignoring the magnetic field here violates symplecticity?

Solution Traversing an RF cavity can cause some transverse effects due to E_r and B_θ. Consider a particle with longitudinal displacement z relative to the synchronous particle — $z > 0$ meaning it is ahead of the synchronous particle. This particle arrives the RF cavity at time $t = -\frac{z}{c}$. For a pillbox cavity, the fields seen by this particle traversing the cavity are

$$\begin{aligned} E_z &= E_0 J_0\left(\frac{\omega_{\rm rf}}{c}r\right)\sin\left(-\frac{\omega_{\rm rf}}{c}z+\phi_s\right), \\ B_\theta &= \frac{E_0}{c}J_1\left(\frac{\omega_{\rm rf}}{c}r\right)\cos\left(-\frac{\omega_{\rm rf}}{c}z+\phi_s\right). \end{aligned}$$

This gives the map traversing the cavity, for a relativistic beam,

$$\begin{aligned} \Delta p_z &= \frac{eE_0\ell}{c}J_0\left(\frac{\omega_{\rm rf}}{c}r\right)\sin\left(-\frac{\omega_{\rm rf}}{c}z+\phi_s\right), \\ \Delta p_x &= -\frac{eE_0\ell}{c}\frac{x}{r}J_1\left(\frac{\omega_{\rm rf}}{c}r\right)\cos\left(-\frac{\omega_{\rm rf}}{c}z+\phi_s\right), \\ \Delta p_y &= -\frac{eE_0\ell}{c}\frac{y}{r}J_1\left(\frac{\omega_{\rm rf}}{c}r\right)\cos\left(-\frac{\omega_{\rm rf}}{c}z+\phi_s\right). \end{aligned}$$

This map (nonlinear) is symplectic.[7]

No edge fields play roles here. For the pillbox cavity, unlike the static cases as shown in Figs. 1.33 and 3.22, the hard-edge fields obey the Maxwell equations and do not stop abruptly at the end walls. The prize we paid is that the particle will have to penetrate a metal wall, which we assume is infinitely thin.

[7]You may want to prove that. To do so, it suffices to show that $\Delta\vec{p}$ can be written as a gradient of a function V, and that function V is $\frac{eE_0\ell}{\omega_{\rm rf}}J_0(\frac{\omega_{\rm rf}}{c}r)\cos(-\frac{\omega_{\rm rf}}{c}z+\phi_s)$.

When linearized, the 4×4 map for the vector (x, p_x, y, p_y, z, p_z) is given by

$$\begin{bmatrix} 1 & 0 & 0 & 0 & 0 & 0 \\ -\frac{1}{2}\frac{eE_0\ell}{c}\frac{\omega_{\rm rf}}{c}\cos\phi_s & 1 & 0 & 0 & 0 & 0 \\ 0 & 0 & 1 & 0 & 0 & 0 \\ 0 & 0 & -\frac{1}{2}\frac{eE_0\ell}{c}\frac{\omega_{\rm rf}}{c}\cos\phi_s & 1 & 0 & 0 \\ 0 & 0 & 0 & 0 & 1 & 0 \\ 0 & 0 & 0 & 0 & -\frac{eE_0\ell}{c}\frac{\omega_{\rm rf}}{c}\cos\phi_s & 1 \end{bmatrix}$$

There is no coupling among the x, y, z coordinates in the linear order, although they are coupled nonlinearly. Linearizing the map, the focal length f, for both x and y planes is given by (focusing if $f > 0$),

$$\frac{1}{f} = \frac{eE_0\ell}{\gamma mc^2}\frac{\pi}{\lambda_{\rm rf}}\cos\phi_s \,.$$

For a typical microwave RF cavity, f might be $\gtrsim 10^3$ m. The effect on the transverse motions is small.

There is of course also the focusing effect on the synchrotron motion due to the RF cavity as seen readily by the linearized map. That longitudinal focusing is in fact why the RF is installed in the first place. As will be explained later, operating a storage ring below transition, we need to choose ϕ_s for longitudinal focusing ($\cos\phi_x > 0$). The RF then will also focus the transverse motions. Operating above transition, ϕ_s is chosen so that the RF defocuses the longitudinal motion ($\cos\phi_s < 0$), then the transverse motions are also slightly defocused.

Homework 5.22

(a) Go through the thinking process and derive Eqs. (5.20) and (5.21).

(b) Consider a storage ring with circumference of 200 m, momentum compaction of $\alpha_c = 0.01$, and a RF harmonic number $h = 100$. In an effort to measure chromaticity down to a level of 0.1 units, how much do you have to swing the RF frequency $\Delta f_{\rm rf}$? Assume your tune measurement accuracy is 10^{-3}.

(c) Is it safe to change the RF frequency as much as required in (b)? Do you run the risk of losing the beam? Assume the maximum dispersion function around the ring to be 1 m.

Homework 5.23 Consider a synchrotron in which a beam is being accelerated by an RF system. The voltage at the RF at each beam passage is $V_0 \sin\phi_s$. The bending magnetic field B is increased proportionally to the beam momentum in such a way that the bending radius ρ, and therefore the beam trajectory, is kept constant during acceleration. Since the beam is relativistic, as a first estimate, let us make the approximation that the particle velocity $v = c$ throughout the acceleration process.

(a) Show that the rate of momentum change is

$$\dot{P} = \frac{eV_0 \sin\phi_s}{C} \,,$$

and the rate of change of the magnetic field must be

$$\dot{B} = \frac{V_0 \sin\phi_s}{\rho C},$$

where C is the accelerator circumference.

(b) The Fermilab Main Injector accelerates a proton beam from 9 GeV/c to 120 GeV/c in 1 sec. and has $\rho = 235$ m and $C = 3319.4$ m. Find the rate of change needed for the bending magnetic field and the needed RF voltage $V_0 \sin\phi_s$.

Solution (b) $V_0 \sin\phi_s = 1.23$ MV, $\dot{B} = 1.6$ T/s.

5.3.2 Synchrotron equation of motion

We mentioned earlier [see Eq. (5.1)] that since the synchrotron motion is slow, the longitudinal coordinates do not change much in one revolution around the ring (unlike the betatron motion), and as a result, we can make drastic simplifications. More specifically, what we can do is to make an *adiabatic approximation* (sometimes also called smooth approximation, as a somewhat trivial application of the sophisticated smooth approximation mentioned in Sec. 3.7) by averaging the particle motion over one revolution. The RF cavity, which normally is rather localized in one area in a storage ring, for example, is approximated as uniformly spread out over the ring. The discrete time step of one revolution is smoothed out into a continuous evolution in time.

The synchronous phase ϕ_s is defined as the RF phase seen by the idealized synchronous particle as it traverses the cavity. This means the accelerating voltage seen by the synchronous particle, turn after turn, is

$$V_s = V_0 \sin\phi_s,$$

where V_0 is the total integrated voltage around the ring. Assuming all cavities are phased and timed correctly, then it is equal to the sum of the voltages of all cavities. Subscript s means synchronous particle.

At this point, we find it convenient to first introduce the coordinates $\Delta E = E - E_s$ = energy of of a particle relative to the synchronous particle, and ϕ = RF phase at the arrival time of the particle. These coordinates change from turn to turn and, as we shall see, they are directly related to the z, δ coordinates.

Energy variation In the adiabatic approximation, the rate of change of ΔE due to the RF acceleration is given by

$$\dot{\Delta E} = eV_0 \frac{\omega_0}{2\pi}(\sin\phi - \sin\phi_s),$$

where ω_0 is the angular revolution frequency of the synchronous particle. Since

$$\delta = \frac{\Delta P}{P_s} = \frac{1}{\beta_s^2}\frac{\Delta E}{E_s}, \tag{5.22}$$

where $\beta_s c, P_s$ and E_s are the velocity, design momentum and design energy of the synchronous particle, we have

$$\dot{\delta} = eV_0 \frac{\omega_0}{2\pi\beta_s^2 E_s}(\sin\phi - \sin\phi_s) . \tag{5.23}$$

We make a comment here regarding the case when the beam energy is accelerated (the jargon is energy is ramped) in a synchrotron. The fact that there is an acceleration means E_s, and therefore β_s if the beam is not ultrarelativistic, will change with time. If the magnet strengths are made to scale properly with E_s (linearly with P_s), then the circumference of the synchrotron can be kept constant. But the increasing β_s will make the revolution period shorter, i.e. ω_0 higher. To maintain synchronism, the RF frequency has to be increased in such a way that $\omega_{\text{rf}} = h\omega_0$ is maintained (or at least approximately maintained if we count on the phase stability principle to help). In our analysis, we assume this has been programmed into the ramping process. In practice, this requires some tuning capability of the RF cavities using either a tuner or some other device. This is a messy requirement; it might be necessary to intentionally install some lossy material such as ferrite into the otherwise pristine RF cavities to lower its Q-value.

Phase variation Equation (5.23) gives the first equation of motion for synchrotron dynamics that describes how δ varies with time. We need another — more complicated — equation that describes how ϕ varies with time.

The arrival time of a particle depends on its instantaneous revolution period T. If T is slightly different from the ideal revolution period T_s for the synchronous particle, then there is a change in ϕ per turn by an amount $\Delta\phi = \omega_{\text{rf}}(T - T_s)$. The rate of change of ϕ in the adiabatic approximation is then

$$\begin{aligned}\dot{\phi} &= \frac{\Delta\phi}{T} = \omega_{\text{rf}}\frac{T - T_s}{T} \\ &\approx \omega_{\text{rf}}\left(\frac{\Delta C}{C_s} - \frac{\Delta v}{v_s}\right),\end{aligned}$$

where we have introduced $T = \frac{C_s+\Delta C}{v_s+\Delta v}$ and $T_s = \frac{C_s}{v_s}$ with C and v the circumference and velocity, respectively.

Two terms contribute to the rate of change in ϕ. Both terms are related to the particle's momentum error δ.

The first term comes from the fact that the circumference of an off-momentum particle is slightly different from the design circumference. In Sec. 4.5, we learned that

$$\frac{\Delta C}{C_s} = \alpha_c \delta ,$$

where α_c is the momentum compaction factor, which typically (but not always) is positive and has a value $\sim \frac{1}{\nu_x^2}$ with ν_x the horizontal betatron tune.

The second term comes from the fact that an off-momentum particle has a velocity slightly different from the design velocity,

$$\frac{\Delta v}{v_s} = \frac{\delta}{\gamma_s^2}, \tag{5.24}$$

with γ_s the relativistic energy factor for the synchronous particle.

It turns out that both contributions are related to $\delta = \frac{\Delta P}{P}$. It should however be pointed out that they actually have different physical origins, and their connection to the same quantity δ is conditional. In reality, $\frac{\Delta v}{v}$ being related to δ is straightforward due to the obvious kinematic connection between v and P. The fact that $\frac{\Delta C}{C}$ relates to δ, on the other hand, is only indirectly through its dependence on the magnetic rigidity $B\rho \propto P$. Homework 5.25 illustrates a consequence of this subtle point by an example of mass spectrometer.

The mass spectrometer case aside, both contributions of $\frac{\Delta C}{C}$ and $\frac{\Delta v}{v}$ can indeed be related to δ. In most cases, these two contributions are of opposite signs. For a particle with $\delta > 0$, for example, its revolution period will lengthen slightly because its circumference is increased, but will shorten slightly because its traveling velocity is also increased. Combining the two contributions, we obtain the second equation for synchrotron motion,

$$\dot{\phi} = \omega_{\rm rf}\eta\delta, \tag{5.25}$$

where a *phase slippage factor* has been defined as

$$\eta = \alpha_c - \frac{1}{\gamma_s^2}.$$

For electron synchrotrons or storage rings, most likely γ_s is so large that

$$\eta \approx \alpha_c.$$

For proton or heavy ion accelerators, however, this is not always the case. In fact, one would have to worry about the effect of a *transition energy*, defined as the beam energy when its corresponding γ is equal to

$$\gamma_t = \frac{1}{\sqrt{\alpha_c}}.$$

Typically, the transition gamma is approximately equal to ν_x.

When the beam energy is below the transition energy, $\eta < 0$, the operation mode is called *below transition*. Similarly, an *above transition* operation has $\eta > 0$. Electron accelerators are most likely operated above transition. For a proton accelerator that requires beam acceleration, sometimes the beam energy crosses transition during energy ramping. This is a moment requiring extreme caution. At the moment of transition, the beam loses its longitudinal focusing

– momentarily losing the phase stability principle condition – and becomes especially vulnerable.[8]

Obviously, the complication of a phase transition occurs only when $\alpha_c > 0$. Special lattices have been designed to give $\alpha_c < 0$ to avoid transition crossing. In these cases, the problem of transition crossing is traded for the problem of a more complex lattice design.

Synchrotron oscillation Combining Eqs. (5.23) and (5.25) eliminates δ,

$$\frac{d^2}{dt^2}(\phi - \phi_s) = \frac{h\omega_0^2 eV_0\eta}{2\pi\beta_s^2 E_s}(\sin\phi - \sin\phi_s)\,.$$

If $\phi - \phi_s$ is small, i.e. if the particle's longitudinal location is close to the synchronous particle and $|\phi - \phi_s| \ll 1$, then we can linearize the equation of motion to obtain

$$\dot{\delta} \approx \left(\frac{eV_0\omega_0}{2\pi\beta_s^2 E_s}\cos\phi_s\right)(\phi - \phi_s)\,,$$

and

$$\frac{d^2}{dt^2}(\phi - \phi_s) - \left(\frac{h\omega_0^2 eV_0\eta}{2\pi\beta_s^2 E_s}\cos\phi_s\right)(\phi - \phi_s) \approx 0\,,$$

which we recognize describes a simple harmonic oscillation. The stability of the motion require that

$$\frac{h\omega_0^2 eV_0\eta}{2\pi\beta_s^2 E_s}\cos\phi_s < 0, \quad \text{or} \quad \eta\cos\phi_s < 0\,,$$

and when satisfied, the simple harmonic oscillation angular frequency is given by

$$\omega_s = \omega_0\sqrt{-\frac{heV_0\eta}{2\pi\beta_s^2 E_s}\cos\phi_s}\,.$$

This quantity is referred to as the *synchrotron frequency*. (The subscript s here, as an exception, does not stand for synchronous.) The ratio

$$\nu_s = \frac{\omega_s}{\omega_0} = \sqrt{-\frac{heV_0\eta}{2\pi\beta_s^2 E_s}\cos\phi_s} \tag{5.26}$$

is called the *synchrotron tune*.

The choice of parameters are such that

above transition	$\eta > 0$, $\cos\phi_s < 0$
below transition	$\eta < 0$, $\cos\phi_s > 0$

Figure 5.28: Linearization of the function $\sin(\phi - \phi_s)$. The solid curves are $\sin\phi - \sin\phi_s$; the dashed lines are $\cos\phi_s(\phi - \phi_s)$; the left figure is when $\phi_s = 0$; the right one is for $\phi_s = \frac{\pi}{4}$.

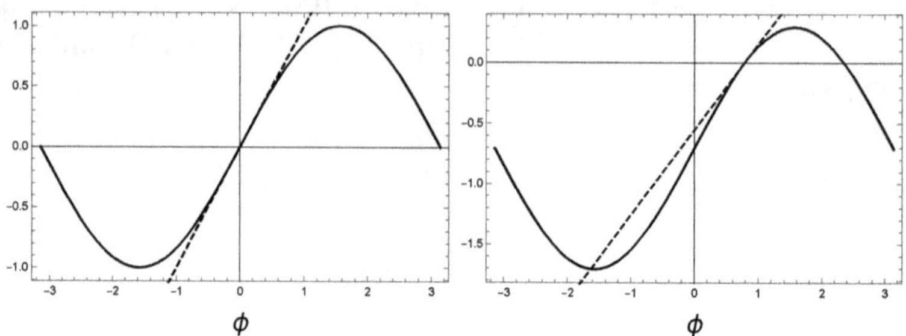

In particular, in the case of no net acceleration, $\sin\phi_s = 0$. The phase stability requires to choose $\phi_s = 0$ if $\eta < 0$ and $\phi_s = \pi$ if $\eta > 0$.

It may be useful to say a few words about the linearization approximation here. The linearization is to approximate $\sin\phi - \sin\phi_s \approx \cos\phi_s(\phi - \phi_s)$, as shown in Fig. 5.28. One should obtain from Fig. 5.28 a feel of how small the oscillation amplitude should be in order for this linearization approximation to hold. An inspection of the case $\phi_s = 0$ brings the following expectations,

- Since the linearization *over-estimates* the restoring force for larger oscillation amplitudes, one expects the effective synchrotron tune to *decrease* with oscillation amplitude.
- If the oscillation amplitude is too large, there should be some value beyond which the restoring force becomes completely ineffective, and the particle motion becomes unstable.

Both expectations will be borne out by later discussions.

Longitudinal dynamical variable We have mentioned a longitudinal coordinate z earlier. But in the above analysis, we used ϕ instead. By definitions of ϕ and z, they are related by

$$\phi - \phi_s \;=\; -\omega_{\rm rf}\frac{z}{\beta_s c}.$$

Note the minus sign on the right hand side, which comes from the convention that a positive (negative) z means a particle is ahead of (behind) the synchronous particle, while a positive (negative) $\phi - \phi_s$ means a particle is behind (ahead of) the synchronous particle.

[8]See, for example, J. Wei, Sec. 4.9, Handbook Accel. Phys. & Eng., 2nd Ed., World Scientific (2013).

The longitudinal coordinate variable can be confusing. Table 5.3 gives a few quantities related to the longitudinal coordinate. Hopefully, it clarifies some of the confusion.

Table 5.3: Possible quantities to be considered as the longitudinal coordinate dynamical variable.

longitudinal coordinate	s	distance traveled by the synchronous particle along its designated trajectory
relative longitudinal coordinate	z	distance of a nonsynchronous particle from the synchronous particle along the longitudinal direction. ($z > 0$ means the nonsynchronous particle is ahead of the synchronous particle.)
time	t	absolute time in the laboratory frame
relative arrival time	τ	time of arrival of a nonsynchronous particle relative to the synchronous particle, $\tau = -\frac{z}{\beta_s c}$. ($\tau > 0$ means the nonsynchronous particle is trailing the synchronous particle.)
RF phase	ϕ	phase of the RF voltage at the time of arrival of the particle, $\phi = \phi_s + \omega_{\text{rf}}\tau$. ($\phi > \phi_s$ means the nonsynchronous particle is trailing the synchronous particle.)

The stability condition requires that $\cos\phi_s$ be chosen to be of the opposite sign to η. Above transition, $\cos\phi_s$ must be negative, and below transition, it must be positive. According to the previous result, we have

$$\dot{\delta} \propto -z\cos\phi_s .$$

A particle with $z > 0$ (i.e. it arrives too early relative to the synchronous particle) will receive more acceleration above transition, and less energy below transition, compared to the synchronous particle. Also according to the result,

$$\dot{z} \propto -\eta\delta$$

A particle with $\delta > 0$ (i.e. it has a momentum slightly higher than that of the synchronous particle) will gain below transition, and lose above transition, in its longitudinal position relative to the synchronous particle.

The Hamiltonian The synchrotron equations of motion can be described in terms of a Hamiltonian

$$H(\phi,\delta,t) = \frac{1}{2}h\omega_0\eta\delta^2 + \frac{eV_0\omega_0}{2\pi\beta_s^2 E_s}[\cos\phi - \cos\phi_s + (\phi-\phi_s)\sin\phi_s]. \tag{5.27}$$

The first term in H is the kinetic energy term, and the second term is the potential well term. The Hamilton equations are

$$\begin{aligned}\dot{\phi} &= \frac{\partial H}{\partial \delta} &&= h\omega_0\eta\delta\,,\\ \dot{\delta} &= -\frac{\partial H}{\partial \phi} &&= \frac{eV_0\omega_0}{2\pi\beta_s^2 E_s}(\sin\phi - \sin\phi_s)\,.\end{aligned} \tag{5.28}$$

Incidentally, for the sake of good practice, note that the phase space is (ϕ, δ), not (δ, ϕ). Write $H(\phi, \delta, t)$, not $H(\delta, \phi, t)$. Also, when writing a Hamiltonian, avoid just writing $H = \cdots$. You should write $H(\phi, \delta, t) = \cdots$. This is because Hamiltonian is not a function.[9] Its meaning is contained in how it depends on the canonical coordinates while its numerical value has no meaning. Only when you clearly specify your canonical coordinates (ϕ, δ) — in the correct order — would your Hamiltonian carry a clear meaning. Other parameters in the Hamiltonian, e.g. V_0, η, are just some input parameters, and don't have the same status as (ϕ, δ). Substituting a numerical value for the other parameters in a Hamiltonian is fine, but substituting numerics for the canonical variables would be meaningless. The time variable to be used in the Hamilton equations should also be explicitly specified when writing the Hamiltonian. In our case, the time variable is clearly specified to be the absolute time t, and is not s for example.

The slow synchrotron motion has permitted us to make the adiabatic approximation. As a result, the equations of motion do not have a time-dependence. We can therefore avoid having to make a Courant-Snyder transformation for synchrotron motion. It follows that H is time-independent and is an invariant of the motion. Particle moves in the phase space (ϕ, δ) along constant-H contours. Note that in this Hamiltonian, we have made the adiabatic approximation; we did not make the linearization approximation.

Homework 5.24 Derive Eqs. (5.22) and (5.24), i.e.

$$\delta = \frac{\Delta P}{P_s} = \frac{1}{\beta_s^2}\frac{\Delta E}{E_s} = \gamma_s^2 \frac{\Delta v}{v_s}\,.$$

Homework 5.25 A slight extension of the discussions in the text yields the expression

$$\frac{\Delta f}{f} = \left(1 - \frac{\gamma^2}{\gamma_t^2}\right)\frac{\Delta v}{v}\,,$$

where f is the revolution frequency. But if we want to accommodate multiple particle species in the storage ring or synchrotron (most likely in an application

[9]This is a mathematician's statement. Hamiltonian does not mathematically fulfill the definition of a function.

for heavy ions), show that there is an additional term to be included, i.e.

$$\frac{\Delta f}{f} = \left(1 - \frac{\gamma^2}{\gamma_t^2}\right)\frac{\Delta v}{v} - \frac{1}{\gamma_t^2}\frac{\Delta(m/q)}{(m/q)}, \tag{5.29}$$

where m/q is the mass-to-charge ratio of the particle species. Note that it is m/q that appears here and not m or q separately. We use m/q instead of q/m to facilitate application to mass spectrometers.

Solution As mentioned in the text, when calculating $\frac{\Delta f}{f} = \frac{\Delta v}{v} - \frac{\Delta C}{C}$, the contribution $\frac{\Delta v}{v}$ comes from the velocity variation, while the contribution $\frac{\Delta C}{C}$ comes from the variation in the magnet rigidity $\frac{m\gamma v}{q}$. Considering two particles with the same v (therefore the same γ) but different m/q, their magnet rigidities are different as if their momentum differ by $\frac{\Delta P}{P} = \frac{\Delta(m/q)}{(m/q)}$. This equivalent variation of effective momentum, however, does not contribute to $\frac{\Delta v}{v}$; it only contributes to $\frac{\Delta C}{C}$. Adding this additional contribution from $\frac{\Delta(m/q)}{(m/q)}$, we obtain the final result (5.29).

As a test of this result, consider the case of a ring with uniform bending magnetic field,

$$\begin{aligned} f &= \frac{v}{2\pi\rho} = \frac{qB}{2\pi m\gamma} \\ \Longrightarrow \quad \frac{\Delta f}{f} &= -\frac{\Delta\gamma}{\gamma} - \frac{\Delta(m/q)}{(m/q)} = -\gamma^2\beta^2\frac{\Delta v}{v} - \frac{\Delta(m/q)}{(m/q)}, \end{aligned}$$

which agrees with Eq. (5.29) because in this case $\gamma_t = 1$.

Homework 5.26 The synchrotron motion is very similar to that of the pendulum when $\phi_s = 0$.

(a) Derive the pendulum equation of motion and its Hamiltonian. (Do not make a small-angle approximation.) Compare your result with the synchrotron equation and Hamiltonian. Obtain insight as which quantities are responsible for the gravity constant g, the pendulum length ℓ and mass m. What corresponds to motion within the RF bucket, and what corresponds to motion outside of the bucket?

(b) Contemplate what arrangement of the pendulum corresponds to $\phi_s \neq 0$. Find the equation of motion for this new setup of pendulum and study its stability limits.

Solution (a) For a pendulum, the Hamiltonian is

$$H(\theta, p_\theta, t) = \frac{p_\theta^2}{2m\ell^2} + mg\ell(1 - \cos\theta).$$

Physical meanings are such that $-\frac{1}{\eta}$ resembles $m\ell^2$, and V_0 resembles $mg\ell$.

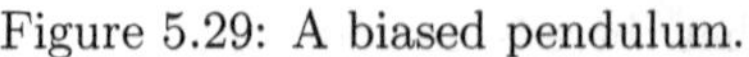

Figure 5.29: A biased pendulum.

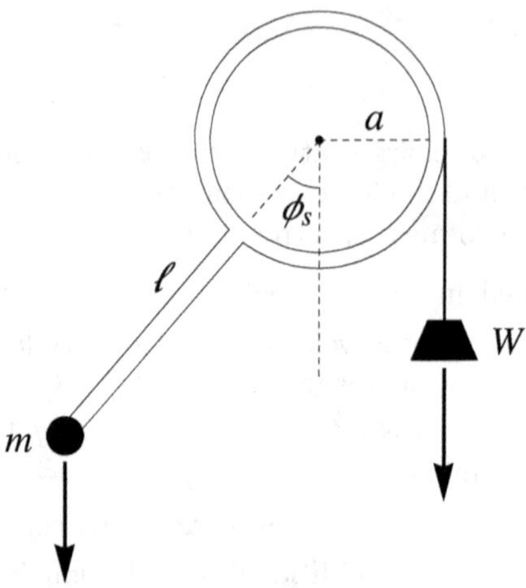

(b) See Fig. 5.29. Equilibrium state is given by $m\ell \sin\phi_s = Wa$. The quantity Wa therefore represents the particle energy gain per turn ΔE_s.

Homework 5.27 The text gave an expression of Hamiltonian in the (ϕ, δ) phase space. Find the Hamiltonian of the longitudinal dynamics in the adiabatic approximation in the $(\tau, \Delta E)$ phase space, where τ is the relative arrival time defined in Table 5.3.

Solution

$$K(\tau, \Delta E, t) = \frac{\beta_s^2 E_s}{\omega_{\rm rf}} H\left(\phi_s + \omega_{\rm rf}\tau, \frac{\Delta E}{\beta_s^2 E_s}, t\right).$$

Homework 5.28 In many accelerator applications, we require a bright beam with small emittance. It would be wonderful if there is a way to damp the synchrotron oscillation. As it stands, the synchrotron equations of motion are

$$\dot{\delta} = -\frac{\omega_s^2}{h\eta\omega_0}\Delta\phi, \qquad \dot{\Delta\phi} = h\eta\omega_0\delta$$
$$\Longrightarrow \qquad \ddot{\delta} + \omega_s^2\delta = 0.$$

One might invent an RF gadget perhaps like shown in Fig. 5.30 to damp this synchrotron oscillation. The gadget introduces an accelerating voltage that has a gradient $G = \frac{dV}{dx}$. If we install this RF at a location with dispersion $D \neq 0$, then the x displacement of the particle contains a contribution of $D\delta$, so the accelerating field contains a term $\Delta\delta = \frac{eG}{E_0} D\delta$. For simplicity, consider $\beta_s = 1$.

Figure 5.30: A possible gadget to induce damping of synchrotron oscillations.

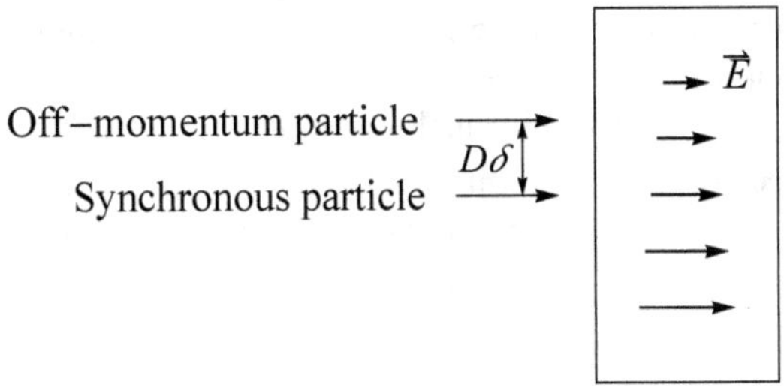

The equations of motion then becomes

$$\begin{aligned} \dot{\delta} &= -\frac{\omega_s^2}{h\eta\omega_0}\Delta\phi + \frac{\omega_0 eGD}{2\pi E_0}\delta, \qquad \dot{\Delta\phi} = h\eta\omega_0\delta \\ \Longrightarrow \qquad \ddot{\delta} &- \frac{\omega_0 eGD}{2\pi E_0}\dot{\delta} + \omega_s^2\delta = 0 . \end{aligned}$$

This is a damped equation if we choose $GD > 0$. This gadget damps synchrotron oscillations — or does it?

Solution The Hamiltonian is

$$H(\Delta\phi, \delta, t) = \frac{h\eta\omega_0}{2}\delta^2 + \frac{\omega_s^2}{2h\eta\omega_0}\Delta\phi^2 - \frac{\omega_0 eGD}{2\pi E_0}\Delta\phi\delta .$$

The combined equation of motion is

$$\ddot{\delta} + \left[\omega_s^2 - \left(\frac{eGD}{2\pi E_0}\right)^2 \omega_0^2\right]\delta = 0 .$$

There is only a frequency shift. No damping.[10]

5.3.3 Transition crossing

The negative mass The fact that a higher momentum particle moves *backward* in its longitudinal position, which occurs when the operation is above transition, and is counter-intuitive because the particle's actual velocity is faster than the synchronous particle, is called the *negative mass effect*. The mass being negative is meant to indicate that the particle moves opposite to the direction it is pushed.

[10]Incidentally, I happen to have known and witnessed this otherwise-clever gadget or its equivalents been invented and shot down a number of times. The use of Hamiltonian and implementation of symplecticity avoids such occasions.

Terminologies coined in past history often turned out inaccurate or even misleading in later days. The "negative mass effect", however, is one relatively rare example of a terminology that remains vividly illuminating. Longitudinal mass of the particle is sometimes negative!

The quantity $-\frac{1}{\eta}$ has the physical meaning of the mass of the dynamical system [noting $\delta \propto (-\frac{1}{\eta})\dot{z}$]. This mass is positive below transition and negative above transition. When crossing transition, this mass changes from $+\infty$ to $-\infty$, as illustrated in Fig. 5.31.

Figure 5.31: The longitudinal mass $-\frac{1}{\eta}$ as a function of beam energy.

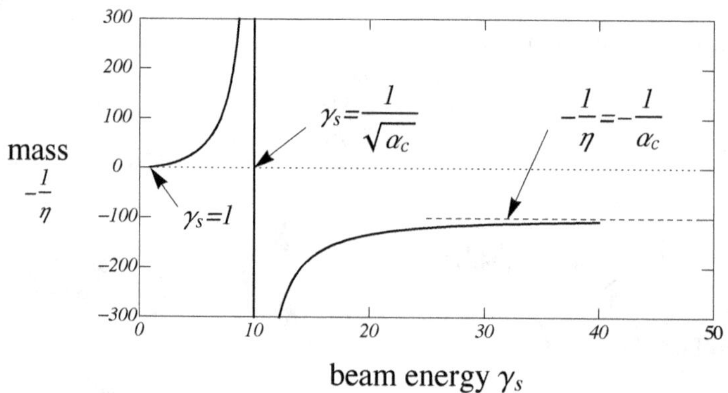

At this point, it may be useful to recall the two essential steps taken to establish the synchrotron equation of motion. Remember that the first step was adiabatic approximation, and the second step was linearization. Our results so far follow from these two steps. As promised, no Courant-Snyder transformation or $\alpha_z, \beta_z, \gamma_z$-functions have been introduced. In particular, in the adiabatic approximation, α_z is always zero, i.e. the phase space ellipse is always upright. We will return to review these two essential steps later.

Transition The linearized motion of a particle in the phase space (z, δ) traces out an upright ellipse with an aspect ratio of $\frac{\omega_s}{|\eta|\beta_s c}$ (Homework 5.29). Illustrated in Fig. 5.32 are the behavior of phase space above, below, and at transition. The rotation in phase space is counterclockwise (negative mass) above transition and clockwise (positive mass) below transition. At transition (for an "isochronous storage ring") with $\eta = 0$, the z-position of the particle is frozen in time. Energy of the particle continues to increase or decrease according to the temporarily-frozen voltage seen by the particle, leading to an instability.

Figure 5.32 intends to sketch the process of transition crossing. Figure (a) represents the motion of a typical particle in the (z, δ) phase space below tran-

Figure 5.32: Longitudinal dynamics in phase space below, at, and above transition. There is a transition jump of RF phase from $\cos\phi_s > 0$ to $\cos\phi_s < 0$ at the moment of transition crossing, as the panels (b) and (c) illustrate.

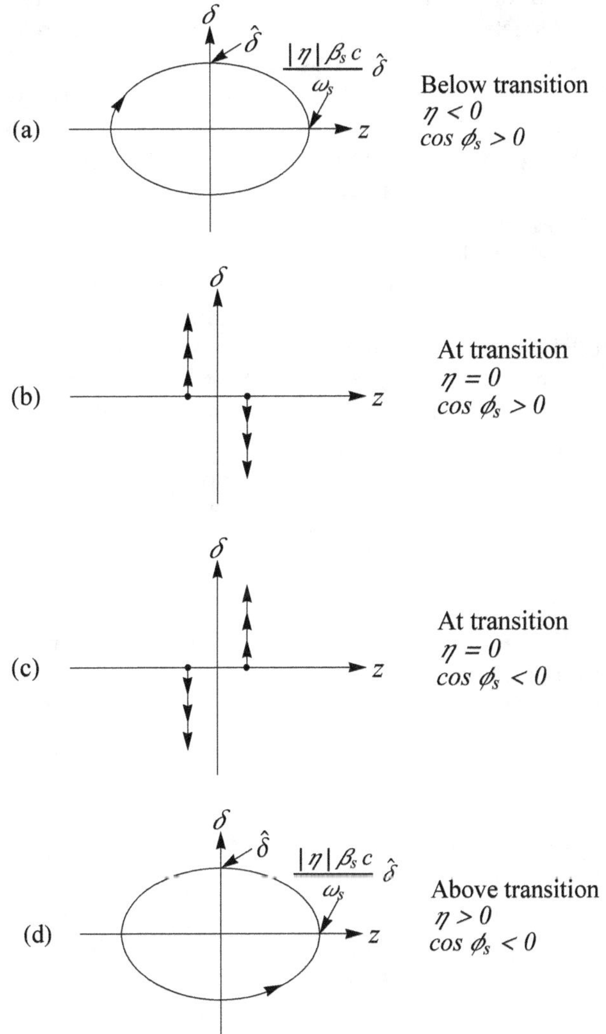

sition. When the beam is accelerated and transition crossing is approaching, the phase space ellipse grows shorter in z and taller in δ [See Eq. (5.30)]. Figure (b) is when the beam is right before crossing with a very short bunch length while particles' energies keep on increasing or decreasing. The beam is most vulnerable at this point.

As the beam continues to be accelerated, the accelerator operator will have to make a sudden jump of RF phase at the moment when transition is crossed. The RF phase seen by the center of the beam bunch is changed from ϕ_s to $\pi - \phi_s$ so that particle motion is stable after crossing. Immediately after crossing, we have figure (c). Particle energies now keep on decreasing or increasing, reversing their motion right before crossing. During the transition and RF phase jump, acceleration continues because the acceleration voltage $\propto \sin\phi_s$ is not affected.

After crossing, the phase space ellipse is reestablished by the phase jump. Initially the ellipse is very short and tall, but as acceleration continues, it flattens out and becomes normal looking, as sketched in (d). Compared with the case before crossing, however, particle motion in phase space has changed from clockwise to counter-clockwise.

Such violent dynamics makes particle motion very vulnerable. To minimize the perturbation to particle motion due to a transition jump, we sometimes not only jump the RF synchronous phase but also expedite the transition crossing speed by controlling the value of $\eta = \alpha_c - \frac{1}{\gamma_s^2}$ at the moment of the transition. The speed η changes with time is nominally determined by the rate of acceleration, but we can increase the speed by temporarily making a sudden time variation of α_c. With this temporary α_c jump at the moment of transition, the resulting η looks like illustrated in Fig. 5.33. The instantaneous speed of changing at the moment of crossing can be substantially increased.

Figure 5.33: Illustration of transition jump by a rapid blip of α_c. Dashed curves represent α_c. Solid curves are η. Plotted are functions of time as the beam is accelerated through the transition. The left and the right cases are without and with the α_c jump.

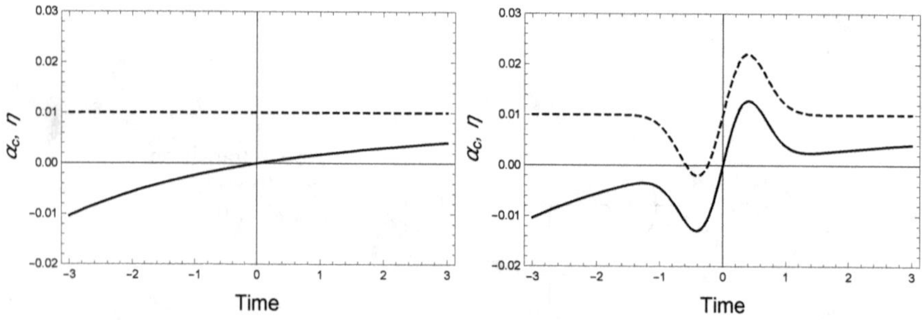

The α_c jump is executed by temporarily powering a set of selected quadrupoles. This requires the quadrupoles be located at positions with nonzero dispersion function, and they need to be powered in such a way that there is little net change in the two betatron tunes. When executed, the process is sometimes called a γ_t-jump. In more sophisticated scenarios, they also make a sign switch of the chromaticities from slightly negative to slightly positive values to put under control of the head-tail instability (Sec. 8.7.4).

Detailed analysis of particle motion follows from a similar analysis that led to Eqs. (5.23) and (5.25), except that because of the acceleration, all the external parameters $E_s, \beta_s, \omega_0, \phi_s, V_0, \eta$ now depend on time t. It is quite an achievement by the accelerator physicists that this problem has been resolved, requiring much careful and clear thinking. The problem is solved under the condition that all these external parameters change with time slowly (compared with synchrotron motion except at the very moment of crossing). In this case, it is found that the appropriate phase space coordinates to use are τ and $E - E_s$ instead of z and δ. It is then found that both τ and $E - E_s$ of a particle still execute synchrotron oscillation with *instantaneous* frequency given by Eq. (5.26), but their amplitudes have a slow time dependence given by[11]

$$\begin{aligned}
(\text{Ampl. of } \tau) &\propto \left(\frac{-\eta}{\beta_s^2 E_s \omega_0^2 e V_0 \cos\phi_s} \right)^{\frac{1}{4}}, \\
(\text{Ampl. of } E - E_s) &\propto \left(\frac{-\eta}{\beta_s^2 E_s \omega_0^2 e V_0 \cos\phi_s} \right)^{-\frac{1}{4}}.
\end{aligned} \tag{5.30}$$

During the acceleration, it has been assumed that the RF frequency $\omega_{\rm rf}$ is made to vary with the revolution frequency ω_0 according to $\omega_{\rm rf} = h\omega_0$ with a fixed harmonic number h.

An alert reader would note that, even though all external parameters vary individually, the amplitudes of τ and $E - E_s$ are such that their product remains a constant during the acceleration. This product of amplitudes of τ and $E - E_s$ of each particle, representing an area covered by the particle motion in the phase space, in units of eV-sec, is called the *normalized longitudinal emittance* of the particle, and it is an invariant during acceleration, including transition crossing.

Had we used other phase space coordinates in our analysis, we would have a phase space area that depends on time. For example, if we use z and δ, then the corresponding longitudinal emittance, in units of mm, would scale with $\frac{1}{\beta_s E_s}$. As the beam energy is accelerated, the apparent beam emittance decreases. This dependence of emittance on acceleration is referred to as a longitudinal adiabatic damping, in analogy to the adiabatic damping we discussed in the transverse case, Sec. 3.4. As pointed out at that time, adiabatic damping is not true damping.

Homework 5.29

(a) Show that, as claimed in the text, the aspect ratio of the upright ellipse in the (z, δ) phase space is $\frac{\omega_s}{|\eta|\beta_s c}$.

(b) Show that the longitudinal emittance ϵ in the (z, δ) coordinates and the normalized emittance ϵ_N in the $(\tau, E - E_s)$ coordinates are related by

$$\epsilon = \frac{c}{\beta_s E_s} \epsilon_N .$$

[11] See D.A. Edwards and M.J. Syphers, "An Introduction to the Physics of High Energy Accelerators". Wiley (1993).

Homework 5.30 Consider the case $\gamma_t \gg 1$ and uniform acceleration with V_0 and $|\cos\phi_s|$ being held constant during the energy ramping process. Apply Eq. (5.30) to show that near the time of transition crossing,

$$A_\tau \propto \frac{1}{A_{\Delta E}} \propto \left(\frac{\left| \frac{1}{(1+g\Delta t)^2} - 1 \right|}{1+g\Delta t} \right)^{1/4},$$

where $g = \frac{1}{\gamma_t}\frac{d\gamma}{dt}$ is given by the acceleration gradient and Δt is the time measured from the moment of transition. Plot this function to gain a feel of the phase space evolution during transition crossing.

5.3.4 How is ϕ_s determined

We now discuss how ϕ_s is determined. It is first to be determined by the acceleration rate for the synchronous particle, i.e.

$$\begin{aligned} eV_0 \sin\phi_s &= \Delta E_s \\ &= \text{desired energy gain per turn for the synchronous particle}. \end{aligned}$$

This condition gives two values of ϕ_s between 0 and 2π. Stability condition is needed to determine which one is the correct (stable) choice of ϕ_s. The result is illustrated in Fig. 5.34. More discussion can be found in Sec. 5.3.8.

Figure 5.34: Illustration of how ϕ_s is determined.

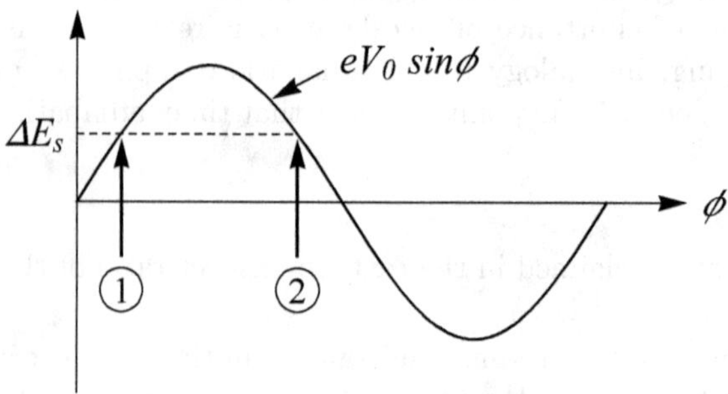

Actually, the accelerator designer does not choose the value for ϕ_s. The beam will make the choice automatically. Once injected into the storage ring, the beam will execute its motion around the correct, stable value of ϕ_s. Particles close to the correct stable ϕ_s will stay in the beam. Particles too far away are unstable and will be lost.

Physically, what happens is as follows. Consider the case when $\eta > 0$. A particle with higher energy ($\delta > 0$) executes a larger circumference. This particle will arrive at the RF cavity a little later in its next turn. If the synchronous phase is chosen correctly, the RF voltage will be dropping in time as the beam arrives. This means this particle with higher energy will receive less energy from the cavity in its next turn because it arrives late. Similarly, a particle with too low an energy will get a boost in energy from the RF cavity. In either case, the energy motion is *focused.* One can go through similar steps for the case $\eta < 0$. This is of course just the principle of phase stability mentioned in Sec. 5.3.1.

When the synchrotron or storage ring is equipped with an RF system and the principle of phase stability applied, it follows that the beam must have a bunched structure, with each beam bunch centered around a stable value of ϕ_s. The spacing between adjacent stable buckets is the RF wavelength.

In case the beam is being accelerated, the synchronous phase ϕ_s will adjust itself to be determined by the rate of acceleration. The rate of acceleration, on the other hand, is determined by the rate at which the bending magnet strengths are being ramped. The magnet ramping rate, therefore, determines the instantaneous stable ϕ_s — the beam will do all the rest thanks to the phase stability principle. All particles then execute synchrotron oscillations about this ϕ_s. Below transition, ϕ_s is between 0 and $\frac{\pi}{2}$. Above transition, ϕ_s is between $\frac{\pi}{2}$ and π. For a smooth acceleration through transition, the RF voltage must go through a sudden phase jump from one value of ϕ_s to the other. (What is being jumped is the RF phase; you cannot jump the beam's longitudinal position.) The phase space rotation slows down as the beam energy approaches transition, and speeds up away from it. The γ_t-jump, if applied, is performed during a short time period (milliseconds) when particles' longitudinal positions are basically frozen.

Homework 5.31 Consider a storage ring with two RF cavities. Let their frequencies be $\omega_{1,2}$ and their voltages be $V_{1,2}$. Let the revolution frequency be ω_0, and the two cavities have harmonic numbers $h_{1,2}$, i.e. $\omega_{1,2} = h_{1,2}\omega_0$. Let $\Delta\phi$ be the difference between their RF phases. Let the required energy gain per turn per particle be U_0.

(a) Find the expression for the synchronous phase of the synchronous particle in terms of the parameters given above.

(b) Find an expression for the synchrotron tune.

(c) The above assumes the synchronization condition is fulfilled with integral values of the harmonic numbers. This necessarily requires ω_1 and ω_2 be related by a rational number $\frac{h_2}{h_1}$. What happens to the beam dynamics if there is a slight error in ω_2 so that their ratio is not the intended value, perhaps even becomes an irrational number?

5.3.5 Negative mass instability

We make a detour here to discuss an effect called "negative mass instability". The first study of this instability was in fact by Maxwell. In 1856, 27 years before his celebrated equations, Maxwell (he was 25) won the Adams Prize when he showed that the Saturn's rings could not possibly be made of a single solid piece, and instead, it could only be composed of many small fragmented satellites.[12]

Consider a circular ring of dusts rotating around Saturn and consider a section of the ring. Let the mass distribution in the ring be nominally uniform, but has accidentally acquired a clustering as shown in Fig. 5.35. A particle at the trailing edge of the clustering, particle A, experiences a gravitational pull in the forward direction of its motion, thus gaining energy, while particle B at the leading edge of the clustering loses some energy.

Figure 5.35: A segment of the Saturn ring and a negative mass instability.

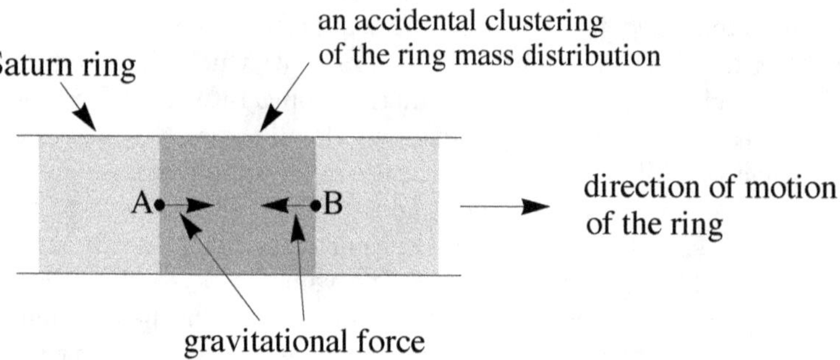

Following the gravitational pull, particle A starts to execute a slightly larger orbit in the gravity field of Saturn and its rotational period becomes longer (see Homework 5.32). As a result, its longitudinal position will trail behind the clustering some more. Conversely particle B's longitudinal position gets further ahead relative to the clustering. This is a negative mass effect in action. The net result is that any effort of condensing the ring into a solid piece would be broken up.

A similar negative mass instability also occurs in a circular accelerator above transition,[13] except that in this case, an accidental clustering in a section of the beam will get further clustered now that the attractive gravitational force is replaced by the repulsive Coulomb force. The net result is that any accidental clustering in the beam's charge distribution will get further clustered, leading to

[12]See "Maxwell on Saturn's Rings", Edited by S.G. Brush, C.W. F. Everitt, and E. Garber, The MIT Press (1983).

[13]C. E. Nielson and A. M. Sessler, Rev. Sci. Instr. 30, 80 (1959); C. E. Nielson, A. M. Sessler, and K. R. Symon, Proc. Int. Conf. High Energy Accel. and Instr., CERN, 1959, p. 239; L. J. Laslett, V. K. Neil, and A. M. Sessler, Rev. Sci. Instr. 32, 276 (1961).

a spontaneous clustering of the beam. Fortunately, this instability is relatively weak, and can be overcome by other damping mechanisms also active in the storage ring. We will discuss collective instabilities in Chapter 8.

Homework 5.32 In the text we claimed that particle A, after a gravitational pull in the forward direction, will make a larger orbit, meaning $\frac{\Delta C}{C} > 0$. Actually, in addition, its speed will slow down as it climbs up the gravitational potential well of the Saturn. In contrast to the case of an accelerator, therefore, both the $\frac{\Delta C}{C}$ and the $\frac{\Delta v}{v}$ terms contribute to negative mass. Find an expression for the change in rotational period after the gravitational pull, and show that this change has $\frac{2}{3}$ contribution from the $\frac{\Delta C}{C}$ term and $\frac{1}{3}$ from the $\frac{\Delta v}{v}$ term. The negative mass should be reflected by the fact that $\eta > 0$.

Solution Let particle A have mass m. Let the ring have radius R and rotational velocity v. Let Saturn mass be M, and G be the gravitational constant. Assume ring motion is nonrelativistic. The total energy of the particle is given by E = K.E. + P.E., where K.E $= \frac{m}{2}v^2$ and P.E. $= -\frac{GMm}{R}$. Balancing the centrifugal force with the gravitational force, we have $\frac{GM}{R} = v^2$, which then requires P.E. $= 2E$ and K.E. $= -E$. The total energy $E < 0$ is negative for a ring bound to Saturn's gravity.

Let particle A's velocity after the gravitation pull changes from v to $v + dv$, which means the total energy of particle A increased by an amount $\Delta E = mv\,dv$. We assume $|\Delta E| \ll |E|$ (so that, in particular, the particle is still bounded after the pull). This incremental energy is to be split between the kinetic and potential energies of particle A in its subsequent motion according to ΔK.E. $= -\Delta E$ and ΔP.E. $= 2\Delta E$. This in turn means the average radius changes by $\frac{\Delta R}{R} = -\frac{\Delta E}{E} = 2\frac{dv}{v}$ and the average velocity changes by $\frac{\Delta v}{v} = \frac{\Delta E}{2E} = -\frac{dv}{v}$. Pulling the particle velocity forward by dv forces it to climb up the potential well and causes it to lose its average velocity by $\Delta v = -dv$. The change in the rotational period is given by $\frac{\Delta T}{T} = \frac{\Delta C}{C} - \frac{\Delta v}{v} = -\frac{3\Delta E}{2E} = 3\frac{dv}{v}$.

5.3.6 Longitudinal strong-focusing analysis

So far, we have done the following: We made an adiabatic approximation, we developed the equations of motion averaged over one revolution, we linearized these equations of motion, and we then derived an expression for the synchrotron tune ν_s in Eq. (5.26). Question arises what if the derived ν_s turns out not $\ll 1$, because then the adiabatic approximation would break down, leading to an internal self-inconsistency. There is an apparent logic blind spot if ν_s is not guaranteed to be $\ll 1$.

This is indeed the case, and we need to examine this closer. To do so, let us back off from the adiabatic approximation, and assume the accelerator contains a single localized RF cavity. The analysis then resembles the transverse Courant-Snyder, and this can be referred to as longitudinal strong-focusing.

When needed, we can still linearize the system and derive an exact expression of $\nu_{s,\text{exact}}$.

The difference equations With a localized RF cavity, the action of the RF cavity is described by the map $(\phi_{\text{before}}, \Delta E_{\text{before}}) \to (\phi_{\text{after}}, \Delta E_{\text{after}})$,

$$\begin{aligned}\Delta E_{\text{after}} &= \Delta E_{\text{before}} + eV_0(\sin\phi_{\text{before}} - \sin\phi_s)\,, \\ \phi_{\text{after}} &= \phi_{\text{before}}\,.\end{aligned}$$

The action of the storage ring for one turn around the RF cavity is described by another map

$$\begin{aligned}\Delta E_{\text{after}} &= \Delta E_{\text{before}}\,, \\ \phi_{\text{after}} &= \phi_{\text{before}} + \frac{2\pi h\eta}{\beta_s^2 E_s}\Delta E_{\text{before}}\,.\end{aligned}$$

Combining the two maps gives the one-turn total map $(\phi_n, \Delta E_n) \to (\phi_{n+1}, \Delta E_{n+1})$ around the entrance of the cavity, where n refers to the n-th turn,

$$\begin{aligned}\Delta E_{n+1} &= \Delta E_n + eV_0\left(\sin\phi_n - \sin\phi_s\right)\,, \\ \phi_{n+1} &= \phi_n + \frac{2\pi h\eta}{\beta_s^2 E_s}\Delta E_{n+1}\,.\end{aligned} \tag{5.31}$$

Equation (5.31) can be used in an iteration procedure, e.g. in a computer simulation program, to obtain the turn-by-turn coordinates of a particle.

The map (5.31) is nonlinear and is symplectic. Its Jacobian matrix

$$\begin{bmatrix} 1 + \frac{2\pi h\eta}{\beta_s^2 E_s} eV_0\cos\phi_n & \frac{2\pi h\eta}{\beta_s^2 E_s} \\ eV_0\cos\phi_n & 1 \end{bmatrix}$$

has determinant of 1.

Note that on the right hand side of the ϕ_{n+1} equation in (5.31), we had ΔE_{n+1}. Had we carelessly written down ΔE_n there, the symplecticity condition would be violated, and that a mistake has been made. Readers beware.

Exact synchrotron tune We cannot solve Eq. (5.31) analytically in general. But we can solve at least the case when $\Delta\phi \equiv \phi - \phi_s$ is small by linearizing it (see Fig. 5.28). After linearization, the equations of motion can be cast in a matrix form,

$$\begin{bmatrix}\Delta\phi \\ \Delta E\end{bmatrix}_{n+1} = M\begin{bmatrix}\Delta\phi \\ \Delta E\end{bmatrix}_n\,,$$

where

$$M = \begin{bmatrix} 1 + \frac{2\pi h\eta}{\beta_s^2 E_s} eV_0\cos\phi_s & \frac{2\pi h\eta}{\beta_s^2 E_s} \\ eV_0\cos\phi_s & 1 \end{bmatrix} \tag{5.32}$$

is the one-turn map matrix around the entrance of the RF cavity.

As we did for the transverse motion, one can now perform a Courant-Snyder analysis on it. The condition for a stable linearized synchrotron motion is given by

$$|\mathrm{tr}M| \leq 2,$$

or

$$0 < -\frac{\pi h\eta}{2\beta_s^2 E_s} eV_0 \cos\phi_s < 1. \tag{5.33}$$

Below transition, $\eta < 0$, stability requires $\cos\phi_s > 0$ or $0 \leq \phi_s < \pi/2$. Above transition, we require $\pi/2 < \phi_s \leq \pi$. This result we already know.

From $\mathrm{tr}M$, we can calculate the exact value of the synchrotron phase advance per turn, which in turn gives an exact expression of the synchrotron tune,

$$\begin{aligned}\nu_{s,\mathrm{exact}} &= \frac{1}{2\pi}\cos^{-1}\left(\frac{\mathrm{Tr}M}{2}\right)\\ &= \frac{1}{2\pi}\cos^{-1}\left(1+\frac{\pi h\eta}{\beta_s^2 E_s} eV_0\cos\phi_s\right).\end{aligned}$$

In terms of ν_s, the synchrotron oscillation tune in the adiabatic approximation found in Eq. (5.26), we have

$$\nu_{s,\mathrm{exact}} = \frac{1}{2\pi}\cos^{-1}\left[1-\frac{(2\pi\nu_s)^2}{2}\right]. \tag{5.34}$$

When $\nu_s \ll 1$, we find

$$\nu_{s,\mathrm{exact}} \approx \nu_s,$$

i.e. the adiabatic approximation holds well. However, the approximation breaks down if ν_s becomes larger. In fact, if one makes the adiabatic approximation, one would miss out on any upper limit of the value of ν_s, while the stability condition (5.33) requires that the adiabatic result must satisfy the condition

$$\nu_s < \frac{1}{\pi}. \tag{5.35}$$

This instability limit results only if one departs from the adiabatic approximation.

To put in a different way: it is not true that to enhance longitudinal focusing, all one has to do is to crank up the RF voltage. If overdone, the system becomes overfocused.

To put in yet another way: Suppose there is one RF station in the ring. If Eq. (5.26) predicts $\nu_s = 0.25$, the actual ν_s is in fact 0.288. If it predicts 0.35, then the design is flawed; the longitudinal dynamics is in fact unstable.

Typically, ν_s is much smaller than $\frac{1}{\pi}$ in practical accelerators, although some accelerator designs have ν_s getting closer to this limit, e.g. $\nu_s \sim 0.1$. Furthermore, this limit on ν_s is derived by having only one RF station in the accelerator. With more RF stations, the upper limit is higher.

Courant-Snyder analysis In writing down Eq. (5.32) and by drawing an analogy to the transverse analysis, it can be realized that what we did was to represent the RF by a thin-lens "quadrupole" and the rest of the ring by a "drift space". The ring is therefore simply consisting of a thin-lens quadrupole plus a drift. The focal length of the equivalent quadrupole is $-\frac{1}{f} = eV_0 \cos\phi_s$, and the equivalent drift length is $\frac{2\pi h\eta}{\beta_s^2 E_s}$.

Here we come to conclude that, if we use $(\Delta\phi, \Delta E)$ as our dynamical variables, then we can draw a longitudinal analogy that

$$\text{Above transition} \quad \Longrightarrow \quad \begin{cases} \text{RF} = \text{focusing quadrupole}, \\ \text{Ring} = \text{positive drift}, \end{cases}$$

$$\text{Below transition} \quad \Longrightarrow \quad \begin{cases} \text{RF} = \text{defocusing quadrupole}, \\ \text{Ring} = \text{negative drift}. \end{cases}$$

Both cases, above and below transition, are of course stable having made the proper choice of ϕ_s.

One subtlety is that here the roles exchange if we use (z, δ) as the dynamical variables, because $\Delta\phi = -\frac{\omega_{\text{rf}}}{\beta_s c} z$. However, this is only a nomenclature issue with no physical consequences.

Following the transverse dynamics, we can perform a Courant-Snyder analysis on the synchrotron motion with one RF station in the storage ring in the $(\Delta\phi, \Delta E)$ phase space. The analysis and the "longitudinal Courant-Snyder functions" at the entrance of the RF cavity are found to be

$$\begin{aligned}
\text{one-turn map} &= \begin{bmatrix} 1 - 4\pi^2\nu_s^2 & \frac{2\pi h\eta}{E_s\beta_s^2} \\ eV_0\cos\phi_s & 1 \end{bmatrix}, \\
\mu_z &= \cos^{-1}(1 - 2\pi^2\nu_s^2), \\
\beta_z &= \frac{h|\eta|}{E_s\beta_s^2\nu_s\sqrt{1-\pi^2\nu_s^2}}, \\
\alpha_z &= -\frac{\pi\nu_s}{\sqrt{1-\pi^2\nu_s^2}}\,\text{sgn}(\eta), \\
\gamma_z &= \frac{|eV_0\cos\phi_s|}{2\pi\nu_s\sqrt{1-\pi^2\nu_s^2}},
\end{aligned} \tag{5.36}$$

where ν_s is the adiabatic synchrotron tune. The subscripts z indicates these Courant-Snyder functions refer to the longitudinal dimension. In most practical cases, ν_s is small. In those weak-focusing cases, the adiabatic approximation holds and we are led to the approximate expressions discussed before. Equation (5.36) however applies to larger values of ν_s, i.e. in the strong-focusing cases.

Beam tilt in phase space We recall that in the conventional adiabatic approximation, the beam distribution in the longitudinal $(\Delta\phi, \Delta E)$ phase space is an upright ellipse, and the same distribution applies at all positions around the accelerator circumference. In comparison, in a synchrotron with a single

RF cavity and with longitudinal strong focusing, the phase space distribution will be tilted ellipses and it changes around the accelerator, much like in the transverse dynamics. In this case, the rms bunch length and the rms energy spread of the beam will vary around the ring.

To quantify the tilt angle in the phase space, perhaps we can introduce the tilt parameter

$$t = \frac{\langle \Delta\phi \Delta E \rangle}{\sqrt{\langle \Delta\phi^2 \rangle \langle \Delta E^2 \rangle}} .$$

In terms of the Courant-Snyder functions, Eq. (2.44) says

$$t = \frac{-\alpha_z}{\sqrt{1+\alpha_z^2}} .$$

Observed at the entrance to the RF cavity, this tilt parameter in the longitudinal phase space becomes $t = \pi \nu_s \operatorname{sgn}(\eta)$.

Beam tilt in x-z space One can also look for a tilt angle of the beam in the physical x-z space when the beam has strong longitudinally focusing. The physical tilt angle is determined by

$$\theta = \frac{1}{2} \tan^{-1} \left(\frac{2\langle zx \rangle}{\langle z^2 \rangle - \langle x^2 \rangle} \right) . \tag{5.37}$$

This beam tilt is a result of discrete distribution of the RF. It will not appear in the adiabatic approximation. Later in Sec. 5.5.5, we will discuss another effect on the beam's closed orbit due to discrete RF distribution.

If we substitute $x = D \frac{1}{\beta_s^2} \frac{\Delta E}{E_s} + x_\beta$ and $z = -\frac{\beta_s c}{\omega_{\rm rf}} \Delta\phi$ into the expression (D is the dispersion at the location of beam tilt observation), and assume no synchro-betatron coupling between the betatron component x_β with the longitudinal coordinates, we obtain

$$\theta = -\frac{1}{2} \tan^{-1} \left(\frac{2 \frac{Dc}{\beta_s E_s \omega_{\rm rf}} \langle \Delta\phi \Delta E \rangle}{(\frac{\beta_s c}{\omega_{\rm rf}})^2 \langle \Delta\phi^2 \rangle - (\frac{D}{\beta_s^2 E_s})^2 \langle \Delta E^2 \rangle - \langle x_\beta^2 \rangle} \right) .$$

The distribution second moments in this expression are readily obtained from the longitudinal Courant-Snyder functions.

In case of a small tilt angle θ, $\beta_s \approx 1$, and $\langle z^2 \rangle \gg \langle x^2 \rangle$, the tilt angle in the x-z space due to a discrete RF distribution oscillates around the ring with an approximate amplitude

$$\theta \approx \pi \nu_s \frac{D\sigma_\delta}{\sigma_z} .$$

In order for this tilt not to change the intended beam orientation in the x-z space (e.g. for the applications in a collider or a free electron laser), we need the condition $\theta \ll \frac{\sigma_x}{\sigma_z}$, or

$$\pi \nu_s D \sigma_\delta \ll \sigma_x .$$

In most practical cases, this is fulfilled unless ν_s approaches $\frac{1}{\pi}$.

Homework 5.33 Consider a linearized longitudinally strong focused ring with a single RF station. The value of ν_s is less than but is close to $\frac{1}{\pi}$ so that the system is stable but only barely.

(a) Now somewhere in the ring is installed a dispersion bump, in such a way that it introduces an extra momentum compaction $\Delta\alpha_c$. How does this affect the stability of the accelerator? How does your answer depend on the sign of $\Delta\alpha_c$?

(b) This time there is installed an extra small RF cavity at the same location as your main cavity. The small cavity has the same ω_{rf} and same ϕ_s, but with a small voltage V_1. What happens to the stability of your accelerator?

(c) Same as (b) but now the extra RF is set to a different ϕ_s.

Solution Case (c) is instructive but requires more thinking. Still linearize the problem. But now you will find a "closed orbit" effect. It is a closed orbit in the $(\Delta\phi, \Delta E)$ phase space. Find this closed orbit and explain its meaning. You are advised to include a stand-alone map for the small RF cavity. Furthermore, you may want to extend your matrices to 3×3.

5.3.7 Phase space topology

We now return to the adiabatic approximation and a Hamiltonian description of the synchrotron motion. Given the Hamiltonian and its phase space, our first task is to examine the phase space topology which gives a qualitative description of how particles move in the phase space.

Consider first the simplest case of a simple harmonic oscillator. The Hamiltonian $H(x,p,t) = \frac{p^2}{2} + \frac{x^2}{2}$, for example, has a simple phase space topology, well known as in Fig. 5.36. The phase space topology consists of a fixed point at the origin and a set of clockwise circles around the fixed point. This case is particularly simple because the equations of motion are linear.

There are three ingredients of a phase space topology,

- stable and unstable fixed points,
- trajectory contours,
- arrows indicating the directions of particle motion along the contours.

Figure 5.36 has illustrated all three ingredients for the case of a simple harmonic oscillator.

For our storage ring, we will continue to keep the adiabatic approximation, but will back off from linearization. Without linearization, we cannot use matrices to analyze the dynamics. We use the Hamiltonian (5.27). With adiabatic approximation, on the other hand, the Hamiltonian is time-independent, and the system is integrable.

The Hamiltonian (5.27) is nonlinear in ϕ.[14] Its phase space topology is more complicated than that of a simple harmonic oscillator. It is a general rule that,

[14] In a Hamiltonian, quadratic dependence is linear, so it is only nonlinear in ϕ.

Figure 5.36: The phase space topology of the simple harmonic oscillator.

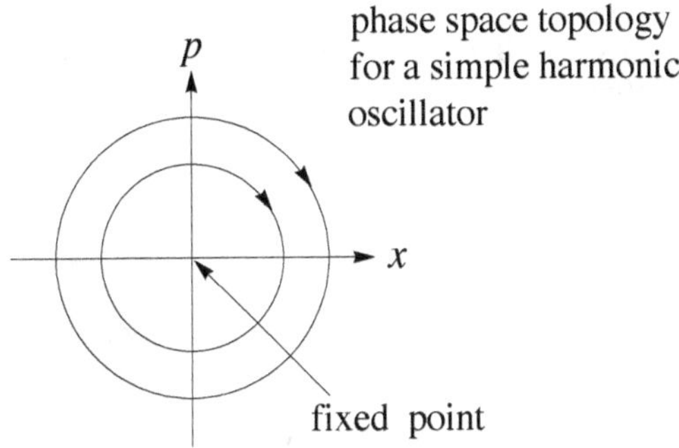

to examine the phase space topology, the first thing to do is to look for the *fixed points*. These are points in phase space which have the property that a particle at those points do not move, i.e. their position in the phase space is fixed.[15]

Obviously, the conditions for a fixed point are $\dot{\phi} = \frac{\partial H}{\partial \delta} = 0$ and $\dot{\delta} = -\frac{\partial H}{\partial \phi} = 0$. For the simple harmonic case, there is one and only one fixed point, namely the origin. For the storage ring, from the equations of motion, the fixed-point conditions are [see Eq. (5.28)]

$$\delta = 0, \qquad \text{and} \quad \sin\phi - \sin\phi_s = 0 .$$

There are an infinite number of points in the (ϕ, δ) phase space that satisfy the fixed point conditions, i.e. there are an infinite number of fixed points, but they divide into two classes,

$$\begin{cases} \phi = \phi_s + 2\pi k \\ \delta = 0 \end{cases} \quad \text{and} \quad \begin{cases} \phi = \pi - \phi_s + 2\pi k \\ \delta = 0 \end{cases} . \tag{5.38}$$

The first class is going to be *stable fixed points* (SFP) and the other is *unstable fixed points* (UFP). In the range $0 \le \phi < 2\pi$ — which we can limit ourselves to because the motion is periodic in ϕ with period of 2π — there are only two fixed points. The one at $\phi = \phi_s$ is the SFP, while the one at $\phi = \pi - \phi_s$ is the UFP. (Be reminded that ϕ_s is chosen so that $\eta \cos\phi_s < 0$.)

Once the fixed points are found, we already know how particle motion in phase space will look like qualitatively near them. At a fixed point, a particle does not move. Near a SFP, particle motions are stably focused, meaning that their phase space trajectories trace out ellipses around the SFP. Near an UFP, trajectories are hyperbolic, and diverge away from the UFP. In this sense, the UFP is a singular point. Although a particle exactly at an UFP will stay there forever, any infinitesimal deviation from it will lead to the eventual departure of the particle. The topology is illustrated in Fig. 5.37.

[15]How to translate fixed points to Chines? 乾脆把 Fixed Point 翻譯作“節骨眼”.

So far we considered one-turn fixed points. When multiturn fixed points are considered, the phase space topology generates islands. The stable SFPs become the multiturn COD and the centers of the islands. The set of m-turn SFPs show up as centers of a string of m islands. See Sec. 3.2.2.

Figure 5.37: Phase space topology near stable and unstable fixed points.

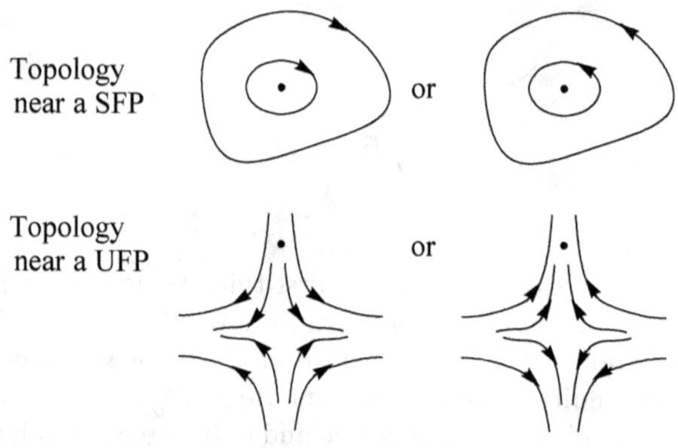

Homework 5.34 Calculate and plot the phase space topology for a dynamical system with Hamiltonian $H(x,p,t) = \frac{p^2}{2} - \frac{x^2}{2}$.

Solution

$$\begin{aligned} x(t) &= x_0 \cosh t + p_0 \sinh t \,, \\ p(t) &= x_0 \sinh t + p_0 \cosh t \,. \end{aligned}$$

Figure 5.38 shows the trajectories (red curves) following nine initial conditions (blue dots) as a function of time from $t = 0$ to $t = 1$. The only fixed point of the system is at the origin, and it is an UFP. From the lengths of the curves, obtain a feel of how rapidly the particle moves for each of the nine initial conditions. Particles A and B moves exactly towards the UFP, while slowing down as the UFP is approached and will never reach the UFP. Particles C and D moves away from the UFP and speed up as they go.

5.3.8 Longitudinal phase space of the storage ring

A moment's reflection then reveals that the longitudinal phase space topology of our Hamiltonian (5.27) must look something like shown in Fig. 5.39. In particular, the ellipses near the SFPs are clockwise when $\eta > 0$ and counter-clockwise when $\eta < 0$. These conclusions were made before in Fig. 5.32 except for two differences. One difference is that here we are not linearizing the motion and as a result no longer limited to the neighborhood of the SFPs, and the

Figure 5.38: Trajectories of a few particles with different initial conditions are plotted in phase space, showing the dynamics near the UFP.

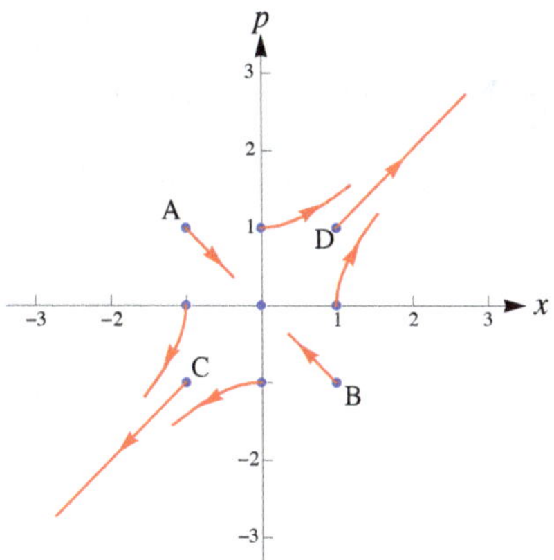

Figure 5.39: Longitudinal phase space of the storage ring, above transition (upper) and below transition (lower).

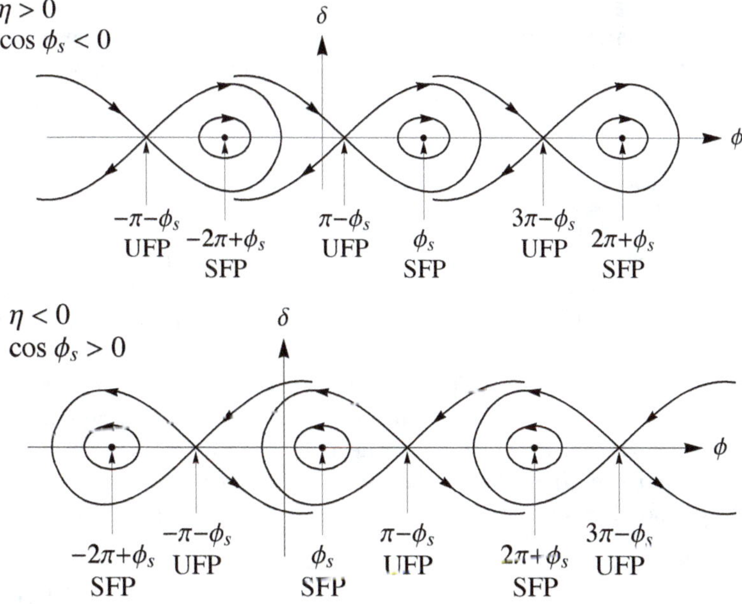

UFPs appear. The other difference is that we are using (ϕ, δ) instead of (z, δ) as the phase space. The clockwise and counterclockwise rotations are switched because z is related to ϕ with a change of sign.

The longitudinal phase space analysis goes in steps. (Steps 1 and 2 were discussed in Sec. 5.3.4, but are included below for completeness.)

1. First, one determines how much energy gain per turn U_0 is needed. Knowing the RF voltage V_0, we use $U_0 = eV_0 \sin\phi_s$ to obtain $\sin\phi_s$. There are two solutions for ϕ_s in the region $0 < \phi_s < 2\pi$. In most applications, we want to accelerate the particle, meaning $U_0 > 0$. In that case, one of the two solutions is in the range $0 < \phi_s < \frac{\pi}{2}$ and the other in the range $\frac{\pi}{2} < \phi_s < \pi$.

2. We then ask if η is positive (above transition) or negative (below transition). If $\eta > 0$, then we choose $\frac{\pi}{2} < \phi_s < \pi$. If $\eta < 0$, we choose $0 < \phi_s < \frac{\pi}{2}$.

3. Once ϕ_s is obtained, the SFPs and the UFPs in phase space are found from Eq. (5.38).

4. The next step is to draw the *separatrices* in the phase space using the Hamiltonian. These are the constant Hamiltonian contours that pass through the UFPs. A separatrix is a curve in the phase space that divides the stable and the unstable regions. We will explicitly calculate the separatrices in the following section.

5. Once the separatrices are drawn, the inside of the separatrices — the side that encloses a SFP, representing stable region in phase space – is called the *bucket*. One then calculates the height, the width, and the area of the bucket.

6. Finally, one asks if his/her beam stays safely inside the bucket.

Since the motion is stable near ϕ_s and unstable near $\pi - \phi_s$, the longitudinal coordinates of particles must cluster around ϕ_s, and stay away from $\pi - \phi_s$. As a result, and as mentioned before, one already makes a nontrivial observation that beams in a synchrotron when RF is applied are necessarily bunched.

Homework 5.35 When analyzing linearized motion in the text in the adiabatic approximation, we considered the stable motion near the SFP. The same analysis applies also to the UFP. Start with the equations of motion, but this time linearize around the UFP. Study the hyperbolic motion of a particle near the UFP. Show that $\Delta\phi = \phi - \pi + \phi_s$ obeys the linearized equation of motion

$$\ddot{\Delta\phi} = \omega_s^2 \Delta\phi \,,$$

where ω_s is the synchrotron frequency defined for the motion around the SFP.

This homework addresses a modification to the phase stability principle, Sec. 5.3.1. The principle addresses only the neighborhood of the SFP. The same principle actually also points out a principle of instability near the UFP. Both the principle of alternating gradient and the principle of phase stability are conditional.

5.3.9 Separatrix

The motion of a particle is described by the motion in the (ϕ, δ) phase space. In the adiabatic approximation, this motion is a continuous tracing along a constant-H contour. Different particles trace out different contours, but for a given particle, all points on the trajectory will have the same value of H. In a 1-D system like the one we are dealing with, the problem has been completely solved (or "integrated" if we use a mathematical jargon). An n-D system is integrated when n constants of the motion are found. In our 1-D case, this constant of the motion is the Hamiltonian itself.

Recall that in the transverse dynamics, the contour that a particle traces out, the Courant-Snyder ellipse, evolves around the storage ring. This complication does not occur for synchrotron motion. Here, the same contour applies around the storage ring due to the adiabatic approximation. In addition, this contour is up-right in the phase space. Had we chosen to analyze the problem with maps without making the adiabatic approximation, then even synchrotron motion will have evolving contours. In that case, the beam distribution will wobble with a tilt in the phase space. The linearized case was discussed in Sec. 5.3.6. For the nonlinear case, one should imagine the phase space plot of Fig. 5.39 wobbling slightly around the ring as well. But in addition, beware, a combination of nonlinearity and nonadiabaticity will cause a fundamental change; it causes chaos. We will return to this later.

The condition $H(\phi, \delta) = H_0$ gives a curve in the (ϕ, δ) space. This curve describes the constant-H contour of a particle whose $H = H_0$. The separatrix is defined to be the particular contour that passes through the UFP. It defines a boundary between bounded and unbounded particle motions in phase space. In this sense, drawing an analogy to particle's transverse dynamics, the separatrix can be considered a "longitudinal dynamic aperture".

Using the expression of the Hamiltonian, the value of H at the UFP is given by[16]

$$\begin{aligned} H_{sx} &= H(\phi = \pi - \phi_s, \delta = 0) \\ &= \frac{eV_0\omega_0}{2\pi\beta_s^2 E_s}[-2\cos\phi_s + (\pi - 2\phi_s)\sin\phi_s] \,. \end{aligned}$$

The separatrix therefore is described by the condition

$$H(\phi, \delta) = H_{sx} \,,$$

or

$$\delta^2 + \frac{eV_0}{\pi\beta_s^2 E_s h\eta}[\cos\phi + \cos\phi_s - (\pi - \phi - \phi_s)\sin\phi_s] = 0 \,.$$

Incidentally, the only contour that passes through the SFP is when $H_0 = 0$, and the contour consists of only a single point, the SPF itself.

[16]In this and the following two sections, we follow some of the development and notations of S.Y. Lee, Accelerator Physics, 3rd Edition, World Scientific (2011).

Figure 5.40: Separatrices in the longitudinal phase space.

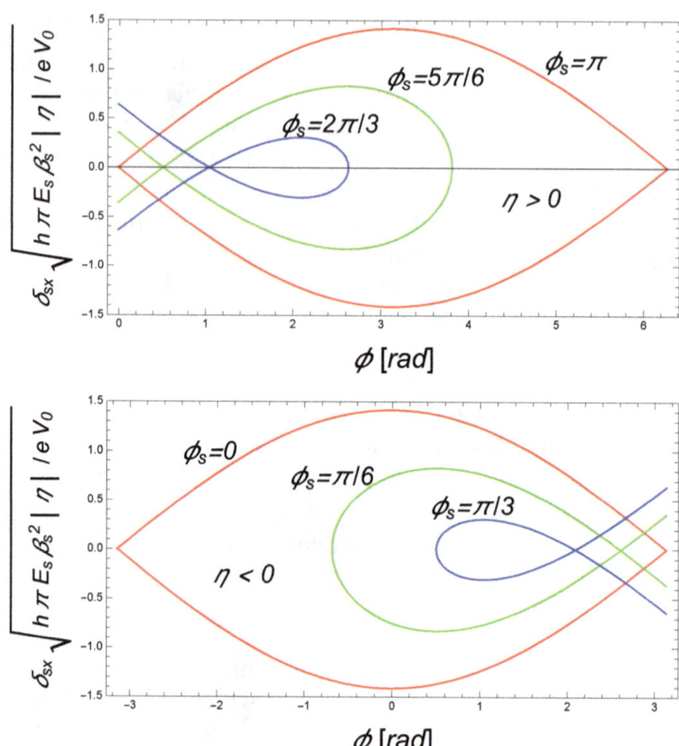

Figure 5.40 shows the separatrices. Each α-shaped curve is one separatrix corresponding to one value of ϕ_s. Each plot of this figure contains three curves for three values of ϕ_s, corresponding to three different storage ring RF settings. (Later we will give other plots when several curves appear in the same phase space. Those later plots correspond to various particles for one value of ϕ_s, i.e. in one storage ring RF setting, and are not to be confused with the separatrix plots shown here.) There are six cases plotted: $\phi_s = \frac{2\pi}{3}, \frac{5\pi}{6}, \pi$ (upper) and $\phi_s = 0, \frac{\pi}{6}, \frac{\pi}{3}$ (lower). Here we assumed the case of acceleration (i.e. $U_0 > 0$). Then the upper figure is for $\eta > 0$, while the lower figure is for $\eta < 0$. For convenience of presentation, the horizontal axis is from 0 to 2π in the upper plot and from $-\pi$ to π in the lower plot.

Recall that SFP is located at $\phi = \phi_s$. Each of the α-shaped stable region be centered around $\phi = \phi_s$.

Consider a particle moving along the separatrix. Let it start from a point very close to (but not exactly on) the UFP. Initially, it very slowly begins to move away from the UFP, but as time goes on, it picks up speed exponentially and moves away along the separatrix. After making one α-shaped loop, it

Figure 5.41: Synchrotron tune of a particle depends on which Hamiltonian contour it lies on. The tune varies from ν_s near the SFP to 0 near the UFP.

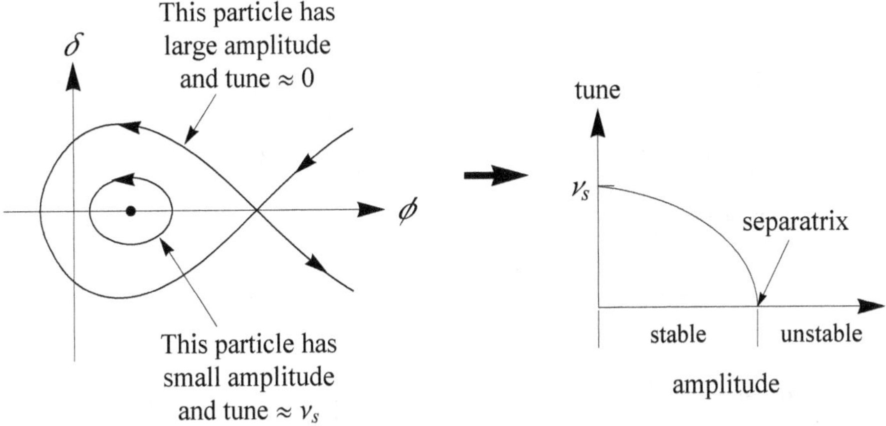

returns towards the UFP. As it approaches the UFP, it slows down, and will take infinitely long time to get to the UFP. Note that the particle does not bounce back and reverse its course and trace out the separatrix again in the opposite direction.

Figure 5.41 illustrates this situation. The left panel shows that a particle close to the SFP rotates continuously with frequency close to ν_s, while a particle close to the separatrix takes infinitely long time to complete one loop in phase space. The right panel then illustrates the dependence of the expected behavior of the synchrotron tune as a function of the particle's synchrotron oscillation amplitude. A particle on the separatrix has vanishing synchrotron tune. Beyond the separatrix, the motion becomes unbounded.

Homework 5.36 In Fig. 5.40, a quick recap,

(a) where are the SFP's and the UFP's? There should be one SFP and one UFP for each of the six curves.

(b) which direction is the particle moving along each of the separatrix curve? Draw arrows on the graphs to indicate the directions.

5.3.10 RF bucket

The separatrix encloses an area in the phase space. This enclosed area represents the stable region in the phase space called the RF bucket which is an important aspect of the longitudinal phase space topology. A particle in the bucket will be bounded by the separatrix, and therefore *trapped* within it. At the center of the bucket lies the SFP. The bucket defines the area allowed by a bunched beam to be stored in a storage ring. Particles outside the bucket are lost from this bunched beam. The reader should be reminded that particles are never allowed to cross the separatrix in phase space. Trapped particles do not leak out of the bucket and outside particles cannot get in.

Bucket width We now examine the bucket parameters closely. We first calculate the bucket width. The width is defined by the two extreme points on the bucket along the $\delta = 0$ axis. These two extreme points can be obtained by setting $\delta = 0$ in the separatrix equation and solving for ϕ. This yields

$$\cos\phi + \phi\sin\phi_s = -\cos\phi_s + (\pi - \phi_s)\sin\phi_s .$$

There are two solutions to this equation. One solution just gives the UFP at

$$\delta = 0, \qquad \phi_1 = \pi - \phi_s .$$

Let the other solution be designated as ϕ_2, i.e. let the other extreme point be located at

$$\delta = 0, \qquad \phi = \phi_2 .$$

The value of ϕ_2 has to be obtained numerically, and has to be chosen in the appropriate quadrant (refer to Fig. 5.40 for the quadrant to choose). For the special case $\phi_s = 0$, we have the two extreme points at $\phi_1 = \pi$ and $\phi_2 = -\pi$.

The bucket width is given by

$$\text{bucket width} = |\phi_2 - \phi_1| = |\phi_2 + \phi_s - \pi| .$$

Bucket area To find the bucket area, we first solve δ in terms of ϕ on the separatrix. This gives, in the region where there is a bucket ($\phi_2 > \phi > \phi_1$ when $\eta > 0$ and $\phi_1 > \phi > \phi_2$ when $\eta < 0$),

$$\delta_{\text{sx}}(\phi) = \left[-\frac{eV_0}{\pi\beta_s^2 E_s h\eta}[\cos\phi + \cos\phi_s - (\pi - \phi - \phi_s)\sin\phi_s]\right]^{1/2} .$$

The bucket area is given by integrating

$$\text{bucket area} = \oint \delta_{\text{sx}}(\phi)d\phi = \frac{16\nu_s}{h|\eta|\sqrt{|\cos\phi_s|}}\alpha(\phi_s) ,$$

where the dimensionless function $\alpha(\phi_s)$ is the bucket area normalized to the case when $\phi_s = 0$,

$$\alpha(\phi_s) = \frac{1}{4\sqrt{2}}\begin{cases} \int_{\phi_2}^{\pi-\phi_s}[\cos\phi + \cos\phi_s - (\pi - \phi - \phi_s)\sin\phi_s]^{1/2}\,d\phi, & \eta < 0, \\ \int_{\pi-\phi_s}^{\phi_2}[-\cos\phi - \cos\phi_s + (\pi - \phi - \phi_s)\sin\phi_s]^{1/2}\,d\phi, & \eta > 0 . \end{cases}$$

When $\phi_s = 0$, we have $\alpha(0) = 1$ and bucket area $= \frac{16\nu_s}{h|\eta|}$. It can be shown that $\alpha(\phi_s) \leq 1$ for all ϕ_s.

Figure 5.42: Quantities ϕ_1 (red), ϕ_2 (green), α (blue) and β (black dashed) as functions of ϕ_s. This plot shows all three bucket properties, the bucket area (α), bucket half-height (β) and width ($|(\phi_2 - \phi_1|$), above and below transition.

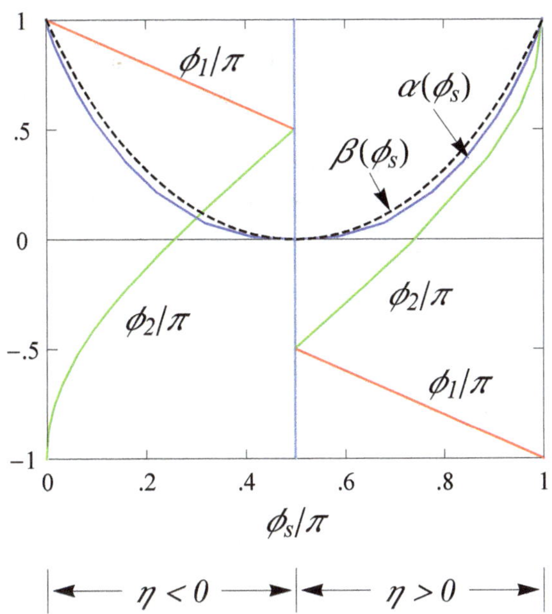

Figure 5.42 plots ϕ_1 (red), ϕ_2 (green), and $\alpha(\phi_s)$ (blue) as functions of ϕ_s. The function $\beta(\phi_s)$ will be mentioned momentarily. Note the jump of $\phi_{1,2}$ by π when ϕ_s crosses π. The two regions are disjoint, one applies to $\eta > 0$, the other $\eta < 0$.

When $\phi_s = \frac{\pi}{2}$, we have $\alpha = 0$. The RF bucket shrinks to vanish in this case. There is no stable region left if one tries to accelerate at the crest of the oscillating RF voltage. In fact, as ϕ_s approaches $\frac{\pi}{2}$, both the bucket length and bucket half-height approach zero, so the bucket area shrinks very quickly.

In our treatment, the bucket area is in units of radians. If we choose different coordinates for our phase space, then different units apply. For example, the longitudinal coordinate can be in radians (like ours), meters, or seconds. The momentum coordinate can be unitless (like ours), eV/c, or eV.

Another common unit of bucket area is eV-seconds, as discussed following Eq. (5.30). The important advantage of using eV-s is that adiabatic damping effect is automatically taken into account, and the bucket area becomes an adiabatically invariant quantity. This is very useful when a given beam is being accelerated by a cascade of accelerators. A bucket area in eV-s units occurs when one uses $(\Delta t, \Delta E)$ or $(\phi, \frac{\Delta E}{\omega_{\mathrm{rf}}})$ for the phase space. With $\Delta E = \beta_s^2 E_s \delta$, we see (see also Homework 5.29)

$$\text{bucket area in eV-s} \;=\; \text{bucket area in radian} \times \frac{\beta_s^2 E_s}{\omega_{\text{rf}}} .$$

Incidentally, sometimes the bucket area in eV-s is applied to the phase space $(\phi, \frac{\Delta E}{\omega_0})$ or $(h\Delta t, \Delta E)$. Then the area will be a factor h larger.

Bucket height To determine the bucket height, we first look for the location where $\frac{\partial \delta_{\text{sx}}}{\partial \phi} = 0$, i.e. where δ_{sx} reaches a maximum. This occurs at $\sin\phi = \sin\phi_s$, or $\phi = \phi_s$. The maximum height along the separatrix occurs at the synchronous phase, i.e. the maximum momentum deviation allowed by the RF bucket occurs at the phase of the SFP – a reasonable result.

The bucket half-height is then found to be

$$\hat{\delta}_{\frac{1}{2}} \;=\; \delta_{\text{sx}}(\phi_s) \;=\; \frac{2\nu_s}{h|\eta|}\left|1 - \left(\frac{\pi - 2\phi_s}{2}\right)\tan\phi_s\right| . \tag{5.39}$$

When $\phi_s = 0$, we have

$$\hat{\delta}_{\frac{1}{2}} \;=\; \frac{2\nu_s}{h|\eta|} .$$

Normalized to the value for the case $\phi_s = 0$, the RF bucket half-height can be written as

$$\hat{\delta}_{\frac{1}{2}} \;=\; \frac{2\nu_s}{h|\eta|}\beta(\phi_s) ,$$

where $\beta(\phi_s)$ has been plotted in Fig. 5.42. The functions $\alpha(\phi_s)$ and $\beta(\phi_s)$ happen to be numerically approximately equal.

For the case $\phi_s = 0$, the bucket half height is $\frac{1}{8}$ of the bucket area, while the full width of the bucket covers from $-\pi$ to π. Up to the bucket height, the bucket therefore covers $\frac{2}{\pi} = 64\%$ of the phase space area, as illustrated in Fig. 5.43.

Homework 5.37 A proton storage ring has a circumference $C = 100$ m, beam energy $\gamma = 10$. There is one RF station in the ring; its peak voltage is 2 MV and the synchronous particle gains 1 MeV per passage. We also know $\gamma_t = 5$ and the harmonic number is $h = 10$. Calculate the values of the following quantities,

$\omega_{\text{rf}}, \nu_s, \nu_{s,\text{exact}}$, longitudinal β-function at the RF,
RF bucket width, bucket half-height $\hat{\delta}_{\frac{1}{2}}$,
bucket area in (ϕ, δ), bucket area in $(\Delta t, \Delta E)$.

The β-function is defined in the (ϕ, δ) phase space, so it should be in the unit of radian.

Figure 5.43: The RF bucket when $\phi_s = 0$.

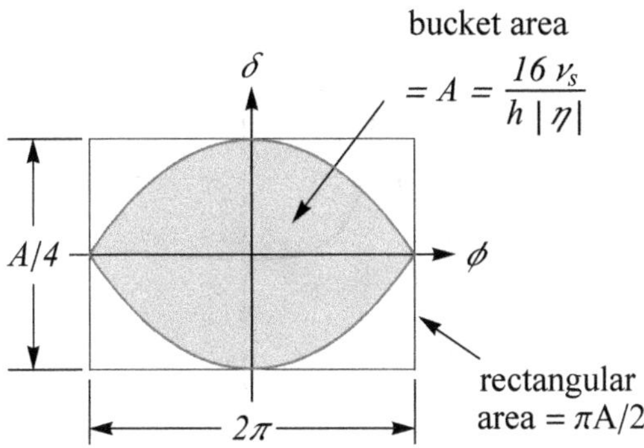

Solution $\omega_{\rm rf} = 1.88 \times 10^8\ {\rm s}^{-1}$, $\eta = 0.03, \phi_s = 2.618$ rad, $\nu_s = 0.00298423$, $\nu_{s,\rm exact} = 0.00298427$, ($\beta$ at RF) $= 101.5$ rad, (bucket width) $= \phi_2 - \phi_1 = 1.048\,\pi$, $\hat{\delta}_{\frac{1}{2}} = 0.0079$, [bucket area in (ϕ, δ) space] $= 0.057$ rad, [bucket area in $(\Delta t, \Delta E)$ space] $= 2.9$ eV-s.

Homework 5.38[17] Consider a synchrotron operated with a fixed RF voltage and fixed synchronous phase while varying the stored beam energy. The RF frequency is adjusted to match the fixed circumference.

(a) How does the RF bucket area [in eV-s unit in $(\Delta t, \Delta E)$ space] behave as a function of beam energy?

(b) Show that the bucket area has a minimum as γ changes, and that when $\gamma_t \gg 1$, this minimum occurs when $\gamma = \sqrt{3}\,\gamma_t$.

Solution

(a)

$$\text{Bucket area} \propto \sqrt{\frac{E_s}{|\eta|}} \propto \sqrt{\frac{\gamma}{\left|\frac{1}{\gamma_t^2} - \frac{1}{\gamma^2}\right|}}.$$

Denote the factor on the right hand side as $F(\gamma)$. It is plotted in Fig. 5.44.

(b) The function $F(\gamma)$ has a minimum at $\gamma = \sqrt{3}\,\gamma_t$. When $\gamma = \sqrt{3}\,\gamma_t$, $F' = 0$, bucket area is a minimum because $F'' \approx \frac{3^{3/4}}{2\sqrt{2\gamma_t}} > 0$. The minimum is rather shallow because F'' is small.

[17]This is a homework given in S.Y. Lee, Accelerator Physics, 3rd Edition, World Scientific (2011).

Figure 5.44: The quantity $F(\gamma)$ plotted as a function of γ for the case $\gamma_t = 5$. The vertical dashed line indicates the location $\gamma = \sqrt{3}\,\gamma_t$.

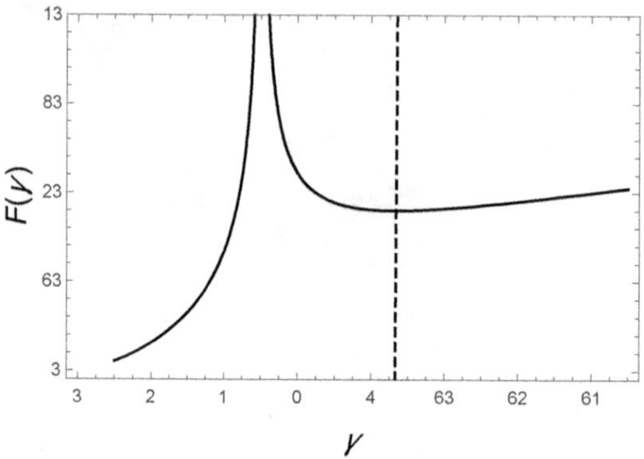

5.3.11 Particle motion in phase space

We have discussed the phase space bucket. We will now look at the motion of individual particles in phase space conforming to the topology of the fixed points and the bucket. In this consideration, we fix the RF system with a fixed value of ϕ_s, and consider the motion of individual particles with various initial conditions. To do that, we return to the Hamiltonian (5.27) and bring attention to its potential well term,

$$\text{Pot. well} = -\frac{eV_0\omega_0}{2\pi\beta_s^2 E_s}\left[\cos\phi - \cos\phi_s + (\phi - \phi_s)\sin\phi_s\right].$$

This potential well is illustrated in the two upper plots in Fig. 5.45 for the cases with $\phi_s = 0$ and $\phi_s = \frac{\pi}{6}$ (with $\eta < 0$). The motion of an individual particle can then be thought of as a particle moving in this potential well.

The red lines in Fig. 5.45 represent the energy level $H_{\rm sx}$ for particles that happen to be moving along the separatrix. Their trajectories in phase space therefore trace out the separatrices as shown by the two lower plots. When the potential well exhausts the available energy, the particle reaches a terminating point, where the phase coordinate is ϕ_2.

For a non-separatrix particle, its motion depends on whether its energy is below or above $H_{\rm sx}$. Below $H_{\rm sx}$, the particle motion is bounded around the SFP. Above $H_{\rm sx}$, the motion becomes unbounded. Two cases are illustrated in the "swimming fish" diagrams in Fig. 5.46 for $\phi_s = \pi$ and $\frac{5\pi}{6}$ (both cases assume $\eta > 0$). Red curves correspond to particles moving along the separatrices.

When $\phi_s = 0$ (or $\phi_s = \pi$ for $\eta > 0$), for example, we have

$$\text{Pot. well} = \frac{eV_0\omega_0}{2\pi\beta_s^2 E_s}(1 - \cos\phi).$$

Figure 5.45: The RF potential well.

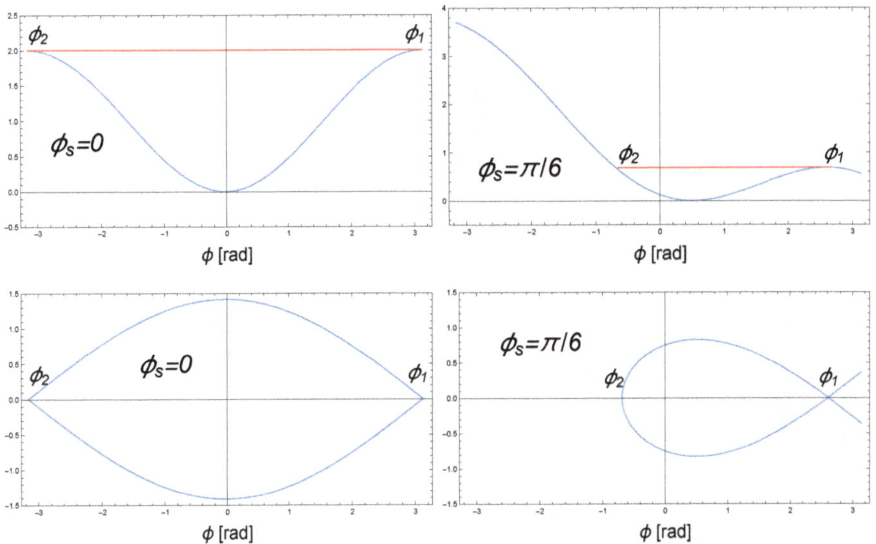

The phase space behaves like the upper panel in Fig. 5.46. This case can be modeled as a swinging pendulum (Homework 5.26). In an oscillation with small amplitude, the pendulum executes a simple harmonic motion. Particles executing small oscillations are illustrated as closed elliptical trajectories in the upper plot of Fig. 5.46. If the oscillation energy exceeds a certain value, the pendulum executes complete revolutions in a rotational motion. Four such trajectories were shown by the wavy non-closed curves. Which directions do the particles flow along their contours?

Small-amplitude oscillation For a beam to be trapped by an RF bucket, its particles will be kept to the neighborhood of the SFP. For small-amplitude particles near the SFP, i.e. $\phi \approx \phi_s$ and $\delta \approx 0$, we can linearize their motion, and the linearized Hamiltonian reads

$$\begin{aligned} H(\Delta\phi, \delta, t) &= \frac{1}{2}h\omega_0\eta\delta^2 - \frac{\omega_0 eV_0\cos\phi_s}{4\beta_s^2 E_s}\Delta\phi^2 \\ &= \frac{1}{2}h\omega_0\eta\delta^2 + \frac{1}{2}\frac{\nu_s^2\omega_0}{h\eta}\Delta\phi^2 , \end{aligned}$$

where $\Delta\phi = \phi - \phi_s$. Solving the equations of motion gives

$$\begin{cases} \dot{\Delta\phi} &= \frac{\partial H}{\partial \delta} \\ \dot{\delta} &= -\frac{\partial H}{\partial \Delta\phi} \end{cases} \Longrightarrow \begin{cases} \Delta\phi &= \hat{\Delta\phi}\cos(\nu_s\omega_0 t + \chi) \\ \delta &= -\hat{\delta}\sin(\nu_s\omega_0 t + \chi) \end{cases} ,$$

Figure 5.46: Particle motion in the longitudinal phase space for two cases with $\phi_s = 0$ and $\phi_s = \frac{5\pi}{6}$.

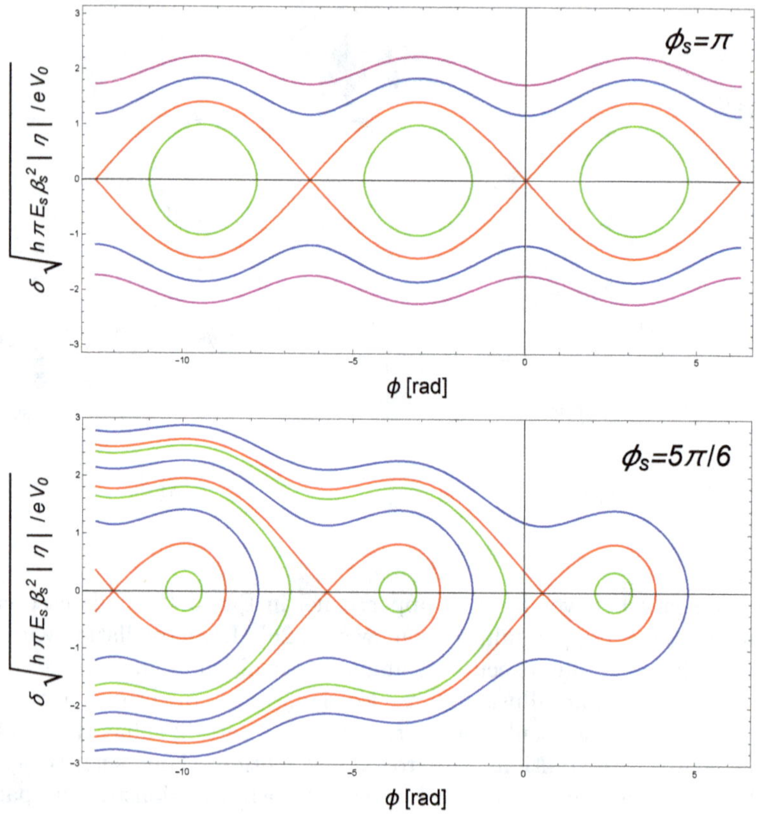

where

$$\hat{\delta} = \frac{\nu_s}{h\eta}\hat{\Delta\phi}.$$

This last expression relates the bunch length and its momentum spread. The z-motion is given by $z = \hat{z}\cos(\nu_s\omega_0 t + \chi)$ with

$$\hat{z} = -\frac{\beta_s c}{h\omega_0}\hat{\Delta\phi} = \frac{\beta_s c\eta}{\nu_s\omega_0}\hat{\delta}. \tag{5.40}$$

These small-amplitude particles trace out ellipses in the $(\Delta\phi, \delta)$ phase space,

$$\frac{\delta^2}{\hat{\delta}^2} + \frac{\Delta\phi^2}{\hat{\Delta\phi}^2} = 1.$$

The phase space area of the ellipse is $\pi|\hat{\delta}\hat{\Delta\phi}|$.

Needless to say, the trapped beam must fit inside the bucket. As mentioned before, in the adiabatic approximation, these ellipses are always upright, and are the same at all positions around the storage ring. When a beam bunch is injected into an RF bucket, its bunch length and its energy spread will need to be *matched* to the RF bucket. Otherwise the motion of the bunch as a whole will tumble inside the bucket, eventually leading to some bunch dilution, or even some loss of particles.

Large-amplitude oscillation For a large-amplitude particle, the period of synchrotron oscillation becomes longer, i.e. the synchrotron frequency becomes lower. This dependence of the oscillation frequency on the amplitude of oscillation is referred to as a *detuning* effect. Given the Hamiltonian (5.27) and the Hamilton's equation $\dot{\phi} = h\omega_0\eta\delta$, the synchrotron oscillation period is given by

$$T = \oint \left(2h\omega_0\eta \left[H_0 - \frac{eV_0\omega_0}{2\pi\beta_s^2 E_s}[\cos\phi - \cos\phi_s + (\phi - \phi_s)\sin\phi_s]\right]\right)^{-1/2} d\phi\,,$$

where H_0 is the invariant value of the Hamiltonian for the particle under consideration. The angular synchrotron frequency is given by $\frac{2\pi}{T}$.

In general, this synchrotron detuning needs to be computed numerically. For the special case of a stationery bucket (i.e. when $\phi_s = 0$), the value of H_0 can be expressed either in terms of the maximum momentum deviation $\hat{\delta}$, or in terms of the maximum phase deviation $\hat{\phi}$. Assuming $\eta < 0$, these expressions are

$$H_0 = -\frac{1}{2}h\omega_0|\eta|\hat{\delta}^2 = -\frac{\omega_0\nu_s^2}{h|\eta|}(1 - \cos\hat{\phi})\,,$$

which also connects the two quantities $\hat{\delta}$ and $\hat{\phi}$,

$$\hat{\delta} = \frac{2\nu_s}{h|\eta|}\left|\sin\frac{\hat{\phi}}{2}\right|.$$

With $\phi_s = 0$, we find

$$T(\hat{\phi}) = \frac{4K\left(\sin\frac{\hat{\phi}}{2}\right)}{\omega_0\nu_s}\,,$$

where

$$K\left(\sin\frac{\hat{\phi}}{2}\right) = \frac{1}{4}\int_{-\hat{\phi}}^{\hat{\phi}} \frac{d\phi}{\sqrt{\sin^2\frac{\hat{\phi}}{2} - \sin^2\frac{\phi}{2}}}\,.$$

By a change of variable from ϕ to w with $\sin\frac{\phi}{2} = \sin\frac{\hat{\phi}}{2}\sin w$, it follows that

$$K(x) = \int_0^{\pi/2} \frac{dw}{\sqrt{1 - x^2\sin^2 w}}$$

is the complete elliptical integral of the first kind. If we define a detuned synchrotron tune $\nu_s(\hat{\phi})$ while keeping the notation $\nu_s = \nu_s(0)$, then

$$\nu_s(\hat{\phi}) = \frac{2\pi}{\omega_0 T(\hat{\phi})} = \frac{\pi \nu_s}{2K\left(\sin\frac{\hat{\phi}}{2}\right)}. \tag{5.41}$$

For small oscillation amplitudes, $\hat{\phi} \ll 1$, we have

$$K\left(\sin\frac{\hat{\phi}}{2}\right) \approx \frac{\pi}{2}\left(1+\frac{\hat{\phi}^2}{16}\right) \quad \Longrightarrow \quad \nu(\hat{\phi}) \approx \nu_s\left(1-\frac{\hat{\phi}^2}{16}\right).$$

As the oscillation amplitude is increased, the synchrotron oscillation slows down. As the amplitude approaches the separatrix with $H_0 \to H_{\rm sx}$, the frequency drops to zero. This behavior of detuning was mentioned in Fig. 5.41.

Figure 5.47 shows the experimental measurement as well as the theoretical calculation of the synchrotron frequency versus the maximum phase deviation $\hat{\phi}$ for the IUCF Cooler.[18] The data were measured by kicking a stored beam and are found in good agreement with the theory.

Figure 5.47: The synchrotron detuning effect in an RF bucket. The solid curve is a theoretical prediction. The data are taken in an IUCF experiment [Courtesy S.Y. Lee, 2019].

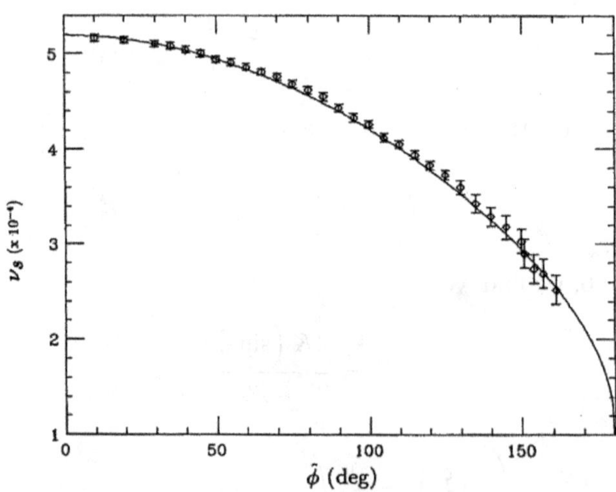

The reason the experiment was carried out up to $\hat{\phi} \approx 160°$ and not all the way to 180° is because the bunch has a finite length. If the beam centroid is kicked too close to 180°, part of the beam will get out of the bucket.

[18]M. Ellison, et al., Phys. Rev. Lett. 70, 591 (1993).

From the data one sees that the bunch length occupies approximately 20° of RF phase.

Homework 5.39 Write a computer simulation program to examine the longitudinal dynamics using the difference equations (without adiabatic approximation).

(a) Observe the adiabatic approximation by taking appropriate parameters when adiabatic approximation applies.

(b) Continue (a) to find $\nu_s(\hat{\phi})$ numerically. Compare with the theoretical expectation (5.41).

(c) Depart from the adiabatic approximation, and observe chaotic behavior near the UFP.

(d) Depart very much away from the adiabatic approximation. See if you might observe chaotic behavior even well inside the RF bucket — this will take a bit of effort and requires a strategy.

Homework 5.40 Consider a synchrotron ramping a beam of injection energy E_i to a final energy E_f. The intension is to complete the ramping process in a time period T. The required energy gain per turn is not just $(E_f - E_i)$ divided by the revolution time because the revolution time is not constant. Find the correct expression of the required energy gain per turn from the RF. Assume ϕ_s and V_0 are held fixed during acceleration. The synchrotron circumference is C.

Solution Energy gaining rate of the synchronous particle is

$$\frac{dE}{dt} = feV_0 \sin\phi_s \, ,$$

where revolution frequency f is

$$f = \frac{\beta c}{C} = \frac{c}{\gamma C}\sqrt{\gamma^2 - 1} \, .$$

We have

$$\frac{dE}{dt} = mc^2 \frac{d\gamma}{dt} = \frac{ceV_0 \sin\phi_s}{\gamma C}\sqrt{\gamma^2 - 1} \, .$$

Solving for $\gamma(t)$ yields

$$\sqrt{\gamma(t)^2 - 1} = \sqrt{\gamma_i^2 - 1} + \left(\frac{eV_0 \sin\phi_s}{Cmc}\right) t \, .$$

The required energy gain per turn from the RF is therefore

$$eV_0 \sin\phi_s = \frac{Cmc}{T}\left(\sqrt{\gamma_f^2 - 1} - \sqrt{\gamma_i^2 - 1}\right) .$$

In the relativistic limit, this reduces to $\frac{C}{cT}(E_f - E_i)$ as expected.

Homework 5.41

(a) Let $\hat{\phi}$ be the maximum synchrotron phase amplitude for a particle inside the RF bucket. Show that the maximum momentum deviation is

$$\hat{\delta} = \frac{\sqrt{2}\nu_s}{h|\eta|}\left[1 - \frac{\cos\hat{\phi}}{\cos\phi_s} - (\hat{\phi} - \phi_s)\tan\phi_s\right]^{1/2}.$$

(b) Find the small amplitude limit of this expression.

(c) In the limit of no acceleration, show that this reduces to

$$\hat{\delta} = \frac{2\nu_s}{h|\eta|}\left|\sin\left(\frac{\hat{\phi} - \phi_s}{2}\right)\right|,$$

where $\phi_s = 0$ when $\eta < 0$ and $\phi_s = \pi$ when $\eta > 0$.

Homework 5.42 In the text, we analyzed the synchrotron motion assuming the RF voltage is given by $V(t) = V_0 \sin(\omega_{\rm rf} t + \phi_s)$. Follow the text to analyze the case when $V(t)$ is kept arbitrary except for the fact that it must be periodic in t with period $\frac{2\pi}{\omega_{\rm rf}}$.

(a) With this arbitrary $V(t)$, find expressions for the two equations of motion, the small-amplitude synchrotron tune, and the Hamiltonian in the adiabatic approximation.

(b) Now specialize $V(t)$ to be a sawtooth as in Fig. 5.48. Sketch the RF bucket in the (ϕ, δ) phase space. Include contours of particle motion both inside and outside the bucket. Find expressions for the synchrotron tune, the bucket half-height and bucket area. What is the RF detuning of this system?

(c) Is this $V(t)$ more appropriate for operation above or below transition? How to make it suitable for the other case?

Figure 5.48: A sawtooth RF waveform.

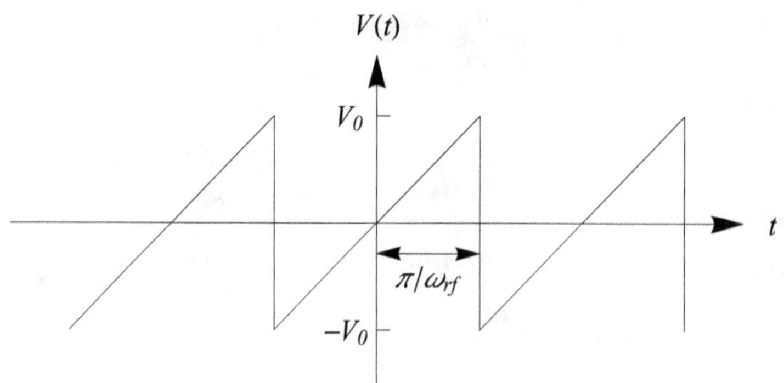

Solution (b) The rate of change of ΔE due to the RF acceleration is given by

$$\dot{\Delta E} = eV(\phi)\frac{\omega_0}{2\pi} = e\phi\frac{dV}{d\phi}\frac{\omega_0}{2\pi} = \phi\frac{eV_0}{\pi}\frac{\omega_0}{2\pi}.$$

With $\delta = \frac{\Delta P}{P_s} = \frac{1}{\beta_s^2}\frac{\Delta E}{E_s}$, we have

$$\dot{\delta} = \phi\frac{eV_0}{\pi\beta_s^2 E_s}\frac{\omega_0}{2\pi}.$$

The second equation for synchrotron motion is

$$\dot{\phi} = \omega_{\rm rf}\eta\delta.$$

The Hamiltonian is

$$H(\phi,\delta,t) = \frac{1}{2}h\omega_0\eta\delta^2 - \frac{\omega_0 eV_0}{4\pi^2\beta_s^2 E_s}\phi^2.$$

The value of H on the separatrix is

$$H_{\rm sx} = \frac{\omega_0 eV_0}{4\beta_s^2 E_s}.$$

The bucket half-height is

$$\hat{\delta}_{\frac{1}{2}} = \sqrt{\frac{eV_0}{2\beta_s^2 E_s h\eta}}.$$

The equation for the separatrix can be written as

$$\frac{\delta^2}{\hat{\delta}^2_{\frac{1}{2}}} + \frac{\phi^2}{\pi^2} = 1.$$

The bucket area is $\pi^2\hat{\delta}_{\frac{1}{2}}$.

5.3.12 Nonlinear phase slippage factor

The phase slippage factor η is a critical parameter in a storage ring design. It controls the synchrotron tune, and it controls the longitudinal phase space topology. In particular, in many applications a short beam bunch is required. To achieve that, a ready trick is to design the ring with a small η. However, the η designed is not strictly a constant; it depends on δ,

$$\eta = \eta_0 + \eta_1\delta + \cdots$$

Nominally, η_0 dominates and the phase space behaves as we discussed. When η_0 is made small, however, we need to take into account of effects due to the higher order terms, particularly η_1.[19]

[19] J. Wei, Ph.D. Thesis, Stony Brook, (1990, rev. 1994); K.Y. Ng, Sec. 2.3.11, Handbook Accel. Phys. & Eng., 2nd Ed., World Scientific (2013); S.Y. Lee, Accelerator Physics, 3rd Ed., World Scientific (2011).

The Hamiltonian of synchrotron motion was given by Eq. (5.27), but now it is to be modified to read

$$H(\phi, \delta, t) = h\omega_0 \left(\frac{1}{2}\eta_0 \delta^2 + \frac{1}{3}\eta_1 \delta^3 \right) + \frac{eV_0\omega_0}{2\pi\beta_s^2 E_s} \left[\cos\phi - \cos\phi_s + (\phi - \phi_s)\sin\phi_s \right]. \tag{5.42}$$

As always, the first thing is to look for the fixed points,

$$\begin{cases} \dot{\phi} = h\omega_0(\eta_0 + \eta_1\delta)\delta = 0 \\ \dot{\delta} = \frac{\omega_0 eV_0}{2\pi\beta_s^2 E_s}(\sin\phi - \sin\phi_s) = 0 \end{cases},$$

$$\Longrightarrow \quad \delta = 0, -\frac{\eta_0}{\eta_1}, \quad \text{and} \quad \phi = \phi_s,\ \pi - \phi_s\,.$$

There are four sets of fixed points,

$$(\phi_s, 0),\ (\pi - \phi_s, 0),\ \left(\phi_s, -\frac{\eta_0}{\eta_1}\right),\ \left(\pi - \phi_s, -\frac{\eta_0}{\eta_1}\right).$$

Figure 5.49: Longitudinal phase space (ϕ, δ) of a storage ring with nonlinear phase slippage factor. Contours of constant Hamiltonian (5.42) is plotted. The left column is the case $\phi_s = 0$. The right column is $\phi_s = \frac{\pi}{4}$. The three rows correspond to $\frac{\eta_1}{\eta_0} = 0, 0.5, 4$ respectively. The parameter $\frac{eV_0}{2\pi\beta_s^2 E_s h\eta_0}$ is set to be an exaggerated value of 0.1, representing a small η_0.

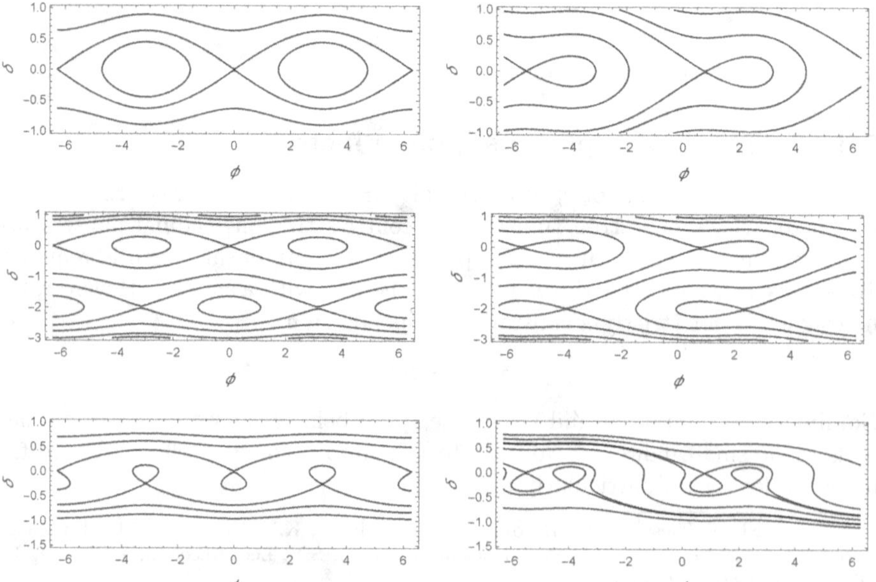

Figure 5.49 shows the phase space of this nonlinear system. The first row is the conventional case when $\eta_1 = 0$. When $\eta_1 \neq 0$, another series of islands are formed. The four sets of fixed points, two stable and two unstable, can be identified. When η_1 is still small, the new island set lies vertically separated from the original set as shown in the two plots in the middle row. When the magnitude of η_1 increases further, the new islands begin to line up horizontally relative to the original set, shown as the plots in the bottom row. When this happens, these buckets are referred to as the "alpha buckets".

In the two middle panels of Fig. 5.49, the two sets of buckets have opposite signs of the sense of rotation, i.e. one clockwise and one counter-clockwise, meaning that one has positive mass and the other has negative mass. In contrast, the lowest two panels have the same sign of mass in both sets of buckets. It is suggested that these conclusions can be drawn by observing the phase space topology shown in Fig. 5.49.

5.4 Nonlinear dynamics and chaos

We have now looked at the longitudinal synchrotron beam dynamics in two ways. We first made the adiabatic approximation, taking advantage of the fact that synchrotron motion is slow when observed once per revolution. In this case, we have a time-independent Hamiltonian, and we analyzed the motion accordingly. Other than the adiabatic approximation, the analysis is exact, even though the Hamiltonian is nonlinear in ϕ.

We also looked at the problem without making adiabatic approximation. In that case, we analyzed the problem using a set of difference equations (maps). We then obtained an exact analysis when we linearized the equations. The matrix formalism that allowed exact treatment was possible only after linearization.

So we have solved the problem only when we made an approximation that the system is either nonlinear but time-independent, or time-dependent but linear. One natural question to ask is then what happens if the system is nonlinear and time-dependent? This would be the case if we examine the difference equation before linearization, i.e. Eq. (5.31). Table 5.4 attempts to summarize the situation. We will discuss chaos in this section, although only very briefly.

Chaotic layer Since there is no reason to doubt the adiabatic approximation as the synchrotron motion is indeed slow, we expect the *numerical* difference between Eq. (5.31) and the RF bucket analysis would not be too great. Since linearization has to be valid at least close to the phase space origin, we expect that whatever the difference is, it will not occur near the center of the bucket. It turns out that, although these expectations turn out to be valid, the "small difference" between Eq. (5.31) and the RF bucket analysis also turns out very drastic, intricate and surprising. What one sees are the *chaotic layers.* Figure 5.50 gives a sketch of the expected difference.

Table 5.4: Where we stand in terms of analysis of synchrotron dynamics.

	With linearization	Without linearization
With adiabatic approximation	• Phase space ellipses are upright and independent of s • Approximate ν_s, Eq. (5.26)	• Hamiltonian (5.27) is constant of motion • RF bucket is not an ellipse but is upright and independent of s • $\nu_s(\hat{\phi})$ detunes with particle amplitude $\hat{\phi}$, Eq. (5.41)
Without adiabatic approximation	• Matrix map (5.32) • Courant-Snyder analysis • Exact ν_s, Eq. (5.34)	• Nonlinear map (5.31) • Not analytically soluble. • Chaos!

Figure 5.50: What is expected to be the slight differences between the adiabatic system and the discrete system. The differences are small but rather drastic.

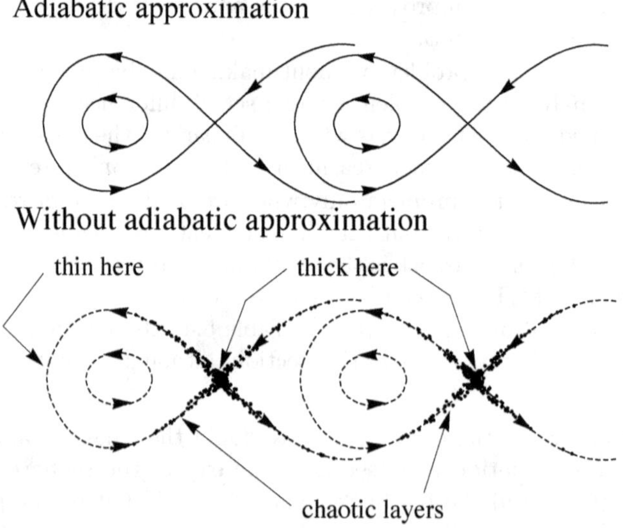

The chaotic layer is generally thin and is unnoticeable unless one is close to the separatrix, and even then, significant layer appears only near the unstable fixed points. In a chaotic layer, particle motion in phase space does not follow

Figure 5.51: Chaotic motion near an UFP by simulation.

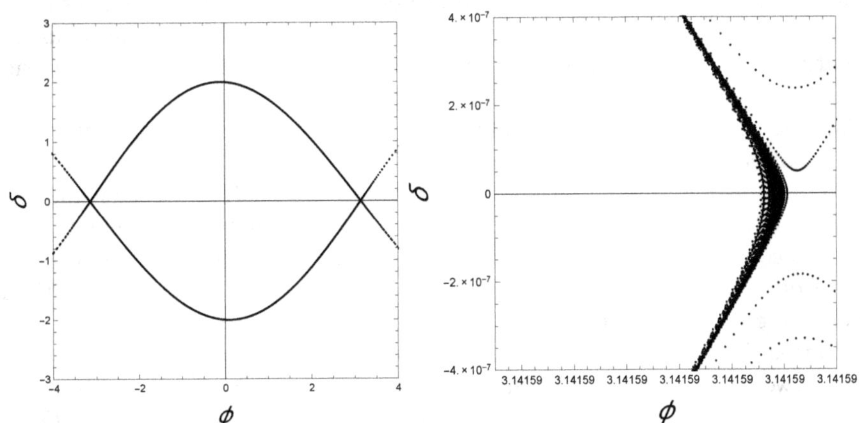

a smooth curve. More accurately, in fact, one should say that a smooth curve contour does not even exist. Instead, it appears to move chaotically within the layer.

Note that the particle motion only *appears* to behave chaotically. There is no diffusion or random walk effects to speak of, all motions are deterministic. Note also that we are no longer speaking of a smoothed dynamics. The system is described by a set of difference equations. We are only entitled to observe the dynamics at each of its iterations. The time variable is not continuous. Each observation gives one dot on the phase space plot. As these dots begin to jump around, they give rise to an appearance of chaotic, random-walk-like motion. You are not allowed to connect the dots.

An actual example of the chaotic layer near the unstable fixed point is illustrated in Fig. 5.51 by a numerical simulation. At a larger scale, the phase space bucket is evident on the left panel. Zooming in to the unstable fixed point, however, chaotic behavior appears on the right panel.

A proper discussion of these intricate effects belongs to the subject of nonlinear dynamics. By making the adiabatic approximation, we have inadvertently excluded this exciting topic in our discussions. Concerning chaos, we shall only make a few sketchy comments below.

Intricate phase space Consider the case of a 1-D system (2-D phase space), and we observe the motion of a particle at a fixed location $s = 0$ in the accelerator. The trajectory of the particle in the phase space consists of the sequence of dots as observed from turn to turn. We mentioned before that there will be no chaos if the Hamiltonian is independent of s (i.e. if adiabatic approximation holds). This is easy to understand, because although we are not observing the trajectory between turns, the particle still follows a well-defined curve (the constant-Hamiltonian contour) from point to point with the same Hamiltonian as if s is continuously defined. There is therefore implicitly a well-defined smooth continuous curve connecting all the discrete dots. Chaos cannot occur in this situation as no constant-Hamiltonian contour curves are allowed to cross each other. When a smooth curve exists that connects the dots, we call the system *integrable*. A 1-D system with s-independent Hamiltonian is integrable. No chaos can occur here.

What happens when the Hamiltonian is s-dependent (i.e. without adiabatic approximation)? In this case, chaos can occur (although it does not have to occur) because there is now not an implicit smooth curve that connects the dots from turn to turn. The dots from different particles can mix without having to cross each other's path.

Indeed, between the times of observation when we are not looking, trajectories of two particles executing chaotic motions do and must cross each other's path in phase space, but they arrive at the alleged point of crossing at *different times*, and that does not count as path crossing, and is not forbidden in a system with time-dependent Hamiltonian.

Can particles' paths pass through the same phase space point at different times in case the Hamiltonian is s-independent? The answer is no. If the Hamiltonian is s-independent, then two paths are not allowed to cross even if they cross at different times.

Indeed in the case of chaos, the dots from different particles fill up the phase space by an infinitely dense intertwined collection of nonconnected dots. It is a fascinating fact that a seemingly nondrastic s-dependence can be so critical in terms of making chaos or not. Many of the comments, for example (3.33), made for the transverse nonlinear resonances applies here as well.

Chaotic layer gets thicker near UFPs We mentioned earlier in the case of smooth RF bucket for Fig. 5.41 that as a particle proceeds along the separatrix toward the UFP, its approaching speed slows down exponentially. This general trend is true also for the nonintegrable case. On the separatrix, the dots from turn to turn also approach the UFP and the speed slows down as it gets closer to the UFP. As it slows down, however, it turns out that the scattering range of the chaotic motion gets wider. This is then reflected by the observation that the chaotic layer gets thicker near the UFP.

This behavior that when many particles crowd up while slowing down is as if the only to conserve phase space is to widen the passageway. This in not unlike the widening of a river as it ends into an ocean. Particle motion in conserved phase space behaves the same as the incompressible water in real space.

On the other hand, away from the UFP, we should mention that there are chaos almost everywhere in the phase space. The statement that they occur near the separatrix was an oversimplification. For example, there are chaos layers also well inside the RF bucket when the detuned ν_s is close to a rational number. When that happens, a chain of islands can be formed well within the bucket, and each chain of islands has its own set of UFPs and its own chaotic layer. In this sense, the RF buckets themselves are simply the biggest chain of islands with the thickest chaotic layer, and we are now speaking of island chains inside an island. In fact, there will also be island chains inside those smaller islands. However, for practical purposes, these chaotic layers are so extremely thin that they are not at all noticeable – until you are so determined to search for them, for example in a computer simulation that prints out many digits.

Chaos is not random walk In addition to the apparent loss of smoothness in the calculations, the existence of chaos has the philosophical implication that there is something indeterministic in a completely deterministic system. The impact of this observation is rather drastic, and has revolutionized our view of classical dynamics. The chaotic appearance, however, can be misleading because there are some fundamental rules, such as symplecticity and phase space volume conservation, that these chaotic behavior must strictly obey.

One testing technique that can be applied to tell whether an observed chaotic behavior is due to the true dynamics or due to noise (or numerical errors in a computer simulation) is to follow the particle motion until chaotic behavior appears, and then reverse the time and follow its motion backwards in time to see if the initial condition is recovered. If recovered, the chaotic behavior is truly due to dynamics. Otherwise, some artifact noise effects have entered.

Adiabatic trapping Adiabatically varying the RF bucket is a useful technique of manipulations of a beam's longitudinal phase space in synchrotrons and storage rings. The mechanism that allows this deserves a brief discussion.

Consider an RF bucket established in the (ϕ, δ) phase space with an RF system. From the meaning of the bucket, we intend to have the particles inside the bucket to stay inside and particles outside to stay outside, i.e. we intend the bucket to have a clean phase space separatrix.

If we now change the RF parameters slightly and slowly, the bucket's size and shape will change accordingly. Let us say the bucket now becomes a bit taller because we have just increased the RF voltage a little. But how could this happen? We know the phase space area is conserved. The slight increase in the bucket area immediately violates the Liouville theorem!

The answer to this question lies in the very thin, almost unnoticeable, chaotic layer of the bucket. During the process of varying the RF voltage, phase space areas permeate through the thin, leaky chaotic layer — overwhelmingly over the region near the UFP — from the outside to the inside of the bucket. As they permeate in, the bucket area grows.

The slower the RF parameters are varied, the slower the outside phase space permeates inside. However, no matter how slowly the parameters are varied, there will always be some phase space permeation taking place. This effect is sometimes called *phase space displacement* in the literature.[20]

[20]K.N. Henrichsen, M.J. de Jonge, CERN report CERN-ISR-RF-MA/74-21 (1974); E.W. Messerschmid, CERN report CERN/ISR-TH/73-31 (1973).

One very useful phase space manipulation technique calls for adiabatically varying the RF parameters. Here, adiabaticity refers to the synchrotron oscillation frequency ν_s. If the parameters vary slowly compared with ν_s, we call the process adiabatic. But ν_s applies only to the bucket center; it vanishes towards the separatrix. When the parameters are varied adiabatically, what happens is that the phase space that permeates into the bucket over the UFPs stays predominately around the outer edge of the bucket, while the interior of the bucket slowly conforms to the changing bucket topology. The result is that although the phase space area of the outer edge of the bucket grows, the interior phase space area remains mostly intact even though its shape changes.

If we now consider a beam that sits in the interior of an initial bucket, and now adiabatically increase the RF voltage from V_0 to V_1, the beam's longitudinal emittance will not change even though the RF bucket has changed. If for example, $V_1 \gg V_0$, then the beam will become much shorter in length and higher in energy spread, keeping their product intact. This adiabatic technique, called *adiabatic trapping* by the RF bucket, allows a very useful way to manipulate the beam's longitudinal distribution without increasing its longitudinal emittance.

5.5 Transverse-longitudinal coupled system

As we demand more and more from accelerators, beam manipulation techniques get more sophisticated, and *phase space gymnastics* has evolved to become a critical topic in accelerator physics. Just like the physical gymnastics, e.g. in the Olympic games, the skills needed in phase space gymnastics are highly technical and precise, while the resulting performance exquisite and beautiful. The ability to manipulate the beam's 6-D phase space offers many precision oriented operations of the beam, and opens up many applications, e.g. for linear colliders, free electron lasers and microbunching schemes. This is a new and fertile R&D field. As a glimpse of this effort, we mentioned briefly the term *lattice adaptor* in Sec. 3.5.3 in connection to a solenoid-immersed gun. Somewhat more content can be found in Sec. 5.6 later on the subject of emittance exchange techniques. Microbunching techniques such as steady-state microbunching also belong to this effort.

As mentioned in Sec. 1.2.3, such phase space gymnastics are all based on the validity of the Liouville theorem in phase space. Early-day phase space gymnastics were mostly applied to the 2-D longitudinal phase space, and took the form of RF manipulations in beam injection, extraction, adiabatic trapping, and phase space displacement acceleration.[21] Many of the newer techniques, however, invoke the fact that the transverse and the longitudinal phase spaces are not stand-alone separated pieces. Often they are coupled, and by taking advantage of their coupling, a surprisingly rich arsenal of techniques have been

[21] See, for example, R. Garoby, Sec. 4.10, Handbook Accel. Phys. & Eng., 2nd Ed., World Scientific (2013).

invented. We shall invest some time to discuss this topic in the remaining of this chapter.

5.5.1 Bunch compression

Our first example application to transverse-longitudinal coupled phase space gymnastics is for bunch compressors in transport lines. In accelerator applications, there are cases when we wish to have a beam with high peak current. However, high peak currents are not always available, and in any case often come associated with a high possibility of beam instability. The trick is then to carry the beam with a long bunch length all along, and only compress it at the last moment before using. Since beam instabilities become less damaging as the beam energy increases, bunch compressions are usually applied when the beam has been sufficiently accelerated.

Although its mechanism involves transverse-longitudinal coupling, the bunch compressor is simple enough to be analyzed purely in the longitudinal dimension. This is possible because in this case, the transverse dimension contributes only a momentum compaction term R_{56}. Once an appropriate momentum compaction R_{56} is provided by the transverse dimensions, we short-circuit the analysis to concentrate on the longitudinal phase space (z, δ).[22]

For a beamline that contains no electric components, δ = constant. The only function of the beamline that concerns us is to introduce R_{56}. Let the beamline length traveled by the synchronous particle be L_0. The path length of an off-momentum particle be $L_0 - R_{56}\delta$. By the time the synchronous particle completes the travel, the off-momentum particle will acquire a longitudinal displacement relative to the synchronous particle by an amount

$$\Delta z = \left(R_{56} + \frac{L_0}{\gamma^2}\right)\delta\,,$$

where the second terms comes from the velocity contribution. In what follows, we will assume a relativistic beam, so the second term can be dropped. The function of the beamline can then be described by a 2×2 transfer matrix

$$\begin{bmatrix} 1 & R_{56} \\ 0 & 1 \end{bmatrix}.$$

In front of the beamline, we now introduce an RF cavity with voltage $V(t) = V_0 \sin \omega_{\rm rf} t$. If we linearize the RF voltage assuming the synchronous particle arrives at the RF cavity at $t = 0$, the transfer map can be written as a matrix,

$$\begin{bmatrix} 1 & 0 \\ k & 1 \end{bmatrix},$$

where $k = -\frac{eV_0\omega_{\rm rf}}{E_0 c}$ with E_0 the design beam energy.

---

[22]Yes, we are returning to the (z, δ) phase space in this section.

Figure 5.52: A process of bunch compression.

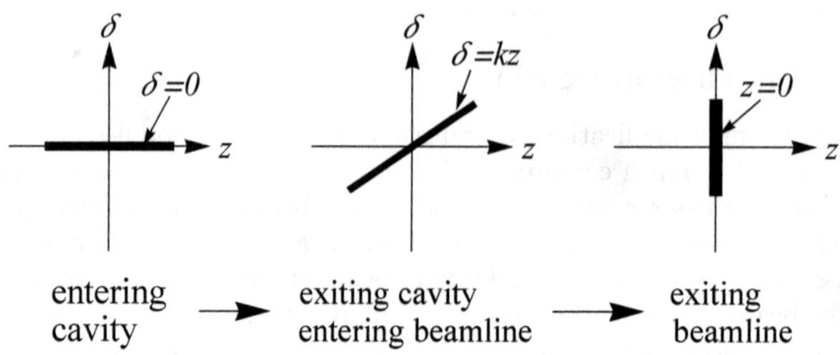

The total map is

$$M_{\text{comp}} = \begin{bmatrix} 1 & R_{56} \\ 0 & 1 \end{bmatrix} \begin{bmatrix} 1 & 0 \\ k & 1 \end{bmatrix} = \begin{bmatrix} 1+kR_{56} & R_{56} \\ k & 1 \end{bmatrix}.$$

Consider a beam entering the compressor with no energy spread, $\delta_0 = 0$, but with a finite bunch length. After it traverses the RF cavity, the beam distribution in the (z, δ) phase space acquires a *chirp*, i.e. the particle's energy increases (or decreases) along the bunch length. After exiting the cavity, the beamline, by applying the momentum compaction R_{56}, translates the chirp into a change in bunch length. A bunch compression occurs as a result if proper parameters are chosen. The condition for a perfect compression occurs when

$$1 + kR_{56} = 0\,. \tag{5.43}$$

With this condition, the bunch distribution makes a perfect rotation in phase space, as illustrated in Fig. 5.52. Compression requires R_{56} to have a particular sign that depends on the polarity of the chirp. The case illustrated has $k > 0$ and $R_{56} < 0$.

For a beam with finite initial energy spread, the situation is basically the same. For example, suppose we want to compress a beam bunch with an initial bi-Gaussian distribution in the (z, δ) phase space. Let the initial Σ-matrix be

$$\Sigma_0 = \begin{bmatrix} \sigma_{z0}^2 & 0 \\ 0 & \sigma_{\delta 0}^2 \end{bmatrix}.$$

After compression, its Σ-matrix becomes

$$\Sigma = M_{\text{comp}} \Sigma_0 \tilde{M}_{\text{comp}} = \begin{bmatrix} R_{56}^2 \sigma_{\delta 0}^2 + (1+kR_{56})^2 \sigma_{z0}^2 & R_{56}\sigma_{\delta 0}^2 + k(1+kR_{56})\sigma_{z0}^2 \\ R_{56}\sigma_{\delta 0}^2 + k(1+kR_{56})\sigma_{z0}^2 & \sigma_{\delta 0}^2 + k^2 \sigma_{z0}^2 \end{bmatrix}.$$

With a perfect compression (5.43), we obtain

$$\Sigma = \begin{bmatrix} R_{56}^2\sigma_{\delta 0}^2 & R_{56}\sigma_{\delta 0}^2 \\ R_{56}\sigma_{\delta 0}^2 & \sigma_{\delta 0}^2 + \frac{\sigma_{z0}^2}{R_{56}^2} \end{bmatrix}.$$

A good compression occurs when $|R_{56}|\sigma_{\delta 0} \ll \sigma_{z0}$. Note that the beam now acquires a tilt in its phase space under the compression condition ($\langle z\delta\rangle \neq 0$). Note also that the compression in bunch length has to pay the price that the energy spread has increased. It is necessary to judge whether that beamline as well as its downstream optics can accommodate the beam with the larger energy spread. A slightly different optimization is considered in Homework 5.43.

Homework 5.43 The text introduced a "perfect" compression with condition (5.43), derived considering an incoming beam with zero energy spread. For a beam with finite energy spread, the optimum for bunch compression is in fact slightly different. As shown in the text, in case of the "perfect" condition, the beam is actually tilted in phase space. Obviously, one can improve the compressor design by removing this tilt. Show that, with k set by the condition $1 + kR_{56} = 0$, the optimum design occurs by adding an additional dispersion section with

$$r_{56} = -R_{56}\left(\frac{R_{56}^2\sigma_{\delta 0}^2}{R_{56}^2\sigma_{\delta 0}^2 + \sigma_{z0}^2}\right).$$

(Therefore the net momentum compaction becomes $R_{56} + r_{56}$.) The Σ-matrix after compression then becomes

$$\Sigma = \begin{bmatrix} R_{56}^2\sigma_{\delta 0}^2\left(\frac{\sigma_{z0}^2}{R_{56}^2\sigma_{\delta 0}^2+\sigma_{z0}^2}\right) & 0 \\ 0 & \sigma_{\delta 0}^2 + \frac{\sigma_{z0}^2}{R_{56}^2} \end{bmatrix}.$$

5.5.2 The 4 × 4 formalism

In Chapter 3, we described a 4×4 formalism for the case when the two transverse dimensions are coupled. In this section, we mainly consider the case with transverse-longitudinal coupling when one of the transverse dimensions, either horizontal or vertical, is coupled to the longitudinal. Obviously, there is the general case when all three dimensions are coupled together, and we will need a 6×6 formalism to describe it. However, we shall not enter the 6×6 formalism here. As we stand presently, most of the phase space gymnastics do not invoke 3-D coupling. It is probably safe for now to leave the 6×6 formalism to applications involving errors and their corrections.

Dipole To describe the transverse-longitudinal coupled system, we will use 4×4 matrices. For most beamline elements, the 4×4 matrices are simply given by the decoupled two diagonal 2×2 submatrices. There are basically two exceptions. One of them is the bending dipoles, which we discuss first.

Let us consider a horizontal bending dipole, and first review what we did in Chapter 4 in connection to the dispersion function. We recall that, for a sector dipole magnet, the matrix map for the vector

$$\begin{bmatrix} x \\ x' \\ \delta \end{bmatrix}$$

was found to be

$$\begin{bmatrix} \cos\frac{s}{\rho} & \rho\sin\frac{s}{\rho} & \rho(1-\cos\frac{s}{\rho}) \\ -\frac{1}{\rho}\sin\frac{s}{\rho} & \cos\frac{s}{\rho} & \sin\frac{s}{\rho} \\ 0 & 0 & 1 \end{bmatrix},$$

where we have replaced the vector $(D, D', 1)$ by (x, x', δ). For example, the horizontal displacement of the particle inside of the dipole is

$$x(s) = x_0 \cos\frac{s}{\rho} + x_0'\rho\sin\frac{s}{\rho} + \left(1-\cos\frac{s}{\rho}\right)\rho\delta\,. \tag{5.44}$$

To describe the transverse-longitudinal coupling properly, however, we need to consider the vector

$$\begin{bmatrix} x \\ x' \\ z \\ \delta \end{bmatrix},$$

and the transfer map should be 4×4. We are missing the component of how z changes across the dipole magnet. To fill that in, we note that the path length change in a dipole magnet leads to the change in z,

$$\begin{aligned} \Delta z(s) &= -\int_0^s \frac{x(s')}{\rho}ds' \\ &= -x_0\sin\frac{s}{\rho} - \rho x_0'\left(1-\cos\frac{s}{\rho}\right) - \left(s-\rho\sin\frac{s}{\rho}\right)\delta\,. \end{aligned}$$

With this addition, we obtain the 4×4 map for a sector dipole of length ℓ,

$$\begin{bmatrix} \cos\frac{\ell}{\rho} & \rho\sin\frac{\ell}{\rho} & 0 & \rho(1-\cos\frac{\ell}{\rho}) \\ -\frac{1}{\rho}\sin\frac{\ell}{\rho} & \cos\frac{\ell}{\rho} & 0 & \sin\frac{\ell}{\rho} \\ -\sin\frac{\ell}{\rho} & -\rho(1-\cos\frac{\ell}{\rho}) & 1 & -\ell+\rho\sin\frac{\ell}{\rho} \\ 0 & 0 & 0 & 1 \end{bmatrix}. \tag{5.45}$$

For a thin-lens dipole with $\theta = \frac{\ell}{\rho} \ll 1$ and $\ell \to 0$, the map becomes

$$\begin{bmatrix} 1 & 0 & 0 & 0 \\ 0 & 1 & 0 & \theta \\ -\theta & 0 & 1 & 0 \\ 0 & 0 & 0 & 1 \end{bmatrix}.$$

This thin-lens expression is the t_3 case of Eq. (2.32). It applies to both sector and rectangular dipoles. Both the exact and the thin-lens expressions are of course symplectic.

Transverse cavity The other exception of a transverse-longitudinal coupled element is the transverse deflecting RF cavities, also called crab cavities, discussed in Sec. 5.1.3. Let the deflecting voltage be $V_\perp(t) = V_{\perp\text{rf}} \sin\omega_{\text{rf}} t \approx -V_{\perp\text{rf}}\omega_{\text{rf}}\frac{z}{c}$, its 4×4 map was found to be

$$\begin{bmatrix} 1 & 0 & 0 & 0 \\ 0 & 1 & k & 0 \\ 0 & 0 & 1 & 0 \\ k & 0 & 0 & 1 \end{bmatrix}. \tag{5.46}$$

where $k = -\frac{eV_{\perp\text{rf}}\omega_{\text{rf}}}{cE_0}$.

A less used 4×4 device is the Wien filter, Sec. 4.3, in the nonrelativistic regime.

It may be useful to recap a bit here. The most common x-y 4×4 coupled systems are the skew quadrupoles and the solenoids. The most common x-z (or y-z as the case maybe) 4×4 coupled systems are the dipoles and the transverse cavities. Many of the phase space gymnastics are based on manipulations and interplays among these four coupling elements together with the noncoupling elements (quadrupoles, drifts, normal RF cavities) arranged in some clever manner to accomplish the desired phase space transformations. As mentioned, this turns out to be a particularly fruitful direction of research towards advanced applications. All 4×4 maps and their corresponding hardware, on the other hand, cannot exceed the ten symplectic building blocks of Eq. (2.32).

Dog-leg Let us first apply the 4×4 formalism to a dog-leg. A dog-leg configuration is shown in Fig. 5.53 (upper).

The map from the entrance of the first dipole to the exit of the second dipole is given by, in thin-lens approximation,

$$\begin{bmatrix} 1 & 0 & 0 & 0 \\ 0 & 1 & 0 & \theta \\ -\theta & 0 & 1 & 0 \\ 0 & 0 & 0 & 1 \end{bmatrix} \begin{bmatrix} 1 & L & 0 & 0 \\ 0 & 1 & 0 & 0 \\ 0 & 0 & 1 & 0 \\ 0 & 0 & 0 & 1 \end{bmatrix} \begin{bmatrix} 1 & 0 & 0 & 0 \\ 0 & 1 & 0 & -\theta \\ \theta & 0 & 1 & 0 \\ 0 & 0 & 0 & 1 \end{bmatrix} = \begin{bmatrix} 1 & L & 0 & -L\theta \\ 0 & 1 & 0 & 0 \\ 0 & -L\theta & 1 & L\theta^2 \\ 0 & 0 & 0 & 1 \end{bmatrix}. \tag{5.47}$$

This expression applies whether the dipoles are sector or rectangular.

Consider for example a particle entering the dog-leg with initial coordinates $x_0 = 0, x_0' = 0, z_0 = 0$ but with a finite δ_0. When it exits the dog-leg, it acquires the coordinates

$$\begin{bmatrix} -L\theta \\ 0 \\ L\theta^2 \\ 1 \end{bmatrix} \delta_0 .$$

Figure 5.53: Layouts of a dog-leg (upper) and a reverse dog-leg (lower).

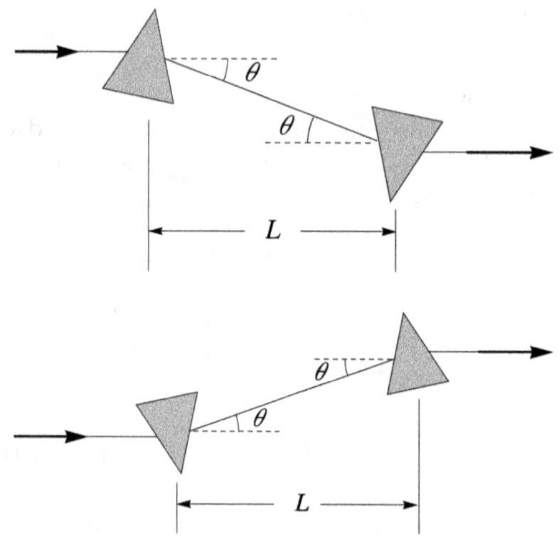

By convention, we designate this result as

$$\begin{bmatrix} R_{16} \\ R_{26} \\ R_{56} \\ 1 \end{bmatrix} \delta_0, \quad \text{i.e.} \quad R_{16} = -L\theta, \ R_{26} = 0, \ R_{56} = L\theta^2 .$$

One can also consider a reverse dog-leg like shown in Fig. 5.53 (lower). It has a 4×4 matrix

$$\begin{bmatrix} 1 & L & 0 & L\theta \\ 0 & 1 & 0 & 0 \\ 0 & L\theta & 1 & L\theta^2 \\ 0 & 0 & 0 & 1 \end{bmatrix}, \tag{5.48}$$

and with coefficients

$$R_{16} = L\theta, \quad R_{26} = 0, \quad R_{56} = L\theta^2 .$$

It might be noted that the two matrix elements L and $L\theta^2$ in Eqs. (5.47) and (5.48) do not affect the symplecticity. The maps are still symplectic if these terms are removed or changed.

When the dog-leg is reversed, the sign of R_{16} is reversed, but the sign of R_{56} is always positive. Higher momentum particles are bent less by the dipole, and its excursion amplitude is less. A simple geometric argument easily shows that the path length becomes shorter. As a result, it gets ahead of the on-momentum particle, and its z-displacement gets a positive gain.

Chicane One can join up a dog-leg and a reverse dog-leg to arrive at a chicane like Fig. 5.54. The 4×4 map for the chicane, from the entrance of the first dipole to the exit of the last one, is

$$\begin{bmatrix} 1 & L & 0 & -L\theta \\ 0 & 1 & 0 & 0 \\ 0 & -L\theta & 1 & L\theta^2 \\ 0 & 0 & 0 & 1 \end{bmatrix}\begin{bmatrix} 1 & L' & 0 & 0 \\ 0 & 1 & 0 & 0 \\ 0 & 0 & 1 & 0 \\ 0 & 0 & 0 & 1 \end{bmatrix}\begin{bmatrix} 1 & L & 0 & L\theta \\ 0 & 1 & 0 & 0 \\ 0 & L\theta & 1 & L\theta^2 \\ 0 & 0 & 0 & 1 \end{bmatrix} = \begin{bmatrix} 1 & 2L+L' & 0 & 0 \\ 0 & 1 & 0 & 0 \\ 0 & 0 & 1 & 2L\theta^2 \\ 0 & 0 & 0 & 1 \end{bmatrix}.$$

With a chicane configuration, the R_{16} terms cancel out, and there is only the R_{56} left, and it is positive. A chicane is a clean device that generates a pure path length change according to the particle momentum, and is a useful tool for phase space gymnastics. Its main (stubborn) limitation is that it contributes an R_{56} that can only be positive.

Figure 5.54: Layout of a chicane.

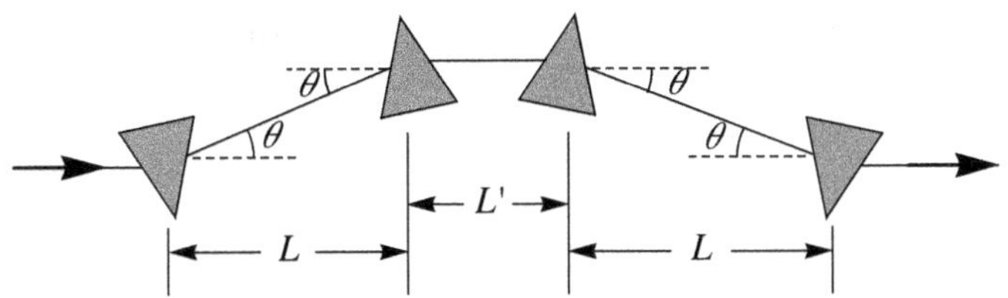

4×4 map in terms of dispersion function So far, the 4×4 formalism has been discussed in terms of the maps of individual elements. For a beamline with a string of elements, we multiply the matrices to obtain the total map. Following the approach of Chapter 2, our next job is to represent this total 4×4 matrix in terms of the Courant-Snyder functions $\alpha_x, \beta_x, \gamma_x, \psi_x$ and the dispersion function D, D' at the entrance and the exit of the beamline.

This effort applies when the beamline consists only of magnetic elements and no electric elements, i.e. the beam energy is preserved. Electric devices such as RF cavities and transverse deflecting cavities are excluded. For one thing, dispersion functions D and D' are undefined if particle energy changes.

In this case the total map in general can be expressed as

$$M = \begin{bmatrix} R_{11} & R_{12} & 0 & R_{16} \\ R_{21} & R_{22} & 0 & R_{26} \\ R_{51} & R_{52} & 1 & R_{56} \\ 0 & 0 & 0 & 1 \end{bmatrix},$$

where the upper-left 2×2 corner is well known,

$$R = \begin{bmatrix} R_{11} & R_{12} \\ R_{21} & R_{22} \end{bmatrix} = \begin{bmatrix} \sqrt{\frac{\beta_{x2}}{\beta_{x1}}}(\cos\psi_x + \alpha_{x1}\sin\psi_x) & \sqrt{\beta_{x1}\beta_{x2}}\sin\psi_x \\ \frac{\alpha_{x1}-\alpha_{x2}}{\sqrt{\beta_{x1}\beta_{x2}}}\cos\psi_x - \frac{1+\alpha_{x1}\alpha_{x2}}{\sqrt{\beta_{x1}\beta_{x2}}}\sin\psi_x & \sqrt{\frac{\beta_{x1}}{\beta_{x2}}}(\cos\psi_x - \alpha_{x2}\sin\psi_x) \end{bmatrix},$$

where subscripts 1 and 2 refer to the beginning and end of the transfer map, and $\psi_x = \psi_x(s_2) - \psi_x(s_1)$ is the betatron phase advance from position s_1 to position s_2.

We now impose the necessary symplecticity condition on the matrix M. First, it requires that R be symplectic, and it already is. Second, it also requires that the R-elements satisfy

$$\tilde{R}S\begin{bmatrix} R_{16} \\ R_{26} \end{bmatrix} + \begin{bmatrix} R_{51} \\ R_{52} \end{bmatrix} = 0\,, \tag{5.49}$$

where $S = \begin{bmatrix} 0 & 1 \\ -1 & 0 \end{bmatrix}$.

Take a note of Eq. (5.49) here. Symplecticity has demanded that the effect of momentum error on horizontal motion must be connected to the effect of horizontal coordinates on the longitudinal arrival time of a particle. Given R_{16}, R_{26}, for example, the values of R_{51}, R_{52} are determined. From a first sight, this looks like coming from nowhere! We mentioned in Chapter 2, in fact, its origin traces back to the Maxwell equations.

To complete the analysis, we still need to find expressions for R_{16}, R_{26}. This is done by invoking the dispersion functions. We have the condition

$$\begin{bmatrix} D_2 \\ D_2' \\ -\Delta C \\ 1 \end{bmatrix} = M \begin{bmatrix} D_1 \\ D_1' \\ 0 \\ 1 \end{bmatrix},$$

where $D_{1,2}, D'_{1,2}$ are the dispersion functions at the entrance and exit of the beamline, and $\Delta C = \int_{s_1}^{s_2} ds \frac{D(s)}{\rho(s)}$ is the change in longitudinal coordinate due to a path difference of an off-momentum particle — do not forget the minus sign in front of ΔC.

This condition can be rewritten,

$$\begin{aligned} \begin{bmatrix} D_2 \\ D_2' \end{bmatrix} &= R\begin{bmatrix} D_1 \\ D_1' \end{bmatrix} + \begin{bmatrix} R_{16} \\ R_{26} \end{bmatrix}, \\ -\Delta C &= R_{51}D_1 + R_{52}D_1' + R_{56}\,. \end{aligned} \tag{5.50}$$

The second equation gives R_{56}. The first equation gives explicit result for R_{16}, R_{26}. We then use Eq. (5.49) to obtain

$$\begin{aligned} \begin{bmatrix} R_{51} \\ R_{52} \end{bmatrix} &= -\tilde{R}S\begin{bmatrix} R_{16} \\ R_{26} \end{bmatrix} = -\tilde{R}S\begin{bmatrix} D_2 \\ D_2' \end{bmatrix} + \tilde{R}SR\begin{bmatrix} D_1 \\ D_1' \end{bmatrix} \\ &= -\tilde{R}S\begin{bmatrix} D_2 \\ D_2' \end{bmatrix} + S\begin{bmatrix} D_1 \\ D_1' \end{bmatrix} = -\tilde{R}S\begin{bmatrix} D_2 \\ D_2' \end{bmatrix} + \begin{bmatrix} D_1' \\ -D_1 \end{bmatrix}. \end{aligned}$$

Substituting the known expression for R, and collecting all the terms, we obtain the final general expression of a 4×4 map of a beamline consisting of magnetic elements, from position s_1 to position s_2,

$$M = \begin{bmatrix} R_{11} & R_{12} & 0 & D_2 - R_{11}D_1 - R_{12}D_1' \\ R_{21} & R_{22} & 0 & D_2' - R_{21}D_1 - R_{22}D_1' \\ A & B & 1 & -\Delta C - AD_1 - BD_1' \\ 0 & 0 & 0 & 1 \end{bmatrix}, \tag{5.51}$$

$$\begin{aligned}
R_{11} &= \sqrt{\frac{\beta_{x2}}{\beta_{x1}}}[\cos\psi_x + \alpha_{x1}\sin\psi_x]\,, \\
R_{12} &= \sqrt{\beta_{x1}\beta_{x2}}\sin\psi_x\,, \\
R_{21} &= \frac{1}{\sqrt{\beta_{x1}\beta_{x2}}}[(\alpha_{x1}-\alpha_{x2})\cos\psi_x - (1+\alpha_{x1}\alpha_{x2})\sin\psi_x]\,, \\
R_{22} &= \sqrt{\frac{\beta_{x1}}{\beta_{x2}}}[\cos\psi_x - \alpha_{x2}\sin\psi_x]\,, \\
A &= R_{21}D_2 - R_{11}D_2' + D_1'\,, \\
B &= -R_{12}D_2' + R_{22}D_2 - D_1\,, \\
\Delta C &= \int_{s_1}^{s_2} ds\frac{D(s)}{\rho(s)}\,,
\end{aligned}$$

where ψ_x is the horizontal phase advance from position s_1 to position s_1.

In case the entrance to the beamline is dispersion-free, i.e. if $D_1 = 0, D_1' = 0$, the map has a simpler expression,

$$\begin{bmatrix} R_{11} & R_{12} & 0 & D_2 \\ R_{21} & R_{22} & 0 & D_2' \\ R_{21}D_2 - R_{11}D_2' & -R_{12}D_2' + R_{22}D_2 & 1 & -\Delta C \\ 0 & 0 & 0 & 1 \end{bmatrix}.$$

Another simpler special case is for the one-turn map. The 4×4 one-turn map is

$$M_{\text{one-turn}} = \begin{bmatrix} R_{11} & R_{12} & 0 & D - R_{11}D - R_{12}D' \\ R_{21} & R_{22} & 0 & D' - R_{21}D - R_{22}D' \\ A & B & 1 & -\Delta C - AD - BD' \\ 0 & 0 & 0 & 1 \end{bmatrix}, \tag{5.52}$$

$$\begin{aligned}
R_{11} &= \cos\mu_x + \alpha_x\sin\mu_x\,, \\
R_{12} &= \beta_x\sin\mu_x\,, \\
R_{21} &= -\gamma_x\sin\mu_x\,, \\
R_{22} &= \cos\mu_x - \alpha_x\sin\mu_x\,, \\
A &= R_{21}D - R_{11}D' + D'\,, \\
B &= -R_{12}D' + R_{22}D - D\,, \\
\Delta C &= \int_0^C ds\frac{D(s)}{\rho(s)} = \alpha_c C\,.
\end{aligned}$$

Incidentally, the reason we have been using subindices 1,2,5,6 and not 1,2,3,4 is because the 3,4 indices are reserved for the y motion.

4 × 4 map for betatron coordinates The quantities x and x' in our discussions are the physical horizontal displacement and its slope. It is sometimes useful to decompose it into a synchrotron part and a betatron part, so that

$$x_\beta = x - D\delta, \qquad x'_\beta = x' - D'\delta \,. \tag{5.53}$$

This is a somewhat abstract decomposition particularly because it is meaningful only when there is no RF and $\delta =$ constant. However, valuable intuition can be gained and the algebra becomes simpler, so we make a detour to discuss it at this junction.

We know we must transform coordinates symplectically. To make the transformation symplectic, Eq. (5.53) is to be supplemented by an additional transformation concerning the longitudinal betatron coordinate,

$$z_\beta \;=\; z + D'x - Dx' \,. \tag{5.54}$$

The reader should keep in mind that betatron coordinate transformation is not given by the more apparent Eq. (5.53) alone; it *must* also include the more hidden Eq. (5.54).

The coordinate transformation then can be described by

$$\begin{bmatrix} x_\beta \\ x'_\beta \\ z_\beta \\ \delta \end{bmatrix} = T \begin{bmatrix} x \\ x' \\ z \\ \delta \end{bmatrix},$$

where

$$T \;=\; \begin{bmatrix} 1 & 0 & 0 & -D \\ 0 & 1 & 0 & -D' \\ D' & -D & 1 & 0 \\ 0 & 0 & 0 & 1 \end{bmatrix} \tag{5.55}$$

is symplectic. See also a discussion around Eq. (2.32).

The 4×4 map for the new betatron coordinates is given by

$$M_\beta \;=\; T_2 M T_1^{-1} \,,$$

where M is the matrix (5.51) and $T_{1,2}$ are the transformation matrices at $s_{1,2}$. It is straightforward to show that

$$M_\beta \;=\; \begin{bmatrix} R_{11} & R_{12} & 0 & 0 \\ R_{21} & R_{22} & 0 & 0 \\ 0 & 0 & 1 & -\Delta C \\ 0 & 0 & 0 & 1 \end{bmatrix} . \tag{5.56}$$

Homework 5.44

(a) The text gave an expression of the 4×4 matrix for a thin-lens dipole. According to that expression, a particle traversing the dipole receives a path length change $\Delta z = -\theta x_0$. Give a physical explanation of this observation using a sketch of the geometry of the particle trajectory. This should be an easy homework.

(b) Similarly explain the expression $R_{56} = L\theta^2$ for a dog-leg of either polarity. Assume thin-lens dipoles. It then is an easy extension to show $R_{56} = 2L\theta^2$ for a chicane.

Homework 5.45 Find the 4×4 map for the zigzag system shown in Fig. 5.55. Assume thin-lens dipoles. Find its R_{16}, R_{26}, R_{56}. Extend your result to the case of a wiggler magnet with N_w wiggle periods.

Figure 5.55: Layout of a zigzag system.

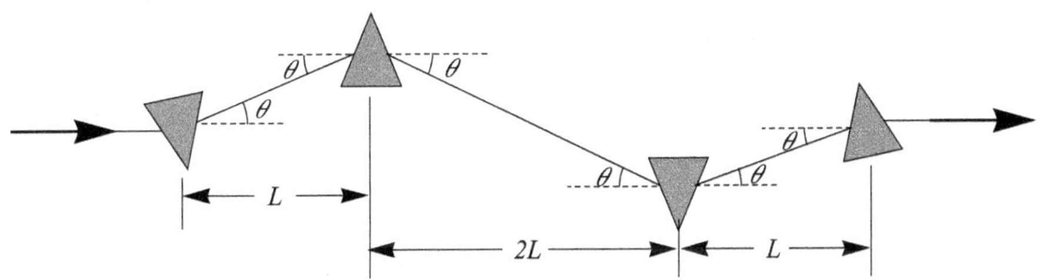

Homework 5.46 The text derived the 4×4 map (5.45) assuming a sector dipole. In particular, it uses Eq. (5.44) to calculate the path length change in the dipole. How are these equations change if we want the 4×4 map for a rectangular dipole?

Solution The difference between a sector dipole and a rectangular dipole occurs only at the two magnet edges.

Homework 5.47

(a) Follow the discussion in the text and show Eq. (5.49).

(b) Also show Eq. (5.50).

Homework 5.48 The map for a dog-leg with thin-lens dipoles was found to be given by Eq. (5.47).

(a) The beam enters the dog-leg with D_1, D_1'. Use the general expression of the 4×4 map to find the dispersion D_2 and D_2' at the end of the dog-leg.

(b) Show that the dispersion is unchanged by the dog-leg if $\theta = D_1'$, not only the transverse dispersion D and D' but also the longitudinal dispersion $\frac{z}{\delta}$. Explain why this happens.

Homework 5.49 The text discusses the canonical transformation from the physical coordinates (x, x') to their betatron components (x_β, x'_β). Although we prefer to consider the physical (x, x') throughout our lectures, this is an important discussion. Go through the derivation to convince yourself of Eq. (5.56).

Homework 5.50 Consider a dog-leg with map (5.47). Nominally it refers to a dog-leg that displaces the beamline axis horizontally from its continuing direction. In case the movement is not horizontal but at a certain angle ϕ around the continuing axis, what is the resulting map of the dog-leg section?

Solution Follow Sec. 3.5.5, but this time we need 6×6 matrices,

$$\begin{bmatrix} \cos\phi & 0 & \sin\phi & 0 & 0 & 0 \\ 0 & \cos\phi & 0 & \sin\phi & 0 & 0 \\ -\sin\phi & 0 & \cos\phi & 0 & 0 & 0 \\ 0 & -\sin\phi & 0 & \cos\phi & 0 & 0 \\ 0 & 0 & 0 & 0 & 1 & 0 \\ 0 & 0 & 0 & 0 & 0 & 1 \end{bmatrix} \begin{bmatrix} 1 & L & 0 & 0 & 0 & -L\theta \\ 0 & 1 & 0 & 0 & 0 & 0 \\ 0 & 0 & 1 & L & 0 & 0 \\ 0 & 0 & 0 & 1 & 0 & 0 \\ 0 & -L\theta & 0 & 0 & 1 & L\theta^2 \\ 0 & 0 & 0 & 0 & 0 & 1 \end{bmatrix} \times$$

$$\begin{bmatrix} \cos\phi & 0 & -\sin\phi & 0 & 0 & 0 \\ 0 & \cos\phi & 0 & -\sin\phi & 0 & 0 \\ \sin\phi & 0 & \cos\phi & 0 & 0 & 0 \\ 0 & \sin\phi & 0 & \cos\phi & 0 & 0 \\ 0 & 0 & 0 & 0 & 1 & 0 \\ 0 & 0 & 0 & 0 & 0 & 1 \end{bmatrix} = \begin{bmatrix} 1 & L & 0 & 0 & 0 & -L\theta\cos\phi \\ 0 & 1 & 0 & 0 & 0 & 0 \\ 0 & 0 & 1 & L & 0 & L\theta\sin\phi \\ 0 & 0 & 0 & 1 & 0 & 0 \\ 0 & -L\theta\cos\phi & 0 & L\theta\sin\phi & 1 & L\theta^2 \\ 0 & 0 & 0 & 0 & 0 & 1 \end{bmatrix}.$$

You might have guessed this result.

Homework 5.51 The one-turn map is given by Eq. (5.52) when we use the physical coordinates as the state vector, and it is given by Eq. (5.56) when use the betatron parts. One difference occurs in their (3, 4)-element,

$$M_{34} = \begin{cases} -\Delta C - AD - BD', & \text{physical coordinates,} \\ -\Delta C, & \text{betatron coordinates.} \end{cases}$$

(a) Show that their difference can be written as

$$-AD - BD' = (\gamma_x D^2 + 2\alpha_x DD' + \beta_x D'^2) \sin\mu_x .$$

(b) Give a physical explanation for this difference.

Solution (b) Consult Eq. (5.54).

Homework 5.52

(a) Find explicitly the eigenvalues and eigenvectors of the 4×4 one-turn map (5.52).

(b) Identify the betatron and synchrotron coordinates from the eigenvectors. Keep in mind that this map does not contain RF, otherwise separation of betatron and synchrotron coordinates is not meaningful.

5.5.3 RF cavity at dispersive location

We have been treating the transverse betatron motions and the longitudinal synchrotron motion separately assuming they are not coupled. The justification is not that they are uncoupled — they are actually pretty strongly coupled and that is where dispersion comes from — but has been that the betatron and synchrotron tunes are so far apart, and since the synchrotron tune is so low, we have averaged out any fast oscillating betatron effects when analyzing the synchrotron motion, taking cover from the adiabatic approximation.

The 4×4 formalism now provides us a tool to examine this more closely. In this and the following few sections, we study the applications of this 4×4 formalism to various examples. The first example applies to the case when an RF cavity is installed at a dispersive location, an issue first raised by Piwinski.[23]

Consider a localized RF cavity at position $s = 0$. Its 4×4 map for the vector (x, x', z, δ), linearized, is

$$M_{\rm cav} = \begin{bmatrix} 1 & 0 & 0 & 0 \\ 0 & 1 & 0 & 0 \\ 0 & 0 & 1 & 0 \\ 0 & 0 & k & 1 \end{bmatrix}, \tag{5.57}$$

where $k = -\frac{eV_0\omega_{\rm rf}}{cE_0}\cos\phi_s$, and we already know that $k\eta C$ must be positive, where η is the phase slippage factor and C the ring circumference. We have also calculated the exact unperturbed synchrotron tune, Eq. (5.34),

$$\cos\mu_s = 1 - \frac{k\eta C}{2}.$$

When the cavity is inserted, the particle motion gets synchrobetatron coupled. The map around the cavity for one turn around the storage ring is given by Eq. (5.52). The total map, observed at the entrance to the cavity, is

$$M = M_{\rm one-turn} M_{\rm cav}.$$

The eigenvalues are determined by $\det(M - \lambda I) = 0$. Let $\lambda + \frac{1}{\lambda} = 2z$, the determinant equation for z is

$$z^2 - \left(\cos\mu_s + \cos\mu_x + \frac{\xi}{2}\sin\mu_x\right) z + \left(\cos\mu_s\cos\mu_x + \frac{\xi}{2}\sin\mu_x\right) = 0, \tag{5.58}$$

where the dimensionless quantity

$$\xi = k(\gamma_x D^2 + 2\alpha_x DD' + \beta_x D'^2)$$

---

[23] A. Piwinski, 11th Int. Conf. High-Energy Accel., 638 (1980); A. Piwinski, Sec. 2.3.4, Handbook Accel. Phys. & Eng., 2nd Ed., World Scientific (2013).

Figure 5.56: Contour plot of the value of μ_+ when the RF is located at a dispersive region. The parameter used is $\xi = 0.1$. The white areas are stopbands, occurring near the resonances $\mu_x - \mu_s = 2\pi n$ (dashed lines). The appearing of a difference resonance stopband is due to the choice of $\xi > 0$ (therefore $k > 0$) corresponding to the case above transition and the longitudinal mass is negative.

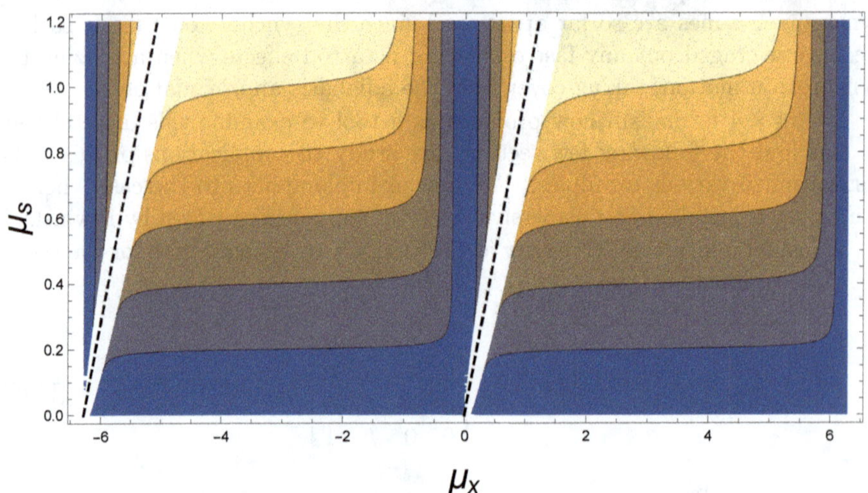

is the driving term of this synchrobetatron coupling effect. The solution is

$$
\begin{aligned}
z_\pm &= \frac{1}{2}\Bigg[\cos\mu_s + \cos\mu_x + \frac{\xi}{2}\sin\mu_x \\
&\quad \pm\sqrt{\left(\cos\mu_s + \cos\mu_x + \frac{\xi}{2}\sin\mu_x\right)^2 - 4\cos\mu_s\cos\mu_x - 2\xi\sin\mu_x}\,\Bigg].
\end{aligned}
$$

If the cavity is inserted at a nondispersive location, $\xi = 0$, the solution reduces to the unperturbed result $z_\pm = \cos\mu_{x,s}$. Stability of the system requires $|z_\pm| \leq 1$. In the stable region, the eigentunes are given by $\cos\mu_\pm = z_\pm$.

Figure 5.56 shows the result of a calculation for μ_+ for an exaggerated case $\xi = 0.1$. It shows a stopband around and on both sides of $\nu_x - \nu_s = n$. This is a clear demonstration of a linear coupled synchrobetatron resonance.

An alert reader would point out the conclusion that there is no sign of instability around the resonance $\nu_x + \nu_y = n$, and that this time, opposite to the transverse case, the motion is unstable near the difference resonance and stable near the sum resonance. The reason of this switching is the negative longitudinal mass, assumed by our choice of $k > 0$. Had we chosen $k < 0$, Fig. 5.56 would show difference resonance stopbands instead.

This rule of stability of the sum resonances and instability of the difference resonances above transition is a general one. It applies also to nonlinear synchrobetatron resonances.

As long as we stay away from the synchrobetatron stopbands (and the betatron integer resonances), the contour lines for the bulk of Fig. 5.56 show a behavior of μ_s = constant. This region of tune space is what we have been assuming when we ignored the synchrobetatron effect. To minimize this synchrobetatron effect, most storage rings install their RF cavities in nondispersive regions although not strictly required.

Homework 5.53 Show that the stopband boundaries of Fig. 5.56 are determined by

$$\cos\mu_s = \cos\mu_x - \frac{\xi}{2}\sin\mu_x \pm 2\sin\frac{\mu_x}{2}\sqrt{\xi\sin\mu_x}\,.$$

This expression applies also when $\xi < 0$.

5.5.4 Crab cavity

An RF cavity excites synchrobetatron resonance as discussed in the previous section. The way it does so is through the coupling effect due to the dispersion at the cavity. The resonance driving is indirect through dispersion. The map for the cavity itself is uncoupled, as seen by Eq. (5.57).

In a crab cavity (transverse deflecting cavity), the coupling is direct. Although dispersion affects the end result, it is not required for it to drive the synchrobetatron resonances. The situation is more analogous to a skew quadrupole driving the x-y coupling resonances discussed in Sec. 3.5.2.

Consider a single crab cavity with map (5.46). The one-turn map around it is given by Eq. (5.52). To provide longitudinal focusing, we will also need an accelerating RF cavity with map (5.57). For simplicity, we consider the crab cavity to be at a nondispersive location. We will also assume $\alpha_x = 0$ at the crab cavity, and locate the acceleration RF to be at the same location as the crab cavity. The total map is

$$T_{\text{one-turn}}T_{\text{cav}}T_{\text{crab}} = \begin{bmatrix} \cos\mu_x & \beta_x\sin\mu_x & \beta_s k\sin\mu_x & 0 \\ -\frac{\sin\mu_x}{\beta_x} & \cos\mu_x & k\cos\mu_x & 0 \\ -k\eta C & 0 & 1-k_0\eta C & -\eta C \\ k & 0 & k_0 & 1 \end{bmatrix},$$

where k_0 is the strength of the accelerating cavity and k is that of the crab cavity. We also have the unperturbed $\cos\mu_s = 1 - \frac{k_0\eta C}{2}$.

The eigentunes satisfy the determinant equation

$$(\cos\mu - \cos\mu_s)(\cos\mu - \cos\mu_x) = -\frac{1}{4}\beta_x k^2\eta C\sin\mu_x\,,$$

Figure 5.57: Contour plot of the value of $\cos\mu_+$ when a crab cavity is installed at a nondispersive location. The parameter used is $\beta_x k^2\eta C = 0.01$. The white areas are stopbands, occurring near the resonances $\mu_x - \mu_s = 2\pi n$ (dashed lines). The integer resonances $\mu_x = 2\pi n$ and $\mu_s = 2\pi n$ are also visible.

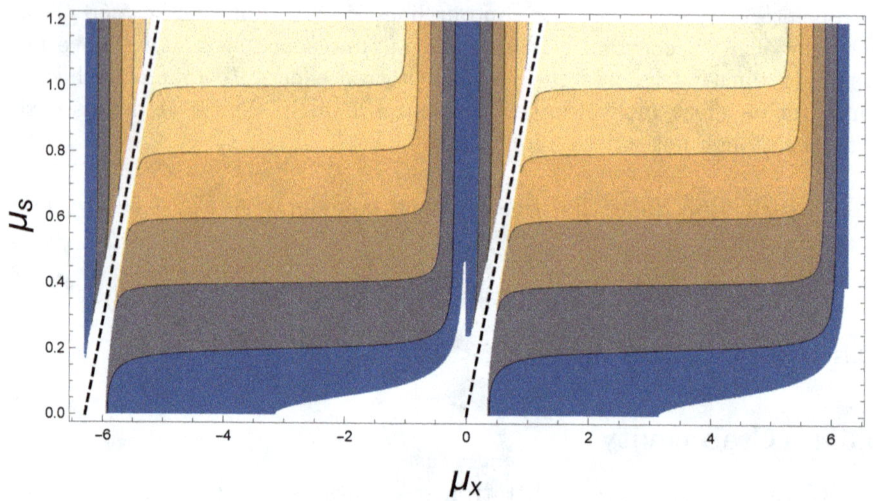

with the solution

$$\cos\mu = \frac{1}{2}\left(\cos\mu_s + \cos\mu_x \pm \sqrt{(\cos\mu_s - \cos\mu_x)^2 - \beta_x k^2\eta C \sin\mu_x}\right). \quad (5.59)$$

The crab cavity also drives synchrobetatron coupling resonance instabilities. It drives the difference resonances if $\eta > 0$ (above transition) and the sum resonances if $\eta < 0$ (below transition). The instability stopbands for an exaggerated case with $\beta_x k^2\eta C = 0.01$ is shown in Fig. 5.57.

Homework 5.54

(a) Derive Eq. (5.59).

(b) Find the stopband width of the synchrobetatron resonance driven by the crab cavity.

5.5.5 Sawtooth orbit

We discussed one beam-tilt effect of a lumped distribution of the RF in Eq. (5.37) deviating from the adiabatic approximation. Another effect due to lumped RF is a sawtooth structure in the beam's closed orbit.

The effect becomes notable for electron storage rings where there are synchrotron radiation losses at all the dipole magnets. The energy loss due to

Figure 5.58: Schematic of energy sawtooth in an electron storage ring. The red curve shows the beam energy deviation from the synchronous value as a function of location around the ring. This energy sawtooth induces an orbit sawtooth.

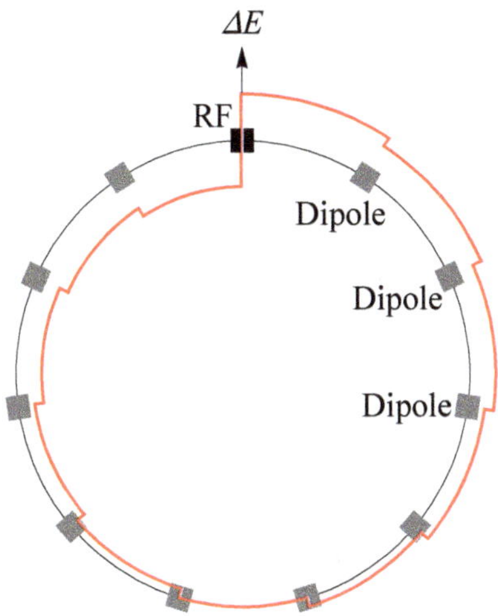

synchrotron radiation is replenished at the RF stations. If the RF is not uniformly distributed around the ring, there will be energy gain at the RF stations and energy losses at the dipoles, resulting in a sawtooth behavior of the beam energy around the ring, as sketched schematically in Fig. 5.58. This sawtooth energy variation in turn induces a sawtooth horizontal closed orbit around the ring.

Using the 4 × 4 formalism To calculate the sawtooth orbit, consider the case of a ring with a single localized RF. Let the synchrotron radiation energy loss at the i-th dipole be u_i. We work in the 4-D space with vector (x, x', z, δ), and our goal is to calculate a 4-D closed orbit. This orbit will contain not only the x, x' components but also z, δ components.[24]

The energy loss at the i-th dipole will be replenished by the RF. We first calculate the sawtooth orbit caused by the energy sawtooth due to the i-th dipole and its RF compensation.

Let the 4-D closed orbit at the entrance of the i-th dipole be

$$X_{0i} = \begin{bmatrix} x_{0i} \\ x'_{0i} \\ z_{0i} \\ \delta_{0i} \end{bmatrix}.$$

[24] Yes, closed orbit does not have to be the 2-D ones we are familiar with.

At the exit of the dipole, the closed orbit acquires an additional energy kick and becomes $X_{0i} - \Delta_i$, where

$$\Delta_i = \begin{bmatrix} 0 \\ 0 \\ 0 \\ u_i/E_s \end{bmatrix}.$$

The quantity E_s is the design energy of the reference synchronous particle, uniform around the ring without sawtooth.

At the RF, the energy u_i is added back. Let T_i be the 4×4 one-turn map around the i-th dipole and let S_i be the map from the RF back to the i-th dipole. Being the closed orbit, we demand that after one revolution around, the vector X_{0i} returns to its original value. This means it satisfies the condition

$$T_i(X_{0i} - \Delta_i) + S_i\Delta_i = X_{0i},$$

or

$$X_{0i} = (T_i - I)^{-1}(T_i - S_i)\Delta_i.$$

Once X_{0i} is obtained, the total sawtooth orbit is just the superposition of contributions from all dipoles.

Exact calculation A sharp eyed reader might notice that we have ignored the RF focusing action at the RF in this treatment. In the previous calculation, the RF functions solely as an energy replenisher, and does not provide longitudinal focusing. In case it is to be included, the way to treat it is to use a 5-D vector

$$\begin{bmatrix} x \\ x' \\ z \\ \delta \\ 1 \end{bmatrix}.$$

If y-motion is also to be included, this leads to 7-D vector. Such extensions are adopted in a technique called SLIM, a subject we shall cover in Sec. 6.5. The extensive use of matrices in accelerator physics was mentioned on page 334. The 4×4 calculation is normally accurate enough in most practical cases.

Calculation using COD formula The same result of the 4×4 technique can be obtained by a more conventional way as well. One first calculates the sawtooth energy E_i at the middle of the i-th dipole. The dipole will kick the closed orbit (this time just the orbit in x) because $E_i \neq E_s$. The kick angle is

$$\Delta\theta_i = \theta_i \frac{E_s - E_i}{E_s}.$$

Applying the closed orbit formula (3.7), and summing over all dipoles, it follows that the sawtooth orbit is given by

$$\Delta x(s) = \frac{\sqrt{\beta_x(s)}}{2\sin\pi\nu_x}\sum_i \sqrt{\beta_{xi}}\,\Delta\theta_i \cos(\nu_x - |\psi_x(s) - \psi_{xi}|)\,. \tag{5.60}$$

It can be shown that this calculation gives the same result as that obtained using the 4×4 formalism.

This approach also yields tune shifts caused by the sawtooth effect. They are given by

$$\Delta\nu_{x,y} = \pm\frac{1}{4\pi}\sum_i \beta_{x,yi}\left(\frac{G\ell}{B\rho}\right)_i \frac{E_s - E_i}{E_s}\,.$$

Homework 5.55

(a) Consider an electron storage ring with many identical dipoles evenly distributed around the ring so that the bending can be considered uniformly distributed. Let there be a single RF station. Let the total synchrotron radiation loss per turn be U_0. By approximating the ring with a β-function uniform around the ring with a value of $\frac{R}{\nu_x}$, give an expression for $\Delta x(s)$ around the ring using Eq. (5.60).

(b) What would the tune shifts be due to the sawtooth effect for this ring?

5.5.6 Feedback for synchrobetatron coupled system

Feedback systems are used extensively to control collective motions of the beam whether it being caused by misinjection or collective instabilities. Basically a feedback works by detecting the displacement of the beam x, and then give the beam a kick $\Delta x' \propto x$ at a location 90° phase from the detecting location in such a way that the subsequent x-motion is reduced. With the feedback signal $\propto x$ and properly phased, the x-motion of the beam is exponentially damped with a certain damping rate.

One condition for this scheme is that the detected signal must be easily accessible to a beam monitor, and the applied kick must be applicable to some kicker device. A complication occurs if the measured coordinate or the applied kick are not in a pure dimension. For example, an easily measurable quantity is the horizontal displacement x of the beam at a beam position monitor, but x may contain both betatron and a synchrotron components. Similarly, an easily applicable feedback is to kick the angle x', but this kick generally excites both betatron and synchrotron motions. A proper analysis of the feedback system will require a more careful analysis of the coupled dynamics. And we have the tool to do just that, the 4×4 formalism.

Among the four coordinates (x, x', z, δ), only x and z are easily detectable by monitors, and only x', δ are applicable for kicks. In combination, this suggests four possible types of feedback systems,

$$
\begin{aligned}
\text{Type } (x, x') : &\quad \text{detecting } x, \text{ kick } x', \\
\text{Type } (x, \delta) : &\quad \text{detecting } x, \text{ kick } \delta, \\
\text{Type } (z, x') : &\quad \text{detecting } z, \text{ kick } x', \\
\text{Type } (z, \delta) : &\quad \text{detecting } z, \text{ kick } \delta\,.
\end{aligned}
$$

If parameters are properly chosen, in principle, one feedback system can damp both horizontal and longitudinal motions of the beam. If not, it might damp one dimension but excite the other.

We analyze only the type (x, δ) as an illustration.[25] We consider a relativistic beam, above transition. Measurement of x is from a beam position monitor. Feedback is applied to δ at the RF cavity. At the RF cavity, the beam receives a kick

$$
\Delta\delta = \frac{(2\pi\nu_s)^2}{\alpha_c C} z + \xi x_{\text{monitor}}\,,
$$

where ξ is the feedback strength. The first term is the RF focusing; the second term is feedback.

Written in terms of a matrix, we have

$$
T_{\text{cav}} = \begin{bmatrix} 1 & 0 & 0 & 0 \\ 0 & 1 & 0 & 0 \\ 0 & 0 & 1 & 0 \\ 0 & 0 & \frac{(2\pi\nu_s)^2}{\alpha_c C} & 1 \end{bmatrix} + \begin{bmatrix} 0 & 0 & 0 & 0 \\ 0 & 0 & 0 & 0 \\ 0 & 0 & 0 & 0 \\ \xi & 0 & 0 & 0 \end{bmatrix} T_{CM}^{-1}\,,
$$

where T_{CM} is the map from the monitor to the cavity. The action of a feedback is not symplectic.

The total map around the exit of the feedback is

$$
T = T_{\text{cav}} T_{\text{one-turn}}\,,
$$

where $T_{\text{one-turn}}$ is given by Eq. (5.52). Substituting the appropriate lattice functions, after some algebra, we obtain

$$
T = \begin{bmatrix} R_{11} & R_{12} & 0 & D_c - R_{11}D_c - R_{12}D_c' \\ R_{21} & R_{22} & 0 & D_c' - R_{21}D_c - R_{22}D_c' \\ A & B & 1 & -\alpha_c C - AD_c - BD_c' \\ \frac{(2\pi\nu_s)^2}{\alpha_c C}A + \xi r_{11} & \frac{(2\pi\nu_s)^2}{\alpha_c C}B + \xi r_{12} & \frac{(2\pi\nu_s)^2}{\alpha_c C} & G \end{bmatrix},
$$

with

$$
R_{11} = \cos\mu_x + \alpha_{xc}\sin\mu_x\,,
$$

[25]In this section, we analyze the type (x, δ). Analysis for the other types, as well as their feedback power requirements and noise considerations can be found in A.W. Chao, P.L. Morton, J. Rees, SLAC Report, PEP-281 (1979).

$$\begin{aligned}
R_{12} &= \beta_{xc} \sin \mu_x \,, \\
R_{21} &= -\gamma_{xc} \sin \mu_x \,, \\
R_{22} &= \cos \mu_x - \alpha_{xc} \sin \mu_x \,, \\
A &= R_{21} D_c - R_{11} D'_c + D'_c \,, \\
B &= -R_{12} D'_c + R_{22} D_c - D_c \,, \\
r_{11} &= \sqrt{\frac{\beta_{xm}}{\beta_{xc}}} [\cos(\mu_x + \Delta\psi) + \alpha_{xc} \sin(\mu_x + \Delta\psi)] \,, \\
r_{12} &= \sqrt{\beta_{xc}\beta_{xm}} \sin(\mu_x + \Delta\psi) \,, \\
G &= 1 - \frac{(2\pi\nu_s)^2}{\alpha_c C} (\alpha_c C + A D_c + B D'_c) + \xi (D_m - r_{11} D_c - r_{12} D'_c) \,, \\
\Delta\psi &= \psi_{xm} - \psi_{xc} \,,
\end{aligned}$$

where subscripts c and m refer to the cavity and the monitor respectively.

We then calculate the eigenvalues λ of T. Replacing $(2\pi\nu_s)^2$ by $2(1-\cos\mu_s)$, we find the determinant equation

$$\begin{aligned}
&(1 - 2\cos\mu_x \lambda + \lambda^2)(1 - 2\cos\mu_s \lambda + \lambda^2) \\
&\quad -\frac{2}{\alpha_c C}(1 - \cos\mu_s) \sin\mu_x (\gamma_{xc} D_c^2 + 2\alpha_{xc} D_c D'_c + \beta_{xc} D_c'^2) \lambda (1 - \lambda)^2 \\
&\quad +\xi (1 - \lambda)(1 - 2\cos\mu_x \lambda + \lambda^2) D_m \\
&\quad -\xi (1 - \lambda)^2 (Q - \lambda Q') \\
&= 0 \,, \qquad (5.61)
\end{aligned}$$

with

$$\begin{aligned}
Q &= \sqrt{\frac{\beta_m}{\beta_c}} \left[D_c \cos\Delta\psi + (\alpha_c D_c + \beta_c D'_c) \sin\Delta\psi \right] \,, \\
Q' &= \sqrt{\frac{\beta_m}{\beta_c}} \left[D_c \cos(\Delta\psi + \mu_x) + (\alpha_c D_c + \beta_c D'_c) \sin(\Delta\psi + \mu_x) \right] .
\end{aligned}$$

The solution to the complete equation requires solving (5.61) numerically for λ. Express the exact solution as

$$\lambda = e^{-\alpha_k \pm i(\mu_k + \Delta\mu_k)} \,, \qquad k = x, s \,.$$

Quantities α_k are the growth or damping constants and $\Delta\mu_k$ are the beam's coherent phase advance shifts.

The product of all eigenvalues is equal to the determinant of the matrix,

$$\begin{aligned}
e^{-2\alpha_x - 2\alpha_s} &= \operatorname{Det} T \\
&= 1 + \xi D_m - \xi Q \,.
\end{aligned}$$

For weak damping, $|\alpha_{x,s}| \ll 1$, we have a sum rule,

$$\alpha_x + \alpha_s = -\frac{\xi}{2}(D_m - Q)\,. \tag{5.62}$$

If the monitor and the cavity are at the same location ($\beta_c = \beta_m, D_c = D_m, \Delta\psi = 0$), then $\alpha_x + \alpha_s = 0$, which means that damping for one mode necessarily means antidamping for the other mode. This is not what a feedback system wants to accomplish.

There are four terms in Eq. (5.61). The first term is the unperturbed contribution. If all other terms are dropped, the solution is $\lambda = e^{\pm i\mu_x}, e^{\pm i\mu_s}$. Each of the other three terms has a clear physical origin. The second term comes from the RF being located at a dispersive location. The third term is induced by the feedback to the synchrotron motion. The fourth term is the effect of the feedback on the horizontal motion.

It is interesting that the three physical effects contribute clearly in three separate terms, linearly added, in the determinant equation (5.61). Had they appear in a factorized manner, it would mean they do not interfere with each other. But here it is not the case. The general solution will require solving Eq. (5.61) numerically. However, to simplify our following discussion, we shall simply examine each of the three terms individually, assuming the coupling among the individual terms is weak.

The second term is the contribution from a dispersive RF. If we drop the remaining terms, we recover Eq. (5.58). This simplification can be justified if the feedback system is sufficiently weak.

The third and the fourth terms, being proportional to ξ, are contributions from the feedback system. Dropping the second and the fourth terms, the solution for λ is

$$\lambda = \begin{cases} e^{\pm i\mu_x}, \\ \cos\mu_s + \frac{\xi D_m}{2} \pm i\sqrt{\sin^2\mu_s + \xi D_m(1-\cos\mu_s) - \frac{\xi^2 D_m^2}{4}}\,. \end{cases} \tag{5.63}$$

This means the x-mode is unperturbed, while the synchrotron mode acquires a feedback damping.

The damping constant for the synchrotron mode is[26]

$$\alpha_s \approx -\frac{\xi D_m}{2}\,. \tag{5.64}$$

For the calculation of $\Delta\mu_s$, see Homework 5.56.

Physically, this feedback damping mode is simply to detect the energy deviation at a x-monitor with a dispersion D_m, and act back on the energy with a

[26]We can calculate α_s and $\Delta\mu_s$ from the exact expression (5.63), but it is simpler to note the algebraic rule that the solution to the equation $\lambda^2 + A\lambda + B = 0$ has the property that $\lambda_1\lambda_2 = B$.

strength ξ. Damping requires ξD_m to be negative. The extra factor of $\frac{1}{2}$ in the expression for α_s is also as expected.

Dropping the second and the third term, we will be working on the x-mode with the feedback. Assuming $\mu_s \ll 1$, Eq. (5.61) becomes

$$1 - 2\cos\mu_x\lambda + \lambda^2 - \xi(Q - \lambda Q') = 0 .$$

We then obtain

$$\alpha_x \approx \frac{\xi Q}{2} . \tag{5.65}$$

The quantity Q determines the feedback damping rate, while the quantity Q' does not contribute.

Note that the feedback damping constants, Eqs. (5.64) and (5.65), satisfy the sum rule (5.62). Feedback damping rates are obtained by properly choosing the beam lattice parameters D_m and Q. To simultaneously damp the horizontal and longitudinal motion, we need both ξ and Q to have opposite sign as D_m.

Take for example $D_m = 1$ m, and if we want to damp an energy oscillation with amplitude $\delta = 10^{-3}$ of a 1 GeV electron beam in 1000 turns by the feedback system, then we need $\xi = -2 \times 10^{-3}$/m, and the RF must provide a feedback voltage up to 2 kV. The same feedback system also can damp horizontal orbital motion. If we choose $\Delta\psi = \pi, \beta_c = \beta_m, D_c = 1$ m, then $Q = -1$ m and the horizontal oscillation is damped also in 1000 turns.

Homework 5.56

(a) We calculated the damping constant α_s for the synchrotron feedback mode with the (x, δ) type of feedback system. Use Eq. (5.63) to calculate the feedback mode phase advance shift $\Delta\mu_s$. Find an approximate expression for $\Delta\mu_s$ when $|\xi D_m| \ll \mu_s \ll 1$.

(b) Similarly find an approximate expression of $\Delta\mu_x$ for the horizontal feedback mode when $|\xi Q| \ll 1, |\xi Q'| \ll 1$.

Solution

(a) $\Delta\mu_s \approx -\frac{1}{4}\mu_s\xi D_m$. This phase advance shift per turn appears as a synchrotron tune shift when this feedback is turned on.

(b) $\Delta\mu_x \approx \frac{\xi}{2\sin\mu_x}(Q' - Q\cos\mu_x)$. This phase advance shift per turn appears as a horizontal betatron tune shift when this feedback is turned on.

Homework 5.57 The text analyzed the case of a coupled synchrobetatron system. A simpler case, e.g. the case when a feedback is applied to the y-motion, involves only a 1-D analysis. Consider a feedback system for the y-motion in a storage ring. The vertical kicker is located 90° betatron phase downstream of the monitor. Analyze the feedback system following the text but using the 2×2 formalism.

5.6 Emittance exchange technique

The fact that the transverse and longitudinal emittances of a typical beam are so different (see Sec. 2.4.4 when we identify the emittance as the internal temperature of the beam motion) has generated various ideas of beam emittance manipulations. What if we can exchange the large emittance in one dimension with the small emittance of another dimension (emittance exchange)? What if we can arbitrarily move the emittance from one dimension to another (emittance partitioning)? What if we can equalize the emittances in different dimensions (thermal equilibrium)? What if we can cool the emittances (beam cooling)? It turns out that the latter three ideas are by no means trivial because they require nonsymplectic processes. Each is an extensive subject of research and applications beyond the scope of these lectures. The first idea, however, the emittance exchanges (EEX), in which the beam emittances in the different dimensions maintain their values, but simply get switched, is quite accessible with our regular symplectic beamline devices. We briefly mention a few of the EEX techniques in this section.

Eigenemittances Earlier we have discussed linearly x-y coupled lattices in which the horizontal and vertical beam emittance can exchange between them (Secs. 3.5.2, 3.5.3, 3.7.7). In those cases, one can exchange any percentage of one emittance into the other as long as one observes a certain emittance sum rule. Recall however, as we cautioned at the time, that we were talking about the *apparent* emittances defined in the original uncoupled coordinates, not the *eigenemittances* defined in the now-coupled coordinates. For a given linear coupled system, no matter how its coordinates are coupled, resonances or not, as long as the couplings are linear and symplectic, a theorem dictates that the eigenemittances cannot be changed. As much as we can change the apparent emittances, we cannot change the eigenemittances.

To establish this theorem, let us first define the Σ matrix of the second moments in the phase space. Eigenemittances are related to the eigenvalues of the matrix $S\Sigma$ where S is the symplectic form. For the 1-D case, for example, we have

$$S\Sigma = \begin{bmatrix} 0 & 1 \\ -1 & 0 \end{bmatrix} \begin{bmatrix} \langle x^2\rangle & \langle xx'\rangle \\ \langle xx'\rangle & \langle x'^2\rangle \end{bmatrix} = \begin{bmatrix} \langle xx'\rangle & \langle x'^2\rangle \\ -\langle x^2\rangle & -\langle xx'\rangle \end{bmatrix},$$

whose eigenvalues are $\pm i\epsilon$, where $\epsilon = \sqrt{\langle x^2\rangle\langle x'^2\rangle - \langle xx'\rangle^2}$ is the eigenemittance in this case.

For a 2-D or 3-D system, the eigenemittances of the beam is also given by the eigenvalues of the matrix $S\Sigma$. To see this, let us first consider a decoupled beam. If the beam is decoupled, its Σ_0 matrix constitutes of diagonal 2×2 blocks. The eigenemittances are defined in an obvious manner, and are easily seen to be related to the eigenvalues of the matrix $S\Sigma_0$. Now consider this beam going through a beamline with some coupling elements that messes up the block diagonal feature of the original Σ_0 matrix. The new Σ however is related to Σ_0 by

$$\Sigma = T\Sigma_0\tilde{T}, \tag{5.66}$$

with some symplectic matrix T. It is easy to show that the matrix $S\Sigma$ is related to $S\Sigma_0$ by a similarity transformation. Its eigenvalues are unchanged. Q.E.D.

It should be mentioned that the theorem of eigenemittance preservation is valid only if we use the true canonical coordinates and momenta of the phase space. For example, if we use our beloved $(x, x', y, y', z, \delta)$ as coordinates, the eigenemittances of the corresponding transport dynamics will appear nonpreserved in a few special circumstance such as when going through a solenoid or a Wien filter. However, the nonpreservations here are only in appearance and is only an artifact of using noncanonical coordinates. Their canonical values do not change.

Emittance exchange Suppose we have a decoupled beamline making a transition into another decoupled beamline, and we are designing the transition beamline to connect the two beamlines. If in the design of the transition beamline, we are allowed only to implement symplectic elements (which may include coupling elements), then the three eigenemittances in the final decoupled beamline are necessarily equal to the three eigenemittances of the initial decoupled beamline, no matter how clever our transition beamline could be. Since both the initial and the final systems are decoupled, their three eigenemittances are simply their own respective horizontal, vertical, and longitudinal emittances.

However, that does not forbid permutations among the three eigenemittances. It opens up the possibility that we can *exchange* emittances, for example ϵ_x and ϵ_z, without violating the eigenemittance theorem. Therefore the idea of "emittance exchange" (EEX) techniques.

To facilitate an emittance exchange, we will need coupling elements in the transition beamline. The map that this transition beamline must provide is of the form

$$T = \begin{bmatrix} 0 & A \\ B & 0 \end{bmatrix}, \tag{5.67}$$

where A, B are 2×2 block matrices. It is easy to show that symplecticity requires that A and B be symplectic.

Consider an uncoupled beam with an initial second-moment matrix $\Sigma_0 = \begin{bmatrix} \Sigma_{x0} & 0 \\ 0 & \Sigma_{z0} \end{bmatrix}$. Going through an EEX with a map T given by (5.67), the beam's Σ matrix becomes, using Eq. (5.66),

$$\Sigma = \begin{bmatrix} A\Sigma_{z0}\tilde{A} & 0 \\ 0 & B\Sigma_{x0}\tilde{B} \end{bmatrix}.$$

It is clear that the emittances have made an $x \leftrightarrow z$ exchange.

Solenoid x-y EEX A simplest EEX system is given by a solenoid. When the solenoid strength $kL = \pi$, the transfer map, including solenoid edge fields, for the (x, x', y, y') vector, according to Eq. (3.30), reads

$$\begin{bmatrix} 0 & 0 & 0 & \frac{2}{k} \\ 0 & 0 & -\frac{k}{2} & 0 \\ 0 & -\frac{2}{k} & 0 & 0 \\ \frac{k}{2} & 0 & 0 & 0 \end{bmatrix}.$$

It is clear that this solenoid will exchange the x- and y-coordinates and thus constitutes an x-y EEX system.

Note that here we continue to use (x, x', y, y') for a solenoid. This is permitted because we only talk about the total solenoid including the edge fields. If we wish to study the particle motion inside the solenoid, then we must use (x, p_x, y, p_y) as our vector. As emphasized in Sec. 3.5.3, the function of the entrance and exit edge fields transform (x, x', y, y') into (x, p_x, y, p_y) and then transform it back.

Double dog-leg z-x EEX While the x-y EEX is straight forwardly given by a solenoid, there has been an evolution of the more involved z-x EEX designs. The key beamline element here will be a transverse RF cavity (crab cavity). These designs are triggered initially to be applied to the free electron lasers when the beam coming out of the photocathode gun has a small longitudinal emittance, while FEL gain requires small transverse emittances. The idea was to implement a z-x EEX for such devices.[27]

For one example, by introducing a transverse RF cavity in a horizontal double-dog-leg, one obtains an x-z EEX system as shown in Fig. 5.59.

Figure 5.59: An EEX design with a double dog-leg configuration.

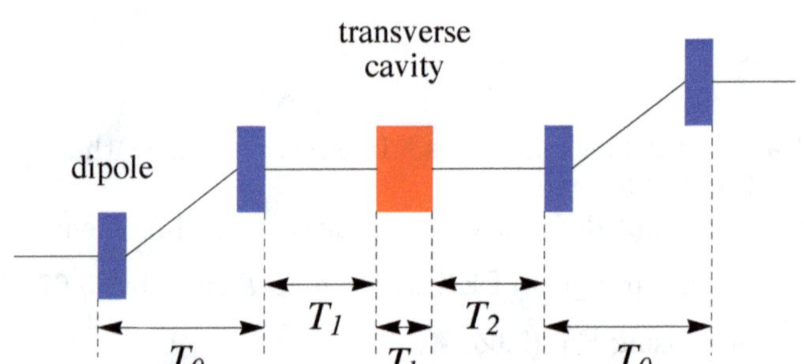

[27]M. Cornacchia, P. Emma, Phys. Rev. ST Accel. Beams 5, 084001 (2002); P. Emma, Z. Huang, K.-J. Kim, and P. Piot, Phys. Rev. ST Accel. Beams 9, 100702 (2006); D. Xiang, A. Chao, Phys. Rev. ST Accel. Beams 14, 114001 (2011).

The dog-leg map is given in Eq. (5.48). The transverse cavity map is given in Eq. (5.46). The total transfer map of the system is given by

$$\begin{aligned} T &= T_0 T_2 T_k T_1 T_0 \\ &= \begin{bmatrix} 1 & L & 0 & L\theta \\ 0 & 1 & 0 & 0 \\ 0 & L\theta & 1 & L\theta^2 \\ 0 & 0 & 0 & 1 \end{bmatrix} \begin{bmatrix} 1 & L_2 & 0 & 0 \\ 0 & 1 & 0 & 0 \\ 0 & 0 & 1 & 0 \\ 0 & 0 & 0 & 1 \end{bmatrix} \begin{bmatrix} 1 & 0 & 0 & 0 \\ 0 & 1 & k & 0 \\ 0 & 0 & 1 & 0 \\ k & 0 & 0 & 1 \end{bmatrix} \\ &\quad \times \begin{bmatrix} 1 & L_1 & 0 & 0 \\ 0 & 1 & 0 & 0 \\ 0 & 0 & 1 & 0 \\ 0 & 0 & 0 & 1 \end{bmatrix} \begin{bmatrix} 1 & L & 0 & L\theta \\ 0 & 1 & 0 & 0 \\ 0 & L\theta & 1 & L\theta^2 \\ 0 & 0 & 0 & 1 \end{bmatrix}, \end{aligned}$$

where L is the length of the dog-leg, θ is the dipole bending angle of each of the dipoles in the dog-leg, $k = -\frac{eV_{\perp \mathrm{rf}}\omega_{\mathrm{rf}}}{cE_0}$, and $L_{1,2}$ are the drift lengths of $T_{1,2}$.

We multiply out the matrices to obtain the total map T. To be an EEX beamline, we want the two diagonal 2×2 blocks of T to vanish. This can be achieved by imposing the condition

$$L\theta k = -1 .$$

When this condition is fulfilled, T becomes

$$T = \begin{bmatrix} 0 & 0 & -\frac{L+L_2}{L\theta} & -L_2\theta \\ 0 & 0 & -\frac{1}{L\theta} & -\theta \\ -\theta & -L_1\theta & 0 & 0 \\ -\frac{1}{L\theta} & -\frac{L+L_1}{L\theta} & 0 & 0 \end{bmatrix} .$$

The EEX system tends to be rather substantial. Take for example a case of 1 GeV electron beam, the condition $L\theta k = -1$ can be fulfilled if we have $V_{\perp \mathrm{rf}} = 10$ MV, $\lambda_{\mathrm{rf}} = 6$ cm, $L = 10$ m, and $\theta = 0.1$ rad.

Chicane z-x EEX The double dog-leg EEX system introduces a sideway step of the beam line, sometimes undesirable for beamline construction or operation. This sideway step can be removed by introducing a $-I$ transformation inside the chicane, as shown in Fig. 5.60, where T_0' is the map (5.47).

In this design, the T_1 drift section of Fig. 5.59 is replaced by a $-I$ section. This reversal of signs allowed the reversal of the second dog-leg.

The total transfer map of the system is given by

$$\begin{aligned} T &= T_0' T_2 T_k (-I) T_0 \\ &= \begin{bmatrix} 1 & L & 0 & -L\theta \\ 0 & 1 & 0 & 0 \\ 0 & -L\theta & 1 & L\theta^2 \\ 0 & 0 & 0 & 1 \end{bmatrix} \begin{bmatrix} 1 & L_2 & 0 & 0 \\ 0 & 1 & 0 & 0 \\ 0 & 0 & 1 & 0 \\ 0 & 0 & 0 & 1 \end{bmatrix} \begin{bmatrix} 1 & 0 & 0 & 0 \\ 0 & 1 & k & 0 \\ 0 & 0 & 1 & 0 \\ k & 0 & 0 & 1 \end{bmatrix} \\ &\quad \times \begin{bmatrix} -1 & 0 & 0 & 0 \\ 0 & -1 & 0 & 0 \\ 0 & 0 & 1 & 0 \\ 0 & 0 & 0 & 1 \end{bmatrix} \begin{bmatrix} 1 & L & 0 & L\theta \\ 0 & 1 & 0 & 0 \\ 0 & L\theta & 1 & L\theta^2 \\ 0 & 0 & 0 & 1 \end{bmatrix} \end{aligned}$$

Figure 5.60: An EEX with a chicane configuration.

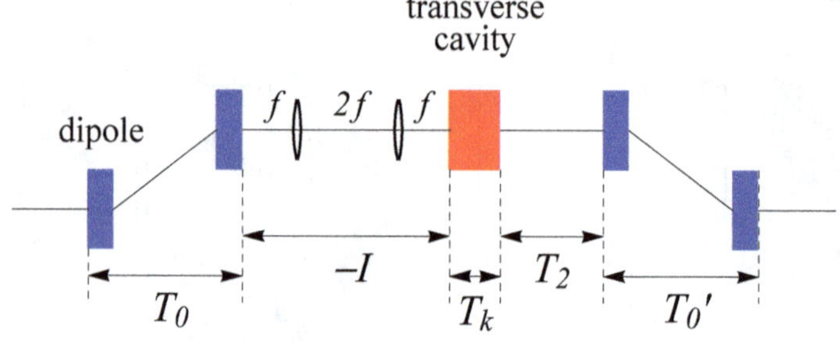

To be an EEX in this configuration, we choose

$$L\theta k = 1 .$$

The total map then becomes

$$\begin{bmatrix} 0 & 0 & \frac{L+L_2}{L\theta} & L_2\theta \\ 0 & 0 & \frac{1}{L\theta} & \theta \\ -\theta & 0 & 0 & 0 \\ -\frac{1}{L\theta} & -\frac{1}{\theta} & 0 & 0 \end{bmatrix} .$$

Application One may ask how such EEX systems could be useful. It should be noted that as accelerator technology advances, the requirements on accelerator beam quality become increasingly demanding. Facing these new demands, the topic of phase space gymnastics is becoming a newly-focused and critical area of R&D. Anticipating this demand, we mention several conceived applications here without elaboration.

- When $\epsilon_z \ll \epsilon_x$, EEX allows small ϵ_x after EEX for FEL applications;
- When $\epsilon_z \gg \epsilon_x$, EEX allows longitudinal bunch compression;
- Observe z-distribution by an x-profile monitor;
- Clean the z- and δ-tails by x-scraper;
- Measure slice energy spread by an x-profile monitor;
- Observe beam microbunching in z by an x-profile monitor;
- Generate z-double bunches by x-wire scraper;
- Generate z-microbunching by modulating the x-profile of a beam;
- Longitudinal phase space linearizer by a sextupole;
- Study coherent synchrotron radiation effect by converting the affected z-δ correlation to x-x' correlation;
- Suppress coherent synchrotron radiation by $\epsilon_z \gg \epsilon_x$ and exchange back afterwards;

- Observe curvature of $z(y)$ by an x-y profile monitor;
- Bunch compression without energy chirp.

Homework 5.58 Given $\langle x^2\rangle, \langle xx'\rangle, \langle x'^2\rangle$, we calculate the emittance as $\epsilon = \sqrt{\langle x^2\rangle\langle x'^2\rangle - \langle xx'\rangle^2}$. To obtain the true value of ϵ, we assume to have removed the centroid values from x and x'. Show that if we have neglected to do so, the resulting value of ϵ — let us designate it as ϵ' — will be an overestimate of the true ϵ, i.e. $\epsilon' > \epsilon$.

Solution

$$\epsilon' - \epsilon = \langle (\langle x\rangle x' - \langle x'\rangle x')^2\rangle\,.$$

Homework 5.59 The text outlined the way to prove the theorem of preservation of eigenemittances for general symplectic systems. It would be instructive to follow the outline and go through the proof in more detail.

Homework 5.60

(a) Consider a decoupled system in the x-y phase space whose Σ-matrix at position $s = 0$ is

$$\Sigma = \begin{bmatrix} \langle x^2\rangle & \langle xx'\rangle & 0 & 0 \\ \langle xx'\rangle & \langle x'^2\rangle & 0 & 0 \\ 0 & 0 & \langle y^2\rangle & \langle yy'\rangle \\ 0 & 0 & \langle yy'\rangle & \langle y'^2\rangle \end{bmatrix}.$$

Find its eigenemittances.

(b) Now add a thin-lens skew quadrupole with strength $k = \frac{\ell}{B\rho}\frac{\partial B_y}{\partial y}$ at position $s = 0$. The system is x-y coupled, and the Σ-matrix becomes Σ'. Calculate the apparent emittances.

(c) Calculate the eigenemittances of Σ' and show that they have not changed.

Solution (b) Both apparent emittances grow by an amount quadratic in k. No resonance effect has been included yet for the sum rule to apply.

$$\begin{aligned} \epsilon^2_{x,\text{apparent}} &= \langle x^2\rangle\langle x'^2\rangle - \langle xx'\rangle^2 + k^2\langle x^2\rangle\langle y^2\rangle \\ \epsilon^2_{y,\text{apparent}} &= \langle y^2\rangle\langle y'^2\rangle - \langle yy'\rangle^2 + k^2\langle x^2\rangle\langle y^2\rangle\,. \end{aligned}$$

Homework 5.61 As illustrated in the text, EEX beamlines manage to switch the beam emittances with a map $\begin{bmatrix} 0 & A \\ B & 0 \end{bmatrix}$. The beam dynamics after the switch, however, have been disrupted by the fact that A and B are not identity maps. Show that it is possible to add a short section in front, and another short section after the EEX section, each of which consisting only of quadrupoles and drifts, to rectify this concern.

Solution The added section in front of the EEX should provide a map $\begin{bmatrix} A^{-1} & 0 \\ 0 & I \end{bmatrix}$.

The section after the EEX should provide a map $\begin{bmatrix} B^{-1} & 0 \\ 0 & I \end{bmatrix}$.

It is in principle also possible to add sections with only RFs and bending magnets to provide $\begin{bmatrix} I & 0 \\ 0 & B^{-1} \end{bmatrix}$ and $\begin{bmatrix} I & 0 \\ 0 & A^{-1} \end{bmatrix}$, but that does not seem a wise alternative to pursue.

Homework 5.62 Verify the claims made in the text that
(a) a double dog-leg becomes an EEX when $L\theta k = -1$.
(b) a chicane becomes an EXX when $L\theta k = 1$.

Homework 5.63 Design an EEX that switches the x and z emittances, and in addition, the exiting beam's energy spread does not depend on the beam's initial spread in x.

Chapter 6

Synchrotron Radiation

It was mentioned in Sec. 1.1.1 that since the advent of Maxwell equations in 1883, it took several evolutions until their true significance, the elegance and the self-contained nature were more fully appreciated. One of the several implications of the Maxwell equations is the possibility of synchrotron radiation, subject of this chapter. (See Sec. 1.1.1 for a mention of the other important implications of the Maxwell equations.) A brief reminder of the history of synchrotron radiation might look as follows.

1883 Maxwell equations;
1887 Hertz observes E&M wave — E&M fields can radiate like waves, and with constant speed c;
1895 Röntgen discovers X-ray;
1897 Thomson discovers electron;
1898 Liénard-Wiechert potential — E&M waves start to be connected to point charges;
1907 Schott's classical analysis of synchrotron radiation;
1946 Schwinger's quantum analysis of synchrotron radiation;
1947 Synchrotron radiation is observed at 70-MeV General Electric electron synchrotron (Fig. 6.1);
1970s First generation of synchrotron radiation facilities;
1980s Second generation of synchrotron radiation facilities;
1990s Third generation of synchrotron radiation facilities;
2010s Fourth generation of synchrotron radiation facilities.

History of synchrotron radiation did not at all come by as readily as implied by the textbooks. It took $\gtrsim 100$ years to evolve, and none of the following steps were easy,

- Maxwell equations
- electromagnetic waves
- concept of point charge
- synchrotron radiation

Figure 6.1: When observed at the GE Electron Synchrotron in 1947, synchrotron radiation had been expected from the circulating electron beam. But it still came as a surprise when actually observed because the radiation would otherwise be expected to be at microwave frequencies comparable to the revolution frequency of the beam. Imagine the excitement that ensued when they observed visible light, and only from one side of the synchrotron (see photo)!

It took 4 years after Maxwell equations to show that these equations implied electromagnetic waves. It took 24 years to realize that they implied synchrotron radiation, and another 41 years to observe it. See Fig. 6.1.[1] A few more following decades were taken to appreciate their deeper significance on gauge invariance.

In the first half of this chapter, we discuss how a point charge radiates synchrotron radiation. The focus is on the radiation, given the prescribed motion of the point charge. Strictly following the prescribed motion, the point charge is considered unperturbed by the action of radiation. In the second half of the chapter, we will return our focus to the point charge and examine how the radiation acts back on the radiating charge.

Let us make a comment here. The image charges and currents on the vacuum chamber walls, although with apparent relativistic motion, unlike the point charges moving in free space with relativistic momentum, do not have and do not follow prescribed motions. Their radiation affects their subsequent motion immediately because they do not have reservoirs of energy and momentum like their free-space counterparts do. We will return to this point later.

This chapter applies mainly to electron storage rings. It also applies to very high energy proton storage rings such as the Large Hadron Collider with a much longer time scales. The emittance and energy spread considerations of this chapter also applies to some cases of electron linacs that contain bending magnets.

[1]John P. Blewett, Synchrotron Radiation — Early History, J. Synch. Rad. 5, 135 (1998).

6.1 Beam producing radiation

This chapter contains two parts. In this first part, we discuss the radiation emitted by a charged particle. The particle motion is assumed sufficiently rigid because it is relativistic, so that the radiation does not affect its prescribed motion. In the second part, we will calculate the various effects when the radiation acts back on the particle and affecting its subsequent motion. The reader can easily conceive the perturbative approach being assumed underneath.

6.1.1 Point charge

The concept of a *point charge* has a long and treacherous history. Looking back, it is almost inconceivable that a concept of an object that has a charge and a mass but has no size could survive a century of scrutinizing.

Indeed it had been the attempt by many prominent 19th century classical physicists to resolve this dilemma, but each time the attempt was met only by more paradoxes, and often, absurdities. Most of the subtleties can be traced to the fact that the field energy of a true point charge is infinite. The pot thickens when considering the spin. A point structure does not have a moment of inertia, so how can it possess an angular momentum?

The resolution of these problems, it turned out, had to wait for the introduction of quantum mechanics and the renormalizable quantum electrodynamics. However, it is most remarkable that part of the concept of point charge survived till today, and what's left becomes standard material treated in textbooks, including the *point charge synchrotron radiation*, the subject of this chapter.

It seems the only real justification of this material lies in the fact that this calculation agrees with what is predicted by quantum electrodynamics up to the leading order in the fine structure constant α_F — and yet almost all electrodynamics textbooks start immediately assuming point charges and develop the material as if naturally given! To some extent, the reader is suggested to take these concepts for granted. Questioning too deeply is likely to be awarded only by unending confusion.

睜一眼閉一眼
眼不見為淨.
Don't ask, don't tell.
孟子: 盡信書則不如無書.
Mencius: Don't be fooled by textbooks.

Classical radius One remnant left over from the decades of struggle was the introduction of the quantity, the *classical radius of the point charge*,[2]

$$r_0 = \frac{e^2}{4\pi\epsilon_0 mc^2}.$$

[2] "Radius" of a "point" particle? Speaking of contradicting of terms!

Today, it has lost the physical meaning at the time when it was introduced, and has now no deeper meaning than just a convenient fundamental quantity with a unit of length. One curious connection between r_0 and the electron spin will be mentioned momentarily.

As mentioned, r_0 does not have a physical meaning, but it does give a dimensional length scale of the system. For completeness, note that there are two other length dimensions of interest, i.e. the Compton wavelength $\lambdabar_C = \frac{r_0}{\alpha_F}$ and the Bohr radius $a_{\text{Bohr}} = \frac{\lambdabar_C}{\alpha_F}$, where $\alpha_F = \frac{1}{137.036}$ is the fine structure constant. These three length dimensions are connected by the scaling factor of α_F. Roughly, we use a_{Bohr} to describe atoms, $\lambdabar_C$ to describe photons, and r_0, the smallest among the trio, to describe our beloved point charges.

Electron-photon system The physical system we will address consists of electrons and photons, the same system that both the classical and quantum electrodynamics dedicate to. Perhaps at this point we could start with a few initial intuition about this e-γ system. For this purpose, we should first keep in mind that electrons and photons interact very easily. They arguably interact more easily than they interact among their own species. If you just push an electron ever so slightly, for example, it would be a good assumption that it will radiate photons. It in fact would be extremely difficult for an electron beam consisting of many electrons not to radiate. Sometimes the radiation becomes coherent because so many photons are emitted simultaneously all in the same manner that the radiation becomes coherent. When that happens, even the coherence becomes so robust that it is difficult to get rid of it. Such is the situation occurring for example, in the mechanism of free electron lasers. In that sense, you might think that FEL is such an advanced scientific instrument that it must be very subtle and touchy. That being true, however, once established, FEL mechanism is surprisingly robust even when it invokes the subtle effect of beam microbunching.

Electron's Poynting angular momentum and spin The Poynting vector $\vec{S} = \frac{1}{\mu_0}\vec{E} \times \vec{B}$ gives the electromagnetic field energy flux and the associated momentum flux. Consider a point electric charge q that also carries a point magnetic dipole moment $\vec{\mu}_B$ — such is the case of an electron — sitting at rest.[3] The charge produces a Coulomb electric field $\vec{E}$ centered around q. The magnetic dipole produces a dipolar magnetic field around $\vec{\mu}_B$, given by Eq. (1.28). The Poynting vector produced by the electric and magnetic fields will be spinning azimuthally around the $(q, \vec{\mu}_B)$ combination. As a result, there is an angular momentum of the system carried by the fields. This is the case in spite of the fact that both q and $\vec{\mu}_B$ are sitting still. This angular momentum was imbedded

[3]This $(q, \vec{\mu})$ combination was briefly mentioned in Feynman Lectures on Physics, Vol. 2, Ch. 27, where it considers this example "rather unimportant". Here we dare to think it is at least intriguing.

Figure 6.2: The electric field (red), magnetic field (blue) and the Poynting vector generated by a combined point charge q and point magnetic dipole moment $\vec{\mu}_B$.

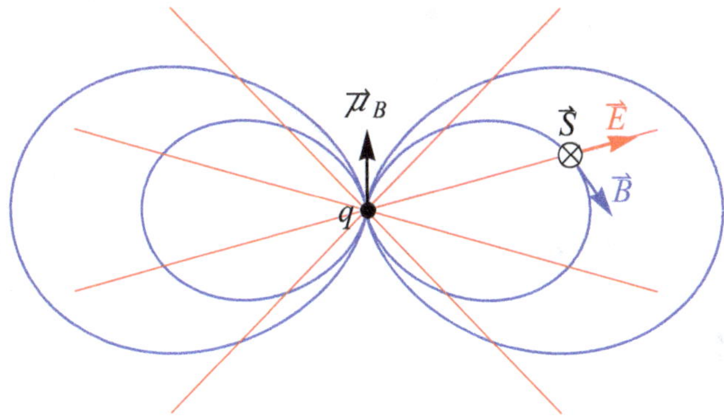

in the original process when q and $\vec{\mu}_B$ were brought together in the first place — whatever the process was, it must have applied a torque to create this system.

Let $\vec{\mu}_B = \mu_B \hat{z}$. We have

$$\begin{aligned} \vec{E} &= \frac{q}{4\pi\epsilon_0 r^2}\hat{r}, \\ \vec{B} &= \frac{\mu_0}{4\pi}\frac{\mu_B}{r^3}(\hat{\theta}\sin\theta + 2\hat{r}\cos\theta), \\ \implies \qquad \vec{S} &= \frac{1}{\mu_0}\vec{E}\times\vec{B} = \frac{q\mu_B}{16\pi^2\epsilon_0}\frac{\sin\theta}{r^5}\hat{\phi}. \end{aligned}$$

The Poynting vector circulates azimuthally around the combined q and $\vec{\mu}_B$ as shown in Fig. 6.2.

The angular momentum diverges at $r = 0$, but if we assume the point structure actually has a finite radius r_0, then the total angular momentum is

$$\begin{aligned} \vec{j} &= \frac{2\pi}{c^2}\int_{r_0}^{\infty} r^2 dr \int_0^{\pi} \sin\theta d\theta\,(r\hat{r}\times\vec{S}) \\ &= \frac{2\pi}{c^2}\frac{q\mu_B}{16\pi^2\epsilon_0}\hat{z}\int_{r_0}^{\infty} r^2 dr \int_0^{\pi} \sin\theta d\theta \left(\frac{\sin\theta}{r^4}\right)\sin\theta \\ &= \frac{q\mu_B}{6\pi c^2\epsilon_0 r_0}\hat{z}. \end{aligned}$$

The field angular momentum is in the same direction as the magnetic dipole moment.

An electron possesses a charge and a magnetic dipole moment, so is just the example of such a system. We mentioned earlier that a point particle has no moment of inertia, so it is questionable how it can have a spin. Here we actually can let the point particle possessing an angular momentum through its electric and magnetic fields.

The electron's magnetic dipole moment and its spin is related by

$$\vec{\mu}_B = \frac{ge\hbar}{2m}\hat{s},$$

where $\hat{s}$ is the unit vector of the spin direction, and g is the gyromagnetic ratio whose value is close to $g \approx 2$. The angular momentum carried by the electron is $\frac{\hbar}{2}\hat{s}$.

Assume an electron is a spherical ball of radius r_0. Ignore any possible interior fields. Applying the calculation relating the angular momentum to the magnetic moment to connect the electron spin to its Poynting angular momentum, we need a condition for r_0,

$$r_0 = \frac{ge^2}{6\pi\epsilon_0 mc^2}.$$

This is very close to the classical radius of the electron. If the electron has the size of its classical radius r_0, therefore, its Poynting angular momentum could explain its spin.

This Poynting field angular momentum lies outside of the electron. Quantum mechanics and quantum field theory should not affect its presence. The Coulomb electric and the dipole magnetic fields outside of the electron are inevitable, so the existence of this angular momentum should not be questioned.

Homework 6.1 The text attempted a classical model to connect the electron spin to its Poynting field angular momentum. The same argument applies also to positrons. A curious reader might try the same for protons. A similar conclusion can also be drawn.

Homework 6.2 The text tried to connect the electron spin to the angular momentum carried by its electric and magnetic fields. We try another model here. The electron possessing both a charge and a magnetic moment can be represented by a charge e (electron charge) and mass m (electron mass) circulating in a loop of radius r_0 (electron classical radius). Show that this model predicts too small a magnetic dipole moment and too small an angular momentum. In other words, this model cannot explain the large values of the magnetic dipole moment and the spin of an electron.

Solution The circulating current is $I = \frac{ev}{2\pi r_0}$, assuming the charge moves with speed v. The magnetic dipole moment can be shown to be $\mu_B = \frac{evr_0}{2}$, which is a factor $\frac{\alpha_F}{g}\frac{v}{c}$ times the electron's magnetic dipole moment, where α_F is the fine structure constant; it is too small even if $v = c$. Similarly, the spinning electron carries only a small fraction $2\alpha_F\frac{v}{c}$ of the electron's spin angular momentum $\frac{\hbar}{2}$. The electron spin is better modeled by its Poynting field than by a spinning charge.

6.1.2 Radiation by a point charge

Electromagnetic radiation emitted from charged particles when they are being accelerated is called *synchrotron radiation.* As we will see soon, this radiation is much more pronounced for electrons than for protons, and is more pronounced for relativistic particles than for nonrelativistic particles. The subject is treated in electrodynamics textbooks in detail. Here we first summarize some of the main results from the textbooks, and then apply these results to the considerations of a relativistic electron storage ring.

Prescribed motion Consider a point charge e tracing out a *prescribed motion* with a relativistic velocity $\vec{\beta}c$. Bear in mind that, in all that will follow, the mechanism of synchrotron radiation makes sense only if the notion of prescribed motion is accepted. What constitutes a prescribed motion? Answer: we first accept the concept of a point charge, and then we are given the position as a function of time $\vec{r}(t)$ for the point charge. Once $\vec{r}(t)$ is given, the velocity and acceleration of the point charge are also known by taking time derivatives. All these quantities are considered given when we discuss synchrotron radiation.

One implicit assumption that the notion of prescribed motion must make is that all we care is that we are given $\vec{r}(t)$. In particular, it does not matter what force is causing the particle's time-dependent trajectory. If a particle's motion follows some curved trajectory $\vec{r}(t)$, it gives out identical synchrotron radiation whether the trajectory bending is due to gravitational, electrical, or magnetic, or any combination of these forces. All systems in Fig. 6.3, for example, emit identical synchrotron radiation under the notion of prescribed motion.

Figure 6.3: Synchrotron radiation of a point charge depends solely on the prescribed motion of the particle. It does not depend on the origin of the force. All these cases produce identical synchrotron radiation.

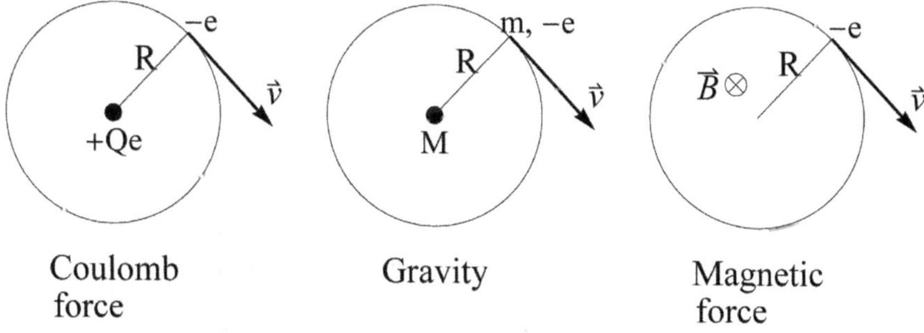

If reminded at this junction that the theory of general relativity is based on Einstein's observation that gravity is intrinsically a geometric phenomenon of the space-time, one might be led to think that this observation on point charge radiation is a consequence that electromagnetic radiation is also a purely geometric manipulation. It does not even care about the force it is undergoing.

It is important to point out here that the notion of prescribed motion is applicable only when the point charge is moving relativistically, or at least sufficiently energetically. In fact, it has to move with sufficient momentum that its trajectory is unperturbed due to the radiation process. In case its momentum is too low, or nonexistent like the case of a moving image charge on a conducting wall, so that the particle does not follow a prescribed motion as it radiates, then the analysis of synchrotron radiation does not apply.

Liénard-Wiechert potential From textbooks, the radiation fields emitted by this point charge can be written as

$$\begin{aligned} \vec{E} &= -\frac{\partial}{\partial t}\vec{A} - \nabla\phi, \\ \vec{B} &= \nabla\times\vec{A}, \end{aligned} \tag{6.1}$$

where the *retarded Liénard-Wiechert potentials* (in Lorentz gauge) at position $\vec{r}$ at time t are given by[4]

$$\begin{aligned} \vec{A}(\vec{r},t) &= \frac{1}{4\pi\epsilon_0 c}\frac{e}{R}\left.\frac{\vec{\beta}}{1-\hat{n}\cdot\vec{\beta}}\right|_r, \\ \phi(\vec{r},t) &= \frac{1}{4\pi\epsilon_0}\frac{e}{R}\left.\frac{1}{1-\hat{n}\cdot\vec{\beta}}\right|_r, \end{aligned} \tag{6.2}$$

where $\vec{R}(t_r)$ is the distance vector from the point charge to the observer with the unit vector $\hat{n}=\frac{\vec{R}}{R}$. All quantities on the right hand sides are taken at the *retarded time* $t_r = t - \frac{R(t_r)}{c}$. The retarded time refers to the earlier time as experienced by the point charge (in the laboratory frame) when its radiation reaches the observer at position $\vec{r}$ at time t.

Doppler factor The quantity

$$\kappa = \left.(1-\hat{n}\cdot\vec{\beta})\right|_r$$

that appears often in the analysis is a *Doppler factor.* Note that it is not a result of a Lorentz transformation as both the moving charge and the observer are in the laboratory frame. To see the physical meaning of κ, consider two emission events by the point charge separated by time $\Delta t_{\rm r}$ at the retarded time — we designate Δt_r to mean the time elapse as observed by the moving electron in the laboratory frame. When the two emissions arrive at the observer, their

[4]A. Liénard, L'Eclaire Electr. 16, 5 (1898); E. Wiechert, Archives Neerlandaises de Sciences Exaktes et Naturelles, The Hague, p. 546 (1900).

Figure 6.4: Calculation of the Doppler factor.

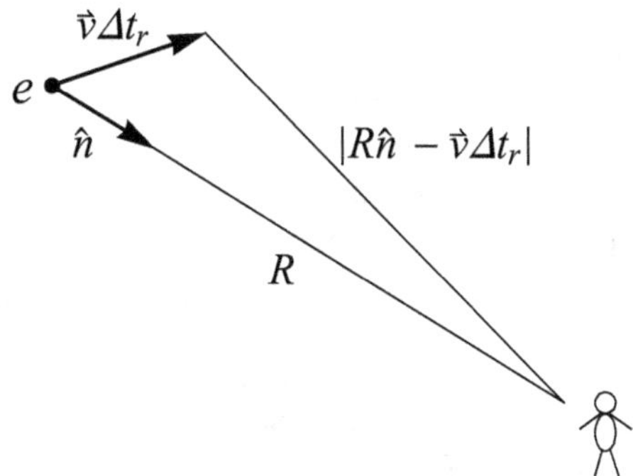

time separation as seen by the observer becomes Δt. The quantity κ is then the Doppler factor

$$\kappa = \frac{\Delta t}{\Delta t_r}.$$

Generally, $\kappa < 1$ when the point charge is moving toward the observer, $\kappa > 1$ when receding from the observer, and $\kappa \approx 1$ when the particle motion is slow, i.e. nonrelativistic.

Figure 6.4 illustrates the calculation of the Doppler factor. The two emissions will travel the distances of R and $|R\hat{n} - \vec{v}\Delta t_r|$, respectively to reach the observer. As a result, the time separation between these two emissions as seen by the observer is

$$\Delta t = \Delta t_r - \left(\frac{R}{c} - \frac{|R\hat{n} - \vec{v}\Delta t_r|}{c} \right) \approx \Delta t_r (1 - \hat{n} \cdot \vec{\beta})$$

$$\Longrightarrow \qquad \kappa = 1 - \hat{n} \cdot \vec{\beta}.$$

Liénard-Wiechert fields Substituting the Liénard-Wiechert potentials (6.2) into the expressions for the radiation fields (6.1), after some complicated but straightforward algebra, we find (see textbooks)

$$\vec{E} = \frac{e}{4\pi\epsilon_0} \frac{\hat{n} - \vec{\beta}}{\gamma^2 \kappa^3 R^2} \bigg|_r + \frac{e}{4\pi\epsilon_0 c} \frac{\hat{n} \times [(\hat{n} - \vec{\beta}) \times \dot{\vec{\beta}}]}{\kappa^3 R} \bigg|_r ,$$

$$\vec{B} = \frac{1}{c} \hat{n} \bigg|_r \times \vec{E}. \qquad (6.3)$$

These expressions are valid in open space, not valid if there are boundaries present. The first term on the right hand side of $\vec{E}$ is a Coulomb field proportional to $\frac{1}{R^2}$, which we will drop in the following. The second term, proportional to $\frac{1}{R}$, is the radiation field we are looking for. The first term is a near-field term; the second term a far-field term. Here we see explicitly that the radiation far-field term comes from acceleration; if $\dot{\vec{\beta}} = \vec{0}$, then there will be no radiation field. In contrast, the Coulomb field term that we are going to drop is independent of the acceleration. The reader should easily check that when $\vec{\beta} = \vec{0}$ and $\dot{\vec{\beta}} = \vec{0}$, we recover the static Coulomb fields $\vec{E} = \frac{e\,\hat{n}}{4\pi\epsilon_0 R^2}$ and $\vec{B} = \vec{0}$.

For the radiation field, $\vec{E}$, $\vec{B}$, and $\hat{n}$ form a right-handed set of orthogonal vectors. For the near-field Coulomb field, $\vec{B}$ is still orthogonal to $\hat{n}$, but $\vec{E}$ points in a different direction.

A cautious reader might correctly question the validity of dropping the Coulomb term. After all, if we drop it, we will no longer be able to maintain the Maxwell equation $\nabla \cdot \vec{E} = \frac{\rho}{\epsilon_0}$. This is evident especially for a charge at rest when dropping the Coulomb term gives a bare point charge stripped of all fields. Indeed this is a very valid and daring question. If we drop the Coulomb term, we have dropped the Maxwell equations. In fact, keeping the far field term only, it can be shown[5] that $\nabla \cdot \vec{E} \neq 0$ everywhere, even in the free space far from the charge — which in turn means we are filling the entire space with fictitious charges. However, this slight error ($\sim R^{-3}$) in the field does not prevent us later from using it to calculate the radiation power.

A comment is in order. Admittedly having to violate so fundamental a law carries with it some discomfort – especially since we made a big deal of having to observe this same physical law, for example when we discussed edge fields of a solenoid. Just remember that, after dropping the Coulomb term, use your result carefully.

As usual, when facing something too difficult to explain: it's an art! It seems that the art of physics lies in how to make the right approximations. The fact is that if physicists always have to carry along exact calculations, most likely they cannot move too many inches 寸步難行.

Homework 6.3 When given the position $\vec{R}$ and time t of an observation, we will have to convert them to the retarded position and time with $t_r = t - \frac{R(t_r)}{c}$. It would be a mess if there were multiple past positions and past times that contribute to the observation at $\vec{R}, t$. Show that, fortunately, this cannot happen, i.e. show that there is a one-to-one correspondence between t and t_r and there is therefore one and only one solution for t_r.

Homework 6.4

(a) Calculate the electric and magnetic fields of a point charge moving relativistically with a uniform velocity using Eq. (6.3).

(b) Sketch the field pattern. Observe Lorentz contraction.

(c) Demonstrate explicitly that Gauss law is obeyed.

---

[5]Albert Hofmann, The Physics of Synchrotron Radiation, Cambridge University Press (2004).

6.1.3 Radiation power

The *Poynting vector* as seen by the observer is given by

$$\vec{S}_{\text{observer}} = \frac{1}{\mu_0}\vec{E}\times\vec{B}\,,$$

which gives, when dropping the contribution from Coulomb field,

$$\vec{S}_{\text{observer}} = \left.\frac{1}{\mu_0 c}\,\vec{E}^2\hat{n}\right|_r\,,$$

and the instantaneous differential power per unit solid angle radiated by the point charge is

$$\begin{aligned}\frac{dP(t_r)}{d\Omega} &= \hat{n}\cdot\vec{S}R^2|_r = (\hat{n}\cdot\vec{S}_{\text{observer}})R^2\frac{dt}{dt_r} = \frac{1}{\mu_0 c}\kappa R^2\vec{E}_r^2\\ &= \frac{r_0 mc}{4\pi}\frac{1}{\kappa^5}\left|\hat{n}\times[(\hat{n}-\vec{\beta})\times\dot{\vec{\beta}}]\right|_r^2\,, \end{aligned}\qquad (6.4)$$

where $\kappa = 1-\beta\cos\theta$ with θ the angle between $\hat{n}$ and $\vec{\beta}$. In the above derivation, $\vec{S}_{\text{observer}}$ appears because we want the Poynting vector as observed by the observer, while the factor $\frac{dt}{dt_r}$ appears because we want to calculate the power radiated by the point charge and as experienced by the point charge.

Equation (6.4) is a general result applicable to any prescribed motion of a point charge. The prescribed motion yields explicitly $\hat{n}, \vec{\beta}, \dot{\vec{\beta}}$ for the formula, and as mentioned, regardless of the physical mechanism that moved the point charge. Figure 6.5 illustrate a few example applications to calculate radiation effects for various physical processes.

Larmor formula We first try out our result on a nonrelativistic charge. With $\beta \ll 1$, we have $\kappa \approx 1$ and

$$\frac{dP}{d\Omega} \approx \frac{r_0 mc}{4\pi}\left|\hat{n}\times(\hat{n}\times\dot{\vec{\beta}})\right|^2.$$

Let Θ be the angle between $\hat{n}$ and $\dot{\vec{\beta}}$ (not to be confused with θ, the angle between $\hat{n}$ and $\vec{\beta}$), then

$$\frac{dP}{d\Omega} \approx \frac{r_0 mc}{4\pi}|\dot{\vec{\beta}}|^2\sin^2\Theta\,.$$

One first sees that radiation power depends on acceleration $\dot{\vec{\beta}}$, regardless of the velocity $\vec{\beta}$.

One also sees that, at least in the nonrelativistic limit, it is the projection of acceleration $\dot{\vec{\beta}}$ perpendicular to the line of sight that gives radiation, as sketched

Figure 6.5: Various cases of radiation can be calculated using the same formula (6.4).

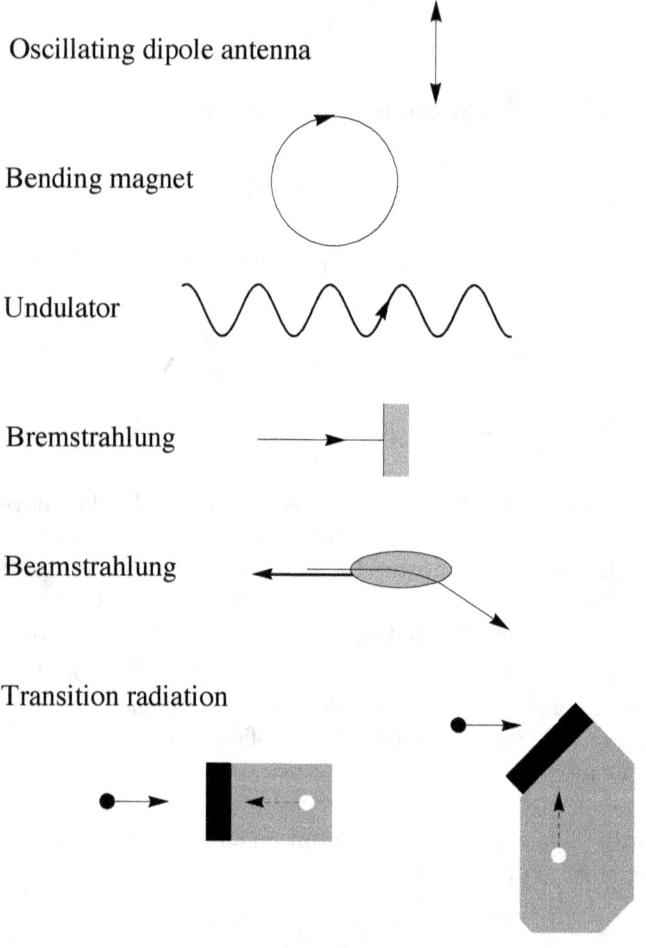

in Fig. 6.6. When $\Theta = 0$, for example, there is no radiation because the observer does not see a projection of the vector $\dot{\vec{\beta}}$ perpendicular to his line of sight. Radiation in a general direction has the protection with the factor $\sin\Theta$ and gives rise to the factor $\sin^2\Theta$ in the radiation power. This projection also tells how the radiation field is polarized.

This observation of radiation by the acceleration projection perpendicular to the line of sight, even though not rigorous [rigorous description is given by Eq. (6.4)], is a very useful one. The reader is suggested to establish an intuition based on it.

Figure 6.6: Radiation by a nonrelativistic point charge under acceleration. The radiation centers around the acceleration vector $\dot{\vec{\beta}}$, not the velocity vector $\vec{\beta}$. In fact, $\vec{\beta}$ has no effect on the radiation.

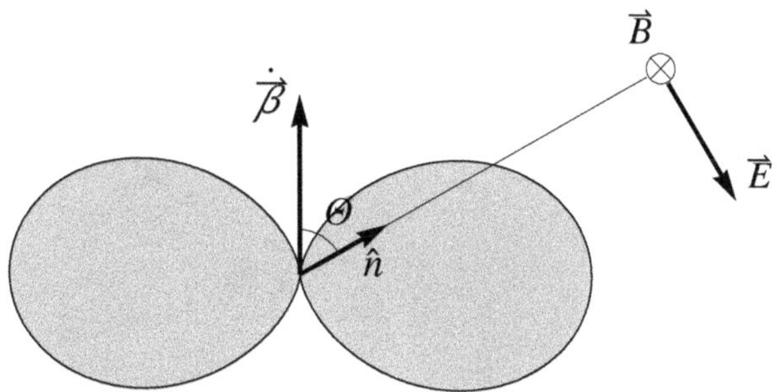

The $\sin^2\Theta$ radiation pattern of a nonrelativistic charge stored in a circular storage ring is illustrated in Fig. 6.7 (left). In this figure, $\dot{\vec{\beta}}$ is large while $\vec{\beta}$ is negligible. Radiation is measured relative to the direction of $\dot{\vec{\beta}}$. The radiation pattern is donut-shaped with its donut-axis lying along the direction of $\dot{\vec{\beta}}$, and is not to be confused as a pattern of two dumb-bell shaped back-to-back cones with the cone axis along the direction of $\vec{\beta}$.

Integrating over the solid angle gives the total radiated power

$$P = 2\pi \int_0^{\pi} \sin\Theta d\Theta \frac{dP}{d\Omega} = \frac{2}{3} r_0 mc|\dot{\vec{\beta}}|^2 . \tag{6.5}$$

This result is due to Larmor. In case the point charge has charge q and mass M, we should replace r_0 by $\frac{q^2}{4\pi\epsilon_0 Mc^2}$. The radiation power is proportional to $\frac{1}{M}$ for given acceleration and proportional to $\frac{1}{M^3}$ given the force.

Relativistic radiation The radiation pattern shown in Fig. 6.7 (left) gives already the basics of what to expect. What needs to be done is to complete a relativistic extension of the calculation. This extension is only kinematic in nature involving basically Lorentz transformations and invoking some factors of the Doppler factor κ, but the resulting differences are so great that, in the relativistic limit, the radiation pattern looks rather qualitatively different from the nonrelativistic case. The radiation pattern in this case is dominated not by $\dot{\vec{\beta}}$ but by $\vec{\beta}$. The donut-shaped radiation will be relativistically boosted into a rather forwardly collimated cigar in the direction of $\vec{\beta}$ with an opening angle $\sim \frac{1}{\gamma}$. The total radiation power will also be much enhanced by a factor $\sim \gamma^4$ compared with the nonrelativistic Larmor power.

If $\gamma = 2000$, $\gamma^4 \approx 10^{13}$! Synchrotron radiation is really a relativistic effect. The relativistic case is illustrated in Fig. 6.7 (right), except that the reader is

Figure 6.7: Comparison of radiation patterns of a nonrelativistic (left) and a relativistic (right) point charge executing circular orbits. The two patterns are vastly different, although described by the same formula.

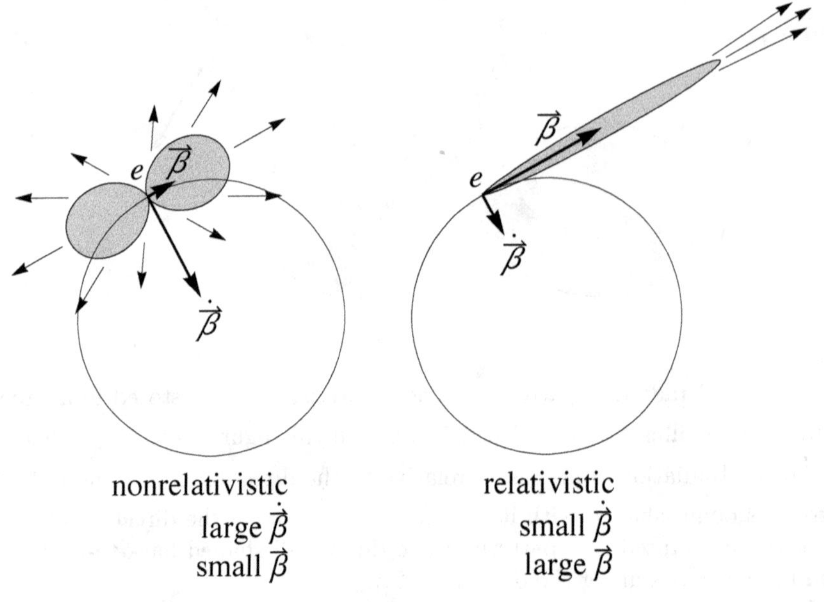

asked to mentally lengthen the cigar region by a factor of 10^{13} or so. A careful reader will also note that the magnitude of $\dot{\vec{\beta}}$ is proportional to β^2 when the bending radius ρ is held fixed.

It may be instructive to examine where does the factor γ^4 come from. We will derive it next, but it basically comes from a Lorentz transformation from the rest frame (nonrelativistic case) to the laboratory frame (relativistic case) yielding

$$\left|\dot{\vec{\beta}}\right|_{\text{rest}} \to \gamma^2 \left|\dot{\vec{\beta}}\right|_{\text{lab}} \quad \text{when} \quad \dot{\vec{\beta}} \perp \vec{\beta},$$

because acceleration involves a double time derivative and therefore doubly time dilated in the Lorentz transformation.

In the relativistic case, we need to start with Eq. (6.4). There are two main factors in Eq. (6.4). We first observe that there is the factor κ^5 in the denominator with an impressive 5th power, indicating extreme sensitivity to the Doppler factor. In the nonrelativistic case, we had $\kappa \approx 1$. At relativistic energies, we have

$$\beta = \left(1 - \frac{1}{\gamma^2}\right)^{1/2} \approx 1 - \frac{1}{2\gamma^2},$$

and letting θ be the angle between the radiation direction and $\vec{\beta}$ and expecting

$\theta \ll 1$, we have

$$\kappa = 1 - \hat{n} \cdot \vec{\beta} = 1 - \beta \cos\theta \approx \frac{\theta^2}{2} + \frac{1}{2\gamma^2}\,.$$

The radiation power therefore contains a factor

$$\frac{1}{\kappa^5} = \frac{32}{\left(\theta^2 + \frac{1}{\gamma^2}\right)^5}\,,$$

and is sharply peaked in the forward direction $\theta = 0$ with an opening angle $\sim \frac{1}{\gamma}$. With $\theta \sim \frac{1}{\gamma}$, we have $\kappa \sim \frac{1}{\gamma^2} \ll 1$. This qualitative change in the relativistic limit comes mainly from the dramatic factor $\frac{1}{\kappa^5} \sim \gamma^{10}$ concentrated in the forward direction. For $\theta \gg \frac{1}{\gamma}$, the $\frac{1}{\kappa^5}$ factor becomes so comparatively smaller (even though nonzero) that we are not too interested in it any more.

The second main factor in Eq. (6.4) is $\left|\hat{n} \times [(\hat{n} - \vec{\beta}) \times \dot{\vec{\beta}}]\right|_r^2$. This is the factor where acceleration $\dot{\vec{\beta}}$ comes into play. In a nonrelativistic limit $|\vec{\beta}| \ll 1$, or when $\dot{\vec{\beta}} || \vec{\beta}$, this term is basically the projection of acceleration $\dot{\vec{\beta}}$ perpendicular to the line of sight of the observer. In those cases, to receive radiation power, the observer must *see* a finite projection of the acceleration vector.

An inspection of the vector $\hat{n} \times [(\hat{n} - \vec{\beta}) \times \dot{\vec{\beta}}]$ also leads to the curious observation that it suppresses the direction $\hat{n}$ in the direction of the velocity $\vec{\beta}$ because the factor $\hat{n} - \vec{\beta}$ then contains a factor of $(1 - \beta) \approx \frac{1}{2\gamma^2} \ll 1$. This suppression acts opposed to the enhancement due to the Doppler factor, partly removing its $\frac{1}{\kappa^5}$ enhancement, although the Doppler effect overwhelms it. See Homework 6.5.

Homework 6.5 The vector $\hat{n} \times [(\hat{n} - \vec{\beta}) \times \dot{\vec{\beta}}]$ in Eq. (6.4) deserves a bit attention. Consider a point charge instantaneously executing a uniform circular motion with curvature ρ and speed βc.

(a) Show that this vector in the far field observed in the direction along the instantaneous forward tangential direction of motion $\hat{\beta}$ is

$$\frac{\beta^2(1-\beta)c}{\rho}(\hat{\beta} \times \hat{z})\,,$$

where $\hat{z}$ is the unit direction along the axis of the circular orbit. Give also the result when observed in the backward tangential direction $-\hat{\beta}$.

(b) Show that observed along the direction $\hat{z}$ or $-\hat{z}$ it is

$$\frac{\beta^2 c}{\rho}(\hat{\beta} \times \hat{z})\,.$$

(c) It might come a bit of a surprise that this vector is much suppressed in the forward tangential direction than either the backward direction or the perpendicular sideways direction by a factor $\sim \frac{1}{\gamma^2}$. But this is before the overwhelming Doppler factor $\frac{1}{\kappa^5}$ is taken into account. Show that when all factors are considered, the radiation power angular spectrum scales as $\sim \gamma^6$. When integrated over the forward solid angle, the total power $\sim \gamma^4$ as also seen in Eq. (6.9) later.

6.1.4 Formation length

As mentioned, the opening angle of this radiation around the direction of motion of the point charge is $\sim \frac{1}{\gamma}$, reflecting very pronounced relativistic effect. Curiously, this simple fact that the radiation is sharply forwardly directed with an opening angle of $\frac{1}{\gamma}$ alone can be used to derive the frequency spectrum of the synchrotron radiation. The result is that synchrotron radiation, again boosted strongly by relativity, must necessarily have extremely high frequencies, all the way toward hard X-rays in many cases. Needless to say, when first realized and observed, these properties of synchrotron radiation were quite striking at the time (Fig. 6.1).

To illustrate the heavy boosting of radiation frequency in the relativistic limit, we first observe that there is a squeeze of the observer's time by a Doppler effect. This squeezing action is illustrated in Fig. 6.8.

Figure 6.8: The simple fact that the synchrotron radiation has a narrow opening angle $\frac{1}{\gamma}$ leads to the conclusion that the radiation frequency must be heavily relativistically boosted by a factor $\sim \gamma^3$. The underlying mechanism is illustrated.

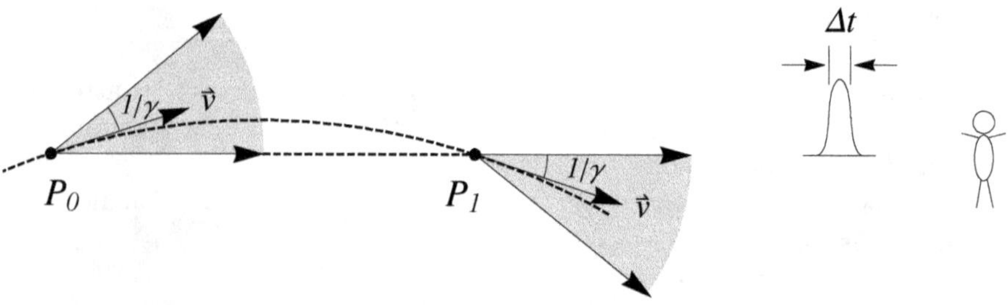

For an observer, the far field synchrotron radiation from a point charge in a storage ring resembles the beacon light emitted from a light house. The radiation appears only when the point charge is moving along the line of sight within an angle $\sim \pm\frac{1}{\gamma}$. From Fig. 6.8, synchrotron radiation emitted by a charge moving along the circular trajectory cannot reach the observer before it reached the point P_0 when those photons emitted on one edge of the radiation cone at an angle $\approx -\frac{1}{\gamma}$ aims directly toward the observer. Similarly, the last photons to reach the observer are emitted from point P_1 at an angle $\approx +\frac{1}{\gamma}$. Between

P_0 and P_1, the emitting electron has gone through a deflection angle $\sim \frac{2}{\gamma}$ and a path length $\sim \frac{2\rho}{\gamma}$ where ρ is the bending radius. The time interval in which the radiation will reach the observer, as recorded by the emitting charge, is therefore

$$\Delta t_r \approx \frac{2\rho}{\gamma\beta c}.$$

The time interval of the radiation as received by the observer, however, is not Δt_r, but is Doppler squeezed by a factor $\approx \kappa$. Indeed,

$$\begin{aligned} \Delta t &= \left(\text{time for particle to travel the arc with path length } \tfrac{2\rho}{\gamma}\right) \\ &\quad - \left(\text{time for light to travel the chord with path length } 2\rho \sin\tfrac{1}{\gamma}\right) \\ &= \frac{2\rho}{\gamma\beta c} - \frac{2\rho\sin\frac{1}{\gamma}}{c} \approx \frac{4\rho}{3c\gamma^3}, \end{aligned}$$

is much shorter than Δt_r by a Doppler factor $\kappa \sim \frac{1}{\gamma^2}$.

The length $\approx \frac{2\rho}{\gamma}$ over which the electron typically emits a photon is called the *formation length* of synchrotron radiation. Another photon will be emitted by the electron over another formation length. We will demonstrate later that these formation lengths for different photons rarely overlap, and each photon radiation is to be considered an independent event.

Although different photons are uncorrected with each other, the radiated photons are coherent within its own wavepacket. The wavepacket has a *coherence length* $\approx \frac{\rho}{\gamma^3}$, which is Doppler compressed from the formation length. In the above discussion, Δt_r is associated with the formation length, Δt connects to the coherence length. Formation length applies to the charge; coherence length applies to the radiation; there is a Doppler factor between them.

The radiation coming out from an emitting electron forms a pattern in space. For an electron circulating in a storage ring, for example, a *snapshot* of its synchrotron radiation forms a spiral pattern as shown in Fig. 6.9. The red region represents the radiation wavefront at a given point of time. Its thickness is given by the coherence length $\sim c\Delta t \sim \frac{\rho}{\gamma^3}$. As seen by a fixed observer, the separation from one wavefront and the following one is the storage ring circumference $2\pi\rho$. The radiation power as received by the observer is also sketched as a function of time t in Fig. 6.9.

Note that the red region in Fig. 6.9 (upper), also represented as simple peaks in radiation power (lower), contains complicated detailed structure within its narrow width $\sim \frac{\rho}{c\gamma^3}$. This detailed structure is not shown in Fig. 6.9, but it will eventually determine the detailed shape of the overall spectrum of the synchrotron radiation. All of our synchrotron radiation analysis of the following sections is just a careful derivation, and examining closely, of this detailed structure.

Figure 6.9: The upper panel shows a snapshot of the synchrotron radiation pattern of an electron circulating in a storage ring. The small black circle represents the storage ring. The lower panel is the radiation power signal received by an observer. Although the electron radiates continuously, the radiation received by the observer is sharply peaked. Each of the red arrows in the upper panel can be traced back to a tangential point off the electron's circular orbit.

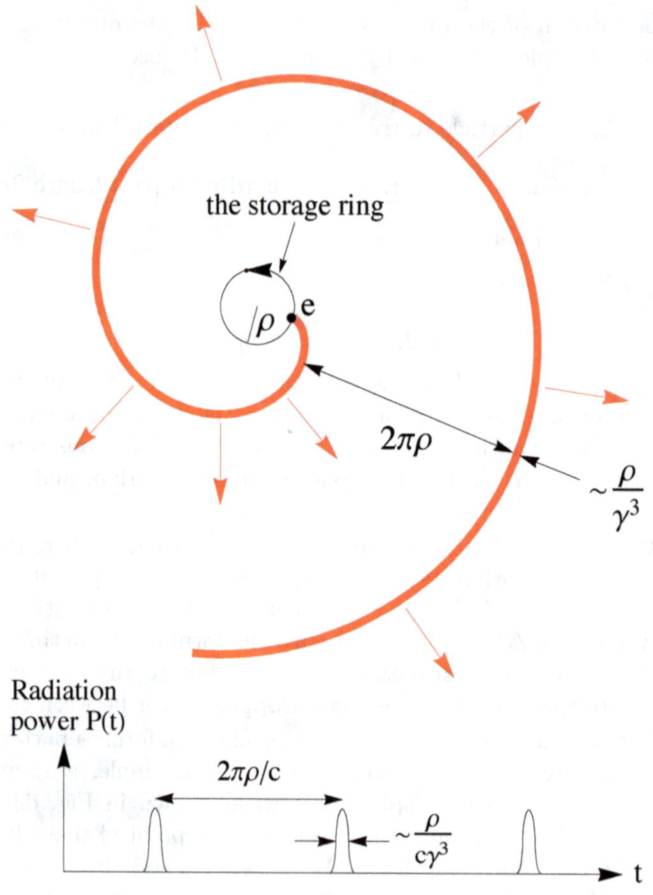

Due to the $\frac{1}{\gamma}$ opening in the radiation angle, the red curve also has a vertical height that gets taller as the radiation spreads out into space. In contrast, the thickness of the red curve in the bending plane does not change.

We have thus painted a picture in which the radiation from the point charge has filled a spiraling thin ribbon of space. This space is filled with a continuous distribution of electric and magnetic fields of the radiation. While outside this space, it remains vacuum. As we shall see later, this picture is to be very substantially altered when we paint a quantum mechanical picture.

Figure 6.10: Illustration to show the apparent electron size for a radiation event.

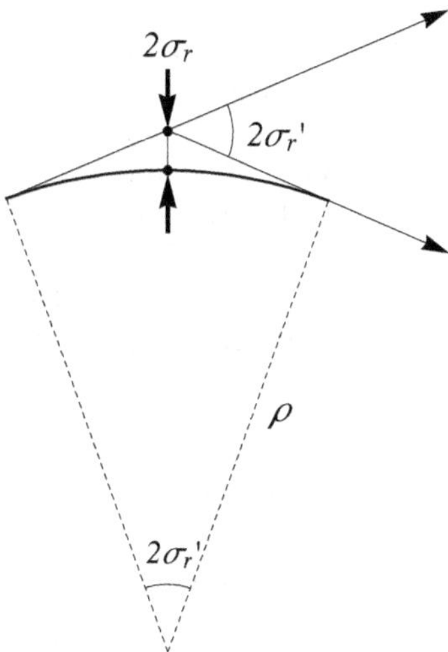

Diffraction limit Due to the finite formation length of synchrotron radiation, there is an apparent angular divergence of the emitted radiation and an apparent size of the radiating electron even though it is considered a point charge.

It is readily seen that the rms radiation angle σ_r' should be $\approx \frac{1}{\gamma}$. As will be shown later [see Eq. (6.17)], however, it also depends on the wavelength of the radiation. The answer is given by

$$\sigma_r' \approx \left(\frac{\lambda}{\rho}\right)^{\frac{1}{3}} .$$

If the emitted photon's frequency is equal to the critical frequency, the typical frequency of the synchrotron radiation, with $\lambda = \lambda_c = \frac{2\rho}{3\gamma^3}$, then it follows that $\sigma_r' \approx \frac{1}{\gamma}$. But for long wavelengths $\lambda > \lambda_c$, we have $\sigma_r' > \frac{1}{\gamma}$, and the opposite for short wavelengths. Formation length is then given by $\sigma_r'\rho$; it is a function of the radiation wavelength.

We then can estimate the apparent electron size by the geometry shown in Fig. 6.10,

$$\sigma_r \approx \rho\frac{\sigma_r'^2}{2} \approx \frac{1}{2}(\rho\lambda^2)^{1/3} .$$

We now make an important observation that the product of the apparent size and the apparent angular divergence of the radiated photon, the intrinsic radiation emittance, is given by

$$\sigma_r\sigma_r' \approx \frac{\lambda}{2} . \tag{6.6}$$

The above discussion applies to the radiation from a point charge. With a beam consisting of a large number of point charge particles, a folding integration is to needed to include the emittance from the beam. The resulting radiation emittance becomes larger than the intrinsic emittance (6.6). A careful design of the accelerator might lead to a beam radiation emittance approaching it, i.e. approaching the *diffraction limit.* Reaching the diffraction limit means the radiation from the beam has reached the smallest possible value in its transverse emittance. Keep in mind, however, that the limit depends on the radiation wavelength of interest. It is much harder to reach the diffraction limit if X-ray is the targeted wavelength. Diffraction limited storage ring design is one focused area of accelerator research.

Longitudinal coherence Diffraction limit addresses a requirement on the transverse beam quality and the transverse coherence of the beam's radiation. There is also an issue concerning the beam's longitudinal quality and its longitudinal coherence. The longitudinal coherence plays the important role to allow the entire beam bunch radiate together, gaining a factor of N to the radiation power, where N is the number of particles within the bunch. In comparison, the influence of transverse coherence alone on the radiation power is rather limited. High power radiation sources care more about the longitudinal coherence.

Different particles in the beam will have different spiral patterns. Radiation within each spiral is internally coherent at all radiation angles, but if the particle motions are uncorrelated among themselves, then different spirals are incoherent relative to one another. If the beam bunch has a length of $\sigma_z \gg \frac{\rho}{c\gamma^3}$, then observed at a fixed angle, the observed radiation is a superimposition of random uncorrelated pulses, each pulse being coherent within itself and lasting a very short time $\sim \frac{\rho}{c\gamma^3}$, and the superimposed total radiation is of length $\sim \sigma_z$. The most interesting case would be when $\sigma_z < \frac{\rho}{c\gamma^3}$. In that case, wonderful things happen because the entire beam bunch radiate coherently. Unfortunately, that is unlikely to happen except at extremely low energies.

On the other hand, ideas have been proposed to microbunch the electron beam in such a way that the microbunches might have a much shorter length to fulfill the coherence requirement.

Homework 6.6 When discussing the formation length, we have implicitly assumed that it is very short. In particular, it is much shorter than the β-function; otherwise the radiation cannot be considered a local event and the prescribed motion of a circle is no longer applicable. Show that this condition gives a requirement on the transverse betatron beam size and divergence in a bending dipole, conditions easily fulfilled,

$$\sigma_x \ll \rho, \quad \text{and} \quad \sigma_x' \ll \gamma\,.$$

Homework 6.7 A schematic of the red spiral of Fig. 6.9 is sketched in Fig. 6.11. Work out the parametric equation for the red spiral. Stay in the horizontal plane and ignore the vertical thickness of the red region.

Figure 6.11: A sketch of the synchrotron radiation pattern.

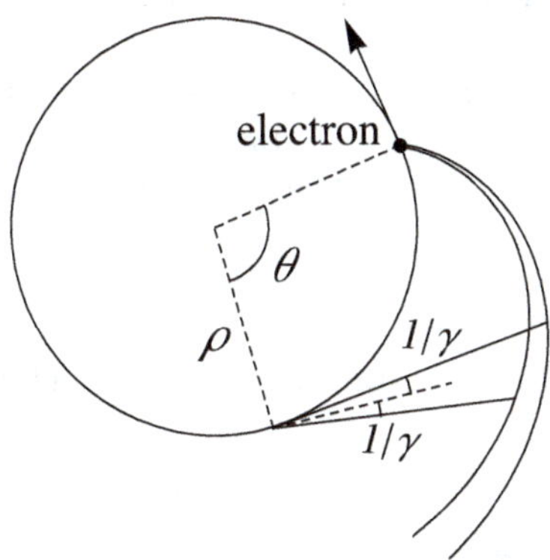

Figure 6.12: Result of calculation of the synchrotron radiation pattern. Figure on the left is a global view. On the right is a close-up view. The black circle is the storage ring. The blue dot is where the radiating electron is located. To exaggerate some of the details, this illustration takes a small value for $\gamma = 2$.

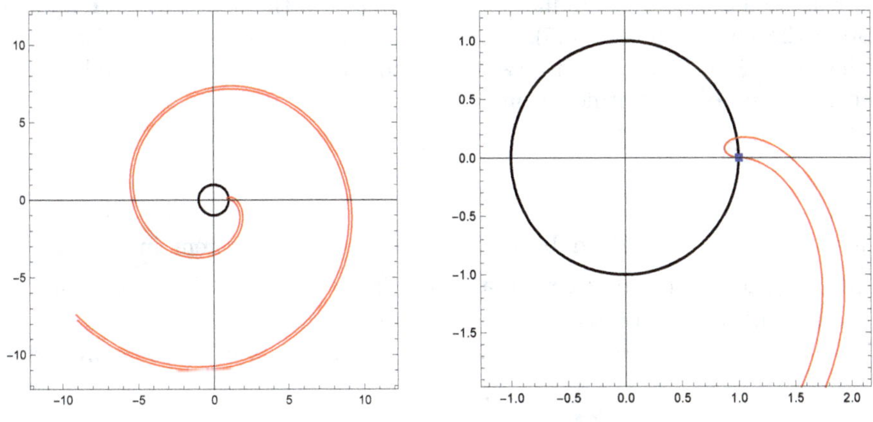

Solution The parametric equations for photons emitted with opening angles of $\pm\frac{1}{\gamma}$ are

$$\begin{aligned} x(\theta) &= \rho\cos(\theta_0 - \theta) - \frac{\rho\theta}{\beta}\sin(\theta_0 - \theta \pm \frac{1}{\gamma}) . \\ y(\theta) &= \rho\sin(\theta_0 - \theta) + \frac{\rho\theta}{\beta}\cos(\theta_0 - \theta \pm \frac{1}{\gamma}) , \end{aligned}$$

where $\beta = \frac{v}{c}$, ρ is the ring radius, θ_0 specifies the present location of the point charge, and $\theta > 0$. An example of the result is shown in Fig. 6.12.

Homework 6.8 Apply the notion of prescribed motion for synchrotron radiation, and assume all motions are nonrelativistic. Consider a point particle with charge q and mass m.

(a) A stationary point particle falls vertically over a distance h under gravity. Find its energy loss due to synchrotron radiation. How does your result depend on h? Does the free falling particle reach an equilibrium terminal velocity (no air resistance)?

(b) The same problem as in (a), except that the particle has a horizontal initial velocity v_0.

(c) Replace gravity by a constant electric field between two parallel plates and repeat (a).

Solution

(a) Radiation power $P = \frac{2r_0mg^2}{3c}$ with g the gravitation acceleration. There is no terminal velocity because when the velocity becomes constant, the radiation ceases and gravity acceleration picks up again.

6.1.5 Relativistic Larmor radiation

Earlier, we integrated the radiation power over Θ in the nonrelativistic limit to obtain the Larmor power (6.5). This can be extended to the relativistic case. Using Eq. (6.4), it is straightforward to show that in general, the relativistic generalization of the Larmor formula is given by

$$P = \frac{2}{3} r_0 mc\gamma^6 \left(|\dot{\vec{\beta}}|^2 - |\vec{\beta} \times \dot{\vec{\beta}}|^2 \right) . \tag{6.7}$$

This result can be derived by choosing a coordinate system in which $\vec{\beta} = |\vec{\beta}|(0,0,1)$, $\dot{\vec{\beta}} = |\dot{\vec{\beta}}|(0, \sin\chi, \cos\chi)$, and $\hat{n} = (\sin\theta\cos\phi, \sin\theta\sin\phi, \cos\theta)$, inserting into Eq. (6.4), and integrating over $\int_0^\pi \sin\theta d\theta \int_0^{2\pi} d\phi$.

With a slight manipulation, the total radiation power can be reexpressed as

$$P = \frac{2}{3} r_0 mc\gamma^2 \left[\left(\frac{d}{dt}(\gamma\vec{\beta}) \right)^2 - \left(\frac{d\gamma}{dt} \right)^2 \right] . \tag{6.8}$$

This result reduces to the nonrelativistic result properly when $\beta \to 0$ and $\gamma \to 1$. The first term comes from time change in momentum, the second term from time change in energy of the radiating charge in its prescribed motion.

It may be curious that the radiation power results from the difference between the rates of momentum change and energy change.[6] Note that, in general,

[6] It becomes less curious if expressed in terms of a Lorentz invariant 4-vector where the power is related to $\frac{dP}{d\tau}$ with τ the covariant time. The same 4-vector approach also provides a

the rate of momentum change is always larger than the rate of energy change; Eq. (6.8)'s validity obviously requires this general rule.

Equation (6.8) can be applied to the special case of a point charge making a circular motion (such as the motion in a uniform magnetic field). The radiation power is found to be

$$P = \frac{2}{3} r_0 m c^3 \frac{\beta^4 \gamma^4}{\rho^2}, \tag{6.9}$$

where ρ is the radius of the circular trajectory of the particle.

Homework 6.9 Follow the outline in the text, derive Eq. (6.7).

Homework 6.10

(a) Derive Eq. (6.8).

(b) Derive Eq. (6.9).

Solution

(a) Use has been made of the identities $\dot{\gamma} = \gamma^3 \vec{\beta} \cdot \dot{\vec{\beta}}$ and $|\vec{\beta} \times \dot{\vec{\beta}}|^2 = |\vec{\beta}|^2 \, |\dot{\vec{\beta}}|^2 - (\vec{\beta} \cdot \dot{\vec{\beta}})^2$.

(b) Use the facts that $\vec{\beta}$ and $\dot{\vec{\beta}}$ are orthogonal, $|\vec{\beta} \times \dot{\vec{\beta}}| = \beta^2 \omega$, $|\dot{\vec{\beta}}| = \omega \beta$, and $\omega = \frac{\beta c}{\rho}$. The total power is proportional to the fourth power of the momentum, not the energy.

Homework 6.11 A point particle with charge q and mass m is made to circulate around a charge-neutral mass $M \gg m$ by gravitational force. Find expressions of the synchrotron radiation power P_γ and radiated energy per turn U_0 by the charge. (a) Consider the motion nonrelativistic. (b) Consider it relativistic.

Homework 6.12

(a) A point charge q in constant motion $\vec{\beta} =$ constant is made to come to a halt from time $t = 0$ to $t = \tau$ by a uniform deceleration. Find the radiation loss of this point charge. Do the nonrelativistic case, and repeat for the relativistic case. What happens in the limit $\tau \to 0$?

(b) Repeat the same problem when a stationary particle suddenly starts to move with a reversed time pattern.

6.1.6 Synchrotron radiation in linac

Consider an electron being accelerated in a linear accelerator with energy gradient $\frac{dE}{ds} = G$.

---

more elegant and concise derivation for Homework 6.9, as developed in J.D. Jackson, Classical Electrodynamics, 3rd Edition, John Wiley (1999). The truth is that we receive the confidence that our lengthy derivation is correct because it conforms to Lorentz covariance at the end.

Incidentally, should a linac acceleration be characterized by a constant spatial gradient of energy, $\frac{dE}{ds} = G$, as we just did, or a constant force, characterized by a constant time gradient of momentum $mc\frac{d}{dt}(\gamma\beta) = F$? In fact, two additional options might be a time derivative of energy, or space gradient of momentum. Which is the correct representation of a linac?

In a linac, we have $\vec{\beta} \parallel \dot{\vec{\beta}}$. Applying Eq. (6.4), we obtain the radiation angular distribution

$$\frac{dP}{d\Omega} = \frac{r_0 mc}{4\pi} \frac{\dot{\beta}^2 \sin^2\theta}{(1-\beta\cos\theta)^5} ,$$

where θ is the angle between $\vec{\beta}$ and $\hat{n}$. Radiation vanishes along the direction of motion with $\theta = 0$. This is expected because there is no projection of electron acceleration perpendicular to the line of sight. The radiation angle θ scales with $\frac{1}{\gamma}$, indicating the radiation becomes increasingly collimated in the forward direction as the electron is accelerated. When $\gamma \gg 1, \theta \ll 1$, we have the familiar Doppler expression $1 - \beta\cos\theta \approx \frac{1}{2}(\frac{1}{\gamma^2} + \theta^2)$.

By integrating over the solid angle, and using $\dot{\beta} = \frac{G}{mc\gamma^3}$, we obtain the radiation power given by

$$P = \frac{2r_0}{3mc} G^2 .$$

This power is independent of γ, i.e. it is a constant throughout the linac. Over a linac of total length L, the total radiated energy per particle is

$$\frac{PL}{c} = \frac{2r_0 G}{3mc^2} E_{\text{final}} ,$$

with E_{final} the final particle energy at the end of the linac. For the SLAC linac, for example, this radiated energy is about 3 meV. The radiation in a linac is negligible.

The rate at which the electron energy is radiated away is smaller than the rate at which it gains from acceleration by a factor $\sim \frac{r_0}{mc^2}G$, which typically $\ll 1$ unless G is so that the electron gains an energy mc^2 in a distance r_0.

Homework 6.13 Go through the steps outlined in the text to derive the synchrotron radiation power of a linac.

Homework 6.14 Consider a beam with N electrons with a sufficiently small beam size to act as a point charge, then in the radiation power expression obtained in the text, we need to replace e^2 by N^2e^2. The radiated power is N^2 times larger. The radiated energy per electron will be a factor N larger. When the linac electron beam bunch is short enough, even a linac synchrotron radiation can be noteworthy. Give your analysis of a criterion when this coherent linac synchrotron radiation becomes significant.

Homework 6.15

(a) Compare the radiation power of a relativistic particle when it is accelerated transversely and longitudinally. Show that their ratio is

$$\frac{P_{\text{long}}}{P_{\text{trans}}} = \frac{1}{\beta^4\gamma^2}\left(\frac{G\rho}{\gamma mc^2}\right)^2,$$

where G specifies the longitudinal acceleration, ρ specifies the transverse acceleration.

(b) Calculate this ratio for the case of electrons, $\gamma mc^2 = 1$ GeV, $\rho = 30$ m, $G = 15$ MeV/m.

Solution (b) 5.3×10^{-8}.

6.1.7 Synchrotron radiation spectrum

Frequency spectrum Now let us return to the case of synchrotron radiation in a storage ring or a bending magnet. The radiation power as a function of time plotted in the lower panel of Fig. 6.9 is related to the radiation spectrum as received by the observer by a Fourier transform. The power as a function of time is sharply peaked pulses appearing periodically at a long interval of $T_0 = \frac{2\pi\rho}{c}$, while each of the sharply peaked pulse has an extremely narrow and yet finite width (these peaks look like δ-functions but are not δ-functions) of $\Delta t \sim \frac{\rho}{c\gamma^3}$. We have assumed $\gamma \gg 1$. Given the power as function of time as an input signal, one asks what is the frequency content, i.e. the *spectrum*, of this signal. The answer is given by its Fourier transform.

It might be counter intuitive that the raw signal as a function of time is less physical than its Fourier transform (i.e. its spectrum). But not to be surprised. It is how nature works. The reason humans might feel more "intuitive" with a time-domain signal is because we have the ability of memory. We have an instinctive feel of the flow of time. Nature does not have, or care to have, a memory, so it only reacts spontaneously and works only in the frequency domain.[7] Unlike humans, an ear, or an electronic detector would feel much more "intuitive" with frequency-domain signals.

Our eyes are a separate matter. Our eyes work not by Fourier transforming the signal at the optical frequencies; they require quantum effects. On the other hand, their function still does not require memory.

What does the Fourier transform of the synchrotron radiation signal, as sketched in the lower panel of Fig. 6.9, look like? The Fourier transform of such a signal is given by a collection of δ-function peaks (these peaks are truly δ-functions) with a small frequency interval between adjacent peaks of $\omega_0 = \frac{2\pi}{T_0} =$

[7]It might be argued that the function of artificial intelligence (AI) is a way to aid decision makings by electronically introducing memory to the otherwise-nature system. Memory and intelligence are helpful if one wants to learn something. Nature does not need to learn; it already knows everything.

$\frac{c}{\rho}$, while these δ-function peaks extends not to infinity, but to a very broad spectrum, the envelope of which extends to a very high frequency $\sim \frac{1}{\Delta t} \sim \frac{c\gamma^3}{\rho}$.

The spectrum is illustrated in Fig. 6.13. It shows the δ-function structure with spacing $\frac{c}{\rho}$ and an envelope that extends up to the frequency $\sim \frac{\gamma^3 c}{\rho}$. The exact shape of the envelope is determined by the exact detailed shape within each of the radiation peaks in the time domain. In fact the detailed shape of the envelope is given by the Fourier transform of the detailed pulse. Since we have not yet calculated the detailed structure of the time-domain pulses, the envelope sketched in Fig. 6.13 is only schematic. Accurate expressions are to be given later. The point, however, is that so much information on the synchrotron radiation spectrum has already been available at this point. As a reminder, all we have as input at this point is the fact that the synchrotron radiation from a relativistic point charge is pointed in the forward direction of motion of the particle with a narrow opening angle $\frac{1}{\gamma}$.

Figure 6.13: A Schematic of the synchrotron radiation frequency spectrum.

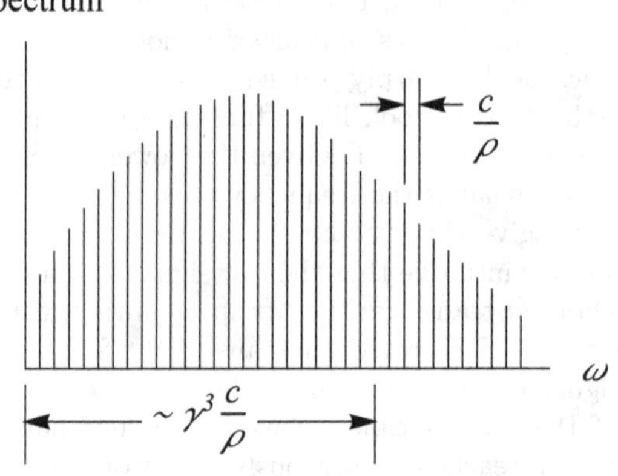

In practice, however, the δ-function structure of the spectrum from a point-charge does not reveal easily because of the various frequency spread mechanisms (such as an energy spread of the beam particles) in the storage ring. These mechanisms will smear out the δ-function structure and make the spectrum into a continuum. The condition for smearing near the upper end of the radiation spectrum due to energy spread is

$$\gamma^3 |\eta| \frac{\Delta\gamma}{\gamma} > 1 ,$$

which is easily fulfilled in practice (η is phase slippage factor). To reveal the δ-function structure of synchrotron radiation, we will have to use a very low energy beam, or an isochronous storage ring, and to limit the examination to the very low frequency part of the spectrum.

Critical frequency We define a *critical frequency*

$$\omega_c = \frac{3}{2}\frac{\gamma^3 c}{\rho} . \tag{6.10}$$

The inclusion of the factor of $\frac{3}{2}$ will become clear later. It is basically the cut-off frequency beyond which the synchrotron radiation spectrum begins to fall off. A photon that carries the critical frequency has a *critical photon energy*

$$u_c = \hbar\omega_c = \frac{3}{2}\frac{\hbar\gamma^3 c}{\rho} = \frac{3\hbar c}{2(mc^2)^3}\frac{E^3}{\rho} .$$

In more practical units, and for electrons, we write

$$u_c\,[\mathrm{keV}] = 2.22\,\frac{E^3\,[\mathrm{GeV}^3]}{\rho\,[\mathrm{m}]} = 0.665\,E^2\,[\mathrm{GeV}^2]\,B\,[\mathrm{T}] .$$

For protons, we would have

$$u_c\,[\mathrm{keV}] = 0.36\,\frac{E^3\,[\mathrm{TeV}^3]}{\rho\,[\mathrm{m}]} = 1.07\times 10^{-4}\,E^2\,[\mathrm{TeV}^2]\,B\,[\mathrm{T}] .$$

Extending the spectrum In the consideration above, the bending magnet is assumed to be much longer than the photon formation length, $\frac{\rho}{\gamma}$, and the natural synchrotron radiation spectrum extends to ω_c. If the magnet length ℓ is shorter than $\frac{\ell}{\gamma^2} < \frac{\rho}{\gamma}$, however, the radiated pulse will be $\frac{\rho}{\gamma^3}$, and the cut-off frequency will be higher than ω_c by a Fourier transformation,

$$\Delta t \sim \frac{\ell}{c\gamma^2} \quad\Longrightarrow\quad \omega_c \sim \frac{c\gamma^3}{\ell} .$$

This idea is, at least in principle, one way to extend the frequency spectrum of synchrotron radiation for a *proton* beam. Although the radiated total power will be smaller, the higher frequency portion of the radiation allows it to be used as a beam diagnostics tool, such as a synchrotron light monitor for a proton beam.

To extend the idea further, one may consider other examples.

- Consider a long bending magnet which contains a very short gap of length $\ell < \frac{\rho}{\gamma}$. This gap will cause a gap of length $\frac{\ell}{\gamma^2}$ in the observed radiation pulse of length $\frac{\rho}{\gamma^3}$, which in turn pushes up the spectrum frequency. The situation is illustrated in Fig. 6.14.

Figure 6.14: One way to extend the synchrotron radiation spectrum to higher frequencies is to introduce a gap in a long dipole magnet with $\ell < \frac{\rho}{\gamma}$.

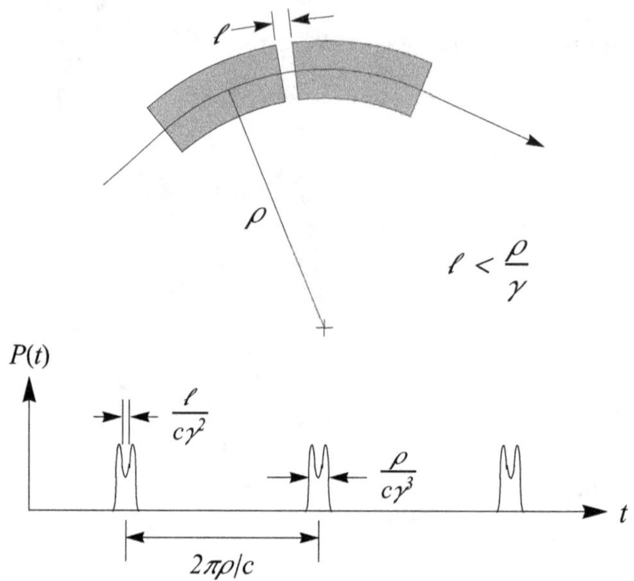

- Consider a long magnet whose magnetic field drops rapidly to zero in a distance $\ell < \frac{\rho}{\gamma}$ at its ends. The radiation at the magnet ends will also contain higher frequencies. This magnet end field radiation has also been used for diagnostics for proton beams.[8]

- One more way to extend the radiation spectrum, at least in principle, is considered in Homework 6.16.

Spectrum calculation So far, our description has been qualitative. To find the radiation spectrum more explicitly, we first consider a calculation of the radiation as seen by an observer, and integrate the power per solid angle over the observer's time t to calculate the energy radiated per solid angle Ω,

$$\begin{aligned}\frac{dW}{d\Omega} &= \int_{-\infty}^{\infty} dt\, \frac{dP(t)}{d\Omega} \\ &= \frac{1}{\mu_0 c}\int_{-\infty}^{\infty} dt\, R^2|\vec{E}_r(t)|^2 \\ &= \frac{1}{2\pi\mu_0 c}\int_{-\infty}^{\infty} d\omega\, R^2|\tilde{\vec{E}}_r(\omega)|^2\,,\end{aligned}$$

[8]R. Coisson, Opt. Commun. 22, 135 (1977); R. Bossart et al., Nucl. Instr. & Meth. 164, 375 (1979).

where we have introduced a Fourier transform

$$\tilde{\vec{E}}_r(\omega) = \int_{-\infty}^{\infty} dt\, \vec{E}_r(t) e^{-i\omega t} ,$$

and applied the *Paseval's theorem*

$$\int_{-\infty}^{\infty} dt\, |\vec{E}_r(t)|^2 = \frac{1}{2\pi} \int_{-\infty}^{\infty} d\omega\, |\tilde{\vec{E}}_r(\omega)|^2 .$$

Note that this integration is over time t, not over the retarded time t_r. Recall that the electron cares about the retarded time t_r, the observer cares about t. Note also that what is just calculated is the total radiated energy by the point charge upon its completion of the whole traversal of the prescribed motion. If the particle repeats the traversal such as in the case of a circular storage ring, its radiated energy needs to be repeated.

Perhaps one should raise a question in passing concerning the Parseval theorem. The quantity $\vec{E}_r(t)$ is a real vector. It is clear that $|\vec{E}_r(t)|$ on the left hand side means the magnitude of the vector $\vec{E}_r(t)$. The quantity $\tilde{\vec{E}}_r(\omega)$ on the right hand side, however, is a complex vector. What does the symbol $|\tilde{\vec{E}}_r(\omega)|$ mean? Does the absolute value mean taking the magnitude of the complex number or the magnitude of the vector? Go over the derivation of the Parseval theorem and make a conclusion what it means.

An inspection of the expression of $\frac{dW}{d\Omega}$ as an integral over ω indicates that what we have obtained is the differential energy spectrum [energy radiated in frequency range $(\omega, d\omega)$ per solid angle $(\hat{n}, d\Omega)$],

$$\frac{d^2W}{d\omega d\Omega} = \frac{2}{2\pi\mu_0 c} R^2 |\tilde{\vec{E}}_r(\omega)|^2 . \tag{6.11}$$

We have multiplied the result by a factor of 2 here because we will consider the spectrum to cover from 0 to ∞ instead of from $-\infty$ to ∞. The factor R^2 is not too significant; it is to be canceled by the factor $\frac{1}{R}$ contained in $\tilde{\vec{E}}$ in the far field. The reader should absorb the fact that basically the differential energy spectrum — not the power spectrum — is simply $|\tilde{\vec{E}}|^2$.

Note the very clever and amazing step toward Eq. (6.11). By a physical argument, we have obtained a double-spectrum $\frac{d^2W}{d\omega d\Omega}$, a spectrum in both angle and frequency, from a calculation of a single spectrum $\frac{dW}{d\Omega}$, a spectrum in angle alone.

Substituting

$$\vec{E}_r(t) = \frac{e}{4\pi\epsilon_0 c} \left. \frac{\hat{n} \times [(\hat{n} - \vec{\beta}) \times \dot{\vec{\beta}}]}{\kappa^3 R} \right|_r ,$$

and using $\mu_0\epsilon_0 = \frac{1}{c^2}$ and $\frac{e^2}{4\pi\epsilon_0} = r_0 mc^2$, we obtain

$$\begin{aligned}\frac{d^2W}{d\omega d\Omega} &= \frac{r_0 mc}{4\pi^2}\left|\int_{-\infty}^{\infty} dt\, e^{-i\omega t}\,\frac{\hat{n}\times[(\hat{n}-\vec{\beta})\times\dot{\vec{\beta}}]}{\kappa^3}\right|_r^2 \\ &= \frac{r_0 mc}{4\pi^2}\left|\int_{-\infty}^{\infty} dt_r\, e^{-i\omega(t_r+\frac{R}{c})}\,\frac{\hat{n}\times[(\hat{n}-\vec{\beta})\times\dot{\vec{\beta}}]}{\kappa^2}\right|^2,\end{aligned}$$

where the last step is to transform the time from the observer's time t to the retarded time t_r. This step is needed because the prescribed motion is given in terms of the retarded time describing the motion of the electron. In doing so, we drop a factor of κ in the denominator. Quantities $\vec{\beta}$ and $\dot{\vec{\beta}}$ as functions of t_r are known through the prescribed motion of the point charge.

The next step involves a subtle step of integration by parts. We will not reproduce this textbook step here. It suffices to give the result,

$$\frac{d^2W}{d\omega d\Omega} = \frac{r_0 mc}{4\pi^2}\omega^2\left|\int_{-\infty}^{\infty} dt_r\, e^{-i\omega(t_r-\frac{\hat{n}\cdot\vec{r}(t_r)}{c})}\left[\hat{n}\times\left(\hat{n}\times\vec{\beta}(t_r)\right)\right]\right|^2, \qquad (6.12)$$

where we have assumed the observation point is sufficiently far from the point charge so that we can replace $e^{i\omega(t_r+\frac{R}{c})}$ by $e^{i\omega(t_r-\frac{\hat{n}\cdot\vec{r}(t_r)}{c})}$ multiplied by an overall phase factor that does not matter, where $\vec{r}(t_r)$ is the position vector of the radiating electron. Given $\vec{r}(t_r)$, $\vec{\beta}(t_r)$ and $\hat{n}(t_r)$ for the point charge, this equation gives the frequency and angular distribution of the radiated energy. Beyond this point, we will drop the subscript r from t_r, trusting that does not cause a confusion.

The reason that the derivation of Eq. (6.12) is a subtle step can be appreciated by observing the fact that it involves only $\vec{\beta}$ and not $\dot{\vec{\beta}}$, as if the radiation comes from the velocity and not the acceleration, and as if there will be radiation by a charge moving with a constant velocity. This apparent paradox is hidden in the step of integration by parts. The physical meaning of Eq. (6.12) is obscured by its apparent mathematical elegance. Although not obvious, had we inserted a constant $\vec{\beta}$ in the above expression, we would obtain zero radiation power. This subtlety also has a practical implication. The reader is warned to apply this otherwise very elegant formula (6.12) with great caution when numerically integrating through the particle trajectory. He/She must make sure that, as a minimum test, in case of constant velocity, the numerical integration routine does give zero radiation accurately. See Homework 6.17.

Homework 6.16 The reasoning given in Fig. 6.8 to explain the radiation pulse compression assumes the electron is moving in vacuum. It is in principle possible to increase the compression factor by adding some weakly refractive material in the radiation formation length. Let the index of refraction be n, with

$n-1 \ll 1$, show that the radiation pulse length as seen by the observer becomes $\frac{\rho}{\gamma c}\left[\frac{4}{3\gamma^2} - 2(n-1)\right]$. This in principle can be another way to extend the maximum frequency reach of synchrotron radiation spectrum, and all is required is to add some dilute gas over a short section in a bending magnet.

Homework 6.17 Use Eq. (6.12) to show explicitly that there is no radiation when a particle moves with a constant velocity. Visualize and appreciate the numerical subtlety of integrating a rapidly oscillating integrand from $-\infty$ to ∞.

Homework 6.18 Consider an electron executing a small oscillating motion with $\vec{r}(t) = a\hat{z}\sin\omega_0 t$. Observe this electron from a far distance in the direction $\hat{n} = (\sin\theta\cos\phi, \sin\theta\sin\phi, \cos\theta)$.

(a) Show that the radiation power spectrum consists of δ-function peaks at the harmonic frequencies $\omega = \nu\omega_0$ where $\nu = 1, 2, 3, \cdots$. The ν-th harmonic contribution to the angular distribution of the radiation power is given by

$$\frac{d^2P_\nu}{d\omega d\Omega} = \frac{r_0 mc}{2\pi}\nu^2\omega_0^2\tan^2\theta\, J_\nu^2(\xi)\,\delta(\omega - \nu\omega_0)\,,$$

where $\xi = \frac{a\nu\omega_0}{c}\cos\theta$. Radiation vanishes in the direction $\theta = 0, \pi$, as expected. Note that here we are calculating the spectrum of the radiated power, not the radiated energy as in the text. There is no divergence at $\theta = \frac{\pi}{2}$ in spite of the factor $\tan^2\theta$.

(b) In the nonrelativistic limit, $a\omega_0 \ll c$, the dipole mode $\nu = 1$ dominates, and we obtain

$$\frac{d^2P_{\nu=1}}{d\omega d\Omega} = \frac{r_0 m\omega_0^4 a^2}{8\pi c}\sin^2\theta\,\delta(\omega - \omega_0)\,.$$

(c) Integrate the result in (b) to obtain the total radiation power

$$P_{\nu=1} = \frac{r_0 m\omega_0^4 a^2}{3c}\,.$$

Compare the result with the Larmor formula.

Solution (a) When calculating $\frac{d^2W}{d\omega d\Omega}$ using Eq. (6.12), one gets

$$\frac{d^2W}{d\omega d\Omega} = \frac{r_0 mc}{4\pi^2}\left(\frac{a\omega_0\omega}{c}\sin\theta\right)^2 |A|^2\,,$$

$$A = \int_{-\infty}^{\infty} dt\, e^{-i\omega t + i\frac{\omega a}{c}\cos\theta\sin\omega_0 t}\cos\omega_0 t\,,$$

$$e^{i\xi\sin\omega_0 t} = \sum_{k=-\infty}^{\infty} J_k(\xi)e^{ik\omega_0 t}, \quad \text{(identity)}\,.$$

One then encounters expressions involving squares of δ-functions in the form of $\delta^2(\omega)$, which formally is divergent. No worry. This is because W, the radiated energy is indeed divergent when integrated over time from $-\infty$ to ∞. What we need here is not W, but the power P. This is achieved by replacing W by P while also replacing $\delta^2(\omega)$ by $\frac{1}{2\pi}\delta(\omega)$. Sketch of a "proof" is as follows,

$$\begin{aligned}
\delta^2(\omega) &= \lim_{T\to\infty}\left|\frac{1}{2\pi}\int_{-T/2}^{T/2} dt\, e^{-i\omega t}\right|^2 \\
&= \lim_{T\to\infty}\frac{1}{\pi^2}\left(\frac{\sin^2\frac{\omega T}{2}}{\omega^2}\right) = \frac{T}{2\pi}\delta(\omega).
\end{aligned}$$

Homework 6.19 Consider an electron executing a motion around a small circle with velocity $\vec{\beta}(t) = \beta(-\sin\frac{\beta ct}{\rho}, \cos\frac{\beta ct}{\rho}, 0)$. Observe this electron from a far distance in the direction $\hat{n} = (\sin\theta\cos\phi, \sin\theta\sin\phi, \cos\theta)$.

(a) Find the radiation power spectrum

$$\frac{d^2P_\nu}{d\omega d\Omega} = \frac{r_0 mc^3\nu^2\beta^2}{2\pi\rho^2}\left[\cot^2\theta J_\nu^2(\xi) + \beta^2 J_\nu'^2(\xi)\right]\delta(\omega-\nu\omega_0),$$

where $\omega_0 = \frac{\beta c}{\rho}$, $\xi = \nu\beta\sin\theta$, and $\nu = 1, 2, 3\cdots$.

(b) Show that in the nonrelativistic limit when $\beta \ll 1$, the dominating mode is $\nu = 1$, and

$$\begin{aligned}
\frac{d^2P_{\nu=1}}{d\omega d\Omega} &= \frac{r_0 mc^3\beta^4}{8\pi\rho^2}(1+\cos^2\theta)\delta(\omega-\omega_0), \\
P_{\nu=1} &= \frac{2r_0 mc^3\beta^4}{3\rho^2}.
\end{aligned} \tag{6.13}$$

Compare with Eq. (6.9).

It should not be unexpected that the dipole radiation power when observed at $\theta = 0$ (seeing the entire circular motion of the electron) is twice that at $\theta = \frac{\pi}{2}$ (seeing only a linear projection of motion).

(c) It would be a useful exercise to obtain the nonrelativistic limit again by resorting directly to Eq. (6.12) instead of treating it as a special case of the general relativistic case.

Solution (a) You may use the identities

$$\begin{aligned}
J_{\nu-1}(\xi)e^{i\phi} - J_{\nu+1}(\xi)e^{-i\phi} &= 2J_\nu'(\xi)\cos\phi + i\frac{2\nu}{\xi}J_\nu(\xi)\sin\phi, \\
J_{\nu-1}(\xi)e^{i\phi} + J_{\nu+1}(\xi)e^{-i\phi} &= \frac{2\nu}{\xi}J_\nu(\xi)\cos\phi - i2J_\nu'(\xi)\sin\phi.
\end{aligned}$$

6.2 Bending magnet radiation

6.2.1 The double spectrum

We now apply what we have learned to a point charge of energy E moving along a circular trajectory in a uniform bending magnetic field B. Knowing $\vec{r}(t), \vec{\beta}(t)$ and $\hat{n}(t)$ from the circular trajectory, we can calculate the synchrotron radiation spectrum. The algebra is somewhat involved, but is straightforward and can be found in textbooks. We just quote the result here,

$$\frac{d^2W}{d\omega d\Omega} = \frac{3r_0mc^2}{4\pi^2c}\gamma^2\left(\frac{\omega}{\omega_c}\right)^2(1+\gamma^2\theta^2)^2\left[K^2_{2/3}(\xi)+\frac{\gamma^2\theta^2}{1+\gamma^2\theta^2}K^2_{1/3}(\xi)\right], \quad (6.14)$$

where θ is the inclination angle between the direction of radiation $\vec{n}$ and the bending plane (it is no longer the angle between $\hat{n}$ and $\vec{\beta}$), ω_c is the critical frequency (6.10), and

$$\xi = \frac{\omega}{2\omega_c}(1+\gamma^2\theta^2)^{3/2}.$$

The modified Bessel functions $K_{1/3}(x)$ and $K_{2/3}(x)$ are plotted in Fig. 6.15.

Figure 6.15: Modified Bessel functions $K_{1/3}(x)$ (red) and $K_{2/3}(x)$ (blue), together with their asymptotic forms (6.16) (dashed black curves $1.075\,x^{-2/3}, 1.688\,x^{-1/3}, \sqrt{\frac{\pi}{2x}}\,e^{-x}$, respectively).

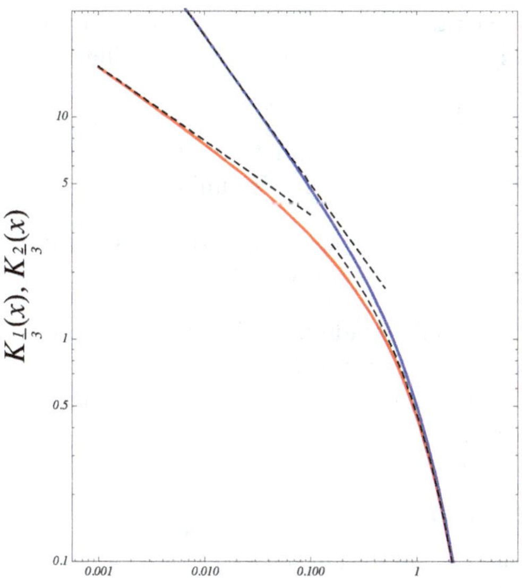

These Bessel functions have a power law behavior when $x \lesssim 1$ and an exponential tail behavior when $x \gtrsim 1$. The exponential tail region is not of interest because their values are too small. The asymptotic values of $K_\nu(x)$ for all ν's are the same when $x \gtrsim 1$.

Perhaps we make a comment on the Bessel functions. The Bessel functions are not something obscure if one is familiar with the sine and cosine functions, which are just sinusoidal functions used to describe waves. But note that they are functions describing waves only when you are looking at the waves in a *Cartesian* coordinate system (x, y, z). What if you look at these same waves in a different coordinate system, in a sense asking you to look at the same object wearing a pair of glasses? Clearly you will have to use a different mathematical functions to describe these waves. If you happen to be in a polar coordinate system (r, θ, z), then they appear as Bessel functions of one kind $J(x)$. If we are in a spherical coordinate system (r, θ, ϕ), you will encounter Bessel functions of another kind $j(x)$. Furthermore, if you replace the stable sinusoidal $\sin x$ and $\cos x$ functions by their twin exponential counterparts e^x and e^{-x}, and repeat the argument above, then you will encounter the Bessel functions $K(x)$ and $I(x)$.

A few comments concerning Eq. (6.14) are in order, especially since we skipped its derivation.

- The solid angle element $d\Omega$ is given by

$$d\Omega = \cos\theta d\theta d\psi , \tag{6.15}$$

  where θ is the angle normal to the bending plane as just mentioned, and ψ is the angle parallel to the bending plane. Note it is $\cos\theta$, not $\sin\theta$, in (6.15), possibly contrary to what is more familiar.

- The energy double spectrum $\frac{d^2W}{d\omega d\Omega}$ depends on ω and θ. The fact that it depends on θ and not on ψ is due to the fact that bending is uniform, and the angular distribution in ψ is also uniform.

- The fact that θ always appears in a combined quantity $\gamma\theta$ indicates the angular distribution in θ simply scales with $\frac{1}{\gamma}$.

- In case the radiation is made by a single magnet and the beam trajectory covers only a fraction of a revolution, ψ does not cover a range of 2π. The result (6.14) still applies to the radiation in the magnet if the magnet length covers a range of $\psi > \frac{1}{\gamma}$.

6.2.2 Polarization

Synchrotron radiation is highly polarized. The polarization can be decomposed into components parallel (σ-mode) and normal (π-mode) to the bending orbital plane. The two terms in Eq. (6.14) correspond to these two polarizations. The first term gives the energy radiated into the σ-mode, the second term gives the π-mode.

For both polarizations, the angular distribution is such that most of the radiated energy occurs near the forward direction with $\theta \lesssim \frac{1}{\gamma}$. To see this more explicitly, note that the asymptotic expressions of the Bessel functions are

$$K_\nu(x) \approx \begin{cases} 2^{\nu-1}\Gamma(\nu)x^{-\nu}, & \text{if } x \ll 1, \\ \sqrt{\frac{\pi}{2x}}e^{-x}, & \text{if } x \gg 1. \end{cases} \tag{6.16}$$

They behave as a power law for small x and become exponentially negligible for large x. This means that most radiation occurs when

$$\xi \lesssim 1,$$

or

$$\omega \lesssim \frac{2\omega_c}{(1+\theta^2\gamma^2)^{3/2}}, \quad \text{or} \quad \theta \lesssim \left(\frac{\omega_c}{\omega}\right)^{1/3}\frac{1}{\gamma}. \tag{6.17}$$

As mentioned, synchrotron radiation spans a wide spectrum up to ω_c. What we see here with the double-spectrum is that the high frequency radiation with $\omega \sim \omega_c$ is confined to $\theta \lesssim \frac{1}{\gamma}$, while lower frequency radiation actually spans a larger angle. The statement that the radiation occurs with an opening angle of $\frac{1}{\gamma}$ is therefore slightly modified here. Conversely, the radiation will red shift as the observation angle is increased. The angular and spectral distribution of the σ- (upper) and the π-modes (lower) are shown in Fig. 6.16.

Note the following features of Fig. 6.16,

- As will soon be shown, if properly normalized, the total volume covered under the surface in the σ-mode (upper figure) is 7 times that of the π-mode (lower figure).

- The σ-mode has a maximum radiation along the bending plane with $\theta = 0$, while the π-mode vanishes there.

- Along the axis $\omega = \omega_c$, the distribution has a width in $\gamma\theta \approx \pm 1$. As ω decreases, the width increases, and the reader can check that the width increases as $\gamma\theta \approx \pm(\omega_c/\omega)^{1/3}$. This behavior is the same for both polarization modes. See also Homework 6.24 for a calculation of the rms angular spread.

- The π-mode radiation is suppressed at $\theta = 0$, i.e. when viewed exactly along the bending plane. This is expected because projected along the line of sight one observes only a linear line motion having only a pure σ-component. When $\theta > 0$, i.e. observing above the bending plane, the observer sees a radiation largely still in the σ-mode but now it contains a slight left-handed circular contribution to the π-mode. Similarly, there is a slight right-handed circular contribution when $\theta < 0$. Figure 6.17 illustrates the situation.

Figure 6.16: Synchrotron radiation frequency-angle spectrum, decomposed into σ-mode (upper) and π-mode (lower), for the case of a uniform bending magnet [Courtesy Helmut Wiedemann (2019)].

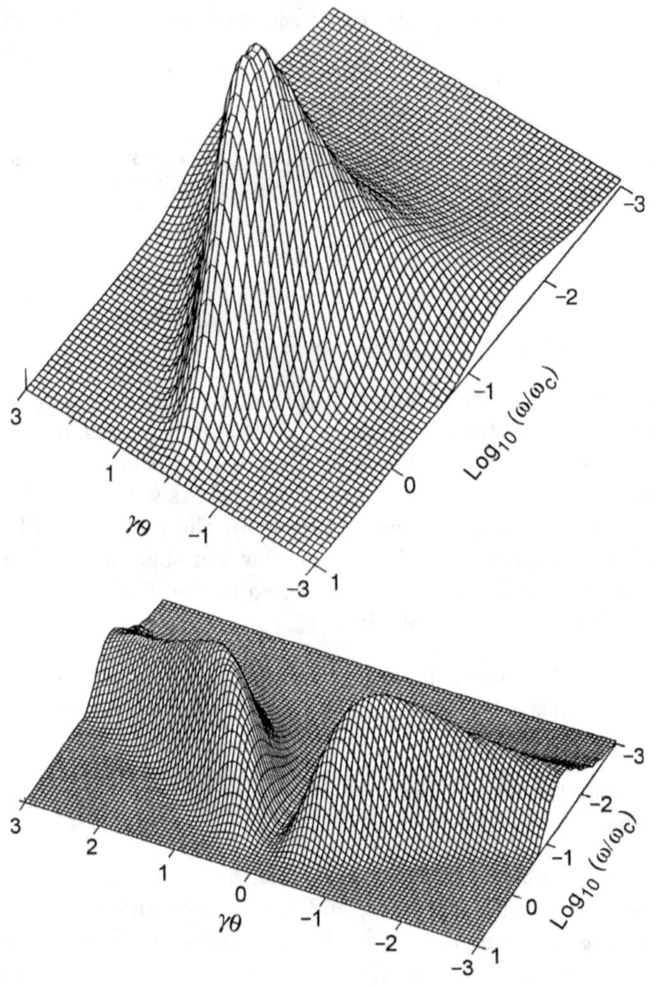

It may be instructive to make a comment on the property that the radiation angle should scale with radiation frequency with the seemingly peculiar scaling (6.17). One way to appreciate this scaling is to consider the radiation of extremely low frequency when $\omega \approx \omega_0$, the revolution frequency. At this frequency, the radiation wavelength is comparable to the storage ring circumference, and one expects the radiated angle would cover all directions, i.e. $\theta \sim 1$. Although, strictly speaking, our approximation $\theta \ll 1$ breaks down here, the scaling still

Figure 6.17: Polarization of the radiation is determined by the electron acceleration as projected along the line of sight.

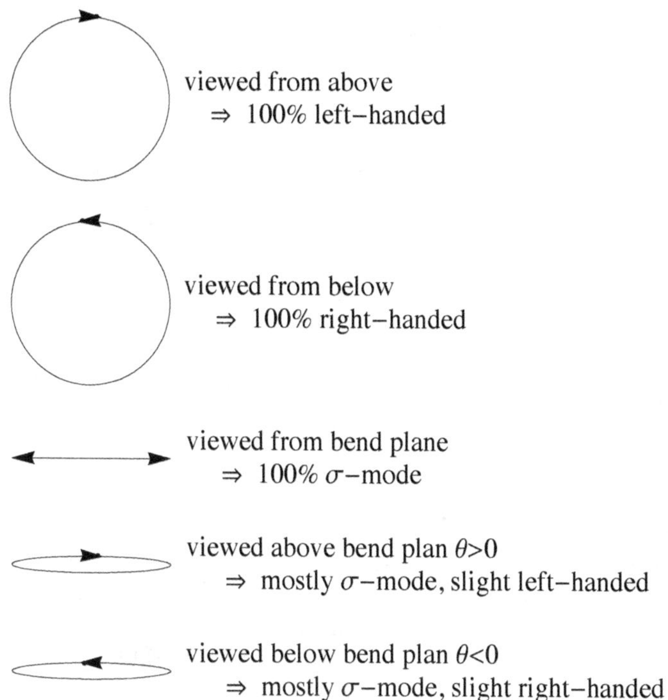

gives quite a reasonable result,

$$\theta \approx \pm \left(\frac{3}{2}\right)^{1/3} .$$

Another way to appreciate this scaling can be seen in Homework 6.20.

Figure 6.18 shows how strongly the radiation is polarized as a function of θ, for four values of the radiation frequency $\frac{\omega}{\omega_c} = 0.001, 0.01, 0.1, 1$. We again see that the radiation is 100% linearly polarized in the σ-mode at $\theta = 0$, and that the angular span widens as ω decreases. For a given ω, as θ increases, polarization evolves from purely linear when $\theta = 0$ to circular (half σ and half π) polarization when $\theta \to \infty$.

It is clear that when $\theta = \frac{\pi}{2}$, we are observing the particle motion from above, and what we see is a circular motion, and not surprisingly, the radiation observed will be circular polarized. However, again, beware that it is not the circular trajectory that causes the circular polarization. It is the circular *acceleration* that causes it.

Figure 6.18: Dependence of polarization on the radiation angle θ [Courtesy Helmut Wiedemann (2019)].

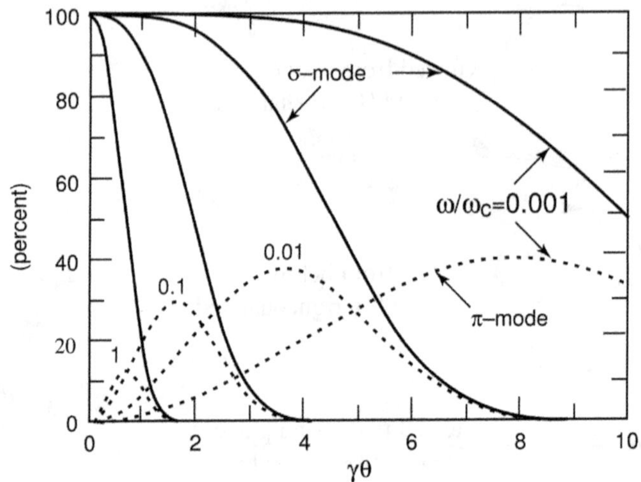

Homework 6.20 The radiation angle scaling (6.17) mentioned in the text has a physical meaning connected to the radiation formation length. Physically, in order to radiate at a general wavelength λ, the condition for radiation is that the Doppler compressed radiation pulse length is comparable to λ — recall their connection by Fourier transform. Show that the scaling $\theta \approx \pm(\omega_c/\omega)^{1/3}(1/\gamma)$ follows from this condition. The peculiar scaling (6.17) has its reason.

Solution This is an instructive homework worth pondering. Refer to Fig. 6.8 as needed. The formation length is $2\rho\theta$ instead of $\frac{2\rho}{\gamma}$ in the general case.

6.2.3 Frequency spectrum

Once we obtain the double-spectrum $\frac{d^2W}{d\omega d\Omega}$, we may integrate it over the solid angle to obtain the frequency spectrum, or integrate over the frequency to obtain the angular distribution. We will do both below.

Integrating over θ gives (derivation omitted[9]) the frequency spectrum of the synchrotron radiation

$$\frac{d^2W}{d\psi d\omega} = \frac{4}{9} r_0 mc\gamma \, S\left(\frac{\omega}{\omega_c}\right). \tag{6.18}$$

[9]To prove it, use the identity

$$\int_{-\infty}^{\infty} dy(1+y^2)^2 \left[K_{2/3}^2(\xi) + \frac{y^2}{1+y^2} K_{1/3}^2(\xi)\right] = \frac{2\pi}{\sqrt{3}x} \int_x^{\infty} dx' K_{5/3}(x'),$$

where $\xi = \frac{x}{2}(1+y^2)^{3/2}$. In the derivation, note that $d\Omega \approx d\theta d\psi$ when $\theta \approx 0$, in contrast to perhaps a more familiar form $d\Omega \approx \theta d\theta d\psi$. See Eq. (6.15).

Figure 6.19: The function $S(x)$.

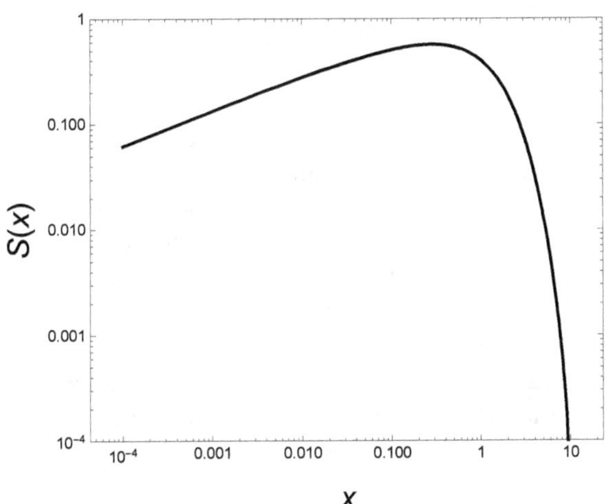

The function $S(x)$ is a universal function, defined as

$$S(x) = \frac{9\sqrt{3}}{8\pi} x \int_x^\infty dx' \, K_{5/3}(x') \,,$$

where $K_{5/3}(x')$ is a Bessel function. Numerically, $S(x)$ is shown in Fig. 6.19.

With Fig. 6.19, Eq. (6.18) says that the dipole radiation has rather wide spectrum, starting low at low frequencies, peaking around $\omega \approx \omega_c$,[10] and then substantially extending beyond ω_c. Note that Fig. 6.19 is a log-log plot.

The quantity W on the left hand side of Eq. (6.18) represents the electromagnetic wave energy radiated by the electron per unit radian along the arc that it traverses. The quantity $\frac{d^2W}{d\psi d\omega}$ represents the frequency spectrum of this radiated energy per radian of bending. It is obviously a function of the frequency ω, and the radiated energy per radian into a frequency range from ω to $\omega + \Delta\omega$ will be $\frac{d^2W}{d\psi d\omega} \cdot \Delta\omega$.

Radiation from a beam bunch So far, we have derived our results having in mind a point charge making a single-pass prescribed circular motion over a section of uniform bending magnet. There are several effects that can alter the results. A few comments are in order.

First of all, the calculation was done considering a single pass of a point charge, not repetitive passages. There is not a δ-function structure in the spectrum as alluded to in Fig. 6.13. This single-pass calculation yields only the

[10]The peak actually occurs at $\omega \approx 0.2858\,\omega_c$. The half height occurs at $\omega \approx 0.01107\,\omega_c$ and $\omega \approx 1.4763\,\omega_c$.

"envelope" of the spectrum. The spectrum lumps up in the case of repetitive passages into δ-function spikes. Due to a finite spread of revolution frequencies in the beam, however, the δ-function peaks are most likely not observable anyway.

The beam traversing the dipole magnet consists of many particles ($N \gg 1$). Collectively their longitudinal distribution has a bunch length σ_z. Inside the distribution, however, particles are randomly positioned relative to each other. In this case, the beam's synchrotron radiation power can be divided into two parts. At high frequencies with $\omega \gtrsim \frac{c}{\sigma_z}$, The radiation spectrum is the same as that obtained for a point-charge. The radiation is incoherent and the total radiation power is simply N times that predicted point-charge result. At low frequencies, $\omega \lesssim \frac{c}{\sigma_z}$, however, the radiation is coherent and the radiated power scales with N^2 instead of N and is therefore much enhanced. This in fact is one way to produce the very useful Terahertz radiation, an active area of research of industrial applications, provided the electron beam bunch can be made sufficiently short.

All radiation is coherent if $\frac{\lambda_c}{2\pi\sigma_z} \gtrsim 1$. All radiation is incoherent if $\frac{\lambda_c}{2\pi\sigma_z} \lesssim \frac{1}{N}$. The coherent radiation power becomes comparable to the incoherent power approximately when $\frac{\lambda_c}{2\pi\sigma_z} \sim \frac{1}{\sqrt{N}}$. See also Fig. 7.31.

An important possible future direction of research in fact is to intentionally generate microbunching structures in the beam distribution, e.g. by a coherent synchrotron radiation instability mechanism or by an intentional laser modulation, so that the bunch structure is tailored to the radiation wavelength and intensity required for specific industrial applications.

Homework 6.21 Reproduce the numerical plot of $S(x)$ shown in Fig. 6.19 (you might want first to exhibit the figure in linear scales instead of log-log to get an intuitive feel of the scales), and then show either analytically or numerically that

(a) the function $S(x)$ is normalized,

$$\int_0^\infty dx\, S(x) \;=\; 1\,.$$

(b) exactly half of the energy radiated has $\omega < \omega_c$, while half has $\omega > \omega_c$, i.e.

$$\int_0^1 dx\, S(x) \;=\; \int_1^\infty dx\, S(x) \;=\; \frac{1}{2}\,.$$

Incidentally, this might be a justification for including a factor of $\frac{3}{2}$ in the definition of ω_c, Eq. (6.10).

Homework 6.22 Find the asymptotic expressions for $S(x)$ when $x \ll 1$ and when $x \gg 1$.

Figure 6.20: The function $S(x)$ (solid) and its asymptotic expressions (dashed).

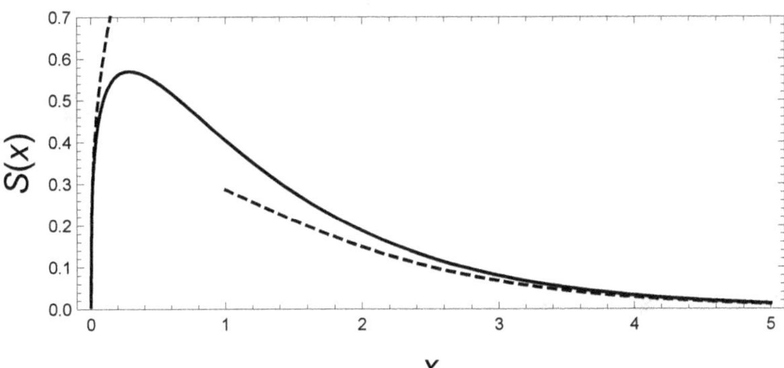

Solution

$$S(x) \approx \frac{9}{8}\sqrt{\frac{3x}{2\pi}}\,e^{-x} = 0.777\sqrt{x}\,e^{-x}, \qquad x \gg 1,$$

$$S(x) \approx \frac{27\sqrt{3}}{8\pi\,2^{1/3}}\Gamma(\frac{5}{3})\,x^{1/3} = 1.333\,x^{1/3}, \qquad x \ll 1.$$

See Fig. 6.20.

Homework 6.23 Synchrotron radiation sometimes is a useful product of an accelerator. For users of synchrotron radiation, it is useful to characterize the radiation intensity in terms of a *photon flux* per solid angle into a frequency bin $\Delta\omega$ around frequency ω and from a circulating beam current I.

(a) Show that this desired quantity is

$$\frac{d^2\dot{N}_{\rm ph}}{d\theta d\psi}(\omega) = \frac{I}{e\hbar}\frac{\Delta\omega}{\omega}\frac{d^2W}{d\omega d\Omega}\,.$$

(b) Integration over θ gives

$$\begin{aligned}\frac{d\dot{N}_{\rm ph}}{d\psi}(\omega) &= \frac{4\alpha_F}{9}\gamma\frac{I}{e}\frac{\Delta\omega}{\omega}\,S\left(\frac{\omega}{\omega_c}\right)\\ &= 3.96\times 10^{19}\left[\frac{\text{photons}}{\text{s rad}}\right]\,E\,[\text{GeV}]\;I\,[\text{A}]\,\frac{\Delta\omega}{\omega}\,S\left(\frac{\omega}{\omega_c}\right),\end{aligned}$$

where α_F is the fine structure constant.

(c) For an electron beam with $E = 1$ GeV and a peak current of $I = 1$ kA, find the photon flux (the number of photons radiated per unit time) accumulated over a bending angle of 1°, and within 0.1% of the critical photon energy. [You may use $S(1) = 0.404$.]

(d) How does the answer in (c) depend on the magnetic field of the bending magnet? What effect does the magnetic field have on the photon flux?

Solution

(c) 2.8×10^{17} photons/s.

(d) For a given beam energy, the magnetic field (or equivalently the bending radius) enters only through ω_c. If we fix the observation at $\omega = \omega_c$, then the photon flux is independent of magnetic field.

6.2.4 Angular spectrum

Integrating over ω gives[11]

$$\begin{aligned}\frac{d^2W}{d\psi d\theta} &= \int_0^\infty d\omega \frac{d^2W}{d\omega d\Omega} \\ &= \frac{1}{16} \frac{r_0 mc^2}{(1+\gamma^2\theta^2)^{5/2}} \left(7 + \frac{5\gamma^2\theta^2}{1+\gamma^2\theta^2}\right) \frac{\gamma^5}{\rho},\end{aligned}$$

where the two terms again refer to contributions from the σ- and the π-modes respectively. See Fig. 6.21. In particular, again we see that the π-mode contribution vanishes along $\theta = 0$.

Figure 6.21: Angular spectrum of synchrotron radiation in a bending magnet, decomposed into the σ- and π-modes.

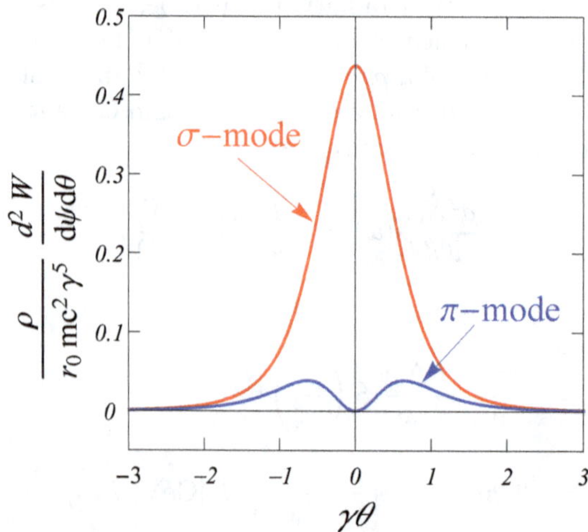

[11]Use has been made of the identity $(\nu + \mu > 0)$

$$\int_0^\infty K_\mu^2(x) x^{2\nu-1} dx = \frac{\sqrt{\pi}}{4} \frac{\Gamma(\nu)\Gamma(\nu+\mu)\Gamma(\nu-\mu)}{\Gamma(\nu+\frac{1}{2})}. \tag{6.19}$$

Homework 6.24

(a) Calculate the rms spread in θ for σ- and π-modes. Calculate the rms spread in θ of the total radiation including both σ- and π-modes. Be reminded that θ is an angle relative to the bending plane.

(b) Use the double spectrum (6.14) to calculate the rms angular spread for the σ- and π-modes as functions of ω. Incentive of this homework is to compare with Fig. 6.16. Show that at low frequencies $\sigma_\theta(\omega) \approx \frac{1}{\sqrt{2}\,\gamma}(\omega_c/\omega)^{1/3}$ and at high frequencies, $\sigma_\theta(\omega) \approx \frac{1}{2\gamma}(\omega_c/\omega)^{1/3}$. Integrating over ω gives the result in (a).

Solution

(a)

$$\gamma^2\sigma_\theta^2 = \begin{cases} \frac{1}{2}, & \sigma\text{-mode}, \\ \frac{3}{2}, & \pi\text{-mode}, \\ \frac{5}{8}, & \text{combined}. \end{cases}$$

For the combined case, show that the σ-mode contributes $\frac{7}{10}$ and π-mode contributes $\frac{3}{10}$ to the rms angular spread.

(b) A numerical calculation for the modes combined gives Fig. 6.22. The angular spread diverges at low frequency.

Figure 6.22: The quantity $\gamma^2\langle\theta^2\rangle$ as a function of $\frac{\omega}{\omega_c}$. The dashed curve is the low-frequency asymptote $\gamma^2\langle\theta^2\rangle \approx \frac{1}{2}(\omega_c/\omega)^{2/3}$. The dotted curve is the high-frequency asymptote $\gamma^2\langle\theta^2\rangle \approx \frac{1}{4}(\omega_c/\omega)^{2/3}$.

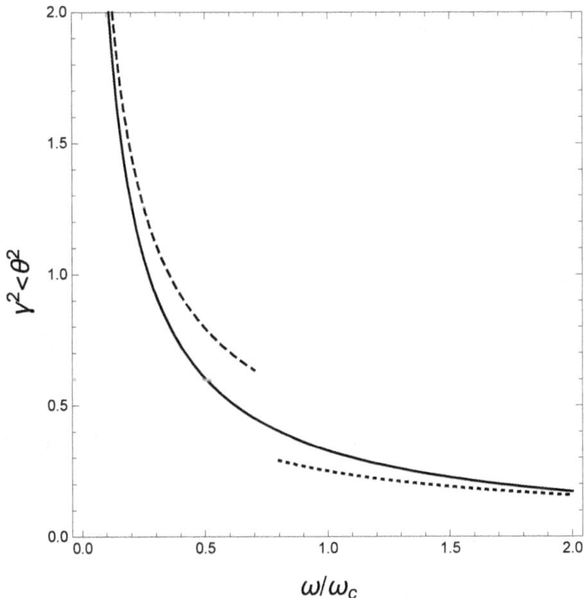

6.2.5 Radiation power

We can integrate $\frac{d^2W}{d\omega d\Omega}$ over both ω and solid angle to obtain the total radiated energy per bending angle ψ. The result is

$$\begin{aligned} \frac{dW}{d\psi} &= \int_{-\infty}^{\infty} d\theta \, \frac{d^2W}{d\psi d\theta} \\ &= \frac{r_0 mc^2\gamma^5}{16\,\rho} \int_{-\infty}^{\infty} d\theta \left[\frac{7}{(1+\gamma^2\theta^2)^{5/2}} + \frac{5\gamma^2\theta^2}{(1+\gamma^2\theta^2)^{7/2}} \right] \\ &= \frac{r_0 mc^2\gamma^4}{12\rho}(7+1)\,. \end{aligned} \tag{6.20}$$

The point charge radiates 7 times more energy into the σ-mode than into the π-mode. This perhaps should not be too surprising.

The *total radiated energy per bending angle* is then

$$\frac{dW}{d\psi} = \frac{2r_0 mc^2\gamma^4}{3\rho}\,.$$

The *total power* of radiation by the point charge, designated as P_γ, is

$$P_\gamma = \frac{dW}{d\psi}\frac{c}{\rho} = \frac{2r_0 mc^3\gamma^4}{3\rho^2}\,, \tag{6.21}$$

and the *total energy radiated per revolution*, designated as U_0, is

$$U_0 = 2\pi\frac{dW}{d\psi} = \frac{4\pi r_0 mc^2\gamma^4}{3\rho}\,. \tag{6.22}$$

In practical units, we have

$$U_0 = \begin{cases} 0.0885\,[\text{MeV}]\,\dfrac{(E[\text{GeV}])^4}{\rho[\text{m}]} & \text{for electrons,} \\[2ex] 0.00778\,[\text{MeV}]\,\dfrac{(E[\text{TeV}])^4}{\rho[\text{m}]} & \text{for protons.} \end{cases} \tag{6.23}$$

Compared with Eq. (6.9), we see that Eq. (6.21) misses a factor β^4 due to a relativistic approximation made here. If we keep this β^4 factor, then U_0 should be given by

$$U_0 = \frac{4\pi r_0 mc^2\beta^3\gamma^4}{3\rho}\,.$$

Compared with Eq. (6.13), we see Eq. (6.21) contains an additional factor γ^4. This is because Eq. (6.13) is a nonrelativistic result.

For practical calculations, it is convenient to introduce a fundamental constant[12]

$$C_\gamma = \frac{4\pi}{3}\frac{r_0}{(mc^2)^3} = \frac{1}{3}\frac{e^2}{\epsilon_0(mc^2)^4}$$
$$= 8.85\times 10^{-5}\frac{\text{m}}{\text{GeV}^3} \qquad \text{(electrons)}$$

Then we have

$$P_\gamma = C_\gamma \frac{cE^4}{2\pi\rho^2}, \quad \text{and} \quad U_0 = C_\gamma \frac{E^4}{\rho}.$$

The reader should note, for given ρ, the strong dependence on the 4th power of the point charge's energy E. Synchrotron radiation is strongly a relativistic effect. One should also note that the factor C_γ is inversely proportional to the 4th power of mass m. This means protons radiate much less than electrons by a factor of

$$\frac{P_{\gamma\text{p}}}{P_{\gamma\text{e}}} = \left(\frac{m_\text{e}}{m_\text{p}}\right)^4 = 8.80\times 10^{-14}.$$

Replacing ρ using, setting $\beta = 1$,

$$\frac{1}{\rho} = \frac{ecB}{E} = 0.2998\frac{B\,[\text{T}]}{E\,[\text{GeV}]},$$

we have

$$P_\gamma = C_\gamma \frac{e^2c^3}{2\pi}E^2B^2, \quad \text{and} \quad U_0 = C_\gamma\, ecE^3B. \tag{6.24}$$

It is useful to remember that the instantaneous power is proportional to E^2B^2. In a given magnetic field, the radiation power is quadratic in the particle energy E.

Energy limit of electron storage ring The sensitive dependence of the synchrotron radiation energy loss U_0 on the beam energy leads to the question on how high can the energy realistically reach for electron storage rings. The synchrotron radiation loss $U_0 \propto \frac{E^4}{C}$. To put it under control, we must have large circumference C. However, we are using the first power of C to fight the fourth power of the beam energy E.

In the history of electron-positron colliders, we had pushed the limits with the example of CERN LEP, now extinct and replaced by the proton-proton collider LHC. LEP was fist pushed to reach as high an energy as practically could be reached by a cost-optimized LEP-I. Later, in an effort to reach as high an energy as practically possible by adding superconducting RF without much optimization of cost, it became LEP-II. These two examples, plus some scaling, can probably be used as sample data to address our question in hand.

[12]Many of our notations follow the classic Matthew Sands, SLAC-121 (1970).

First let us consider the design of a high energy electron-positron collider with a cost-optimized manner. The total cost of the collider contains two contributions, one proportional to the expensive RF system, the other for the rest of the collider. The RF system cost is proportional to U_0. The remaining of the collider is proportional to the circumference C. The total cost of the collider is therefore

$$\text{Cost} = C + \frac{E^4}{C}. \tag{6.25}$$

It follows from this expression that the total cost is minimum when $C = E^2$, and the minimum cost is $\text{Cost}_{\text{min}} = 2E^2$. Both the circumference and the total cost of the collider scale with E^2.

LEP-I was designed to minimize the total cost. Its parameters were $E = 55$ GeV and $C = 27$ km. It follows that a cost-optimized electron-positron collider would have a circumference approximately $C = 27\left(\frac{E}{55\text{ GeV}}\right)^2$ km, and a cost that scales quadratically with E from that of LEP-I. We will return to the cost estimate in Sec. 6.4.14 when we discuss a scaling law from the point of view of beam dynamics. It will be seen that the cost optimization scaling here and the beam dynamics scaling in Secs. 6.4.7 and 6.4.14 happen to be in good agreement much to the comfort of the accelerator designers.

If minimum total cost is not the issue but the total synchrotron radiation power is, then we should scale by holding the total synchrotron radiation power fixed, i.e. the scaling is $C \propto E^4$. Since LEP-II was designed with synchrotron radiation power as the limit, we now scale from LEP-II. The LEP-II parameters were $E = 104.5$ GeV, $C = 27$ km. It follows that this collider would have a circumference approximately $C = 27\left(\frac{E}{104.5\text{ GeV}}\right)^4$ km, and a cost that scales with E^4 from that of LEP-II, adding the fixed cost of the non-RF systems.

Figure 6.23 shows the parameter choices of the storage rings. The solid line indicates the scaling to reach maximum beam energy. The dashed line is the result with cost optimization. Added to the figure are the data of two other electron storage rings SPEAR-II and PEP-I for reference.

Homework 6.25 Derive Eq. (6.23).

Homework 6.26 Calculate U_0 for the case of a fully stripped uranium ion ($A = 238, Q = 92$) with ion kinetic energy of 100 MeV/u, and circulates in a storage ring of radius $\rho = 10$ m. Because of the nonrelativistic nature of the case, make appropriate decisions on the factors of γ and β in the formula you use.

Homework 6.27 The cosmic ray seen on earth consists $\sim 90\%$ protons and $\sim 1\%$ electrons. The highest energy protons may have 10^{20} eV, electrons maybe 2 TeV. Assume the intergalaxial magnetic field is $\sim 10\ \mu$G, find the synchrotron radiation losses of these high energy protons and electrons as they move along

Figure 6.23: Electron storage ring parameters plotted on the C versus E parameter space, where C is the ring circumference and E is the design beam energy. Solid line is when E is pushed to a practical limit. Dashed line is when cost is optimized.

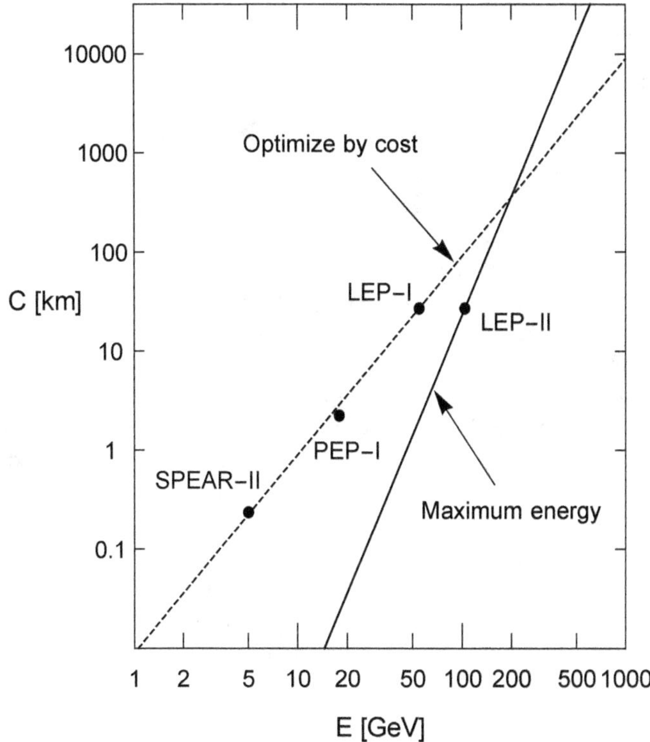

the intergalaxial space. Would synchrotron radiation be the limit on the highest energies these particles can reach in cosmic rays?[13]

Homework 6.28 The solar wind containing approximately equal number of electrons and protons is emitted from the surface of the sun toward the earth. The velocities of these particles are equivalent to a temperature of about 5000°K at the sun's surface. Their flux can be estimated by the fact that the sun emits about 1 million tons per second of material into solar wind. The earth's magnetic field is about 1 Gauss at surface of the earth.

(a) Consider the synchrotron radiation of the solar wind particles as observed on the earth. Design a crude estimate for the radiation frequency and the radiated power per invading particle.

[13]The well-known Greisen-Zatsepin-Kuzim cutoff in cosmic ray energies is due to particles interacting with the cosmic microwave background, not their synchrotron radiation.

(b) Does synchrotron radiation explain the phenomenon of aurora? Whether your answer is yes or no, why do you think aurora is observed only near the polar regions?

6.3 Quantum mechanical picture

6.3.1 Photon energy spectrum

In the classical picture, synchrotron radiation is described as a continuous emission of electromagnetic waves. In quantum mechanics, however, we understand that the radiation consists really of a large number of discrete photons, each carrying an energy $u = \hbar\omega$.

The radiation spectrum is then represented by the distribution density of the radiation as a function of the photon energy u. The classical frequency spectrum now becomes quantum mechanical photon energy spectrum. In this quantum mechanical picture, neither the exact time of each photon emission nor the exact amount of energy carried by each photon is predictable. All we know from quantum electrodynamics is a statistical description of these emission events. What is most amazing is the fact that the calculation of the radiated spectrum, based on quantum electrodynamics, averaged over the statistics and to the first order in the fine structure constant α_F, gives exactly the same result as that predicted by the classical picture. In particular, it invokes exactly the same distribution function $S(\frac{\omega}{\omega_c})$. This is a highly nontrivial conclusion, obtained after extensive development of quantum electrodynamics and calculations based on it, but in our treatment, we shall reverse the order of its logic and simply start with it a priori, and short-circuit the elaborate quantum electrodynamical derivations.

Now we make this quantum jump.

In the quantum mechanical picture, what is being radiated by the electron is not a continuous E&M wave, but many quantized photons. The energy carried by each photon depends on ω, and is given by

$$u = \hbar\omega,$$

where $\hbar$ is the reduced Planck constant.

As just mentioned, it turns out that the expression of radiated energy per radian of arc emitted by the electron, given in Eq. (6.18), happens to be valid even in the quantum mechanical picture. (It is actually only valid to the leading order in α_F, but let's not worry about the higher order effects here.) Armed with this powerful piece of information, we can now translate Eq. (6.18) to "derive" its quantum mechanical counterpart. In this quantum mechanical picture, we are interested in an expression for $\dot{n}(u)$ which is the statistically averaged number of photons emitted per unit time in the energy range between $u = \frac{\omega}{\hbar}$ and $u+du = \frac{\omega}{\hbar}+\frac{d\omega}{\hbar}$ as the electron travels around its circular trajectory. Obviously, $\dot{n}$ is a function of photon energy u.

It should be easy to see that, in the quantum mechanical picture, the radiated energy spectrum per unit time should just be $u \cdot \dot{n}$. We then, by equating the quantum mechanical and the classical spectra of radiated energy, to obtain

$$
\begin{aligned}
u\frac{d\dot{n}}{d\theta}du &= \frac{d^2W}{d\theta d\psi d\omega}\frac{d\psi}{dt}d\omega = \frac{c}{\rho}\frac{d^2W}{d\theta d\psi d\omega}d\omega\,. \\
\Longrightarrow \quad \frac{d\dot{n}}{d\theta} &= \frac{c}{\rho\hbar u}\frac{d^2W}{d\theta d\psi d\omega}\,. \qquad (6.26)
\end{aligned}
$$

where we have identified $\frac{d\psi}{dt} = \frac{c}{\rho}$.

Integrating over θ then gives

$$\dot{n} = \frac{c}{\rho\hbar u}\frac{dW}{d\psi d\omega}\,,$$

or

$$\dot{n}(u) = \frac{P_\gamma}{u_c^2}\frac{u_c}{u}S\left(\frac{u}{u_c}\right)\,,$$

where $u_c = \hbar\omega_c$ is the critical energy of the photons. The same amount of energy is being radiated away on the average, except now it is carried away by discrete quantized photons instead of continuous electromagnetic waves.

In Fig. 6.24, $S(x)$ is again plotted (red curve) together with the function $\dot{n}(x)$ (the blue curve). The three dashed curves are the asymptotic forms $1.333\,x^{1/3}$, $1.333\,x^{-2/3}$ and $0.777\,x^{1/2}\,e^{-x}$, respectively.

The above seemingly innocent change in the overall physical picture, from classical to quantum mechanical, drastically changes our results, as we will describe next. Figure 6.25 is a schematic of the two pictures, classical picture on the left and quantum picture on the right. Each red arrow in the right figure represents a statistically emitted photon. Spaces actually occupied by photons are relatively sparse in the quantum picture, while spaces filled with radiation in the classical picture form a continuum (the red spiral region). We will soon learn that, just from the rough quantum picture sketched here in the right figure, we can tell that the electron storage ring's energy is approximately 70 MeV or 130 GeV if it is a proton ring.

The $S(\frac{u}{u_c})$ function has a power law behavior $\sim u^{1/3}$ for small u. This means the energy carried by low-energy photons in the synchrotron radiation is small. However, there will need to be many of these low-energy photons to carry that small amount of energy, and that is reflected by the fact that $\dot{n}(u) \sim u^{-2/3}$ for small u. This divergence at small u, however, does not prevent a calculation of the *total number of photons radiated per unit time*,

$$
\begin{aligned}
\dot{\mathcal{N}} &= \int_0^\infty \dot{n}(u)du \\
&= \frac{15\sqrt{3}}{8}\frac{P_\gamma}{u_c} = \frac{5}{2\sqrt{3}}\alpha_F\frac{c\gamma}{\rho}\,,
\end{aligned}
$$

Figure 6.24: Functions $S(x)$ (red) and $\dot{n}(x)$ (blue). dashed curves are their respective asymptotic expressions.

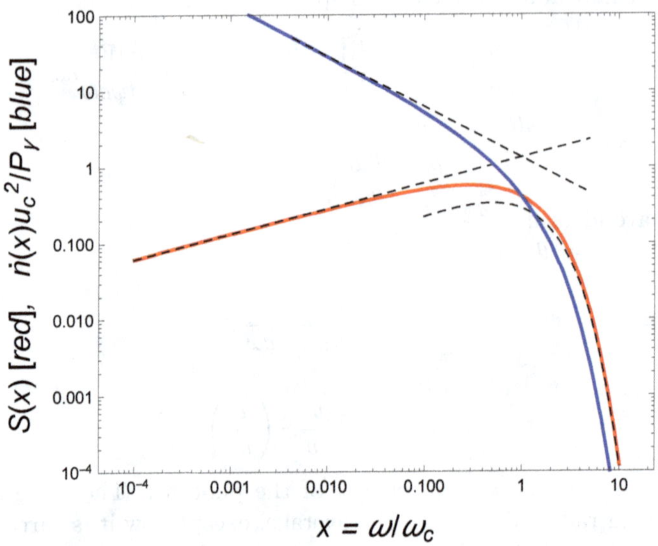

Figure 6.25: The classical and the quantum mechanical pictures of synchrotron radiation are very different. Storage rings are indicated by the two small black circles. All arrows are carefully drawn so that they all trace back to a tangential direction from the respective storage rings.

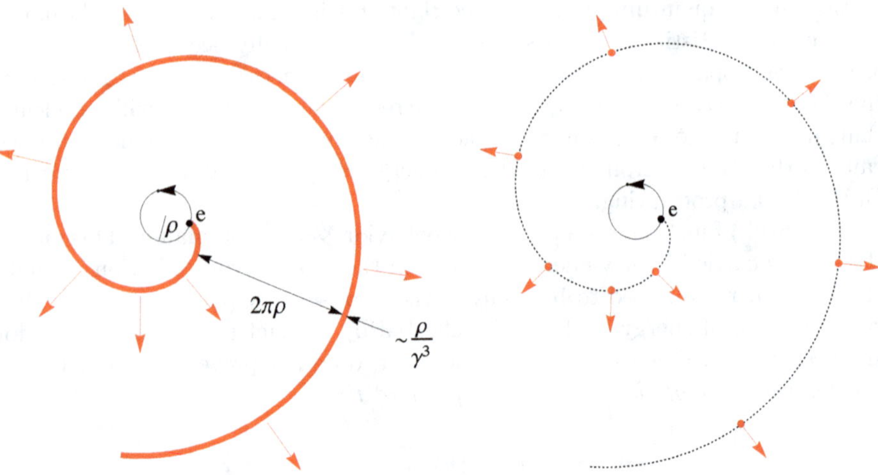

where α_F is the fine structure constant. It should be emphasized again that $\dot{\mathcal{N}}$ is a statistical quantity in the quantum description. The exact number of photons emitted per second will fluctuate around this expected statistical value.

Homework 6.29 In terms of radiated energy, we showed in Homework 6.21 that $\frac{1}{2}$ of the energy is radiated below and $\frac{1}{2}$ above u_c. Show that, in terms of the number of radiated photons, about 91% of the photons have energies below the critical energy, while 9% are above it.

Solution This exercise asks you to develop an intuition that synchrotron radiation really consists of many soft photons. Although they do not carry too much energy, there is a large number of them. This is also somewhat reflected by the fact that the maximum of $S(x)$ occurs not near $x = 1$ but near $x = 0.2$.

6.3.2 Each emission is statistically independent event

A more interesting presentation of $\dot{\mathcal{N}}$ is to calculate the total number of photons emitted per revolution, and it is

$$\mathcal{N}_0 = \dot{\mathcal{N}} \frac{2\pi\rho}{c} = \frac{5\pi}{\sqrt{3}} \alpha_F \gamma \approx \frac{\gamma}{15} . \tag{6.27}$$

This is a surprisingly simple expression. The expected number of photons emitted per revolution by a point charge, whether it is an electron or a proton, is equal to its γ, divided by 15. It is even independent of the bending radius. For example, a 2-GeV electron in a storage ring will radiate ~ 250 photons per turn. If each turn is made by 10 bending magnets, then there is 25 photons as the electron passes through each of the magnets. Incidentally, this is also how we can tell the ring sketched in Fig. 6.25 has an energy of 70 MeV.

Furthermore, we also know that each photon emission takes place over a small piece of arc as the point charge is being bent – we called it the formation length. The arc subtends an angle $\sim \frac{2}{\gamma}$ for each emission. The total bending angle per turn over which the electron is executing an emission is therefore

$$\sim \frac{\gamma}{15} \times \frac{2}{\gamma} = \frac{2}{15} \text{ radians} ,$$

which is simply a constant! Regardless of all details of the storage ring, including whether it is for electrons or for protons, what is the particle's energy, or how large or small the ring is, therefore, a particle is radiating a photon over about $\frac{2/15}{2\pi} \approx 2\%$ of its bending trajectory, and is *doing nothing* over the remaining 98% of the time. Radiation events are rather sparse.

We obtained this result of the 2% effect by connecting to the classical electrodynamics, which strictly speaking is valid only to leading order in the fine structure constant α_F. Higher orders are to be obtained using quantum electrodynamics. However, when and how the first-order consideration breaks down can be anticipated by this 2% effect. The overlap of two photon emission events, with probability $(2\%)^2$, breaks our classical analysis, and this overlap probability is of the order α_F^2. The chance of a two-photon process (second order quantum electrodynamics) is rare, of the order of 2% of all radiation events.

This 2% value depends on $\hbar$. If $\hbar \to 0$ or is much smaller than it already is, then 2% value becomes 100%, and we recover the classical picture. The reader ought to appreciate how a seemingly infinitesimal change (after all $\hbar$ is very small) of the fundamental principles can affect the physical outcome in such a drastic manner.

Two important conclusions can be drawn following the above discussion,

- Each emission can be considered an independent event.
- Synchrotron radiation events can be considered a statistical process.

A few statistical quantities We can also calculate other statistical quantities. The mean energy carried by each photon is

$$\langle u \rangle = \frac{1}{\dot{\mathcal{N}}} \int_0^\infty u \dot{n}(u) du = \frac{8}{15\sqrt{3}} u_c ,$$

with mean square spread

$$\begin{aligned} \langle u^2 \rangle &= \frac{1}{\dot{\mathcal{N}}} \int_0^\infty u^2 \dot{n}(u) du = \frac{11}{27} u_c^2 , \\ \langle u^2 \rangle - \langle u \rangle^2 &= \frac{1297}{18225} u_c^2 . \end{aligned}$$

We see that u_c is too high an estimate for a typical energy of a radiated photon. The mean radiated energy per photon is only 30% of u_c. The fluctuation around this mean value is also large, about 27% of u_c. The reason of this effect again lies in the fact that there are a very large number of low energy photons.

Another quantity that will be used in later discussions is the product of $\dot{\mathcal{N}}$ and $\langle u^2 \rangle$,

$$\begin{aligned} \dot{\mathcal{N}} \langle u^2 \rangle &= \int_0^\infty u^2 \dot{n}(u) du \\ &= \frac{55}{24\sqrt{3}} u_c P_\gamma \\ &= \frac{55}{24\sqrt{3}} \hbar r_0 m c^4 \frac{\gamma^7}{\rho^3} \\ &= \frac{55}{24\sqrt{3}} \alpha_F \hbar^2 c^3 \frac{\gamma^7}{\rho^3} . \end{aligned} \tag{6.28}$$

Note the 7th power dependence on energy E for fixed ρ.

The $\hbar$ involved indicates that any effect that involves the quantity $\dot{\mathcal{N}} \langle u^2 \rangle$ is necessarily of a quantum mechanical nature. As we shall see later, although $\hbar$

is small, the 7th power of γ will make this statistical effect stand out macroscopically.

Contributions from σ- and π-modes Each of the synchrotron radiation spectral quantities can be decomposed into contributions from the σ- and the π-modes. As mentioned earlier, the radiation power has $\frac{7}{8}$ from σ-mode and $\frac{1}{8}$ from π-mode. Homework 6.35 shows that for the quantity $\dot{\mathcal{N}}\langle u^2\rangle$, $\frac{10}{11}$ comes from the σ-mode and $\frac{1}{11}$ is from the π-mode. For other quantities, the results are given in Table 6.1. Homework 6.24 has dealt with the case for $\langle\theta^2\rangle$. The quantity $\dot{\mathcal{N}}\langle u^2\theta^2\rangle$ will be used in Sec. 6.4.12.

Table 6.1: Contributions from the σ- and the π-modes to the various radiation spectral quantities.

Quantity	Contribution from σ-mode	Contribution from π-mode
$\dot{\mathcal{N}}$	$\frac{4}{5}$	$\frac{1}{5}$
$P_\gamma = \dot{\mathcal{N}}\langle u\rangle$	$\frac{7}{8}$	$\frac{1}{8}$
$\langle\theta^2\rangle$	$\frac{7}{10}$	$\frac{3}{10}$
$\dot{\mathcal{N}}\langle u^2\rangle$	$\frac{10}{11}$	$\frac{1}{11}$
$\dot{\mathcal{N}}\langle u^2\theta^2\rangle$	$\frac{20}{23}$	$\frac{3}{23}$

Homework 6.30 You have a bending magnet in a proton beam line. Magnetic field is 1 T and magnet length is 1 m. Proton momentum is $P = 10$ GeV/c. How many photons on the average does each proton radiate going through your magnet? What is the typical energy of these photons? What is the rms fluctuation of the photon energies around the average value you estimated?

Solution Each proton radiates on the average 0.0034 photons. Average energy of each photon has 3.37μeV with rms fluctuation of 2.9 μeV.

Homework 6.31 In the text we showed that the expected number of photons emitted by a point charge e per revolution in a circular accelerator is about equal to $\frac{\gamma}{15}$. Show that the same result can be obtained by identifying $\mathcal{N}_0$ to be

$$\mathcal{N}_0 = \frac{U_0}{\langle u\rangle}.$$

Homework 6.32 The text found a value of 2% duty factor of actual radiation formation. If you have not yet been driven by this curiosity, now is the time to trace the reason why this duty factor is numerically so small. Connect this parameter to the fine structure constant α_F. As alluded to in the text, the smallness of this parameter is the underlying reason why quantum electrodynamics deviates from classical electrodynamics. Fortunately, however, for some reason, nature has decided to compensate for this by dictating the equivalence between the two pictures to first order in α_F.

Homework 6.33 Show the following practical formulae for an electron beam in a storage ring ($\gamma \gg 1$),

$$\begin{aligned} \text{radiation power} &= 6.03\times 10^{-9}[\text{W}]\,\frac{\gamma^4\,R[\text{m}]\,I[\text{A}]}{(\rho[\text{m}])^2}\,, \\ \text{photon flux} &= 4.14\times 10^{17}[\text{photons/s}]\,\frac{\gamma\,R[\text{m}]\,I[\text{A}]}{\rho[\text{m}]}\,, \end{aligned}$$

where $C = 2\pi R$ is the storage ring circumference, I is the average circulating beam current.

Homework 6.34 (a) Use the identity[14]

$$\int_0^\infty dx\, x^\mu K_\nu(x) = 2^{\mu-1}\Gamma\left(\frac{1+\mu+\nu}{2}\right)\Gamma\left(\frac{1+\mu-\nu}{2}\right),$$

and an integration by parts to show that

$$\begin{aligned} \langle u^n\rangle &= \frac{1}{\dot{\mathcal{N}}}\int_0^\infty u^n \dot{n}(u)du \\ &= \frac{3}{5\pi}\frac{2^n}{n+1}\Gamma\left(\frac{n}{2}+\frac{11}{6}\right)\Gamma\left(\frac{n}{2}+\frac{1}{6}\right)u_c^n\,. \end{aligned}$$

Check the validity when $n = 0, 1, 2$.

(b) Show that the radiation moments diverges badly for large n,

$$\frac{\langle u^n\rangle}{u_c^n} \approx \frac{3}{5}n^n e^{-n}\,.$$

n	0	1	2	3	4	5	6	7	8
$\frac{\langle u^n\rangle}{u_c^n}$	1	0.3079	0.4074	0.958	3.232	14.19	76.70	492.0	3652
$\frac{3}{5}n^n e^{-n}$	-	0.2207	0.3248	0.807	2.813	12.63	69.39	450.6	3377

Homework 6.35 Show that the quantity $\dot{\mathcal{N}}\langle u^2\rangle$ has $\frac{10}{11}$ from σ-mode and $\frac{1}{11}$ from π-mode radiation.

[14]Gamma function $\Gamma(x)$ is not too mysterious. We all know the factorial function $n!$. But the factorial function is defined only for integers n. Gamma function is just an attempt to extrapolate it to arbitrary x. It is defined in such a way that $\Gamma(n) = (n-1)!$.

Solution Carry out the calculation of $\dot{\mathcal{N}}\langle u^2\rangle$ from scratch using Eqs. (6.14) and (6.26). Keep track of the two polarization contributions. Change variables from θ to $y = \gamma\theta$ and u to $\xi = \frac{u}{2u_c}(1+y^2)$ to obtain

$$\dot{\mathcal{N}}\langle u^2\rangle = \frac{3r_0 mc^2\gamma u_c^2}{4\pi^2\rho\hbar}\int_{-\infty}^{\infty} dy \int_0^{\infty} d\xi \frac{16\,\xi^3}{(1+y^2)^4}\left[K_{2/3}^2(\xi) + \frac{y^2}{1+y^2}K_{1/3}^2(\xi)\right].$$

Then apply Eq. (6.19). Other quantities in Table 6.1 can be obtained similarly.

Homework 6.36 Consider a particle of mass m and charge e executing a cyclotron motion in a plane perpendicular to a uniform, static magnetic field B. Find the trajectory of the particle. Take into account the fact that its energy is being radiated away. In particular, find the time constant for the shrinking orbital radius. In the old days of atomic physics, it was a puzzle to be addressed why the atoms do not just collapse due to the synchrotron radiation of the atomic electrons.

Homework 6.37 In the text, we assumed that all bending magnets have the same strength and there is only one value of ρ. Consider a storage ring that contains two bending radii ρ_1 and ρ_2, each contributes to a bending of π as shown in Fig. 6.26. Quantities $u_c, P_\gamma, \dot{\mathcal{N}}$ can still be defined locally with local values of ρ, but global quantities defined for one revolution will be modified. Find expressions of $U_0, \mathcal{N}_0, \langle u\rangle, \langle u^2\rangle$, and $\dot{\mathcal{N}}\langle u^2\rangle$.

Figure 6.26: A storage ring containing two strengths of bending magnets.

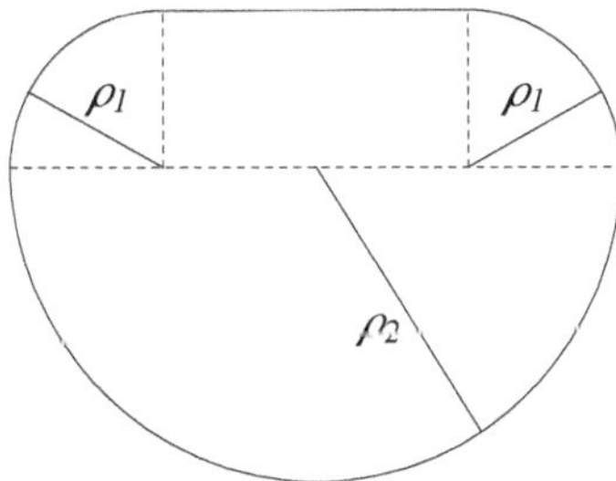

Solution To do this homework, it is necessary to understand that the averaging refers to the averaging over the radiated photon counts. The number of radiated photons in the two ρ_1-arcs is equal to that in the ρ_2-arc. It follows that $U_0 = \frac{1}{2}C_\gamma E^4(\frac{1}{\rho_1}+\frac{1}{\rho_2}), \mathcal{N}_0 = \frac{5\pi}{3}\alpha_F\gamma, \langle u\rangle = \frac{2}{5\sqrt{3}}\gamma^3 c\hbar(\frac{1}{\rho_1}+\frac{1}{\rho_2}), \langle u^2\rangle = \frac{11}{12}\gamma^6 c^2\hbar^2(\frac{1}{\rho_1^2}+\frac{1}{\rho_2^2})$. The quantity $\dot{\mathcal{N}}\langle u^2\rangle$ requires more thinking. Its physical meaning refers to a summation $\sum_i u_i^2$ over all photon emission events, divided by the revolution period $T_0 = \frac{1}{c}[\pi(\rho_1+\rho_2)+2(\rho_2-\rho_1)]$. The result is $\dot{\mathcal{N}}\langle u^2\rangle = \frac{55\pi}{36}\frac{\alpha_F\gamma^7 c^2\hbar^2}{T_0}(\frac{1}{\rho_1^2}+\frac{1}{\rho_2^2})$.

Homework 6.38 Extending Homework 6.37 further, sometimes bending can be made with an opposite sign, i.e. $\rho < 0$. Find expressions for u_c, P_γ, $\dot{\mathcal{N}}$, U_0, $\mathcal{N}_0$, $\langle u \rangle$, $\langle u^2 \rangle$, and $\dot{\mathcal{N}}\langle u^2 \rangle$,

(a) for the figure-8 storage ring shown in Fig. 6.27.

Figure 6.27: A figure-8 storage ring.

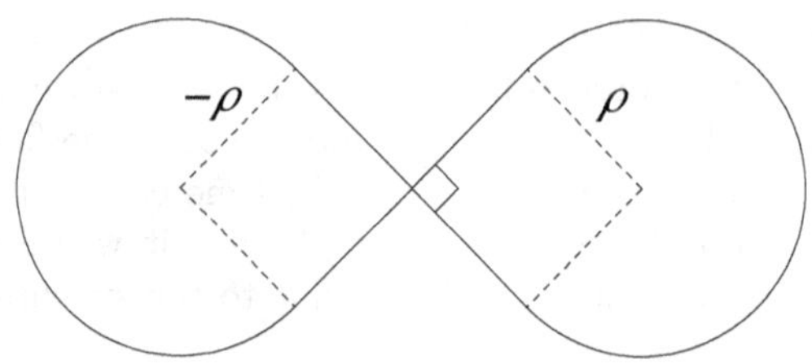

Figure 6.28: A storage ring containing wigglers.

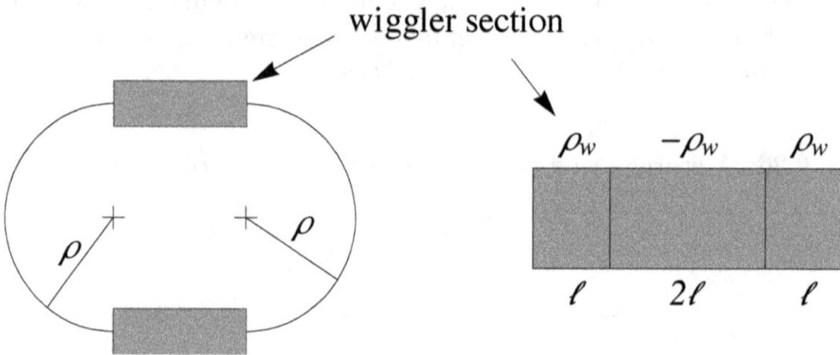

(b) for a storage ring containing wiggler sections as shown in Fig. 6.28.

Solution

(a) Quantities $\mathcal{N}_0, \langle u \rangle, \langle u^2 \rangle$ are the same as a circular ring. Also, $U_0 = C_\gamma \frac{E^4}{|\rho|}, \dot{\mathcal{N}}\langle u^2 \rangle = \frac{55}{24\sqrt{3}} \alpha_F \frac{\hbar^2 c^3 \gamma^7}{|\rho|^3} \frac{3\pi}{3\pi+4}$. (An alert reader will point out the unnatural occurrence of the term $3\pi + 4$.)

(b) Keeping in mind that we are averaging over the photon counts,

$$U_0 = \frac{C_\gamma}{2\pi} E^4 \left(\frac{2\pi}{\rho} + \frac{8\ell}{\rho_w^2} \right),$$

$$\mathcal{N}_0 = \frac{5}{2\sqrt{3}} \alpha_F \gamma \left(2\pi + \frac{8\ell}{|\rho_w|} \right),$$

$$
\begin{aligned}
\langle u\rangle &= \frac{4}{5\sqrt{3}}\frac{\hbar c\gamma^3}{\rho}\left(\frac{1+\frac{4\ell\rho}{\pi\rho_w^2}}{1+\frac{4\ell}{\pi|\rho_w|}}\right),\\
\langle u^2\rangle &= \frac{11}{12}\left(\frac{\hbar c\gamma^3}{\rho}\right)^2\left(\frac{1+\frac{4\ell\rho^2}{\pi|\rho_w|^3}}{1+\frac{4\ell}{\pi|\rho_w|}}\right),\\
\dot{\mathcal{N}}\langle u^2\rangle &= \frac{55}{24\sqrt{3}}\frac{\alpha_F\hbar^2c^3\gamma^7}{\rho^3}\left(\frac{1+\frac{4\ell\rho^2}{\pi|\rho_w|^3}}{1+\frac{4\ell}{\pi\rho}}\right).
\end{aligned}
$$

6.3.3 Beamstrahlung

Electron and positron bunches collide at the interaction point of a linear or circular collider. As an electron encounters the positron beam and sees its electromagnetic field, it radiates and this radiation is called *beamstrahlung*.[15] As colliders demand increasingly high luminosity, the beam intensity is pushed high and the transverse beam size at the collision point becomes extremely small. The result is a high luminosity but also a strong beamstrahlung, meaning that energetic photons can be emitted during the collision. If the emitted photon is sufficiently energetic, the emitting electron or positron will be lost in its subsequent motion because its energy falls out of the stable momentum aperture of the accelerator. For very high energy electron-positron storage ring colliders, beamstrahlung constitutes one of the leading performance limitations.[16]

Assume that the colliding bunches each has a cylindrical shape with uniform charge density and radius a and length ℓ. For both electron and positron beams, the particle energy is E_0 (assumed ultrarelativistic for a collider), and the number of particles per bunch is N. An electron traversing the positron bunch at its center $r = 0$ will see no beam-beam force and will not radiate. But an electron traversing a finite displacement r (consider $r < a$) from the center of the positron beam will experience beamstrahlung.

During each passage, the electric and magnetic fields experienced by the electron at transverse position r are

$$
E_r = cB_\theta = \frac{Ne}{2\pi\epsilon_0\ell a^2}r\,.
$$

The Lorentz force experienced by the electron is

$$
m\gamma c^2 r'' = -\frac{Ne^2}{\pi\epsilon_0\ell a^2}r\,.
$$

[15] J.E. Augustin et al., Proc. Workshop on Possibilities and Limitations of Accelerators and Detectors (1978); P. Chen, D. Schulte, Handbook Accel. Phys. & Eng., 2nd Ed., Sec.2.5.3, World Scientific, Singapore (2013).

[16] V.I. Telnov, arXiv:1203.6563 and arXiv:1307.3915v1 [physics.acc-ph] (2013).

The beam bunches pass through each other over a distance $\frac{\ell}{2}$ (the factor of 2 is because both beams move with speed c toward each other). As a result, the electron is bent by an angle

$$\theta = r''\frac{\ell}{2} = -\frac{2Nr_0}{\gamma a^2}r\,.$$

The peak bending angle occurs for electrons at $r = a$, with bending angle

$$\hat{\theta} = \frac{2Nr_0}{\gamma a}\,.$$

The bending radius ρ during the collision is given by

$$\frac{1}{\rho} = r'' = \frac{4Nr_0}{\gamma\ell a^2}r\,.$$

During the collision, the beamstrahlung critical photon energy is

$$u_c = \frac{3}{2}\hbar c\frac{\gamma^3}{\rho} = \frac{6\hbar c\gamma^2 Nr_0}{\ell a^2}r\,.$$

For an electron skimming over the edge of the positron beam with $r = a$, it has the highest value of u_c,

$$\hat{u}_c = \frac{6\hbar c\gamma^2 Nr_0}{\ell a}\,.$$

In the above analysis, we have assumed the beam-beam bending angle is small, so that the initial transverse displacement r does not change much throughout the collision. This requires the transverse beam size at the collision point $a \gg \theta\frac{\ell}{2}$ for all electrons, i.e.

$$a \gg \sqrt{\frac{Nr_0\ell}{\gamma}}\,.$$

We also assumed the interaction length is longer than the formation length for radiating a photon. This requires $\theta > \frac{2}{\gamma}$, or

$$a \ll Nr_0\,.$$

The instantaneous beamstrahlung radiation power is

$$P_\gamma = \frac{2}{3}\frac{\gamma^4 mc^3 r_0}{\rho^2} = \frac{32}{3}\frac{N^2\gamma^2 r_0^3 mc^3}{\ell^2 a^4}r^2\,.$$

The radiated energy per beam-beam collision of this electron is therefore

$$\Delta E_\gamma = P_\gamma\frac{\ell}{2c} = \frac{16}{3}\frac{N^2\gamma^2 r_0^3 mc^2}{\ell a^4}r^2\,.$$

The expected number of photons emitted per electron per collision is

$$\mathcal{N}_\gamma = \frac{5\pi}{\sqrt{3}}\alpha_F\gamma \times \frac{\theta}{2\pi} = \frac{5}{\sqrt{3}}\alpha_F \frac{Nr_0}{a}\frac{r}{a}.$$

Typically $\mathcal{N}_\gamma \ll 1$ but that is not a requirement.

Here we have assumed that the emitted beamstrahlung photons have energies $\ll E_0$. Otherwise, energy conservation will come in to severely limit the radiation, and our results will require substantial modifications. We shall here avoid that regime for simplicity. That requires $\hat{u}_c \ll E_0$, or

$$a \gg \frac{6N\gamma r_0^2}{\alpha_F \ell}.$$

The average radiated energy by the electrons is obtained by averaging over the beam distribution,

$$\Delta E_{\gamma,\text{ave}} = \frac{2}{a^2}\int_0^a rdr\, \Delta E_\gamma = \frac{8}{3}\frac{N^2\gamma^2 r_0^3}{\ell a^2} mc^2.$$

The average number of radiated photons per collision is

$$\mathcal{N}_{\gamma,\text{ave}} = \frac{2}{a^2}\int_0^a rdr\, \mathcal{N}_\gamma = \frac{10}{3\sqrt{3}}\alpha_F \frac{Nr_0}{a}.$$

Physically this quantity contains a fine structure constant $\alpha_F \ll 1$ that specifies the duty factor of quantum emissions and another factor $\frac{Nr_0}{a} \gg 1$, which is the number of formation lengths during the collision. It is expected that the number of radiated photons be given by their product.

The radiated photons cover energies up to the level of $\hat{u}_c$. However, some most energetic photons will exceed much beyond this level, and it is those photon emissions that cause most damage due to the beamstrahlung.

Beamstrahlung lifetime We now proceed to give a rough estimate of the beam lifetime due to beamstrahlung for the case of an electron-positron storage ring collider. Let us assume the storage ring has a momentum aperture given by A_δ in such a way that any particle with energy deviation $\delta > A_\delta$, even briefly before radiation damping takes control of it, will be lost.

An electron is lost due to beamstrahlung when it emits an energetic photon whose energy exceeds $A_\delta E_0$. A crude estimate of beam lifetime is therefore given by the loss rate

$$\tau_{\text{bs}}^{-1} = \frac{\mathcal{P}(A_\delta E_0)}{T_0},$$

where T_0 is the revolution period, assuming there is one beam-beam collision per revolution, and $\mathcal{P}(E)$ is the probability when the electron emits a photon of energy larger than E during each collision. Obviously we must have $\mathcal{P}(A_\delta E_0) \ll 1$, which requires $A_\delta E_0 \gg \hat{u}_c$.

The expected differential emission rate of the number of photons is given by

$$\dot{n}(u) = \frac{P_\gamma}{u_c u} S\left(\frac{u}{u_c}\right) = \frac{16}{9} \frac{N\alpha_F r_0 c}{\ell a^2} r \frac{1}{u} S\left(\frac{u}{u_c}\right).$$

The probability of radiating an energetic photon is approximately given by

$$\begin{aligned}
\mathcal{P}(E) &= \frac{\ell}{2c} \int_E^\infty du\, \dot{n}(u) \\
&= \frac{8}{9} \frac{N\alpha_F r_0}{a^2} r \int_E^\infty \frac{du}{u} S\left(\frac{u}{u_c}\right) \\
&= \frac{8}{9} \frac{N\alpha_F r_0}{a^2} r \int_{E/u_c}^\infty \frac{dx}{x} S(x).
\end{aligned}$$

If $E \gg \hat{u}_c$, we can approximate

$$S(x) \approx \frac{9}{8} \sqrt{\frac{3}{2\pi}} \sqrt{x} e^{-x}, \qquad x \gg 1,$$

which yields

$$\mathcal{P}(E) \approx \sqrt{\frac{3}{2\pi}} \frac{N\alpha_F r_0}{a^2} r \frac{e^{-\xi}}{\sqrt{\xi}},$$

where $\xi = \frac{E}{u_c}$ is a function of r. We then have the loss rate for electrons at radial position r,

$$\tau_{\rm bs}^{-1} \approx \sqrt{\frac{3}{2\pi}} \frac{N\alpha_F r_0}{a^2 T_0} r \frac{e^{-\xi}}{\sqrt{\xi}},$$

with

$$\xi = \frac{\alpha_F A_\delta \ell a^2}{6N\gamma r_0^2} \frac{1}{r}.$$

The beamstrahlung loss rate of the electron beam is obtained by averaging over the transverse electron beam distribution,

$$\begin{aligned}
\tau_{\rm bs,ave}^{-1} &= \frac{2}{a^2} \int_0^a r dr\, \tau_{\rm bs}^{-1} \\
&\approx \frac{36}{\sqrt{\pi}} \frac{\gamma^{3/2} N^{5/2} r_0^4}{\alpha_F^{1/2} a^{5/2} A_\delta^{3/2} \ell^{3/2} T_0} \exp\left(-\frac{\alpha_F A_\delta \ell a}{6N\gamma r_0^2}\right).
\end{aligned}$$

As a numerical example, contemplating an electron-positron storage ring collider with $E_0 = 125$ GeV, circumference of 100 km, $a = 3$ μm, $\ell = 5$ mm, $N = 8 \times 10^{10}$, we obtain $\hat{\theta} = 0.6$ mrad, which barely justifies the approximation that r basically remains constant during the collision. We obtain $\hat{u}_c = 1.05$ GeV. We then obtain the average radiated energy per electron per collision $\Delta E_{\gamma,\rm ave} = 0.25$ GeV, the average number of photons emitted is $N_{\gamma,\rm ave} = 1.05$.

Figure 6.29: Beamstrahlung lifetime versus the momentum aperture A_δ for a conceived electron-positron collider. Upper figure is when the beams have uniform round transverse cross-sections with $a = 3\ \mu$m. Middle figure is round beams with $a = 10\ \mu$m. Lower figure shows the case of a flat beam.

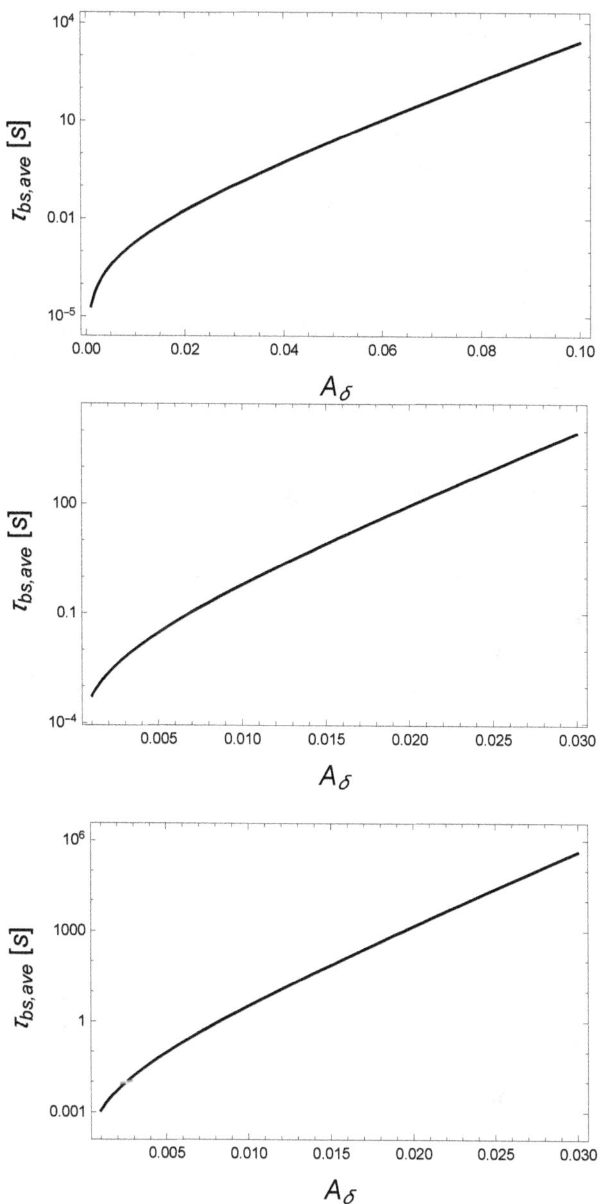

The beamstrahlung lifetime is then calculated as a function of the momentum aperture A_δ. As one would expect, the lifetime is a sensitive function of A_δ, as shown in Fig. 6.29 (upper). It is seen that in order to have a reasonable beam lifetime, the required A_δ would be about 10%, which is very demanding.

Our model assumed a round beam while the actual design may have a flat beam at the collision point. The above value of $a = 3$ μm was chosen to maintain the nominal luminosity by taking $a = 2\sqrt{\sigma_x\sigma_y}$. If we increase a from 3 μm to 10 μm, the plot becomes more relaxed as shown in Fig. 6.29 (middle). For a reasonable lifetime, the required A_δ becomes 2.5%. The demanding requirement on momentum aperture is relaxed, but the luminosity decreases by approximately a factor of 10.

Flat beam The beamstrahlung difficulty is to be relieved by replacing the round beam by a very flat beam at the collision point. We now model the beam at the collision point as a flat rectangle with uniform distribution of width $2w$ and height $2h$ and $w \gg h$. To relate to a bi-Gaussian distribution, we may take $w = 2\sigma_x$ and $h = 2\sigma_y$. The above analysis can be repeated to find the beamstrahlung lifetime. The result is

$$\begin{aligned}
E_y &= -cB_x = \frac{Ne}{4\epsilon_0 \ell wh} y\,, \\
\theta_y &= \frac{\pi N r_0}{\gamma wh} y, \quad \hat{\theta}_y = \frac{\pi N r_0}{\gamma w}\,, \\
\frac{1}{\rho} &= \frac{2\pi N r_0}{\gamma wh\ell} y\,, \\
u_c &= \frac{3\pi\hbar c\gamma^2 N r_0}{wh\ell} y, \quad \hat{u}_c = \frac{3\pi\hbar c\gamma^2 N r_0}{w\ell}\,, \\
h &\gg \frac{\pi N r_0 \ell}{2\gamma w}\,, \\
w &\ll \frac{\pi}{2} N r_0\,, \\
P_\gamma &= \frac{8\pi^2}{3} \frac{N^2\gamma^2 r_0^3 mc^3}{\ell^2 w^2 h^2} y^2\,, \\
\Delta E_\gamma &= \frac{4\pi^2}{3} \frac{N^2\gamma^2 r_0^3 mc^2}{\ell w^2 h^2} y^2\,, \\
\mathcal{N}_\gamma &= \frac{5\pi}{2\sqrt{3}} \alpha_F \frac{N r_0}{wh} y\,, \\
w &\gg \frac{3\pi N\gamma r_0^2}{\alpha_F \ell}\,, \\
\Delta E_{\gamma,\text{ave}} &= \frac{4\pi^2}{9} \frac{N^2\gamma^2 r_0^3}{\ell w^2} mc^2\,, \\
\mathcal{N}_{\gamma,\text{ave}} &= \frac{5\pi}{4\sqrt{3}} \alpha_F \frac{N r_0}{w}\,,
\end{aligned}$$

$$
\begin{aligned}
\frac{P_\gamma}{u_c} &= \frac{8\pi}{9}\frac{\alpha_F N r_e c}{\ell w h} y\,, \\
\mathcal{P}(E) &= \sqrt{\frac{3\pi}{8}\frac{N\alpha_F r_e}{wh} y}\,\frac{e^{-\xi}}{\sqrt{\xi}}, \quad \xi = \frac{E}{u_c}\,, \\
\tau_{\rm bs}^{-1} &= \sqrt{\frac{3\pi}{8}\frac{N\alpha_F r_e}{whT_0} y}\,\frac{e^{-\xi}}{\sqrt{\xi}}, \quad \xi = \frac{\alpha_F A_\delta \ell w h}{3\pi\gamma N r_e^2}\frac{1}{y}\,, \\
\tau_{\rm bs,ave}^{-1} &= \frac{9\pi^2}{2\sqrt{2}}\frac{\gamma^{3/2}N^{5/2}r_e^4}{\alpha_F^{1/2}w^{5/2}A_\delta^{3/2}\ell^{3/2}T_0}\exp\left(-\frac{\alpha_F A_\delta \ell w}{3\pi N\gamma r_e^2}\right).
\end{aligned}
$$

We now consider an electron-positron storage ring collider with $E_0 = 125$ GeV, circumference of 100 km, $\ell = 5$ mm, $N = 8 \times 10^{10}$ again, but this time with a flat beam with $w = 20$ μm, $h = 0.4$ μm. The result of beamstrahlung lifetime is then shown in Fig. 6.29 (lower). Nominal luminosity is maintained, while the required A_δ is about 2%.

Homework 6.39 It may be constructive to fill in the derivation for the flat beam case. The results have been given in the text.

6.4 Radiation acting back on beam

6.4.1 Shifting focus to the radiating electron

So far, we have been concentrating on the radiation, i.e. the photons. In doing so, as we emphasized, we have assumed the radiating particle has a rigid, unperturbed trajectory, and has a prescribed motion as a function of time. The radiation does not act back on this prescribed particle motion. This has been the basic assumption of the classical theory of synchrotron radiation. One way to appreciate the tentative nature of this picture is to note two aspects.

1. Energy and momentum conservation has been compromised. The radiating particle, as it stands, does not receive any reduction in energy or a recoil in momentum as it radiates.

   You may rightfully question how could we possibly ignore so fundamental two laws, but we did! Sometimes we are not allowed to make approximations by dropping some higher-order terms – when it all seem so reasonable to do so – because it ever so subtly violates a hidden principle. While at some other times, we seem to ignore fundamental principles so readily.[17]

2. Another compromise made by this classical picture is that the discrete and random nature — together with the *noise* effects that come with it — of photon emissions have been ignored.

[17] We know devils are in the details but for physicists, devils are in the approximations.

The first part so far of this chapter is what we communicate to a synchrotron radiation user. The remaining of the chapter is what an accelerator physicist would focus on.

The approach we take is a *perturbative* one. We assume the radiation is determined by the unperturbed particle motion and calculate its properties accordingly – as we have done so far – as the zeroth order description. With this zeroth order result of the radiation, we calculate its effects on the particle motion as a first order perturbation correction. We can continue the iteration in principle and modify the radiation using the perturbed electron motion as the new prescribed motion, but we stop our calculation here. Higher order effects are ignored. The topics covered in this second part of this chapter (first order effects) are necessarily more subtle compared with the previous part of this chapter (zeroth order effects).

As a prelude to what will follow in the remaining of this chapter, from the outset, there are at least the following effects of synchrotron radiation on the motion of the radiating particle.

1. The very least we need to do is to recover the energy conservation. The energy carried away by radiation has to be replenished by the acceleration at the RF cavity. This requires

$$U_0 = eV_0 \sin\phi_s ,$$

as already discussed in Chapter 5. However, this is far from the only effect to pay attention to.

2. The expression (6.22) for U_0 was derived in the classical picture, which, as mentioned, ignores energy and momentum conservation. The real expression will be slightly different. Let E be the energy of a particle before radiation, then after the radiation, its expected energy is $E - \langle u \rangle$. One could perhaps get a rough estimate of the real expression of U_0 by substituting $E - \frac{\langle u \rangle}{2}$ for E in the classical expression for U_0. Since U_0 is proportional to E^4 (holding bending curvature fixed), we expect that the real U_0 would be slightly reduced from its classical value, by a factor $\sim (1 - 4\frac{\langle u \rangle}{2E})$. This estimate turns out not too badly. A quantum mechanical calculation by Schwinger gives

$$U_{0,\text{qm}} = U_{0,\text{cl}} \left(1 - \frac{55}{16\sqrt{3}} \frac{u_c}{E}\right) .$$

This slightly reduced radiation power is not too problematic for our purpose. We will not worry about it in what follows.

3. When an electron beam is stored in a storage ring, as the electrons synchrotron radiate, the beam's spin polarization slowly aligns with the bend-

ing magnetic field.[18] Such a polarization effect, as all physical effects involving spin, is necessarily quantum mechanical in origin. In a planar ideal storage ring, the level of final polarization is $P_0 = \frac{8}{5\sqrt{3}} = 92\%$. The beam slowly gets polarized with a surprisingly high level of polarization. The polarization time is given by

$$\tau_{\text{pol}}^{-1} = \frac{5\sqrt{3}}{8} \frac{r_0 \gamma^5 \hbar}{m|\rho|^3}. \tag{6.29}$$

Although most interesting, this subject is considered beyond our scope. We shall not cover it further in the present lectures.

4. Radiation damping.

5. Quantum excitation.

The last two effects 4 and 5 are what we will discuss in detail in the remaining of this chapter. Here we just mention the fact that radiation damping results from the attempt to recover the momentum conservation, while quantum excitation comes from energy conservation and the noise effect,

$$\begin{aligned} &\text{Momentum conservation} \quad \Longrightarrow \quad \text{radiation damping}, \\ &\text{Energy conservation} \quad \Longrightarrow \quad \text{quantum excitation}. \end{aligned}$$

Radiation damping will not involve $\hbar$, and is a purely classical effect. Quantum excitation does involve $\hbar$, and is quantum mechanical in nature. Effects 1, 2, and 4 are classical, while effects 3 and 5 are quantum mechanical.

Both radiation damping and quantum excitation are mainly relevant for electrons. For protons, the related time scales are typically much longer than the time of beam storage unless for the very few ultra-high energy proton storage rings.

Topics to be covered in the remaining of this chapter are

- radiation damping,
- quantum excitation,
- quantum lifetime,
- empirical scaling laws.

6.4.2 Radiation damping — rough calculation

Synchrotron damping Synchrotron radiation provides a natural damping mechanism to synchrotron oscillation of a particle. To see this mechanism, consider a particle with an energy slightly higher than the synchronous energy E_s by an amount ΔE. Having a slightly higher energy, this particle radiates with

[18] A.A. Sokolov and I.M. Ternov, Sov.Phys. Dokl. 8, No.12, 1203 (1964); D.P. Barber and G. Ripken, Handbook Accel. Phys. & Eng., 2nd Ed., Sec. 2.6.6, World Scientific (2013).

a slightly more power than the synchronous particle. As a result, its ΔE in its next turn will be slightly reduced, which contributes to a damping mechanism.

For a more quantitative feel of this damping mechanism, consider a rough calculation below. A more careful calculation will be given in the next section. The radiation power for a particle with energy deviation ΔE in a given magnetic field B is, according to Eq. (6.24),

$$P_\gamma(\Delta E) = C_\gamma \frac{e^2 c^3}{2\pi}(E_s + \Delta E)^2 B^2 .$$

Note that we are holding the magnetic field B fixed. This is a subtle point. Had we held the bending curvature ρ constant, we would obtain $P_\gamma(\Delta E) \propto (E_s + \Delta E)^4$, yielding a different result. The reader is suggested to resolve why we choose to hold B fixed in this consideration, and we assume that the reader is satisfied with his/her conclusion.[19]

The energy radiated away by this particle per turn is higher than that of the synchronous particle by an amount

$$\begin{aligned} \text{change of } \Delta E \text{ per turn} &= -[P_\gamma(\Delta E) - P_\gamma(0)]T_0 \\ &= -2\frac{P_\gamma(0)\Delta E}{E_s}T_0 = -2\frac{U_0}{E_s}\Delta E , \end{aligned} \tag{6.30}$$

where the overall minus sign is because the particle energy is *reduced* due to radiation; T_0 is the period of revolution. It is now clear that this gives a damping effect on ΔE, and the origin of this damping is that higher energy particles radiate more. Soon we will do a more careful calculation of the synchrotron radiation damping time. For now, we only note that the damping decrement for synchrotron oscillation is $\sim \frac{2U_0}{E_s}$ per turn. The damping time is therefore $\sim \frac{E_s}{2U_0}$ turns. Obviously, $\frac{E_s}{U_0} \gg 1$.

Betatron damping Synchrotron radiation also provides a natural damping mechanism to the two transverse betatron oscillations of the particles. It is marvelous and fortunate that synchrotron radiation provides natural damping to all three degrees of freedom. Mother nature is sometimes kind to us, at least to the accelerator physicists.

The transverse radiation damping mechanism is very similar to the mechanism of *adiabatic damping* mentioned in Sec. 3.4. Radiation damping we are going to discuss occurs in a storage ring. Its counterpart that occurs in a linac is just the adiabatic damping.

When an electron emits synchrotron photons, the photons are emitted in its instantaneous direction of motion. Each time it radiates (in a bending magnet, for example), its momentum vector retards a little and becomes smaller, but points in the same direction as before the radiation, leaving the electron's

[19]This subtlety, however, is not to be confused with another subtlety we will mention at the end of this section.

Figure 6.30: Sketches illustrating the mechanism of radiation damping at the RF cavity.

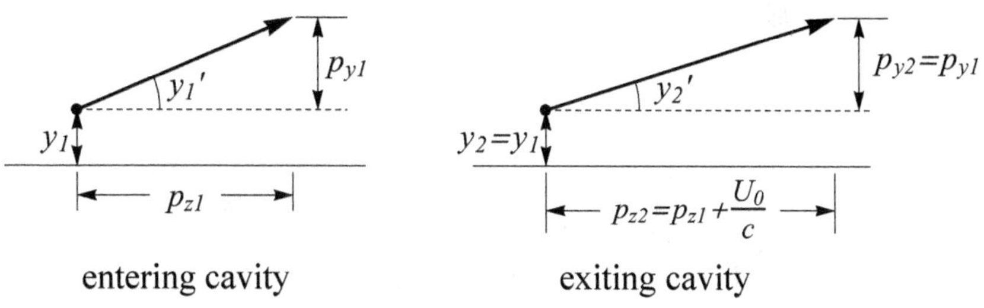

betatron coordinates x, x', y, y' (and therefore its x- and y-Courant-Snyder invariants) unchanged. These emission events do not damp the betatron motion of the electrons.

However, as the electron makes more and more radiation in one revolution, its momentum vector gets shorter and shorter. By the time the electron reaches the RF cavity to replenish its energy by an amount U_0, the RF acceleration is not along the instantaneous direction of motion but along the z-axis. All electrons are accelerated in the same z-direction as dictated by the RF cavity. Because of this acceleration, the electron's slopes x' and y' are reduced as it passes through the RF cavity, as sketched in Fig. 6.30.

The transverse momentum p_y is the same before and after the cavity. So is the transverse displacement y. The changes in the slope y' is given by

$$\begin{aligned} y_2' &= \frac{p_y}{p_z + \frac{U_0}{c}} \approx \frac{p_y}{p_z}\left(1 - \frac{U_0}{p_z c}\right) \\ &= y_1'\left(1 - \frac{U_0}{p_z c}\right) \approx y_1'\left(1 - \frac{U_0}{E_s}\right), \end{aligned}$$

and similarly

$$x_2' \approx x_1'\left(1 - \frac{U_0}{E_s}\right).$$

One sees that both betatron oscillations are damped with the same damping decrement per turn which is half that of the synchrotron motion, and the damping time twice long as the synchrotron radiation damping time. One also sees that no $\hbar$'s have been invoked. The mechanism is purely classical.

Note the statement "all electrons are accelerated in the same z-direction at the RF cavity" above. The betatron radiation damping comes from the fact that all electrons are accelerated in the same direction. It is in fact not important that this same direction has to be the z-direction. Had the RF cavity been misaligned, the same radiation damping will still apply as long as all electrons are accelerated in that misaligned direction. The reader is suggested to consider the question what are the undesirable effects of a misaligned RF cavity.

Note also the qualitative difference between the synchrotron damping and the two betatron dampings, in addition to a difference of a factor of 2 in the damping rates. The synchrotron damping occurs each time when the particle passes through a bending magnet, while the two betatron dampings occur only at the RF cavity. In a sense, the term *radiation* damping arguably applies more for the synchrotron than to the betatron degrees of freedom.

In all three degrees of freedom, the radiation damping times are of the order of $\frac{E_s}{U_0}$ turns, or the time it takes to radiate away all the energy of the electron if not replenished. We have of course assumed that $U_0 \ll E_s$. Even then, however, one should be impressed by the fact that the damping rates are really very fast. Typically $\frac{E_s}{U_0}$ is of the order of 10^3, meaning that the three damping rates are of the order of 1000 turns. This is of course welcome by accelerator designers.

A subtlety But an alert reader will notice in the above that when we calculated the synchrotron damping rate

$$\text{change of } \Delta E \text{ per turn} = -2\frac{U_0}{E_s}\Delta E\,,$$

the factor of *two* on the right comes from the fact that we calculated the energy change using the formula for P_γ, which is proportional to the *square* of E.

But if we have a storage ring that simply consists of a uniform magnetic field as in the case of a weak focusing ring, then we could directly calculate the change in ΔE per turn by using the formula for U_0 (without resorting to P_γ). For a given magnetic field, we know that U_0 is proportional to the *cube* of E. So at least for the special case of a uniform magnetic field, we should obtain a damping rate

$$\text{change of } \Delta E \text{ per turn} = -3\frac{U_0}{E_s}\Delta E\,,$$

which immediately says our analysis is at least incomplete, if not wrong.

Indeed this subtlety points in a direction that something is seriously incomplete with the rough calculations here. We will now perform the calculations more carefully in the following section. We will then see that the careful calculations will modify the synchrotron and the horizontal betatron damping rates, while keeping the vertical betatron damping rate.

6.4.3 Synchrotron radiation damping – detailed calculation

We want to calculate carefully the change in ΔE per turn due to synchrotron radiation for a particle with energy $E_s + \Delta E$. This change will be designated as $U_0(\Delta E)$, with $U_0(0)$ given by the familiar nominal expression for the synchronous particle. The derivation follows that by Sands.[20] We have, for $\gamma_s \gg 1$,

$$\begin{aligned} U_0(\Delta E) &= \oint P_\gamma(\Delta E)dt = \oint P_\gamma(\Delta E)\frac{dt}{ds}ds \\ &= \frac{1}{c}\oint P_\gamma(\Delta E)\left(1+\frac{x}{\rho}\right)ds = \frac{1}{c}\oint P_\gamma(\Delta E)\left(1+\frac{D\Delta E}{\rho E_s}\right)ds\,, \end{aligned}$$

where $P_\gamma(\Delta E)$ is the radiation power, s is the coordinate that runs along the design orbit, and $x = D\frac{\Delta E}{E_s}$ is the horizontal displacement due to the off-energy motion. Betatron motions are ignored.

The replacement $x = D\frac{\Delta E}{E_s}$ has made a few assumptions: (i) The synchrotron oscillation is very slow per revolution; (ii) the ring has a planar design; (iii) no x-y coupling; and (iv) the tunes are chosen to be away from synchrobetatron resonances. In case these assumptions are not valid, a more general approach using a 6×6 formalism will be used to calculate the three fully-coupled radiation damping rates. We will mention a SLIM algorithm in Sec. 6.5 for that purpose.

We now need to find $P_\gamma(\Delta E)$. Recognizing $P_\gamma \propto E^2B^2$, the deviation of $P_\gamma(\Delta E)$ from the nominal $P_\gamma(0)$ has two contributions. One comes from the E^2 dependence that we treated before, and that gives

$$P_\gamma(\Delta E) - P_\gamma(0) = 2\frac{P_\gamma(0)}{E_s}\Delta E\,.$$

The B^2 dependence also makes a contribution if the bending magnets also contain some gradients, i.e. if the magnets are combined-functioned. To see that, let $B = B + Gx = B + GD\frac{\Delta E}{E_s}$ with G the field gradient, then B^2 will contribute another term

$$P_\gamma(\Delta E) - P_\gamma(0) = 2\frac{P_\gamma(0)}{B}GD\frac{\Delta E}{E_s}\,.$$

Combining these two contributions, we have, keeping to first order in ΔE,

$$U_0(\Delta E) - U_0(0) = \frac{\Delta E}{cE_s}\oint P_\gamma(0)\left(2 + 2\frac{GD}{B} + \frac{D}{\rho}\right)ds\,,$$

where the first term is the result of our rough calculation earlier for the dipole magnets, the second term comes from the contribution of radiation in combined-function magnets with dispersion, and the third term comes from the path length dependence on ΔE in dipoles with dispersion. The first term does not depend on dispersion, and integrates to give $2U_0\frac{\Delta E}{E_s}$.

We then have

$$\text{change of } \Delta E \text{ per turn} = -U_0(\Delta E) + U_0(0) = -U_0(0)(2+\mathcal{D})\frac{\Delta E}{E_s}\,,$$

[20]Matthew Sands, SLAC report SLAC -121 (1970).

where $\mathcal{D}$ is a dimensionless *radiation damping partition number*,

$$\mathcal{D} = \frac{1}{cU_0} \oint dsP_\gamma(0)\left(2\frac{G}{B}+\frac{1}{\rho}\right)D = \frac{\oint \frac{D\,ds}{\rho}\left(2k+\frac{1}{\rho^2}\right)}{\oint \frac{ds}{\rho^2}}, \tag{6.31}$$

where $k = \frac{G}{B\rho}$. In most (but not all) cases, $\mathcal{D}$ is positive. It is a property of the lattice design, and is independent of any beam properties.

It should perhaps be mentioned that the quantity ρ in Eq. (6.31) can have opposite signs when the bending is in the reversed direction. In those cases, the quantity $B\rho$ in the definition of $k = \frac{G}{B\rho}$ should use the local values of the magnetic rigidity $\pm\frac{P}{e}$ including its sign.

Uniform bend ring The more careful calculation therefore introduced a new term involving $\mathcal{D}$ compared with the rough calculation (6.30). At this point, one urgent thing to do is to see what happens to the case of a uniform bending storage ring – the subtlety we mentioned at the end of the previous section. In that case, we have $G = 0$, and $D = \rho$. Indeed, we see that our formula gives $\mathcal{D} = 1$, and therefore satisfactorily gives $2 + \mathcal{D} = 3$. The subtlety issue is resolved.

Strong-focusing separate-function ring For strong focusing, separated-function storage rings, on the other hand, usually $\mathcal{D} \ll 1$, and the rough calculation is approximately valid. To see this, we note that, with separated-function magnets, $\frac{k}{\rho} = 0$, so

$$\mathcal{D} = \frac{\oint \frac{D\,ds}{\rho^3}}{\oint \frac{ds}{\rho^2}}.$$

If all bending magnets have the same ρ (isomagnetic), then

$$\mathcal{D} = \frac{1}{2\pi\rho}\oint \frac{D\,ds}{\rho} = \frac{\alpha_c R}{\rho},$$

where $2\pi R$ is the storage ring circumference and α_c is the momentum compaction factor. In this case, one notes that $\mathcal{D}$ comes solely from the path length contribution. Since for a strong focusing ring, $\alpha_c \ll 1$, the partition number is usually $\ll 1$. As an order of magnitude estimate, in fact, we have

$$\mathcal{D} \approx \alpha_c \approx \frac{1}{\nu_x^2}.$$

Combined-function ring In the other extreme, for a storage ring with combined-function magnets, the partition number plays an important role. Consider an isomagnetic storage ring,[21] for example, the dispersion function D satisfies

$$D'' + \left(\frac{1}{\rho^2} + k\right) D = \frac{1}{\rho}.$$

Using this expression, we have

$$\begin{aligned}
\mathcal{D} &= \frac{\oint \frac{D\,ds}{\rho^3} + 2\oint \frac{ds}{\rho} Dk}{2\pi\rho/\rho^2} \\
&= \frac{\oint \frac{D\,ds}{\rho^3} + 2\oint \frac{ds}{\rho}\left(-D'' - \frac{D}{\rho^2} + \frac{1}{\rho}\right)}{2\pi/\rho} \\
&= -\alpha_c \frac{R}{\rho} + \frac{1}{\pi}\oint \left(\frac{1}{\rho} - D''\right) ds \\
&= -\alpha_c \frac{R}{\rho} + 2\,.
\end{aligned}$$

For a strong-focusing combined-function storage ring, $\alpha_c \ll 1$, we have $\mathcal{D} \approx 2$. We will see later that the fact that $\mathcal{D}$ can be as large as 2 creates a serious problem. It in fact rules out the possibility of building a strong-focusing electron storage ring exclusively of combined-function magnets. The AGS, for example, works fine for storing protons and heavy ions, but will not do so for electrons.

For a weak-focusing combined-function ring, it will be shown in Homework 6.50 that $\mathcal{D} = \frac{1-2n}{1-n}$ with n the field index $(1 > n > 0)$.

To summarize,

$$\mathcal{D} \begin{cases} = 1, & \text{uniform bend,} \\ \approx \frac{1}{\nu_x^2}, & \text{strong focusing, separated-function,} \\ \approx 2, & \text{strong focusing, combined-function,} \\ = \frac{1-2n}{1-n}, & \text{weak focusing.} \end{cases} \tag{6.32}$$

Damping rate We have now obtained the change rate of ΔE per turn. We still have to compute the synchrotron radiation damping rate. For this, we need to return to the equations of motion. In terms of the variables of ΔE and RF phase ϕ, we have from Chapter 5 that

$$\begin{aligned}
\dot{\Delta E} &= \frac{\omega_0}{2\pi}\left[eV_0 \sin\phi - U_0(\Delta E)\right], \\
\dot{\phi} &= h\omega_0\eta \frac{\Delta E}{E_s}, \\
U_0(0) &= eV_0 \sin\phi_s\,.
\end{aligned}$$

[21]The term isomagnetic applies to the bending magnets, i.e. it refers to $\rho =$ constant around the storage ring. It does not apply to quadrupoles; so $k(s) \neq$ necessarily constant. There are therefore two kinds of isomagnetic rings. One is weak focusing with $k(s)$ = constant (e.g. ZGS) and one is strong focusing with $k(s)$ alternating gradient (e.g. AGS).

Combining these equations and the equation for $U_0(\Delta E)$, and linearize with respect to $\Delta\phi = \phi - \phi_s$, we obtain

$$\ddot{\Delta\phi} + 2\bar{\alpha}_z\dot{\Delta\phi} + \omega_s^2\Delta\phi = 0,$$

which we recognize is an equation describing a damped simple harmonic oscillator. The first and third terms are the familiar equation for small-amplitude synchrotron oscillation. The second term is the new radiation damping term, with the synchrotron oscillation damping decrement

$$\bar{\alpha}_z = \frac{U_0(0)}{E_s}\left(1+\frac{\mathcal{D}}{2}\right),$$

and the e-folding radiation damping time

$$\tau_z = \frac{T_0}{\bar{\alpha}_z} = \frac{E_s T_0}{U_0(0)(1+\frac{\mathcal{D}}{2})},$$

where $T_0 = \frac{2\pi}{\omega_0}$ is the revolution period. We recover the result of our rough calculation when $\mathcal{D} = 0$.

The solution for the damped motion, when $\bar{\alpha}_z \ll \omega_s$, is

$$\Delta\phi(t) = Ae^{-\bar{\alpha}_z t/T_0}\cos(\omega_s t + \phi_0).$$

Control $\mathcal{D}$ by RF frequency The fact that radiation damping partition number $\mathcal{D}$ changes sensitively on combined-function magnets also implies its sensitivity to the RF frequency. When the RF frequency is varied, the beam has to adjust its trajectory circumference to maintain the harmonic condition $\omega_{\rm rf} = h\omega_0$. The way the beam does it is to find a new central beam energy with

$$\Delta\delta = -\frac{1}{\eta}\frac{\Delta\omega_{\rm rf}}{\omega_{\rm rf}}, \tag{6.33}$$

where η is the phase slippage factor. In the mean time, the beam moves on an orbit with displacements $D\Delta\delta$ in all quadrupoles, making all quadrupoles behaving as combined-function magnets even in a separate-function ring.

The effective bending field is $B = GD\Delta\delta$ with a bending curvature $\frac{1}{\rho_k} = kD\Delta\delta$ with $k = \frac{G}{B\rho}$. It follows that $\mathcal{D}$ changes by

$$\Delta\mathcal{D} = \frac{2\oint ds\frac{kD}{\rho_k}}{\oint\frac{ds}{\rho^2}} = \frac{2\oint dsD^2k^2}{\oint\frac{ds}{\rho^2}}\Delta\delta.$$

For a separate-function ring, the coefficient $\frac{\Delta\mathcal{D}}{\Delta\delta}$ is always positive. The dependence of $\mathcal{D}$ on the RF frequency is then obtained by relating $\Delta\delta$ to $\frac{\Delta\omega_{\rm rf}}{\omega_{\rm rf}}$. See also Eqs. (5.20) and (5.21) when a similar technique is used to measure the dispersion function and the chromaticities.

Homework 6.40

(a) Show that a separate-function FODO cell contributes

$$\frac{\Delta\mathcal{D}}{\Delta\delta} = \frac{L}{\ell}\left(\frac{8}{\sin^2\frac{\Phi}{2}}+2\right),$$

where L is the half-cell length, ℓ is the length of each thin quadrupole, and Φ is the horizontal phase advance per cell.

(b) Consider a ring consisting of N_{cell} FODO cells. Give an estimate of its $\frac{\Delta\mathcal{D}}{\Delta\delta}$. Note that $\frac{L}{\ell}$ can be large (but do not set $\ell \to 0$ even though for thin magnets). Convince yourself that $\mathcal{D}$ depends sensitively on RF frequency variation especially for large rings.

Solution (b) The quantity $\frac{\Delta\mathcal{D}}{\Delta\delta}$ for the whole ring is the same as that for the one FODO cell given in (a). And yet, the effect of RF frequency on $\mathcal{D}$ is stronger for larger rings, basically quadratically with the ring size.

6.4.4 Vertical betatron radiation damping – detailed calculation

Consider an electron executing vertical betatron oscillation with an emittance

$$\epsilon_y = A_y^2 = \frac{1}{\beta_y}[y^2+(\alpha_y y+\beta_y y')^2],$$

where A_y is the vertical betatron amplitude (in units of $\sqrt{\text{m}}$). When this electron emits a photon, it gets a recoil in the direction opposite to its instantaneous direction of motion. As discussed earlier, this action does not change its displacement y or its slope y'. As a result, the emission does not affect its emittance.

Here we have ignored the fact that the emission does not exactly align with the direction of motion of the electron. There is in fact an opening angle $\sim \frac{1}{\gamma}$ (Fig. 6.31). This opening angle *does* generate a sideway additional excitation of the vertical betatron motion, and should in principle be taken into account. This excitation, however, does not affect the radiation damping. Instead, it belongs to a treatment of quantum excitation which will be dealt with later.

As the electron reaches and passes the RF cavity, as illustrated in Fig. 6.30, its displacement y remains the same, but its slope is reduced,

$$\Delta y' = -\frac{U_0}{E_s}y'.$$

Although this damping is localized at the RF cavity, it is a weak effect on a turn-by-turn basis, so it can be considered to be spread out over the whole circumference C, with a decrement of $\frac{U_0}{E_s C}$ per unit distance. We then obtain

Figure 6.31: Illustration of the effect on vertical betatron motion due to synchrotron radiation with an opening angle.

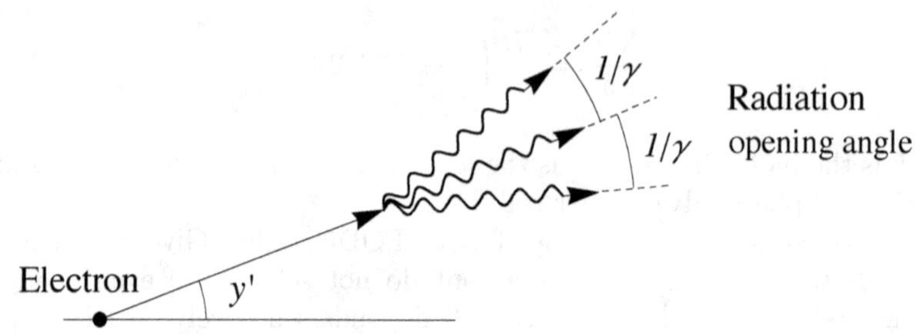

the equations of motion

$$\begin{aligned}\frac{dy}{ds} &= y', \\ \frac{dy'}{ds} &= -K(s)y - \frac{U_0}{E_s C}y',\end{aligned}$$

which are combined to yield

$$\frac{d^2y}{ds^2} + \frac{U_0}{E_s C}\frac{dy}{ds} + K(s)y = 0.$$

It is clear that the motion is a damped oscillator with its amplitude damped exponentially with a damping decrement

$$\bar{\alpha}_y = \frac{U_0}{2E_s},$$

(we put a bar over α_y to avoid possible confusion with the Courant-Snyder function) and an e-fold damping time

$$\tau_y = \frac{T_0}{\bar{\alpha}_y} = 2\frac{E_s T_0}{U_0}.$$

The vertical betatron amplitude behaves like

$$A_y \sim e^{-\bar{\alpha}_y t/T_0},$$

and its change per turn is

$$\frac{\Delta A_y/\text{turn}}{A_y} = \frac{e^{-\bar{\alpha}_y(t+T_0)/T_0} - e^{-\bar{\alpha}_y t/T_0}}{e^{-\bar{\alpha}_y t/T_0}} \approx -\bar{\alpha}_y = -\frac{U_0}{2E_s}.$$

This is the same result as the rough calculation given earlier.

6.4.5 Horizontal betatron radiation damping – detailed calculation

It might first seem that the effect of synchrotron radiation on the horizontal and the vertical betatron motions would be the same, so the results in the last section would apply here too, e.g. $\bar{\alpha}_x = \frac{U_0}{2E_s}$. This is almost so, except for one complication, i.e. there is a dispersion horizontally and no dispersion vertically (in an ideal planar storage ring). This horizontal dispersion complicates the physics, and we have to deal with some subtle effects.

To analyze these subtle effects correctly requires some patience. One might wonder what is the reason that motivates one to invest such an endeavor in the first place. In other words, there must be some reason to suspect that the answer obtained by the rough calculation is somehow wrong. Indeed there has been a reason for suspicion. The reason that the rough calculation $\bar{\alpha}_x = \frac{U_0}{2E_s}$ must be wrong comes from a powerful theorem called the *Robinson sum rule*, to be discussed in Sec. 6.4.6,

$$\bar{\alpha}_x + \bar{\alpha}_y + \bar{\alpha}_z = \frac{2U_0}{E_s}.$$

Since $\bar{\alpha}_y = \frac{U_0}{2E_s}$ and $\bar{\alpha}_z = (2 + \mathcal{D})\frac{U_0}{2E_s}$, this sum rule demands that $\bar{\alpha}_x = (1 - \mathcal{D})\frac{U_0}{2E_s}$, and if true, the rough calculation for the horizontal betatron damping must be modified. We shall do that next. It is interesting to note that, although wrong for other reasons, the rough calculation does obey the sum rule.

Let us make another comment at this point. It concerns the assumptions being made of our storage ring. In our derivations so far, we have assumed the ring has a planar design. It also assumes no x-y or synchrobetatron coupling. In addition, the use of dispersion function also assumes the synchrotron oscillation is slow, i.e. $\nu_s \ll 1$. Our results obtained painstakingly apply only when all these assumptions are valid. It is inevitable to ask the question why we invest so much effort to get a result for such a singular special case. What in case the ring is nonplanar, or if it is coupled, or if ν_s is not very small, as a real storage ring must be when various error effects are included? Our analysis will all have to be reformulated somehow. The way to proceed involves a 3-D treatment. Later in this chapter, we shall discuss a SLIM formalism. We shall see that if we do not aim for closed-form expressions, the actual calculations of the 3-D radiation damping and quantum excitation effects can be done in a rather straight forward manner.

We now return to the conventional treatment of uncoupled planar ring. Taking into account of the dispersion term, the horizontal displacement of a particle can be written as

$$\begin{aligned} x &= x_\beta + x_\epsilon, & x_\epsilon &= D\delta, \\ x' &= x'_\beta + x'_\epsilon, & x'_\epsilon &= D'\delta, \end{aligned}$$

Figure 6.32: The closed-orbit gets shifted when a synchrotron radiation photon is emitted but the instantaneous displacement of the particle is not.

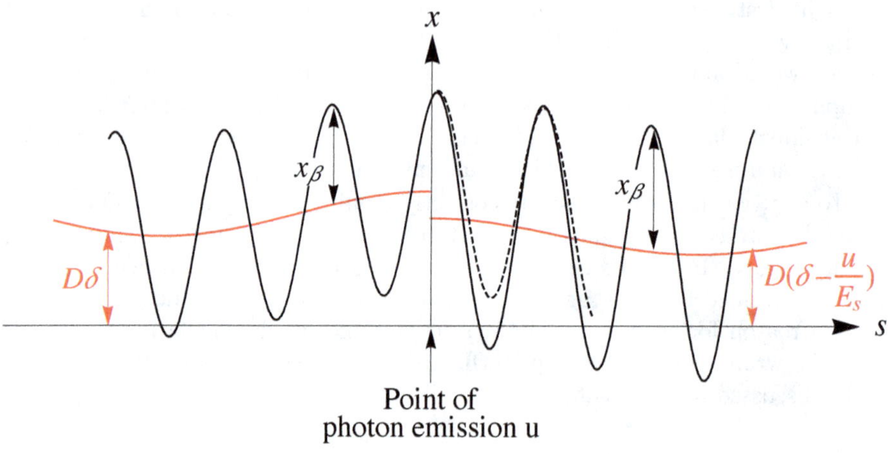

where x_β and x_ϵ are the betatron and energy contributions respectively, D is dispersion function. The energy term represents an equilibrium orbit for the off-momentum particle, while the betatron term oscillates around it. When a photon of energy u is emitted, the electron energy changes by $-u$. The energy term of the orbit changes by $-D\frac{u}{E_s}$. After the emission, therefore, the particle begins to oscillate around a different, shifted closed orbit $x_\epsilon - D\frac{u}{E_s}$. It is important to know that the photon emission can shift the closed orbit but it cannot shift the instantaneous position of the particle. The situation is sketched in Fig. 6.32. The red curves are the off-energy dispersion orbits before and after emission. On top of it is the betatron oscillation. A photon emission excites an additional component of betatron oscillation. The betatron oscillation without the additional excitation is depicted by the dashed curve. Although the red closed-orbit gets shifted, the instantaneous position of the electron is not changed at the emission point.

Since the radiation does not change x, the change in x_ϵ must be accompanied by an opposite change in x_β, at the point of emission,

$$\Delta x_\beta = -\Delta x_\epsilon = D\frac{u}{E_s},$$

$$\Delta x'_\beta = -\Delta x'_\epsilon = D'\frac{u}{E_s}.$$

We now introduce the horizontal betatron amplitude A_x by

$$A_x^2 = \frac{x_\beta^2}{\beta_x} + \frac{(\beta_x x'_\beta + \alpha_x x_\beta)^2}{\beta_x}.$$

This amplitude is before radiation. After radiation, it becomes

$$(A_x+\Delta A_x)^2=\frac{1}{\beta_x}\left\{\left(x_\beta+\frac{uD}{E_s}\right)^2+\left[\beta_x\left(x'_\beta+\frac{uD'}{E_s}\right)+\alpha_x\left(x_\beta+\frac{uD}{E_s}\right)\right]^2\right\}.$$

Subtracting the above two expressions gives

$$\begin{aligned}(A_x+\Delta A_x)^2-A_x^2 &= \frac{u}{\beta_x E_s}[2Dx_\beta+2(\beta_x x'_\beta+\alpha_x x_\beta)(\beta_x D'+\alpha_x D)]\\ &\quad+\frac{u^2}{\beta_x E_s^2}[D^2+(\beta_x D'+\alpha_x D)^2]\,.\end{aligned}\qquad(6.34)$$

The first term on the right hand side is proportional to $\frac{u}{E_s}$. The second term is proportional to $(\frac{u}{E_s})^2$. It might be tempting to drop the second term because $\frac{u}{E_s}$ is small. And indeed that is what we will do next to perform our careful calculation of the damping rate. However, the second term is also important — one indication of which is the fact that it is always positive, and, although very small, it can accumulate over time. The second term will become the source of *quantum excitation*, as will be picked up again in the next section.

It may be useful to have an order of magnitude feel of the two terms in Eq. (6.34). Take a 5-GeV electron storage ring for example, with bending radius $\rho=25$ m. The radiation loss per turn is $U_0=2.2$ MeV. The number of photons emitted per turn is $\frac{\gamma}{15}=650$. The mean value $\frac{\langle u\rangle}{E_s}$ that enters the first term therefore is about 0.68×10^{-6}. The second term then is of the order of $\frac{\langle u\rangle^2}{E_s^2}=0.46\times10^{-12}$, a truly small quantity.

Dropping the second term for now (to be picked up again later), and keeping only the linear term in ΔA_x, we obtain

$$A_x\Delta A_x = \frac{u}{\beta_x E_s}[Dx_\beta+(\beta_x x'_\beta+\alpha_x x_\beta)(\beta_x D'+\alpha_x D)]\,.\qquad(6.35)$$

Since we expect synchrotron radiation to affect particle motion only weakly and slowly, our next step is to average the right hand side over time. A rapidly oscillating term on the right hand side would not contribute to a net change in A_x. At this point, it is tempting to drop the whole term altogether because we expect the two conjugate coordinates

$$\begin{aligned}x_\beta &\sim \cos\omega_x t\,,\\ \beta_x x'_\beta+\alpha_x x_\beta &\sim -\sin\omega_x t\,,\end{aligned}$$

where ω_x is the rapidly oscillating horizontal betatron frequency. The square brackets in Eq. (6.35) contains one x_β term and one $(\beta_x x'_\beta+\alpha_x x_\beta)$ term. If we average over t at this point, we would obtain zero.

But here comes yet another subtlety. The synchrotron radiation, although largely occurring in bending magnets, and the radiated photon energy u is largely independent of x_β, there *is* a small part of u that does correlate with x_β. If we filter out this small part of u, keeping to first order in x_β, then this small contribution will resonate with the x_β term in the square brackets and will no longer average to zero.

The drive, at least partly, behind this persistent searching, as mentioned earlier, is so that the Robinson sum rule gets to be obeyed.[22]

Our next job is therefore to filter out this small x_β-dependent term in u. Note that u will not contain a $(\beta_x x'_\beta + \alpha_x x_\beta)$-dependent term, so the $(\beta_x x'_\beta + \alpha_x x_\beta)$-term in the square brackets can be dropped because it does average to zero.

We showed before that the synchrotron radiation power $P_\gamma \propto E^2 B^2$, and the synchrotron radiation damping came mostly from its E^2-dependence. Here, the x_β-dependence of u comes from the B^2-dependence of P_γ. We have

$$P_\gamma(x_\beta) - P_\gamma(0) \;=\; 2P_\gamma(0)\frac{Gx_\beta}{B}\,.$$

Following an analysis similar to that used in Sec. 6.4.3, and consider an element ds along the design orbit, the radiated photon in a distance ds is

$$\begin{aligned} u(x_\beta) \;&=\; P_\gamma(x_\beta)\left(1+\frac{x_\beta}{\rho}\right)\frac{ds}{c} \\ \Longrightarrow\quad u(x_\beta) - u(0) \;&=\; P_\gamma(0)x_\beta\left(\frac{2G}{B}+\frac{1}{\rho}\right)\frac{ds}{c}\,. \end{aligned}$$

As promised, we have found a small term in u proportional to x_β.

Combining results so far, the change of A_x per turn is then given by

$$\frac{A_x \Delta A_x}{\text{turn}} \;=\; \oint \frac{D}{E_s}\frac{\langle x_\beta^2\rangle}{\beta_x}P_\gamma(0)\left(\frac{2G}{B}+\frac{1}{\rho}\right)\frac{ds}{c}\,,$$

where $\langle x_\beta^2\rangle$ means taking a time averaging over x_β^2. Although x_β oscillates rapidly, $\langle x_\beta^2\rangle$ assumes a well-defined finite value. By the definition of A_x^2, we have in fact

$$\frac{\langle x_\beta^2\rangle}{\beta_x} \;=\; \frac{A_x^2}{2}\,.$$

[22]If you pause at this step, a dose of Chinese philosophy may help,

> 格物致知
> The path to knowledge is through inquisitive cultivations.
> (Philosopher Yangming Wang 1472–1529)
> 湊答案
> Do anything to get the answer you want.

Therefore, with $k = \frac{G}{B\rho}$,

$$\begin{aligned}
\frac{\Delta A_x/\text{turn}}{A_x} &= \frac{1}{2E_s c}\oint ds\, DP_\gamma(0)\left(\frac{2G}{B}+\frac{1}{\rho}\right) \\
&= \frac{U_0}{2E_s}\frac{\oint ds\,\frac{D}{\rho}\left(2k+\frac{1}{\rho^2}\right)}{\oint\frac{ds}{\rho^2}} = \frac{U_0}{2E_s}\mathcal{D}\,.
\end{aligned}$$

This is quite a remarkable result. The complicated expression seemingly miraculously turns out exactly to be the partition number defined earlier. This is particularly remarkable as the synchrotron radiation damping (that mostly occurs in the bends) and the betatron radiation damping (that occurs at the RF cavity)[23] are quite different physical mechanisms — at least superficially — and requires very different calculations, and yet after two complicated, lengthy, seemingly unrelated calculations are completed, they both invoke the same complicated quantity $\mathcal{D}$. However, as mentioned, this is inescapable due to the Robinson sum rule.

When $\mathcal{D}$ is positive, as usually is the case, this new subtle contribution of synchrotron radiation causes the horizontal betatron amplitude to grow, i.e. it acts against radiation damping.

On top of this $\mathcal{D}$ contribution, there is also the main contribution from the RF cavity,

$$\frac{\Delta A_x/\text{turn}}{A_x} = -\frac{U_0}{2E_s}\,.$$

Adding two contributions together gives

$$\frac{\Delta A_x/\text{turn}}{A_x} = -\frac{U_0}{2E_s}(1-\mathcal{D})\,.$$

The first term comes from the RF cavities. The second term comes from the magnets. This then gives the damping decrement

$$\bar{\alpha}_x = \frac{U_0}{2E_s}(1-\mathcal{D})\,,$$

and the damping time

$$\tau_x = \frac{T_0}{\bar{\alpha}_x} = \frac{2E_s T_0}{U_0(1-\mathcal{D})}\,.$$

The reader should note that indeed the Robinson sum rule is observed.

In passing it should be noted that we have derived $\frac{\Delta A_x/\text{turn}}{A_x}$ by adding two contributions: one unit from the RF cavity, and $-\mathcal{D}$ units from the magnets.

[23] Strictly speaking, where exactly the damping occurs depends on which dynamical variables we use. Here, we use the variables $(x, x', y, y', z, \delta)$. This observation is to be changed had we used $(x, p_x, y, p_y, z, \delta)$ as the dynamical variables.

These two contributions have completely different physical origins. In a sense, only the RF contribution is truly a damping effect. The $-\mathcal{D}$ contribution, strictly speaking, is a synchrobetatron *coupling* effect. When it damps one dimension, it necessarily antidamps the other dimension. For both dimensions to be damped, we have the stability condition that

$$1 > \mathcal{D} > -2 .$$

The part $1 > \mathcal{D}$ is for stability of horizontal betatron motion; the part $\mathcal{D} > -2$ is for stability of synchrotron motion. The vertical betatron motion on the other hand is free from this synchrobetatron coupling.

The fact that the rough calculation of the radiation damping obeys the Robinson sum rule in fact is not accidental. The reason is that the rough calculation does include all the main *damping* effects. What it ignored was coupling effects, which do not affect the sum rule.

The reader is reminded of the statement made earlier that, to assure radiation damping stability, a strong-focusing electron storage ring cannot be made exclusively of combined-function magnets (such as the case of the AGS synchrotron). The reason is that in that case, we have shown that $\mathcal{D} \approx 2$. As a result, the horizontal betatron motion is radiation *antidamped*, and is therefore unstable. This is of course not allowed. In a sense, this is unfortunate; a strong focusing lattice using combined-function magnets should constitute the most compact lattice design. But as we just saw, this does not work for electrons.

6.4.6 Robinson sum rule

Although we may claim having proven the Robinson sum rule by our derivations above, one might find our proof somewhat unsatisfactory because it seems the sum rule comes about by accident. In this section, we give Robinson's own proof, in the mean time addressing its physical origin.[24]

In this context, all it involves are linear dynamics, so we can use matrices to describe the particle motion. In general, we have a 3-D motion with a 6-D phase space $(x, x', y, y', z, \delta)$. Around the storage ring, each beamline element therefore has a map given by a 6×6 matrix.

We multiply all these 6×6 matrices together to form a one-turn map

$$M_{\text{tot}} = M_1 M_2 \cdots M_{\text{cav}} \cdot\cdot M_n ,$$

where M_{cav} is the map for the RF cavity, and some of the M_i-matrices are for the bending magnets. All other matrices remain unspecified, including the possibility that they may couple all three degrees of freedom (including coupling to the vertical dimension which is so far a bystander in our treatment).

[24] K.W. Robinson, Phys. Rev. 111, 373 (1958); Iu. F. Orlov and E. K. Tarasov, J. Exp. Theor. Phys. 34, 651 (1958) [Sov. Phys. JETP 34, 3 (1958)]; R. Talman and A.W. Chao, Phys. Rev. Special Topics - Accel. and Beams, 12, 084901 (2009).

We learned now that in a bending magnet, say beamline element M_5, a particle radiates more when its $\delta > 0$, and as a result, the (6,6)-element of M_5 is not equal to 1 as we have always assumed. Instead it is to be changed into

$$(M_5)_{66} = 1 - 2\frac{U_5}{E_s},$$

with U_5 the radiated energy in the magnet. The factor of 2 should be clear.

We also learned that in the RF cavity, x' and y' get changed due to radiation damping, we then have

$$(M_{\text{cav}})_{22} \;=\; (M_{\text{cav}})_{44} \;=\; 1 - \frac{U_0}{E_s},$$

where U_0 is the total radiated energy per turn.

We are now ready to prove the Robinson sum rule. Let us take the determinant of M_{tot}. We know it is given by

$$\det M_{\text{tot}} \;=\; \text{product of determinants of all beamline maps}.$$

Normally, because of the symplecticity condition of beam dynamics, all the beamline element maps have unit determinants, and we simply obtain $\det M_{\text{tot}} = 1$, a well-known result. But now, having made the modifications due to radiation damping, the determinant of the M_5 bending magnet becomes

$$\det M_5 \;=\; 1 - 2\frac{U_5}{E_s},$$

while the determinant of the RF cavity becomes

$$\det M_{\text{cav}} \;=\; \left(1 - \frac{U_0}{E_s}\right)^2.$$

As a result, now we have

$$\begin{aligned}\det M_{\text{tot}} \;&=\; \left(1 - \frac{U_0}{E_s}\right)^2 \prod_{i=\text{bends}} \left(1 - 2\frac{U_i}{E_s}\right) \\ &\approx\; 1 - 2\frac{U_0}{E_s} - 2 \sum_{i=\text{bends}} \frac{U_i}{E_s} \\ &=\; 1 - 4\frac{U_0}{E_s},\end{aligned}$$

where we have used the property that $\sum_{i=\text{bends}} U_i = U_0$.

In our present considerations, there are two operations that involve nonsymplectic processes, the radiation damping and the quantum excitation. In the Robinson rum rule analysis, only radiation damping is involved.

Perhaps a recap would be useful here. We learned before that the actions of radiation damping occur in the bends and in the RF cavities. These actions are true damping actions, and they modify the beam dynamics nonsymplectically. As to all the other elements, they may couple the 6-D phase space beam dynamics, but will not render their maps nonsymplectic. As a result, they can cause redistribution of the damping decrements among the three dimensions – just like what the partition number $\mathcal{D}$ does, but $\mathcal{D}$ is far from being the only example – but they cannot create additional damping action by affecting the sum rule.

The one-turn map $M_{\rm tot}$ can also be eigenanalyzed. Let its six eigenvalues be

$$\lambda_k = e^{-\bar{\alpha}_k \pm i\mu_k}, \quad k = 1, 2, 3,$$

where $\bar{\alpha}_k$ is the damping decrement per turn and μ_k is the oscillation phase per turn for the k-th eigenmode. The fact that the eigenvalues form complex conjugate pairs is a consequence of the fact that the matrix $M_{\rm tot}$ is real.

We then have

$$\begin{aligned} \det M_{\rm tot} &= \text{product of all six eigenvalues} \\ &= \prod_{k=1}^{3} e^{-2\bar{\alpha}_k} \approx 1 - 2\sum_{k=1}^{3} \bar{\alpha}_k\,. \end{aligned}$$

Comparing the two expressions, we conclude that

$$\sum_{k=1}^{3} \bar{\alpha}_k = 2\frac{U_0}{E_s},$$

which we recognize is the Robinson sum rule applied to the general case with arbitrary 3-D linear coupling.

Note that nowhere in this derivation concerns the details of the calculation for the partition number $\mathcal{D}$. This again indicates that the origin of $\mathcal{D}$ is due to a symplectic coupling effect, and not a more fundamental modification to the symplecticity condition of the beamline elements.

Partition number The Robinson sum rule states that the three dimensions partition a total damping decrement of $2\frac{U_0}{E_s}$. In the general case, we define *partition numbers*,

$$\bar{\alpha}_{x,y,z} = J_{x,y,z}\frac{U_0}{2E_s}\,.$$

For a conventional planar storage ring, for example, we have

$$J_x = 1 - \mathcal{D}, \quad J_y = 1, \quad J_z = 2 + \mathcal{D}, \quad J_x + J_y + J_z = 4\,.$$

Radiation damping of a long arc As a variation of the storage ring considerations up to now, consider an electron beam being transported by a long arc, as was the case for the SLAC Linear Collider (SLC, now extinct). In this case, since there is no RF in this arc, the partition of the radiation becomes[25]

$$J_x = -\mathcal{D}, \quad J_y = 0, \quad J_z = 2 + \mathcal{D}.$$

In this case, the sum rule reads

$$J_x + J_y + J_z = 2.$$

Assume the arc lattice is combined-function strong focusing, as is the case of the SLC arc, which implies $\mathcal{D} \approx 2$, according to Eq. (6.32). Due to the radiation damping effect, the horizontal emittance will grow by

$$\frac{\Delta\epsilon_x}{\epsilon_x} = 2\mathcal{D}\frac{U_0}{2E_s} = \frac{2U_0}{E_s},$$

where account is taken for ϵ_x is quadratic in x, so it grows twice the rate of x. For the SLC at 50 GeV, $\frac{U_0}{E_s} = 2.8\,\%$, we find the emittance grows by 5.6 %. Since there is no acceleration, there is no adiabatic damping.

At the same time, the energy spread damps by

$$\frac{\Delta\sigma_\delta}{\sigma_\delta} = -2(2+\mathcal{D})\frac{U_0}{2E_s} = -\frac{4U_0}{E_s}.$$

This result of energy damping is twice what is expected if the arc is not part of a storage ring. Also because there is no RF, all the damping is applied to δ and without having to be shared with z. The energy spread shrinks by 11 % for the SLC at 50 GeV.

6.4.7 Quantum excitation of synchrotron oscillation

Beam does not damp to zero-size So far we discussed in detail the radiation damping effects in the three degrees of freedom. The three damping times are all of the order of $\frac{E_s T_0}{U_0}$. These are all classical effects, indicated by the fact that no expressions so far involve $\hbar$.

According to the picture so far, a beam of particles, once injected into a storage ring, will eventually damp to a zero size. This is of course absurd and we must ask what effects will emerge as the beam size gets smaller, and function to prevent the beam size to shrink to zero. A long list, limited only by lack of imagination, follows, e.g.

- lattice nonlinearities;
- space charge effect;
- intrabeam scattering;

[25] A. Chao, R. Helm, SLAC Report CN-201 (1982).

- collective instability;
- beam-beam interaction;
- magnet power supply ripple;
- continuous random vibration of the ground;
- residual gas scattering;
- scattering with black body radiation in the vacuum chamber;
- quantum excitation.

The surprise is that what comes first to limit the shrinking beam size is synchrotron radiation itself. And this leading offender is a *quantum* effect, proportional to $\hbar$. When taken into account of quantum excitations, the beam sizes (length, height, width, energy spread) all become nonzero. And furthermore, in spite of the fact that this effect depends on $\hbar$, it dominates over all the other mechanisms listed above which do not depend on $\hbar$.

In the discussion earlier, we compared a classical picture and a quantum mechanical picture of synchrotron radiation, Fig. 6.25. A classical picture describes synchrotron radiation as a continuous process, while the quantum mechanical picture views the radiation consisting of many stochastic sudden emission events. It turns out that nature has chosen the quantum mechanical picture, and needless to say, there are important consequences — consequences involving $\hbar$.

Excitation of synchrotron motion One consequence is the quantum excitation of synchrotron oscillations. Consider an ideal synchronous particle that at time $t = 0$ emits a sudden photon with energy u. Its energy gets a recoil $-u$, and starting from $t = 0$, it begins to execute a synchrotron oscillation with frequency ω_s. Because of radiation damping, however, this oscillation damps out in time τ_z. At a certain time $t = t_1$ later, this same electron emits another photon, uncorrelated to the previous emission. Another excitation-damping cycle resumes. The process repeats with the photon energies and the emission times basically random except for some average behavior. The situation is depicted in Fig. 6.33.

In Fig. 6.33, the photon energy of each emission event is random, but they are all of the order of (in most cases somewhat smaller than) the critical energy u_c. The emission times are also random, but with an average duration t_1.

Time scale hierarchy We know that synchrotron oscillation is slow compared with the revolution frequency. However, it is fast compared with radiation damping. This means in Fig. 6.33,

$$\tau_z \gg \frac{2\pi}{\omega_s},$$

as reflected in the figure.

Figure 6.33: Schematic of quantum excitation process of an electron emitting synchrotron radiation photons. Each photon is emitted with a random energy at a random time, obeying only a constraint on the average radiation power. Quantities u, ω_s, t_1, τ_z are indicated.

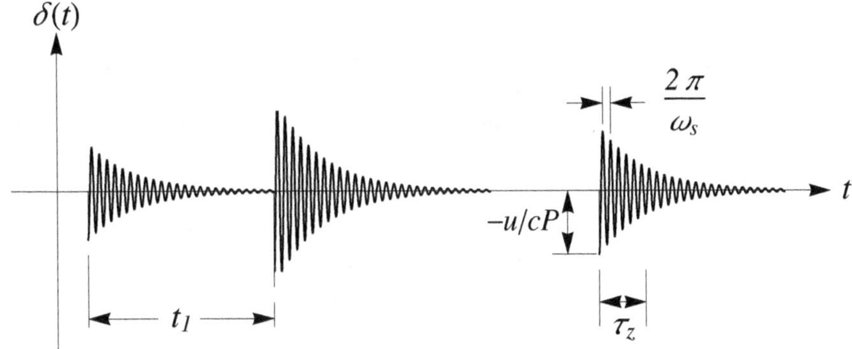

The average duration t_1 between emissions, however, has been greatly exaggerated in Fig. 6.33. An estimate of t_1 is

$$t_1 \sim \frac{1}{\dot{\mathcal{N}}} = \frac{T_0}{\mathcal{N}_0} = \frac{T_0}{\gamma/15},$$

where $\mathcal{N}_0 = \frac{\gamma}{15}$ is the expected number of emissions per turn, Eq. (6.27). This is an extremely short time, and that means the randomly timed excitation-damping cycles overlap grossly, and not at all as cleanly separated as sketched in Fig. 6.33.

There is a hierarchy of time scales in an electron storage ring as follows,

$$\tau^{(\mathrm{q})}_{x,y,z} \gg \tau_{\mathrm{pol}} \gg \tau_{x,y,z} \gg \frac{2\pi}{\omega_s} \gg T_0 \gg \frac{2\pi}{\omega_{x,y}} \gg t_1 \gg \frac{\rho}{c\gamma}, \qquad (6.36)$$

where τ_{pol} is the spin polarization time (6.29) we did not cover, $\tau^{(\mathrm{q})}_{x,y,z}$ are quantum lifetimes to be discussed later, $\frac{\rho}{c\gamma}$ is the photon emission formation time. The reader should appreciate that our treatments all along implicitly assume this time hierarchy. If any of these time scales switch around, for example, it is likely to substantially affect the validity of some of our treatments.

Protons would have less number of relevant time scales. This is in fact one fundamental difference between electrons and protons.

The reader should appreciate the fact that, as a general rule of thumb, every time a new time scale or a length scale is reached, a new platform becomes available for the access to new physics. We shall mention a length hierarchy in accelerator physics later.

Rms energy spread As grossly overlapped as these excitation-damping cycles, however, if all emission events are uncorrelated, we can still imagine a superposition of them piling up. The net result is that the particle will be constantly being excited and damped, and an equilibrium will be reached. At this equilibrium, the particle will acquire an rms energy spread from a balance between quantum excitation on one hand and radiation damping on the other.

Roughly, the number of excitation-damping cycles going on at any given time is $\sim \frac{\tau_z}{t_1}$, which easily reaches the level of 10^5. Since each excitation gives an amplitude $\sim u_c$, we expect that the rms equilibrium energy spread of the particle to be

$$\sigma_E \sim u_c\sqrt{\frac{\tau_z}{t_1}} ,$$

where the square root is of course characteristic of the random walk process. Each excitation is small, of the order of u_c, but randomly accumulating a large number of them can give a sizable σ_E. Note that u_c is proportional to $\hbar$, and t_1 actually also contains $\hbar$ in it through the fine structure constant. The reader is suggested to check how σ_E will scale with $\hbar$ at this junction.

The energy deviation as a function of time of a particle is

$$\delta(t) = \sum_{i=-\infty\,(t>t_i)}^{\infty} \left(\frac{-u_i}{E_s}\right) e^{-(t-t_i)/\tau_z} \cos\omega_s(t-t_i) ,$$

where u_i is the energy of the i-th photon.

The rms equilibrium energy spread is calculated as follows,

$$\begin{aligned}\sigma_\delta^2 &= \langle \delta^2(t) \rangle \\ &= \left\langle \sum_{i,j\ (t>t_i,t>t_j)} \left(\frac{u_i u_j}{E_s^2}\right) e^{-(t-t_i)/\tau_z} \cos\omega_s(t-t_i) e^{-(t-t_j)/\tau_z} \cos\omega_s(t-t_j) \right\rangle ,\end{aligned}$$

where the brackets $\langle\ \rangle$ is somewhat loosely defined as averaging over photon emission statistics. First, since the emission times are uncorrelated — we showed before that this statement was 98% accurate — we replace

$$\langle \cos\omega_s(t-t_i) \cos\omega_s(t-t_i) \rangle \rightarrow \frac{1}{2}\delta_{ij} ,$$

because unless $i=j$, the averaging contributes zero, and when $i=j$, the term averages to $\frac{1}{2}$. We then have

$$\sigma_\delta^2 = \left\langle \sum_{i\ (t>t_i)} \frac{u_i^2}{2E_s^2} e^{-2(t-t_i)/\tau_z} \right\rangle .$$

We next make the replacement

$$\left\langle \sum_{i\ (t>t_i)} e^{-2(t-t_i)/\tau_z} \right\rangle \rightarrow \dot{\mathcal{N}} \int_{-\infty}^{t} dt_i e^{-2(t-t_i)/\tau_z} = \dot{\mathcal{N}} \frac{\tau_z}{2} , \tag{6.37}$$

where $\dot{\mathcal{N}}$ is the rate of photon emissions. We then obtain the final result for the rms equilibrium energy spread of a particle,

$$\sigma_\delta^2 = \frac{\dot{\mathcal{N}}\langle u^2\rangle\tau_z}{4E_s^2}. \tag{6.38}$$

A few comments are in order.

- Although we started by averaging $\delta^2(t)$, which formally is a function of time t, the final result is independent of t. This is because although the instantaneous value of δ of a particle fluctuates in time, its rms expectation value is constant in time. In case we have a beam of particles, this rms energy spread will in fact be the rms energy spread of the whole beam at all times although individual particles' instantaneous energies fluctuate.
- Watch out for a trick being played just now. We started with taking a *time average* of the instantaneous energy of a single particle to find σ_δ^2 of the particle. We then quietly made a transition to assign it the meaning of an *ensemble average* of 10^{10} particles at a snap-shot of time 偷天換日. The justification for this is due to the ergodic theorem: in a truly statistical system, time-averaging is equal to ensemble-averaging. For example, the averaging over tossing a coin 100 times (time averaging) is assumed to yield the same result as tossing 100 coins once (ensemble averaging).
- Also in the above expression, we dropped the subscript i in u_i with the understanding that the last brackets $\langle\ \rangle$ mean averaging over the synchrotron radiation spectrum. The reader might get the feeling that the "time averaging" we just performed here is rather extensive, hopefully well guided by physical intuition (Homework 6.42).
- Observing $t_1 \sim \frac{1}{\dot{\mathcal{N}}}$, we have established the earlier expectation that $\sigma_E \sim u_c\sqrt{\frac{\tau_z}{t_1}}$ by Eq. (6.38).

The quantity $\dot{\mathcal{N}}\langle u^2\rangle$ in Eq. (6.38) is a key quantity in quantum excitation effects. We introduced it in Eq. (6.28), reproduced below,

$$\dot{\mathcal{N}}\langle u^2\rangle = \frac{55}{24\sqrt{3}}\frac{e^2\hbar c^2}{4\pi\epsilon_0}\frac{\gamma^7}{\rho^3}.$$

This quantity is proportional to $\hbar$.

In spite of being proportional to the small quantity $\hbar$, however, σ_δ^2 contains a large factor γ^7. For relativistic beams, therefore, the rms energy spread *can* be macroscopic. The rms energy spread and rms beam sizes in an electron storage ring are therefore some of the relatively rare examples of macroscopic – visible to naked eyes – quantum effects. The macroscopic nature is also helped by the fact that the rms beam sizes are proportional to $\sqrt{\hbar}$, not $\hbar$.

There is another subtle point to make here. The reader might wonder how come we have been using exclusively the quantity $\langle u^2\rangle$ in our calculations, and not the quantity $\langle (u-\langle u\rangle)^2\rangle = \langle u^2\rangle - \langle u\rangle^2$ more commonly seen in other statistical analysis for the rms values. This is curious especially because synchrotron radiation has only one sign, i.e. u is always positive and therefore the initial kick in each excitation-damping cycle is always a negative one, and $\langle u\rangle \neq 0$.

The answer to this question lies in the fact that a photon-emission excites a rapid oscillation, and its excitation amplitude is u, not $u - \langle u\rangle$. This is true even with the fact that the average energy loss $\langle u\rangle$ will be replenished at the RF cavity. To help consideration of this point, imagine a synchrotron radiation in which all emitted photons have the same energy. In such a case, $u - \langle u\rangle = 0$ and yet the particle energy will still be quantum excited by these emissions. The reader is encouraged to think through this subtlety. The time hierarchy (6.36) should be kept in mind.

Substituting in expressions of $\dot{\mathcal{N}}\langle u^2\rangle$ and noting

$$\begin{aligned} \tau_z^{-1} &= \left(1+\frac{\mathcal{D}}{2}\right)\frac{cC_\gamma}{2\pi}\frac{E_s^3}{\rho^2}, \qquad C_\gamma = \frac{4\pi}{3}\frac{r_0}{(mc^2)^3} \\ \Longrightarrow \quad \tau_z^{-1} &= (2+\mathcal{D})\frac{cr_0}{3\rho^2}\gamma^3, \end{aligned}$$

we obtain

$$\begin{aligned} \sigma_\delta^2 &= \frac{55}{32\sqrt{3}}\frac{\hbar}{mc}\frac{\gamma^2}{2+\mathcal{D}}\frac{\oint \frac{ds}{|\rho|^3}}{\oint \frac{ds}{\rho^2}} \\ &= \left[3.83\times 10^{-13}\,\mathrm{m}\right]\frac{\gamma^2}{2+\mathcal{D}}\frac{\oint \frac{ds}{|\rho|^3}}{\oint \frac{ds}{\rho^2}}, \end{aligned}$$

where we have introduced an averaging over the circumference, taking into account now that ρ is actually a function of s. Considering the possibility of some reversed bending, we also introduced an absolute value for $\rho(s)$. The numerator $\oint \frac{ds}{|\rho|^3}$ comes from quantum excitation; the denominator $\oint \frac{ds}{\rho^2}$ comes from radiation damping. Their ratio, indicating a balancing between their opposing effects, determines the rms equilibrium energy spread.

Putting this comment explicitly, when these two opposing effects are in balance,

$$\text{equilibrium beam emittance} = \frac{\text{quantum excitation}}{\text{radiation damping}} \propto \frac{\dot{\mathcal{N}}\langle u^2\rangle}{\tau_{x,y,z}^{-1}}.$$

This formula applies to all three beam dimensions.

For an isomagnetic ring, we have

$$\sigma_\delta^2 = \frac{55}{32\sqrt{3}}\frac{\hbar}{mc}\frac{\gamma^2}{(2+\mathcal{D})\rho}. \tag{6.39}$$

Take for example, an electron storage ring with $E_s = 5$ GeV, $\rho = 30$ m, and ignore $\mathcal{D}$, we find $\sigma_\delta = 0.78 \times 10^{-3}$. This energy spread is large enough to accommodate a reasonable beam within it. It is also small enough that it is not too painstaking to design a storage ring optics whose momentum aperture can accommodate the beam. In a sense, we lucked out again! As mentioned, σ_δ assumes a macroscopic value in spite of the factor of $\sqrt{\hbar}$.

The reader is alerted to one observation: Regardless of the fact that we are dealing with the longitudinal dynamics, the energy spread we obtained is independent of the RF system. It is independent of the RF voltage, the synchronous phase, the synchrotron frequency, the harmonic number, or even the non-RF phase slippage factor. Energy spread is therefore an intrinsic property to the lattice optics design. As we vary the RF voltage in a storage ring, the prime player of longitudinal dynamics, its energy spread remains unchanged. We shall continue this discussion in the next section discussing the bunch length.

Limit to storage ring energy In Eq. (6.39), the left-hand-side contains the rms of the *relative* energy spread instead of the energy spread itself. In practice, indeed, it is the relative energy spread that often matters in the performance of an accelerator. We then note that the right-hand-side contains mainly $\frac{\gamma^2}{\rho}$. After a storage ring is built, ρ is fixed, we find that as we vary the beam energy in this given accelerator (as long as γ is less than some maximum value $\hat{\gamma}$ determined by the magnet strengths), its relative energy spread will scale like

$$\sigma_\delta \propto \gamma \qquad (\gamma \leq \hat{\gamma}),$$

and the maximum relative energy spread $\hat{\sigma}_\delta$ occurs when the storage ring is operated at its maximum design energy.

Pursuing along this line further, when we design an electron storage ring that aims for a given highest reachable energy $\hat{\gamma}$, we should choose an optimal bending radius that scales with

$$\rho \propto \hat{\gamma}^2,$$

so that the energy spread of the beam stays within a practical tolerable level (most likely $\sim 10^{-3}$). It follows that, once this optimal ring radius is chosen, $\hat{\sigma}_\delta$ of the beam is almost a universal constant, regardless of the design energy of the ring, and regardless of the size of the storage ring. This interesting scaling property will be taken on again later in Sec. 6.4.14.

It is also a curious fact that this scaling of ring size with energy obtained from a practical consideration of acceptable beam energy spread coincides with that obtained by cost optimization consideration introduced in Eq. (6.25).

Homework 6.41 Show that the rms energy spread σ_E can also be expressed as the geometric mean between the beam energy E_s and the critical photon energy u_c. This exercise reemphasizes the fact that typically $E_s \gg \sigma_E \gg u_c$, and we have here actually an energy hierarchy, except that it is small scale compared with the time and length hierarchies mentioned in the text.

Solution Ignoring $\mathcal{D}$,

$$\sigma_E^2 \approx \frac{55}{96\sqrt{3}} u_c E_s .$$

Homework 6.42 Convince yourself of the replacement (6.37) and then verify (6.38). A careful reader will note that the brackets in Eq. (6.37) represents a four-fold averaging: (i) averaging over the fast synchrotron oscillation with frequency ω_s, (ii) averaging over the slower damping time τ_z, (iii) averaging over the synchrotron radiation spectrum, and finally (iv) averaging over the circumference around the storage ring. In comparison, the brackets in Eq. (6.38) represents averaging only in (iii) and (iv) after the other two averaging have been taken. We have done some hand-waving along the way.

Homework 6.43 Show that, for an isomagnetic storage ring, and ignoring $\mathcal{D}$,

$$\frac{\sigma_E^2}{U_0} = \frac{165}{256\pi\sqrt{3}} \frac{mc^2}{\alpha_F} = 16.23\, mc^2 ,$$

where $\sigma_E = \sigma_\delta E_s$, and α_F is the fine structure constant.

Solution What physics message is behind this homework? Consider a beam-line that contains dipoles evenly distributed along the way without RF. As an electron propagates along its path, it radiates, and its radiated energy U_0 is proportional to the path length it traverses. On the other hand, its energy spread σ_E, being a random walk process, grows as the square root of the path length. If we apply this picture to a storage ring, it leads to a question of what determines the ratio σ_E^2 to U_0. Indeed, in this homework, one finds that the ratio is a fundamental constant.

Homework 6.44 We had a figure-8 storage ring and a wiggler storage ring suggested in Homework 6.38. Find $\tau_x, \tau_y, \tau_z, \sigma_\delta$ for these designs. Set $\mathcal{D}$ to zero but first examine its behavior in these rings.

6.4.8 Bunch length

Rms bunch length We mentioned that the energy spread σ_δ basically is a universal constant, or at least it does not change with the RF parameters. What changes with the RF parameters is the rms bunch length – the other longitudinal coordinate – according to [e.g. Eq. (5.40)]

$$\sigma_z = \frac{\beta_s c|\eta|}{\omega_s}\sigma_\delta \approx \frac{c|\eta|}{\omega_s}\sigma_\delta . \tag{6.40}$$

This relationship holds because it is obeyed by each of the excitation-damping cycles. These cycles are triggered by δ-deviations, and then carried over to z-deviations through synchrotron oscillation.

The limit when $|\eta| \to 0$ Equation (6.40) gives the rms bunch length to a good approximation in most practical cases. With $\omega_s \propto \sqrt{|\eta|}$, it then seems to indicate that by reducing the slippage factor $|\eta|$ of the storage ring lattice — i.e. to make the storage ring *isochronous* — the bunch length can be reduced to as small a value as desired. This however, is not the case.

There are many reasons why the bunch length does not become zero when $|\eta| \to 0$. One of them is due to nonlinear phase slippage mentioned in Sec. 5.3.12. In this section, however, we ignore any nonlinear effects. Even in that case, we will show that the bunch length does not vanish when $|\eta| \to 0$.

The reason that the nominal bunch length reduces to zero when $\eta = 0$ is that for a particle executing synchrotron oscillation, its motion is being analyzed using the adiabatic approximation when the time unit is considered to be the turn numbers and the motion within each turn is ignored. When completing one revolution, and when $\eta = 0$, the particle's longitudinal movement vanishes, and the bunch length becomes zero. In reality, however, $\eta = 0$ only applies when the particle motion is integrated over one complete turn. Within each revolution, the particle's longitudinal position does move. We need to analyze this *partial momentum compaction* effect more carefully.[26]

Consider an observation point at s_2 where it has a horizontal β-function β_2 and dispersion function D_2. Let the ring have a momentum compaction factor $R_{56} = -\eta C$. Let there be a photon emission of energy u at a location s_1 with horizontal β-function β_1 and dispersion D_1. Let us assume $D_1' = D_2' = 0, \alpha_1 = \alpha_2 = 0$ for simplicity. See Homework 6.45 for the general case.

Consider an electron that moves along the design orbit, and it now emits a photon at s_1. After the photon emission, the electron's subsequent motion is disturbed, and we calculate its z-coordinate as it circulates around the ring turn after turn.

We first ignore the RF, i.e. no synchrotron focusing. The z-coordinate in the first turn is, using Eq. (5.51),

$$z_0 = -\frac{u}{E_s}\left(-\Delta C + \frac{D_1 D_2}{\sqrt{\beta_1 \beta_2}} \sin\psi\right),$$

where ψ is the horizontal betatron phase advance from s_1 to s_2, and

$$\Delta C = \int_{s_1}^{s_2} ds\, \frac{D(s)}{\rho(s)}$$

is the partial momentum compaction.

[26] Y. Shoji, H. Tanaka, M. Takao, and K. Soutome, Phys. Rev. E 54, R4556 (1996); Y. Shoji, Nucl. Instr. Meth. Phys. Res. A637, S120 (2011); X.J. Deng et al., Phys. Rev. Accel. & Beams, 23, 044002 (2020).

In the subsequent revolutions, it is easy to show that, for the n-th revolution,

$$z_n = -\frac{u}{E_s}\left(nR_{56} - \Delta C + \frac{D_1 D_2}{\sqrt{\beta_1\beta_2}}\sin(n\mu+\psi)\right), \tag{6.41}$$

where $\mu = 2\pi\nu$ is the phase advance per revolution.

The first term describes the conventional synchrotron oscillation when RF is added. It contributes to the conventional bunch length (6.40). The second term makes an additional contribution due to the partial momentum compaction between s_1 and s_2. The third term comes from a betatron oscillation coupled to path length. Usually the first term dominates. But that is no longer true when $R_{56} = 0$ for an isochronous ring.

The dominating first term is proportional to n, i.e. it continues to drift turn after turn and requires the RF to focus it back. The second term is a static shift. It does not require the RF to focus it. But its value is different depending on where the photon is emitted and depending on the size of the emitted photon. A spread in z, after summing over all photon emission locations and photon spectrum, will result in a finite bunch length. The third term oscillates in time. It averages to zero and will make a significant contribution only near a synchrobetatron resonance.

Consider the case of perfect isochronous ring with $R_{56} = 0$. The z coordinate of the electron sits at $z = \frac{u}{E_s}\Delta C$ plus an oscillatory term with amplitude $\frac{u}{E_s}\frac{D_1 D_2}{\sqrt{\beta_1\beta_2}}$. Even when $R_{56} = 0$, there is therefore a finite excitation of the longitudinal length. Piling up random noises over n revolutions, even with $R_{56} = 0$, we obtain a bunch length diffusion at the observation point with

$$\frac{d}{dn}(\Delta\sigma_z^2) = \frac{\langle u^2\rangle}{E_s^2}\left(\Delta C^2 + \frac{D_1^2 D_2^2}{2\beta_1\beta_2}\right),$$

where $\langle\ \rangle$ means averaging over the synchrotron radiation spectrum. It is understood that there is another averaging over the photon emission locations around the ring.

The first term containing ΔC^2 comes from a tilt in the phase space and can be removed by an artificial $r_{56} = \Delta C$. It is not intrinsic in that sense. However, photons are emitted at different locations with different values of ΔC. So only its average value can be removed, and there will remain a residual contribution even after the average tilt is removed.

There is also the second term containing $\frac{D_1^2 D_2^2}{2\beta_1\beta_2}$. It oscillates from turn to turn, and cannot be removed by an optics design. However, its value is typically much smaller than ΔC^2, and in any case can be removed if $D_2 = 0$ at the observation point.

When radiation damping is included, following a step similar to Eq. (6.38), a balance between radiation damping and quantum excitation leads to

$$\sigma_{z,\mathrm{tot}}^2 = \sigma_z^2 + \left(\Delta C^2 + \frac{D_1^2 D_2^2}{2\beta_1\beta_2}\right)\sigma_\delta^2 .$$

where σ_z^2 is the nominal bunch length given by Eq. (6.40), and σ_δ^2 is the nominal energy spread given by (6.38).

At equilibrium, this additional ΔC contribution to rms bunch length is typically small compared with the nominal rms bunch length. An order of magnitude estimate gives their typical ratio,

$$\frac{\sigma_\delta \Delta C}{\sigma_z} \sim \frac{2\pi\nu_s}{\sqrt{3}} , \tag{6.42}$$

but again this is no longer the case if $\eta \to 0$.

When RF and synchrotron oscillation are added, the bunch length will be carried by synchrotron oscillation into a contribution to the equilibrium energy spread $\sigma_{\delta,\mathrm{tot}}$ according to

$$\sigma_{\delta,\mathrm{tot}} = \frac{\omega_s}{c|\eta|}\sigma_{z,\mathrm{tot}} ,$$

causing it to increase beyond the conventional prediction. In particular, when $\eta = 0$ for a truly isochronous ring, the energy spread becomes infinite. Physically, this is because the partial momentum compaction generates a finite bunch length at the RF cavity. When $\eta = 0$, the longitudinal location of each electron is frozen; the energy gain or loss at the RF cavity for each particle continues turn after turn without synchrotron focusing, leading to a diverging energy spread. In this case, nonlinear phase slippage becomes critical to control the energy spread.

A self-consistent calculation of both σ_z and σ_δ can be obtained using the 3-D SLIM formalism of Sec. 6.5. Figure 6.34 shows an example calculation that compares the result using a computer simulation according to the discussions of this section, the result using the nominal formulae (6.40) and (6.38) based on the Sands formalism, and that using the SLIM algorithm.

Homework 6.45

(a) Derive Eq. (6.41).

(b) Extend Eq. (6.41) to the case when $\alpha_{1,2}$ and $D'_{1,2}$ are nonzero.

Solution (b)

$$\begin{aligned} z_n &= -\frac{u}{E_s}\left(nR_{56} - \Delta C + \frac{Z}{\sqrt{\beta_1\beta_2}}\right) , \\ Z &= D_1 D_2 \sin(n\mu+\psi) - (\alpha_1 D_1 + \beta_1 D_1')D_2\cos(n\mu+\psi) \\ &+ (\alpha_2 D_2 + \beta_2 D_2')\cos(n\mu+\psi) - (\alpha_1 D_1 + \beta_1 D_1')(\alpha_2 D_2 + \beta_2 D_2')\sin(n\mu+\psi) . \end{aligned}$$

Figure 6.34: Calculations of the rms bunch length and energy spread in a storage ring in the neighborhood of $|\eta| = 0$. The green curves are predictions using the nominal Sands formalism. The simulation and the SLIM results are marked as blue and red dots [Courtesy Xiujie Deng (2019)]. The sample storage ring uses that of the Metrology Light Source at Berlin. The nonzero bunch length as well as the divergent energy spread at $\eta = 0$ are explained in the text.

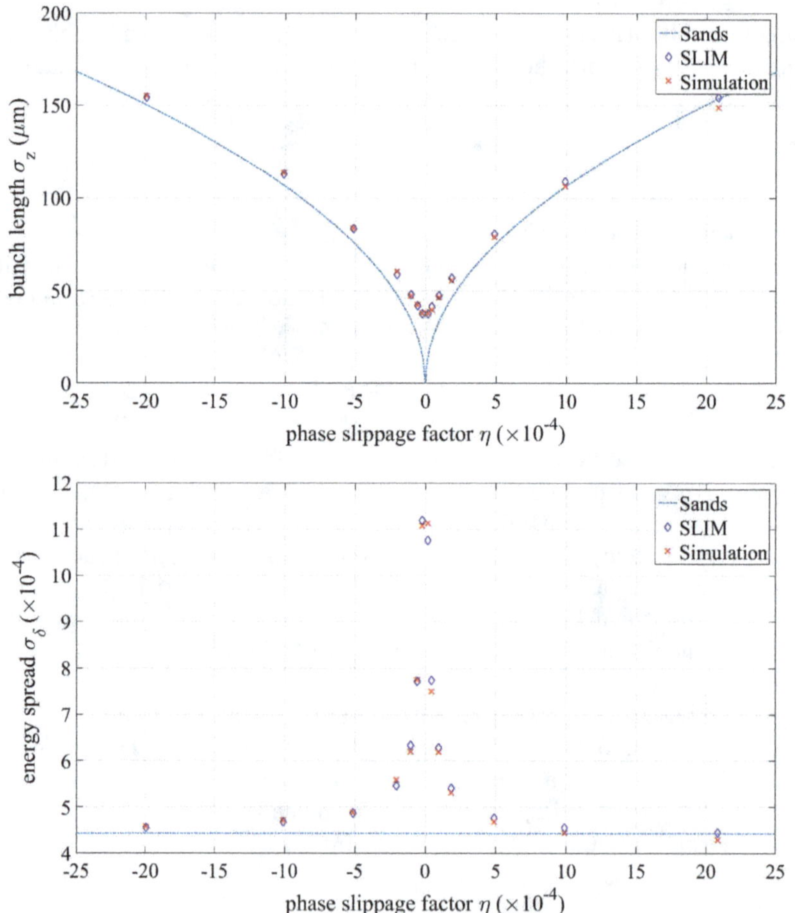

Homework 6.46

(a) Convince yourself of the statement made in the text that typically

$$\Delta C \gg \frac{D_1 D_2}{\sqrt{\beta_1 \beta_2}}\,.$$

(b) Fill in the steps to show the isomagnetic estimate Eq. (6.42).

6.4.9 Central limit theorem

The intention here is not to prove the *central limit theorem*, but to discuss its implication for our beam size and energy spread calculations.

We have learned that the synchrotron motion of a particle in a storage ring is driven by a large number of uncorrelated random quantum excitations. Central limit theorem says that, in this case, and considering linear optics only, the expected beam distribution in the (z, δ) phase space is going to be Gaussian.

The central limit theorem is quite amazing. It says that, regardless of how wild the spectrum of the single events may be or actually is, one invariably obtains a Gaussian distribution when sufficiently many events are accumulated. This theorem can be realized by thinking about the process of a pin-ball machine, that of tossing coins, or quantum excitation by synchrotron radiation photons.

In the absence of damping, the expected rms spreads σ_z and σ_δ of this Gaussian beam will grow with time as $\sqrt{t}$. With damping, the growths will be limited and they will reach an equilibrium. We have in the previous section calculated these equilibrium rms values. What the central limit theorem now tells us in addition is that the whole distribution in the (z, δ) space is Gaussian. This is a powerful theorem, because knowing the rms values, we now know everything there is to know about the distribution. Each distribution has one and only one parameter to control it,

$$\begin{aligned} \psi(z,\delta) &= N\psi_z(z)\psi_\delta(\delta)\,, \qquad (6.43)\\ \psi_z(z) &= \frac{1}{\sqrt{2\pi}\,\sigma_z}e^{-z^2/2\sigma_z^2}\,,\\ \psi_\delta(\delta) &= \frac{1}{\sqrt{2\pi}\,\sigma_\delta}e^{-\delta^2/2\sigma_\delta^2}\,,\\ \sigma_z &= \frac{c|\eta|}{\omega_s}\sigma_\delta\,, \end{aligned}$$

where N is the number of particles in the beam. Normalization is

$$\int_{-\infty}^{\infty} dz \int_{-\infty}^{\infty} d\delta\, \psi(z,\delta) = N\,.$$

In general, a distribution can be characterized by all its distribution moments. The rms values σ_δ and σ_z are only two of its second moments. There are an infinitely many more higher-order moments to characterize completely a distribution in general. For a Gaussian distribution here, once the first and the second moments are given, all higher moments are automatically specified.

The reader should now note that the second moments of this Gaussian distribution is solely determined by *one* distribution parameter $\dot{\mathcal{N}}\langle u^2\rangle$. In other words, although synchrotron radiation has a complicated radiation spectrum (which we painstakingly calculated), the one single parameter of this spectrum that determines the beam distribution is just $\dot{\mathcal{N}}\langle u^2\rangle$.

Had synchrotron radiation had a completely different spectrum, it will give the same Gaussian beam distribution as long as this different spectrum has the same $\dot{\mathcal{N}}\langle u^2\rangle$. And this is why in computer simulations, it is not critical how exactly the photon energies are generated or at what locations the photons are emitted. The reader should imagine the pain if we had to generate a simulation program that follows in detail the photon spectrum and simulate where each photon is to be radiated. We now avoid all of that, thanks to the central limit theorem.

With some exaggeration, a uniform square photon spectrum emitted at a single appropriately chosen location with appropriate β-functions and dispersion function in the ring, for example, is often just as good as any other more sophisticated choices as long as it gives the right $\dot{\mathcal{N}}\langle u^2\rangle$. This somewhat surprising conclusion testifies the power of the theorem.

Perhaps a word of caution however should be said here. If we indeed choose to simulate the synchrotron radiation events by lumping the radiation to few locations around the ring, it would actually be advisable not to simulate using the true synchrotron radiation spectrum. An attempt to do a "better job" using a true synchrotron radiation spectrum could yield misleading results because the synchrotron radiation spectrum allows accidentally very strong photons be emitted with nonnegligible probability. The result of imposing a true synchrotron radiation spectrum together with lumping the emission events will lead to a gross over-estimate of very large photons, which in turn could exaggerate a quantum lifetime problem, even though the rms distribution remains correct.

Having stated that, one might still wonder what effects do the higher moments of the radiation spectrum such as $\dot{\mathcal{N}}\langle u^n\rangle$ have on the beam distribution. In fact, they do potentially affect the beam distribution and distort it from a strict Gaussian. Suffice it to say here, however, that these effects occur only for large amplitude particles. The distortion from Gaussian appears only in the tails of the distribution, and the distortion is small. On the other hand, there are effects, such as the quantum lifetime effects to be discussed later, that are sensitive to beam tail distributions. For those effects, it might be necessary to examine the higher moments and deviation from central limit theorem more carefully if some accuracy is desired. A way to analyze the problem using Fokker-Planck equation is outside the scope of these lectures.

Longitudinal beam distribution The bi-Gaussian distribution (6.43) is an upright elliptical distribution in the (z,δ) phase space. The reason it is upright is because we made an adiabatic approximation that smoothed out the RF focusing. In general, the elliptical Gaussian is tilted, and there is another second moment $\langle z\delta\rangle \neq 0$. We have ignored this tilt here.

It is sometimes useful to transform the phase space from (z,δ) to (ϕ, A) as illustrated in Fig. 6.35. The synchrotron oscillation amplitude A is defined with

$$A^2 = \delta^2 + \left(\frac{\omega_s}{c\eta}\right)^2 z^2 .$$

Figure 6.35: Coordinate transformation from (z, δ) to (ϕ, A).

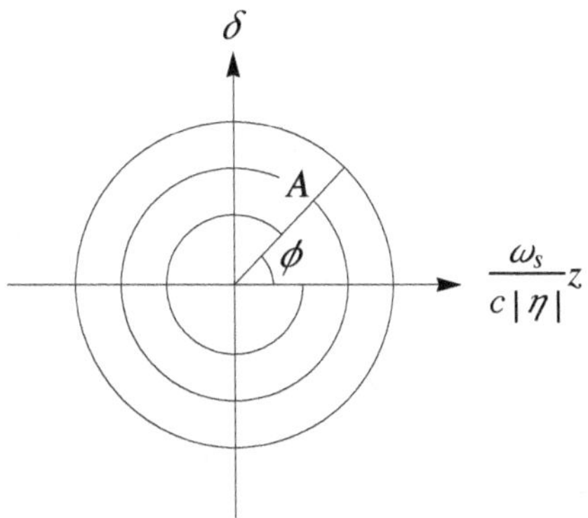

The equilibrium Gaussian distribution in the (A, ϕ) phase space is (using $dzd\delta = \frac{c|\eta|}{\omega_s} AdAd\phi$)

$$\psi(A, \phi) = \frac{N}{2\pi\sigma_\delta^2} e^{-A^2/2\sigma_\delta^2},$$

which is independent of ϕ, and integrates with normalization

$$\int_0^{2\pi} d\phi \int_0^\infty AdA\, \psi(A, \phi) = N.$$

We can also integrate over ϕ to obtain a distribution in A,

$$\psi(A) = \frac{N}{\sigma_\delta^2} e^{-A^2/2\sigma_\delta^2},$$

with normalization

$$\int_0^\infty AdA\, \psi(A) = N.$$

We can calculate the rms values of the amplitude,

$$\begin{aligned}
\langle \delta^2 \rangle &= \frac{1}{N} \int_{-\infty}^{\infty} dz \int_{-\infty}^{\infty} d\delta\, \delta^2 \psi(z, \delta) = \sigma_\delta^2, \\
\langle z^2 \rangle &= \frac{1}{N} \int_{-\infty}^{\infty} dz \int_{-\infty}^{\infty} d\delta z^2 \psi(z, \delta) = \sigma_z^2, \\
\langle A^2 \rangle &= \frac{1}{N} \int_0^\infty AdAA^2 \psi(A) = 2\sigma_\delta^2 = \langle \delta^2 \rangle + \left(\frac{\omega_s}{c\eta}\right)^2 \langle z^2 \rangle.
\end{aligned}$$

The amplitude A is an invariant of the motion. We point out that any equilibrium distribution of the beam (i) must not depend on time explicitly, and (ii) must be a function of the invariants of the motion. In this 1-D case, there is only one invariant, namely A, which means the equilibrium distribution has to be a function solely of A. Central limit theorem then takes over and dictates $\psi(A)$ to be Gaussian.

6.4.10 Quantum excitation of horizontal betatron oscillation

Quantum excitation Quantum emissions excite horizontal betatron oscillations. This was already analyzed earlier in Eq. (6.34) when we analyzed the horizontal betatron radiation damping effect, but at the time we dropped the second term. It is this second term that we now return to. We have, due to the emission of a photon of energy u, a change in amplitude A_x,

$$\Delta A_x^2 = \frac{u^2}{E_s^2}\mathcal{H},$$

where, following Sands, we have defined a function

$$\mathcal{H} = \frac{D^2 + (\beta_x D' + \alpha_x D)^2}{\beta_x},$$

with the lattice functions β_x, α_x, D, D' evaluated at the point where photon emission takes place. This $\mathcal{H}$-function resembles the Courant-Snyder invariant, except that it takes on the dispersion coordinates instead of the betatron coordinates. Quantum excitation of the beam emittance by synchrotron radiation comes about through the dispersion functions D and D'.

When discussing Eq. (6.34), we gave a numerical estimate, and found that its second term was ~six orders of magnitude smaller than the first term. It should be made a cautionary remark here when we drop the first linear term and keep only the second quadratic term in calculating the emittance quantum excitation in this section (linear and quadratic in reference to the photon energy). The linear term is dropped because it is oscillatory and does not accumulate from turn to turn. The quadratic term, although small, is kept because it accumulates. If we perform a simulation of the radiation process of a beam distribution going through a bending magnet, a section of beamline, or a revolution of a storage ring, and compute the apparent emittance in that passage, we are likely to find a large apparent "emittance growth". It must be kept in mind that this apparent growth is not a real growth. Most of it comes from the linear contribution and will wash out after accumulating many revolutions, and will not contribute to the equilibrium emittance of the beam distribution. Only the very small second quadratic term is of interest here. Furthermore, it takes many revolutions for the first linear term to average out before the second quadratic term begins to emerge. Simulators beware.

As it explicitly involves the horizontal dispersion function D, clearly it is applied when there is no vertical dispersion. The vertical dispersion can in fact be included by extending the analysis. However, a straightforward extension strictly does not apply when the vertical dispersion originates from coupling to the horizontal dimension. Furthermore, the application of $\mathcal{H}$ breaks down anyway when ν_s is not small or near a coupling resonance. How to extend this conventional analysis to the more general 3-D coupled cases is a very practical issue. We shall revisit this issue in Sec. 6.5.

The rate of change of the beam average $\langle A_x^2 \rangle$ is given by including a rate of emissions $\dot{\mathcal{N}}$ and averaging over the circumference $2\pi R$,

$$\frac{d\langle A_x^2 \rangle}{dt} = \frac{1}{2\pi R E_s^2} \oint ds \, \mathcal{H}\dot{\mathcal{N}} \langle u^2 \rangle \, .$$

Here again, only one parameter, i.e. $\dot{\mathcal{N}}\langle u^2 \rangle$ determines the entire horizontal betatron distribution from synchrotron radiation spectrum, and the central limit theorem says the equilibrium distribution shall be Gaussian.

The $\mathcal{H}$-function The function $\mathcal{H}$ has the following properties.

- It depends only on the storage ring lattice, and is a periodic function of s.
- It has dimensionality of length.
- It is always positive.
- There are two terms in $\mathcal{H}$. Usually the first term dominates. For a rough estimate, we have

$$\mathcal{H} \approx \frac{D^2}{\beta_x} \approx \frac{(R/\nu_x^2)^2}{R/\nu_x} = \frac{R}{\nu_x^3} \, .$$

- It has a physical meaning of a *coupling coefficient* that specifies how much does the recoil of a photon emission excite the horizontal betatron oscillation of the particle. Since photon emission recoils occur in particle's energy, the coupling coefficient requires the photon emission to occur in a location with finite dispersion. If $D = 0$ and $D' = 0$, then $\mathcal{H} = 0$ and photon emissions at this location will not couple to horizontal betatron motion. For most applications, this coupling is to be made the smaller the better.

Length scale hierarchy There is a hierarchy of lattice functions, in descending magnitudes of length, in a storage ring, as follows,

$$\begin{aligned} &R \, , \\ \beta_x &\approx \frac{R}{\nu_r} \, , \\ D &\approx \frac{R}{\nu_x^2} \, , \\ \mathcal{H} &\approx \frac{R}{\nu_x^3} \, , \end{aligned}$$

$$R \gg \beta_{x,y} \gg D \gg \mathcal{H} \, .$$

This hierarchy of length scales is to be read together with the hierarchy of time scales (6.36).

Rms horizontal betatron beam size When radiation damping is also taken into account, we have

$$\frac{d\langle A_x^2\rangle}{dt} = -\frac{2}{\tau_x}\langle A_x^2\rangle + \frac{1}{2\pi RE_s^2}\oint ds\,\mathcal{H}\dot{\mathcal{N}}\langle u^2\rangle\,. \tag{6.44}$$

This equation is useful to determine the time evolution of beam emittance in a storage ring under the influence of radiation damping and quantum excitation. For example, when a beam is injected into a storage ring with an emittance mismatch, then this equation can be used to calculate the time evolution of the beam emittance after injection as it oscillates and transients toward the final equilibrium value. If there are other mechanisms (such as those listed near the beginning of Sec. 6.4.7) that affect beam emittance, they can be included at this stage. For our purpose below, we will only look for the equilibrium emittance, and only for the case when radiation damping and quantum excitation are taken into consideration.

Equilibrium emittance is reached when the damping term and the diffusion term balance each other,

$$\frac{d\langle A_x^2\rangle}{dt} = 0 \qquad \Longrightarrow \quad \langle A_x^2\rangle = \frac{\tau_x}{4\pi RE_s^2}\oint ds\,\mathcal{H}\dot{\mathcal{N}}\langle u^2\rangle\,.$$

With

$$\tau_x^{-1} = (1-\mathcal{D})\frac{cr_0}{3\rho^2}\gamma^3\,,$$

the rms horizontal betatron beam size $\sigma_{x\beta}$ is found to be

$$\begin{aligned}\frac{\sigma_{x\beta}^2}{\beta_x} &= \frac{\langle A_x^2\rangle}{2} = \frac{\tau_x}{8\pi RE_s^2}\oint ds\,\mathcal{H}\dot{\mathcal{N}}\langle u^2\rangle \\ &= \frac{55}{32\sqrt{3}}\frac{\hbar}{mc}\frac{\gamma^2}{1-\mathcal{D}}\frac{\oint ds\frac{\mathcal{H}}{|\rho|^3}}{\oint\frac{ds}{\rho^2}} \\ &= \left[3.83\times10^{-13}\text{ m}\right]\frac{\gamma^2}{1-\mathcal{D}}\frac{\oint ds\frac{\mathcal{H}}{|\rho|^3}}{\oint\frac{ds}{\rho^2}}\,.\end{aligned} \tag{6.45}$$

The right hand side is independent of s. This means the rms beam size is proportional to $\sqrt{\beta_x(s)}$ as a function of s, as one would expect.

For an isomagnetic storage ring, we find

$$\begin{aligned}\frac{\sigma_{x\beta}^2}{\beta_x} &= \frac{55}{32\sqrt{3}}\frac{\hbar}{mc}\frac{\gamma^2\langle\mathcal{H}\rangle_{\text{dipole}}}{(1-\mathcal{D})\rho}\,, \\ \langle\mathcal{H}\rangle_{\text{dipole}} &= \frac{1}{2\pi\rho}\oint_{\text{dipole}} ds\,\mathcal{H}\,.\end{aligned}$$

where a subscript "dipole" means integrating only over the dipoles. Physically this of course is due to the fact that radiation occurs only there. Numerically it is also important because if the integration is incorrectly taken over the entire circumference, one is likely to obtain an answer a factor 2-3 off.

It is instructive to relate the horizontal betatron beam size to the beam energy spread as, for an isomagnetic storage ring,

$$\frac{\sigma_{x\beta}^2}{\beta_x} = \frac{2+\mathcal{D}}{1-\mathcal{D}} \langle \mathcal{H} \rangle_{\text{dipole}} \, \sigma_\delta^2 \,.$$

This expression illustrates clearly again that $\mathcal{H}$ has the physical meaning of a coupling coefficient from δ to x.

We showed that roughly $\mathcal{H} \approx \frac{R}{\nu_x^3}$. If we ignore $\mathcal{D}$, and let $\rho \approx R$, then we obtain a rough formula

$$\frac{\sigma_{x\beta}^2}{\beta_x} \approx \frac{55}{32\sqrt{3}} \frac{\hbar}{mc} \frac{\gamma^2}{\nu_x^3} = \left[3.83 \times 10^{-13} \text{ m}\right] \frac{\gamma^2}{\nu_x^3} \,.$$

Under the same conditions, we also have

$$\frac{\sigma_{x\beta}^2}{\beta_x} \approx \frac{2R}{\nu_x^3} \sigma_\delta^2 \,. \tag{6.46}$$

If we further take $\beta_x \approx \frac{R}{\nu_x}$, then

$$\sigma_{x\beta} \approx \sqrt{2} \frac{R}{\nu_x^2} \sigma_\delta \,.$$

In an earlier numerical example following Eq. (6.39) with $R = 30$ m and $E_s = 5$ GeV, we had $\sigma_\delta = 0.78 \times 10^{-3}$. If $\nu_x \approx 5$, then $\sigma_{x\beta} \approx 1.3$ mm. Like σ_δ, this beam size is again macroscopic on one hand and reasonably small to handle by an accelerator on the other hand.

The fact that the beam emittance is proportional to γ^2 in electron storage rings poses another limitation on the energy of a storage rings for synchrotron radiation or high energy collider applications.

FODO cell For synchrotron radiation facilities, it is desirable to design a special storage ring lattice that provides as bright a beam as possible, which translates to making the beam emittance as small as possible. Special lattices of this type include the *Chasman-Green lattice* (also called *double-bend-achromat*), *triple-bend-achromats*, etc. (Sec. 4.6) constitute an important research subject. Basically the idea is to design a lattice cell that contains a special arrangement of dipoles and quadrupoles in such a way that $\langle \mathcal{H} \rangle_{\text{dipole}}$ is minimized.

In the following, however, we will discuss only the simplest lattice design we have been familiar with, i.e. a storage ring lattice consisting of FODO cells. In particular, we assume thin-lens quadrupoles and thin-lens dipoles, from Sec. 4.2.2, with L the drift space length (total cell length $= 2L$), f the

Figure 6.36: The quantity $\langle\mathcal{H}\rangle_{\text{dipole}}$ as a function of the FODO-cell's horizontal betatron phase advance Φ. Dashed line indicates the value 1.

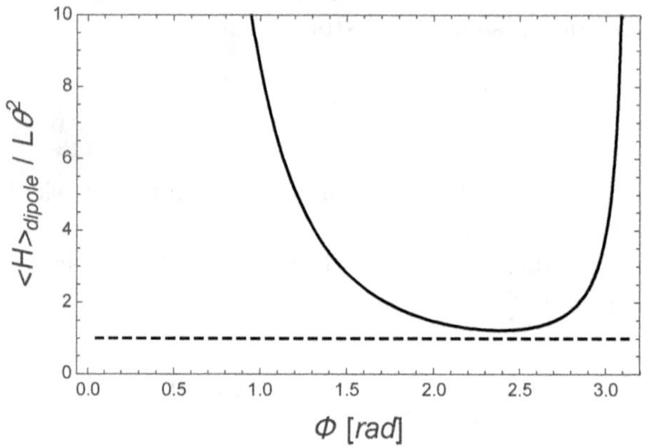

focal length of both the QF and QD quadrupoles, and Φ the betatron phase advance per cell. We ignore the weak focusing of the bending magnets.

Using the expressions of $D(s), D'(s), \alpha_x(s), \beta_x(s)$ from Sec. 4.2.2, we obtain, for the region $L > s > 0$,

$$\mathcal{H}_{\text{dipole}}(s) = \frac{L\theta^2}{4}\left[\left(1-\frac{s}{L}\right)^2\frac{2s^2}{L^2} - \frac{1-\frac{6s^2}{L^2}+\frac{4s^3}{L^3}}{\sin\frac{\Phi}{2}} - \frac{3}{\sin^2\frac{\Phi}{2}} + \frac{4}{\sin^4\frac{\Phi}{2}}\right]\tan\frac{\Phi}{2}.$$

We then obtain

$$\langle\mathcal{H}\rangle_{\text{dipole}} = 2L\theta^2\left(\frac{1-\frac{3}{4}\sin^2\frac{\Phi}{2}+\frac{1}{60}\sin^4\frac{\Phi}{2}}{\sin^2\frac{\Phi}{2}\sin\Phi}\right). \tag{6.47}$$

As a check, if we take $\Phi \ll 1, \Phi = \frac{2\pi\nu}{N}, \theta = \frac{2\pi}{2N}, L = \frac{2\pi R}{2N}$, then we confirm the rough value $\langle\mathcal{H}\rangle_{\text{dipole}} \approx \frac{R}{\nu^3}$.

Figure 6.36 shows $\frac{\langle\mathcal{H}\rangle_{\text{dipole}}}{L\theta^2}$ as a functions of Φ. A minimum of $\langle\mathcal{H}\rangle_{\text{dipole}}$, therefore a minimum of beam emittance, occurs around $\Phi = 137°$. (Note that this is nominally Φ_x, not Φ_y, although the FODO cell here assumed $\Phi_x = \Phi_y$.) In practice, such a large phase advance per cell may encounter difficulties in chromaticity corrections because the needed sextupole strengths become large.

Homework 6.47 Show that the $\mathcal{H}$ function is an invariant over any stretch of beamline that contains no bending magnets. In particular, it does not change going through a pure quadrupole.

Solution First show that

$$\frac{d\mathcal{H}}{ds} = \frac{2}{\rho}(\beta_x D' + \alpha_x D).$$

This means $\mathcal{H}$ is constant unless going through a dispersive bending magnet where $\frac{1}{\rho} \neq 0$ *and* either D or D' is nonzero.

Homework 6.48 Continue the previous homework. Show that

$$\frac{d\mathcal{H}}{ds} = -\frac{\sqrt{\beta_x(s)}}{\rho(s)\sin\pi\nu_x}\int_s^{s+C} ds'\frac{\sqrt{\beta_x(s')}}{\rho(s')}\sin[\pi\nu_x - \psi_x(s') + \psi_x(s)]\,.$$

Homework 6.49

(a) Derive Eq. (6.47).

(b) Plot the result to confirm Fig. 6.36.

(c) Show that, for the FODO cell lattice mentioned in the text,

$$\langle\mathcal{H}\rangle_{\text{dipole}} = \theta^2 L \times \begin{cases} \frac{781}{60\sqrt{3}}, & \Phi = 60^\circ, \\ \frac{151}{60}, & \Phi = 90^\circ. \end{cases}$$

Homework 6.50 This is a continuation from Homework 4.27. Show that for a weak focusing synchrotron with field index n $(0 < n < 1)$,

$$\begin{aligned}
\langle\mathcal{H}\rangle &= \frac{\rho}{(1-n)^{3/2}}, \\
\mathcal{D} &= \frac{1-2n}{1-n}, \\
\frac{\Delta\mathcal{D}}{\Delta\delta} &= \frac{2n^2}{(1-n)^2}, \\
\frac{\sigma_{x\beta}^2}{\beta_x} &= \frac{55}{32\sqrt{3}}\frac{\hbar}{mc}\frac{\gamma^2}{n\sqrt{1-n}}, \\
\sigma_\delta^2 &= \frac{55}{32\sqrt{3}}\frac{\hbar}{mc}\frac{\gamma^2}{\rho}\frac{1-n}{3-4n}, \\
\sigma_x^2 &= \sigma_{x\beta}^2 + D^2\sigma_\delta^2 = \frac{55}{32\sqrt{3}}\frac{\hbar}{mc}\frac{3\rho\gamma^2}{n(3-4n)}\,.
\end{aligned}$$

To fulfill the stability condition $1 > \mathcal{D} > -2$, we require

$$\frac{3}{4} > n > 0\,.$$

Homework 6.51 Denote a time dependent $\langle A_x^2\rangle$ by $\langle A_x^2\rangle(t)$. Denote the equilibrium value of $\langle A_x^2\rangle$ by $\langle A_x^2\rangle_{\text{eq}}$.

(a) Show that Eq. (6.44) can be rewritten as

$$\frac{d}{dt}\langle A_x^2\rangle(t) = -\frac{2}{\tau_x}\left(\langle A_x^2\rangle(t) - \langle A_x^2\rangle_{\text{eq}}\right).$$

(b) For a beam injected into the storage ring at $t = 0$ with $\langle A_x^2\rangle(0) = \langle A_x^2\rangle_{\text{in}}$, solve the above equation to find the time evolution of its horizontal beam emittance $\langle A_x^2\rangle(t)$.

(c) Show that when $t \to \infty$, $\langle A_x^2\rangle(\infty) = \langle A_x^2\rangle_{\text{eq}}$.

Solution This problem is particularly relevant for the consideration of a damping ring.

(b) $\langle A_x^2\rangle(t) \;=\; \langle A_x^2\rangle_{\text{in}}\, e^{-2t/\tau_x} + \langle A_x^2\rangle_{\text{eq}}\left(1 - e^{-2t/\tau_x}\right)$.

Homework 6.52 In Chapter 2, we mentioned a lattice scaling law in Sec. 2.4.5. Find out how do the following quantities scale with λ: $\mathcal{D}$, $\langle\mathcal{H}\rangle_{\text{dipole}}$, beam sizes $\sigma_{x,y}, \sigma_{x',y'}, \sigma_z, \sigma_\delta$.

6.4.11 Horizontal beam distribution

The equilibrium beam distribution in the horizontal betatron (x_β, x'_β) phase space is obliged to be a function of the one and only invariant of the 1-D system, i.e. the Courant-Snyder invariant

$$A_x^2 \;=\; \frac{1}{\beta_x}[x_\beta^2 + (\beta_x x'_\beta + \alpha_x x_\beta)^2]\,.$$

Furthermore, according to the central limit theorem, the equilibrium distribution should be Gaussian in x_β and x'_β, i.e. it must be of the form

$$\psi(x_\beta, x'_\beta) \;=\; \frac{N\beta_x}{2\pi\sigma_{x\beta}^2}\, e^{-\frac{x_\beta^2 + (\beta_x x'_\beta + \alpha_x x_\beta)^2}{2\sigma_{x\beta}^2}}\,,$$

$$\psi(A_x) \;=\; \frac{N\beta_x}{\sigma_{x\beta}^2}\, e^{-\frac{\beta_x A_x^2}{2\sigma_{x\beta}^2}}\,,$$

$$\psi(x_\beta) \;=\; \frac{N}{\sqrt{2\pi}\,\sigma_{x\beta}}\, e^{-\frac{x_\beta^2}{2\sigma_{x\beta}^2}}\,,$$

where we have imposed the normalization

$$\int_{-\infty}^{\infty} dx_\beta \int_{-\infty}^{\infty} dx'_\beta\, \psi(x_\beta, x'_\beta) \;=\; \int_0^\infty A_x dA_x \psi(A_x) \;=\; N\,.$$

This distribution refers to the horizontal betatron motion of the beam. The physical total horizontal displacement of a particle contains an additional contribution from synchrotron motion, i.e.

$$x \;=\; x_\beta + D\delta\,.$$

It follows that the rms of the total x-coordinate is determined by

$$\begin{aligned}\sigma_x^2 &= \sigma_{x\beta}^2 + D^2\sigma_\delta^2 \\ &= \frac{55}{32\sqrt{3}}\frac{\hbar}{mc}\frac{\gamma^2}{\oint \frac{ds}{\rho^2}}\left[\frac{\beta_x}{1-\mathcal{D}}\oint ds\frac{\mathcal{H}}{|\rho|^3} + \frac{D^2}{2+\mathcal{D}}\oint\frac{ds}{|\rho|^3}\right].\end{aligned}$$

For an isomagnetic ring, we have

$$\sigma_x^2 = \frac{55}{32\sqrt{3}}\frac{\hbar}{mc}\frac{\gamma^2}{\rho}\left[\frac{\beta_x\langle\mathcal{H}\rangle_{\text{dipole}}}{1-\mathcal{D}} + \frac{D^2}{2+\mathcal{D}}\right].$$

For a rough estimate, we substitute $\beta_x \approx \frac{R}{\nu_x}$, $D \approx \frac{R}{\nu_x^2}$, $\mathcal{H} \approx \frac{R}{\nu_x^3}$, $\rho \approx R$, and ignoring $\mathcal{D}$ (i.e. $|\mathcal{D}| \ll 1$), yielding

$$\begin{aligned}\sigma_x^2 &= \frac{55}{32\sqrt{3}}\frac{\hbar}{mc}\frac{\gamma^2 R}{\nu_x^4}\left[1+\frac{1}{2}\right] \\ &= \left(3.83\times 10^{-13}\ \text{m}\right)\frac{\gamma^2 R}{\nu_x^4}\left[1+\frac{1}{2}\right],\end{aligned}$$

where the two terms in the square brackets come from betatron and synchrotron contributions respectively. We see that

- The beam size is proportional to γ for a given storage ring lattice.
- The beam size is inversely proportional to ν_x^2. A strongly focused lattice is a sensitive way to minimize the beam size.
- The synchrotron contribution to beam size is found to be comparable to, and slightly less than the betatron contribution.
- The rough expression can also be written as

$$\sigma_x^2 = \sigma_\delta^2\frac{R^2}{\nu_x^4}[2+1] = \sigma_\delta^2 D^2[2+1]$$

  by relating it to the energy spread σ_δ.

Homework 6.53

(a) Compute the distribution $\psi(x_\beta')$.

(b) The text readily gives $\langle x_\beta^2\rangle = \sigma_{x\beta}^2$. Compute the other two second moments $\langle x_\beta x_\beta'\rangle$ and $\langle {x_\beta'}^2\rangle$.

Solution (a) $\psi(x_\beta') = \sqrt{\frac{\beta_x}{2\pi\gamma_x}}\frac{N}{\sigma_{x\beta}}\exp\left[-\frac{\beta_x {x_\beta'}^2}{2\gamma_x\sigma_{x\beta}^2}\right]$.

(b) $\langle x_\beta x_\beta'\rangle = -\frac{\alpha_x}{\beta_x^2}\sigma_{x\beta}^2$, $\langle {x_\beta'}^2\rangle = \frac{\gamma_x}{\beta_x}\sigma_{x\beta}^2$.

6.4.12 Quantum excitation of vertical betatron oscillation

Synchrotron radiation excites directly energy oscillation in the motion of a stored particle. This energy oscillation then couples to a horizontal betatron oscillation through the horizontal dispersion function. A stored electron beam therefore acquires natural longitudinal and horizontal beam sizes as discussed so far.

The beam will also acquire a natural vertical beam size. The mechanism is somewhat more subtle than that producing the longitudinal and horizontal sizes. This is because in an ideal storage ring, there is no vertical dispersion and there is therefore not a direct coupling to the energy oscillation due to emissions of synchrotron photons.

So far in the discussion of quantum excitations, we have considered photons emitted strictly along the instantaneous direction of motion of the emitting electron. If this is indeed strictly true, then there will be no natural excitation of vertical beam size. However, the photons are emitted with a small, but finite, opening angle $\sim \frac{1}{\gamma}$. This small opening angle allows direct excitation of a natural vertical beam size, as illustrated in Fig. 6.37.

Figure 6.37: A sideways emission of synchrotron photon excites a vertical betatron oscillation directly.

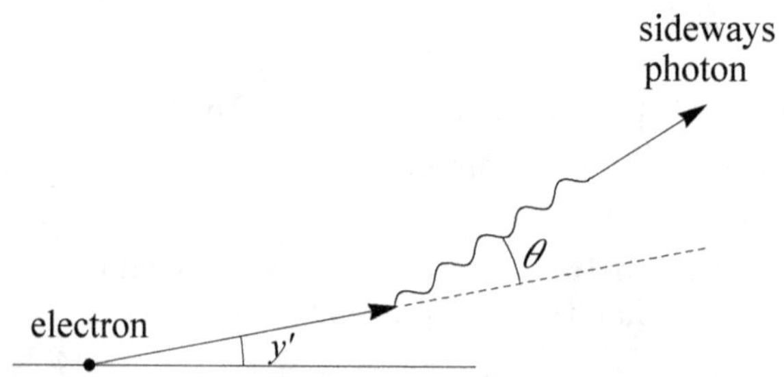

This same mechanism of course also exists to excite the horizontal beam size. However, this effect is much weaker than the coupling excitation induced by the horizontal dispersion, and was ignored when we discussed the horizontal quantum excitation effect. But for vertical quantum excitation, we can no longer do so.

When an electron emits a photon of energy u, it receives a recoil momentum $-\frac{u}{c}$. When a photon is emitted not strictly along the direction of motion of the electron, but with an opening angle θ relative to the bending plane of its orbit, the recoil momentum has a vertical component $-\frac{u}{c}\theta$ that is available to excite its subsequent vertical betatron oscillation with

$$\Delta y' = -\frac{u}{E_s}\theta\,.$$

The change in vertical betatron amplitude due to this emission event is

$$\Delta A_y^2 = \beta_y \Delta y'^2 = \beta_y \frac{u^2}{E_s^2}\theta^2 .$$

A quantum excitation analysis then leads to

$$\frac{d\langle A_y^2\rangle}{dt} = \frac{1}{2\pi R E_s^2}\oint ds\, \beta_y \dot{\mathcal{N}}\langle u^2\theta^2\rangle .$$

As is by now familiar, we then add the contribution from radiation damping,

$$\frac{d\langle A_y^2\rangle}{dt} = -2\frac{\langle A_y^2\rangle}{\tau_y} + \frac{1}{2\pi R E_s^2}\oint ds\, \beta_y \dot{\mathcal{N}}\langle u^2\theta^2\rangle ,$$

and conclude that the equilibrium vertical betatron beam size is given by

$$\langle A_y^2\rangle = \frac{\tau_y}{4\pi R E_s^2}\oint ds\, \beta_y \dot{\mathcal{N}}\langle u^2\theta^2\rangle .$$

The rms vertical betatron beam size σ_y is therefore

$$\frac{\sigma_y^2}{\beta_y} = \frac{\langle A_y^2\rangle}{2} = \frac{\tau_y}{8\pi R E_s^2}\oint ds\, \beta_y \dot{\mathcal{N}}\langle u^2\theta^2\rangle .$$

We note one qualitative difference from the previous quantum excitation calculations. Here again, one and only one synchrotron radiation spectrum parameter determines the entire vertical beam distribution, except that this time this key parameter is no longer $\dot{\mathcal{N}}\langle u^2\rangle$, but is $\dot{\mathcal{N}}\langle u^2\theta^2\rangle$.

Using the detailed spectrum, it can be calculated (Homework 6.54) that

$$\dot{\mathcal{N}}\langle u^2\theta^2\rangle = \frac{23}{48\sqrt{3}} r_0 \hbar m c^4 \frac{\gamma^5}{|\rho|^3} , \tag{6.48}$$

which can also be written as

$$\dot{\mathcal{N}}\langle u^2\theta^2\rangle = \frac{23}{110}\frac{\dot{\mathcal{N}}\langle u^2\rangle}{\gamma^2} ,$$

which looks reasonable because θ is expected to $\sim \frac{1}{\gamma}$.

The smallness of the value $\frac{23}{110}$ indicates an anticorrelation between θ and u, which is what is expected physically. More explicitly, indeed

$$\frac{\dot{\mathcal{N}}\langle u^2\theta^2\rangle}{\dot{\mathcal{N}}\langle u^2\rangle\langle\theta^2\rangle} = \frac{92}{275}$$

is a relatively small number, an indication of anticorrelation between u and θ.

With

$$\tau_y^{-1} = \frac{U_0}{2E_sT_0} = \frac{cr_0}{3\rho^2}\gamma^3 ,$$

we obtain

$$\frac{\sigma_y^2}{\beta_y} = \frac{23}{64\sqrt{3}} \frac{\hbar}{mc} \frac{\oint ds \frac{\beta_y}{|\rho|^3}}{\oint \frac{ds}{\rho^2}} .$$

For an isomagnetic ring, we have

$$\frac{\sigma_y^2}{\beta_y} = \frac{23}{64\sqrt{3}} \frac{\hbar}{mc} \frac{\langle\beta_y\rangle_{\text{dipole}}}{\rho} .$$

To obtain a rough estimate, let $\beta_y \approx \frac{R}{\nu_y}$ and $\rho \approx R$, we find

$$\begin{aligned}\sigma_y^2 &\approx \frac{23}{64\sqrt{3}} \frac{\hbar}{mc} \frac{R}{\nu_y^2} \\ &= (0.801 \times 10^{-13} \text{ m}) \frac{R}{\nu_y^2} .\end{aligned}$$

Take for example, $R = 30$ m and $\nu_y \approx 5$, we have $\sigma_y = 0.31$ μm.

We now compare this result with the natural horizontal beam size. If ν_x and ν_y are comparable, then we see

$$\frac{\sigma_y}{\sigma_{x\beta}} \sim \sqrt{\frac{23}{110}} \frac{\nu_y}{\gamma} .$$

The factor ν_y in the numerator came from $\frac{\beta_y}{D}$, while the γ in the denominator came from the sideways angle. The small parameter $\sqrt{\frac{23}{110}}$ is due to the anticorrelation between photon energy u and the photon emission angle θ. For typical high energy electron storage rings, the natural σ_y is then easily much smaller than the natural σ_x.

Furthermore, one also notes that this natural vertical beam size is independent of the beam energy. This curious effect comes from the fact that $\dot{\mathcal{N}}\langle u^2\rangle\tau_y \propto \gamma^2$ while $\theta^2 \propto \frac{1}{\gamma^2}$, and as a result, these γ-dependences cancel. An electron with higher energy would radiate more, but at the same time, it radiation is more collimated in the forward direction.

Homework 6.54 Verify Eq. (6.48) from synchrotron radiation spectrum.

Solution Follow similar steps as Homework 6.35.

Homework 6.55 Consider a beam in a storage ring without synchrotron radiation (e.g. a proton beam). A feedback system with damping time τ_d is turned on for its vertical betatron motion. At the same time the kicker of this feedback system contains an electrical power noise in such a way that at each turn of beam passage, it gives a component of random kicking angle whose rms average is $\theta_{\rm rms}$. Find an expression for the rms spread of the beam centroid position under the influence of this feedback system. Observed by a multiturn sampling scope, this centroid position contribution will appear as an increased beam spot size.

6.4.13 Vertical beam size due to coupling

The natural vertical beam size calculated in Sec. 6.4.12 is, however, often too small to be of practical significance. In reality, the vertical beam size comes mainly from error effects. It turns out that in addition to the natural contribution, there are three more contributions to the betatron beam size, all due to error effects.

Contribution from coupling to longitudinal First, the storage ring will inevitably contain some spurious vertical dispersion D_y. This small vertical dispersion will directly couple vertical betatron oscillations to energy excitations just like in the horizontal case. Following similar steps as the horizontal betatron excitation, we obtain

$$\begin{aligned}
\frac{\sigma_{y\beta}^2}{\beta_y} &= \frac{55}{32\sqrt{3}}\frac{\hbar}{mc}\gamma^2\frac{\oint ds\frac{\mathcal{H}_y}{|\rho|^3}}{\oint\frac{ds}{\rho^2}} \\
&= \left[3.83\times10^{-13}\ \text{m}\right]\gamma^2\frac{\oint ds\frac{\mathcal{H}_y}{|\rho|^3}}{\oint\frac{ds}{\rho^2}}, \qquad (6.49)
\end{aligned}$$

where

$$\mathcal{H}_y = \frac{D_y^2+(\beta_y D_y'+\alpha_y D_y)^2}{\beta_y}.$$

This contribution dominates over the natural quantum excitation if

$$\frac{D_y}{D} \gtrsim \frac{\nu_y}{\gamma}.$$

Typically, this condition is fulfilled.

Contribution from energy spread The contribution (6.49) is due to a direct excitation to the betatron part of the vertical beam size. When $D_y \neq 0$, there is a contribution $D_y\sigma_\delta$ to the synchrotron part of the vertical beam size. The net vertical beam size is a quadratic sum of these two contributions. On the other hand, this $D_y\sigma_\delta$ term contributes only to the beam size, and not the emittance.

Contribution from coupling to horizontal The beam also gains vertical betatron emittance by coupling to the horizontal betatron oscillation through a small x-y coupling. If this coupling results from a distribution of skew quadrupoles, we have derived in the smooth approximation, Eq. (3.62), a complex x-y linear *coupling coefficient* g. Here, consistent with notations used in Sec. 3.7.7, we introduce

$$\epsilon = 2g = \frac{1}{2\pi}\oint ds \frac{1}{B\rho}\frac{\partial B_y}{\partial y}\sqrt{\beta_x\beta_y}e^{i\psi_x - i\psi_y - i(\nu_x-\nu_y-n)\frac{s}{R}} .$$

We showed in Sec. 3.7.7 that when the betatron tunes $\nu_{x,y}$ are close to a difference resonance $\nu_x - \nu_y = n$, there is a slow but systematic transfer of betatron oscillation emittances back and forth between $\epsilon_x = \frac{\sigma_{x\beta}^2}{\beta_x}$ and $\epsilon_y = \frac{\sigma_y^2}{\beta_y}$ in such a way that

$$\epsilon_x + \epsilon_y = \epsilon_{x0} ,$$

where ϵ_{x0} is the unperturbed horizontal beam emittance in the absence of coupling. We also showed by Eq. (3.68) that, when the emittance coupling exchange is undergoing, the emittance beating frequency (in tune unit) is

$$\nu_{\text{beat}} = \sqrt{\Delta\nu^2 + |\epsilon|^2} ,$$

where $\Delta\nu = \nu_x - \nu_y - n$.

Now if we are considering an electron beam in a storage ring, its emittances cannot beat forever. Quantum excitation and radiation damping will force them to reach an equilibrium. In this case, the natural horizontal beam emittance we calculated earlier is now to be split between ϵ_x and ϵ_y. The question is how is the splitting to be executed and what are their respective equilibrium values.

Obviously the analysis of this problem will have to invoke quantum excitation and radiation damping in the x-y coupled environment. This analysis can be done using the Fokker-Planck equation, which is out of our scope. We will only quote the result here.[27] It can be shown that at the equilibrium, we have

$$\begin{aligned}
\epsilon_x &= \frac{1}{1+\kappa}\epsilon_{x0} ,\\
\epsilon_y &= \frac{\bar{\alpha}_x}{\bar{\alpha}_y}\frac{\kappa}{1+\kappa}\epsilon_{x0} ,\\
\kappa &= \frac{\bar{\alpha}_y(\bar{\alpha}_x+\bar{\alpha}_y)|\epsilon|^2}{\bar{\alpha}_x\left[(\bar{\alpha}_x+\bar{\alpha}_y)|\epsilon|^2 + 4\bar{\alpha}_y\Delta\nu^2\right]} ,
\end{aligned}$$

where $\bar{\alpha}_{x,y}$ are the radiation damping decrements obeying the sum rule $\bar{\alpha}_x + \bar{\alpha}_y + \bar{\alpha}_z = \frac{2U_0}{E_s}$, and ϵ_{x0} is the horizontal nominal emittance before splitting due to the coupling. The emittances obey a sum rule

$$\bar{\alpha}_x\epsilon_x + \bar{\alpha}_y\epsilon_y = \bar{\alpha}_x\epsilon_{x0} .$$

[27] A.W. Chao and M.J. Lee, J. Appl. Phys. 47, No. 10, 4453 (1976).

Figure 6.38: The betatron emittances ϵ_x (solid) and ϵ_y (dashed) in the presence of linear x-y coupling near a $\nu_x - \nu_y = n$ resonance. Left panel shows the case $\bar{\alpha}_x = \bar{\alpha}_y$. Right panel shows $\bar{\alpha}_x = \frac{1}{2}\bar{\alpha}_y$.

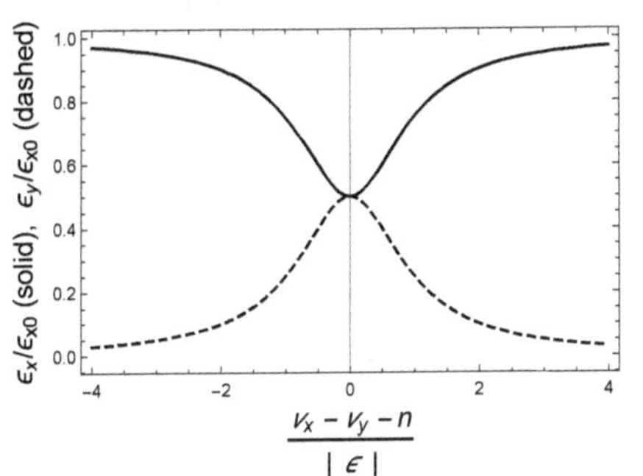

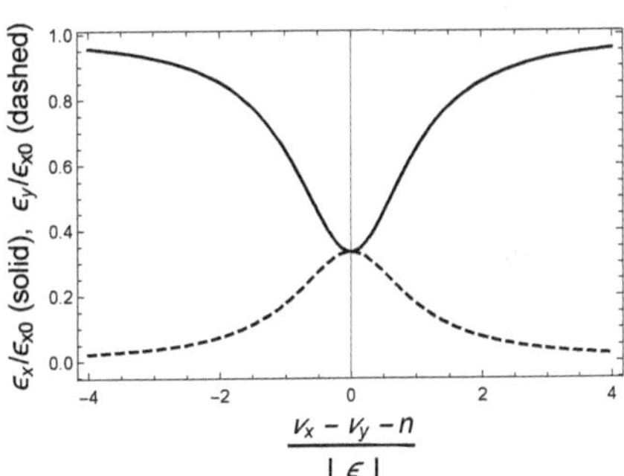

In general, $\bar{\alpha}_x \neq \bar{\alpha}_y$. In case they are equal (e.g. ignoring the contribution from the partition number $\mathcal{D}$), the split is made according to

$$\begin{aligned} \epsilon_x &= \frac{1}{1+\kappa}\,\epsilon_{x0}\,, \\ \epsilon_y &= \frac{\kappa}{1+\kappa}\,\epsilon_{x0}\,, \\ \kappa &= \frac{|\epsilon|^2}{|\epsilon|^2 + 2\Delta\nu^2}\,. \end{aligned}$$

The emittance sum rule becomes

$$\epsilon_x + \epsilon_y = \epsilon_{x0}\,.$$

How the emittances split ϵ_{x0} depends on the ratio of the distance to the resonance $\Delta\nu$ to the coupling strength $|\epsilon|$. Figure 6.38 (left) shows the result when $\bar{\alpha}_x = \bar{\alpha}_y$. When the tunes are exactly on resonance, $\kappa = 1$, the natural horizontal emittance is split equally between ϵ_x and ϵ_y. We call this a *round beam* case – even though the beam may not be physically round unless $\beta_x = \beta_y$. Away from resonance, meaning $|\Delta\nu| \gg |\epsilon|$, we regain the uncoupled case with $\epsilon_x = \epsilon_{x0}$ and $\epsilon_y \approx 0$.

Figure 6.38 (right) shows the result when $\bar{\alpha}_x = \frac{1}{2}\bar{\alpha}_y$ (or $\mathcal{D} = \frac{1}{2}$). When exactly on resonance, the splitting is done according to ratio $\bar{\alpha}_x/\bar{\alpha}_y$.

As mentioned in Eq. (3.53), the coupling coefficient ϵ is a complex quantity depending on the location where it is evaluated. But it was shown that its magnitude $|\epsilon|$ is an invariant, and only its phase varies depending on the observation location. The fact that the emittance splitting is made according to $|\epsilon|$ and is independent of the phase of ϵ is therefore expected because the emittances must be invariant around the ring. The phase of ϵ comes in when we calculate the tilt angle of the beam's profile in its x-y projection, and that depends on the point of observation.

A source of confusion exists here. Sometimes κ, instead of ϵ, is referred to as the coupling coefficient. We prefer to call ϵ the coupling coefficient and κ the *emittance coupling coefficient*, noting that for the case $\bar{\alpha}_x = \bar{\alpha}_y$,

$$\frac{\epsilon_y}{\epsilon_x} = \kappa\,.$$

Needless to say, ϵ is a more fundamental quantity than κ.

The vertical beam size due to x-y coupling dominates over the natural contribution if

$$\kappa \gtrsim \sqrt{\frac{\nu_y}{\gamma}}\,.$$

Lower limit to the vertical beam size In many applications, it is desirable to minimize the vertical beam size by reducing various error effects. It should be noted that as the error sources are reduced, one will approach the ultimate limit of the natural beam size. For example, it would not be too meaningful to reduce the vertical dispersion to values much smaller than $\frac{\nu_y}{\gamma}D$, or to reduce the emittance coupling coefficient to values much smaller than $\sqrt{\frac{\nu_y}{\gamma}}$. The reader is suggested to stay alert to bold claims of successful minimization of coupling effects when the vertical beam size becomes smaller than the natural contribution. We shall show momentarily in the next section that the natural contribution of vertical beam size is in fact $\approx 0.283\,\mu$m. So remember to express your skepticism when someone boasts to possess such a powerful correction algorithm to reduce his/her vertical beam size down to $< 0.283\,\mu$m!

6.4.14 Empirical scaling laws

It is instructive to examine deeper into the rough expressions of the beam sizes obtained so far. To summarize,

$$\begin{aligned}
\sigma_\delta &\approx \gamma\sqrt{\frac{C_q}{2R}}\,, \\
\sigma_z &= \frac{c|\eta|}{\omega_s}\sigma_\delta\,, \\
\sigma_{x\beta} &\approx \sqrt{2}\,\frac{R}{\nu_x^2}\,\sigma_\delta\,, \\
\sigma_{x\delta} &\approx \frac{R}{\nu_x^2}\sigma_\delta\,, \\
\sigma_x &\approx \sqrt{3}\,\frac{R}{\nu_x^2}\,\sigma_\delta\,,
\end{aligned}$$

$$\sigma_y \approx \begin{cases} \frac{1}{\nu_y}\sqrt{\frac{23}{110}C_q R} & \text{natural} \\ \sqrt{3}\,D_y\sigma_\delta & \text{vertical dispersion} \\ \frac{|\epsilon|}{\sqrt{|\epsilon|^2+2\Delta\nu^2}}\sqrt{\frac{\nu_x}{\nu_y}}\sigma_{x\beta} & x\text{-}y \text{ coupling} \end{cases},$$

where we have used the approximate expressions $\beta_{x,y} \approx \frac{R}{\nu_{x,y}}, D \approx \frac{R}{\nu_x^2}, \rho \approx R$, and

$$C_q = \frac{55}{32\sqrt{3}}\frac{\hbar}{mc} = 3.83 \times 10^{-13} \text{ m}.$$

For a given storage ring with fixed R, ν_x, ν_y, we see that all beam sizes (except for the natural vertical beam size) scales linearly with γ. As we raise the beam energy in a storage ring, we will therefore encounter a practical limit (in addition to limits on the magnet strengths), and that is the beam size will become too big to be contained by a suitable vacuum chamber or within the available dynamic apertures. We briefly mentioned these observations by a numerical example when we discussed the beam energy spread (6.39) and horizontal beam size (6.46). We now want to explore a bit what was really being said.

In practice, this will mean that, for a typical electron storage ring appropriately optimized, its transverse beam size should be $\sigma_{x,y} \sim 1$ mm or so at the design top beam energy. At the same time its lattice design should accommodate a beam with energy spread $\sigma_\delta \sim 0.1\%$ or so. The fact that a beam size ~ 1 cm and $\sigma_\delta \sim 1\%$ are too big is due to a quantum lifetime consideration, to be discussed later in this chapter.

If we take $\sigma_{x\delta} = 1$ mm and $\sigma_\delta = 0.1\%$, and apply the above rough formulae, we obtain some universal empirical scaling laws. First of all, we see that at the design top energy, the beam sizes are

$$\sigma_\delta = 10^{-3}, \quad \sigma_{x\delta} = 1 \text{ mm}, \quad \sigma_{x\beta} = \sqrt{2} \text{ mm}, \quad \sigma_x = \sqrt{3} \text{ mm}.$$

These amazing simple scaling laws apply regardless of all details of the beam or the storage ring other than the fact that this ring is for electrons!

We then note that the storage ring radius R, the design top energy E_s, and betatron tune ν_x must satisfy

$$\begin{aligned} \sigma_{x\delta} = 1 \text{ mm} &\implies R \text{ [m]} = \nu_x^2 \\ \sigma_\delta = 10^{-3} &\implies \gamma = 2300\,\nu_x \implies E_s \text{ [GeV]} = \nu_x. \end{aligned}$$

What this means is that all practical electron storage rings (not insertion-dominated) will approximately follow another empirical scaling law,

$$E_s \text{ [GeV]} = \sqrt{R \text{ [m]}} = \nu_x.$$

It is indeed an amazing scaling law with its simplicity. For example, if we know the storage ring has a design energy of 5 GeV, then its betatron tune is likely to be about 5, and its radius is going to be approximately 25 m.

Although the arguments used in "deriving" these scaling laws assume an electron storage ring, it is interesting to note that the E [GeV] $= \nu_x$ law holds approximately also for proton storage rings. The scaling law breaks down however relating to the proton storage ring radius R because in the proton case, determination of R is magnetic-field dominated, not synchrotron-radiation dominated.

Figure 6.39 shows the data illustrating this scaling law. The lower figure for E_s versus R shows some deviation from the scaling for small rings. This

Figure 6.39: Design data of electron storage rings demonstrating the empirical scaling laws. Synchrotron radiation facilities dominated by insertion devices are not included.

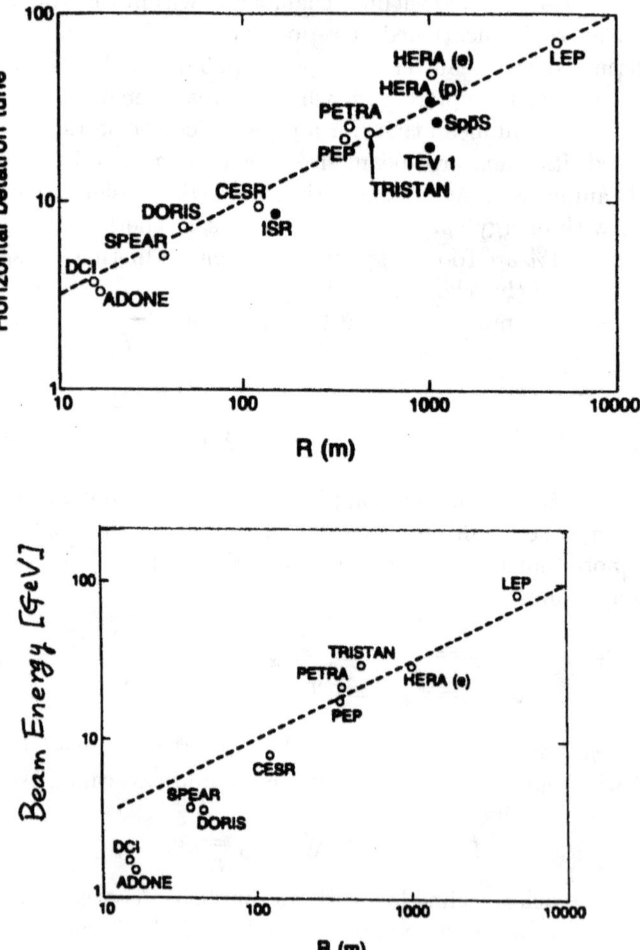

is because the additional various overhead functions such as injection and RF sections become significant, and R becomes longer than otherwise needed for beam dynamics reasons.

Furthermore, the magnetic field averaged over the circumference can be written as

$$\langle B \rangle \ [T] \ = \ \frac{3.3}{E_s \ [\text{GeV}]} \, .$$

It clearly shows the effort to decrease the bending magnetic field with increasing beam energy to reduce the synchrotron radiation. For the 55-GeV LEP-I, for example, $\langle B \rangle$ is only 0.06 T.

One can also make a rough estimate of the storage ring cost by scaling to LEP-I. We find that the scaling $R \propto E_s^2$ obtained for cost optimization as discussed around Eq. (6.25) agrees with our scaling obtained using beam dynamics arguments. With LEP-I energy of 55 GeV, we obtain the cost scaling,

$$\text{Cost} \ = \ (\text{LEP-I cost}) \times \left(\frac{E_s[\text{GeV}]}{55} \right)^2 ,$$

once the LEP-I cost is known.

We are not finished yet. It further follows from these scaling laws also various interesting scaling properties such as

$$\begin{aligned}
\beta_x \ [\text{m}] \ &\approx \ \frac{R \ [\text{m}]}{\nu_x} \ = \ \nu_x \, , \\
D \ [\text{m}] \ &\approx \ \frac{R \ [\text{m}]}{\nu_x^2} \ = \ 1 \, , \\
\alpha_c \ &\approx \ \frac{1}{\nu_x^2} \ = \ \frac{1}{R \ [\text{m}]} \, , \\
E_s \ [\text{GeV}] \ &\approx \ \beta_x \ [\text{m}] \, .
\end{aligned}$$

In particular, we have found that all electron storage rings will have an average dispersion of the order of 1 m.

Furthermore, we also find that the natural vertical beam size has a universal value of

$$\sigma_y \ (\text{natural}) \ = \ \frac{\nu_x}{\nu_y} \sqrt{\frac{23}{110} C_q} \, .$$

In case $\nu_y \approx \nu_x$, this gives

$$\sigma_y \ (\text{natural}) \ = \ 0.283 \ \mu\text{m} \, .$$

For advanced synchrotron radiation facilities, one tries to minimize σ_y by correcting all kinds of error effects, and are typically approaching this natural limit. What we learn here is that trying harder will meet with diminishing returns.

Homework 6.56 In the spirit of the empirical scaling laws, show that once the beam energy is given, the synchrotron radiation loss per turn U_0 is readily estimated by

$$U_0[\text{MeV}] \approx 0.09(E_s[\text{GeV}])^2 = 0.09\,R[\text{m}]\,,$$

and the radiation damping times $\tau_{x,y,z}$ are

$$\tau_{x,y}\,[\text{ms}] = 2\tau_z\,[\text{ms}] = 0.5\,E_s\,[\text{GeV}]\,.$$

The critical photon energy is

$$u_c[\text{keV}] = 2\,E_s[\text{GeV}]\,.$$

Homework 6.57 Show two more scaling laws:

$$\begin{aligned}\text{natural } x\text{-emittance } \epsilon_{x0} &= \frac{1.5}{\nu_x}\ \text{mm-mrad}\,,\\ \text{normalized natural emittance } \epsilon_{N,x0} &= 3\ \text{mm-rad}\,.\end{aligned}$$

The storage ring normalized emittance is enormous compared with an electron beam coming out of a high quality gun, for which 1 μm-rad is typical. Quantum excitation has very significantly enlarged the beam's emittances in synchrotron-radiation dominated rings.

Homework 6.58 Show that according to the scaling law, the value of $\frac{\langle u\rangle}{E_s}$ is a universal constant,

$$\frac{\langle u\rangle}{E_s} = 0.7\times 10^{-6}\,.$$

Each emission event thus carries away the beam energy by this universal fractional value independent of the beam energy or the ring circumference. Does this expression apply to a proton beam?

Homework 6.59 This homework attempts to pull various considerations together for a zeroth order design effort of an electron storage ring. Design a 10-GeV ring. Your RF system frequency is 400 MHz. Ignore special-purpose insertion sections in your considerations. In this effort, 10 GeV and 400 MHz are the two main input parameters you will receive.

(a) According to the empirical scaling law, how long should the storage ring circumference be? What should its betatron tune ν_x be?

(b) Start with a crude lattice design. Let the lattice consist purely of FODO cells, each with phase advance of approximately 90° in both x and y. Assume thin-lens quadrupoles. How many FODO cells do you need? How long is the cell length? What is the bending magnetic field? What is the focal length of the quadrupoles? If the quadrupole pole tip field is 1 Tesla and pole tip is at radius 10 cm, how long are these quadrupoles? Is the thin-lens approximation justified? What are the maximum and minimum values of the $\beta_{x,y}$-functions in the lattice and where are their locations? What are the maximum and minimum values of the dispersion function and where are they located? What value is the momentum compaction factor and the phase slippage factor?

(c) Is your design away from major low-order resonances? If not, make a proposal to adjust your lattice slightly to avoid them.

(d) Estimate the sextupole strengths $T_{F,D}$ needed for chromaticity compensations. If the sextupole pole tip field is 0.5 Tesla and pole tip radius is 10 cm, how long are these sextupoles? Where do you located them?

(e) Assuming the quadrupoles and sextupoles are iron magnets whose cross-sections are as shown in Figs. 1.26 and 1.27, calculate the needed winding currents nI for these magnets.

(f) Estimate synchrotron radiation loss per turn U_0 per electron. You are now asked to install a total of 50 copper RF cavities, each with voltage $V_0 = 0.5$ MV and frequency 400 MHz. All cavities are tuned so that they all contribute the same voltage at the time of beam arrival. Using pillbox model for the cavities, give reasonable dimensions of the pillbox. How does the cavity voltage compare with the Kilpatrick breakdown limit? Estimate the Q-value and shunt impedance per RF cavity. Are you above or below transition? Calculate the synchronous phase ϕ_s, the harmonic number h, the synchrotron tune ν_s, and the rms bunch length σ_z. What is the rms energy spread σ_δ? What is the uncoupled horizontal emittance ϵ_{x0} and the normalized emittance $\epsilon_{N,x0}$? What are your estimated values of the three radiation damping times (with FODO cells, you have assumed separate-function magnets)?

(f) If the beam contains a single bunch of 10^{11} electrons, what is the average beam current? What is the peak beam current? How much fraction of stored RF field energy is extracted by the beam each turn as the beam passes? Estimate the fraction of stored field energy lost on heating the cavity walls per revolution of the beam. How much RF power must you feed to each of the 50 cavities in order to maintain beam storage? Is that acceptable?

(g) Estimate the RF bucket half-height. Estimate the synchrotron quantum lifetime as limited by the RF bucket. Is it acceptable?

(h) Finally, if you have an idea how much LEP-I cost, estimate the cost to your storage ring.

Summarize the above results into a parameters list in Table 6.2. You have now (almost) designed a storage ring. Congratulations!

Homework 6.60 A more venturous homework is to apply the empirical scalings to design an electron storage ring at 150 GeV perhaps for a collider. Due to the high RF power required, it is advisable to replace the room temperature RF by superconducting ones. This crude design will also have to be added insertion sections for interaction regions.

Table 6.2: Parameters list of a zeroth order design of an electron storage ring.

Beam energy E	10 GeV
Circumference C	
Number of FODO cells	
Betatron phase advance/cell $\Phi_{x,y}$ (avoiding resonances)	
Betatron tunes $\nu_{x,y}$ (avoiding resonances)	
FODO cell length	
Maximum and minimum values of $\beta_{x,y}$	
Maximum and minimum values of dispersion function	
Momentum compaction factor α_C	
Phase slippage factor η	
Bending angle per dipole magnet	
Bending magnetic field	
Quadrupole focal length	
Quadrupole length	
Quadrupole winding current nI	
Sextupole strengths $T_{F,D}$	
Sextupole lengths $\ell_{F,D}$	
Sextupole winding current nI	
Synchrotron radiation energy loss per turn per electron U_0	
Critical synchrotron photon energy u_c	
Total RF voltage per turn	
Ratio of RF acceleration gradient to Kilpatrick limit	
RF frequency $\omega_{\rm rf}/2\pi$	400 MHz
Length and radius of each pillbox RF cavity	
Quality Q-value of RF cavities	
Shunt impedance R_s per cavity	
Field energy stored in each cavity	
Percentage of stored energy extracted by beam per passage	
Power consumed per RF cavity by beam	
Power consumed per RF cavity by wall heating	
Synchronous phase ϕ_s	
Harmonic number h	
Synchrotron tune ν_s	
RMS energy spread σ_δ	
RMS bunch length σ_z	
Uncoupled horizontal emittance ϵ_{x0}	
Normalized $\epsilon_{x0,N}$	
Radiation damping times $\tau_{x,y,z}$	
Average beam current $I_{\rm ave}$	
Peak beam current $\hat{I}$	
RF bucket half-height	
Synchrotron quantum lifetime	
Cost	

6.4.15 Vertical quantum lifetime

We insisted earlier, by applying the central limit theorem, that the equilibrium beam distribution in phase space shall be Gaussian. We mentioned at the time that the bulk of the distribution depends on one and only one parameter $\dot{\mathcal{N}}\langle u^2\rangle$ of the radiation spectrum, but the distribution will deviate slightly from Gaussian with its tails slightly modified by the higher moments of synchrotron radiation spectrum.

This topic aside, there is another reason that the beam distribution tail must not be Gaussian, namely, the vacuum chamber has a finite size and simply cannot accommodate the infinite tail of a Gaussian. For one thing, the Gaussian tail has to be cut off by the vacuum chamber.

But the brutal removal of a beam tail is intrinsically inconsistent with an equilibrium Gaussian distribution which has an infinite tail. As the tail is removed, quantum excitation will try to reestablish it. As particles are pumped into the tail in the process, they are removed from the beam. The beam, therefore, cannot have an equilibrium distribution. Instead, the distribution will have to decrease in time, leading to a finite lifetime of the beam. This beam lifetime is called *quantum lifetime*. Since there are three degrees of freedom, and therefore three potential aperture limits, we have three quantum lifetimes $\tau^{(\mathrm{q})}_{x,y,z}$, corresponding to horizontal, vertical, and synchrotron dimensions respectively.

Let us first consider the y-dimension, and consider a vacuum chamber limit in the form of a vertical aperture at $y = A$. In practice, we always have the condition $A \gg \sigma_y$. Whenever a particle acquires a vertical displacement $y > A$, it is removed from the beam. We first note that since betatron oscillation is much faster than radiation damping and quantum excitation [Eq. (6.36)], this limit on y translates to a limit on the critical betatron amplitude

$$A_y = \sqrt{\frac{y^2 + (\alpha_y y + \beta_y y')^2}{\beta_y}},$$

and the aperture limit becomes

$$A_y \le \frac{A}{\sqrt{\beta_y}},$$

as illustrated in Fig. 6.40. Note that the unit of A is mm, the unit of A_y is $\sqrt{\mathrm{mm}}$.

When the beam distribution is described by an equilibrium Gaussian before imposing the aperture limit, this distribution is a steady state only in its overall appearance. Underneath, particles are streaming along rapidly. Represented in phase space, the motion of each particle is dominated by an elliptical trajectory along its respective Courant-Snyder ellipse. Radiation damping and quantum excitation cause the motion to deviate slightly from this elliptical motion. The time scale of the effect due to these deviations are slow compared with the

Figure 6.40: Aperture limit on the vertical amplitude imposes a vertical quantum lifetime of a stored beam. Particles in the shaded regions are removed from the beam.

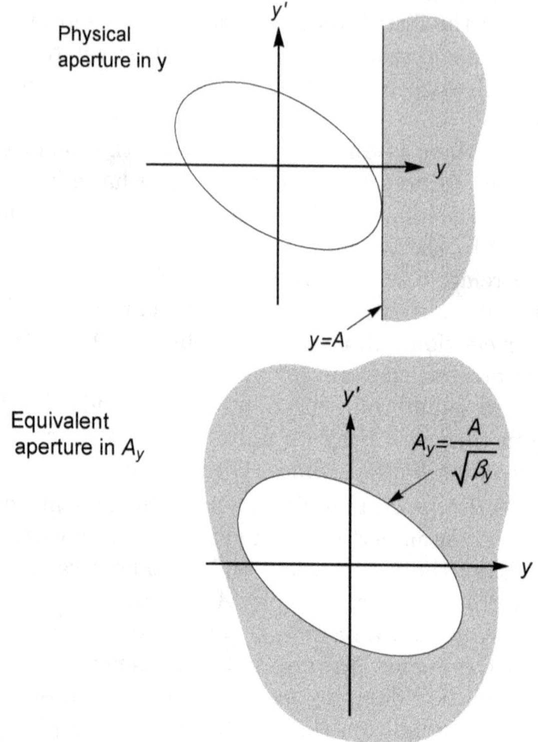

dominating elliptical motion, but they are fast by the time scale of the quantum lifetime. The fast elliptical motion is the reason why the physical aperture $y < A$ becomes the effective aperture $A_y < \frac{A}{\sqrt{\beta_y}}$.

Imagine the situation when an aperture is suddenly inserted into an initially happy Gaussian beam with infinite tails. Three steps happen to the beam distribution after the aperture is inserted: (i) the tail is cut off, the time scale is immediate or within a few turns; (ii) particles close to the aperture diffuse quickly to the aperture and get lost, generating a sharp drop of beam distribution from a normal Gaussian to a truncated zero at the aperture in a distance $\lesssim \sigma$, the time scale of this step is ~radiation damping time; (iii) While maintaining the overall truncated Gaussian distribution in a steady-state, the entire beam intensity slowly decreases with a long time scale, the quantum lifetime.

Step (i) is obvious. As to step (ii), the detailed analysis is skipped here. Suffice it to say that a proper analysis is to apply the Fokker-Planck equation, and

Figure 6.41: Illustration of the steady-state distribution when the beam is subject to an aperture limit at $y = A \gg \sigma_y$. The plot is meant to be semi-logarithmic, and the illustration is for the case $A = 5\sigma_y$. The red dashed curve is the nominal Gaussian distribution. The black dashed curve indicates the cutoff of the tails in step (i) of the text. Effect of step (ii) gives a steady-state distribution shown as the blue solid curve. The slope of the blue curve at the aperture location determines the quantum lifetime.

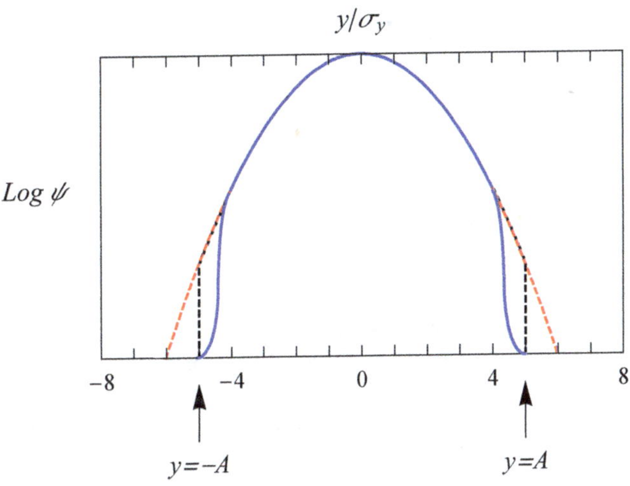

the resulting truncated steady-state distribution has basically the unperturbed Gaussian shape in the beam core, cleanly truncated beyond $y = A$, and then near the aperture, the distribution rises from zero at $y = A$ and grows rapidly to match the normal Gaussian within a distance $\lesssim \sigma_y$. The situation is sketched in Fig. 6.41.

For step (iii), we can calculate the quantum lifetime three ways. Below, we give first very crude estimate, and then a second, improved estimate. A third way, an exact calculation applying a Fokker-Planck equation with boundary condition, is outside our scope, and will not be included.[28]

Crude estimate The number of particles residing in the beam tail cut away by the aperture is

$$N \int_{A/\sqrt{\beta_y}}^{\infty} A_y dA_y \psi(A_y) = \frac{N\beta_y}{\sigma_y^2} \int_{A/\sqrt{\beta_y}}^{\infty} A_y dA_y e^{-\frac{\beta_y A_y^2}{2\sigma_y^2}} = N e^{-\frac{A^2}{2\sigma_y^2}} ,$$

where N is the total number of particles in the beam. In most practical cases, the aperture limit A is much larger than σ_y. The fraction of particles residing

[28] A.W. Chao, Part. Accel. Conf., Chicago (1977), IEEE Trans. Nucl. Sci. NS-24, p. 1889.

in the beam tail and thus removed by the aperture is small. For example, if $A = 6\sigma_y$, then there is only $e^{-18} = 1.5 \times 10^{-8}$ of the beam particles in the tail.

Quantum excitation and radiation damping will try to reestablish this beam tail. The time scale for this reestablishment is of the order of the radiation damping time τ_y. In other words, in every time interval of τ_y, a fresh beam tail is to be reestablished, and each time it is wiped away by the aperture limit. The particle loss rate is therefore roughly

$$\frac{dN}{dt} \approx \frac{N}{\tau_y} e^{-\frac{A^2}{2\sigma_y^2}} .$$

This then leads to a rough estimate of vertical quantum beam lifetime as

$$\tau_y^{(q)} \approx \tau_y e^{\frac{A^2}{2\sigma_y^2}} , \tag{6.50}$$

which of course is recognized to be much longer than the radiation damping time τ_y.

Improved estimate This expression (6.50) however is not as accurate as we would like. When we took τ_y to be the characteristic time to reestablish the beam tail, we have been too crude. What happens is that indeed the beam distribution would reestablish itself in a time interval of $\sim \tau_y$ for the *core* of the beam. But the *tail* particles diffuse much faster. Quantum lifetime (6.50) is too optimistic.

What one needs to calculate is the rate at which particles diffuse into the tail, i.e. the rate $\frac{dN}{dt}$ at which particles cross the aperture outward. We now apply a heuristic argument to obtain $\left.\frac{dN}{dt}\right|_{\text{aperture}}$ as follows.[29]

Without an aperture, at equilibrium, the rates of particle crossing the aperture inward and outward are equal. When an aperture is inserted, there are of course no particles damped back into the beam from the tail, but the same rate still applies to the crossing rate outward, at least approximately. Now that unperturbed crossing rate outward is equal to the unperturbed crossing rate inward when the beam is in the equilibrium Gaussian state. So all we need to calculate is the rate particles are damped from tail back into the beam through the aperture for an equilibrium distribution without aperture, and that result shall be identified to be $\left.\frac{dN}{dt}\right|_{\text{aperture}}$. What an insight to visualize such a heuristic physical argument!

The equation of motion for A_y^2 is

$$\frac{dA_y^2}{dt} = -2\frac{A_y^2}{\tau_y} + (\text{diffusion term}) .$$

The first term on the right hand side comes from radiation damping. The second term comes from quantum diffusion. At equilibrium, the two terms are equal

[29]M. Sands, SLAC Report SLAC-121 (1970).

Figure 6.42: The vertical betatron phase space to obtain the rate of particle crossing inward through the aperture.

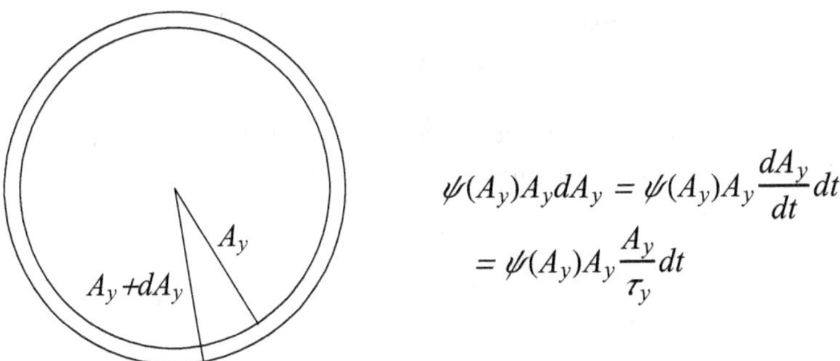

throughout the distribution, core or tail. As seen by the factor A_y^2 in the first term, it is clear that both the damping and the diffusion rates become large in the tail. Particle motion is more agitated in the tail.

In a time interval of dt, the number of particles damped back into the beam is equal to the number of particles contained between the amplitude $\frac{A}{\sqrt{\beta_y}}$ and $\frac{1}{\sqrt{\beta_y}}(A + \frac{dA_y}{dt}dt)$, and that is given by $NA_y\psi(A_y)\frac{dA_y}{dt}dt$ evaluated at $A_y = \frac{A}{\sqrt{\beta_y}}$. The rate at which particles damp back into the beam through the aperture is therefore

$$NA_y\psi(A_y)\frac{dA_y}{dt}\bigg|_{A_y = \frac{A}{\sqrt{\beta_y}}} = - NA_y\psi(A_y)\frac{A_y}{\tau_y}\bigg|_{A_y=\frac{A}{\sqrt{\beta_y}}} .$$

See Fig. 6.42.

If we insert the equilibrium Gaussian distribution for $\psi(A_y)$ in the above expression, we obtain the particle loss rate,

$$\begin{aligned}\frac{dN}{dt} &= - NA_y\psi(A_y)\frac{A_y}{\tau_y}\bigg|_{A_y=\frac{A}{\sqrt{\beta_y}}} \\ &= - \frac{N\beta_y A_y^2}{\tau_y\sigma_y^2} e^{-\frac{\beta_y A_y^2}{2\sigma_y^2}}\bigg|_{A_y=\frac{A}{\sqrt{\beta_y}}} = -\frac{NA^2}{\tau_y\sigma_y^2} e^{-\frac{A^2}{2\sigma_y^2}} .\end{aligned}$$

Compared with the more optimistic expression (6.50), this loss rate is substantially larger by a factor $\frac{A^2}{\sigma_y^2}$. The quantum lifetime is correspondingly shorter,

$$\tau_y^{(q)} = \left(\frac{\sigma_y}{A}\right)^2 \tau_y e^{\frac{A^2}{2\sigma_y^2}} . \tag{6.51}$$

With a 6-σ aperture, the quantum lifetime is approximately $\frac{1}{36}e^{18} = 2\times 10^6$ times the radiation damping time. If radiation damping time is 10 ms, then quantum lifetime with a 6-σ aperture is 6 hrs, according to this formula. The reader should note the extreme sensitivity to the choice of the aperture. For a 7-σ aperture, for example, the quantum lifetime would increase to $\frac{1}{49}e^{49/2} = 9\times 10^8$ times τ_y.

In practice, however, we often demand an aperture significantly larger than 6 or 7 times σ. In addition to plain safety margin including possible orbit distortion, it has also another important reason that quantum lifetime, as mentioned, is extremely sensitive to the tail of the beam distribution. A tail distribution that falls off slower than Gaussian for whatever reason will have a large impact on the quantum lifetimes.

Exact calculation There are two reasons to question the above estimates of quantum lifetime. First of all, vertical betatron oscillation is not the only contribution to the vertical displacement of the beam particles. We mentioned that a particle's vertical displacement has mainly four contributions (Sec. 6.4.13),

- a natural cause of quantum excitation that occurs when the radiation photons are emitted sideways,
- quantum excitation due to a spurious vertical dispersion in the dipoles,
- coupling to the horizontal beam size through x-y coupling by error fields,
- contribution of $D_y\sigma_\delta$ from coupling to the synchrotron dimension.

Strictly speaking, each contribution requires a different analysis, and Eq. (6.51) applies only when the quantum excitation comes from the first two contributions. The other contributions coming from couplings to the other dimensions are more involved. However, in most practices, the same formula is usually applied anyway although exact solution exists.

The second reason refers to the approximation made in the derivation of Eq. (6.51). An exact calculation without the heuristic argument can be made using the Fokker-Planck equation with a boundary condition. The same analysis can also be extended to treat cases with coupled motion. It is beyond our present scope. We shall say a few more words in the next two sections on the cases of horizontal and synchrotron quantum lifetimes.

A careful reader might note a third reason to question Eq. (6.51). In its derivation, one might conclude that the quantum lifetime is proportional to $\psi(\frac{A}{\sqrt{\beta_y}})$ at the aperture, and Eq. (6.51) was obtained by inserting Gaussian distribution for the quantity $\psi(\frac{A}{\sqrt{\beta_y}})$. However, when the aperture is inserted, we know $\psi(\frac{A}{\sqrt{\beta_y}})$ must become zero. So what is the justification to insert the finite Gaussian value for it? The answer to this question lies in the fact that in the exact derivation, the quantum lifetime is actually related to the slope

$\psi'(\frac{A}{\sqrt{\beta_y}})$ at the aperture. The heuristic argument has swept this subtlety under the rug and yet gives the correct approximate result.

6.4.16 Horizontal quantum lifetime

Consider a horizontal vacuum chamber aperture at $x = A$ in a storage ring. If the lattice is nondispersive, i.e. $D = 0$ at the aperture limit, the aperture imposes a limit only on amplitudes of the particles' horizontal betatron oscillations. The analysis for the vertical betatron aperture limit would apply, and the corresponding horizontal quantum lifetime reads

$$\tau_x^{(\mathrm{q})} = \left(\frac{\sigma_{x\beta}}{A}\right)^2 \tau_x \, e^{\frac{A^2}{2\sigma_{x\beta}^2}} .$$

When the horizontal aperture occurs at a dispersive location with $D \neq 0$, however, the aperture limit is imposed on the total x,

$$x = x_{x\beta} + D\delta < A . \tag{6.52}$$

In this case, the dynamics involves coupled horizontal betatron and synchrotron motions. The quantum lifetime calculation becomes more complicated and our heuristic argument no longer apply. It requires setting up a Fokker-Planck diffusion equation and solving it with a boundary condition at $x = A$. The derivation is omitted here; the result when $r = \frac{D^2\sigma_\delta^2}{\sigma_x^2}$ is not too close to 0 or 1 and when the damping partition number $\mathcal{D} \approx 0$ (therefore $\tau_z = \frac{1}{2}\tau_x$), is given by[30]

$$\tau_x^{(\mathrm{q})} = \frac{e^{n^2/2}}{\sqrt{2\pi}\, n^3} \frac{\tau_x}{(1+r)\sqrt{r(1-r)}} ,$$

where $n = \frac{A}{\sigma_x} \gg 1$.

Compared with the case when $D = 0$ with $\tau_x^{(\mathrm{q})} \sim \frac{e^{n^2/2}}{n^2}$, we see that this quantum lifetime $\sim \frac{e^{n^2/2}}{n^3}$ is shorter by a factor $\sim n$ while all other factors are of the order of unity. This is because the aperture limit (6.52) cuts closer to the core than the apertures by either $x_{x\beta} < A$ or $D\delta < A$.

We mentioned earlier that the vertical quantum lifetime is more involved in the presence of x-y coupling. For this case, a similar involved calculation can be performed to obtain the corresponding quantum lifetime.

Homework 6.61 To appreciate the shorter quantum lifetime in a coupled case when the aperture limit occurs at a location with finite dispersion, consider the

[30] A.W. Chao, US/CERN Accel. School, S. Padre Island, 1986, Lecture Notes Phys., V. 296, Springer, p. 51.

case when the total beam size σ_x is given, while $r = \frac{D^2\sigma_\delta^2}{\sigma_x^2}$ is being varied. Show that the shortest quantum lifetime occurs when

$$r = \frac{1+\sqrt{17}}{8} = 0.64, \quad \frac{D\sigma_\delta}{\sigma_x} = 0.80, \quad \frac{\sigma_{x\beta}}{\sigma_x} = 0.60\,.$$

The shortest quantum life time is

$$\tau_{x,\min} = \frac{64}{\sqrt{2\pi}(9+\sqrt{17})(\sqrt{6\sqrt{17}-10})}\frac{e^{n^2/2}}{n^3}\tau_x = 0.507\frac{e^{n^2/2}}{n^3}\tau_x\,,$$

which is a factor $\frac{0.507}{n}$ shorter than that predicted by a nominal formula.

6.4.17 Longitudinal quantum lifetime

In Eq. (5.27), synchrotron oscillation has been described by Hamiltonian

$$H(\phi,\delta,t) = \frac{1}{2}h\omega_0\eta\delta^2 + \frac{eV_0\omega_0}{2\pi E_s}[\cos\phi - \cos\phi_s + (\phi-\phi_s)\sin\phi_s]\,.$$

We also derived in Eq. (5.39) an expression for the RF bucket half height,

$$\hat{\delta}_{\frac{1}{2}} = \frac{2\nu_s}{h|\eta|}\left|1-\left(\frac{\pi-2\phi_s}{2}\right)\tan\phi_s\right|\,.$$

The Hamiltonian on the separatrix is

$$H(\phi,\delta) = H_{\rm sx} = \frac{1}{2}h\omega_0\eta\hat{\delta}^2_{\frac{1}{2}}\,.$$

In the absence of radiation damping and quantum diffusion, the Hamiltonian H is a constant of the motion of a particle, and the RF bucket provides a stable region for particle motion in the phase space. With radiation damping and quantum diffusion, particles can slowly change their H, and sometimes accidentally cross the separatrix and get out of the bucket. In the picture we adopt, we will consider a particle lost from the beam when it crosses the separatrix. In this picture, synchrotron quantum lifetime is given by the time for particles to diffuse out of the bucket.

This is of course a crude picture because

- As a particle approaches the separatrix, its synchrotron oscillation frequency slows down towards zero. The approximation we made in the transverse analyses that the betatron oscillations are fast compared with radiation damping and diffusion does not apply.

- A particle out of the bucket does not immediately mean that it is lost. It still stays in the storage ring until its subsequent dynamics carries it to some aperture limit due to a large momentum deviation or a large transverse displacement. During this time, it actually is possible to diffuse or damp back into the bucket.
- Chaotic layer has not been considered.

However, for our purpose here for an order of magnitude estimate, we shall ignore these complications. The main reason is that an appropriate and more dependable theory — probably based on Fokker-Planck equation — has not yet been developed.

We then draw the analogy to the vertical quantum lifetime calculation by

- replacing the Courant-Snyder invariant A_y by the Hamiltonian H for the synchrotron motion,
- replacing the $A_y = \frac{A}{\sqrt{\beta_u}}$ aperture limit by the separatrix,
- replacing the cutting aperture A by the RF bucket half-height $\hat{\delta}_{\frac{1}{2}}$.

We then write down the synchrotron quantum lifetime as

$$\tau_z^{(q)} \approx \left(\frac{\langle H\rangle}{2H_{\rm sx}}\right) \tau_z \, e^{\frac{H_{\rm sx}}{\langle H\rangle}} ,$$

where $\langle H\rangle = h\omega_0\eta\sigma_\delta^2$ is the rms value of H of the unperturbed beam (which has a Gaussian distribution in δ), and $H_{\rm sx}$ acts as the aperture limit in H. We can also write the synchrotron quantum lifetime as

$$\tau_z^{(q)} = \left(\frac{\sigma_\delta}{\hat{\delta}_{\frac{1}{2}}}\right)^2 \tau_z \exp\left[\frac{\hat{\delta}_{\frac{1}{2}}^2}{2\sigma_\delta^2}\right] .$$

In case the momentum aperture is not limited by the RF bucket half height but by some other means such as a momentum aperture of the storage ring lattice design, then $\hat{\delta}_{\frac{1}{2}}$ should be replaced by the corresponding momentum aperture.

6.5 SLIM formalism

6.5.1 Going beyond the 1-D Courant-Snyder formalism

In Chapter 2, we covered the subject of Courant-Snyder formalism extensively. That formalism, as elegant as it is, we also mentioned the fact that it is not unique, i.e. it is an artifact. We also emphasized the fact that it has many limitations such as limitation to uncoupled 1-D systems (2-D phase space). The concept of dispersion function, for example, must assume quasistatic energy motion, therefore seriously limiting synchrotron motions to $\nu_s \ll 1$. Even more questions arise when a 3-D coupled system is to be considered in general, particularly when errors are introduced. In this section, we introduce a SLIM formalism as a way to proceed.

Courant-Snyder formalism — a recap Consider the linear optics of a circular accelerator. Following Courant-Snyder, we analyze particle motion using transport matrices. We first form a phase space state vector

$$X = \begin{bmatrix} x \\ x' \end{bmatrix} .$$

Particle motion from position s_1 to position s_2 in the accelerator beamline is described by a 2×2 transport matrix $M(s_2|s_1)$,

$$\begin{bmatrix} x(s_2) \\ x'(s_2) \end{bmatrix} = M(s_2|s_1) \begin{bmatrix} x(s_1) \\ x'(s_1) \end{bmatrix} .$$

To calculate $M(s_2|s_1)$, one multiplies the transport matrices element by element from position s_1 to position s_2. The intimate connection of matrices to accelerator physics was emphasized in Sec. 2.2.10.

In this formalism, it is important to recognize that *all* the physics associated with linear optics are contained in the transport matrices. In other words, these matrices should give us everything we want to know, and all physical results must be derivable from them without having to seek additional information. Our job is therefore to analyze these transport matrices to extract all the physics information they contain to the fullest possible extent. Question, is how.

The way Courant-Snyder does it was to introduce the Courant-Snyder parameters $\alpha(s), \beta(s), \gamma(s), \psi(s)$, as well as the dispersion functions $D(s), D'(s)$ to deal with the off-momentum particles. We shall refer to these functions generally as "auxiliary lattice functions". By calling them "auxiliary", there is an implicit implication that they are not fundamental. The reason of doing so will become clear momentarily,

$$\text{Auxiliary functions}: \quad \alpha(s), \beta(s), \gamma(s), \psi(s), D(s), D'(s), \mathcal{H}(s) , \qquad (6.53)$$

where we have included the function $\mathcal{H}(s)$ to the list for synchrotron radiation dominated storage rings.

The goal, of course, is to calculate all the physical quantities associated with the beam. By physical quantities, it mean quantities directly related to the beam's measurable physical properties, such as closed orbit distortion, momentum compaction factor, betatron and synchrotron tunes, rms beam sizes, bunch length, and energy spread, etc. Note that this long list does not contain the auxiliary functions themselves. In the Courant-Snyder formalism, these functions play an auxiliary role helping us to calculate the physical quantities, but they themselves are not physical.

So, in the Courant-Snyder tradition, we have been doing accelerator physics in three steps,

1. Find the transport matrices $M(s_2|s_1)$ by multiplying element matrices.
2. Compute the auxiliary lattice functions using the transport matrices.
3. Compute beam's physical properties, what we really want to compute, using the auxiliary lattice functions.

Again, step 1 starts with the calculation of the transport matrices which contain all the physics. No more information in addition should be required once the transport matrices are obtained. Step 2 contains no physics but, for a 1-D uncoupled system, the set of auxiliary functions provide sufficient flexibility to allow all physical quantities to be calculated in step 3. The 3-step process is a self-contained set for a linear 1-D uncoupled system.

The main weakness of this formalism is that it applies only to 1-D uncoupled dynamics. Just for two immediate examples, β_x and β_y lose their meanings when x- and y-motions are coupled, and D and D' become obsolete when the synchrotron tune becomes sizable, or when a synchrobetatron resonance is nearby. Trying to retrofit the Courant-Snyder formalism along the way of *generalized β-functions* to deal with these coupled cases is often awkward if not impossible.

To establish Step 3 above, we are accustomed to derive and then apply formulae explicitly involving the auxiliary functions. But these textbook formulae work only for the 1-D uncoupled case.[31] In actual applications, we often have to ask what replaces these formulae when they break down? What if there is a skew quadrupole in the storage ring? What if there is a crab cavity? What happens if the tunes are near, very close to, or exactly on a resonance?

Facing these weaknesses, we look for another way to calculate the beam's physical parameters directly from the transport matrices (6×6 with general coupling) without resorting to the auxiliary functions, and in effect bypass the Courant-Snyder formalism altogether. After all, as mentioned, these transport matrices contain all the information we need for the arbitrarily coupled systems under study,[32] so auxiliary functions, step 2, should not be needed if we can directly link them to the physical quantities we want to compute.

The SLIM alternative We just mentioned a three-step process of accelerator physics and suggested that the middle step was an artifact. At this point, it is natural to contemplate the possibility of skipping the middle step and all the auxiliary functions. A moment's reflection shows that the way to proceed is to resort to the eigenvalues and eigenvectors of the transport matrices.[33] Schematically, we look for an alternative scheme as follows.

[31] Examples are so abundant that explicit enumerating is not useful. In retrospect, in fact, it is amazing that we would go so far just to maintain a formalism that has so limited an applicability.

[32] More accurately, we are talking about transport matrices for linear optics, and transport maps for nonlinear optics. Here in the SLIM formalism, we only deal with the linear cases.

[33] Alexander W. Chao, J. Appl. Phys. 50, 595 (1979); A.W. Chao, Nucl. Inst. Meth. 180, 29 (1981).

Traditional scheme:

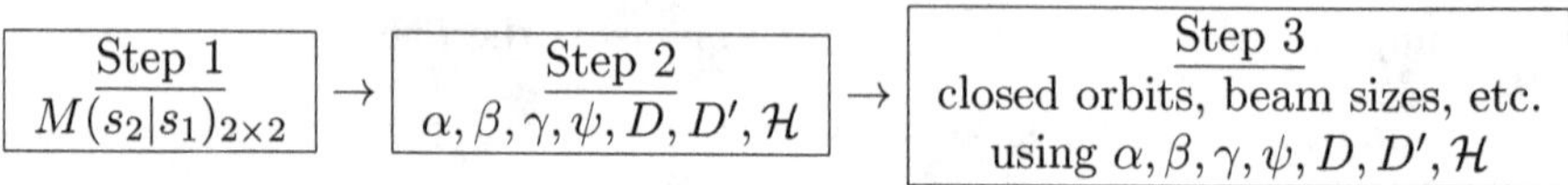

Eigenvector (SLIM) scheme:

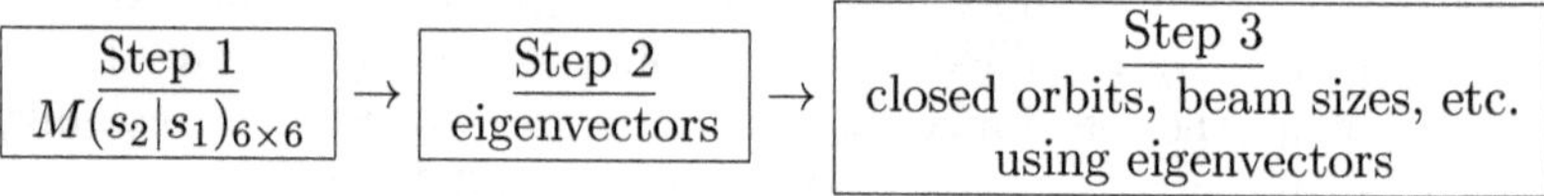

Note that $M(s_2|s_1)$'s in the eigenvector scheme now refer to general, coupled, 6×6 transport matrices, while the traditional scheme is limited to 2×2 uncoupled cases. Note also that the set of all eigenvalues and eigenvectors contains all the information contained in a transport matrix. There is no loss (and of course also no gain) of information by resorting to the eigen-quantities.

The two schemes are very different in their required scales of effort. In the conventional scheme, accelerator physics textbooks spend more than half of its volume to develop the auxiliary functions and then to derive painstakingly the analytical expressions one by one of all its physical quantities valid only in the 1-D uncoupled cases. In the eigenvector scheme, in contrast, step 2 is just a one-line statement of calling an eigenanalysis subroutine. Step 3 is to calculate the physical quantities directly from the eigenvalues and eigenvectors. Since all physics are contained in the eigenvalues and eigenvectors, this scheme offers the calculation of all desired physical quantities rather directly.

A program called SLIM was written based on the SLIM formalism. One sign of its conciseness is seen by the fact that the code has only 1000 lines. For readers who are used to the conventional scheme, it perhaps is worthwhile to give a few specific examples of what the SLIM code does beyond the conventional scheme.

- There is no need of calculation of the auxiliary functions.
- There is no need to derive analytical formulae of physical quantities in terms of the auxiliary functions — those formulae are valid only approximately and only for uncoupled 1-D cases anyway.
- The accelerator lattice, and therefore its transport matrices, can be fully coupled as needed. Coupling elements such as skew quadrupoles, solenoids, RF cavities at dispersive locations, crab cavities are treated naturally on equal footing as quadrupoles, dipoles and drifts.
- Synchrotron and betatron motions are treated on equal footing. Synchrotron oscillation does not have to be slow. For a ring with strong longitudinal focusing as discussed in Sec. 5.3.6, for example, the adiabatic approximation and the traditional analysis breaks down. SLIM remains applicable.

- All linear resonances (i.e. $\nu_x = k, \nu_y = k, \nu_s = k, 2\nu_x = k, 2\nu_y = k, 2\nu_s = k, \nu_x \pm \nu_y = k, \nu_x \pm \nu_s = k, \nu_y \pm \nu_s = k$, where k covers all integers) are covered simultaneously and exactly without approximation of weak resonances.
- One example application of SLIM was already mentioned in Fig. 6.34 when we calculated the bunch length in the limit $|\eta| \to 0$. The conventional formula predicts zero bunch length. SLIM calculates the bunch length and energy spread accurately and in agreement with a computer simulation.
- The core of SLIM is a single call to the eigenanalysis. When the 6-D phase space dynamics is studied, it treats $(x, x', y, y', z, \delta)$ on equal footing; its transport matrices are 6×6. If spin dynamics is to be included, the same single call of eigenanalysis also applies to the spin dynamics; only the transport matrices now become 8×8.[34] The SLIM code basically stays at 1000 lines when spin dynamics is included.

Flowchart Figure 6.43 shows a flow chart of the original SLIM code. The input is the storage ring lattice consisting of the various beamline elements and various coupling elements. The calculation steps will be discussed in the remaining of this section. In the SLIM input, only when the flag ISPIN is set to 1 will the polarization calculations be carried out.

Calculating physical quantities using eigenvectors We claimed that all physical quantities can be obtained from the eigenvalues and eigenvectors of the 6×6 transport matrices (8×8 if including spin) without the need of auxiliary functions or the derivation of analytical closed forms for a myriad of physical quantities. In the following sections, we shall elaborate on that claim.

To set up the problem, let us ignore spin dynamics in the following discussions. In the SLIM formalism, each linear element in the storage ring lattice is represented by a 6×6 transport matrix which transforms the state vector

$$X = \begin{bmatrix} x \\ x' \\ y \\ y' \\ z \\ \delta \end{bmatrix}$$

as an electron passes through the element. Nonlinear elements are approximated by linearizing their electromagnetic fields around the origin of coordinate system. Knowing the transport matrices around the storage ring, the distribution parameters can be obtained from their eigenvalues and eigenvectors.

Ignoring nonlinear perturbations, by the central limit theorem, we know the equilibrium particle distribution in an electron storage ring is Gaussian in the

[34] Alexander W. Chao, Nucl. Instr. Meth. 180, 29 (1981).

Figure 6.43: Flow chart of the SLIM program. The solid boxes are the main outputs. The dashed boxes are for reference only for the purpose of comparing with conventional codes, and are not used in subsequent calculations.

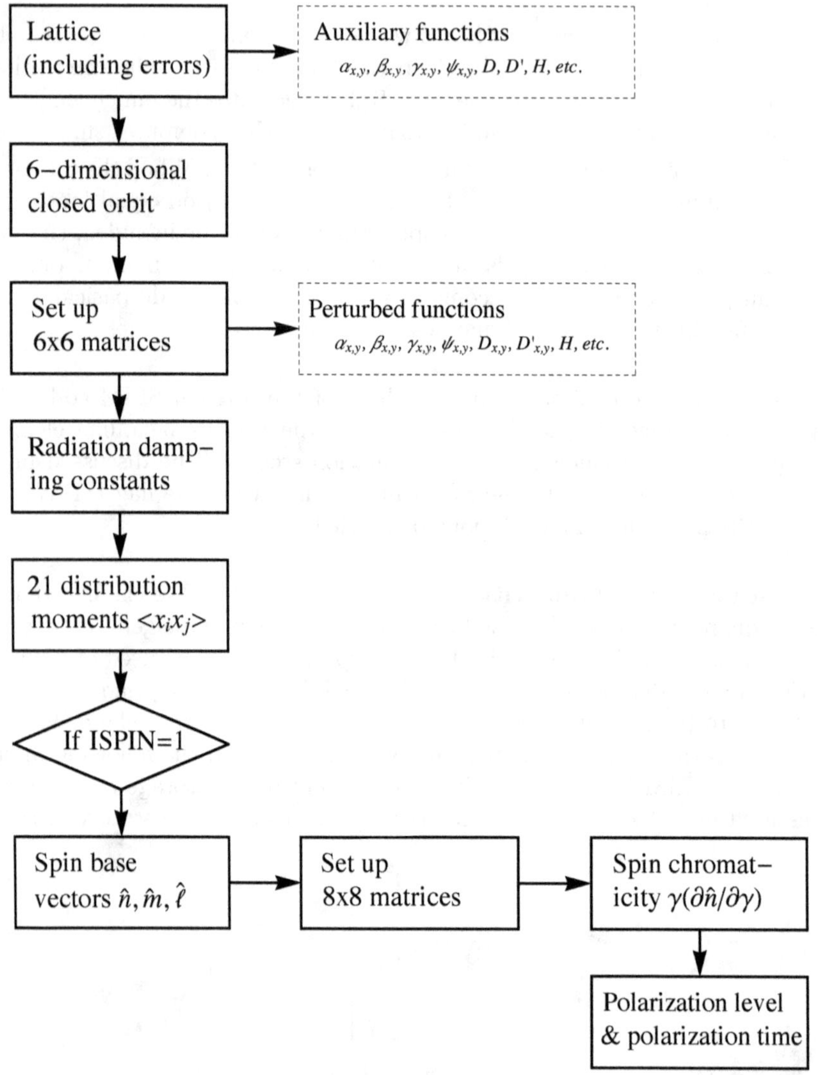

coordinates of X. To describe this particle distribution, it is necessary and sufficient to specify the distribution parameters which include the six first-moments $\langle X_i \rangle$ and the 21 second-moments $\langle X_i X_j \rangle, i, j = 1, 2, \cdots, 6$. These 27 parameters uniquely determine the beam distribution. The SLIM formalism serves as a straightforward method of obtaining the beam distribution parameters under a wide variety of storage ring operating conditions.

By adding the spin coordinates to the analysis and treating them all on equal footing, SLIM also calculates all the spin polarization related physical quantities without changing the calculation framework and using the same SLIM code.

Homework 6.62 This is an elementary homework aiming to illustrate the point that for the 1-D case, the Courant-Snyder parameters are contained implicitly in the eigenvalues and eigenvectors of the one-turn matrix maps. SLIM is just an extension of this observation to higher dimension phase space.

The one-turn Courant-Snyder map around a position s can be written as

$$\begin{bmatrix} \cos\mu + \alpha\sin\mu & \beta\sin\mu \\ -\frac{1+\alpha^2}{\beta}\sin\mu & \cos\mu - \alpha\sin\mu \end{bmatrix}.$$

(a) Find its eigenvalues and eigenvectors.

(b) Observe that the betatron phase advance μ enters the eigenvalues, while the Courant-Snyder functions α, β enter the eigenvectors.

(c) Observe that μ is invariant when transformed to another position s', while α and β change.

You are familiar with these results already.

6.5.2 Closed orbit

Relative to the ideal design trajectory, the motion of an electron can be decomposed into two components, X_0 and X, with X_0 describing the trajectory of the beam distribution center and X being the oscillatory deviation from X_0. Here, X_0 and X are 6-D vectors.

The fact that $X_0 \neq 0$ could be due to several causes. Some of them are well-known, e.g. a quadrupole magnet misaligned in its transverse position, or a dipole magnet misaligned in its roll angle will kick the beam transversely. Additional dipole magnets may exist in the lattice for orbit correction or beam injection purposes. A lesser-known example could be the sawtooth effect due to the discrete energy losses/gains at the dipoles and the RF cavities, as discussed in Sec. 5.5.5. In SLIM, these orbit distortion effects are treated all inclusively in one 6-D calculation including the coupling effects present in the lattice.

Once all the lattice elements and their errors are specified in the lattice, the closed orbit is calculated from the condition that the resulting 6-D phase space closed-orbit vector X_0 closes onto itself after one complete turn. To do so, we first form a seven-dimensional vector V by adding to X_0 a seventh component which is always given by unity, i.e.

$$V(s) = \begin{bmatrix} x_0 \\ x_0' \\ y_0 \\ y_0' \\ z_0 \\ \delta_0 \\ 1 \end{bmatrix}.$$

The vector $V(s)$ describes the trajectory of beam distribution center as a function of s. Individual particles in the beam in general oscillate around this 6-D center with three eigenfrequencies in three eigenmodes.

The vector V_f at the exit of a given lattice element is linearly related to the vector V_i at the entrance by $V_f = MV_i$. We first find these 7×7 matrices M for every element in the storage ring lattice. In general, the M matrix looks like

$$M = \left[\begin{array}{c|c} 6 \times 6 & \\ \text{transport} & \vec{d} \\ \text{matrix} & \\ \hline 0\,0\,0\,0\,0\,0 & 1 \end{array}\right] \tag{6.54}$$

The upper-left 6×6 corners of these M matrices are the usual transport matrices. The 7th row of all M matrices is always filled with [0 0 0 0 0 0 1]. The remaining upper-right 6×1 vector $\vec{d}$ is the cause of orbit distortion.

For simplicity, we assume that all lattice elements aside from the drift spaces and solenoids are short compared with the focal lengths so that thin-lens approximations apply. Generalization to the thick-lens case is straightforward. Under the thin-lens condition, the diagonal elements of M are always equal to unity and off-diagonal elements except those listed in Table 6.3 vanish. Solenoids, being intrinsically thick elements, are not listed in Table 6.3. Although not included here, they are included in the SLIM program.

In Table 6.3, we have defined ℓ = length of the lattice element, $B\rho$ = particle rigidity, $2\pi R$ = circumference of the storage ring, E_0 = design particle energy, $C_\gamma = \frac{4\pi r_0}{3(mc^2)^3}$. For the RF cavities, we define $\hat{V}$ = peak applied voltage, h = harmonic number, ϕ_s = synchronous phase. For the crab cavities, $\hat{V}_\perp$ is the peak transverse voltage applied to a traversing electron, $h_\perp$ is the harmonic number, and we assume the beam crosses the crab cavity at zero synchronous phase.

In Table 6.3, as mentioned, the upper left 6×6 corners of these matrices are the thin-lens version of the usual transport matrices. The element lengths ℓ are used only in the calculation of synchrotron radiation energy losses. They do not enter any lattice calculations.

More detailed discussions of Table 6.3 are given below.

- For DRF, QUA, SKQ elements, $\vec{d} = \vec{0}$.
- For a HB with bending angle θ, a particle going through the HB with initial coordinates $x_i, x'_i, y_i, y'_i, z_i, \delta_i$ will come out with x_f, y_f, y'_f unchanged from their initial values. The change in z_f is simply that determined by the usual transport matrix, i.e. $z_f = z_i - \theta x_i$, and this yields the m_{51} element. The change in δ_f is caused by an energy loss due to synchrotron radiation in the HB. This energy loss is given by $\frac{C_\gamma E_0^4 \theta^2}{2\pi\ell}$, leading to $\delta_f = \delta_i - \frac{C_\gamma E_0^3 \theta^2}{2\pi\ell}$. Written in the 7×7 matrix language, this gives the m_{67} term, this time

Table 6.3: Nonvanishing off-diagonal matrix elements of the 7×7 transport matrices in the thin-lens approximation. All diagonal elements are 1 for these elements.

Lattice element	Matrix elements
Drift space (DRF)	$m_{12} = m_{34} = \ell$
Horizontal bending magnet (HB)	$m_{26} = -m_{51} = \frac{B_y \ell}{B\rho}$ $m_{67} = -\frac{C_\gamma E_0^3}{2\pi\ell}\left(\frac{B_y \ell}{B\rho}\right)^2$ $m_{27} = \frac{m_{67}}{2}\frac{B_y \ell}{B\rho}$
Vertical bending magnet (VB)	$-m_{46} = m_{53} = \frac{B_x \ell}{B\rho}$ $m_{67} = -\frac{C_\gamma E_0^3}{2\pi\ell}\left(\frac{B_x \ell}{B\rho}\right)^2$ $m_{47} = -\frac{m_{67}}{2}\frac{B_x \ell}{B\rho}$
Quadrupole magnet (QUA)	$-m_{21} = m_{43} = \frac{\ell}{B\rho}\frac{\partial B_y}{\partial x}$
Skew quadrupole magnet (SKQ)	$-m_{23} = -m_{41} = \frac{\ell}{B\rho}\frac{\partial B_y}{\partial y}$
RF cavity (CAV)	$m_{65} = -\frac{e\hat{V} h \cos\phi_s}{RE_0}, \quad m_{67} = \frac{e\hat{V}\sin\phi_s}{E_0}$
Crab cavity (x-CRAB)	$m_{25} = m_{61} = -\frac{e\hat{V}_\perp h_\perp}{RE_0}$
Crab cavity (y-CRAB)	$m_{45} = m_{63} = -\frac{e\hat{V}_\perp h_\perp}{RE_0}$
Horizontal dipole kicker (HK)	$m_{26} = -m_{51} = -m_{27} = \frac{B_y \ell}{B\rho}$ $m_{67} = -\frac{C_\gamma E_0^3}{2\pi\ell}\left(\frac{B_y \ell}{B\rho}\right)^2$ $m_{27} = \frac{m_{67}}{2}\frac{B_y \ell}{B\rho}$
Vertical dipole kicker (VK)	$-m_{46} = m_{53} = m_{47} = \frac{B_x \ell}{B\rho}$ $m_{67} = -\frac{C_\gamma E_0^3}{2\pi\ell}\left(\frac{B_x \ell}{B\rho}\right)^2$ $m_{47} = -\frac{m_{67}}{2}\frac{B_x \ell}{B\rho}$

in the $\vec{d}$ column. To find x'_f, we note that the change in x' in a bending magnet is given by $\theta\bar{\delta}$, where $\bar{\delta}$ is the relative energy deviation of the particle in the middle of the HB. We take $\bar{\delta} = \frac{1}{2}(\delta_i + \delta_f) = \delta_i - \frac{C_\gamma E_0^3 \theta^2}{4\pi\ell}$. This then gives $x'_f = x'_i + \theta\delta_i - \theta\frac{C_\gamma E_0^3\theta^2}{4\pi\ell}$, yielding the m_{26} and m_{27} elements. Table 6.3 then sums up these calculations for the HB element.

- Calculations for VB, HK, VK elements are similar to those for HB. The difference between kickers and bends comes from the choice of reference design orbit.
- For CAV, the only change occurs in δ_f. We note that $\delta_f = \delta_i + \frac{e\hat{V}}{E_0}\sin(\phi_s + \frac{hz_i}{R})$. Linearizing with respect to z_i then gives $\delta_f = \delta_i + \frac{e\hat{V}\sin\phi_s}{E_0} +$

$\frac{e\hat{V}h\cos\phi_s}{RE_0}z_i$. The m_{67} term at the RF together with the m_{67} terms at all the HBs, VBs, HKs, VKs drive the RF sawtooth orbit distortion.

The trajectory of the particle distribution center, described by $V(s)$ with s indicating the position around the storage ring, must satisfy the following set of conditions,

$$\begin{aligned} x_0, x_0', y_0, y_0'(s+2\pi R) &= x_0, x_0', y_0, y_0'(s)\,, \\ \text{net energy change around one turn} &= 0\,, \\ \text{net path length traveled around one turn} &= \text{exactly } 2\pi R\,. \end{aligned}$$

where the last condition is imposed by the fact that the RF cavity frequency is accurately fixed.

To find $V(s)$, we first look for the transformation matrix for one revolution, $W(s)$, by multiplying the M matrices of all lattice elements successively from s to $s+2\pi R$. The closed orbit vector $V(s)$ is then simply given by the condition that $V(s)$ must return to its starting value when the particle completes one full turn, i.e.

$$W(s)V(s) = V(s)\,.$$

This equation, we note, is also saying that $V(s)$ is an eigenvector of $W(s)$ with eigenvalue 1. That one of the eigenvalues of $W(s)$ must be equal to 1 can be easily proved. The trajectory of the particle distribution center obtained here, $X_0(s)$, is the closed orbit of the beam. See Homework 6.64.

The closed-orbit obtained this way should not be taken too lightly. It should be emphasized that what one finds here is an orbit in the 6-D phase space. In the nominal uncoupled case, its 1st and 3rd components give the usual x_0 and y_0 obtained in conventional theories. But what is obtained here contains more, such as x-y coupled closed orbit, synchrobetatron coupled orbit, the RF sawtooth orbit, etc. It also yields the transverse orbit slopes x_0', y_0', the longitudinal closed orbit z_0 and the energy shift closed orbit δ_0 around the ring. Note that no analytical expressions in terms of auxiliary functions have been invoked.

Iteration for nonlinear elements So far, the closed orbit calculation takes into account only the linear lattice elements as listed in Table 6.3. Effects on closed orbit due to nonlinear elements can also be treated by iteration. A sextupole, for example, is represented by a horizontal and a vertical kicker with nonzero transformation matrix elements

$$\begin{aligned} \text{SEXT:} \qquad m_{26} = -m_{51} = -m_{27} &= \frac{1}{2}\lambda(x_0^2 - y_0^2)\,, \\ -m_{46} = m_{53} = m_{47} &= \lambda x_0 y_0\,, \end{aligned} \tag{6.55}$$

where $\lambda = \frac{\ell}{B\rho}\frac{\partial^2 B_y}{\partial x^2}$ is the strength of the sextupole. Nonlinear effects due to radiation energy losses in quadrupole, skew quadrupole, sextupole magnets, and crab cavities can also be included by inserting the following additional matrix elements,

$$\begin{aligned}
\text{SEXT:} \quad m_{67} &= -\frac{C_\gamma E_0^3 \lambda^2}{8\pi\ell}(x_0^2+y_0^2)^2\,, \\
\text{SKQ:} \quad m_{67} &= -\frac{C_\gamma E_0^3}{2\pi\ell}\left(\frac{\ell}{B\rho}\frac{\partial B_y}{\partial y}\right)^2 (x_0^2+y_0^2)\,, \\
\text{QUA:} \quad m_{67} &= -\frac{C_\gamma E_0^3}{2\pi\ell}\left(\frac{\ell}{B\rho}\frac{\partial B_y}{\partial x}\right)^2 (x_0^2+y_0^2)\,, \\
\text{CRAB:} \quad m_{67} &= -\frac{C_\gamma E_0^3}{2\pi\ell}\left(\frac{e\hat{V}_\perp h_\perp}{RE_0}\right)^2 z_0^2\,.
\end{aligned} \tag{6.56}$$

The last term comes from synchrotron radiation loss due to the transverse bending at the crab cavity.

The additional matrix elements listed in Eqs. (6.55) and (6.56) are small. Unlike the elements listed in Table 6.3, they also depend on the closed orbits themselves, so a calculation of the closed orbit when they are taken into account can only be obtained by iteration. The iteration steps are as follows: (i) calculate V without nonlinearities, (ii) modify transformation matrices for nonlinear elements according to Eqs. (6.55) and (6.56), (iii) recalculate V with the new matrices, (iv) iterate. One or two iterations should be enough for convergence.

As a by-product of the closed orbit calculation, the total energy loss per turn due to synchrotron radiation can be obtained from

$$U_0 = \sum_{\text{CAV}} e\hat{V}\left(\sin\phi_s - \frac{h}{R}z_0\cos\phi_s\right),$$

where z_0 is the longitudinal component of the closed orbit. As seen in the matrix elements listed, this U_0 includes the radiation losses in all linear and nonlinear magnets as well as crab cavities. It also is calculated with the presence of general coupling.

Homework 6.63 Derive some of the elements listed in Table 6.3 to gain familiarity with these matrices.

Homework 6.64 The 7×7 one-turn matrix $W(s)$ is obtained by multiplying the element matrices, each of which has the form (6.54).

(a) Show that the end product $W(s)$ also has the same form (6.54).

(b) Show that $W(s)$ always has an eigenvector with eigenvalue 1. As mentioned in the text, this eigenvector is just the 6-D closed orbit at position s.

(c) The proof of (b) indicates that in this coupled linear system, a closed orbit always exists. However, this statement is not 100% water tight. What are the conditions when an eigenvector of eigenvalue 1, and therefore the closed orbit, does not exist?

Solution

(b) Let the one-turn transport matrix be expressed as (6.54). An eigenvector with eigenvalue 1 means $W_{7\times7}\vec{V}_7 = \vec{V}_7$, where the subscripts have obvious meanings. Rewritten in a 6×6 form, we have $W_{6\times6}\vec{V}_6 = \vec{V}_6 - \vec{d}$ where $\vec{V}_6$ is just the closed orbit X_0. A solution exists and is given by

$$\vec{V}_6 = (W_{6\times6} - I)^{-1}\vec{d}.$$

(c) The conditions when the closed orbit does not exist occur when determinant of $W_{6\times6} - I$ is zero. It is easy to show that these singular conditions occur when any one of the six eigenvalues of $W_{6\times6}$ is equal to 1, or equivalently, when any of the three eigentunes is equal to an integer.

This seemingly purely mathematical exercise concludes what you knew already, i.e. there is no closed orbit if any one of the tunes is on an integer resonance (Fig. 3.3). Note that here we include all three tunes. A synchrotron tune of $\nu_s = 0$, as would be the case for an unbunched beam, for example, would not have a 3-D closed orbit, although it does have a 2-D one.

6.5.3 Coupling effects

Once the closed orbit X_0 is obtained, we next calculate the transport matrices around it. A particle's motion is described by the state vector X, which is the deviation of the particle's coordinates from the closed orbit X_0. The 6×6 transport matrices describing the motion for X are given by linearizing around X_0. For linear beamline elements, they are just the upper left 6×6 corners of the 7×7 matrices in Table 6.3. For nonlinear elements, such as for sextupoles, by linearizing around the closed orbit, a sextupole becomes a combination of HK, VK, SKQ and QUA due to feed-down effects,

$$\begin{bmatrix} 1 & 0 & 0 & 0 & 0 & 0 \\ -\lambda x_0 & 1 & \lambda y_0 & 0 & 0 & \frac{\lambda}{2}(x_0^2 - y_0^2) \\ 0 & 0 & 1 & 0 & 0 & 0 \\ \lambda y_0 & 0 & \lambda x_0 & 1 & 0 & -\lambda x_0 y_0 \\ -\frac{\lambda}{2}(x_0^2 - y_0^2) & 0 & \lambda x_0 y_0 & 0 & 1 & 0 \\ 0 & 0 & 0 & 0 & 0 & 1 \end{bmatrix}. \tag{6.57}$$

The purpose of introducing 7×7 formalism is to calculate the closed-orbit. After the closed orbit is obtained, from here on, we will concentrate on these 6×6 transport transformations. By successively multiplying all these matrices from s to $s+2\pi R$, one obtains the transformation for one revolution, $T(s)$, which will be used extensively next. The matrix $T(s)$ depends on s, the location where the one-turn matrix is evaluated.

All linear 6-D coupling effects are now contained in the matrices $T(s)$. Because of its symplecticity, the six eigenvalues of $T(s)$ can be expressed by

$$\lambda_k = e^{i2\pi\nu_k}, \qquad k = \pm I, \pm II, \pm III, \tag{6.58}$$

with $(I, -I), (II, -II), (III, -III)$ forming three pairs,

$$\nu_{-k} = -\nu_k, \qquad k = I, II, III.$$

Calculation of these eigenvalues then immediately gives three eigentunes $\nu_{I,II,III}$. In the nominal case, they are the horizontal, vertical, and synchrotron tunes. In case of an arbitrary coupled system, they are the tunes of the three eigenmodes. The three eigenmodes are treated on equal footing without distinction among themselves.

If any one of the eigentunes turns out imaginary, the storage ring is unstable. This may happen even when all the unperturbed motions are stable but becomes unstable due to 3-D coupling, e.g. x-y coupling near a sum resonance or x-z coupling near a difference synchrobetatron resonance.

Each eigenmode also has a pair of eigenvectors. The six eigenvectors are denoted $E_{I,II,III}$ and $E_{-I,-II,-III}$. We will discuss more about these eigenvectors later. As mentioned, all beam dynamics contained in the transport matrices are now contained in these eigenvalues and eigenvectors. Our next job is then to construct all remaining physical quantities out of these eigenvectors, without resorting to other auxiliary lattice functions.

Linear resonances, whether x-y or synchrobetatron in nature, are treated on equal footing. When we calculate the beam parameters later, these coupling effects are included automatically and no approximations such as weak coupling are necessary. The calculation is exact, even very close to the resonances. To be more specific, the calculation as given here is valid if the tunes stay away from the resonances by a distance of the order of the radiation damping constants, whose effect — nonsymplectic — has yet to be discussed in Sec. 6.5.4. Since the radiation damping time constants are typically very small, the sacrifice of parameter space around the resonances is negligible.

Homework 6.65

(a) Show that the map (6.57) is symplectic.

(b) The text has not included the case of skew sextupoles. Find the necessary additions to include skew sextupoles in SLIM. Start with additions in Table 6.3 and include its linearized feeddown map.

(c) Extend again to include octupoles.

6.5.4 Radiation damping

Radiation damping originates from two effects, one at the magnets and one at the cavities, as discussed below.

1. The first effect is that in a bending element, a particle with higher (lower) energy will lose more (less) energy due to synchrotron radiation. This

effect leads to a longitudinal radiation damping. This results in the following changes of the diagonal m_{66} elements:

$$\begin{aligned}
\text{HB, VB, HK, VK:} \qquad m_{66} &= 1 - \frac{C_\gamma E_0^3}{\pi \ell}\left(\frac{\ell B_{x,y}}{B\rho}\right)^2, \\
\text{QUA, SKQ:} \qquad m_{66} &= 1 - \Lambda_q (x_0^2 + y_0^2), \\
\text{SEXT:} \qquad m_{66} &= 1 - \frac{\Lambda_s}{2}(x_0^2 + y_0^2),
\end{aligned}$$

where $\Lambda_q = \frac{C\gamma E_0^3}{\pi\ell}\left(\frac{\ell}{B\rho}\frac{\partial B_y}{\partial x}\right)^2$ or $\frac{C\gamma E_0^3}{\pi\ell}\left(\frac{\ell}{B\rho}\frac{\partial B_y}{\partial y}\right)^2$, and $\Lambda_s = \frac{C\gamma E_0^3}{2\pi\ell}\lambda^2(x_0^2 + y_0^2)$. Derivation of these m_{66} elements should be straightforward. But note that we are now modifying the *diagonal* elements of the transport matrices, and such modifications, otherwise forbidden, affect the symplecticity of the matrices.

In addition, energy radiated in quadrupole, skew quadrupole, and sextupole magnets also depends on the transverse displacements of an electron. This contributes to the mixing among the transverse and longitudinal damping effects. Upon linearization, it yields

$$\begin{aligned}
\text{QUA, SKQ:} \qquad m_{61} = -\Lambda_q x_0, \quad m_{63} = -\Lambda_q y_0\,, \\
\text{SEXT:} \qquad m_{61} = -\Lambda_s x_0, \quad m_{63} = -\Lambda_s y_0\,.
\end{aligned}$$

These changes will be the cause of the damping partition number $\mathcal{D}$ of Eq. (6.31) for the ideal case of a planar ring.[35] For the general coupled 3-D cases, this calculation gives the three damping constants without further ado.

To derive this result for a QUA, for example, we note that the radiated energy loss in a quadrupole is given by $\frac{C_\gamma E_0^2 E^2}{2\pi\ell}\left(\frac{\ell \vec{B}}{B\rho}\right)^2$. In a quadrupole, the magnetic field seen by a particles is given by $\vec{B} = \frac{\partial B_y}{\partial x}[(x_i + x_0)\hat{y} + (y_i + y_0)\hat{x}]$, where x_0 and y_0 are the closed orbit displacements at the quadrupole. Keeping linear terms in x_i, y_i and δ_i (but not linearizing with respect to x_0 and y_0), we obtain

$$\begin{aligned}
\delta_f &= \delta_i - \frac{\Lambda_q}{2}(1+\delta_i)^2[(x_0 + x_i)^2 + (y_0 + y_i)^2] \\
&\approx \delta_i[1 - \Lambda_q(x_0^2 + y_0^2)] - \frac{\Lambda_q}{2}(x_0^2 + y_0^2) - \Lambda_q[x_0 x_i + y_0 y_i]\,,
\end{aligned}$$

which reproduces the m_{67}, m_{66} terms already obtained before, but also gives two new coupling terms m_{61} and m_{63}.

[35] Actually, the nominal derivation for $\mathcal{D}$ takes into account only the contributions from the quadrupoles. It does not consider the effects of the other magnets.

2. The other effect is that as an electron passes through an RF cavity, it receives a boost in its longitudinal momentum, while keeping the transverse momentum unchanged. Consequently, the slopes x' and y' of the electron's trajectory are reduced in magnitude by a factor of $1 - \frac{u}{E_0}$, where $u = e\hat{V}[\sin\phi_s - \frac{h}{R} z_0 \cos\phi_s]$ is the energy gain of the electron at the RF cavity. This effect provides the transverse radiation damping to the motion of an electron. In the matrix formalism, it is taken into account by replacing the diagonal m_{22} and m_{44} elements in the CAV matrices,

$$m_{22} = m_{44} = 1 - \frac{e\hat{V}}{E_0}\left(\sin\phi_s - \frac{h}{R} z_0 \cos\phi_s\right).$$

After these modifications, the diagonal elements are no longer given by 1's, even in the thin-lens approximation. The symplecticity condition is violated — as it should in this calculation.

After making these modifications, one calculates once again the 6×6 transformation matrix for one revolution, $D(s)$, by multiplying matrices from s to $s + 2\pi R$. The eigenvalues of $D(s)$ are expressed as

$$\lambda_k = e^{-\alpha_k + i2\pi\nu_k}, \qquad k = \pm I, \pm II, \pm III,$$

with

$$\nu_{-k} = -\nu_k, \qquad \alpha_{-k} = \alpha_k,$$

where ν_k remains the tune of the k-th mode, and the new parameters α_k's are the radiation damping decrements.

In particular, if anyone of the α_k's is negative, motion of the electron will be unstable. As to the eigentunes obtained in this new calculation, they will differ from the values obtained without the radiation damping modifications. The difference, however, is exceedingly small, second order in the α_k's, and can be safely ignored.

By evaluating the eigenvalues of the one-turn map $D(s)$, we have thus derived a method of calculating the radiation damping constants in the presence of various coupling effects. As mentioned, the contribution of the partition number $\mathcal{D}$ is automatically included.

Furthermore, using the property that the product of all six eigenvalues of a matrix is equal to its determinant, we obtain

$$\prod_{k=\pm I, \pm II, \pm III} \lambda_k = \exp[-2(\alpha_I + \alpha_{II} + \alpha_{III})] = \det[D(s)].$$

For weak damping with $|\alpha_k| \ll 1$, the determinant of the nearly-symplectic matrix $D(s)$ is equal to the product of all its diagonal elements, all of them are

1's except for those we just modified for the calculation of radiation damping constants. The Robinson sum rule, Sec. 6.4.6, then follows,

$$\alpha_I + \alpha_{II} + \alpha_{III} = \frac{2U_0}{E_0}.$$

All the couplings do not affect the validity of the Robinson sum rule.

Homework 6.66 For the special case of no orbit distortion and no coupling, the SLIM formalism can be performed analytically, and the result agrees with the conventional calculation. A more curious reader may try this instructive exercise. Set up the SLIM framework for an ideal uncoupled planar storage ring, calculate the three radiation damping times. Show that they agree with the results obtained by the conventional approach, including the contribution of the partition number $\mathcal{D}$.[36]

Homework 6.67 We derived a radiation damping partition number $\mathcal{D}$ earlier, making the conventional assumptions of 1-D uncoupled motion and slow synchrotron oscillation. The result is particularly applied to note the important influence of combined-function dipoles to the radiation damping mechanism. Does a crab cavity also have a similar role in affecting the partition number?

Solution Although a crab cavity, like a bending magnet, also bends the beam trajectory transversely and is a radiation source, its effect is to introduce a x-z coupling and it does not directly affect the radiation damping partition number. Multiply a few SLIM matrices to show this.

Homework 6.68 In Sec. 5.5.3, we analyzed the coupling effect when an RF cavity location is dispersive. In particular, we found the perturbed eigentunes and the corresponding stopbands. In this section, we described how radiation damping can be included in the matrix formulation. Combine these efforts to find analytical expressions of the perturbed radiation damping constants $\alpha_{x,z}$ when the RF cavity is dispersive. Do this to first order in $\frac{U_0}{E_0}$. It suffices to consider a 4-D phase space for this problem.

Solution to simplify the analysis, you may simply replace the cavity map (5.57) by

$$M_{\rm cav} = \begin{bmatrix} 1 & 0 & 0 & 0 \\ 0 & 1-\frac{U_0}{E_0} & 0 & 0 \\ 0 & 0 & 1 & 0 \\ 0 & 0 & k & 1-\frac{2U_0}{E_0} \end{bmatrix}.$$

Homework 6.69 A similar problem as Homework 6.68 applies to the case when a skew quadrupole is present to couple the x-y motion. Find the effect of coupling on the two radiation damping constants $\alpha_{x,y}$.

[36] Boaz Nash, Ph. D. Thesis, Stanford University, 2006.

6.5.5 Equilibrium beam size and shape

SLIM can be used to determine the beam size and shape quantities, such as the horizontal and vertical beam sizes, the tilt angle in the x-y plane, the rms bunch length and energy spread, etc. What SLIM calculates is the 21 distribution moments $\langle X_i X_j \rangle$. To do this, let us consider a photon being emitted at position s_0 with energy u. Let $T(s_0)$ be the coupled 6×6 transport matrix for one revolution obtained earlier without radiation damping. The eigenvalues, λ_k, and eigenvectors, $E_k(s_0)$, of $T(s_0)$ are defined by

$$T(s_0)E_k(s_0) = \lambda_k E_k(s_0), \qquad k = \pm I, \pm II, \pm III\,. \tag{6.59}$$

It can be shown, using the fact that $T(s_0)$ is real and symplectic, that we can choose the eigenvectors in such a way that

$$\begin{aligned} E_k^*(s_0) &= E_{-k}(s_0)\,, \\ \lambda_{\pm k} &= e^{\pm i2\pi\nu_k}\,. \end{aligned}$$

There are three eigenmodes. The six eigenvalues and six eigenvectors form three pairs, each corresponding to one of the eigenmodes. In the absence of coupling, these they reduce to the horizontal betatron, vertical betatron and synchrotron modes. With coupling in general, there are still three eigenmodes; they are referred to as modes I, II, and III.

In passing, the reader should note that we have explicitly included the s_0 dependence in $E_k(s_0)$ but not in λ_k because λ_k is independent of s_0.

When a photon of energy u is emitted, due to energy conservation, the emitting electron suffers a sudden quantum excitation on its state vector,

$$\Delta X(s_0) = \begin{bmatrix} 0 \\ 0 \\ 0 \\ 0 \\ 0 \\ -u/E_0 \end{bmatrix}. \tag{6.60}$$

It should be pointed out here that the quantities x and y considered here in the state vector X are the total transverse displacements of a particle, not just the "betatron parts". As noted before, the distinction among the betatron and the synchrotron motions in the conventional theory, valid only when ν_s is small and in the absence of synchrobetatron coupling, is not adopted in the SLIM language. The physical coordinates x, x', y, y', z are not excited at the instant of photon emission.

Following the photon emission, the subsequent motion of the electron is described by a decomposition into the six eigenmodes according to

$$\Delta X(s) = \sum_{k=\pm I,\pm II,\pm III} A_k E_k(s), \quad s \geq s_0\,, \tag{6.61}$$

where $E_k(s)$ is the eigenvector of $T(s)$ obtained from $E_k(s_0)$ by the matrix transformation from s_0 to s. Since $\Delta X(s)$ is real, and $E_{-k} = E_k^*$, it follows that

$$A_{-k} \;=\; A_k^* \,.$$

Equation (6.61) satisfies the initial condition

$$\begin{bmatrix} 0 \\ 0 \\ 0 \\ 0 \\ 0 \\ -u/E_0 \end{bmatrix} \;=\; \sum_k A_k E_k(s_0)\,. \tag{6.62}$$

We need to calculate the decomposition coefficients A_k's from Eq. (6.62). There are six equations in (6.62) to be used to solve for the six unknowns A_k's.

From the symplecticity condition of $T(s)$, i.e.,

$$\tilde{T}(s)ST(s) \;=\; S\,,$$

where a tilde means taking the transpose of a matrix and S is the symplectic form, one can prove that

$$\tilde{E}_j S E_i \;=\; 0 \qquad \text{unless} \qquad j \;=\; -i\,. \tag{6.63}$$

We will normalize the eigenvectors so that

$$\tilde{E}_k^* S E_k \;=\; \begin{cases} i, & k = I, II, III, \\ -i, & k = -I, -II, -III. \end{cases} \tag{6.64}$$

This normalization condition is preserved as a function of s due to the symplecticity of $T(s)$. It then follows that Eq. (6.62) yields the solution

$$A_k \;=\; \begin{cases} i\frac{u}{E_0}E_{k5}^*(s_0), & k = I, II, III, \\ -i\frac{u}{E_0}E_{k5}^*(s_0), & k = -I, -II, -III, \end{cases} \tag{6.65}$$

where E_{ki} means the i-th component of the vector E_k.

Assuming all photon emission events are uncorrelated, one obtains the quantum diffusion rate of $|A_k|^2$ by averaging Eq. (6.65) over the synchrotron radiation spectrum and averaging around the storage ring,

$$\frac{d}{dt}\langle |A_k|^2\rangle \;=\; \frac{1}{2\pi R}\oint ds \left\langle \dot{\mathcal{N}}\frac{u^2}{E_0^2}\right\rangle |E_{k5}(s)|^2\,,$$

where $\dot{\mathcal{N}}$ is the average number of photons emitted per unit time and, using Eq. (6.28),

$$\left\langle \dot{\mathcal{N}}\frac{u^2}{E_0^2}\right\rangle \;=\; \frac{2C_L\gamma^5}{|\rho(s)|^3}\,,$$

where

$$C_L = \frac{55}{48\sqrt{3}} \frac{r_0 \hbar}{m}.$$

So far we have ignored the radiation damping which, when taken into account, gives an additional contribution

$$\frac{d}{dt}\langle |A_k|^2 \rangle = -\frac{2\alpha_k}{T_0}\langle |A_k|^2 \rangle,$$

with T_0 the revolution time and α_k the radiation damping constants found in Sec. 6.5.4. The equilibrium values of $\langle |A_k|^2 \rangle$ is given by a balance between quantum diffusion and radiation damping, which yields

$$\langle |A_k|^2 \rangle = \langle |A_{-k}|^2 \rangle = C_L \frac{\gamma^5}{c\alpha_k} \oint ds \frac{|E_{k5}(s)|^2}{|\rho(s)|^3}. \tag{6.66}$$

It follows from Eq. (6.61) that the particle distribution second moments at position s are given by

$$\langle X_i X_j \rangle(s) = 2 \sum_{k=I,II,III} \langle |A_k|^2 \rangle \mathrm{Re}[E_{ki}(s) E_{kj}^*(s)]. \tag{6.67}$$

In general, all three eigenmodes contribute to all the 21 distribution moments. Only in a 1-D case, and only when synchrotron motion is adiabatically slow, would the modes be identified cleanly with horizontal betatron, vertical betatron, and longitudinal synchrotron dimensions.

Equations (6.66) and (6.67) are our final expressions. In a linear system, the beam distribution is strictly Gaussian, and the 21 quantities in (6.67), together with the six first moments $\langle X \rangle = X_0$ given by the closed orbit, completely specify the equilibrium distribution, and consequently the sizes and shapes, of the beam in the 6-D phase space.

The tilt angle θ_{xy}, for example, of the x-y beam profile relative to the horizontal axis can be found from

$$\tan 2\theta_{xy} = \frac{2\langle xy \rangle}{\langle x^2 \rangle - \langle y^2 \rangle}, \tag{6.68}$$

and the rms transverse beam area in the x-y plane (e.g. for luminosity calculation) is given by

$$A_{xy} = \pi\sqrt{\langle x^2 \rangle \langle y^2 \rangle - \langle xy \rangle^2}. \tag{6.69}$$

Similar expressions can be given for example of the beam tilt in the x-z or the y-z planes, etc.

The conventional theory basically has an expression similar to Eq. (6.66) but with $\mathcal{H}$ replacing the eigenvector $|E_{k5}|^2$, and the resulting expression gives the unperturbed horizontal beam size $\langle x^2 \rangle$ for a 1-D beam. In contrast, Eq. (6.66) gives the three eigenmode emittances for a 3-D, arbitrarily coupled beam, and Eq. (6.67) gives the 21 beam distribution moments.

Homework 6.70 Following Eq. (6.59), it was stated that the eigenvectors $E_k(s_0)$ depend on the position s_0, while the eigenvalues λ_k do not depend on s_0. Prove this statement.

Homework 6.71

(a) Prove the orthogonality condition (6.63).

(b) Establish the normalization condition (6.64). Explain why the normalization conditions (6.64) are not simply normalized to 1 on the right hand side, and why they form complex conjugate pairs for each of the three modes.

(c) Show that both conditions (6.63) and (6.64) are preserved as functions of s. In other words, once it is established at one location s, the same condition will hold automatically at other locations.

(d) Prove Eq. (6.65).

Solution

(a) Consider $\tilde{E}_j S E_i = \widetilde{\tilde{E}_j S E_i} = \tilde{E}_j \tilde{T} S T E_i$.

(b) Note that $\tilde{E}_k^* S E_k = -(\tilde{E}_k^* S E_k)^*$.

Homework 6.72 Prove Eqs. (6.68) and (6.69).

Homework 6.73 The SLIM formalism can be applied to a simpler case of a feedback system. Consider the 1-D x-motion of a proton beam (no synchrotron radiation). Let the one-turn transfer matrix around position s_0 be

$$T(s_0) = \begin{bmatrix} \cos\mu & \beta\sin\mu \\ -\frac{1}{\beta}\sin\mu & \cos\mu \end{bmatrix},$$

where β is the β-function at s_0 and μ is phase advance per turn, and we have assumed α-function vanishes at s_0.

Now consider a feedback device installed at s_0. The feedback action is modeled such that it detects the value of x' at s_0 and immediately reduces it by an amount $\Delta x' = -\alpha x'$. Here x and x' refer to the coordinates of the beam centroid. This will provide a feedback action on the beam's x-motion. However, this feedback system also has a noise when kicking the beam each time it comes by, and the rms kicking angle of this noise is $\sigma_{\Delta\theta}$.

(a) The two eigenvalues λ_x, λ_{-x} and eigenvectors E_x, E_{-x} of $T(s_0)$ were found in Homework 6.62.

(b) Check the orthonormal conditions (6.63) and (6.64).

(c) Following the SLIM technique to find the feedback damping and the noise diffusion rates on the emittance, $\langle|A_x^2|\rangle$, of the beam centroid trajectory.

(d) Find the equilibrium emittance $\langle|A_x^2|\rangle$ of the beam centroid motion by balancing the damping and the diffusion rates.

(e) Find the feedback orbit distribution moments $\langle x^2\rangle, \langle xx'\rangle, \langle x'^2\rangle$ when the beam is observed at s_0.

(f) Remove the simplifying assumption that α-function vanishes. What do you expect to be the result for $\langle x^2\rangle, \langle xx'\rangle, \langle x'^2\rangle$?

6.5.6 Sideways photon

We just calculated the effect on beam distribution moments due to the quantum noise (6.60). Clearly, the same formalism can be used to calculate the effects due to other random noise sources. Here we give a few examples.

- A particle is scattered randomly when it hits the residual gas molecules in the vacuum chamber. In that case, we calculate this effect by replacing the right hand side of Eq. (6.60) by the appropriate scattering angles in x' and y'.
- Intrabeam scattering, i.e. the binary Coulomb scatterings among the beam particles, can be treated as a noise source.[37]
- Phase or amplitude noise of the RF voltage can be considered a noise source at the location of the RF with noises occurring in z and δ respectively.
- Power supply noises can be treated similarly, as illustrated for example in Homework 6.73.
- When generalized to 8×8, the spin diffusion rate due to synchrotron radiation can be calculated.

In this section, we consider yet another example, the effect of sideways synchrotron radiation noise.

Recall that Eq. (6.60) assumes the photon is emitted exactly along the direction of motion of the electron. Although a good approximation, in reality the photon can be radiated at a small opening angle $\sim \frac{1}{\gamma}$ relative to the electron's direction of motion, and these sideways emissions contribute to additional quantum diffusion effects, as illustrated in Fig. 6.37.

When an electron emits a synchrotron radiation photon not exactly in the direction of its motion, the electron receives a sideways angular kick. Let θ_x and θ_y be the direction of emission projected to the x- and y-directions respectively, the transverse recoil to the electron is given by $\Delta x' = -\frac{u}{E_0}\theta_x$ and $\Delta y' = -\frac{u}{E_0}\theta_y$.

The physical meaning of θ_x is obscured by the horizontal bending of the particle trajectory, but the noise to y' has a clean meaning. In any case, this small additional contribution is particularly relevant to the vertical dimension because the vertical beam dimension would otherwise be zero at least in the ideal planar case.

[37] J.D. Bjorken and S.K. Mtingwa, Part. Accel. 13, 115 (1983).

Consider the vertical motion, a sideways emission contributes to the initial sudden impulse, in addition to (6.60),

$$\Delta X(s_0) = \begin{bmatrix} 0 \\ 0 \\ 0 \\ -\frac{u}{E_0}\theta_y \\ 0 \\ 0 \end{bmatrix}.$$

Following similar derivation as before, we decompose the subsequent motion of the electron in terms of eigenvectors according to Eq. (6.61). This time, the decomposition coefficients are found to be

$$A_k = \begin{cases} -i\frac{u}{E_0}\theta_y E_{k3}^*, & k = I, II, III, \\ i\frac{u}{E_0}\theta_y E_{k3}^*, & k = -I, -II, -III. \end{cases}$$

By balancing these quantum diffusion contributions with radiation damping, we obtain additional contributions to the eigenemittances,

$$\langle |A_k|^2 \rangle = \langle |A_{-k}|^2 \rangle = \frac{1}{2c\alpha_k} \oint ds \left\langle \dot{\mathcal{N}} \frac{u^2}{E_0^2} \theta_y^2 \right\rangle |E_{k3}(s)|^2 .$$

We then use the expression (6.48),

$$\left\langle \dot{\mathcal{N}} \frac{u^2}{E^2} \theta_y^2 \right\rangle = \frac{23}{55} \frac{C_L \gamma^3}{|\rho(s)|^3}$$

to obtain

$$\langle |A_k|^2 \rangle = \langle |A_{-k}|^2 \rangle = \frac{23}{110} \frac{C_L \gamma^3}{c\alpha_k} \oint ds \frac{|E_{k3}(s)|^2}{|\rho(s)|^3} .$$

These contributions are to be added to the contributions (6.66) to calculate the beam's second moments (6.67).

6.5.7 Gaussian beam distribution in the 6-D phase space

We have now calculated the 6 first moments and the 21 second moments of the beam distribution in its 6-D phase space. Together with the knowledge that the beam in equilibrium has a Gaussian distribution, we have now the complete knowledge of the beam distribution. However, it would still be desirable to be able to express the beam distribution function $\psi(X)$ in the 6-D phase space explicitly in terms of the 6 first-moments and the 21 second-moments.

Let us observe the beam at a position s, which has a 6×6 one-turn transport matrix $T(s)$. The inclusion of the six first moments, the closed orbit X_0, is easy. All we have to do is to shift the center of the 6-D Gaussian by the closed orbit.

So we will concentrate on finding an expression for the Gaussian distribution around the closed orbit. The result is readily given by

$$\psi(X) = \frac{1}{(2\pi)^3\sqrt{\det\Sigma}}\exp\left(-\frac{1}{2}\tilde{X}\Sigma^{-1}X\right), \tag{6.70}$$

where Σ is the distribution's second moment matrix, whose ij-th element is $\langle X_i X_j \rangle$. But we shall also examine its internal structure in terms of three invariants as follows.

Quadratic forms of the three invariants The particle motion as observed at s has three invariants, each can be expressed as a quadratic form. Let these invariants be written as

$$g_k = \tilde{X}G_kX, \qquad k = I, II, III,$$

where G_k is a symmetric, positive definite, 6×6 matrix yet to be found.

We now define a matrix U. The six columns of U are the eigenvectors of $T(s)$ as follows,

$$U = [E_I \quad -iE_{-I} \quad E_{II} \quad -iE_{-II} \quad E_{III} \quad -iE_{-III}]. \tag{6.71}$$

This matrix U is symplectic, and it can easily be shown to have the property

$$U^{-1}TU = \begin{bmatrix} e^{i2\pi\nu_I} & 0 & 0 & 0 & 0 & 0 \\ 0 & e^{-i2\pi\nu_I} & 0 & 0 & 0 & 0 \\ 0 & 0 & e^{i2\pi\nu_{II}} & 0 & 0 & 0 \\ 0 & 0 & 0 & e^{-i2\pi\nu_{II}} & 0 & 0 \\ 0 & 0 & 0 & 0 & e^{i2\pi\nu_{III}} & 0 \\ 0 & 0 & 0 & 0 & 0 & e^{-i2\pi\nu_{III}} \end{bmatrix} \equiv \Lambda,$$

where Λ is the diagonal eigenvalue matrix.

The three invariants g_k's are determined by the matrices G_k's. Let us first prove that g_k's would be invariants if

$$G_k = -iSUH_k\tilde{U}S, \tag{6.72}$$

where

$$H_I = \begin{bmatrix} 0 & 1 & 0 & 0 & 0 & 0 \\ 1 & 0 & 0 & 0 & 0 & 0 \\ 0 & 0 & 0 & 0 & 0 & 0 \\ 0 & 0 & 0 & 0 & 0 & 0 \\ 0 & 0 & 0 & 0 & 0 & 0 \\ 0 & 0 & 0 & 0 & 0 & 0 \end{bmatrix},$$

$$H_{II} = \begin{bmatrix} 0 & 0 & 0 & 0 & 0 & 0 \\ 0 & 0 & 0 & 0 & 0 & 0 \\ 0 & 0 & 0 & 1 & 0 & 0 \\ 0 & 0 & 1 & 0 & 0 & 0 \\ 0 & 0 & 0 & 0 & 0 & 0 \\ 0 & 0 & 0 & 0 & 0 & 0 \end{bmatrix},$$

$$H_{III} = \begin{bmatrix} 0 & 0 & 0 & 0 & 0 & 0 \\ 0 & 0 & 0 & 0 & 0 & 0 \\ 0 & 0 & 0 & 0 & 0 & 0 \\ 0 & 0 & 0 & 0 & 0 & 0 \\ 0 & 0 & 0 & 0 & 0 & 1 \\ 0 & 0 & 0 & 0 & 1 & 0 \end{bmatrix}.$$

First note that these expressions of G_k's are symmetric, i.e. $\tilde{G_k} = G_k$, as they must.

To prove that g_k would be invariant under the transformation T, we need to show that

$$\tilde{T}G_kT = G_k, \qquad k = I, II, III.$$

This is proven as follows,

$$\begin{aligned} \tilde{T}G_kT &= -i\tilde{T}SUH_k\tilde{U}ST \\ &= -iST^{-1}UH_k\tilde{U}\tilde{T}^{-1}S \\ &= -iSU\Lambda^{-1}H_k\Lambda^{-1}\tilde{U}S \\ &= -iSUH_k\tilde{U}S = G_k, \qquad \text{Q.E.D.} \end{aligned}$$

The factor $-i$ in front of the definition (6.72) is to make the invariant a real and positive quantity. To show that, we introduce the matrix K given by

$$K = \frac{1}{\sqrt{2}} \begin{bmatrix} 1 & i & 0 & 0 & 0 & 0 \\ i & 1 & 0 & 0 & 0 & 0 \\ 0 & 0 & 1 & i & 0 & 0 \\ 0 & 0 & i & 1 & 0 & 0 \\ 0 & 0 & 0 & 0 & 1 & i \\ 0 & 0 & 0 & 0 & i & 1 \end{bmatrix}.$$

Matrix K is symplectic. We then consider the change of coordinates from X to the coordinates V,

$$X = UKV. \tag{6.73}$$

The transformation matrix UK is real and symplectic. This makes all the components of V real because X components are all real. The columns of UK are in fact $\sqrt{2}$ times the real and imaginary parts of the eigenvectors E_k^*. In terms of the V coordinates, the quadratic form g_k becomes

$$\begin{aligned} g_k &= \tilde{X} G_k X \\ &= -i\tilde{V}\tilde{K}\tilde{U} G_k U K V \\ &= -i\tilde{V}\tilde{K} S H_k S K V \\ &= -i\tilde{V}\tilde{K} H_k K V \\ &= \tilde{V} I_k V\,, \end{aligned}$$

Where I_k is a matrix that is all zero's except for a 2×2 identity matrix in the k-th diagonal 2×2 block. Thus,

$$\begin{aligned} g_I &= V_1^2 + V_2^2\,, \\ g_{II} &= V_3^2 + V_4^2\,, \\ g_{III} &= V_5^2 + V_6^2\,. \end{aligned} \tag{6.74}$$

are real and positive quantities. It is then straightforward to express the invariants g_k's in terms of the coordinates of X.

Equilibrium distribution Once the quadratic forms g_k's are given, the equilibrium Gaussian beam distribution can be expressed as

$$\psi(X) = \frac{1}{\pi^3 \langle g_I\rangle\langle g_{II}\rangle\langle g_{III}\rangle} \exp\left(-\frac{g_I}{\langle g_I\rangle} - \frac{g_{II}}{\langle g_{II}\rangle} - \frac{g_{III}}{\langle g_{III}\rangle}\right), \tag{6.75}$$

where $\langle g_k\rangle$'s are the expectation values of the three invariants. The beam distribution is therefore known explicitly once the values of $\langle g_k\rangle$'s are specified.

To establish Eq. (6.75), we need to show

$$\begin{aligned} \int \psi(X) d^6X &= 1\,, \\ \int \psi(X) g_k\, d^6X &= \langle g_k\rangle, \qquad k = I, II, III\,. \end{aligned} \tag{6.76}$$

Applying the change of variables (6.73) to (6.76), and noting that since UK is symplectic, we have $d^6X = d^6V$. Therefore,

$$\begin{aligned} \int \psi(X) d^6X &= \int d^6V \frac{1}{\pi^3\langle g_I\rangle\langle g_{II}\rangle\langle g_{III}\rangle} e^{-\frac{V_1^2+V_2^2}{\langle g_I\rangle} - \frac{V_3^2+V_4^2}{\langle g_{II}\rangle} - \frac{V_5^2+V_6^2}{\langle g_{III}\rangle}} \\ &= 1\,, \end{aligned}$$

$$\begin{aligned} \int \psi(X) g_{I,II,III}\, d^6X &= \int d^6V \frac{V_{1,3,5}^2 + V_{2,4,6}^2}{\pi^3\langle g_I\rangle\langle g_{II}\rangle\langle g_{III}\rangle} e^{-\frac{V_1^2+V_2^2}{\langle g_I\rangle} - \frac{V_3^2+V_4^2}{\langle g_{II}\rangle} - \frac{V_5^2+V_6^2}{\langle g_{III}\rangle}} \\ &= \langle g_{I,II,III}\rangle\,, \end{aligned}$$

after performing the various decoupled 1-D integrals. Q.E.D.

The second moments With Eq. (6.70), the equilibrium beam distribution (6.75) says the beam's second moment matrix is given by

$$\Sigma^{-1} = \frac{2G_I}{\langle g_I \rangle} + \frac{2G_{II}}{\langle g_{II} \rangle} + \frac{2G_{III}}{\langle g_{III} \rangle} . \tag{6.77}$$

The second moment matrix Σ is obtained by inverting Eq. (6.77), and the result is found to be

$$\Sigma = -\frac{1}{2}\langle g_I \rangle S G_I S - \frac{1}{2}\langle g_{II} \rangle S G_{II} S - \frac{1}{2}\langle g_{III} \rangle S G_{III} S . \tag{6.78}$$

To prove this, let us first consider G_k and G_ℓ for some $k, \ell = I, II, III$, and calculate the quantity

$$\begin{aligned} S G_k S G_\ell &= -S(SUH_k\tilde{U}S)S(SUH_\ell\tilde{U}S) \\ &= -UH_k\tilde{U}SUH_\ell\tilde{U}S \\ &= -UH_kSH_\ell SU^{-1} \\ &= -\delta_{k\ell}\, UI_kU^{-1} . \end{aligned} \tag{6.79}$$

By multiplying Eqs. (6.77) and (6.78) together, and applying the formula (6.79) repeatedly, one shows that the product is equal to I, the identity matrix. Q.E.D.

Using Eq. (6.72), we can further express (6.78) as

$$\begin{aligned} \langle X_i X_j \rangle &= \Sigma_{ij} \\ &= -\frac{1}{2} \sum_{k=I,II,III} \langle g_k \rangle (S G_k S)_{ij} \\ &= \sum_{k=I,II,III} \langle g_k \rangle \mathrm{Re}(E_{ki} E_{kj}^*) , \end{aligned} \tag{6.80}$$

where the last step has used $SG_kS = -iUH_k\tilde{U}$, the ij-th element of which can be found to be $-2\,\mathrm{Re}(E_{ki}E_{kj}^*)$.

Expectation values of the three invariants So far in this section, we have been discussing completely general mathematics; no physics has been introduced. The only input to the discussions has been the transport matrix $T(s)$ at a given position s. Without physics inputs, (6.75) can be a feasible equilibrium beam distribution for any values of $\langle g_{I,II,III} \rangle$. The expression (6.80) that relates the beam distribution's second moments to the eigenvectors of $T(s)$ is still missing the values of $\langle g_{I,II,III} \rangle$.

To do so, we connect the general discussion to the SLIM calculation. Due to a balance between radiation damping and quantum excitation, it was found that the beam's second moments are determined by Eqs. (6.66) and (6.67). These calculations can be summarized simply by stating

$$\langle g_k \rangle = 2\langle |A_k|^2 \rangle = C_L \frac{2\gamma^5}{c\alpha_k} \oint ds \frac{|E_{k5}(s)|^2}{|\rho(s)|^3} .$$

A similar expression can be derived for the contributions from sideways photon emissions.

Homework 6.74 Show that the matrix U defined in Eq. (6.71) is symplectic.

Homework 6.75 Show that the matrix UK defined in Eq. (6.73) is real and symplectic.

Homework 6.76 Suppose you are given the explicit expression of the one-turn matrix,[38] i.e., $T(s) =$

$$\begin{bmatrix} R_{11} & R_{12} & 0 & 0 & A_{\rm rf}R_{16} & R_{16} \\ R_{21} & R_{22} & 0 & 0 & A_{\rm rf}R_{26} & R_{26} \\ 0 & 0 & R_{33} & R_{34} & 0 & 0 \\ 0 & 0 & R_{43} & R_{44} & 0 & 0 \\ D'-D'R_{11}+DR_{21} & -D+DR_{22}-D'R_{12} & 0 & 0 & 1+A_{\rm rf}R_{56} & R_{56} \\ 0 & 0 & 0 & 0 & A_{\rm rf} & 1 \end{bmatrix},$$

where

$$\begin{aligned}
R_{11} &= \cos\mu_x + \alpha_x \sin\mu_x\,, \\
R_{12} &= \beta_x \sin\mu_x\,, \\
R_{16} &= D - DR_{11} - D'R_{12}\,, \\
R_{21} &= -\gamma_x \sin\mu_x\,, \\
R_{22} &= \cos\mu_x - \alpha_x \sin\mu_x\,, \\
R_{26} &= D' - D'R_{22} - DR_{21}\,, \\
R_{33} &= \cos\mu_y + \alpha_y \sin\mu_y\,, \\
R_{34} &= \beta_y \sin\mu_y\,, \\
R_{43} &= -\gamma_y \sin\mu_y\,, \\
R_{44} &= \cos\mu_y - \alpha_y \sin\mu_y\,, \\
R_{56} &= (\gamma_x D^2 + 2\alpha_x DD' + \beta_x D'^2)\sin\mu_x - C\alpha_c\,, \\
A_{\rm rf} &= \frac{2\pi e\hat{V}h\cos\phi_s}{CE_0}\,,
\end{aligned}$$

and $\alpha_{x,y}, \beta_{x,y}, \gamma_{x,y}$ are the Courant-Snyder functions, D, D' are the dispersion functions (all functions are evaluated at position s), $\mu_{x,y}$ are the betatron phase advances per turn, α_c is momentum compaction factor, C is circumference, $\hat{V}, h$ are the peak voltage and harmonic number of the RF cavity, ϕ_s is the synchronous phase, E_0 is the design beam energy. Follow the SLIM formalism

[38] This expression is different from Eq. (5.52) because RF must be added to provide a longitudinal focusing.

to do the following,

(a) Compute the eigenvalues λ_k and eigenvectors E_k, $k = \pm I, \pm II, \pm III$.

(b) Find the three quadratic invariants g_k, $k = I, II, III$.

(c) Find the three expectation values $\langle g_k \rangle$ due to synchrotron radiation without sideways photons.

(d) Find the 21 distribution moments $\langle X_i X_j \rangle$.

(e) Find an expression for the beam distribution function $\psi(X)$ in the 6-D phase space.

Compare your results with that obtained in conventional theory. Note that the conventional theory further assumes the synchrotron tune is very small (i.e. $0 < -A_{\rm rf} R_{56} \ll 1$). In the absence of x-y coupling as assumed here, the problem reduces to two independent parts: a 2×2 case for the y-motion alone, and a 4×4 case for the coupled x- and z-motions.

Chapter 7

Undulator and free electron laser

7.1 Four generations

We are familiar with the subject of synchrotron radiation learned in classical electrodynamics. We have just dedicated Chapter 6 to the subject. As a reminder, in textbooks, the subject of synchrotron radiation is typically learned in the following steps.

1. Synchrotron radiation from a single electron — a point charge — in a prescribed circular motion such as that occurs in a bending magnet.

2. Same as Step 1, but now allow the radiated photons to act back on the single electron. The perturbation affects the radiating electron, and not so much the radiation itself. The electrons as a result experience additional physical effects of quantum excitation and radiation damping.

3. Same as Step 1, but replace the simple circular prescribed motion in a bending magnet by a more sophisticated case when the prescribed motion is an oscillatory motion such as that occurs in an undulator magnet.

4. Same as Step 3, but allow the radiation to act back on the electron. In addition, replace the electron point charge by an electron beam with finite size.

The first half of Chapter 6 dealt with item 1. The second half of Chapter 6 dealt with item 2. In this chapter, we will deal with items 3 and 4.

It turns out that these four steps follow closely the history of development of synchrotron radiation facilities over the past few decades. Indeed, there are four generations of radiation facilities,

- 1st generation: parasitic use of synchrotron radiation on storage rings for high energy physics;

- 2nd generation: dedicated storage rings with beam properties optimized for producing synchrotron radiation;
- 3rd generation: dedicated storage rings with undulator insertions;
- 4th generation: free electron lasers.

Each step of our learning corresponds to one generation of the synchrotron radiation facilities. The first generation pays attention only to the synchrotron radiation emitted from the dipole magnets. Step 1 of the textbooks suffice. However, as we explore deeper into the first generation storage rings, we must also pay attention to the stored electron beam, which suffers from emitting the radiation in the first place. Only when the effects of radiation back to the electrons are carefully analyzed and controlled as done in Step 2 can we design a storage ring as a dedicated synchrotron radiation facility of the second generation.

It then was realized that one can significantly enrich and empower the radiation if we replace the simple dipole magnet radiator by a periodic array of dipoles of alternating polarities called undulators. Significant investments went into devising storage rings equipped with undulators as radiators, thus a third generation was launched.

An unexpected turn of events then led to the realization, pioneered by John Madey in 1971,[1] that when radiation acts back on the radiating electrons in the case of undulators, a marvelous thing happens. The electron beam under the effect of its own undulator radiation align themselves up into a *microbunched* pattern, which in turn enhances the undulator radiation, which in turn enhances the microbunching. The process feedback onto itself leading to an exponential growth of both the microbunching and the radiation. If the undulator is long enough, the final radiation power can be extremely high. This mechanism is called free electron laser (FEL). The same exponentiation process is not applicable to dipole magnet radiators. When the first hard X-ray FEL was made to lase in 2010,[2] for example, its radiation brightness exceeded the then-existing X-ray sources by 10 orders of magnitude overnight, highlighting the accomplishment of the fourth generation facilities.

To discuss FEL, we will need to get familiar with the undulator radiation. In this chapter, we will first discuss the undulator radiation, and then introduce the subject of FEL in a second part.

The perturbative approach It is necessary to include a discussion of a perturbative nature of accelerator physics here. Hidden behind the development of the four generations of synchrotron radiation facilities is the fundamental assumption that a perturbative approach is applicable. This perturbative nature

[1] J. M. Madey, J. App. Phys. 42, 1906 (1971).
[2] P. Emma et al., Nature Photonics, Vol. 4, p. 641 (2010).

runs rather deep in accelerator physics in general, so let us discuss it a bit further.

Let us first consider the context of collective effects, a subject we devote Chapter 8. The way we treat the subject is to first assume a given, unperturbed, prescribed motion of the beam — we call this the "rigid beam approximation" in Chapter 8. We use Maxwell equations to calculate the electromagnetic fields assuming this unperturbed beam. These fields then are inserted in the Vlasov equation to solve the beam motion under their influence. These two steps constitute one iteration of the accelerator treatment of collective instability effects.

In principle, we ought to continue to recalculate the fields using the now-perturbed beam motion and obtain the new fields. The new fields then are used to recalculate the new beam motion, etc. However, for all practical applications, one iteration suffices for accelerators, and that is what we do.

A similar situation occurs in the subject of synchrotron radiation in accelerator physics. In this case, a charged particle is first assigned an unperturbed, prescribed motion — in a bending magnet, for example, the particle shall follow a perfect circle. This prescribed motion is used to calculate the radiation fields, as done in all electrodynamics textbooks. This radiation is then used to calculate how the particle motion is affected when it acts back on the particle. These two steps constitutes one iteration of our treatment. Again, ideally, we ought to continue the iteration because the perturbed particle motion must change the radiation. But we stop the iteration here. Indeed that is what we did in Chapter 6.

There is however one very important exception to the rule of convergence at first iteration in accelerator physics, namely for the case of free electron laser. In the case of FEL, the beam starts with an unperturbed, prescribed, oscillatory motion in an undulator. Radiation is calculated accordingly. The radiation acts back on the beam, affecting its motion, as we would do in the perturbative scheme. For the case of an FEL, iteration does not stop here. One more iteration is needed. It turns out that the perturbed beam motion has a pronounced resonance effect, and as a result, the perturbation to beam motion is so significant that the radiation is greatly affected, and a second iteration is required, yielding a surprising exponential growth of the radiation power. The key to the FEL as an exception to the first-order-perturbation rule is due to this resonance effect. We shall discuss this effect of the FEL in the second part of this chapter.

7.2 Undulator

7.2.1 Electromagnetic undulator

There are basically two types of undulator designs, electromagnetic and permanent magnetic. Consider the electromagnetic arrangement of Fig. 7.1. The

Figure 7.1: Sketch of an electromagnetic undulator.

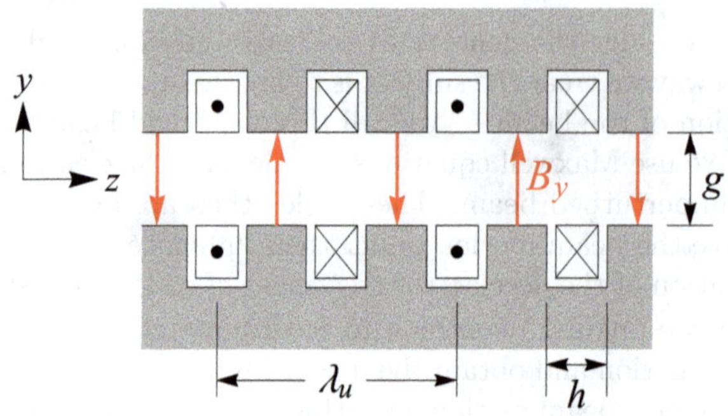

arrangement is assumed to be 2-D planar. Let I be the current in each coil package. The shaded region is iron with $\mu = \infty$.

The vertical magnetic field depends on y and z only,

$$B_y = \frac{8\mu_0 I}{\pi h} \sum_{m=1,\text{odd}}^{\infty} \frac{\sin\left(\frac{mk_u h}{2}\right)}{m \sinh\left(\frac{mk_u g}{2}\right)} \sin(mk_u z)\cosh(mk_u y)\,, \tag{7.1}$$

where $k_u = \frac{2\pi}{\lambda_u}$. The fundamental harmonic $m = 1$ contribution is

$$B_{y,m=1} = \frac{8\mu_0 I}{\pi h} \frac{\sin\left(\frac{k_u h}{2}\right)}{\sinh\left(\frac{k_u g}{2}\right)} \sin(k_u z)\cosh(k_u y)\,. \tag{7.2}$$

Figure 7.2: On-axis magnetic field $\frac{B_y}{\mu_0 k_u I}$ for an electromagnetic undulator as a function of $k_u z$ for the case when $k_u h = k_u g = \frac{\pi}{2}$.

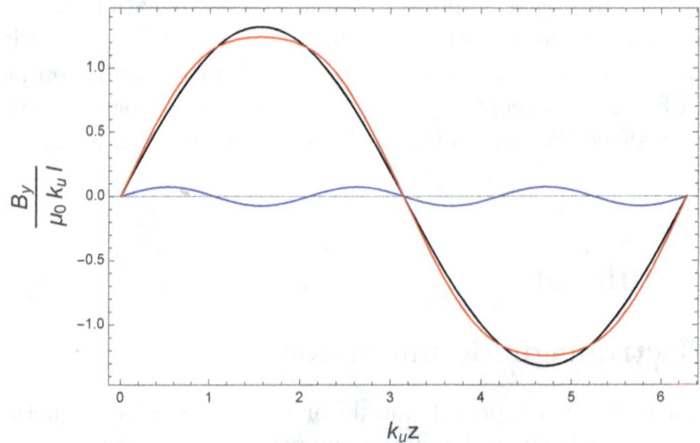

The magnetic field at the magnet center $y = 0$ is shown in Fig. 7.2 for the case when $k_u h = k_u g = \frac{\pi}{2}$. The black, blue, and red curves represent the $m = 1$ harmonic, the $m = 3$ harmonic, and the total field, respectively. The case is chosen so that the $m = 3$ contribution is more visible. Optimized choice would try to reduce the higher order contributions. When $k_u h = k_u g = 2$, for example, the $m = 3$ contribution is basically invisible.

Take for example, $I = 10$ kA, $g = h = \frac{\lambda_u}{4} = 3$ cm, we find a peak on-axis undulator field of about 0.8 T.

For reference, at the pole tip $y = \frac{g}{2}$, the field $\frac{B_y}{\mu_0 k_u I}$ versus $k_u z$ is shown in Fig. 7.3. There is a logarithmic divergence at the edge of the pole tips due to the sharp corner of the idealized $\mu = \infty$ material.

Figure 7.3: Pole-tip magnetic field $\frac{B_y}{\mu_0 k_u I}$ for an electromagnetic undulator as a function of $k_u z$ for the case when $k_u g = k_u h = \frac{\pi}{2}$.

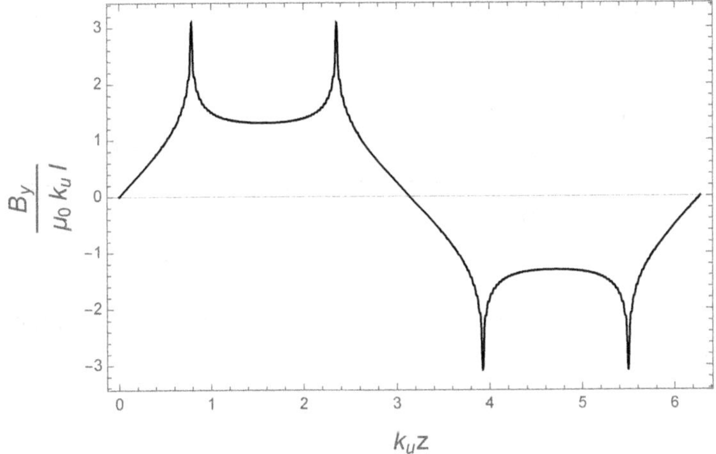

7.2.2 Permanent magnet undulator

Consider the arrangement as shown in Fig. 7.4. The magnetic field is given by[3]

$$
\begin{aligned}
B_y + iB_z &= \frac{8B_0}{\pi} \sum_{m=1,5,9,13,}^{\infty} \frac{1}{m} e^{-mk_u g/2} \left(1 - e^{-mk_u h}\right) \\
&\quad \times \sin\frac{m\pi}{4} \cos[mk_u(z + iy)] , \qquad (7.3)
\end{aligned}
$$

[3] K. Halbach, Sec. 7.2.8, Handbook of Accel. Phys. & Eng., 2nd Ed., World Scientific (2013).

Figure 7.4: Sketch of a permanent magnet undulator.

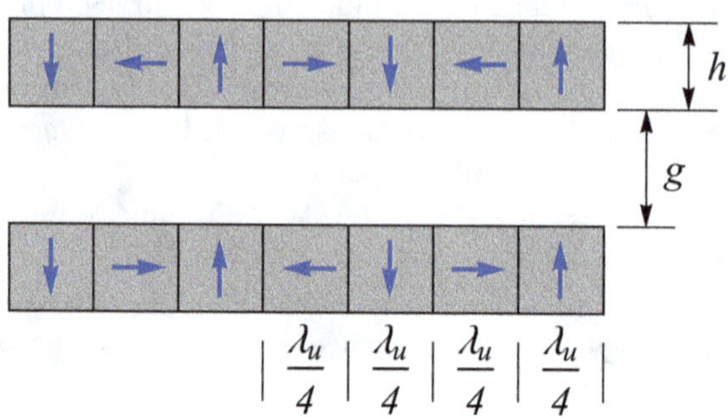

where B_0 is the remnant field of a permanent magnet when it stands alone, and $\cos[mk_u(z+iy)] = \cos mk_u z \cosh mk_u y - i \sin mk_u z \sinh mk_u y$. Typically, $B_0 \sim 1$ T.

The fundamental harmonic $m = 1$ contribution on axis ($y = 0$) is

$$B_y(y=0, m=1) \;=\; \frac{4\sqrt{2}}{\pi} B_0 e^{-k_u g/2} \left(1 - e^{-k_u h}\right) \cos k_u z\,. \tag{7.4}$$

Figure 7.5 shows the contributions from $m = 1$ (black), $m = 5$ (blue), and sum over all m's (red) for the case $k_u g = k_u h = 1$. Again, the parameters are chosen so that higher order contribution is more visible.

Figure 7.5: On-axis magnetic field $\frac{B_y}{B_0}$ for a permanent magnet undulator as a function of $k_u z$ for the case when $k_u g = k_u h = 1$.

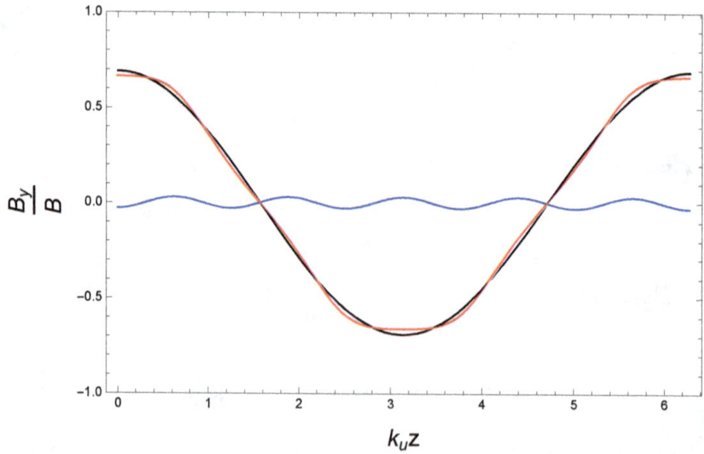

Figure 7.6: Pole-tip magnetic field $\frac{B_y}{B_0}$ for a permanent magnet undulator as a function of $k_u z$ for the case when $k_u g = k_u h = 1$.

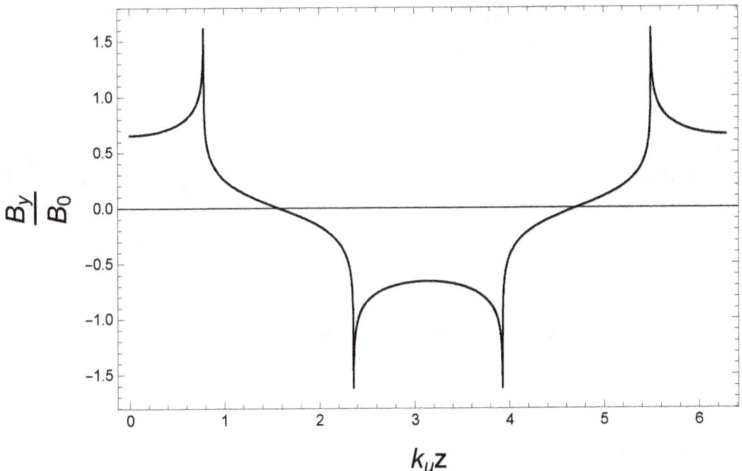

The pole tip field at $y = \pm\frac{g}{2}$ is as seen in Fig. 7.6 for a plot of $\frac{B_y}{B_0}$ as a function of $k_u z$. Again there is a logarithmic divergence at the junctions of the permanent magnet blocks due to the idealized assumptions made.

Homework 7.1 Do this homework as a basic exercise.

(a) Show that Eq. (7.3) obeys the Maxwell equation $\nabla \cdot \vec{B} = 0$.

(b) Figure 7.4 is missing the pattern of magnetic fields. Fill in the magnetic field pattern. It may not be as trivial as it sounds.

Homework 7.2 The peak $m - 1$ field on axis is given by

$$\frac{4\sqrt{2}}{\pi} B_0 e^{-k_u g/2} \left(1 - e^{-k_u h}\right) .$$

Try to choose a set of undulator parameters to reach a peak on-axis field as high as possible.

Solution Ideally, we want $h \to \infty, g \to 0$, which should be expected intuitively. We then have a peak field of $\frac{4\sqrt{2}}{\pi} B_0 \sim 1.8\, B_0$. In practice, we back off from it. Choose $h > 0.36\, \lambda_u$ so that $(1 - e^{-k_u h}) > 0.9$. Also, $e^{-k_u g/2} > 0.9$ would need $g < \frac{\lambda_u}{30}$, but we back off to $g = \frac{\lambda_u}{\pi}$ (the conventional rule of thumb), which gives a unfavorable factor of $e^{-k_u g/2} = 0.37$. The peak on-axis field reached is then $0.6\, B_0$, or about 0.6 T.

7.2.3 Prescribed motion in undulator

We emphasized in Sec. 6.1.2 that the power spectrum of synchrotron radiation by a point charge is solely determined by the trajectory of the electron $\vec{r}(t)$. Once $\vec{r}(t)$ is known at all times t, the radiation power spectrum, as well as its polarization properties, are determined, regardless of what causes the electron to accelerate in the first place. It was also observed that the physical mechanism of radiation is geometric in nature. Based on this notion of radiation by prescribed motion, Chapter 6 then proceeded to calculate the synchrotron radiation in bending magnets assuming a circular prescribed motion.

By the same token, to discuss FELs in this Chapter, we start with a discussion of a single electron radiation in an undulator. To do so, we first calculate the prescribed motion $\vec{r}(t)$ of an electron when it traverses an undulator.

The undulator on-axis magnetic fields are in the vertical y-direction approximately given by Eqs. (7.2) and (7.4). Let us then consider

$$B_y \approx B_u \cos k_u z \,,$$

where

$$B_u = \begin{cases} \frac{8\mu_0 I}{\pi h} \frac{\sin(k_u h/2)}{\sinh(k_u g/2)}, & \text{electromagnetic,} \\ \frac{4\sqrt{2}}{\pi} B_0 e^{-k_u g/2}(1-e^{-k_u h}), & \text{permanent magnet.} \end{cases}$$

Here we have dropped all field components of high harmonics. Homework 1.46 gives the magnetic pole face shapes required to produce a pure sinusoidal undulator field.

Prescribed trajectory Given the magnetic field, we now calculate the prescribed motion $\vec{r}(t)$ of a point charge traversing the undulator. This requires solving the equation of motion

$$m\gamma\dot{\vec{v}} = e\vec{v}\times\vec{B} \,.$$

The particle moves mainly in the z-direction, with a small-but-significant x-displacement but no y-displacement. The x-displacement is sinusoidal, out of phase for electrons (in phase for positrons) with B_y. A sketch is shown in Fig. 7.7.

The solution is found to be, for relativistic particle with $\gamma \gg 1$,

$$\begin{aligned} v_x &= -\frac{\sqrt{2}a_u c}{\gamma} \sin k_u z \,, \\ x &= \frac{\sqrt{2}a_u}{\gamma k_u} \cos k_u z \,, \\ v_z &= c - \frac{c}{2\gamma^2} - \frac{a_u^2 c}{\gamma^2} \sin^2 k_u z \,, \\ z &= \bar{v}_z t + \frac{a_u^2}{4\gamma^2 k_u} \sin 2k_u z \,, \end{aligned} \tag{7.5}$$

Figure 7.7: A sketch of the prescribed motion of an electron in an undulator.

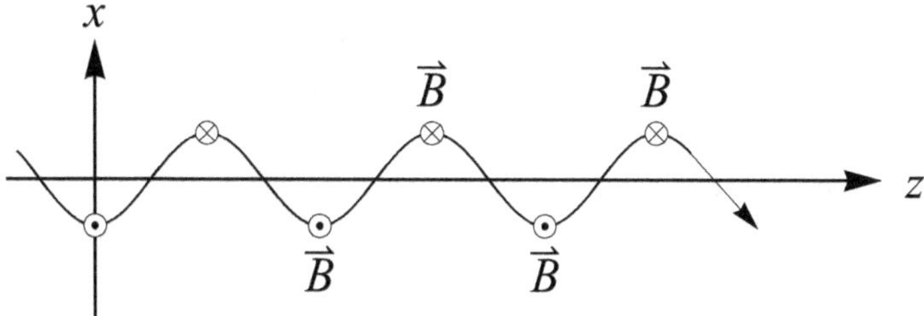

where a dimensionless parameter a_u is defined as

$$a_u = \frac{eB_u}{\sqrt{2}\,mck_u}.$$

A more practical formula would be

$$a_u = 0.66\,B_u[\mathrm{T}]\,\lambda_u[\mathrm{cm}].$$

Typical undulators have a_u in the range of 1 to 10. Sometimes the undulator strength parameter is defined by a quantity K, which is related to a_u by

$$K = \sqrt{2}\,a_u.$$

Note that the field polarity chosen in Fig. 7.7 is for an electron. By convention, strictly, the parameter a_u is negative for electrons. Field polarity will be the opposite for a positron. In the following, however, we will not worry about the sign of a_u as it does not affect any consideration other than the field polarity itself.

Equation (7.5) constitutes the prescribed motion of a particle traversing the undulator. By taking the time derivative of $\vec{v}(t)$ with respect to t, we have also included the acceleration $\ddot{\vec{r}}$ in the prescribed motion.

Equation (7.5) assumes the undulator field $B_y = B_u \cos k_u z$ is purely sinusoidal in z. If the field contains higher harmonics such as those given by Eqs. (7.1) or (7.3), the higher harmonics of the prescribed motion will be affected. This in turn will affect the higher harmonics of the calculated undulator spectrum. The effect tends to be small, however. Pursuing along this line to try to enhance or suppress the high harmonic spectrum by shaping the undulator field seems not very efficient — although more detailed effort is not yet available.

As we will learn in the later discussions of the subject, the longitudinal velocity v_z is of particular importance because it enters directly the calculation of the *phase of the radiation wave* and it has to be calculated most accurately.

This is reflected in the fact that in Eq. (7.5) we have calculated it to second order in $\frac{1}{\gamma}$, while the transverse velocity v_x is calculated to first order. In fact, as the subject evolves further, it is important to recognize that it is the longitudinal dynamics and the longitudinal beam structure that will play the most significant and rich role of the FEL physics.

As seen from Eq. (7.5), the longitudinal velocity is oscillatory in time, with a time-average velocity

$$\bar{v}_z = c - \frac{c}{2\gamma^2}(1 + a_u^2)\,. \tag{7.6}$$

The actual v_z oscillates about $\bar{v}_z$ with an oscillation magnitude of the order of $\frac{1}{\gamma^2}$ and with a frequency twice that of the undulator field.

The transverse velocity $\beta_x = \frac{v_x}{c}$ oscillates with amplitude $\frac{\sqrt{2}a_u}{\gamma}$. It is of the order $\frac{1}{\gamma}$. The electron moves basically with a uniform velocity $\bar{v}_z$ in the z-direction. In addition, it executes a figure-8 motion in the x-z plane. Since z is one order of $\frac{1}{\gamma}$ smaller than x, however, this figure-8 pattern is very tall and slender. This figure-8 motion constitutes the prescribed motion for the undulator radiation, and is our result to be entered into later undulator calculations.

Figure 7.8 (left panel) shows the transverse and longitudinal displacements of the particle relative to a reference particle that moves uniformly in the z-direction with velocity $\bar{v}_z$. The vertical axis is $k_u x$, horizontal axis is $k_u(z-\bar{v}_z t)$, for three cases of $\frac{a_u}{\gamma} = 0.01, 0.03, 0.05$. Note the different scales of the two axes. Note also that it is the motion relative to a reference particle in uniform motion with velocity $\bar{v}_z$, not to be confused to be the motion as observed by the reference particle — no Lorentz transformation is invoked.

The right panel of Fig. 7.8 shows the figure-8 motion in the velocity space. It shows $\frac{v_x}{c}$ (vertical axis) versus $\frac{v_z-\bar{v}_z}{c}$ (horizontal axis) for the same three values of $\frac{a_u}{\gamma}$.

Homework 7.3

(a) Verify Eq. (7.5) in the relativistic limit $\gamma \gg 1$.

(b) It would be instructive to reproduce Fig. 7.8 using information contained in Eq. (7.5) to gain a feel of the figure-8 motion. The motion contains three components: a zero-th order straight ahead motion, a $\sim \frac{1}{\gamma}$ component of sideways motion, and a small $\sim \frac{1}{\gamma^2}$ wobbling back-and-forth motion. We shall see later that each component has its contribution to the undulator radiation power.

(c) Understand why at both the top and the bottom of the figure-8 trajectory, the particle will necessarily be moving forward, not backward.

(d) Plot x versus t of the electron motion in the undulator.

Figure 7.8: The figure-8 motion of an electron traversing an undulator. The left panel shows the spatial trajectory, and the right panel shows the velocity trajectory of the figure-8 motion. The vertical and horizontal scales are very different.

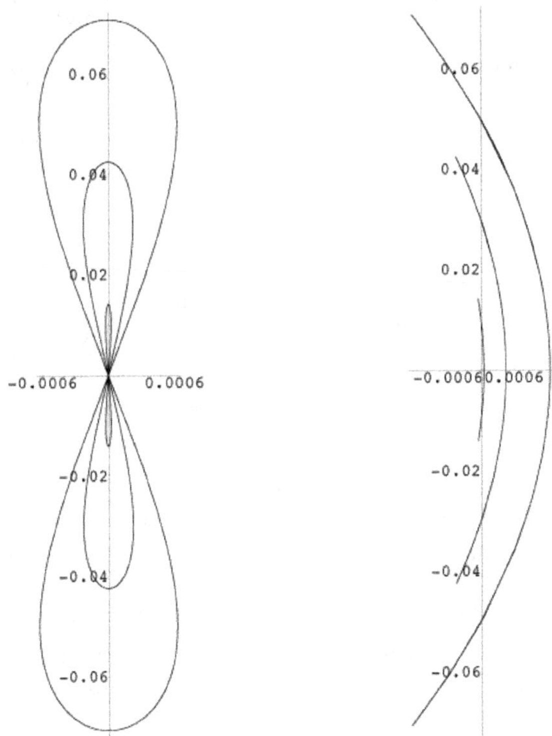

7.3 Undulator radiation

7.3.1 Radiation spectrum — part I

Once the prescribed motion is known, we apply Eq. (6.12),

$$\frac{d^2W}{d\Omega d\omega} = \frac{e^2\omega^2}{16\pi^3\epsilon_0 c^3}\left|\int_{-\infty}^{\infty} dt\,\hat{n}\times(\hat{n}\times\vec{v})\,e^{-i(\omega/c)(ct-\hat{n}\cdot\vec{r})}\right|^2 . \tag{7.7}$$

Equation (7.7) gives the undulator radiation spectrum, i.e. the radiated energy per solid angle per frequency interval, by a point charge executing a prescribed motion (note that $\omega > 0$). As mentioned, this radiation spectrum is determined once the prescribed motion, $\vec{r}(t)$ and $\dot{\vec{r}}(t)$, is specified. The quantity $\frac{d^2W}{d\Omega d\omega}$ is a double spectrum of radiation energy in radiation angle and in radiation frequency. Integrated over the radiation angle Ω, it yields the frequency spectrum

Table 7.1: A brief comparison of the radiation spectra of bending magnet and undulator radiations.

	Bending magnet	Undulator
Brightness	high	higher
Opening angle	$\sim \frac{1}{\gamma}$ narrow collimated	$\sim \frac{1}{\gamma\sqrt{N_u}}$ narrower more collimated
Frequency	up to $\omega_c = \frac{3}{2}\frac{\gamma^3 c}{\rho}$ high no need to be tunable	$\omega_r = \frac{2\omega_u\gamma^2}{1+a_u^2}$ can be high tunable
Frequency bandwidth	$\Delta\omega \sim \omega_c$ wide	$\Delta\omega \sim \frac{\omega_r}{N_u}$ narrow

of the radiation energy. Integrated over the frequency ω, it gives the angular spectrum. Integrated over both Ω and ω, it gives the total radiated energy by the electron upon completion of the traversal of the prescribed motion over the length of the undulator.

As discussed with Eq. (6.12), Eq. (7.7) is a subtle result as it involves only $\vec{\beta}$ and not $\dot{\vec{\beta}}$, as if the radiation comes from the velocity and not the acceleration. The physical meaning of Eq. (7.7) is therefore obscured by its apparent mathematical elegance.

We shall return to the derivation of undulator radiation spectrum using Eq. (7.7) to a later Sec. 7.3.4. Here, let us bring up the result and compare it with the radiation spectrum of a dipole magnet. A brief comparison can be seen in Table 7.1. We give a few comments about Table 7.1 below.

- N_u is the total number of undulator periods. Typically $N_u \gtrsim 100$.
- The bending magnet radiation is pushed to high frequencies because of the relativistic Doppler factor of γ^3. The undulator radiation also contains a relativistic factor, but it is only γ^2. However, $\lambda_u \ll \rho$. So the undulator radiation can also reach X-ray. The price to pay is that a good effort will be needed in the undulator design. One might take the viewpoint that the undulator design has become an important effort because of the need to gain back the extra factor of γ.
- The frequency bandwidth of dipole radiation is 100% while that of the undulator is $\frac{1}{N_u}$. The narrowness of the bandwidth for an undulator is due to a resonance effect to be discussed in Sec. 7.3.2. The resonant frequency is $\omega_r = \frac{2\omega_u\gamma^2}{1+a_u^2}$, where $\omega_u = 2\pi\frac{c}{\lambda_u}$ is the undulator period frequency.

- A connection between ω_c of the dipole radiation and ω_r of the undulator radiation can be established by replacing the dipole formation time $\frac{\rho}{\gamma c}$ by its undulator counterpart $\frac{\lambda_u(1+a_u^2)}{c}$, which is $\frac{1}{N_u}$ times the undulator formation time.

7.3.2 The undulator resonance

As mentioned in Table 7.1, one big difference between a dipole magnet radiation and undulator radiation is that the dipole radiation is very broad-band, while the undulator radiation is narrow-band due to a resonance effect. We also mentioned towards the end of Sec. 7.1 that this resonance effect is the key to the undulator and the free electron laser radiation. In particular, lacking this resonance effect, dipole radiation cannot be made into a free electron laser. The power spectrum calculated using Eq. (7.7) with the prescribed motion (7.5) will automatically and prominently contain this resonance effect in its result. But before we derive the spectrum in Sec. 7.3.4, it is instructive to add a discussion to illustrate how this resonance effect comes about and how it is expected to affect the radiation spectrum. The actual details of the calculation will be less emphasized instead.

Consider a relativistic electron moving down the undulator. The velocity changes as the electron moves along, but in a magnetic field, the speed of the particle stays constant, given by βc, where

$$\beta = \sqrt{1-\frac{1}{\gamma^2}} \approx 1-\frac{1}{2\gamma^2},$$

i.e. the electron moves slightly slower than c by a factor $1-\frac{1}{2\gamma^2}$.

But β_z in the z-direction moves slightly slower than β because the electron moves in an sinusoidal trajectory instead of a straight line,

$$\begin{aligned}\beta_z &= \sqrt{\beta^2-\beta_x^2} \approx \beta\left(1-\frac{\beta_x^2}{2\beta^2}\right) \approx \beta\left(1-\frac{\beta_x^2}{2}\right)\\ &\approx \left(1-\frac{1}{2\gamma^2}\right)\left(1-\frac{\beta_x^2}{2}\right) \approx 1-\frac{1}{2\gamma^2}-\frac{\beta_x^2}{2}.\end{aligned}$$

Since $v_x = -\frac{\sqrt{2}a_u c}{\gamma}\sin k_u z$, we obtain

$$\langle\beta_x^2\rangle = \frac{a_u^2}{\gamma^2}.$$

Substituting into β_z, we obtain the average $\bar{v}_z$,

$$\bar{v}_z \approx c\left(1-\frac{1+a_u^2}{2\gamma^2}\right),$$

which we recognize is simply Eq. (7.6).

As mentioned, the undulator radiation is dominated by a resonance effect. To understand the resonance, consider the electron in the prescribed undulating motion radiating along its path. Consider the radiation in the forward z-direction. Consider a photon radiated at a position A. The detailed radiation process changes according to the instantaneous condition of the prescribed motion, but exactly the same condition will occur again when the electron reaches position B one undulator period λ_u later. A resonance occurs when the radiation wavelength is such that the photon radiated at A and the photon radiated at B add constructively in phase, because when that happens, not only photons radiated from A and B, but also photons radiated at all repetitive positions like A and B will add. If the undulator has N_u periods, then radiation from all N_u positions add constructively, yielding a strong radiation in the forward direction.

Let the electron radiate a photon at A, and then it moves on down to B. The time it takes for the electron to move from A to B is $t_1 = \frac{\lambda_u}{\bar{v}_z}$. Since $\bar{v}_z$ is less than c, by the time it reaches B, the photon emitted at A has moved further past B and is at a distance

$$(c - \bar{v}_z)t_1 \;\approx\; \lambda_u \frac{1 + a_u^2}{2\gamma^2}$$

ahead of the electron.

Although v_z changes along the passage, we used $\bar{v}_z$ in the above calculation. This is justified because positions A and B are separated by the undulator period. Over one undulator period, the average v_z is given by $\bar{v}_z$.

Now, the photons are really waves. So the photons will be in phase and add constructively if

$$\lambda_u \frac{1 + a_u^2}{2\gamma^2} \;=\; \lambda_r \,, \tag{7.8}$$

where λ_r is the resonant wavelength of the radiation. Since this constructive radiation occurs only at specified wavelength, the strong radiation is necessarily narrow banded.

We now extend the above analysis of resonance condition to the case when the radiation is not in the forward direction but with a radiation angle θ. The effect is illustrated in Fig. 7.9.

Radiation at positions A and B separated by λ_u will add coherently if

$$c\frac{\lambda_u}{\bar{v}_z} - \lambda_u \cos\theta \;=\; \lambda_r \,.$$

If we then note $\cos\theta \approx 1 - \frac{\theta^2}{2}$, the resonance condition becomes

$$\lambda_r \;=\; \frac{\lambda_u}{2\gamma^2}\left(1 + a_u^2 + \gamma^2\theta^2\right),$$

Figure 7.9: Illustration of the undulator resonance at an angle θ of radiation.

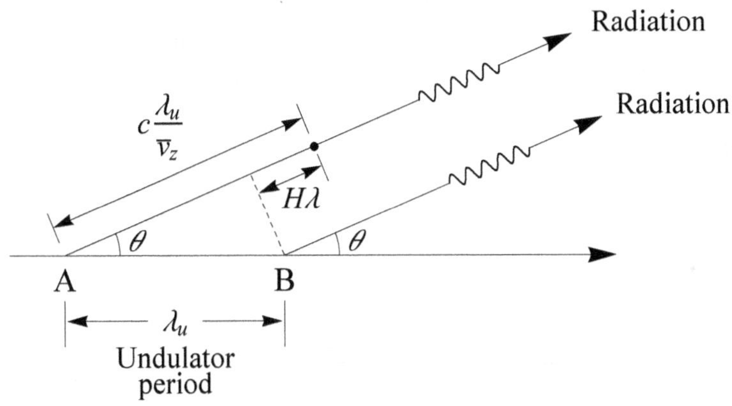

or,

$$\frac{1}{\omega_r} = \frac{1 + a_u^2 + \gamma^2\theta^2}{2ck_u\gamma^2} . \tag{7.9}$$

For later use, we also define the fundamental radiation wavelength in the forward direction, $\theta = 0$,

$$\lambda_0 = \frac{\lambda_u}{2\gamma^2}(1 + a_u^2), \qquad \frac{1}{\omega_0} = \frac{1 + a_u^2}{2ck_u\gamma^2} , \tag{7.10}$$

We make a few observations,

- The resonance frequency ω_r defined by Eq. (7.9) depends on θ. As θ increases from 0, the resonant frequency decreases. We thus expect the radiation spectrum to *red shift* as the observation angle deviates from along the undulator axis.
- The angle θ in Fig. 7.9 was drawn in the x-z plane, but the resonant condition occurs when θ represents the polar angle $\sqrt{\theta_x^2 + \theta_y^2}$ in a 3-D configuration. The resonance condition therefore is fulfilled by a cone distribution, not a planar distribution although this is a planar undulator. Keep this point in mind as it might be a bit counterintuitive.
- Resonance also occurs at shorter wavelengths $\frac{\lambda_r}{H}$ for H = integers. So an undulator also offers higher harmonics in its radiation. Each of the harmonics is narrow-banded around its resonant frequency $H\omega_r$.
- There is a suppression of coherent radiation at wavelengths when H = half-integers (complete suppression if N_u is even); the waves now superimpose destructively. The radiation is basically drops to zero power at the wavelengths $2\lambda_r, \frac{2}{3}\lambda_r, \frac{2}{5}\lambda_r, \cdots$. See Fig. 7.10.

- Less obvious is the fact that Fig. 7.9 has an important counterpart that illustrates the resonant condition for systematic exchange of energies between a laser and an undulating electron. This point will be discussed in detail in Sec. 7.4.3 and homework 7.29.

Figure 7.10: Constructive and destructive pattern of undulator radiation. Horizontal axis is $\frac{\lambda}{\lambda_r}$. Red bars indicate destructive interference. Green bars are constructive. The thick green bar indicates the fundamental mode.

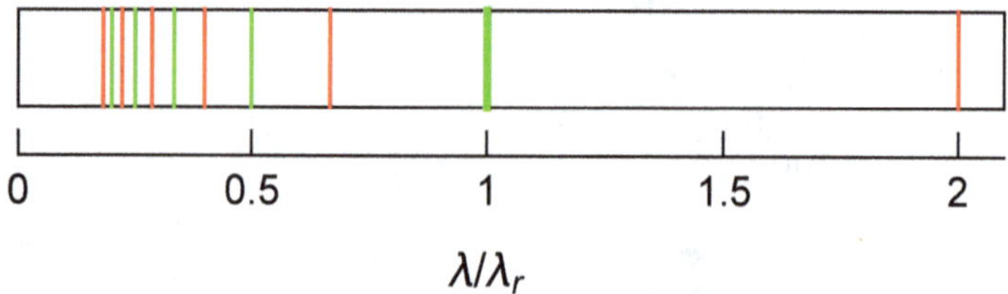

Homework 7.4 Fill in the derivation that leads to Eq. (7.9).

Homework 7.5 Interference, either constructive or destructive, of a wave is the key to the physics of all wave phenomenon. This is true also for sound waves.

(a) If two singers on stage, shown in Fig. 7.11, sing exactly one note at a fixed frequency, does the audience below suffer from an interference pattern, i.e. the intensity of the music an audience hears depends on the position where he/she sits?

(b) What if the frequencies of the two singers are different but one is an integral multiple of the other?

7.3.3 Expected frequency spectrum of undulator radiation

Single electron grows a nose The discussion of the resonance effect gives rise to a physical picture, at least approximately, of the radiation emitted by an electron without too much of mathematical derivations. As the electron propagates down the undulator, its radiation moves with it basically in the same forward direction. However, the radiation moves ever so slightly faster than the electron, with the result that the radiation gets ahead by a distance of λ_r each time the electron propagates a distance of λ_u. This is of course a tiny amount because λ_u is in centimeters while λ_r is most likely submicrons.

As the electron moves another distance of λ_u, the radiation would accumulate by another distance of λ_r. Each time the electron goes one period of the undulator, the radiation accumulates another headway distance of λ_r in front of it, as if the electron is growing a nose, and the nose gets longer each time the

Figure 7.11: Two singers interfering each other.

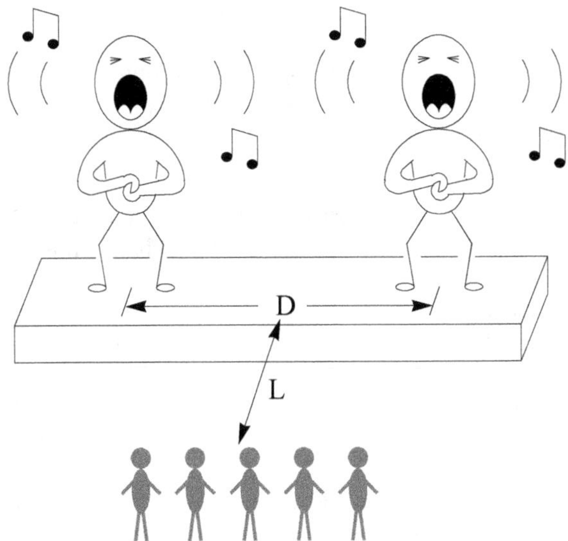

electron moves one undulator period. Each time the electron propagates one centimeter, it grows a tiny radiation nose of one submicron. By the time the electron completes the traversal of the undulator with N_u periods, the radiation has grown a radiation nose with a total length of $N_u\lambda_r$. As the end of the undulator, the electron and its radiation get separated as the electron is bent off by a magnetic field while its nose continues to fly ahead.

This growing-nose picture allows a good grip of the frequency spectrum of the undulator radiation. Indeed, via this picture, we expect the radiation to take a wavepacket form more or less like this,

$$E(t) = E_0 \times \begin{cases} e^{-i\omega_r t}, & \text{if } 0 < t < \frac{N_u\lambda_r}{c}, \\ 0, & \text{otherwise.} \end{cases} \tag{7.11}$$

In other words, the radiation is like a free space radiation with frequency $\omega_r = \frac{2\pi c}{\lambda_r}$, but it has grown only N_u oscillations long.

To obtain the frequency content of the undulator radiation, we make a Fourier transform on this radiation wavepacket (7.11),

$$\tilde{E}(\omega) = E_0 \int_0^{\frac{N_u\lambda_r}{c}} dt\, e^{i\omega t - i\omega_r t} = \frac{\sin(\frac{\omega-\omega_r}{2}\frac{N_u\lambda_r}{c})}{(\frac{\omega-\omega_r}{2})}\, e^{i(\frac{\omega-\omega_r}{2}\frac{N_u\lambda_r}{c})},$$

which gives

$$|\tilde{E}(\omega)|^2 = E_0^2 \left(\frac{\sin N_u\pi\epsilon}{N_u\pi\epsilon}\right)^2, \tag{7.12}$$

where

$$\epsilon = \frac{\omega - \omega_r}{\omega_r}, \qquad \text{with } |\epsilon| \ll 1$$

and ω_r given by Eq. (7.9). The same spectrum applies at an angle $\theta \neq 0$. We have assumed $N_u \gg 1$. This frequency spectral factor will enter prominently in the exact calculation to be discussed in Sec. 7.3.4.

Formation length It may be instructive to add a general discussion of radiation processes here. Regardless of the radiation mechanism, a radiation process requires an internal coherence during a certain formation time of radiation. For an undulator radiation by an electron, the radiation is fully coherent within each wavepacket (7.11). Different electrons radiate different wavepackets. In an undulator, there is no coherence among the different wavepackets, so the radiation is said to be *incoherent.* It is not to be confused with the fact that the radiation wavepacket internally is — and indeed must be, in order to radiate in the first place — fully coherent.

In the case of free electron laser, to be discussed later, it is found that under some favorable conditions, these wavepackets from different electrons can align themselves so that they become locked in phase and become coherent. When this happens, we say the FEL radiation is *coherent.*

One then defines two quantities, the *formation length* and the *coherence length.* The formation length of the radiation by an electron traversing an undulator is the undulator length $L_u = N_u\lambda_u$. The radiation's coherence length is $N_u\lambda_r$. Although referring to the same radiation event, they differ numerically by a large factor. The formation length emphasizes the electron, while the coherence length focuses on the laser. Their ratio is a Doppler factor. In this case, it is given by $\frac{\lambda_r}{\lambda_u}$.

For the case of a bending dipole, a similar situation happens when each electron radiates synchrotron radiation. As discussed in Sec. 6.1.4, the wavepacket emitted by an electron is radiated within the formation length $\sim \frac{2\rho}{\gamma}$, and when Doppler compressed, the radiated photons have a coherence length $\sim \frac{\rho}{\gamma^3}$. Unlike the case of an undulator, the pattern of this synchrotron radiation wavepacket is rather complicated, not simple like (7.11). On the other hand, each of these synchrotron radiation wavepackets, as complicated as they are, is identical and is fully internally coherent. If this complicated wavepacket is Fourier transformed, we obtain the synchrotron radiation frequency spectrum, just like what we did for the case of an undulator. A brief table gives a comparison below.

	Formation length	Coherence length	Doppler factor
Bending magnet	$\frac{2\rho}{\gamma}$	$\frac{\rho}{\gamma^3}$	$\frac{2}{\gamma^2}$
Undulator	$N_u\lambda_u$	$N_u\lambda_r$	$\frac{\lambda_r}{\lambda_u}$

What a bending magnet differs from an undulator is the fact that there is no way for an electron to align with its radiation systematically and repetitively in a bending magnet so that a resonance effect occurs. Consequently, there is no free electron laser using a bending magnet.

Microbunched beam So far we addressed the expected radiation spectrum of a single electron. What happens if the electron is replaced by a microbunched beam? What do we expect its radiation spectrum to behave?

Consider a microbunched electron beam. Let its microbunches be spaced by distance λ_0 given by Eq. (7.10). Let the k-th microbunch at location $k\lambda_0$ have density n_k. The Fourier transform of the beam distribution is

$$\rho(z) = \sum_k n_k \delta(z - k\lambda_0) \quad \Longrightarrow \quad \tilde{\rho}(\omega) = \sum_k n_k \, e^{ik(\frac{\omega-\omega_0}{c})\lambda_0} \, .$$

As we shall see in Sec. 7.4.1, this $\tilde{\rho}(\omega)$ is just the bunching factor of our beam at frequency ω near $\omega = \omega_0$.

In case the beam has an over-all Gaussian distribution,

$$n_k = e^{\frac{-k^2\lambda_0^2}{2\sigma_z^2}} \, ,$$

with $\sigma_z \gg \lambda_0$, we have

$$\tilde{\rho}(\omega) = \int_{-\infty}^{\infty} dk \, e^{\frac{-k^2\lambda_0^2}{2\sigma_z^2}} \, e^{ik(\frac{\omega-\omega_0}{c})\lambda_0} = \sqrt{2\pi}\,\frac{\sigma_z}{\lambda_0} e^{-\frac{(\omega-\omega_0)^2}{2c^2}\sigma_z^2} \, .$$

The radiation distribution is given by the convolution of $\rho(z)$ and the wave-packet E_0 of Eq. (7.11). Its Fourier transform is the product $\tilde{E}(\omega)$ and $\tilde{\rho}(\omega)$. The radiation spectrum is then given by

$$|\tilde{E}(\omega)\tilde{\rho}(\omega)|^2 \propto E_0^2 \left(\frac{\sin N_u \pi \epsilon}{N_u \pi c} \right)^2 e^{\frac{-(\omega-\omega_0)^2\sigma_z^2}{c^2}} \, , \tag{7.13}$$

with $\epsilon = \frac{\omega-\omega_r}{\omega_r}$, $|\epsilon| \ll 1$.

The first factor in Eq. (7.13) is the same as that of a single electron centered at the frequency ω_r. The main effect of a microbunched beam is reflected by the second factor, i.e. the bunching factor, which centers around ω_0. The two factors center around different frequencies. The difference between ω_r and ω_0 lies in the red-shift with radiation angle θ.

In particular, in case $\frac{\sigma_z}{\lambda_0} > N_u$, i.e. if the number of microbunches in the beam is larger than N_u, there will be a spectrum-narrowing effect of the radiation spectrum. The frequency bandwidth is then determined by $\frac{\Delta\omega}{\omega} \sim \frac{\lambda_0}{\sigma_z}$ instead of the natural bandwidth $\sim \frac{1}{N_u}$ as the case of a single electron.

Furthermore, the bunching factor also significantly suppresses the off-axis, red shifted, radiation of a single electron. This is because the bunching factor does not red shift. As a result, the red shifted part of the spectrum is significantly suppressed if $\frac{\sigma_z}{\lambda_0} > N_u$. The frequency content of the microbunched beam forces the radiation spectrum to concentrate around the frequency ω_0 and in the forward direction if the beam contains sufficient number of microbunches.

Homework 7.6

(a) In the text, we wrote down the expected electric field of the radiation nose in an off-axis direction θ. Convince yourself that the result is given by Eq. (7.11). In particular, we know that the nose has a length $N_u\lambda_0$ in the forward direction. What is the length of the nose at an angle θ? Give a physical explanation of the result.

(b) Fill in the derivation that leads to Eq. (7.12). Show that the factor $F(\epsilon) = \left(\frac{\sin(\pi N_u \epsilon)}{\pi\epsilon}\right)^2$ is a function sharply peaked at $\epsilon = 0$ with $F(0) = N_u^2$ and its FWHM is $\Delta\epsilon = \frac{0.8859}{N_u}$. The area under the curve $F(\epsilon)$ is equal to N_u.

(c) Derive Eq. (7.13).

7.3.4 Radiation spectrum — part II

The calculation using the prescribed motion (7.5) and Eq. (7.7) is a bit tedious but straightforward. We omit the detailed derivation, but bring a few of the more technical points to attention here.

- We first need to define the coordinate system. We choose it according to Fig. 7.12. In particular, the electron prescribed motion is in the x-z plane and the observation unit vector is chosen to be

$$\hat{n} = \hat{x}\sin\theta\cos\phi + \hat{y}\sin\theta\sin\phi + \hat{z}\cos\theta\,.$$

Note that $\hat{n}$ is not assumed to be along the instantaneous direction of motion of the electron. As it will turn out, the angle $\theta \sim \mathcal{O}(\frac{1}{\gamma})$, while ϕ is not restricted.

- The quantity $\hat{n}\times(\hat{n}\times\vec{v})$ in Eq. (7.7) needs to be calculated to accuracy $\mathcal{O}(\frac{1}{\gamma})$,

$$\begin{aligned}\hat{n}\times(\hat{n}\times\vec{v}) &= \hat{n}(\hat{n}\cdot\vec{v}) - \vec{v} \\ &\approx (c\theta\cos\phi - v_x)\hat{x} + c\theta\sin\phi\,\hat{y}\,.\end{aligned} \tag{7.14}$$

The $\hat{x}$-component will give rise to a σ-mode polarization in the undulating plane, while the $\hat{y}$-component gives rise to the π-mode radiation. Note that the π-mode vanishes when $\theta = 0$ (observing in the forward direction), or when (even if $\theta \neq 0$) $\sin\phi = 0$ (observing in the x-z plane). This is as expected by the line-of-sight argument.

Figure 7.12: Coordinate system chosen to calculate the undulator radiation spectrum. The undulator magnetic field is in the y-direction. The red wiggly curve for the electron trajectory is in the x-z plane.

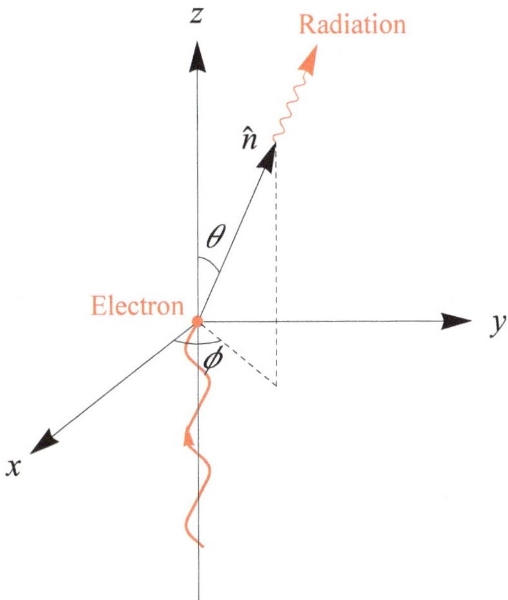

- More important, the phase $(ct - \hat{n} \cdot \vec{r})$ needs to be calculated to a much higher accuracy, order $\mathcal{O}(\frac{1}{\gamma^2})$,

$$ct - \hat{n}\cdot\vec{r} \approx \frac{ct}{2}\left(\theta^2 + \frac{1+a_u^2}{\gamma^2}\right) - \frac{a_u^2}{4\gamma^2 k_u}\sin(2k_u\bar{v}_z t) - \frac{\sqrt{2}a_u\theta\cos\phi}{\gamma k_u}\cos(k_u\bar{v}_z t)\,. \tag{7.15}$$

- As seen by Eq. (7.14), the term $\hat{n} \times (\hat{n} \times \vec{v})$ does not contain higher harmonic contributions. The higher harmonics $H > 1$ originate solely from the phase factor $e^{-ik(ct-\hat{n}\cdot\vec{r})}$. As given in Eq. (7.15), the quantity $ct - \hat{n} \cdot \vec{r}$ contains a term $\propto \sin(2k_u\bar{v}_z t)$ from the v_z-modulation and a term $\propto \cos(k_u\bar{v}_z t)$ from the x-modulation. The v_z-modulation frequency is twice that of the x-modulation, resulting from the figure-8 motion of the electron.

  Along the forward $\theta = 0$ direction, only the v_z-modulation contributes. The higher harmonics therefore originate from the v_z-modulation of the figure-8 electron motion. Since the v_z-modulation tends to be small $\sim \mathcal{O}(\frac{1}{\gamma^2})$, high harmonics are typically much weaker in power compared with the main $H = 1$ component.

Details omitted (see Homework 7.7), the radiation energy spectrum is found, near the frequency $\omega = \omega_r(H+\epsilon)$ (H is the harmonic number — an integer — and $|\epsilon| \ll 1$), to be

$$\frac{d^2W}{d\Omega d\omega}(\theta,\phi,\epsilon) \;=\; \frac{2e^2\gamma^2}{\pi\epsilon_0 c}G(\theta,\phi)F(\epsilon)\,. \tag{7.16}$$

The spectrum contains two factors; basically the factor $F(\epsilon)$ gives the frequency spectrum, the factor $G(\theta,\phi)$ gives the angular distribution.

The frequency ϵ-dependence of the spectrum comes from

$$F(\epsilon) \;=\; \left(\frac{\sin\pi N_u\epsilon}{\pi\epsilon}\right)^2,$$

which one recognizes to be what was anticipated from Eq. (7.12), while the angular (θ,ϕ)-dependence comes from

$$\begin{aligned}
G(\theta,\phi) &= G_\sigma(\theta,\phi) + G_\pi(\theta,\phi)\,,\\
G_\sigma(\theta,\phi) &= \left[\frac{H\left(a_u\mathcal{D}_1 + \frac{\gamma\theta}{\sqrt{2}}\mathcal{D}_2\cos\phi\right)}{1+a_u^2+\gamma^2\theta^2}\right]^2,\\
G_\pi(\theta,\phi) &= \frac{1}{2}\left(\frac{H\gamma\theta\,\mathcal{D}_2\sin\phi}{1+a_u^2+\gamma^2\theta^2}\right)^2,\\
\mathcal{D}_1 &= -\frac{1}{2}\sum_{m=-\infty}^{\infty} J_{H+2m-1}(H\alpha)\left[J_m(H\zeta)+J_{m-1}(H\zeta)\right],\\
\mathcal{D}_2 &= \sum_{m=-\infty}^{\infty} J_{H+2m}(H\alpha)J_m(H\zeta)\,,\\
\alpha &= \frac{2\sqrt{2}a_u\gamma\theta\cos\phi}{1+a_u^2+\gamma^2\theta^2}, \qquad \zeta = \frac{a_u^2/2}{1+a_u^2+\gamma^2\theta^2}\,.
\end{aligned}$$

As noted earlier, the physical meaning of the frequency form factor $F(\epsilon)$ is a result of $|\tilde{E}(\omega)|^2$ of Eq. (7.12). Some properties of $F(\epsilon)$ were already mentioned in Homework 7.6.

Subscripts σ and π refer to the radiation polarization parallel and perpendicular to the undulating plane respectively. Since the electron's prescribed motion lies in the undulator's midplane, one expects that the radiation will be predominantly in the σ polarization. Only when one observes with an inclined angle relative to the undulating plane will be observed some π polarization power.

As we skipped the detailed derivation, we make it up by making a few closer observations of the result (7.16) as follows.

- Frequency spectrum is *the same* for both σ- and π-radiations and is given by $F(\epsilon)$.

- The σ-radiation dominates. The π-mode vanishes when $\theta \sin\phi = 0$. Note that it is the inclined angle, $\theta \sin\phi$, relative to the undulating plane, instead of the polar angle θ, that enters here — see expression for $G_\pi(\theta, \phi)$. This is of course expected by the familiar line-of-sight argument.

- It might appear that the ω- and the Ω-dependences factorize cleanly because $\frac{d^2W}{d\Omega d\omega}$ factorizes cleanly into $F(\epsilon)$ and $G(\theta, \phi)$. But this is not quite true because ϵ depends on θ, implying a red-shift effect,

$$\epsilon = \frac{\omega}{\omega_r} - H = \frac{\omega}{2ck_u\gamma^2}(1 + a_u^2 + \gamma^2\theta^2) - H . \qquad (7.17)$$

  On the other hand, $G(\theta, \phi)$ does not explicitly depend on ω.

- The spectrum $\frac{d^2W}{d\Omega d\omega}$ is measured, aside from the dimensionless form factors F and G, in units of of

$$\frac{2e^2\gamma^2}{\pi\epsilon_0 c} = 8r_0\gamma^2 mc = 3.8 \times 10^{-17}\gamma^2 \frac{\text{eV}}{\text{rad}^2\text{-particle-s}^{-1}} .$$

  Aside from the dimensionless form factors F, G_σ, G_π, this "natural unit" specifies the order of magnitude of the total radiated energy per electron.[4]

  The "natural unit" $\frac{2e^2\gamma^2}{\pi\epsilon_0 c}$, however, is missing an expected factor of N_u, which is hidden in the area under the form factor $F(\epsilon)$. In addition, as will be mentioned later, the form factors $G_{\sigma,\pi}$ are typically (disappointingly) of the order of a few percent. So we expect an order of magnitude

$$\frac{d^2W}{d\Omega d\omega} \text{ in the forward direction} \sim \text{(a few percent)} \times \frac{2e^2\gamma^2}{\pi\epsilon_0 c} N_u . \quad (7.18)$$

Homework 7.7 It is suggested that a practitioner go through the derivation of Eq. (7.16) at least once in lifetime. The steps are a bit tedious but doable.

Solution It would be useful to Fourier decompose the quantity

$$\exp\left[i\frac{\omega}{c}\frac{a_u^2}{4\gamma^2 k_u}\sin 2s + i\frac{\omega}{c}\theta\cos\phi\frac{\sqrt{2}a_u}{\gamma k_u}\cos s\right] = \sum_{r=-\infty}^{\infty} \beta_r e^{-irs} ,$$

where $s = k_u ct$. It can then be shown that

$$\beta_r = i^r \sum_{m=-\infty}^{\infty} J_{r-2m}(H\alpha)J_m(H\zeta) .$$

Another trick is to observe that, when k is an integer and $|\epsilon| \ll 1$, we can write

$$\int_0^{2\pi N_u} ds\, e^{-i(k+\epsilon)s} \approx \left(\frac{1 - e^{-i2\pi N_u\epsilon}}{i\epsilon}\right)\delta_{k0} .$$

[4]Note the use of s^{-1}, instead of Hz, as the unit of ω. Do not use Hertz as the unit for ω. See Sec. 1.1.2.

7.3.5 The form factor $F(\epsilon)$

The form factor $F(\epsilon)$ is sharply peaked at $\epsilon = 0$. Its peak value is N_u^2. The spectral width (half width between ± 1 nodes) is

$$\Delta\epsilon = \frac{1}{N_u}.$$

The area under the curve is N_u. Two cases $N_u = 5$ (red) and 10 (blue) are shown in Fig. 7.13.

Figure 7.13: The frequency spectrum form factor $F(\epsilon)$ as a function of ϵ. Red curve is for $N_u = 5$; blue curve is for $N_u = 10$.

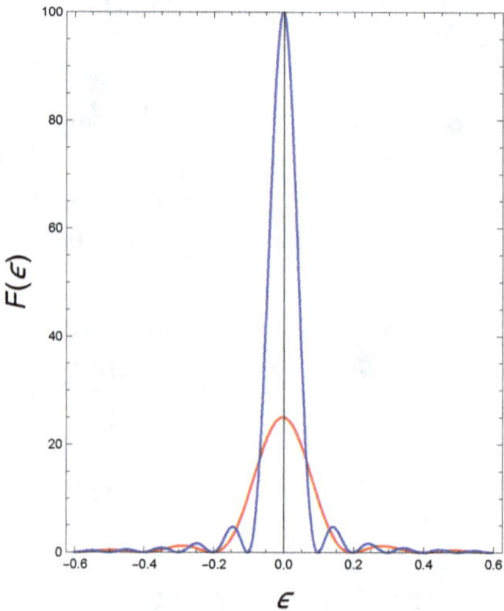

$F(\epsilon)$ as frequency spectrum We have just shown $F(\epsilon)$ as a function of ϵ. But ϵ contains both ω and θ, Eq. (7.17), and we really want to look at its dependences on θ and ω separately. Let us examine $F(\epsilon)$'s ω-dependence, i.e. its spectral distribution, first.

The spectral distribution depends on θ. As θ increases, i.e. moving away from the straight-ahead direction, the frequency spectrum is red shifted. To show this red shift effect clearer, let us write

$$\begin{aligned} \epsilon &= Hx(1+y^2) - H\,, \\ x &= \frac{\omega}{H\omega_0}, \qquad y = \frac{\gamma\theta}{\sqrt{1+a_u^2}}, \end{aligned}$$

where

$$\omega_0 = \frac{2ck_u\gamma^2}{1+a_u^2}$$

Figure 7.14: The frequency spectrum form factor $F(\epsilon)$ as a function of x for three values of off-axis angles with $y = 0, 1, 2$ and $H = 1$. The red curves are for $N_u = 5$; blue curves are for $N_u = 10$.

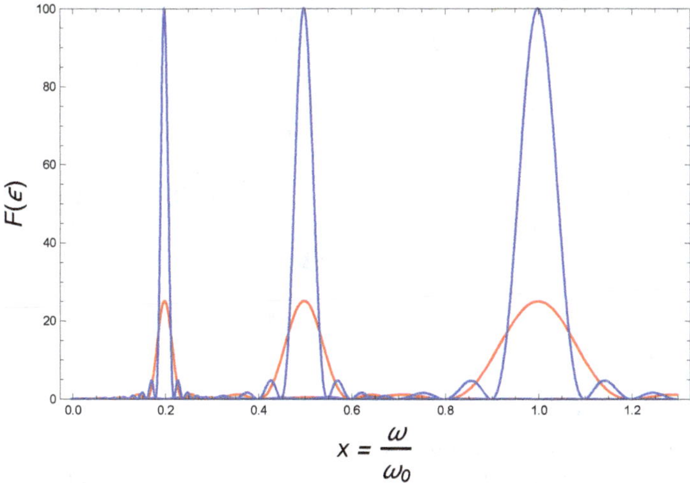

is the fundamental resonant frequency at the forward direction $\theta = 0$. Figure 7.14 then shows $F(\epsilon)$ as a function of x for $H = 1$. The red and blue curves are for $N_u = 5, 10$ respectively. The three clusters correspond to $y = 0$ (right), 1 (middle), and 2 (left), respectively.

As illustrated in Fig. 7.14, the undulator radiation spectrum in the forward direction $\theta = 0$ is centered around the frequency $H\omega_0$ with a half width between ± 1-nodes given by $\Delta\omega = \frac{H\omega_0}{N_u}$. When θ increases (radiation away from the forward direction, either up-down or left-right), the center frequency shifts downward (red shifted) and at the same time becomes narrower although the height of $F(\epsilon)$ remains N_u^2 for all y. The half width between the first zeros is in general given by $\Delta\omega = \frac{H\omega_0}{N_u(1+y^2)}$ and becomes narrower as y increases. In terms of $\frac{\Delta\omega}{\omega}$, however, the width is constant independent of y,

$$\frac{\Delta\omega}{\omega} = \frac{1}{N_u}.$$

All these features can be confirmed in Fig. 7.14.

The resonant frequency ω_0 corresponds to a resonant wavelength in the forward direction,

$$\lambda_0 = \frac{\lambda_u}{2\gamma^2}(1 + a_u^2). \tag{7.19}$$

Take the undulator of the Linac Coherent Light Source for example, $\lambda_u = 3$ cm, $B_y = 1.25$ T, $\gamma = 2.7 \times 10^4$, we obtain $a_u = 2.5$ and $\lambda_0 = 1.5$ Angstrom.

For X-ray FEL users, it is desirable to have shorter radiation wavelength λ_0. It might appear that a shorter undulator period λ_u would help. But that makes it more difficult in the undulator construction with required quality, and it also has the tendency of reducing the value of a_u.

If one integrates over all angles θ, based on Fig. 7.14 and a moment's reflection, one expects a frequency spectral distribution covering x from 0 to ~ 1. For small x, one expects a vanishing spectral distribution. As x increases, one expects a slow power-law climb of the radiation power spectrum until it reaches a maximum in the neighborhood $x \approx 1$. The spectrum then extends beyond $x = 1$ only by a small tail, with $x - 1 \sim \frac{1}{N_u}$. A quantitative calculation of this integrated power spectrum will be given later in Sec. 7.3.10 and Fig. 7.24, but we now already know what to expect qualitatively.

$F(\epsilon)$ as angular spectrum The angular distribution of undulator radiation comes from a product of $F(\epsilon)$ and $G_{\sigma,\pi}(\theta,\phi)$. The angular dependences of these two form factors, however, are quite different.

Let us first look at the angular structure of $G_{\sigma,\pi}(\theta,\phi)$ first. The repeated occurrence of the combination of parameters into $1+a_u^2+\gamma^2\theta^2$ in the expressions for $G_{\sigma,\pi}(\theta,\phi)$ reflects the fact that a typical value of θ of the radiation would scale as

$$\theta \sim \frac{\sqrt{1+a_u^2}}{\gamma}.$$

One might loosely — and only loosely — consider this angular spread comes from a convolution of the electron's oscillating direction of motion (which contributes $\sim \frac{a_u}{\gamma}$) and the natural divergence of the radiation relative to the tangential direction of motion (which contributes $\sim \frac{1}{\gamma}$).

It should be reminded that θ is the polar angle relative to the undulator axis. In case, as we will need later, we want to use $\theta_{x,y}$, they are related to θ by $\theta^2 = \theta_x^2 + \theta_y^2$. The above convolution picture should apply only to x, not to y, but the scaling holds for both. We shall return to this point momentarily when discussing Figs. 7.16 and 7.17. As we will then see, what this convolution picture fails to consider is the interference effect due to the electron's undulating motion, i.e. what the undulator was introduced to do in the first place.

The expression $\theta \sim \frac{\sqrt{1+a_u^2}}{\gamma}$ can be recast into the form $\theta \sim \sqrt{\frac{\lambda_0}{\lambda_u}}$. We said before that λ_u is in centimeters and λ_r is in submicrons. The radiation angular spread is the square root of their ratio, i.e. the square root of the Doppler factor.

Now consider $F(\epsilon)$. Compared with the more modest $G_{\sigma,\pi}(\theta,\phi)$, $F(\epsilon)$ is much more sharply peaked. To see that, we need to plot $F(\epsilon)$ as a function of θ. Figure 7.15 now plots $F(\epsilon)$ as a function of $y = \frac{\gamma\theta}{\sqrt{1+a_u^2}}$ for different values of $x = \frac{\omega}{H\omega_0}$.

Figure 7.15: $F(\epsilon)$ as a function of $y = \frac{\gamma\theta}{\sqrt{1+a_u^2}}$. The columns from left to right correspond to $H = 1, 2, 3$. The rows correspond to $x = \frac{\omega}{H\omega_0} = 0.5, 0.7, 0.95, 1, 1.05$. Red curves are for $N_u = 5$. Blue curves are for $N_u = 10$.

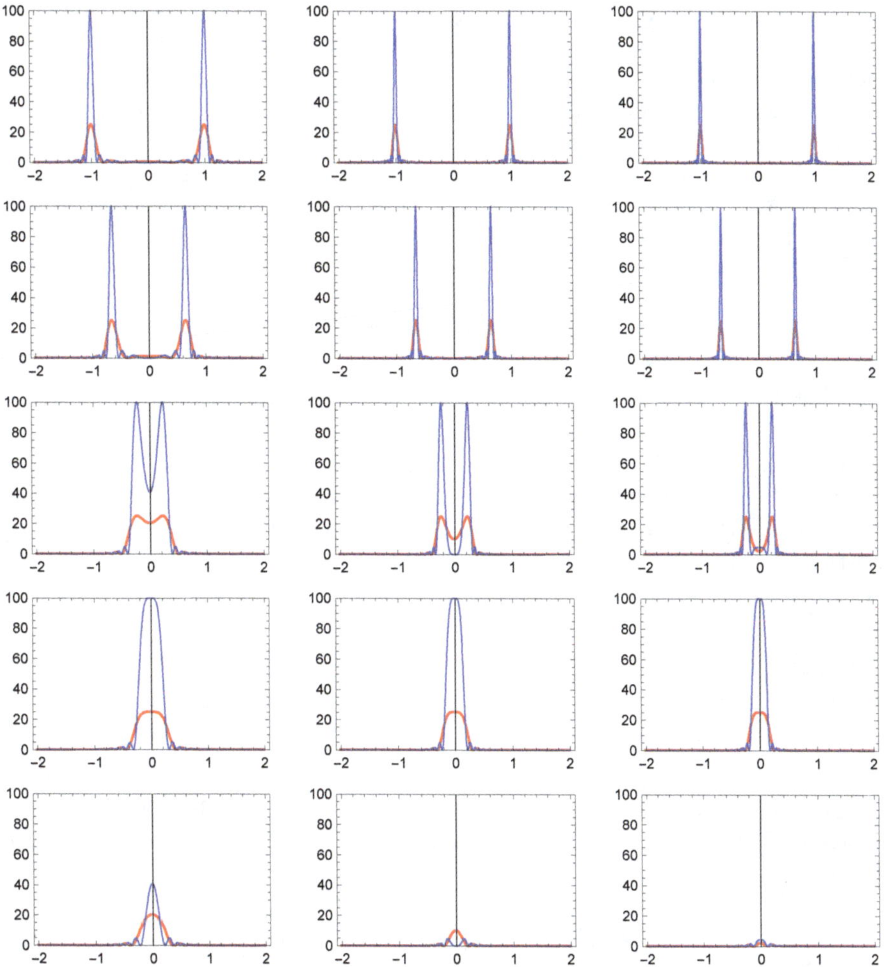

From Fig. 7.15, we see that when ω is below $H\omega_0$, and $x = \frac{\omega}{H\omega_0} < 1$, $F(\epsilon)$ has two peaks at

$$(y^2)_{\text{peak}} = \frac{1}{x} - 1\,.$$

The width around each peak is

$$\Delta(y^2) \approx \frac{1}{HxN_u}\,.$$

There are two peaks as a $\pm$ pair. As x is increased toward 1, these two peaks approach each other. Cases for $x = 0.95$ show double-humped distributions in y because the two peaks begin to overlap. Further increase of x toward 1 causes the two peaks to merge. The two-hump structure becomes noticeable when $(y^2)_{\rm peak} \approx \Delta(y^2)$, or when x is just slightly less than 1, with $1 - x \approx \frac{1}{HN_u}$. The cases $x = 1$, i.e. when ω is exactly equal to $H\omega_0$, these two peaks merge into one and they show a bell-shaped single-hump distribution. The cases $x = 1.05$, i.e. when ω is just above ω_0, $F(\epsilon)$ diminishes very quickly and they show a single-hump suppressed distribution.

The angular sharpness of $F(\epsilon)$ compared with that of $G_{\sigma,\pi}(\theta,\phi)$ can be seen easily. When $x = 1$, i.e. $\omega = H\omega_0$, for example, $F(\epsilon)$ is significant when $y = 0$ within a bandwidth $\Delta y \sim \frac{1}{\sqrt{HN_u}}$, or equivalently,

$$\theta \sim \frac{1}{\sqrt{HN_u}} \frac{\sqrt{1+a_u^2}}{\gamma} . \tag{7.20}$$

This angular bandwidth is much narrower than that of $G_{\sigma,\pi}(\theta,\phi)$ by a factor $\sqrt{HN_u} \gg 1$. See Fig. 7.23 for another illustration.

Another way to appreciate how wide the form factors $G_{\sigma,\pi}$ are relative to the sharp $F(\epsilon)$ is to note that when the radiation is emitted with the wide angle at the edge of the $G_{\sigma,\pi}$ (i.e. when $y \gtrsim 1$), the radiation has already cause a very substantial red shift (i.e. with $x \lesssim \frac{1}{2}$).

Homework 7.8

(a) Figure 7.15 shows $F(\epsilon)$ as a function of $y = \frac{\gamma\theta}{\sqrt{1+a_u^2}}$ has two peaks, and the peak locations move when $x = \frac{\omega}{H\omega_0}$ is varied — the red shift effect. For $0 < x < 1$ and staying away somewhat from $x = 0$, the peaks are located at positions $y_{\rm peak} = \pm\sqrt{\frac{1}{x} - 1}$.

(b) Figure 7.15 also shows that these peaks have finite widths. Show that if x stays away from 0 and 1, the node-to-node half-width of these peaks is given by

$$\Delta y = \frac{1}{2\sqrt{x}} \left[\sqrt{1 + \frac{1}{HN_u} - x} - \sqrt{1 - \frac{1}{HN_u} - x} \right] . \tag{7.21}$$

If $HN_u \gg 1$,

$$\Delta y \approx \frac{1}{2HN_u\sqrt{x(1-x)}} .$$

7.3.6 The form factors $G_{\sigma,\pi}(\theta,\phi)$

Figures 7.16 and 7.17 show $G_{\sigma,\pi}(\theta,\phi)$ for various values of a_u and H as functions of

$$u_x = \gamma\theta\cos\phi, \qquad u_y = \gamma\theta\sin\phi .$$

Figure 7.16: The form factor $G_\sigma(u_x, u_y)$ plotted in 3-D view in the (u_x, u_y) space. The three columns from left to right are for $a_u = 0.5, 1, 2$. The three rows from top to bottom are for $H = 1, 2, 3$.

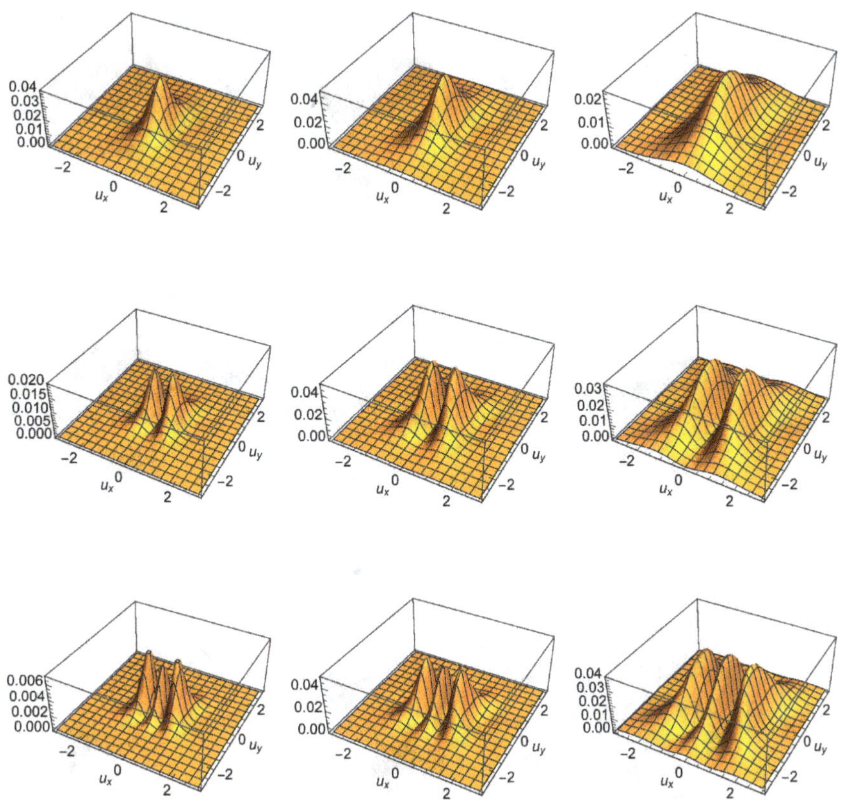

Several observations should be made from Figs. 7.16 and 7.17. For example, G_σ is seen to be typically $\sim$ a few times 0.01, while G_π is typically an order of magnitude less than G_σ. The π-radiation is, although not negligible, much weaker than the σ-radiation. It is also seen that $G_\pi = 0$ when $\theta = 0$ or when $\sin\phi = 0$, as pointed out earlier.

The form factors exhibit interference patterns. In particular, the u_x-distribution of G_σ contains H humps. If $H = 1$, it resembles a Gaussian. The u_y-distribution of G_σ is always Gaussian-like with a single hump for all values of H. The lack of symmetry between u_x and u_y is not surprising because, after all, the electron is making its undulating motion in x, not in y. As a result, interesting interference pattern occurs in x, and nothing as sophisticated can occur in y.

Figure 7.17: The form factor $G_\pi(u_x, u_y)$ plotted in 3-D view in the (u_x, u_y) space. The three columns from left to right are for $a_u = 0.5, 1, 2$. The three rows from top to bottom are for $H = 1, 2, 3$.

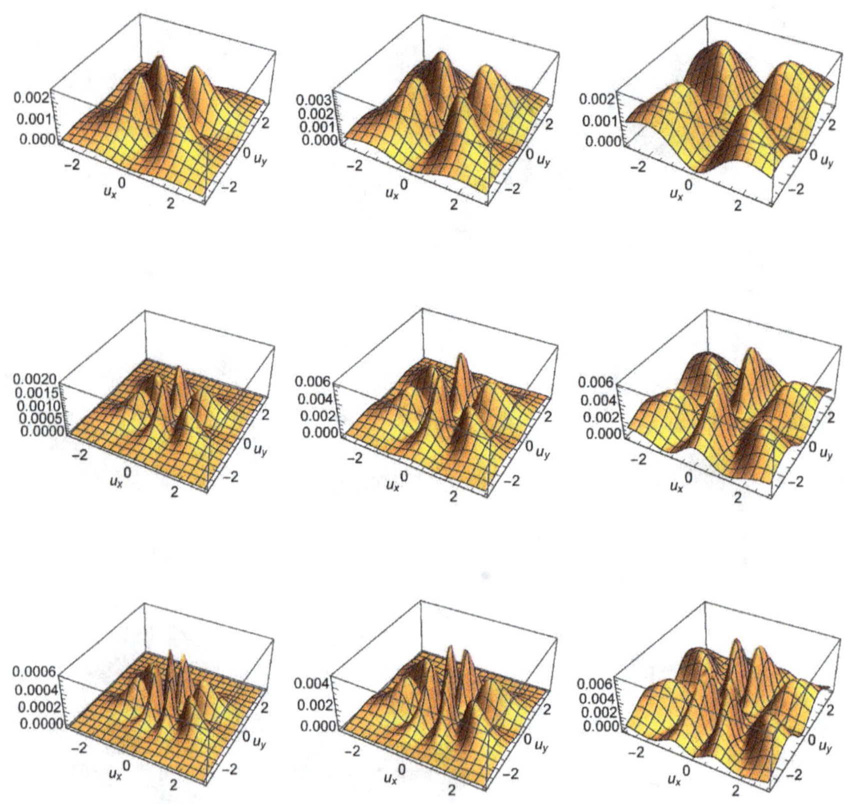

There is another observation to be made with the angular distribution of the form factors. For the $H = 1, \sigma$-mode, its Gaussian-like distribution is narrower in x than in y. What we might have expected (incorrectly) is to see a wider distribution in x than in y because the orbital modulation is in x. This somewhat counter intuitive observation is to be explained as follows. Interference certainly plays a role to this observation. In addition, it is to be noted that the $H = 1$ mode dominates only when a_u is small. When a_u increases, higher harmonics begin to contribute. When $a_u \gtrsim 1$, the $H = 1$ harmonic no longer dominates. The higher harmonics, with their interference patterns widen only in x, not in y, then widens the total angular distribution. As a result, for $a_u \gtrsim 1$, the total angular spread is wider in x than in y, as one would expect — but only when all harmonics are summed together and only when $a_u \gtrsim 1$.

Similarly, the u_x-distribution of G_π has $H+1$ humps, while its u_y-distribution always has two humps. In contrast, the dependence of the form factors on a_u is less drastic.

All angular spreads in G_σ and G_π are limited to a region ($|u_x| \lesssim 1, |u_y| \lesssim 1$), or $\gamma\theta \lesssim 1$. The angular distribution spreads out as a_u increases according to $\gamma\theta \propto \sqrt{1+a_u^2}$. Note that for higher harmonics, the total angular spread $\theta \sim \frac{\sqrt{1+a_u^2}}{\gamma}$ does not depend much on the harmonic number H. It is the fine structures within the total spread that becomes narrower as H is increased.

An alert reader might note that it is more physical to consider the angular form factors as functions of u_x and u_y than as functions of θ and ϕ — as indicated for example by how the distribution humps occur in the higher harmonics. The angular form factors are better described in Cartesian coordinates (u_x, u_y) than the polar coordinates $(\gamma\theta, \phi)$. This is in contrast to the spectral function $F(\epsilon)$, which depends on θ and is independent of ϕ, indicating it is more physical to describe $F(\epsilon)$ in the polar coordinates — this point was made in our discussion of the physical origin of the undulator resonance in Sec. 7.3.2. It is a bit unfortunate that these two factors prefer different coordinates.[5] For our discussions, we think it is perhaps overall more convenient to compromise on the angular form factors, and to consider them as functions of (θ, ϕ).

ACO observation A nice actual-color photo taken at the ACO storage ring, shown in Fig. 7.18, is an excellent illustration of the undulator spectrum $\frac{d^2W}{d\Omega d\omega}$. The radiation happens to be in the visible light range. The orientation of the photo is such that horizontal is x and vertical is y, with the planar undulator modulating the x motion of the electron beam.

The fundamental undulator frequency is that of blue color. Referring to Figs. 7.16 and 7.17, one first identifies the central blue region of Fig. 7.18 to be the $H = 1, \sigma$-mode with a central hump. Being the contribution from only the lowest $H = 1$ radiation, it is narrower in x than in y. As the radiation angle increases, this σ-radiation gets red-shifted and becomes the central yellow and the red regions of the photo.

The $H = 1, \pi$-mode radiation shows up as the four blue spots in the corners. As the radiation angle increases, these four blue spots become the four faint red spots in the far corners. If observed closer, one may even see a hint of yellow regions between the four blue and the four red π-mode spots.

All the higher harmonics $H > 1$ are out of the visible light range and not seen by the camera.

Rough expressions for $G_{\sigma,\pi}(\theta, \phi)$ Sometimes a rough estimate of the form factors $G_{\sigma,\pi}(\theta, \phi)$ is needed, at least for low harmonics and modest values of $a_u \lesssim 1$. It turns out that for small to modest values of a_u and for $H = 1, 2$, it

[5]Had we considered a helical undulator, this would be less of an issue.

Figure 7.18: An actual-color photo taken at the forward direction for an undulator at the ACO storage ring. [Courtesy Helmut Wiedemann, Ives Petrov (2019)]

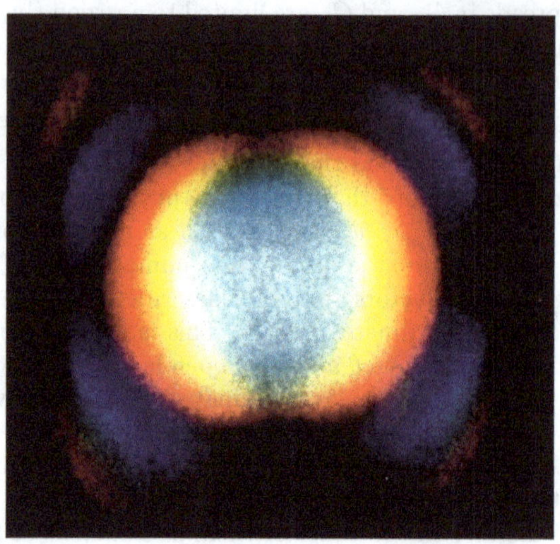

is sometimes adopted in the literature to approximate G_σ by ignoring $\mathcal{D}_2$ and taking only the $m=0$ component in $\mathcal{D}_1$, i.e.

$$\begin{aligned} \mathcal{D}_1 &\approx \begin{cases} -\frac{1}{2}J_0(\alpha)[J_0(\zeta)-J_1(\zeta)], & \text{for } H=1, \\ -\frac{1}{2}J_1(2\alpha)[J_0(2\zeta)-J_1(2\zeta)], & \text{for } H=2, \end{cases} \\ \Longrightarrow \quad G_\sigma &\approx \begin{cases} \frac{a_u^2}{4(1+a_u^2+\gamma^2\theta^2)^2} J_0^2(\alpha)[J_0(\zeta)-J_1(\zeta)]^2, & \text{for } H=1, \\ \frac{a_u^2}{(1+a_u^2+\gamma^2\theta^2)^2} J_1^2(2\alpha)[J_0(2\zeta)-J_1(2\zeta)]^2, & \text{for } H=2. \end{cases} \end{aligned} \tag{7.22}$$

On the other hand, for G_π, we need to keep $\mathcal{D}_2$. For sufficiently small a_u, and for $H=1$, this can be done again by taking only the $m=0$ term in the summation,

$$\begin{aligned} \mathcal{D}_2 &\approx J_1(\alpha)J_0(\zeta), \\ \Longrightarrow \quad G_\pi &\approx \frac{\gamma^2\theta^2\sin^2\phi}{2(1+a_u^2+\gamma^2\theta^2)^2} J_1^2(\alpha)J_0^2(\zeta). \end{aligned} \tag{7.23}$$

Further consider only the σ-radiation with $H=1$, ignoring $\mathcal{D}_2$, taking only the $m=0$ contribution, and letting $\gamma\theta \ll 1$, we have $\alpha \approx 0, \zeta \approx \frac{a_u^2/2}{1+a_u^2}$, and from Eq. (7.22),

$$G_\sigma \approx \frac{a_u^2}{4(1+a_u^2+\gamma^2\theta^2)^2}[JJ]^2,$$

Figure 7.19: The $[JJ]$ factor as a function of a_u. Dashed line marks the asymptotic value $[JJ] = 0.6962$.

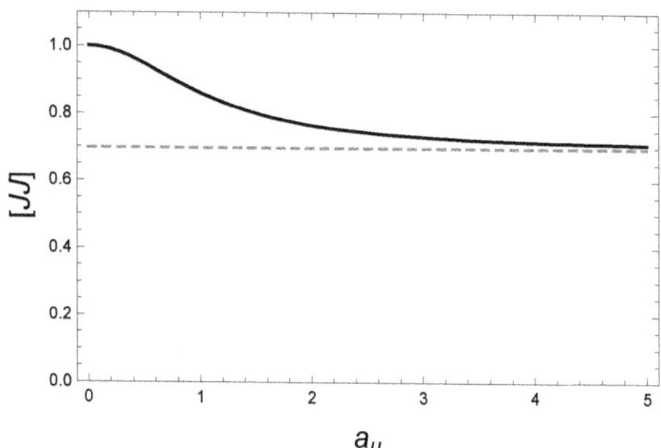

where, following the literature, we have introduced an abbreviation,

$$[JJ] = J_0\left(\frac{a_u^2/2}{1+a_u^2}\right) - J_1\left(\frac{a_u^2/2}{1+a_u^2}\right). \tag{7.24}$$

The presence of the quantity $[JJ]$ originates from the fact that v_z has a figure-8 sinusoidal modulation, i.e. $v_z \approx \bar{v}_z + \frac{a_u^2 c}{\gamma^2}\cos 2k_u z$. (See also Homework 7.28 later.) Figure 7.19 shows the function $[JJ]$. Although significant in its physical meaning, it is seen that its value lies between 1 and 0.7, and is not a very critical numerical factor in practice.

A word of caution is in order here. In the derivation of the rough expressions for the σ-radiation, we have assumed the contribution from $\mathcal{D}_2$ is small and can be dropped. This not quite so. When $\mathcal{D}_2$ contribution is included, the rough expressions become more complicated. We will show, for example in Eq. (7.39) and Homework 7.22 later that the inclusion of $\mathcal{D}_2$ can change the numerical results by a factor of 2 or so.

Forward radiation when $\theta = 0$ In the forward direction, $\theta = 0$, we have $u_x = u_y = 0, \alpha = 0, G_\pi = 0$. There is only σ-radiation, and this radiation is only in odd harmonics ($H =$ odd). Returning to the exact expression Eq. (7.16), we have

$$\begin{aligned} \frac{d^2W}{d\Omega d\omega}(\theta=0) &\rightarrow \frac{2e^2\gamma^2}{\pi\epsilon_0 c} G_\sigma(\theta=0) F\left(\frac{\omega}{\omega_0} - H\right), \\ G_\sigma(\theta=0) &= \left(\frac{H a_u}{2(1+a_u^2)}\right)^2 \left[J_{\frac{H-1}{2}}\left(\frac{H a_u^2/2}{1+a_u^2}\right) - J_{\frac{H+1}{2}}\left(\frac{H a_u^2/2}{1+a_u^2}\right)\right]^2. \end{aligned} \tag{7.25}$$

We have dropped the ϕ-dependence as it becomes not relevant.

Figure 7.20: The forward form factor $G_\sigma(\theta = 0)$ as a function of a_u. Six curves are for harmonic number $H = 1, 3, 11, 51, 201$ successively.

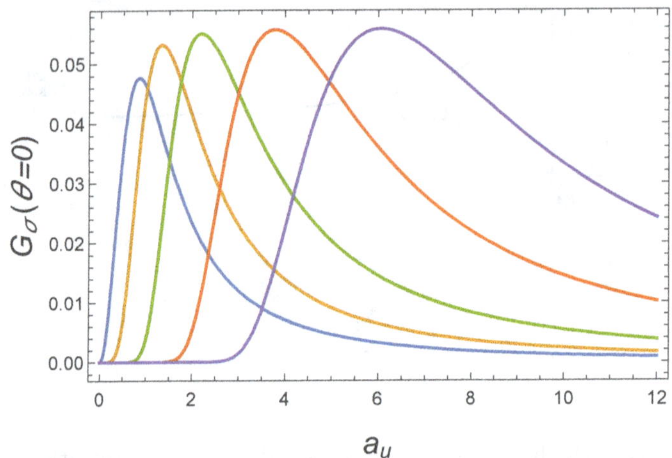

Figure 7.20 shows $G_\sigma(\theta = 0)$ as a function of a_u for $H = 1, 3, 11, 51, 201$. When $a_u = 0$, there is no undulator radiation. For small a_u, the radiation is dominated by the fundamental harmonic. When $a_u \approx 1$, a few low harmonics begin to contribute. When a_u is increased beyond 1, high harmonics very quickly join; for example, 201-harmonic is very significant when $a_u = 6$.

The factor within the expression for $G_\sigma(\theta = 0)$ can be written as $[JJ]_H$ for abbreviation as a generalization of Eq. (7.24),

$$[JJ]_H = J_{\frac{H-1}{2}}\left(\frac{Ha_u^2/2}{1+a_u^2}\right) - J_{\frac{H+1}{2}}\left(\frac{Ha_u^2/2}{1+a_u^2}\right).$$

However, it is meaningful only when $H =$ odd.

Homework 7.9 We mentioned that radiation is completely suppressed when $\omega = \frac{\omega_r}{2}, \frac{3\omega_r}{2}, \cdots$ if $N_u =$ even. Show that explicitly by proving $F(\epsilon) = 0$ in those cases.

Homework 7.10 Follow the text to show that even harmonics vanish in the forward direction $\theta = 0$. Explain physically why this is the case.

Solution Physically this is because (i) higher harmonics come from v_z in its figure-8 motion, and (ii) v_z has no line-of-sight component in the forward direction.

Homework 7.11 When a_u switches sign, meaning the undulator field switches polarity or replacing electrons by positrons, the radiation spectrum is unchanged.

Demonstrate this by showing that both $F(\epsilon)$ and $G(\theta, \phi)$ are even functions of a_u.

Homework 7.12 Show that an alternative expression for $\mathcal{D}_1$ is

$$\mathcal{D}_1 = -\frac{1}{2} \sum_{m=-\infty}^{\infty} J_m(H\zeta)[J_{H+2m-1}(H\alpha) + J_{H+2m+1}(H\alpha)] .$$

Homework 7.13 Make a plot of G_σ and G_π using the approximate Eqs. (7.22) and (7.23). Compare your results with the appropriate cases of the exact calculations shown in Figs. 7.16 and 7.17, at least for the case $H = 1$. You should get a reasonable comparison at least for the cases when a_u is not too large.

Homework 7.14

(a) Integrate the radiation spectrum $\frac{d^2W}{d\Omega d\omega}$ of Eq. (7.25) over ω to obtain the total angular density in the forward direction,

$$\frac{dW}{d\Omega}(\theta = 0) = \frac{2e^2\gamma^2}{\pi\epsilon_0 c} N_u\omega_0 \left(\frac{Ha_u}{2(1+a_u^2)}\right)^2 [JJ]_H^2 .$$

(b) A beam of 10^{10} electrons and energy of 1 GeV traverses an undulator with $L_u = 0.6$ m, $\lambda_u = 3$ cm, $B_y = 1$ T. A detector with a circular opening radius of 3 mm is located at the forward $\theta = 0$ direction a distance 5 m downstream of the center of the undulator. First find the undulator radiation's fundamental resonant wavelength λ_0. Then calculate the total radiation energy received by the detector in the fundamental frequency as well as for the harmonics $H = 2$ and $H = 3$. Give the result in unit of picoJoule.

(c) Compare with the expected behavior Eq. (7.18).

Solution (b) First confirm the detector accepting angle is less than $\frac{1}{\gamma}\sqrt{\frac{1+a_u^2}{HN_u}}$, Eq. (7.20), before applying the formula. Watch out for the case $H = 2$.

7.3.7 Radiation spectrum — part III

As mentioned, the frequency spectrum of the undulator radiation depends on the form factor $F(\epsilon)$, while the angular spectrum depends on both $F(\epsilon)$ and $G_{\sigma,\pi}(\theta, \phi)$. It was also mentioned that the angular distribution of $F(\epsilon)$ is much more sharply peaked than that of $G_{\sigma,\pi}(\theta, \phi)$, although it does not impose any restrictions on ϕ. In this section, we want to find the angular distribution of the product $F(\epsilon)G_{\sigma,\pi}(\theta, \phi)$ from Eq. (7.16) for given ω.

Because $F(\epsilon)$ is sharply peaked, the angular distribution can be approximated as

$$G(\theta, \phi)F(\epsilon) \approx G(\theta, \phi)\big|_{y^2=\frac{1-x}{x}} F(\epsilon), \qquad (0 < x < 1) .$$

where $x = \frac{\omega}{H\omega_0}, y = \frac{\gamma\theta}{\sqrt{1+a_u^2}}$. Because $G_{\sigma,\pi}(\theta,\phi)$ is a smooth function of θ within the much narrower $F(\epsilon)$, we ignore the variation of $G(\theta,\phi)$ within the peak region of $F(\epsilon)$. The peak of $F(\epsilon)$ is located at $y^2 = \frac{1-x}{x}$, or

$$\gamma^2\theta^2 = (1+a_u^2)\left(\frac{H\omega_0}{\omega} - 1\right).$$

As one example, take the case $H = 1$ and σ-polarization, and if we apply the approximate expression (7.22), the angular distribution becomes

$$\begin{aligned} G_\sigma(\theta,\phi)F(\epsilon) &\approx \frac{a_u^2}{4(1+a_u^2)^2(1+y^2)^2} J_0^2(\alpha)\,[J_0(\zeta) - J_1(\zeta)]^2 \Big|_{y^2=\frac{1-x}{x}} \left(\frac{\sin \pi\epsilon_1 N_u}{\pi\epsilon_1}\right)^2 \\ &= \frac{a_u^2 x^2}{4(1+a_u^2)^2} J_0^2\left(\frac{2a_u\sqrt{2x(1-x)}\cos\phi}{\sqrt{1+a_u^2}}\right) \\ &\quad \times \left[J_0\left(\frac{a_u^2 x/2}{1+a_u^2}\right) - J_1\left(\frac{a_u^2 x/2}{1+a_u^2}\right)\right]^2 \left(\frac{\sin \pi\epsilon_1 N_u}{\pi\epsilon_1}\right)^2, \end{aligned} \tag{7.26}$$

where $x = \frac{\omega}{\omega_0}$ and $\epsilon_1 = x(1 + \frac{\gamma^2\theta^2}{1+a_u^2}) - 1$ is the value of ϵ when $H = 1$.

For $H = 2$ and σ-mode, we have from Eq. (7.22),

$$\begin{aligned} G_\sigma(\theta,\phi)F(\epsilon) &\approx \frac{a_u^2 x^2}{(1+a_u^2)^2} J_1^2\left(\frac{4a_u\sqrt{2x(1-x)}\cos\phi}{\sqrt{1+a_u^2}}\right) \\ &\quad \times \left[J_0\left(\frac{a_u^2 x}{1+a_u^2}\right) - J_1\left(\frac{a_u^2 x}{1+a_u^2}\right)\right]^2 \left(\frac{\sin 2\pi\epsilon_1 N_u}{2\pi\epsilon_1}\right)^2, \end{aligned} \tag{7.27}$$

where $\epsilon_2 = 2\epsilon_1$ is the value of ϵ when $H = 2$.

Similarly, using Eq. (7.23), the π-polarization and $H = 1$ is found to have the spectrum

$$G_\pi(\theta,\phi)F(\epsilon) \approx \frac{x(1-x)\sin^2\phi}{2(1+a_u^2)} J_1^2\left(\frac{2a_u\sqrt{2x(1-x)}\cos\phi}{\sqrt{1+a_u^2}}\right) J_0^2\left(\frac{a_u^2 x/2}{1+a_u^2}\right)\left(\frac{\sin \pi\epsilon_1 N_u}{\pi\epsilon_1}\right)^2. \tag{7.28}$$

These results Eqs. (7.26), (7.27) and (7.28) give approximately the angular distribution for arbitrary radiation frequency ω. In particular, they include the red shift effect when the polar angle increases from 0. If we further specialize to the case $x = 1$ (i.e. $\omega = \omega_0$), the angular distribution (7.26) simplifies to read

$$G_\sigma(\theta,\phi)F(\epsilon) \approx \frac{a_u^2}{4(1+a_u^2)^2} [JJ]^2 \left(\frac{\sin(\frac{\pi\gamma^2\theta^2}{1+a_u^2} N_u)}{(\frac{\pi\gamma^2\theta^2}{1+a_u^2})}\right)^2. \tag{7.29}$$

Due to the sharpness of $F(\epsilon)$ only the $\theta = 0$ part of G_σ contributes, and the resulting angular distribution does not depend on ϕ.

Rms angular spread A word of caution is to be made with these results when we let $F(\epsilon)$ to define the angular spectral shape while letting $G_{\sigma,\pi}$ only to determine the height of the spectrum peak. The bell-shaped Sinc-function distribution (7.29) has a long power-law tail in θ. Sometimes it will require the weak angular dependence in $G(\theta,\phi)$ to help the convergence. The angular distribution therefore contains a sharp peak due to $F(\epsilon)$ (angular spread $\Delta y \sim \frac{1}{\sqrt{N_u}}$) *and* a long tail weakly confined by $G(\theta,\phi)$ (angular spread $\Delta y \sim 1$).

The situation can be demonstrated if one calculates the rms angular spread σ_θ. If we ignore the weak cut-off due to $G_{\sigma,\pi}(\theta,\phi)$, then the rms angular spread in θ logarithmically diverges.

In spite of weakness in the definition of the rms σ_θ, however, one might still concentrate on the peaked portion of the angular distribution defined solely from $F(\epsilon)$, and connect an equivalent rms through the FWHM sense. When this is done, for the case $H = 1$, we obtain

$$\sigma_\theta^2 \approx 0.3 \times \frac{1+a_u^2}{\gamma^2 N_u}.$$

With the fundamental radiation wavelength λ_0 given by Eq. (7.19), and the undulator total length $L_u = N_u\lambda_u$, we obtain (for $\omega = \omega_0$)

$$\sigma_\theta \approx 0.8 \times \sqrt{\frac{\lambda_0}{L_u}},$$

and since θ represents the polar angle,

$$\sigma_{\theta x} = \sigma_{\theta y} \approx \frac{0.8}{\sqrt{2}} \times \sqrt{\frac{\lambda_0}{L_u}}. \tag{7.30}$$

7.3.8 Radiation phase space

We discussed a diffraction limit for the case of synchrotron radiation in a dipole magnet given by Eq. (6.6). We pointed out that such a diffraction limit is intrinsic and applies to all radiation mechanisms. Here let us discuss how it also applies to the undulator radiation.

As mentioned earlier, the undulator radiation by a single electron is like an electron growing a coherent nose as it propagates down the undulator. The wavepacket radiated occupies a certain area of phase space. We need to calculate or estimate the phase space area occupied by this wavepacket, and this phase space area will be the intrinsic radiation emittance by a point-charge electron.

Crude mechanical model We will contend with a simple mechanical model as follows. Consider an electron traversing an undulator of length L_u. To illustrate the radiation emittance, we model the undulator radiation by a uniform line source of radiation extending from the region $-\frac{L_u}{2} \le z \le \frac{L_u}{2}$. The model is not meant to be rigorous as it ignores the wave nature of the radiation and does not take into account of diffraction, but it is useful to illustrate the phase space effect of undulator radiation. Admittedly, as will be seen, it would be a bit awkward trying to explain a diffraction limit using a mechanical model.

Consider a small section of the undulator at position $z = z_0$. The radiation emitted by the electron in this short section occupies an angular spread $-\frac{\Delta\theta}{2} \leq \theta \leq \frac{\Delta\theta}{2}$ as illustrated in Fig. 7.21. In the phase space (x, θ), it is represented as the red segment shown in Fig. 7.22(a).

Figure 7.21: A model to calculate the undulator radiation phase space. The red region indicates the interior of the undulator.

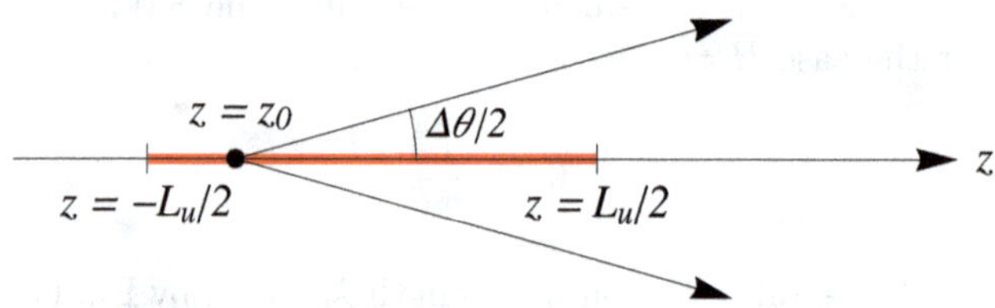

The wavepacket is built up from radiations from all short sections with z_0 running from $-\frac{L_u}{2}$ to $\frac{L_u}{2}$. To calculate the phase space occupied by the entire wavepacket, we first transform the phase space segment of Fig. 7.22(a) from $z = z_0$ to $z = 0$. The result will look like Fig. 7.22(b) and (c), with (b) when $z_0 > 0$ and (c) when $z_0 < 0$. The phase space segments in (b) and (c) are still uniform straight lines although the density is stretched thinner compared to that of (a). Figure (d) then is an integration of radiation from the entire extended source by integrating over z_0. The red region of Fig. 7.22(d) gives the intrinsic radiation phase space in this simplified model.

The distribution in Fig. 7.22(d) can be calculated (Homework 7.15). The phase space origin ($\theta = 0, x = 0$) is logarithmically singular. The θ-projection of the phase space is uniform distribution extended the range $-\frac{\Delta\theta}{2} \leq \theta \leq \frac{\Delta\theta}{2}$. The x-projection is more complicated,

$$\psi(x) = \begin{cases} 2\ln\left(\frac{\Delta\theta L_u}{4|x|}\right), & |x| < \frac{\Delta\theta L_u}{4}, \\ 0, & \text{otherwise.} \end{cases}$$

One can also calculate the rms values of the radiation phase space distribution. The result is

$$\begin{aligned} \langle\theta^2\rangle &= \frac{\Delta\theta^2}{12}, \\ \langle\theta x\rangle &= 0, \\ \langle x^2\rangle &= \frac{L_u^2\Delta\theta^2}{144}. \end{aligned} \tag{7.31}$$

Figure 7.22: Calculation of the intrinsic phase space occupied by the wavepacket of undulator radiation.

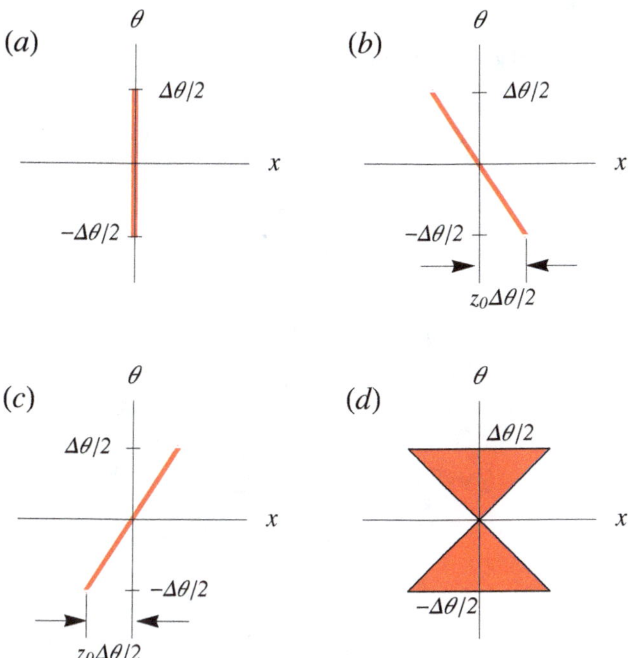

The corresponding radiation emittance and the equivalent β-function at the middle of the undulator are

$$\begin{aligned} \epsilon_{\text{rad}} &= \sqrt{\langle\theta^2\rangle\langle x^2\rangle} = \frac{L_u \Delta\theta^2}{12\sqrt{12}}, \\ \beta^*_{\text{rad}} &= \sqrt{\frac{\langle x^2\rangle}{\langle\theta^2\rangle}} = \frac{L_u}{\sqrt{12}}. \end{aligned}$$

In this model, the radiation from each of the source segments has a zero slice emittance, but when integrated over the extended source, it acquires a finite total projected emittance.

To connect to the undulator radiation, we now use Eq. (7.30) to assign

$$\begin{aligned} \Delta\theta &= \frac{0.8}{\sqrt{2}}\sqrt{\frac{12\,\lambda_0}{L_u}}, \\ \Longrightarrow \quad \sqrt{\langle\theta^2\rangle} &= \frac{\Delta\theta}{\sqrt{12}} = \frac{0.8}{\sqrt{2}}\sqrt{\frac{\lambda_0}{L_u}}, \\ \sqrt{\langle x^2\rangle} &= \frac{L_u\Delta\theta}{12} = \frac{0.8}{\sqrt{24}}\sqrt{\lambda_0 L_u}, \end{aligned}$$

which then yields

$$\epsilon_{\text{rad}} = \frac{(0.8)^2}{\sqrt{48}}\lambda_0 .$$

For later use, we also need to find out the radiation beam size averaged over the length of the undulator,

$$\langle\bar{x^2}\rangle = \int_{-L_u/2}^{L_u/2} \frac{ds}{L_u}(\langle x^2\rangle + \langle\theta^2\rangle s^2) = \frac{L_u^2\Delta\theta^2}{72} .$$

or

$$\langle\bar{x^2}\rangle = \frac{(0.8)^2}{12}\lambda_0 L_u .$$

The result is not too different from what is expected,

$$\begin{aligned}
\epsilon_{\text{rad}} &= \frac{\lambda_0}{4\pi} , \\
\sigma_\perp^2 &= \langle\bar{x^2}\rangle = \frac{\lambda_0 L_u}{4\pi} , \\
\sigma_\theta^2 &= \frac{\lambda_0}{4\pi L_u} .
\end{aligned} \tag{7.32}$$

The model does not take into account of wave nature of the laser. Given the crude nature of the model, we consider it reasonable. We shall use Eq. (7.32) in our later discussions. When doing so, $\sigma_\perp$ is meant to represent the radiation's transverse radius over the undulator, not only at its center.

Incidentally, for synchrotron radiation in a dipole, we may apply this model by replacing the undulator length L_u by the synchrotron radiation coherence length $\frac{2\rho}{\gamma}$, and replacing $\Delta\theta$ by $\frac{2}{\gamma}$, and obtain

$$\epsilon_{\text{rad}} \approx \frac{\rho}{3\sqrt{3}\,\gamma^3} = \frac{\lambda_c}{4\pi\sqrt{3}} ,$$

consistent with expectation.

Coherence — recap This radiation wavepacket from a single electron is coherent within itself. As mentioned earlier, each radiation event by a single electron is necessarily intrinsically coherent, regardless of the radiation mechanism. The radiation wavepacket from each electron, whether it is the long nose (7.11) being piled up during the passage through an undulator or the relativistically compressed synchrotron radiation wavepacket emitted in a bending magnet, remains fully coherent within itself, and such an internal coherence is in fact a requirement for the radiation to be possible in the first place. When making a statement like "synchrotron radiation is incoherent", the incoherence being referred to is among the radiation wavepackets from different electrons or different emission events by the same electron. When one says "FEL radiation is coherent", it means all the noses from different electrons in this FEL undulator turn out to be in phase relative to one another. In this context, undulator radiation is incoherent, while FEL as we will see is coherent.

Let us continue this discussion a bit further. Let us compare the synchrotron radiation in a bending magnet with the undulator radiation. Both are internally coherent, and each has a total length. The undulator wavepacket has a length of $N_u\lambda_0$, while the synchrotron radiation wavepacket has a relativistically compressed length $\frac{\rho}{\gamma^3}$. Inside a undulator wavepacket is a quasi-sinusoidal pattern (7.11), while inside a synchrotron radiation wavepacket is a complicated (but well-defined) pattern (not shown). Fourier transform of each wavepacket yields the frequency spectrum of the corresponding radiation. The Fourier transform of the undulator wavepacket is a narrow banded $F(\epsilon)$. The Fourier transform of the synchrotron radiation wavepacket yields the well-known wide-banded synchrotron radiation spectrum.

In most practical undulators, there are five lengths involved: the undulator length L_u, the electron bunch length σ_z, the undulator period λ_u, the coherent length $\ell_{\rm coh} = N_u\lambda_0$, and the radiation wavelength λ_0. Usually, we have the length hierarchy,

$$L_u \gg (\sigma_z, \lambda_u) \gg \ell_{\rm coh} \gg \lambda_0 \,.$$

Radiation emittance from electron beam All individual radiation events obey the diffraction limit. What still needs to be done is to produce an electron beam that has an emittance smaller than this diffraction limit in order to fully use the coherence of the intrinsic radiation. In case the electron beam has its own transverse size $\sigma_{e,x}$ and divergence $\sigma'_{e,x}$, then the total radiation has

$$\sigma_{\rm tot}\sigma'_{\rm tot} \approx \sqrt{\sigma_x^2 + \sigma_{e,x}^2}\,\sqrt{{\sigma_\theta}^2 + {\sigma'_{e,x}}^2} \,.$$

The radiation will then be transversely coherent if

$$\sigma_{e,x} \lesssim \sigma_x, \qquad \text{and} \qquad \sigma'_{e,x} \lesssim \sigma_\theta \,, \tag{7.33}$$

where σ_x, σ_θ are given in Eq. (7.32). That in turn puts a requirement on the emittance of the electron beam

$$\epsilon_e \lesssim \sigma_x\sigma_\theta = \frac{\lambda_0}{4\pi} \,. \tag{7.34}$$

It should be noted that the diffraction limit is reached by imposing two conditions (7.33). Satisfying their combined condition (7.34) is not sufficient. However, following the conventional claims, we shall not insist here, understanding that electron beam optics will provide the exchange between $\sigma_{e,x}$ and $\sigma'_{e,x}$ as needed.

Take LCLS for example, $\lambda_0 = 1.5$ Angstrom requires $\epsilon_e \lesssim 1.2\times 10^{-11}$ m-rad. At 15 GeV, the required normalized emittance of the electron beam is $\epsilon_{e,N} \lesssim 0.5$ mm-mrad, which is what LCLS achieved.

Wiggler or undulator? We mentioned earlier in Eq. (7.5) that the prescribed rms orbital angular oscillation of an individual electron in the undulator is

$$\sigma_x' \approx \frac{a_u}{\gamma}.$$

Since we know about the natural opening angle $\frac{1}{\gamma}$ of radiation in dipoles, and noting that an undulator consists simply of a series of dipoles, one may have expected (incorrectly) the radiation angular spread to be given by adding in quadrature the contributions from this $\frac{1}{\gamma}$ contribution and a contribution from the orbital angular spread $\frac{a_u}{\gamma}$.

We now know that the angular spread of radiation from a single electron is given by [ignoring the numerical factor and consider $\omega = \omega_0$ only, see Eq. (7.20)]

$$\sigma_\theta \approx \frac{\sqrt{1+a_u^2}}{\sqrt{N_u}\gamma}.$$

So the initial expectation is missing a large factor $\frac{1}{\sqrt{N_u}}$.

The additional factor of $\frac{1}{\sqrt{N_u}}$ shows how strong the interference effect introduced by the undulator has been. The undulator narrows the angular spread by a large factor $\sqrt{N_u}$, whether the angular spread comes originally from the $\frac{1}{\gamma}$ radiation opening angle or the $\frac{a_u}{\gamma}$ electron oscillation angle if we focus on a single harmonic.

It is customary to call the device an "undulator" if the radiation angular spread is dominated by the radiation opening angle, and a "wiggler" if it is dominated by the electron's orbital motion. It then follows that an undulator would have $a_u \lesssim 1$ while a wiggler would have $a_u \gg 1$.

However, although there is every reason to accept this customary definition, one should note that the reasoning leading to this definition is flawed. This can be seen by noting that there is no y-oscillation and yet the radiation angular spread is given by Eq. (7.20) even in the y-direction. The reason of an extra contribution $\frac{a_u}{\gamma}$ in angular spread actually comes from the longitudinal velocity modulation of the figure-8 motion. It does not come from the x-oscillation of the electron.

One more point along this line of reasoning: It might be argued that if we increase a_u in a flat undulator, all we have to do in terms of the magnet aperture is to widen the horizontal width of the vacuum chamber because the extra space needed for both the electron beam and the radiation will be only in the x-direction. This is incorrect. The electron will need more space in the x-dimension, but both x- and y-dimensions need to take into account of a contribution from $\frac{a_u}{\gamma}$.

Furthermore, there does not seem much physical meaning to be assigned to a dominance by radiation angle or orbital angle. In all cases, there will be interference and there will be the undulator resonance effect. Only that when a_u is large, the resonance occurs at a long wavelength, and higher harmonics make increasingly important contributions. A criterion like $a_u \approx 1$ does not define a cutoff.

Nevertheless, in a crude manner, a difference is being made between undulators and wigglers. To the extent that is being done, all modern radiation sources are of the undulator type — although they may refer to a_u values as high as ~ 30 and in principle can go higher. Wigglers these days are still used, but mostly not as radiation sources. For example, they are used as damping wigglers whose function is mainly to force the beam to radiate as hard as possible without actual bending of the net trajectory.

Homework 7.15 Set up the model as suggested in the text.

(a) Show that the phase space distribution is normalized by $\int dx d\theta\, \psi(x,\theta) = L_u \Delta\theta$.

(b) Derive Eq. (7.31).

7.3.9 Angular spectrum

Equation (7.20) refers to the angular spread when ω is fixed at $\omega = H\omega_0$. As emphasized, it is an angular spread much narrower than the expected $\frac{\sqrt{1+a_u^2}}{\gamma}$ when no radiation interference is taken into account.

The situation changes when ω is integrated over. In Fig. 7.15, we have seen how sharply the form factor $F(\epsilon)$ is as a function of θ, and how the sharp peaks red shift as the angle increases. Equation (7.20) is a statement about how sharp the peaks are — more accurate expression is given by Eq. (7.21). However, if we integrate over frequency ω, the sharp peaks are integrated over, and we obtain a wider angular distribution that extends over a range $\sim \frac{\sqrt{1+a_u^2}}{\gamma}$. A graphic illustration is shown in Fig. 7.23.

Indeed, integrating $\frac{d^2W}{d\Omega d\omega}$ over ω gives

$$\begin{aligned}
\frac{dW}{d\Omega} &= \int_0^\infty d\omega \frac{d^2W}{d\Omega d\omega} \\
&= \frac{2e^2\gamma^2}{\pi\epsilon_0 c} G(\theta,\phi) \int_0^\infty d\omega \left(\frac{\sin \pi\epsilon N_u}{\pi\epsilon}\right)^2 \\
&= \frac{4e^2\gamma^4 k_u}{\pi\epsilon_0} \frac{N_u G(\theta,\phi)}{1+a_u^2+\gamma^2\theta^2}, \qquad (7.35) \\
G(\theta,\phi) &= G_\sigma(\theta,\phi) + G_\pi(\theta,\phi).
\end{aligned}$$

The sharp $F(\epsilon)$ has been integrated over. The total angular spread is mainly determined by the more smooth form factors $G_{\sigma,\pi}(\theta,\phi)$. Dependence on the harmonic number H is hidden in the form factors $G_{\sigma,\pi}(\theta,\phi)$.

Figure 7.23: Upper panel shows the angular distribution of undulator radiation at a certain fixed frequency $\omega < H\omega_0$. Lower panel is the angular distribution when ω is integrated over.

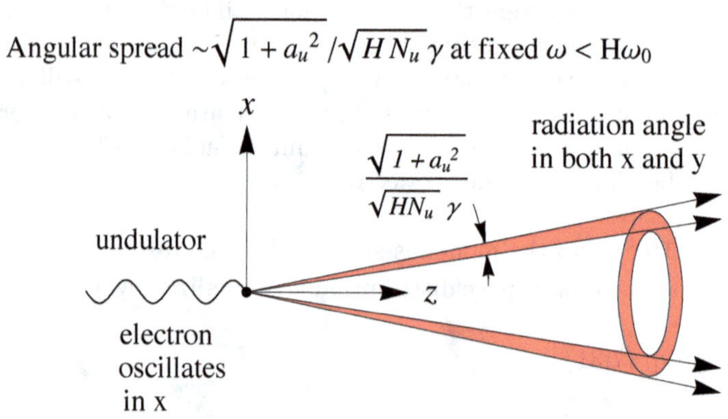

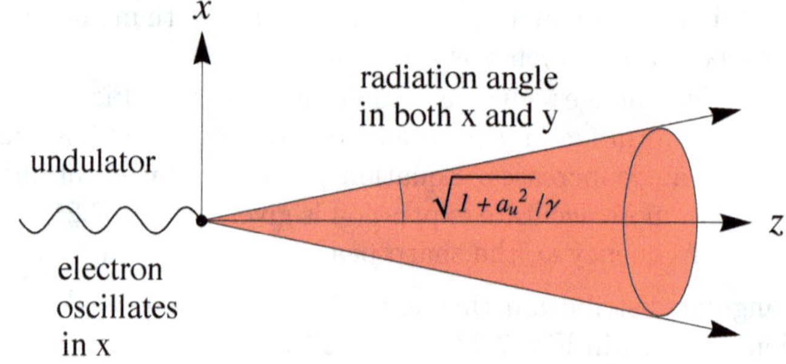

7.3.10 Frequency spectrum

One can also obtain the frequency spectrum by integrating $\frac{d^2W}{d\Omega d\omega}$ over the solid angle Ω. The result $\frac{dW}{d\omega}$ is to be expressed as a function of $x = \frac{\omega}{H\omega_0}$ when H, a_u, N_u are given.

The sharpness of the form factor $F(\epsilon)$ comes handy to help this calculation. As we observed in Fig. 7.15, The result $\frac{dW}{d\omega}$ has significant values only when $0 < x < 1$. When x exceeds 1, the value of $\frac{dW}{d\omega}$ drops precipitously within a distance $x - 1 \sim \frac{1}{HN_u}$. In this section, we shall confine ourselves to $0 < x \leq 1$.

For a given x with $0 < x \leq 1$, there is only one value of θ that contributes significantly to the integration over Ω,

$$\theta_1 = \frac{1}{\gamma}\sqrt{(1+a_u^2)(\frac{1}{x}-1)}\,.$$

We then have, for $0 < x = \frac{\omega}{H\omega_0} < 1$,

$$\begin{aligned}
\frac{dW}{d\omega} &= \int_0^\infty \theta d\theta \int_0^{2\pi} d\phi \frac{d^2W}{d\Omega d\omega} \\
&\approx \frac{e^2}{\pi\epsilon_0 c}\int_0^{2\pi} d\phi\, G(\theta_1,\phi)\int_0^\infty d(\gamma\theta)^2\left(\frac{\sin\pi\epsilon N_u}{\pi\epsilon}\right)^2 \\
&= \frac{2e^2}{\epsilon_0 c}\bar{G}(\theta_1)\int_{(x-1)H}^\infty \frac{1+a_u^2}{xH}d\epsilon\left(\frac{\sin\pi\epsilon N_u}{\pi\epsilon}\right)^2 \\
&\approx \frac{2e^2}{\epsilon_0 c}\bar{G}(\theta_1)\frac{1+a_u^2}{xH}N_u\,. \qquad (7.36)
\end{aligned}$$

The apparent divergence when $x \to 0$ is overcome by the factor $\bar{G}(\theta_1) \to 0$, and we have defined

$$\bar{G}(\theta) = \frac{1}{2\pi}\int_0^{2\pi} d\phi\, G(\theta,\phi)\,.$$

Figure 7.24 plots $\frac{\epsilon_0 c}{2N_u e^2}\frac{dW_{\sigma,\pi}}{d\omega}$ as a function of $x = \frac{\omega}{H\omega_0}$ over the range $0 < x < 1$ for $a_u = 0.5$ (red), 1 (green), 2 (blue). As mentioned, an exponential tail of the spectrum beyond $x = 1$ has been cut-off. The σ-radiation is generally an order of magnitude stronger than the π-radiation. High harmonic radiations are sensitive to a_u. A large a_u means a large excursion in terms of the figure-8 electron motion.

It is incorrect to expect the radiation into the harmonic H is peaked at the frequency $\omega = H\omega_0$. This statement is approximately correct only for $H = 1$ and only for the σ-mode. Red shift effect is rather significant when the solid angle Ω is integrated over. Even for the $H = 1, \sigma$ mode, in fact, the radiation spectrum is peaked at $\omega = \omega_0$ but has a triangular shape with a wide spread covering from $\omega = 0$ up to ω_0, as seen by the upper-left case in Fig. 7.24.

Red shift effect is even more pronounced for the other radiation modes. One observes from Fig. 7.24 that, for the σ-mode, the spectral peaks are located at $\omega \approx \omega_0$ for the $H = 1$ mode, $\omega \approx 1.7\,\omega_0$ for the $H = 2$ mode, and for the $H = 3$ mode, there are two peaks located at $\omega \approx 3\,\omega_0$ and $\omega \approx 2.4\,\omega_0$. For the π-mode, the peaks are located at $0.4\,\omega_0, 0.8\,\omega_0, 1.2\,\omega_0$ for the $H = 1, 2, 3$ respectively.

Asymptotic $a_u \to 0$ Expression (7.36) can be used to find the frequency dependence of $\frac{dW}{d\omega}$ when $a_u \ll 1$. When $a_u \ll 1$, we have for the case $H = 1$,

$$\begin{aligned}
\alpha &\approx 2\sqrt{2}a_u\sqrt{x(1-x)}\cos\phi\,, \\
\zeta &\approx 0\,, \\
\mathcal{D}_1 &\approx -\frac{1}{2}\,, \\
\mathcal{D}_2 &\approx \frac{\alpha}{2}\,,
\end{aligned}$$

Figure 7.24: Frequency spectrum of undulator radiation $\frac{\epsilon_0 c}{2N_u e^2}\frac{dW_{\sigma,\pi}}{d\omega}$ as a function of $x = \frac{\omega}{H\omega_0}$ over the range $0 < x < 1$. The three rows are for $H = 1, 2, 3$ respectively. The two columns are for the σ- (left) and the π-radiation (right) components. Red, green, and blue curves are for $a_u = 0.5, 1, 2$ respectively.

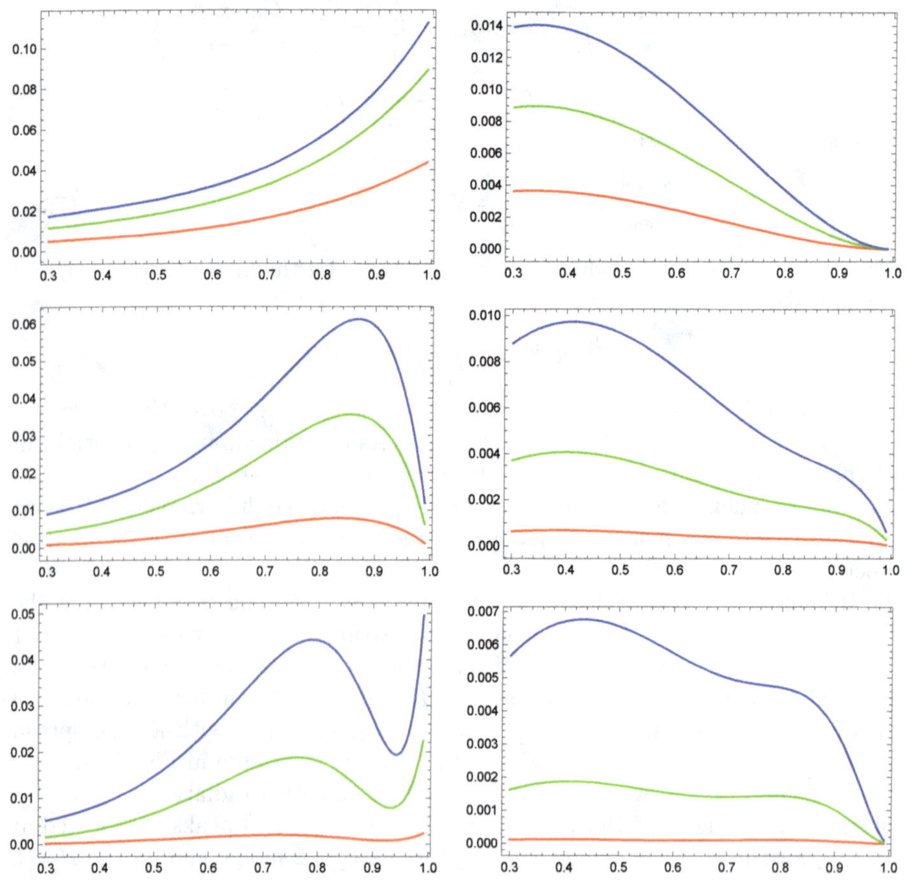

$$
\begin{aligned}
G_\sigma(\theta_1, \phi) &\approx \frac{a_u^2 x^2}{4}[1 - 2(1-x)\cos^2\phi]^2 \,, \\
\bar{G}_\sigma(\theta_1) &\approx \frac{a_u^2 x^2}{8}(1 - 2x + 3x^2) \,, \\
G_\pi(\theta_1, \phi) &\approx \frac{a_u^2 x^2}{4}(1-x)^2 \sin^2 2\phi \,, \\
\bar{G}_\pi(\theta_1) &\approx \frac{a_u^2 x^2}{8}(1-x)^2 \,,
\end{aligned}
$$

where $x = \frac{\omega}{\omega_0}$. We then obtain, for $a_u \ll 1$,

$$\begin{aligned}\frac{dW_\sigma}{d\omega} &\approx \frac{e^2 N_u}{4\epsilon_0 c}\, a_u^2\, x(1-2x+3x^2)\,, \\ \frac{dW_\pi}{d\omega} &\approx \frac{e^2 N_u}{4\epsilon_0 c}\, a_u^2\, x(1-x)^2\,. \end{aligned} \tag{7.37}$$

These spectra (7.37) agree with the $H = 1, a_u = 0.5$ cases shown in Fig. 7.24. In particular, it says that the red shift is very substantial for the π-mode, in such a way that there is little radiation into the π-mode unless ω is substantially below ω_0.

Expressions can also be obtained for $H \neq 1$. Homework 7.18 gives the result for the case $H = 2$. For $a_u \ll 1$, we have $\frac{dW_{\sigma,\pi}}{d\omega} \propto a_u^{2H}$. For small a_u, it is then observed that the radiated energy lies mostly in the fundamental $H = 1$ mode. As the harmonic number H increases, the radiated energy monotonically decreases. Note that here we are speaking of the spectrum $\frac{dW}{d\omega}$ after the radiation angle Ω is integrated. This observation is, for example, not to be confused with the earlier observation that only odd harmonics are radiated in the forward $\theta = 0$ direction.

Spectral flux The quantity $\frac{dW}{d\omega}$ is sometimes more conveniently expressed in terms spectral flux, $\mathcal{F}(\omega)$, the number of photons per unit time in a given bandwidth $\frac{\Delta\omega}{\omega}$ when $\frac{\Delta\omega}{\omega} \ll \frac{1}{N_u}$. Solid angle Ω is integrated over. In that case, we have, using Eq. (7.36),

$$\begin{aligned}\mathcal{F}(\omega) &= \frac{dW}{d\omega}(\omega)\frac{\Delta\omega}{\hbar\omega}\frac{I}{e} \\ &\approx 8\pi\alpha_F \left(\frac{I}{e}\right) N_u \frac{\Delta\omega}{\omega} \bar{G}(\theta_1)\frac{1+a_u^2}{xH}\,,\end{aligned}$$

where α_F is fine structure constant, I is the electron beam current and $x = \frac{\omega}{H\omega_0}$. This spectral flux is for radiation from an incoherent beam. It does not apply to the case of FEL.

In many cases, we are interested in the flux when $H = 1$ and when ω is within the neighborhood $\Delta\omega$ of $\omega = \omega_0$. In that case, $x = 1, \theta_1 = 0, \sigma$-radiation dominates, and we find

$$\mathcal{F}(\omega = \omega_0) = 2\pi\alpha_F \left(\frac{I}{e}\right) N_u \frac{\Delta\omega}{\omega}\frac{a_u^2}{1+a_u^2}[JJ]^2\,. \tag{7.38}$$

This expression depends linearly on I and N_u. It does not depend explicitly on the undulator wavelength λ_u, although of course $\omega = \omega_0$ depends on λ_u. Note that $\frac{\Delta\omega}{\omega} \ll \frac{1}{N_u}$ is also required condition to apply Eq. (7.38).

Take for example a third generation synchrotron radiation storage ring (not an FEL), we might have $I = 100$ mA, $N_u = 100$, Figure 7.25 shows $\mathcal{F}(\omega = \omega_0)$

Figure 7.25: An example calculation of undulator photon flux for a storage ring facility. Plotted is $\mathcal{F}(\omega = \omega_0)$ as a function of a_u for the case $I = 100$ mA, $N_u = 100$.

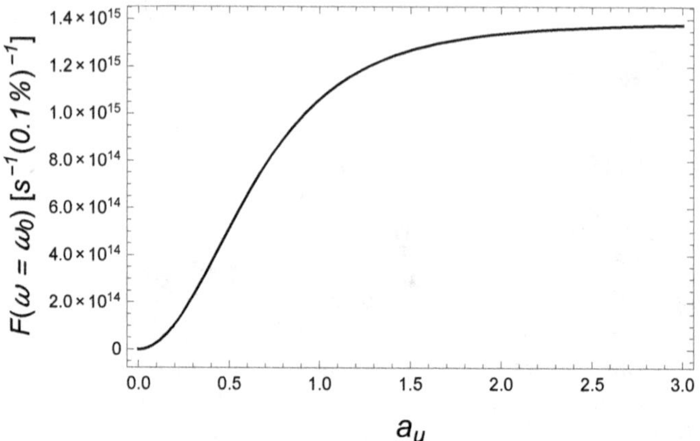

(in units of number of photons per second per 0.1% bandwidth) as a function of a_u.

As $a_u \to \infty$, the spectral flux reaches a plateau value

$$\mathcal{F}(\omega = \omega_0, a_u = \infty) = 0.0222 \left(\frac{I}{e}\right) N_u \frac{\Delta\omega}{\omega}.$$

From the point of view of undulator radiation, there is not too much to gain in terms of the photon flux when $a_u > 1.5$ or so. This is no longer the case with the FEL.

Brightness Brightness is defined as the spectral flux normalized by the emittances,

$$\mathcal{B} = \frac{\mathcal{F}}{(2\pi\Sigma_x\Sigma'_x)(2\pi\Sigma_y\Sigma'_y)},$$

where the quantities in the denominator represent the folded rms emittances of the undulator radiation and the electron beam. In the present consideration, the beam is radiating incoherently. The folded emittances are most likely dominated by those of the electron beam. Take $\Sigma_x\Sigma'_x = 10^{-2}$ mm-mrad, $\Sigma_y\Sigma'_y = 10^{-4}$ mm-mrad for a typical 3rd generation storage ring, then in units of number of photons per second per 0.1% bandwidth per (mm-mrad)^2, we have the brightness as a function of a_u shown in Fig. 7.26 when the interested frequency is in the neighborhood of ω_0.

Homework 7.16 We calculated $\frac{dW}{d\omega}$ by integrating over the solid angle. The resulting $\frac{dW}{d\omega}$ covers a wide range of ω from 0 to $\omega \approx H\omega_0$. Consider the case

Figure 7.26: An example calculation of undulator radiation brightness as a function of a_u for a storage ring facility for the case $I = 100$ mA, $N_u = 100$, $\Sigma_x\Sigma'_x = 10^{-2}$ mm-mrad, $\Sigma_y\Sigma'_y = 10^{-4}$ mm-mrad.

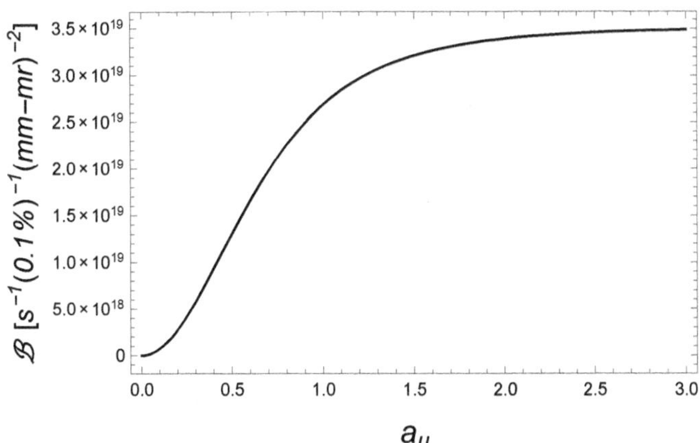

when the radiation is made to go through a pinhole on an otherwise opaque screen. Let the pinhole extend over $\theta_1 \pm \Delta\theta$ with $\delta\theta \ll \theta_1$ and a small range $\phi_1 \pm \Delta\phi$. What would the frequency spread $\frac{\Delta\omega}{\omega}$ of the radiation be after the pinhole? If we demand the radiation after the pinhole to be quasi-monochromatic with $\frac{\Delta\omega}{\omega}$ not to exceed $\frac{1}{N_u}$ by more than a factor of 2, how narrow do we have to choose $\Delta\theta$?

Solution With a pinhole at $(\theta_1 \pm \Delta\theta, \phi_1 \pm \Delta\phi)$, we have

$$\frac{dW}{d\omega} = \theta_1\Delta\theta\Delta\phi\frac{d^2W}{d\omega d\Omega}(\theta_1, \phi_1) = \theta_1\Delta\theta\Delta\phi\,\frac{2e^2\gamma^2}{\pi\epsilon_0 c}G(\theta_1, \phi_1)F(\epsilon)\,.$$

Frequency narrowing comes from $F(\epsilon)$, yielding $\frac{\Delta\omega}{\omega} \approx \frac{1}{N_u}$ even with infinitesimal pinholes.

To require $\frac{\Delta\omega}{\omega} \lesssim \frac{2}{N_u}$, $G(\theta, \phi)$ must not vary as much as $\frac{1}{N_u}$ in the pinhole range. Near the forward direction $\theta_1 = 0$, for example, this requires

$$\frac{\gamma^2\Delta\theta^2}{1 + a_u^2} \lesssim \frac{1}{N_u}\,.$$

With $\lambda_0 = \frac{\lambda_u(1+a_u^2)}{(2\gamma^2)}$, we obtain the condition

$$\Delta\theta \lesssim \sqrt{\frac{2\lambda_0}{L_u N_u}}\,.$$

Homework 7.17 Fill in the derivations of the asymptotic expressions (7.37) of $\frac{dW_{\sigma,\pi}}{d\omega}$ when $a_u \to 0$.

Homework 7.18 Expression (7.36) can be used to find the frequency dependence of $\frac{dW_{\sigma,\pi}}{d\omega}$ when $a_u \ll 1$. The text gives the case for $H = 1$. Extend the calculation to find the asymptotic expressions for the case $H = 2$.

Solution

$$\begin{aligned}
\alpha &\approx 2\sqrt{2}a_u\sqrt{x(1-x)}\cos\phi\,, \\
\zeta &\approx \frac{xa_u^2}{2}\,, \\
\mathcal{D}_1 &\approx -\frac{\alpha}{2}\,, \\
\mathcal{D}_2 &\approx \frac{\alpha^2}{2} - \frac{xa_u^2}{2}\,, \\
G_\sigma(\theta_1,\phi) &\approx 32\,a_u^4x^3(1-x)\cos^2\phi\left[\frac{5}{8}-(1-x)\cos^2\phi\right]^2, \\
\bar{G}_\sigma(\theta_1) &\approx \frac{5a_u^4}{4}\,x^3(1-x)(1-4x+8x^2)\,, \\
G_\pi(\theta_1,\phi) &\approx 32\,a_u^4x^3(1-x)\sin^2\phi\left[(1-x)\cos^2\phi-\frac{1}{8}\right]^2, \\
\bar{G}_\pi(\theta_1) &\approx \frac{a_u^4}{4}x^3(1-x)(5-12x+8x^2)\,, \\
\frac{dW_\sigma}{d\omega} &\approx \frac{5e^2N_u}{4\epsilon_0c}\,a_u^4\,x^2(1-x)(1-4x+8x^2)\,, \\
\frac{dW_\pi}{d\omega} &\approx \frac{e^2N_u}{4\epsilon_0c}\,a_u^4\,x^2(1-x)(5-12x+8x^2)\,.
\end{aligned}$$

This result agrees with the $a_u = 0.5$ case of Fig. 7.24.

Homework 7.19 Expression (7.36) can also be used to find the frequency dependence of $\frac{dW_{\sigma,\pi}}{d\omega}$ when $a_u \gg 1$. Show that the result for $H = 1$ is given by

$$\begin{aligned}
\alpha &\approx 2\sqrt{2x(1-x)}\cos\phi\,, \\
\zeta &\approx \frac{x}{2}\,, \\
\mathcal{D}_1 &\approx -\frac{1}{2}\sum_{m=-\infty}^{\infty} J_{2m}(\alpha)[J_m(\zeta)+J_{m-1}(\zeta)]\,, \\
\mathcal{D}_2 &\approx -\sum_{m=\infty}^{\infty} J_{2m+1}(\alpha)J_m(\zeta)\,,
\end{aligned}$$

$$\begin{aligned} \frac{dW_\sigma}{d\omega} &\approx \frac{e^2 N_u}{2\pi\epsilon_0 c} \int_0^{2\pi} d\phi \left(\sqrt{x}\mathcal{D}_1 + \sqrt{\frac{1-x}{2}}\,\mathcal{D}_2 \cos\phi \right)^2 , \\ \frac{dW_\pi}{d\omega} &\approx \frac{e^2 N_u}{\pi\epsilon_0 c}(1-x) \int_0^{2\pi} d\phi\, \mathcal{D}_2^2 \sin^2\phi \,. \end{aligned}$$

The result can be plotted and compared with the $a_u = 2$ case shown in Fig. 7.24.

Homework 7.20 Consider an undulator in a third generation storage ring. With its configuration and its magnetic field fixed, one may vary the storage ring energy. As the beam energy varies, how do the following quantities depend on it? (a) ω_0, (b) $\mathcal{F}(\omega_0)$, (c) $\mathcal{B}(\omega_0)$?

Homework 7.21

(a) Prove Eq. (7.38).

(b) Reproduce numerically the plot in Fig. 7.25.

7.3.11 Total radiated energy

Exact results To obtain the total radiated energy, we can integrate $\frac{dW}{d\Omega}$, Eq. (7.35), over the solid angle Ω,

$$\begin{aligned} W_{\sigma,\pi} &= \int d\Omega\, \frac{dW_{\sigma,\pi}}{d\Omega} = \int_0^\infty \theta d\theta \int_0^{2\pi} d\phi\, \frac{dW_{\sigma,\pi}}{d\Omega} \\ &= \frac{2e^2 N_u \omega_0}{\epsilon_0 c} \int_0^\infty d(\gamma\theta)^2\, \frac{\bar{G}_{\sigma,\pi}(\theta)}{1 + \frac{(\gamma\theta)^2}{1+a_u^2}} \,. \end{aligned}$$

One can also obtain the same result by integrating $\frac{dW}{d\omega}$, Eq. (7.36), over frequency ω,

$$\begin{aligned} W_{\sigma,\pi} &= \int_0^\infty d\omega\, \frac{dW_{\sigma,\pi}}{d\omega} \\ &\approx \frac{2e^2 N_u \omega_0}{\epsilon_0 c}(1+a_u^2) \int_0^1 dx\, \frac{\bar{G}_{\sigma,\pi}(\theta_1)}{x} , \\ \theta_1 &= \frac{1}{\gamma}\sqrt{(1+a_u^2)(\frac{1}{x}-1)} \,. \end{aligned}$$

The two calculations give the same result. Let us define two form factors $g_{\sigma,\pi}$ by [compare with Eq. (7.18)]

$$W_{\sigma,\pi} = \frac{2e^2 N_u \omega_0}{\epsilon_0 c}\, g_{\sigma,\pi} \,.$$

Figure 7.27 shows $g_{\sigma,\pi}$ as functions of a_u for $H = 1, 2, 3$ using the exact expressions of $G_{\sigma,\pi}(\theta,\phi)$, Eq. (7.16), in a numerical calculation.

Figure 7.27: The total radiated energy by a single electron going through an undulator. Upper panel shows g_σ. Lower panel shows g_π. The red, green, blue curves are for $H = 1, 2, 3$ respectively.

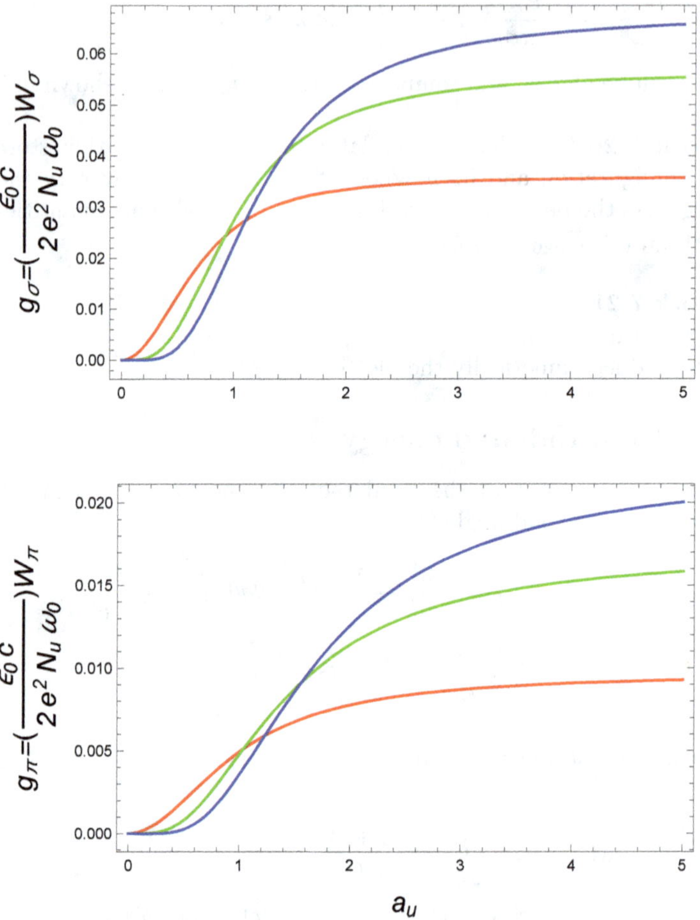

In Fig. 7.27, we see that typically W_σ is several times larger than W_π. Most radiation has a σ-polarization. For small $a_u \ll 1$ and $H = 1$, we will show in Eq. (7.40) that the energy radiated in σ-mode is seven times that in the π-mode. When $a_u \gtrsim 1$, radiation energies into higher harmonics become more than the $H = 1$ radiation energy. As a_u continues to increase, the quantities $g_{\sigma,\pi}$ approach constant values as a_u becomes $\gtrsim 2$. It should be noted however that the radiated energy is spread out over an angular spread proportional to a_u as a_u becomes large.

Figure 7.27 can be read off to give the total undulator radiated energy. The result depends only on three parameters N_u, ω_0 and a_u. The other parameters such as beam energy, undulator period, magnetic field, etc. are contained in these three parameters.

An approximate expression To gain a bit more insight, we can also apply the approximate Eqs. (7.22) and (7.23) to calculate the total radiated energy. For the case of $H = 1, \sigma$-mode, for example, we obtain (Homework 7.22),

$$
\begin{aligned}
g_\sigma &= (1+a_u^2)\int_0^1 \frac{dx}{x}\bar{G}_\sigma(\theta_1)\,,\\
G_\sigma(\theta_1,\phi) &\approx \frac{x^2}{(1+a_u^2)^2}\left(a_u\mathcal{D}_1 + \frac{\sqrt{(1+a_u^2)(\frac{1}{x}-1)}}{\sqrt{2}}\cos\phi\,\mathcal{D}_2\right)^2,\\
\mathcal{D}_1 &= -\frac{1}{2}J_0(\alpha_1)[J_0(\zeta_1)-J_1(\zeta_1)]\,,\\
\mathcal{D}_2 &= J_1(\alpha_1)J_0(\zeta_1)\,,\\
\alpha_1 &= \frac{2\sqrt{2}\,a_u\cos\phi}{\sqrt{1+a_u^2}}\sqrt{x(1-x)}\,,\\
\zeta_1 &= \frac{a_u^2 x}{2(1+a_u^2)}\,.
\end{aligned}
\tag{7.39}
$$

A word of caution is in order. In deriving Eq. (7.22), we made a crude approximation that $\mathcal{D}_2$ does not contribute much to the σ-radiation. This is not quite so. Here in Eq. (7.39), we have compensated for that crudeness and included the contribution from $\mathcal{D}_2$ while still taking only the $m = 0$ contributions in $\mathcal{D}_{1,2}$ under this approximation.

Asymptotic $a_u \to 0$ Asymptotic expressions of $\bar{G}_{\sigma,\pi}(\theta_1)$ for $H = 1$ and $a_u \ll 1$ were given by Eq. (7.37). Integrating over x then gives

$$
\begin{aligned}
W_\sigma &\approx \frac{7}{12}\frac{e^2N_u\omega_0}{4\epsilon_0 c}a_u^2\,,\\
W_\pi &\approx \frac{1}{12}\frac{e^2N_u\omega_0}{4\epsilon_0 c}a_u^2\,,
\end{aligned}
$$

or equivalently,

$$
g_\sigma \approx \frac{7}{96}a_u^2, \quad g_\pi \approx \frac{1}{96}a_u^2\,. \tag{7.40}
$$

These expressions agree well with the red curves exhibited in Fig. 7.27 when $a_u \ll 1$.

The radiated power lies $\frac{7}{8}$ in the σ-mode, and $\frac{1}{8}$ in the π-mode. This is the same as in the case of synchrotron radiation in a bending dipole, as seen in Eq. (6.20), and is what is anticipated.

The form factors $g_{\sigma,\pi}$ should scale approximately as $[JJ]^2$ for $H = 1$ over a range of modest a_u. From Fig. 7.27, and consistent with Eq. (7.40), it appears that a reasonable fit can be made to obtain

$$(g_\sigma + g_\pi)\big|_{H=1} \approx \frac{a_u^2[JJ]^2}{12(1+a_u^2)} .$$

In addition, it appears that the total radiation form factor, summed over all H's is approximately

$$\sum_H (g_\sigma + g_\pi) \approx \frac{a_u^2(1+a_u^2)}{12} . \tag{7.41}$$

The ratio of energy radiated into the fundamental mode relative to the total is therefore

$$\frac{(g_\sigma + g_\pi)\big|_{H=1}}{\sum_H (g_\sigma + g_\pi)} = \frac{[JJ]^2}{(1+a_u^2)^2} . \tag{7.42}$$

We shall denote this quantity $r(a_u)$ — we will use it later. We have expected $r(0) = 1$, and $r(a_u) \to 0$ when $a_u \to \infty$.

Radiation efficiency is limited by $\boldsymbol{\alpha_F}$ We just found from Fig. 7.27 that $g_{\sigma,\pi}$ are typically of the order of a few times 0.01, a disappointingly low value, for a fixed harmonic number H. The total radiated energy per electron is typically a few percent of the value

$$\frac{2e^2 N_u \omega_0}{\epsilon_0 c} = \frac{16\pi N_u \gamma^2 mc^2 k_u r_0}{1+a_u^2} .$$

The radiated energy per electron per undulator period is

$$\frac{W_{\sigma,\pi}}{N_u} = \frac{2e^2\omega_0}{\epsilon_0 c} g_{\sigma,\pi} .$$

The number of photons radiated per electron per undulator period is

$$\frac{W_{\sigma,\pi}}{N_u H \hbar\omega_0} = \frac{8\pi\alpha_F}{H} g_{\sigma,\pi} ,$$

where $\alpha_F \approx \frac{1}{137}$ is the fine structure constant. This number approaches the value of α_F when $8\pi g_{\sigma,\pi} \sim H$.

We see that, as a product of three small numbers $8\pi\alpha_F$, $\frac{1}{H}$, and $g_{\sigma,\pi}$, this is typically a very small fractional number indeed. The electron moving down the undulator basically rarely radiates. This situation has been familiar when we discussed the quantum mechanical picture of synchrotron radiation in bending magnets. We see that this feature is not much improved in an undulator. Situation will change in an FEL, however.

Take $N_u = 100, \gamma = 10^4$ (5 GeV beam), $\lambda_u = 10$ cm, $B_u = 0.2$ T. We then have $a_u = 1.32$, $\omega_0 = 1.4 \times 10^{18}$ s^{-1}, and $\lambda_0 = \frac{2\pi c}{\omega_0} = 1.4$ nm. The radiated photon energy is $H\hbar\omega_0$ with $\hbar\omega_0 = 0.9$ keV. We also have $g_\sigma = 0.03, g_\pi = 0.006$ from Fig. 7.27 for $H = 1$. The total radiated energy per electron is found to be 0.6 keV. On the average, about $\frac{2}{3}$ of the electrons get to radiate one photon in the 100-period undulator.

Radiation based on single passages are not very efficient. After painstakingly preparing an electron beam, only a fraction of them actually radiate. This is why 2nd and 3rd generation facilities use storage rings so that the electron beam is used multiple times in multiple passages. In an FEL, however, single passages are again adopted. The efficiency of radiation is improved using a completely different mechanism.

It is remarkable that even with such sparse radiation, the picture that there is a distinct resonance condition and a distinct pattern produced by interference between radiations from adjacent undulator periods actually occur. It can perhaps be viewed as yet another accomplishment of quantum mechanics.

Compared with bending magnet radiation The total radiated energy by an electron is obtained by integrating $\frac{d^2W}{d\omega d\Omega}$ over ω and Ω. As we have seen, the radiated frequency spectrum and angular distribution from an undulator is distinctly different from those from a bending magnet. However, the total radiated energy from an undulator can be very simply, at least approximately, obtained from the known result of the bending magnet when the undulator strength is moderately weak ($a_u \lesssim 1$). The sensitive structure in frequency spectrum and angular distribution of an undulator all wash out when integrated, conspiring to the net effect that the total radiated energy is approximately the same as that of a bending magnet.

The radiation power in a bending magnet is

$$P_\gamma = \frac{2r_0 mc^3\gamma^4}{3\rho^2} .$$

By equating $\frac{1}{\rho} = \frac{eB}{m\gamma c}$, we have

$$P_\gamma = \frac{2r_0\gamma^2 e^2 B^2 c}{3m} .$$

For an undulator, all we have to do is to replace B^2 by the average value of the undulator field, i.e. $B^2 \to \frac{B_u^2}{2}$,

$$P_\gamma = \frac{r_0\gamma^2 e^2 B_u^2 c}{3m} .$$

Replacing B_u by a_u, we have

$$P_\gamma = \frac{2}{3} r_0 m\gamma^2 a_u^2 k_u^2 c^3 . \tag{7.43}$$

We will return to Eq. (7.43) later in Sec. 7.4.2.

The radiated energy per electron passing through the undulator is then

$$W = P_\gamma \frac{N_u \lambda_u}{c} = \frac{e^2 \gamma^2 a_u^2 k_u N_u}{3\epsilon_0}.$$

This is in fact the same as the predicted total radiated energy of an undulator

$$W = \left(\frac{2e^2 N_u \omega_0}{\epsilon_0 c}\right) \sum_H (g_\sigma + g_\pi).$$

if we observe Eq. (7.41).

Undulator immersed in medium? We observed earlier that one problem with undulator radiation is that the form factors $G_{\sigma,\pi}$ (and also form factors $g_{\sigma,\pi}$) are too small. These form factors involve dimensionless quantities, which nominally should be of the order of unity, and yet their factors conspire to make it typically only a few times 0.01. It would indeed be desirable if one can think of a way to substantially increase the form factors $G_{\sigma,\pi}$.

It might be that the origin of the problem comes from the fact that the lowest order harmonic resonance occurs at $H = 1$. Had it been possible to radiate in the $H = 0$ mode, the form factor might become an order of magnitude larger.

Unfortunately, the physical origin of undulator radiation resonance as discussed in Sec. 7.3.2 indicates that it is not possible to radiate in the $H = 0$ mode, because the photon definitively travels faster than the electron. There is however, at least in principle, one way to overcome this limitation. Consider an undulator immersed in a medium of index of refraction n instead of in vacuum.[6] Let n be slightly larger than 1 so that the propagation speed of the radiation is slightly slowed down from c. Then the resonance condition is rederived to give

$$\lambda_u \left[-(n-1) + \frac{1}{2\gamma^2} + \frac{\langle v_x^2 \rangle}{2c^2} + \frac{\theta^2}{2}\right] = H\lambda,$$

which will allow $H = 0$ as a permissible resonance radiation. For example, if we choose

$$n - 1 = \frac{1}{2\gamma^2} + \frac{\langle v_x^2 \rangle}{2c^2} = \frac{1 + a_u^2}{2\gamma^2},$$

then resonance radiation can occur for $H = 0$ in the forward direction $\theta = 0$. A rederivation of the form factor should then be performed. Note that when a $H = 0$ resonance is made to occur, the radiation is in resonance for arbitrary wavelengths λ.

Homework 7.22

(a) Go through the algebra to derive Eq. (7.39).

[6] One will of course have to deal with other problems such as the electron ionizing the medium and also its additional Cerenkov radiation.

(b) Perform a numerical calculation of Eq. (7.39) to obtain g_σ as a function of a_u and compare your result with the $H = 1$ case in Fig. 7.27.

Solution (b) Figure 7.28 shows the result. The agreement is reasonable. It is confirmed here that dropping the terms with $m \neq 0$ in the summation for the exact expressions of $G_{\sigma,\pi}$ in Eq. (7.16) is acceptable, but for some accuracy, it is advisable to include both $\mathcal{D}_1$ and $\mathcal{D}_2$. You may show that the result changes by a factor of 2 or so depending on whether $\mathcal{D}_2$ is or is not included.

Figure 7.28: $\left(\frac{\epsilon_0 c}{2e^2 N_u \omega_0}\right) W_\sigma$ as calculated by the approximate expression Eq. (7.39) as a function of a_u.

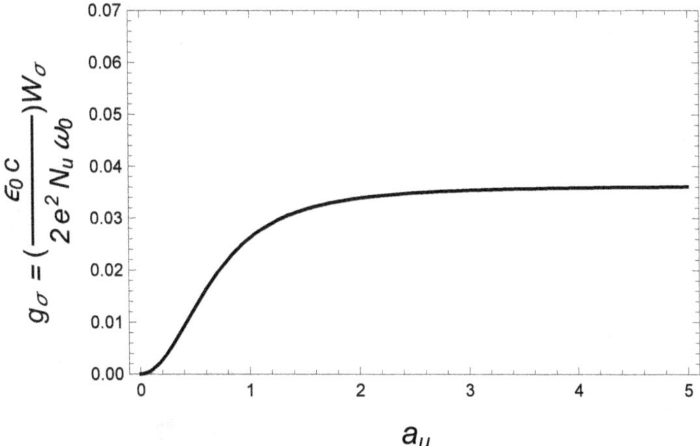

Homework 7.23 The text derived $g_{\sigma,\pi}$ for the fundamental harmonic $H = 1$ when $a_u \ll 1$. Use the result of Homework 7.18 to calculate $g_{\sigma,\pi}$ for the second harmonic $H = 2$.

Solution $g_\sigma = \frac{3}{16} a_u^4$, $g_\pi = \frac{1}{48} a_u^4$. This time the π-radiation energy is $\frac{1}{9}$ that of the σ-radiation. Good agreement with Fig. 7.27 is found when $a_u \lesssim 0.3$.

Homework 7.24 You are asked to design an FEL that provides a radiation at a given wavelength λ_0. Undulator has been given to you with a given λ_u. The electron beam is also given with a given γ. You have only the value of a_u under your control by varying the undulator magnetic field. You of course know how to produce the λ_0 radiation by a $H = 1$ mode by choosing an appropriate a_u. But you can also choose another a_u' to produce the λ_0 radiation by choosing the second harmonic $H = 2$.

(a) Find the connection between a_u' and a_u.

(b) Compare their radiation energy. Which radiates more? When the undulator design and the electron beam are fixed, is it more advantageous to operate your FEL with the fundamental $H = 1$ mode or the $H = 2$ mode?

7.4 Low gain free electron laser

Conventional lasers use atomically *bound* electrons. The laser mechanism invokes electrons bound either in an atom or a molecule. As a result, photon energies of typical lasers are limited to ∼eV level and wavelength ∼atomic size. It is difficult to generate X-rays this way because there is no way to bound these high-energy photons between atomic energy levels. In fact it was for a long while considered impossible to ever be able to generate X-ray lasers.

A brilliant idea proposed by John Madey[7] (Fig. 7.29) was then to invent a way to use *free* electrons, circumventing the required use of bound electrons. The proposed concept is a linear-accelerator based free electron laser. The challenge is to somehow manipulate the electrons in an accelerator so that they align themselves in such a way to radiate coherently, and to radiate at the desired wavelength. By getting free electrons to radiate coherently, skipping the step of having them bounded between two well-defined energy levels, the dream of X-ray lasers has become a possibility after all.

Figure 7.29: John Madey and his FEL. [Courtesy Stanford News Service (2019)]

It is important to appreciate that a good coherence has two very important consequences,

1. The radiation becomes laser.
2. The radiation power is much enhanced by an additional factor of N, where N is the number of electrons in coherence.

[7]John M. Madey, J. Appl. Phys. 42, 5, 1906 (1971); D.A.G. Deacon et al., Phys. Rev. Lett. 38, 892 (1977).

Both consequences are critical, both are based on the coherence, and they come together hand-in-hand. For LCLS, for example, the radiation constitutes a laser because of the coherence. The coherence also contributes to the huge enhancement that amounts to a factor of $N \sim 10^6$. Together with a few other factors, LCLS provides X-ray brightness that increased by 10 orders of magnitude overnight compared with the third generation synchrotron radiation sources.[8]

There is one price to pay in reaching free electron lasers. Not invoking well-defined energy levels of bound states of electrons, the FEL spectrum is not sharply defined. A typical FEL bandwidth is $\sim 10^{-2}$, while conventional atomic lasers might have easily a bandwidth $\sim 10^{-6}$ or better.

7.4.1 Bunching factor

We have now discussed in detail the radiation from a single point charge with a prescribed oscillatory motion in an undulator. To prepare for the discussion of free electron lasers, we will need to consider two more aspects. One is to replace the single point charge by a beam with a finite distribution. The other is to let go of the motion being prescribed and allow the motion be affected by the radiation itself. As we shall see, the first consideration leads to a low-gain FEL; the second leads to a high-gain FEL. Here in this section, we start with a short discussion of a finite beam distribution and a low-gain variety of the FEL.

Diffraction limit versus microbunching First, the beam would have a transverse distribution. We shall let the beam have a very tight transverse distribution so that its two transverse emittances are diffraction limited, i.e.

$$\epsilon_{x,y} \lesssim \frac{\lambda}{4\pi} .$$

This puts a tight constraint on the beam quality. Typical 3rd generation storage ring light sources do not meet the diffraction limit for X-rays. Coherence will require frequency filtering or the application of pin-holes, both of which induce substantial reduction of radiation intensity. On the other hand, a high-quality linac reaches the diffraction limit more readily.

Reaching the diffraction limit makes the radiation transversely coherent. Once achieved, all particles in the same longitudinal slice of the beam radiate coherently not too differently from a beam with zero transverse sizes. However, different longitudinal slices of the beam still radiate separately. To make the entire beam bunch radiate coherently, in addition to the diffraction limit, we still have to demand a favorable longitudinal beam distribution $\rho(z)$. The condition on $\rho(z)$ is given in terms of a quantity called the *bunching factor*.

The reason bunching factor becomes so important is traced to the fact that the radiation phase $e^{-i(\omega/c)(ct-\hat{n}\cdot\vec{r})}$ in the forward direction depends sensitively

[8] P. Emma et al., Nature Photonics, Vol. 4, p. 641 (2010).

on the z-positions of the radiating particles, and does not depend sensitively on their transverse positions.

We have concentrated on discussing the diffraction limit from the point of view in the transverse coherence. In principle, there is also a longitudinal diffraction limit. More specifically, a longitudinal diffraction limit occurs when the electron beam is shorter than λ. However, exceedingly short bunches is not the only way to reach coherence. The proper definition of a longitudinal diffraction limit is given in terms of the bunching factor, and even a long bunch can have a high degree of bunching factor if the bunch is microbunched.

Incidentally, there is also a transverse counterpart of microbunching. In that case, $\epsilon_{x,y} \lesssim \frac{\lambda}{4\pi}$ is not the only way to reach coherence. Fresnel zone plate is an example, although a near-field effect.

At present, the design of a transversely diffraction limited storage ring is a concentrated research effort. These efforts focus on the transverse coherence and pay attention to the transverse dynamics of the electron beam. It should be pointed out however that once the emittances are made to reach the diffraction limit, there is no more games to play in the transverse dimension. In comparison, an observation is to be made that much richer and more to be gained is the research to manipulate the beam's longitudinal distribution to meet various specific needs of its radiation through the wide-open longitudinal dynamics. This observation in turn forecasts an enhanced research future to tailor make the longitudinal beam distribution than the effort to make a storage ring of small transverse emittances. Free electron lasers is one, but unlikely to be the only one, prominent candidate in this critical direction of research. Major advances are yet to be discovered in longitudinal dynamics and 6-D phase space gymnastics 海闊天空.

In what follows, we calculate the bunching factor and the corresponding radiation power for four cases — a smooth beam, a shot noise beam, a DC beam, and a microbunched beam. The fourth case, being our focused case, is moved into a section of its own in the next section.

Smooth beam Bunching factor is defined for a beam bunch that contains a large number of electrons and specifically aims for the far field radiation of this beam bunch at a given frequency ω. A large bunching factor at a frequency ω indicates the potential of a high level of coherent radiation from the beam at frequency ω. As we shall see later, in a high-gain FEL, we aim to produce a beam with high bunching factor at the undulator resonant frequency.

Consider a beam with longitudinal charge distribution $Ne\rho(z)$ entering an undulator (or any other radiating device, such as a bending magnet, or a metallic or dielectric foil), with N the total number of electrons in the beam, and

$$\int_{-\infty}^{\infty} dz\, \rho(z) \;=\; 1\,. \tag{7.44}$$

Figure 7.30: Illustration of the additional phase factor in the bunching factor.

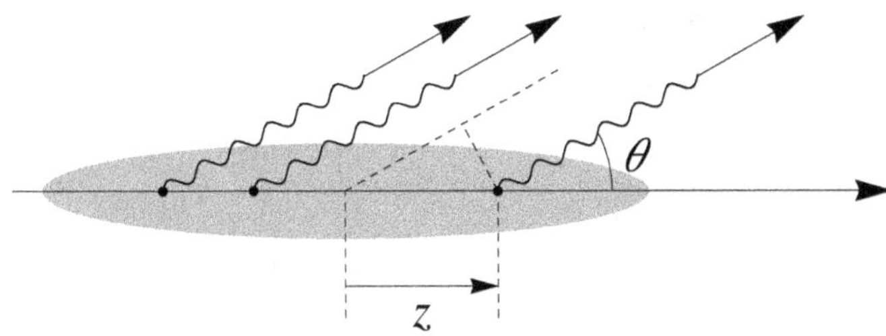

To calculate the radiation spectrum of this beam, we apply Eq. (7.7), which so far has been used for the case of a point charge, but now the radiation from different parts of the beam must include an additional phase factor. More specifically, a slice of the beam at position z will require a replacement in Eq. (7.7),

$$\int_{-\infty}^{\infty} dt\, \hat{n} \times (\hat{n} \times \vec{v})\, e^{-i(\omega/c)(ct-\hat{n}\cdot\vec{r})}$$

$$\rightarrow \quad e^{-i(\omega/c)z\cos\theta} \int_{-\infty}^{\infty} dt\, \hat{n} \times (\hat{n} \times \vec{v})\, e^{-i(\omega/c)(ct-\hat{n}\cdot\vec{r})}\,,$$

where θ is the polar angle of the observation direction $\hat{n}$ relative to the forward z-direction of beam motion, and the additional phase factor $e^{-i(\omega/c)z\cos\theta}$ comes from the phase of radiation relative to the bunch center. The inclusion of the additional phase factor is illustrated in Fig. 7.30.

This phase factor, and therefore the concept of bunching factor, relies on a far field approximation. The radiation being considered must be far from the radiation source. Near field consideration would lead to concepts like Fresnel zone plates, outside of our scope.

Recall that, when we calculated the prescribed motion in an undulator, we had kept the longitudinal coordinates z to order $\frac{1}{\gamma^2}$ while the transverse coordinates only to order $\frac{1}{\gamma}$. The reason of requiring much more accurate longitudinal coordinates is particularly reflected in the bunching factor. Because the radiation lies mainly in the forward (small angle θ) direction, the phase information is much more sensitive to z than to x and y.

When radiation from all parts of the beam are combined together, the replacement acquires an overall factor

$$\int_{-\infty}^{\infty} dt\, \hat{n} \times (\hat{n} \times \vec{v})\, e^{-i(\omega/c)(ct-\hat{n}\cdot\vec{r})}$$

$$\rightarrow \quad N\mathcal{B}(\omega) \int_{-\infty}^{\infty} dt\, \hat{n} \times (\hat{n} \times \vec{v})\, e^{-i(\omega/c)(ct-\hat{n}\cdot\vec{r})}\,,$$

where

$$\mathcal{B}(\omega) = \int_{-\infty}^{\infty} dz\, \rho(z)\, e^{-i(\omega/c)z\cos\theta} \tag{7.45}$$

is referred to as the bunching factor of the beam at frequency ω.

The bunching factor generally depends on the observation polar angle θ. In case of forward direction, $\theta = 0$, the bunching factor

$$\mathcal{B}(\omega)\big|_{\theta=0} = \int_{-\infty}^{\infty} dz\, \rho(z)\, e^{-i(\omega/c)z}$$

is basically the Fourier transform of $\rho(z)$,

$$\mathcal{B}(\omega)\big|_{\theta=0} = \tilde{\rho}(\omega)\,.$$

When $\theta \neq 0$, the bunching factor is blue-shifted, with

$$\mathcal{B}(\omega)\big|_{\theta} = \mathcal{B}(\omega\cos\theta)\big|_{\theta=0}\,.$$

The bunching factor has a dimensionality of the (fractional) number of particles in the beam, i.e. dimensionless in the usual convention. It is a complex number with magnitude always ≤ 1.

The new radiation spectrum is then given by

$$\left.\frac{d^2W}{d\Omega d\omega}\right|_{\text{beam}} = N^2|\mathcal{B}(\omega)|^2 \left.\frac{d^2W}{d\Omega d\omega}\right|_{\text{pt.charge}},$$

with $\left.\frac{d^2W}{d\Omega d\omega}\right|_{\text{pt.charge}}$ given by Eq. (7.7).

Shot noise beam The bunching factor discussed so far refers to a beam with a smooth distribution $\rho(z)$. In reality, a beam contains a large number N of discrete point charges. The actual beam distribution reads

$$\rho(z) = \frac{1}{N}\sum_{i=1}^{N} \delta(z - z_i)\,,$$

where z_i is the location of the i-th particle. The normalization $\int_{-\infty}^{\infty} dx\rho(z) = 1$ is maintained. Distribution of z_i's obeys a probability given by a smooth function $\rho_0(z)$, but otherwise stochastic. What we know in reality is only $\rho_0(z)$. We do not know the exact values of z_i's. Such a distribution would be applicable to a beam entering a radiator without any pre-manipulations, i.e. it represents the shot noise of the beam distribution.

When this beam radiates, its bunching factor depends on the exact locations z_i's. The exact expression (for $\theta = 0$)

$$\mathcal{B}(\omega) = \frac{1}{N}\sum_{i=1}^{N} e^{-i(\omega/c)z_i}$$

is stochastic. However, we can calculate its *expectation value*. It is given by

$$\langle \mathcal{B}(\omega) \rangle = \int_{-\infty}^{\infty} dz\, \rho_0(z) e^{-i(\omega/c)z} ,$$

which is the same as (7.45) as one would expect.

But $\langle \mathcal{B}(\omega) \rangle$ is not the important quantity. More important is the calculation of the radiation spectrum by finding the expectation value of $|\mathcal{B}(\omega)|^2$ as follows,

$$\begin{aligned}
|\mathcal{B}(\omega)|^2 &= \frac{1}{N^2} \sum_{i=1}^{N} \sum_{j=1}^{N} e^{-i(\omega/c)(z_i - z_j)} \\
&= \frac{1}{N^2} \left[N + \sum_{i=1}^{N} \sum_{j \neq i} e^{-i(\omega/c)(z_i - z_j)} \right] , \\
\Longrightarrow \quad \langle |\mathcal{B}(\omega)|^2 \rangle &= \frac{1}{N} + \frac{N-1}{N} \int_{-\infty}^{\infty} dz_i \int_{-\infty}^{\infty} dz_j\, \rho_0(z_i) \rho_0(z_j) e^{-i(\omega/c)(z_i - z_j)} \\
&= \frac{1}{N} + \frac{N-1}{N} \left| \int_{-\infty}^{\infty} dz\, \rho_0(z) e^{-i(\omega/c)z} \right|^2 \\
&= \frac{1}{N} + \frac{N-1}{N} |\langle \mathcal{B}(\omega) \rangle|^2 \\
&= \frac{1}{N} + \frac{N-1}{N} |\tilde{\rho}_0(\omega)|^2 .
\end{aligned}$$

The expectation value of the radiation spectrum is therefore

$$\left\langle \frac{d^2W}{d\Omega d\omega} \right\rangle \Bigg|_{\text{beam}} = \left[N + N(N-1) |\tilde{\rho}_0(\omega)|^2 \right] \frac{d^2W}{d\Omega d\omega} \Bigg|_{\text{pt.charge}} . \tag{7.46}$$

The first term is the stochastic contribution proportional to N. The second term is a coherent radiation from the collective radiation from the beam, and is proportional to $N(N-1)$. If the beam contains only one particle, the result reduce to the trivial case.

In a 3rd generation synchrotron radiation facility, the radiation frequency of interest $\omega \gg \frac{c}{\sigma_z}$, there is no significant coherent radiation because the bunching factor is exponentially small. But the incoherent radiation still makes them useful tools. On the other hand, if σ_z is small enough and there is interest in the radiation at low frequencies $\lesssim \frac{c}{\sigma_z}$, then the radiation becomes coherent, and the radiation power available becomes very high due to the $N(N-1) \approx N^2$ dependence. A schematic view of the situation is shown in Fig. 7.31.

For a point beam with N particles ($\sigma_z = 0$), it radiates an energy N^2 times that of a single particle. Strictly speaking, only $\frac{N-1}{N}$ portion of it is truly coherent. The remaining $\frac{1}{N}$ is incoherent.

Figure 7.31: A schematic of incoherent and coherent radiation power behavior for a 3rd generation synchrotron radiation facility as a function of the frequency of the radiation. The figure is meant to be a log-log plot.

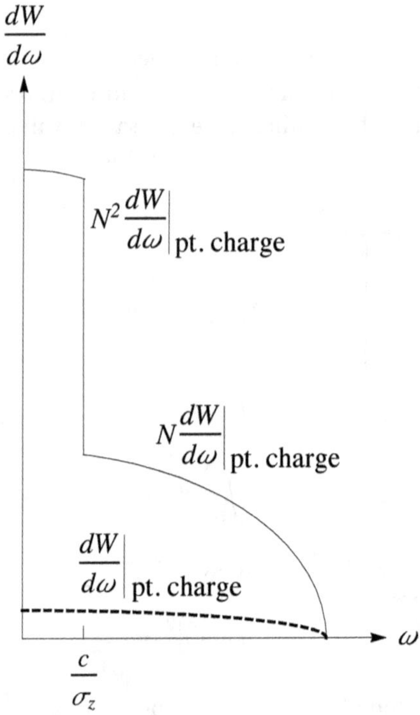

DC beam A true DC beam will not radiate because $\mathcal{B} = 0$ except for the static Coulomb field. Mathematically, it is easily seen, for a DC beam,

$$\langle \mathcal{B} \rangle \;=\; \rho_0 \int_{-\infty}^{\infty} dz\, e^{-i(\omega/c)z} \;=\; 2\pi c \rho_0\, \delta(\omega)\,.$$

The only radiation would be at $\omega = 0$. This is a very general result and holds for any kind of far field radiation. It means a true DC beam does not radiate to far distances except for the static Coulomb field, which is in any case a near-field effect.

However, there is no true DC beam. When stochastic contribution is included, Eq. (7.46) says that a DC beam has a radiation given by N times the point charge contribution.

Homework 7.25 With the normalization (7.44), prove mathematically that, for any $\rho(z)$ and any ω,

$$|\mathcal{B}(\omega)| \;\leq\; 1\,.$$

Homework 7.26 In case the beam distribution has been manipulated so that now it contains an oscillating component at a certain frequency ω_r. For example, let us assume

$$\rho(z) = \frac{1}{\sqrt{2\pi}\sigma_z} e^{-\frac{z^2}{2\sigma_z^2}} \left(1 + \epsilon \sin \frac{\omega_r z}{c}\right), \quad (7.47)$$

with $|\epsilon| < 1$. Calculate the bunching factor of this beam in the forward direction $\theta = 0$.

As we will see, in an FEL, we manipulate the beam distribution so that its longitudinal distribution will contain a frequency component resonant with the undulator resonance. Low-gain FELs will have a small bunching factor. High-gain FELs can have bunching factor approaching 1. Pay special attention to the case when $\frac{\omega_r \sigma_z}{c} \gg 2\pi$ because that will be the parameters relevant to the FELs.

Solution First show that the normalization (7.44) is obeyed. The bunching factor is given by

$$\mathcal{B}(\omega) = e^{-\frac{\sigma_z^2\omega^2}{2c^2}} \left[1 - i\epsilon\, e^{-\frac{\sigma_z^2\omega_r^2}{2c^2}} \sinh\left(\frac{\sigma_z^2 \omega\omega_r}{c^2}\right)\right].$$

The special case when $\epsilon = 0$ deserves some attention. A Gaussian beam's bunching factor is

$$\mathcal{B}(\omega) = e^{-\frac{\sigma_z^2\omega^2}{2c^2}}.$$

The bunching factor is significant only if $\omega \lesssim \frac{c}{\sigma_z}$ as one would have expected.

Plotted in Fig. 7.32 is the value of $|\mathcal{B}(\omega)|$ as a function of $x = \frac{\sigma_z\omega}{c}$ for three spectra $y = \frac{\sigma_z\omega_r}{c} = 10\pi, 15\pi, 20\pi$, respectively. Each spectrum contains two peaks, one located at $x = 0$, and the other at $x = y$. The peak at $x = 0$ represents the unmodulated component $e^{-\frac{\sigma_z^2\omega^2}{2c^2}}$. The peak at $x = y$ says the modulated beam will resonantly radiate at its modulated frequency ω_r. Show that the height of the resonant peak $\approx c^2$. Find an estimate of its width.

Homework 7.27 This homework is similar to Homework 7.26, but might be closer to the case of a high-gain FEL. Assuming that you now have managed to produce a microbunched beam with

$$\rho(z) = \begin{cases} \frac{1}{2M\lambda_0}\left(1 + \cos\frac{2\pi z}{\lambda_0}\right), & \text{if } -M\lambda_0 < z < M\lambda_0, \\ 0, & \text{otherwise,} \end{cases}$$

where M is an integer $\gg 1$ because the original bunch is long.

(a) Show that the bunching factor at the forward direction $\theta = 0$ is given by ($k = \frac{\omega}{c}$)

$$\mathcal{B}(k) = \frac{2(k^2\lambda_0^2 - 2\pi^2)\sin(Mk\lambda_0)}{Mk\lambda_0(k^2\lambda_0^2 - 4\pi^2)}.$$

Figure 7.32: Bunching factor $|\mathcal{B}(\omega)|$ of a modulated beam (7.47) with $\epsilon = 0.5$. The red, green, blue curves are for $y = \frac{\sigma_z \omega_r}{c} = 10\pi, 15\pi, 20\pi$, respectively. Horizontal axis is $x = \frac{\sigma_z \omega}{c}$.

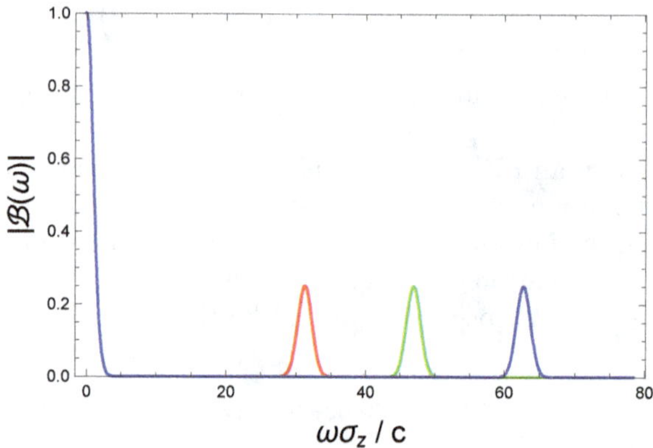

(b) Bunching factor is sharply peaked at $k = 0$ with $\mathcal{B} = 1$ and at $k = \pm\frac{2\pi}{\lambda_0}$ with $\mathcal{B} = \frac{1}{2}$.

(c) Plot $\mathcal{B}$ as a function of $x = \frac{k\lambda_0}{2\pi}$ for the case $M = 10$ and convince yourself that this microbunched beam has a possible strong coherent radiation at wavelength λ_0.

7.4.2 Radiation by a microbunched beam

We just discussed the bunching factor and how it affects the radiation of a beam in various beam conditions. One more case to be discussed is the radiation of a microbunched beam in an undulator. Because of its particular relevance, however, we decide to treat it in a dedicated section. We assume the microbunching wavelength is equal to the resonant wavelength of the undulator.

It is possible to derive rigorously in detail to obtain the radiation power in this case. But we prefer to obtain the result using a series of heuristic arguments — together with their physical meanings — as follows. We will then compare our heuristic result with that of the detailed derivation at the end of this section.

First consider a single electron. We made earlier a heuristic argument connecting the radiation power in an undulator to that in a bending magnet, and derived Eq. (7.43),

$$P_{\text{single}} = \frac{2}{3} r_0 m \gamma^2 a_u^2 k_u^2 c^3 .$$

We will now assign this same result to be the total radiation power by a point electron in the undulator to find an expression for the power of the radiation in the $H = 1$ mode. We will use a series of physical arguments to do so.

This expression (7.43) represents the power radiated by the electron over a distance of $L_u = N_u\lambda_u$. It is not the power of the radiation itself. The radiation itself has a length of $N_u\lambda$. The radiation's instantaneous peak power therefore is compressed as

$$P_{\text{single,rad}} = \frac{\lambda_u}{\lambda} P_{\text{single}} .$$

This radiation power is what enters into all undulator radiation modes. We want only the portion that enters the fundamental $H = 1$ mode. Using Eq. (7.42), the peak power becomes

$$P_{\text{single,rad,fund.mode}} = P_{\text{single,rad}}\, r(a_u) ,$$

where

$$r(a_u) \approx \frac{[JJ]^2}{(1+a_u^2)^2} .$$

The factor $r(a_u)$ gives the fraction of radiated energy into the fundamental mode for a single point charge. For a Gaussian beam, the higher order modes are suppressed more due to the smaller bunching factors if $\sigma_z \gtrsim \lambda$. For a long microbunched beam, it radiates at the frequency where the microbunching occurs and all other harmonics are suppressed. Here we assume the beam is microbunched at the fundamental harmonic of the undulator. Unlike the case of a single particle, here all higher harmonics are suppressed and therefore ignored.

We have one more factor to consider. The fundamental mode radiation has a frequency red shift as the radiation angle increases from a straight-ahead forward direction. The total radiation angle, including all frequencies, is $\Delta\theta \sim \frac{\sqrt{1+a_u^2}}{\gamma}$. The radiation of interest is un-red-shifted frequency with wavelength λ. The angular spread of that is $\Delta\theta' \sim \frac{\sqrt{1+a_u^2}}{\gamma\sqrt{N_u}}$. Therefore, there is another reduction factor of $(\frac{\Delta\theta'}{\Delta\theta})^2 = \frac{1}{N_u}$ due to frequency filtering (or equivalently, angle filtering),

$$\begin{aligned} P_{\text{single,rad,fund.mode,w/o red shift}} &= \frac{1}{N_u} P_{\text{single,rad,fund.mode}} \\ &= \frac{1}{N_u} \frac{[JJ]^2}{(1+a_u^2)^2} \frac{\lambda_u}{\lambda} \frac{2}{3} r_0 m\gamma^2 a_u^2 k_u^2 c^3 . \end{aligned}$$

Below diffraction limit The radiation from a single electron covers a coherence region over a distance, the coherence length, of $N_u\lambda$. Below the diffraction limit, all electrons within one longitudinal slice of the beam radiate coherently. Furthermore, all slices within a distance $N_u\lambda$ also radiate coherently. This means the total number of electrons radiating coherently is

$$N_{\text{coh}} = \left(\frac{N_{\text{tot}}}{L_z}\right) N_u\lambda , \tag{7.48}$$

where N_{tot} is the total number of electrons in the electron macrobunch, L_z is the total length of the macrobunch. The radiation power is proportional to N_{coh}^2. The radiation power by this group of coherent electrons then becomes

$$N_{\text{coh}}^2 P_{\text{single,rad,fund.mode,w/o red shift}} \,.$$

This expression represents the radiation power within one coherence length $N_u\lambda$ by one group of coherent electrons. Other groups occupying different longitudinal units of $N_u\lambda$ will radiate at different longitudinal positions. There is no overlap among the different groups, so this expression also represents the peak power of the radiation.

In this derivation, we have assumed to be below diffraction limit, and

$$4\pi\epsilon_{x,y} \lesssim \lambda < N_u\lambda < L_z \,.$$

Finally, we include the longitudinal bunching factor $\mathcal{B}$ to obtain our final result for the peak power of the radiation into the undulator fundamental mode by a microbunched beam, below the diffraction limit,

$$\hat{P}_{\text{below diff. limit}} \approx \frac{4\pi^2}{3} r_0 mc^3 |\mathcal{B}|^2 \frac{a_u^2}{1+a_u^2} [JJ]^2 N_u \left(\frac{N_{\text{tot}}}{L_z}\right)^2 \,. \tag{7.49}$$

Above diffraction limit So far we discussed the case if the transverse electron beam size is small, $\sigma_\perp^2 \lesssim \frac{\lambda}{4\pi} L_u$, i.e. below the diffraction limit, where $\frac{\lambda}{4\pi}$ is the radiation emittance — see Eq. (7.32). In this case, all electrons inside the same longitudinal slice of the beam radiate coherently. In case the beam size is larger, $\sigma_\perp^2 \gtrsim \frac{\lambda}{4\pi} L_u$, i.e. if the beam is above diffraction limit, this is no longer true.

As just discussed, the radiation from a single electron covers a coherence volume. The longitudinal length of this volume is $N_u\lambda$. Nominally, the transverse radius of this coherence volume is $\sqrt{\frac{\lambda}{4\pi} L_u}$. The coherence volume is therefore $N_u\lambda \cdot \pi \frac{\lambda}{4\pi} L_u$

Below the diffraction limit, the transverse beam size is smaller than the diffraction radius. The coherence volume becomes $N_u\lambda \cdot \pi\sigma_\perp^2$. The number of electrons in coherence is given by Eq. (7.48).

Above diffraction limit, not all electrons in each longitudinal slice are in coherence. The number of coherent electrons becomes

$$N_{\text{coh}} = \left(\frac{N_{\text{tot}}}{L_z}\right) N_u\lambda \frac{\lambda L_u}{4\pi\sigma_\perp^2} \,.$$

This means the radiation power (7.49) acquires an additional factor of $\left(\frac{\lambda L_u}{4\pi\sigma_\perp^2}\right)^2$.

However, the radiation obtained is by the coherent electrons within the group under consideration. There are other groups occupying the same longitudinal space and also contribute to the same peak power due to the overlapping. The number of these coherent groups is $\frac{4\pi\sigma_\perp^2}{\lambda L_u}$. Taking all factors into consideration, we obtain the final result, above diffraction limit,

$$\hat{P}_{\text{above diff. limit}} \approx \frac{4\pi^2}{3} r_0 mc^3 |\mathcal{B}|^2 \frac{a_u^2}{1+a_u^2} [JJ]^2 N_u \left(\frac{N_{\text{tot}}}{L_z}\right)^2 \frac{\lambda L_u}{4\pi\sigma_\perp^2}. \tag{7.50}$$

Equation (7.50) differs from (7.49) by a factor $\frac{\lambda L_u}{4\pi\sigma_\perp^2}$.

General case The two results Eqs. (7.49) and (7.50) can be combined,

$$\hat{P}_{\text{general}} \approx \frac{4\pi^2}{3} r_0 mc^3 |\mathcal{B}|^2 \frac{a_u^2}{1+a_u^2} [JJ]^2 N_u \left(\frac{N_{\text{tot}}}{L_z}\right)^2 \begin{cases} 1, & \text{if } r < 1 \\ \frac{1}{r}, & \text{if } r > 1 \end{cases}. \tag{7.51}$$

where r is the ratio of the transverse area of the electron beam to that of the laser,

$$r = \frac{4\pi\sigma_\perp^2}{\lambda L_u}.$$

This result, however, is based on heuristic physical arguments. A more accurate expression can be obtained from a more careful derivation.[9] The result is

$$\hat{P} \approx 2\pi^2 r_0 mc^3 |\mathcal{B}|^2 \mathcal{F}_\perp(r) \frac{a_u^2}{1+a_u^2} [JJ]^2 N_u \left(\frac{N_{\text{tot}}}{L_z}\right)^2, \tag{7.52}$$

where a transverse form factor $\mathcal{F}_\perp$ has been introduced,

$$\mathcal{F}_\perp(r) = \frac{2}{\pi}\left[\tan^{-1}\left(\frac{1}{r}\right) + \frac{r}{2}\ln\left(\frac{r^2}{r^2+1}\right)\right].$$

We define the diffraction limit to be when $r = 1$. Below diffraction limit, we take $r \ll 1$. Above diffraction limit, we take $r \gg 1$.

The function $\mathcal{F}_\perp(r)$ is plotted in Fig. 7.33. Its asymptotic values are as follows,

$$\mathcal{F}_\perp(r) \approx \begin{cases} 1, & r \ll 1, \text{ below diffraction limit}, \\ \frac{1}{2} - \frac{\ln 2}{\pi} \approx 0.28, & r = 1, \text{ at diffraction limit}, \\ \frac{1}{\pi r}, & r \gg 1, \text{ above diffraction limit}. \end{cases}$$

It can be seen that our heuristic derivation (7.49) differs from the final result (7.52) by a factor of $\frac{2}{3}$, and our (7.50) differs by a factor of $\frac{2\pi}{3}$. This is considered reasonable given the approximate nature of our derivations.

[9] E.L. Saldin, E.A. Schneidmiller, and M.V. Yurkov, Nucl. Instru. Meth. A 539, 499 (2005).

Figure 7.33: Transverse form factor $\mathcal{F}_\perp(r)$ as a function of $r = \frac{4\pi\sigma_\perp^2}{\lambda L_u}$.

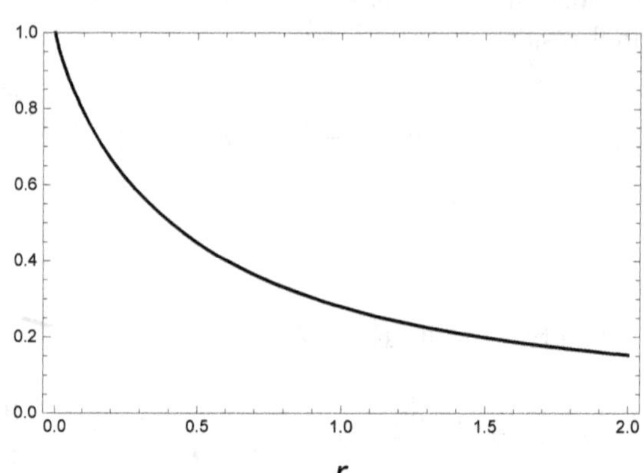

7.4.3 Adding a copropagating laser

We now add a complication to the undulator configuration. Consider the case when an electron is sent down an undulator together with a copropagating laser. Let the laser be planar polarized with electric field

$$\vec{E} = \hat{x} E_0 \cos(\omega t - kz) ,$$

where $k = \frac{\omega}{c}$.

As the electron propagates down the undulator, its motion is affected by the laser. However, we assume for now that the laser field is weak enough that the motion of the electron is still dominated by the undulator field. In other words, let us assume the electron motion still has the prescribed oscillatory motion in the undulator, Eq. (7.5). We will also very safely consider only the case when $k_u \ll k$.

The electron energy $\mathcal{E}_0$ changes according to

$$\begin{aligned} \frac{d\mathcal{E}_0}{dt} &= e\vec{v}\cdot\vec{E} \\ &= -\frac{\sqrt{2}a_u ceE_0}{\gamma} \sin k_u z \, \cos(\omega t - kz) \\ &= -\frac{\sqrt{2}a_u ceE_0}{\gamma} \frac{1}{2} \left[\sin(k_u z + \omega t - kz) + \sin(k_u z - \omega t + kz)\right] . \end{aligned}$$

Note that the undulator is necessary in order for $\frac{d\mathcal{E}_0}{dt} \neq 0$. Laser field alone does not exchange energy with the electrons because the laser field is perpendicular to the direction of motion of the electron in the absence of the undulator modulation.

The next step is to extract the nonoscillatory contribution to obtain the systematic energy transfer from the laser to the electron. There are two sine terms in the $\frac{d\mathcal{E}}{dt}$ expression. Nonoscillatory contributions can only occur when their arguments stay around constant values. Since z progresses in time systematically with an average speed of $\bar{v}_z$, in order for the first sine term to deliver or extract energy systematically from the electron, we need

$$k_u\bar{v}_z + kc - k\bar{v}_z = 0, \quad \text{or} \quad \frac{\bar{v}_z}{c} = \frac{k}{k-k_u},$$

which is not possible because it requires $\bar{v}_z > c$. As a result, the first sine term always oscillates and does not accumulate a systematic energy transfer to or from the electron.

So only the second sine term should be kept, and we obtain

$$\frac{d\mathcal{E}_0}{dt} = -\frac{a_u ceE_0}{\sqrt{2}\,\gamma}\sin(k_u z - \omega t + kz)\,. \tag{7.53}$$

This term is nonoscillatory only when

$$\frac{\bar{v}_z}{c} = \frac{k}{k+k_u}\,.$$

This condition can be rearranged a bit, and we find that, in order for the copropagating laser to systematically exchange energy with the undulating electron, the laser wavelength must be chosen to be (with $k \gg k_u$)

$$\lambda = \frac{1+a_u^2}{2\gamma^2}\lambda_u\,.$$

Under this condition, we obtain a systematic energy transfer from the laser to the electron with

$$\frac{d\mathcal{E}_0}{dt} = -\frac{a_u ceE_0}{\sqrt{2}\gamma}\sin\frac{\omega z_0}{\bar{v}_z},$$

where we have set $z = z_0 + \bar{v}_z t$ for the particle under consideration. We will use this result later.

The rate of laser acceleration of an electron depends on its initial launching longitudinal position z_0. The acceleration rate oscillates as a function of z_0. For a particle with the launching phase $\sin\frac{\omega z_0}{\bar{v}_z} = -1$, it will have a maximum energy gain from the laser, with

$$\frac{d\mathcal{E}_0}{dt} = \frac{a_u ceE_0}{\sqrt{2}\gamma}\,. \tag{7.54}$$

Particles with other phases will receive positive and negative energy exchange rates.

There is a subtlety to be noted, however. In the above derivation, we have argued to set $z = \bar{v}_z t$, while z in fact contains an additional very small oscillatory term yielding a figure-8 motion. Although the figure-8 is extremely thin, meaning that the forward-backward z-motion $\ll$ the sideways x-motion, this z motion makes a meaningful contribution. When $z = \bar{v}_z t + \frac{a_u^2}{4k_u\gamma^2}\sin 2k_u\bar{v}_z t$ is used in the previous analysis, both sine terms will contribute. The first term contributes a term proportional to $-J_1(\xi)$ and the second term contributes a term proportional to $J_0(\xi)$, where $\xi = \frac{a_u^2/2}{1+a_u^2}$. The final result then becomes

$$\frac{d\mathcal{E}_0}{dt} = \frac{a_u ceE_0}{\sqrt{2}\gamma}[JJ], \tag{7.55}$$

where $[JJ] = J_0(\xi) - J_1(\xi)$ was defined in Eq. (7.24).

The reader would note that we have in fact designed a laser accelerator — it is sometimes called the inverse free electron laser.[10] By choosing the right laser wavelength and properly phasing the laser with respect to the electrons, we have designed a laser accelerator. It is not a very efficient accelerator concept because of the diminishing factor of $\frac{1}{\gamma}$ in its acceleration rate. But in any case that is not what we are after in the present discussion.

More relevant to us, attention is to be drawn to the observation that this magic value of the laser wavelength $\frac{1+a_u^2}{2\gamma^2}\lambda_u$, which will allow constant systematic energy exchange between the laser and the electron in the undulaor, happens to coincide with the resonant wavelength $\lambda_r(\theta=0)$ radiated by the point charge in the undulator in the absence of the laser as discussed in Eq. (7.8). This apparent coincidence turns out to be the source of the wonder of the FEL mechanism. Hidden of course behind this apparent coincidence is the fact that they both originate from the one and the same resonant behavior we emphasized multiple times. We will return to this point again later.

Figure 7.9, while playing a key role illustrating the undulator radiation resonance, has a counterpart (homework 7.29) that illustrates the resonance mechanism for a systematic energy transfer of a laser-electron system in an undulator.

Homework 7.28 The text gave a qualitative reasoning in going from Eq. (7.54) to Eq. (7.55). Fill in the necessary derivation.

Solution Use is to be made of the identity

$$e^{iA\sin x} = \sum_{n=-\infty}^{\infty} J_n(A)e^{inx}.$$

Homework 7.29 It would be instructive to modify Fig. 7.9 to illustrate the systematic energy transfer mechanism discussed in this section. Identify the

[10] R.B. Palmer, J. Appl. Phys. 43, 3014 (1972); P. Sprangle and C.M. Tang, IEEE Trans. Nucl. Sci. NS-28, 3346 (1981); E.D. Courant, C. Pellegrini and W. Zakowicz, Phys. Rev. A32, 2813 (1985).

resonance condition. Assume the laser amplitude stays constant during the undulator passage.

7.4.4 Laser-undulator pendulum

The pendulum Before proceeding to the FEL analysis, let us briefly review our classical mechanics on a pendulum motion as well as its phase space (θ, p_θ). We only summarize the results here. See Fig. 7.34.

angular coordinate	θ
angular momentum p_θ	$mL^2\dot{\theta}$
phase space	(θ, p_θ)
equation of motion	$\begin{cases} \dot{\theta} = \frac{1}{mL^2}\, p_\theta \\ \dot{p}_\theta = -mLg\sin\theta \\ \ddot{\theta} + \frac{g}{L}\sin\theta = 0 \end{cases}$
Hamiltonian	$H(\theta, p_\theta) = mgL(1-\cos\theta) + \dfrac{1}{2mL^2}p_\theta^2$ $= \text{constant of the motion}$
small amplitude oscillation	$\begin{cases} \ddot{\theta} + \frac{g}{L}\theta = 0 \\ \theta(t) = \theta_0\cos\omega_0 t + \frac{\dot{\theta}_0}{\omega_0}\sin\omega_0 t \\ \omega_0 = \sqrt{\frac{g}{L}} \\ H = \frac{mgL}{2}\theta^2 + \frac{1}{2mL^2}p_\theta^2 \\ \text{trajectory in phase space is ellipse} \end{cases}$
separatrix	$\begin{cases} \text{defines the boundary of the ``bucket''} \\ H = 2mgL \\ \frac{p_\theta^2}{2mL^2} = mgL(1+\cos\theta) \end{cases}$
large amplitude gyration	$\begin{cases} H > 2mgL \\ \text{trajectory in phase space is outside the bucket} \\ \text{pendulum continues to gyrate in one direction} \end{cases}$

Laser-undulator equation of motion Returning to our system of an electron going through an undulator with a copropagating laser. Let the electron energy $= \gamma mc^2$, undulator period $= \lambda_u$, undulator strength $= a_u$, laser wavelength $= \lambda$. Choose the laser $\vec{E} = \hat{x}E_0\cos(\omega t - kz)$. In this section, we establish

Figure 7.34: A simple frictionless nonrelativistic pendulum (left) and its dynamics in phase space (θ, p_θ) (right). The green closed curves (approximately ellipses) represent small amplitude oscillations. Blue curves are large amplitude gyration motion. Red curves are the separatrix, which we refer to as the "bucket". Note the arrows indicating the direction of motion of the pendulum.

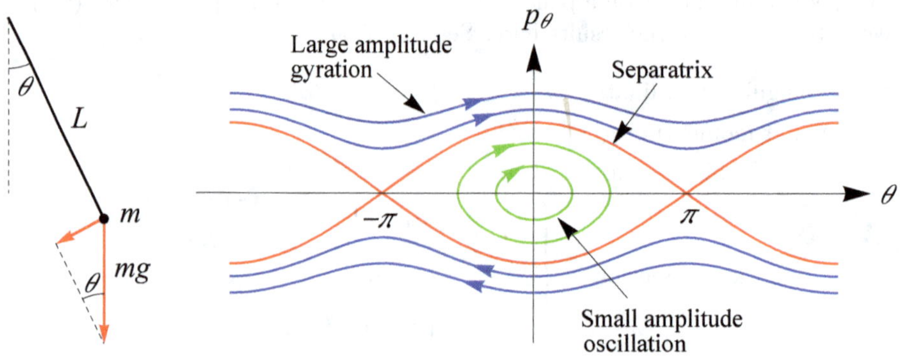

the equation of motion of the electron by assembling the various pieces we have learned so far.

First note that the laser wavelength is chosen nominally to fulfill the resonant condition at γ_r. When the electron energy changes due to the electron-laser-undulator interaction, its γ changes from γ_r, and the electron slips off resonance slightly. This means the phase

$$\psi = (k_u \bar{v}_z - \omega + k\bar{v}_z)t$$

cannot be zero always. It is now a *dynamic* variable. The earlier analysis of laser acceleration still applies but now the phase is no longer fixed.

This dynamical quantity ψ is sometimes referred to as the *ponderomotive phase*. Physically it is equal to kz, where z is the instantaneous longitudinal deviation of an individual electron from the center of the electron bunch. Note that in the definition of ψ, $\bar{v}_z$ refers to the average longitudinal velocity of the particle under consideration, so it depends on δ, but k and ω refer to the resonant values.

We know that, as always, the first step is to define the phase space of our dynamical system. Since γ is not expected to slip off γ_r too much, we define

$$\delta = \frac{\gamma - \gamma_r}{\gamma_r}, \qquad |\delta| \ll 1 .$$

Quantities δ, ψ are the two dynamical variables of the system, and (ψ, δ) forms our phase space; ψ is the coordinate, δ is the momentum.

The equation of motion for the system will consist of an expression for $\frac{d\psi}{dt}$ and an expression for $\frac{d\delta}{dt}$. The rate of change of ψ is

$$\frac{d\psi}{dt} \approx k_u \bar{v}_z - \omega + k\bar{v}_z \,.$$

With $\gamma \approx \gamma_r$, it follows that $\bar{v}_z \approx c\left[1 - \frac{1+a_u^2}{2\gamma_r^2}(1-2\delta)\right]$, and with $k \gg k_u$,

$$\frac{d\psi}{dt} \approx 2k_u c\delta \,.$$

This $\frac{d\psi}{dt}$ equation says the ponderomotive phase ψ changes slowly because it is proportionally to δ which is $\ll 1$. In addition, it is small also because of the factor k_u, which is much smaller than k. This means the ponderomotive phase of a particle in the laser-undulator system is a slowly varying quantity. Note that what is small is $\frac{d\psi}{dt}$. The phase ψ itself may slowly wonder away significantly from 0 because t is large for a long undulator.

The rate of change of electron energy is, according to Eq. (7.53) with an additional factor of $[JJ]$,

$$\frac{d\delta}{dt} = \frac{1}{\gamma_r mc^2}\frac{d\mathcal{E}}{dt} = -\frac{a_u ceE_0}{\sqrt{2}\,\gamma_r^2 mc^2}[JJ]\,\sin\psi \,,$$

where we have dropped the oscillatory term. The term kept is slowly oscillating with $\sin\psi$.

This $\frac{d\delta}{dt}$ equation says the systematic energy transfer from the laser to the electron is proportional to $\sin\psi$. The laser systematically transfers energy to the electron if $\sin\psi < 0$, and systematically absorbs energy from the electron if $\sin\psi > 0$. Both ψ and δ change slowly.

Combining what we have so far,

$$\begin{aligned}
\frac{d\psi}{ds} &= 2k_u\delta \,, \\
\frac{d\delta}{ds} &= -\epsilon\,\sin\psi \,, \\
\epsilon &= \frac{a_u eE_0}{\sqrt{2}\,\gamma_r^2 mc^2}[JJ] \,,
\end{aligned} \tag{7.56}$$

where $s \approx ct$ is the distance traveled by the electron along the undulator length. Equation (7.56) describes the evolution of the electron motion in the laser-undulator system in the phase space (ψ, δ). The system is symplectic.

Laser-undulator bucket One recognizes that Eq. (7.56) is a familiar pendulum equation, and is also the same as the synchrotron equation of motion in a storage ring with adiabatic approximation.

The bucket in the (ψ, δ) phase space covers a range of 2π in the ψ-coordinate. Had we used Δz as the coordinate, a bucket will cover a range of the laser wavelength λ. As we will soon see, the electron beam will be modulated by the buckets into bunchlets spaced by the laser wavelength λ. Recognizing that λ is very small, this phenomenon is called "microbunching".

An electron close to the bucket center executes small amplitude oscillation around the center with frequency

$$\Omega = c\sqrt{2k_u|\epsilon|} = c\sqrt{\frac{\sqrt{2}\,k_u|a_u|eE_0}{\gamma_r^2 mc^2}[JJ]}\,. \tag{7.57}$$

We loosely refer to Ω as the FEL synchrotron oscillation frequency. During one oscillation period, the electron has traveled a distance

$$D = \frac{2\pi c}{\Omega} = \frac{2\pi}{\sqrt{2k_u|\epsilon|}}\,. \tag{7.58}$$

Bucket half-height is given by

$$\hat{\delta}_{\frac{1}{2}} = \sqrt{\frac{2|\epsilon|}{k_u}} = \sqrt{\frac{\sqrt{2}}{\pi}\,\frac{eE_0\lambda_r}{mc^2}[JJ]\,\frac{|a_u|}{1+a_u^2}}\,. \tag{7.59}$$

The bucket half-height is related to the synchrotron oscillation frequency by

$$\hat{\delta}_{\frac{1}{2}} = \frac{\Omega}{ck_u}\,.$$

As for all existing lasers, $eE_0\lambda_r \ll mc^2$ (the Schwinger limit), the bucket height $\hat{\delta}_{\frac{1}{2}} \ll 1$ — otherwise our analysis breaks down anyway.

Length hierarchy There are several length scales in an FEL: the total undulator length L_u, together with λ_u, D, λ_r, and the electron microbunch length σ_z. Later we will add another quantity called the gain length L_G. It may be useful here to consider an approximate hierarchy of their magnitudes. Typically, for a low-gain FEL, we might have

$$L_G \gg L_u \sim D \gg \lambda_u \sim \sigma_z \gg \lambda_r\,.$$

For a high-gain FEL, we have, at least for the last gain length of the long undulator,

$$L_u \gg D \sim L_G \gg \lambda_u \gg \sigma_z \gg \lambda_r\,.$$

Homework 7.30 Derive Eq. (7.57) for the FEL synchrotron frequency and Eq. (7.59) for the FEL bucket half-height.

7.4.5 Low-gain FEL

The low-gain FEL consists of an electron beam passing through an undulator together with a copropagating injected laser. We learned that the electron beam will radiate at the undulator's resonant wavelength $\frac{(1+a_u^2)}{2\gamma^2}\lambda_u$. We also learned that systematic energy exchange between the copropagating laser and the electrons occurs when the laser wavelength λ is equal to $\frac{(1+a_u^2)}{2\gamma^2}\lambda_u$. These two effects are separate physical effects — the former has nothing to do with the injected laser, the later does not care how the electron radiates — but their wavelength requirements coincide. This apparent coincidence lays the key to the FEL mechanism. As mentioned, underlying their coincidence is the one and only undulator resonance effect.

We now consider a low-gain FEL, an undulator-laser-electron-beam system, with the condition

$$\lambda = \frac{(1+a_u^2)}{2\gamma^2}\lambda_u \,.$$

This condition connects the three systems: γ comes from the electron beam, λ comes from the injected laser, λ_u and a_u come from the undulator.

Radiation by the electron has the same frequency as the injected laser. By energy conservation, the energy gain/loss of the electron becomes the energy loss/gain of the radiation energy in the laser, so the laser field strength will change in time. We will come to this point later for the high-gain FEL. For now, we assume the radiation is dominated by the injected laser throughout the undulator passage. The reader is advised not to apply this energy conservation argument to a case of bending magnet radiation. Lacking the underlying resonance effect and therefore lacking the apparent frequency coincidence just mentioned, bending magnets cannot be made into an FEL.

If the laser strength dominates the electron beam's radiation, the motion of each electron can be considered independently of the other electrons. Its motion will obey the pendulum equation (7.56), and its phase space (ψ, δ) will look like that shown for the pendulum, Fig. 7.34. In the phase space, a single particle follows one of the contour trajectories of the plot. Some particles gain energy, some lose energy, and there is a bucket determining the phase space topology. Individual electrons move in the bucket just like in a pendulum motion or in a synchrotron oscillation of a storage ring below transition.

With the magic of coincident frequencies, we have already a low-gain FEL. Consider a monoenergetic electron beam injected with $\delta = \delta_0$. Since typical electron beam is much longer than the laser wavelength, we approximate the beam as uniform in ψ at injection. The beam's initial distribution in phase space is therefore a uniform horizontal line at $\delta = \delta_0$. Once injected, this distribution begins to evolve.

Figure 7.35 shows the phase space evolution for three cases with different initial δ_0. In each case, a monoenergetic beam is injected, and then its evolution is followed as the beam propagates down the undulator for a total distance of D

Figure 7.35: Phase space evolution of a low-gain FEL for three cases of injected beam. The upper panel is the case when the injected beam is on-momentum, $\delta_0 = 0$. The middle case is when the beam is injected with an energy slightly above that of the resonant value, $\delta_0 > 0$. The lower panel is when $\delta_0 < 0$.

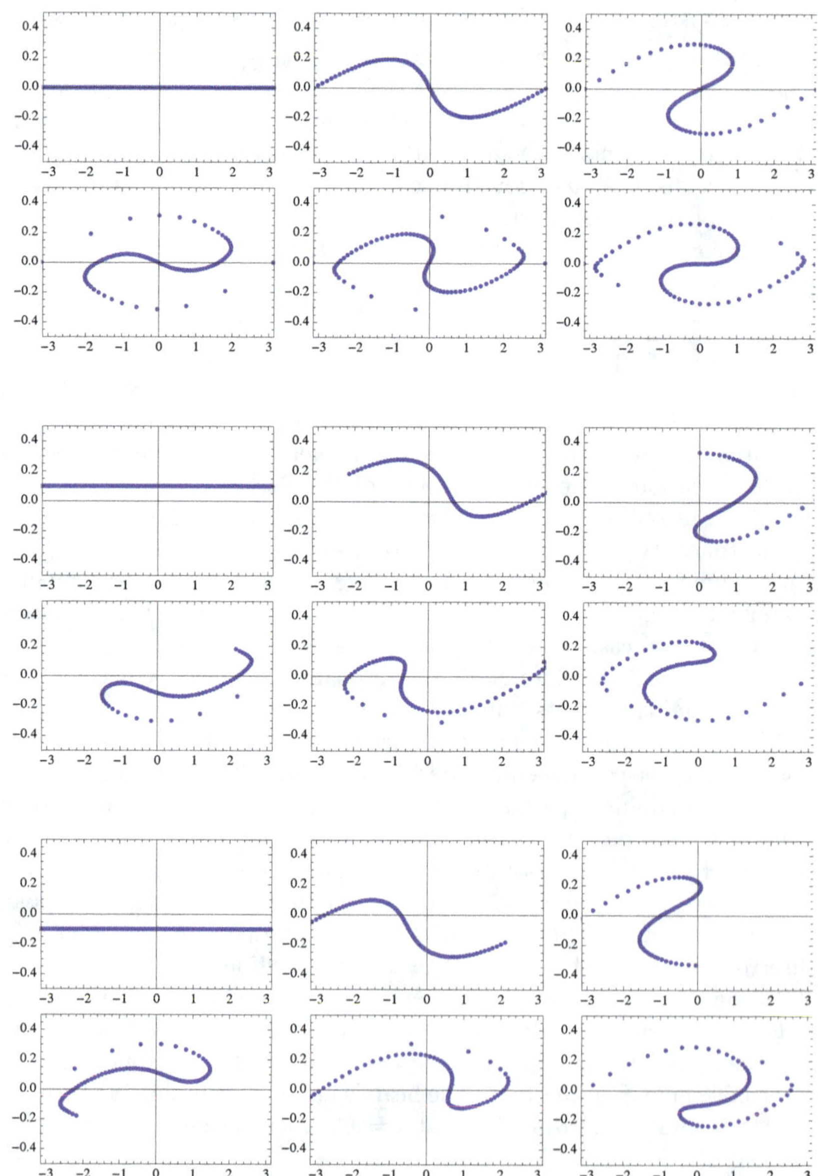

given by Eq. (7.58) — one complete synchrotron oscillation of small-amplitude particles. The shape of the bucket is apparent. The width of the bucket $\Delta\psi = 2\pi$ represents one laser wavelength λ. The electron bunch length is assumed to be much longer than λ.

From Fig. 7.35 we see that the beam with $\delta_0 = 0$ has no net energy change as it evolves down the undulator. Some particles gain energy while others lose energy, the net change of beam energy is zero. By energy conservation, this means the laser radiation does not grow or dissipate. One also observes that the beam with $\delta_0 > 0$ loses net energy, which means the laser grows, while the beam with $\delta_0 < 0$ would mean the laser dissipates.

In general, for the laser to grow, the beam should be injected slightly above the bucket center but not too much higher than the bucket height so that most of the particles are still inside the bucket, i.e. its energy slightly higher than $\gamma_r mc^2$ where γ_r is the resonant value. Or equivalently, the undulator is tuned so that its resonant radiation frequency is slightly higher than the injected laser frequency.

The total length of the undulator should not be too long. Particles inside the bucket execute synchrotron oscillations and complete one synchrotron period in a distance D. When $\delta_0 > 0$, more particles lose energy than gain energy initially, so the laser grows. If the undulator length is $> \frac{D}{2}$, the particles gaining and losing energy are exchanged, and the laser will begin to lose energy to the electron beam. The optimum undulator length is expected to be around $L_u \approx \frac{D}{2}$.

Note in Fig. 7.35 the phenomenon of microbunching. Although injected uniformly in ψ, the beam develops a microbunching in its longitudinal distribution due to the presence of the buckets. The microbunching wavelength is very short, equal to the laser wavelength. Microbunching is a passive outcome in a low-gain FEL, but as we will see, it becomes the dynamical agent for the high-gain FEL.

In this scenario, however, the degree of microbunching and the laser growth are rather limited. The buckets confine the electron motion, and the laser gain is low. This is a low-gain FEL. It can be used in a storage ring with FEL radiation produced at each passage of the beam. The radiation gain at each passage is low, but is made up by the high revolution rate of the storage ring. The gain must not be too high per passage; otherwise the beam is too disrupted to complete its next revolution.

Estimate of gain Consider a uniform monoenergetic beam at $\delta = \delta_0$ again. By the pendulum equation, it is possible to calculate exactly how δ behaves as a function of s for a particle that starts initially at δ_0 and ψ_0 at $s = 0$. By integrating over ψ_0, we obtain the total beam energy gain or loss, ΔE as a function of s as the beam propagates down the undulator. This ΔE is a function of s and δ_0.

We look for the values of s^* and δ^* that give the largest possible beam energy loss ΔE. We then choose the undulator length to be given by s^*. We also choose the injection offset δ_0 to be δ^*. This combination will give the optimum design of the low-gain FEL.

We will not elaborate on this exact calculation. We consider a rough estimate sufficient. The optimum can be anticipated as follows. First, the undulator length is expected to be approximately $s^* = \frac{D}{2}$, where D is the small amplitude synchrotron oscillation length, Eq. (7.58). Second, the optimum energy offset is expected to be close to but slightly below the bucket half-height. For the optimum low-gain FEL, we therefore choose our parameters,

$$L_u = \frac{D}{2}, \quad \delta_0 = \frac{1}{2}\hat{\delta}_{\frac{1}{2}} .$$

This choice of parameters gives

$$\delta_0 k_u L_u = \frac{1}{4}\hat{\delta}_{\frac{1}{2}} k_u D = \frac{\pi}{2} ,$$

or equivalently a simple criterion for the optimum energy offset,

$$\delta_0 = \frac{1}{4N_u} .$$

The choice $L_u = N_u \lambda_u = \frac{D}{2}$ can also be rewritten in terms of a condition for the laser strength parameter ϵ,

$$|\epsilon| = \frac{\pi}{4N_u L_u} ,$$

which one notes is a small quantity — because N_u is large and L_u is long — as expected. The injected copropagating laser strength is therefore optimized to be

$$eE_0 = \frac{\sqrt{2}\pi\gamma_r^2 mc^2}{4N_u L_u a_u [JJ]} .$$

The laser power can be calculated accordingly. If the laser has an electric field E_0 (assumed uniform) over a cross-sectional area A, then the Injected laser energy contained in one wavelength is given by

$$E_{\text{laser}} = \frac{\epsilon_0}{2} E_0^2 \lambda A .$$

For the laser to cover the length of the undulator within a Rayleigh range, we set $A = L_u \lambda$.

With these choices, all particles execute half a synchrotron oscillation going through the undulator. Approximating the synchrotron oscillation to be that in the linear regime, on the average the particles oscillate from $+\delta_0$ to $-\delta_0$, and the energy change of the beam is approximately given by

$$\Delta E \approx Nmc^2\gamma_r \cdot 2\delta_0 = \frac{Nmc^2\gamma_r}{2N_u} ,$$

where N is the number of electrons within the laser wavelength.

By energy conservation, the laser has gained the energy ΔE. By taking the ratio of ΔE to that of the required laser energy in one laser wavelength, one then defines a laser gain of the device per passage,

$$G = \frac{\Delta E}{E_{\rm laser}} = \frac{64NN_u^2 r_0}{\pi\gamma_r\lambda}\frac{a_u^2[JJ]^2}{1+a_u^2}\,.$$

Expressed this way, it is independent of the undulator length L_u. When L_u is increased, for example, the laser spot size has to increase to maintain the Rayleigh condition, and the gain remains unchanged.

As G specifies a percentage gain, not an incremental gain, it hints at a possibility of exponential growth, not a linear growth, when pushed for performance, as we will discuss in the following section.

Homework 7.31 A low-gain FEL has the parameters $\lambda_u = 5$ cm, $\gamma_r = 2\times 10^3$, $a_u = 3$ and $L_u = 1$ m. The electron beam has 10^{10} electrons and a length (assume uniform longitudinal distribution) of 1 mm. What would be the optimum choice for eE_0, the optimum energy offset, the energy gained by the laser, and the laser gain per passage?

Solution $B_y = 0.9$ T, $\lambda = 0.0625\,\mu$m, $N_u = 20, \delta_0 = 0.0125, \epsilon = 0.039\,{\rm m}^{-1}$, $[JJ] = 0.73, E_0 = 52$ GV/m, $N = 0.625\times 10^6$, $\Delta E = 2.6\,\mu$J, $A = 6.25\times 10^{-8}\,{\rm m}^2$. Injected laser energy in one wavelength is 47 μJ. The laser gain per passage is about 5.5%.

7.4.6 Anticipating high gain

In the previous analysis, the laser radiation that copropagates with the electron beam has a constant electric field strength E_0. The microbunching buckets have a fixed height. As mentioned, the laser gain is low, limited by the bucket height.

However, we also pointed out there was a certain degree of microbunching occurring in the process. If the undulator gets longer and/or the laser growth rate gets faster, the laser field E_0 will increase along the undulator, and the bucket height increases. The increasing bucket height expedites the microbunching of the electron beam, which in turn enhances the radiation from the beam, and the laser growth speeds up. This mechanism leads to an exponential growth of the microbunching as well as the laser radiation. Several orders of magnitude are available for the laser power to grow, leading to high-gain FELs, to be discussed in more detail in the following few sections.

As mentioned, the electron motion inside the bucket is analogous to synchrotron oscillation in a storage ring. For a low-gain FEL, the FEL mechanism requires the electrons to lose energy in its synchrotron oscillation. It is easy to see that an effective low-gain FEL demands the electrons to execute approximately half a synchrotron oscillation. An effective low-gain FEL, therefore, might have a total undulator length $L_u \approx \frac{D}{2}$ where D is the synchrotron oscillation distance. The undulator is typically short (in meters). In contrast, the high-gain FEL undulators are long (~100 meters). On the other hand, as we will see, the integrated synchrotron phase advance over the 100-m undulator remains around π.

As a preview of the beam dynamics in a high-gain FEL, we return to the $\delta_0 = 0$ case of Fig. 7.35 and repeat the simulation, except that this time we artificially introduce an exponential growth in the laser-beam coupling coefficient ϵ. The exponential growth rate is such that ϵ has grown substantially within the distance D.

Figure 7.36: Phase space evolution of a mimic case of a high-gain FEL. The case run is the $\delta_0 = 0$ case of a low-gain FEL, but when ϵ is artificially and somewhat exaggeratedly made to grow exponentially.

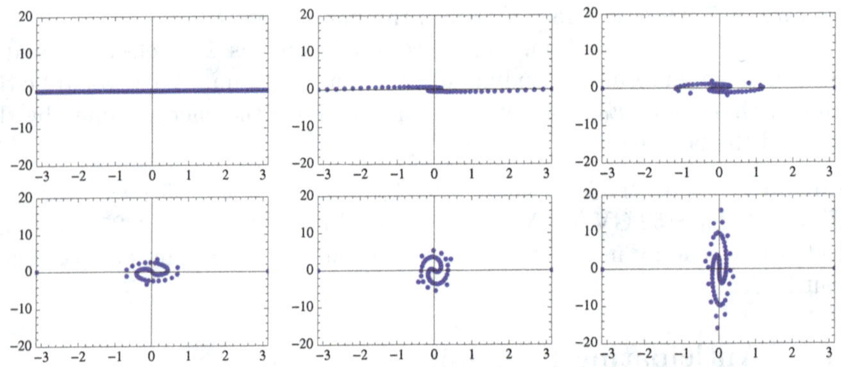

The result of simulation is shown in Fig. 7.36. The $\delta_0 = 0$ beam is simulated over a distance D as before. A phase space bucket is provided by an initial value of ϵ determined by the initial low laser strength, and the beam begins to microbunch accordingly with a low bucket height as in the case of low-gain FEL. As the beam propagates down the undulator, ϵ increases, and so does the bucket height. The microbunch conforms to its shape and becomes more microbunched. The more microbunched beam then radiates more strongly into the laser strength, making ϵ to grow even more. The process fuels the exponential growth of ϵ, and the exponentiation process continues. A comparison of Fig. 7.36 with the upper panel of Fig. 7.35 would indicate the more pronounced microbunching effect, which in turn plays the critical role in the high-gain FEL mechanism, particularly noting the much exaggerated scale of the vertical axis in Fig. 7.36.

Along the line of the above discussion, note the mechanism of RF trapping, as mentioned in Sec. 5.4. The beam trapped well inside the bucket conforms to the bucket. Ideally, the trapping would be adiabatic, which means the exponential growth rate is slower than the synchrotron frequency. But doing so reduces to the low-gain limit. An optimum choice is likely to be when the growth distance is comparable to the distance D. Again, this will be confirmed in Sec. 7.5.5. We will see that at least in the last few gain lengths, D approaches the value of the gain length, and the FEL gain process stops due to saturation.

One might ask where does the radiation energy come from in this case because according to Fig. 7.36, the electrons have an average zero energy change. The answer is that the entire beam will have to lower its energy slowly, i.e. the whole microbunch loses energy to radiation. Another way to say the same thing is to redefine the resonant energy as the beam propagates down the undulator. In that case, one sees the bucket shifting to the left, giving rise to a beam that is constantly decelerated.

7.5 High gain free electron laser

As elaborated in Sec. 7.4.6, the low-gain picture foresees a very important opportunity of a high-gain FEL. Let us recap it here. Let us start with a low-gain FEL,[11] but now increase the laser gain to the extent that the constant-laser-strength approximation begins to break. In addition, we further make the undulator length much longer so it covers many gain lengths. This then potentially becomes a high-gain FEL. The physical process can be viewed in steps as follows.

1. The undulator radiation from the electron beam is sharply peaked at the same frequency as the injected laser. The radiated energy therefore feeds directly to the laser.

2. Initially, E_0 increases by the low-gain mechanism. In the mean time, the electron beam microbunches, its bunching factor at the laser frequency grows, and its radiation increases, making E_0 to increase more.

3. As E_0 increases, the pendulum buckets get taller. The electron beam gets increasingly microbunched by the RF trapping mechanism. With increasing degree of microbunching, the enhanced radiation adds even faster to the laser field E_0.

4. The process exponentiates. The bunching factor and the radiation field E_0 both grow exponentially with s. The laser power $\propto E_0^2$. A typical high-gain FEL allows > 5-6 orders of magnitude growth in laser power.

The radiation from an FEL, whether it is low-gain or high-gain, is coherent. We mentioned before that coherence refers to coherence among the radiation wavepackets from different electrons. The wavepacket by each individual electron is coherent within itself. The situation is sketched in Fig. 7.37.

[11] An important variant called SASE free electron laser deviates from these steps. We shall discuss SASE as well later.

Figure 7.37: Upper panel shows the radiation wavepacket radiated by a single electron as it traverses an undulator. The wavepacket is like a nose attached to the electron when it exits the undulator. This same wavepacket applies to both the undulator and the FEL radiations in the two panels below. Middle panel is a schematic of the beam at the end of an undulator; the wavepackets of different electrons are incoherent, their phases are stochastic. Lower panel shows the coherence of radiation at the end of an FEL. With microbunching, the radiation becomes coherent.

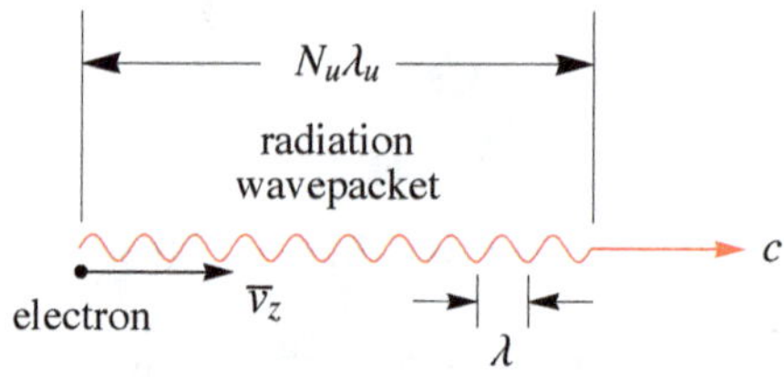

Undulator radiation

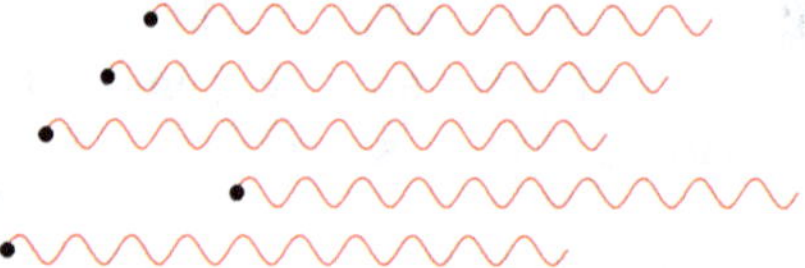

FEL radiation

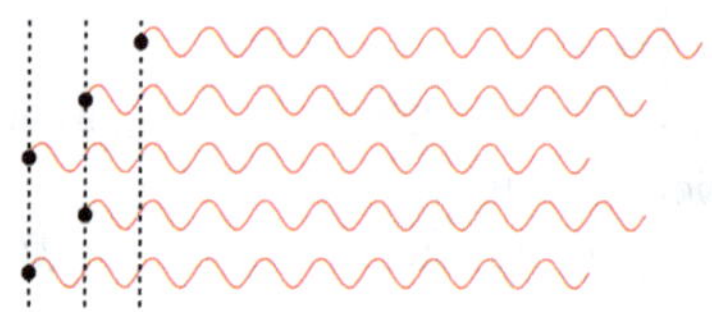

7.5.1 Maxwell-pendulum equations

To describe the laser-electron-undulator system in the high-gain regime, we need to establish a set of coupled Maxwell-pendulum equations. We first need to establish how the laser field is driven by the microbunched beam with a varying (exponentiating) bunching factor. This requires the Maxwell equations. We then need to establish how the increasing laser field continues to microbunch the beam, and that requires the pendulum equation. Together, our framework consists of a set of coupled Maxwell-pendulum equations.

In the high-gain physical system, there are three dynamical quantities,

$$\begin{aligned} \text{bunching factor } \mathcal{B} &= \langle e^{-i\psi} \rangle \,, \\ \text{energy bunching factor } \mathcal{D} &= \langle \delta e^{-i\psi} \rangle \,, \\ \text{laser field amplitude } \ \mathcal{E} \,, \end{aligned}$$

where a bracket $\langle \ \rangle$ means taking an ensemble average of the electron beam. All three quantities are slowly changing along the undulator distance. The phase factor $e^{-i\psi}$ singles out the microbunching component at the radiation wavelength λ.

It is quite amazing that the dynamics of the system of a high-gain FEL, as complex as it is, can be described by just three dynamical quantities. Hidden behind has been distilled insights accumulated by pioneers,[12] and we shall follow their footsteps in our analysis below.

A comment should be made on what we are planning to do with this analysis. The quantity $\psi = (k_u + k)\bar{v}_z t - \omega t$ represent the phase between a particle motion relative to the laser field. The quantities $\langle \cdots e^{-i\psi} \rangle$ represent a frequency filtering of the dynamical quantities with the resonant frequency. The dynamics we will study addresses the time evolution of these frequency-filtered quantities, i.e. only their filtered Fourier components in the narrow bandwidth of the resonant frequency are considered relevant and are being studied in the proposed dynamical system.

The procedure adopted offers a deep insight of the FEL high-gain dynamics. However, it is also a subtle step as it addresses the time evolution of a Fourier component, and as such necessarily convolutes a time-domain and a frequency-domain analysis. The analysis must be done with extreme care. Computer simulation results, as an example of a word of caution, based on these analysis must be interpreted accordingly. Taking a result of simulation based on such filtered dynamics too seriously to examine detailed particle distribution other than the resonant harmonics would be unfounded.

Later in Chapter 8, we will discuss another subject of collective instabilities. We will discuss a time-domain treatment using wake functions and a frequency-domain treatment using impedances. However, in that case, we can choose one treatment or another yielding the same results. The situation here is different. The time-domain and frequency-domain treatments are convoluted into one, and subtleties can be generated when doing so.

Physically $\mathcal{E}$ quantifies the laser field strength, while $\mathcal{B}$ and $\mathcal{D}$ characterizes the electron beam properties — as mentioned, all after filtering out the appropriate Fourier components. Let us elaborate on their dynamics below.

Bunching factor $\mathcal{B}$ The quantity $\mathcal{B}$ has been discussed in Sec. 7.4.1. It quantifies the degree of microbunching in the beam's density ψ-distribution. It is a dynamical quantity that evolves as the beam propagates down the high-gain undulator with its growing degree of microbunching.

Given the pendulum equation, it is easy to establish the dynamics of $\mathcal{B}$,

$$\begin{aligned} \frac{d\mathcal{B}}{ds} &= \left\langle -ie^{-i\psi}\frac{d\psi}{ds} \right\rangle \\ &= \langle -ie^{-i\psi} 2k_u\delta \rangle = -2ik_u\mathcal{D} \,. \end{aligned}$$

[12] R. Bonifacio, C. Pellegrini, L.M. Narducci, Opt. Commun. Vol. 50, p. 373 (1984).

Figure 7.38: Sketch of the beam distribution when $\mathcal{B} \neq 0$ and when $\mathcal{D} \neq 0$. The three panels are for the cases $\mathcal{B} = 0, \mathcal{D} = 0$ (upper), $\mathcal{B} \neq 0$ (middle), and $\mathcal{D} \neq 0$ (lower). The upper two panels show the distribution in real space. The third panel shows the distribution in phase space.

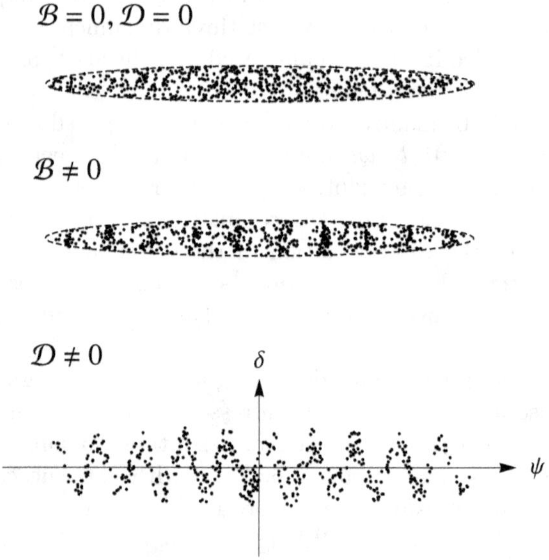

$\mathcal{B} = 0, \mathcal{D} = 0$

$\mathcal{B} \neq 0$

$\mathcal{D} \neq 0$ δ ψ

Energy bunching factor $\mathcal{D}$ It is easy to visualize that the bunching factor $\mathcal{B}$ specifies the degree of microbunching in the beam density's ψ-distribution. When $\mathcal{B} \neq 0$, it means the beam's longitudinal density distribution is microbunched. The quantity $\mathcal{D}$ specifies the degree of modulation in the beam energy's ψ-distribution. When $\mathcal{D} \neq 0$, it means the beam's energy deviation along the bunch length is modulated at the resonance wavelength. In the (ψ, δ) phase space, the beam distribution might have a sinusoidal structure. See illustration Fig. 7.38.

Again from the pendulum equation, we establish the dynamics of $\mathcal{D}$,

$$\begin{aligned}
\frac{d\mathcal{D}}{ds} &= \left\langle e^{-i\psi}\frac{d\delta}{ds} - ie^{-i\psi}\frac{d\psi}{ds}\delta \right\rangle \\
&= \langle e^{-i\psi}(-\epsilon\sin\psi) - ie^{-i\psi}2k_u\delta^2 \rangle \\
&\approx \langle e^{-i\psi}(-\epsilon\sin\psi) \rangle \qquad \text{(dropping the second order term } \propto \delta^2\text{)} \\
&\approx \frac{-\epsilon}{2i} \qquad \left(\text{with } \sin\psi = \frac{e^{i\psi} - e^{-i\psi}}{2i} \text{ and dropping the oscillatory term}\right) \\
&= i\frac{a_u e}{2\sqrt{2}\gamma_r^2 mc^2}[JJ]\,\mathcal{E}\,.
\end{aligned}$$

Laser field amplitude $\mathcal{E}$ We still need a third equation for $\frac{d\mathcal{E}}{ds}$. Physically this change in laser field comes from radiation of the microbunched electron beam. As the microbunched beam undulates, its radiation field adds to the laser. The additional radiation field, and therefore $\frac{d\mathcal{E}}{dt}$, is expected to be proportional to the bunching factor $\mathcal{B}$. Unlike the case of low-gain FEL, this laser gain comes from the undulating beam directly, and is not a result of the laser field copropagating with the electron beam.

One can work through a detailed Maxwell equations to derive an expression for $\frac{d\mathcal{E}}{dt}$. But we will instead use a physical argument starting with Eq. (7.52), the radiation power of a single electron into the fundamental mode, reproduced below,

$$\hat{P} \approx 2\pi^2 r_0 mc^3 |\mathcal{B}|^2 \frac{a_u^2}{1+a_u^2} [JJ]^2 N_u \left(\frac{N_{\text{tot}}}{L_z}\right)^2 \begin{cases} 1, & \text{below diffraction limit} \\ \frac{\lambda L_u}{4\pi^2\sigma_\perp^2}, & \text{above diffraction limit} \end{cases},$$

Let us consider the case above diffraction limit. The electron beam has a transverse cross-sectional area $\pi\sigma_\perp^2$. The radiation field E, together with its magnetic counterpart, has an energy density $\frac{\epsilon_0}{2}E^2$, where E is the field amplitude, not the oscillating instantaneous field. By connecting E to the peak radiation power $\hat{P}$, we have

$$\frac{\epsilon_0}{2} E^2 \pi\sigma_\perp^2 = \frac{\hat{P}}{c}.$$

We then take the square root on both sides to find $|E|$,

$$|E| = \frac{e}{2\sqrt{2}\epsilon_0\gamma} |\mathcal{B}| a_u [JJ] \left(\frac{N_{\text{tot}}}{\pi\sigma_\perp^2 L_z}\right) L_u,$$

or

$$\frac{d|E|}{ds} = \frac{e}{2\sqrt{2}\epsilon_0\gamma} |\mathcal{B}| a_u [JJ] \left(\frac{N_{\text{tot}}}{\pi\sigma_\perp^2 L_z}\right).$$

The phase of E is of course lost in our derivation because we derived it via the expression of radiation power. However, we know the radiation is generated by the microbunched beam. The phase of the incremental radiation electric field is locked to the phase of the bunching factor $\mathcal{B}$ of the radiating beam. To find this lost phase factor, we resort to the Maxwell equations.

Consider a beam with a longitudinal distribution

$$\rho \propto e^{-i(\omega t - kz)}.$$

The bunching factor $\mathcal{B}$ has the same phase as ρ. Our task is to find the phase of the electric field it generates. This beam first generates a current density $J_z \propto \rho \propto e^{-i(\omega t-kz)}$. By Maxwell equation $\nabla\times\vec{B} \propto \vec{J}$, we find $ik\vec{B} \propto e^{-i(\omega t-kz)}$. Again by Maxwell equation $\nabla\times\vec{E} + \frac{\partial\vec{B}}{\partial t} = \vec{0}$, we know E and B are in phase. We therefore conclude that E has a phase of $-i$ relative to ρ. As mentioned, the phase of the incremental of electric field is locked to that of $\mathcal{B}$; the locking relative phase is therefore $-i$.

Concluding the results, we then obtain

$$\frac{d\mathcal{E}}{ds} = -i\frac{e}{2\sqrt{2}\epsilon_0\gamma}\mathcal{B}\,a_u[JJ]\left(\frac{N_{\text{tot}}}{\pi\sigma_\perp^2 L_z}\right). \tag{7.60}$$

We have now derived an expression of $\frac{d\mathcal{E}}{ds}$ for the case above diffraction limit. In our later discussions, including the cubic equation (Sec. 7.5.2), we are going to address only the case above diffraction limit.

The case below diffraction limit is more subtle. Along our present approach, we would proceed by taking the square root of the exact general expression (7.52). The dependence of $\frac{d\mathcal{E}}{ds}$ on $r = \frac{4\pi\sigma_\perp^2}{\lambda L_u}$ is a bit complicated through the form factor $\mathcal{F}_\perp(r)$. In the literature, however, sometimes one replaces $\sigma_\perp^2$ in the expression, in a simplified way to take into account of the effect of diffraction limit, by

$$\sigma_\perp^2 \to \sigma_\perp^2 + \frac{\lambda L_u}{4\pi},$$

which is equivalent to assigning $\mathcal{F}_\perp(r) \to \frac{1}{\pi(1+r)}$. If adopting this replacement, then one replaces $\sigma_\perp^2$ by $2\sigma_\perp^2$ in the expression (7.60) to represent the case at the diffraction limit.

Homework 7.32 Figure 7.37 showed the radiation in the case of an undulator. The radiation consists of a collection of N sinusoidal wave packets, all of the same wavelength and equal wave energy but linearly superimposed with random phases.

(a) Calculate the net wave energy. Show that this net energy fluctuates, but its average value $\langle E\rangle$ is N times the wave energy of each individual wave.

(b) Calculate the rms spread $\langle(E-\langle E\rangle)^2\rangle$. Find how much the instantaneous net beam energy fluctuates around its average expectation value.

(c) Perform a numerical test to add up N sinusoidal wave packets and test your results obtained in (a) and (b).

Homework 7.33

(a) Estimate the values of $\mathcal{B}$ in the middle panel of Fig. 7.38. Refer to Homework 7.27.

(b) Estimate the value of $\frac{\mathcal{D}}{\sigma_\delta}$ in the bottom panel.

7.5.2 Cubic equation

We finally obtain the coupled linear Maxwell-pendulum equations for the dynamics of a high-gain FEL,

$$\frac{d\mathcal{B}}{ds} = -2ik_u\mathcal{D},$$

$$\begin{aligned}
\frac{d\mathcal{D}}{ds} &= i\frac{a_u e}{2\sqrt{2}\gamma^2 mc^2}[JJ]\,\mathcal{E}\,, \\
\frac{d\mathcal{E}}{ds} &= -i\,\frac{e}{2\sqrt{2}\epsilon_0\gamma}a_u[JJ]\left(\frac{N_{\rm tot}}{\pi\sigma_\perp^2 L_z}\right)\mathcal{B}\,.
\end{aligned} \tag{7.61}$$

The rate of change of $\mathcal{B}$ is given by $\mathcal{D}$. The rate of change of $\mathcal{D}$ is given by $\mathcal{E}$. The rate of change of $\mathcal{E}$ is given by $\mathcal{B}$.

Let any one of the three quantities be designated by X. Equation of motion for X is

$$\begin{aligned}
\frac{d^3X}{ds^3} &= (-2ik_u)\left(i\frac{a_u e}{2\sqrt{2}\gamma^2 mc^2}[JJ]\right)\left(-i\frac{e}{2\sqrt{2}\epsilon_0\gamma}a_u[JJ]\,\frac{N_{\rm tot}}{\pi\sigma_\perp^2 L_z}\right)X \\
&= -i\,\frac{r_0 a_u^2 k_u N_{\rm tot}}{\gamma^3 L_z\sigma_\perp^2}[JJ]^2\,X\,.
\end{aligned}$$

Define a quantity L_G as

$$\begin{aligned}
\frac{r_0 a_u^2 k_u N_{\rm tot}}{\gamma^3 L_z\sigma_\perp^2}[JJ]^2 &\equiv \frac{1}{(\sqrt{3}\,L_G)^3}\,, \\
\Longrightarrow \qquad L_G &= \left(\frac{\gamma^3 L_z\sigma_\perp^2}{3\sqrt{3}\,r_0 a_u^2 k_u N_{\rm tot}[JJ]^2}\right)^{1/3}\,,
\end{aligned} \tag{7.62}$$

then

$$\frac{d^3X}{ds^3} = \frac{-i}{(\sqrt{3}\,L_G)^3}X\,.$$

This is the cubic equation for a 1-D high-gain FEL.[13]

In the definition (7.62) of gain length, we intend to use the rms electron beam size for $\sigma_\perp$. Keep in mind we are above the diffraction limit when the electron size dominates. In the case near the diffraction limit when the laser also contributes equally, for example, it sometimes is done to replace $\sigma_\perp$ by $\sqrt{2}\sigma_\perp$ — see discussion following Eq. (7.60).

The cubic equation, a third order differential equation, has three independent solutions. All solutions in general can be expressed as a linear superposition of these three solutions. These three independent solutions are

$$\begin{aligned}
X_1 &= \exp\left(i\frac{s}{\sqrt{3}\,L_G}\right), && \text{oscillatory solution}\,, \\
X_2 &= \exp\left(-i\frac{s}{2\sqrt{3}\,L_G} - \frac{s}{2L_G}\right), && \text{damped solution}\,, \\
X_3 &= \exp\left(-i\frac{s}{2\sqrt{3}\,L_G} + \frac{s}{2L_G}\right), && \text{exponentially growing solution}\,.
\end{aligned} \tag{7.63}$$

[13] R. Bonifacio, C. Pellegrini, L.M. Narducci, Opt. Commun. Vol. 50, p. 373 (1984).

After a beam is injected into the system, starting with initial conditions $\mathcal{B}(0), \mathcal{D}(0), \mathcal{E}(0)$, the system evolves with time. The system is decomposed into three components, represented by a linear superposition of the three solutions. After traveling some distance, the components in X_1 and X_2 are dominated by the exponentially growing X_3 component. As a result, all three quantities have the behavior

$$\mathcal{B}, \mathcal{D}, \mathcal{E} \sim \exp\left(\frac{s}{2L_G}\right),$$

which means the laser power grows exponentially as

$$\text{Laser power} \sim |\mathcal{E}|^2 \sim \exp\left(\frac{s}{L_G}\right).$$

The quantity L_G is the *gain length* of the laser power.

Furthermore, in the exponential regime, the three quantities are interconnected. Their connections, including the relative amplitudes and phases, can be easily obtained from the Maxwell-pendulum equations. The result is

$$\mathcal{B} : \mathcal{D} : \mathcal{E} = \left(3e^{i\frac{\pi}{3}} \frac{ea_u k_u L_G^2}{\sqrt{2}\gamma^2 mc^2}[JJ]\right) : \left(\sqrt{3}e^{i\frac{2\pi}{3}} \frac{ea_u L_G}{2\sqrt{2}\gamma^2 mc^2}[JJ]\right) : 1. \tag{7.64}$$

Gain length We have defined a gain length (7.62). It is clear that a high-gain FEL aims for a short gain length. A high peak current of the electron beam is important even though the dependence is rather weak, $L_G \propto \left(\frac{N_{\rm tot}}{L_z}\right)^{-1/3}$. A tight transverse beam size is also important, with $L_G \propto \sigma_\perp^{2/3}$ — recall our formulae apply above diffraction limit with $\sigma_\perp^2 \gtrsim \frac{\lambda L_u}{4\pi}$. The LCLS free electron laser at SLAC, the first hard-X-ray FEL facility ever built, achieved during commissioning an emittance smaller than expected and thus benefitted from a substantially shorter L_G. Higher beam energy helps reaching hard X-rays, but it hurts the gain length, $L_G \propto \gamma\sigma_\perp^{2/3}$, even when adiabatic damping is considered.

For a high-gain X-ray SASE FEL (see Sec. 7.6), the total length of the undulator L_u must be many (typically ~20) times L_G to secure saturation (see later Sec. 7.5.5). Since the FEL laser power grows exponentially, most of the laser power from a long FEL in fact comes from the very last gain length. One way to view a high-gain SASE FEL device is to consider the beginning 19 gain lengths to be a way to fully microbunch the beam, and then let the microbunched beam radiate in the last gain length of the undulator — and that radiation turns out to be extremely bright.

Along this line of consideration, research efforts in several directions aim for producing a relativistic electron beam that is already microbunched before entering a radiation device, thus saving the first 19 gain lengths of undulator. One example of these efforts tries to microbunch the electron beam in a storage ring in a steady state, and is called steady-state microbunching.[14] If accomplished, as the electron beam circulates around the storage ring, it is always microbunched as it traverses the radiating undulator turn by turn. In this case, the undulator can be rather weak and short. The device does not aim to produce the high peak power of a high-gain FEL but produces high average power.

Homework 7.34 The gain length (7.62) derived in text is for the case above the diffraction limit. Find the expression of gain length when the beam is below the diffraction limit.

Homework 7.35 Confirm that $X_{1,2,3}$ in Eq. (7.63) are the solution to the cubic equation.

Homework 7.36 Derived Eq. (7.64).

7.5.3 FEL impedance

The connection between $\frac{d\mathcal{E}}{ds}$ and $\mathcal{B}$ in Eq. (7.61) contains information for the FEL impedance. The concept of impedance will be discussed in Chapter 8 in detail, but for completeness, let us briefly mention it in connection to the FEL dynamics.

To proceed, we first need to connect the laser field $\mathcal{E}$, which is mainly transverse, to the longitudinal voltage seen by the beam due to its modulated motion in the undulator. The analysis was given in Sec. 7.4.3. The voltage gain $\mathcal{V}$ along the undulator by applying Eq. (7.56), satisfies

$$\frac{d\mathcal{V}}{ds} = i\frac{a_u[JJ]}{2\sqrt{2}\,\gamma}\mathcal{E}\,.$$

In the exponential growing regime, we also have

$$\frac{d\mathcal{V}}{ds} = \frac{1}{\sqrt{3}\,L_G}\,e^{-i\pi/6}\,\mathcal{V}\,.$$

Relating $\mathcal{V}$ to $\mathcal{E}$, and then apply Eq. (7.61), we obtain the voltage gain per unit length along the undulator,

$$\frac{\mathcal{V}}{L} = \frac{\sqrt{3}\,a_u^2[JJ]^2}{8\pi\epsilon_0\gamma^2\sigma_\perp^2 c}\,L_G\,e^{i\pi/6}\left(\frac{eN_{\rm tot}c}{L_z}\,\mathcal{B}\right)\,.$$

The FEL impedance is defined as the ratio of the beam-radiation induced voltage $\mathcal{V}$ to the beam current $\frac{eN_{\rm tot}c}{L_z}\mathcal{B}$. We then obtain the value of the

[14]D.F. Ratner and A.W. Chao, Phys. Rev. Lett. 105, 154801 (2010); D. Ratner and A. Chao, Proc. FEL Conf., Shanghai (2011); A.W. Chao, et al., 7th Int. Part. Accel. Conf., Busan, Korea, 2016, p. 1048; C.X. Tang, et al., Proc. 60th ECFA Advanced Beam Dynamics Workshop on Future Light Sources, Shanghai (2018).

impedance per unit length at the resonant frequency $\omega_0 = \frac{2ck_u\gamma^2}{1+a_u^2}$,

$$\frac{Z(\omega_0)}{L} = \frac{\sqrt{3}\,a_u^2[JJ]^2}{8\pi\epsilon_0\gamma^2\sigma_\perp^2 c}\,L_G\,e^{i\pi/6}\,.$$

This impedance per unit length applies over the length of the undulator.

As saturation is reached, the radiation power is dominated by the last gain length of the undulator. The total impedance as seen by the electron beam is then given by $Z_{\rm tot} \approx \frac{Z(\omega_0)}{L}\,L_G$. The radiation power is

$$P_{sat} = {\rm Re}Z_{\rm tot}\,\hat{I}^2 = \frac{3a_u^2[JJ]^2 r_0 mc^3}{4\gamma^2\sigma_\perp^2}L_G^2\left(\frac{N_{\rm tot}}{L_z}\right)^2, \qquad (7.65)$$

where $\hat{I}$ is the electron beam current at saturation when $\mathcal{B}=1$, i.e. $\hat{I} = \frac{eN_{\rm tot}c}{L_z}$. It is straightforward to show that this expression is exactly the same as Eq. (7.67) to be derived later.

For the LCLS, $a_u = 2.5, \gamma = 2.8\times 10^4, \lambda_u = 3$ cm, $\sigma_\perp = 25\,\mu$m, $\hat{I} = 3.4$ kA, $L_G = 2.7$ m, we find an impressive $\frac{{\rm Re}Z(\omega_0)}{L} = 400\,\Omega$/m over the length of the LCLS. The peak radiation power, occurring in the last gain length when $\mathcal{B}\approx 1$, is $P_{\rm sat} = 1.3$ GW.

This impedance is peaked at ω_0 and has a bandwidth $\frac{\Delta\omega}{\omega_0} = \frac{1}{N_u}$. If we approximate it as an LRC circuit impedance (Sec. 8.2.4), then we have

$$\frac{R_S}{L} = \frac{3a_u^2[JJ]^2 L_G}{16\pi\epsilon_0\gamma^2\sigma_\perp^2 c}, \qquad \omega_R = \frac{2ck_u\gamma^2}{1+a_u^2}, \qquad Q = N_u\,, \qquad (7.66)$$

where $\frac{R_S}{L}$ is the shunt impedance per unit length along the undulator.

FEL sideband instability As mentioned, the three coupled FEL equations (7.61) refer to the dynamics of the beam parameters filtered at the resonant frequency of the undulator and the radiation. On the other hand, we know that the dynamics also involves a synchrotron oscillation of the microbunches inside the microbuckets. The beam dynamics therefore necessarily involves the synchrotron sidebands around the resonant frequency, whose effects have been filtered out by the cubic equation.

Considering the dynamics of these sidebands leads to potential FEL instabilities especially as the FEL approaches the point of saturation. Study of this beam dynamics goes beyond the scope of these lectures. Suffice it to mention that the instability dynamics would be an FEL counterpart to the Robinson instability in storage rings, a subject we will discuss in Sec. 8.7.2. The corresponding impedance can be modeled by (7.66) and operation is below transition.

7.5.4 Transfer matrix for the FEL cubic equation

Our previous analysis singled out the exponentially growing solution X_3 and claimed that the other two components projected onto $X_{1,2}$ could be ignored in comparison. This is a good approximation in the high-gain regime. In case the gain is small, or if an undulator is installed in a storage ring where the beam passes the undulator repetitively, the same FEL quantities $\mathcal{B}, \mathcal{D}, \mathcal{E}$ can still be applied to describe the dynamics of the beam-laser system. In such a situation, it becomes useful to solve the complete dynamics without ignoring the X_1 and X_2 terms.

The Maxwell-pendulum equations are linear. The dynamics can be described by a matrix map. In this section, we provide the 3×3 complex map T for the beam-laser system to go through the undulator,

$$\begin{bmatrix} \mathcal{B} \\ \mathcal{D} \\ \mathcal{E} \end{bmatrix}_{\text{exit}} = T \begin{bmatrix} \mathcal{B} \\ \mathcal{D} \\ \mathcal{E} \end{bmatrix}_{\text{entrance}} .$$

The algebra is somewhat involved but straightforward. We skip the derivation and give the result. The nine matrix elements of T are given by

$$\begin{aligned}
t_{11} &= \frac{1}{3} e^{-\frac{\lambda_+ t}{\sqrt{3}}} \left(1 + e^t + e^{i\lambda_- t}\right), \\
t_{12} &= \frac{\alpha L_G}{\sqrt{3}} e^{-\frac{\lambda_+ t}{\sqrt{3}}} \left(i\lambda_- - i\lambda_+ e^t - e^{i\lambda_- t}\right), \\
t_{13} &= \alpha\beta L_G^2 e^{-\frac{\lambda_+ t}{\sqrt{3}}} \left(-i\lambda_+ + i\lambda_- e^t - e^{i\lambda_- t}\right), \\
t_{21} &= \frac{1}{3\sqrt{3}\alpha^2\beta L_G^3} t_{13}, \\
t_{22} &= t_{11}, \\
t_{23} &= -\frac{\beta}{\alpha} t_{12}, \\
t_{31} &= \frac{1}{3\sqrt{3}\,\alpha^2\beta L_G^3} t_{12}, \\
t_{32} &= -\frac{1}{3\sqrt{3}\,\alpha\beta^2 L_G^3} t_{13}, \\
t_{33} &= t_{11},
\end{aligned}$$

where we have defined $\alpha = 2k_u, \beta = \frac{ea_u[JJ]}{2\sqrt{2}\,\gamma^2 mc^2}, \lambda_+ = e^{i\frac{\pi}{6}}, \lambda_- = e^{-i\frac{\pi}{6}}, t = \frac{s}{L_G}$. The end of the undulator has $s = L_u$.

Storage ring based SASE coherent radiation Given the initial values $\mathcal{B}(0), \mathcal{D}(0), \mathcal{E}(0)$, the matrix T gives $\mathcal{B}(s), \mathcal{D}(s), \mathcal{E}(s)$. Consider an electron beam stored in a storage ring interacting with its own radiation of previous revolutions as is the case of a FEL oscillator (see later) or in the steady-state microbunching scheme. The three dynamical quantities $\mathcal{B}, \mathcal{D}, \mathcal{E}$ interact at the undulator by the map T once per revolution. After the interaction, the laser and the electron beam are separated. The laser carries the quantity $\mathcal{E}$ and transforms it by the laser cavity and the mirrors. The electron beam carries $\mathcal{B}$ and $\mathcal{D}$ and transforms them by the storage ring longitudinal map. After these transformations, the laser and the electron beam meet at the undulator and the three quantities interact by the map T again. The process then repeats.

This beam-laser system can be set up to resonate and grow exponentially to provide a storage ring based, either external or self-generated, source of coherent radiation. If self-generated, the mechanism parallels that of a SASE FEL except that it occurs in a storage ring. The challenge for such a beam-laser system in a storage ring is to maintain the phase lock between the microbunched electron beam and the laser on a turn-by-turn basis.

Homework 7.37 Calculate the three eigenvalues of the 3×3 matrix T. You should just recover the three solutions of the cubic equation. The eigenvector of the matrix T as functions of s are just $X_{1,2,3}$. Also, the determinant of T should be 1.

Homework 7.38 When $t \to \infty$, the matrix elements of the T matrix are dominated by the X_3 terms. Find the asymptotic expression of T and understand its dynamics.

Solution

$$\begin{aligned} T &\approx T_0\, e^{\frac{t}{2} - i\frac{t}{2\sqrt{3}}}\,, \\ T_0 &= \begin{bmatrix} \frac{1}{3} & -\frac{i\lambda_+ \alpha L_G}{\sqrt{3}} & i\lambda_- \alpha\beta L_G^2 \\ \frac{i\lambda_-}{3\sqrt{3}\alpha L_G} & \frac{1}{3} & \frac{i\lambda_+ \beta L_G}{\sqrt{3}} \\ \frac{-i\lambda_+}{9\alpha\beta L_G^2} & -\frac{i\lambda_-}{3\sqrt{3}\beta L_G} & \frac{1}{3} \end{bmatrix}. \end{aligned}$$

All three dynamical quantities $\mathcal{B}, \mathcal{D}, \mathcal{E}$ grow with $e^{\frac{t}{2} - i\frac{t}{2\sqrt{3}}}$ in time with fixed ratios given by a time-independent matrix T_0. Determinant of T_0 is zero, eigenvalues are 0, 0, 1. The eigenvector with eigenvalue 1 is

$$\begin{bmatrix} 3\lambda_+^2 \alpha\beta L_G^2 \\ \sqrt{3} i\lambda_+ \beta L_G \\ 1 \end{bmatrix},$$

which is just the result that appeared in Eq. (7.64).

7.5.5 Saturation

The Maxwell-pendulum system of equations are linear in $\mathcal{B}, \mathcal{D}, \mathcal{E}$. The cubic equation predicts that they grow exponentially in time in lock step. Such an exponential growth has to cease and the laser power saturates at some point. That cease point occurs when the bunching factor reaches 1,

$$|\mathcal{B}|_{\text{sat}} = 1\,,$$

At this point, the beam is 100% microbunched and cannot go beyond. The system becomes nonlinear when described by the dynamical quantities $\mathcal{B}, \mathcal{D}, \mathcal{E}$. The cubic equation breaks down. The beam begins to exchange energy back and forth with the laser. The laser power saturates and transients.

Field amplitude An estimate of the laser saturation power can be obtained as follows. From Eq. (7.64), we see that when $|\mathcal{B}| = 1$, we have

$$|\mathcal{E}|_{\rm sat} = \frac{\sqrt{2}\,\gamma^2 mc^2}{3ea_u k_u L_G^2 [JJ]}\,.$$

Power The saturation laser power is then

$$\begin{aligned} P_{\rm sat} &= \frac{c\epsilon_0}{2}|\mathcal{E}|^2 \pi\sigma_\perp^2 = \frac{c\epsilon_0}{2}\pi\sigma_\perp^2 \left(\frac{\sqrt{2}\,\gamma^2 mc^2}{3ea_u k_u L_G^2 [JJ]}\right)^2 \\ &= \frac{1}{4\sqrt{3}}\frac{\gamma mc^3}{k_u L_G}\frac{N_{\rm tot}}{L_z}\,. \end{aligned} \tag{7.67}$$

We have again assumed the case above diffraction limit, $\sigma_\perp^2 \gtrsim \frac{\lambda L_u}{4\pi}$. As mentioned, this result reproduces the result (7.65) derived from the FEL impedance.

If we define the beam power

$$P_{\rm beam} = \gamma mc^3 \frac{N_{\rm tot}}{L_z}\,,$$

then

$$P_{\rm sat} = \frac{1}{4\sqrt{3}}\frac{1}{k_u L_G} P_{\rm beam}\,.$$

Obviously we require $k_u L_G \gg 1$, i.e. the gain length is much longer than the undulator period length — otherwise some of the approximations we made would have broken down. In addition, we have assumed $L_G \ll L_u$, otherwise the X_3 component does not dominate.

One can express the result in terms of the Pierce parameter commonly used in the literature,

$$\rho = \frac{1}{2\sqrt{3}}\frac{1}{k_u L_G} \ll 1\,.$$

We then have, above the diffraction limit $\sigma_\perp^2 \gtrsim \frac{\lambda L_u}{4\pi}$,

$$P_{\rm sat} = \frac{\rho}{2} P_{\rm beam}\,.$$

Pierce parameter ρ has the physical meaning of energy transfer efficiency from the electron beam to the FEL laser. Although ρ tends to be small, it is desirable to have it as large as possible to optimize the radiation power.

Energy spread The energy spread of the electron beam at the point when saturation is reached can be found from Eq. (7.64),

$$|\mathcal{D}|_{\rm sat} \;=\; \frac{1}{2\sqrt{3}\,k_u L_G} \;=\; \rho\,.$$

Pierce parameter also has the physical meaning of being the final relative energy spread of the electron beam as it exits the high-gain FEL. Its value tends to be small, but should be kept small enough if the spent electron beam is to be reused.

Bucket height The microbunching bucket half-height at the saturation point is given by, using Eq. (7.59),

$$\hat{\delta}_{\frac{1}{2},\,\rm sat} \;=\; \sqrt{\frac{2}{3}}\,\frac{1}{k_u L_G} \;=\; 2\sqrt{2}\,\rho\,.$$

This bucket half-height $\hat{\delta}_{\frac{1}{2},\,\rm sat}$ is a factor $2\sqrt{2}$ times the beam energy spread $|\mathcal{D}|_{\rm sat}$ at the saturation point.

Synchrotron frequency The synchrotron frequency at the saturation point is, using Eq. (7.57),

$$\Omega_{\rm sat} \;=\; \sqrt{\frac{2}{3}}\,\frac{c}{L_G}\,.$$

The synchrotron oscillation frequency grows exponentially along the length of the undulator with $\sqrt{|\mathcal{E}|}$. That means it is approximately given by, at a distance $s > 0$ before reaching the saturation point,

$$\Omega(s) \;\sim\; \Omega_{\rm sat}\, e^{-\frac{s}{4L_G}}\,.$$

The total integrated synchrotron oscillation phase accumulated over the entire undulator is

$$\Phi \;=\; \int_0^\infty \frac{ds}{c}\,\Omega(s) \;=\; 4\sqrt{\frac{2}{3}} \;\approx\; \pi\,.$$

The entire passage does not complete even a single synchrotron oscillation. It constitutes only approximately half a synchrotron period. Out of this total phase advance, about a fraction $\int_0^{L_G} / \int_0^\infty = 1 - e^{-1/4} \approx 22\%$ occurs within the last gain length. In contrast, the laser power, scaling as $|\mathcal{E}|^2$, grows $1 - e^{-1} \approx 63\%$ in the last gain length.

The fact that $\Phi \approx \pi$ deserves a comment here. First note that the same condition results from optimizing the low-gain FEL, where the undulator length is chosen such that the incident beam executes approximately half a synchrotron oscillation so that a maximum amount of beam energy is extracted — see Sec. 7.4.5. What happens here for the high-gain FEL is somewhat similar. As the beam radiates so substantially, its energy decreases. The pendulum bucket lowers as the beam propagates down the undulator. As the bucket lowers, the beam tries to follow it by its instantaneous synchrotron oscillation. The optimum occurs when the beam arrives to the bottom of the final bucket, and that corresponds to the case when the integrated phase $\Phi \approx \pi$.

Synchrotron oscillation period The corresponding synchrotron oscillation period length D shortens exponentially. At the saturation point, it has shortened to the value

$$D_{\rm sat} \approx 2\pi\sqrt{\frac{3}{2}}\,L_g\,.$$

The fact that $\frac{D_{\rm sat}}{2\pi} \approx L_G$ towards the end of the high-gain FEL is the optimum choice was anticipated in Sec. 7.4.6.

It is interesting to note that when saturation is reached at $|\mathcal{B}|_{\rm sat} = 1$, the two other optimal conditions occur at that same moment,

$$\Phi \approx \pi, \qquad \frac{D_{\rm sat}}{2\pi} \approx L_G\,.$$

After reaching the saturation point, the bucket more or less stays constant for a while. The beam continues to execute synchrotron oscillation in a transient motion, but then at the constant saturated frequency $\Omega_{\rm sat}$. The beam breaks up after the transient. It is possible to extract more of the beam energy into the laser if the undulator parameters are tapered to guide the residual synchrotron motion. Tapered FELs are a research subject considered beyond our scope.

Homework 7.39 The LCLS-I FEL has the following parameters: $a_u = 2.5, \lambda_u = 3$ cm, $B_y = 1.25$ T, $N_u = 4000, \gamma = 28000$, peak current $= \hat{I} = 3.4$ kA, $\sigma_\perp = 25$ μm (including the contribution from diffraction), $\epsilon_N = 0.5$ mm-mrad, $\lambda_0 = 1.4$ Angstrom. Find the gain length L_G and the Pierce parameter ρ. Find also the electron beam's energy spread $|\mathcal{D}|_{\rm sat}$, the synchrotron oscillation period length $D_{\rm sat}$, the bucket half-height $\hat{\delta}_{\frac{1}{2},\,{\rm sat}}$, and the radiation power at saturation $P_{\rm sat}$.

Solution $\frac{N_{\rm tot}}{L_z} = \frac{\hat{I}}{ec} = 7.1 \times 10^{13}$/m, $L_G = 2.7$ m, $\rho = 5.2 \times 10^{-4}$, $P_{\rm sat} = 12.6$ GW, $|\mathcal{D}|_{\rm sat} = 5.2 \times 10^{-4}$, $D_{\rm sat} = 2\pi\sqrt{\frac{3}{2}}\,L_G = 20.4$ m, $\hat{\delta}_{\frac{1}{2},\,{\rm sat}} = 0.0015$.

Homework 7.40 Verify that Eqs. (7.65) and (7.67) give the same result for the radiation power at saturation.

Homework 7.41 Consider the envisioned 20-TeV proton beam in the ill-fated Superconducting Super Collider. The SSC circumference is $C = 87$ km, bending radius is $\rho = 10.1$ km, bending magnetic field is 6.6 T. Other parameters include $\alpha_c = 2.2 \times 10^{-4}, \nu_x \approx 80, \nu_s = 0.0019$. An SSC beam bunch contains $N =$

7.5×10^9 protons. The betatron and synchrotron radiation damping times are 26 and 13 hours respectively.

(a) Find the equilibrium rms relative energy spread and rms bunch length of the stored proton beam.

(b) If the beam is stored a few days after injection, it potentially will reach the fully damped transverse emittances of $\epsilon_x = \epsilon_y = \frac{1}{2}\epsilon_{x0}$ when operated at full coupling. Find this equilibrium emittance. If the aim is to produce a radiation at $\lambda_0 = 1$ Angström, is this beam diffraction limited?

(c) Undulator is an issue. Find a set of undulator parameters λ_u, B_u and a_u for the beam to radiate at λ_0.

(d) If this radiation is based on SASE mechanism, find the gain length, an appropriate length of the undulator L_u, and the radiation energy per passage of the undulator you designed. This is potentially a SASE free proton laser. However, its radiation power is low and it fires once every few days unless a steady-state mode is implemented.

(e) Compared with an FEL, a free proton laser has the potential advantage of approaching a coherent γ-ray laser, but its radiation power is weak. Contemplate a possibility of a free muon laser as an intermediate case between an FEL and an FPL.

Solution (a, b) $U_0 = 0.123$ MeV, $\sigma_\delta = 2.2 \times 10^{-6}$, $\sigma_z = 3.5$ mm, $\epsilon_x = \epsilon_y = 0.9 \times 10^{-13}$ mm-mrad.

7.6 Seeding

The exponential growth of a high-gain FEL, however, requires an initial condition to set it off. Had $\mathcal{B}(0), \mathcal{D}(0), \mathcal{E}(0)$ been all zero, there will be no FEL growth. There are three ways to initiate a high-gain FEL,

- $\mathcal{B}(0) \neq 0 \Longrightarrow$ the injected beam is initially microbunched in its longitudinal distribution, even if only slightly.
- $\mathcal{D}(0) \neq 0 \Longrightarrow$ the injected beam is initially energy modulated, even if only slightly.
- $\mathcal{E}(0) \neq 0 \Longrightarrow$ there is an injected laser at the resonant frequency, even if weakly.

The process of preparing the initial conditions for a high-gain FEL is called *seeding*. Many ingenious ideas have been brought forward, leading to various seeding and beam manipulation concepts. This is an exciting area of research.

In parallel, there are basically three schemes of FELs according to their seeding scenarios, the oscillator type, the amplifier type, and the SASE type, as sketched in Fig. 7.39. But more specifically, there can be several variations as follows.

Figure 7.39: Sketch of three FEL schemes, the FEL oscillator, the seeded FEL amplifier, and SASE FEL. The electron beam wiggles in the undulator in the x-plane but is shown in the y-plane for illustration. The SASE undulator is much longer than those of the other two schemes. The oscillator FEL requires mirrors to store the laser. The amplifier and SASE FELs are single-pass configurations.

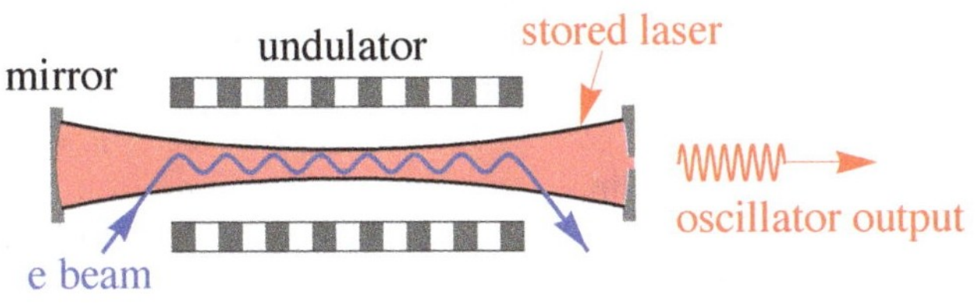

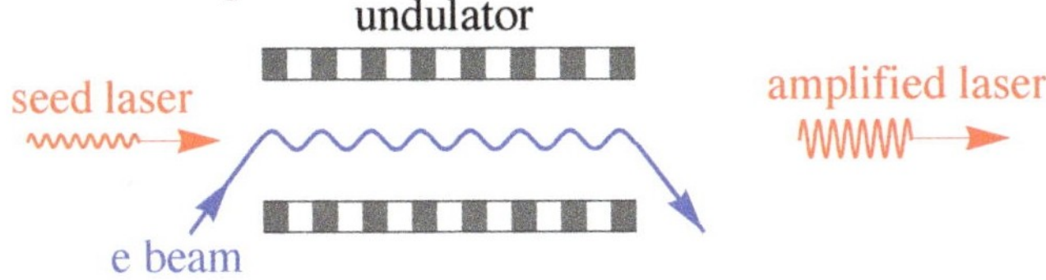

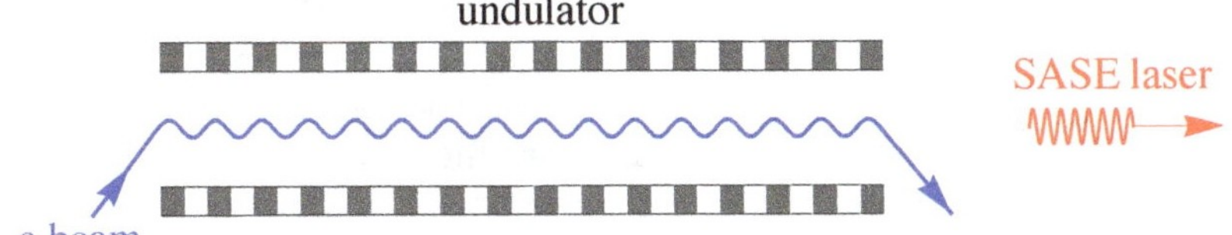

1. When $\mathcal{E}(0) \neq 0$, a laser is injected copropagating with a nonmicrobunched electron beam. This is the case of a low-gain FEL. But if the undulator is long enough, it becomes high-gain. The injected laser gets amplified by the FEL. The scheme acts as an FEL amplifier, shown as the middle panel in Fig. 7.39.

   Drawback of this scheme is that one needs to have a seed laser at the desired wavelength to begin with. For X-ray, this is not easily available.

2. An alternative to the amplifier is when the injected electron beam is microbunched already with the resonant wavelength. This is a case of $\mathcal{B}(0) \neq 0$, and there is no need to inject a copropagating laser. The laser will grow by the FEL mechanism.

   Drawback of this scheme is it is not easy initially to microbunch an electron beam at the X-ray wavelength.

3. One can also inject a beam that is micro-modulated in its energy distribution. This is a case with $\mathcal{D}(0) \neq 0$. To produce the needed energy modulation, however, one needs an X-ray laser to start with. This case is in fact the same as case 1 above because the purpose of the copropagating laser is just to produce a $\mathcal{D}(0) \neq 0$ in any case.

4. A low-gain FEL has the drawback that the laser with the same frequency not only needs to preexist but also needs to be powerful, and the amplification is weak. However, by adding a pair of high-quality mirrors, the laser can be stored in the laser cavity. By repetitive passages, either in a storage ring or in a repetitive linac, the low gains accumulate into a significant growth of the laser power. This is the oscillator type illustrated in the upper panel of Fig. 7.39. A small seed laser can get the process started, or it can be self-triggered by the spontaneous radiation of the electron beam.

   The electron beam is not microbunched at its entrance to the undulator, but its pulse needs to synchronize with respect to the laser pulse. A certain fraction of the laser is extracted from the laser cavity for the user through a small opening in one of the mirrors.

5. An important special case with $\mathcal{B}(0) \neq 0$ is when a nonmicrobunched beam is injected without a laser. A nonzero value of $\mathcal{B}(0)$ comes from a statistical noise in the beam distribution. The fact that a stochastic beam possesses a nonzero bunching factor was analyzed in Sec. 7.4.1. This amazing possibility, easiest to achieve because it has no prerequisites on injected laser or premicrobunched beam, yielding a robust facility of brilliant X-ray laser source, is what is employed at all the presently conceived hard X-ray FELs. This scheme is called Self-Amplified Spontaneous Emission (SASE),[15] illustrated as the lower panel in Fig. 7.39.

   Starting with shot noise, the SASE X-ray laser is necessarily noisy. The "long noses" of radiation, each of length $N_u\lambda$, generated by neighboring electrons do sometimes add coherently, but it is not possible to predict which parts of the electron beam will generate these coherent long noses. As a result, SASE X-ray laser has a poor temporal coherence. A laser pulse at the end of a SASE FEL might look like Fig. 7.40 in one particular shot, and the pattern changes in the next shot.

6. Yet another concept is to seed the electron beam at a low-frequency f_0, then manipulate the electron beam in such a way that the beam becomes microbunched in such a way that its bunching factor becomes large at a higher frequency kf_0 with $k \gg 1$. This scheme is called "harmonic

[15] A.M. Kondratenko and S.L. Saldin, Part. Accel. 10, 207 (1980); R. Bonifacio, C. Pellegrini, and L.M. Narducci, Opt. Commun. 50, 373 (19984); P. Emma et al., Nat. Photon. 4, 641 (2010); T. Ishikawa et al., Nat. Photon. 6, 540 (2012).

Figure 7.40: A typical laser pulse output from a SASE FEL in a computer simulation. The pulse consists of a random collection of bursts, each approximately of the coherence length $\sim N_u\lambda$. Within each burst is internally coherent but is uncorrelated to the next burst. The peak power inside each burst can reach the saturation level if the undulator is sufficiently long. Typically, the FEL is designed to be about $\sim$20 gain lengths, so that most bursts, once formed, have a good chance to grow from shot noise to saturation.

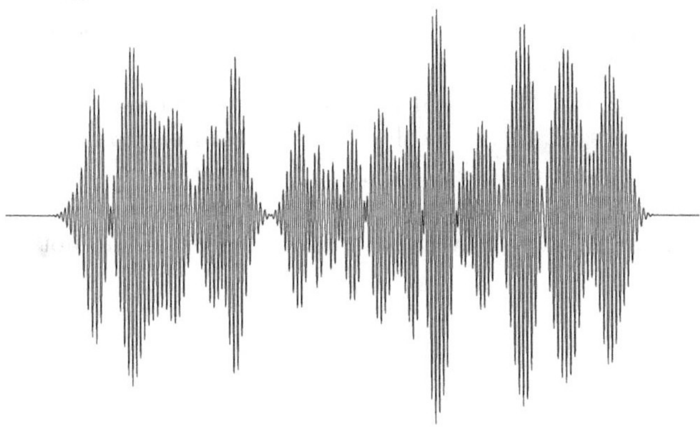

generation". Once made available, one can generate a high-gain X-ray FEL using a weak low-frequency laser seed and a short undulator.

Presently conceived schemes along this line, but beyond our scope, include (not intended to be exhaustive)

- high gain high harmonic (HGHG)[16]
- echo-enhanced high harmonic (EEHG)[17]
- self-seeding[18]
- enhanced SASE
- transverse-gradient undulator
- phase-merging enhanced harmonic generation PEHG
- steady-state microbunching (SSMB)
- chirp-compress harmonic generation
- sawtooth seeding

[16] L.-H. Yu, Phys. Rev. A44, 5178 (1991); L.-H. Yu et al., Science 289, 932 (2000); L.-H. Yu and J.H. Wu, Nucl. Instr. Meth. Phys. Res., A483, 493 (2002).

[17] G. Stupakov, Phys. Rev. Lett. 102, No. 7, 074801 (2009); D. Xiang and G. Stupakov, Phys. Rev. ST Accel. Beams, 12, No. 3, 030702 (1009).

[18] J. Feldhaus et al., Opt. Commun. 140, 341 (1997); G. Geloni, V. Kocharyan and E. Saldin, J. Mod. Opt. 58, 1391 (2011); J. Amann et al., Nat. Photon. 6, 693 (2012).

The subject of free electron laser, together with its variants and seeding schemes constitutes an important subject. Aside from the basic principle discussed in this chapter, the reader is encouraged to study more with further readings.[19]

Homework 7.42 When speaking of SASE, we have considered the case when the initial shot noise comes from the bunching factor $\mathcal{B}(0)$ in the stochastic longitudinal distribution of the electrons as it enters the undulator. Another possible SASE source is when the electron beam enters the undulator with stochastic energy distribution.

(a) Develop the analysis of an FEL that starts with a nonzero $\mathcal{D}(0)$.

(b) Apply the result of (a) to the case when an electron beam is injected into the undulator with a stochastic energy distribution.

(c) Which contribution is more important, the noise in $\mathcal{B}(0)$ or the noise in $\mathcal{D}(0)$? Why do we consider only $\mathcal{B}(0)$ in a SASE process and not $\mathcal{D}(0)$?

[19] See for example, Kwang-Je Kim, Zhirong Huang, and Ryan Lindberg, "Synchrotron Radiation and Free-Electron Lasers", Cambridge Univ. Press (2017).

Chapter 8

Collective Instability

Accelerators are devices that control and manipulate the motion of charged particles. To design an accelerator, one first starts by considering the motion of a single particle. To describe the dynamics of a beam of particles, one then regards the beam as a collection of noninteracting single particles moving in the environment prescribed by the accelerator design. The environment is defined by the electric and magnetic fields of the various accelerator components specified in the design. Different particles differ in their motions exclusively due to their different initial conditions. Given these fields, and depending on the initial conditions of the particles, intricate effects of linear and nonlinear dynamics are studied in detail.

Many accelerator applications, however, require beams of medium or high intensities. As the beam intensity is increased, the electromagnetic fields self-generated by the beam, including the fields carried by the direct Coulomb and Ampere fields and the fields generated by the beam interacting with its immediate vacuum chamber surroundings, will perturb the external prescribed fields. When the perturbation becomes sufficiently strong, the beam motion becomes unstable. To describe this aspect of the beam dynamics, the single-particle picture breaks down and a multiparticle picture is introduced. This multiparticle picture takes into consideration the important self-generated fields, but for the sake of simplicity — or lack of sophistication — usually omits the detailed nonlinear aspects included in the single-particle picture.

To be more specific, consider an intense particle beam moving in a metallic vacuum chamber of an accelerator. The beam interacts electromagnetically with its surroundings to generate an electromagnetic field, known as the *wakefield*. This field then acts back on the beam, perturbing its motion. Under unfavorable conditions, the perturbation on the beam generates new fields constructively to the existing wakefield and therefore further enhances it. The process exponentiates and the beam-surroundings interaction leads to an instability, known as a *collective instability*, and often a subsequent beam loss. The beam and its surroundings form a self-consistent dynamical system, and it is this system that we will study in this chapter. Thus,

$$\begin{aligned} \text{dynamical system} &= \text{beam} + \text{surroundings}\,, \\ \text{mediator of interaction} &= \text{wakefield}\,. \end{aligned}$$

The subject of collective instabilities in high energy accelerators has been studied since the late 1950s and early 1960s. The importance of the subject lies in the fact that it is one of the main factors that determine the ultimate performance of the accelerator. The advancement of this subject over the years can be evidenced by the discovery and curing of a large number of collective instability mechanisms. Each accelerator, when pushed for performance, will encounter some intensity limit. After this limit is analyzed, understood and possibly cured, a new limit emerges. The process repeats, and the end result is the improved understanding and higher performance of the accelerator. The confidence gained in turn provides a basis for ever more daring proposals for new advanced accelerators.

Over the years, accelerator physicists have observed, explained, and (mostly) cured several intricate instability mechanisms. An incomplete list follows below.

- negative mass instability 1959
- resistive wall instability 1960
- Robinson instability 1964
- beam breakup instability 1966
- head-tail instability 1969
- microwave instability 1969
- Landau damping 1969
- beam-beam limit in colliders 1971
- potential well distortion 1971
- Sacherer formalism 1972
- anomalous bunch lengthening 1974
- transverse mode coupling instability 1980
- hose instability 1987
- coherent synchrotron radiation instability 1990
- sawtooth instability 1993
- electron beam-ion instability 1996
- electron cloud instability 1997
- microbunching instability 2005
- interplay of multiple instability mechanisms 2013

As accelerator technology and demand progresses, instability mechanisms have been uncovered at a rate of approximately one every 3 years. Today, the subject has grown into a large collection of activities. Each activity constitutes an important research area; each needs to be understood or implemented on the accelerators of the future. This is an active area of accelerator physics research.

In contrast, our scope of these lectures is much more limited. To proceed with our discussions, two issues first need to be addressed,

1. How wakefields are generated: We will mention a fundamental theorem called Panofsky-Wenzel theorem, which provides the backbone of the concept of wake functions and impedances and the framework of the theory of collective instabilities.

2. How the wakefields act back to the beam: Here we introduce a few mechanisms leading to the various instabilities observed in accelerators under various conditions. We will limit ourselves to simplified models to describe these instability mechanisms.

We will focus our attention only on these two steps in this chapter. Step 1 mainly involves electromagnetics. The key is Maxwell equations. Step 2 involves beam dynamics. The key is Vlasov equation.

8.1 Wake function

8.1.1 The Gauss law

Collective instabilities in accelerators mostly come from an intense charged-particle beam electromagnetically interacting with its vacuum chamber environment. As the beam interacts with its environment, it generates a wakefield, and the wakefield acts back on the beam, disturbing its motion, and if the disturbance is strong enough, the beam becomes unstable.

To discuss the wakefields, we must start with its ultimate origin, the *Gauss law*, which states that each charged particle always is attached with it a definite amount of electric field lines. We can distort these field lines but we can never cut them loose from the charge under any circumstances. Furthermore, the amount of field lines attached to each charge cannot be changed, neither increased nor decreased.[1] Since the field lines do not disappear, they can only bounce around in the vacuum chamber, and in doing so, wakefields are generated.

Gauss law is an amazing law. Mathematically, it reads

$$\nabla \cdot \vec{E} = \frac{\rho}{\epsilon}.$$

Physically it reads: Electric field lines are absolutely attached to the charges. Gauss law is extremely robust. In contrast, for example, Newton's law is not so robust; it breaks down when relativity sets in.

[1] Gauss law is extremely geometric in its nature.

高斯定律: 丟不開 甩不掉 揮之不去的電場線

Gauss law: The annoyingly forever attached electric field lines.

We also have an integral form of the Gauss law: Integration of the normal component of electric field over the surface S of any 3-D volume V is equal to $\frac{4\pi Q}{\epsilon}$,

$$\oint_S \vec{E} \cdot d\vec{S} = \frac{Q}{\epsilon},$$

where Q is the total charge inside the volume enclosed by the surface S. It is amazing that this law holds no matter how the charges are moving – nonrelativistic, relativistic, or under acceleration, or whether they are embedded in any type of material. It also does not matter how close the charges might be immediately next to the surface S. The field integral will make a sudden change when a charge crosses the surface even infinitesimally.

8.1.2 A moving charge

If the charge is stationary and if it is in a free space, its Coulomb field lines radiate radially outwards isotropically, as in Fig. 8.1(a). For a moving charge, we see Lorentz contraction as shown in Fig. 8.1(b). When v approaches c, the field lines get contracted into a thin "pancake". The pancake is attached to the particle and moves with it with velocity v. When we take the ultrarelativistic limit $v = c$, then the pancake reduces to an infinitely thin sheet, as shown in Fig. 8.1(c). In all cases, integration of the field lines yields the same value of $\frac{q}{\epsilon_0}$, obeying the Gauss law. Since in accelerators, the particles typically move with speed very close to c, we will now think of the picture in (c).

In addition to the electric field as shown in Fig. 8.1, when the particle moves, it also generates a magnetic field. This magnetic field also contracts to a thin pancake when the particle's velocity approaches c, and into an infinitely thin sheet when $v = c$. But although their distributions are the same, their directions are different; direction of the electric field is radial outward; direction of the magnetic field is azimuthal (right-hand rule),

$$\begin{aligned} E_r &= \frac{q}{2\pi\epsilon_0 r}\,\delta(z-ct)\,, \\ B_\theta &= \frac{q}{2\pi\epsilon_0 c r}\,\delta(z-ct)\,. \end{aligned} \tag{8.1}$$

The electric field E_r and the magnetic field cB_θ have the same magnitude.

However, the magnetic field differs from the electric field on another important point. When $v = 0$ as in Fig. 8.1(a), there is electric field, but no magnetic field. When v increases, the magnitude of the magnetic field increases, but still weaker than the electric field. Only when $v = c$, the magnetic field increases to become the same magnitude as the electric field. The fact that the magnitudes of the electric and magnetic fields are equal when $v = c$ has important consequences, see later.

Figure 8.1: Field lines attached to a point charge q. (a) When the charge is stationary. (b) When charge is moving with uniform velocity v. (c) What happens in the ultrarelativistic limit $v = c$.

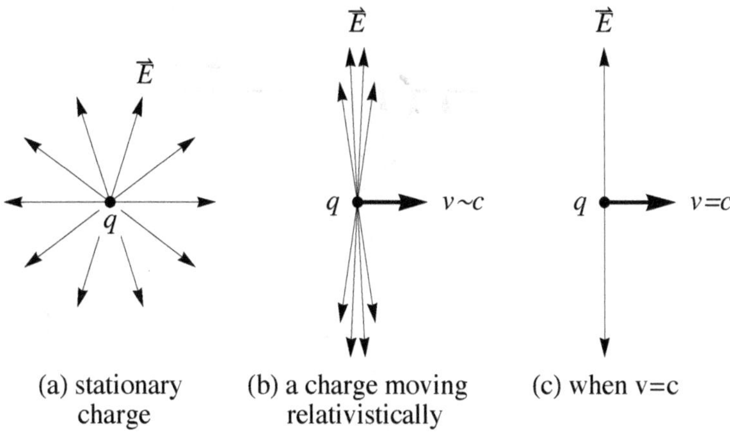

Homework 8.1 It is easy to derive the fields (8.1) given the knowledge that in the ultrarelativistic limit, the fields contract to an infinitely thin pancake. Show that the electric field follows immediately from Gauss law and the magnetic field follows from Ampere's law.

8.1.3 The vacuum chamber

So far we have discussed a particle in free space. We now need to add the vacuum chamber. Consider a very smooth vacuum chamber beam pipe.[2] For now, let us also consider the smooth pipe wall to be perfectly conducting.

The ultrarelativistic beam going down the axis of the pipe, together with its electromagnetic field and the vacuum chamber looks like that sketched in Fig. 8.2.

Figure 8.2 depicts a complicated arrangement. Note first that the electromagnetic fields are perfectly and cleanly terminated on the pipe wall. No fields penetrate into the wall because it is perfect conductor. The image charge on the wall is exactly equal and opposite to that of the beam, and it moves also with $v = c$ in the same forward direction. As the beam moves forward, the entire field pattern moves with it. In particular, there are no electromagnetic fields left behind this pattern. There is no "wakefield" generated in this configuration.

We just mentioned the image charge moves with $v = c$. But note that this is phase velocity. Basically, the image charges do not carry momentum as real

[2]How smooth does a vacuum chamber have to be? It depends on the circumstance. Typically, a 1-mm discontinuity on the pipe wall is going to be significant. In some circumstances, the small 1-μm roughness on the wall surface can have an appreciable effect.

Figure 8.2: Sketch of an ultrarelativistic beam moving down the axis of a smooth perfectly conducting beam pipe. The beam produces an image charge and current on the pipe wall. A test charge e copropagates with the beam.

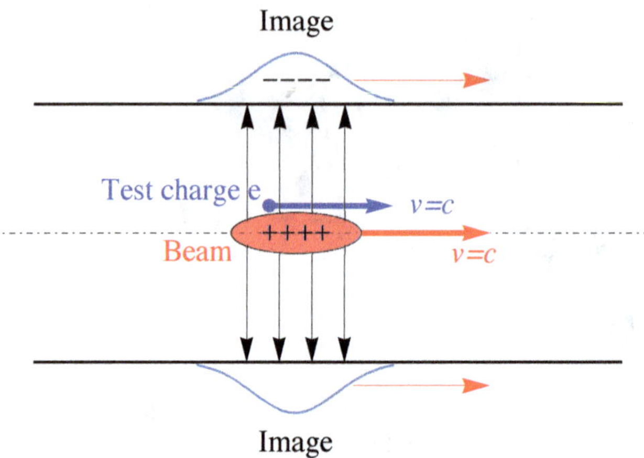

charges do, therefore they do not carry a reservoir of relativistic energy and are easily perturbed. The notion of prescribed motion does not apply to the image charges — they do not synchrotron radiate, for example. This is an important point, more discussion of this later.

Is this beam stable? To study the beam instability problem, we need to examine what effect does the electromagnetic field carried by the beam have on the motion of the particles in the beam. So, let us now consider a particular particle in the beam, the blue charge e in Fig. 8.2, which of course moves with $v = c$ as a member of the beam. We call this particle the "test particle". This test particle will see an electric force $e\vec{E}$ due to the electric field of the beam. This force is easily seen to push e towards the vacuum chamber wall because the test charge e has the same sign as the charges of the beam.

But there is also a magnetic force. The magnetic field is in the azimuthal direction (right hand rule). The magnetic force is $e\vec{v} \times \vec{B}$. It is easily seen that this magnetic force is pointing towards the pipe axis.

We mentioned that when $v = c$, the magnitude of E and magnitude of cB are equal. In the ultrarelativistic limit, therefore, the electric and the magnetic forces exactly cancel. The particles in the ultrarelativistic beam do see electric force and magnetic force, but they do not see a net force because they exactly cancel each other. The collective electromagnetic fields carried by the beam do not influence particle motion. A moment's reflection then says that there cannot be any collective instabilities!

This cancellation effect between the electric and magnetic forces due to the beam's self fields is very fortunate and very important. Without this fortunate cancellation, no modern accelerators would have worked.

Let us make a conclusion by stating the following theorem: There are no collective instabilities if the following conditions are all fulfilled: (a) the beam is ultrarelativistic, (b) the vacuum chamber is smooth, (c) the vacuum chamber wall is perfectly conducting.

Note that the pipe has to be smooth only in the z-direction. The pipe's cross section in the transverse plane can have an arbitrary shape as long as this shape is maintained along the length of the accelerator, and the theorem remains valid.

Putting the theorem in another way, we may say that there are three possible ways for a collective instability to occur,

1. the beam is not relativistic enough;
2. the vacuum chamber is too resistive;
3. the vacuum chamber is not smooth enough.

If any one of these conditions occurs, the exact cancellation of the electric and magnetic forces is disrupted, and the beam can encounter an instability if its intensity is too high.

In reality, we try to avoid these conditions as much as possible. We invariably design and construct our accelerators with the aim to be as close to the cancellation condition as possible. In fact, as we will elaborate later, we generally do such a good job in the design and the construction of accelerators that the electric and magnetic forces generally get to cancel to a high accuracy.

However, the cancellation is never perfect. The vacuum chamber is generally made of copper or aluminum, which are good conductors but not perfect. There will be many small necessary discontinuities along the vacuum chamber pipe, such as beam position monitors, vacuum pumping ports, etc. There are also those very big discontinuities known as RF cavities. As to the condition of $v = c$, it is never satisfied completely, especially for proton and heavy ion accelerators. So the cancellation of electric and magnetic forces are not perfect. And what is left over after the cancellation leads to collective instabilities.

8.1.4 Discontinuity

When a charged particle beam traverses a discontinuity structure in the conducting vacuum chamber, an electromagnetic wakefield is generated by a beam-structure interaction, as illustrated in Fig. 8.3.

The reason a wakefield is generated when there is a wall discontinuity is because the beam's image charges moving along the pipe now have to move around a corner. We all know that when a charge is bent in its trajectory, it radiates. Although they do not move like real ballistic particles, image charges are real charges. Their turning around a corner still radiates. Wakefields are really the radiation fields of the image charges when their trajectories are bent.

An immediate comment needs to be made about the radiation from image charges and currents. Image charges on the surface of a metal move mainly in

Figure 8.3: Wakefields are generated when a beam traverses a discontinuity on the vacuum chamber wall.

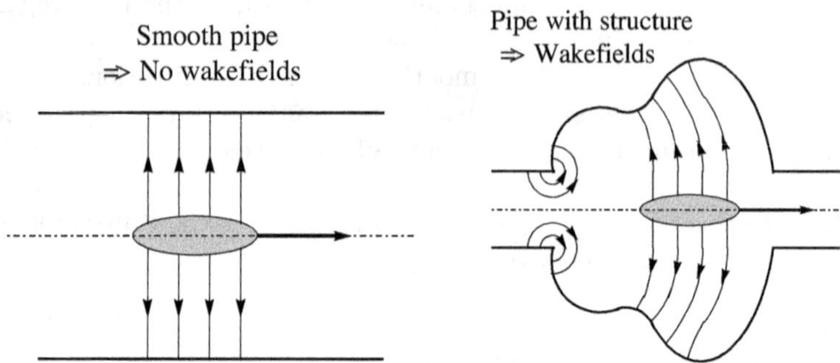

the transverse direction penetrating into or emerging out of the metal surface instead of longitudinally cruising parallel along the surface. This is not dissimilar to the motion of water waves on the surface of a lake. This motion does not allow the image charges to carry relativistic momentum. Formulation of synchrotron radiation based on prescribed motion therefore does not apply here. Any radiation immediately perturbs the motion of the image charges. Image charges are much softer objects than relativistic particles.

Once stated that these wakefields are a result of radiation, it is natural to ask what is the frequency content of these wakefields? The answer is that it depends on the details of the beam and the detailed geometry of the discontinuity. In general, it covers a very wide range, from micron wavelengths to long microwaves. There will hardly be any short wavelength radiation like X-rays, because again, there is no relativistic boost or Doppler effect, like in synchrotron radiation. To describe the frequency content of these wakefields, we will later introduce a quantity called the *impedance*. Impedance is essentially the Fourier transform of wakefield.

8.1.5 Resistive wall

As mentioned, a discontinuity structure in the vacuum chamber generates wakefields even when the vacuum chamber is perfectly conducting. On the other hand, when the vacuum chamber is smooth but is resistive, there are also wakefields generated, by the following physical process.[3]

1. When the beam's image charges flow on the vacuum chamber wall, the electric field carried by the beam will be terminated immediately by the

[3] Alexander Wu Chao, Physics of Collective Beam Instabilities in High Energy Accelerators, Wiley, 1993; Karl L. F. Bane, SLAC Report AP-87 (1991); H. Henke and O. Napoly, Proc. Euro. Part. Accel. Conf., Nice, 1990, p. 1046; O. Henry and O. Napoly, Part. Accel. 35, 235 (1991).

image charges on the wall surface — charges are forbidden in metal, see the definition of metal, Eq. (5.8). On the other hand, the magnetic field carried by the beam is not stopped immediately at the wall surface, because the wall image current is not a strictly a surface current but penetrates into the wall by a skin depth.

2. As the image current, pushed a skin depth by the beam into the wall, slowly resurfaces after the beam departs, a resurfacing image current drives new magnetic fields in the vacuum chamber. These new magnetic fields occur behind the passing beam.

3. The resurfacing current and magnetic field will execute some transient behavior, and appear to oscillate a few times. After the initial transient, the resurfacing current and magnetic field decay away but only very slowly.

4. The resurfacing magnetic field, changing with time in its transient, now drives an electric field by Maxwell equation's induction. This yields some electric field inside the resistive wall after all, but it is very weak.

5. Outside the wall material and inside the vacuum chamber is now generated the resistive wall wakefields.

In the case of a resistive wall pipe with circular cross section, an ultrarelativistic point charge q going down its axis deposits a wakefield shown in Fig. 8.4, which shows the electric component of the wakefield inside the vacuum chamber (outside of the wall material). Note that there is a matching magnetic field pattern following the leading point charge just like the electric field pattern does, and that both the electric and magnetic field patterns follow the leading point charge as a frozen pattern, indicating a phase velocity of c, but it is important to know that the Poynting vector (obtained by $\vec{E} \times \vec{B}$) does not indicate the field energy flows in the z-direction with the speed of light. The field energy does not propagate down the pipe. New field energy is deposited by the point charge as it moves down the pipe, while old energy gets transformed to heat locally by the resistivity on the pipe wall.

The parameter χ in Fig. 8.4 is a small dimensionless parameter defined by

$$\chi = \frac{1}{\mu\sigma_c bc}, \tag{8.2}$$

with b the vacuum pipe radius, σ_c the conductivity of the pipe material. Conductivity of aluminum at room temperature is about 3.6×10^7 ohm^{-1}m^{-1}. If the beam pipe is made of aluminum and $b = 5$ cm, we have $\chi = 1.5 \times 10^{-9}$.

As seen by Fig. 8.4, there is no wakefield ahead of the point charge, as causality would dictate. The wakefield pattern following the point charge is measured in distance z in units of $s_0 = (2\chi)^{1/3}b$. Since $\chi \ll 1$, the resistive wall wakefield decays very quickly — initially, that is — following the passage of the point charge. The span of Fig. 8.4 covers only about 0.7 mm longitudinal

Figure 8.4: Pattern of electric field following a driving ultrarelativistic point charge in a cylindrically symmetric resistive wall pipe. The field line density to the left of the heavy dashed line has been magnified by a factor of 40. The horizontal coordinate is $\frac{z}{s_0}$, the vertical coordinate is $\frac{y}{s_0}$, where $s_0 = (2\chi)^{1/3}b$. Note the different horizontal and vertical scales. Arrows indicate the directions of the electric fields. [Courtesy Karl Bane 2016]

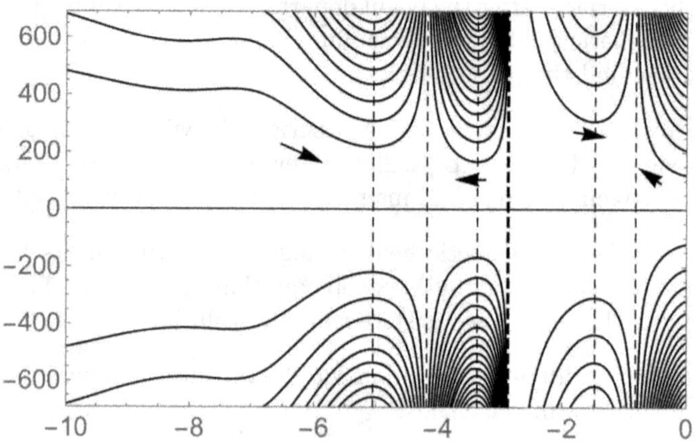

length. Note that the polarity of the electric field immediately following the driving point charge q is retarding. During this short distance of ∼0.7 mm, the electric field has oscillated twice between retarding and accelerating polarities and attempts to make a third one.

On the other hand, after the quick initial decay, at long distances, the residual resistive wall wake starts to decay very slowly. This means the resistive wall wakefield has a long tail after its initial sharp peak. An intense beam bunch, for example, can leave a resistive wall wakefield that lasts long enough to affect its motion when the bunch returns after making one or more complete circulations around a circular accelerator.

As will be shown later, the fact that the resistive wall generates both short-range and long-range wakefields is reflected by the fact that its corresponding impedance has an exceptionally wide spectrum, covering from very high to very low frequencies. Resistive wall is one of the more annoying features of an accelerator in practice.

8.1.6 Particle motion in presence of wakefield

We now have introduced wakefields due to both vacuum chamber discontinuities and resistive wall. We still need to explain how the wakefields affect particle's motion. To address the question of beam instability, the particles we are interested in are those moving with $v \approx c$ together with the beam. One such particle was shown as the test particle e in Fig. 8.2, except that we discussed only the case when no wakefields were generated. We now need to examine the effect on the test particle when wakefields are present.

To a large degree, accelerator physics and plasma physics are quite similar. Both involve nonlinear dynamics (single-particle effects) and collective instabilities (multiparticle effects). However, there is an important difference,

$$\begin{aligned} &\text{beam self fields} > \text{external applied fields} \qquad (\text{plasma})\,, \\ &\text{beam self fields} \ll \text{external applied fields} \qquad (\text{accelerators})\,. \end{aligned}$$

This difference means *perturbation techniques* are applicable to accelerators with

$$\begin{aligned} \text{unperturbed motion} &= \text{external fields (magnets, RF cavities)}, \\ \text{perturbation} &= \text{self fields, wakefields}\,. \end{aligned}$$

In fact, in accelerator physics, a first order perturbation often suffices. This makes accelerators much cleaner physical system to study compared to typical plasma systems, although the mathematics and physics tools are quite similar. Their similarity and difference can perhaps be analogous to the case between QED (quantum electrodynamics) and the QCD (quantum chromodynamics) prior to the standard model.

As will be illustrated, the concepts of wakefields and impedances are unequivocally based on the validity of this perturbation technique as applied to high energy accelerators – and the words "high energy" are critical, as we will explain.

Consider a beam with distribution ψ in phase space $(\vec{q}, \vec{p})$. The dynamics of the evolution of ψ is described by the Vlasov equation (Sec. 8.5),

$$\frac{\partial \psi}{\partial t} + \frac{\vec{p}}{m} \cdot \frac{\partial \psi}{\partial \vec{q}} + \vec{f} \cdot \frac{\partial \psi}{\partial \vec{p}} = 0\,,$$

where

$$\vec{f} = e\left(\vec{E} + \vec{v} \times \vec{B}\right)$$

is the Lorentz force. In case the beam is intense, the EM fields contain two contributions,

$$\begin{aligned} \vec{E} &= \vec{E}_{\text{ext}} + \vec{E}_{\text{wake}}\,, \\ \vec{B} &= \vec{B}_{\text{ext}} + \vec{B}_{\text{wake}}\,, \end{aligned}$$

where $\vec{E}_{\text{ext}}, \vec{B}_{\text{ext}}$ are the external fields determined by the accelerator design. The wakefields (N is beam intensity)

$$\begin{aligned} (\vec{E}, \vec{B})_{\text{wake}} &\propto N\,, \\ (\vec{E}, \vec{B})_{\text{wake}} &\ll (\vec{E}, \vec{B})_{\text{ext}}\,, \end{aligned}$$

are determined by the Maxwell equations where the source terms ρ and $\vec{j}$ are determined by the beam distribution ψ,

$$\rho = \int d^3p\, \psi, \qquad \vec{j} = \int d^3p\, \vec{v}\psi\,.$$

We therefore have the situation when the beam distribution is described by the Vlasov equation whose force terms are given by the electromagnetic fields, while the electromagnetic fields are described by the Maxwell equations whose source terms are given by the beam distribution. It is clear that a full treatment of the beam-wakefield system requires solving a coupled "Vlasov-Maxwell equation". As if these were not complicated enough, the fields also have to obey necessary boundary conditions which in general are 3-D.

Beam-structure interaction is a difficult problem in general. Its solution often involves numerical solution using particle-in-cell (PIC) codes. At this point, it seems the only way to proceed to solve this hopelessly complicated beam-environment problem will have to be based on PIC simulations. However, it is also obvious that applying PIC codes is reasonable only for small devices such as electron guns and klystrons, but it becomes impractical or impossible for large accelerators.

So, can we simplify the problem for our purpose while maintaining sufficiently accurate and trust-worthy results? The answer is yes. For *high energy* accelerators, this complication can be largely mitigated due to two simplifying approximations. These simplifications will pave the way to the concepts of "wake function" and "impedance".

Let us make a comment here before discussing these two simplifications. In retrospect, the reader is to be amazed of how much progress we have made as a result of several decades of accumulated efforts by a large number of accelerator physicists on the subject of collective instability. By analyzing and decomposing the problem piece by piece and step by step, the field has evolved a concise and accurate treatment of this otherwise intractable subject. We will return to this comment later.

Rigid beam approximation The first simplification is the rigid beam approximation. At high energies, beam motion is affected little during the passage of a structure. This means one can calculate the wakefields assuming the beam shape is rigid and its motion is ultrarelativistic with $v = c$ during the passage. In fact, we only need to calculate the wakefields generated by a "rigid $\cos m\theta$ ring beam" as shown in Fig. 8.5, where $m = 0$ is monopole moment (net charge), $m = 1$ is dipole moment, etc. Wakefield of a general beam can be obtained by superposition of wakefields due to the ring beams with different m's and different ring radii.

It should be noted that the rigid beam simplification is applied only during the passage of the beam through the structure in the effort to calculate the

Figure 8.5: An ultrarelativistic $\cos m\theta$ ring beam goes down the beam pipe in the rigid-beam approximation.

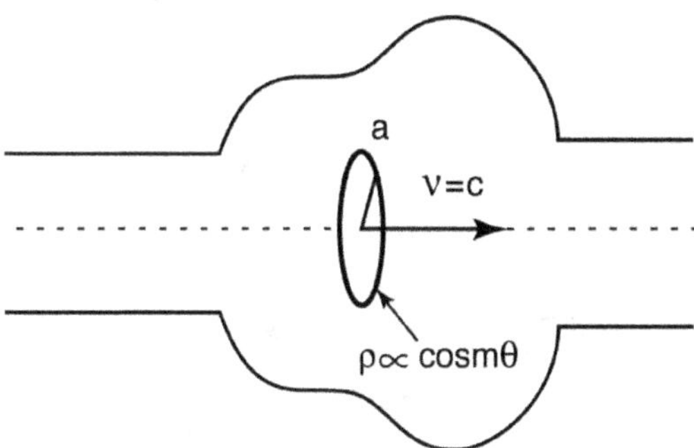

wakefields. After the passage, the beam is not to be held rigid any more, and the beam dynamics determines its subsequent motion.

Impulse approximation The second simplification is the impulse approximation. First, let us note that we do not need to know the instantaneous $\vec{E}$ or $\vec{B}$ separately. We need only to know the Lorentz force $\vec{f}$. Second, for high energies, we do not even need the instantaneous $\vec{f}$. We only need the integrated impulse

$$\Delta\vec{p} = \int_{-\infty}^{\infty} dt\ \vec{f},$$

where the integration over t is performed along the *unperturbed* trajectory of the test charge e, holding D fixed, where D is the longitudinal separation between the test charge and the ring beam. The justification to integrate only along the unperturbed trajectory is the rigid-beam approximation.

Figure 8.6 shows the configuration of a ring beam and a test charge that follows it. The ring beam generates a wakefield. The test charge receives an integrated wake-induced impulse in the impulse approximation.

Note that it does not make sense to ask for $\Delta\vec{p}$ until the test charge has completed its entire passage. It is the total impulse that the test charge receives that is of interest. The integration over time is from $-\infty$ to $+\infty$.

8.1.7 Panofsky-Wenzel theorem

The instantaneous wakefields are complicated, but as we will soon see, $\Delta\vec{p}$ is much simpler and it is $\Delta\vec{p}$ that we need. Mother nature has been very kind to allow this crack opening for the accelerator physicists. The quantity $c\Delta\vec{p}$ is sometimes called the "wake potential". As mentioned, although the beam is considered to be rigid *during* the passage, the impulse will affect the subsequent

Figure 8.6: A test charge e following a $\cos m\theta$ ring beam illustrates the impulse approximation.

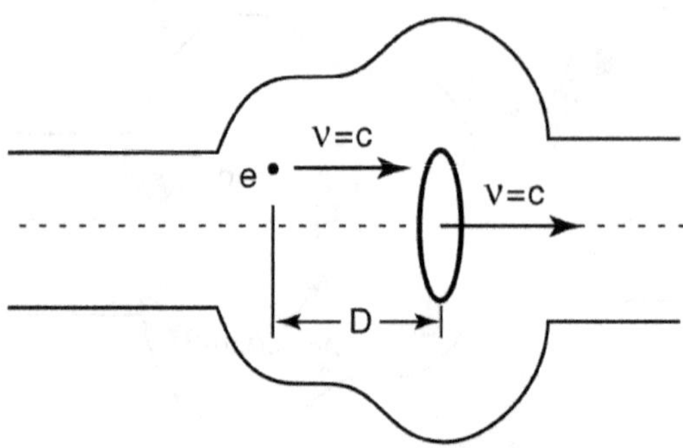

beam motion after the passage. Instabilities, if any, occur as subsequent after-facts of the beam-structure interaction, and not during each passages of the structure discontinuity.

We derive the Panofsky-Wenzel theorem[4] in this section. Details of the derivation are different from the original paper. Reasoning along this line turns out to be quite fruitful, allowing the theorem to be generalized in different system configurations and applications. Such generalizations are beyond our scope. As it will turn out, the Panofsky-Wenzel theorem is the basis of all beam instability analyses in high energy accelerators.

In the present context, Maxwell equations read

$$\begin{aligned} \nabla \cdot \vec{E} &= \frac{\rho}{\epsilon_0}, \\ \nabla \times \vec{B} - \frac{1}{c^2}\frac{\partial \vec{E}}{\partial t} &= \mu_0 \beta c \rho \hat{z}, \\ \nabla \cdot \vec{B} &= 0, \\ \nabla \times \vec{E} + \frac{\partial \vec{B}}{\partial t} &= 0, \end{aligned}$$

where we have made the important rigid beam approximation $\vec{j} = \rho\vec{v}$ (the beam shape is rigid) and $\vec{v} = \beta c\hat{z}$ (the beam motion is rigid). On the other hand, ρ is left unspecified and remains general. The Panofsky-Wenzel theorem will be valid for arbitrary beam distribution as long as its motion is rigid. The beam velocity v is assumed to be $\approx c$, but we delay the declaration of $v = c$ until later.

We assume the fields are in a vacuum environment. These fields include contributions from the wakefields as intended. However, they also include the external fields such as those from magnets and RF cavities. In the region of

[4] W. K. H. Panofsky and W. A. Wenzel, Rev. Sci. Instr. 27, 967 (1956).

interest, those external fields satisfy the source-free Maxwell equations. So the Panofsky-Wenzel theorem will apply to those external-field cases as well.

The Lorentz force as seen by the rigid test charge e is given by

$$\vec{f} = e(\vec{E} + \beta c\hat{z} \times \vec{B}),$$

which leads to

$$\begin{aligned}
\nabla \cdot \vec{f} &= e\left[\frac{\rho}{\epsilon_0} + \beta c\nabla \cdot (\hat{z} \times \vec{B})\right] \\
&= e\left[\frac{\rho}{\epsilon_0} - \beta c\hat{z} \cdot \nabla \times \vec{B}\right] \\
&= e\left[\frac{\rho}{\epsilon_0} - \frac{\beta}{c}\hat{z} \cdot \frac{\partial \vec{E}}{\partial t} - \beta^2 \frac{\rho}{\epsilon_0}\right] \\
&= \frac{e\rho}{\epsilon_0 \gamma^2} - \frac{e\beta}{c}\frac{\partial E_z}{\partial t}, \\
\nabla \times \vec{f} &= e\left[-\frac{\partial \vec{B}}{\partial t} + \beta c\nabla \times (\hat{z} \times \vec{B})\right] \\
&= -e\left(\frac{\partial}{\partial t} + \beta c\frac{\partial}{\partial z}\right)\vec{B}.
\end{aligned}$$

As mentioned, we are only interested in the impulse due to the Lorentz force. To be more specific, we want to calculate the net kick received by a test charge e which has a transverse position (x, y) and longitudinal position D relative to the moving beam. Both the beam and the test charge move with $\vec{v} = \beta c\hat{z}$. The impulse is then written more precisely as

$$\Delta\vec{p}(x, y, D) = \int_{-\infty}^{\infty} dt\ \vec{f}(x, y, D + \beta ct, t). \tag{8.3}$$

We have adopted the convention that $D < 0$ if the test charge trails the beam.

So far, we have not assumed any detailed shape of the beam. Neither have we made any assumptions of the pipe boundary. As mentioned, we have kept β in the derivation, postponing setting $\beta = 1$ until later.

One more point is to be made concerning the impulse equation (8.3). In the picture, what is moving down the pipe is the particle beam (both the driving beam and the test charge). However, the wakefields are localized and are not moving. The impulse, integrated from $-\infty$ to ∞, is to be applied locally at the point of the discontinuity. This important point was emphasized when we discussed the resistive wall wakefields in Sec. 8.1.5. The physical picture that the wakefields are localized just like the magnet and RF fields, should be kept in mind.

We now have

$$\begin{aligned}\nabla \times \Delta\vec{p} &= \int_{-\infty}^{\infty} dt \left[\nabla' \times \vec{f}(x,y,z,t)\right]_{z=D+\beta ct} \\ &= -e\int_{-\infty}^{\infty} dt \left[\left(\frac{\partial}{\partial t} + \beta c\frac{\partial}{\partial z}\right)\vec{B}(x,y,z,t)\right]_{z=D+\beta ct} \\ &= -e\vec{B}(x,y,D+\beta ct,t)\Big|_{t=-\infty}^{t=\infty}.\end{aligned}$$

In this equation, ∇ refers to taking derivative with respect to coordinates (x,y,D), while ∇' refers to taking derivative with respect to coordinates (x,y,z).

If the wakefield $\vec{B}$ vanishes far away from the region of interest, we have

$$\nabla \times \Delta\vec{p} = \vec{0},$$

which is the (generalized) Panofsky-Wenzel theorem.

One might ask if the Panofsky-Wenzel theorem exhausts all the useful information contained in the Maxwell equations regarding the wake impulse. The answer is no. Panofsky-Wenzel theorem in this sense, is necessary but not sufficient. The theorem implies that $\Delta\vec{p}$ can be written as the gradient of another quantity W,

$$\Delta\vec{p}(x,y,D) = -e\nabla W(x,y,D),$$

but it does not say what W is. On the other hand, additional information from the Maxwell equations sometimes allow useful information on W.

The Panofsky-Wenzel theorem is stated as a vector equation. One can decompose it into a component parallel to $\hat{z}$ and a component perpendicular to $\hat{z}$ by taking $\hat{z}\cdot$ or $\hat{z}\times$ operations to it. We then obtain

$$\nabla\cdot(\hat{z}\times\Delta\vec{p}) = 0, \tag{8.4}$$

$$\frac{\partial}{\partial D}\Delta\vec{p}_\perp = \nabla_\perp \Delta p_z. \tag{8.5}$$

Equation (8.4) says something about the transverse components of $\Delta\vec{p}$. Equation (8.5) says that the transverse gradient of the longitudinal wake potential is equal to the longitudinal gradient of the transverse wake potential.

A slight detour on Eq. (8.5) might be in order. In an accelerator design, its optimization typically asks for a decoupled operation as cleanly as possible. The theorem (8.5) says that if after some effort, the longitudinal impact delivered to the beam is made to decouple from the beam's transverse motion, then automatically you have decoupled the transverse impact from the beam's longitudinal motion, and vice versa. This is of course a welcomed result from the Panofsky-Wenzel theorem.

An alert reader also connects Eq. (8.5) to the symplecticity of the Hamiltonian system under consideration. Maxwell equations assure symplecticity through the Panofsky-Wenzel theorem.

A further approximation by setting $\beta = 1$ for an ultrarelativistic case yields additional conditions following the Panofsky-Wenzel theorem. If we observe the following with $\beta = 1$,

$$\begin{aligned}
\nabla \cdot \Delta\vec{p} &= \int_{-\infty}^{\infty} dt \left[\nabla' \cdot \vec{f}(x,y,z,t)\right]_{z=D+ct} \\
&= -\frac{e}{c}\int_{-\infty}^{\infty} dt \left(\frac{\partial E_z}{\partial t}\right)_{z=D+ct} = e\int_{-\infty}^{\infty} dt \left(\frac{\partial E_z}{\partial z}\right)_{z=D+ct} \\
&= \int_{-\infty}^{\infty} dt \left(\frac{\partial f_z}{\partial z}\right)_{z=D+ct} = \frac{\partial \Delta p_z}{\partial D},
\end{aligned}$$

which leads to a neat result (when $\beta = 1$)

$$\nabla_\perp \cdot \Delta\vec{p}_\perp = 0. \tag{8.6}$$

In terms of the function $W(x, y, D)$, it follows that

$$\nabla_\perp^2 W = 0, \qquad \text{or} \qquad \frac{\partial^2 W}{\partial x^2} + \frac{\partial^2 W}{\partial y^2} = 0.$$

It is clear that the Panofsky-Wenzel theorem imposes strong constraints on the impulses received by a test charge from a rigid ultrarelativistic beam.

Homework 8.2 Follow the outline in the text and prove Eqs. (8.4), (8.5), and (8.6).

8.1.8 Lawson-Woodward theorem

In the previous section, we derived the Panofsky-Wenzel theorem based on the assumption that the electric and magnetic fields seen by the particle are in free space. This means we have assumed that the particle does not see any boundaries or materials along its path, but it does not forbid the fields to have boundary conditions. As long as the particle's path lies completely in free space, we are free to have various boundary conditions that modify the fields.

In fact, the boundary conditions that modify the fields play a critical role here. Without boundaries, the fields can be decomposed into a superposition of plane waves. It turns out that in this environment, and under the rigid-beam and impulse approximations, the particle will not receive any net impulse from these oscillating electromagnetic fields. An ultrarelativistic particle does not interact with a laser or RF wave in free space. This is called the Lawson-Woodward theorem.[5]

[5]J.D. Lawson, IEEE Trans. Nucl. Sci. 26, 4217 (1979); P.M. Woodward, J.I.E.E. 93 (IIIA): 1554 (1947); R. Palmer, AIP Conf. Proc. 335, 90 (1995).

In free space without boundaries, electromagnetic fields in general can be expressed as a superposition of plane waves,

$$\begin{aligned}\vec{E}(\vec{x},t) &= \int_{-\infty}^{\infty} d^3\vec{k}\ \vec{e}_0(\hat{k},k)e^{-ik(ct-\hat{k}\cdot\vec{x})}\,,\\ \vec{B}(\vec{x},t) &= \int_{-\infty}^{\infty} d^3\vec{k}\ \hat{k}\times\vec{e}_0(\hat{k},k)e^{-ik(ct-\hat{k}\cdot\vec{x})}\,,\end{aligned}$$

where $\vec{e}_0(\hat{k},k)\cdot\hat{k}=0$.

For a particle moving with constant velocity, its trajectory is given by $\vec{x}=ct\vec{\beta}$ with $\vec{\beta}=$ constant. As seen by this particle,

$$\begin{aligned}\vec{E}(t) &= \int_{-\infty}^{\infty} d^3\vec{k}\ \vec{e}_0(\hat{k},k)e^{-ikct(1-\hat{k}\cdot\vec{\beta})}\,,\\ \vec{B}(t) &= \int_{-\infty}^{\infty} d^3\vec{k}\ \hat{k}\times\vec{e}_0(\hat{k},k)e^{-ikct(1-\hat{k}\cdot\vec{\beta})}\,.\end{aligned}$$

Lawson-Woodward theorem refers to the energy gain by this high-energy particle as it traverses the laser field. So the next step is to integrate over t to find the particle's energy gain. But in our derivation, we aim to generalize the result somewhat. For our purpose, we will calculate the momentum kick by integrating the vector impulse (recalling the impulse approximation),

$$\begin{aligned}\Delta\vec{p} &= \int_{-\infty}^{\infty} dt\ \left[e\vec{E}(t)+e\vec{\beta}\times\vec{B}(t)\right]\\ &= \int_{-\infty}^{\infty} dt\ e\int_{-\infty}^{\infty} d^3\vec{k}\left[\vec{e}_0+\vec{\beta}\times(\hat{k}\times\vec{e}_0)\right]_{\hat{k},k} e^{-ikct(1-\hat{k}\cdot\vec{\beta})}\,. \qquad (8.7)\end{aligned}$$

We make a few remarks here concerning Eq. (8.7),

- It is more general than an equation for the energy gain. For the energy gain, we can obtain it by

$$\Delta\mathcal{E} = c\Delta\vec{p}\cdot\vec{\beta}\,.$$

  It is easy to see that the magnetic contribution drops out in $\Delta\mathcal{E}$ but not in $\Delta\vec{p}$. Magnetic field does not change the particle's energy, although it does change its momentum.

- Integration over time is from $-\infty$ to ∞. This means we have assumed that, throughout time, (a) the plane waves are in free space, and (b) the particle is moving in straight line with constant velocity. If either of these conditions are violated, Lawson-Woodward theorem will not apply. Any energy exchange from a laser or an RF will have to violate either condition (a) or condition (b). An RF cavity acceleration chooses to violate condition (a); a free-electron-laser undulator chooses to violate (b).

- As mentioned, there are two possibilities when the velocity of the particle can be constant, at least approximately: (a) the particle motion is ultra-relativistic, and (b) the plane wave field is weak. For our applications, it is condition (a) that interests us most.

We now continue with Eq. (8.7). Integration over t can actually be performed; it gives a δ-function in k,

$$\begin{aligned}\Delta\vec{p} &= 2\pi e\int_{-\infty}^{\infty} d^3\vec{k}\left[\vec{e}_0+\vec{\beta}\times(\hat{k}\times\vec{e}_0)\right]_{\hat{k},k}\delta[kc(1-\hat{k}\cdot\vec{\beta})]\\ &= 2\pi e\int_{-\infty}^{\infty} d^3\vec{k}\left[\vec{e}_0+\vec{\beta}\times(\hat{k}\times\vec{e}_0)\right]_{\hat{k},0}\frac{\delta(k)}{c(1-\hat{k}\cdot\vec{\beta})},\end{aligned}$$

where we have used the fact that $1-\hat{k}\cdot\vec{\beta}$ never vanishes. Indeed this impulse $\Delta\vec{p}$ is nonzero. But because of the factor $\delta(k)$, it contains contributions from only the $k=0$ components from the electric and magnetic fields. In other words, only the DC part of the spectrum can deliver a net impulse to the particle.

But DC components of the fields are not of interest to us here — of course if there is a DC electric field, the particle will gain net energy, and DC magnetic field from which a particle receives a transverse kick. We now assume that there is no DC component in the laser or RF field, i.e. $\vec{e}_0(\hat{k},0)=\vec{0}$. It then follows

$$\Delta\vec{p} = \vec{0},$$

which is the generalized Lawson-Woodward theorem: an ultrarelativistic particle cannot gain or lose a net energy or momentum from a laser or an RF in free space.

In particular, we have just proved that in order to accelerate a beam with RF, the RF must be confined and stored in some cavity structure.

The Lawson-Woodward theorem states $\Delta\vec{p}=\vec{0}$. The Panofsky-Wenzel theorem states $\nabla\times\Delta\vec{p}=\vec{0}$.

It follows from the Lawson-Woodward theorem that any net impulse, longitudinal or transverse, a laser can deliver to a particle must come from either

1. the particle's velocity is not constant, or
2. laser field is not in free-space.

The various concepts of laser acceleration will have to implement either 1 or 2 without violating the Lawson-Woodward theorem. We shall not elaborate on laser acceleration concepts.[6] Suffice it to mention here that as it turns out, the price to pay for either 1 or 2 is a large reduction of the net available acceleration gradient. Since these conditions are violated only slightly, the Lawson-Woodward reduction factor becomes necessarily very large, and therein lies the challenge of the various laser acceleration schemes.

[6] T. Tajima and J.M. Dawson, Phys. Rev. Lett. 43, 267 (1979); J. Rosenzweig, Handbook Accel. Phys. & Eng., Sec. 1.6.10 (2013); R.B. Palmer, AIP Conf. Proc. 91, 179 (1982).

8.1.9 Cylindrically symmetric pipe

We now return to the discussion of the Panofsky-Wenzel theorem. The special case when the vacuum chamber environment is cylindrically symmetric has a particularly simple form. We discuss it in this section. In cylindrical coordinates, Eq. (8.4) gives

$$\begin{aligned} \nabla \cdot [\hat{z} \times (\Delta p_r \hat{r} + \Delta p_\theta \hat{\theta})] &= 0 \\ \Longrightarrow \qquad \frac{\partial}{\partial r}(r\Delta p_\theta) &= \frac{\partial}{\partial \theta}\Delta p_r \,. \end{aligned}$$

Equation (8.5) gives

$$\begin{aligned} \frac{\partial}{\partial D}(\Delta p_r \hat{r} + \Delta p_\theta \hat{\theta}) &= \left(\hat{r}\frac{\partial}{\partial r} + \frac{\hat{\theta}}{r}\frac{\partial}{\partial \theta}\right)\Delta p_z \\ \Longrightarrow \qquad & \begin{cases} \frac{\partial}{\partial D}\Delta p_r = \frac{\partial}{\partial r}\Delta p_z, \\ \frac{\partial}{\partial D}\Delta p_\theta = \frac{1}{r}\frac{\partial}{\partial \theta}\Delta p_z. \end{cases} \end{aligned}$$

Equation (8.6) gives

$$\begin{aligned} \frac{1}{r}\frac{\partial}{\partial r}(r\Delta p_r) + \frac{1}{r}\frac{\partial}{\partial \theta}\Delta p_\theta &= 0 \\ \Longrightarrow \qquad \frac{\partial}{\partial r}(r\Delta p_r) &= -\frac{\partial}{\partial \theta}\Delta p_\theta \qquad (\beta = 1)\,. \end{aligned}$$

These results are surprisingly simple. They do not contain any beam source terms. Exact shape or distribution of the beam does not matter. Neither do they depend on the boundary conditions. The boundary can be of any shape, and can be perfectly conducting or resistive metal, or it can be dielectric. The boundary does not have to be a sharply defined surface; it can for example be a gradually fading plasma surface. The boundary also does not have to consist of a single piece. The only inputs needed for the Panofsky-Wenzel theorem are the free-space Maxwell equations, the Lorentz force equation, and the rigid-beam and the impulse approximations.

We are now ready to consider a $\cos m\theta$ ring beam with $\vec{v} = c\hat{z}$ as we set out to do by solving the three equations above. It follows that the solution can be expressed in terms of a function $W_m(D)$ such that[7]

$$\begin{aligned} c\Delta\vec{p}_\perp &= -eI_m W_m(D) m r^{m-1}(\hat{r}\cos m\theta - \hat{\theta}\sin m\theta)\,, \\ c\Delta p_z &= -eI_m W_m'(D) r^m \cos m\theta\,, \end{aligned} \tag{8.8}$$

[7]This definition of wake functions was originally chosen for the convenience of the cgs unit system. For the SI system chosen for these lectures, it is better to replace the factor of c on the left hand side by $4\pi\varepsilon_0$. However, we follow the adopted convention and will maintain the original definition (8.8).

where I_m is the m-th multipole moment of the ring beam. Here a prime denotes $\frac{d}{dD}$, $W_m(D)$ is called the transverse wake function and $W'_m(D)$ the longitudinal wake function. The longitudinal wake function is simply the D-derivative of the transverse wake function.

For a simplest case of a ultrarelativistic monopolar beam, $m = 0, I_0 = Q$, the wakefield impulse gives the energy change of the test charge,

$$\Delta E = c\Delta p_z = -eQW'_0(D)\,.$$

Equations (8.8) contain a wealth of information. First note the fact that they are proportional to e and I_m, as expected. It then can be checked explicitly that the Panofsky-Wenzel theorem is obeyed by these expressions. Finally, and most importantly, note that the solution (8.8) contains explicit dependences of r and θ. The fact that we can go so far without any specific details is surprising and shows the power of this line of reasoning. The dependence on D is through the wake function $W_m(D)$ which can be obtained only if explicit boundary conditions are introduced. By the Panofsky-Wenzel theorem, the 3-D Maxwell-Vlasov system is now decomposed into a job of finding the 1-D functions $W_m(D)$.

For completeness, the corresponding $W(r, \theta, D)$ function that yields Eq. (8.8) is

$$W = I_m W_m(D) r^m \cos m\theta\,.$$

When the beam pipe is cylindrically symmetric, each m-multipole component of the beam excites a wake pattern according to (8.8). Different m's do not mix.

Dimensionalities of the wake functions are such that

$$[4\pi\epsilon_0 W_m] = \mathrm{m}^{-2m}, \qquad [4\pi\epsilon_0 W'_m] = \mathrm{m}^{-2m-1}\,.$$

With $\left[\frac{1}{4\pi\epsilon_0}\right] = \Omega\mathrm{ms}^{-1}$ (Homework 1.8), we have

$$[W_m] = \Omega\mathrm{m}^{\,2m+1}\mathrm{s}^{-1}, \qquad [W'_m] = \Omega\mathrm{m}^{-2m}\mathrm{s}^{-1}\,.$$

It follows that the most important longitudinal wake function has $[W'_0] = \Omega\mathrm{s}^{-1}$, and the most important transverse wake function has $[W_1] = \Omega\mathrm{s}^{-1}\mathrm{m}^{-1}$.

Recap A recap might be instructive at this point concerning what has been developed here. At the beginning, it seemed that there was not too much handle on the wakefields because they seemed to have to depend on all kinds of details such as the geometry of the discontinuity, the properties of the wall material, or the shape and distribution of the driving electron beam. So it is quite amazing that on a very general ground, there is such a guideline like the rigid-beam and impulse approximations, and following these approximations, the Panofsky-Wenzel theorem, which relates the longitudinal and the transverse components of the impulse received by a test charge influencing its subsequent motion. The solution (8.8) explicitly solved all the r and θ dependences, leaving only the longitudinal D dependence to be solved, and furthermore, it constrains the transverse and longitudinal components of the impulse in a definitive manner.

As we will see in the second half of this chapter, the subject of collective effects requires another step when these wakefields are applied to accelerators. In that second step, we will leave electromagnetics based on the Maxwell equations and our treatments will turn to beam dynamics.

Homework 8.3 Show explicitly that Eq. (8.8) obeys the Panofsky-Wenzel theorem.

8.1.10 Decomposing wakefield into modes

Armed with the Panofsky-Wenzel theorem, to analyze the problem of instability, we proceed as follows. The problem at hand is what is the impulse received by the test charge e when it integrates the wakefield left behind by a particle beam — both the test charge and the beam are moving down the pipe ultrarelativistically and rigidly. To do so, we first consider the beam to be a δ-function in z, i.e. it is infinitely short in length. If the beam has a finite length, the result we obtain with the δ-function beam will serve as a Green's function, and a beam with a general longitudinal distribution can be obtained by linear superposition.

The beam now reduces to an infinitely short beam with arbitrary transverse distribution. To break down the problem further, we next decompose the transverse distribution into "modes". Any general transverse distribution can be decomposed into a summation of transverse modes, the mode index is designated by m. We then consider a single transverse mode m. A general transverse distribution can be obtained again by superposition with a summation over m.

So the problem is now reduced to finding the impulse integrated by a test charge that is a distance z behind another larger beam; the beam is infinitely short in z, has a transverse m-th moment I_m, and is moving along the pipe axis. The configuration is shown in Fig. 8.7. In this configuration, I_m is the "driving beam" (driving the wakefields), e is the test charge (integrating the wakefields), z is the longitudinal distance that e is trailing behind I_m, and (r, θ) is the transverse displacement of the test charge relative to the pipe axis. The impulse calculated by this configuration is going to be used as a Green's function when we analyze the beam instability problem. Note that I_m does not need to be off axis; it is considered to be on axis. The test charge e however is off axis to probe the wakefield impulses.

Polar coordinate For the case with a cylindrically symmetric vacuum chamber pipe, the driving beam distribution can be decomposed into transverse modes as given in Table 8.1.

Figure 8.7: Configuration of a driving beam I_m, a test charge e to illustrate the wakefield mode decomposition.

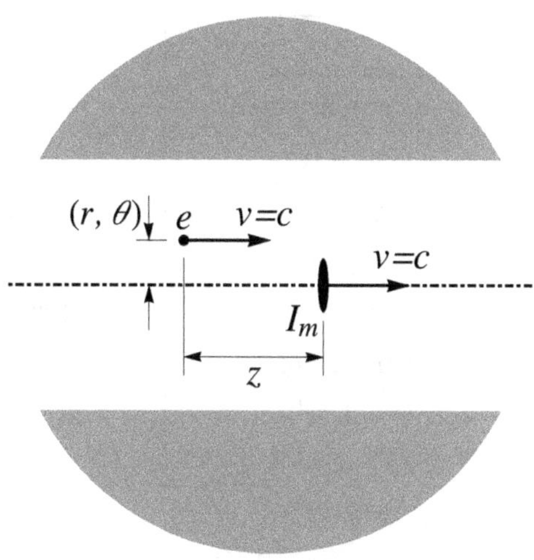

Table 8.1: Mode decomposition for a cylindrically symmetric wakefield.

m	mode	transverse distribution of driving beam	transverse moment of the driving beam I_m
0	monopole	1	q
1	dipole	$\cos\theta$	$q\langle x\rangle$
	skew dipole	$\sin\theta$	$q\langle y\rangle$
2	quadrupole	$\cos 2\theta$	$q\langle x^2 - y^2\rangle$
	skew quadrupole	$\sin 2\theta$	$q\langle 2xy\rangle$
3	sextupole	$\cos 3\theta$	$q\langle x^3 - 3xy^2\rangle$
	skew sextupole	$\sin 3\theta$	$q\langle 3x^2y - y^3\rangle$

For a cylindrically symmetric pipe, the m-th multipole wakefield is driven when and only when the driving beam has an m-th moment. For example, if the beam is transversely displaced from the pipe axis, then it contains an $m = 1$ dipole moment in addition to its original $m = 0$ moment due to its net charge. (If the beam is horizontally displaced, it contains a dipole moment. When it is vertically displaced, it contains a skew dipole moment.) The displaced beam therefore drives both an $m = 0$ wakefield and an $m = 1$ wakefield. Each moment then drives its corresponding wakefield. A beam with no skew quadrupole moment (i.e. a beam with $q\langle 2xy\rangle = 0$), for example, will not drive a $\sin 2\theta$ wakefield. A beam with m-th distribution moment I_m will generate a wakefield in the m-th mode that is proportional to I_m.

Table 8.2: Mode decomposition of a cylindrically symmetric wakefield in Cartesian coordinates.

m	Distribution moments of beam	Longitudinal wake impulse $c\Delta p_z$	Transverse wake impulse $c\Delta\vec{p}_\perp$
0	q	$-eq\, W_0'(z)$	0
1	$\begin{cases} q\langle x\rangle \\ q\langle y\rangle \end{cases}$	$-eq\langle x\rangle x W_1'(z)$ $-eq\langle y\rangle y W_1'(z)$	$-eq\langle x\rangle W_1(z)\hat{x}$ $-eq\langle y\rangle W_1(z)\hat{y}$
2	$\begin{cases} q\langle x^2-y^2\rangle \\ q\langle 2xy\rangle \end{cases}$	$-eq\langle x^2-y^2\rangle(x^2-y^2)W_2'(z)$ $-eq\langle 2xy\rangle 2xy\, W_2'(z)$	$-2eq\langle x^2-y^2\rangle W_2(z)(x\hat{x}-y\hat{y})$ $-\,2eq\langle 2xy\rangle W_2(z)(y\hat{x}+x\hat{y})$
3	$\begin{cases} q\langle x^3-3xy^2\rangle \\ q\langle 3x^2y-y^3\rangle \end{cases}$	$-eq\langle x^3-3xy^2\rangle$ $\times(x^3-3xy^2)W_3'(z)$ $-eq\langle 3x^2y-y^3\rangle$ $\times(3x^2y-y^3)W_3'(z)$	$-3eq\langle x^3-3xy^2\rangle W_3(z)$ $\times[(x^2-y^2)\hat{x}-2xy\hat{y}]$ $-3eq\langle 3x^2y-y^3\rangle W_3(z)$ $\times[2xy\hat{x}+(x^2-y^2)\hat{y}]$

Cartesian coordinate In Cartesian coordinates, Table 8.2 lists the two moments (first the normal moment and then the skewed moment) and the associated transverse and longitudinal wake impulses $c\Delta\vec{p}_\perp$ and $c\Delta p_z$ seen by a test charge e with transverse coordinates x, y that follows, at a distance $|z|$ behind, a beam which possesses an m-th moment. The question being asked is what kick is being received by the test charge as the beam and the test charge complete the traversal of the chamber discontinuity. We have the convention that $z < 0$ if the test charge trails the drive beam. A bracket $\langle\ \rangle$ means averaging over the transverse distribution of the drive beam; $\hat{x}$ and $\hat{y}$ are the unit vectors in the x- and y-directions.

Table 8.2 contains rich and detailed information about the wake effects and should be studied with attention. In most applications, we care mostly about the $m = 0$ monopole mode when discussing longitudinal collective instabilities, and mostly about $m = 1$ dipole and skew dipole modes when discussing transverse collective instabilities. Therefore, for each discontinuity, we just ask for two functions, $W_0(z)$ and $W_1(z)$, and we just calculate these two functions for most applications. Modes driven by higher moments ($m > 1$) and affecting higher mode motions are often — but not always — less important in terms of causing collective beam instabilities.

The reason the monopole wake function $W_0(z)$ is not relevant for transverse instabilities is because the transverse impulse vanishes when $m = 0$ (see Table 8.2). So the leading transverse contribution comes from $m = 1$. However, as mentioned, this is valid only when the vacuum chamber has a cylindrical symmetry.

Wake impulse pattern The left panel of Fig. 8.8 is a sketch of wake impulse pattern for the $m = 0$ case. A monopole charge $I_0 = q$ drives the wakefield seen by test charges at a distance $|z|$ behind. The impulses received by the test charges are indicated as red arrows. Table 8.2 indicates that these impulses have zero transverse component, i.e. $\Delta \vec{p}_\perp = \vec{0}$. Furthermore, all test charges in the slice at the same distance $|z|$ behind the drive charge receive the same longitudinal impulse. For a short beam bunch, this longitudinal impulse is retarding, i.e. $\Delta p_z < 0$.

The right panel of Fig. 8.8 is a sketch for the $m = 1$ case. A dipole moment I_1 (shown as a displaced point charge, although its monopole contribution is ignored in the figure) drives the wakefield seen by test charges at a distance $|z|$ behind. Table 8.2 says that the transverse impulses received by these test charges are all equal. The longitudinal impulses depends linearly on the transverse displacement of the test charge. For a short bunch, the polarities of the transverse and longitudinal impulses are as indicated.

Figure 8.8: The wakefield modes have rigid patterns dictated by the Panofsky-Wenzel theorem. The $m = 0$ and $m = 1$ patterns are shown. The red arrows indicate the integrated impulses received by the test charges. The lengths and the polarities of these red arrows are not drawn casually. See text.

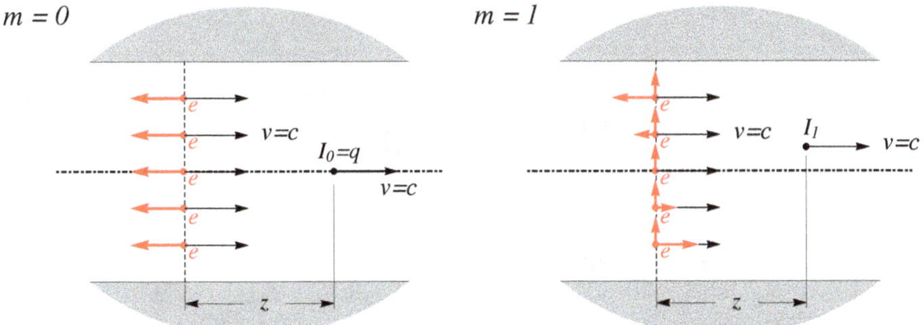

Wakefield pattern Figure 8.8 shows the patterns of the wake impulses for the decomposed modes. They represent the patterns of the wakefields integrated by the test charges. It may be useful also to show the physical field patterns of these wakefields, for example inside of a cavity structure when excited by the beam. One example of a cylindrically symmetric cavity is sketched in Fig. 8.9 for the $m = 0$ and the $m = 1$ modes.

Homework 8.4 Figure 8.9 are meant to be snapshots of the patterns of the electric and magnetic fields for the two modes $m = 0$ and $m = 1$. For simplicity, the patterns have ignored the entrance and exit ports of the cavity.

Figure 8.9: Wakefield patterns excited in a cylindrically symmetric cavity structure in the $m = 0$ (upper) and $m = 1$ (lower) modes.

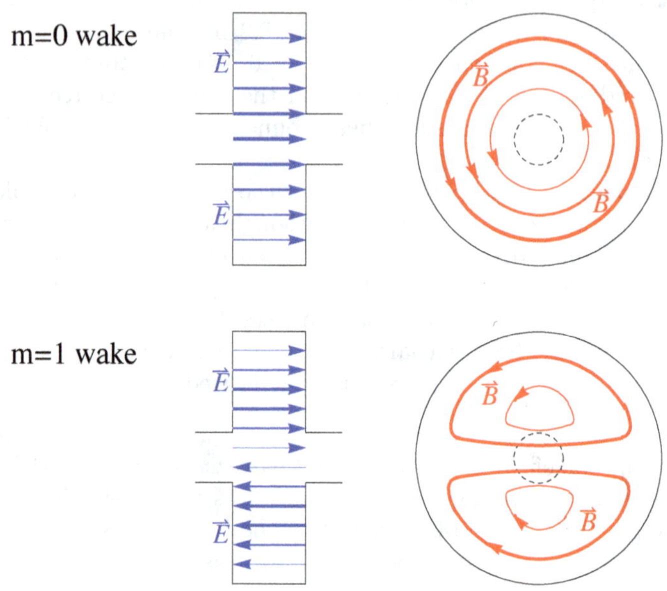

(a) The electric field and the magnetic field snapshots are taken at times when they reach their maximum strength respectively. They are not taken at the same time. Resolve the time sequence of these snapshots.

(b) Convince yourself that the thickness of the sketched field lines is meant to indicate the strength of the fields at least qualitatively.

(c) For each case, relative to the moment when the snapshots of the electric fields are taken, when did the driving beam move across the cavity? Which direction did the beam move? How was the beam displaced, up or down if at all, to drive these wakefields?

Homework 8.5 Consider a linac section with a cylindrically symmetric vacuum chamber whose wake functions integrated over the section are given by $W_0(z)$ and $W_1(z)$. Ignore all higher order wake functions, i.e. assume $W_m(z) = 0$ for $m \geq 2$.

Consider an intense point charge q moving down the linac section ultra-relativistically with $z = ct$ and having a transverse displacement given by $x = x_0, y = 0$, i.e. it is displaced horizontally. Now consider an infinitesimal test charge e trailing the intense point charge q by a distance $|D|$. Let the test charge follow a trajectory given by $x = x_1, y = 0, z = ct + D$ ($D < 0$ for the test charge to be trailing the driving charge q). This test charge is also displaced horizontally. Let the momentum of the test particle be P_0.

(a) Find the expression for the parasitic energy loss experienced by the test charge e, due to the wakefield left behind by the intense point charge q, by the time it reaches the end of the linac section.

(b) Find the transverse kicking angles $\Delta x'$ and $\Delta y'$ experienced by the test charge at the end of the linac section.

(c) Repeat (a) and (b) but this time assume the test charge e is displaced vertically with $x = 0, y = y_1$.

Solution

(a) $\Delta E = -eqW_0'(-D) - eqx_0x_1W_1'(-D)$.
(b) $\Delta x' = -\frac{eqx_0}{cP0}W_1(-D), \Delta y' = 0$.
(c) $\Delta E = -eqW_0'(-D), \Delta x' = -\frac{eqx_0}{cP_0}W_1(-D)$, $\Delta y' = 0$.

8.1.11 Noncylindrical symmetry

Equation (8.8) and Tables 8.1 and 8.2 apply to the case when the vacuum chamber has a cylindrical symmetry. In this case, a beam with an m-th distribution moment will drive the m-th mode wakefield and only the m-mode wakefield. The m-th mode wakefield in turn drives the m-th mode beam motion and only the m-th mode beam motion. A beam with dipole moment drives a dipole wakefield that deflects the transverse dipole motion of the trailing test charge. The m-th mode beam distribution and the m-th mode wakefield form an eigenmode of the beam-environment system.

Consider a case with only the $m = 0$ and $m = 1$ modes, ignoring all higher modes, and consider first a cylindrically symmetric pipe. The wake impulse is given by, from Table 8.2,

$$\begin{aligned}
\Delta p_x &= -eq\langle x_d\rangle W_1(z)\,, \\
\Delta p_y &= -eq\langle y_d\rangle W_1(z)\,, \\
\Delta p_z &= -eqW_0'(z) - eq(x\langle x_d\rangle + y\langle y_d\rangle)W_1'(z)\,,
\end{aligned} \tag{8.9}$$

where we have added subscripts d referring to the coordinate of the driving beam. The transverse impulses clearly indicate dipole kicks driven by the dipole moments of the drive beam.

The dipole kicks driven by the dipole moment cause the beam to oscillate. The beam oscillation produces additional dipole moment, which in turn produces additional dipole wakefields. The system therefore closes in a feedback loop, leading possibly to exponential growth and instabilities under unfavorable conditions. Before instability occurs, the system causes tune shifts. Note that these tune shifts are not the tune shifts we are familiar with, for example, due to a quadrupole magnetic field. No quadrupolar fields are involved here. These tune shifts have the property that both the horizontal and vertical betatron tunes shift down with beam intensity because this wakefield is necessarily defocusing, at least for short beam bunches. We will return to this result later in Eq. (8.38).

The situation changes when the vacuum chamber loses its cylindrical symmetry. In that case, a beam with m-th distribution moment will drive not only the m-th but also other modes of wakefields. Equation (8.8) and Tables 8.1 and 8.2 no longer describe an eigenmode system. On the other hand, we can still decompose the beam distribution and the wakefield in terms of these modes as described in this section. We will then see for example a monopole beam driving a quadrupolar wakefield and a quadrupole beam driving a monopole wakefield, etc.

In case of a noncylindrically symmetric environment, in addition to this dipole-induced dipole kicks, there will be additional monopole-induced quadrupole kicks. This time the wakefield is quadrupolar. Just like the case of a quadrupole magnetic field, this component of wakefield has the property that it contributes to the two betatron tune shifts with opposite signs, causing one tune shifting up and the other shifting down.

On the other hand, unlike the dipole-induced dipole wakefields, the monopole-induced quadrupolar wakefields cause a head-tail tune split of the beam, but they do not directly cause a feedback exponential type of collective instability. More of concern is the complications in the interpretation of tune measurements when diagnosing the impedance of the accelerator — e.g. using Eq. (8.38) — implicitly assuming cylindrical symmetry of the impedances. This particularly becomes an issue because typically each accelerator contains many noncylindrically symmetric objects in its vacuum chamber.

There is another view of the difference between these two contributions of wakefields. The dipole-induced dipole wakefield is a dynamical quantity. It oscillates from one bunch passage to the next because the dipole moments of the driving beam bunches change from bunch to bunch and from turn to turn. The monopole-induced quadrupole wakefield, on the other hand, is driven by the monopole moment, which does not vary in time. It is a static effect.

For a noncylindrically symmetric pipe, the wake impulse can be written, for a structure without cylindrical symmetry but maintaining an up-down symmetry, generalizing Eq. (8.9), as[8]

$$\begin{aligned}
\Delta p_x &= -eq(a'\langle x_d\rangle + ax)W_1(z)\,, \\
\Delta p_y &= -eq(b'\langle y_d\rangle + by)W_1(z)\,, \\
\Delta p_z &= -eqa_0 W_0'(z) - eq\left(\frac{a}{2}x^2 + a'x\langle x_d\rangle + d\langle x_d^2\rangle + \frac{b}{2}y^2 + b'y\langle y_d\rangle + e\langle y_d^2\rangle\right)W_1'(z)\,,
\end{aligned}$$

where a, a', b, b', a_0, d, e are numerical coefficients.

Panofsky-Wenzel theorem in the ultrarelativistic limit and in Cartesian coordinates reads

$$\frac{\partial \Delta p_x}{\partial y} = \frac{\partial \Delta p_y}{\partial x}, \qquad \frac{\partial \Delta p_x}{\partial x} = -\frac{\partial \Delta p_y}{\partial y},$$

[8]V. Balbekov, Proc. Euro. Part. Accel. Conf., Berlin, 792 (1992); J. Laslett, Proc. Summer Study on Storage Rings, Brookhaven, 324 (1993); A.W. Chao, S. Heifets, and B. Zotter, Phys. Rev. Special Topics - Accel. & Beams, 5, 111001 (2002).

$$\frac{\partial \Delta p_x}{\partial z} = \frac{\partial \Delta p_z}{\partial x}, \qquad \frac{\partial \Delta p_y}{\partial z} = -\frac{\partial \Delta p_z}{\partial y}.$$

which dictates the condition

$$a = -b.$$

In general, these coefficients need to be calculated from the boundary conditions of the pipe. The coefficients in the two analytically soluble cases, one of the cylindrical pipe and one for a pipe modeled as two parallel plates, are given in Table 8.3.

Table 8.3: The $m = 0$ and $m = 1$ mode wakefield coefficients for a cylindrical beam pipe and a beam pipe consisting of two parallel plates.

Geometry	a'	a	b'	b	a_0	d	e
Cylindrical pipe of radius b	1	0	1	0	1	0	0
Two parallel plates with gap $2b$	$\frac{\pi^2}{24}$	$-\frac{\pi^2}{24}$	$\frac{\pi^2}{12}$	$\frac{\pi^2}{24}$	1	$-\frac{\pi^2}{48}$	$\frac{\pi^2}{48}$

Resistive wall For the case of resistive wall with conductivity σ_c, for example, the wake functions are given by [see also Eq. (8.14)], for $z < 0$,

$$W_0(z) = -\frac{c}{2\pi b}\sqrt{\frac{Z_0}{\pi\sigma_c}}\frac{C}{|z|^{1/2}}, \qquad W_0'(z) = -\frac{c}{4\pi b}\sqrt{\frac{Z_0}{\pi\sigma_c}}\frac{C}{|z|^{3/2}},$$

$$W_1(z) = -\frac{c}{\pi b^3}\sqrt{\frac{Z_0}{\pi\sigma_c}}\frac{C}{|z|^{1/2}}, \qquad W_1'(z) = -\frac{c}{2\pi b^3}\sqrt{\frac{Z_0}{\pi\sigma_c}}\frac{C}{|z|^{3/2}},$$

where $Z_0 = \sqrt{\frac{\mu_0}{\epsilon_0}}$ (Homework 1.8), C is the circumference of the circular accelerator, b is the pipe radius if the pipe is circular, and the half-gap size if the pipe consists of two parallel plates.

8.1.12 Property of wake function

So, for each vacuum chamber discontinuity, no matter how complicated its geometry, we have now reduced the wakefield problem to finding the wake functions $W_m(z)$. For each discontinuity along the vacuum chamber, we just ask the question what are the wake functions of this discontinuity? When these wake functions are calculated or measured, by superposition, we will know the impulse each particle in the beam receives from the collective wakefields generated by the beam. And when that is known, we can analyze the stability of the beam.

In this section, we discuss the general properties of the wake functions. The first property to note is the causality condition. It dictates that the wake functions must vanish in front of the drive beam,

$$W_m(z) = 0, \qquad \text{if } z > 0 .$$

The causality condition follows from our ultrarelativistic assumption. A beam's wakefields cannot affect any test charges in front of the beam.

Immediately following the beam, we expect to see a longitudinal electric field that retards the beam, regardless of vacuum chamber properties. This is because the beam must not gain energy as it propagates down the pipe; otherwise we can create a perpetual moving machine. This means the quantity $j_z F_\parallel$ must be negative, which by our convention implies

$$W'_m(z) > 0 \qquad \text{for} \quad z \to 0^- .$$

Once $W'_m(0^-) > 0$ is established, and using the fact that $W_m(z > 0) = 0$, it follows that $W_m(0^-) < 0$, i.e. the transverse wake function $W_m(z)$ is negative for $z < 0$, and its magnitude increases monotonically with $|z|$, at least initially, starting from $W_m = 0$ at $z = 0^+$. In other words, in general, the longitudinal wake function $W'_m(z)$ is cosine-like and the transverse wake function $W_m(z)$ is sine-like, as sketched in Fig. 8.10.

In Fig. 8.10, both W_m and W'_m vanish if $z > 0$. The value of W'_m at $z = 0$ is indicated by a solid dot and it sits exactly midway between the values of $W'_m(0^+) = 0$ and $W'_m(0^-) > 0$, according to the fundamental theorem of beam loading, which we will elaborate in Sec. 8.1.13. The lower curve is of course related to the upper one by taking derivative with respect to z. The dotted lines connect the extrema points of $W_m(z)$ with the points when $W'_m(z)$ crosses zero.

It then follows readily from this discussion a nontrivial observation that a point charge does not experience a deflecting force due to its own transverse wake ($W_m(0) = 0$). This is in contrast to the longitudinal dimension, in which a point charge does see its own retarding wake impulse ($W'_m(0) \neq 0$). This property leads to a general preference for short beams when the transverse wake effects dominate, while longer bunches are preferred when the main concern is the longitudinal wake effects.

There are many interesting and very general properties of the wake function $W_m(z)$. A summary of some properties, some already mentioned, is given below.

- $W_m(z) = 0$, $W'_m(z) = 0$ for $z > 0$ (causality).
- $W_m(z) \leq 0$, $W'_m(z) \geq 0$ for $z \to 0^-$.
- $W_m(0) = 0$ (in most cases, except space charge).
- $W'_m(0) = \frac{1}{2} W'_m(0^-)$ (fundamental theorem of beam loading).
- $W'_m(0^-) \geq |W'_m(z)|$ for all z.
- $\int_{-\infty}^{0} W'_m(z) dz \geq 0$.

Figure 8.10: Sketches of the longitudinal wake function $W_m(z)$ and the transverse wake function $W'_m(z)$. The value of W'_m at $z = 0$ is indicated by a solid dot and it sits exactly midway between the values of $W'_m(0^+) = 0$ and $W'_m(0^-) > 0$. The two plots are connected because one is the derivative of the other.

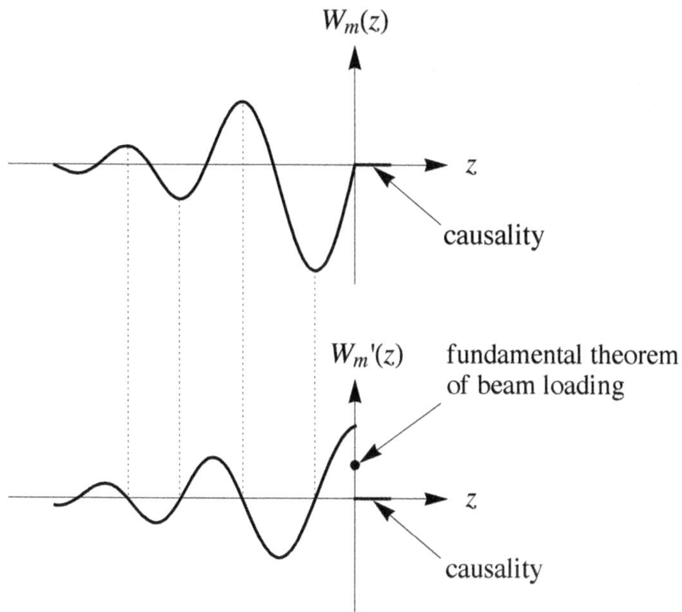

Here let us demonstrate how property that $W'_m(0^-) \geq |W'_m(z)|$ for all z can be established. Consider a point charge q followed by another point charge q at a distance $|z|$ behind. After traversing the discontinuity, the first q loses an energy $\frac{1}{2}q^2 W'_0(0^-)$ due to the wake generated by itself, where the factor of $\frac{1}{2}$ is due to the fundamental theorem of beam loading. The trailing q loses an energy $\frac{1}{2}q^2 W'_0(0^-) + q^2 W'_0(z)$, where the second term is due to the wake left by the leading q. Physically, the two-charge system can never gain energy; this means $W'_0(z) \geq -W'_0(0^-)$ for any z. Similarly, if the second charge is $-q$ rather than q, one proves $W'_0(z) \leq W'_0(0^-)$. The property is thus established for $m = 0$. From this property, it follows that $W'_m(0^-)$ is the maximum value the function $|W'_m(z)|$ ever reaches, and $W'_m(0^-) = 0$ necessarily gives the trivial case when the entire wake potential vanishes.

Polarity of wakefield The other properties of wake functions listed above can be shown similarly. But below, let us make one more comment about the polarity of the transverse wake impulses. This we do using the sketch in Fig. 8.11.

Figure 8.11: The polarity of the wakefield always hurts a short beam. For $m = 0$, the longitudinal wake impulse is retarding. For $m = 1$, the transverse wake impulse further deflects the test charge. For $m = 2$, the tail portion of an elliptical beam becomes further elongated. The arrows indicate the wake impulses.

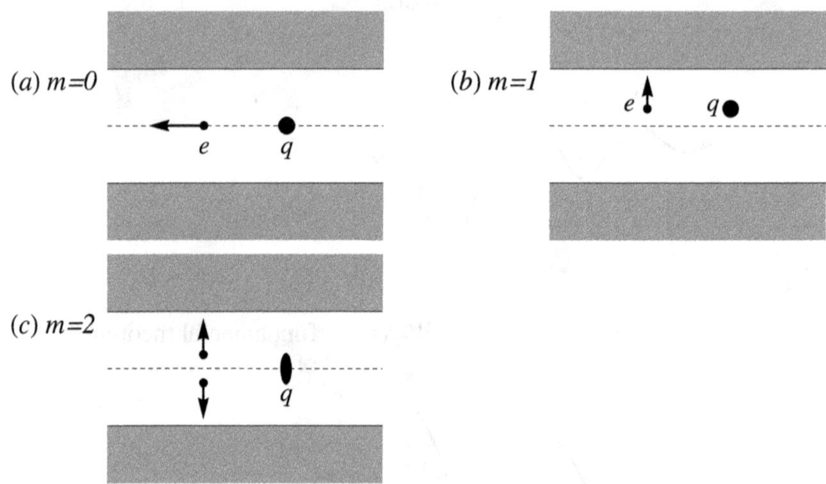

Imagine a short beam traveling down the accelerator with a displacement in x as shown in case (b). The head of the beam bunch will generate a wake impulse that kicks the particles that are in the tail further away from the accelerator axis, since $W_1(z) < 0$ if $|z|$ is short enough. Similarly, if the beam has an elliptical shape in its transverse distribution and thus possesses a quadrupole moment as shown in case (c), the transverse wake impulse is such that it elongates the ellipse further in the bunch tail. Case (a) shows what happens for $m = 0$, i.e. the longitudinal wake is retarding. In general, one finds that the polarity of the wake forces is such that it always *hurts* a short beam.

As $|z|$ increases, on the other hand, W_m' and W_m may change signs and the wakefields may become beneficial. In particular, W_0' may become negative at some finite distance behind the head of the beam. Therefore, if one injects two beam bunches into the accelerator and if the separation of the two bunches is chosen strategically, the trailing bunch can be *accelerated* by the wakefield of the leading bunch. This leads to the idea of *wakefield accelerators*.[9]

Transformer ratio limit We will not discuss the wakefield accelerators, but as an illustration of the property of the wake functions, we will show below that a straightforward application of the wakefield acceleration idea in which a short

[9]J. Rosenzweig, Handbook Accel. Phys. & Eng., Sec. 1.6.10 (2013); E. Esarey and C. Schroeder, Handbook Accel. Phys. & Eng., Sec. 7.3.13 (2013); G. -A. Voss and T. Weiland, DESY Report 82-074 (1982).

leading bunch is followed by a short trailing bunch traveling down the *same* path would not work very well. This is possibly counter intuitive as one might expect that by having an intense leading bunch, leaving a strong wakefield behind to accelerate a small trailing bunch, one should get a strong acceleration of the trailing bunch. Although true, there is more to it than this expectation.

Consider a wakefield accelerator consisting of a short leading bunch of charge N_1e and a short trailing bunch of charge N_2e a distance $|z|$ behind. After a distance L, the leading bunch loses an energy

$$\Delta\mathcal{E}_1 = -\frac{1}{2}N_1^2e^2W_0'(0^-)\,.$$

To maximize the energy gain by the trailing bunch, we design the wake potential and choose the relative position of the two bunches in such a way that $W_0'(z) = -W_0'(0^-)$ — there is no way to do better than that. The energy of the trailing bunch then changes by

$$\Delta\mathcal{E}_2 = \left(-\frac{1}{2}N_2^2 + N_1N_2\right)e^2W_0'(0^-)\,,$$

where the first term is the energy loss due to self-field, and the second term comes from the accelerating wakefield left behind by the leading bunch. For a given N_2, the acceleration rate of the trailing bunch can be made arbitrarily large by having a large N_1. However, a large N_1 also means the leading bunch is decelerated rapidly. In fact, let E_1 and E_2 be the energies of a single particle in the leading and trailing bunches, then the leading bunch will come to a full stop when it has exhausted all its energy, i.e., when $\Delta\mathcal{E}_1 = -N_1E_1$. At this point, the energy of a particle in the trailing bunch has increased by an amount

$$\Delta E_2 = \frac{\Delta\mathcal{E}_2}{N_2} = \left(2-\frac{N_2}{N_1}\right)E_1 < 2E_1\,. \tag{8.10}$$

An intense leading bunch indeed induces a large acceleration rate for the trailing bunch, but the total energy gained by each particle in the trailing bunch cannot exceed $2E_1$, no matter how intense the leading bunch is; most of the energy contained in the leading beam would be wasted. The *transformer ratio* $\frac{\Delta E_2}{E_1}$ cannot exceed 2.[10] To accelerate the trailing beam by 2 GeV, for instance, one has to prepare a 1 GeV leading beam first. Such a wakefield accelerator is not very efficient. Note that this conclusion applies regardless of details of the wakefield mechanism, whether it is provided by a dielectric wall, cavity structures, or a plasma medium.

To improve the transformer ratio, one can consider two beams in *different* paths. In these devices, the wakefield is generated in a special-purpose chamber, optimized, and sent by waveguides to accelerate another beam in a different

[10] J. Seeman, IEEE Trans. Nucl. Sci. NS-30, 3180 (1983); R. D. Ruth, A. W. Chao, P. L. Morton, and P. B. Wilson, Part. Accel. **17**, 171 (1985).

vacuum chamber. The transformer ratio can then be made larger than 2. This is the approach of klystrons, as well as some of the wakefield accelerator concepts.

Homework 8.6 The discussion leading to Eq. (8.10) assumes a short leading bunch. It is in fact possible, in principle at least, to defeat this limit on the transformer ratio by using a longer leading bunch. The wake potential of a long bunch can be obtained by superposition. Consider a wake function $W_0'(z<0) = W_0 \cos kz$ and a leading bunch with a triangular-shaped longitudinal charge density $\rho = \frac{Ne}{2\ell}(1-\frac{z}{\ell})$ if $|z|<\ell$ and 0 otherwise. Let the bunch length 2ℓ be n times the wavelength of the wake field, i.e., $k\ell = \pi n$. Show that by locating a short trailing bunch optimally behind the leading bunch, one can achieve a transformer ratio of $2\pi n$.[11] Figure 8.12 shows the longitudinal wake potential as a function of z for the case of $n=2$.

Figure 8.12: The longitudinal wake potential (solid curve) produced by the beam as a function of $\frac{z}{\ell}$, as discussed in Homework 8.6. The dashed curve gives $\rho(z)$. The vertical scales are unspecified.

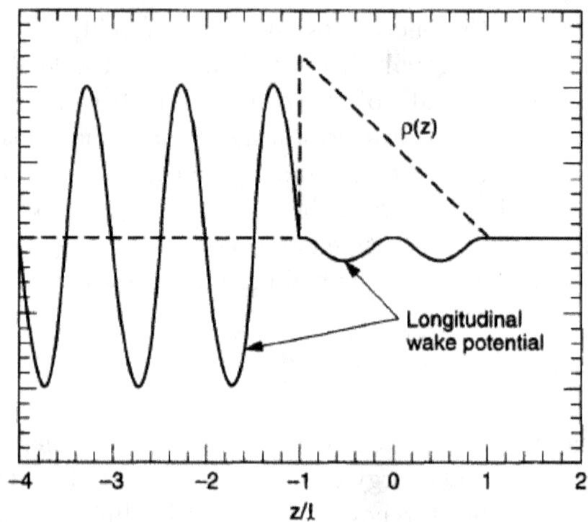

Homework 8.7 Consider a cavity structure with wake function $W_0'(z) = W_0 \cos kz$ with $z<0$. This wake function describes the cavity response to a δ-function beam. Now consider a step-function beam with line charge density λ and beam length ℓ.

(a) Give an expression for the energy change $\Delta E(z)$ for a test particle at location z within the beam.

(b) Calculate the total energy change of the beam.

[11] K. L. F. Bane, P. B. Wilson, and T. Weiland, AIP Proc. 127, Phys. High Energy Part. Accel., BNL/SUNY, 1983, p.875

(c) Is there a possibility that after the beam leaves, it deposits no wakefield, therefore the beam has no net energy loss? When that happens, the head of the beam loses energy while the tail recovers it.

Solution (a) $\Delta E(z) = \frac{e\lambda W_0}{k}\sin kz$ with $0 > z > -\ell$.

8.1.13 Fundamental theorem of beam loading

As mentioned earlier, there is also the important property that $W'_m(0)$, evaluated exactly at $z=0$, is always equal to $\frac{1}{2}$ times the value $W'_m(0^-)$ evaluated at z slightly less than 0. This is referred to as the *fundamental theorem of beam loading.*[12] The factor $\frac{1}{2}$ in the theorem comes from the fact that charges in a beam see the wake produced only by those charges in front of it and as a result see on the average only half of the total beam charge.

To prove the theorem in general, consider a beam with short but finite length that has an otherwise arbitrary longitudinal charge density $\rho(z)$. The beam loses energy at a rate

$$\frac{d\mathcal{E}}{ds} = -\int_{-\infty}^{\infty} dz'\,\rho(z')\int_{z'}^{\infty} dz\,\rho(z)\,E_z(z'-z)\,,$$

where $E_z(z'-z)$ is the longitudinal electric wakefield produced by a unit point charge and seen by another point charge a distance $|z'-z|$ behind in an arbitrary vacuum chamber environment. If the bunch length is short enough that E_z behaves like a step function within the bunch distribution, it becomes

$$\frac{d\mathcal{E}}{ds} \approx -E_s(0^-)\int_{-\infty}^{\infty} dz'\,\rho(z')\int_{z'}^{\infty} dz\,\rho(z).$$

By a change of variable from z to $u=\int_z^\infty dz'\rho(z')$, we have

$$\frac{d\mathcal{E}}{ds} \approx -E_s(0^-)\int_0^q u du = -\frac{q^2}{2}\,E_s(0\ )\,,$$

which proves the theorem. The derivation assumed nothing but causality.

A thought experiment This proof of the theorem may sound a bit too abstract. Perhaps we include an example application of the theorem by a thought experiment as follows.

Consider a setup as shown in Fig. 8.13. A point charge q traverses a cavity which has one and only one mode with wavelength λ. As the point charge leaves the cavity, a wakefield is excited in the cavity. Now send a second point charge q, at a distance $\frac{\lambda}{2}$ behind, through the cavity. It also excites a wakefield, but the two wakefields cancel each other by linear superposition. The net result is

[12] Perry B. Wilson, AIP Proc. 87, Fermilab, 1981, p. 450.

Figure 8.13: A thought experiment arrangement concerning the fundamental theorem of beam loading.

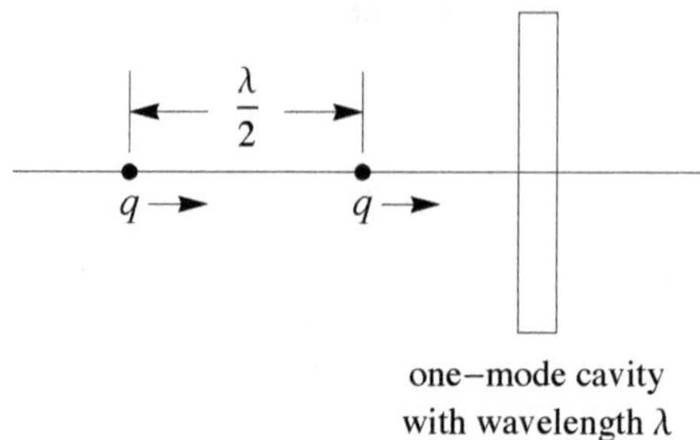

that the wakefield energy left behind by the first particle is completely absorbed by the second particle and there is no energy loss by the two particles together. It is easy to show (Homework 8.8) that in order for this to happen, one must have $W_0'(0) = \frac{1}{2} W_0'(0^-)$.

Another thought experiment Another more sophisticated thought experiment might also be helpful. Consider an infinite train of point charges, each with charge q, propagating with $v = c$ down the axis of a vacuum chamber pipe. Let the time spacing between adjacent point charges be T_s. Let there be an RF cavity structure with perfectly conducting walls at position $z = 0$ in the vacuum chamber. The wakefield of the RF has one and only one resonant frequency ω.

A steady state is established of this beam-cavity system. In this steady state, there is a wakefield generated by the beam and is always oscillating in the RF cavity without attenuation. Consider now a particular point charge traversing the cavity at time $t = 0$. Let the wakefield before the passage of the point charge $(t < 0)$ be

$$W(t < 0) = W_0 \cos(\omega t + \phi) ,$$

where W_0 and ϕ are yet to be determined by a steady state condition. After the passage, a new wakefield is generated by the point charge, so the wakefield becomes

$$W(t > 0) = W_0 \cos(\omega t + \phi) + W_1 \cos \omega t ,$$

where the second term is the new wakefield generated by the passing point charge. We know W_1 is positive, $W_1 > 0$.

Because this is a steady state, the wakefield must repeat its pattern at each passage of the point charges of the beam. This means $W(t > 0)$ must be the same as $W(t < 0)$ if we replace its t by $t - T_s$. In other words, we must have

$$W_0 \cos[\omega(t - T_s) + \phi] = W_0 \cos(\omega t + \phi) + W_1 \cos \omega t .$$

This expression has to be valid for all t within the range $0 < t < T_s$. By expanding in terms of $\sin(\omega t + \phi)$ and $\cos(\omega t + \phi)$, we obtain two necessary conditions,

$$\begin{aligned} W_0 \cos \omega T_s &= W_0 + W_1 \cos \phi \,, \\ W_0 \sin \omega T_s &= W_1 \sin \phi \,. \end{aligned}$$

The solution for steady state is found to be specified by

$$W_0 = \frac{W_1}{2 \sin \frac{\omega T_s}{2}}, \qquad \phi = \frac{\pi}{2} + \frac{\omega T_s}{2} \,.$$

We now calculate the energy change by the point charge as it traverses the RF cavity. Since this system is lossless, the point charge must not receive or deposit any net energy into the RF cavity in the steady state. This means the energy the point charge deposits into the cavity must be equal to the energy it receives from the wakefield at the time when it traverses the cavity. The energy it receives from the wakefield is equal to

$$\Delta \mathcal{E} = -W_0 \cos \phi \,.$$

Substituting W_0 and ϕ into this expression, we obtain

$$\Delta \mathcal{E} = -\frac{W_1}{2} \,.$$

The factor $\frac{1}{2}$ is just what the theorem demands.

When ωT_s approaches the value of 2π, the point charges arrive at the RF cavity in synchronism to the wakefield oscillation, a resonant condition is fulfilled. One expects the steady state wakefield to grow very high. Indeed this is reflected by the fact that $W_0 \to \infty$. However, even in that case, the phase ϕ is self-adjusted so that the energy gain due to the accumulated high wakefield is fixed to the value $\Delta \mathcal{E} = -\frac{W_1}{2}$, and as a result, there is no net energy change by each of the point charges as they interact with the RF cavity, assured by the fundamental theorem of beam loading.

Homework 8.8 Fill in the derivation of the fundamental theorem of beam loading for the first thought experiment.

Homework 8.9 Extend the second thought experiment to the case when the single mode contains a damping (finite Q value). The fundamental theorem of beam loading should still be fulfilled.

8.2 Impedance

8.2.1 Definition

We talked about frequency content of the wakefields and mentioned that its wavelengths cover a wide range from $\sim$1 μm to $\sim$1 m. The quantities that characterize the frequency content of the wakefields (more accurately, the frequency content of the impulses caused by the wakefields) are the impedances, which are just the Fourier transforms of the wake functions,

$$
\begin{aligned}
Z_m^{\parallel}(\omega) &= \int_{-\infty}^{\infty} \frac{dz}{c} e^{-i\omega z/c}\, W_m'(z)\,, \\
Z_m^{\perp}(\omega) &= i \int_{-\infty}^{\infty} \frac{dz}{c} e^{-i\omega z/c}\, W_m(z)\,.
\end{aligned}
$$

Since we have already discussed the wake functions, we consider these equations the definitions of impedances. The idea of representing the accelerator environment by an impedance was introduced by Sessler and Vaccaro.[13]

Strictly speaking, there is a difference between the frequency content in the impedance and the frequency content in the wakefield. The impedance contains only the information in the integrated impulse, while the wakefield contains much more. As a result, if you see a frequency in the impedance, it must be also a frequency contained in the wakefield, but the reverse is not necessarily true.

When an impedance as a function of ω contains sharp peaks, we refer to it as "narrow band". When its behavior in ω is smooth and wide, we refer to it as "broad band". By the property of Fourier transforms, it follows that a narrow band impedance gives rise to a wake function oscillating in z over a long range in z, while a broad band impedance would have a short-ranged rapidly-damped wake function.

To be more specific, for a narrow band impedance with a sharp peak at frequency ω_0 with a width $\Delta\omega \ll \omega_0$, its wake function oscillates with wave number $\frac{\omega_0}{c}$ and damps slowly with range $\frac{c}{\Delta\omega}$. For a broad-band impedance with a width of $\Delta\omega \gg \omega_0$, its wake function basically damps out quickly in a range $\frac{c}{\Delta\omega}$ without a chance to oscillate.

Instead of asking about wake functions, an accelerator designer therefore could alternatively ask about the impedances of the accelerator. The impedance is the quantity most directly related to the maximum beam current that can be accepted by the accelerator. This connection is easy to anticipate because the wakefield impulses can be expressed in units of volts, while the voltage is given by multiplying the beam current in amperes with the impedance in ohms.

It is perhaps useful to make a point about the factorization of physics at this junction. As mentioned in Sec. 1.1.7, useful definitions of quantities in physics usually try to factorize their physical origins. In accelerator physics, a quantity is most usefully defined so that it is either purely dependent of the accelerator properties, or purely dependent of the beam properties. A mixed quantity, if defined, will usually not be very useful, or serves only as an intermediate step,

[13] A. Sessler and V. Vaccaro, CERN Report ISR-RF/67-2 (1967).

waiting to be improved. Here, when we mentioned

$$\text{volt} = \text{ampere} \times \text{ohm},$$

the ampere is a beam quantity, the ohm is an accelerator quantity; their product, the volt, the end product that accelerator physicists want to know as they try to design an accelerator to match the envisioned beam, is factorized successfully. In the present language, we should insist that the impedances are pure property of the accelerator. So are the wake functions.

Inverting the Fourier transforms, we have

$$W'_m(z) = \frac{1}{2\pi}\int_{-\infty}^{\infty} d\omega\, e^{i\omega z/c}\, Z_m^{\parallel}(\omega),$$

$$W_m(z) = \frac{-i}{2\pi}\int_{-\infty}^{\infty} d\omega\, e^{i\omega z/c}\, Z_m^{\perp}(\omega).$$

By the nature of Fourier transforms, it follows that a long-range wake function is contributed by a narrow band impedance, while a short-range wake function comes from a broad band impedance.

It may be instructive to make a point about the time-domain picture (the wake functions) and the frequency-domain picture (the impedances). We mentioned before that humans seem to feel more "intuitive" with time-domain because we have memories. Nature, however, does not care to have a memory, and so she cares about the frequency domain much more – that is how most detectors work. Mathematically, however, time-domain and frequency-domain are exactly equivalent. There is no real reason one treatment is superior to the other. The same applies to the impedances and the wake functions.

The Panofsky-Wenzel theorem, which relates the longitudinal wake function to the derivative of the transverse wake function, also gives a relationship between the longitudinal and transverse impedances for a given m,

$$Z_m^{\parallel}(\omega) = \frac{\omega}{c} Z_m^{\perp}(\omega). \tag{8.11}$$

Note that this expression applies for a fixed mode number m. Transverse and longitudinal impedances at different m's do not necessarily have a rigorous connection.

Dimensionalities of the wake functions were discussed in Sec. 8.1.9. The dimensionalities of the impedances are

$$[Z_m^{\parallel}] = \Omega L^{-2m}, \qquad [Z_m^{\perp}] = \Omega L^{-2m+1}.$$

For example, $Z_0^{\parallel}$ would be in ohms; $Z_1^{\perp}$ would be in ohms per meter.

Homework 8.10 The wake functions and the impedances are connected by Fourier transforms. We know that it is difficult to calculate wake functions at small distances. The same difficulty also appears in the calculation of impedances to high frequencies. An often adopted technique is to perform a computer simulation of a relativistic Gaussian beam with some small but finite rms bunch length σ_z going through a vacuum chamber component and integrate the wakefield to find the wake function. By choosing σ_z (together with the matching mesh size and the time-step size in the simulation) to be small, at the limit of what the computer can handle, the wake function obtained approaches that of the true wake function. Once that is obtained, we make a Fourier transformation to obtain the impedance.

There is an error associated with this approach, however. Suppose $\bar{W}(z)$ is the wake function found numerically by using a Gaussian beam of rms length σ_z, and $\bar{Z}(\omega)$ is the impedance obtained by Fourier transforming $\bar{W}(z)$. Show that the actual impedance is given by

$$Z(\omega) = \bar{Z}(\omega)\, e^{\omega^2\sigma_z^2/2c^2} .$$

This equation means in principle one can obtain the true impedance $Z(\omega)$ by a simulation using a Gaussian beam and then apply the correction factor $e^{\omega^2\sigma_z^2/2c^2}$ to calculate the true impedance. In fact, after making this correction, the impedance can be extended beyond the expected limit of $\omega \ll \frac{c}{\sigma_z}$ and reach $\omega \sim \frac{c}{\sigma_z}$. However, extension to high frequencies using this result becomes doubtful due to the very strong divergence when $\omega > \frac{c}{\sigma_z}$.

8.2.2 Property of impedance

Just like the wake functions, there are some general properties of impedances which we describe below.

- $Z_m^{\parallel}(\omega) = \dfrac{\omega}{c} Z_m^{\perp}(\omega)$, (Panofsky-Wenzel theorem in frequency domain).

- $\left\{ \begin{array}{rcl} Z_m^{\parallel\,*}(\omega) &=& Z_m^{\parallel}(-\omega), \\ Z_m^{\perp\,*}(\omega) &=& -Z_m^{\perp}(-\omega), \end{array} \right.$ (reality of wake functions).

- $\left\{ \begin{array}{rcl} \int_0^\infty d\omega\, \mathrm{Im} Z_m^{\perp}(\omega) &=& 0, \\ \int_0^\infty d\omega\, \dfrac{\mathrm{Im} Z_m^{\parallel}(\omega)}{\omega} &=& 0, \\ \mathrm{Re} Z_m^{\parallel}(0) &=& 0, \end{array} \right.$ ($W_m(0) = 0$, in most cases except space charge).

- $$\left\{ \begin{array}{rcl} \mathrm{Re} Z_m^{\parallel}(\omega) &=& \dfrac{1}{\pi}\,\mathrm{P.V.} \displaystyle\int_{-\infty}^{\infty} d\omega' \, \frac{\mathrm{Im} Z_m^{\parallel}(\omega')}{\omega' - \omega}, \\ \mathrm{Im} Z_m^{\parallel}(\omega) &=& -\dfrac{1}{\pi}\,\mathrm{P.V.} \displaystyle\int_{-\infty}^{\infty} d\omega' \, \frac{\mathrm{Re} Z_m^{\parallel}(\omega')}{\omega' - \omega}, \end{array} \right.$$
  (causality, Hilbert transform).

The same expressions apply to $Z_m^\perp$. P.V. means taking the principal value of the integral. We will explain its meaning later after Eq. (8.43).

- $\begin{cases} \text{Re}Z_m^\parallel(\omega) \geq 0 \text{ for all } \omega, \\ \text{Re}Z_m^\perp(\omega) \geq 0 \text{ if } \omega > 0, \quad \leq 0 \text{ if } \omega < 0, \end{cases}$
(arbitrary beam can only lose energy to impedance).

- $Z_1^\perp \approx \frac{2c}{b^2\omega} Z_0^\parallel$, $Z_{m\neq 0}^\perp \approx \frac{2c}{b^{2m}\omega} Z_0^\parallel$, $Z_{m\neq 0}^\parallel \approx \frac{2}{b^{2m}} Z_0^\parallel$.
These are approximate expressions relating transverse and longitudinal impedances — not to be confused with the exact Panofsky-Wenzel theorem (8.11), b is the pipe radius. On the other hand, these expressions are exact for the case of resistive round pipe.

In a previous discussion, we associated physical meaning to the wake functions and discussed their properties in detail, particularly using some physical arguments of energy conservation and causality principle. Since we simply took the impedances as the Fourier transforms of the wake functions, we shall not elaborate again on their physical properties as listed above, with perhaps one side comment to be made concerning the Hilbert transform.

Mathematically, the fact that the real and the imaginary parts of the impedance form Hilbert transform pairs is a result of the causality principle, stated above without proof. This theorem dictates that if an impedance has a real part, it must also necessarily have an imaginary part, and vice versa. As a result, strictly speaking, a pure resistance $Z_m^\parallel(\omega) = R$, a pure inductance $Z_m^\parallel(\omega) = -i\omega L$ and a pure capacitance $Z_m^\parallel(\omega) = i\frac{C}{\omega}$ are all unphysical. The only physical representation with circuit elements is the LRC circuit with nonzero L, R, C, Sec. 8.2.4. See also Homework 8.12.

The reason of highlighting this comment is that sometimes one is tempted to test a theory of collective instability by applying it to simplified special cases such as pure resistance/inductance/capacitance — as occasionally found in the literature. This step can be unjustified as it may contain hidden divergences at $\omega \to 0$ or $\omega \to \infty$. On the other hand, we ourselves will sweep aside these divergences and violate this cautionary comment when we discuss the space charge impedance in the next section.

Homework 8.11 You probably already realized that the imaginary part of any impedance $Z_m^\parallel(\omega)$ must be an odd function of ω. This obviously means $\text{Im}Z_m^\parallel(\omega)$ must not be all positive or all negative over the entire range $-\infty < \omega < \infty$. Have you realized that $\text{Im}Z_m^\parallel(\omega)$ also must not be all positive or all negative over half of the range, i.e. the range $0 < \omega < \infty$? Convince yourself of this second realization.

Homework 8.12 Later we will introduce a transverse wake function model (8.34), i.e. $W_1(z < 0) = -W_0$, a constant over some distance. Show that its corresponding impedance is

$$Z_1^{\perp}(\omega) = W_0 \left[\frac{1}{\omega} - i\pi\delta(\omega) \right] .$$

Check the Hilbert transformation property.

8.2.3 Analytical example of impedance and wake function

Even armed with the powerful Panofsky-Wenzel theorem, to find the wake function and impedance for a given vacuum chamber discontinuity, one needs to solve Maxwell equations for the electromagnetic fields produced in the vacuum chamber. Over the years, a large arsenal of techniques had been developed to do these calculations. Most results involve numerically solving the associated boundary value problems, sometimes with a PIC code, sometimes assuming rigid beams. Relatively few special cases allow analytic solutions.

We mentioned earlier that there are three ways when wakefields are generated:

1. the beam is not relativistic;
2. the vacuum chamber is resistive or dielectric;
3. the vacuum chamber is not smooth.

Three cases, each representing one of these three ways, that permit analytical expressions are given in this section. For each example, we give the longitudinal and transverse wake functions and the corresponding impedances (derivations omitted). More examples can be found in the Handbook.[14]

Direct space charge We discussed the space charge effects in Chapter 1. The space charge wakefield and impedance come about when the beam is not sufficiently relativistic. In particular, we showed in Sec. 3.3.4 that the space charge effect induces incoherent horizontal and vertical betatron tune shifts, and these shifts are both negative. It also induces an incoherent longitudinal synchrotron tune shift that is positive above transition and negative below transition. We also discussed the fact that space charge force introduces not only the tune shifts for small-amplitude particles, but also tune spreads over the entire beam.

Figure 8.14 shows the space charge wakefields in the x-y plane driven by a ring-shaped, infinitely thin, $\cos m\theta$-distribution beam.

The z-dependence of the wake functions is δ-function in z. The space charge wakefield is δ-function localized as a pancake distribution. With a beam of

[14] K.Y. Ng, K. Bane, Sec. 3.2.4, Handbook Accel. Phys. & Eng., 2nd Ed., World Scientific (2013).

Figure 8.14: Patterns of space charge wakefields driven by a $\cos m\theta$ ring beam in the x-y plane for the cases $m = 0, 1, 2$.

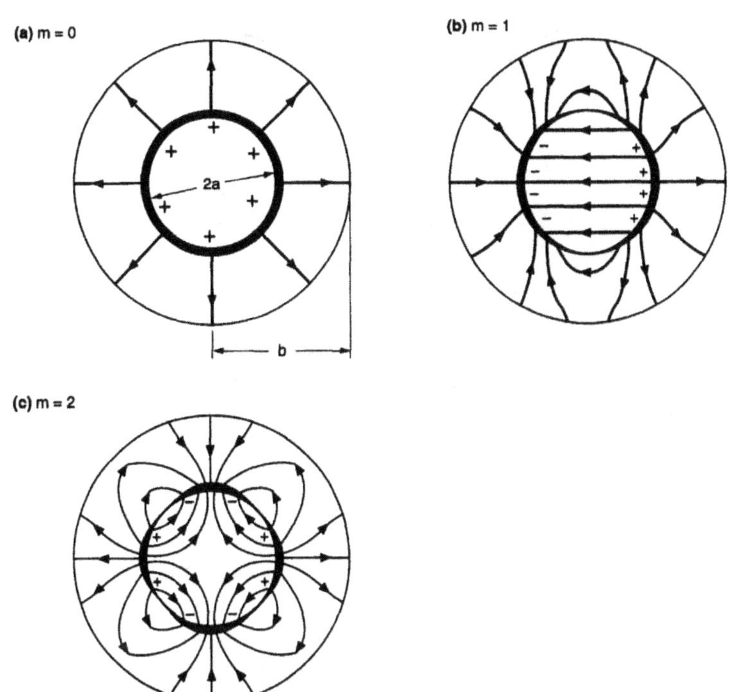

radius a in a perfectly conducting round pipe of radius b and length L, we have (derivation omitted)[15]

Impedances	Wake functions
$Z_0^{\parallel} = i\dfrac{Z_0 L\omega}{4\pi c\gamma^2}\left(1+2\ln\dfrac{b}{a}\right)$	$W_0' = \dfrac{Z_0 cL}{4\pi\gamma^2}\left(1+2\ln\dfrac{b}{a}\right)\delta'(z)$
$Z_{m\neq 0}^{\perp} = i\dfrac{Z_0 L}{2\pi\gamma^2 m}\left(\dfrac{1}{a^{2m}}-\dfrac{1}{b^{2m}}\right)$	$W_{m\neq 0} = \dfrac{Z_0 cL}{2\pi\gamma^2 m}\left(\dfrac{1}{a^{2m}}-\dfrac{1}{b^{2m}}\right)\delta(z)$

(8.12)

where $Z_0 = \sqrt{\mu_0/\epsilon_0} \approx 377\ \Omega$ is the free-space impedance, ϵ_0 and μ_0 are the free-space dielectric constant and magnetic permeability. Because of the factor $\frac{1}{\gamma^2}$, space charge effects are most significant for low-to-medium energy proton or heavy ion accelerators.

[15]See, for example, Alexander Wu Chao, Physics of Collective Beam Instabilities in High Energy Accelerators, Wiley (1993); K.Y. Ng, Physics of Intensity Dependent Beam Instabilities, World Scientific (2006).

The space charge impedance is purely imaginary, and is proportional to $i\omega$. Its ω-dependence is as if it is a pure inductance. However, its sign is as if it is a capacitance. By convention, we call the space charge impedance "capacitive".

Space charge effect is a very special case in collective instability analysis, and often becomes subtle when trying to study. Some of the subtleties are described below.

- As mentioned earlier, a true impedance must have both imaginary and real parts to satisfy the Hilbert causality condition. A purely imaginary impedance of the space charge as given here is strictly speaking illegitimate. However, we adopt this result here, with a caution to its possible divergences at high frequencies in some applications.

- The Panofsky-Wenzel theorem applies when the beam is ultrarelativistic, with $\gamma \to \infty$. Here with finite γ, the wake function and impedance concepts do not strictly apply. Yet, it turns out that insisting on formulating space charge in the language of wake functions and impedances almost fit perfectly – as long as we allow some degree of compromise on their physical properties. One of the properties to compromise is causality.[16]

- From the surface, the space charge wake functions as listed are δ-functions, i.e. there is no wake impulses ahead of the driving source, so it seems that causality is obeyed. However, the sign of the wake functions is such that a beam with finite bunch length will *gain* energy at the bunch head. The energy gain can only come from wake impulses exerted from the back of the bunch, violating causality after all. The reason this is even allowed is because $v < c$.

- Also as mentioned, impedances in general should be a property of the accelerator, and have nothing to do with the beam. The space charge impedances, however, depend on the beam energy and the transverse beam size.

- Of particular importance to the subtlety of space charge dynamics is the fact that the space charge force moves with the beam. This is in contrast to the other wall-induced impedances that are fixed in location relative to the vacuum chamber wall. As a result for example, a dipole oscillation — when the entire beam moves together and the wakefields move with it — under the influence of space charge is qualitatively different from the case of the other wall impedances. This is an exceptional property of the space charge effects. Directly applying the space charge impedance in existing formulae of collective instabilities may lead to wrong conclusions. Users beware!

[16] As if this is not a serious compromise.

- Because the space charge impedance is purely imaginary, it does not directly induce collective instabilities (see later). Often, however, because of its prevalent presence, it often combines with other instability effects (e.g. due to wall impedances, beam-beam perturbation, etc.) to generate a more complicated effect of an interplay. As a result, the net effect becomes involved and subtle.

Resistive wall Another case soluble analytically is for a round resistive pipe with radius b, conductivity σ_c, and length L. Defining the skin depth for the beam pipe [see Eq. (5.9)],

$$\delta_{\text{skin}} = \sqrt{\frac{2c}{|\omega| Z_0 \sigma_c}}, \qquad \delta_{\text{skin}}\,[\text{mm}] = \frac{1}{\sqrt{f\,[\text{MHz}]}} \times \begin{cases} 0.066, & \text{copper}, \\ 0.086, & \text{aluminum}, \\ 0.43, & \text{stainless steel}, \end{cases} \tag{8.13}$$

one finds

Impedances ($|\omega| \ll \chi^{-1/3}\frac{c}{b}$)	Wake functions ($0 > z$ and $|z| \gg \chi^{1/3} b$)
$Z_m^{\parallel} = \frac{\omega}{c} Z_m^{\perp}$	$W_m = -\frac{c}{\pi b^{2m+1}(1+\delta_{m0})}\sqrt{\frac{Z_0}{\pi\sigma_c}\frac{L}{\|z\|^{1/2}}}$
$Z_m^{\parallel} = \frac{1-\text{sgn}(\omega) i}{1+\delta_{0m}} \frac{L}{\pi\sigma_c \delta_{\text{skin}} b^{2m+1}}$	$W_m' = -\frac{c}{2\pi b^{2m+1}(1+\delta_{m0})}\sqrt{\frac{Z_0}{\pi\sigma_c}\frac{L}{\|z\|^{3/2}}}$

(8.14)

where χ was defined in Eq. (8.2).

The impedance is proportional to $(1-i)$, i.e. one might say that it is half resistive and half inductive. Note that the skin depth δ_{skin} is a frequency domain quantity. It appears only in the impedances, not in the wake functions.

These expressions of resistive wall wake functions are for the long-range case only, i.e. they have excluded the very short ranges with $|z| \lesssim \chi^{1/3} b$ (see Fig. 8.4), and therefore excluded the corresponding impedances at very high frequencies. An alert reader might note that the sign of $W_m'(0-)$ is apparently negative, violating the fundamental property of wake functions. However, the range of validity of Eq. (8.14) forbids setting $z \to 0^-$, and indeed a careful calculation of the resistive wall wake function shows that it flips sign in the very short range from $z = 0$ to $z \sim -\chi^{1/3} b$ as shown in Fig. 8.4.

There are two particularly notorious sources of collective instabilities due to long-range wakefields. One is due to the RF cavities. The other is due to resistive wall. The resistive wall impedance, having a power-law behavior in ω, diverges (or at least does not converge easily) at both high and low frequencies.

Furthermore, since the resistive wall impedance applies whenever there is a vacuum chamber wall, it is basically proportional to the circumference of the storage ring. The bigger the storage ring, the larger its resistive wall impedance. Resistive wall instabilities therefore tend to become increasingly prominent as the size of the accelerator increases.

Slowly varying wall boundary The third way to generate wakefields and impedances is when the vacuum chamber has discontinuities, even if perfectly conducting. Consider a case when the vacuum chamber wall varies along the accelerator slowly, a perturbation technique can be used to calculate the impedances. Specify the wall variation by $h(z)$ (cylindrically symmetric bump). At low frequencies $k = \frac{\omega}{c} < 1/(\text{bump length or width})$, the wake functions are[17]

$$W_m(z) = \begin{cases} -\frac{cZ_0}{4\pi}\delta(z)\int_{-\infty}^{\infty} d\zeta\,[h'(\zeta)]^2, & m=0, \\ -\frac{m}{m+1}\frac{cZ_0}{\pi}\delta(z)\int_{-\infty}^{\infty} d\zeta\left[\frac{h'(\zeta)}{h^m(\zeta)}\right]^2, & m>0, \end{cases}$$

and the impedance is purely inductive,

$$Z_m^{\parallel} = \frac{\omega}{c}Z_m^{\perp} = \begin{cases} \frac{-i\omega Z_0}{4\pi c}\int_{-\infty}^{\infty} d\zeta\,[h'(\zeta)]^2, & m=0, \\ -\frac{i\omega Z_0}{\pi c}\frac{m}{m+1}\int_{-\infty}^{\infty} d\zeta\left[\frac{h'(\zeta)}{h^m(\zeta)}\right]^2, & m>0. \end{cases}$$

When the boundary varies rapidly, this impedance formula breaks down, and most likely numerical calculation has to be applied. This formula is mostly applicable to shallow single smooth bumps on the vacuum chamber. A wavy structure with a wave length L like a bellows will not be applicable because the wave number being limited to $k \ll \frac{1}{L}$ automatically rules out the resonance frequencies of interest.

Homework 8.13 Apply the result of slowly varying wall boundary to calculate the wake functions $W_0'(z), W_1(z)$ and the impedances $Z_0^{\parallel}(\omega), Z_1^{\perp}(\omega)$ for the vacuum chamber with $h(z) = d\cos^2\frac{\pi z}{L}$ in the region $|z| < \frac{L}{2}$ and zero otherwise.

Solution One of the results is $Z_0^{\parallel}(\omega) = -i\frac{\pi d^2 Z_0}{8cL}\omega$. Stretching its applicability, take $d = \frac{c}{\omega} = L = b$ for a cavity structure of the size of the vacuum chamber pipe radius b, we see $Z_0^{\parallel}(\omega = \frac{c}{b}) \sim -i\frac{\pi}{8}Z_0$, consistent with what is expected of this cavity in the broad band resonator model to be discussed later.

Homework 8.14

(a) Repeat Homework 8.13 for the two vacuum chamber geometries shown in Fig. 8.15. Give your results in appropriate units.

(b) An alert reader might question what about the three kinks in the function $h(z)$. Strictly speaking, their contributions diverge. In reality, these kinks are rounded off somehow. Investigate an argument to estimate their contributions, either by rounding off the corners or by frequency shielding, and give the condition when they can be ignored.

[17] K. Yokoya, CERN SL/90-88 (AP) (1990).

Figure 8.15: Impedances of two slowly varying beam pipes.

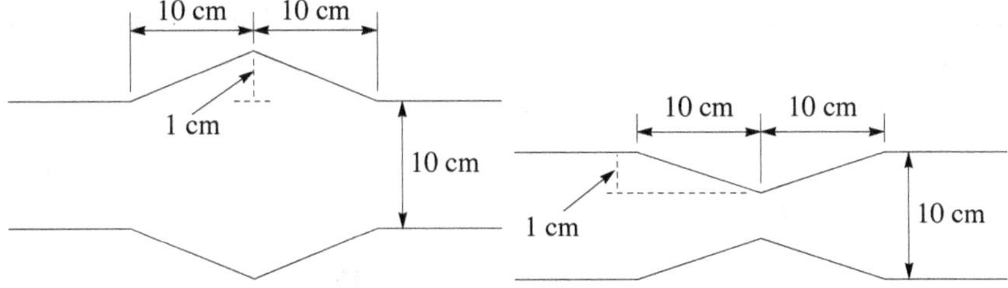

8.2.4 Resonator model

Sometimes, lacking explicit analytical results, one can proceed using analytical models, which do not necessarily represent concrete physical cases but are analytically accessible and obey all the necessary properties of the true physical wake functions and impedances.

Figure 8.16: An *LRC* circuit for resonator impedance.

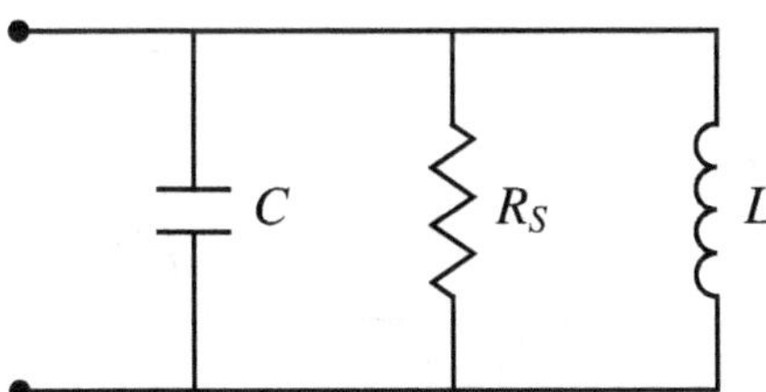

The *LRC* circuit One example of such is the resonator model, modeled after an equivalent parallel *LRC* resonator circuit, shown in Fig. 8.16.[18]

The impedance of the circuit is given by

$$\frac{1}{Z_m^{\|}} = \frac{1}{R_S} + \frac{i}{\omega L} - i\omega C ,$$

or

$$Z_m^{\|} = \frac{R_S}{1 + iQ\left(\frac{\omega_R}{\omega} - \frac{\omega}{\omega_R}\right)} , \tag{8.15}$$

[18] A. Hofmann, K. Hübner, and B. Zotter, IEEE Trans. Nucl. Sci. NS-26, 3514 (1979); P. B. Wilson et al., IEEE Trans. Nucl. Sci. NS-24, 1211 (1977); A. Hofmann, Proc. 11th Int. Conf. High Energy Accel., Geneva, 1980, p. 540.

Figure 8.17: The resonator impedances. The upper panel is the longitudinal impedance of the circuit for $Q = 10$. The lower panel is the transverse impedance of the circuit for $Q = 10$. The solid curves give the real parts and the dashed curves give the imaginary parts of the impedances.

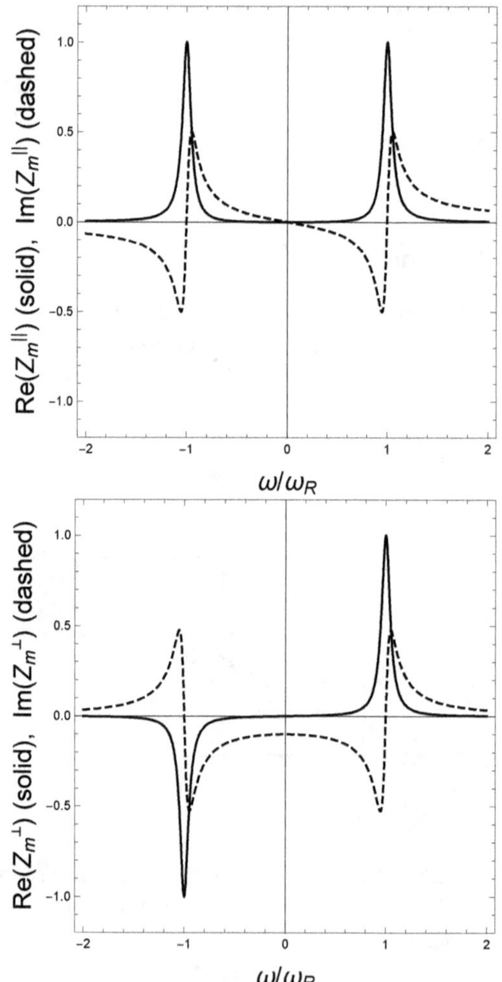

where $Q = R_S\sqrt{C/L}$ is the quality factor and $\omega_R = 1/\sqrt{CL}$ is the resonant frequency of the circuit, R_S has the dimension of ΩL^{-2m}. The width (half width at half maximum) of the resonance peak of $\mathrm{Re}Z_m^{\|}(\omega)$ is about $\Delta\omega \approx \frac{\omega_R}{2Q}$ if $Q \gg 1$. The quality factor Q is typically ~ 1 for a broad-band resonator, $\sim 10^4$ for a fine-tuned sharp resonator, or $\sim 10^{9-10}$ for a superconducting structure. Figure 8.17 shows the resonator impedances $Z_m^{\|}(\omega)$ and $Z_m^{\perp}(\omega)$.

The wake function $W'_m(z)$ can be obtained by performing a Fourier transformation on the impedance (Homework 8.16),

$$W'_m(z) = \begin{cases} 0 & \text{if } z > 0 \\ \alpha R_S & \text{if } z = 0 \\ 2\alpha R_S \, e^{\alpha z/c} \left(\cos\frac{\bar\omega z}{c} + \frac{\alpha}{\bar\omega}\sin\frac{\bar\omega z}{c}\right) & \text{if } z < 0 \end{cases}, \tag{8.16}$$

where $\alpha = \frac{\omega_R}{2Q}$ and $\bar\omega = \sqrt{\omega_R^2 - \alpha^2}$.

At low frequencies $|\omega| \ll \omega_R$, the LRC resonator impedance $Z_m^{\parallel}(\omega) \approx -i\omega L$ is inductive. For $|\omega| \gg \omega_R$, we have $Z_m^{\parallel}(\omega) \approx \frac{i}{\omega C}$, which is capacitive. Around the resonant frequency ω_R, the impedance $Z_m^{\parallel}(\omega) \approx R_S$ is mostly resistive. By convention, one calls an impedance *inductive* or *capacitive* according to the sign of $\mathrm{Im}Z_m^{\parallel}$ in the region $\omega > 0$; an inductive or capacitive impedance then means $\mathrm{Im}Z_m^{\parallel} < 0$ or > 0, respectively. Therefore, the resistive-wall impedance is partly inductive and partly resistive, and the space charge impedance, in spite of the fact that it is proportional to ω like that of an inductance, is referred to as purely capacitive.

The Panofsky-Wenzel theorem requires that the same resonator also gives a transverse impedance,

$$Z_m^{\perp} = \frac{c}{\omega}\frac{R_S}{1 + iQ\left(\frac{\omega_R}{\omega} - \frac{\omega}{\omega_R}\right)}.$$

This impedance has been shown in Fig. 8.17. The corresponding transverse wake function is (for $z < 0$)

$$W_m(z) = \frac{cR_S\omega_R}{Q\bar\omega}\, e^{\alpha z/c} \sin\frac{\bar\omega z}{c}.$$

Sharp resonators A sharply peaked impedance can sometimes be modeled by taking the limit $Q \to \infty$ and $R_S \to \infty$ but holding $\frac{R_S}{Q}$ fixed, i.e.

$$\mathrm{Re}Z_m^{\parallel} = \frac{\pi}{2}\frac{R_S\omega_R}{Q}\left[\delta(\omega - \omega_R) + \delta(\omega + \omega_R)\right],$$

$$\mathrm{Im}Z_m^{\parallel} = \frac{R_S\omega_R}{2Q}\left(\frac{1}{\omega - \omega_R} + \frac{1}{\omega + \omega_R}\right).$$

The corresponding wake function is

$$W'_m(z) = \frac{R_S\omega_R}{Q}\cos\frac{\omega_R z}{c} \qquad \text{for} \quad z < 0.$$

The reader might note that this particularly simple wake function model has been used in Homeworks 8.6 and 8.7, and has been used in the two thought experiments in Sec. 8.1.13.

Similarly, for the transverse resonator impedance, with $Q \to \infty$, $R_S \to \infty$, and fixed $\frac{R_S}{Q}$, we have

$$\begin{aligned}
\mathrm{Re} Z_m^\perp &= \frac{\pi c R_S}{2Q} \left[\delta(\omega - \omega_R) - \delta(\omega + \omega_R) \right], \\
\mathrm{Im} Z_m^\perp &= \frac{c R_S \omega_R}{2Q\omega} \left(\frac{1}{\omega - \omega_R} + \frac{1}{\omega + \omega_R} \right),
\end{aligned}$$

and

$$W_m(z) = \frac{c R_S}{Q} \sin \frac{\omega_R z}{c}.$$

These sharp resonator model wake functions oscillate indefinitely without damping and with infinitely long range.

Broad-band resonator impedance For the study of wakefield effects at short ranges, it is often useful to consider a broad-band resonator model. It gives ball park estimates of the impedances and their effects on the beam sometimes just from the rough geometrical dimensions of a vacuum chamber object.

For example, if one considers the case of a discontinuity of a cavity structure whose size and depth are of the order of b, the pipe radius, by an energy loss argument, it can be shown that the $m = 0$ short range wakefield can be approximated by a broad-band resonator with parameters

$$R_S \approx 60\,\Omega, \qquad Q \approx 1, \qquad \omega_R \approx \frac{c}{b}. \tag{8.17}$$

The bulk of the impedance can be represented as a broad-band resonator. The Q value is always about equal to 1, the shunt impedance is always about 60 Ω as long as the cavity structure has a size similar to the vacuum pipe radius b, details of cavity shape aside. The only thing that depends on b is resonator frequency ω_R. The value of $60\,\Omega$ can also be represented as

$$\frac{Z_0}{2\pi} \approx 60\,\Omega.$$

The same cavity structure also has a transverse impedance. It can be obtained by relating it to the longitudinal impedance. For $m = 1$, we have for $\omega \lesssim \frac{c}{b}$,

$$Z_1^\perp \approx \frac{Z_0}{2\pi b} \approx 60\,\Omega \times \frac{1}{b}.$$

A deep cavity in a $b = 5$ cm pipe, therefore, contributes approximately $Z_1^\perp \approx$ 1.2 kΩ/m in the neighborhood of $\frac{\omega}{2\pi} \lesssim 1$ GHz. In terms of the resonator parameters, we have for $m = 1$,

$$R_S \approx 60\,\Omega \times \frac{1}{b^2}, \qquad Q \approx 1, \qquad \omega_R \approx \frac{c}{b}. \tag{8.18}$$

The broad-band resonator model does not give an accurate description of the impedance at high frequencies $\omega \gg \frac{c}{b}$; it predicts a purely capacitive impedance, while a more accurate model — the diffraction model[19] — predicts an impedance half capacitive and half resistive [See Eq. (8.31)]. For this reason, the model does not give accurate wake functions at very short ranges $|z| \ll b$.

The broad-band resonator model ignores the possibility of the cavity responding resonantly to sharply defined frequencies, i.e. it ignores the possible existence of trapped cavity modes. These cavity modes occur at frequencies below the cutoff frequency $\sim \frac{c}{b}$, and they give rise to wake functions which ring for long periods of time. Neglecting these long-range contributions, the model also does not give accurate wake functions at very long ranges $|z| \gg b$.

In spite of its limitations, however, the broad-band resonator model gives a reasonable estimate of the impedance near $\omega \sim \frac{c}{b}$ and wake functions near $|z| \sim b$. For applications where the beam bunch length is comparable to the cavity size, it provides a simple and handy tool.

Broad-band resonator wake function The broad-band impedance (8.17) for $m = 0$ and (8.18) for $m = 1$ have their corresponding wake functions. Figure 8.18 shows the longitudinal wake function $\frac{W_0'(z)}{c}$ and the transverse wake function $\frac{W_1(z)}{c}$ for the cases $b = 5$ and 10 cm.

The longitudinal wake functions at short range have the properties that

$$\begin{aligned} W_0'(0^-) &= c\,\frac{60\,\Omega}{b}\,, \\ W_1(0^-) &= 0\,, \\ W_1 \text{ peak value} &\approx -c\,\frac{60\,\Omega}{b^2}\,, \quad \text{occurring at } z \approx -b\,, \end{aligned}$$

as are reflected in Fig. 8.18.

However, as mentioned, at short range, there are high frequency contributions neglected by the broad-band resonator model, meaning that the calculated wake functions are not applicable when $z \to 0$. These expressions are only to be read as

$$\begin{aligned} W_0'(z \sim -b) &\approx c\,\frac{60\,\Omega}{b} \quad \text{or} \quad \frac{cZ_0}{2\pi b}\,, \\ W_1(z \sim -b) &\approx -c\,\frac{60\,\Omega}{b^2} \quad \text{or} \quad -\frac{cZ_0}{2\pi b^2}\,. \end{aligned}$$

Take the SLAC linac for example, if we take $b \approx 1$ cm, then $W_0'(z \sim -b) \approx 6\,c\,k\Omega\,\mathrm{m}^{-1}$. When high frequency contributions are taken into account, the wake function at very short range increases to about $W_0'(0^-) \approx 21\,c\,k\Omega\,\mathrm{m}^{-1}$.[20]

[19] J. D. Lawson, Rutherford Lab. Report RHEL/M 144 (1968), and Part. Accel. 25, 107 (1990); K. Bane and M. Sands, Part. Accel. 25, 73 (1990); R. B. Palmer, Part. Accel. 25, 97 (1990).

[20] K. Bane and P.B. Wilson, Proc. 11-th Int. Conf. High Energy Accel., Geneva, p. 592 (1980).

Figure 8.18: The broad-band resonator wake functions $\frac{W_0'(z)}{c}$ and $\frac{W_1(z)}{c}$ as functions of z ($z < 0$) for the two cases $b = 10$ cm (solid) and $b = 5$ cm (dashed). Wake functions vanish for $z > 0$.

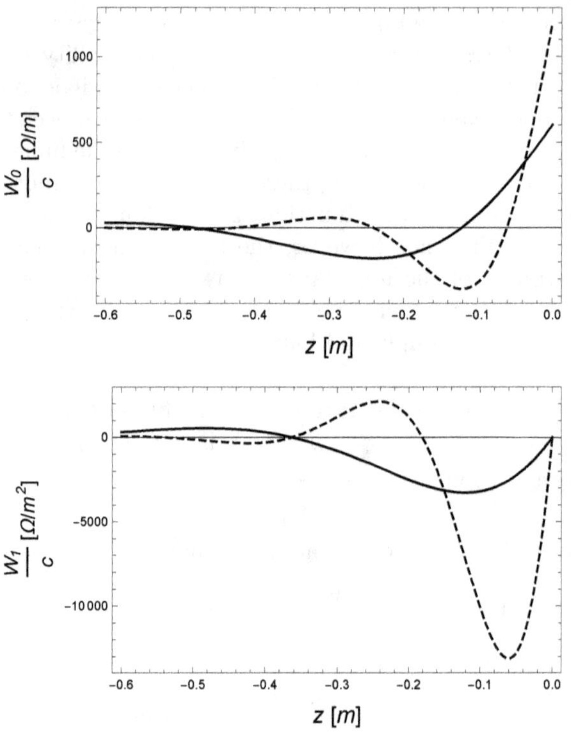

Since W_m and W_m' are always multiplied by e^2 in applications, we add a practical formula here,

$$e^2 c \times 60\,\Omega \;\approx\; e^2 c \times \frac{Z_0}{2\pi} \;=\; 2mc^2 r_0\,, \tag{8.19}$$

where r_0 is the classical radius of the particle under consideration.

Homework 8.15 The text introduced the *LRC* circuit with Fig. 8.16. An alternative view of the same resonator impedance can be represented by a circuit sketched in Fig. 8.19. Confirm that the voltage at the output is related to the input voltage by an impedance that is given by Eq. (8.15). The output voltage is the one across the cavity in this representation. We have $\frac{V_{out}}{V_{in}} = \frac{Z_m^{\parallel}}{R_s}$.

Homework 8.16 Given the impedance (8.15), perform a Fourier transform to obtain the wake function (8.16).

Figure 8.19: An alternative LRC circuit for resonator impedance.

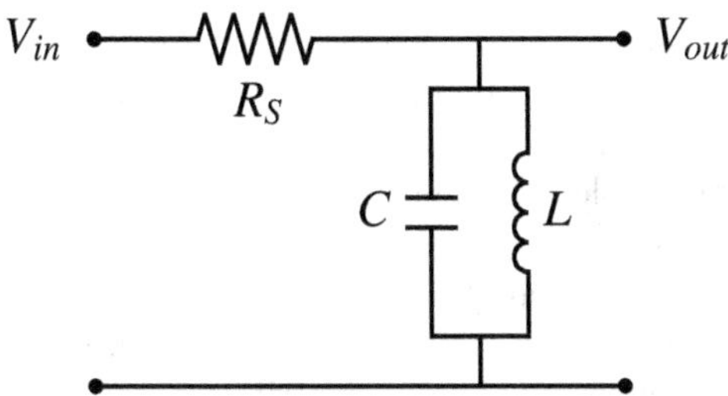

Solution It should be warned that the Fourier transform is not to be performed too casually. Simply inserting (8.15) into Mathematica, for example, might yield a wrong answer. The correct answer must use the correct integration contours in the complex ω-plane by insisting on the causality condition that the wake function vanishes for $z > 0$.

8.2.5 The quantity $\frac{Z}{n}$ for circular accelerator

We mentioned before that for longitudinal instabilities, it is the $m = 0$ effects that dominate. This means we must pay most attention to $Z_0^{\parallel}$. Furthermore, in some accelerators (most likely the earlier ones, such as an older synchrotron), beam bunches are relatively long. When the bunches are longer than the vacuum chamber pipe radius, often it is the lower frequency part of the impedance that dominates the collective instabilities.

From the properties of impedances, we know that $Z_0^{\parallel}(\omega = 0) = 0$ — take the LRC impedance for example. This, together with the fact that $\mathrm{Re}Z_0^{\parallel}(\omega)$ is an even function of ω and $\mathrm{Im}Z_0^{\parallel}(\omega)$ is odd, says that the impedance must be purely imaginary near $\omega \approx 0$ with $\mathrm{Im}Z_0^{\parallel}(\omega) \propto \omega$. For older synchrotrons, the instability therefore must be mainly specified by the low-frequency slope of $\mathrm{Im}Z_0^{\parallel}(\omega)$, i.e. the main quantity of focus is going to be $\frac{\mathrm{Im}Z_0^{\parallel}(\omega)}{\omega}$ at $\omega \to 0$. Such a quantity is called $\frac{Z_0^{\parallel}}{n}$, or $\frac{Z}{n}$ for short. We therefore ask the question, "what is the value of $\frac{Z}{n}$ of your synchrotron?" By $\frac{Z}{n}$, it means

$$\frac{Z}{n} = \left[\text{the low-frequency slope of Im}\frac{Z_0^{\parallel}(\omega)}{\omega}\right] \times \omega_0 \,,$$

where ω_0 is the revolution frequency of the synchrotron. As we shall see later, the quantity $\frac{Z}{n}$ plays a critical role in the evaluation of many — but not all — collective instability effects in circular accelerators.

An alert reader would have noted that the quantity $\frac{Z}{n}$ actually is extracting the inductance out of the impedance. Indeed, this is why sometimes $\frac{Z}{n}$ is quoted in units of henries instead of ohms.

The same alert reader may point out that an inductance is purely imaginary. As we alluded to earlier and will elaborate more later, to first order of the impedance, instabilities were caused only by the real parts of the impedance, and imaginary parts only induce collective frequency shifts. Indeed this is a correct observation as well. Instabilities occur either when the real part of the impedance also plays a role, or when nonlinear effects such as mode coupling enter the picture. In either case, $\frac{Z}{n}$ remains a key parameter.

As an example, consider a single cavity of the size comparable to the vacuum chamber radius b. As mentioned, this cavity can be represented by a broad-band resonator impedance. In terms of $\frac{Z}{n}$, it contributes

$$\frac{Z}{n}(\text{per cavity}) \approx 60\ \Omega \times \frac{\omega_0}{\omega_R} = 60\ \Omega \times \frac{b}{R},$$

where $\omega_R = \frac{c}{b}$ and $\omega_0 = \frac{c}{R}$ and $2\pi R$ is circumference of the circular accelerator.

Careless limit If one imagines a carelessly built circular accelerator in which the vacuum chamber is filled with all sorts of cavities and discontinuities of approximately the same size as the pipe radius, the total impedance around the circumference is $Z_0^{\parallel} \approx n_{\text{cav}} \times 60\ \Omega$, where $n_{\text{cav}} \approx 2\pi R/2b = \pi R/b$ is the total number of cavities, assuming each cavity occupies a longitudinal space of $2b$, and ignoring any interference effects of adjacent cavities. We then find that this carelessly designed accelerator has a total $\frac{Z}{n}$ of

$$\frac{Z}{n}\ (\text{total}) \approx 60\ \Omega \times \frac{b}{R} \times n_{\text{cav}} \approx 60\pi\ \Omega = \frac{Z_0}{2},$$

where $Z_0 = 377\ \Omega$ is the vacuum impedance.[21] This *careless limit* is independent of the accelerator size R and the pipe size b. Basically, one cannot do worse than this limit.

In case a fraction f of the accelerator is filled with cavities, one has

$$\frac{Z}{n} \approx f \times \frac{Z_0}{2}. \tag{8.20}$$

In a typical modern accelerator, attempts are made to make $\frac{Z}{n}$ substantially less than 1 Ω or so. This means the vacuum chamber has to be sufficiently smooth (and conducting) to suppress the impedance by a factor of at least a few hundred compared with the careless limit.

---

[21] Z_0 is the impedance of the vacuum. Yes, vacuum has impedance. An oscillating electromagnetic source will readily radiate into the vacuum. In fact, vacuum impedance is very large, meaning vacuum is an excellent (probably the best possible, considering its infinite band width) radiator. A well designed accelerator will have impedance only a small fraction of the vacuum impedance. Accelerator vacuum chambers are poor radiators purposely by design.

A recap A recap of what we have developed so far might be instructive at this point. Note that $\frac{Z}{n}$ is just a single quantity, in units of ohms (or henries). In comparison, the impedance is a much more complicated quantity because it is a function of frequency. By using $\frac{Z}{n}$, the entire problem of beam instability is reduced to a question of how large is the value of a single quantity $\frac{Z}{n}$.

Recall that we started by introducing the Panofsky-Wenzel theorem, which drastically reduced a 3-D Vlasov-Maxwell problem to the concepts of wake functions and impedances. We have now introduced $\frac{Z}{n}$, which distills the beam dynamics in terms of a single quantity and thus represents another step of an impressive accomplishment by the accelerator physicists over the decades to address the complex problem of collective beam instability.

抽絲剝繭 滴水穿石的功夫
Dripping water drills through rocks.

The fact that these drastic simplifications are even possible is a lucky blessing from Mother Nature plus generations of hard efforts.

For the more modern accelerators, however, particularly when the beam bunches get shorter and peak intensities get higher, the single value $\frac{Z}{n}$ no longer provides a complete picture. The collective instability problem becomes more involved. For these applications, we need not only $\frac{Z}{n}$ but also the entire impedance functions, particularly at high frequencies. Unfortunately, high frequency is also where impedances are difficult to measure or to calculate.

Homework 8.17 Use the expression for $Z_0^{\|}(\omega)$ for the case of space charge, find an expression of $\frac{Z}{n}$ for a storage ring made purely of a perfectly conducting pipe of radius b and a round beam with radius a and $\gamma \neq \infty$. Convince yourself that in this case as well, its $\frac{Z}{n}$ is a large factor smaller than the careless limit.

8.3 Macroparticle model in linear accelerator

We have now described wakefields and impedances. We still need to describe how to use these quantities to calculate beam instabilities. For example, given the impedance, we would want to know if the beam of a certain intensity and beam bunch length is stable, and if it is unstable, what is its growth rate?

As mentioned before, there are a large number of instability mechanisms found over the years. The beam can be stable or unstable depending on which instability mechanism it encounters. For some mechanisms, the beam is stable below a certain intensity threshold, while unstable above it. For some other mechanisms, the beam can be always unstable without a threshold, but is fortunately stabilized by some stabilizing counter-mechanisms such as radiation damping or Landau damping. Each instability occurs under some different conditions and at different beam intensities. But there are many of them, so one would actually encounter a whole list of all kinds of possible instabilities. After encountering one instability mechanism and curing it, one is likely to meet the next one at a higher intensity.

In the remaining of this chapter, we will briefly describe a few of these instability mechanisms or collective effects,

- collective effects in a linac,
- parasitic heating,
- potential-well distortion
- Robinson instability,
- strong head-tail instability,
- head-tail instability.

8.3.1 Beam energy spread in a linac

Consider a bunch of charged particles traveling down the accelerator in a vacuum chamber pipe. The $m = 0$ wakefield excited by the beam produces a longitudinal force on particles in the beam. The main effect of this longitudinal force is a retarding voltage, causing energy changes of individual particles. As a result, there is a net energy loss of the beam to the wakefields. Furthermore, since not all particles in the bunch lose the same amount of energy, the wakefield also causes the beam to acquire an energy spread.

One-particle model Consider first a one-particle model in which the beam bunch is represented by a macroparticle of charge Ne that consists of N particles of charge e. As this macroparticle beam travels down the linac, it experiences the self-generated retarding longitudinal field and loses energy accordingly. The parasitic energy loss per particle of charge e is

$$\Delta E = -\frac{1}{2}Ne^2 W_0'(0^-),$$

where the factor $\frac{1}{2}$ is due to the fundamental theorem of beam loading.

Take the SLAC linac for example. We have $W_0'(0^-) = 21\,c\,k\Omega\,\mathrm{m}^{-1} \times L_0/L$, where $L_0 = 3$ km is the total length of the linac, and $L = 3.5$ cm is the length of the cavity period. We then estimate the parasitic loss of 2.2 GeV per electron for $N = 5 \times 10^{10}$.

Two-particle model This estimate can be improved by using a two-particle model. The beam bunch is then represented by two macroparticles, one leading and another trailing at a distance $|z|$ behind, each with charge $\frac{Ne}{2}$. The parasitic loss per particle in the leading macroparticle is half of that for a one-particle model, i.e., 1.1 GeV, because the leading macroparticle contains only half the beam population. The trailing macroparticle loses, in addition to the 1.1 GeV due to self-field, an energy of

$$\Delta E = -e\left(\frac{Ne}{2}\right) W_0'(z)$$

due to the wakefield left behind by the leading macroparticle. If we take $z = -\sigma_z = -1$ mm, $N = 5 \times 10^{10}$, and $W_0'(-1 \text{ mm}) = 13.5\, c\, k\Omega\, \text{m}^{-1} \times L_0/L$, each particle in the trailing macroparticle loses an additional 1.4 GeV. The net loss of a trailing particle is therefore 2.5 GeV.

Comparing the one- and the two-particle model results, the one-particle model estimates a parasitic loss per particle of 2.2 GeV; the two-particle model estimates an average loss of $(1.1 + 2.5)/2 = 1.8$ GeV. The results agree reasonably well, but the two-particle model offers the additional information that the wakefield has introduced an energy split between the bunch head and the bunch tail, and its magnitude is approximately 1.4 GeV by the time the bunch reaches the end of the linac. If the beam energy at the end of the linac is 50 GeV, this energy spread across the bunch is approximately 3%.

Energy extraction efficiency For linear collider applications, this energy spread makes it difficult to focus the beam to a small spot at the collision point in a final focus system, and is to be avoided.

One concern for a high-intensity linear collider can be described as follows. The energy spread at the end of the linac scales as

$$\frac{\Delta E}{E} \approx \frac{\frac{1}{2}Ne^2 W_0'}{GL_0} \approx \frac{Nmc^2 r_0}{2Gb^2},$$

where G is the acceleration gradient, and $W_0' \approx \frac{cL_0}{b^2} \times 30\,\Omega$ (setting the cell length $L \approx 2b$) is the longitudinal wake function, where b is the vacuum chamber radius characterizing the size of the accelerating cavities. We have used Eq. (8.19) here.

On the other hand, the efficiency of energy extraction by the beam from the field energy U stored in the accelerating cavities $[U \approx \frac{\epsilon_0}{2}(\frac{G}{e})^2 \times \pi b^2 L_0]$ is

$$\text{extraction efficiency} \approx \frac{NE}{U} \approx \frac{8Nmc^2 r_0}{Gb^2},$$

which is equal to 16 times the energy spread. In other words, to improve the uncompensated energy spread of the beam at the end of the linac would require sacrificing the energy extraction efficiency.

One way to ameliorate this problem is to compensate $\frac{\Delta E}{E}$ by phasing the RF voltage. But to do so requires sacrificing acceleration gradient in a rather significant way. Another way is to send a *train* of M bunches per filling of the RF cavities. This will increase the energy extraction efficiency by a factor of M, although at the cost of having to deal with the multibunch interactions due to the long range wakefields.

Distribution of particles We now depart from the simplified models and consider a bunch with a general longitudinal charge density distribution $\rho(z)$. The energy change for a test charge e at longitudinal position z can be written as $eV(z)$, where

$$V(z) = -\int_z^\infty dz' \rho(z') W_0'(z - z'), \tag{8.21}$$

or equivalently

$$V(z) = -\frac{1}{2\pi} \int_{-\infty}^{\infty} d\omega \, e^{i\omega z/c} Z_0^{\parallel}(\omega) \tilde{\rho}(\omega), \tag{8.22}$$

where

$$\tilde{\rho}(\omega) = \int_{-\infty}^{\infty} dz \, e^{-i\omega z/c} \rho(z).$$

A Gaussian beam has $\rho(z) = \frac{q}{\sqrt{2\pi}\sigma_z} e^{-z^2/2\sigma_z^2}$ and $\tilde{\rho}(\omega) = q e^{-\omega^2\sigma_z^2/2c^2}$.

A negative $V(z)$ means the test charge loses energy from the wakefield. An additional integration of $V(z)$ over the bunch then gives the total parasitic loss,

$$\Delta E = \int_{-\infty}^{\infty} \rho(z) V(z) dz \,.$$

For a bunch with Gaussian longitudinal distribution and uniform disk transverse distribution, for example, the energy spread due to space charge effect is, using Eqs. (8.12) and (8.21),

$$\frac{V(z)}{L} = \frac{1}{4\pi\epsilon_0} \sqrt{\frac{2}{\pi}} \frac{q}{\gamma^2 \sigma_z^2} \left(\ln \frac{b}{a} + \frac{1}{2} \right) \frac{z}{\sigma_z} e^{-z^2/2\sigma_z^2}, \tag{8.23}$$

where q is the total charge of the Gaussian bunch.

Generally, particles in the front of the bunch ($z > 0$) lose energy to wakefields, while particles in the back of the bunch ($z < 0$) may gain or lose energy depending on the length of the bunch. This is not true for the special case of the space charge effect, for which particles in the front of the bunch gain energy, and particles in the back of the bunch lose energy. This is not unexpected as it would be a natural consequence of a Coulomb repulsion. In fact, other than a Lorentz factor of $\frac{1}{\gamma^2}$ and a numerical geometric form factor, Eq. (8.23) represents simply a Coulomb repulsion force.

For the space charge effect, the energy gained by the bunch head is necessarily given up by the bunch tail so that the net energy of the bunch is unchanged. As mentioned earlier, the fact that the bunch head *gains* energy from the bunch tail is not considered a violation of causality because $v < c$ in this case.

Consider a numerical example of a 1 GeV (kinetic energy) proton transport line. If we take $q = 10^{11}\, e$, $\sigma_z = 10$ cm, $a = 2$ cm, and $b = 5$ cm, we obtain a longitudinal space charge force of $\pm 2.3\,\mathrm{kV/m}$ for particles located at $z = \pm\sigma_z$. The net energy change of these particles after traveling 100 m of this transport line is ± 0.23 MeV. The space charge induced beam momentum spread is $\frac{\Delta P}{P} = \frac{1}{\beta^2}\frac{\Delta E}{E} = \pm 1.6 \times 10^{-4}$.

For a resistive wall, we have, using Eqs. (8.14) and (8.22),[22]

$$\begin{aligned} \frac{V(z)}{L} &= \frac{qc}{8\pi b\sigma_z^{3/2}}\sqrt{\frac{Z_0}{2\sigma_c}}\, f\left(\frac{z}{\sigma_z}\right), \\ f(u) &= -|u|^{3/2} e^{-u^2/4}\left[(I_{-1/4} - I_{3/4})\,\mathrm{sgn}(u) - I_{1/4} + I_{-3/4}\right], \end{aligned} \tag{8.24}$$

with the Bessel functions $I_{\pm 1/4}$ and $I_{\pm 3/4}$ evaluated at $u^2/4$. The function $f(u)$ is plotted in Fig. 8.20.

Figure 8.20: The function $f(u)$ for the resistive wall energy loss.

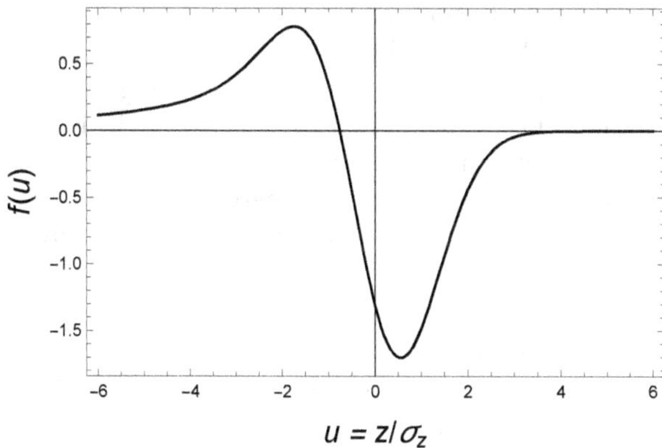

Integrating the resistive wall wake (8.24) over the bunch, we obtain

$$\frac{\Delta E}{L} = -\frac{q^2 c}{4\pi^2 b\sigma_z^{3/2}}\Gamma\left(\frac{3}{4}\right)\sqrt{\frac{Z_0}{2\sigma_c}}, \qquad \Gamma\left(\frac{3}{4}\right) = 1.225, \tag{8.25}$$

with $\Gamma(x)$ the Gamma-function.

Continuing the above numerical example, assuming an aluminum pipe $\sigma_c = 3.6 \times 10^7\ \Omega^{-1}\mathrm{m}^{-1}$, a particle located at $0.5\,\sigma_z$ ahead of bunch center loses an energy of 47 eV after traveling 100 m, and a particle located at $1.8\,\sigma_z$ behind the bunch center gains 22 eV. The average energy loss is 22 eV per particle.

Homework 8.18 Prove Eq. (8.23). Then show that the total parasitic loss of the bunch is $\Delta E = 0$.

[22] A. Piwinski, DESY Report HERA 92-11 (1992).

Homework 8.19

(a) Prove Eq. (8.24).

(b) Then derive Eq. (8.25) for parasitic loss for resistive all.

Solution

(a) This is a more subtle problem than homework 8.18. A straight application of Eq. (8.21) diverges. Use Eq. (8.22) instead. First show

$$f(u) = -\frac{2}{\pi}\int_{-\infty}^{\infty} dx\, e^{-x^2/2+ixu}[1-i\,\mathrm{sgn}(x)]\,|x|^{1/2}\,.$$

(b) First show the identity $\int_{-\infty}^{\infty} du\, e^{-u^2/2} f(u) = -\frac{4}{\sqrt{2\pi}}\,\Gamma(\frac{3}{4})$.
It is advised to find the reason why direct application of Eq. (8.21) is a wrong thing to do while use of Eq. (8.22) is permitted.

8.3.2 Beam breakup

In the previous section, the beam was considered to be centered in a cylindrically symmetric vacuum chamber pipe. There were no transverse wakefields. In case the beam is off-centered, for example due to its executing a betatron oscillation, an $m=1$ dipole wakefield will be excited by the head of the bunch, which causes transverse deflection of the bunch tail. For a high-intensity beam, the betatron motion of the bunch tail can be seriously perturbed, leading to a transverse breakup of the beam. The first observation of this beam breakup effect was made on the SLAC linac.[23]

Coasting beam Let us consider first a coasting beam when the beam is not being accelerated in the linac. To proceed with a simplified macroparticle model, we first note that a one-particle model is not useful here because, unlike the longitudinal $m=0$ case, a point charge does not exert a transverse wake force on itself. We have to start with a two-particle model.

In the two-particle model, the leading macroparticle, unperturbed by its own transverse wakefield, executes a free betatron oscillation

$$y_1(s) = \hat{y}\cos k_\beta s\,,$$

where s is the distance coordinate along the linac and k_β is the betatron wave number. The trailing macroparticle, at a distance $|z|$ behind, sees a deflecting wakefield left behind by its leading partner. According to Table 8.2, we have

$$\begin{aligned} y_2'' + k_\beta^2 y_2 &= -\frac{Ne^2 W_1(z)}{2EL}\, y_1 \\ &= -\frac{2\pi\epsilon_0 N r_0 W_1(z)}{\gamma L}\,\hat{y}\cos k_\beta s\,, \end{aligned} \qquad (8.26)$$

[23]R. B. Neal and W. K. H. Panofsky, Science **152**, 1353 (1966); W. K. H. Panofsky and M. Bander, Rev. Sci. Instr. 39, 206 (1968).

where $E = \gamma m_0 c^2$ is the beam particle energy, r_0 is the classical radius of the particle, $W_1(z)$ is the transverse wake function per cavity period, and L is the cavity period. In writing down this equation of motion, we have assumed smooth betatron focusing (i.e., k_β is independent of s) and $k_\beta L \ll 1$, so that the wakefield can be averaged over cavity periods when describing the particle motion. We have also ignored acceleration of the beam energy for simplicity. For the SLAC linac, $k_\beta \approx 0.06\ \text{m}^{-1}$ and $k_\beta L \approx 0.002$.

Equation (8.26) describes a driven oscillator. It describes an oscillator driven exactly on resonance. The solution to Eq. (8.26) is

$$y_2(s) = \hat{y}\left[\cos k_\beta s - \frac{\pi\epsilon_0 N r_0 W_1(z)}{k_\beta \gamma L} s \sin k_\beta s\right],$$

in which the first term describes the free oscillation and the second term is the resonant response to the driving wakefield. The amplitude of the second term grows linearly with s. The fact that the trailing macroparticle is being driven exactly on resonance by the wakefield is the reason why beam breakup instability is so potentially damaging.

At the end of the linac, the oscillation amplitude of the bunch tail relative to the bunch head is characterized by the dimensionless growth parameter

$$\Upsilon = -\frac{\pi\epsilon_0 N r_0 W_1(z) L_0}{k_\beta \gamma L}, \tag{8.27}$$

where L_0 is the total linac length. For short bunches, $W_1(z) < 0$, the parameter Υ is positive.

One may continue to extend the analysis to a three-particle beam. In this case, the third macroparticle is driven by the wakefields left behind by both leading macroparticles. In particular, driven by the second macroparticle, it will acquire a resonant amplitude that grows quadratically as s^2. In general, the m-th macroparticle will have an amplitude $\propto s^{m-1}$. It is easy to envision an exponentiation of the beam tail's response to the transverse wakefields, leading to the expectation that the intense beam will acquire a banana shape as it propagates down the linac. The net effect goes slower than a full-blown exponentiation like $e^{\alpha s}$ because only the $(m-1)$-th macroparticle, and not the entire beam, contributes to the m-th power in the growth rate. When proper analysis is performed, the actual exponentiation behaves as $e^{(\alpha s^{1/3})}$.[24]

The behavior of the banana beam is as sketched in Fig. 8.21. The sequence of snapshots shown reflects the fact that the motion of the bunch head is described by $\cos k_\beta s$, while the deviation of the bunch tail relative to the bunch head is described by $s \sin k_\beta s$. In particular, when the bunch head is at a maximum displacement, the tail lines up with the bunch head, but when the bunch head displacement is zero, the tail swing is maximum. As the beam propagates down the linac, the swing amplitude of the flapping tail increases with s until the tail

[24] A.W. Chao, B. Richter, and C.Y. Yao, Nucl. Instr. Meth. 178, 1 (1980).

Figure 8.21: Sketch of the beam breakup behavior viewed at four consecutive moments. The dashed curves indicate the trajectory of the bunch head. The phase of the flapping tail is not drawn arbitrarily.

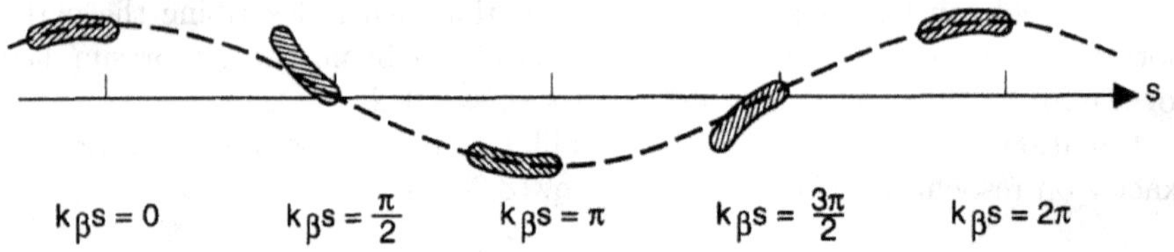

Figure 8.22: Two possible snapshot pictures of a propagating beam.

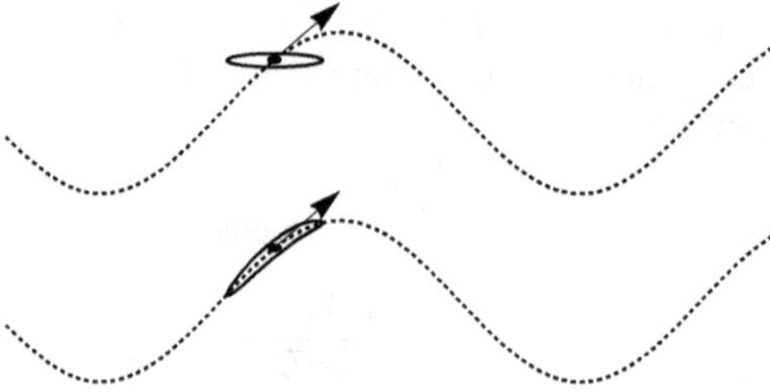

breaks up and particles are lost. Note that the sign of the tail swing shown is not arbitrary, because $\Upsilon > 0$.

Incidentally, how does a particle bunch move in an accelerator? If we take a snapshot of the beam bunch, which one of the pictures in Fig. 8.22 is correct? You may not care about this question if the bunch length is short, but you must care if the bunch is long.

Figure 8.23 shows four transverse beam profiles observed at the end of the SLAC linac with a beam intensity of $N = 2 \times 10^{10}$.[25] The leftmost profile was observed when the beam was carefully steered so that its trajectory was well centered in the beam pipe. When the beam was injected off center by 0.2, 0.5, and 1 mm, the beam profiles are as shown successively to the right. One observes that a tail develops as the injection offset is increased. The curling of the tail indicates the offset has both horizontal and vertical components. The beam sizes σ_x and σ_y are about 120 μm.

Consider a beam coasting down the SLAC linac at 1 GeV without acceleration. The tail swing is magnified by a factor of $\Upsilon \approx 180$ compared to the bunch head if we take $N = 5 \times 10^{10}$, $W_1(-1 \text{ mm}) = -0.21c\,\text{M}\Omega\ \text{m}^{-2}$, $L_0 = 3$ km, $L = 3.5$ cm, and $k_\beta = 0.06\ \text{m}^{-1}$. To preserve the beam emittance, it is

[25] J.T. Seeman, K.L.F. Bane, T. Himel, and W L. Spense, Part. Accel. 30, 97 (1990).

Figure 8.23: Transverse beam profile observed at the end of the SLAC linac when a beam breakup occurs. [Courtesy John Seeman, 1991]

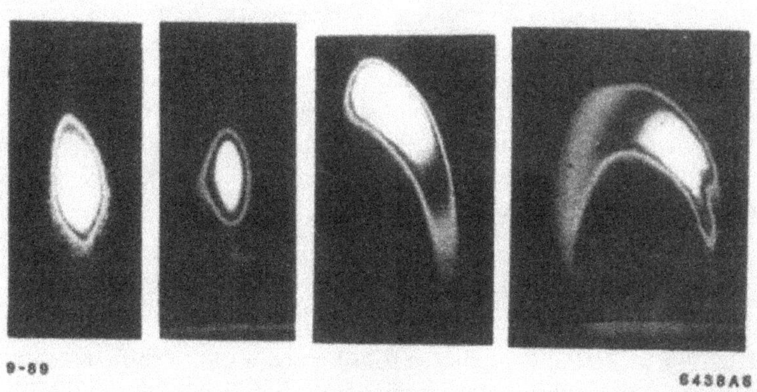

necessary to have $|\Upsilon \hat{y}| \ll$ transverse beam size. This means the beam must be injected onto the linac axis with an accuracy better than a fraction of a per cent of the beam size, which is difficult to achieve.

Accelerated beam So far we have ignored beam acceleration, which has an important stabilizing effect because, as its energy increases, the beam becomes more rigid and less vulnerable to the wakefields. Repeating a similar analysis but taking into account of acceleration and adiabatic damping yields the growth parameter

$$\Upsilon = -\frac{\pi \epsilon_0 N r_0 W_1(z) L_0}{k_\beta \gamma_f L} \ln \frac{\gamma_f}{\gamma_i},$$

which is basically simply replacing the factor L_0/γ by its integral counterpart $\int_0^{L_0} \frac{ds}{\gamma(s)}$. Due to acceleration, the tail amplitude thus grows logarithmically rather than linearly with s, and the growth parameter is much reduced. If the beam is accelerated in the SLAC linac from 1 to 50 GeV, the factor Υ becomes 14, instead of 180, which would be the case for a beam coasting at 1 GeV.

Ion-induced beam breakup So far we have been considering the wakefield generated by a vacuum chamber wall impedance. In this discussion, we consider a special type of beam breakup occurring in a long linac. Assuming in the vacuum chamber there is some residual gas. As the beam bunch passes through, the head of the bunch ionizes the residual gas molecules, rendering them into ions, and the ions perturb the motion of particles in the tail of the bunch. If the linac is long, this may potentially cause a beam breakup.[26]

[26] A.W. Chao, SLAC Report CN-121 (1981).

Let ions be produced at a rate of R per unit distance along the linac per single particle in the beam in air at atmospheric pressure $P_{\rm atm}$. Consider a particle in the very tail of the bunch. Assuming the ionization electrons are immediately repelled by the beam, the tail particle sees only naked ions, assumed to have charges Qe each. The number of ions per distance seen by this particle is $NR\frac{P}{P_{\rm atm}}$, where N is the number of particles in the beam bunch, P is the vacuum pressure. Assume the beam has a round transverse cross-section with radius a, it follows that the equation of motion for the tail particle with charge e, due to the ion effect alone, is

$$x'' = \left(\frac{2NQr_0R}{\gamma a^2}\frac{P}{P_{\rm atm}}\right)x\,.$$

This ion perturbation contributes to a betatron phase shift to the tail particle. At the exit of the linac, the accumulated betatron phase is

$$\Delta\psi = \frac{1}{2}\int_0^L ds\,\beta(s)\left(\frac{2NQr_0R}{\gamma(s)a(s)^2}\frac{P}{P_{\rm atm}}\right)\,.$$

where we have integrated over the entire linac of length L, and $\beta(s), a(s), \gamma(s)$ take their local values. It turns out that the quantity $\frac{\gamma(s)a^2(s)}{\beta(s)}$ is just the normalized emittance ϵ_N, a quantity independent of s. Then we find

$$\Delta\psi = \frac{NQr_0RL}{\epsilon_N}\frac{P}{P_{\rm atm}}\,.$$

Take a long electron linac for example, and take $R = 1000\,{\rm m}^{-1}, P = 10^{-6}$ Torr, $Q = 1, \epsilon_N = 10^{-6}$ m-rad, $N = 5\times 10^{10}$, $L = 3$ km, we find $\Delta\psi = 0.55$ rad.

This phase shift accumulated over the long linac due to ion-induced beam breakup does not seem a serious concern. However, when such a beam disruption is applied in a storage ring, the seemingly small effect can accumulate quickly turn after turn to cause an effect called the *fast ion instability*.[27] We shall not discuss this instability as it is outside of our intended scope.

8.3.3 BNS damping

The beam breakup instability is quite severe even with acceleration. To control it, the beam has to be tightly focused, rapidly accelerated, and carefully injected, and its trajectory carefully steered down the linac. It turns out, however, that there is another interesting and effective method to ameliorate the situation. This method, known as the *BNS damping*[28] is described next.

[27]T. Raubenheimer, F. Zimmermann, Phys. Rev. E52, 5487 (1995).

[28]V. Balakin, A. Novokhatsky, and V. Smirnov, Proc. 12th Int. Conf. High Energy Accel., Fermilab, 1983, p. 119.

We proceed with a two-particle model. Consider first the case without acceleration, where the leading macroparticle executes a free betatron oscillation. The idea of BNS damping requires introducing a slightly stronger betatron focusing of the bunch tail than the bunch head. The equation of motion of the tail particle can be written as

$$y_2'' + (k_\beta + \Delta k_\beta)^2 y_2 = -\frac{2\pi\epsilon_0 N r_0 W_1(z)}{\gamma L}\hat{y}\cos k_\beta s\,.$$

The solution, assuming $|\Delta k_\beta / k_\beta| \ll 1$, is

$$y_2(s) = \hat{y}\cos(k_\beta + \Delta k_\beta)s + \frac{\pi\epsilon_0 N r_0 W_1(z)}{k_\beta \Delta k_\beta \gamma L}\,\hat{y}\big[\cos(k_\beta + \Delta k_\beta)s - \cos k_\beta s\big]\,.$$

Compared with the previous case, one observes that, by introducing a slightly different focusing strength for the bunch tail, the beam breakup mechanism of the bunch head resonantly driving the bunch tail is removed. A further inspection shows that there exists a magical condition for the bunch tail to follow the bunch head exactly for all s, namely

$$\frac{\pi\epsilon_0 N r_0 W_1(z)}{k_\beta \Delta k_\beta \gamma L} = -1\,,$$

or equivalently,

$$\frac{\Delta k_\beta}{k_\beta} = -\frac{\pi\epsilon_0 N r_0 W_1(z)}{k_\beta^2 \gamma L} = \frac{\Upsilon}{k_\beta L_0}\,,$$

where Υ was defined in Eq. (8.27), and $k_\beta L_0$ is the total betatron phase advance of the linac. For short bunches, Υ and Δk_β are positive; the betatron focusing required to fulfill the BNS condition is therefore stronger at the bunch tail than at the bunch head.

The mechanism of BNL damping is not just to shift the frequencies of the driven particles away from resonance. Under the BNS condition, $y_2(s) = y_1(s) = \hat{y}\cos k_\beta s$. The trailing macroparticle executes an apparent free betatron oscillation but at the *unperturbed* frequency k_β, not $k_\beta + \Delta k_\beta$, and the beam no longer breaks up. Physically, this happens because the additional external focusing force introduced for the bunch tail has compensated for the defocusing dipole deflection force due to the wakefield left behind by the bunch head. Note that the BNS focusing has to be adjusted according to the beam intensity.

There are different ways to provide the BNS focusing. One is to introduce a radio frequency quadrupole whose strength changes as the bunch passes by so that the head and tail of the bunch see different quadrupole strengths. Another is to choose the location of the bunch relative to the acceleration RF voltage. By properly choosing the phase of the RF voltage relative to the beam bunch, the betatron focusing required by the BNS condition can be obtained. A third way is to invoke the longitudinal wakefield taking advantage of the fact that its polarity is in fact in line with what is needed for the BNS damping, i.e. in suppressing particle energy at the bunch tail. The energy spread across the bunch then causes a spread in betatron focusing according to

$$\frac{\Delta k_\beta}{k_\beta} = \xi \frac{\Delta E}{E},$$

where ξ is the chromaticity, which is determined by the linac lattice design.

8.4 Parasitic heating

When a beam bunch of charge q and line density $\rho(t)$ traverses an impedance region in the vacuum chamber, it suffers some energy loss to the impedance. Let the longitudinal impedance be $Z_0^{\|}(\omega)$. This *parasitic energy loss* (sometimes also called HOM heating, HOM means "higher order mode") by the beam bunch is

$$\Delta E = -\kappa^{\|} q^2,$$

where $\kappa^{\|}$ is the *loss factor*, often expressed in units of V/pC,

$$\kappa^{\|}(\sigma) = \frac{1}{\pi} \int_0^\infty d\omega \, \mathrm{Re} Z_0^{\|}(\omega) \, |\tilde{\rho}(\omega)|^2, \tag{8.28}$$

where $\sigma = \frac{\sigma_z}{\beta c}$ is the rms beam bunch length in sec. For a Gaussian bunch, $\rho(t) = \frac{1}{\sqrt{2\pi}\,\sigma} e^{-t^2/2\sigma^2}$, $\tilde{\rho}(\omega) = e^{-\omega^2\sigma^2/2}$.

Incidentally, Eq. (8.28), together with the fact that parasitic energy loss must be positive for arbitrary beam distribution $\rho(t)$, provides the proof that the real part of any impedance $Z_0^{\|}(\omega)$ must be positive for all ω.

One can also write the loss factor in terms of the wake function. For a Gaussian bunch,

$$\kappa^{\|} = \frac{1}{2\sqrt{\pi}\,\sigma} \int_{-\infty}^0 \frac{dz}{c} \, W_0'(z) \, e^{-z^2/4\sigma^2 c^2}.$$

Taking the limit $\sigma \to 0$, this equation proves (again) that any wake function must have the property that $W_0'(0^-) > 0$, and $\kappa(\sigma \to 0) = \frac{1}{2} W_0'(0^-)$.

Parasitic loss gives rise to heating of the vacuum chamber wall where there are impedances. For example, in high intensity electron storage rings, the beam position monitors or bellows can heat up and sometimes get burned. This is especially serious when short bunches are required. As a bunch gets shorter, the loss to the low frequency impedances stays the same, but more and more of the higher frequency impedances begin to participate, and the loss increases monotonically as the bunch length shortens. Furthermore, high-frequency wakefields penetrate into finer structures in various corners of the vacuum chamber so are more difficult to identity.

Open structures Different impedances contribute to varying degrees of parasitic losses. Examples of particularly lossy structures include RF cavities, unshielded vacuum pump ports, and vacuum chamber crotches, for which the vacuum chamber contains large open regions for the wakefields to radiate. The latter two are sketched in Fig. 8.24 — such structures are ideally suited to scrape off wakefields to form HOM losses. To reduce the parasitic loss, vacuum pump ports should be shielded. In comparison, a crotch, sometimes used when two beams are made to merge, cannot be shielded, so is typically a location where special cooling will be needed.

Figure 8.24: Two examples of particularly lossy vacuum chamber geometries, an unshielded vacuum pump port and a pipe crotch.

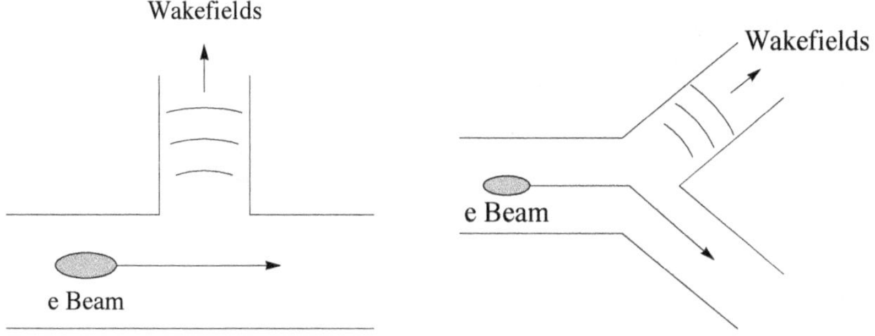

Imaginary impedance Only the real (resistive) part of the impedance contributes to the parasitic loss. Imaginary impedances, such as the space charge or the slowly varying wall impedances, are purely imaginary and therefore do not introduce a net energy loss to the beam. However, this only means the energy loss by particles at the head of the bunch is recovered by particles in the tail of the bunch, so that there are energy transfers but no net energy loss of the entire beam bunch. As mentioned before, the bunch head loses energy and bunch tail recovers the energy in the case of slowly varying wall. Space charge acts the opposite and thus puts causality in question.

Resistive wall In general, it happens that the parasitic loss becomes large for short bunches. Parasitic heating is mainly a problem for high intensity, and short, bunches. For example, Eq. (8.25) says for a cylindrically symmetric resistive wall,

$$\frac{\kappa^{\|}(\sigma)}{L} = \frac{\Gamma(\frac{3}{4})c}{4\pi^2 b\sigma_z^{3/2}}\left(\frac{Z_0}{2\sigma_c}\right)^{1/2}. \tag{8.29}$$

Cavity structure Most of the parasitic loss occurs as the beam encounters a discontinuous structure in the vacuum chamber pipe. Part of the wakefield gets trapped by the structure if the structure is cavity-like and if the wakefield frequency is below the cutoff frequency of the pipe. This trapped field energy is eventually deposited as heat on the cavity walls. The rest of the wakefield, with frequency above cutoff, propagates down the pipe and eventually deposits its energy on lossy material elsewhere in the vacuum chamber.[29]

For a bunch traversing a cavity structure, $\kappa^{\|}$ is given by a sum over cavity modes below cut-off, plus a contribution above cut-off. Each of the cavity modes below cut-off contributes a resonator impedance, and each resonator impedance contributes

$$\kappa^{\|} \approx \begin{cases} \dfrac{\omega_R R_S}{2Q} e^{-\omega_R^2 \sigma^2}, & \text{high-}Q \text{ resonator}, \\ \dfrac{\omega_R R_S}{2Q}, & \text{low-}Q \text{ resonator, short bunch } \omega_R \sigma \ll 1, \\ \dfrac{R_S}{4\sqrt{\pi}\, Q^2 \omega_R^2 \sigma^3}, & \text{low-}Q \text{ resonator, long bunch } \omega_R \sigma \gg 1. \end{cases} \tag{8.30}$$

Above cut-off, the impedance per cavity can be represented by the diffraction model,[30]

$$Z_0^{\|}(\omega) = [1 + \text{sgn}(\omega) i] \frac{Z_0}{2\pi^{3/2}} \frac{1}{b} \sqrt{\frac{cg}{|\omega|}}, \tag{8.31}$$

where g is the gap size of the cavity. This impedance has both real and imaginary parts. The corresponding wake function is

$$W_0'(z < 0) = \frac{Z_0 c}{2\pi^2} \frac{\sqrt{2g}}{b} |z|^{-1/2}.$$

Circular accelerator So far we have considered single passages of a beam traversing an impedance. For a single bunch in a circular accelerator, the integral in Eq. (8.28) is replaced by an infinite sum,

$$\kappa^{\|}(\sigma) = \frac{\omega_0}{2\pi} \sum_{p=-\infty}^{\infty} Z_0^{\|}(p\omega_0) \, |\tilde{\rho}(p\omega_0)|^2 .$$

For short bunches in large accelerators ($\omega_0 \ll \frac{1}{\sigma}$), the sum can be replaced by an integral, and the difference between single passes and multiple passes disappears as it should. On the other hand, when there is a sharp impedance peak resonating at or near a frequency $p\omega_0$, there can be a large enhanced local heating in the case of a storage ring.

[29] Next time you put a potato in a microwave oven, think of it as the impedance.

[30] J.D. Lawson, Rutherford Lab. Report RHEL/M 144 (1968); Karl Bane and Matthew Sands, Part. Accel. 25, 73 (1990).

Parasitic loss is the culprit The parasitic energy lost by the beam goes into wakefields. Typically, only a small fraction of the particle energy is depleted to produce the wakefields, and most of the energy stored in the wakefields ends up as heat on the vacuum chamber walls. But under unfavorable conditions, a small portion of the wakefield energy can be transferred systematically back to beam motion, causing beam instabilities. The parasitic loss, therefore, is ultimately responsible for the various collective beam instabilities. How the wakefields affect the beam dynamics and what are the mechanisms of the various collective beam instabilities are subjects to which we will have to study. The parasitic energy loss itself, of course, will have to be supplied back to the beam by an RF accelerating voltage.

Homework 8.20 Equation (8.29) in the text was derived using the expression of the resistive wall impedance. The parasitic heating per unit length of resistive wall pipe is given by $\frac{\kappa^{\parallel}}{L}$ which is found to be proportional to $\frac{1}{b\sigma_z^{3/2}\sqrt{\sigma_c}}$. Construct a physical explanation of the resistive wall loss by considering Fig. 8.25.

Figure 8.25: Schematic illustration of the resistive wall loss mechanism.

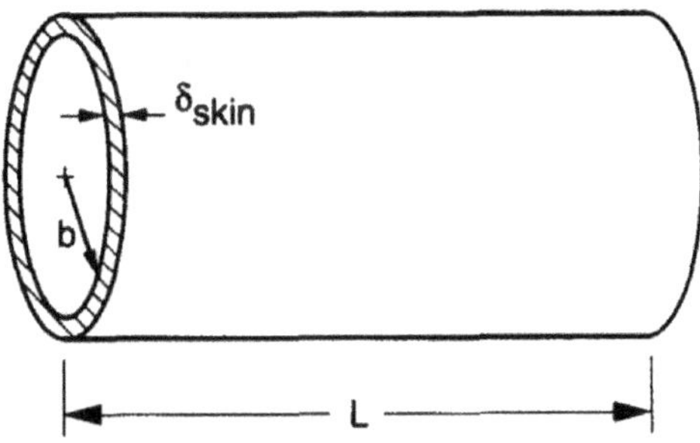

Solution Wall current flows around the pipe surface with a penetration depth of $\delta_{\rm skin} = \sqrt{\frac{2c}{Z_0\sigma_c|\omega|}}$ where $\omega \approx \frac{c}{\sigma_z}$. The total resistance is $R = \frac{L}{\sigma_c 2\pi b\delta_{\rm skin}}$. The heating, or the energy deposited to the wall per unit time, is given by RI^2 with $I \approx \frac{qc}{\sigma_z}$. This quick derivation should go a long way to deriving Eq. (8.29).

Homework 8.21 Derive Eq. (8.30).

Homework 8.22 Consider a storage ring with smooth resistive pipe. Let circumference $C = 100$ m, $b = 5$ cm, $N = 10^{11}$ protons (ignore synchrotron radiation), $\sigma_z = 1$ cm, $\sigma_c = 3.6 \times 10^7\ \Omega^{-1}\text{m}^{-1}$ (aluminum). Let there be one single cavity structure in the ring with cavity gap size $g = 10$ cm, which has 6 trapped HOMs with $\frac{R_s}{Q} = 60\,\Omega$ each.

(a) Estimate the parasitic heating power due to resistive wall.

(b) Estimate the parasitic heating power due to the cavity HOMs below cut-off.

(c) Estimate the parasitic heating power due to the cavity above cut-off.

(d) How are the above heating powers distributed around the accelerator?

(e) Do we need water cooling for the pipe? What if the pipe is made of stainless steel? Do we need water cooling for the RF cavity due to the parasitic heating?

Solution

(b) Since $\omega_{\text{cut-off}} \approx \frac{c}{b}$, the trapped modes will have $\omega_R < \frac{c}{\sigma_z}$. This means we should use the short bunch formula in Eq. (8.30). Let the 6 trapped modes have $\omega_R = (1, \frac{5}{6}, \frac{4}{6}, \frac{3}{6}, \frac{2}{6}, \frac{1}{6})\frac{c}{b}$, add up the heating due to the 6 modes. This heating is trapped by the cavity.

(c) Use diffraction model. This power propagates down the two directions from the cavity.

(d) First estimate the attenuation length of the untrapped modes in (c). Let this power dissipation be given by $e^{-2\beta z}$, then roughly, $\beta \approx \frac{c}{4\pi b \sigma_c \delta_{\text{skin}}}$. Is the untrapped wakefield absorbed near the cavity, or does it propagate around the ring more or less evenly?

(e) Without water cooling, the heated area will have to cool by black body radiation. The black body radiation heat loss is given by

$$\frac{dQ}{dt} = 2\pi b L \frac{\pi^2 k_B^4 T^4}{15\hbar^3 c^2},$$

where k_B is the Boltzmann constant. At equilibrium between HOM heating and black body radiation, the equilibrium temperature T is determined by

$$\frac{dQ_{\text{HOM}}}{dt} = 2\pi b L \frac{\pi^2 k_B^4 (T^4 - T_0^4)}{15\hbar^3 c^2},$$

where T_0 is the room temperature. Estimate the temperature T at the cavity and on the beam pipe around the ring to see if water cooling is needed.

8.5 Vlasov equation

The Vlasov equation describes the collective behavior of a multiparticle system under the influence of long-range electromagnetic forces (Fig. 8.26).[31] It was proposed as a simplified version of many-body statistical mechanics when only long-range effects are important, and as such ignores much of the subtleties of individual particle correlations. For our purpose, we shall be content limiting our discussion to the Vlasov equation.

[31] A.A. Vlasov, J. Phys. USSR 9, 25 (1945).

Figure 8.26: Anatoly Vlasov (1908–1975). [Science Photo Library]

The continuum model Before we introduce the Vlasov equation, perhaps let us make a comment to put our discussion in context. A beam of particles might have, say, 10^{12} particles. Our job is to describe the motion of this collection of particles in a systematic manner. One may choose to describe the motion of each and every one of the 10^{12} particles with its 6-D phase space coordinates, yielding a total of 6×10^{12} functions of time t. This is exactly what we do when considering single-particle dynamics. However, it is obviously not practical when considering collective beam dynamics. So instead, we must look for simplifications. The dilemma is illustrated in Fig. 8.27.[32]

Figure 8.27: The actual beam contains 10^{12} particles. Do we model the actual system as a single particle, or do we model it as a continuum? In other words, is 10^{12} closer to 1 or closer to ∞?

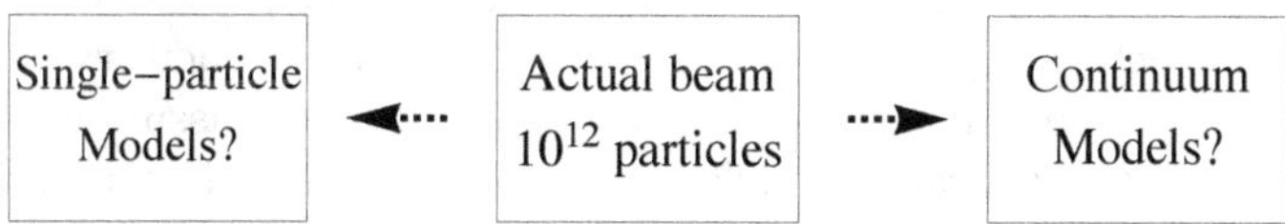

There are several ways to proceed. One possibility is to describe the beam in frequency domain. However, in this case, we effectively will be trading 6×10^{12} functions of time by 3×10^{12} modes, each with a specific frequency. The two pictures are exactly equivalent. Nothing is gained to make the job easier.

Another approach is the one-particle model. This macroparticle with 10^{12} charges generates electromagnetic wakefields, and the fields act back on it to

[32]A. Chao, CERN Accel. School, Thessaloniki, 2018; SLAC report SLAC-PUB-17370 (1918).

Figure 8.28: Schematics of models used to describe a 10^{12}-particle beam.

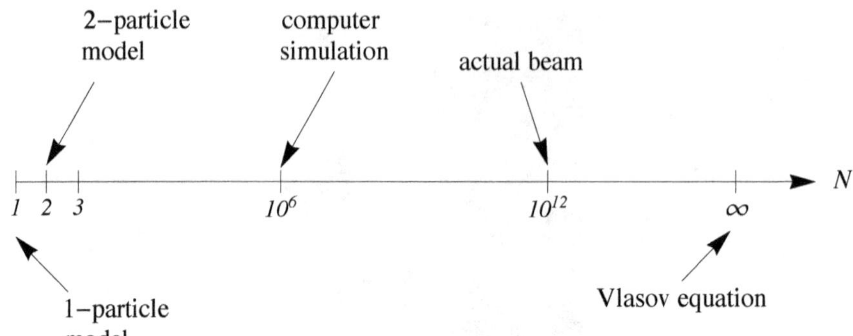

exhibit instabilities. The end result is to reduce the problem to a study of the collective motion of three lowest modes of the beam, one for each dimension. What is sacrificed is all the internal motions of particles within the macroparticle, and all the higher order modes are short-circuited. With this drastic simplification, we obtain a crude, but qualitatively valid description of the gross behavior of the beam under collective effects.

One-particle models can be extended to two-particle models. This time, each macroparticle contains $\frac{1}{2} \times 10^{12}$ charges. Each macroparticle then interacts with its own wakefields as well as wakefields from its companion. There will then be 6 modes, allowing the lowest order of the higher order modes involving internal motions of the beam. Two-particle models obviously are more accurate than the one-particle models, and in many cases still allow analytical results.

One can extend the picture to more macroparticles, but the problem rapidly becomes cumbersome. However, computers can still be used to handle, perhaps up to 10^6 macroparticles, numerically. This leads to the approach of computer simulations; they are special cases of macroparticle models treated numerically.

Failing to reach 10^{12} particles for the actual system, we then try to approach from the other extreme, i.e. we try to analyze the problem assuming an infinite number of particles with each particle having an infinitesimal charge. The beam then becomes a continuum in the 6-D phase space. Discrete "particles" lose its meaning. We shall designate the continuum distribution in phase space by ψ. Vlasov equation then comes in, as it describes the behavior of ψ. The situation is depicted schematically in Fig. 8.28.

It should be emphasized at the start that the continuum being discussed by the Vlasov formalism is in the 6-D phase space, not in the 3-D physical space. This important difference reflects again the importance of the phase space.

Binary model Let us continue the discussion on our context. The Vlasov equation assumes the interaction among particles is long-ranged. Short-range interactions affecting only neighboring particles are ignored (more accurately, they are smoothed out as part of the long-range effect). This is not always a good approximation for the accelerator beam environment, particularly noting the fact that the discrete particles are really point charges with very singular and divergent Coulomb and Liénard-Wiechert fields.

We shall ignore these short-range effects in our discussions in terms of Vlasov continuum model. But here let us mention the other efforts to supplement the Vlasov model by introducing the binary models. One such binary model is used in the treatment of *intrabeam scattering effect*[33] and the *Touschek effect*,[34] when two individual particles scatter within the beam distribution. Sometimes scattering can occur at close proximity and the singular Coulomb fields lead to rare-but-violent scattering and the subsequent particle loss. When this occurs, one scattering causes loss of two particles. Such interactions are excluded in the Vlasov model.

The binary models serve as an important tool to fill the gap left behind by the single-particle model and the continuum model. On the other hand, they are themselves simplifications. In a binary model, short-range interactions between a pair of two point particles are taken into account, but intermediate-range interactions among multiple particles, such as tertiary interactions, are still excluded.

The difference between a discrete model and a continuum model has potentially more subtleties than can be remedied by binary models. Consider Fig. 8.29 in the (x, z) space. The discreteness in the beam distribution can have a significant effect because the wake function $W'_m(z)$ is discontinuous at $z = 0$. When two particles have almost the same z position, the subsequent beam evolution can be different depending on which particle happens to be ahead, even ever so slightly. This sensitivity leads to the fact that the longitudinal microwave instability is very sensitive to any change of parameters in computer simulations. It also leads to a sluggish convergence in longitudinal microwave instability and spreading to involve several higher modes. Such a sensitivity does not occur in the transverse instabilities because $W_m(z)$ is continuous at $z = 0$.

To cast in different words, the fact that longitudinal instabilities typically converge poorly compared with their transverse counterparts comes from the short-range wakefields. In the time domain, it comes from the short-range discontinuity of the wake functions at $z = 0$, as just mentioned. In the frequency domain, it comes from the fact that $Z^{\parallel}_m(\omega) = \frac{\omega}{c} Z^{\perp}_m(\omega)$; the extra factor of ω in $Z^{\parallel}_0(\omega)$ means it is more vulnerable to poor convergence at high frequencies.

The situation is more drastic if one considers the beam-beam interaction. In this case the profiles in Fig. 8.29 is in the (x, y) space. Each discrete particle has around it a singular Coulomb field. It is at least questionable if one can treat

[33]V. Lebedev, Sec. 2.4.12, Handbook Accel. Phys. & Eng., 2nd Ed., World Scientific (2013); J.D. Bjorken and S.K. Mtingwa, Part. Accel. 13, 115 (1983).

[34]C. Bernardini et al., Phys. Rev. Lett. 10, 407 (1963); A. Piwinski, DESY Report DESY-98–179 (1998).

Figure 8.29: Schematics of a discrete and a continuum beam distribution.

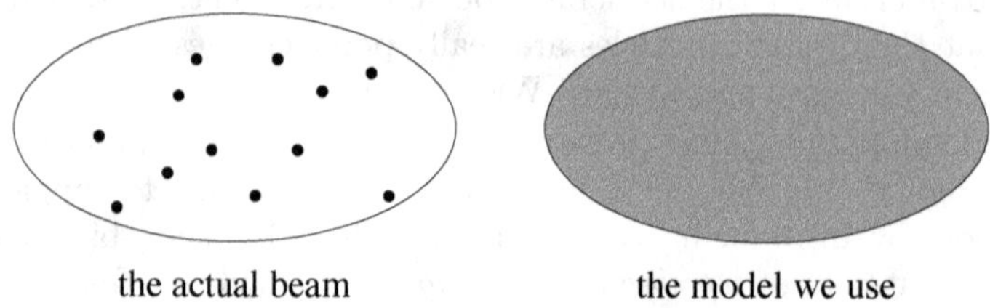

the discrete beam distribution as a continuum as we often do (Sec. 3.7.4).[35] At present, this remains an open question; we do not have binary models for beam-beam effect similar to the case of intrabeam and Toushek effects.

Derivation of Vlasov equation To construct the Vlasov equation, one invariably starts with the single-particle equations of motion (assume 1-D for simplification)

$$\begin{aligned} \dot{q} &= f(q, p, t), \\ \dot{p} &= g(q, p, t). \end{aligned}$$

The motion of a particle is described by the motion of its representative point in phase space (q, p).

In a conservative deterministic system, the particle trajectory in phase space is completely determined by the initial conditions (q_0, p_0) at time $t = t_0$. Two particles having the same initial conditions have exactly the same trajectory in phase space. It follows that the only way for two trajectories to meet in phase space at a given time is for them to coincide at all times. In other words, trajectories either completely coincide or never intersect.

Consider a distribution of particles occupying an area in the phase space. Because they cannot intersect with particles on the boundary of the distribution as the distribution evolves in time, particles inside cannot exit and particles outside cannot enter the distribution. The number of particles inside the area of phase space is therefore conserved.

If the system is conservative, then

$$f = \frac{\partial H}{\partial p} \quad \text{and} \quad g = -\frac{\partial H}{\partial q},$$

where H is the Hamiltonian. It follows that

$$\frac{\partial f}{\partial q} + \frac{\partial g}{\partial p} = 0.$$

[35] A diffusion model was proposed, A.W. Chao, AIP Proc. 127, p. 201 (1983).

Figure 8.30: Phase space evolution of a Vlasov system.

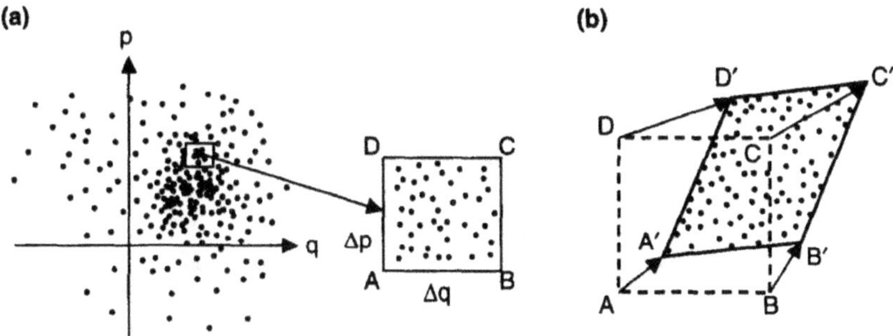

As will be seen later, this condition leads to an area conservation property. As the particle distribution evolves in the phase space, its shape may be distorted but its area remains constant. In fact, in a nonconservative system, $\frac{\partial f}{\partial q} + \frac{\partial g}{\partial p}$ has the physical meaning of the rate of area shrinkage.

Consider a distribution of a group of particles in the phase space at time t as sketched in Fig. 8.30(a). A rectangular $\Delta q \, \Delta p$ box is drawn,

$$\begin{aligned} &A(q,p), \\ &B(q+\Delta q, p), \\ &C(q+\Delta q, p+\Delta p), \\ &D(q, p+\Delta p). \end{aligned}$$

At a later time, $t + dt$, the box moves and deforms into a parallelogram $A'B'C'D'$ with the same area as $ABCD$, as shown in Fig. 8.30(b). All particles inside the box move with the box. Let the number of particles enclosed by the box be

$$\psi(q,p,t) \, \Delta q \, \Delta p \, ,$$

where ψ is the phase space distribution density normalized by

$$\int_{-\infty}^{\infty} dq \int_{-\infty}^{\infty} dp \, \psi(q,p,t) \;=\; N \, .$$

The vertices of the parallelogram are

$$\begin{aligned} &A'[q + f(q,p,t)\,dt, p + g(q,p,t)\,dt], \\ &B'[q + \Delta q + f(q+\Delta q, p, t)\,dt, p + g(q+\Delta q, p, t)\,dt], \\ &C'[q + \Delta q + f(q+\Delta q, p+\Delta p, t)\,dt, p + \Delta p + g(q+\Delta q, p+\Delta p, t)\,dt], \\ &D'[q + f(q, p+\Delta p, t)\,dt, p + \Delta p + g(q, p+\Delta p, t)\,dt]. \end{aligned}$$

The condition that no particles leak into or out of the box gives

$$\psi(q,p,t)\ \text{area}(ABCD) \;=\; \psi(q+f dt, p+g dt, t+dt)\ \text{area}(A'B'C'D')\,.$$

For a Hamiltonian system, the area of the box is conserved,

$$\begin{aligned}
\text{area}(A'B'C'D') &= |\overrightarrow{A'B'} \times \overrightarrow{A'D'}| \\
&= \Delta q\,\Delta p\left[1+\left(\frac{\partial f}{\partial q}+\frac{\partial g}{\partial p}\right)dt\right] \\
&= \Delta q\,\Delta p \;=\; \text{area}(ABCD), \qquad \text{Q.E.D.}
\end{aligned}$$

We then have

$$\begin{aligned}
\psi(q,p,t) &= \psi(q+f\,dt, p+g\,dt, t+dt) \\
&= \psi+\frac{\partial\psi}{\partial q} f\,dt+\frac{\partial\psi}{\partial p} g\,dt+\frac{\partial\psi}{\partial t}\,dt\,,
\end{aligned}$$

or, after canceling out ψ on both sides, the *Vlasov equation*,

$$\frac{\partial\psi}{\partial t}+f\frac{\partial\psi}{\partial q}+g\frac{\partial\psi}{\partial p} \;=\; 0\,.$$

Vlasov equation can also be put in a seemingly elegant but actually vague form,

$$\frac{d\psi}{dt} = 0, \quad \text{or} \quad \psi \;=\; \text{const in time.}$$

Sometimes loosely referred to as the Liouville theorem, it states that the local particle density does not change if (an important if) the observer moves with the flow of boxes, but in this form it does not tell how the boxes flow. The Vlasov form, in contrast, does not have this ambiguity, since it contains explicitly the single-particle information f and g.

Strictly speaking, f and g are given by external forces. Collisions among discrete particles in the system, for example, are excluded. However, if a particle interacts more strongly with the *collective* fields of the other particles than with its nearest neighbors, the Vlasov equation still applies if one treats the collective fields on the same footing as the external fields. This in fact forms the basis of treating the collective instabilities using the Vlasov technique. As mentioned, in this context, short- and intermediate-range effects are ignored (see page 828).

One special case where the Vlasov equation can be solved exactly is when the system is described by a Hamiltonian $H(q,p)$ which does not have an explicit time dependence. A stationary solution is found to be

$$\psi(q,p) \;=\; \text{any function of } H(q,p)\,.$$

In this system, individual particles stream along constant-H contours in the phase space in such a way that the overall distribution is stationary.

In the derivation of the Vlasov equation, we have assumed there are no diffusion or external damping effects. This is usually a good approximation for proton beams. For electron beams, synchrotron radiation contributes to both damping and diffusion, and one needs to apply the *Fokker-Planck equation.*[36] However, when the instability occurs faster than the damping or diffusion times, the Vlasov treatment applies also to electrons.

8.6 Potential-well distortion

8.6.1 Longitudinal potential-well distortion

As a first application of the Vlasov technique, we study the effect of longitudinal wakefield on a distortion of the equilibrium shape of a beam bunch.[37] The mechanism is a static one; no part of the beam bunch is executing collective oscillation. The extent of distortion depends on the beam intensity; higher beam intensities cause larger distortions.

Consider a bunched beam that travels along the axis of a cylindrically symmetric vacuum chamber pipe in a circular accelerator. We assume the beam does not have any transverse dimension, i.e., the beam is a thin thread. Such a beam does not generate transverse wakefields; only the $m = 0$ wake is excited.

Consider a particle in the beam executing longitudinal synchrotron oscillation. The phase space coordinates q and p are

$$q = z \quad \text{and} \quad p = -\frac{\eta c}{\omega_s}\delta\,,$$

where η is the phase slippage factor defined by the accelerator lattice, ω_s is the synchrotron oscillation frequency.

The single-particle equations of motion are

$$z' = -\eta\delta \quad \text{and} \quad \delta' = K(z)\,.$$

We leave $K(z)$ arbitrary for now, except that we do know it cannot depend on δ, because the system is conservative, and its mathematics is symplectic.

The Vlasov equation reads

$$\frac{\partial\psi}{\partial s} - \eta\delta\frac{\partial\psi}{\partial z} + K(z)\frac{\partial\psi}{\partial\delta} = 0\,,$$

[36] A.D. Fokker, Ann. Phys. 348, 810 (1914); M. Planck, Sitzungsberichte der Preussischen Akademie der Wissenschaften zu Berlin, 24, 324 (1917); S. Chandrasekhar, Rev. Mod. Phys. 15, 1 (1943).

[37] C. Pellegrini and A.M. Sessler, Nuovo Cimento 3A, 116 (1971); B. Zotter, Proc. 4th Advanced ICFA Beam Dynamics Workshop on Collective Effects in Short Bunches, Tsukuba, 1990, KEK Report 90-21.

where we will set $\frac{\partial \psi}{\partial s} = 0$ since we are looking for a stationary distribution. The general stationary solution is

$$\begin{aligned}\psi(z,\delta) &= \text{any function of the Hamiltonian } H\,, \\ H &= \frac{\eta^2 c^2}{\omega_s}\left[\frac{\delta^2}{2} + \frac{1}{\eta}\int_0^z K(z')\,dz'\right].\end{aligned}$$

The second integral term in the Hamiltonian is the potential-well term. A simple harmonic system would have a parabolic potential well.

RF bucket If the potential well is provided by an external RF voltage $V_{\rm rf}(z)$, we have

$$K(z) = \frac{eV_{\rm rf}(z)}{CE} = \frac{\omega_s^2}{c^2\eta V'_{\rm rf}(0)} V_{\rm rf}(z)\,,$$

where ω_s is defined as the synchrotron frequency for a particle moving around the synchronous position $z = 0$ with infinitesimal amplitude.

A practical case is given by $V_{\rm rf} = \hat{V}\sin(\frac{\omega_{\rm rf} z}{c})$. The deviation of $V_{\rm rf}(z)$ from a linear dependence on z is a cause of potential-well distortion. The general stationary distribution is given by any function of the Hamiltonian, which in this case is given by

$$H = \frac{\eta^2 c^2}{2\omega_s}\delta^2 + \frac{\omega_s c^2}{\omega_{\rm rf}^2}\left[1 - \cos\left(\frac{\omega_{\rm rf} z}{c}\right)\right].$$

This Hamiltonian also describes the form of the RF bucket. A stationary distribution must conform to the contours of constant Hamiltonian inside the bucket. For small oscillation amplitudes, we have $K = \frac{\omega_s^2}{\eta c^2}\,z$, the case of simple harmonic motion.

One noteworthy special case of the stationary beam distribution is that given by $\exp(-\text{const} \times H)$. This distribution is always Gaussian in δ. In case the bunch length is much shorter than the RF wavelength, ($|z| \ll \frac{c}{\omega_{\rm rf}}$) the familiar quadratic form of the Hamiltonian is reestablished, and the distribution is also Gaussian in z. As the bunch length increases, the bunch shape deviates from Gaussian; the potential well is distorted by the RF bucket, although the distribution remains Gaussian in δ.

Wakefield There is another reason for the Hamiltonian to deviate from the quadratic form, and thus to cause potential-well distortion, namely, the wakefield. Consider a bunch that is short compared with the RF wavelength. Let the wake function be $W_0'(z)$ integrated over the accelerator circumference, and assume that the wake has dissipated before the beam completes one revolution,

$$K(z) = \frac{\omega_s^2}{\eta c^2} z - \frac{4\pi\epsilon_0 r_0}{\gamma C}\int_z^\infty dz'\,\rho(z')W_0'(z-z')\,.$$

The corresponding Hamiltonian is

$$H = \frac{\eta^2 c^2}{2\omega_s}\delta^2 + \frac{\omega_s}{2}z^2 - \frac{4\pi\epsilon_0\eta c^2 r_0}{\omega_s\gamma C}\int_0^z dz'' \int_{z''}^{\infty} dz'\, \rho(z')W_0'(z''-z')\,,$$

where C is storage ring circumference.

The stationary solution to the Vlasov equation must be a function of H. The complication here is that the complicated z-dependence of H now involves the beam density ρ, which in turn is determined by the stationary distribution itself. Clearly some self-consistency requirement is to be imposed.

8.6.2 Haissinski distribution

Continuing the Gaussian example, the stationary distribution maintains its Gaussian distribution in δ, so we can write

$$\psi(z,\delta) = \frac{1}{\sqrt{2\pi}\sigma_\delta}\exp\left(-\frac{\delta^2}{2\sigma_\delta^2}\right)\rho(z)\,.$$

The Gaussian form and the value of σ_δ are arbitrary if the collective behavior is governed by the Vlasov equation, as in the case of a proton beam. However, if the beam behavior is governed, as for an electron beam, by the Fokker-Planck equation, then this Gaussian distribution with a specific value for σ_δ will be the unique solution of the stationary beam distribution. The Gaussian form is of particular interest because physically it is the natural equilibrium distribution for electron beams, and mathematically it is easiest to handle due to the factorization in the dependences on z and δ.

This distribution matches the stationary solution

$$\psi(z,\delta) \propto \exp\left(-\frac{\omega_s}{\eta^2 c^2\sigma_\delta^2}H\right)\,.$$

Self-consistency then imposes a transcendental equation for $\rho(z)$, referred to as the *Haissinski equation*,[38]

$$\rho(z) = \rho(0)\exp\left[-\frac{1}{2}\left(\frac{\omega_s z}{\eta c\sigma_\delta}\right)^2 + \frac{4\pi\epsilon_0 r_0}{\eta\sigma_\delta^2\gamma C}\int_0^z dz''\int_{z''}^{\infty} dz'\,\rho(z')W_0'(z''-z')\right]\,.$$

In the limit of zero beam intensity, the solution reduces to the bi-Gaussian form, where $\sigma_z = \frac{\eta c\sigma_\delta}{\omega_s}$. For high beam intensities, $\rho(z)$ deforms from Gaussian. The Haissinski equation is solved numerically for $\rho(z)$ once $W_0'(z)$ is known and σ_δ specified. Figure 8.31 shows the result for the electron damping ring for the

[38] J. Haissinski, Nuovo Cimento 18B, 72 (1973).

Figure 8.31: Potential well distortion of bunch shape for various beam intensities for the SLC damping ring. The horizontal axis is $x = -\frac{z}{\sigma_{z0}}$, where σ_{z0} is the unperturbed rms bunch length. The vertical scale gives $y = \frac{4\pi e\rho(z)}{V'_{\rm rf}(0)\sigma_{z0}}$. [Courtesy Karl Bane, 1992]

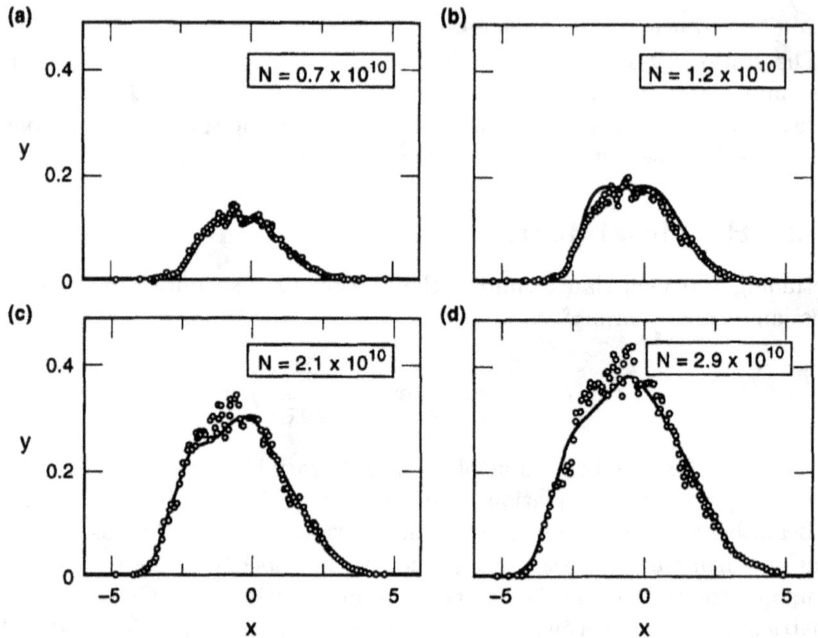

SLAC Linear Collider.[39] The bunch shape is Gaussian at low beam intensities, and it distorts as the beam intensity is increased. The calculations agree with the measurements.

Note that the distribution centroid leans forward ($z > 0$) as the beam intensity increases. This effect comes from the parasitic loss of the beam bunch, and is a consequence of the real (resistive) part of the impedance. Since the SLC damping ring is operated above transition, the bunch moves forward so that the parasitic energy loss can be compensated by the RF voltage.

Note also that the bunch length increases as the beam intensity increases. This bunch shape distortion comes mainly from the imaginary part of the impedance. Above transition, an inductive impedance causes bunch lengthening and a capacitive impedance shortens the bunch. Below transition, it is the reverse. In the present case, that the bunch lengthens is a consequence of the fact that the imaginary part of the impedance is mostly inductive.

[39] K.L.F. Bane and Ronald D. Ruth, Proc. IEEE Conf. Part. Accel., Chicago, 1989, p. 789; L. Rivkin et al., Proc. Euro. Part. Accel. Conf., Rome, 1988, p. 634.

Table 8.4 summarizes the situation. Most storage rings have inductive imaginary impedances, $\mathrm{Im}\, Z_0^{\|} < 0$, while all have positive resistive real impedances, $\mathrm{Re}\, Z_0^{\|} > 0$.

Table 8.4: Behavior of bunch distribution in various circumstances due to potential-well distortion.

	above transition	below transition
bunch centroid	move forward	move backward
bunch length	lengthen if inductive shorten if capacitive	shorten if inductive lengthen if capacitive

Retrieving the impedance from Haissinski distribution It would be an interesting exercise to try to retrieve the impedance given the Haissinski distribution. Assuming that one has measured the equilibrium beam distribution $\rho(z)$ accurately over the range of z covering the entire beam distribution, can we retrieve the impedance $Z_0^{\|}(\omega)$ from the measurement? If applicable, it may be one way to obtain the impedance of an electron storage ring once an accurate measurement of the longitudinal beam distribution is established. The answer to this question is yes, in principle. The derivation is straightforward using the Haissinski equation,

$$Z_0^{\|}(\omega) = \frac{\eta\sigma_\delta^2\gamma Z_0 C}{4\pi r_0}\frac{1}{\tilde{\rho}(\omega)}\int_{-\infty}^{\infty} dz\, e^{-i\omega z/c}\left[\frac{\rho'(z)}{\rho(z)} + \left(\frac{\omega_s}{\eta c\sigma_\delta}\right)^2 z\right], \tag{8.32}$$

where $\tilde{\rho}(\omega)$ is the Fourier transform,

$$\tilde{\rho}(\omega) = \int_{-\infty}^{\infty} dz\, e^{-i\omega z/c}\rho(z).$$

See Homework 8.23, however.

Homework 8.23

(a) Derive Eq. (8.32).

(b) Consider the case when $\rho(z)$ is measured accurately only in the main bulk of the beam and not so accurately at the tails. It is clear that one cannot extract the entire $Z_0^{\|}(\omega)$. What range of ω would this technique allow you to measure $Z_0^{\|}(\omega)$?

Solution (b) Note the $\tilde{\rho}(\omega)$ in the denominator. Only the bulk frequencies contained in the bunch distribution can be trusted. The high frequency tail diverges very rapidly by inverse Gaussian.

8.6.3 Transverse potential well distortion

The potential well distortion discussed in the previous section refers to a static distortion of the longitudinal RF bucket and is strictly a longitudinal phenomenon. There is in fact a transverse counterpart when the beam's transverse shape is statically distorted by wakefields. We call that the transverse potential well distortion. The mechanism is the same as the beam breakup mechanism in linacs, except that in the repetitive system of a storage ring, the beam does not breakup; it instead follows a transversely distorted closed orbit. With the bunch head and bunch tail following different closed orbits, the beam is then tilted in its x-z or y-z distribution.[40]

For simplicity, let us again consider a two-particle model. Consider the vertical motion only. Let the beam have a closed orbit distortion y_0 at a position where there is a transverse impedance with a wake function $W_1(z)$. Unaffected by its own wakefield, the leading macroparticle traverses the impedance with a displacement y_0. The trailing macroparticle at a distance $|z|$ behind, however, sees an additional deflecting kick

$$\Delta y' = -\frac{2\pi \epsilon r_0 N}{\gamma} W_1(|z|) y_0$$

due to the wakefield left behind by the leading macroparticle. We ignore multiturn wakefields in this consideration.

As the beam circulates around the storage ring, the bunch head will always traverse the impedance with the same displacement y_0. The bunch tail is in a steady state receiving the same kick $\Delta y'$ every turn. This constitutes a closed orbit condition, yielding a COD for the bunch tail, Eq. (3.7),

$$y(s) = \Delta y' \frac{\sqrt{\beta_0 \beta(s)}}{2 \sin \pi\nu} \cos(\pi\nu - |\psi(s) - \psi(s_0)|) \, .$$

This additional COD for the bunch tail then contributes to a y-z tilt of the beam distribution around the ring.

A question might be raised what happens if the bunch head and tail execute longitudinal synchrotron oscillation. The answer is that synchrotron oscillation will not change the conclusion as long as one stays away from synchrobetatron resonances.

Summing over all impedance sources, this beam tilt can be important for applications such as a collider or a synchrotron radiation source because these applications care about the projected effective transverse beam profile when pushed for performance.

---

[40] A.W. Chao and S. Kheifets, SLAC PEP Note 365 (1981).

8.7 Macroparticle model in circular accelerator

As elaborated in Sec. 8.5, the analysis of collective instabilities is carried out mainly either by simplified models or by a continuum model via the Vlasov equation. In this section 8.7, we concentrate only on the simplified one- or two-particle models, and we continue the same in Sec. 8.9 when Landau damping mechanism is added to the analysis. The continuum approach applying the Vlasov equation, a larger formal subject, is considered outside of our scope.

8.7.1 Instability mechanism

Here we first make a general remark concerning the instability mechanisms. In circular accelerators, we often categorize instabilities phenomenologically,

- transverse or longitudinal,
- short-range/single-bunch or long-range/multibunch.

This is natural and it categorizes instabilities into four distinct cases. We follow this categorization in our lectures.

Alternatively, in this section, we mention another way to categorize focusing on the basic physical instability mechanisms. In this categorization, whether transverse or longitudinal and whether short-range or long-range become not the issue. There are two distinctly different instability mechanisms,

1. One is the "swing-set" type, like Robinson instability (Sec. 8.7.2) and head-tail (HT, Sec. 8.7.4) instability — the former is longitudinal long-range, the latter is transverse short-range. Other examples include resistive wall instability, longitudinal head-tail instability (Sec. 8.7.5), and negative mass instability (Sec. 5.3.5).
2. The other is "mode coupling" type, like transverse-mode coupling instability (TMCI, Sec. 8.7.3) and longitudinal microwave instability — the former is transverse short-range, the latter is longitudinal short-range.

Swing-set type The swing-set instability is the more common type. It comes about when the swing-set is driven by a force that has a phase lag relative to the displacement just like the mechanism of a swing-set, Fig. 8.32. Stability or instability depends on the sign of the relative phase lag,

$$\begin{aligned}
\theta(t) &\propto \sin\omega t\,, \\
\dot{\theta}(t) &\propto +\cos\omega t\,,
\end{aligned}$$

$$\text{Driving force} \propto \begin{cases} +\cos\omega t & \Longrightarrow \quad \text{antidamping} \quad \Longrightarrow \quad \text{instability}, \\ -\cos\omega t & \Longrightarrow \quad \text{damping} \quad \Longrightarrow \quad \text{stability}. \end{cases}$$

Antidamping occurs when the driving force is in phase with velocity, or 90° phase lag with displacement. Damping occurs when the force is 180° out of phase

Figure 8.32: A swing set. If driven, its stability depends on the phase of the driving force relative to its instantaneous velocity.

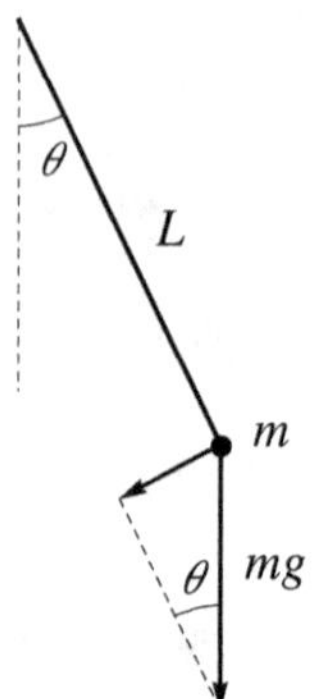

with velocity, or 90° phase lead with displacement. This type of instabilities do not have a threshold. Sign of the phase lag/lead determines the stability. The sign typically is determined by the sign of a certain key parameter,

Instability	Key parameter
Robinson	(RF resonant frequency) - (RF drive frequency)
head-tail	head-tail phase
negative mass	phase slippage factor
longitudinal head-tail	η_1/η_0

Mode coupling type The less common mode coupling type of instabilities are originally stable. But if the perturbation is too strong, the stability can be broken. This type of instabilities has a threshold. Instability occurs when the beam intensity exceeds a threshold.

There is a tendency that the swing-set type of instabilities were discovered earlier in history because they are unstable at low beam currents. The mode coupling type occurred only when accelerator technology evolved and the required beam currents became substantial.

The realization of the mode coupling instabilities was due to Frank Sacherer's pioneering efforts as late as 1977 to explain the longitudinal microwave instability and the bunch lengthening effect in electron storage rings.[41] Although both are mode coupling type, the transverse TMCI is more straight forward than the longitudinal microwave instability — the sluggish convergence in the longitudinal dynamics compared with the transverse dynamics was emphasized on

[41] F. Sacherer, IEEE Trans. Nucl. Sci. NS-24, No. 3, 1393 (1977). A bunch length scaling law was derived earlier based on but without identifying the mode coupling mechanism, A.W. Chao and J. Gareyte, SLAC Report SPEAR-197 (1976) and Part. Accel. 25, 229 (1990); P. B. Wilson et al., IEEE Trans. Nucl. Sci. NS-24, 1211 (1977).

page 829. It was only when the transverse counterpart, the TMCI, was discovered in 1980s when the mode coupling as an instability mechanism was firmly established. Unfortunately Sacherer (Fig. 8.33) did not live to see it happening.

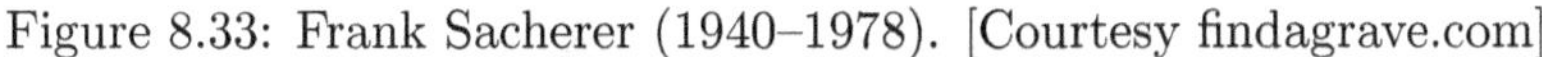
Figure 8.33: Frank Sacherer (1940–1978). [Courtesy findagrave.com]

8.7.2 Robinson instability

Robinson instability is one of the most basic instability mechanisms of the swing-set type. It is a longitudinal instability that occurs in circular accelerators. The main contributor to this instability is the longitudinal impedance due to the RF accelerating cavities. These cavities are tuned to have a resonant frequency ω_R for its fundamental accelerating mode. This mode is where the klystrons feed into, but at the same time, it is also a big source of wakefield and impedance. Since we must have RF cavities to accelerate the beam, we must accept the existence of these big impedance sources and try to live with them.

The RF fundamental mode is the biggest $m = 0$ mode impedance source in the accelerator. The real part of this impedance peaks at ω_R with a narrow width. The width is approximately given by $\frac{\Delta\omega}{\omega_R} \approx \pm\frac{1}{Q}$, where Q is the Q-value of the RF cavity's accelerating mode. Typically, $Q \sim 10^4$ (or 10^{9-10} for superconducting cavities). So this impedance is sharply peaked.

By design, ω_R is very close to an integer multiple of the revolution frequency ω_0 of the beam. This necessarily means that the wakefield excited by the beam in the cavities contains a major frequency component near $\omega_R \approx h\omega_0$ or equivalently, the impedance $Z_0^{\parallel}(\omega)$ has a sharp peak at $\omega_R \approx h\omega_0$, where h is the RF harmonic number.

As we will soon show, the exact value of ω_R relative to $h\omega_0$ is of critical importance for the stability of the beam. Above the transition energy, the beam will be unstable if ω_R is slightly above $h\omega_0$ and stable if slightly below.

Figure 8.34: Kenneth W. Robinson (1925–1979). [Courtesy Ewan Paterson, 2019]

Below transition, it is the other way around. This instability mechanism was first analyzed by Robinson in 1964 (Fig. 8.34).[42]

To simplify the physical picture, let us adopt a one-particle model to discuss the Robinson instability. The advantage of this model is that it allows simple analytical results. We will apply two-particle models to describe other instabilities, those for which internal structures play a role, in later sections.

Since Robinson instability is a longitudinal effect, we now consider the longitudinal motion of this one-particle beam. Let z_n be the longitudinal displacement of the beam at the accelerating RF cavity in the n-th revolution, measured relative to the center of an idealized bunch unaffected by wakefields. The rate of change of z_n is related to the relative energy error $\delta_n = \frac{\Delta E}{E}$ of the beam in the same n-th revolution by

$$\frac{d}{dn} z_n = -\eta C \delta_n \,,$$

where η is the phase slippage factor, C is the accelerator circumference. A positive z_n means the beam arrives the cavity earlier than the idealized reference beam. The storage ring is above transition if $\eta > 0$ and below transition if $\eta < 0$.

The energy error also changes with time. In the absence of wakefields, its equation of motion is

$$\frac{d}{dn} \delta_n = \frac{(2\pi\nu_s)^2}{\eta C} z_n \,,$$

[42]K.W. Robinson, Cambridge Electron Accel. Report CEAL-1010 (1964).

where ν_s is the synchrotron tune. If we combine these two equations, we get a simple harmonic oscillation for both z_n and δ_n,

$$\frac{d^2 z_n}{dn^2} + (2\pi\nu_s)^2 z_n = 0 \,.$$

The oscillation has a phase advance of $2\pi\nu_s$ per revolution, and is just the synchrotron oscillation of the macroparticle beam. Typically, $\nu_s \ll 1$, i.e., synchrotron oscillation is slow and the beam does not execute much synchrotron motion during the time it completes one revolution.

Wakefields perturb the simple harmonic oscillation. For an intense beam, the energy variation also depends on the wakefield generated by the beam. The $\frac{d\delta_n}{dn}$ equation then acquires an additional term,

$$\begin{aligned} \frac{d}{dn}\delta_n &= \frac{(2\pi\nu_s)^2}{\eta C} z_n + \frac{eV(z_n)}{E} \\ &= \frac{(2\pi\nu_s)^2}{\eta C} z_n - \frac{4\pi\epsilon_0 N r_0}{\gamma} \sum_{k=-\infty}^{n} W_0'(kC - nC + z_n - z_k) \,, \end{aligned}$$

where W_0' is the longitudinal wake function accumulated over one turn of the accelerator. The summation over k is over the wakefields left behind by the beam from all revolutions prior to the n-th — this is necessary because as mentioned, Robinson instability is a multi-turn effect. The argument of the wake function, $kC - nC + z_n - z_k$, is the longitudinal separation of beam positions between the n-th and the k-th revolutions. The equation of motion now becomes

$$\frac{d^2 z_n}{dn^2} + (2\pi\nu_s)^2 z_n = \frac{4\pi\epsilon_0 N r_0 \eta C}{\gamma} \sum_{k=-\infty}^{n} W_0'(kC - nC + z_n - z_k) \,.$$

In case the beam bunch has an oscillation amplitude much shorter than the wavelength of the fundamental cavity mode, one can expand the wake function,

$$W_0'(kC - nC + z_n - z_k) \approx W_0'(kC - nC) + (z_n - z_k) W_0''(kC - nC) \,.$$

The first term on the right hand side is a static term independent of the motion of the beam. It describes the parasitic loss effect discussed earlier and can be taken care of by a constant shift in z_n. We will drop this term altogether. The second term, on the other hand, does involve the dynamics of the beam. The quantity $z_n - z_k$ is the difference of z's and — although we will not make such an approximation — resembles a time derivative $\frac{dz}{dn}$, which in turn suggests an instability since a $\frac{dz}{dn}$ term in a $\frac{d^2 z}{dn^2}$ equation indicates a possible exponential growth of z.

We need now to solve this equation for z_n as a function of turn number n. To do so, we let z_n be written in an ansatz form,

$$z_n \propto e^{-in\Omega T_0} \,,$$

where $T_0 = \frac{C}{c} = \frac{2\pi}{\omega_0}$ is the beam revolution period, and Ω is the mode frequency of the beam oscillation and is a key quantity yet to be determined. This ansatz holds because the equation of motion has now been linearized in z_n.

Substituting into the equation of motion, we find an algebraic equation for Ω,

$$\Omega^2 - \omega_s^2 = -\frac{4\pi\epsilon_0 N r_0 \eta c}{\gamma T_0} \sum_{k=-\infty}^{\infty} (1 - e^{-ik\Omega T_0}) W_0''(kC) ,$$

where $\omega_s = \nu_s \omega_0$ is the synchrotron frequency. The differential equation of motion has now been reduced to an algebraic equation, thanks to the correct choice of the ansatz which in turn results from the linearization process.

Now the wake function can be expressed in terms of the longitudinal impedance by a Fourier transform. This yields

$$\Omega^2 - \omega_s^2 = -i \, \frac{4\pi\epsilon_0 N r_0 \eta}{\gamma T_0^2} \sum_{p=-\infty}^{\infty} \left[p\omega_0 Z_0^{\parallel}(p\omega_0) - (p\omega_0 + \Omega) Z_0^{\parallel}(p\omega_0 + \Omega) \right] .$$

Given the impedance, this equation can in principle be solved for Ω. Note that Ω appears on both sides of the equation. Here, however, we will take a perturbative approach and assume Ω does not deviate much from ω_s for modest beam intensities. We thus replace Ω by ω_s on the right hand side of the equation. Quantity Ω is then easily solved.

In general, Ω is complex. The real part of Ω is the perturbed synchrotron oscillation frequency of the collective beam motion, and the imaginary part gives the growth rate (or damping rate if negative) of the motion. We then obtain a *mode frequency shift*,

$$\begin{aligned} \Delta\Omega &= \mathrm{Re}(\Omega - \omega_s) \\ &= \frac{2\pi\epsilon_0 N r_0 \eta}{\gamma T_0^2 \omega_s} \sum_{p=-\infty}^{\infty} \left[p\omega_0 \mathrm{Im} Z_0^{\parallel}(p\omega_0) - (p\omega_0 + \omega_s) \mathrm{Im} Z_0^{\parallel}(p\omega_0 + \omega_s) \right] , \end{aligned}$$

and an *instability growth rate*,

$$\tau^{-1} = \mathrm{Im}(\Omega - \omega_s) = \frac{2\pi\epsilon_0 N r_0 \eta}{\gamma T_0^2 \omega_s} \sum_{p=-\infty}^{\infty} (p\omega_0 + \omega_s) \mathrm{Re} Z_0^{\parallel}(p\omega_0 + \omega_s) .$$

In the expression for τ^{-1}, use has been made of the fact that $\mathrm{Re}Z_0^{\parallel}(\omega)$ is an even function of ω.

It is the imaginary part of the impedance that contributes to the collective frequency shift and the real part that contributes to the instability growth rate. When we measure the synchrotron frequency in a real operation, the frequency we measure is not ω_s, but the shifted mode frequency Ω.

Note that this statement is valid only when $|\Omega - \omega_s| \ll \omega_s$. Otherwise imaginary and real parts of the impedance have mixed contributions. In particular, for an unbunched beam, there is no longitudinal focusing and therefore

Figure 8.35: Robinson stability criterion requires the fundamental RF mode be slightly detuned. (a) above transition. (b) below transition.

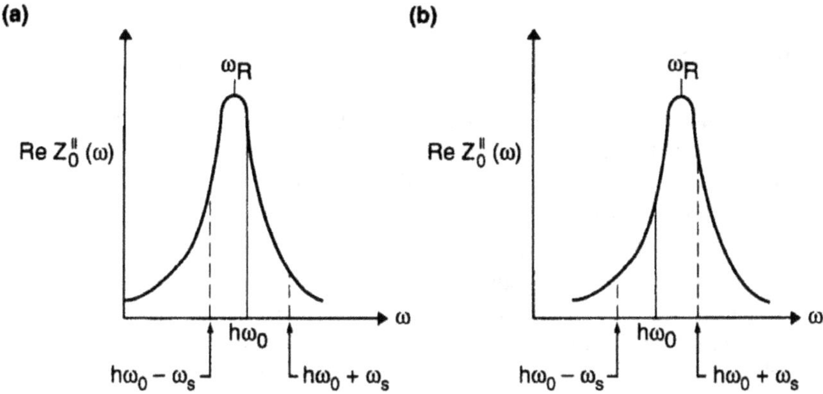

$\omega_s = 0$, and this statement clearly does not apply. We will return to this point in Sec. 8.9.5.

So far our results holds for arbitrary impedance. The Robinson instability nominally was applied to the fundamental mode of the RF impedance. But its mechanism is much more general. It can be applied also to high order RF modes, and to arbitrary longitudinal impedance $Z_0^{\|}$. In fact, it can also be generalized to transverse instabilities, leading to a transverse Robinson instability.

We now focus on the resonator impedance for the fundamental RF cavity mode. The only significant contributions to the growth rate come from two terms in the summation, namely $p = \pm h$ because the real part of the impedance is sharply peaked there. This gives

$$\tau^{-1} \approx \frac{N r_0 \eta h \omega_0}{2\gamma T_0^2 \omega_s} \left[\mathrm{Re} Z_0^{\|}(h\omega_0 + \omega_s) - \mathrm{Re} Z_0^{\|}(h\omega_0 - \omega_s) \right] .$$

Let us make a side comment here in passing. The fact that only the $p = \pm h$ terms dominate applies to the calculation of τ^{-1} when $\mathrm{Re}Z_0^{\|}$ is involved. The same is not applicable to a calculation of $\Delta\Omega$ because $\mathrm{Im}Z_0^{\|}$, with its long tails, does not converge as rapidly.

Beam stability requires $\tau^{-1} \leq 0$. That is, the real part of the impedance must be lower at frequency $h\omega_0 + \omega_s$ than at frequency $h\omega_0 - \omega_s$ if $\eta > 0$ (above transition), and the other way around if $\eta < 0$ (below transition). This condition gives the important Robinson stability criterion that, above transition, the resonant frequency ω_R of the fundamental cavity mode should be slightly detuned downwards from an exact integral multiple of ω_0. Below transition, stability requires ω_R be slightly higher than $h\omega_0$. See Fig. 8.35.

Physically, Robinson instability comes from the fact that the revolution frequency of an off-momentum beam is not given by ω_0 but by $\omega_0(1 - \eta\delta)$. To illustrate the physical origin of the Robinson instability mechanism, consider a beam executing synchrotron oscillation (dipole mode) above transition. Due to the energy error of the beam, the impedance samples the beam signal at a frequency slightly below $h\omega_0$ if $\delta > 0$, and slightly above $h\omega_0$ if $\delta < 0$. In order to damp this synchrotron oscillation, we need to let the beam lose energy when $\delta > 0$ and gain energy when $\delta < 0$. This can be achieved by having an impedance that decreases with increasing frequency in the neighborhood of $h\omega_0$. The Robinson stability criterion then follows.

When $\tau^{-1} > 0$, the beam is unstable because any accidental small synchrotron oscillation of the beam would grow exponentially with the instability growth rate, eventually leading to the loss of the beam. On the other hand, when $\tau^{-1} < 0$, the Robinson mechanism leads to exponential damping of any center-of-mass dipole-mode synchrotron oscillations. Due to the very strong RF fundamental mode impedance, the Robinson damping can be rather strong. When the Robinson criterion is met, the synchrotron oscillation of the beam is "Robinson damped" and this damping will help stabilize the beam against similar instabilities due to other impedance sources.

8.7.3 Transverse mode coupling instability

The next topic is a transverse instability, called transverse mode coupling instability (TMCI), also called the strong head-tail instability or transverse microwave instability. It is to be discussed using a two-particle model.[43] It was first observed and analyzed at the PEP storage ring,[44] where it was found that when intensity is above a certain threshold, the beam becomes transversely unstable. Below it, the beam motion is perturbed but remains stable.

In a sense the TMCI and the Robinson instability are two instabilities that are as different as instabilities can be. A brief comparison, Table 8.5, can perhaps illustrate this.

Two-particle model The physical mechanism of this instability is closely related to the beam breakup instability of the linacs. To illustrate the mechanism of the TMCI, consider a beam with two macroparticles, each with charge of $\frac{Ne}{2}$ and each executing synchrotron oscillation. We assume their synchrotron oscillations have equal amplitude but opposite phases. During time $0 < \frac{s}{c} < \frac{T_s}{2}$, where $T_s = \frac{2\pi}{\omega_s}$ is the synchrotron oscillation period, particle 1 leads particle 2; the equations of motion for the two macroparticles are

$$\begin{aligned} y_1'' + \left(\frac{\omega_\beta}{c}\right)^2 y_1 &= 0\,, \\ y_2'' + \left(\frac{\omega_\beta}{c}\right)^2 y_2 &= \frac{2\pi\epsilon_0 N r_0 W_0}{\gamma C} y_1\,, \end{aligned} \tag{8.33}$$

[43] R. D. Kohaupt, DESY Report M-80/19 (1980); R. Talman, CERN Report ISR-TH/81-17 (1981); R. Talman, Nucl. Instru. Meth. 193, 423 (1982).

[44] PEP Group, Proc. 12th Int. Conf. High Energy Accel. Fermilab, 1983, p. 209.

Table 8.5: A brief comparison between the Robinson instability and the TMCI.

	Robinson instability	TMCI
Dimension	longitudinal	transverse
Mode	$m = 0$	$m = 1$
Wakefield	long-range	short-range
Effect	multibunch/multiturn	Single bunch/single turn
Impedance	sharply peaked	broad-band
Model	one-particle	two-particle
Mechanism	swing-set	mode coupling
Threshold	no	yes

where ω_β is the unperturbed betatron oscillation frequency, whether it is horizontal or vertical. Similarly, during $\frac{T_s}{2} < \frac{s}{c} < T_s$, we have the same equations (8.33) with indices 1 and 2 exchanged. Then during $T_s < \frac{s}{c} < \frac{3T_s}{2}$, Eq. (8.33) applies again, etc. The quantities y_1 and y_2 are considered to be infinitesimal initially. Whether the beam is stable depends on their behavior in time. If they remain infinitesimal, the beam is stable. If they grow exponentially, the beam is unstable.

We have been using s as the time variable. Synchrotron oscillation, for example, is divided into half oscillation periods according to $0 < \frac{s}{c} < \frac{T_s}{2}$ and $\frac{T_s}{2} < \frac{s}{c} < T_s$, and not according to $0 < t < \frac{T_s}{2}$ and $\frac{T_s}{2} < t < T_s$. This is in fact an important subtlety. We will return to this point later in Sec. 8.7.4.

In writing down Eq. (8.33), we have assumed for simplicity that the wake function (integrated over the accelerator circumference C), $W_1(z)$, is a constant, and yet it vanishes before the beam completes one revolution, i.e.,

$$W_1(z) = \begin{cases} -W_0 & \text{if } 0 > z > -(\text{bunch length}), \\ 0 & \text{otherwise.} \end{cases} \tag{8.34}$$

The property of wake functions requires that $W_0 > 0$. This short range wake function corresponds to an impedance which is broad-banded (Homework 8.12). In contrast with the Robinson instability, the TMCI is essentially a single-bunch phenomenon.

We now analyze the stability condition of the two-particle beam system. From Eq. (8.33), the solution for y_1 is simply a free betatron oscillation,

$$\tilde{y}_1(s) = \tilde{y}_1(0)e^{-i\omega_\beta s/c},$$

where

$$\tilde{y}_1 = y_1 + i\frac{c}{\omega_\beta}y_1'.$$

Both the real and imaginary parts are meaningful in the representation.

Substituting this $\tilde{y}_1(s)$ into the equation for y_2 yields the solution

$$\tilde{y}_2(s) = \tilde{y}_2(0)e^{-i\omega_\beta s/c} + i\frac{\pi\epsilon_0 N r_0 W_0 c}{\gamma C\omega_\beta}\left[\frac{c}{\omega_\beta}\tilde{y}_1^*(0)\sin\frac{\omega_\beta s}{c} + \tilde{y}_1(0)se^{-i\omega_\beta s/c}\right]. \tag{8.35}$$

There are three terms in Eq. (8.35). The first two terms describe the free betatron oscillation; the third term, proportional to s, is a resonantly driven response — the same analysis occurred in Sec. 8.3.2.

Equation (8.35) can be simplified if $\omega_\beta T_s \gg 1$, or equivalently, $\omega_\beta \gg \omega_s$. In that case, the second term on the right hand side can be dropped because it is much smaller than the third term, and we can write the solution during the period $0 < \frac{s}{c} < \frac{T_s}{2}$ in a matrix form,

$$\begin{bmatrix}\tilde{y}_1\\ \tilde{y}_2\end{bmatrix}_{s=cT_s/2} = e^{-i\omega_\beta T_s/2}\begin{bmatrix}1 & 0\\ i\Upsilon & 1\end{bmatrix}\begin{bmatrix}\tilde{y}_1\\ \tilde{y}_2\end{bmatrix}_{s=0},$$

where we have defined a positive, dimensionless parameter

$$\Upsilon = \frac{\pi^2\epsilon_0 N r_0 W_0 c^2}{\gamma C\omega_\beta\omega_s}. \tag{8.36}$$

It may be instructive to compare this expression with the growth parameter (8.27) introduced for the linac beam breakup. Note for example that here we have $\Upsilon \propto \frac{1}{\omega_s}$. First of all, it says that synchrotron oscillation is a stabilizing mechanism in TMCI; the faster is synchrotron oscillation, the weaker is the instability. Secondly, it explains the reason why crossing transition is such a big deal; during transition, $\omega_s \to 0$.

The time evolution during $\frac{T_s}{2} < \frac{s}{c} < T_s$ can be obtained by exchanging indices 1 and 2 in the above analysis. The total transformation for one full synchrotron period is therefore

$$\begin{aligned}\begin{bmatrix}\tilde{y}_1\\ \tilde{y}_2\end{bmatrix}_{cT_s} &= e^{-i\omega_\beta T_s}\begin{bmatrix}1 & i\Upsilon\\ 0 & 1\end{bmatrix}\begin{bmatrix}1 & 0\\ i\Upsilon & 1\end{bmatrix}\begin{bmatrix}\tilde{y}_1\\ \tilde{y}_2\end{bmatrix}_0\\ &= e^{-i\omega_\beta T_s}\begin{bmatrix}1-\Upsilon^2 & i\Upsilon\\ i\Upsilon & 1\end{bmatrix}\begin{bmatrix}\tilde{y}_1\\ \tilde{y}_2\end{bmatrix}_0.\end{aligned}$$

As time evolves, the vector formed by the phasors $\tilde{y}_1$ and $\tilde{y}_2$ is repeatedly transformed by the 2×2 matrix of this map. Stability of the system is thus determined by the eigenvalues of this matrix. The two eigenvalues for the two modes (a + mode and a − mode) are

$$\lambda_\pm = e^{\pm i\phi}, \quad \sin\frac{\phi}{2} = \frac{\Upsilon}{2}, \tag{8.37}$$

and the eigenvectors are

$$V_\pm = \begin{bmatrix} \pm e^{\pm i\phi/2} \\ 1 \end{bmatrix}.$$

Stability requires $\phi =$ real, which is fulfilled if $|\sin\frac{\phi}{2}| \le 1$, or

$$\Upsilon \le 2.$$

When $\phi = 0$, we have $V_\pm = \begin{bmatrix} \pm 1 \\ 1 \end{bmatrix}$. When $\phi = \pi$, we have both $V_\pm$ become equal to $\begin{bmatrix} i \\ 1 \end{bmatrix}$. For weak beams, $\Upsilon \ll 1$, we have $\phi \approx \Upsilon$. Near the instability, ϕ approaches π as Υ approaches 2.

A moment of reflection indicates that the instability that occurs when $\Upsilon > 2$ (or even when Υ approaches but stays below 2) causes a rather severe disruption of the beam, as seen by the fact that, during half a synchrotron period, the motion of the trailing particle has grown by an amount more than twice the amplitude of the free-oscillating leading macroparticle. During this half synchrotron period, the trailing macroparticle is driven hard by the wakefield just like the case of beam breakup in a linac. For $\Upsilon \le 2$, however, the growths made during the half synchrotron periods when the particle is trailing do not accumulate and the beam is stable. As the beam intensity increases so that $\Upsilon > 2$, the growths of the particles then do accumulate and bootstrap into an instability.

This *threshold* behavior is very different from the linac case in which the beam — at least its tail — is always unstable. One can imagine that, by periodically exchanging the roles of leading and trailing particles, the two-particle beam is made more stable. The more frequently they are exchanged, the more stable is the beam. As mentioned, this shows up in $\Upsilon \propto \frac{1}{\omega_s}$. Strong betatron focusing and a high beam energy also help, as indicated by $\Upsilon \propto \frac{1}{\gamma\omega_\beta}$. The factor $\frac{1}{\omega_\beta}$ in Υ is related to the β-function. If the β-function β_Z at the location of the impedance is known, a better expression of Υ would be obtained by replacing $\frac{c}{\omega_\beta}$ by β_Z. The factor $\frac{1}{C}$ appearing in Υ is because the effect is driven by the wakefield per unit distance and the physical quantity is really $\frac{W_1(z)}{C}$.

Beam observation under TMCI In an accelerator, the most readily available beam signal comes from the beam position monitors that measure the center of charge of the beam. In the two-particle model, the center of charge signal is given by $y_1 + y_2$. In particular, it will be useful to examine its frequency spectrum. To do that, consider a two-particle beam that is in a pure eigenstate $V_\pm$ at time $\frac{s}{c} = 0$. In the stable region, it is straightforward to show that the subsequent motion of the beam center of charge is given by[45]

$$(\tilde{y}_1 + \tilde{y}_2)(s) = \exp\left[-i(\omega_\beta \mp \frac{\phi\omega_s}{2\pi})\frac{s}{c}\right] \sum_{\ell=-\infty}^{\infty} C_\ell e^{-i\ell\omega_s s/c},$$

$$C_\ell = 2i\Upsilon \frac{1 \pm (-1)^\ell}{(2\pi\ell \mp \phi)^2}(1 \mp e^{\pm i\phi/2}).$$

The $\pm$ modes as observed by a beam position monitor therefore contain the following frequencies, as shown in Fig. 8.36,

$$+ \text{ mode} \quad : \quad \omega_\beta + \ell\omega_s - \frac{\phi}{2\pi}\omega_s \,, \quad \ell = \text{even}\,,$$

$$- \text{ mode} \quad : \quad \omega_\beta + \ell\omega_s + \frac{\phi}{2\pi}\omega_s \,, \quad \ell = \text{odd}\,.$$

Each frequency line in Fig. 8.36 has its own signal strength that varies with the beam intensity. At low intensities, the only significant line is the $\ell = 0$ spectral line. All other lines are weakly excited. As the beam intensity increases towards $\Upsilon = 2$, all other lines become significant, and at the instability threshold, all lines have equal strength.

For weak beams, the two macroparticles cleanly move up and down in phase in the + mode and out of phase in the − mode. As Υ increases, the mode frequencies shift and the particle motions become more complicated; each mode then contains a combination of in-phase and out-of-phase motions. At the stability limit $\Upsilon = 2$, the frequencies of the two modes merge into each other and become imaginary, at which moment the beam becomes unstable.

It should be mentioned that although the two-particle model exhibits higher order modes ($\ell \geq 2$) as shown in Fig. 8.36, they are poorly represented quantitatively due to the limitation of having only two macroparticles. This is particularly the case if the wake function $W_1(z)$ oscillates within the bunch length. If the wake function oscillates m times, then a reasonable accurate quantitative description will require at least $m + 2$ macroparticles. For the two-particle model, it is better to be applied to the case of a constant wake function with no oscillation, as we did.

Head-tail oscillation observed at CERN LEP As mentioned, the most readily available signal from the beam is that of its center of charge motion observed by a beam position monitor. Using special instruments such as a streak camera, however, it is possible to observe the motion of the beam across the length of the beam. One such observation, made on the electron storage

[45]Derivation is somewhat lengthy. It takes into account of the arrival times of the two macroparticles as observed by a BPM at a fixed location. As a result, Fig. 8.36 shows an infinite number of spectral lines even though there are only two eigenmodes — each eigenmode exhibits an infinite number of frequencies.

Figure 8.36: Frequency spectrum of the center-of-charge signal of a two-particle beam in the stable region $\Upsilon < 2$. Solid curves are the spectrum of the $+$ mode, dashed curves the $-$ mode. Instability occurs where the mode frequencies merge.

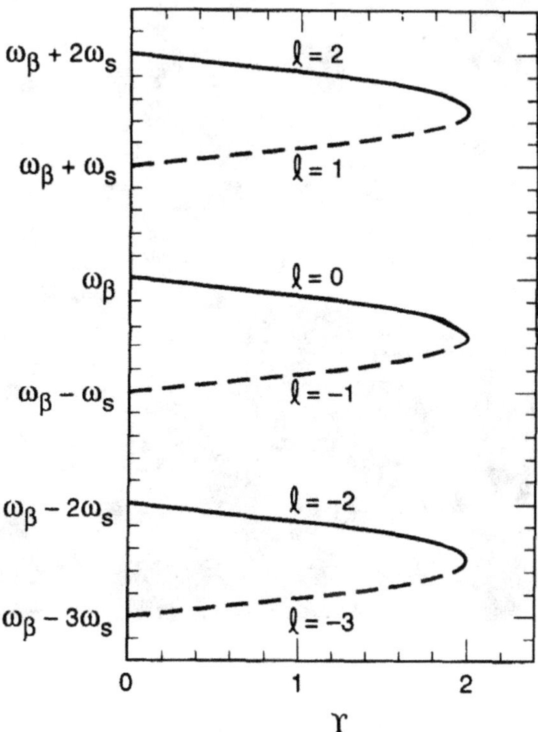

ring LEP at CERN, is shown in Fig. 8.37.[46] It shows the turn-by-turn snapshot pictures of a beam executing a transverse head-tail oscillation. The bunch is seen from the side and one observes a vertical head-tail oscillation (with mode $m = 1$ and $\ell = 1$). The horizontal scale is 500 ps for the total image while the vertical scale is not calibrated. The figure shows the same bunch each turn going from top to bottom.

Coherent betatron tune shift The TMCI is one of the cleanest instabilities to observe in electron storage rings. In particular, one may measure the threshold beam intensity when the beam becomes unstable transversely and associate the observation with $\Upsilon = 2$. Another approach is to measure the "betatron frequency" (what is measured is the frequency of the $m = 1, \ell = 0$ spectral line) as the beam intensity is varied. From our two-particle analysis, the initial slope

[46] E. Rossa et al., Proc. Euro. Part. Accel. Conf. Berlin, 1992, p. 144.

Figure 8.37: The turn-by-turn pictures, taken by a streak camera, of a beam executing a transverse head-tail oscillation in the LEP storage ring. [Courtesy Albert Hofmann and Edouard Rossa, 1992]

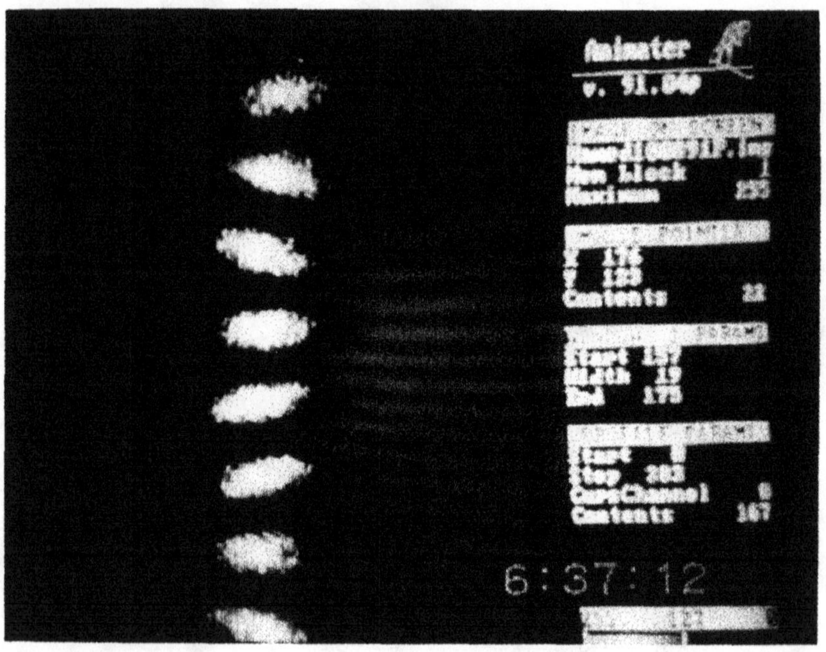

of this frequency with respect to the beam intensity is found to be

$$\left(\frac{d\omega_\beta}{dN}\right)_{N=0} = -\frac{\omega_s}{2\pi}\left(\frac{d\phi}{dN}\right)_{N=0} = -\frac{\pi\epsilon_0 r_0 W_0 c^2}{2\gamma C\omega_\beta}. \tag{8.38}$$

By measuring the instability threshold or by measuring the initial slope of the betatron frequency, information on the short-range wakefield or broad-band impedance can be obtained.

According to the two-particle model, at the instability threshold, the measured betatron frequency has shifted by $\frac{\omega_s}{2}$. The measured value of $(\frac{d\omega_\beta}{dN})_{N=0}$ can be used to predict the instability threshold $N_{\rm th}$ by

$$N_{\rm th} = -\frac{\omega_s}{\pi}\frac{1}{(d\omega_\beta/dN)_{N=0}}.$$

This is useful when the beam intensity is limited for reasons other than the TMCI. By measuring ω_β at low beam intensities, the eventual TMCI threshold can be estimated. However, as mentioned in Sec. 8.1.11, this scenario can be obscured if the vacuum chamber does not have cylindrical symmetry.

The two-particle model predicts that the $\ell = 0$ frequency always shifts *down* as the beam intensity is increased. Physically, this is because, for short bunches, the sign of the wake force is such that the bunch tail is always deflected further away from the vacuum chamber axis if the beam is transversely displaced (Fig. 8.11). With the head and the tail moving together in the $\ell = 0$ mode, the wake force acts as a defocusing effect and the mode frequency shifts down.

We mentioned TMCI was one of the easiest instability phenomena to observe. As mentioned before, although TMCI and longitudinal microwave instabilities are both of the mode-coupling type, the longitudinal microwave turns out to be one of the most ambiguous to observe. The difference is traced to the fact that the longitudinal wake function $W_0'(z)$ is discontinuous at $z = 0$ while the transverse wake function $W_1(z)$ is continuous there. This continuity property is also traced to the frequency domain with the extra factor of ω in the connection $Z_m^{\|}(\omega) = \frac{\omega}{c} Z_m^{\perp}(\omega)$. The longitudinal impedance converges slower than the transverse impedance at high frequencies. Furthermore, this lack of convergence also indicates that a few-particle model would not work as well for the longitudinal microwave instability.[47]

Simulation and observation for PEP The center of charge signal of the beam as a function of time after the beam receives an initial transverse kick can also be studied with a numerical simulation. Figure 8.38 shows results of the numerical simulation of a two-particle model as compared with the experimental observation at the electron storage ring PEP.

Although the exact values of Υ used in the simulation have been slightly adjusted to make the comparison (see figure caption), the agreement is quite reasonable, indicating that the two-particle model describes this instability mechanism remarkably well. The fact that the signal exhibits damping is due to the radiation damping effect in electron storage rings, which was included in the simulation.

In Fig. 8.38, one also observes that the beating frequency decreases as the beam intensity increases. This is expected because the mode frequencies approach each other as beam intensity approaches the instability threshold.

The instability threshold observed at PEP occurred in a single-bunch experiment when the beam intensity was $N_{\text{th}} = 6.4 \times 10^{11}$ with betatron tune $\frac{\omega_\beta}{\omega_0} = 18.19$, synchrotron tune $\frac{\omega_s}{\omega_0} = 0.044$, $E = 14.5$ GeV, $C = 2200$ m, and $\omega_0 = 0.86 \times 10^6\ \text{s}^{-1}$. By relating these parameters to $\Upsilon = 2$, one obtains an estimate of the wake function for PEP of $W_0 = c\,17.4\,\text{M}\Omega\,\text{m}^{-2}$. This translates into an effective angular kick at the bunch tail of 18 μrad per millimeter of bunch head displacement per revolution.

If we draw an equivalence to N_{cav} cavities in the vacuum chamber, each contributing a wake function $W_0 \sim c\frac{60\,\Omega}{b^2}$ as estimated by the broad-band

[47] A two-particle model for the longitudinal microwave instability is ruled out on a fundamental reason. The very minimum model requires a three-particle model. Even then, its quantitative value is not to be trusted because higher modes are compromised.

Figure 8.38: Comparison of a two-particle simulation and the experimental observation for the case of PEP storage ring. Each panel shows the beam-position-monitor signal as a function of time after the beam is kicked. On the left are the signals observed at the PEP storage ring: (a) is when the beam intensity N is 0.86 times the threshold intensity N_{th}, (b) $\frac{N}{N_{\text{th}}} = 0.93$, and (c) $\frac{N}{N_{\text{th}}} = 0.988$. On the right are the results of simulation using a two-particle model with (a) $\frac{\Upsilon}{2} = 0.77$, (b) $\frac{\Upsilon}{2} = 0.96$, and (c) $\frac{\Upsilon}{2} = 0.99$.

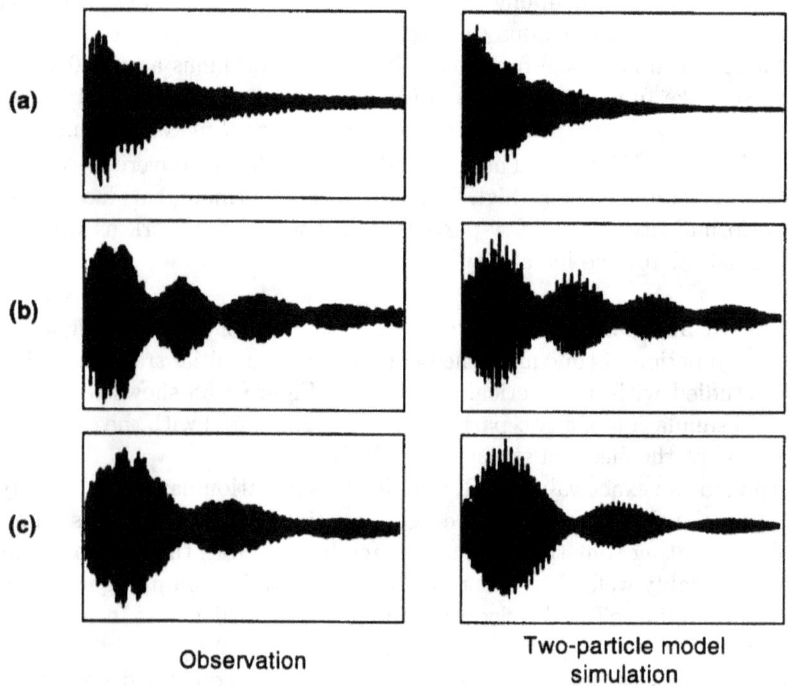

resonator model, it follows that there are an equivalent of $N_{\text{cav}} = 725$ cavities of dimension $b = 5$ cm. They occupy a total of about 1.6% of the circumference if each cavity occupies a length of b.

These data can also be used to estimate the impedances. Using the property of impedance, we have $Z_1^\perp \approx \frac{bW_0}{c}$. Taking a beam pipe radius $b = 5$ cm, we find $Z_1^\perp = 0.88$ MΩ/m.[48]

Using the approximate connection between $Z_1^\perp$ and $Z_0^\parallel$ (Sec. 8.2.2),

$$Z_1^\perp \approx \frac{2c}{b^2\omega} Z_0^\parallel \quad \Longrightarrow \quad \frac{Z_0^\parallel}{n} \approx \frac{b^2\omega_0}{2c} Z_1^\perp = \frac{\pi b^2}{C} Z_1^\perp ,$$

[48]Strictly speaking, transverse impedance effects should be linearly weighted by their local β-functions, and the estimate of $Z_1^\perp$ should take that into account. But we ignore that complication here, assuming the impedances are spread more or less evenly around the ring.

then gives $\frac{Z_0^{\|}}{n} \approx 3.2\ \Omega$. This value of $\frac{Z_0^{\|}}{n}$ indicates that, using Eq. (8.20), about 1.6% of the accelerator circumference is effectively occupied by cavities or their equivalents. The two estimates of the percentage of circumference occupied by equivalent cavity structures are consistent.

8.7.4 Head-tail instability

In our analysis of the TMCI, we assumed that the betatron and the synchrotron motions are decoupled. In doing so, we have ignored an important source of instability known as the *head-tail instability*, to which we now turn.

The head-tail instability is another clean instability to be observed experimentally.[49] Although it involves a mechanism of head-tail phase that is more subtle than that of the TMCI, this instability can occur at a much lower beam intensities. This explains the fact that it was actually observed and explained earlier than the TMCI.[50] It also explains the fact that this instability has preempted the name "head-tail instability" although almost any other collective instability could justify the same name.

Head-tail phase The betatron oscillation frequency of a particle in a circular accelerator depends on the momentum error $\delta = \frac{\Delta P}{P}$ of the particle. If we denote that betatron frequency of an on-momentum particle as ω_β, the betatron frequency for an off-momentum particle can be written as

$$\omega_\beta(\delta) \;=\; \omega_\beta + \omega_0 \xi \delta\,,$$

or equivalently,

$$\nu_\beta(\delta) \;=\; \nu_\beta + \xi \delta\,,$$

where ξ is the chromaticity parameter determined by the accelerator design.

To assure that the beam has a small betatron frequency spread due to a spread in δ, the absolute value of ξ must not be too large. A consequence of the head-tail consideration, as we will soon see, is that in addition to this requirement, ξ must also have a definite sign. The main reason for introducing sextupoles in circular accelerators is, in fact, to control ξ.

Up to now, we have used s, the longitudinal coordinate along the accelerator, as the independent variable and time t is related to s innocently simply by $s = ct$. It is no longer so simple here because now we have to consider synchrotron motions and the varying time-of-arrival confounds the connection between s and t, and it turns out to be simpler to use s as the independent variable, as will be done below.

---

[49]The SPEAR Group, Proc. 9th Int. Conf. High Energy Accel., SLAC, 1974, p. 338; J. Gareyte and F. Sacherer, Proc. 9th Int. Conf. High Energy Accel., SLAC, 1974, p. 341; Y. Miyahara and K. Takata, Part. Accel. 10, 125 (1980).

[50]C. Pellegrini, Nuovo Cimento 64A, 447 (1969); M. Sands, SLAC Reports TN-69-8 and TN-69-10 (1969).

A comment is in order concerning using s or t as the time variable. We have been careful in choosing s; basically the physical reason is that all accelerator components, magnets and RF cavities, act on the beam according to its s position. If we use t instead, the time-of-arrival will confound the dynamics, and the description of beam motion becomes rather cumbersome. Take particularly the synchrotron oscillation for example. In general, the single particle synchrotron oscillation is given by

$$z = \hat{z} \sin \frac{\omega_s s}{c}, \qquad \delta = \hat{\delta} \cos \frac{\omega_s s}{c}.$$

It is important to appreciate the fact that it is *not* given by

$$z = \hat{z} \sin \omega_s t, \qquad \delta = \hat{\delta} \cos \omega_s t.$$

Readers beware!

Let us first examine the free betatron oscillation in the absence of wakefields. The accumulated betatron phase is given by an integration of the betatron frequency, i.e.,

$$\begin{aligned} \phi_\beta(s) &= \int \omega_\beta(\delta) \frac{ds}{c} = \omega_\beta \frac{s}{c} + \omega_0 \xi \int \delta \, \frac{ds}{c} \\ &= \omega_\beta \frac{s}{c} - \frac{\omega_0 \xi}{c\eta} z(s) \end{aligned}$$

where η is the phase slippage factor, and we have used $z' = -\eta\delta$.

This is already a remarkable result. It says that the deviation of the betatron phase of a particle from the nominal value $\omega_\beta \frac{s}{c}$ is determined by its longitudinal position, i.e. the chromatic modulation of the betatron phase depends on z. The modulation is slow and weak.

Observe that η appears in the denominator. It appears by the integration of head-tail phase over synchrotron motion. When η is small, the head-tail phase gets to integrate over a long time and therefore gets bigger. A flag is being raised when one considers an isochronous $\eta \approx 0$ configuration.

Also observe that we are dealing with a modulation of the betatron phase by z, and we shall soon show that this leads to an instability. Same analysis says a modulation of the betatron frequency by δ will also lead to an instability. In contrast, modulation of the betatron phase by δ or the betatron frequency by z does not lead to an instability — they only contribute to mode frequency shift. In other words,

$$\left\{ \begin{array}{l} \text{Modulation of } \phi_\beta \text{ by } z \\ \text{Modulation of } \omega_\beta \text{ by } \delta \end{array} \right. \quad \Longrightarrow \quad \text{Instability},$$

$$\left\{ \begin{array}{l} \text{Modulation of } \phi_\beta \text{ by } \delta \\ \text{Modulation of } \omega_\beta \text{ by } z \end{array} \right. \quad \Longrightarrow \quad \text{Tune shift}.$$

Figure 8.39: The synchrotron oscillations of a two-particle beam observed in the longitudinal phase space. The quantity $\Delta\phi_\beta = \phi_{\beta 1} - \phi_{\beta 2}$ is the difference of the betatron phases of the two particles; it is modulated by the synchrotron motion as shown. The sense of rotation of particle motion in the phase space is for the case above transition, i.e. $\eta > 0$.

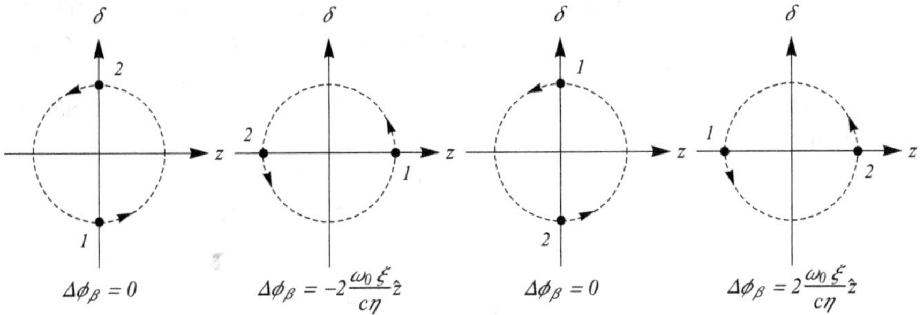

We now consider two macroparticles whose synchrotron oscillations are given by

$$z_1 = \hat{z}\sin\frac{\omega_s s}{c} \quad \text{and} \quad z_2 = -z_1 .$$

Particle 1 leads particle 2 during $0 < \frac{s}{c} < \frac{\pi}{\omega_s}$ and trails it during $\frac{\pi}{\omega_s} < \frac{s}{c} < \frac{2\pi}{\omega_s}$. The free betatron oscillations of the two particles are described by

$$\begin{aligned} y_1(s) &= \tilde{y}_1 e^{-i\phi_{\beta 1}(s)} = \tilde{y}_1 \exp\left(-i\omega_\beta \frac{s}{c} + i\,\frac{\omega_0\xi}{c\eta}\,\hat{z}\sin\frac{\omega_s s}{c}\right) \\ y_2(s) &= \tilde{y}_2 e^{-i\phi_{\beta 2}(s)} = \tilde{y}_2 \exp\left(-i\omega_\beta \frac{s}{c} - i\,\frac{\omega_0\xi}{c\eta}\,\hat{z}\sin\frac{\omega_s s}{c}\right) . \end{aligned} \tag{8.39}$$

As the particles exchange their leading and trailing roles, the betatron phases are such that the leading particle always lags in phase relative to the trailing particle if $\frac{\xi}{\eta} > 0$ and the situation reverses if $\frac{\xi}{\eta} < 0$, as illustrated in Fig. 8.39. This phase leading/lagging, as it will turn out, is the physical origin of the head-tail damping/antidamping. A swing-set type instability is waiting to happen.

The factor

$$\chi = \frac{\omega_0\xi}{c\eta}\hat{z}$$

is called the *head-tail phase*. As a numerical example, one may have an electron accelerator with $\eta = 0.003$, $\xi = 1$, $\hat{z} = 3$ cm, and $\omega_0 = 0.28 \times 10^7$ s^{-1}, which gives a head-tail phase of $2\pi \times 0.016$.

It should be emphasized here that, in spite of its apparent subtlety or intricate nature of the head-tail phase, it is a robust concept. It affects the spectrum of the collective motion of the beam in a fundamental manner. It is a real phenomenon.

Instability growth rate Recalling the TMCI, the trailing particle is always unstable due to the resonant driving by the wakefield of the leading particle; the growths of the trailing particle during the half synchrotron periods are strong, but below a certain threshold the synchrotron oscillation washes away the growths and the net result is that the beam becomes stable. The additional chromatic term that we are considering now does not have this fortunate property. As we will see, the weak growths associated with chromaticity do accumulate persistently from one half synchrotron period to the next, and thus slowly build up an instability without a threshold even though the growths per half synchrotron period are small.

Let us look at the motion of particle 2 during $0 < s < \frac{\pi c}{\omega_s}$ in the presence of the wakefield. The wake function, we assume, is that given by Eq. (8.34). The equation of motion is

$$\begin{aligned} y_2'' + \left[\frac{\omega_\beta(\delta_2)}{c}\right]^2 y_2 &= \frac{2\pi\epsilon_0 N r_0 W_0}{\gamma C} y_1 \,, \\ \omega_\beta(\delta_2) &= \omega_\beta + \frac{\omega_0 \xi \hat{z} \omega_s}{c\eta} \cos\frac{\omega_s s}{c} \,. \end{aligned}$$

The y_1 on the right hand side is given by the free oscillation result. If we let y_2 also be given by Eq. (8.39), but allow $\tilde{y}_2$ to be slowly varying in time, we obtain an equation for $\tilde{y}_2$,

$$\tilde{y}_2'(s) \approx i\frac{\pi\epsilon_0 N r_0 W_0 c}{\gamma C \omega_\beta} \tilde{y}_1(0) \exp\left(2i\frac{\xi\omega_\beta\hat{z}}{c\eta}\sin\frac{\omega_s s}{c}\right) .$$

For most practical cases, the head-tail phase $|\chi| \ll 1$ (unless for an isochronous ring), the exponential factor in the expression can be Taylor expanded and y_2 can be integrated to yield

$$\tilde{y}_2(s) = \tilde{y}_2(0) + i\frac{\pi\epsilon_0 N r_0 W_0 c}{\gamma C \omega_\beta}\tilde{y}_1(0)\left[s + i\frac{2\xi\omega_\beta\hat{z}}{\eta\omega_s}\left(1 - \cos\frac{\omega_s s}{c}\right)\right] .$$

The first term in the square brackets is the resonant response already studied before and is responsible for the TMCI. The second, chromatic term is small, because it is proportional to the head-tail phase and also because it is not a resonant response. On the other hand, the important fact here is that the chromatic term is 90° out of phase from the resonant term.

The transformation from $s = 0$ to $s = \frac{\pi c}{\omega_s}$ is thus described by

$$\begin{bmatrix} \tilde{y}_1 \\ \tilde{y}_2 \end{bmatrix}_{\pi c/\omega_s} = \begin{bmatrix} 1 & 0 \\ i\Upsilon_{HT} & 1 \end{bmatrix} \begin{bmatrix} \tilde{y}_1 \\ \tilde{y}_2 \end{bmatrix}_0 ,$$

where

$$\Upsilon_{HT} = \frac{\pi^2\epsilon_0 N r_0 W_0 c^2}{\gamma C \omega_\beta \omega_s}\left(1 + i\,\frac{4\chi}{\pi}\right) .$$

This Υ_{HT}, of course, reduces to the TMCI Υ, Eq. (8.36), if $\xi = 0$, but now it has acquired an imaginary part if $\xi \neq 0$. A similar procedure applied to the period $\frac{\pi c}{\omega_s} < s < \frac{2\pi c}{\omega_s}$ leads to the transformation

$$\begin{bmatrix} \tilde{y}_1 \\ \tilde{y}_2 \end{bmatrix}_{2\pi c/\omega_s} = \begin{bmatrix} 1 & i\Upsilon_{HT} \\ 0 & 1 \end{bmatrix} \begin{bmatrix} \tilde{y}_1 \\ \tilde{y}_2 \end{bmatrix}_{\pi c/\omega_s} .$$

Stability of the system is determined by the total transformation matrix

$$\begin{bmatrix} 1 & i\Upsilon_{HT} \\ 0 & 1 \end{bmatrix} \begin{bmatrix} 1 & 0 \\ i\Upsilon_{HT} & 1 \end{bmatrix} = \begin{bmatrix} 1-\Upsilon_{HT}^2 & i\Upsilon_{HT} \\ i\Upsilon_{HT} & 1 \end{bmatrix} .$$

The eigenvalues of this matrix are obtained by Eq. (8.37) with Υ replaced by Υ_{HT}. For a weak beam intensity, $|\Upsilon_{HT}| \ll 1$, the two eigenvalues are

$$\lambda_\pm \approx e^{\pm i\Upsilon_{HT}} .$$

The + mode (− mode) is the mode when the two macroparticles oscillate in phase (out of phase) in the limit of weak beam intensity. The imaginary part of Υ_{HT} then gives a growth rate of the betatron oscillations,

$$\tau_\pm^{-1} = \mp \frac{2\epsilon_0 N r_0 W_0 c \xi \omega_0 \hat{z}}{\gamma C \eta \omega_\beta} . \tag{8.40}$$

When the + mode is unstable, the − mode is stable; the transverse displacement of the beam center of charge grows with time but the projected transverse size of the beam essentially remains constant. When the − mode is unstable, the + mode becomes stable; the beam center of charge does not oscillate, but the projected beam size grows exponentially.

The + mode is damped if $\frac{\xi}{\eta} > 0$ and antidamped if $\frac{\xi}{\eta} < 0$. The − mode is damped if $\frac{\xi}{\eta} < 0$ and antidamped if $\frac{\xi}{\eta} > 0$. We conclude from this that the only value of ξ that assures a stable beam is $\xi = 0$. However, it can be shown by using a Vlasov equation technique, the two-particle model has overestimated the growth rate of the − mode. This consideration, together with the presence of some stabilizing mechanisms (such as Landau damping, or radiation damping in the case of circular electron accelerators) leads us to choose slightly positive values for ξ for operation above transition, and slightly negative ξ below transition.

The growth rate is proportional to N and ξ, and inversely proportional to γ as one would expect. The linear dependence on the bunch length $\hat{z}$, however, is a consequence of the constant wake model. Had we assumed a different wake model, the dependence on $\hat{z}$ would change.

It is important to note that the *same* transverse wakefield is responsible for both the TMCI and the head-tail instability. Continuing the PEP example mentioned earlier, we find the head-tail growth rate ∓0.6 ms at the TMCI threshold, $N = 6.4 \times 10^{11}$.[51] The head-tail damping or antidamping can be rather fast.

[51] Strictly speaking, Eq. (8.40) applies only when $|\Upsilon_{HT}| \ll 1$. We apply it here, even though $\mathrm{Re}\Upsilon_{HT}=2$, to obtain an order of magnitude estimate.

Impedance measurement In addition to the methods mentioned earlier concerning TMCI, the head-tail growth rate provides another way to measure the transverse wake function and the impedance of an accelerator. To do so, ξ is made slightly positive (above transition), a beam center-of-charge motion (in a + mode) is excited by a kicker, and its subsequent damped motion is observed. The linear dependence of the damping rate on ξ allows the extraction of the wake function information. The various methods of measuring the wake function are not expected to give identical values for the transverse impedance $Z_1^{\perp}$ due to the simplifying models used, but the results should at least be comparable.

Transverse mode coupling and head-tail instabilities together The head-tail and the transverse mode coupling instabilities are both transverse, short-range effects affecting mainly single bunches. They both are driven by the one and the same impedance, and as a result, they occur together. When one is present, so will be the other. Two key parameters, the chromaticity and the synchrotron tune, are at play. When chromaticity dominates, it is head-tail. When synchrotron tune dominates, it is TMCI.

Linacs do not have swing-set or the mode-coupling instability mechanisms. However, the beam breakup instability is the TMCI in linacs. The BBU, TMCI and head-tail instabilities are driven by the same physical cause.

As such, the TMCI and the head-tail instabilities really should be analyzed together. When combined, the stability conditions depends on both Υ and χ, where Υ is given by Eq. (8.36). For the two-particle model, the instability is given by the shaded region as illustrated in Fig. 8.40.

Figure 8.40: TMCI and head-tail should be considered together because the same impedance drives both. The shaded region indicates the unstable region in the (Υ, χ) parameter space.

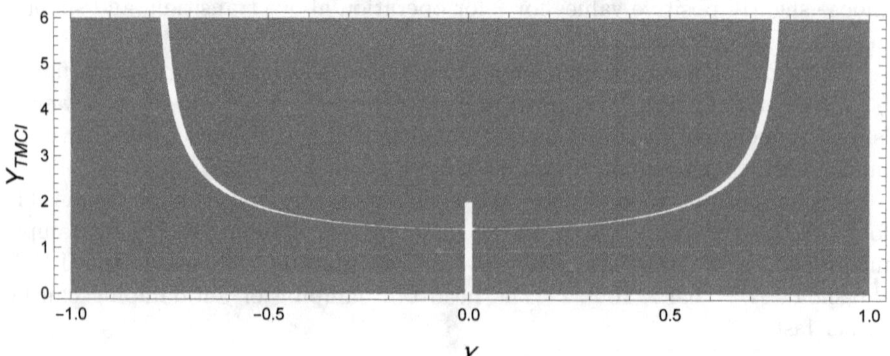

8.7.5 Longitudinal head-tail instability

Yet another swing-set instability is the longitudinal head-tail instability triggered by a high-order chromaticity.[52] An observation at the CERN SPS synchrotron is shown in Fig. 8.41.

Figure 8.41: The possible observed longitudinal head-tail instability at CERN SPS. Observed is the output from the RF phase detector. [Courtesy Trevor Linnecar 2019]

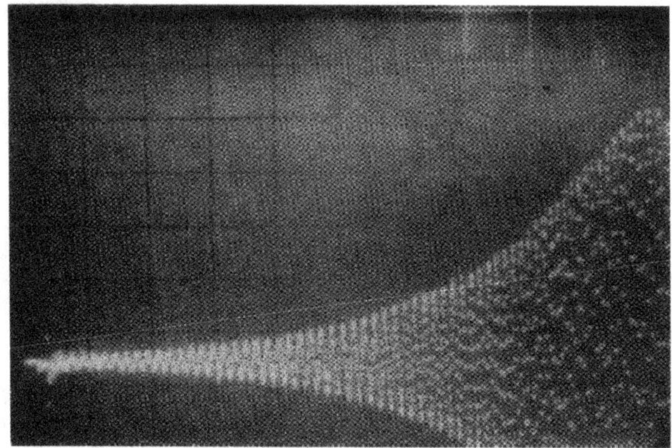

Consider a circular accelerator with a high order phase slippage factor. Its Hamiltonian contains a term proportional to δ^3 as

$$\begin{aligned} H(z,\delta,s) &= -\left(\frac{1}{2}\eta_0\delta^2 + \frac{1}{3}\eta_1\delta^3\right) - \frac{1}{2\eta_0}\left(\frac{2\pi\nu_{s0}}{C}\right)^2 z^2\,, \\ \Longrightarrow \quad \frac{dz}{ds} &= -\eta_0\delta - \eta_1\delta^2\,, \\ \frac{d\delta}{ds} &= \frac{1}{\eta_0}\left(\frac{2\pi\nu_{s0}}{C}\right)^2 z\,. \end{aligned}$$

Motion of particles in phase space follows constant Hamiltonian contours. When $\eta_1 \neq 0$, the contours are deformed from ellipses. When $\eta_1 > 0$, it makes an up-down asymmetry, as illustrated in the left panel of Fig. 8.42.

When a beam bunch executes a dipole synchrotron oscillation, the bunch length is different at the top and the bottom points as illustrated. When $\eta_1 > 0$,

[52] D. Boussard and T. Linnecar, Proc. 2nd Euro. Part. Accel. Conf., France, p. 1560 (1990); Bo Chen, The Longitudinal Collective Instabilities of Nonlinear Hamiltonian Systems in a Circular Accelerator, Ph.D. thesis, Univ. Texas, Austin, 1995.

Figure 8.42: Longitudinal phase space distorted due to a nonlinear phase slippage factor (left) and due to a nonlinear potential-well (right). A beam shown by the shaded regions executes a synchrotron dipole oscillation in the phase space.

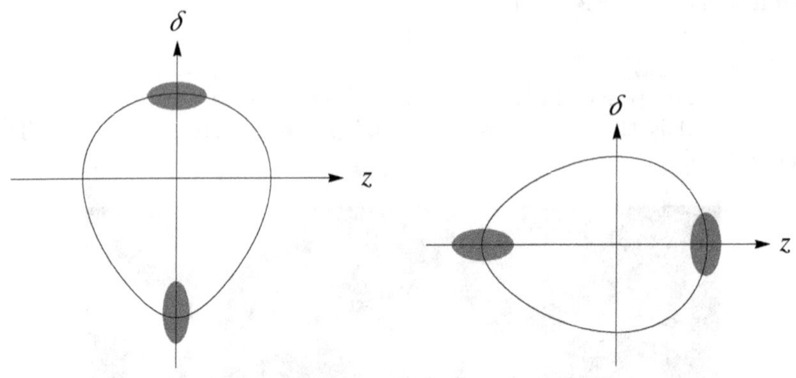

the bunch is longer in the upper half plane and shorter in the lower half plane. The bunch length at the top and at the bottom of the phase space contour have the ratio

$$\frac{\hat{z}_+}{\hat{z}_-} = \left|\frac{(dz/ds)_+}{(dz/ds)_-}\right| = \left|\frac{\eta_0\delta_+ + \eta_1\delta_+^2}{\eta_0\delta_- + \eta_1\delta_-^2}\right| ,$$

where $\delta_\pm \approx \pm\delta_0 - \frac{\eta_1}{3\eta_0}\delta_0^2$, and we have noted the instantaneous bunch length $\hat{z}$ scales by its dipole oscillation according to $\hat{z} \propto \frac{dz}{ds}$. It follows

$$\frac{\hat{z}_+}{\hat{z}_-} \approx \frac{1+\frac{2\eta_1}{3\eta_0}\delta_0}{1-\frac{2\eta_1}{3\eta_0}\delta_0} \approx \frac{1+\frac{2\eta_1}{3\eta_0}\delta_+}{1+\frac{2\eta_1}{3\eta_0}\delta_-} .$$

This means the bunch length is modulated according to

$$\hat{z} \propto 1 + \frac{2\eta_1}{3\eta_0}\delta ,$$

where δ is the instantaneous energy displacement of the centroid of the bunch as it executes a collective dipole synchrotron oscillation.

Now consider the inclusion of impedance. The parasitic loss $\Delta\mathcal{E}$ of the beam per revolution depends on its bunch length. Since we define $\Delta\mathcal{E}$ to be negative to represent energy loss, and usually the energy loss is larger with shorter bunch length, we usually have

$$\frac{d\Delta\mathcal{E}}{d\hat{z}} > 0 .$$

When parasitic loss is included, the equations of motion become

$$\begin{aligned}
\frac{dz}{ds} &= -\eta_0\delta - \eta_1\delta^2 \,, \\
\frac{d\delta}{ds} &= \frac{1}{\eta_0}\left(\frac{2\pi\nu_{s0}}{C}\right)^2 z + \frac{1}{CE}\left[\Delta\mathcal{E}\Big|_{(1+\frac{2\eta_1}{3\eta_0}\delta)\hat{z}} - \Delta\mathcal{E}\Big|_{\hat{z}}\right] \\
&= \frac{1}{\eta_0}\left(\frac{2\pi\nu_{s0}}{C}\right)^2 z + \frac{2\eta_1}{3\eta_0}\frac{\hat{z}}{CE}\frac{d\Delta\mathcal{E}}{d\hat{z}}\delta \,.
\end{aligned}$$

Combining the two equations, and dropping a nonlinear term proportional to η_1,

$$\frac{d^2\delta}{ds^2} - \frac{2\eta_1}{3\eta_0}\frac{\hat{z}}{CE}\frac{d\Delta\mathcal{E}}{d\hat{z}}\frac{d\delta}{ds} + \left(\frac{2\pi\nu_{s0}}{C}\right)^2 \delta = 0 \,.$$

It is easy to recognize this as the equation of motion for a damped/antidamped oscillator. The growth rate is

$$\tau^{-1} = \frac{2\eta_1}{3\eta_0}\frac{c\hat{z}}{2CE}\frac{d\Delta\mathcal{E}}{d\hat{z}} \,.$$

The beam motion is unstable when $\frac{\eta_1}{\eta_0}\frac{d\Delta\mathcal{E}}{d\hat{z}} > 0$, as is the case illustrated in Fig. 8.42 (left), where the bunch loses energy when it has already a low energy and it loses less energy when it already has too high an energy.

Estimate of growth rate Generally the longitudinal head-tail is a weak instability. However, it might become more significant when η_0 becomes small for isochronous designs. If we take broad band resonator model for the longitudinal impedance, and if we use long bunch expression $\Delta\mathcal{E} = -\frac{Ne^2R_Sc^3}{4\sqrt{\pi}\,Q^2\omega_R^2\sigma_z^3}\,n_{\text{cav}}$ with $\hat{z} = \sqrt{2}\sigma_z \gg b$, and broad band resonator model $R_S = \frac{Z_0}{2\pi}, Q = 1, \omega_R = \frac{c}{b}$, we obtain

$$\Delta\mathcal{E} \approx -\sqrt{\frac{2}{\pi}}\frac{Nb^2r_0}{\hat{z}^3} \times n_{\text{cav}} \times mc^2 \,.$$

This gives a longitudinal HT growth rate

$$\tau^{-1} = \sqrt{\frac{2}{\pi}}\frac{\eta_1}{\eta_0}\frac{c}{C\gamma}\frac{Nb^2r_0}{\hat{z}^3} \times n_{\text{cav}} \,.$$

If $n_{\text{cav}} = 10, b = \hat{z} = 5$ mm, $C = 50$ m, $\frac{\eta_1}{\eta_0} = 10^{-2}$, $N = 10^{10}$, $E = 250$ MeV, then we find $\tau \sim 1.8$ s. These inputs corresponds to a single bunch current $I = \frac{Nec}{C} = 10$ mA. We also use $n_{\text{cav}} = \frac{R}{b}\frac{|Z/n|}{60\ \Omega}$ with $|\frac{Z}{n}| = 3.8\ \Omega$.

Potential well distortion Longitudinal head-tail instability has a close cousin, namely the potential well distortion (Sec. 8.6.1). So far we studied the effects in the $\frac{dz}{ds}$ equation with a nonlinear term in δ. Naturally we could also inquire into the counterpart problem when the $\frac{d\delta}{ds}$ equation contains a term nonlinear in z. Such a system describes a potential well distortion effect. Let the Hamiltonian be

$$H(z,\delta,s) = -\frac{1}{2}\eta_0\delta^2 - \frac{1}{2\eta_0}\left(\frac{2\pi\nu_{s0}}{C}\right)^2 z^2(1+\epsilon z)\,,$$

where ϵ characterizes the potential well distortion.

The phase space now has a left-right asymmetry and looks like that shown in the right panel of Fig. 8.42.

Following similar steps as above,

$$\frac{\hat{z}_+}{\hat{z}_-} = \left|\frac{(d\delta/ds)_-}{(d\delta/ds)_+}\right| = \left|\frac{2z_- + 3\epsilon z_-^2}{2z_+ + 3\epsilon z_+^2}\right| \approx \frac{1-\epsilon z_0}{1+\epsilon z_0}\,,$$

where the $\pm$ subscripts refer now to the moments when the bunch oscillation has $z > 0$ and $z < 0$ respectively, $z_\pm = \pm z_0 - \frac{\epsilon}{2}z_0^2$, and we have noted $\hat{z} \propto \left(\frac{d\delta}{ds}\right)^{-1}$. It follows that the bunch length is modulated according to the longitudinal displacement of the oscillating beam,

$$\hat{z} \propto 1 - \epsilon z\,.$$

When parasitic loss is included, the equations of motion are

$$\begin{aligned}
\frac{dz}{ds} &= -\eta_0\delta\,, \\
\frac{d\delta}{ds} &= \frac{1}{\eta_0}\left(\frac{2\pi\nu_{s0}}{C}\right)^2 z + \frac{1}{CE}\left[\Delta\mathcal{E}\Big|_{(1-\epsilon z)\hat{z}} - \Delta\mathcal{E}\Big|_{\hat{z}}\right] \\
&= \frac{1}{\eta_0}\left(\frac{2\pi\nu_{s0}}{C}\right)^2 z - \frac{\epsilon\hat{z}}{CE}\frac{d\Delta\mathcal{E}}{d\hat{z}}z\,.
\end{aligned}$$

This perturbation contributes not to an instability. Instead, it contributes a coherent dipole-mode frequency shift,

$$\Delta\nu_s \approx -\epsilon\frac{\eta_0 C\hat{z}}{8\pi^2\nu_{s0}E}\frac{d\Delta\mathcal{E}}{d\hat{z}}\,.$$

Tangential shifting in phase space leads to frequency shift. Radial shift leads to damping/antidamping.

8.8 Landau damping

8.8.1 Physical origin

There are many collective instability mechanisms acting on a high intensity beam in an accelerator, demanding a wide range of often conflicting stability conditions. Yet the beam as a whole seems basically stable, as evidenced by the

Figure 8.43: Lev Davidovich Landau (1908–1968). [Wikipedia]

existence of a wide variety of working accelerators. One of the reasons for this fortunate outcome is *Landau damping*,[53] which provides a natural stabilizing mechanism against collective instabilities if particles in the beam have a small spread in their natural (synchrotron or betatron) frequencies. The theory of Landau damping of unbunched beams in accelerators was first formulated using the Vlasov technique by Neil and Sessler[54] for the longitudinal case and by Lasslet, Neil and Sessler[55] for the transverse case.

The spread in ω_β has several sources. A dependence of ω_β on the energy of the particle, together with an energy spread in the beam, leads to a spread in ω_β. Nonlinearities in the focusing system cause a dependence of ω_β on the particle's betatron amplitude. A spread in betatron amplitudes then leads also to a spread in ω_β.

The source of spread in ω_s depends on whether the beam is bunched or unbunched. For bunched beams, a spread can result from nonlinearity in the RF focusing voltage. For unbunched beams, there is no synchrotron oscillation and $\omega_s = 0$, but a dependence of the revolution frequency on the particle energy fills in the role.

The physical origin of Landau damping comes from the fact that nothing, no instabilities, existed when $t \to -\infty$. All instabilities will have to start and grow from zero. We will establish its mechanism below.

Anticipating a collective force playing the role of driving the motion of all the particles in a beam, let us consider a single simple harmonic oscillator with

[53] L.D. Landau, J. Phys. USSR 10, 25 (1946).

[54] V.K. Neil and A.M. Sessler, Rev. Sci. Instr. 36, 429 (1965).

[55] L.J. Laslett, V.K. Neil and A.M. Sessler, Rev. Sci. Instr. 36, 436 (1965).

natural frequency ω driven by a sinusoidal force of frequency Ω,

$$\ddot{x} + \omega^2 x = A\cos\Omega t\,,$$

with initial conditions $x(0) = 0$ and $\dot{x}(0) = 0$.[56] The solution is

$$x(t > 0) = -\frac{A}{\Omega^2 - \omega^2}(\cos\Omega t - \cos\omega t)\,. \tag{8.41}$$

The $\cos\Omega t$ term gives the main term responding to the driving force at frequency Ω; the $\cos\omega t$ term comes from matching the initial conditions.

The explicit inclusion of the initial conditions plays an important role. Following Hereward,[57] Eq. (8.41) is our starting point of the Landau damping discussion. Otherwise, one could have carelessly written the solution

$$x(t) = -\frac{A}{\Omega^2 - \omega^2}\cos\Omega t, \quad \text{or} \quad x(t) = -\frac{A}{\Omega^2 - \omega^2}e^{-i\Omega t}\,. \tag{8.42}$$

Equation (8.42) contains only a $\cos\Omega t$ term and no $\cos\omega t$ term, and it contains a singularity at $\Omega = \omega$ while (8.41) is well behaved there. This singularity is the source of subtleties and at this point is to be avoided. As we will see later, by applying some mathematical tricks, it is possible to bypass the explicit inclusion of the initial conditions and go straight to (8.42), but at this point, we stay with (8.41).

Consider now an ensemble of oscillators (each oscillator represents a single particle in the beam) which has a spectrum $\rho(\omega)$ satisfying $\int_{-\infty}^{\infty} d\omega\rho(\omega) = 1$. Now subject this ensemble of particles to the driving force $A\cos\Omega t$ with all particles starting with initial conditions $x(0) = 0$ and $\dot{x}(0) = 0$. The ensemble average response is

$$\langle x\rangle(t > 0) = -\int_{-\infty}^{\infty} d\omega\rho(\omega)\frac{A}{\Omega^2 - \omega^2}(\cos\Omega t - \cos\omega t)\,.$$

For simplicity, let us consider a narrow beam spectrum around a frequency ω_x and a driving frequency near the spectrum, i.e. $\omega \approx \Omega \approx \omega_x$. The beam response is then

$$\langle x\rangle(t) = -\frac{A}{2\omega_x}\int d\omega\rho(\omega)\frac{1}{\Omega - \omega}(\cos\Omega t - \cos\omega t)\,.$$

Changing variable from ω to $u = \omega - \Omega$ leads to

$$\begin{aligned}
\langle x\rangle(t) &= \frac{A}{2\omega_x}\int_{-\infty}^{\infty} du\frac{\rho(u+\Omega)}{u}[\cos\Omega t - \cos(\Omega t + ut)] \\
&= \frac{A}{2\omega_x}\left[\cos\Omega t\int_{-\infty}^{\infty} du\rho(u+\Omega)\frac{1-\cos ut}{u} + \sin\Omega t\int_{-\infty}^{\infty} du\rho(u+\Omega)\frac{\sin ut}{u}\right].
\end{aligned}$$

All integrals are well behaved at $u = 0$.

---

[56] Our main result will not change if we have a general case $x(0) = x_0, x'(0) = x_0'$.

[57] H.G. Hereward, CERN Report 65-20 (1965).

The beam response contains a $\cos \Omega t$ term and a $\sin \Omega t$ term, but their coefficients are time dependent. The next step is to show that those coefficients approach well behaved limits. To do so, one first observes

$$\begin{aligned}
\lim_{t\to\infty} \frac{\sin ut}{u} &= \pi\delta(u) \,, \\
\lim_{t\to\infty} \frac{1-\cos ut}{u} &= \text{P.V.} \left(\frac{1}{u}\right) .
\end{aligned}$$

The proof is illustrated in Fig. 8.44.

If we are not interested in the transient effects immediately following the onset of the driving force, we obtain

$$\langle x\rangle(t) = \frac{A}{2\omega_x} \left[\cos\Omega t \ \text{P.V.} \int d\omega \frac{\rho(\omega)}{\omega - \Omega} + \pi\rho(\Omega) \sin\Omega t \right] , \tag{8.43}$$

where P.V. means taking the principal value of the integral. This expression contains explicitly a $\cos \Omega t$ term and a mysterious $\sin \Omega t$ term. The reason the $\sin \Omega t$ term is considered mysterious is traced to Eq. (8.42).

The sign of the $\cos \Omega t$ term relative to the driving force depends on the sign of P.V. $\int d\omega \frac{\rho(\omega)}{\omega-\Omega}$. A system is referred to as "capacitive" or "inductive" based on whether its sign is positive or negative.

The $\sin \Omega t$ term has a definite sign relative to the driving force because $\rho(\Omega)$ is always positive. In particular, $\frac{d\langle x\rangle}{dt}$ is always in phase with the force, indicating work is being done on the system. The system always reacts to the force "resistively." As we will see, Landau damping comes from this mysterious $\sin \Omega t$ term.

Incidentally, the clear specification of principal value P.V. is important and cannot be skipped. The integral in Eq. (8.43) is undefined without specifying P.V. because of the divergence at $\omega = \Omega$. The trick of P.V. is to utilize the property that the divergences on the side $\omega < \Omega$ and the side $\omega > \Omega$ are of opposite signs and, if the integration is taken *symmetrically* about the singularity so that the divergences on the two sides cancel each other, the integral is actually well defined. Algebraically, this leads to

$$\text{P.V.} \int_{-\infty}^{\infty} dx \, \frac{f(x)}{x-a} = \int_{0}^{\infty} du \, \frac{f(a+u) - f(a-u)}{u} \,,$$

where the expression on the right is well behaved at $u = 0$. The fact that we need to, and are justified to take P.V. specifically in this symmetric manner is reflected in Fig. 8.44 by the way the singularity is avoided by the factor $1-\cos ut$ and by the antisymmetry of the function $\frac{1}{u}$.

The Landau damping effect is to be distinguished from a transient *decoherence* (also called phase mixing, or filamentation) effect that occurs when the beam has nonzero initial conditions. Had we included an initial condition

Figure 8.44: The functions $\frac{\sin ut}{u}$, $\frac{1-\cos ut}{u}$ are shown as functions of u in (a), (b) for two values $t = 3$ and 10. The dashed curves in (b) are for the function $\frac{1}{u}$. The sole function of $(1-\cos ut)$ in (b) is to suppress the singularity at $u = 0$, and as the plots show, the suppression is done in an antisymmetric manner around $u = 0$ — this point will become important later.

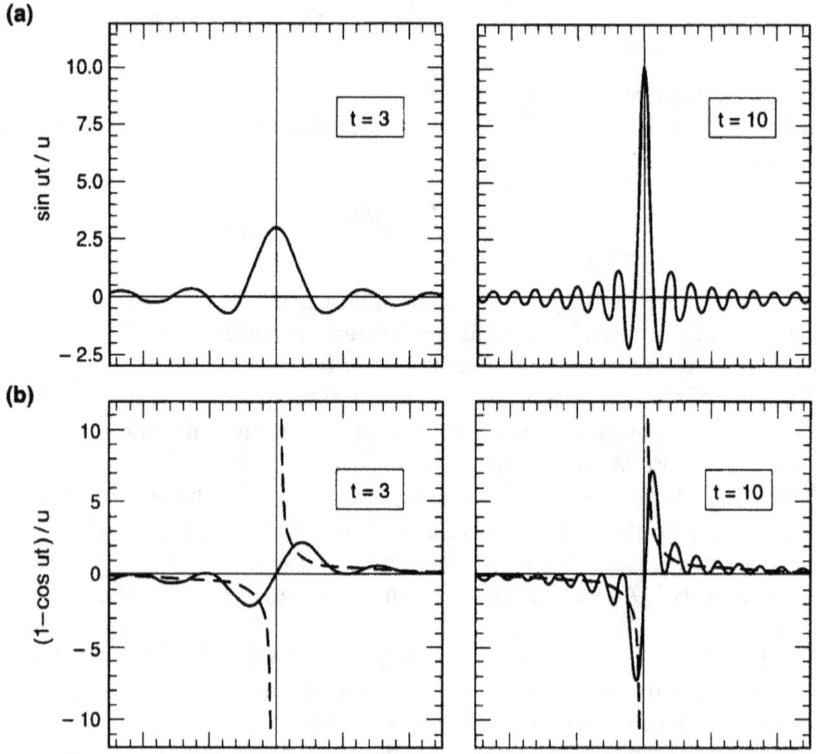

$x(0) = x_0$ and $\dot{x}(0) = \dot{x}_0$, we would have introduced two additional terms into the ensemble response,

$$x_0 \int d\omega\, \rho(\omega) \cos\omega t + \dot{x}_0 \int d\omega\, \rho(\omega) \frac{\sin\omega t}{\omega}\,.$$

These terms do not participate in the *dynamics* of the beam particles and are not very interesting for our purpose here. In this decoherence effect, individual particles continue to execute oscillations of constant amplitude, but the total beam response $\langle x \rangle$ decreases with time.

As mentioned, work is continuously being done on the system. However, the amplitude of $\langle x \rangle$, as given in Eq. (8.43), does not increase with time. Where did the energy go? To investigate this, let us first identify the energy of a particle as the square of its oscillation amplitude. From Eq. (8.41), the amplitude of a particle is given by the slowly varying envelope,

$$\text{amplitude} = \frac{A}{\omega_x(\Omega-\omega)} \sin\frac{(\Omega-\omega)t}{2} .$$

This leads to a total oscillation energy of

$$\begin{aligned} \mathcal{E} &= N\int d\omega\ \rho(\omega)\left[\frac{A}{\omega_x(\Omega-\omega)} \sin\frac{(\Omega-\omega)t}{2}\right]^2 \\ &= \frac{NA^2}{\omega_x^2}\int du\ \rho(u+\Omega)\frac{\sin^2\frac{ut}{2}}{u^2} , \end{aligned}$$

where N is the total number of oscillators. We then note that the behavior of $\frac{\sin^2(ut/2)}{u^2}$ is such that as t increases, the region where this function assumes significant values narrows around $u=0$. The range of the region decreases as $\frac{1}{t}$, but the height of the function increases quadratically as t^2, leading to an area under the function that increases linearly with t. In fact,

$$\lim_{t\to\infty} \frac{\sin^2\frac{ut}{2}}{u^2} = \frac{\pi t}{2}\delta(u) .$$

We then have

$$\mathcal{E} = \frac{\pi}{2}\frac{NA^2}{\omega_x^2}\rho(\Omega)t ,$$

which increases linearly with time. The system therefore absorbs energy from the driving force indefinitely while holding the ensemble beam response within bounds.

An analogy occurs in the decoherence behavior. In that case, the single-particle energies and thus the total beam energy are constant in time, but the ensemble signal $\langle x\rangle$ decreases with time. The analogy connects decoherence to Landau damping, but as mentioned, they remain distinct in terms of their dynamics. The decoherence effect contains no dynamics.

The stored energy $\mathcal{E}$ is incoherent in the sense that the energy is contained in the individual particles, but it is not to be regarded as heat in the system. This is because the stored energy is not distributed more-or-less uniformly in all particles, but is selectively stored in particles with continuously narrowing range of frequencies around the driving frequency. Figure 8.45 shows the driving force and the single-particle responses (8.41) for different particle frequencies. One sees that a particle with $\omega=\Omega$, being resonantly driven, continues to increase in amplitude as t increases. A particle with ω away from Ω gets out of resonance after a time approximately $\frac{\pi}{|\omega-\Omega|}$, and at time $t=\frac{2\pi}{|\omega-\Omega|}$, the particle returns all its energy back to the driving force in a beating process.

It is mainly those particles with $|\omega-\Omega|<\frac{1}{t}$ that contribute to the $\sin\Omega t$ response, and those particles with $|\omega-\Omega|>\frac{1}{t}$ that contribute to the $\cos\Omega t$

Figure 8.45: Single-particle response $x(t)$ to a sinusoidal driving force $f(t) = A\cos\Omega t$. (a) $\frac{f(t)}{A}$ versus $u = \frac{\Omega t}{2\pi}$. (b) $\frac{\Omega^2 x(t)}{A}$ versus u. The solid curve is for a particle on resonance with $\omega = \Omega$. The dashed curve is for a particle with $\omega = 1.1\ \Omega$.

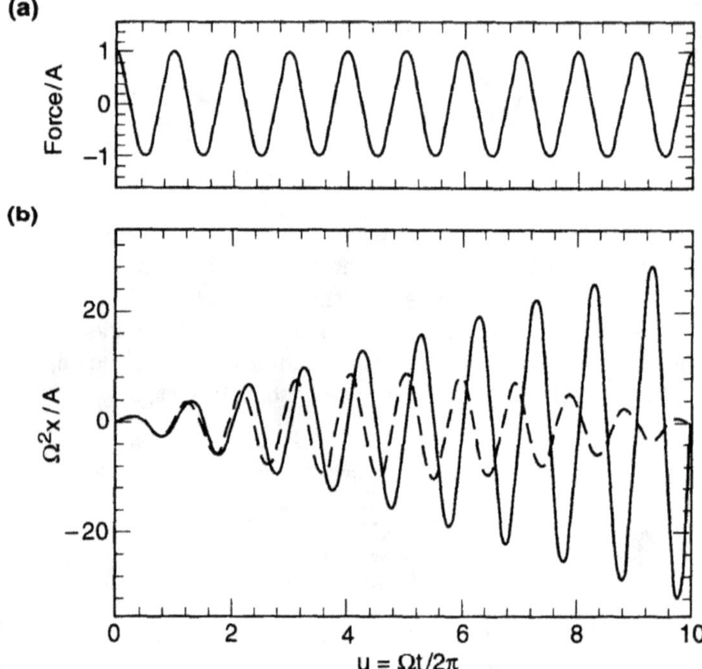

response. Since the number of particles with $|\omega - \Omega| < \frac{1}{t}$ decreases with time as $\frac{1}{t}$ while their amplitude increases as t, the net $\sin\Omega t$ contribution to $\langle x \rangle$ is constant in time.

The asymptotic behavior (8.43) applies if one waits for a time longer than $\frac{1}{\Delta\omega}$ where $\Delta\omega$ is the frequency spread of the beam spectrum. For $t < \frac{1}{\Delta\omega}$, the beam response is confounded by transient terms. Furthermore, the $\sin\Omega t$ term is proportional to $\rho(\Omega)$. If the spectrum is such that there are no particles around frequency Ω to continuously absorb energy, Landau damping will cease and a beating phenomenon takes over. Since a beam consists of a finite number of particles, Landau damping will cease when t is larger than $\frac{1}{\delta\omega}$, where $\delta\omega$ is the frequency spacing between two nearest particles. The range of time for Eq. (8.43) to be applicable is therefore

$$\frac{1}{\delta\omega} \gg t \gg \frac{1}{\Delta\omega}.$$

With N particles in the beam, one might have $\delta\omega \approx \frac{\Delta\omega}{N}$. Taking $N = 10^{11}$ and $\Delta\omega = 10^3$ s^{-1} for example, the time is limited to the range between 1 ms and 10^8 s.

Of course the upper limit can be exceeded long before 10^8 s if the resonant particles are lost when their amplitudes exceed b, the radius of the vacuum chamber, and this occurs when $\frac{At}{2\omega_x} > b$. The applicable time for Landau damping then becomes

$$\frac{2b\omega_x}{A} > t \gg \frac{1}{\Delta\omega}.$$

An alternative way to consider the beam response that avoids the consideration of transient effects is to pretend that the driving force has been in existence since $t \to -\infty$, except that the force contains an extra exponential factor $e^{\epsilon t}$ which goes to 0 as $t \to -\infty$, i.e., the force has been turned on adiabatically starting from zero. This extra exponential factor effectively takes into account the initial conditions while removing the transient effects. The price to pay is that the force then grows indefinitely as $t \to +\infty$. Fortunately it is just the slightly growing solutions that most interest us, because they correspond to the situation when the beam is at the edge of instability. These considerations will be elaborated upon in the following sections.

8.8.2 Examples of frequency spectrum

we now give several explicit examples of beam frequency spectrum for later applications, the first of which is when all particles in the beam have the same natural frequency, i.e.,

$$\rho(\omega) = \delta(\omega - \omega_x).$$

In this trivial case, we have the following asymptotic beam response to a driving force $A\cos\Omega t$,

$$\langle x\rangle = \frac{A}{2\omega_x}\left[\frac{\cos\Omega t}{\omega_x - \Omega} + \pi\delta(\Omega - \omega_x)\sin\Omega t\right].$$

The fact that this spectrum, without a frequency spread, has no Landau damping will become clear in Sec. 8.9.

The first nontrivial case to be considered is for a Lorentz spectrum

$$\rho(\omega) = \frac{\Delta\omega}{\pi}\frac{1}{(\omega - \omega_x)^2 + \Delta\omega^2}.$$

We have

$$\text{P.V.}\int d\omega \frac{\rho(\omega)}{\omega - \Omega} = \frac{\omega_x - \Omega}{(\omega_x - \Omega)^2 + \Delta\omega^2}.$$

Substituting into Eq. (8.43) yields the beam response

$$\langle x\rangle = \frac{A}{2\omega_x}\frac{(\omega_x - \Omega)\cos\Omega t + \Delta\omega\sin\Omega t}{(\omega_x - \Omega)^2 + \Delta\omega^2}. \tag{8.44}$$

The system is purely resistive if driven on resonance with $\Omega = \omega_x$, capacitive if $\Omega < \omega_x$, and inductive if $\Omega > \omega_x$.

To proceed, for the case of a general spectrum $\rho(\omega)$, write the beam response in a complex notation,

$$\begin{aligned} \text{driving force} &= Ae^{-i\Omega t}, \\ \langle x \rangle &= \frac{A}{2\omega_x \Delta\omega} e^{-i\Omega t} \left[f(u) + ig(u) \right], \end{aligned}$$

where $u = \frac{\omega_x - \Omega}{\Delta\omega}$ with $\Delta\omega$ the width of the spectral spread, and

$$\begin{aligned} f(u) &= \Delta\omega \text{ P.V.} \int d\omega \frac{\rho(\omega)}{\omega - \Omega}, \\ g(u) &= \pi \Delta\omega \, \rho(\omega_x - u\Delta\omega). \end{aligned}$$

The dimensionless complex quantity $f + ig$ is the *beam transfer function.*

For the δ-function spectrum,

$$f(u) = \frac{1}{u}, \quad \text{and} \quad g(u) = \pi\delta(u).$$

For the Lorentz spectrum,

$$f(u) = \frac{u}{1+u^2}, \quad \text{and} \quad g(u) = \frac{1}{1+u^2}.$$

A few more examples of different frequency spectra are given below. For a rectangular spectrum

$$\rho(v) = \frac{1}{2\Delta\omega} H(1 - |v|), \qquad v = \frac{\omega_x - \omega}{\Delta\omega},$$

where $H(x)$ is the step function $H(x) = 1$ if $x > 0$ and 0 if $x < 0$, the asymptotic beam response is described by

$$f(u) = \frac{1}{2} \ln \left| \frac{u+1}{u-1} \right| \quad \text{and} \quad g(u) = \frac{\pi}{2} H(1 - |u|).$$

For a parabolic spectrum, we have

$$\begin{aligned} \rho(v) &= \frac{3}{4\Delta\omega}(1 - v^2) H(1 - |v|), \\ f(u) &= \frac{3}{4} \left[(1 - u^2) \ln \left| \frac{u+1}{u-1} \right| + 2u \right], \\ g(u) &= \frac{3\pi}{4}(1 - u^2) H(1 - |u|). \end{aligned}$$

For an elliptical spectrum,

$$\begin{aligned}
\rho(v) &= \frac{2}{\pi\Delta\omega}H(1-|v|)\sqrt{1-v^2}\,, \\
f(u) &= 2\left[u-\operatorname{sgn}(u)H(|u|-1)\sqrt{u^2-1}\right]\,, \\
g(u) &= 2H(1-|u|)\sqrt{1-u^2}\,.
\end{aligned}$$

For a bi-Lorentz spectrum,

$$\begin{aligned}
\rho(v) &= \frac{2}{\pi\Delta\omega(v^2+1)^2}\,, \\
f(u) &= \frac{u(u^2+3)}{(u^2+1)^2}\,, \\
g(u) &= \frac{2}{(u^2+1)^2}\,.
\end{aligned}$$

And, finally, for a Gaussian spectrum,

$$\begin{aligned}
\rho(v) &= \frac{1}{\sqrt{2\pi}\Delta\omega}e^{-v^2/2}\,, \\
f(u) &= \sqrt{\frac{2}{\pi}}e^{-u^2/2}\int_0^\infty \frac{dy}{y}e^{-y^2/2}\sinh(uy)\,, \qquad (8.45) \\
g(u) &= \sqrt{\frac{\pi}{2}}e^{-u^2/2}\,.
\end{aligned}$$

The function $f(u)$ for the Gaussian spectrum is approximately equal to u for small u and to $\frac{1}{u}$ for large u.

These results will be used in later sections. Figure 8.46 shows the f and g functions for the various spectra. The function g is simply related to the spectrum by $\rho = \frac{1}{\pi\Delta\omega}g$, and the function f is related to the principle value integral over the spectra. An inspection of Figure 8.46 indicates that $f(u)$ resembles, but not identical to $-g'(u)$. The reason for this behavior should become clear on inspecting the principle value integral expression for $f(u)$.

Homework 8.24 Derive the beam response (8.44) for a beam with Lorentz frequency spectrum being driven by a sinusoidal force. This problem can be solved exactly without taking the asymptotic limit $t \to \infty$, and including initial conditions x_0, x_0'.

Homework 8.25 For a practice of taking principle values, try to derive the result for one of the spectra mentioned in the text, for example, that of the Gaussian distribution (8.45).

Figure 8.46: The functions $f(u)$ (solid curves) and $g(u)$ (dashed curves) for various spectral distributions. The shape of the frequency spectra is the same as the dashed curves.

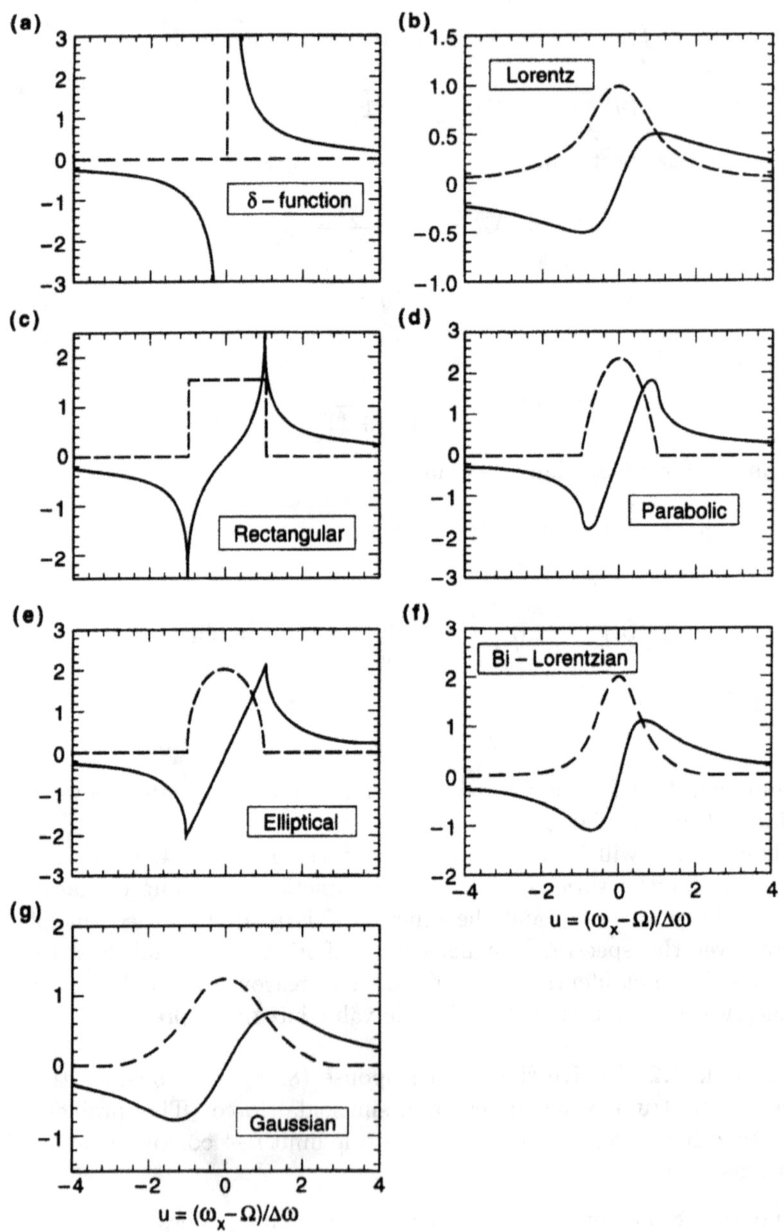

8.8.3 Mathematical trick

We now introduce a mathematical trick. It turns out that one can "derive" the same result by venturing with (8.42). In complex notation, Eq. (8.42) can be written as

$$\langle x \rangle = \frac{A}{2\omega_x} e^{-i\Omega t} \int d\omega \frac{\rho(\omega)}{\omega - \Omega} .$$

Our detailed examinations provides a well-defined way to deal with the otherwise undefined integral, i.e.

$$\int d\omega \frac{\rho(\omega)}{\omega - \Omega} \quad \to \quad \text{P.V.} \int d\omega \frac{\rho(\omega)}{\omega - \Omega} + i\pi\rho(\Omega) , \tag{8.46}$$

or more symbolically

$$\frac{1}{\omega - \Omega} \quad \to \quad \text{P.V.} \left(\frac{1}{\omega - \Omega} \right) + i\pi\delta(\omega - \Omega) .$$

Again, it is necessary to include a mysterious out-of-phase term — with a definite sign — as evidenced by the imaginary term $i\pi\rho(\Omega)$, even though the expression on the left hand side seems to be for a real quantity.

The right hand side of Eq. (8.46) in fact is equal to the left hand side provided one pretends the integration is to be executed in the complex ω-plane and the contour of integration, C, is as illustrated in Fig. 8.47(a). The connection (8.46) now reads

$$\int d\omega \frac{\rho(\omega)}{\omega - \Omega} \quad \to \quad \int_C d\omega \frac{\rho(\omega)}{\omega - \Omega} .$$

The straight line portion of C gives the principal value term in $\langle x \rangle$ and the semicircular portion gives the pole contribution $i\pi\rho(\Omega)$. The semicircular section assures the antisymmetric integration of the P.V. operation; its circular shape is not arbitrarily chosen.

Figure 8.47: Contours in the complex ω-plane.

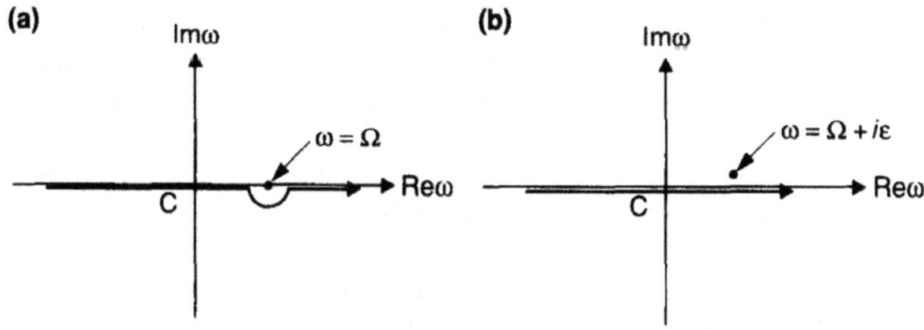

Equivalently one could consider the integration along the real axis of the ω-plane, but move the pole at $\omega = \Omega$ up by an infinitesimal amount, as shown in Fig. 8.47(b),

$$\int d\omega \frac{\rho(\omega)}{\omega - \Omega} \quad \to \quad \int_{-\infty}^{\infty} d\omega \frac{\rho(\omega)}{\omega - \Omega - i\epsilon},$$

or

$$\frac{1}{\omega - \Omega} \quad \to \quad \frac{1}{\omega - \Omega - i\epsilon},$$

or simply

$$\Omega \quad \to \quad \Omega + i\epsilon. \tag{8.47}$$

The replacement $\Omega \to \Omega + i\epsilon$, it is noted, is what constitutes an infinitesimally growing driving force that avoids the requirement to explicitly include the initial conditions. It is also what we need in the search for the condition of the edge of instability.

It is now a matter of taste whether to regard our main conclusion (8.46) as a result of a simple derivation starting with Eq. (8.42) and then make a profound connection (8.47), or to regard it as a result of a detailed calculation which takes into account of initial conditions.

8.9 Collective instability with Landau damping

Results obtained in the previous section applied to circular accelerators lead to Landau damping of collective instabilities. We will demonstrate this in this section. We will do so using simplified macroparticle models.

8.9.1 One-particle model for bunched beam – transverse

The first case concerns transverse instability of a bunched beam. Consider again a one-particle model, except now the N individual particles have a spread in their natural frequencies. This one-particle model is also referred to as the rigid-beam model for obvious reason.

An alert reader might ask the question how this beam can maintain assembled in its motion as a macroparticle, taking into account the fact that the beam has a natural frequency spread and therefore it necessarily decoheres as it evolves in time. The answer lies in the fact that the macroparticle here represents the bunch executing a collective dipole motion, the subject of our study. Here we consider a collective motion of this rigid, one-particle beam at the outset, and ask for the condition when this dipole motion might become unstable.

The driving force on the individual particles comes from the center-of-charge displacement of the beam as a whole, $\langle y \rangle$, through the wakefield. For a single particle whose betatron frequency is ω,

$$y''(s) + \left(\frac{\omega}{c}\right)^2 y(s) = -\frac{4\pi\epsilon_0 N r_0}{\gamma C} \sum_{k=1}^{\infty} \langle y \rangle (s - kC)\, W_1(-kC)\,.$$

Consider the situation when y-motion of the macroparticle is just at the edge of exponential growth due to a collective instability. We have

$$\langle y \rangle (s) = B e^{-i\Omega s/c}\,, \tag{8.48}$$

where Ω carries an imaginary part $i\epsilon$, where ϵ is infinitesimally positive.

Equation (8.48) requires a pause to consider. It is not very interesting to search for damped, stable solutions. Finding stable solutions does not assure beam stability, but finding one unstable solution reveals the beam to be unstable.

We now have

$$y''(s) + \left(\frac{\omega}{c}\right)^2 y(s) = -\frac{4\pi\epsilon_0 B N r_0}{\gamma C} \mathcal{W} e^{-i\Omega s/c}\,,$$

where

$$\mathcal{W} = \sum_{k=1}^{\infty} W_1(-kC) e^{i\omega_\beta k T_0}\,,$$

or in terms of impedance,

$$\mathcal{W} = -\frac{i}{T_0} \sum_{p=-\infty}^{\infty} Z_1^\perp(p\omega_0 + \omega_\beta)\,.$$

We have assumed the mode frequency shift is small so that $\Omega \approx \omega_\beta$, where ω_β is the center of the beam's betatron frequency spectrum.

The beam is driven by a sinusoidal driving force. Our analysis of Landau damping gives the beam response,

$$\langle y \rangle = -\frac{4\pi\epsilon_0 B N r_0 \mathcal{W} c}{2\omega_\beta \gamma T_0} e^{-i\Omega s/c} \left[\text{P.V.} \int d\omega \frac{\rho(\omega)}{\omega - \Omega} + i\pi\rho(\Omega) \right]. \tag{8.49}$$

The reader may recall a comment made earlier that a one-particle beam does not see its own transverse wake impulse because $W_1(0) = 0$. This means in the present consideration, the instability can only come from the wakefields left by the one-particle beam in previous revolutions. In case the wakefield is short-ranged, there should not be an instability in this one-particle model. To check this statement in the present context, consider a short-ranged wake with a broad-band impedance $Z_1^\perp(\omega)$. Because of its broad band width, the summation in $\mathcal{W}$ can be replaced by an integral,

$$\sum_{p=-\infty}^{\infty} Z_1^\perp(p\omega_0 + \omega_\beta) \approx \int_{-\infty}^{\infty} dp\, Z_1^\perp(p\omega_0 + \omega_\beta) = \int_{-\infty}^{\infty} \frac{d\omega}{\omega_0} Z_1^\perp(\omega) = \frac{2\pi i}{\omega_0} W_1(0)\,,$$

which vanishes. For the one-particle rigid beam to have a collective instability, the wake function must be long-ranged.

Dispersion relation We now have obtained Eq. (8.49). But we had already assumed that the collective beam motion is given by Eq. (8.48). This means the mode frequency Ω is not arbitrary. In order for the beam motion to be nontrivial, Ω must satisfy a self-consistency condition, the *dispersion relation*,

$$1 = -\frac{2\pi\epsilon_0 N r_0 \mathcal{W} c}{\omega_\beta \gamma T_0} \left[\text{P.V.} \int d\omega \frac{\rho(\omega)}{\omega - \Omega} + i\pi\rho(\Omega) \right],$$

or

$$-\frac{2\pi\epsilon_0 N r_0 \mathcal{W} c}{\omega_\beta \gamma T_0 \Delta\omega} = \frac{1}{f(u) + ig(u)},$$

where $f(u) + ig(u)$ is the beam transfer function with $u = \frac{\omega_\beta - \Omega}{\Delta\omega}$. For several beam frequency spectra, it has been explicitly calculated in Sec. 8.8.2.

In case the beam does not have a natural frequency spread, we have $f(u) = \frac{1}{u}$, $g(u) = 0$. The complex mode frequency shift is found to be

$$(\Omega - \omega_\beta)_{\text{no Landau damping}} = \frac{2\pi\epsilon_0 N r_0 c \mathcal{W}}{\omega_\beta \gamma T_0}.$$

We shall designate this quantity as ξ_1; it contains essentially the beam intensity, multiplied by the impedance, divided by the focusing strength and the magnetic rigidity, and has the dimensionality of a frequency shift.

For a beam with natural frequency spread, the dispersion relation can be written as

$$-\frac{\xi_1}{\Delta\omega} = \frac{1}{f(u) + ig(u)}. \tag{8.50}$$

The left hand side of Eq. (8.50) contains information about the beam intensity and the impedance. The right hand side contains information about the beam frequency spectrum. For a given impedance, the LHS is obtained by calculating the complex mode frequency shift ξ_1 in the absence of Landau damping. Without Landau damping, the stability condition is simply Im $\xi_1 < 0$.

Stability diagram Once its LHS is obtained, Eq. (8.50) can in principle be used to determine Ω in the presence of Landau damping when the beam is at the edge of instability. However, the exact value of Ω is not too useful. The useful question to ask is under what conditions the beam becomes unstable regardless of the exact value of Ω. Eq. (8.50) can be used in a reversed manner to address this question.

To do so, consider the real parameter $u = \frac{\omega_\beta - \Omega}{\Delta\omega}$ and observe the locus traced out in the complex $\mathcal{D}_1$-plane as u is scanned from ∞ to $-\infty$, where

$$\mathcal{D}_1 = \frac{1}{f(u) + ig(u)}.$$

This locus defines a *stability boundary diagram.* The LHS of Eq. (8.50), a complex quantity, is then plotted in the complex $\mathcal{D}_1$-plane as a single point.

If this point lies on the locus, it means the solution of Ω for Eq. (8.50) is real — with an infinitesimal positive imaginary part, and this ξ_1 value is such that the beam is just at the edge of instability. If it lies on the inside of the locus (the side which contains the origin of the $\mathcal{D}_1$-plane), the beam is stable. If it lies on the outside of the locus, the beam is unstable.

Figure 8.48 gives the stability diagrams for the several example frequency spectra calculated in Sec. 8.8.2.

The dispersion relation is particularly simple for the Lorentz spectrum [case (b) in Fig. 8.48, see also Homework 8.24],

$$\Omega = \omega_\beta + \xi_1 - i\Delta\omega\,.$$

The stability condition $\text{Im}\Omega < 0$ therefore becomes

$$\text{Im}\xi_1 < \Delta\omega\,.$$

The finite frequency spread $\Delta\omega$ therefore enlarges the stability region. This is true not only for the Lorentz spectrum but also all other spectra shown in Fig. 8.48.

The fact that the stable region is always *enlarged* by the frequency spread demonstrates the Landau damping mechanism. Its origin can be traced back to the fact that $g(u)$ is always positive, which in turn comes from the fact that the beam continues to absorb energy from the driving force without having to let $\langle y \rangle$ grow.

For a given spectral shape, the tolerable $\xi_1 \propto \Delta\omega$; the larger the frequency spread, the stronger the Landau damping. On the other hand, for a given $\Delta\omega$, the effectiveness of Landau damping is different for different spectral shapes. The Lorentz spectrum, having a long distribution tail, is most forgiving, while the δ-function spectrum is not effective. This is also obvious from Fig. 8.48. Generally the tighter is the spectral shape, the less is its Landau damping effectiveness.

For practical accelerator operations, there may be information on the value of the half-width-at-half-height $\Delta\omega_{\frac{1}{2}}$, but not enough detailed information on the shape of the frequency spectrum. For those applications, we introduce a simplified stability criterion

$$|\xi_1| = \frac{2\pi\epsilon_0 N r_0 c}{\omega_\beta \gamma T_0^2}\left|\sum_{p=-\infty}^{\infty} Z_1^\perp(p\omega_0+\omega_\beta)\right| < \frac{1}{\sqrt{3}}\Delta\omega_{\frac{1}{2}}\,, \tag{8.51}$$

where the factor $\frac{1}{\sqrt{3}}$ is chosen so that it coincides with the semicircular portion of the boundary for the elliptical spectrum, Fig. 8.48(e). Stability diagram for this simplified criterion is shown in Fig. 8.48(h).

Equation (8.51) says that if the mode frequency shift or growth rate, calculated without Landau damping, is larger than the frequency spread of the beam, Landau damping most likely will not rescue the beam from instability.

Figure 8.48: The stability boundary diagrams for various spectra. Shaded regions are unstable. The coordinates labeled refer to (Re$\mathcal{D}_1$, Im$\mathcal{D}_1$; u). The value of u can be used to obtain Ω. (a) δ-function spectrum, no Landau damping. (h) is the simplified criterion (8.51).

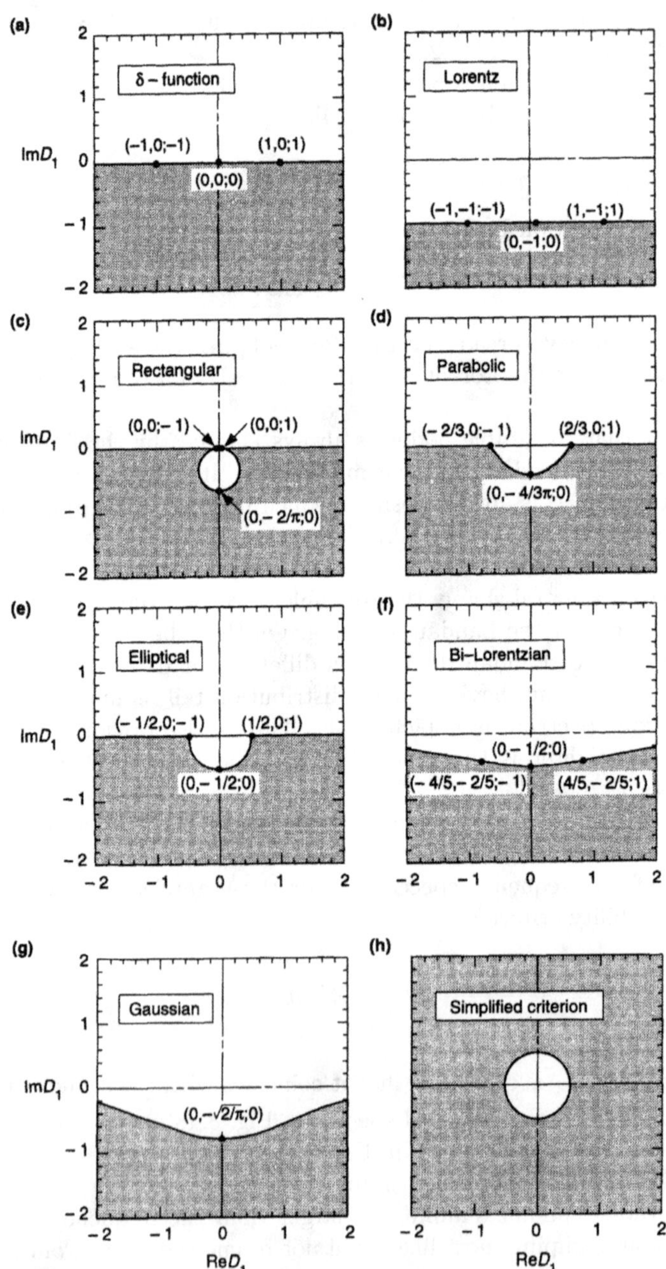

Homework 8.26 It is explained in the text that a one-particle beam does not have transverse instability when the wake function is short-ranged. Consider now the case of a resistive wall.[58] This time the wake function is long-ranged and the corresponding instability is a multiturn effect.

(a) Apply Eq. (8.51) to find an expression for the threshold beam intensity N_{th}.

(b) Consider a storage ring with a single bunch of 1-GeV electrons. Let the storage ring circumference $C = 100$ m, aluminum resistive wall $\sigma_c = 3.6 \times 10^7\,\Omega^{-1}\text{m}^{-1}$, betatron tune $\nu_\beta = 10.87$, beam pipe radius $b = 5$ cm, and betatron tune spread $\frac{\Delta\omega_{\frac{1}{2}}}{\omega_0} = 10^{-4}$. Calculate the value of N_{th}.

Solution

(a) You may use either the wake function or the impedance expressions for $\mathcal{W}$, although the wake function expression is simpler. Assume the resistive wall covers the entire circumference of the storage ring, it can be shown that

$$\begin{aligned}\xi_1 &= -\frac{2Nr_0c}{\gamma\omega_\beta T_0 b^3}\sqrt{\frac{c}{\sigma_c\omega_0 Z_0}}\,f(\Delta_\beta)\,,\\ f(\Delta_\beta) &= \sum_{k=1}^{\infty}\sqrt{\frac{2}{k}}\,e^{i2\pi k\Delta_\beta}\,,\end{aligned}$$

where Δ_β is the fractional part of the betatron tune $\nu_\beta = \frac{\omega_\beta}{\omega_0}$ chosen to be between $-\frac{1}{2}$ and $\frac{1}{2}$. The threshold intensity N_{th} is determined by the condition $|\xi_1| = \frac{1}{\sqrt{3}}\Delta\omega_{\frac{1}{2}}$.

Figure 8.49 shows the function $f(\Delta_\beta)$. It should be concluded from the plot that, in the absence of Landau damping, the beam is stable when $0 < \Delta_\beta < \frac{1}{2}$, i.e. the betatron tune is above an integer, and is unstable when the tune is below an integer. Physically, this happens because of the swing-set instability mechanism at work.

(b) Using $f(-0.13) = -0.092 - 1.72i$, it follows that $N_{\text{th}} = 1.8 \times 10^{14}$. Landau damping easily suppresses this resistive wall instability.

8.9.2 Feedback system

A feedback system can be modeled as an impedance. Not really a passive device as an impedance is supposed to be, modeling it as an impedance will necessarily compromise on some of its basic properties, such as its real part may become negative — in fact one might say that the purpose of the feedback is to artificially introduce a negative real part in its impedance.

Consider a transverse feedback system. In this feedback model, the center-of-charge displacement $\langle y\rangle$ is detected and a kick proportional to $\langle y\rangle$ is applied

[58] E.D. Courant and A.M. Sessler, Rev. Sci. Instru. 37, 1579 (1966).

Figure 8.49: The function $f(\Delta_\beta)$. Dashed curve is the imaginary part. Solid is the real part. The imaginary part gives the instability growth rate if negative and damping rate if positive. The real part gives the mode frequency shift.

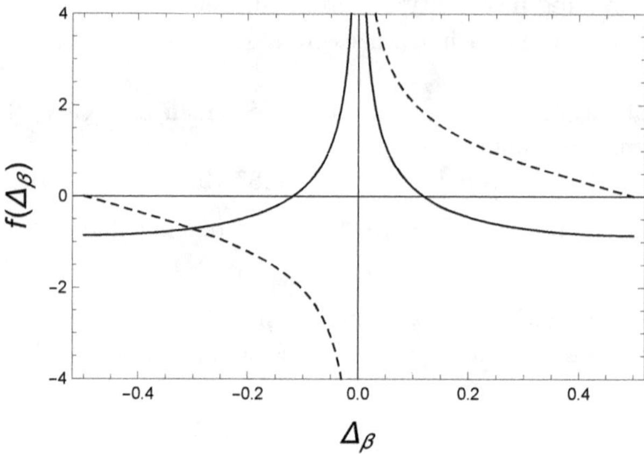

to beam particles at a later time with a delay $\frac{s_0}{c}$. A particle with betatron frequency $\omega \approx \omega_\beta$ has the equation of motion

$$y''(s) + \left(\frac{\omega}{c}\right)^2 y(s) \;=\; -g\,\langle y\rangle(s - s_0)\,,$$

where g is a feedback gain factor. In this model, g is a real quantity.

Let the center-of-charge motion be

$$\langle y\rangle(s) \;=\; Be^{-i\Omega s/c}\,.$$

The equation of motion of the particle reads

$$\begin{aligned} y''(s) + \left(\frac{\omega}{c}\right)^2 y(s) \;&=\; -gBe^{-i\Omega(s-s_0)/c} \\ \Longrightarrow \qquad y(s) \;&\approx\; -\frac{gBc^2}{2\omega_\beta(\omega-\Omega)}e^{-i\Omega(s-s_0)/c}\,. \end{aligned}$$

Without frequency spread, i.e. without Landau damping, we obtain

$$\Omega - \omega_\beta \;\approx\; \frac{gc^2}{2\omega_\beta}e^{i\omega_\beta s_0/c}\,.$$

The feedback growth rate is given by the imaginary part of Ω, i.e.

$$\tau^{-1} \;=\; \mathrm{Im}\Omega \;\approx\; \frac{gc^2}{2\omega_\beta}\sin\frac{\omega_\beta s_0}{c}\,.$$

Maximum feedback damping occurs when τ^{-1} is as negative as possible, i.e. when $g < 0$, $\frac{\omega_\beta s_0}{c} = \frac{\pi}{2}$. The feedback is purely resistive. This is consistent with the requirement for negative feedback of a swing set. It becomes purely reactive when $\frac{\omega_\beta s_0}{c} = 0$ or π.

When there is a frequency spread, the frequency spread will interfere with the feedback. The dispersion relation depends on whether the feedback is implemented to passively suppress instabilities or to actively damp a beam's betatron oscillation. The former asks for the condition when the beam is at the edge of exponential growth, the latter for the edge of exponential damping. The frequency spread prevents the instability to grow in the former, and it prevents the feedback to damp the beam oscillation in the latter.

When the feedback is used to actively damp a betatron oscillation, the required dispersion relation is for the beam to be barely damped, we need to change the beam transfer function from $f(u) + ig(u)$ to $f(u) - ig(u)$,

$$-\frac{gc^2}{2\omega_\beta} e^{i\omega_\beta s_0/c} \int d\omega \frac{\rho(\omega)}{\omega - \Omega} = 1\,,$$

$$\Longrightarrow \quad -\frac{gc^2}{2\omega_\beta \Delta\omega} e^{i\omega_\beta s_0/c} = \frac{1}{f(u) - ig(u)}, \qquad u = \frac{\omega_\beta - \Omega}{\Delta\omega}\,.$$

Take a Lorentz spectrum for example, this becomes

$$\frac{1}{f - ig} = i + \frac{\omega_\beta - \Omega}{\Delta\omega}$$

$$\Longrightarrow \quad \Omega = \omega_\beta + i\Delta\omega + \frac{gc^2}{2\omega_\beta} e^{i\omega_\beta s_0/c}\,.$$

For the feedback to function as intended, we require

$$\mathrm{Im}\Omega < 0 \quad \Longrightarrow \quad \frac{gc^2}{2\omega_\beta} \sin\frac{\omega_\beta s_0}{c} < -\Delta\omega\,. \tag{8.52}$$

Homework 8.27 The text gave the requirement of a feedback system to actively damp the betatron oscillation of a beam with a Lorentz betatron frequency spread. Suppose now this same beam is suffering from a transverse impedance in such a way that in the absence of feedback, its dispersion relation is given by Eq. (8.50). Find the condition for the feedback system so that the impedance-induced instability is passively suppressed.

Homework 8.28 This model of feedback can also be cast in terms of wake function and impedance. The corresponding wake function is a δ-function in z, with

$$\frac{4\pi\epsilon_0 N r_0}{\gamma C} W_1(z) = gC\delta(z + s_0)\,.$$

Find its corresponding impedance. Examine which fundamental properties of a passive impedance have been violated.

8.9.3 One-particle model for bunched beam – longitudinal

A similar analysis can be performed for the longitudinal Robinson instability using a one-particle model,

$$\begin{aligned} z''(s) + \left(\frac{\omega_s}{c}\right)^2 z(s) &= \frac{4\pi\epsilon_0 N r_0 \eta}{\gamma C} \sum_{k=1}^{\infty} \left[\langle z\rangle(s) - \langle z\rangle(s-kC)\right] W_0''(-kC) \\ &= \frac{4\pi\epsilon_0 N r_0 \eta}{\gamma C} B e^{-i\Omega s/c} \mathcal{W}, \end{aligned}$$

where we have introduced

$$\langle z\rangle(s) = B e^{-i\Omega s/c},$$

and

$$\begin{aligned} \mathcal{W} &= \sum_{k=1}^{\infty} \left(1 - e^{i\omega_s k T_0}\right) W_0''(-kC) \\ &= \frac{i}{C} \sum_{p=-\infty}^{\infty} \left[p\omega_0 Z_0^{\parallel}(p\omega_0) - (p\omega_0 + \omega_s) Z_0^{\parallel}(p\omega_0 + \omega_s)\right]. \end{aligned}$$

Self-consistency then gives rise to a dispersion relation

$$\frac{2\pi\epsilon_0 N r_0 \eta \mathcal{W} c^2}{\omega_s \gamma C \Delta\omega} = \frac{1}{f(u) + i g(u)},$$

similar to the transverse case except that the frequency spectrum now refers to synchrotron frequency, and the complex mode frequency shift in the absence of Landau damping is

$$\xi_1 = -\frac{2\pi\epsilon_0 N r_0 \eta \mathcal{W} c^2}{\omega_s \gamma C}.$$

The simplified stability criterion reads

$$|\xi_1| = \frac{2\pi\epsilon_0 N r_0 \eta c^2}{\omega_s \gamma C^2} \left| \sum_{p=-\infty}^{\infty} \left[p\omega_0 Z_0^{\parallel}(p\omega_0) - (p\omega_0 + \omega_s) Z_0^{\parallel}(p\omega_0 + \omega_s)\right] \right| < \frac{\Delta\omega_{\frac{1}{2}}}{\sqrt{3}},$$

where $\Delta\omega_{\frac{1}{2}}$ refers to the half-width-half-height spread of the synchrotron frequency of the beam.

The conclusion that the longitudinal Landau damping behaves analogously to the transverse case, however, is valid only for *bunched* beams for which $\omega_s \neq 0$. The analysis depends on the assumption that the mode frequency shift $|\Omega|$ is small compared with ω_s. For unbunched beams, $\omega_s = 0$, longitudinal analysis gives results very different from its transverse counterpart. In fact, in case of an unbunched beam, both the transverse and the longitudinal cases require a closer look, as discussed in the next two sections.

8.9.4 Transverse instability of unbunched beam

Clearly we will not be able to apply one-particle or two-particle models to an unbunched beam. To set up the problem here, consider an unbunched beam in a circular accelerator with a rigid uniform round cross section with radius a. Consider an infinitesimal transverse displacement of the beam that behaves in time as $e^{-i\Omega t}$ and in angular coordinate as $e^{ins/R}$, where n is a mode index. Observed at a fixed location s, the beam oscillates with frequency Ω. Observed at a fixed time t as a snapshot, the beam makes n oscillations around the accelerator. This perturbation induces a wakefield; the field acts back on the beam, leading to a reduction or an enhancement of the initial beam perturbation, and the beam is stable or unstable accordingly.

Figure 8.50: Snapshots of an unbunched beam executing a collective mode with mode index $n = 4$. (a) Transverse, (b) Longitudinal.

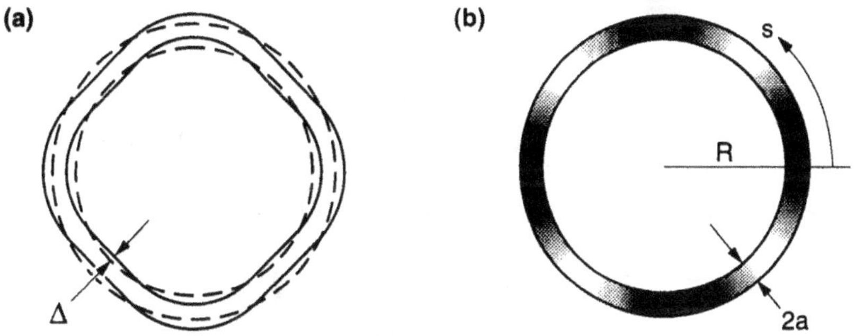

We consider the transverse case in this section. The longitudinal case is postponed to the next section. The snapshot beam distribution is sketched in Fig. 8.50(a). The perturbing part of the oscillating beam has a ring-beam distribution

$$\rho = \frac{D}{\pi a^2}\,\delta(r-a)\cos\theta\,,$$

where D is the dipole moment per unit length; it is related to the maximum beam displacement Δ by

$$D(s,t) = \frac{Ne\Delta}{2\pi R}\exp\left(in\frac{s}{R} - i\Omega t\right). \tag{8.53}$$

The transverse force F_y on a test charge, at location s and time t, is determined by the dipole moment of the ring beam at the same location s as it past by at an earlier time t'.[59] By superposition, the wake force is obtained by

[59]This statement is carefully worded. It contains much of the physics of wakefield impulses

integrating the wake field over t'. This gives

$$F_y(s,t) = -\frac{e}{2\pi R}\int_{-\infty}^{t} cdt'\, W_1(ct'-ct)\, D(s,t')\,.$$

Substituting the expression for D and expressing the result in terms of the impedance $Z_1^\perp$ yield

$$F_y(s,t) = ecD(s,t)\, i\frac{Z_1^\perp(\Omega)}{2\pi R}\,.$$

Note that the impedance is evaluated at frequency Ω. The impedance responds to the oscillating beam signal observed at a fixed location. It does not have any information on, and therefore does not respond to, the mode index n.

Consider now a specific particle in the beam that passes position S at time 0. It is located at position $s = S + ct$ at time t, and experiences a transverse force $F_y(S+ct,t)$ at any time t. Modeling the impedance as evenly distributed around the accelerator circumference, the equation of motion of the particle is

$$\begin{aligned}\ddot{y} + \omega_\beta^2 y &= \frac{F_y(S+ct,t)}{m_0\gamma} \\ &= \frac{4\pi\epsilon_0 N r_0 c^2}{\gamma T_0}\, i\frac{Z_1^\perp(\Omega)}{2\pi R}\Delta \exp\left[in\frac{S}{R} - i(\Omega - n\omega_0)t\right]\,.\end{aligned}$$

Although the impedance responds only to Ω, the moving particle sees a force with a driving frequency $\omega_d = \Omega - n\omega_0$.

The solution is given by, with $s = S + ct$,

$$y(s,t) = \frac{4\pi\epsilon_0 N r_0 c^2}{\gamma T_0}\, i\frac{Z_1^\perp(\Omega)}{2\pi R}\frac{\Delta \exp\left(in\frac{s}{R} - i\Omega t\right)}{\omega_\beta^2 - (\Omega - n\omega_0)^2}\,.$$

Self-consistency requires that the beam at (s,t) have a displacement of $y(s,t) = \Delta\exp(in\frac{s}{R} - i\Omega t)$. This yields the dispersion relation

$$1 = \frac{4\pi\epsilon_0 N r_0 c^2}{\gamma T_0}\, i\frac{Z_1^\perp(\Omega)}{2\pi R}\frac{1}{\omega_\beta^2 - (\Omega - n\omega_0)^2}\,,$$

which is to be solved for Ω. The fact that the s and t dependences nicely cancel in obtaining the dispersion relation is a consequence of the correct choice of the ansatz (8.53).

There are two solutions for Ω, one close to $n\omega_0 - \omega_\beta$, the other close to $n\omega_0 + \omega_\beta$. We consider the solution close to $n\omega_0 + \omega_\beta$. The other solution leads to an identical instability criterion. The solution is

$$\Omega = n\omega_0 + \omega_\beta + \xi_1\,,$$

---

and how they act on the beam. In particular, the wakefields do not propagate; they stand still oscillating or transient at the same location where they were generated.

where

$$\xi_1 = -\frac{2\pi\epsilon_0 N r_0 c^2}{\omega_\beta \gamma T_0} i \frac{Z_1^\perp(\Omega)}{2\pi R}$$

is the complex mode frequency shift in the absence of Landau damping.

The quantity ξ_1 involves the impedance evaluated at frequency Ω, but it can be approximately evaluated at the unperturbed frequency $n\omega_0 + \omega_\beta$. The instability growth rate and the mode frequency shift are then given by

$$\begin{aligned} \tau^{-1} &= \mathrm{Im}\xi_1 = -\frac{2\pi\epsilon_0 N r_0 c^2}{\omega_\beta \gamma T_0} \frac{\mathrm{Re}Z_1^\perp(n\omega_0 + \omega_\beta)}{2\pi R}, \\ \Delta\Omega &= \mathrm{Re}\xi_1 = \frac{2\pi\epsilon_0 N r_0 c^2}{\omega_\beta \gamma T_0} \frac{\mathrm{Im}Z_1^\perp(n\omega_0 + \omega_\beta)}{2\pi R}. \end{aligned}$$

The real part of the impedance appears in the growth rate, the imaginary part in the mode frequency shift.

As an example, the resistive wall impedance gives

$$\xi_1 = -\frac{4\pi\epsilon_0 N r_0 c^2}{\omega_\beta \gamma T_0 b^3} \frac{1}{\sqrt{2\pi\sigma |n\omega_0 + \omega_\beta|}} \left[1 + i\ \mathrm{sgn}(n\omega_0 + \omega_\beta)\right].$$

The dominating mode occurs when n is equal to the nearest integer to $-\frac{\omega_\beta}{\omega_0}$. Let the betatron tune $\nu_\beta = \frac{\omega_\beta}{\omega_0}$ be written as $N_\beta + \Delta_\beta$, we have

$$\Omega - \Delta_\beta \omega_0 = -\frac{4\pi\epsilon_0 N r_0 c^2}{\omega_\beta \gamma T_0 b^3} \frac{1 + i\ \mathrm{sgn}(\Delta_\beta)}{\sqrt{2\pi\sigma\omega_0 |\Delta_\beta|}}.$$

The beam is stable if the tune is above an integer and unstable below an integer. The same conclusion was obtained in Homework 8.26 for a bunched beam.

So far we have not included Landau damping. For a beam with spectrum $\rho(\omega)$ in its betatron frequency, the dispersion relation reads

$$1 = \frac{4\pi\epsilon_0 N r_0 c^2}{\gamma T_0} i \frac{Z_1^\perp(\Omega)}{2\pi R} \int d\omega \frac{\rho(\omega)}{\omega^2 - (\Omega - n\omega_0)^2}.$$

There are two poles at $\omega = \pm(n\omega_0 - \Omega)$. For a given mode n, usually only one of the poles dominates. Consider the $\omega = -n\omega_0 + \Omega$ pole. The dispersion relation can be written as

$$1 = -\xi_1 \int d\omega \frac{\rho(\omega)}{\omega - \omega_d},$$

where $\omega_d = \Omega - n\omega_0$ is the frequency of the driving wake force.

From our Landau damping analysis, we have

$$1 = -\xi_1 \left[\mathrm{P.V.} \int d\omega \frac{\rho(\omega)}{\omega - |\omega_d|} + i\pi\ \mathrm{sgn}(\omega_d)\ \rho(|\omega_d|)\right],$$

or

$$1 = -\frac{\xi_1}{\Delta\omega}\left[f(u) + ig(u)\right],$$

where $u = \frac{\omega_\beta - |\omega_d|}{\Delta\omega}$.

Beam stability is determined by whether $-\frac{\xi_1}{\Delta\omega}$ lies in the stable region in the complex $\mathcal{D}_1 = (f + ig)^{-1}$ plane. When unstable, the beam is said to have a *transverse microwave instability.*

For example, a Lorentz spectrum gives

$$\Omega = n\omega_0 + \omega_\beta + \xi_1 - i\Delta\omega,$$

with beam stability condition

$$\mathrm{Im}\xi_1 = -\frac{4\pi\epsilon_0 N r_0 c^2}{2\omega_\beta \gamma T_0}\frac{\mathrm{Re}Z_1^\perp(n\omega_0 + \omega_\beta)}{2\pi R} < \Delta\omega.$$

For resistive wall, and for a beam with Lorentz spectrum, stability condition is

$$-\frac{4\pi\epsilon_0 N r_0 c^2}{\omega_\beta \gamma T_0 b^3}\frac{\mathrm{sgn}(\Delta_\beta)}{\sqrt{2\pi\sigma\omega_0|\Delta_\beta|}} < \Delta\omega.$$

In case the detailed spectral distribution is not available, one could apply the simplified criterion,

$$|\xi_1| = \frac{2\pi\epsilon_0 N r_0 c}{\omega_\beta \gamma T_0^2}|Z_1^\perp(n\omega_0 + \omega_\beta)| < \frac{1}{\sqrt{3}}\Delta\omega_{\frac{1}{2}}.$$

Landau damping results from a finite $\Delta\omega_{\frac{1}{2}}$ which is the HWHM spread of $n\omega_0 + \omega_\beta$. One source of this spread is an energy spread of the particles in the beam, which contributes due to the dependence of ω_β on δ (through the chromaticity ξ) and due to a dependence of ω_0 on δ (through the phase slippage factor η),

$$\Delta\omega_{\frac{1}{2}} = \left|-n\omega_0\eta + \xi\omega_0\right|\Delta\delta_{\frac{1}{2}}.$$

It follows the stability condition becomes

$$|Z_1^\perp(n\omega_0 + \omega_\beta)| < Z_0\frac{R\gamma\,|-n\eta + \xi|}{2\sqrt{3}\,\epsilon_0 N r_0 \beta_Z}\Delta\delta_{\frac{1}{2}},$$

where β_Z is the β-function at the location of the impedance.

8.9.5 Longitudinal instability of unbunched beam

This section deals with the longitudinal case for unbunched beams. We will see that the analysis becomes quite different from the previous cases because there is no external focusing, i.e. $\omega_s = 0$. Landau damping in this case does not come directly from a spread in the natural focusing frequencies, but indirectly from a spread in the revolution frequencies.

Because of this change, what we derived before, including the beam transfer function $f + ig$ (Fig. 8.46) and the stability diagram (Fig. 8.48), will need to be rederived and replaced in this section.

Consider a relativistic unbunched beam with a uniform longitudinal line density

$$\lambda_0 = \frac{N}{2\pi R}.$$

Let the perturbation be an infinitesimal longitudinal density wave given by

$$\Delta\lambda(s,t) = \Delta\hat{\lambda}\, \exp\left(in\frac{s}{R} - i\Omega t\right).$$

This perturbation was illustrated in Fig. 8.50(b) for the case $n = 4$. The mode $n = 0$ is excluded because it violates charge conservation. The beam interacts with the vacuum chamber environment through a longitudinal impedance. We assume the impedance is evenly distributed around the circular accelerator taking advantage of the fact that the longitudinal motion is slow compared with beam revolution even in the presence of the wakefields.

The longitudinal wake force F_s on a test charge at position s and time t is determined by the beam density as it passes by position s at an earlier time $t' < t$. By superposition, we have

$$F_s(s,t) = -\frac{e^2}{2\pi R}\int_{-\infty}^{t} cdt'\; W_0'(ct' - ct)\Delta\lambda(s,t'),$$

which leads to

$$F_s(s,t) = -e^2 c\Delta\lambda(s,t)\frac{Z_0^{\|}(\Omega)}{2\pi R}.$$

Consider now a specific particle that passes position S at time $t = 0$. It experiences a longitudinal force $F_s(S + ct, t)$ at time t. Let the energy deviation and longitudinal coordinate of this particle be designated as δ_S and z_S; the equations of motion are

$$\begin{aligned}
\dot{z}_S &= -\eta c\delta_S,\\
\dot{\delta}_S &= -\frac{4\pi\epsilon_0 r_0 c}{\gamma T_0}\Delta\hat{\lambda}\exp\left[in\frac{S}{R} - i(\Omega - n\omega_0)t\right] Z_0^{\|}(\Omega).
\end{aligned}$$

We first ignore Landau damping, the solution is

$$\begin{aligned}
\delta_S(t) &= -i\frac{4\pi\epsilon_0 r_0 c}{\gamma T_0}\,\frac{\Delta\hat{\lambda}}{\Omega - n\omega_0}\exp\left[in\frac{S}{R} - i(\Omega - n\omega_0)t\right] Z_0^{\|}(\Omega),\\
z_S(t) &= -\frac{4\pi\epsilon_0\eta r_0 c^2}{\gamma T_0}\,\frac{\Delta\hat{\lambda}}{(\Omega - n\omega_0)^2}\exp\left[in\frac{S}{R} - i(\Omega - n\omega_0)t\right] Z_0^{\|}(\Omega).
\end{aligned}$$

The quantities δ_S and z_S describe the motion of a single particle in the beam. A moment's reflection considering the whole beam shows that

$$\Delta\lambda(s,t) = -\lambda_0 \frac{\partial z_{s-ct}}{\partial s},$$

which in turn yields

$$\Delta\lambda(s,t) = \frac{8\pi^2\epsilon_0 N r_0 \eta}{\gamma T_0^3} \frac{in}{(\Omega - n\omega_0)^2} \Delta\hat{\lambda} \exp\left(in\frac{s}{R} - i\Omega t\right) Z_0^{\parallel}(\Omega).$$

Self-consistency requires that the beam at (s,t) have a density perturbation of $\Delta\hat{\lambda}\exp\left(in\,\frac{s}{R} - i\Omega t\right)$. This yields the dispersion relation

$$1 = \frac{\xi_2}{(\Omega - n\omega_0)^2},$$

where

$$\xi_2 = i\frac{8\pi^2\epsilon_0 N r_0 n\eta}{\gamma T_0^3} Z_0^{\parallel}(n\omega_0).$$

The real and imaginary parts of $\Omega - n\omega_0$ give

$$\begin{aligned}\Delta\Omega &= \mathrm{Re}(\Omega - n\omega_0) = \mp\mathrm{sgn}(\mathrm{Im}\xi_2)\sqrt{\frac{|\xi_2| + \mathrm{Re}\xi_2}{2}} \\ &= \mp\mathrm{sgn}(n\eta)\sqrt{\frac{4\pi^2\epsilon_0 N r_0}{\gamma T_0^3}\left(|n\eta Z_0^{\parallel}| - n\eta\mathrm{Im}Z_0^{\parallel}\right)},\end{aligned}$$

and

$$\begin{aligned}\tau^{-1} &= \mathrm{Im}(\Omega - n\omega_0) = \mp\sqrt{\frac{|\xi_2| - \mathrm{Re}\xi_2}{2}} \\ &= \mp\sqrt{\frac{4\pi^2\epsilon_0 N r_0}{\gamma T_0^3}\left(|n\eta Z_0^{\parallel}| + n\eta\mathrm{Im}Z_0^{\parallel}\right)},\end{aligned}$$

where the impedance $Z_0^{\parallel}$ is evaluated at $n\omega_0$, and the $\pm$ signs refer to two solutions for the mode.

Recall that, in the transverse case, only the real part of the impedance appears in the growth rate and only the imaginary part appears in the mode frequency shift; this is no longer true here. For stability, τ^{-1} must not be positive for both of the $\pm$ modes. This is possible only when ξ_2 is real and positive, in which case the instability growth rate is zero. Away from the positive real axis, the $+$ mode grows and the $-$ mode damps; the beam is unstable.

The new dispersion relation has an important difference from the previous one, namely, it is the *square* of the complex mode frequency shift that is related to the impedance. This feature can be traced back to the fact that there is no external focusing.

This quadratic dependence has an important consequence. In all previous cases, half of the complex ξ_1-plane is stable even without Landau damping, whereas in the present case, the stable region is restricted to the positive real ξ_2-axis. Landau damping is absolutely required against the longitudinal collective instabilities for unbunched beams.

This quadratic behavior is also reflected in the stability diagram, which will be shown later in Fig. 8.52. The unstable (shaded) region of the longitudinal unbunched case can be obtained from the unstable region of the previous cases by "folding over" its upper half plane, i.e. by rotating the negative real axis clockwise around the origin toward the positive real axis, stretching elastically the unstable region behind it — a procedure equivalent of squaring of the complex plane. In the case without Landau damping, the only stable region after folding is the positive real axis.

Landau damping for the longitudinal instabilities of an unbunched beam comes from a spread in the revolution frequency ω_0, which could result from an energy spread of the beam. Let $\rho(\omega_0)$ be the spectrum centered around $\bar{\omega}_0$; the dispersion relation reads

$$1 = \xi_2 \int d\omega_0 \, \frac{\rho(\omega_0)}{(n\omega_0 - \Omega - i\epsilon)^2},$$

where the impedance is to be evaluated at $n\bar{\omega}_0$. We have attached an infinitesimal positive imaginary part to Ω.

An integration by parts gives

$$\begin{aligned} 1 &= \frac{\xi_2}{n} \int d\omega_0 \, \frac{\rho'(\omega_0)}{n\omega_0 - \Omega - i\epsilon} \\ &= \frac{\xi_2}{n} \left[\text{P.V.} \int d\omega_0 \, \frac{\rho'(\omega_0)}{n\omega_0 - \Omega} + \frac{i\pi}{|n|} \rho' \left(\frac{\Omega}{n} \right) \right]. \end{aligned}$$

The absence of external focusing gives rise to the consequence that longitudinal Landau damping of unbunched beams involves the *derivative* of $\rho(\omega_0)$.

Denoting the spectrum width by $\Delta\omega$, the dispersion relation can be written as

$$\frac{\xi_2}{n^2 \Delta\omega^2} = \frac{1}{f(u) + i \, \text{sgn}(n) g(u)} \equiv \mathcal{D}_2,$$

where $u = \frac{n\bar{\omega}_0 - \Omega}{n\Delta\omega}$ and

$$\begin{aligned} f(u) &= n\Delta\omega^2 \, \text{P.V.} \int d\omega_0 \frac{\rho'(\omega_0)}{n\omega_0 - \Omega}, \\ g(u) &= \pi \Delta\omega^2 \rho' \left(\frac{\Omega}{n} \right). \end{aligned}$$

Figure 8.51 shows the functions $f(u)$ and $g(u)$ for various frequency spectra.

Figure 8.51: Functions $f(u)$ (solid curves) and $g(u)$ (dashed curves). Dotted curves are $\pi\Delta\omega\rho(v)$.

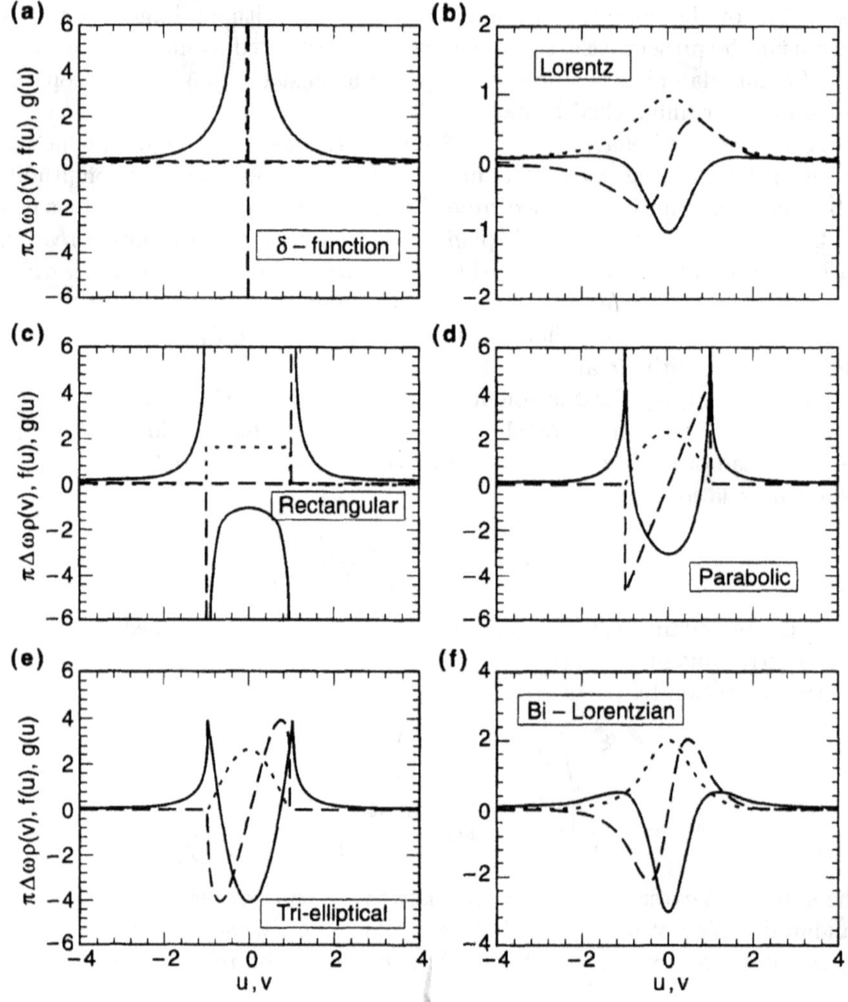

The case of Lorentz spectrum is particularly simple. The mode frequency shift is the same as the case without Landau damping. The growth rate is found to be

$$\tau^{-1} = \mp\sqrt{\frac{4\pi^2\epsilon_0 N r_0}{\gamma T_0^3}\left(|n\eta Z_0^{\|}| + n\eta \mathrm{Im} Z_0^{\|}\right)} - |n|\Delta\omega\,,$$

containing an extra negative term $-|n|\Delta\omega$, demonstrating Landau damping. Note that the damping term, unlike its transverse counterpart, is proportional to n. The higher the mode index, the more strongly the mode is Landau damped.

Figure 8.52 shows the stability diagrams for various spectra, i.e. it shows the locus of $\mathcal{D}_2$ as u is scanned from $-\infty$ to ∞. The beam is unstable if the value of $\frac{\xi_2}{n^2\Delta\omega^2}$ lies in the shaded region. In the unstable region, the beam has a *longitudinal microwave instability*.

If one does not have detailed spectral information and is interested in a rough estimate, a simplified stability criterion, called *Keil-Schnell criterion*,[60] can be used,

$$|\xi_2| = \frac{8\pi^2\epsilon_0 N r_0}{\gamma T_0^3}|n\eta Z_0^{\parallel}(n\omega_0)| < \frac{1}{4}n^2\Delta\omega^2 ,$$

where the factor $\frac{1}{4}$ is such that the condition is exact for the tri-elliptical spectrum.

If the spread in revolution frequency ω_0 comes from the energy spread of the beam, we have

$$\Delta\omega = \bar{\omega}_0|\eta|\Delta\delta .$$

We then obtain the stability condition

$$\left|\frac{Z_0^{\parallel}(n\bar{\omega}_0)}{n}\right| < Z_0\frac{|\eta|\gamma R}{16\,\epsilon_0 N r_0}\Delta\delta^2 . \tag{8.54}$$

Observe that, instead of imposing a condition on the impedance $Z_0^{\parallel}$ directly, Eq. (8.54) imposes a condition on $\frac{Z_0^{\parallel}}{n}$. In the present context, $\frac{Z_0^{\parallel}}{n}$ is a consequence of balancing the instability strength $|\xi_2| \propto nZ_0^{\parallel}$, and the Landau damping effect $\propto n^2\Delta\omega^2$. The $\frac{Z_0^{\parallel}}{n}$ results as their ratio is being taken.

One may take another viewpoint that the final stability condition should involve not the impedance but the impedance per unit length. By identifying $\frac{N}{2\pi R}$ of Eq. (8.54) as the line density of the beam, and recognizing that $\frac{Z_0^{\parallel}}{n}$ has the meaning of the impedance per unit length, one may anticipate the form of Eq. (8.54).

In the case of a bunched beam, if the microwave instability growth rate is much faster than the synchrotron oscillation frequency, one may modify Eq. (8.54) by replacing the unperturbed uniform density $\frac{N}{2\pi R}$ by the peak density $\frac{N_B}{\sqrt{2\pi}\sigma_z}$ across the length of the bunch. We then obtain the *Boussard criterion*,[61]

$$\left|\frac{Z_0^{\parallel}(n\bar{\omega}_0)}{n}\right| < \frac{1}{16\sqrt{2\pi}}Z_0\frac{|\eta|\gamma}{\pi\epsilon_0 N_B r_0}\Delta\delta^2\sigma_z .$$

[60] E. Keil and W. Schnell, CERN Report TH-RF/69-48 (1969); V. K. Neil and A. M. Sessler, Rev. Sci. Instr. 36, 429 (1965).

[61] D. Boussard, CERN Lab II/RF/Int 75-2 (1975); J. M. Wang and C. Pellegrini, Proc. 11th Int. Conf. High Energy Accel., Geneva, 1980, p. 554.

Figure 8.52: Stability boundary diagrams for various spectra in the complex $\mathcal{D}_2$-plane. The numbers in parentheses are $(\mathrm{Re}\mathcal{D}_2, \mathrm{Im}\mathcal{D}_2; \mathrm{sgn}(n)u)$. (a) without Landau damping, the stability region is only along the positive real axis.

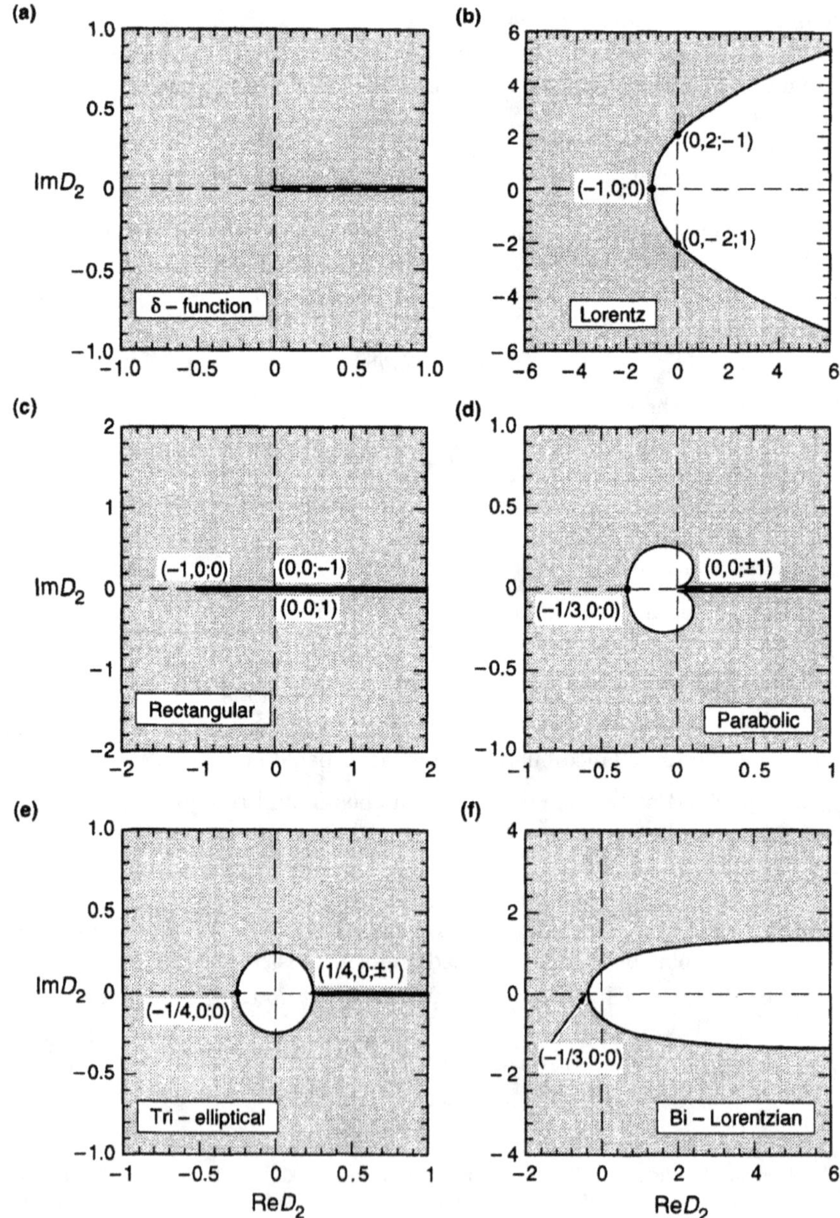

Subject Index

www.ingramcontent.com/pod-product-compliance
Lightning Source LLC
Chambersburg PA
CBHW052114230326
41598CB00079B/3666